Lowell S. Thomas
1813 Wilmington Dr.
Midland, MI 48642

SPI Plastics Engineering Handbook

of the Society of the Plastics Industry, Inc.

fourth edition

edited by **Joel Frados**

VNR VAN NOSTRAND REINHOLD
New York

Published by Van Nostrand Reinhold
115 Fifth Avenue
New York, New York 10003

Van Nostrand Reinhold International Company Limited
11 New Fetter Lane
London EC4P 4EE, England

Van Nostrand Reinhold
480 La Trobe Street
Melbourne, Victoria 3000, Australia

Macmillan of Canada
Division of Canada Publishing Corporation
164 Commander Boulevard
Agincourt, Ontario M1S 3C7, Canada

15 14 13 12 11

Library of Congress Cataloging in Publication Data

Society of the Plastics Industry.
    Plastics engineering handbook of the Society of the
Plastics Industry, inc.

    First ed. published in 1947 under title: SPI
handbook; 2d-3d editions published under title: SPI
plastics engineering handbook.
    Includes bibliographical references and index.
    1. Plastics—Handbooks, manuals, etc. I. Frados,
Joel. II. Title.
TP1130.S58 1976 668.4'1 75-26508
ISBN 0-442-22469-9

FOREWORD

It is a great pleasure to present herewith the Fourth Edition of the Plastics Engineering Handbook. This new edition reflects substantial revisions and updating throughout. We are indeed indebted to the many contributors to this edition, especially to Joel Frados, Publisher of *Plastics Focus*, who served as Editor, and authored several sections.

When the Handbook was first published in 1947 the total reported production of plastics was 1,252,000,000 pounds. Currently, in 1976, we are likely to exceed 26 billion pounds. It is reasonable to believe that the earlier editions contributed in no small way to that fantastic growth through its emphasis on quality, sound engineering, and good design.

Today, plastics products are being used substantially in every industry, and the need for an authoritative and reliable Handbook like this has never been greater. It is essential that the designers and users of plastics recognize not only the potential capabilities but also the limitations of our materials.

Looking ahead, we can anticipate continuing rapid growth for plastics. There will inevitably be new and improved materials and dramatic innovations in production techniques, as well as thousands of new applications. Thus, plans must be made immediately for the publication of the Fifth Edition a few years hence.

All of us in plastics today owe a special acknowledgment to all of the men who worked on the earlier editions of this handbook. The most notable and most unsung of these is Charles L. Condit, now Staff Vice President of SPI, whose untiring devotion to our industry's progress continues to be an inspiration for all of us.

RALPH L. HARDING, JR.
President of SPI

PREFACE

The plastics industry has changed in many ways since the last edition of the *SPI Engineering Handbook* was published—so much so that it becomes difficult to make comparisons. Where thermosets dominated the Third Edition, the thermoplastics move to the fore in this one. The blow molding process, as another example, had only a few paragraphs of description in the Third Edition—it has a chapter of its own in this volume. And no one had even heard of rotational molding or structural foam molding when the Third Edition was published.

In fact, to keep pace with the fast-moving industry, virtually every chapter from the Third Edition has either been completely rewritten or revised extensively and a number of new chapters have been added.

However, the basic format and coverage that has made the *SPI Engineering Handbook* so important to the plastics industry has been retained. The general flow of the book continues to duplicate the general flow of plastics through the manufacturing operation—from original materials selection to processing to secondary finishing to final use.

Chapter 1 provides an up-to-date glossary of the words and expressions in common use in the plastics industry (including a special illustrated section on injection molding and extrusion nomenclature).

Chapters 2 and 3 form a basic guide to plastics materials highlighting their chemistry, their characteristics, and their applications. These two chapters are new to the *SPI Engineering Handbook*, but because the family of plastics has become so diverse over the years, this information is essential to a full understanding of the various manufacturing operations covered in the Handbook.

Chapters 4 through 18 are devoted to the most popular methods of plastics processing. The three major processing techniques—injection molding, extrusion, and thermoset processing—are covered first (Chapters 4 through 11). Next, thermoforming, including the relatively new concept of "cold stamping" (Chapter 12), and blow molding (Chapter 13) are reviewed. Chapters 14 through 18 cover those processing techniques that involve the use of plastics in various powder, paste, and liquid grades: rotational molding (Chapter 14), calendering (Chapter 15), processing vinyl dispersions (Chapter 16), powder coating (Chapter 17), and casting (Chapter 18).

In most instances, each chapter covers all aspects of the individual process: machinery and equipment, molds or dies, processing variables, etc. However, for injection molding and thermoset processing the subjects of mold design are discussed in separate chapters (Chapters 6 and 10, respectively). The reasons for this lie both in the complexity of the subjects and in the fact that many of the basic rules and principles outlined in these two chapters serve as a good starting point for understanding mold design as it applies to all the processes covered in subsequent chapters. Similarly, we have also devoted a separate chapter (5) to the subject of controls for injection molding. Again, this is intended as a review of the basic principles of process control as it applies to all processes (each subsequent chapter, however, does cover the type of controls used for the particular techniques being discussed).

Chapters 19 and 20 are devoted to two forms of plastic—reinforced plastics and foamed or cellular plastics—that are so unique and so widely used that they have virtually spawned entire industries of their own. Chapter 21 on radiation processing is another subject that is entirely new to an Engineering Handbook.

The next three chapters—mold making and materials (Chapter 22), designing molded products (Chapter 23), and standards for tolerances of molded articles (Chapter 24)—should more logically have been placed in the front of the Handbook, since they cover procedures that are generally undertaken after the basic plastic has been selected and before processing begins. However, we have carried them in the middle of the Handbook, after processing and before secondary finishing, because we feel that the success of these activities, especially design, will depend to a large extent on a complete understanding of the process to be used. As the reader will quickly note, when working with plastics, there are strong interrelationships between the basic material, the process, the design, and even finishing and assembly. This latter subject is covered in the four chapters that follow the sections on design—Chapter 25 through Chapter 28.

Finally, there is a chapter on Compounding and Materials Handling (Chapter 29) and one on Performance Testing of Molded Products (Chapter 30).

ACKNOWLEDGMENTS

Previous editions of the SPI Engineering Handbook were prepared by SPI-appointed sub-committees who were responsible for reviewing and up-dating each of the chapters of the book under the very able direction of Charles L. Condit (now Staff Vice President of SPI). Their efforts have provided the current Editors with an invaluable base of information with which to work—and we are deeply indebted to them.

As will be evident to readers of this edition, we have also been fortunate in having the assistance of a number of leading industry experts in compiling this Handbook. Wherever possible, we have acknowledged their contributions directly with a credit line, but since we are very much aware of the effort that went into the preparation of this material, we wish to extend to them our personal thanks.

Acknowledgment should also be made to the many materials suppliers in the plastics industry who have built up over the years a sizable body of literature relating to the use of plastics and who granted permission to use various sections in this Handbook. It would be impossible to list them all individually, since they span the entire gamut of current resin suppliers, but again, we have provided appropriate credit in the Handbook for their contributions.

Because of my own personal association with the publishing industry, I also found numerous occasions to call on my friends at the various plastics magazines for their advice and assistance on various sections of the Handbook. Since the plastics industry is serviced by a number of high-caliber publications, I was fortunate in obtaining permission to use editorial material here that originally appeared in those publications. I am especially indebted to Malcolm W. Riley and Lowell L. Scheiner of *Plastics Technology,* Sid Gross of *Modern Plastics,* Len Berringer and Bernie Miller of *Plastics World*, George Smoluk (ex-Editor) and Abe Schoengood (current Editor) of *Plastics Engineering*, Charles Cleworth of *Plastics Machinery & Equipment*, and Don Swanson of *Plastics Design and Processing*. And another note of thanks to J. O'Rinda Trauernicht from our own *Plastics Focus* staff, particularly for permission to use excerpts from her award-winning series on the bonding and decorating of plastics (as it originally appeared in *Plastics Technology*).

And finally, my very special gratitude to Julia Frados and Robin Frados for their assistance and encouragement.

JOEL FRADOS

CONTENTS

Plastics Engineering Handbook

of the Society of the Plastics Industry, Inc.

1
GLOSSARY

Over the years, the plastics industry has built up a language and a terminology of its very own. In this chapter, the most commonly used words and expressions are classified and defined. Definitions of other terms can also be found in the text, and can be located by means of the index at the back of the book.

In many instances, the words being defined are peculiar to the plastics industry and the way in which it manufactures its products. In other cases, the expression used by the industry may derive from words commonly used in other branches of manufacturing (e.g., the concept of *forging* plastics derives from metalworking terminology); as applied to plastics, however, these definitions may differ from common usage.

Readers are especially referred to the illustrations that accompany the definitions of extrusion terminology, injection molding terminology, and mold terminology.

The Editor wishes to extend special thanks and appreciation to the following for permission to reprint various definitions: *Modern Plastics Encyclopedia* (published by McGraw-Hill, Inc.); Phillips Petroleum Co.'s "Glossary of Plastics Terms"; SPI's Machinery Division's "Standard Nomenclature for Single Screw Extruders"; SPI's Machinery Division's "Standard Nomenclature and Specifications for Plastics Injection Molding Machines"; SPI's Plastics Pipe Institute's "A Glossary of Plastics Piping Terms"; SPI's Bottle Division's "Bottle Gloss-

ary"; SPI's Custom Molders Division's "Standards and Practices of Plastics Custom Molders."

A-stage—An early stage in the reaction of certain thermosetting resins, in which the material is still soluble in certain liquids and fusible. Sometimes referred to as resol. (See **B-stage, C-stage**.)

ablative plastics—This description applies to a material which absorbs heat (while part of it is being consumed by heat) through a decomposition process known as pyrolysis, which takes place in the near surface layer exposed to heat. This mechanism essentially provides thermal protection of the subsurface materials and components by sacrificing the surface layer.

accumulator—An auxiliary cylinder and piston (plunger) mounted on injection molding or blowing machines. It is used to provide extremely fast molding cycles. In blow molding, the accumulator cylinder is filled (during the time between parison deliveries or "shots") with melted plastic coming from the main (primary) extruder. The molten plastic is stored or "accumulated" in this auxiliary cylinder until the next parison is required. At that time the piston in the accumulator cylinder forces the molten plastic into the dies that form the parison.

acid-acceptor—A compound which acts as a stabilizer by chemically combining with acid which may be initially present in minute quantities in a plastic, or which may be formed by the decomposition of the resin.

activation, n.–The process of inducing radio-activity in a specimen by bombardment with neutrons or other types of radiation.

adiabatic–An adjective used to describe a process or transformation in which no heat is added to or allowed to escape from the system under consideration. It is used, somewhat incorrectly, to describe a mode of extrusion in which no external heat is added to the extruder although heat may be removed by cooling to keep the output temperature of the melt passing through the extruder constant. The heat input in such a process is developed by the screw as its mechanical energy is converted to thermal energy.

adsorption, n.–A concentration of a substance at a surface or interface of another substance.

aging, n.–(1) The effect of exposure of plastics to an environment for an interval of time. (2) The process of exposing plastics to an environment for an interval of time, resulting in improvement or deterioration of properties.

air ring–A circular manifold used to distribute an even flow of the cooling medium, air, onto a hollow tubular form passing through the center of the ring. In blown tubing, the air cools the tubing uniformly to provide uniform film thickness.

air-slip forming–A variation of snap-back forming in which the male mold is enclosed in a box in such a way that when the mold moves forward toward the hot plastic, air is trapped between the mold and the plastic sheet. As the mold advances, the plastic is kept away from it by the air cushion formed as described above, until the full travel of the mold is reached, at which point a vacuum is applied, destroying the cushion and forming the part against the plug.

air vent–Small outlet, usually a groove, to provide a path for air to flow out of a mold cavity as the material enters.

ambient temperature–Temperature of the medium surrounding an object.

anchorage, n.–Part of the insert that is molded inside of the plastic and held fast by the shrinkage of the plastic.

angle press–A hydraulic molding press equipped with horizontal and vertical rams, and specially designed for the production of complex moldings containing deep undercuts.

anneal–(1) To heat a molded plastic article to a predetermined termperature and slowly cool it, to relieve stresses. (2) To heat steel to a predetermined temperature above the critical range and slowly cool it, to relieve stresses and reduce harness. (Annealing of molded or machined parts may be done dry as in an oven or wet as in a heated tank of mineral oil.)

antioxidant–A chemical substance that can be added to a plastic resin to minimize or prevent the effects of oxygen attack on the plastic (e.g., yellowing or degradation). Such chemical attack by oxygen may render a plastic brittle or cause it to lose desired mechanical properties.

antistatic agent–A chemical substance that can be applied to the surface of a plastic article, or incorporated in the plastic from which the article is to be made. Its function is to render the surface of the plastic article less susceptible to accumulation of electrostatic charges which attract and hold fine dirt or dust on the surface of the plastic article.

arc resistance–Time required for a given electrical current to render the surface of a material conductive because of carbonization by the arc flames. Ref.: Standard Method of Test for High-Voltage, Low-Current Arc Resistance of Solid Electrical Insulating Materials (ASTM Designation: D 495).

artificial aging (see also **aging**)–The exposure of the plastic to conditions which "accelerate" the effects of time. Such means include heating, exposure to cold, flexing, exposure to chemicals, ultraviolet lights, etc. Typically the conditions chosen for such testing reflect the conditions under which the plastic article will be used. The length of time the article is exposed to these test conditions is generally relatively short. Properties such as dimensional stability, mechanical fatigue, chemical resistance, stress crack resistance, etc., are evaluated.

autoclave–A closed vessel for conducting a chemical reaction or other operation under pressure and heat.

autoclave molding–As used in reinforced plastic molding. After lay-up, entire assembly is placed in steam autoclave at 50 to 100 psi. Additional pressure achieves higher reinforcement loadings and improved removal of air. (Modification of pressure bag method.)

automatic mold—A mold for injection, compression or transfer molding that repeatedly goes through the entire molding cycle, including ejection, without human assistance.

average molecular weight—Plastics (polymers) are long, chain-like structures. The number of units which comprise an individual chain varies from chain to chain. Average Molecular Weight indicates chain length of the most typical chain in a given plastic; it is neither the longest chain nor the shortest.

B-stage—An intermediate stage in the reaction of a thermosetting resin in which the material softens when heated and swells in contact with certain liquids but does not entirely fuse or dissolve. Resins in thermosetting molding compounds are usually in this stage. See also **A-stage** and **C-stage**.

back pressure—Resistance of a material, because of its viscosity, to continue flow when mold is closing.

back-pressure-relief port—An opening from an extrusion die for escape of excess material.

back taper—Reverse draft used in mold to prevent molded article from drawing freely. (See **undercut**.)

backing plate—In injection molding equipment, a heavy steel plate that is used as a support for the cavity blocks, guide pins, bushings, etc. In blow molding equipment, it is the steel plate on which the cavities (i.e., the bottle molds) are mounted.

baffle—A device used to restrict or divert the passage of fluid through a pipe line or channel. In hydraulic systems the device, which often consists of a disc with a small central perforation, restricts the flow of hydraulic fluid in a high pressure line. A common location for the disc is in a joint in the line. When applied to molds, the term is indicative of a plug or similar device located in a stream or water channel in the mold and designed to divert and restrict the flow to a desired path.

bag molding—A method of applying pressure during bonding or reinforced plastics molding in which a flexible cover, usually in connection with a rigid die or mold, exerts pressure on the material being molded, through the application of air pressure or drawing of a vacuum.

Bakelite—A proprietary name for phenolic and other plastics materials, often used indiscriminately to describe any phenolic molding material or molding. The name is derived from that of Dr. Leo Hendrik Baekeland (1863–1944), a Belgian who developed phenolic resins in the early 1900's.

Banbury—An apparatus for compounding materials composed of a pair of contra-rotating rotors which masticate the materials to form a homogeneous blend. This is an internal type mixer which produces excellent mixing.

barrel—See **extruder**.

barrier plastics—A general term applied to a group of lightweight, transparent, and impact-resistant plastics, usually rigid copolymers of high acrylonitrile content. The barrier plastics are generally characterized by gas, aroma, and flavor barrier characteristics approaching those of metal and glass.

binder, n.—A component of an adhesive composition which is primarily responsible for the adhesive forces which hold two bodies together. (See **extender**; **filler**; **matrix**).

biscuit—See **cull** and **preform**

blanking—The cutting of flat sheet stock to shape by striking it sharply with a punch while it is supported on a mating die. Punch presses are used. Also called **die cutting**.

bleed—**(1)** To give up color when in contact with water or a solvent. **(2)** Undesired movement of certain materials in a plastic (e.g. plasticizers in vinyl) to the surface of the finished article or into an adjacent material. Also called "Migration." **(3)** An escape passage at the parting line of a mold, like a vent but deeper, which allows material to escape or bleed out.

blind hole—Hole that is not drilled entirely through.

blister, n.—Undesirable rounded elevation of the surface of a plastic, whose boundaries may be indefinitely outlined, somewhat resembling in shape a blister on the human skin. A blister may burst and become flattened.

blocking, n.—An adhesion between touching layers of plastic, such as that which may develop under pressure during storage or use.

bloom–(1) A non-continuous surface coating on plastic products that comes from ingredients such as plasticizers, lubricants, anti-static agents, etc., which are incorporated into the plastic resin. It is not always visible. "Bloom" is the result of ingredients coming out of "solution" in the plastic and migrating to the surface of the plastic. (2) Also used to describe an increase in diameter of the parison as it comes from the extruder die(s) in the blow molding process.

blow molding–A method of fabrication in which a warm plastic parison (hollow tube), is placed between the two halves of a mold (cavity) and forced to assume the shape of that mold cavity by use of air pressure. The air pressure is introduced through the inside of the parison and thereby forces the plastic against the surface of the mold that defines the shape of the product.

blow pin–Part of the tooling used to form hollow objects or containers by the blow molding process. It is a tubular tool through which air pressure is introduced into the parison to create the air pressure necessary to form the parison into the shape of the mold. In some blow molding systems, it is a part of, or an extension of, the core pin.

blow pressure–The air pressure required to form the parison into the shape of the mold cavity, in a blow molding operation.

blow rate–The speed or rate at which the air enters or the time required for air to enter the parison during the blow molding cycle.

blow-up ratio–In blow molding, the ratio of the diameter of the product (usually its greatest diameter) to the diameter of the parison from which the product is formed. In blown film extrusion, the ratio between the diameter of the final film tube and the diameter of the die orifice.

blown film extrusion–Technique for making film by extruding the plastic through a circular die, followed by expansion (by the pressure of internal air admitted through the center of the mandrel), cooling, and collapsing of the bubble.

blown tubing–A thermoplastic film which is produced by extruding a tube, applying a slight internal pressure to the tube to expand it while still molten and subsequent cooling to set the tube. The tube is then flattened through guides and wound up flat on rolls. The size of blown tubing is determined by the flat width in inches as wound rather than by the diameter as in the case of rigid types of tubing.

blueing–A mold blemish in the form of a blue oxide film on the polished surface of a mold due to abnormally high mold temperatures is termed blueing.

bolster–Space or filler in a mold.

boss–Projection on a plastic part designed to add strength, to facilitate alignment during assembly, to provide for fastenings, etc.

bottom blow–A specific type of blow molding technique which forms hollow articles by injecting the blowing air into the parison from the bottom of the mold (as opposed to introducing the blowing air at a container opening.)

bottom plate–Part of the mold which contains the heel radius and the push-up.

breakdown voltage–The voltage required, under specific conditions, to cause the failure of an insulating material. See **dielectric strength** and **arc resistance**

breaker plate–A perforated plate located at the rear end of an extruder or at the nozzle end of an injection cylinder. It often supports the screens that prevent foreign particles from entering the die, and is used to keep unplasticized material out of nozzle and to improve distribution of color particles.

breathing–The opening and closing of a mold to allow gases to escape early in the molding cycle. Also called degassing. When referring to plastic sheeting, "breathing" indicates permeability to air.

brinell hardness–Similar to Rockwell Hardness (q.v.).

bubble–A spherical, internal void, globule of air or other gas trapped within a plastic. See **void**.

bubbler–A device inserted into a mold force, cavity or core, which allows water to flow deep inside the hole into which it is inserted and to discharge through the open end of hole. Uniform cooling of the molds and of isolated mold sections can be achieved in this manner.

bulk density–The density of a molding material in loose form (granular, nodular, etc.) expressed as a ratio of weight to volume (e.g., g/cm^3 or lb/ft^3).

bulk factor—Ratio of the volume of loose molding powder to the volume of the same weight of resin after molding.

burned—Showing evidence of thermal decomposition through some discoloration, distortion, or localized destruction of the surface of the plastic.

burning rate—A term describing the tendency of plastics articles to burn at given temperatures.

burst strength—The internal pressure required to break a pipe or fitting. This pressure will vary with the rate of build-up of the pressure and the time during which the pressure is held.

butt fusion—A method of joining pipe, sheet, or other similar forms of a thermoplastic resin wherein the ends of the two pieces to be joined are heated to the molten state and then rapidly pressed together.

C-stage—The final stage in the reaction of certain thermosetting resins, in which the material is relatively insoluble and unfusible. The resin in a fully cured thermosetting molding is in this stage. Sometimes referred to as resite. (See **A-stage**; **B-stage**).

calender (v.)—To prepare sheets of material by pressure between two or more counter-rotating rolls. (n.)—The machine performing this operation.

carbon black—A black pigment produced by the incomplete burning of natural gas or oil. It is widely used as a filler, particularly in the rubber industry. Because it possesses useful ultraviolet protective properties, it is also much used in molding compounds intended for outside weathering applications.

case harden—To harden surface of a piece of steel to a relatively shallow depth.

cast film—A film made by depositing a layer of liquid plastic onto a surface and stabilizing this form by evaporation of solvent, by fusing after deposition, or by allowing a melt to cool. Cast films are usually made from solutions or dispersions.

cast—(1) To form a "plastic" object by pouring a fluid monomer-polymer solution into an open mold where it finishes polymerizing. (2) Forming plastic film and sheet by pouring the liquid resin onto a moving belt or by precipitation in a chemical bath.

casting (n.)—The finished product of a casting operation; should not be used for molding, q.v.

catalyst—A substance which markedly speeds up the cure of a compound when added in minor quantity as compared to the amounts of primary reactants. (See **hardener, inhibitor, promoter.**)

cavity—Depression in mold, which usually forms the outer surface of the molded part; depending on number of such depressions, molds are designated as a single cavity or multi-cavity.

cavity retainer plate—Plates in a mold which hold the cavities and forces. These plates are at the mold parting line and usually contain the guide pins and bushings. Also called "Force Retainer Plate."

cell—A single cavity formed by gaseous displacement in a plastic material. (See **cellular plastic.**)

cellular plastic—A plastic whose density is decreased substantially by the presence of numerous cells disposed throughout its mass. (See **cell; foamed plastic.**)

center gated mold—An injection or transfer mold wherein the cavity is filled with molding material through a sprue or gate directly into the center of the part.

centipose—A unit of viscosity, conveniently and approximately defined as the viscosity of water at room temperature.

centrifugal casting—A method of forming thermoplastic resins in which the granular resin is placed in a rotatable container, heated to a molten condition by the transfer of heat through the walls of the container, and rotated so that the centrifugal force induced will force the molten resin to conform to the configuration of the interior surface of the container. Used to fabricate large diameter pipes and similar cylindrical items.

chalking—Dry, chalk-like appearance or deposit on the surface of a plastic. See **haze** and **bloom.**

charge—The measurement or weight of material used to load a mold at one time or during one cycle.

chase—An enclosure of any shape, used to: (a) shrink-fit parts of a mold cavity in place; (b) prevent spreading or distortion in hobbing; (c) enclose an assembly of two or more parts of a split cavity block.

chill roll—A cored roll, usually temperature controlled with circulating water, which cools the web before winding. For chill roll (cast) film, the surface of the roll is highly polished. In extrusion coating, either a polished or a matte surface may be used depending on the surface desired on the finished coating.

chill roll extrusion (or cast film extrusion)—The extruded film is cooled while being drawn around two or more highly polished chill rolls cored for water cooling for exact temperature control.

chromium plating—An electrolytic process that deposits a hard film of chromium metal onto working surfaces of other metals where resistance to corrosion, abrasion, and/or erosion is needed.

C.I.L. (flow test)—A method of determining the rheology or flow properties of thermoplastic resins developed by Canadian Industries Limited. In this test, the amount of the molten resin which is forced through a specified size orifice per unit of time when a specified, variable force is applied gives a relative indication of the flow properties of various resins.

clamping—See injection molding machines.

clamping area—The largest rated molding area an injection or transfer press can hold closed under full molding pressure.

clamping plate—A plate fitted to a mold and used to fasten mold to a molding machine.

clamping force—In injection molding and in transfer molding, the pressure which is applied to the mold to keep it closed, in opposition to the fluid pressure of the compressed molding material, within the mold cavity (cavities) and the runner system. In blow molding, the pressure exerted on the two mold halves (by the locking mechanism of the blowing table) to keep the mold closed during formation of the container. Normally, this pressure or force is expressed in tons.

coefficient of expansion—The fractional change in dimension (sometimes volume) specified, of a material for a unit change in temperature. Values for plastics range from 0.01 to 0.2 mils/in., °C. Ref: Standard Method of Test for Coefficient of Linear Thermal Expansion of Plastics (ASTM Designation D 696).

co-extrusion—Process of combining two or more layers of extrudate to produce a multiple layer product in a single step.

cold-cure foams—See high-resiliency flexible foams.

cold drawing—Technique for using standard metalworking equipment and systems for forming thermoplastic sheet (e.g., ABS) at room temperature.

cold flow—See creep.

cold molding—A procedure in which a composition is shaped at room temperature and cured by subsequent baking.

cold parison blow molding—Technique in which parisons are extruded or injection molded separately and then stored for subsequent transportation to the blow molding machine for blowing.

cold runner molding—Sprue and runner system (the manifold section) is insulated from the rest of the mold and temperature-controlled to keep the plastic in the manifold fluid. This mold design eliminates scrap loss from sprues and runners.

cold slug—The first material to enter an injection mold; so called because in passing through sprue orifice it is cooled below the effective molding temperature.

cold slug well—Space provided directly opposite the sprue opening in an injection mold to trap the cold slug.

cold stretch—A pulling operation with little or no heat, usually on extruded filaments to increase tensile properties.

color concentrate—A measured amount of dye or pigment incorporated into a predetermined amount of plastic. This pigmented or colored plastic is then mixed into larger quantities of plastic material to be used for molding. The "concentrate" is added to the bulk of plastic in measured quantity in order to produce a precise, predetermined color of the finished articles to be molded.

combination mold—See family mold.

compound—The plastic material to be molded or blown into final form. Included are the resin itself, along with modifiers, pigments, antioxidants, lubricants, etc., needed to process the resin efficiently and to produce the desired properties in the finished article.

compression mold—A mold which is open when the material is introduced and which shapes the material by heat and by the pressure of closing.

compression molding–A technique of thermoset molding in which the molding compound (generally preheated) is placed in the heated open mold cavity, mold is closed under pressure (usually in a hydraulic press), causing the material to flow and completely fill the cavity, pressure being held until the material has cured.

compression ratio–In an extruder screw, the ratio of volume available in the first flight at the hopper to the last flight at the end of the screw.

compressive strength–Crushing load at the failure of a specimen divided by the original sectional area of the specimen. Ref: Tentative Method of Test for Compressive Properties of Rigid Plastics (ASTM Designation D 695).

concentrate–A measured amount of additive (e.g., dye, pigment, foaming agent, anti-static agent, flame retardant, glass reinforcement, etc.) is incorporated into a predetermined small amount of plastic. This "concentrate" can then be mixed into larger quantities of plastic to achieve a desired color or end-property.

condensation–A chemical reaction in which two or more molecules combine with the separation of water or some other simple substance. If a polymer is formed, the condensation process is called Polycondensation. (See also **polymerization**.)

condensation resin–A resin formed by polycondensation, e.g., the alkyd, phenol-aldehyde, and urea formaldehyde resins.

conditioning–The subjection of a material to a stipulated treatment so that it will respond in a uniform way to subsequent testing or processing. The term is frequently used to refer to the treatment given to specimens before testing. (Standard ASTM test methods that include requirements for conditioning are indexed in Index of ASTM standards.)

cooling channels–Channels or passageways located within the body of a mold through which a cooling medium can be circulated to control temperature on the mold surface. May also be used for heating a mold by circulating steam, hot oil or other heated fluid through channels as in molding of the thermosetting and some thermoplastic materials.

cooling fixture–Block of metal or wood holding the shape of a molded piece which is used to maintain the proper shape or dimensional accuracy of a molding after it is removed from the mold until it is cool enough to retain its shape without further appreciable distortion. Also called: "Shrink Fixture."

copolymer–See **polymer**.

core–(1) Male element in die which produces a hole or recess in part. (2) Part of a complex mold that molds undercut parts. Cores are usually withdrawn to one side before the main sections of the mold open. (Usually called Side Cores). (3) A channel in a mold for circulation of a heat-transfer medium. Also called "Force."

core pin–Pin used to mold a hole.

core pin plate–Plate holding core pins.

coring–The removal of excess material from the cross section of a molded part to attain a more uniform wall thickness.

cratering–Depressions on coated plastic surfaces which are caused by excessive lubricant. Cratering results when paint thins excessively and later ruptures leaving pin holes and other voids. Use of less thinners in the coating can reduce or eliminate cratering as can less lubricant on the part.

crazing–Fine cracks which may extend in a network on or under the surface or through a layer of plastic material.

creep–The dimensional change with time of a material under load, following the initial instantaneous elastic deformation. Creep at room temperature is sometimes called "Cold Flow." Ref: Recommended Practices for Testing Long-Time Creep and Stress-Relaxation of Plastics Under Tension or Compression Loads at Various Temperatures (ASTM Designation: D 674).

crystallinity–A state of molecular structure in some resins which denotes uniformity and compactness of the molecular chains forming the polymer. Normally can be attributed to the formation of solid crystals having a definite geometric form.

cull–Material remaining in a transfer chamber after mold has been filled. Unless there is a slight excess in the charge, the operator cannot be sure cavity is filled. Charge is generally regulated to control thickness of cull.

cure–To change the physical properties of a

material by chemical reaction, which may be condensation, polymerization, or vulcanization; usually accomplished by the action of heat and catalysts, alone or in combination, with or without pressure.

curing temperature—Temperature at which a cast, molded, or extruded product, a resin-impregnated reinforcing material, an adhesive, etc., is subjected to curing.

curing time—In the molding of plastics, the interval of time between the instant of cessation of relative movement between the moving parts of a mold and the instant that pressure is released. Also called "Molding Time."

curtain coating—A method of coating which may be employed with low viscosity resins or solutions, suspensions, or emulsions of resins in which the substrate to be coated is passed through and perpendicular to a freely falling liquid "curtain" (or "waterfall"). The flow rate of the falling liquid and the linear speed of the substrate passing through the curtain are co-ordinated in accordance with the thickness of coating desired.

cut-off—The line where the two halves of a mold come together. Also called "Flash Groove" and "Pinch-Off."

cycle—The complete, repeating sequence of operations in a process or part of a process. In molding, the cycle time is the period, or elapsed time, between a certain point in one cycle and the same point in the next.

daylight—See **injection molding machines.**

debossed—An indented or depressed design or lettering that is molded into an article so as to be below the main outside surface of that article.

deflashing—Any technique or method which removes excess, unwanted material from a molded article. Specifically, the excess material is removed from those places on the article where parting lines of the mold that formed the article may have caused the excess material to be formed.

deflection temperature—The temperature at which a specimen will deflect a given distance at a given load under prescribed conditions of test. See ASTM D 648. Formerly called heat distortion.

degassing—See **breathing.**

degradation—A deleterious change in the chemical structure of a plastic. See **deterioration.**

delamination—The splitting of a plastic material along the plane of its layers. Physical separation or loss of bond between laminate plies. See **laminate.**

deliquescent—Capable of attracting moisture from the air.

density—Weight per unit volume of a substance, expressed in grams per cubic centimeter, pounds per cubic foot, etc.

desiccant—Substance which can be used for drying purposes because of its affinity for water.

destaticization—Treatment of plastic materials which minimizes the effects of static electricity on the surface of the articles. This treatment can be accomplished either by treating the surface with specific materials, or incorporating materials in the molding compound. Minimization of surface static electricity prevents dust and dirt from being attracted to and/or clinging to the surface of the article.

deterioration—A permanent change in the physical properties of a plastic evidenced by impairment of these properties.

diaphragm gate—Gate used in molding annular or tubular articles. Gate forms a solid web across the opening of the part.

die adaptor—That part of an extrusion die which holds the die block.

die cutting—(1) Blanking (q.v.); (2) Cutting shapes from sheet stock by striking it sharply with a shaped knife edge known as a "steel-rule die." Clicking and Dinking are other names for die cutting of this kind.

dielectric constant—Normally the relative dielectric constant; for practical purposes, the ratio of the capacitance of an assembly of two electrodes separated solely by a plastics insulating material to its capacitance when the electrodes are separated by air. (ASTM Designation: D 150.)

dielectric heating (electronics heating or R.F. heating) —The plastic to be heated forms the dielectric of a condenser to which is applied a high-frequency (20 to 80 mc.) voltage. Dielectric loss in the material is the basis. Process used for sealing vinyl films and preheating thermoset molding compounds.

dielectric strength—The electric voltage gradient at which an insulating material is broken down or "arced through," in volts per mil of thickness. Ref. Standard Methods of Test for Dielectric Breakdown Voltage and Dielectric Strength of Electrical Insulating Materials at Commercial Power Frequencies (ASTM Designation: D 149).

diffusion—The movement of a material, such as a gas or liquid, in the body of a plastic. If the gas or liquid is absorbed on one side of a piece of plastic and given off on the other side, the phenomenon is called permeability. Diffusion and permeability are not due to holes or pores in the plastic but are caused and controlled by chemical mechanisms.

diluent—In an organosol, a liquid component which has little or no solvating action on the resin, its purpose being to modify the action of the dispersant.

dimensional stability—Ability of a plastic part to retain the precise shape in which it was molded, fabricated, or cast.

dip coating—Applying a plastic coating by dipping the article to be coated into a tank of melted resin or plastisol, then chilling the adhering melt.

disc gate—See **diaphragm gate.**

discoloration—Any change from the original color, often caused by overheating, light exposure, irradiation, or chemical attack.

dished—Showing a symmetrical distortion of a flat or curved section of a plastic object, so that, as normally viewed, it appears concave, or more concave than intended. See **warp.**

dispersion—Finely divided particles of a material in suspension in another substance.

dissipation factor—See **power factor.**

doctor roll; doctor bar—A device for regulating the amount of liquid material on the rollers of a spreader.

domes—Showing a symmetrical distortion of a flat or curved section of a plastic object, so that, as normally viewed, it appears convex, or more convex than intended. See **warp.**

double-shot molding—Means of producing two-color parts and/or two different thermoplastic materials by successive molding operations.

dowel—Pin used to maintain alignment between two or more parts of a mold.

draft—The degree of taper of a side wall or the angle of clearance designed to facilitate removal of parts from a mold.

drape forming—Method of forming thermoplastic sheet in which the sheet is clamped into a movable frame, heated, and draped over high points of a male mold. Vacuum is then pulled to complete the forming operation.

draw down ratio—The ratio of the thickness of the die opening to the final thickness of the product.

drawing—Drawing is the process of stretching a thermoplastic to reduce its cross sectional area thus creating a more orderly arrangement of polymer chains with respect to each other.

dry blend—Refers to a molding compound, containing all necessary ingredients mixed in a way that produces a dry-free-flowing, particulate material. This term is commonly used in connection with polyvinyl chloride molding compounds.

dry coloring—Method commonly used by fabricators for coloring plastic by tumble blending uncolored particles of the plastic material with selected dyes and pigments.

dry strength—The strength of an adhesive joint determined immediately after drying under specified conditions or after a period of conditioning in the standard laboratory atmosphere. See **wet strength.**

duplicate cavity plate—Removable plate that retains cavities, used where two-plate operation is necessary for loading inserts, etc.

durometer hardness—The hardness of a material as measured by the Shore Durometer. Ref: Tentative Method of Test for Indentation Hardness of Rubber and Plastics by Means of a Durometer (ASTM Designation: D 2240).

dwell—A pause in the application of pressure to a mold, made just before the mold is completely closed, to allow the escape of gas from the molding material.

dyes—Synthetic or natural organic chemicals that are soluble in most common solvents. Characterized by good transparency, high tinctorial strength, and low specific gravity.

ejection mark—A surface mark on the part caused by the ejector pin as it pushes the part out of the mold cavity.

ejector pin (Or ejector sleeve)—A rod, pin or sleeve which pushes a molding off a force or out of a cavity of a mold. It is attached to an ejector bar or plate which can be actuated by the ejector rod(s) of the press or by auxiliary hydraulic or air cylinders.

ejector pin retainer plate—Retainer into which ejector pins are assembled.

ejector return pins—Projections that push the ejector assembly back as the mold closes; also called safety pin, and position pushbacks.

ejector rod and bar—Bar that actuates the ejector assembly when mold is opened.

elastic deformation—The part of the deformation of an object under load which is recoverable when the load is removed.

elasticity—That property of a material by virtue of which it tends to recover its original size and shape after deformation. If the strain is proportional to the applied stress, the material is said to exhibit Hookean or ideal elasticity.

elastomer—A material which at room temperature stretches under low stress to at least twice its length and snaps back to the original length upon release of stress. See **rubber**.

electric discharge machining (EDM)—A metal working process applicable to mold construction in which controlled sparking is used to erode away the work piece.

electronic treating—A method of oxidizing a polyolefin part or film to render it printable by passing it between electrodes and subjecting it to a high voltage corona discharge.

electroformed molds—A mold made by electroplating metal on the reverse pattern on the cavity. Molten steel may be then sprayed on the back of the mold to increase its strength.

electroplating—Deposition of metals on certain plastics and molds for finish.

elongation—The fractional increase in length of a material stressed in tension.

embossing—Techniques used to create depressions of a specific pattern in plastics film and sheeting. Such embossing in the form of surface patterns can be achieved on molded parts by the treatment of the mold surface by photoengraving or other process.

encapsulating—Enclosing an article (usually an electronic component or the like) in a closed envelope of plastic, by immersing the object in a casting resin and allowing the resin to polymerize or, if hot, to cool. See **potting**.

environmental stress cracking (ESC)—The susceptibility of a thermoplastic article to crack or craze formation under the influence of certain chemicals, or aging, or weather, or stress.

exotherm—(1) The temperature/time curve of a chemical reaction giving off heat, particularly the polymerization of casting resins. (2) The amount of heat given off. The term has not been standardized with respect to sample size, ambient temperature, degree of mixing, etc.

expandable plastic—A plastic which can be made cellular by thermal, chemical or mechanical means.

extender—A substance, generally having some adhesive action, added to a plastic composition to reduce the amount of the primary resin required per unit area.

extrudate—The product or material delivered by an extruder, such as film, pipe, the coating on wire, etc.

extrusion coating—The resin is coated on a substrate by extruding a thin film of molten resin and pressing it onto or into the substrates, or both, without the use of an adhesive.

extrusion—The compacting of a plastic material and the forcing of it through an orifice in more or less continuous fashion.

extruder—Basically, a machine (Fig. 1-1) which accepts solid particles (pellets or powder) or liquid (molten) feed, conveys it through a surrounding barrel by means of a rotating screw and pumps it, under pressure, through an orifice. Nomenclature used applies to the barrel, the screw, and other extruder elements.

> **barrel**—A cylindrical housing in which the screw rotates, including a replaceable liner, if used, or an integrally formed special surface material. Nomenclature covers:
>
> > **feed openings**: A hole through the feed section for introduction of the feed material into the barrel. Variations (see Fig. 1-2) include vertical, tangential, side feed openings.
> >
> > **feed section**: A separate section, located at the upstream end of the barrel, which contains the feed opening into the barrel.
> >
> > **grooved liner** (or barrel): A liner whose bore is provided with longitudinal grooves.

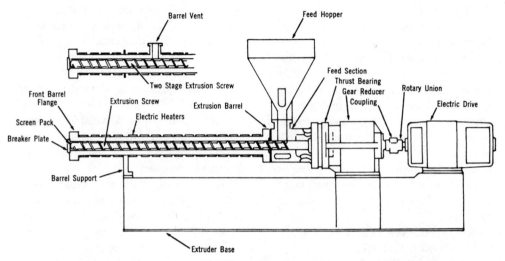

Fig. 1-1. Typical extruder cross-section. (All illustrations on extrusion, *courtesy SPI Machinery Division*).

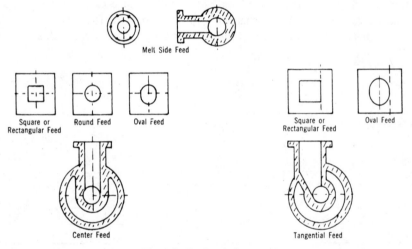

Fig. 1-2. Feed variations.

barrel heaters: The electrical resistance or induction heaters mounted on or around the barrel.

barrel jacket: A jacket surrounding the outside of the barrel for circulation of a heat transfer medium.

barrel heating zone: A portion of barrel length having independent temperature control.

barrel vent: An opening through the barrel wall, intermediate in the extrusion process, to permit the removal of air and volatile matter from the material being processed.

screw—A helically flighted screw which when rotated within the barrel mechanically works and advances the material being processed. Types of screws include:

    constant lead screw, uniform pitch screw: A screw with a flight of constant helix angle.

    constant taper screw: A screw of constant lead and a uniformly increasing root diameter over the full flighted length.

    cored screw: A screw with a hole in the center of the root for circulation of a heat transfer medium, or installation of a heater.

decreasing lead screw: A screw in which the lead decreases over the full flighted length (usually of constant depth).

metering type screw: (Fig. 1-3): A screw which has a metering section.

multiple-flighted screw: A screw having more than one helical flight such as: double flighted, double lead, double thread, or two starts, and triple flighted, etc.

multiple-stage screw (Fig. 1-4): A two or more stage screw with introduction of special mixing sections, choke rings or torpedoes.

single-flighted screw (Fig. 1-3): A screw having a single helical flight.

two-stage screw (Fig. 1-5): A screw constructed with an initial feed section followed by a restriction section, and then an increase in the flight channel volume to release the pressure on the material while carrying it forward, such as a screw used for venting at an intermediate point in the extruder.

vented screw (special): A two-stage screw with a screw vent in the second stage. (For internal venting through core of screw.)

water-cooled screw: A cored screw suitable for the circulation of cooling water.

Nomenclature applicable to screw design and construction includes:

screw channel: With the screw in the barrel, the space bounded by the surfaces of flights, the root of the screw and the bore of the barrel. This is the space through which the stock is conveyed and pumped.

feed section of screw: The portion of a screw which picks up the material at the feed opening (throat) plus an additional portion downstream. Many screws have

A —Feed Section Flight Depth
B —Metering Section Flight Depth
C —Lead
Ø —Helix Angle (Pitch)
D —Screw Outside Diameter

L —Effective Screw Length
OA—Screw Overall Length
R —Screw Root Diameter
W —Screw Flight Width

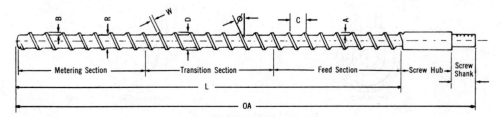

Fig. 1-3. Single flight, single-stage extrusion screw.

A —Feed Section Flight Depth
B —Metering Section Flight Depth
C —Feed Section Flight Depth
D —Metering Section Flight Depth
E —Lead
Ø —Helix Angle (Pitch)

F —Screw Outside Diameter
L —Effective Screw Length
OA —Screw Overall Length
R —Screw Root Diameter
W —Screw Flight Width

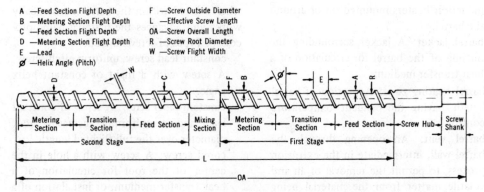

Fig. 1-4. Single flight, two-stage extrusion screw with mixing section.

A —Feed Section Flight Depth
B —Metering Section Flight Depth
C —Feed Section Flight Depth
D —Metering Section Flight Depth
E —Lead
∅ —Helix Angle (Pitch)
F —Screw Outside Diameter
L —Effective Screw Length
OA —Screw Overall Length
R —Screw Root Diameter
W —Screw Flight Width

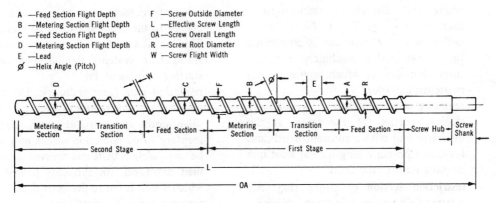

Fig. 1-5. Single flight, two-stage extrusion screw.

an initial constant lead and depth section, all of which is considered the feed section.

screw flight: The helical metal thread of the screw.

flight land: The surface at the radial extremity of the flight constituting the periphery or outside diameter of the screw.

hardened flight land: A screw flight having its periphery harder than the base metal by flame hardening, induction hardening, depositing of hard facing metal, etc.

helix angle: The angle of the flight at its periphery relative to a plane perpendicular to the screw axis.

screw hub: The portion immediately behind the flight which prevents the escape of the material.

length to diameter ratio (L/D ratio) (Fig. 1-6): The length to diameter ratio (L/D) can be expressed in two ways: (A) L/D ratio based on the barrel length—the distance from the forward edge of the feed opening to the forward end of the barrel bore divided by the bore diameter and expressed as a ratio wherein the diameter is reduced to 1 such as 20 : 1 or 24 : 1; and (B) L/D ratio based on the screw flighted length—The distance from the rear edge of the feed opening to the forward end of the barrel bore divided by the bore diameter and expressed as a ratio

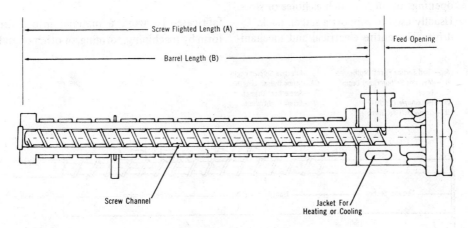

Use A for L/D ratio based on barrel length
Use B for L/D ratio based on screw flighted length

Fig. 1-6. Barrel cross-section.

wherein the diameter is reduced to 1, such as 20 : 1 or 24 : 1. Note: Either definition of L/D ratio can be considered to be correct, but machinery manufacturer should state which applies to his particular extruder.

metering section of screw: A relatively shallow portion of the screw at the discharge end with a constant depth and lead, and having a length of at least one or more turns of the flight.

restriction section or choke ring: An intermediate portion of a screw offering a resistance to the forward flow of material.

torpedo (Fig. 1-7): An unflighted cylindrical portion of the screw usually located at the discharge end but can be located in other sections, particularly in multiple stage screws.

transition section of screw (compression section) (Fig. 1-3): The portion of a screw between the feed section and metering section in which the flight depth decreases in the direction of discharge.

screw vent: A vent opening through the stem of the screw into the core for purpose of venting back through the core of the screw.

Nomenclature applicable to other extruder parts includes:

breaker plate: A metal plate installed across the flow of the stock between the end of the screw and the die with openings through it such as holes or slots. Usually used to support a screen pack.

drive: The entire electrical and mechanical system used to supply mechanical energy to the input shaft of gear reducer. This includes motor, constant or variable speed belt system, flexible couplings, starting equipment, etc.

reduction gear (gear reducer): The gear device used to reduce speed between drive motor and extruder screw. Supplementary speed reduction means may also be used, such as belts and sheaves, etc.

melt extrusion: An extrusion process in which a melt is fed to the extruder.

screens: A woven metal screen or equivalent device which is installed across the flow of stock between the tip of the screw and the die and supported by a breaker plate to strain out contaminants or to increase the back pressure or both.

stock or melt temperature: The temperature of the stock or melt as sensed by stock or melt thermocouple and indicated on a compatible meter. The location of measurement must be specified.

surging: A pronounced fluctuation in output over a short period of time without deliberate change in operating conditions.

thrust bearing: The bearing used to absorb the thrust force exerted by the screw.

thrust: The total axial force exerted by the screw on the thrust bearing. (For practical purposes equal to the extrusion pressure times the cross section area of the bore.)

fabricate—To work a material into a finished form by machining, forming, or other operation.

A —Feed Section Flight Depth
B —Metering Section Flight Depth
C —Lead
∅ —Helix Angle (Pitch)
D —Screw Outside Diameter

L —Effective Screw Length
OA—Screw Overall Length
R —Screw Root Diameter
W —Screw Flight Width

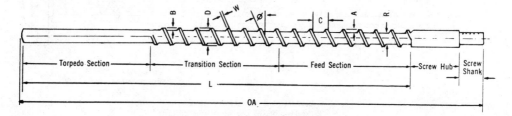

Fig. 1-7. Single flight, single-stage, with torpedo, extrusion screw.

fadeometer—An apparatus for determining the resistance of resins and other materials to fading.

family mold—A multi-cavity mold wherein each of the cavities forms one of the component parts of the assembled finished object. The term often applied to molds wherein parts from different customers are grouped together in one mold for economy of production. Sometimes called "combination mold."

fan gate—A shallow gate somewhat wider than the runner from which it extends.

female—In molding practice, the indented half of a mold designed to receive the male half.

fiber—A reinforcing filler characterized by a high ratio of length to diameter (e.g., glass, asbestos, synthetic, metal, etc.). In practice, this term usually refers to thin fibers of glass which are used to reinforce both thermoplastic and thermosetting materials. One-inch long fibers are occasionally used, but the more commonly used fiber lengths are $\frac{1}{2}''$ and $\frac{1}{4}''$, or less.

filament winding—Roving or single strands of glass, metal, or other reinforcement are wound in a predetermined pattern onto a suitable mandrel. The pattern is so designed as to give maximum strength in the directions required. The strands can either be run from a creel through a resin bath before winding or pre-impregnated materials can be used. When the right number of layers have been applied, the wound mandrel is cured at room temperatures or in an oven.

fill-and-wipe—Parts are molded with depressed designs; after application of paint, surplus is wiped off, leaving paint remaining only in depressed areas. Sometimes called "wipe-in."

filler—A cheap, inert substance added to a plastic to make it less costly. However, fillers may also improve physical properties, particularly hardness, stiffness, dimensional stability and impact strength. The particles are usually small, in contrast to those of reinforcements (q.v.); but there is some overlap between the functions of the two.

fillet—A rounded filling of the internal angle between two surfaces.

film—Sheeting having a nominal thickness not greater than 0.010 inch.

fin—Excess material left on a molded object at those places where the molds or dies mated. Also, the web of material remaining in holes or openings in a molded part which must be removed in finishing.

fines—Very small particles (usually under 200 mesh) accompanying larger grains, usually of molding powder.

finish—To complete the secondary work on a molded part so that it is ready for use. Operations such as filing, deflashing, buffing, drilling, tapping, degating are commonly called finishing operations. See **surface finish**.

fisheye—Small globular mass which has not blended completely into the surrounding material; particularly evident in a transparent or translucent material.

fixture—Means of holding a part during a machining or other operation.

flake—Used to denote the dry, unplasticized base of plastics.

flight—See **extruder**.

flame retardant resin—A resin which is compounded with certain chemicals to change its burning characteristics.

flame spraying—Method of applying a plastic coating in which finely powdered fragments of the plastic, together with suitable fluxes, are projected through a cone of flame onto a surface.

flame treating—A method of rendering inert thermoplastic objects receptive to inks, lacquers, paints, adhesives, etc., in which the object is bathed in an open flame to promote oxidation of the surface of the article.

flash—Extra plastic attached to a molding along the parting line; under most conditions it would be objectionable and must be removed before the parts are acceptable.

flash gate—Usually a long gate extending from a runner which runs parallel to an edge of a molded part along the flash or parting line of the mold.

flash line—A raised line appearing on the surface of a molding and formed at the junction of mold faces. See **parting line**.

flash mold—A mold in which the mold faces are perpendicular to the clamping action of the press, so that the higher the clamping force, the tighter the mold seam.

flexibilizer—An additive that makes a resin or rubber more flexible, i.e., less stiff. Also a plasticizer.

flexible molds—Molds made of rubber or elastomeric plastics used for casting plastics. They can be stretched to remove cured pieces with undercuts.

flexural modulus, psi—Flexural modulus is the ratio of stress to strain for a given material within its proportional limit under bending load conditions. (ASTM test method D 790).

flexural strength—Ability of a material to flex without permanent distortion or breaking. Ref: Standard Method of Test for Flexural Properties of Plastics (ASTM Designation: D 790).

floating chase—Mold member, free to move vertically, which fits over a lower plug or cavity, and into which an upper plug telescopes.

flow—A qualitative description of the fluidity of a plastic material during the process of molding.

flow line—A mark on a molded piece made by the meeting of two flow fronts during molding. Also called "weld line."

flow marks—Wavy surface appearance on a molded object caused by improper flow of the material into the mold. See **splay marks**.

flow molding—Technique of producing leather-like materials by placing a die-cut plastic blank (either solid or expanded vinyl or vinyl-coated substrate) in a mold cavity (usually silicone rubber molds) and applying power via a high-frequency RF generator to melt the plastic so that it flows into the mold to the desired shape and with the desired texture.

fluidized bed coating—A method of applying a coating of a thermoplastic resin to an article in which the heated article is immersed in a dense-phase fluidized bed of powdered resin and thereafter heated in an oven to provide a smooth, pin-hole-free coating.

foamed plastics—Plastics with internal voids or cells. The foam may be flexible or rigid, the cells closed or connected, the density anything from slightly below that of the solid parent resin down to, in some cases, 2 lb./cu. foot, or less.

foaming agents—Chemicals added to plastics and rubbers that generate inert gases on heating, causing the resin to assume a cellular structure.

foam-in-place—Refers to the deposition of foams which requires that the foaming machine be brought to the work which is "in place" as opposed to bringing the work to the foaming machine.

foil decorating—Molding paper, textile, or plastic foils printed with compatible inks directly into a plastic part so that the foil is visible below the surface of the part as integral decoration.

force plate—The plate that carries the plunger or force plug of a mold and guide pins or bushings.

force—That portion of the mold which forms the inside of the molded part. Sometimes called a core or a plunger.

forging—See **solid phase forming**.

form-and-spray—Technique for thermoforming plastic sheet into an end-product and then backing up the sheet with spray-up reinforced plastics.

forming—A process in which the shape of plastic pieces such as sheets, rods, or tubes is changed to a desired configuration. See also **thermoforming**. Note.—The use of the term "forming" in plastics technology does not include such operations as molding, casting, or extrusion, in which shapes or pieces are made from molding materials or liquids.

friction welding—A method of welding thermoplastics materials whereby the heat necessary to soften the components is provided by friction. See **spin welding** and **vibration welding**.

frost line—In the extrusion of polyethylene lay-flat film, a ring-shaped zone located at the point where the film reaches its final diameter. This zone is characterized by a "frosty" appearance to the film caused by the film temperature falling below the softening range of the resin.

frothing—Technique for applying urethane foam in which blowing agents or tiny air bubbles are introduced under pressure into the liquid mixture of foam ingredients.

fusion—In vinyl dispersions, the heating of a dispersion to produce a homogeneous mixture.

gamma transition; glassy transition—The change in an amorphous polymer or in amorphous regions of a partially crystalline polymer from (or to) a viscous or rubbery condition to (or from) a hard and relatively brittle one. This transition generally occurs over a relatively narrow temperature region and is similar to the solidification of a liquid to a glassy state; it is not a phase transition. Not only do hardness and brittleness undergo rapid changes in this temperature region, but other properties such as thermal expansibility and specific heat also change rapidly. This phenomenon has been called second-order transition, glass transition, rubber transition, and rubbery transition. The word transformation has also been used instead of transition.

gamma-transition temperature; glassy-transition temperature—The temperature region in which the gamma or glassy transition occurs. (See **gamma** or **glassy** transition.) The measured value of gamma- or glassy-transition temperature depends to some extent on the details of the method of test.

gate—The short, usually restricted section of the runner at the entrance to the cavity of an injection or transfer mold.

gate mark—A gate mark is a surface discontinuity on the part caused by the presence of the mold orifice through which material enters the cavity.

gel—(1) A semisolid system consisting of a network of solid aggregates, in which liquid is held. (2) The initial jelly-like solid phase which develops during the formation of a resin from a liquid. Both types of gel have very low strengths and do not flow like a liquid. They are soft and flexible, and will rupture under their own weight unless supported externally.

gel point—The stage at which a liquid begins to exhibit pseudo-elastic properties. This stage may be conveniently observed from the inflection point on a viscosity-time plot. (See **gel**.)

gelatin—(1) Formation of a gel. (2) In vinyl dispersions, formation of gel in the early stages of fusion.

glass finish—A material applied to the surface of a glass reinforcement to improve its effect upon the physical properties of the reinforced plastic.

glassy transition—See **gamma transition; glassy transition**.

gloss—The shine or lustre of the surface of a material.

granular structure—Nonuniform appearance of

finished plastic material due to retention of, or incomplete fusion of, particles of composition, either within the mass or on the surface.

grinding-type resin—A vinyl resin which requires grinding to effect dispersion in plastisols or organosols.

grit blasted—A surface treatment of a mold in which steel grit or sand materials are blown on the walls of the cavity to produce a roughened surface. Air escape from mold is improved and special appearance of molded article is often obtained by this method.

guide pins—Devices that maintain proper alignment of force plug and cavity as mold closes. Also called "leader pins."

guide-pin bushing—A guiding bushing through which the leader pin moves.

grid—Channel-shaped mold-supporting members.

halocarbon plastics—Plastics based on resins made by the polymerization of monomers composed only of carbon and a halogen or halogens.

hardener—A substance or mixture of substances added to a resin or adhesive to promote or control the curing reaction by taking part in it. The term is also used to designate a substance added to control the degree of hardness of the cured film. (See **catalyst**.)

hardness—The resistance of a plastic material to compression and indentation. Among the most important methods of testing this property are Brinell hardness, Rockwell hardness and Shore hardness, q.v.

haze—Indefinite cloudy appearance within or on the surface of a plastic, not describable by the terms chalking or bloom.

head block; retainer board—A thick (3 to 5 in.) large piece of laminated lumber, usually with veneer crossings, used for bottom and top of a bale of plywood, during pressing and clamping.

heat-distortion point—The temperature at which a standard test bar deflects 0.010 in. under a stated load of either 66 or 264 psi. Ref: Standard Method of Test for Deflection Temperature of Plastics Under Load. (ASTM Designation: D 648).

heat forming—See **thermoforming**.

heat-treat—Term used to cover annealing, hardening, tempering, etc., of metals.

heater-adapter—That part of an extrusion die around which heating medium is held.

heating chamber—In injection molding, that part of the machine in which the cold feed is reduced to a hot melt. Also called "heating cylinder."

heat-sealing—A method of joining plastic films by simultaneous application of heat and pressure to areas in contact. Heat may be supplied conductively or dielectrically.

helix—See **extruder**.

high-frequency heating—The heating of materials by dielectric loss in a high-frequency electrostatic field. The material is exposed between electrodes, and by absorption of energy from the electrical field is heated quickly and uniformly throughout.

high-pressure laminates—Laminates molded and cured at pressures not lower than 1000 psi and more commonly in the range of 1200 to 2000 psi.

high-resiliency flexible foams—Urethane foams that offer low hysteresis and modulus, can be made more flame retardant, and process at lower oven temperatures and with shorter molding cycles. In trade parlance, these foams have a sag factor of 2.7 and above (i.e., better cushioning). Can be produced "cold cure" (no additional heat over that supplied by the exothermic reaction of the foaming process) or with heated molds and heat cures.

hob—A master model used to sink the shape of a mold into a soft steel block.

hobbing—A process of forming a mold by forcing a hob of the shape desired into a soft steel blank.

hold-down groove—A small groove cut into the side wall of the molding surface to assist in holding the molded article in that member while the mold opens.

holder block—See **chase**.

honeycomb—Manufactured product consisting of sheet metal or a resin impregnated sheet material (paper, fibrous glass, etc.) which has been formed into hexagonal-shaped cells. Used as core material for sandwich constructions.

hopper—Feed reservoir into which molding powder is loaded and from which it falls into a molding machine or extruder, sometimes through a metering device.

hopper dryer—A combination feeding and drying device for extrusion and injection molding of thermoplastics. Hot air flows upward through the hopper containing the feed pellets.

hopper loader—A curved pipe through which molding powders are pneumatically conveyed from shipping drums to machine hoppers.

hot gas welding—A technique of joining thermoplastic materials (usually sheet) whereby the materials are softened by a jet of hot air from a welding torch, and joined together at the softened points. Generally a thin rod of the same material is used to fill and consolidate the gap.

hot heated manifold mold—A thermoplastic injection mold in which the portion of the mold which contains the runner system has its own heating elements which keep the molding material in a plastic state ready for injection into the cavities, from which the manifold is insulated.

hot-runner mold—A thermoplastic injection mold in which the runners are insulated from the chilled cavities and remain hot so that the center of the runner never cools in normal cycle operation. Runners are not, as is the case usually, ejected with the molded pieces. Sometimes called "insulated runner mold."

hot-stamping—Engraving operation for marking plastics in which roll leaf is stamped with heated metal dies onto the face of the plastics.

impact, Izod—A specific type of impact test made with a pendulum type machine. The specimens are molded or extruded with a machined notch in the center.

impact resistance—Relative susceptibility of plastics to fracture by shock, e.g., as indicated by the energy expended by a standard pendulum-type impact machine in breaking a standard specimen in one blow.

impact strength—(1) The ability of a material to withstand shock loading. (2) The work done in fracturing, under shock loading, a specified test specimen in a specified manner. (3) Molded plastics are usually given a value on an Izod scale. An Izod impact test is designed to determine the resistance of a plastics material to a shock loading. It involves the notching of a specimen, which is then placed in the jaws of the machine and struck with a weighted pendulum. Ref: Standard Methods of Test for Impact Resistance of Plastics and Electrical Insulating Materials (ASTM Designation: D 256).

impact, tup—A falling weight (tup) impact test developed specifically for pipe and fittings. There are several variables that can be selected.

impregnation—The process of thoroughly soaking a material such as wood, paper or fabric, with a synthetic resin so that the resin gets within the body of the material. The process is usually carried out in an impregnator.

impulse sealing—A heat sealing technique in which a pulse of intense thermal energy is applied to the sealing area for a very short time, followed immediately by cooling. It is usually accomplished by using an RF heated metal bar which is cored for water cooling or is of such a mass that it will cool rapidly at ambient temperatures.

inhibitor—A substance which prevents or retards a chemical reaction.

injection blow molding—A blow molding process in which the parison to be blown is formed by injection molding.

injection mold—A mold into which a plasticated material is introduced from an exterior heating cylinder.

injection molding—A molding procedure whereby a heat-softened plastic material is forced from a cylinder into a cavity which gives the article the desired shape. Used with both thermoplastic and thermosetting materials.

injection molding machine—There are three essential parts to an injection molding machine: the mold clamping device, the mold (see **mold**), and the injection unit. Accepted terminology is as follows:

> **Clamping system terminology:**
>
> **clamping unit**: That portion of an injection molding machine in which the mold is mounted, and which provides the motion and force to open and close the mold and to hold the mold closed during injection. When the mold is closed in a horizontal direction, the clamp is referred to as a **horizontal clamp**. When closed in a

vertical direction, the clamp is referred to as a **vertical clamp**. This unit can also provide other features necessary for the effective functioning of the molding operation.

daylight, open—Fig. 1-8: The maximum distance that can be obtained between the stationary platen and the moving platen when the actuating mechanism is fully retracted without ejector box and/or spacers.

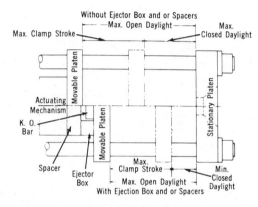

Fig. 1-8. Clamp die space nomenclature. (All illustrations on injection molding, *courtesy SPI Machinery Division*).

daylight, closed or minimum mold thickness—Fig. 1-8: The distance between the stationary platen and the moving platen when the actuating mechanism is fully extended, with or without ejector box and/or spacers. Minimum mold thickness will vary, depending upon the size and kind of ejector boxes and/or spacers used.

daylight, maximum closed—Fig. 1-8: That distance between the stationary platen and the moving platen when the actuating mechanism is fully extended without ejector box and/or spacers.

daylight, minimum closed—Fig. 1-8: That distance between the stationary platen and the moving platen when the actuating mechanism is fully extended with standard ejector box and/or spacers.

ejector (knockout): A provision in the clamping unit which actuates a mechanism within the mold to eject the molded part(s) from the mold. The ejection actuating force may be applied hydraulically or pneumatically by a cylinder(s) attached to the moving platen or mechanically by the opening stroke of the moving platen.

full hydraulic clamp—Fig. 1-9: A clamping unit actuated by a hydraulic cylinder which is directly connected to the moving platen. Direct fluid pressure is used to open and close the mold, and to provide the clamping force to hold the mold closed during injection.

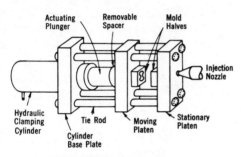

Fig. 1-9. Hydraulic clamp.

moving platen—Figs. 1-9 and 1-10: That member of the clamping unit which is moved toward a stationary member. The moving section of the mold is bolted to this moving platen. This member usually includes the ejector (knockout) holes and mold mounting pattern of bolt holes or "T" slots. A standard pattern is recommended by SPI Standards Testing Method (Injection Machinery Division Standards, September 11, 1958).

stationary platen—Figs. 1-9 and 1-10: The fixed member of the clamping unit on which the stationary section of the mold is bolted. This member usually includes a mold mounting pattern of bolt holes or "T" slots. A standard pattern is recommended by SPI Standards Testing Method (Injection Machinery Division Standards, September 11, 1958). In addition, the stationary platen usually includes provision for locating the mold on the platen and aligning the sprue bushing of the mold with the nozzle of the injection unit.

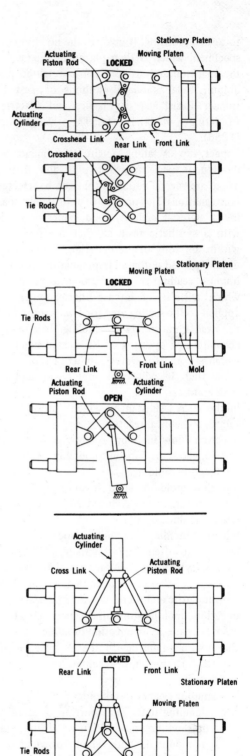

Fig. 1-10. Types of toggle clamps.

tie rods or beams—Figs. 1-9 and 1-10: Those members of the clamping unit which join and align the stationary platen with the clamping force actuating mechanism and which serve as the tension members of the clamp when it is holding the mold closed.

toggle clamp (hydraulic actuated, mechanical actuated)—Fig. 1-10: A clamping unit with a toggle mechanism directly connected to the moving platen. A hydraulic cylinder, or some mechanical force device, is connected to the toggle system to exert the opening and closing force and hold the mold closed during injection.

Injection system terminology

injection plasticizing (plasticating) unit: That portion of an injection molding machine which converts a plastic material from a solid phase to a homogeneous semi-liquid phase by raising its temperature. This unit maintains the material at a moldable temperature and forces it through the injection unit nozzle into a mold.

plunger unit—Fig. 1-11: a combination injection and plasticizing device in which a heating chamber is mounted between the plunger and the mold. This chamber heats the plastic material by conduction. The plunger, on each stroke, pushes unmelted plastic material into the chamber, which in turn forces plastic melt at the front of the chamber out through the nozzle.

prepack: Prepacking, also called "stuffing," is a method which can be used to increase the volumetric output per shot of the injector plunger unit by prepacking or stuffing additional material into the heating cylinder by means of multiple strokes of the injector plunger. (Applies only to plunger unit type injection machines.)

reciprocating screw—Fig. 1-12: A combination injection and plasticizing unit in which an extrusion device with a reciprocating screw is used to plasticize the material. Injection of material into a mold can take place by direct extrusion into the mold, or by reciprocating the screw as an injection plunger, or by a combination of the two. When the screw serves as an injection plunger, this unit acts as a holding, measuring, and injection chamber.

two-stage plunger unit—Fig. 1-13: An injection and plasticizing unit in which the plasticizing is performed in a separate

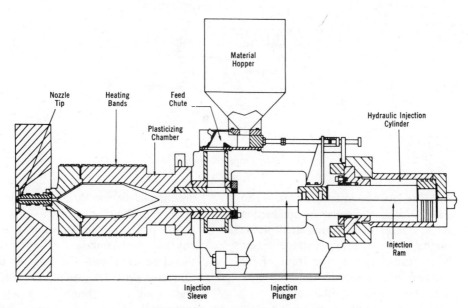

Fig. 1-11. Plunger unit.

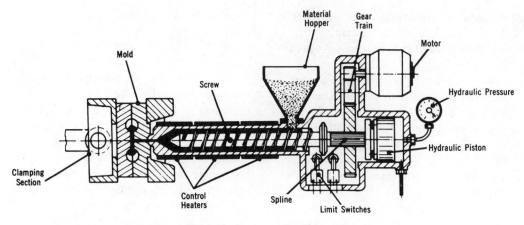

Fig. 1-12. Reciprocating screw unit.

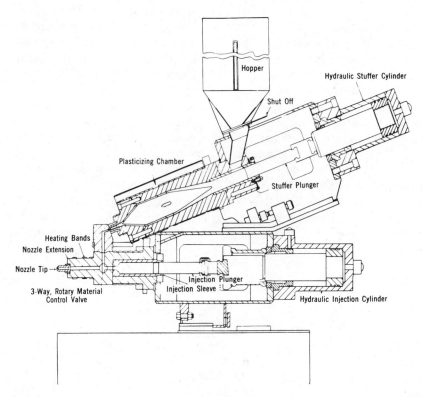

Fig. 1-13. Two-stage plunger unit.

unit. The latter consists of a chamber to heat the plastic material by conduction and a plunger to push unmelted plastic material into the chamber, which in turn forces plastic melt at the front of the chamber into a second stage injection unit. This injection unit serves as a combination holding, measuring, and injection chamber. During the injection cycle the shooting plunger forces the plastic melt from the injection chamber out through the nozzle.

two-stage screw unit—Fig. 1-14: An injection and plasticizing unit in which the plasticizing is performed in a separate unit which consists of a screw extrusion

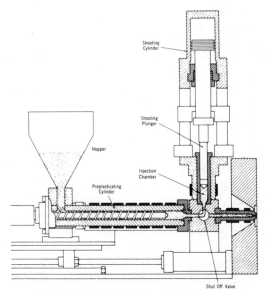

Fig. 1-14. Two-stage screw unit.

device to plasticize the material and force it into a second stage injection unit. This injection unit serves as a combination holding, measuring, and injection chamber. During the injection cycle a plunger forces the plastic melt from the injection chamber out through the nozzle.

inorganic pigments—Natural or synthetic metallic oxides, sulfides, and other salts, calcined during processing at 1200 to 2100°F. They are outstanding in heat- and light-stability, weather-resistance, and migration-resistance.

insert—An integral part of a plastic molding consisting of metal or other material which may be molded into position or may by pressed into the molding after the molding is completed.

in situ foaming—The technique of depositing a foamable plastic (prior to foaming and generally in the form of a liquid mix) into the place where it is intended that foaming shall take place.

integral-skin foams—As applied to urethane or structural foams, the description covers molded foams that develop their own integral surface skins. The surface skin is generally "solid," as contrasted to the cellular construction in the interior of the part.

iridescence—Loss of brilliance in metallized plastics and development of multi-color reflec-

tance. Iridescence is caused by cold flow of plastic or coating and from extra heat during vacuum metallizing.

irradiation (atomic)—As applied to plastics, refers to bombardment with a variety of subatomic particles, generally alpha-, beta-, or gamma-rays. Atomic irradiation has been used to initiate polymerization and copolymerization of plastics and in some cases to bring about changes in the physical properties of a plastic material.

isotactic—A chain of unsymmetrical molecules combined head to tail with their methyl groups occupying the same relative position in space along the chain.

jet molding—Processing technique characterized by the fact that most of the heat is applied to the material as it passes through the nozzle or jet, rather than in a heating cylinder as is done in conventional processes.

jetting—Turbulent flow of resin from an undersized gate or thin section into a thicker mold section, as opposed to laminar flow of material progressing radially from a gate to the extremities of the cavity.

jig—Means of holding a part and guiding the tool during machining or assembly operation.

joint—The location at which two adherends are held together with a layer of adhesive. (See **bond**).

joint, butt—A type of edge joint in which the edge faces of the two adherends are at right angles to the other faces of the adherends. (See **joint; joint, edge; joint, scarf.**)

joint, edge—A joint made by bonding the edge faces of two adherends with adhesives. (See **joint; joint, scarf; joint, butt.**)

joint, lap—A joint made by placing one adherend partly over another and bonding together the overlapped portions. (**See joint, scarf.**)

joint, scarf—A joint made by cutting away similar angular segments of two adherends and bonding the adherends with the cut areas fitted together. (See **joint, lap.**)

joint, starved—A joint which has an insufficient amount of adhesive to produce a satisfactory bond. This condition may result from too thin a spread to fill the gap between the adherends, excessive penetration of the adhesive into the adherend, too short an assembly time, or the use of excessive pressure.

kirksite—An alloy of aluminum and zinc used for the construction of blow molds; it imparts high degree of heat conductivity to the mold.

kiss-roll coating—This roll arrangement carries a metered film of coating to the web; at the line of web contact, it is split with part remaining on the roll, the remainder of the coating adhering to the web.

knife coating—A method of coating a substrate (usually paper or fabric) in which the substrate, in the form of a continuous moving web, is coated with a material whose thickness is controlled by an adjustable knife or bar set at a suitable angle to the substrate.

knit line—See weld line.

knockout—See injection molding machines.

knockout bar—A bar or plate in a knockout frame used to back up a row or rows of knockout pins.

knockout-pin plate—See ejector plate.

knockout pin—See ejector pin.

laminar flow—Laminar flow of thermoplastic resins in a mold is accompanied by solidification of the layer in contact with the mold surface that acts as an insulating tube through which material flows to fill the remainder of the cavity. This type of flow is essential to duplication of the mold surface.

laminated plastics (synthetic resin-bonded laminate, laminate)—A plastics material consisting of superimposed layers of a synthetic resin-impregnated or -coated substrate (paper, glass mat, etc.) which have been bonded together, usually by means of heat and pressure, to form a single piece.

land—(1) The horizontal bearing surface of a semipositive or flash mold by which excess material escapes. See cutoff. (2) The bearing surface along the top of the flights of a screw in a screw extruder; (3) the surface of an extrusion die parallel to the direction of melt flow.

lay-up—(n.) As used in reinforced plastics, the reinforcing material placed in position in the mold; also the resin-impregnated reinforcement. —(v.) The process of placing the reinforcing material in position in the mold.

L/D ratio—See extruder.

leach—To extract a soluble component from a mixture by the process of percolation.

light-resistance—The ability of a plastics material to resist fading after exposure to sunlight or ultraviolet light. Ref: Tentative Recommended Practice for Exposure of Plastics to Fluorescent Sunlamp (ASTM Designation: D 1501).

liquid injection molding (LIM)—(1) A process that involves an integrated system for proportioning, mixing, and dispensing two-component liquid resin formulations and directly injecting the resultant mix into a mold which is clamped under pressure. Generally used for the encapsulation of electrical and electronic devices. (2) Variation on reaction injection molding. Involves mechanical mixing rather than the high-pressure impingement mixing used with reaction injection molding. However, unlike mechanical mixing in other systems, the mixer here does not need to be flushed, since a special feed system automatically dilutes the residue in the mixer with part of the polyol needed for the next shot, thereby keeping the ingredients from reacting.

loading tray—A device in the form of a specially designed tray which is used to load the charge of material or metal inserts simultaneously into each cavity of a multi-cavity mold by the withdrawal of a sliding bottom from the tray. Also called "charging tray."

locating ring—A ring which serves to align the nozzle of an injection cylinder with the entrance of the sprue bushing and the mold to the machine platen.

loss factor—The product of the power factor and the dielectric constant.

low-pressure laminates—In general, laminates molded and cured in the range of pressures from 400 psi down to and including pressures obtained by the mere contact of the plies.

low-profile resins—Applied to special polyester resin systems for reinforced plastics. These systems are combinations of thermoset resins and thermoplastic resins. Although "low-profile" and "low-shrink" are sometimes used interchangeably, there is a difference. Low-shrinks contain up to 30% thermoplastic polymer by weight of total resin, while low-profiles contain from 30 to 50%. Low-shrink offers minimum surface waviness in the molded part (as low as 1 mil per inch mold shrinkage); low-profile offers no surface waviness (from 0.5 to 0 mil/inch mold shrinkage).

low-shrink resins—See **low-profile resins.**

lubricant bloom—An irregular, cloudy, greasy film on a plastic surface.

lug—(1) A type of thread configuration, usually thread segments disposed equidistantly around a bottle neck (finish). The matching closure has matching portions that engage each of the thread segments. (2) A small indentation or raised portion on the surface of a container, provided as a means of indexing the container for operations such as multi-color decoration or labeling.

luminescent pigments—Special pigments available to produce striking effects in the dark. Basically there are two types: one is activated by ultra-violet radiation, producing very strong luminescence and, consequently, very eye-catching effects; the other type, known as phosphorescent pigments, does not require any separate source of radiation.

macerate—(v.) To chop or shred fabric for use as a filler for a molding resin.—(n.) The molding compound obtained when so filled.

machine shot capacity—Refers to the maximum weight of thermoplastic resin which can be displaced or injected by the injection ram in a single stroke.

mandrel—(1) In blow molding, part of the tooling that forms the inside of the container neck and through which air is forced to form the hot parison to the shape of the molds. (2) In extrusion, the solid, cylindrical part of the die that forms tubing or pipe. (3) In filament winding of reinforced plastic, the form (usually cylindrical) around which the filaments are wound.

manifold—A term used mainly with reference to blow molding and sometimes with injection molding equipment. It refers to the distribution or piping system which takes the single channel flow output of the extruder or injection cylinder and divides it to feed several blow molding heads or injection nozzles.

masterbatch—A plastics compound which includes a high concentration of an additive or additives. Masterbatches are designed for use in appropriate quantities with the basic resin or mix so that the correct end concentration is achieved. For example, color masterbatches for a variety of plastics are extensively used as they provide a clean and convenient method of obtaining accurate color shades.

mat—A randomly distributed felt of glass fibers used in reinforced plastics lay-up molding.

material distribution—A term which describes the variation in thickness of various parts of a product, i.e., body, wall, shoulder, heel, base, etc.

material well—Space provided in a compression mold to care for bulk factor of the material load.

matched metal molding—Method of molding reinforced plastics between two close-fitting metal molds mounted in a press.

matrix—(n.) The continuous phase of a composite material. The resin component in a reinforced plastics material.

matte finish—A type of dull, non-reflective finish. See **surface finish.**

mechanically foamed plastic—A cellular plastic whose structure is produced by physically incorporated gases.

melt fracture—An instability in the melt flow through a die, starting at the entry to the die. It leads to surface irregularities on the finished article like a regular helix or irregularly-spaced ripples.

melt index—The amount, in grams, of a thermoplastic resin which can be forced through a 0.0825 inch orifice when subjected to 2160 gms. force in 10 minutes at 190°C. Ref: Tentative Method of Measuring Flow Rates of Thermoplastics by Extrusion Plastometer (ASTM Designation: D 1238).

melt strength—The strength of the plastic while in the molten state.

metering screw—An extrusion screw which has a shallow constant depth, and constant pitch section over, usually, the last 3 to 4 flights.

migration of plasticizer—Loss of plasticizer from an elastomeric plastic compound with subsequent absorption by an adjacent medium of lower plasticizer concentration.

metalizing—Applying a thin coating of metal to a nonmetallic surface. May be done by chemical deposition or by exposing the surface to vaporized metal in a vacuum chamber.

metallic pigments—A class of pigments consisting of thin opaque aluminum flakes (made by ball milling either a disintegrated aluminum foil or a rough metal powder and then polishing to obtain a flat, brilliant surface on each particle) or copper alloy flakes (known as bronze pigments). Incorporated into plastics, they produce unusual silvery and other metal-like effects.

metering screw—An extrusion screw which has a shallow constant depth, and constant pitch section over, usually, the last 3 to 4 flights.

modulus of elasticity—The ratio of stress to strain in a material that is elastically deformed. Ref: Standard Method of Test for Flexural Properties of Plastics (ASTM Designation D 790).

moisture vapor transmission—The rate at which water vapor permeates through a plastic film or wall at a specified temperature and relative humidity. Ref: Standard Methods of Test for Water Vapor Transmission of Materials in Sheet Form (ASTM Designation: E 96).

mold—(n.) A medium or tool designed to form desired shapes and sizes. For a guide to mold terminology see Fig. 1-15. —(v.) To process a plastics material using a mold.

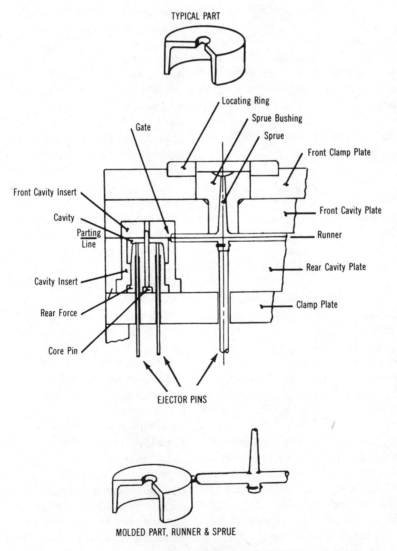

Fig. 1-15. Mold terminology. (*Courtesy, E. I. du Pont de Nemours & Co., Inc.*).

mold base—The assembly of all parts making up an injection mold, other than the cavity, core, and pins.

mold insert (removable)—Part of a mold cavity or force which forms undercut or raised portions of a molded article.

mold number—The number assigned to each mold or set of molds for identification purposes. The number is usually placed in that part of the container mold that forms the base of the container.

mold release—See **parting agent**.

molding cycle—See **cycle**.

molding material—Plastic material in varying stages of granulation, often comprising resin, filler, pigments, plasticizers and other ingredients, ready for use in the molding operation. Also called "molding compound" or "powder."

molding pressure—The pressure applied directly or indirectly on the compound to allow the complete transformation to a solid dense part.

molding shrinkage—The difference in dimensions, expressed in inches per inch, between a molding and the mold cavity in which it was molded, both the mold and the molding being at normal room temperature when measured. Also called "mold shrinkage," "shrinkage," and "contraction."

mold mark—Identifying symbol of the molder who produced the part. Usually molded into an unobtrusive area.

mold release—A lubricant used to coat a mold cavity to prevent the molded piece from sticking to it, and thus to facilitate its removal from the mold. Also called "release agent."

mold seam—A line formed by mold construction such as removable members in cavity, cam slides, etc. (Not to be confused with mold parting line).

molecular weight—The sum of the atomic masses of the elements forming the molecule.

molecular weight distribution—The ratio of the weight average molecular weight to the number average molecular weight gives an indication of the distribution.

monomer—A relatively simple compound which can react to form a polymer.

movable platen—The moving platen of an injection or compression molding machine to which half of the mold is secured during operation.

This platen is moved either by a hydraulic ram or a toggle mechanism.

multi-cavity mold—A mold with two or more mold impressions, i.e., a mold which produces more than one molding per molding cycle.

multiple-screw extruders—As contrasted to conventional single-screw extruders, these machines involve the use of two or four screws (conical or constant depth). Types include machines with intermeshing counter-rotating screws, intermeshing co-rotating screws, and nonintermeshing counter-rotating screws.

neck-in—In extrusion coating, the difference between the width of the extruded web as it leaves the die and the width of the coating on the substrate.

needle blow—A specific blow molding technique where the blowing air is injected into the hollow article through a sharpened hollow needle which pierces the parison.

nonrigid plastic—A plastic which has a lack of stiffness or apparent modulus of elasticity of not over 50,000 psi at $25^\circ C$ when determined according to Standard Method of Test for Stiffness of Plastics by Means of a Cantilever Beam (ASTM Designation: D 747).

notch sensitivity—The extent to which the sensitivity of a material to fracture is increased by the presence of a break in the homogenity of the surface, such as a notch, a sudden change in section, a crack, or a scratch. Low notch sensitivity is usually associated with ductile materials, and high notch sensitivity with brittle materials.

nozzle—The hollow cored metal nose screwed into the extrusion end of (a) the heating cylinder of an injection machine or (b) a transfer chamber where this is a separate structure. A nozzle is designed to form under pressure a seal between the heating cylinder or the transfer chamber and the mold. The front end of a nozzle may be either flat or spherical in shape.

olefins—A group of unsaturated hydrocarbons of the general formula $C_n H2_n$, and named after the corresponding paraffins by the addition of "ene" to the stem. Examples are ethylene and propylene.

one-shot molding—In the urethane foam field, indicates a system whereby the isocyanate, polyol, catalyst, and other additives are mixed together directly and a foam is produced immediately (as distinguished from prepolymer).

opaque—Descriptive of a material or substance which will not transmit light. Opposite of transparent, q.v. Materials which are neither opaque nor transparent are sometimes described as semi-opaque, but are more properly classified as translucent, q.v.

open-cell foam—A cellular plastic in which there is a predominance of inter-connected cells.

orange peel—Uneven leveling of coating on plastic surfaces, usually because of high viscosity. Simple spray gun adjustments and/or addition of high boiling solvent to coating for a wetter spray is helpful.

organic pigments—Characterized by good brightness and brilliance. They are divided into toners and lakes. Toners, in turn, are divided into insoluble organic toners and lake toners. The insoluble organic toners are usually free from salt-forming groups. Lake toners are practically pure, water-insoluble heavy metal salts of dyes without the fillers or substrates of ordinary lakes. Lakes, which are not as strong as lake toners, are water-insoluble heavy metal salts or other dye complexes precipitated upon or admixed with a base or filler.

organosol—A suspension of a finely-divided resin in a volatile organic liquid. The resin does not dissolve appreciably in the organic liquid at room temperature, but does at elevated temperatures. The liquid evaporates at the elevated temperature and the residue on cooling is a homogeneous plastic mass. Plasticizers may be dissolved in the volatile liquid.

orientation—The alignment of the crystalline structure in polymeric materials so as to produce a highly uniform structure. Can be accomplished by cold drawing or stretching during fabrication.

orifice—The opening in the extruder die formed by the orifice bushing (ring) and mandrel.

orifice bushing—The outer part of the die in an extruder head.

outgassing—Devolatilization of plastics or applied coatings during exposure to vacuum in vacuum metallizing. Resulting parts show voids or thin spots in plating with reduced and spotty brilliance. Additional drying prior to metallizing is helpful, but outgassing is inherent to plastic materials and coatings ingredients, including plasticizer and volatile components.

parallels—The support spacers placed between the mold and press platen or clamping plate. Also called "risers."

parison swell—In blow molding, the ratio of the cross-sectional area of the parison to the cross-section area of the die opening.

part—In its proper literal meaning, a component of an assembly. However, the word is widely misused to designate any individual manufactured article, even when (like a cup, a comb, a doll) it is complete in itself, not part of anything.

parting agent—A lubricant, often wax, used to coat a mold cavity to prevent the molded piece from sticking to it, and thus to facilitate its removal from the mold. Also called "release agent."

parting line—Mark on a molding or casting where halves of mold met in closing.

partitioned mold cooling—See **bubbler**.

pearlescent pigments—A class of pigments consisting of particles that are essentially transparent crystals of a high refractive index. The optical effect is one of partial reflection from the two sides of each flake. When reflections from parallel flakes reinforce each other, the result is a silvery luster. Effects possible range from brilliant high-lighting to moderate enhancement of the normal surface gloss.

permanence—Resistance of a plastic to appreciable changes in characteristics with time and environment.

permanent set—The increase in length, expressed in a percentage of the original length, by which an elastic material fails to return to original length after being stressed for a standard period of time.

permeability—(1) The passage or diffusion of a gas, vapor, liquid, or solid through a barrier without physically or chemically affecting it. (2) The rate of such passage.

pigment—A pigment is a coloring agent mixed with plastic material prior to processing to provide a uniform color.

pill—See **preform**.

pinhole—A very small hole in a plastic container, film, etc.

pinch-off—A raised edge around a cavity in the mold, which seals off the part and separates the excess material as the mold closes around the parison in the blow molding operation.

pinpoint gate—A restricted orifice of 0.030 in. or less in diameter through which molten resin flows into a mold cavity.

pipe—A hollow cylinder of a plastic material in which the wall thicknesses are usually small when compared to the diameter and in which the inside and outside walls are essentially concentric. See **tubing**.

pipe train—A term used in extrusion of pipe which denotes the entire equipment assembly used to fabricate the pipe, e.g., extruder, die, cooling bath, haul-off and cutter.

pitch—The distance from any point on the flight of a screw line to the corresponding point on an adjacent flight, measured parallel to the axis of the screw line or threading.

plastic—(n.) One of many high-polymeric substances, including both natural and synthetic products, but excluding the rubbers. At some stage in its manufacture, every plastic is capable of flowing, under heat and pressure, if necessary, into the desired final shape. (a.) Made of plastic; capable of flow under pressure or tensile stress.

plastic, rigid—A plastic with a stiffness or apparent modulus of elasticity greater than 100,000 psi at 23°C. (ASTM D 747.)

plastic, semi-rigid—A plastic with a stiffness or apparent modulus of elasticity between 10,000 and 100,000 psi at 23°C. (ASTM D 747.)

plastic deformation—A change in dimensions of an object under load that is not recovered when the load is removed; opposed to elastic deformation.

plastics tooling—Tools, e.g., dies, jigs, fixtures, etc., for the metal forming trades constructed of plastics, generally laminates or casting materials.

plasticate—To soften by heating or kneading. Synonyms are: plastify, flux, and (imprecisely) plasticize (q.v.).

plasticity—A property of plastics which allows the material to be deformed continuously and permanently without rupture upon the application of a force that exceeds the yield value of the material.

plasticize—To soften a material and make it plastic or moldable, either by means of a plasticizer or the application of heat.

plasticizer—A material incorporated in a plastic to increase its workability and its flexibility or distensibility; normally used in thermoplastics. The addition of the plasticizer may lower the melt viscosity, the temperature of the glassy transition, or the elastic modulus of the plastic.

plastify—See **plasticate**.

plastigel—A plastisol exhibiting gel-like flow properties; one having an effective yield value.

plastisols—Mixtures of vinyl resins and plasticizers which can be molded, cast, or converted to continuous films by the application of heat. If the mixtures contain volatile thinners also, they are known as "organosols."

plate dispersion plug—See **breaker plate**.

platens—The mounting plates of a press to which the entire mold assembly is bolted. See **injection molding machines**.

plug forming—A thermoforming process in which a plug or male mold is used to partially preform the part before forming is completed using vacuum or pressure.

plunger—That part of a transfer or injection press that applies pressure on the unmelted plastic material to push it into the chamber, which in turn forces plastic melt at the front of the chamber out the nozzle. See **ram**.

plunger injection molding machines—See **injection molding machines**.

polishing roll(s)—A roll or series of rolls, which have a highly polished chrome plated surface, that are utilized to produce a smooth surface on sheet as it is extruded.

polyliner—A perforated longitudinally ribbed sleeve that fits inside the cylinder of an injection molding machine; used as a replacement for conventional injection cylinder torpedos.

polymer—A high-molecular-weight organic compound, natural or synthetic, whose structure can be represented by a repeated small unit, the mer; e.g., polyethylene, rubber, cellulose. Synthetic polymers are formed by addition or condensation polymerization of monomers. If two or more monomers are involved, a copolymer is obtained. Some polymers are elastomers, some plastics.

polymerization—A chemical reaction in which the molecules of a monomer are linked together to form large molecules whose molecular weight is a multiple of that of the original substance. When two or more monomers are involved, the process is called copolymerization or heteropolymerization.

porosity—Porosity is the existence in a plastic material of very small voids.

porous molds—Molds which are made up of bonded or fused aggregate (powdered metal, coarse pellets, etc.) in such a manner that the resulting mass contains numerous open interstices of regular or irregular size through which either air or liquids may pass through the mass of the mold.

positive mold—A mold designed to trap all the molding material when it closes.

postforming—The forming, bending, or shaping of fully cured, C-stage thermoset laminates that have been heated to make them flexible. On cooling, the formed laminate retains the contours and shape of the mold over which it has been formed.

pot—Chamber to hold and heat molding material for a transfer mold.

pot life—See **working life**.

pot plunger—A plunger used to force softened molding material into the closed cavity of a transfer mold.

potting—Similar to encapsulating (q.v.) except that steps are taken to insure complete penetration of all the voids in the object before the resin polymerizes.

powder molding—General term used to denote several techniques for producing objects of varying sizes and shapes by melting plastic powder, usually against the inside of a mold. The techniques vary as to whether the molds are stationary (e.g., as in variations on slush molding techniques) or rotating (e.g., as in variations on rotational molding).

preform (n.)—A compressed tablet or biscuit of plastic composition used for efficiency in handling and accuracy in weighing materials. (v.)—To make plastic molding powder into pellets or tablets.

preheating—The heating of a compound prior to molding or casting in order to facilitate the operation, reduce cycle, and improve product.

preplastication—Technique of premelting injection molding powders in a separate chamber, then transferring the melt to the injection cylinder. Device used for preplastication is commonly known as a preplasticizer.

prepolymer—A chemical structure intermediate between that of the monomer or monomers and the final polymer or resin.

prepolymer molding—In the urethane foam field, indicates a system whereby a portion of the polyol is pre-reacted with the isocyanate to form a liquid propolymer with viscosity range suitable for pumping or metering. This component is supplied to end-users with a second premixed blend of additional polyol, catalyst, blowing agent, etc. When the two components are mixed together, foaming occurs. (See **one-shot molding**).

prepreg—A term generally used in reinforced plastics to mean the reinforcing material containing or combined with the full complement of resin before molding.

preprinting—In sheet thermoforming, the distorted printing of sheets before they are formed. During forming the print assumes its proper proportions.

press polish—A finish for sheet stock produced by contact, under heat and pressure, with a very smooth metal which gives the plastic a high sheen.

pressure forming—A thermoforming process wherein pressure is used to push the sheet to be formed against the mold surface as opposed to using a vacuum to suck the sheet flat against the mold.

pressure pads—Reinforcements distributed around the dead areas in the faces of a mold to help the land absorb the final pressure of closing without collapsing.

primary plasticizer—Has sufficient affinity to the polymer or resin so that it is considered compatible and therefore it may be used as the sole plasticizer.

primer—A coating applied to a surface, prior to the application of an adhesive or lacquer, enamel or the like, to improve adhesion or finishing.

printed circuit—An electrical or electronic circuit produced mainly from copper clad laminates.

programming—The extrusion of a parison which differs in thickness in the length direction in order to equalize wall thickness of the blown container. It can be done with a pneumatic or hydraulic device which activates the mandrel shaft and adjusts the mandrel position during parison extrusion (parison programmer, controller, or variator). It can also be done by varying extrusion speed on accumulator-type blow molding machines or, in some "parison reheat" systems by varying the amount of heat applied.

promoter—A chemical, itself a feeble catalyst, that greatly increases the activity of a given catalyst.

prototype mold—A simplified mold construction often made from a light metal casting alloy, an epoxy resin, or an RTV silicone rubber, in order to obtain information for the final mold and/or part design.

pulp molding—Process by which a resin-impregnated pulp material is preformed by application of a vacuum and subsequently oven cured or molded.

pultrusion—Automated method for producing continuous reinforced plastics shapes by pulling pre-impregnated reinforcing fibers through a heated die where the resin is cured.

purging—Cleaning one color or type of material from the cylinder of an injection molding machine or extruder by forcing it out with the new color or material to be used in subsequent production. Purging materials are also available.

quench (thermoplastics)—A process of shock cooling thermoplastic materials from the molten state.

quench bath—The cooling medium used to quench molten thermoplastic materials to the solid state.

quench-tank extrusion—The extruded film is cooled in a quench-water bath.

radio frequency (r.f.) preheating—A method of preheating molding materials to facilitate the molding operation and/or reduce the molding cycle. The frequencies most commonly used are between 10 and 100 mc/sec.

radio frequency welding—A method of welding thermoplastics using a radio frequency field to apply the necessary heat. Also known as high frequency welding.

ram—The press member that enters the cavity block and exerts pressure on the plastic compound. It is designated as the "top force" or "bottom force" by position in the assembly.

ram travel—Distance ram moves when operating a complete molding cycle.

reaction injection molding (RIM)—A process that involves the high-pressure impingement mixing of two (or more) reactive liquid components; after mixing, the liquid stream is injected into a closed mold at low pressure. The finished parts can be cellular or solid elastomers, with a wide range of hardness and modulus. Also known as high-pressure impingement mixing (HPIM). Used especially with urethanes.

reaming—A method used to trim and size plastic bottle finishes. A special rotating cutting tool trims the sealing surface smooth and simultaneously reams (bores) the bottle opening to desired size.

reciprocating screw injection molding—A combination injection and plasticizing unit in which an extrusion device with a reciprocating screw is used to plasticize the material. Injection of material into a mold can take place by direct extrusion into the mold, or by reciprocating the screw as an injection plunger, or by a combination of the two. When the screw serves as an injection plunger, this unit acts as a holding, measuring, and injection chamber. See **injection molding machines.**

recycle—Material from flash, trimmings, scrap, rejects, etc., which can be ground-up or re-pelletized and fed back into the processing machine.

reinforced molding compound—A material reinforced with special fillers or fibers to meet specific requirements (glass, synthetic fibers, minerals, etc.).

reinforced plastics—Plastics with some strength properties greatly superior to those of the base resin, resulting from the presence of high-strength fillers embedded in the composition. The reinforcing fillers are usually fibers, fabrics, or mats made of fibers.

reinforcement—A strong inert material bound into a plastic to improve its strength, stiffness, and impact resistance. Reinforcements are usually long fibers of glass, sisal, cotton, etc.—in woven or nonwoven form.

release agent—See **mold release.**

resin—Any of a class of solid or semi-solid organic products of natural or synthetic origin, generally of high molecular weight with no definite melting point. Most resins are polymers (q.v.).

resin, liquid—An organic, polymeric liquid which, when converted to its final state for use, becomes a solid.

resin pocket—An apparent accumulation of excess resin in a small localized section visible on cut edges of molded surfaces. Also called "resin segregation."

resistivity—The ability of a material to resist passage of electrical current either through its bulk or on a surface. The unit of volume resistivity is the ohm-cm., of surface resistivity, the ohm.

restricted gate—A very small orifice between runner and cavity in an injection or transfer mold. When the piece is ejected, this gate breaks cleanly, simplifying separation of runner from piece.

restrictor ring—A ring-shaped part protruding from the torpedo surface which provides increase of pressure in the mold to improve welding of two streams.

retainer plate—The plate on which demountable pieces, such as mold cavities, ejector pins, guide pins, and bushings are mounted during molding; usually drilled for steam or water.

retarder—See **inhibitor.**

reverse-roll coating—The coating is premetered between rolls and then wiped off on the web. The amount of coating is controlled by the metering gap and also by the speed of rotation of the coating roll.

rib—A reinforcing member of a fabricated or molded part.

rigid resin—See **plastic, rigid.**

Rockwell hardness—A common method of testing material for resistance to indentation in which a diamond or steel ball, under pressure, is used to pierce the test specimen. Ref: Standard Method of Test for Rockwell Hardness of Plastics and Electrical Insulating Materials (ASTM Designation: D 785).

roll mill—Two rolls placed in close relationship to one another used to admix a plastic material with other substances. The rolls turn at different speeds to produce a shearing action to the materials being compounded.

roller coating—Used for applying paints to raised designs or letters.

rotating spreader—A type of injection torpedo which consists of a finned torpedo which is rotated by a shaft extending through a tubular cross section injection ram behind it.

rotational casting (or molding)—A method used to make hollow articles from thermoplastic materials. Material is charged into a hollow mold capable of being rotated in one or two planes. The hot mold fuses the material into a gel after the rotation has caused it to cover all surfaces. The mold is then chilled and the product stripped out.

roving—A form of fibrous glass in which spun strands are woven into a tubular rope. The number of strands is variable but 60 is usual. Chopped roving is commonly used in preforming.

rubber—An elastomer capable of rapid elastic recovery after being stretched to at least twice its length at temperatures from 0 to 150°F at any humidity. Specifically, Hevea or natural rubber, the standard of comparison for elastomers.

runner (refers to mold)—In an injection or transfer mold, the channel that connects the sprue with the gate to the cavity.

runner system (refers to plastic)—The term usually applied to all the material in the form of sprues, runners and gates which lead material from the nozzle of an injection machine or the pot of a transfer mold to the mold cavity.

rupture strength, psi—The true value of rupture strength is the stress of a material at failure based on the ruptured cross-sectional area itself.

sag—The extension locally (often near the die face) of the parison during extrusion by gravitational forces. This causes necking-down of the parison. Also refers to the flow of a molten sheet in a thermoforming operation.

sag-streaks—Uneven plastic surface due to heavy coating application and poor flow-out. Can be eliminated by thinning the coating, adjusting the spray gun, or changing the stroke.

sandwich constructions—Panels composed of a lightweight core material—honeycomb, foamed plastic, etc., q.v.—to which two relatively thin, dense, high strength faces or skins are adhered.

sandwich heating—A method of heating a thermoplastic sheet prior to forming which consists of heating both sides of the sheet simultaneously.

scrap—Any product of a molding operation that is not part of the primary product. In compression molding, this includes flash, culls, runners, and is not reusable as a molding compound. Injection molding and extrusion scrap (runners, rejected parts, sprues, etc.) can usually be reground and remolded.

screws—See extruders.

screw plasticating injection molding—A technique in which the plastic is converted from pellets to a viscous melt by means of an extruder screw which is an integral part of the molding machine. Machines are either single stage (in which plastication and injection are done in the same cylinder) or double stage in which the material is plasticated in one cylinder and then fed to a second for injection into a mold. See injection molding machines.

segregation—A separation of components in a molded article usually denoted by wavy lines and color striations in thermoplastics. In thermosettings, usually meaning segregation of resin and filler on surface.

semi-automatic molding machine—A molding machine in which only part of the operation is controlled by the direct action of a human. The automatic part of the operation is controlled by the machine according to a predetermined program.

semipositive mold—A mold whose principle is: As the two halves of the mold begin to close, the mold acts much like a flash mold, as the excess material is allowed to escape around the loose-fitting plunger and cavity. As the plunger telescopes further into the cavity, the mold becomes a positive mold with very little clearance and full pressure is exerted on the material, producing a part of maximum density.

This type of mold takes advantage of the free flow of material in a flash mold and the quality of producing dense parts in the positive mold.

set—To convert into a fixed or hardened state by chemical or physical action, such as condensation, polymerization, oxidation, vulcanization, gelation, hydration, or evaporation of volatile constituents. (See cure.)

shear strength, psi—Shear strength is the stress at which a material fails under a shear loading condition. (ASTM test method D 732).

shear stress—The stress developing in a polymer melt when the layers in a cross section are gliding along each other or along the wall of the channel (in laminar flow).

$$\text{shear stress} = \frac{\text{force}}{\text{area sheared}} = \text{psi}$$

sheet (thermoplastic)—A flat section of a thermoplastic resin with the length considerably greater than the width and 10 mils or greater in thickness.

shelf life—An expression to describe the time a molding compound can be stored without losing any of its original physical or molding properties.

Shore hardness—A method of determining the hardness of a plastic material using a scelroscope. This device consists of a small conical hammer fitted with a diamond point and acting in a glass tube. The hammer is made to strike the material under test and the degree of rebound is noted on a graduated scale. Generally, the harder the material the greater will be the rebound. A single indentor, without hammer, can be used to obtain Shore A or Shore D durometer measurements. Ref: Tentative Method of Test for Indentation Hardness of Rubber and Plastics by Means of a Durometer (ASTM Desgination: D 2240).

short or short shot—A molded part produced when the mold has not been filled completely.

shot—The yield from one complete molding cycle, including cull, runner, and flash.

shot capacity—The maximum weight of material which a machine can produce from one forward motion of the plunger or screw.

shrink wrapping—A technique of packaging in which the strains in a plastics film are released by raising the temperature of the film thus causing it to shrink over the package. These shrink characteristics are built into the film during its manufacture by stretching it under controlled temperatures to produce orientation, q.v., of the molecules. Upon cooling, the film retains its stretched condition, but reverts toward its original dimensions when it is heated.

shrinkage—The change in dimension (decrease) which a molded article undergoes after being molded. Shrinkage is caused by cooling and subsequent contraction of the plastic material. See **mold shrinkage.**

shrink fixture—See **cooling fixture.**

side coring or side draw pins—Projections used to core a hole in a direction other than the line of closing of a mold, and which must be withdrawn before the part is ejected from the mold.

silk screen printing—This printing method, in its basic form, involves laying a pattern of an insoluble material, in outline, on a finely woven fabric, so that when ink is drawn across it, it is able to pass through the screen only in the desired areas. Also called "screen process decorating."

silicone—Chemical derived from silica; used in molding as a release agent and general lubricant.

single cavity mold (injection)—An injection mold having only one cavity in the body of the mold, as opposed to a multiple cavity mold or family mold which have numerous cavities.

sink mark—A depression or dimple on the surface of an injection molded part due to collapsing of the surface following local internal shrinkage after the gate seals. May also be an incipient short shot.

sintering—In forming articles from fusible powders, e.g., nylon, the process of holding the pressed-powder article at a temperature just below its melting point for about $\frac{1}{2}$ hour. Particles are fused (sintered) together, but the mass, as a whole, does not melt.

sizing (n.)—The process of applying a material to a surface to fill pores and thus reduce the absorption of the subsequently applied adhesive or coating or to otherwise modify the surface. Also, the surface treatment applied to glass fibers used in reinforced plastics, for improving the bond between glass and plastic. The material used is sometimes called size.

slip additive—A modifier that acts as an internal lubricant which exudes to the surface of the plastic during and immediately after processing. In other words, a non-visible coating blooms to the surface to provide the necessary lubricity to reduce coefficient of friction and thereby improve slip characteristics.

slip forming—Sheet forming technique in which some of the plastic sheet material is allowed to slip through the mechanically operated clamping rings during a stretch-forming operation.

slot extrusion—A method of extruding film sheet in which the molten thermoplastic compound is forced through a straight slot.

slurry preforming—Method of preparing reinforced plastics preforms by wet processing techniques similar to those used in the pulp molding (q.v.) industry.

slush molding—Method for casting thermoplastics, in which the resin in liquid form is poured into a hot mold where a viscous skin forms. The excess slush is drained off, the mold is cooled, and the molding stripped out.

softening range—The range of temperature in which a plastic changes from a rigid to a soft state. Actual values will depend on the method of test. Sometimes erroneously referred to as softening point.

solid phase forming—Using metal working techniques to form thermoplastics in a solid phase. Procedure begins with a plastic blank that is then heated and fabricated (i.e., forged) by bulk deformation of the materials in constraining dies by application of force.

solvent—Any substance, usually a liquid, which dissolves other substances.

solvent molding—Process for forming thermoplastic articles, by dipping a male mold in a solution or dispersion of the resin and drawing off the solvent to leave a layer of plastic film adhering to the mold.

specific gravity—The density (mass per unit volume) of any material divided by that of water. Ref: Standard Methods of Test for Specific Gravity and Density of Plastics by Displacement (ASTM Designation: D 792).

spider gate—Multi-gating of a part through a system of radial runners from the sprue.

spider—(1) In a molding press, that part of an ejector mechanism which operates the ejector pins. (2) In extrusion, a term used to denote the membranes supporting a mandrel within the head/die assembly.

spin welding—A process of fusing two objects together by forcing them together while one of the pair is spinning, until frictional heat melts the interface. Spinning is then stopped and pressure held until they are frozen together.

spiral flow test—A method for determining the flow properties of a thermoplastic resin in which the resin flows along the path of a spiral cavity. The length of the material which flows into the cavity and its weight gives a relative indication of the flow properties of the resin.

spiral mold cooling—A method of cooling injection molds or similar molds wherein the cooling medium flows through a spiral cavity in the body of the mold. In injection molds, the cooling medium is introduced at the center of the spiral, near the sprue section, as more heat is localized in this section.

splay marks—Lines found in part after molding, usually due to flow of material in mold. Sometimes called silver streaking.

split cavity—Cavity made in sections.

split-ring mold—A mold in which a split cavity block is assembled in a chase to permit the forming of undercuts in a molded piece. These parts are ejected from the mold and then separated from the piece.

spray coating—Usually accomplished on continuous webs by a set of reciprocating spray nozzles traveling laterally across the web as it moves.

sprayed metal molds—Mold made by spraying molten metal onto a master until a shell of predetermined thickness is achieved. Shell is then removed and backed up with plaster, cement, casting resin, or other suitable material. Used primarily as a mold in sheet-forming processes.

spray-up—Covers a number of techniques in which a spray gun is used as the processing tool. In reinforced plastics, for example, fibrous glass and resin can be simultaneously deposited in a mold. In essence, roving is fed through a chopper and ejected into a resin stream which is directed at the mold by either of two spray systems. In foamed plastics, very fast-reacting urethane foams or epoxy foams are fed in

liquid streams to the gun and sprayed on the surface. On contact, the liquid starts to foam.

spreader—A streamlined metal block placed in the path of flow of the plastics material in the heating cylinder of extruders and injection molding machines to spread it into thin layers, thus forcing it into intimate contact with the heating areas.

sprue—Feed opening provided in the injection or transfer mold; also, the slug formed at this hole. Spur is a shop term for the sprue slug.

sprue-bushing—A hardened steel insert in an injection mold which contains the tapered sprue hold and has a suitable seat for the nozzle of the injection cylinder. Sometimes called an adapter.

sprue gate—A passageway through which molten resin flows from the nozzle to the mold cavity.

sprue lock or puller—In injection molding, a portion of the plastic composition which is held in the cold slug well by an undercut; used to pull the sprue out of the bushing as the mold is opened. The sprue lock itself is pushed out of the mold by an ejector pin. When the undercut occurs on the cavity block retainer plate, this pin is called the sprue ejector pin.

stabilizer—An ingredient used in the formulation of some plastics to assist in maintaining the physical and chemical properties of the compounded materials at their initial values throughout the processing and service life of the material.

stationary platen—The plate of an injection or compression molding machine to which the front plate of the mold is secured during operation. This platen does not move during normal operation.

steam molding (expandable polystyrene)—Used to mold parts from pre-expanded beads of polystyrene using steam as a source of heat to expand the blowing agent in the material. The steam in most cases is contacted intimately with the beads directly or may be used indirectly to heat mold surfaces which are in contact with the beads.

steam plate—Mounting plate for molds, cored for circulation of steam.

stir-in resin—A vinyl resin which does not require grinding to effect dispersion in a plastisol or an organosol.

storage life—The period of time during which a liquid resin or packaged adhesive can be stored under specified temperature conditions and remain suitable for use. Storage life is sometimes called shelf life.

stress-crack—External or internal crack in a plastic caused by tensile stresses less than that of its short-time mechanical strength. The development of such cracks is frequently accelerated by the environment to which the plastic is exposed. The stresses which cause cracking may be present internally or externally or may be combinations of these stresses. The appearance of a network of fine cracks is called crazing.

striation—A separation of colors resulting in a linear effect of color variation.

stripper-plate—A plate that strips a molded piece from core pins or force plugs, the stripper-plate is set into operation by the opening of the mold.

structural foams—Expanded plastics materials having integral skins and outstanding rigidity. Structural foams involve a variety of thermoplastics resins as well as urethanes.

submarine gate—A type of edge gate where the opening from the runner into the mold is located below the parting line or mold surface as opposed to conventional edge gating where the opening is machined into the surface of the mold. With submarine gates, the item is broken from the runner system on ejection from the mold.

surface finish—Finish of molded product.

surface resistivity—The electrical resistance between opposite edges of a unit square of insulating material. It is commonly expressed in ohms. Ref: Standard Methods of Test for D-C Resistance or Conductance of Insulating Materials (ASTM Designation: D 257).

surface treating—Any method of treating a material so as to alter the surface and render it receptive to inks, paints, lacquers, and adhesives such as chemical, flame, and electronic treating.

surging—Unstable pressure build-up in an extruder leading to variable throughput and waviness of the parison.

sweating—Exudation of small drops of liquid, usually a plasticizer or softener, on the surface of a plastic part.

synergism—A term used to describe the use of two or more stabilizers in an organic material where the combination of such stabilizers improves the stability to a greater extent than could be expected from the additive effect of each stabilizer.

T-die—A term used to denote a center-fed, slot extrusion die for film which in combination with the die adapter resembles an inverted T.

tab gated—A small removable tab of approximately the same thickness as the mold item, usually located perpendicular to the item. The tab is used as a site for edge gate location, usually on items with large flat areas.

temper, n.—To reheat after hardening to some temperature below the critical temperature, followed by air cooling to obtain desired mechanical properties and to relieve hardening strains.

tensile bar (specimen)—A compression or injection molded specimen of specified dimensions which is used to determine the tensile properties of a material.

tensile modulus, psi—Tensile modulus is the ratio of stress to strain for a given material within its proportional limit under tensile loading loading conditions. (ASTM test method D 638).

tensile strength—The pulling stress, in psi, required to break a given specimen. Area used in computing strength is usually the original, rather than the necked-down area. Ref: Tentative Method of Test for Tensile Properties of Plastics (ASTM Designation: D 638).

Therimage—A trademark for a decorating process for plastic which transfers the image of a label or decoration to the object under the influence of heat and light pressure.

thermal conductivity—Ability of a material to conduct heat; physical constant for quantity of heat that passes through unit cube of a substance in unit of time when difference in temperature of two faces is 1°. Ref: Standard Method of Test for Thermal Conductivity of Materials by Means of the Guarded Hot Plate (ASTM Designation: C 177).

thermal expansion—See coefficient of expansion.

thermal stress cracking (TSC)—Crazing and cracking of some thermoplastic resins which results from over-exposure to elevated temperatures.

thermoelasticity—Rubber-like elasticity exhibited by a rigid plastic and resulting from an increase of temperature.

thermoforming—Any process of forming thermoplastic sheet which consists of heating the sheet and pulling it down onto a mold surface.

thermoforms—The product which results from a thermoforming operation.

thermoplastic—(a.) Capable of being repeatedly softened by heat and hardened by cooling—(n.) A material that will repeatedly soften when heated and harden when cooled. Typical of the thermoplastic family are the styrene polymers and copolymers, acrylics, cellulosics, polyethylenes, polypropylene, vinyls, nylons, and the various fluorocarbon materials.

thermoset—A material that will undergo or has undergone a chemical reaction by the action of heat and pressure, catalysts, ultra-violet light, etc., leading to a relatively infusible state. Typical of the plastics in the thermosetting family are the aminos (melamine and urea), most polyesters, alkyds, epoxies, and phenolics.

thixotropic—Said of materials that are gel-like at rest but fluid when agitated. Liquids containing suspended solids are apt to be thixotropic.

thread plug or ring or core—A part of a mold that shapes a thread and must be unscrewed from the finished piece.

tie rods—See injection molding machines.

thrust—See extruder.

toggle action—A mechanism which exerts pressure developed by the application of force on a knee joint. It is used as a method of closing presses and also serves to apply pressure at the same time.

tolerance—A specified allowance for deviations in weighing, measuring, etc., or for deviations from the standard dimensions or weight. Ref: SPI Standards and Practices of Plastics Custom Molders.

torpedo—See extruder.

tracking—See arc resistance.

transfer molding—A method of molding thermosetting materials, in which the plastic is first softened by heat and pressure in a transfer chamber, then forced by high pressure through suitable sprues, runners, and gates into closed mold for final curing.

translucent—Descriptive of a material or substance capable of transmitting some light, but not clear enough to be seen through.

transparent—Descriptive of a material or substance capable of a high degree of light transmission (e.g., glass).

tubing—A particular size of plastics pipe in which the outside diameter is essentially the same as that of copper tubing. See pipe.

tumbling—Finishing operation for small plastic articles by which gates, flash, and fins are removed and/or surfaces are polished by rotating them in a barrel together with wooden pegs, sawdust, and polishing compounds.

tunnel gate—See submarine gate.

twin-screw extrusion—See multiple-screw extruders.

twin-sheet thermoforming—Technique for thermoforming hollow objects by introducing high-pressure air in-between two sheets and blowing the sheets in the mold halves (vacuum is also applied).

two-stage machines—See injection molding machines and extruders.

ultrasonic sealing or bonding—A sealing method in which sealing is accomplished through the application of vibratory mechanical pressure at ultrasonic frequencies (20 to 40 kc.). Electrical energy is converted to ultrasonic vibrations through the use of either a magnetostrictive or piezoelectric transducer. The vibratory pressures at the interface in the sealing area develop localized heat losses which melt the plastic surfaces effecting the seal.

ultrasonic insertion—The inserting of a metal insert into a thermoplastic part by the application of vibratory mechanical pressure at ultrasonic frequencies.

undercut—(a.) Having a protuberance or indentation that impedes withdrawal from a two-piece, rigid mold. Flexible materials can be ejected intact even with slight undercuts. (n.) Any such protuberance or indentation; depends also on design of mold.

unit mold—Mold designed for quick changing interchangeable cavity parts.

uv stabilizer (ultraviolet)—Any chemical compound which, when admixed with a thermoplastic resin, selectively absorbs UV rays.

vacuum forming—Method of sheet forming in which the plastic sheet is clamped in a stationary frame, heated, and drawn down by a vacuum into a mold. In a loose sense, it is sometimes used to refer to all sheet forming techniques, including drape forming (q.v.), involving the use of vacuum and stationary molds.

vacuum metalizing—Process in which surfaces are thinly coated with metal by exposing them to the vapor of metal that has been evaporated under vacuum (one millionth of normal atmospheric pressure).

valley printing—Ink is applied to the high points of an embossing roll and subsequently deposited in what becomes the valleys of the embossed plastic material.

vehicle—The liquid medium in which pigments, etc., are dispersed in coatings such as paint, q.v., and which enable the coating to be applied.

vent—In a mold, a shallow channel or minute hole cut in the cavity to allow air to escape as the material enters. Also called "breathers." (See also **air vent**).

vented screw—See **extruder**.

venturi dispersion plug—A plate having an orifice with a conical relief drilled therein which is fitted in the nozzle of an injection molding machine to aid in the dispersion of colorants in a resin.

vertical flash ring—The clearance between the force plug and the vertical wall of the cavity in a positive or semi-positive mold; also the ring of excess material which escapes from the cavity into this clearance space.

vibration welding—Assembly technique in which frictional heat is generated by pressing the surfaces of parts together and vibrating the parts through a small relative displacement. This displacement can be either linear or angular. Vibration welding machines operate at relatively low frequencies, 90 to 120 Hz.

Vicat softening temperature—Measurement of the heat distortion temperature of a plastic material. Also called "Heat Deformation Point." Ref: Tentative Method of Test for Vicat Softening Point of Plastics (ASTM Designation: D 1525).

viscoelastic—A term that refers to plastics which "store" and dissipate energy during mechanical deformation. The term explains "flow" (q.v.). of plastic materials under stress.

viscosity—Internal friction or resistance to flow of a liquid. The constant ratio of shearing stress to rate of shear. In liquids for which this ratio is a function of stress, the term "apparent viscosity" is defined as this ratio.

void—A void or bubble occurring in the center of a heavy thermoplastic part usually caused by excessive shrinkage.

volume resistivity—The electrical resistance between opposite faces of a 1-cm cube of insulating material. It is measured under prescribed conditions using a direct current potential after a specified time of electrification. It is commonly expressed in ohm-centimeters. Also called "Specific Insulation Resistance." Ref: Standard Methods of Test for D-C Resistance or Conductance of Insulating Materials (ASTM Designation: D 257).

warpage—Dimensional distortion in a plastic object after molding.

water absorption—The ability of a thermoplastic material to absorb water from an environment. Ref: Standard Method of Test for Water Absorption of Plastics (ASTM Designation: D 570).

weatherometer—An instrument which is utilized to subject articles to accelerated weathering conditions, e.g., rich UV source and water spray.

web—A thin sheet in process in a machine. The molten web is that which issues from the die. The substrate web is the substrate being coated.

weld lines (also, weld marks or flow lines)—A mark on a molded plastic piece made by the meeting of two flow fronts during the molding operation.

welding—Joining thermoplastic pieces by one of several heat-softening processes. In hot-gas welding, the material is heated by a jet of hot air or inert gas directed from a welding "torch" onto the area of contact of the surfaces which are being welded. Welding operations to which this method is applied normally require the use of a filler rod. In spin-welding (q.v.), the heat is generated by friction. Welding also includes heat sealing and the terms are synonymous in some foreign countries, including Britain.

wet strength—The strength of paper when saturated with water, especially used in discussions of processes whereby the strength of paper is increased by the addition, in manufacture, of plastics resins. Also, the strength of an adhesive joint determined immediately after removal from a liquid in which it has been immersed under specified conditions of time, temperature, and pressure.

wheelabrating—Deflashing molded parts by bombarding with small particles at a high velocity.

window—A defect in a thermoplastics film, sheet or molding, caused by the incomplete "plasticization" of a piece of material during processing. It appears as a globule in an otherwise blended mass. See also **fisheye**.

working life—The period of time during which a liquid resin or adhesive, after mixing with catalyst, solvent, or other compounding ingredients, remains usable.

yield value (yield strength)—The lowest stress at which a material undergoes plastic deformation. Below this stress, the material is elastic; above it, viscous.

Young's modulus of elasticity—The modulus of elasticity in tension. The ratio of stress in a material subjected to deformation.

2
POLYMER CHEMISTRY

The chemical structure and nature of plastics materials have a significant relationship not only to the properties of the plastic but to the ways in which it can be processed, designed, or otherwise translated into an end-product. Throughout this volume the reader will find various references to polymer chemistry (e.g., the section entitled "Theory of injection molding," in Chapter 4 on "Injection Molding").

This chapter therefore will provide a basic review of polymer chemistry, with emphasis on the distinctions between various structures and their influence on the engineering of plastics products.

Forming Polymers

Basically, all polymers are formed by the creation of chemical linkages between relatively small molecules or monomers to form very large molecules or polymers; the same idea as connecting boxcars on a railroad to form a train; the boxcars being monomers and the train being formed a polymer. Like a boxcar the molecules must at least have the power to be coupled at either end and to continue coupling on added boxcars after one end is connected.

Actually, in polymer formation, the process

* by George Smoluk, former Editor, *SPE Journal*. As adapted from "Polymer Chemistry", appearing in the *SPE Journal*, June, 1970, Volume 26. (*SPE Journal* is now known as *Plastics Engineering*).

is more like forming many, many trains in a railroad yard simultaneously from the boxcars available in the yard in a competitive fashion so that the switching engine which moved the fastest would form the longest train while the slowest switching engine would form the shortest train, due to depletion of the available rolling stock by the concurrent train-forming process. The train-forming process of polymerization comes to a stop when factors prevent any additional boxcars from being added to any of the trains being assembled. Thus, we ultimately end up with trains having a variety of lengths, yet all composed of the essentially identical boxcars.

The railroad analogy described above is basically what happens in an "addition" type polymerization.

The above process is characterized by the simple combination of molecules without the generation of any by-products formed as the result of the combination. The molecules which combine do not decompose to produce fission products which then remain as part of the reaction debris or need to be removed from the reaction to either allow it to continue the molecule-building process or to insure the formation of a pure polymer.

In reality the addition-type process can occur in several ways. One way simply involves the external chemical activation of molecules that cause them to start combining with each other in a chain reaction type fashion (by the bonding of atoms directly within the reacting molecule). Another way for an addition polymerization to occur is through a rearrangement of atoms within both reacting molecules, but still without the net loss of any atoms from the polymer molecule. And still a third way for addition polymerization to occur is for a molecule composed of a ring of atoms to open up and connect with other ring type molecules being opened up under the influence of the proper catalytic activators; once again with no net loss of any atoms from the polymer structure.

In another type of polymerization reaction which has been called "condensation" polymerization, the chemical union of two molecules can only be achieved by the splitting out of a molecule (usually small) formed by the atoms which must be removed from the two molecules being joined to allow the coupling process or polymerization to continue. This is the type of polymerization which is involved in the formation of some nylons, phenolics, amino resins, and polyester pre-polymers.

Normally the reaction by-product in a condensation type of polymerization must be immediately removed from the reacting polymer since it may either inhibit further polymerization or appear as an undesirable impurity in the finished polymer.

There is yet another method by which polymers may be formed but it is in reality simply a sequential combination of the previous two processes. Such a process is used in the formation of plastics such as the polyesters and the polyurethanes.

In such a polymerization, a condensation reaction is usually carried out first to form a relatively small polymer which is then capable of undergoing further reaction by addition polymerization to form larger polymer molecules with a third ingredient. This is what is done when a polyester is first formed by a condensation reaction and the then still-active polyester is reacted with styrene to form what is essentially a polyester-styrene copolymer.

The various types of polymerization reactions are shown in Fig. 2-1a, Fig. 2-1b, and Fig. 2-1c.

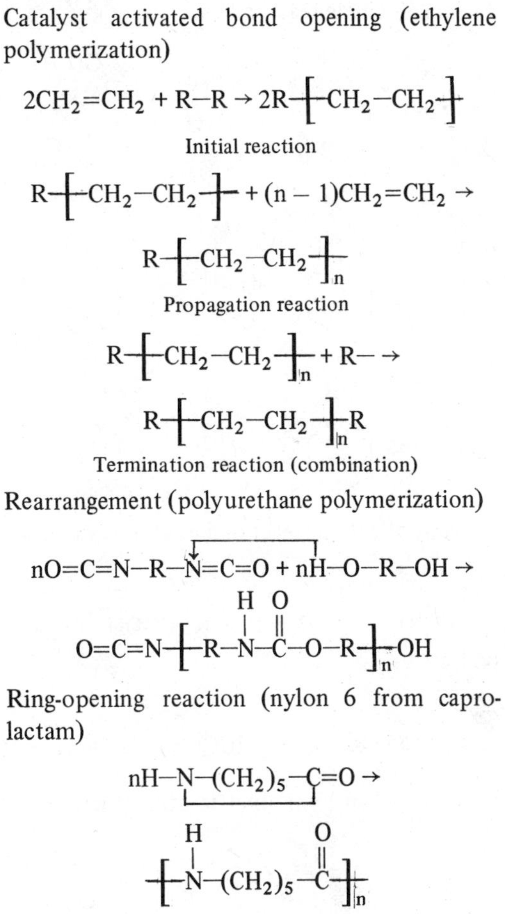

Catalyst activated bond opening (ethylene polymerization)

$$2CH_2{=}CH_2 + R{-}R \rightarrow 2R{-}\!\!\left[CH_2{-}CH_2\right]\!\!-$$

Initial reaction

$$R{-}\!\!\left[CH_2{-}CH_2\right]\!\!- + (n-1)CH_2{=}CH_2 \rightarrow$$

$$R{-}\!\!\left[CH_2{-}CH_2\right]_n$$

Propagation reaction

$$R{-}\!\!\left[CH_2{-}CH_2\right]_n + R{-} \rightarrow$$

$$R{-}\!\!\left[CH_2{-}CH_2\right]_n\!\!R$$

Termination reaction (combination)

Rearrangement (polyurethane polymerization)

$$nO{=}C{=}N{-}R{-}N{=}C{=}O + nH{-}O{-}R{-}OH \rightarrow$$

$$O{=}C{=}N{-}\!\!\left[R{-}N{-}C{-}O{-}R\right]_n\!\!OH$$
$$\qquad\qquad\;\; \overset{H}{|}\;\; \overset{O}{\|}$$

Ring-opening reaction (nylon 6 from caprolactam)

$$nH{-}N{-}(CH_2)_5{-}C{=}O \rightarrow$$

$$\left[N{-}(CH_2)_5{-}C\right]_n$$
$$\;\overset{H}{|}\qquad\qquad\;\; \overset{O}{\|}$$

Fig. 2-1a. Typical addition polymerizations (no by-products).

Polymerization Techniques

In actual practice there are many different techniques used to carry out polymerization reactions; however, most involve one of four general methods of polymerization. These include the polymerization of the monomer or reactants in bulk, in solution, in suspension, and in emulsion forms. The bulk and solution methods are used for the formation of both addition and condensation type polymers, whereas suspension and emulsion techniques are largely used for addition polymerizations.

Bulk polymerization. This type of polymerization involves the reaction of monomers or reactants among themselves without placing them in some form of extraneous media such as is done in the other types of polymerizations.

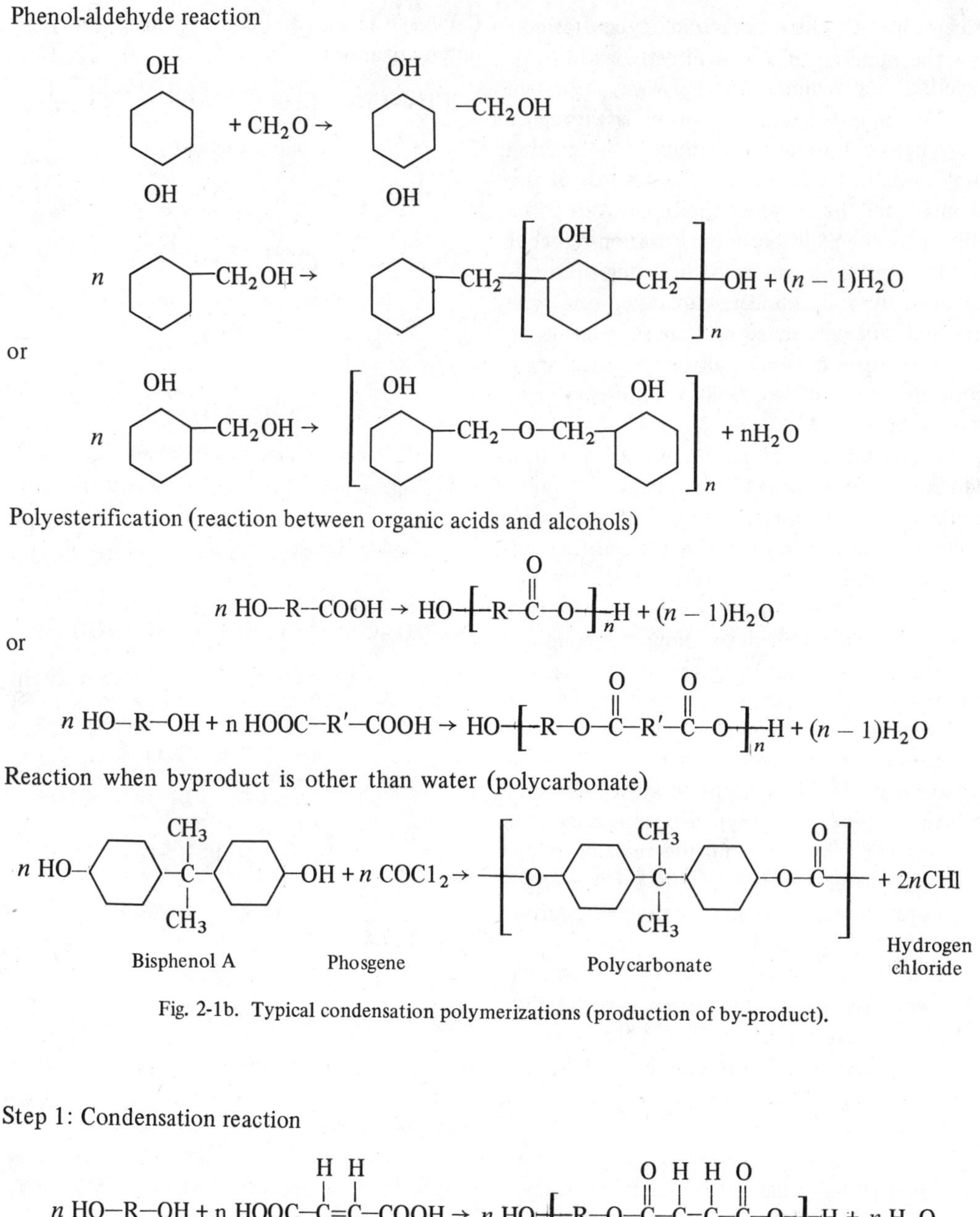

Phenol-aldehyde reaction

or

Polyesterification (reaction between organic acids and alcohols)

$$n\ HO{-}R{-}COOH \rightarrow HO{-}\!\left[R{-}\overset{\displaystyle O}{\overset{\|}{C}}{-}O\right]_n\!{-}H + (n-1)H_2O$$

or

$$n\ HO{-}R{-}OH + n\ HOOC{-}R'{-}COOH \rightarrow HO{-}\!\left[R{-}O{-}\overset{\displaystyle O}{\overset{\|}{C}}{-}R'{-}\overset{\displaystyle O}{\overset{\|}{C}}{-}O\right]_n\!{-}H + (n-1)H_2O$$

Reaction when byproduct is other than water (polycarbonate)

Bisphenol A   Phosgene   Polycarbonate   Hydrogen chloride

Fig. 2-1b. Typical condensation polymerizations (production of by-product).

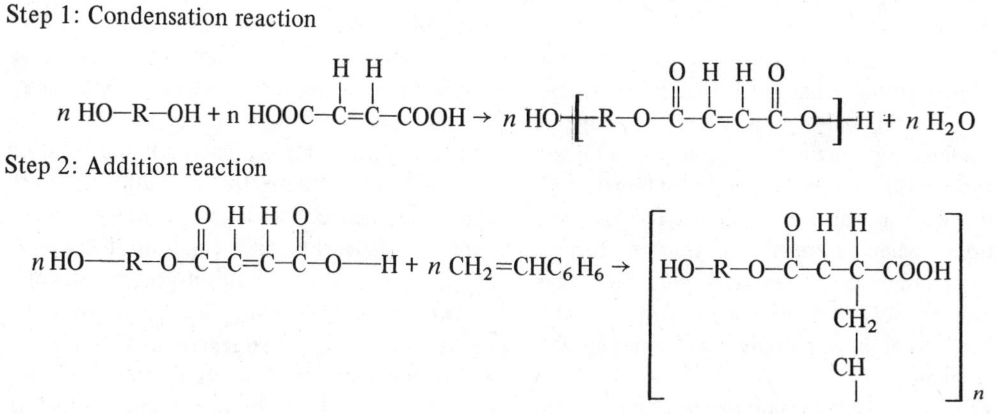

Step 1: Condensation reaction

$$n\ HO{-}R{-}OH + n\ HOOC{-}\overset{H}{\overset{|}{C}}{=}\overset{H}{\overset{|}{C}}{-}COOH \rightarrow n\ HO{-}\!\left[R{-}O{-}\overset{O}{\overset{\|}{C}}{-}\overset{H}{\overset{|}{C}}{=}\overset{H}{\overset{|}{C}}{-}\overset{O}{\overset{\|}{C}}{-}O\right]{-}H + n\ H_2O$$

Step 2: Addition reaction

Fig. 2-1c. Combination polymerizations (curing of polyesters).

Two types of behaviour are observed in bulk polymerizations. In one case, the polymer is soluble in the monomer during all stages of the polymerization and a monomer-soluble initiator is used. As polymerization progresses, viscosity increases significantly, and chain growth takes place in the monomer or polymers dissolved in the monomer until all of the monomer is consumed.

In the second case, the polymer is insoluble in the monomer system. In such systems, the polymerization is believed to occur within the growing polymer chains since very high molecular weights are formed even though the polymer chain drops out of the monomer solution.

One of the disadvantages of carrying out a polymerization in bulk is the fact that the rise in viscosity can interfere with keeping reaction conditions under control due to the difficulty of maintaining proper agitation and removing heat from exothermic polymerization reactions which give off heat. However, the process is widely used.

Solution polymerization. Solution polymerization is similar to bulk polymerization except that whereas the solvent for the forming polymer in bulk polymerization is the monomer, the solvent in solution polymerization is usually a chemically inert medium. The solvents used may be complete, partial, or nonsolvents for the growing polymer chain.

When monomer and polymer are both soluble in the solvent, initiation and propagation of the growing polymer chains take place in the oil or organic phase. Because of the mass-action law, rates of polymerization in solvents are slower than in bulk polymerizations and the molecular weight of the polymers formed is decreased.

In another case when the monomer is soluble in the solvent but the polymer is only partially soluble or completely insoluble in the solvent, initiation of the polymerization takes place in the liquid phase. However, as the polymer molecules grow, some of the propagation of polymers takes place within monomer swollen molecules which are beginning to precipitate from the reaction. When this occurs it again becomes possible to build up molecular weights because of the decreased dilution within the polymers. Thus, molecular weights as high as those possible with bulk polymerizations can also be achieved in solution polymerizations provided the polymer precipitates out of solution as it is formed and creates a propagation site.

In the third case, in which the polymer is completely insoluble in the solvent and the monomer is only partially soluble in the solvent, rates of reaction are reduced and lower molecular weights, below those possible in bulk polymerizations, are formed. However, the formation of relatively high molecular weight polymers is still possible in such a system.

In addition to the relative solubilities of monomer, polymer, and solvent in the system, the way in which the ingredients are fed to the system can also have a significant effect on how the polymerization proceeds, and hence the structure of the finished polymer.

Suspension polymerization. Often called "pearl" polymerization, this technique is normally used only for catalyst-initiated or free-radical addition polymerizations. The monomer is mechanically dispersed in a liquid, usually water, which is a nonsolvent for the monomer as well as for all sizes of polymer molecules which form during the reaction.

The catalyst initiator is dissolved in the monomer and it is preferable that it does not dissolve in the water so that it remains with the monomer. The monomer and polymer being formed from it remain within the beads of organic material dispersed in the phase. Actually suspension polymerization is essentially a finely divided form of bulk polymerization. The advantage of the suspension polymerization over bulk is that it allows the operator to effectively cool exothermic polymerization reactions and thus maintain closer control over the chain building process. Other behavior is the same as bulk polymerization.

By controlling the degree of agitation, monomer-to-water ratios, and other variables, it is also possible to control the particle size of the finished polymer, thus eliminating the need for reforming the material into pellets from a melt such as is usually necessary with bulk polymerizations.

Emulsion polymerization. This is a technique in which addition polymerizations are carried out in a water medium containing an emulsifier (a soap) and a water-soluble initiator. It is used because emulsion polymerization is much more rapid than bulk or solution polymerization at the same temperatures and produces polymers with molecular weights much greater than those obtained at the same rate in bulk polymerizations. The polymerization reaction in emulsion polymerization involves causing the reaction to take place within a small hollow sphere composed of a film of soap molecules, called a *micelle*. Monomer diffuses into these micelles and control of the soap concentration, overall reaction-mass recipe, and reaction conditions provide additional controls over the reaction.

Polymerization techniques can have a significant effect on the number, size, and characteristics of the polymer molecules formed and will thus have a significant effect on the properties of the polymer. Thus, batches of a polymer such as polystyrene, which can be made by any of the four polymerization techniques described above, will differ depending on which type of polymerization method was used to make the material.

Weight, Size and Distribution of Molecules

Because there is such diversity among polymer molecules, a number of techniques for defining and quantifying these characteristics are in use by the industry—and are also of value to processors and end-users as a determinant of polymer properties.

One such parameter relates to the size of the molecules in the polymer and is known as *molecular weight* (MW). MW refers to the average weight of the molecules in the mixture of different size molecules that make up the polymer. It is expressed either as a number average based on the sum of the number fractions of the weight of each species or size of molecule present or as a weight average based on the weight fractions of each species or size of molecule present in the polymer.

The molecular weight of a polymer has a

significant effect on its properties. Thus, higher molecular weight polymers tend to be tougher and more chemically resistant, whereas low molecular weight polymers tend to be weaker and more brittle. In the polyethylene family, for example, low molecular weight polyethylenes are almost wax-like in characteristics, whereas ultra-high molecular weight polyethylenes offer outstanding chemical resistance and toughness (although, conversely, the higher the molecular weight goes, the more energy in the form of temperature and pressure is required to process the material).

Another expression applicable to molecular weight is the *degree of polymerization* (DP). This refers to the number of monomer molecules that combine to form a single polymer molecule and is estimated by dividing the number-average molecular weight of the polymer by the molecular weight of the monomer. The relationship can be expressed as so:

MW of polymer = DP x MW of monomer

And finally, it is important to know something about the molecular weight distribution within the polymer, that is, the relative proportions of molecules of different weight. Obviously, if one could create a mono-disperse polymer, all of the molecules would be of a single size. This has not been achieved commercially, however, and so another parameter of definition used to describe polymers is the distribution of the various sizes of molecules within a poly-disperse polymer; that is, the breadth of distribution or ratio of large, medium, and small molecular chains in the resin. If the resin is made up of chains close to the average length, it is called narrow; if it is made up of chains of a wider variety of lengths, it is called broad.

In general, narrow-range resins have better mechanical properties than broad-range resins, although, as with the case of the higher molecular weight materials, they are somewhat more difficult to process.

The molecular weights of polymers and molecular-weight distribution are determined indirectly by measuring the properties of the polymers themselves or of their solutions and correlating this information with the type of molecular weight believed to correspond to the type of property measured.

For example, chemical analysis of the end-groups present in polymer molecules, studies of the boiling points and freezing points of solutions of polymers, and osmotic pressure studies on polymer solutions yield data on the number-average molecular weight of the polymer. Light scattering in polymer solutions and sedimentation methods in the ultra-centrifuge yield data related to the weight-average molecular weights of polymers. Such methods yield the overall average molecular weight of the polymer samples. To get direct data on molecular-weight distributions, the polymers must first be separated by fractionation methods into rather sharp fractions of samples of relatively mono-disperse molecular weight.

Polymer Structure

In addition to the size of molecules and the distribution of sizes in a polymer, the shapes or structures of individual polymer molecules also play a major role in determining the properties of a plastic.

Earlier, in the discussion of the analogy between the connecting of railroad boxcars to form a train and the formation of polymers, it was implied that polymers form by aligning themselves into long chains of molecules without any side protrusions or branches, or lateral connections, between molecules. Some polymers do largely this and nothing more; however, it is also possible for polymers to form more complex structures. Thus, polymer molecules may form in the shape of branched molecules, in the form of giant three-dimensional networks, in the form of linear molecules with regular lateral connections to form "ladder-type" polymers, in the form of two-dimensional networks or platelets, and so forth, depending on how many connections or bonds can exist between the mono-disperse monomeric molecules which were used to form the polymer and between sites on the forming or already formed polymer molecules. See Fig. 2-2.

Because of the geometry of such molecules, some can come closer together than others in which the structure prevents more intimate contact. The structural obstruction to close approach is called *steric hindrance*. Thus, polymers which can be packed closely or exhibit little steric hindrance can ordinarily more easily form crystalline structures in which the molecules are aligned in some regular pattern. Others, such as polymers which are cross-linked prior to crystallization, are prohibited from aligning themselves in crystals due to the hindrance created by the multiple interconnections and hence tend to be amorphous, or noncrystalline.

Amorphous polymers do not have melting points, but rather softening ranges, are normally transparent, and undergo only small volume changes when solidifying from the melt, or when the solid softens and becomes fluid. Crystalline polymers, on the other hand, have considerable order to the molecules in the solid state, indicating that many of the other atoms are regularly spaced, have a true melting point with a latent heat of fusion associated with the melting and freezing process, and a relatively large volume change during the transition from melt to solid.

Thus, many different structures are possible with plastics—and each will affect the basic properties of the polymer. For example, linear polymers, like high-density polyethylene, are made of molecules which resemble spaghetti in a bowl and are relatively free to slide over one another or to pack more closely together (in the absence of steric hindrances due to large pendant side groups). Branched polymers, like low-density polyethylene, on the other hand, have side appendages and interconnections that cause the molecules to resemble clumps of tree branches that cannot be easily compressed or compacted. Thus, branched polymers (with more voids) are more permeable to gases and solvents than linear polymer, lower in density (since the molecules are not compacted together), and more flexible. Linear polymers, on the other hand, are higher in tensile, stiffness and softening temperatures.

Cross-linked structures, in which the individual chain segments are strongly bound together by chemical unions, also have special characteristics (as in the family of thermo-

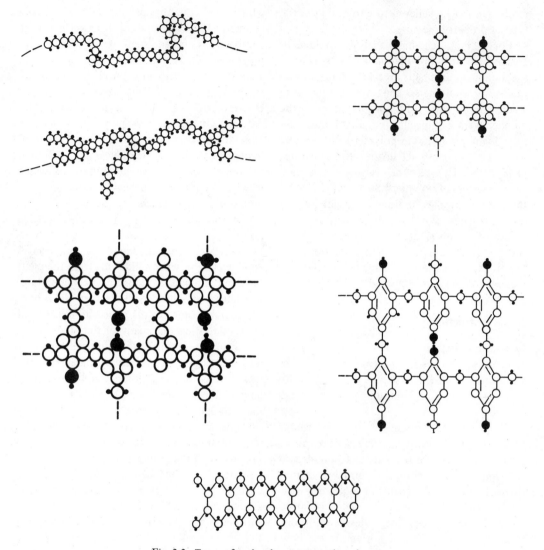

Fig. 2-2. Types of molecular structures in polymers.

setting plastics). They do not exhibit creep or relaxation unless such primary bonds actually are broken by continually applied stress or by elevated temperatures high enough to cause chemical decomposition of the polymer. Cross-linked polymers are also fairly resistant to solvent attack; solvents may swell such polymers, but seldom cause complete rupture or dissolution.

Ladder structures have unusual stability and have become important in terms of the new heat-resistant plastics. Aromatic compounds (such as benzene) and heterocyclic compounds (such as benzimidazole) that have semi-ladder, ladder, or spiro structures offer heat stabilities in excess of 900°F.

Effect of Time and Temperature

It should also be noted that whereas the effect of molecular weight and molecular-weight distribution on properties is relatively fixed and stable with temperature (barring decomposition of the polymer), the arrangement of the molecules within the structure of a polymer mass is in most cases relatively sensitive to temperature. Thus, the structure of any given polymer can be significantly changed by exposing it to different temperatures and thermal treatments.

For example, heating a crystalline-type polymer above its melting point and then quenching it can produce a polymer that is far more amorphous or noncrystalline in structure than the original polymeric sample. Such a quenched material can have properties that are significantly different from the properties of a sample that is cooled slowly and allowed to re-crystallize.

The effects of time on a polymer structure are similar to those of temperature in the sense that any given polymer has a "most preferred" or equilibrium structure in which it would prefer to arrange itself but is prevented from doing so instantaneously on short notice by steric hindrances. However, given enough time, the molecules in a polymer ultimately migrate to arrange themselves in this form. Elevating the temperature and making the molecules more mobile or spreading them apart allows them to accomplish this in a shorter time and vice versa.

Thus over an extended period of time, the properties of a polymer can become significantly different from those measured earlier if the structure of the polymer was in an unstable form when the properties were initially evaluated.

Chemical Composition

In addition to all the variations in the make-up of the polymer discussed above and having to do largely with extra- or intermolecular phenomena, polymer properties are also heavily dependent on such factors as the intramolecular chemical composition of the polymer.

For example, as shown in Fig. 2-3, polyethylene consists (except for catalyst or extraneous impurities) completely of atoms of carbon and hydrogen. This internal make-up of the polymer molecule affects in turn all the previously discussed variables and hence contributes its own basic characteristics to the overall properties of the polymer.

Going further, if one takes every fourth hydrogen atom occurring in polyethylene and replaces it with a methyl (CH_3) group in regular intervals along the length of the chain-like molecule as shown in Fig. 2-3, the polyethylene

is transformed into polypropylene. In such a case, even though the degree of polymerization existing and the molecular-weight distribution remain roughly the same, the spacing of the molecules in the polymer matrix or the morphology of the polymer changes and hence so do its macroscopic physical properties.

Similarly, if every fourth hydrogen atom of the polyethylene were substituted with a benzene ring structure (C_6H_6), or with a chlorine atom (Cl), as shown in Fig. 2-3, the ethylene would become either polystyrene or polyvinyl chloride, respectively.

The simple substitution of a single hydrogen atom with another atom or chemical group can cause a drastic change in the properties of a polymer. Whereas polyethylene is translucent, flexible, and crystalline, polystyrene is transparent, brittle, and amorphous.

Copolymers

In addition to making changes in the basic repeating unit by substitution as illustrated above, it is also possible to change the chemical composition and hence the morphology and properties of a polymer by mixing the types of structural groups or basic repeating units within the chain of a polymer. This is done by the process called *copolymerization*. In such a reaction, for example, monomers of styrene and acrylonitrile can be reacted to form styrene-acrylonitrile copolymers; or the styrene may be reacted with butadiene to form styrene-butadiene copolymers; or the acrylonitrile, butadiene, and styrene may all be reacted simultaneously to form the now familiar ABS (acrylonitrile-butadiene-styrene) copolymers. Any of the polymerization techniques previously discussed can be used.

In addition to varying the types of starting ingredients used to form copolymers, the relative amounts of each monomer used in the reaction may also be varied to produce a literally unlimited number of possible permutations and combinations of types and amounts of monomers. It is this ability of the polymer chemist to react different monomers in different amounts that has given rise to the concept of "tailor-making" plastic materials.

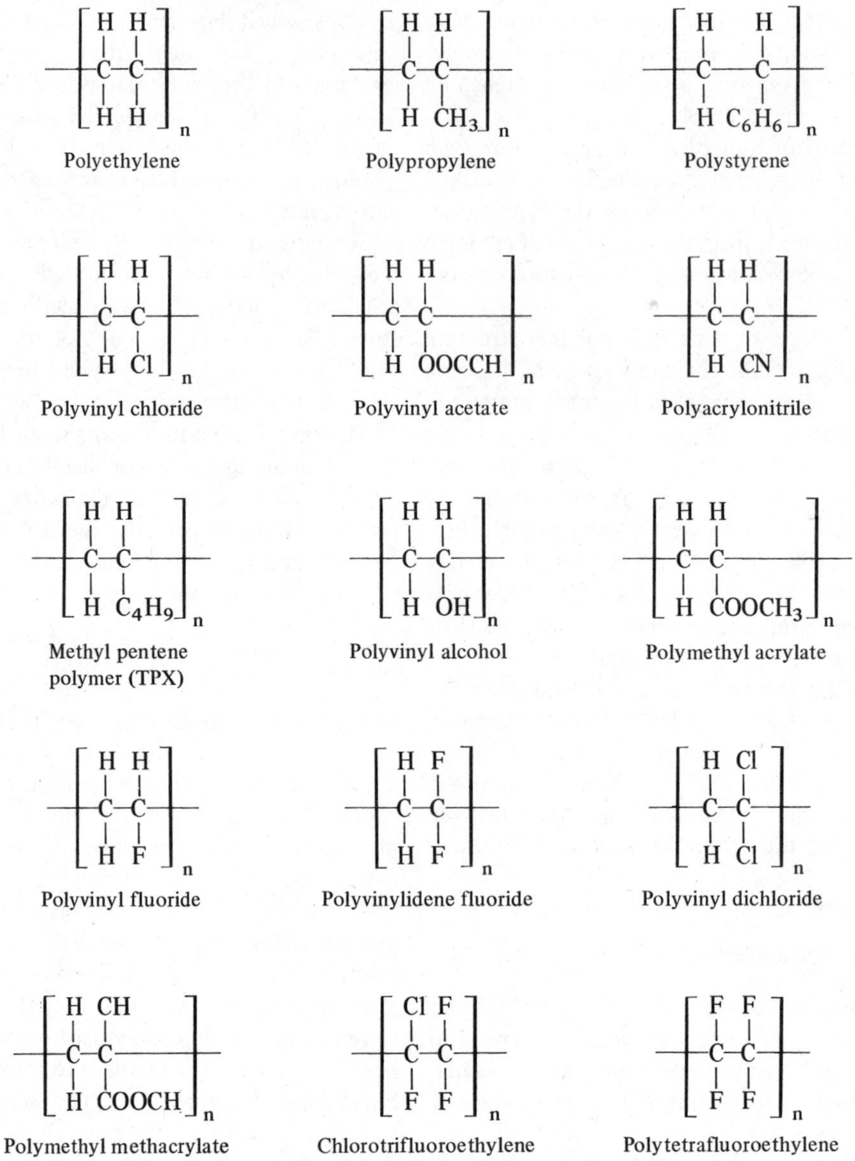

$$\left[\begin{array}{cc} H & H \\ | & | \\ -C - C - \\ | & | \\ H & H \end{array}\right]_n$$

Polyethylene

$$\left[\begin{array}{cc} H & H \\ | & | \\ -C - C - \\ | & | \\ H & CH_3 \end{array}\right]_n$$

Polypropylene

$$\left[\begin{array}{cc} H & H \\ | & | \\ -C - C - \\ | & | \\ H & C_6H_6 \end{array}\right]_n$$

Polystyrene

$$\left[\begin{array}{cc} H & H \\ | & | \\ -C - C - \\ | & | \\ H & Cl \end{array}\right]_n$$

Polyvinyl chloride

$$\left[\begin{array}{cc} H & H \\ | & | \\ -C - C - \\ | & | \\ H & OOCCH \end{array}\right]_n$$

Polyvinyl acetate

$$\left[\begin{array}{cc} H & H \\ | & | \\ -C - C - \\ | & | \\ H & CN \end{array}\right]_n$$

Polyacrylonitrile

$$\left[\begin{array}{cc} H & H \\ | & | \\ -C - C - \\ | & | \\ H & C_4H_9 \end{array}\right]_n$$

Methyl pentene
polymer (TPX)

$$\left[\begin{array}{cc} H & H \\ | & | \\ -C - C - \\ | & | \\ H & OH \end{array}\right]_n$$

Polyvinyl alcohol

$$\left[\begin{array}{cc} H & H \\ | & | \\ -C - C - \\ | & | \\ H & COOCH_3 \end{array}\right]_n$$

Polymethyl acrylate

$$\left[\begin{array}{cc} H & H \\ | & | \\ -C - C - \\ | & | \\ H & F \end{array}\right]_n$$

Polyvinyl fluoride

$$\left[\begin{array}{cc} H & F \\ | & | \\ -C - C - \\ | & | \\ H & F \end{array}\right]_n$$

Polyvinylidene fluoride

$$\left[\begin{array}{cc} H & Cl \\ | & | \\ -C - C - \\ | & | \\ H & Cl \end{array}\right]_n$$

Polyvinyl dichloride

$$\left[\begin{array}{cc} H & CH \\ | & | \\ -C - C - \\ | & | \\ H & COOCH \end{array}\right]_n$$

Polymethyl methacrylate

$$\left[\begin{array}{cc} Cl & F \\ | & | \\ -C - C - \\ | & | \\ F & F \end{array}\right]_n$$

Chlorotrifluoroethylene

$$\left[\begin{array}{cc} F & F \\ | & | \\ -C - C - \\ | & | \\ F & F \end{array}\right]_n$$

Polytetrafluoroethylene

Fig. 2-3. Polymers based on the ethylene chain.

Figure 2-4 illustrates some of the more common copolymers now available commercially. The subscripts of x, y, and z denote differing amounts of each of the structural units in the polymer chain. It should be noted that the structural formulas in Fig. 2-4 are somewhat fictional in the sense that there is an implication that a long chain of one group is connected to a long chain of another group, and successively to a long chain of still a third type of polymer. Actually, the distribution of each of the species within the chain will depend both on the amount present and how the polymerization was carried out.

For example, an acrylonitrile-styrene copolymer might be made by first reacting a good portion of the acrylonitrile to form polyacrylonitrile and then adding the styrene and remaining acrylonitrile to complete the polymerization of the desired end product. Or, in another

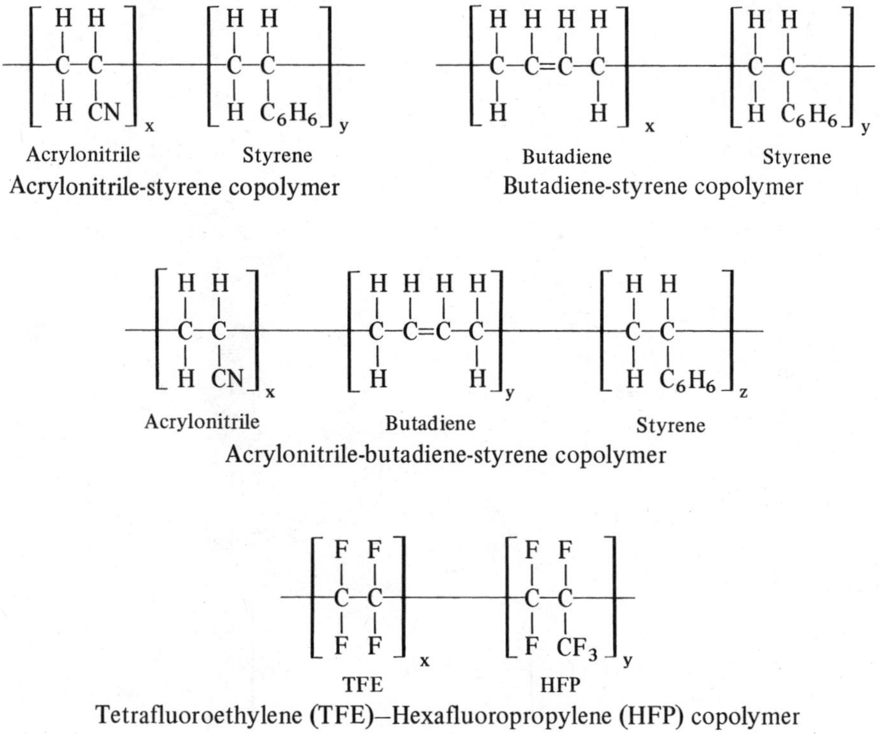

Fig. 2-4. Carbon "backbone" copolymers.

case, both ingredients might be added to a reactor simultaneously and allowed to react concurrently.

Other Polymers

From a study of Figs. 2-3 and 2-4, it will be noted that all of the polymers listed there are formed by combination of polymerization made possibly by the reaction of the double bonds that exist between adjacent carbon atoms in the monomers used as starting materials. However, as mentioned earlier, this is not the only way in which polymers are formed. When polymers are formed by rearrangements or condensation reactions other types of polymeric chains can be formed. Figure 2-5 lists the structural formulas of several commercially important polymers in which oxygen (O) is an integral part of bonds holding the polymer together. Similarly, Fig. 2-6 lists several types of polyamides and a polyurethane in which nitrogen (N) atoms form

the bonds between portions of the polymer molecule. In some cases the polymers are formed by molecular rearrangements (in the case of polyurethanes and epoxies) and in other cases by condensation reactions (some nylons).

Other condensation polymers are shown in Fig. 2-7 which shows some of the complex structures proposed for the more common types of thermosets. Note that the thermosets always have more than two linkages connecting the various structural units in the polymer and are commonly referred to as cross-linked materials involving a network-like molecular structure.

There are at least 30 to 40 different families of thermoplastics now offered and about 10 different basic families of thermosets. And this takes into account only the more common types of polymers. It should also be noted that copolymers, mixtures, and chemically modified versions of the polymers listed make the total list of polymeric materials which are theoreti-

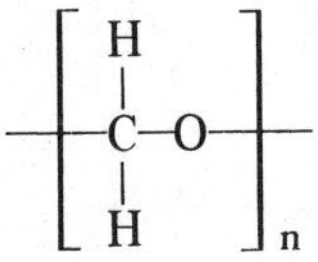

Polyacetal resin

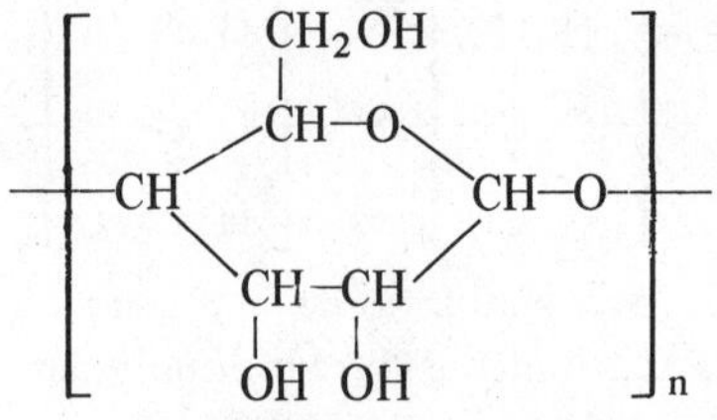

Cellulose (natural polymer)

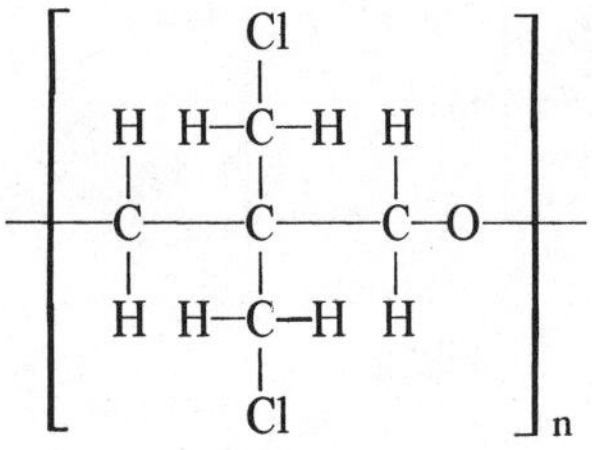

Chlorinated polyether

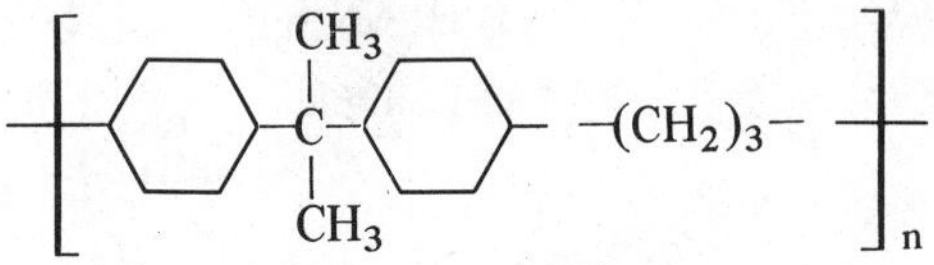

Phenoxy resin (polyhydroxyether)

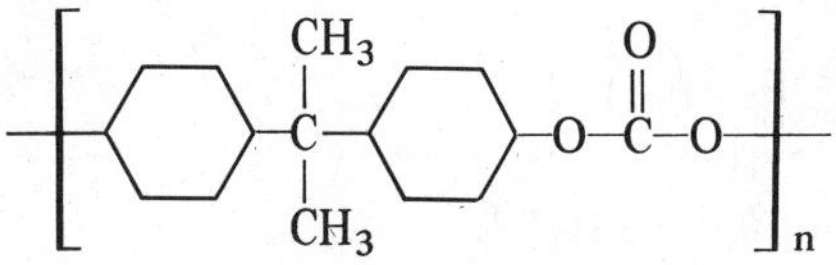

Polycarbonate

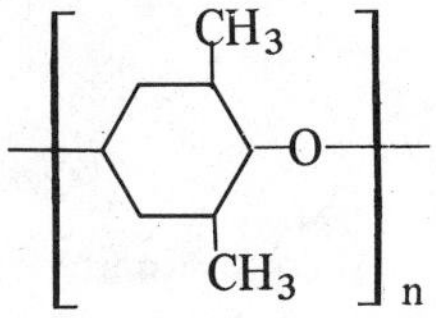

Polyphenylene oxide

Fig. 2-5. Polymers with oxygen in the chain.

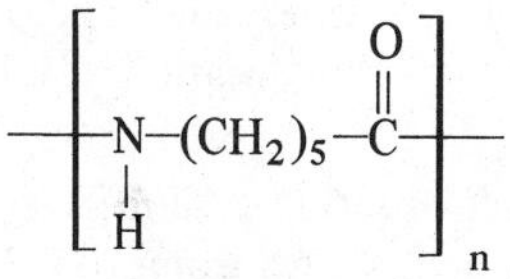

Nylon 6

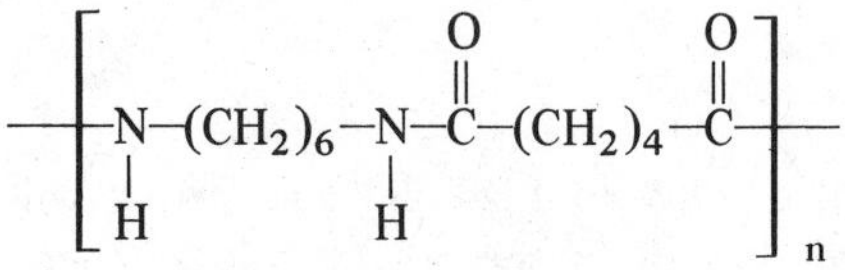

Nylon 6/6

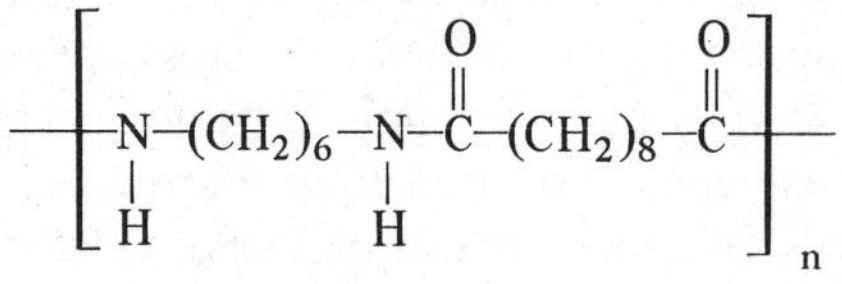

Nylon 6/10

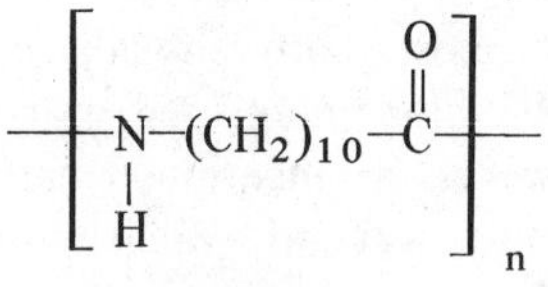

Nylon 11

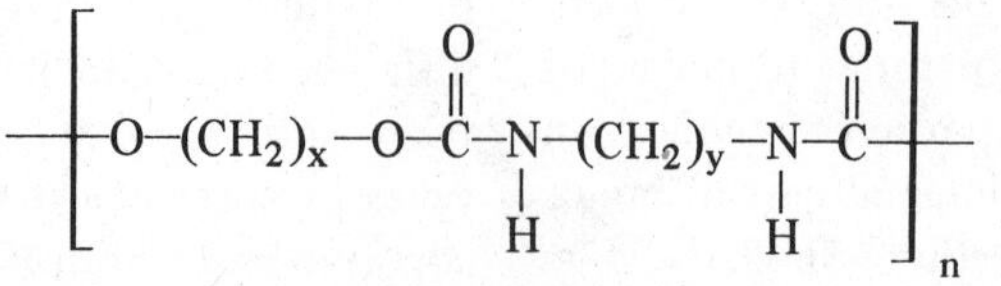

Polyurethanes

Fig. 2-6. Polymers with nitrogen in the chain.

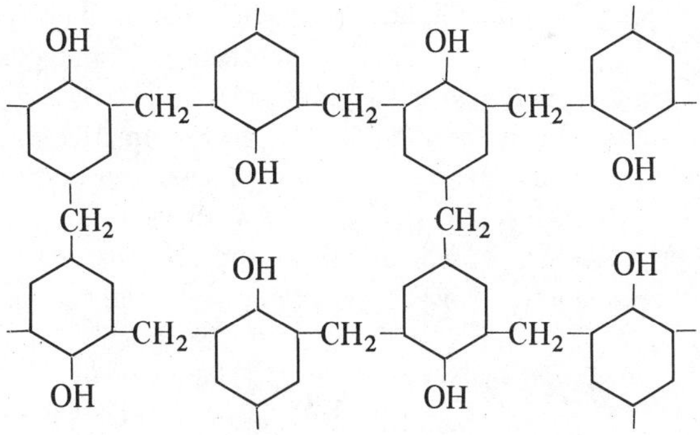

Phenol-formaldehyde resin

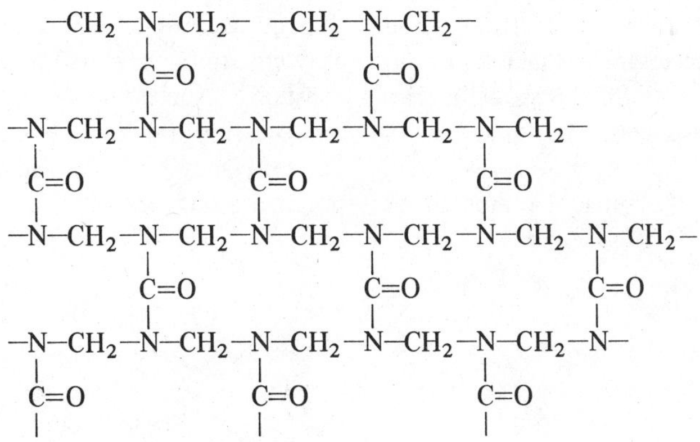

Urea-formaldehyde

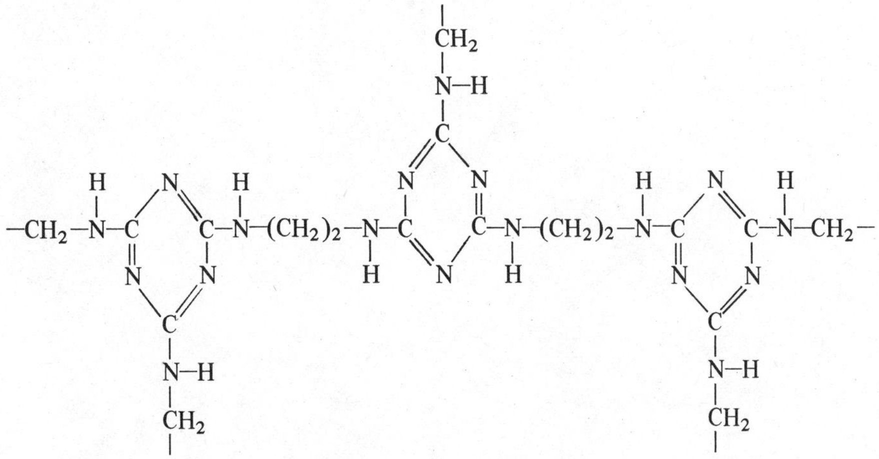

Melamine-formaldehyde

Fig. 2-7. Thermoset structures.

cally possible, entirely too long to enumerate in any one place.

In addition, the properties of a polymer may be radically altered by mixing it with non-polymeric materials and chemicals. In fact, some polymers would be worthless to the molder or extruder if additives were not used to modify them before processing into finished products.

For example, phenolic, urea, and melamine resins as they come out of the polymerization kettle are largely brittle, frangible solids with low impact strength. To make them usable, such materials must be mixed with a filler of some kind to reinforce the strength of the polymer. In the case of the three thermosets mentioned, cellulose fillers made from wood flour or cotton are necessary to make the resins commercially moldable. Similarly, cellulosics and vinyl resins must also be mixed with semi-solvents, called *plasticizers*, to soften them to some degree, so they may be molded or extruded, or to modify their properties (to make them soft and pliable, rather than hard and stiff).

In actual practice the number of different types of chemical additives and fillers that are mixed with the fundamental polymer is so large that the chemistry of plastic compounds becomes extremely complex.

It is also possible to vary existing polymers by any number of chemical, mechanical, or irradiation techniques to change the structure of the polymer. During processing, for example, it is possible to align or orient the molecules in a polymer to significantly increase the strength of the polymer in the direction of the orientation. Chemical or irradiation methods can also be used to cross-link a plastic, like polyethylene, to improve its toughness and chemical resistance. Nucleation techniques and chemical modification (as in the case of chlorinated polyethylene) are further adaptable to changing polymer structures.

3

ENGINEERING AND
DESIGN GUIDE TO PLASTICS

As indicated in Chapter 2 on Polymer Chemistry, the family of plastics is extraordinarily varied and complex. There are, however, some fairly broad and basic guidelines that can be followed when engineering or designing a product to be made out of plastics. In this chapter, the major groupings of plastics materials will be analyzed in terms of their engineering properties and characteristics. This will be followed by a review of the basic structural design considerations to be followed when working with plastics.

The Family of Plastics

Plastics generally are organic high polymers (i.e., they consist of large chain-like molecules containing carbon) which are formed in a plastic state either during or after their transition from a small-molecule chemical to a solid material. Stated very simply, the large chain-like molecules are formed by hooking together short-chain molecules of chemicals (monomers) in a reaction known as polymerization. When chemically similar monomers are hooked together, the resulting plastic is a polymer, such as polyethylene made from the ethylene monomer. When chemically dissimilar monomers are hooked together, the process is called copolymerization and the resulting plastic is a copolymer, such as PVC copolymers made from monomeric vinyl chloride and vinyl acetate.

In the broad classification of plastics, there are two generally accepted categories: thermoplastic resins and thermosetting resins.

Thermoplastic resins consist of long molecules, each of which may have side chains or groups that are not attached to other molecules (i.e., are not cross-linked). Thus, they can be repeatedly softened and hardened by heating and cooling. Usually, thermoplastic resins are purchased as a granular polymer. Heat softens the material so that it can be formed; subsequent cooling hardens it in the final desired shape. No chemical change generally takes place during forming.

In thermosetting resins, reactive portions of the molecules form cross-links between the long molecules during polymerization; and thus, once "polymerized" or "cured," the material cannot be softened by heating. Thermosets are usually purchased as a liquid monomer-polymer mixture or as a partially polymerized molding compound. In this uncured condition, they can be formed to the finished shape with or without pressure and polymerized with chemicals or heat.

As technology progresses, the line between thermoplastics and thermosets has become less distinct. On the one hand, more and more processes are evolving which make use of the economic processing of thermoplastics and then convert the material to essentially a thermoset, e.g., extruded polyethylene wire coating which is cross-linked after extrusion (either chemically or by irradiation) to form what is actually a thermoset material that cannot be subsequently melted by heating. On the other hand, materials and machinery have become modified to provide the economics of thermoplastic processing to thermosetting materials, e.g., injection molding of phenolics and other thermosetting molding materials.

Some plastics materials even have members on both sides of the fence. There are, for example, both thermoset polyesters and thermoplastic polyesters.

The distinction between thermoplastics and thermosetting resins, however, is still widely made in the industry and will be adapted to this discussion.

It should be noted that many of the data figures used are averages and are not to be taken as a guide to selection of any material. Since plastics are so adaptable to modification, various reinforced or modified grades are on the market with properties well above those discussed here. The use of data in this chapter is intended primarily to place the various plastics in perspective within the overall family of plastics. For a quick broad comparison, the reader is referred to Table 3-1.

Thermoplastic Resins

ABS resins—ABS materials contain acrylonitrile, butadiene, and styrene monomers in varying proportions, both as terpolymers and blends. Originally an outgrowth of polystyrene modification, the materials have properties so distinct from those of polystyrene that they are classified separately.

ABS materials provide a balanced combination of mechanical toughness (Izod impacts can range from 2 ft-lb/in. of notch to 10 ft-lb/in. of notch), wide service temperature range (−40 to 240°F), good dimensional stability, good chemical resistance, good electrical insulating properties and ease of fabrication.

ABS materials are generally subdivided into five major classifications as follows:

1. *Medium impact.* A hard, rigid tough material used for parts that must have high strength, good fatigue resistance, surface hardness, and gloss.

2. *High impact.* Used where additional impact strength is required and can be justified at the expense of rigidity and hardness.

3. *Extra hard impact.* With further decreases in rigidity, hardness and strength, this material has the highest impact resistance.

4. *Low temperature impact.* This material is designed for high impact strength at temperatures as low as −40°F. Strength and rigidity again suffer, as does heat resistance.

5. *High strength, heat resistant.* A widely used ABS, this type provides maximum heat resistance (maximum recommended continuous service temperature is about 200°F; at 264 psi, heat distortion point is about 185-215°F). Its impact strength is comparable to that of the high impact type, but it has higher tensile and flexural strengths, stiffness, and hardness.

Like other amorphous polymers, ABS materials have a relatively flat stress-temperature curve. Thus, an increase from 66 to 264 psi drops the heat distortion point only 12°F. In selecting a specific type of ABS, remember that the medium impact materials offer better resistance to creep, abrasion, and heat than the higher impact grades.

Tensile yield strength and ultimate tensile strength for ABS materials are the same—about 3000 to 9000 psi. Strain at yield is typically 2 to 5%, while ultimate elongation is quite variable. Compressive strength for these thermoplastics is higher than their tensile strengths.

ABS materials have relatively good electrical insulating properties which make them suitable for secondary insulating applications.

In general, ABS materials have very good resistance to a wide range of chemicals. They are very good in weak acids, and both weak and strong alkalis; generally good in strong acids; and poor in solvents such as esters, ketones, aldehydes, and some chlorinated hydrocarbons. These plastics also have low water absorption.

ABS materials are available as compounds for injection molding, blow molding, extrusion and calendering, as sheet for thermoforming or cold forming, and in expandable grades for

Table 3-1. Typical property ranges for plastics[*]

THERMOSETS[a]	Specific gravity	Transparency	Ten str, 1000 psi	Ten mod, 10⁶ psi	Impact str, Izod, ft-lb[c]	Dielec constant at 60 cps	Dielec str, v/mil	Max use temp, F (no load)	HDT at 264 psi[d]	Weather res[e]	Weak acid[e]	Strong acid[e]	Weak alkali[e]	Strong alkali[e]	Solvents[e]
Alkyds															
Glass filled	2.12-2.15	No	4-9.5	20-28	0.6-10	5.7	250-530	450	400-500	R	A	A	A	A	A
Mineral filled	1.60-2.30	No	3-9	5-30	0.3-0.5	5.1-7.5	350-450	300-450	350-500	R	A	A	A	D	A
Asbestos filled	1.65	No	4.5-7	-	0.4-0.5	-	380	450	315	R	R	S	R	S	R
Syn. fiber filled	1.24-2.10	No	4.5-7	20	0.5-4.5	3.8-5.0	365-500	300-430	245-430	R	R	S	R	S	A
Allyl diglycol carbonate	1.30-1.40	Yes	5-6	3.0	0.2-0.4	4.4	380	212	140-190	R	R	A[k]	R	R-S	R
Diallyl phthalates															
Glass filled	1.61-1.78	No	6-11	14-22	0.4-15	4.3-4.6	400-450	300-400	330-540	R	R	S	R-S	S	R
Mineral filled	1.65-1.68	No	5-9	12-22	0.3-0.5	5.2	400-420	300-400	320-540	R	R	S	R-S	S	R
Asbestos filled	1.55-1.65	No	7-8	12-22	0.4-0.5	4.6-6.2	400-450	300-400	320-540	R	R	S	R-S	S	R
Epoxies (bis A)															
No filler	1.06-1.40	Yes	4-13	2.15-5.2	0.2-1.0	3.2-5.0	400-650	250-500	115-500	R	R	A	R	S	R-S
Graphite fiber reinf.	1.37-1.38	No	185-200	118-120	-	-	-	-	-	S	R	R	R	R	R-S
Mineral filled	1.6 -2.0	No	5-15	-	0.3-0.4	3.5-5.0	300-400	300-500	250-500	S	R	R	R	R	R-S
Glass filled	1.7 -2.0	No	10-30	30	10-30	3.5-5.0	300-400	300-500	250-500	S	R	R-S	R	R	R-S
Epoxies (novolac)															
No filler	1.12-1.24	No	5-11	2.15-5.2	0.3-0.7	3.11-4.0	360-600	400-500	450-500	R	R	R	R	R	R
Epoxies (cycloaliphatic)															
No filler	1.12-1.18	Yes	10-17.5	5-7	-	3.6	-	480-550	500-550	R	R	R-A	R	R-A	R
Melamines															
Cellulose filled	1.45-1.52	No	5-9	11	0.2-0.4	6.2-7.6	350-400	250	270	S	R-S	D	R	D	R
Flock filled	1.50-1.55	No	7-9	-	0.4-0.5	-	300-330	250	270	S	R-S	D	R	D	R-S
Asbestos filled	1.70-2.0	No	5-7	20	0.3-0.4	6.4-10.2	410-430	250-400	265	S	R-S	D	S	S	R
Fabric filled	1.5	No	8-11	14-16	0.6-1.0	7.6-12.6	250-350	250	310	S	R	D	R	A	R-S
Glass filled	1.8 -2.0	No	5-10	24	0.6-18	9.7-11.1	170-300	300-400	400	S	R	D	R	R-S	R
Phenolics															
Woodflour filled	1.34-1.45	No	5-9	8-17	0.2-0.6	5-13	260-400	300-350	300-370	S	R-S	S-D	S-D	A	R-S
Asbestos filled	1.45-2.00	No	4.5-7.5	10-30	0.2-0.4	5-20	260-360	350-500	300-500	S	R-S	S-D	S-D	A	R-S
Mica filled	1.65-1.92	No	5.5-7	25-50	0.3-0.4	4.7-6	350-400	250-300	300-350	S	R-S	S-D	S-D	A	R-S
Glass filled	1.69-1.95	No	5-18	19-33	0.3-18	5-7.1	140-400	350-550	300-600	S	R-S	S-D	S-D	A	R-S
Fabric filled	1.36-1.43	No	3-9	9-14	0.8-8	5.2-21	200-400	220-250	250-330	S	R-S	S-D	S-D	A	R-S
Polybutadienes															
Very high vinyl (no filler)	1.00	No	8	2	1.1	2.4	1800	500	-	S	R	R	R	R	R
Polyesters															
Glass filled BMC	1.7 -2.3	No	4-10	16-25	1.5-16	5.3-7.3	300-420	300-350	400-450	R-E	R-A	S-A	S-A	S-D	A-D
Glass filled SMC	1.7 -2.1	No	8-20	16-25	8-22	-	320-400	300-350	400-450	R-E	R-A	S-A	S-A	S-D	A-D
Glass cloth reinf.	1.3 -2.1	No	25-50	19-45	5-30	4.1-5.5	350-500	300-350	400-450	R-E	R-A	S-A	S-A	S-D	A-D
Silicones															
Glass filled	1.7 -2.0	No	4-6.5	10-15	3-15	3.3-5.2	200-400	600	600	R-S	R-S	R-S	S	S-A	R-A
Mineral filled	1.8 -2.8	No	4-6	13-18	0.3-0.4	3.5-3.6	200-400	600	600	R-S	R-S	R-S	S	S-A	R-A
Ureas															
Cellulose filled	1.47-1.52	No	5.5-13	10-15	0.2-0.4	7.0-7.5	300-400	170	260-290	S	R-S	A-D	S-A	D	R-S
Urethanes															
No filler	1.1 -1.5	No & Yes	0.2-10	1-10	5-NB	4.0-7.5	400-500	190-250	-	R-S	S	A	S	S-A	R-S

THERMOPLASTICS		Spec ific Grav ity	Trans paren cy	Ten. str., 1000 psi	Ten. Mod., 10⁶ psi	Imp. str. Izod[c] ft.-lb.	Dielec. con. at 60 cps	Dielec. str., v/mil	Max use temp, F (no load)	HDT at 66 psi	HDT at 264 psi	Weath-er res	Chemical resistance[e]				
													W ac	S ac	W al	S al	Solv
ABS	GP	1.05-1.07	No	5.9	3.1	6	2.8-3.2	385	160-200	210-225	190-206	R-E	R	A[k]	R	R	A[m]-R
	Hi. imp.	1.01-1.06	No	4.8	2.4	7.5	2.8-3.5	300-375	140-210	210-225	188-211	R-E	R	A[k]	R	R	A[m]-R
	Ht. res.	1.06-1.08	No	7.4	3.9	2.2	2.7-3.5	360-400	190-230	225-252	226-240	R-E	R	A[k]	R	R	A[m]-R
	Trans.	1.07	Yes	5.6	2.9	5.3	—	—	130	180	165	R-E	R	A[k]	R	R	A[m]-R
		1.20	No	6.0	3.2	2.5	2.4-5.0	370-400	130-180	210-220	195	R-E	R	A[k]	R	R	A[m]-R
Acetals	Homo	1.42	No	10	5.2	1.4	3.7	320	195	338	255	R[j]	R	A	R	A-D	R
	Copol	1.41	No	8.8	4.1	1.2-1.6	3.7	500	212	316	230	R[j]	R	A	R	R	R
Acrylics	GP	1.11-1.19	Yes	5.6-11.0	2.25-4.65	0.3-2.3	3.0-3.7	450-500	130-230	175-225	165-210	R	R	A[k]	R	A	A[m]-R
	Hi. imp.	1.12-1.16	No	5.8-8.0	2.3-3.3	0.8-2.3	3.5-3.9	400-500	140-195	180-205	165-190	R	R	A[k]	R	R	A[m]-R
		1.21-1.28	No	8.0-12.5	3.5-4.8	0.3-0.4	3.5-5.1	400-440	125-200	170-200	155-205	R	R	A[k]	R	A	A[m]-R
	Cast	1.18-1.28	Yes	9.0-12.5	3.7-5.0	0.4-1.5	3.5-5.1	400-530	140-200	165-235	160-215	R	R	A[k]	R	A	A[m]-R

[*]Reprinted from the May 1972 issue of *Materials Engineering* with permission of the publishers.

Table 3-1. Typical property ranges for plastics. (continued)

THERMOPLASTICS		Spec. ific Grav- ity	Trans- paren- cy	Ten. str., 1000 psi	Ten. Mod., 10^5 psi	Imp. str. Izod^c ft.-lb.	Dielec. con. at 60 cps	Dielec. str., v/mil	Max use temp, F (no load)	HDT at 66 psi	HDT at 264 psi	Weath- er res	W ac	S ac	W al	S al	Solv
	Multi poly- mer	1.09- 1.14	Yes	6-8	3.1- 4.3	1-3	3.3- 3.5	495	165- 175	—	185- 195	E	R	A^k	R	S	A^m
Cellulosics	Acetate	1.23- 1.34	Yes	3.0- 8.0	1.05- 2.55	1.1- 6.8	3.5- 7.5	250- 600	140- 220	120- 209	111- 195	S	S	D	S	D	D-S
	Butyrate	1.15- 1.22	Yes	3.0- 6.9	0.7- 1.8	3.0- 10.0	3.5- 6.4	250- 400	140- 220	130- 227	113- 202	S	S	D	S	D	D-S
	E. cellu- lose	1.10- 1.17	Yes	3-8	0.5- 3.5	1.7- 7.0	3.2	350- 500	115- 185	—	115- 190	S	S	D	R	S	D
	Nitrate	1.35- 1.40	Yes	7-8	1.9- 2.2	5-7	7.0- 7.5	300- 600	140	—	140- 160	E	S	D	S	D	D
	Propionate	1.19- 1.22	Yes	4.0- 6.5	1.1- 1.8	1.7- 9.4	3.7- 4.0	300- 450	155- 220	147- 250	111- 228	S	S	D	S	D	D-S
Ch. polyether		1.4	No	5.4	1.5	0.4	3.0	400	290	285	—	R-S	R	A^k	R	R	R
Eth. copolymers	EEA	0.93	Yes	2.0	0.05	NB	2.7	550	190	—	—	S	R	A^k	R	R	A-D
	EVA	0.94	Yes	3.6	0.02- 0.12	NB	2.50- 3.16	525	—	140- 147	93	S	R	A	R	R	A-D
Fluoropolymers	FEP	2.14- 2.17	No	2.5- 3.9	0.5- 0.7	NB	2.1	500- 600	400	158	—	—	R	R	R	R	R
	PTFE	2.1- 2.3	No	1-4	0.38- 0.65	2.5- 4.0	2.1	400- 500	550	250	—	—	R	R	R	R	R
	CTFE	2.10- 2.15	Yes- No	4.6- 5.7	1.8- 2.0	3.5- 3.6	2.6- 2.7	530- 600	350- 390	258	—	—	R	R	R	R	S^n
	PVF$_2$	1.77	No	7.2	1.7	3.8	10.0	260	300	300	195	S	R	A^l	R	R	R
	ETFE & ECTFE	1.68- 1.70	No	6.5- 7.0	2-2.5	NB	2.4- 2.6	400	300	220	160	R	R	R	R	R	R
Methylpentene		0.83	Yes	3.3- 3.6	1.3- 1.9	0.95- 3.8	—	700	275	—	—	E	R	A^k	R	R	A
Nylons	6/6	1.13- 1.15	No	9-12	3.85	2.0	4.0	385	180- 300	360- 470	150- 220	R	R	A	R	R	R-D^o
	6	1.14	No	12.5	—	1.2	4.0- 5.3	385	180- 250	300- 365	140- 155	R	R	A	R	R	R-A^o
	6/10	1.07	No	7.1	2.8	1.6	3.9	470	180	300	—	R	R	A	R	R	R-A^o
	8	1.09	No	3.9		>16	9.3	340				R	R	A	R	R	R-A^o
	12	1.01	No	6.5- 8.5	1.7- 2.1	1.2- 4.2	3.6	840	175- 260	—	120- 130	R	R	A	R	R	R-A^o
	Copoly- mers	1.08- 1.14	No	7.5- 11.0		1.5-19	3.2- 4.5	440- 450	180- 250	—	130- 350	R	R	A	R	R	R-A^o
Polyesters	PET	1.37	No	10.4	—	0.8	3.65	—	175	240	185	R	R	A^k	R	A	R-A^o
	PBT	1.31	No	8.0- 8.2	3.6	1.2- 1.3	3.3- 3.7	590- 700	280	310	130	R	R	R	R	A	R ^f
	PTMT	1.31	No	8.2	—	1.0	3.16	420	270	302	122	R	R	R	R	A	R
	Copol.	1.2	Yes	7.3	—	1.0	—	—	—	—	154	—	—	—	—	—	—
Polyaryl ether		1.14	No	7.5	3.2	10	3.14	430	250	320	300	E	R	R	R	R	A
Polyaryl sulfone		1.36	No	13	3.7	2	3.94	350	500	—	525	Darkens	R	R	R	R	R
Polybutylene		0.910	No	3.8	0.26	NB	2.25	—	225	215	130	E	R	A^k	R	R	—
Polycarbonate		1.2	Yes	9	3.45	12-16	3.17	380	250	270- 290	265- 285	R	R	A^k	A	A	A
PC/ABS		1.14	No	8.2	3.7	10	2.74	500	220	235	220	R-E	R	A^k	R	S	A
Polyethylenes	LD	0.91- 0.93	No	0.9- 2.5	0.20- 0.27	NB	2.3	480	180- 212	100- 120	90-105	E	R	A^k	R	R	R
	HD	0.95- 0.96	No	2.9- 5.4		0.4- 14	2.3	480	175- 250	140- 190	110- 130	E	R	R-A^k	R	R	R
	HMW	0.945	No	2.5	1	NB	2.3	480	—	155- 180	105- 180	E	R	A^k	R	R	R
Ionomer		0.94- 0.95	Yes	3.4- 4.5	0.3- 0.7	6-NB	2.4	1000	160- 180	110	100- 120	E	A	A^k	R	R	R
Phenylene oxide based mtls.		1.06- 1.10	No	7.8- 9.6	3.5- 3.8	5.0	2.65- 2.69	400- 500	175- 220	230- 280	212- 265	R	R	R	R	R	R-A
Polyphenylene sulfide		1.34	No	10	4.8	0.3		595	500	—	278	R	R	A^k	R	R	R
Polyimide		1.43	No	5-7.5	5.4	5-7	4.12	310	500	—	680	—	R	A	A	A	R
Polypropylenes	GP	0.90- 0.91	No	4.8- 5.5	1.6- 2.2	0.4- 2.2	2.20- 2.28	650	225- 300	200- 230	125- 140	E	R	A^k	R	R	R
	Hi. imp.	0.90- 0.91	No	3-5	1.3	1.5- 12	2.20- 2.28	450- 650	200- 250	160- 200	120- 135	E	R	A^k	R	R	A
Propylene copolymer		0.91	No	4	1.0- 1.7	1.1	2.25- 2.30	450- 600	190- 240	185- 230	115- 140	E	R	A^k	R	R	R
Polystyrenes	GP	1.04- 1.07	Yes	6.0- 7.3	4.5	0.3	2.45- 2.65	400- 600	150- 170	—	180- 220	S	R	A^k	R	R	D
	Hi. imp.	1.04- 1.07	No	2.8- 4.6	2.9- 4.0	0.7- 1.0	2.45- 4.75	400- 500	140- 175	—	175- 210	S	R	A^k	R	R	D

^aAll values at room temperature unless otherwise listed. ^bPer ASTM. ^cNotched samples. ^dHeat deflection temperature. ^eAc is acid and Al is alkali; R is resistant. A is attacked, S is slight effects, E is embrittles and D is decomposes. ^fChalks slightly. ^kBy oxidizing acids. ^lBy fuming sulfuric. ^mBy Ketones, esters, and chlorinated and aromatic hydrocarbons. ^nHalogenated solvents cause swelling. ^oDissolved by phenols and formic acid.

Table 3-1. Typical property ranges for plastics. (continued)

THERMOPLASTICS	Specific Gravity	Transparency	Ten. str., 1000 psi	Ten. Mod., 10⁵ psi	Imp. str. Izod ft.-lb.	Dielec. con. at 60 cps	Dielec. str., v/mil	Max use temp, F (no load)	HDT at 66 psi	HDT at 264 psi	Weather res	Chemical resistance				
												W ac	S ac	W al	S al	Solv
Polysulfone	1.24	No	10.2	3.6	1.2	3.14	425	300	360	345	S	R	R	R	R	R-A
Polyurethanes	1.11-1.25	No	4.5-8.4	0.1-3.5	NB	5.4-7.6	460	190	—	—	R-S	S-D	S-D	S-D	S-D	R
Vinyl Rigid	1.3-1.5	No-Yes	5-8	3-5	0.5-20	3.2-3.6	425-1300	150-175	135-180	130-175	R	R	R-S	R	R	R-A
Vinyl Flexible	1.2-1.7	Yes & No	1-4	—	0.5-20	5-6	250-800	140-175	—	—	S	R	R-S	R	R	R-A
Rigid CPVC	1.49-1.58	Yes-No	7.5-9.0	3.6-4.7	1.0-5.6	2.8-3.6	—	230	215-245	200-235	R	R	R	R	R	R
PVC/acrylic	1.30-1.35	No	5.5-6.5	2.75-3.35	15	3.9-4.0	400	—	180	170	R	R	S	R	R	A
PVC/ABS	1.10-1.21	No	2.6-6.0	0.8-3.4	10-15	—	600	—	—	—	S	R	R-S	R	R	R-D
SAN	1.08	Yes	10-12	5.0-5.6	0.4-0.5	3.0-3.8	1775	140-200	—	190-220	S-E	R	A	R	R	A

[a]All values at room temperature unless otherwise listed. [b]Per ASTM. [c]Notched samples. [d]Heat deflection temperature. [e]Ac is acid and Al is alkali; R is resistant, A is attached, S is slight effects, E is embrittles and D is decomposes. [j]Chalks slightly. [k]By oxidizing acids. [l]By fuming sulfuric. [m]By ketones, esters, and chlorinated and aromatic hydrocarbons. [n]Halogenated solvents cause swelling. [o]Dissolved by phenols and formic acid.

foam molding. Although most ABS materials are opaque, a transparent grade has been introduced.

Typical applications for ABS include: automobile dash-boards and grilles, pump impellers, appliance housings and parts, pipe and fittings, luggage, helmets, telephones, electrical connectors, refrigerator door liners, shoe heels, knobs and handles, and business machine housings (Fig. 3-1).

Fig. 3-1. Main housing, cover, and keyboard cover for a computer-type printing system are formed from ABS sheet. (*Courtesy Borg-Warner Chemicals*)

Acetals—One of the strongest (tensile strength: 10,000 psi) and stiffest (modulus in flexure: 410,000 psi) thermoplastics, acetal is also characterized by excellent fatigue life and dimensional stability. Other outstanding properties include low friction coefficients, exceptional solvent resistance and high heat resistance for extended use up to 220°F.

At present there are two basic types of acetal: a homopolymer (du Pont's Delrin) and a copolymer (Celanese's Celcon).

Acetals are highly crystalline thermoplastics which accounts for their excellent properties and predictable long-range performance under load. In creep resistance, acetal is one of the best thermoplastics; however, apparent modulus falls off consistently with long-term loading.

For practical design purposes, yield point and ultimate tensile strength for acetal homopolymer are the same at room temperature; however, a well defined yield point with considerable elongation is exhibited by the copolymer at room temperature and by both types at elevated temperature.

Acetals have an excellent fatigue endurance limit; at 100% RH it is 5000 psi at 77°F and is still 3000 psi at 150°F. Furthermore, lubricants and water have little effect on the fatigue life. Although the effects of frequency appear negligible in the 60 to 1800 cycles per sec range, higher frequencies affect properties due to the generation of heat.

The impact strength of an acetal does not fall off abruptly at subzero temperature as does that of many other thermoplastics.

Hardness of acetals is only slightly reduced by moisture absorption or temperature below 215°F.

While not as good as that of nylons, abrasion resistance of acetals is better than that of many other thermoplastics. Like nylons, acetals have a slippery feel.

Acetals are especially notable among thermoplastics because of their resistance to organic solvents. However, in contact with strong acids, acetals will craze.

Combined with the good mechanical properties of an acetal are its good electrical properties. Its dielectric constant and dissipation factor are uniform over a wide frequency range and up to temperatures of 250°F. Aging also has little effect on its electrical properties.

Acetals are available as compounds for injection molding, blow molding, and extrusion. Grades reinforced with glass fibers (higher stiffness, lower creep, improved dimensional stability) or TFE-fluorocarbon fibers (improved frictional and wear properties) are also on the market.

Many applications involve replacement of metals where the higher strength of metals is not required and costly finishing and assembly operations can be eliminated. Typical parts include pump impellers; gears; appliance cases and housings; automobile turn signals and carburetors; conveyor belt sections; bearings; plumbing components; pipe and fittings; machinery parts; and aerosol containers.

Acrylics—Acrylics have outstanding resistance to long-term exposure to sunlight and weathering. Polymethyl methacrylate (PMMA), a hard, rigid and transparent material, is the most widely used member of the acrylic family. Cast PMMA sheet has excellent optical properties (it transmits about 92% total light) and is more resistant to impact than glass. It is not as resistant to surface scratching as glass, however, although new surface coatings are being made available to overcome this limitation.

In addition to excellent optical properties, acrylics have low water absorption, good electrical resistivity, and fair tensile strength. The heat resistance of acrylics is on the order of 200°F.

More recently, modified acrylics and acrylic multipolymers that offer high impact strength and toughness in addition to the standard acrylic properties have been made available. These grades incorporate elastomeric or alloying constituents that impart the added strength (up to 10 to 20 times as much as general-purpose acrylic crystal). Acrylics are available as compounds for extrusion, injection molding, blow molding, and casting. Extruded or cast sheet, and film are also marketed.

Typical applications include: outdoor signs, glazing, aircraft canopies, washbasins (formed from sheet and backed up with spray-on reinforced plastics), lighting applications, knobs, handles, escutcheons, safety shields, and machine covers. Some of the transparent acrylic multipolymers have found application in the drug and food packaging industry (Fig. 3-2).

Acrylonitrile-based resins—Sometimes called *barrier resins*, this family of plastics is intended primarily for use in packaging (i.e., bottles) carbonated beverages. Although exact formulation of these materials has not been revealed, it is generally agreed that they all incorporate acrylonitrile materials. One supplier (Vistron) describes its barrier resin (Barex) as an acrylic thermoplastic made from acrylonitrile and methacrylonitrile. Major characteristics of the materials are barrier properties relating to the transmission of gas, aroma, or flavor of the package's contents. The resins are similar in melt processability to rigid PVC and some grades of ABS and can be extruded, blow molded, and thermoformed.

Cellulosics—The family of cellulosics includes cellulose acetate, cellulose acetate butyrate, cellulose propionate, and ethyl cellulose. There are other cellulosics but these are the most widely used. Cellulosics are characterized by good strength, toughness and transparency, and high surface gloss. In addition, they have good chemical resistance. Generally, these thermoplastics should not be used at temperatures much above 170-220°F.

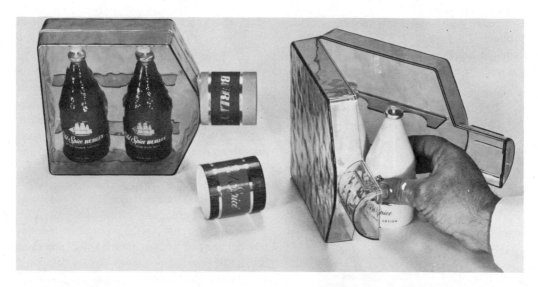

Fig. 3-2. Transparent package with excellent impact strength is molded of an acrylic multipolymer. (*Courtesy American Cyanamid*)

Cellulose acetate is the lowest cost cellulosic material. It has good toughness and rigidity. (Izod impact strength: 0.4-5.2 ft-lb/in. notch). This easily molded material is available in a variety of grades ranging from "soft" to "hard."

Cellulose acetate butyrate, although a little more expensive than the straight acetate, is somewhat tougher, with a horn-like quality, and has lower moisture absorption (0.9-2.2% in 24 hr vs. 1.9-7% for acetate). It has relatively good weatherability and excellent transparency.

Cellulose propionate is similar to cellulose acetate butyrate in both cost and properties, but it has somewhat higher tensile strength, modulus and impact strength (1.5 to 11.5 ft-lb/in. notch).

Ethyl cellulose is what might be called the impact grade of the cellulosics. The excellent toughness of this material (Izod impact strengths range from 1.7 to 8.5 ft-lb/in. notch depending on the specific resin) is maintained over a wide temperature range. Ethyl cellulose also has moderately low moisture absorption (1-2% in 24 hr). In addition, this cellulosic is available in self-lubricating grades.

Nearly all cellulosics are noted for their toughness, but none of these materials is generally recommended for applications involving anything more than relatively low loads.

The main feature of cellulosics is their excellent moldability which results in a brilliant, high-gloss finish.

Cellulose acetate is subject to dimensional changes due to cold flow, extreme heat, and moisture absorption. Cellulose acetate butyrate is a slightly more stable material, but still not outstanding in comparison to some other thermoplastics. This material, however, is one of the few thermoplastics that resists weathering. This property, combined with its good optical properties, makes cellulose acetate butyrate an excellent material for outdoor signs. Cellulose propionate is similar in properties to butyrate. Ethyl cellulose has several outstanding properties. It withstands heavy abuse and has good environmental resistance: heat distortion point at 264 psi is up to 190 F.; water absorption is 1.8% in 24 hr; weathering resistance is excellent. Like other cellulosics, however, it is not primarily a load-bearing material.

Cellulosic compounds are available for extrusion, injection molding, blow molding, and rotational molding. Cellulosics are also widely used in the form of film and sheet.

Typical applications include: cellulose acetate—knobs, appliance housings and handles; cellulose acetate butyrate—tool handles, knobs, dials, appliance housings, steering wheels, signs, light globes; cellulose propionate—automobile arm rests, pen and pencil barrels, appliance housings and toys (Fig. 3-3); ethyl cellulose—flashlight housings, tool handles, roller wheels, and refrigerator breaker strips.

Fig. 3-3. High impact resistance, clarity, and economy were major reasons for using cellulose propionate in a new toy. (*Courtesy Eastman Chemical Products*)

Fluorocarbons—Outstanding properties of fluorocarbons (or fluoroplastics as they are also called) include inertness to most chemicals and resistance to high temperatures. Fluorocarbons have a rather waxy feel, extremely low coefficients of friction, and excellent dielectric properties which are relatively insensitive to temperature and power frequency. Mechanical properties are normally low, but this changes dramatically when the fluorocarbons are reinforced with glass fibers or molybdenum disulfide fillers.

There are numerous fluorocarbons available, but the two most widely used types are tetrafluoroethylene (TFE) and chlorotrifluoroethylene (CTFE).

TFE is extremely heat resistant (up to 500°F) and has outstanding chemical resistance, being inert to most chemicals. Its coefficient of friction is lower than that of any other plastic and it can be used unlubricated. TFE has a tensile strength on the order of 1500 to 5000 psi and an impact strength of 2.5 to 3.0 ft-lb/in. of notch. TFE also has outstanding low temperature characteristics and will remain flexible at even very low temperatures.

CTFE has a heat resistance of up to 390°F, a tensile strength of 4500 to 6000 psi, and an impact strength of 2.5 to 2.7 ft-lb/in. of notch. It is chemically resistant to all inorganic corrosive liquids, including oxidizing acids and is resistant to most organic solvents except certain halogenated materials and oxygen-containing compounds (which cause a slight swelling).

In terms of processability, however, CTFE can be molded and extruded by conventional thermoplastic processing techniques while TFE (since it does not soften like other thermoplastics do) must be formed by processes similar to powder metallurgy (i.e., powders are compacted to the desired shape and sintered). More recently, a copolymer of TFE and hexafluoropropylene known as FEP (fluorinated ethylene-propylene) has been made available. FEP shows essentially some of the same properties as TFE (i.e., toughness, excellent dimensional stability, and outstanding electrical insulating characteristics over a wide range of temperatures and humidities), but it exhibits a melt viscosity low enough to permit it to be molded by conventional thermoplastic processing techniques.

Other members of the family of fluoroplastics include: vinylidene fluoride, a high-molecular weight material which has a high tensile strength (6500 to 6800 psi), low cold flow characteristics, and is thermally stable (650°F for short periods of time, 300°F for longer durations); ethylene trifluoroethylene (ETFE), a melt-processable fluorocarbon with good high temperature and chemical resistance characteristics (Fig. 3-4); and ethylene-chlorotrifluoroethylene (E-CTFE), a strong, highly impact resistant material that can also be molded or extruded on conventional thermoplastics equipment.

In the unfilled condition, fluorocarbons, in general, have only fair resistance to deformation under load. When reinforced with such materials as molybdenum disulfide, graphite, asbestos or glass, the fluorocarbons show increased stiffness, hardness and compressive strength, and reduced elongation and deformation under load.

Fig. 3-4. A variety of parts with excellent mechanical, chemical, and electrical characteristics can be easily molded from both plain and glass-reinforced ETFE fluoropolymer resins. (*Courtesy E. I. du Pont de Nemours & Co., Inc.*)

Stress-strain data on fluorocarbons show these materials to be much better in compression than in tension. Depending on crystallinity, TFE, for example, has a tensile yield strength at 73°F of about 1800 psi (30% strain); in compression, however, this value can be as high as 3700 psi (25% strain).

Typical applications include: TFE—nonlubricated bearings, chemical-resistant pipe and pump parts, high temperature electronic parts, packings, gaskets, seals and rings; CTFE—coil forms, chemical-resistant pipe and pump parts, connectors and valve diaphragms, fuel sight lenses, electrical insulators and inserts; FEP—wire insulation and jacketing, high-frequency connectors and microwave components, coils, gaskets, electrical terminals, tube sockets and terminal insulators; vinylidene fluoride—electrical insulation, seals and gaskets, diaphragms, pipe and other chemical process equipment; ETFE—molded labware, valve liners, electrical connectors, coil bobbins, electrical insulation; E-CTFE—wire and cable coatings (for applications requiring high-performance wire), chemi-

cally resistant linings and coatings, molded labware, and film for laminates used in aircraft interiors, flexible circuitry, and medical packaging.

Phenylene oxide based resins—These resins are formulated from a patented process for oxidative coupling of phenolic monomers (Noryl, General Electric). The family is characterized by outstanding dimensional stability at elevated temperatures; broad temperature use range; outstanding hydrolytic stability; and excellent dielectric properties over a wide range of frequencies and temperatures.

An outstanding property of these resins is stability under long-term loading. Strain as low as 0.3% occurs even at 150°F under 3000 psi for 500 hr. The resin also has a very low water absorption rate (0.006% in 24 hr at 73°F and 0.14% at equilibrium). Dimensional change is negligible even in boiling water, detergents, and weak and strong acids and bases. Izod impact strength is 5.0 ft-lb/in. of notch, measured on a $\frac{1}{8}$ by $\frac{1}{2}$ in. sample. Unnotched specimens indicate an impact strength in excess of 60 ft-lb/in.

The tensile yield stress at 73°F is 9600 psi. Compressive strength reaches 16,400 psi at about 10% deformation. Tensile yield stress and tensile modulus show relatively little change with temperature to over 200°F.

Based on the grades chosen, heat deflection temperatures can range from 212 to 300°F at 264 psi. Temperatures beyond these may be used for occasional short-time exposure.

The resin will soften or dissolve in many halogenated or aromatic hydrocarbons. If an application requires exposure to or immersion in a given environment, stressed samples should be tested under operating conditions.

Molding and extrusion grades are available. Typical applications include: automotive dashboards, electrical connectors, TV tuners and deflection yoke components, pumps, plumbing fixtures, small appliance and business machine housings, automotive grilles and trim.

Polyallomers—As a member of the olefins family (e.g., polyethylene, polypropylene), polyallomers offer excellent chemical resistance. They fall between high-density polyethylene and polypropylene in many properties. Some grades have a brittleness temperature as low as −40°F, a heat distortion temperature at 66 psi of 200°F and at 264 psi of 140°F, and impact strengths as high as 6 ft-lb/in. of notch.

In general, the material can be expected to perform much like a polypropylene with respect to creep, stress-strain, and fatigue. In all cases, the low temperature performance is usually better. Although surface hardness is slightly less than that of polypropylene, resistance to abrasion is greater.

Polyallomers are available as extrusion, injection molding, and blow molding compounds. Film and sheet are also on the market.

Typical applications include: pipe fittings, bowling ball bags and tool boxes with integral hinges, textile bobbins, and detergent bottles.

Polyamides (nylons)—Nylons have high tensile strengths (up to 8000 to 11,000 psi in some grades), high modulus (in flexure: 210,000 to 500,000 psi), and good impact strength (up to 3 ft-lb/in. of notch and higher in the Izod test). Abrasion resistance is high, resistance to heat can go to 250°F and elec-

trical properties and chemical resistance are good. Nylons resist nonpolar solvents, including aromatic hydrocarbons, esters, and essential oils. They are softened by and absorb polar materials such as alcohols, glycols, and water.

Several different types of nylon are on the market, but the two most widely used types are nylon 6/6 (hexamethylene diamine adipic acid) and nylon 6 (polycaprolactam).

Nylon 6/6 is the most common polyamide molding material. Special grades of nylon 6/6 include: (1) heat-stabilized grades for molding electrical parts, (2) hydrolysis-stabilized grades for parts to be used in contact with water, (3) light-stabilized grades for weather-resistant moldings, and (4) higher-melt-viscosity grades for molding of heavy sections and for better extrudability.

Special grades of nylon 6 include: (1) grades with higher flexibility and impact strength, (2) heat and light-stabilized grades for resistance to outdoor weathering, (3) grades for incorporating nucleating agents to promote consistent crystallinity throughout sections, thereby providing better load-bearing characteristics, and (4) higher viscosity grades for extrusion of rod, film, pipe, large shapes, and blown bottles.

Other types of nylon include: nylon 8, an alcohol-soluble type used for impregnating and coating synthetic and natural fabrics, rubber goods, paper, leather, metal, etc.; and nylon 6/10, similar to type 6/6 but with lower moisture absorption, lower heat resistance, lower yield strength, and greater flexibility; and types 12 and 11 which have substantially lower moisture absorption and high strength.

Creep rates for nylons at various stress levels under both tension and compression show only a small deformation within the initial 24-hour period and increase from this point in a linear manner. This means that long-term deformation can usually be predicted accurately on the basis of short-term tests.

Stress-strain data on nylon show that the material reacts in a linear manner up to loads of about 300 psi. At higher stresses, elongation becomes proportionately greater until a yield point of 8500 psi is reached at 25% elongation. In spite of the reduction in area, the further drawing of nylon strengthens it so that the ultimate tensile strength is as high as 11,000 psi (at 300% elongation).

The high moisture absorption of nylons causes dimensional change with increasing moisture content. While type 6 absorbs moisture more rapidly than type 6/6, both eventually reach equilibrium at about 2.7% moisture content in 50% RH air, and about 9 to 10% in water.

Other key engineering characteristics of nylons are excellent wear, abrasion and chemical resistance.

Nylons are resistant to most solvents and are especially resistant to petroleum oils and greases, alkalis, lactic acids, and photographic solutions. They are generally soluble in molten phenol and hot formic acid and are attacked by concentrated solutions of mineral acids.

The dielectric strength of nylon is on the order of 385 to 470 v per mil, volume resistivity is 45×10^{13} ohm-cm, and the dissipation factor is 0.01 to 0.09. However, in many of the electrical applications in which it is used, its mechanical strength and resistance to oils and greases are also important factors.

General-purpose nylon molding materials are available for extrusion, injection molding, blow molding, rotational molding, and (for the nylon 6 materials) casting or anionic polymerization. Nylon sheet and film are also marketed.

For specific engineering applications, a number of specialty nylons has been developed, including: molybdenum-disulfide filled nylons (to improve wear and abrasion resistance, frictional characteristics, flexural strength, stiffness, and heat resistance); glass fiber-filled nylons (to improve tensile strength, heat distortion temperatures, and, in some cases, impact strength), and sintered nylons. Sintered nylons are fabricated by processes similar to powder metallurgy (the same as used for TFE fluorocarbons). The resulting materials have improved frictional and wear characteristics, as well as higher compressive strength.

Typical applications include: gears, cams and other sliding contact devices, slide fasteners, door hinges, gaskets, wire insulation, coil forms, high-pressure flexible tubing, boat propellors, chemical containers, belting, wear pads, and appliance parts (Fig. 3-5).

Polyaryl ether—Another relatively new engineering plastic (Arylon, Uniroyal, Inc.) that offers: a high heat-deflection temperature of 300°F measured at a 264 psi fiber stress on a ¼ by ½ by 5 in. bar; an Izod impact of 8 ft-lb/in. of notch at 72°F and 2.5 ft-lb/in. of notch at −20°F—among the highest of plastics; and excellent chemical resistance to organic solvents (except chlorinated aromatics, esters, and ketones).

Fig. 3-5. A fan blade and blower wheel molded of mineral-reinforced nylon. (*Courtesy Monsanto Co.*)

Polyaryl ether can be molded in reciprocating screw injection molding machines. It can also be plated in conventional preplate systems without annealing or solvent conditioning prior to etching.

The material is recommended for use in the automotive (as a painted exterior automotive trim or in under-the-hood applications requiring outstanding heat resistance), appliance, plumbing, and electrical industries.

Polyaryl sulfone—This material (Astrel, Carborundum Co.) consists of phenyl and biphenyl groups linked by thermally stable ether and sulfone groups. It is distinguished from polysulfone polymers by the absence of aliphatic groups, which are liable to oxidative attack.

Polyaryl sulfone is characterized by a very high heat-deflection temperature, 525°F at 264 psi. At 500°F it maintains a tensile strength in excess of 4000 psi and a flexural modulus of over 300,000 psi.

The resistance to oxidative degradation is indicated by the ability of polyaryl sulfone to retain its tensile strength after 2000 hr exposure to 500°F air-oven aging.

Its unique high-temperature performance characteristics are achieved without sacrificing room temperature mechanical or electrical properties. At normal ambient temperatures, polyaryl sulfone is a strong, stiff, tough material that in general offers properties comparable to any other engineering thermoplastic. At −320°F polyaryl sulfone still has 6% elongation, indicating that it has some toughness and utility in cryogenic applications.

Polyaryl sulfone has good resistance to a wide variety of chemicals including acids, bases, and common solvents. It is unaffected by practically all fuels, lubricants, hydraulic fluids, and cleaning agents used on or around electrical components. Highly polar solvents such as N,N-dimethylformamide, N,N-dimethylacetamide, and N-methylpyrrolidone are solvents for the material.

Polyaryl sulfone may be injection molded on conventional equipment with sufficient injection pressure and temperature capabilities. Generally, the cylinder and nozzle must be equipped to reach temperatures of 800°F and the injection system should be capable of providing from 20,000 to 30,000 psi injection pressure. Reciprocating screw machines are preferred. Polyaryl sulfone can also be extruded on equipment of varying design.

Polybutylene—Polybutylene is a flexible, linear polyolefin having a density of 0.91 and offering a unique combination of properties. Some of its important characteristics are: excellent resistance to creep at room and elevated temperatures—no failures after long-term loading at 90% of yield strength; higher long-term temperature resistance than other polyolefins; good toughness (high yield strength, high impact strength, high tear strength, high puncture resistance); exceptional resistance to environmental stress cracking (no failures after 2000+ hours per ASTM D-1693-60T); high filler loadings—up to 75% (300 PHR); and good flexibility similar to medium density polyethylenes.

Other important advantages include good moisture barrier properties, excellent electrical insulation characteristics, and resistance to most chemical environments.

Evaluations of the material indicate that it is a versatile resin adaptable to a wide variety of processes and applications. Successful evaluations to date include: film and sheeting, hot melt coatings and adhesives, flexible pipe and tubing, color compounding of polyolefins, electrical insulation, flexibilizer for polypropylene, shrink film, and processing aid.

Polybutylene can be processed on the same equipment used for low density polyethylene, but it should be noted that after cooling the polymer undergoes a crystalline transformation. During this change, tensile strength, hardness, specific gravity, and other crystallinity dependent properties attain their ultimate values. This behavior presents opportunities for post-forming techniques such as cold forming of sheeting or molded parts. This delay in aging does not pose any unusual processing problems.

Polycarbonates—Polycarbonates are among the stronger, tougher and more rigid thermoplastics. In addition, they have a ductility normally associated with the softer, lower modulus thermoplastics.

These properties, together with excellent electrical insulating characteristics, are maintained over a wide range of temperatures (−60 to 279°F) and loading rates. Although there may be a loss of toughness with heat aging, the material still remains stronger than many thermoplastics.

Polycarbonates are transparent materials and resistant to a variety of chemicals. They are, however, attacked by solvents.

The creep resistance of these materials is one of the best for thermoplastics. With polycarbonates, as with other thermoplastics, creep at a given stress level increases with increasing temperature. Yet, even at temperatures as high as 250°F, creep resistance is good.

Tensile yield strength at room temperature is about 8000 to 9500 psi. Yield occurs at a strain of about 5%. Elongation is in the range of 60 to 100%, with ultimate tensile strength falling in the same range as tensile yield strength.

The characteristic ductility of polycarbonate provides it with very high impact strength. Typical values are about 14 ft-lb/in. of notch on a $\frac{1}{8}$ in. thick specimen, although grades are available as high as 18 ft-lb. Unnotched specimens show impact resistance greater than 60 ft-lb. Fatigue resistance is also very good.

Moisture absorption for polycarbonates is low and equilibrium is reached rapidly. The materials, however, are adversely affected by weathering (slight color change and slight embrittlement can occur on exposure to ultraviolet rays).

Polycarbonate molding compounds are available for extrusion, injection molding, blow molding, and rotational molding. Film and sheeting with excellent optical and electrical properties are also available. Among the specialty grades, glass-reinforced polycarbonates have proved especially popular by virtue of their improved ultimate tensile strength, flexural modulus, tensile modulus, and chemical resistance.

Typical applications include: safety shields, lenses, glazing, gears, electrical relay covers, helmets, pump impellers, sight gages, cams and gears, aircraft components, automotive fender extensions, bezels, telephone switchgear, snowmobile components, boat propellors, water bottles, and housings for hand-held power tools and small appliances (Fig. 3-6).

Fig. 3-6. Frameless design mixer housing molded of polycarbonate provides double insulation and sole support for current-carrying internal components. (*Courtesy General Electric Co.*)

Polyester, high-temperature aromatic—Introduced in the early 1970's, this material (Ekonol, Carborundum Co.) shows excellent stability in air at over 600°F and combines inherent self-lubricating properties with high stiffness, thermal conductivity, electrical insulating properties, and solvent resistance. Because of its toughness and high heat resistance, it was originally fabricated by metallurgical-type processes. However, an injection moldable version (Ekcel) was introduced in 1974. This grade can be processed with conventional equipment.

The structure is characterized as a linear repeating chain of *p*-oxybenzoyl units. The polymer is highly crystalline, but no melting point has been observed even at 900 to 1000°F, where the polymer decomposes rapidly. However, the material undergoes a crystal-crystal transition at temperatures of 325 to 360°C.

The polymer displays excellent high temperature stability in air. Its flexural strength appears to be limited by thermal stresses induced during fabrication, but incorporation of 18% of a finely divided Al_2O_3 filler tends to reduce thermal stresses. The modulus is extremely high and the polymer possesses high compressive strength with little or no tendency to cold flow.

The resin displays a thermal conductivity three to five times higher than values reported for other polymers. This property is significant for high-temperature bearing or electrical applications where localized high spots can lead to premature breakdown. The thermal expansion coefficient is relatively low.

It displays a very high dielectric strength combined with a low dissipation factor. Tested against a variety of organic solvents, the polymer had no tendency to dissolve or even swell up to reflux conditions. With acids and bases, it displays good resistance in the cold, but hot caustic and hot concentrated H_2SO_4 tend to attack the polymer.

Typical applications include: self-lubricating bearings; seals, rotors, or vanes of process pumps; high temperature circuit boards; insulation components; encapsulation of diodes, transistors, and integrated circuits, and handles and slip-free coatings for frying pans.

Polyesters, thermoplastic—These linear polyesters are highly crystalline, with a melting point near 435°F. They are hard, strong, and extremely tough. Other characteristics include:

high resistance to abrasion, a low coefficient of friction, high resistance to cold flow, good chemical resistance, good dielectric properties, and low moisture absorption. Thermoplastic polyesters can be extruded, injection molded, or blow molded.

Reinforced compounds (with glass, talc, or asbestos) are available. The reinforced compounds retain most of the easy processing characteristics of the unmodified resin.

Applications include: gears, bearings, housings for pumps and appliances, impellers, pulleys, switch parts, furniture, fender extensions, and other products in competition with metals or high-performance plastics. The resin is also recommended for packaging applications, especially for products (i.e., food and medical goods) that are to be sterilized by radiation. It is also being used for blow molded carbonated beverage bottles.

Polyethylenes—Polyethylenes are characterized by toughness, near-zero moisture absorption, excellent chemical resistance, excellent electrical insulating properties, low coefficient of friction, and ease of processing.

Polyethylenes are classified by density as follows: Type I—0.910-0.925 gm/cu cm (called low density); Type II—0.926-0.940 gm/cu cm (medium density); and Type III—0.941-0.965 gm/cu cm (high density, also called linear polyethylene). The primary differences among the three types are in rigidity, heat resistance, chemical resistance, and ability to sustain loads. In general, as density increases, hardness, heat resistance, stiffness, and resistance to permeability increase. As density decreases, tensile strength and impact strength increase.

Low density polyethylene is quite flexible (stiffness modulus: 13,000 to 30,000 psi), with high impact strength and relatively low heat resistance (maximum recommended service temperature is 140 to 175°F), although grades are available with heat resistance up to 200°F.

Medium density polyethylene has a stiffness modulus of 31,000 to 150,000 psi and maximum continuous service temperature of about 180 to 250°F. Although the low and medium density polyethylenes are so flexible that they do not completely break in Izod impact tests, the high density types have values up to

20 ft-lb/in. of notch, depending on the resin.

Other interesting types of polyethylenes include: cross-linked (chemically or by irradiation) polyethylene, an infusible thermosetting material which has high heat resistance; olefin copolymers, consisting of ethylene and materials like vinyl acetate or ethyl acrylate, which offer a combination of low temperature flexibility with non-migratory ingredients, toughness, good stress-crack resistance, and ease of processing; and high-molecular weight polyethylene (72,000 and over mol. wt.) which has extremely high impact strength, excellent low temperature properties, and outstanding wear and abrasion resistance.

Polyethylenes, in general, are not outstanding load-bearing materials, but high density polyethylene can be used for some short-term light loads.

Since a polyethylene part under load can deform a great deal, it is usually considered to have failed before the actual breaking point is reached; yield strength is much more meaningful for design purposes. Although high density polyethylene has a good low temperature yield strength (11,000 psi at −40°F), it does not perform as well as some other thermoplastics as temperature is raised (3700 psi at 73°F, 1100 psi at 200°F).

The fatigue endurance limit of high density polyethylene is 1800 psi (50 to 100% RH, 70°F); low and medium density materials are better. This is also true for impact strength.

Few thermoplastics have the excellent chemical resistance and dielectric properties of polyethylenes. Soluble in some organic solvents above 140°F, polyethylenes resist bases and acids at room temperature. Resistivity (both volume and surface) and dielectric strength are high.

Forms available include polyethylene compounds that can be extruded, injection molded, blow molded, rotational molded, etc. Film and sheet are available, as are systems filled with carbon black (for ultraviolet stability), asbestos (for improved high temperature performance), and glass fiber (for improved mechanical properties). Cellular polyethylene, a foam-like material, has improved mechanical and electrical properties, which make it an excellent electrical insulator. Rigid polyethylene foam is also available.

Typical applications include: housings, pipe, heater ducts, housewares, toys, automobile interior side panels, battery parts, containers, wire and cable insulation, bottles, film and other packaging products, and materials handling (Fig. 3-7).

Fig. 3-7. Dunnage tray molded of high-density polyethylene weighs 30 lb and can carry a payload of about 6000 lb. (*Courtesy Monsanto Co.*)

Polyimides—Generally, this class of materials is based on an aromatic dianhydride and an aromatic diamine. However, although they share the general linear structure of a thermoplastic, melting points are so high that conventional processing techniques cannot be used; rather polyimides are available as fabricated parts (by machining, punching, or specialized direct forming techniques), coatings for wire and fabrics, and as a binder for diamond abrasive wheels. It should be noted, however, that in the early 1970's, polyimides were introduced that could be molded using conventional thermoset conditions and molding equipment, as well as thermoplastic conditions and molding equipment. These will not be discussed in this section.

The outstanding physical properties of the polyimides make them valuable in applications involving severe environments such as high temperature and radiation, heavy load, and high rubbing velocities. In terms of heat resistance, the continuous service of polyimide in air is on the order of 500°F. At elevated temperatures, polyimide parts retain an unusually high degree of strength and stiffness. However, prolonged exposure at high temperatures can cause a gradual reduction in tensile strength.

Polyimides are resistant to most dilute or weak acids and organic solvents. They are not attacked by aliphatic or aromatic solvents, ethers, esters, alcohols, and most hydraulic fluids. But they are attacked by bases such as strong alkali and aqueous ammonia solutions. The resin is also not suitable for prolonged exposure to hydrazine, nitrogen dioxide, or to primary or secondary amine components.

The minimum tensile strength for polyimides at 73°F is 10,500 psi. In tension, ultimate elongation at 73°F is about 5 to 6%. In compression, parts fabricated from polyimide resin can be permanently distorted, but will not break until a strain level of over 50% is reached. The parts are stiff, with a modulus of elasticity at 73°F of 450,000 psi.

Parts fabricated from unfilled polyimide resin have unusual resistance to ionizing radiation and good electrical properties.

Typical applications include: components such as gears, covers, bushings, turbo-fan engine backing rings, insulators, washers, piston rings, valve seats, etc., for such industries as aircraft and aerospace, nuclear power, office equipment, and electrical/electronic.

Polyphenylene sulfide—This resin is produced by a proprietary process (Ryton, Phillips Petroleum Co.) and is characterized by outstanding chemical resistance, good retention of mechanical properties, and high stiffness at elevated temperatures.

The tensile strength of molded parts is on the order of 21,000 psi and glass-filled polyphenylene sulfide parts provide a 2.2 million psi flexural modulus. Thermal properties are also exceptional—excellent long-term aging characteristics, high deflection temperatures greater than 425°F, and Vicat softening point near 600°F. In terms of chemical resistance, polyphenylene sulfide has hydrolytic stability at elevated temperatures and is resistant to a broad range of solvents, mineral and organic acids, and alkalis.

Coatings of polyphenylene sulfide may be applied in a variety of ways. Liquid slurries and dry powders may be applied by spraying and dry powders by the fluidized bed technique. Each technique requires a post-baking operation and multiple coats may be applied provided each coat is thoroughly baked before adding the next layer.

As with most coating operations, surface preparations (i.e., degreasing and grit blasting) is essential. But aluminum, steel, cast iron, brass, bronze, titanium, and a host of other metals have been successfully coated. Only those substrates which cannot stand a baking temperature of 700°F for at least 30 minutes should be restricted.

Molding grades of polyphenylene sulfide exhibit excellent affinity for reinforcing fillers such as glass or asbestos. These filled compositions may be processed by either injection or compression molding. Molded parts are hard, stiff, strong, and retain useful properties at temperatures as high as 500°F.

The injection molding grade containing 40% glass is markedly different from the unfilled resin. It is much more rigid and has a higher tensile strength. Of significance for high temperature use, the heat deflection temperature of the glass-filled resin is greater than 425°F, compared to 278°F for the unfilled resin.

Applications include: corrosive-resistant, thermally stable, protective coatings for oil-field pipe, valves, fittings, couplings, thermowells, probes and other equipment in both the petroleum and chemical processing industries; release coatings for molds, cookware and industrial containers; moldings of electrical components, valves, flex-rings, pump housings, pump impellers, conveyor rollers, spur gears, etc.; and machined parts for sleeves, pistons, compressor rings, and engineered components requiring excellent chemical resistance at elevated temperatures.

Polypropylenes—Interest in polypropylenes lies in a balance of many desirable charac-

teristics. Fairly rigid materials (flexural modulus: 150,000 to 240,000 psi), polypropylenes have better resistance to heat (heat distortion temperature at 66 psi: 200 to 250°F) and to more chemicals than do other thermoplastic materials at the same cost level. Polypropylenes have negligible water absorption, excellent electrical properties, and they are easy to process.

In much the same manner as density is important in determining the mechanical properties of polyethylenes, the isotactic index (related to the repeated units in the stereoregular molecular chain) of a polypropylene very often determines the characteristics of the material. An increase in the isotactic index of a polypropylene will sharply increase the yield strength of the material. Hardness, stiffness, and tensile strength also increase. On the other hand, as the isotactic index decreases, elongation and impact strength increase.

The ability to carry light loads for long periods and over wide temperature ranges is one of the properties that make polypropylenes valuable in engineering applications. For example, tensile yield strength (up to 5500 psi at 73°F) and tensile modulus (an average of 170,000 psi at 73°F) are still quite good at 212°F: 1800 psi and 40,000 psi, respectively. In addition, polypropylenes tested at 140°F react in a linear manner to stresses up to about 1000 psi; in this range there should be no permanent deformation.

Polypropylenes do not have outstanding long-term creep resistance, but fatigue endurance limit is excellent. In fact, it is often referred to as the "living hinge" thermoplastic.

One of the limitations most often mentioned for polypropylenes is their low temperature brittleness (−4°F). However, polypropylene copolymers have been developed with brittleness points of about −20°F.

Like all other polyolefins, polypropylenes have excellent resistance to water and to water solutions, such as salt and acid solutions that are destructive to metals. They are also resistant to organic solvents and alkalis. Above 175°F, polypropylene is soluble in aromatic substances such as toluene and xylene, and chlorinated hydrocarbons such as trichlorethylene.

Polypropylenes have excellent electrical resistivity (both volume and surface) and their dielectric strength is high.

Polypropylenes are available as molding compounds for extrusion, injection molding, blow molding, and rotational molding. Film and sheet are on the market, as are several different types of filled and reinforced molding compounds (Fig. 3-8). Polypropylene foams are available.

Fig. 3-8. Glass-reinforced polypropylene is used in this injection molded bakery tray that weighs less than metal trays and eliminates the problem of baked goods sticking to the tray. (*Courtesy Fiberfil Div., Dart Industries*)

For the most part, the filled polypropylenes use asbestos. These materials offer excellent impact strength and stiffness at temperatures up to 250°F. Other favorable properties include: low coefficient of friction, low heat absorption, excellent chemical resistance, and good dielectric properties.

Typical applications include: portable radio and television cabinets, pipe and fittings, automotive interior parts, housewares, bottles, carrying cases with integral hinges, pump impellers, fibers, coil forms, luggage, electrical connectors, and packaging.

Polystyrenes—The name polystyrene designates a large family of thermoplastics derived from the basic styrene monomer. In general, polystyrenes are characterized by low cost, ease of processing, hardness, and excellent dielectric properties. An important characteristic of polystyrenes is that within this family of materials there are many types specifically tailored to provide good mechanical, thermal, and chemical properties.

General-purpose polystyrene is the low cost member of the family. Properties include: hardness, rigidity, optical clarity, dimensional stability and excellent processability.

Modified or impact polystyrene (rubber reinforced) extends the uses of polystyrenes into those areas where high impact strength and good elongation are required. Most products of this type have the rubber molecularly grafted to the polystyrenes; however, some are produced by mechanical mixing.

Light-stable polystyrene meets IES, NEMA and SPI standards for non-yellowing service life up to five years.

Styrene-acrylonitrile (SAN) is the most chemical resistant polystyrene. It has excellent resistance to acids, bases, salts, and some solvents. It also has a high heat distortion point (195°F at 264 psi) and a higher tensile modulus (400,000 to 600,000 psi) than most other thermoplastics.

Another type of polystyrene commercially available is a low-temperature grade.

High impact polystyrene is generally recognized in the field as the best styrene material for load-bearing applications. The other types of polystyrene offer various improvements in chemical, thermal, and optical performance, usually at the sacrifice of mechanical properties.

High impact polystyrene is a very rigid material (modulus of elasticity in tension: 300,000 psi); however, it does not withstand long-term tensile loads as well as some other thermoplastics.

When loads are to be held longer than 500 hours, most high impact polystyrenes have design strengths between 800 and 1500 psi. Stress-strain data show high impact polystyrene to be very stable at low loads (15% strain at

300 psi for 1000 hr). However, strain is as much as 35% for greater loads (1000 psi) and for less time (100 hr).

Extra-high impact polystyrene is available with an impact strength as high as 4.7 ft-lb/in. notch (Izod). Extrusion, injection molding, compression molding, and blow molding polystyrene compounds are available, as are polystyrene film, sheet, and foam (Fig. 3-9). Glass-fiber-reinforced polystyrenes have improved mechanical and thermal properties.

Fig. 3-9. Foamed sheet extruded from polystyrene is used for deep-draw thermoformed packaging applications, such as egg cartons. (*Courtesy Monsanto Co.*)

Typical applications include: pipe and fittings, refrigerator parts, automobile interior parts, appliance housings, battery cases, TV picture masks, dials and knobs, washing machine filters, packaging, toys, and housewares.

Polysulfone—Polysulfone is a rigid, strong, thermoplastic that is both stable and self-extinguishing. It can be molded, extruded, or thermoformed (in sheets) into a wide variety of shapes. Characteristics of special significance to the design engineer are its heat-deflection temperature of 345°F at 264 psi and long-term use temperature of 300 to 345°F. Adding glass fibers in 5 to 10% concentration greatly improves its environmental stress-crack resistance.

Polysulfone has high tensile strength—10,200 psi at yield—and stress-strain behavior typical for rigid, ductile materials.

Flexural modulus of elasticity is high—nearly 400,000 psi at room temperature. With in-creases in temperature, flexural modulus stays above 300,000 psi to about 320°F.

Polysulfone is highly resistant to mineral acid, alkali, and salt solutions. Its resistance to detergents, oils, and alcohols is good even at elevated temperatures under moderate levels of stress. For example, specimens have withstood 500 hr in 2% detergent solution at 140°F and 1500 psi stress.

Polysulfone will be attacked by such polar organic solvents as ketones, chlorinated hydrocarbons, and aromatic hydrocarbons.

Its electrical properties are retained over a wide temperature range up to 350°F and after immersion in water or exposure to high humidity. Tests of electrical properties of polysulfone, polycarbonates, and polyacetals indicate two areas of significant difference. First, the dissipation factor of polyacetals is higher by a factor of 10, compared to polysulfone and polycarbonates. Second, good electrical properties are retained up to 350 to 375°F for polysulfone, 275 to 300°F for polycarbonates, and 210 to 250°F for polyacetals.

The value of polysulfone as an engineering material is best measured by studying creep or elongation under load. At room temperature for 10,000 hr (over a year) at 3000 psi, polysulfone shows about 1% total strain after a year. At 210°F, the creep of polysulfone at 3000 psi is higher than at room temperature, but total strain is still well under 2% after nearly a full year of such exposure.

Typical applications include: integrated circuit carriers, connectors, coil bobbins, housings for meters, switches and electronic components, light-fixture sockets and shades, under-the-hood switch and relay bases, dome-light bezels.

Vinyls—Vinyls are very versatile, ranging from extremely flexible, plasticized grades to very rigid vinyls (e.g., a modulus in flexure of about 540,000 psi). Chemical resistance of the rigid material is excellent. In addition, vinyls are easy to process by a number of different techniques.

Vinyls are used mainly for their chemical and weathering resistance, high dielectric properties or abrasion resistance. The most widely used vinyls are polyvinyl chloride (PVC) and PVC copolymers (e.g., vinyl chloride-vinyl acetate).

PVC and PVC acetate copolymers are formulated to provide such a wide and diverse range of properties that characterization by performance is difficult. In general, PVC materials are tough, with high strength and abrasion resistance, exceptional resistance to a wide variety of chemicals, and excellent dielectric characteristics. The recommended temperature range for PVC is −65 to 175°F.

Rigid PVC is essentially unplasticized, has a hardness similar to that of hard rubber, an impact strength of 0.25 to 1.2 ft-lb/in. notch (Izod), and modulus in flexure of 380,000 to 540,000 psi.

Flexible PVC is a plasticized material which often contains, in addition to plasticizers, other ingredients such as stabilizers and pigments in various proportions. Typical property data include: tensile strength: 1000 to 3500 psi; elongation: 200 to 450%; 100% modulus: 1200 to 2800 psi. A cold flexure temperature of about −70°F is obtainable.

Other types of vinyls include: polyvinyl acetate, a widely used thermoplastic adhesive; polyvinyl alcohol, a water-soluble film; polyvinyl acetals (formals and butyrals), widely used as additives to improve adhesion and mechanical properties of other plastics; and polyvinylidene chloride, a tough, abrasion resistant vinyl with better than ordinary heat resistance (212°F maximum service temperature).

Vinyls can be extruded, cast, injection molded, blow molded, rotationally molded, and calendered. Film and sheet are available, as are both rigid and flexible foams.

Typical applications include: pipe and fittings, electrical wire insulation, valve seats, chemical storage tanks, adhesives, packaging, upholstery, shoe components, flooring, and outerwear.

Thermosetting Resins

Among plastics materials, thermosetting materials generally provide one or more of the following advantages: (1) high thermal stability, (2) resistance to creep and deformation under load and high dimensional stability, and (3) high rigidity and hardness. These advantages are coupled with the light weight and excellent electrical insulating properties common to all plastics. The compression and transfer molding methods by which the materials are formed, together with the more recent evolution of thermoset injection molding techniques, offer low processing cost and mechanized production.

Thermosetting molding compounds consist of two major ingredients: (1) a resin system which generally contains such components as curing agents, hardeners, inhibitors and plasticizers, and (2) fillers and/or reinforcements, which may consist of mineral or organic particles, inorganic or organic fibers, and/or inorganic or organic chopped cloth or paper.

The resin system usually exerts the dominant effect, determining to a great extent cost, dimensional stability, electrical qualities, heat resistance, chemical resistance, decorative possibilities, and flammability. Fillers and reinforcements affect all these properties to varying degrees, but their most dramatic effects are seen in strength and toughness, and sometimes, electrical qualities.

Following is a summary on each of the eight generic families of molding compounds. Descriptions cannot be all-inclusive in the space allowed; the attempt here is more to characterize the materials in a brief and general manner.

Alkyds−Primarily electrical materials, alkyds combine good insulating properties with low cost. They are available in granular and putty form, permitting incorporation of delicate complex inserts.Moldability is excellent, cure time short, and pressures are low. In addition, their electrical properties in the rf and uhf ranges are relatively heat stable up to a maximum use temperature of 250 to 300°F.

General-purpose grades are normally mineral-filled; compounds filled with glass or synthetic fibers provide substantial improvements in mechanical strength, particularly impact strength. Short fibers and mineral fillers give lower cost and good moldability; longer fibers give optimum strength.

Although the term "alkyd" in paint terminology refers to a fatty-acid-modified polyester, in molding compound parlance the term merely means a dry polyester molding compound, usually cross-linked with a diallyl phthalate (DAP) monomer.

Typical uses for alkyds include circuit breaker insulation, coil forms, capacitor and resistor encapsulation, cases, housings and switchgear components.

Allylics—Of the allylic family, diallyl phthalate (DAP) is the most commonly used molding material. A relatively high priced, premium material, its two outstanding features are excellent dimensional stability and a high insulation resistance (5×10^6 megohms) which is retained to an extremely high degree after exposure to moisture (e.g., same value is retained after 30 days at 100% RH and 80°F).

Both DAP and DAIP (diallyl isophthalate) compounds are available, the latter providing primarily improved heat resistance (maximum continuous service temperatures of 350 to 450°F vs. 300 to 350°F for DAP). Compounds are available with a variety of reinforcements, e.g., glass, asbestos, acrylic and polyester fibers. Physical and mechanical properties are good, and the materials are resistant to acids, alkalis, and solvents.

DAP compounds are available as electrically conductive and magnetic molding grades. This is achieved by incorporating either carbon black or precious metal flake. In electrically conductive DAP, volume resistivity is on the order of 0.4 ohm-cm (i.e., electrical conductivity is 2.5 mhos-cm). The magnetic grade has a residual flux density of 1000 to 1200 gausses, coercive force is 860 to 875 oersteds, and energy product $(BH)_{max}$ is 258,000 gauss-oersteds.

Epoxies—These materials offer excellent electrical properties and dimensional stability, coupled with high strength and low moisture absorption.

Epoxies are used by the plastics industry in several ways. One is in combination with glass fibers (i.e., impregnating fiber with epoxy resin) to produce high-strength composites (or "reinforced plastics") that provide heightened strength, electrical and chemical properties, and heat resistance. Typical uses for epoxy-glass RP

is in aircraft components, filament wound rocket motor casings for missiles, pipes, tanks, pressure vessels, and tooling jigs and fixtures.

Epoxies are also used in the encapsulation or casting of various electrical and electronic components and in the powder coating of various metal substrates. Epoxies can further be transfer or injection molded. Transfer presses, for example, are being used today to encapsulate electronic components in epoxy or to mold epoxy electrical insulators. Injection molding presses are also available for liquid epoxy molding.

Epoxy molding compounds are available in a wide range of mineral or glass-reinforced versions. The low glass-content (up to about 12 to 50%) or mineral-reinforced compounds are primarily used for electrical applications and can be molded at relatively low pressures and short cycles. The high glass content compounds (over 60%) offer extremely high mechanical strengths, but are more difficult to process.

Melamines—Melamine molding compounds are best known for their extreme hardness, excellent and permanent colorability, arc-resistant nontracking characteristics and self-extinguishing flame resistance. Dishware and household goods plus some electrical uses have been primary applications.

General-purpose grades for dishes or kitchenware are usually alpha-cellulose-filled. Mineral-filled grades provide improved electrical properties; fabric and glass fiber reinforcements provide higher shock resistance and strength.

Chemical resistance is relatively good, although the material is attacked by strong acids and alkalis. Melamines are also tasteless, odorless, and are not stained by pharmaceuticals and many foodstuffs.

Like the other member of the amino family, urea, the melamines find extensive use outside of the molding area in the form of adhesives, coating resins, and laminating resins.

Phenolics—Phenolics are the workhorse of the thermosetting molding compound family. With the number and variety of types available, it is difficult to characterize them as a whole. But, in general, they provide low cost, good electrical properties, excellent heat resistance, and good mechanical properties, coupled with excellent moldability. They are generally limited in color (usually black or dark brown) and color stability.

General-purpose grades are usually wood-flour and flock-filled. They provide a good all-around combination of moderately good mechanical, electrical, and physical properties at low cost. They are generally suitable for use at temperatures up to 300°F.

In chemical resistance, the materials are severely attacked by strong acids and alkalis; effects of dilute acids and alkalis and organic solvents vary with the reagents and with the resin formulation.

Impact grades vary with the reinforcement. In order of increasing impact strength, paper, chopped fabric or cord, and glass fibers are used. Glass fiber grades also provide substantial improvement in strength and rigidity.

Glass-containing grades can also be combined with heat resistant resin binders to provide a combination of impact and heat resistance. Dimensional stability is also substantially improved by glass.

Electrical grades are generally mineral- or flock-filled materials designed for improved retention of electrical properties at high temperatures and high humidities.

Heat resistant grades are usually mineral- or glass-filled compounds designed for long-term retention of properties at temperatures in the 400 to 500°F range. Special heat resistant phenolics (such as the phenyl silanes) provide good long-term stability at 500 to 550°F and shorter term stability at 600°F Some types are usable for up to 1 hr as high as 1000°F.

Most ablative plastics consist of phenolic as the resin binder. Here, of course, periods of use are measured in seconds or minutes; temperatures are on the order of 3000 to 7000°F, or upward to about 25,000°F. Molding compounds used include phenolics reinforced with high silica glass, fused silica fibers, and graphite fabrics, usually in macerated form.

Many specialty-grade phenolics are also available. These include chemical-resistant grades in which the resin is formulated particularly for improved stability to certain chemicals; self-lubricating grades incorporating dry lubricants such as graphite or molybdenum disulfide;

grades with improved resistance to moisture and detergents for washing machine parts; rubber-modified phenolics for improved toughness, resilience and resistance to repeated impact; and ultra high strength grades (glass content over 60% by weight) designed to provide mechanical strengths comparable to those of cloth-reinforced plastics laminates.

Polyesters—Another plastics family that is extremely versatile in terms of the forms in which it finds use in the industry. Polyester resins can be formulated to be brittle and hard, tough and resilient, or soft and flexible. In combination with a reinforcement like glass fibers, they offer outstanding strength, a high strength-to-weight ratio, chemical resistance, and other excellent mechanicals.

Although polyesters are finding a growing market as a liquid resin for casting furniture parts (e.g., simulated wood doors, drawer fronts, lamp bases, wall plaques, etc.) their prime outlet is in combination with glass fibers in high-strength composites (or "reinforced plastics"). Using various lay-up, spray-up, and matched metal molding techniques (see Chapter 19 on Reinforced Plastics Processing), polyester-glass is being used for such products as automotive body parts, boat hulls, building panels (Fig. 3-10), housings, bathroom components, tote boxes, appliances, and electronic and electrical applications.

Special equipment is also available to injection mold polyester-glass formulations. Most recently, one-component molding sytems, combining resin, reinforcement, and other additives, have been made available to molders. These materials, known as SMC (sheet molding compounds) are easy to handle and afford a great deal of automation in feeding into a molding press.

Another popular technique of molding involves premix compounds, which are dough-like materials prepared by the molder by combining the premix constituents (e.g., polyester, glass, catalyst, etc.) in a sigma blade mixer or similar piece of equipment prior to molding. This technique is currently being used to make automotive heater housings and air conditioner components.

Also available for use with polyester resins is

Fig. 3-10. Fascia panel for building construction is molded of polyester-glass reinforced plastics. Panel measures 8 by 14 ft, yet weighs only 120 lb. (*Courtesy Arco/Polymers*)

an extrusion-like technique known as pultrusion in which glass fibers are pulled through a resin bath to make continuous profiles, pipe, or other shapes.

Silicones—Unparalleled long-term heat resistance, excellent electrical insulating properties stable under both frequency and temperature changes, good moisture and chemical resistance, exceptionally high cost is about the description of silicone molding compounds. Being premium materials, they are not often competitive with other compounds either in cost or performance.

The two major types are: mineral-filled for maximum heat resistance and optimum moldability, and glass-filled for high mechanical strength and toughness.

The silicones are also available in the form of a room temperature vulcanizing (RTV) rubber that is used in making molds for the casting of various plastics (e.g., polyester furniture parts).

Ureas—Excellent colorability, moderately good strength and low cost are the primary attributes of urea molding materials. Dimensional stability and impact strength are poor.

The materials are used for such products as decorative housings, jewelry casings, lighting fixtures, closures, wiring devices, and buttons.

Structural Design Characteristics of Plastics

The purpose here is to describe some of the basic behavior characteristics of plastics materials as they affect the structural design of plastics parts.[1]

Mechanical Properties—The Stress-Strain Curve. In the tensile testing of plastics, three distinct types of stress-strain curves may be observed. These are illustrated in Fig. 3-11 (a, b, and c). Stress-strain curves, Type A and B, illustrate materials which have gradual and abrupt yielding. The Type C diagram shows a material which fails before yielding occurs.

The designer is usually concerned with two distinct regions of the stress-strain curve. In the first region, designated by the line OA, elastic design principles may be applied. The second region around Point B, which will be referred to as the yield point, can be used when fracture or failure due to large deformation is of prime concern. In the following discussion, both of these regions are considered in detail.

Proportional limit and elastic limit are two terms used to describe the point of termination of the linear portion of a stress-strain curve, the limit of Hooke's law. This is about Point A. However, a careful study of stress-strain curves for thermoplastics has shown that there is no linear section. Rather, there is deviation from linearity from the origin. The deviation is very slight below 0.5% strain and most published figures of elastic modulus, E_o, are the slope of a line tangent to the low strain portion of the stress-strain curve.

The secant modulus, E_s, is the slope of a line drawn from the origin to a point on the curve, and the secant modulus decreases with higher strains. Plotting E_s/E_o emphasizes that application of the initial modulus becomes less accurate with increasing strain. This is shown in Fig. 3-12 for three different plastics. This means a part in service may deflect more than calculations indicate since linearity is assumed in deriving equations for deformation.

With thermoplastics there is no "limit"; rather it is a matter of degree of deviation with

[1] Portions of this section adapted from "Application Design" published by E. I. du Pont de Nemours & Co., Inc.

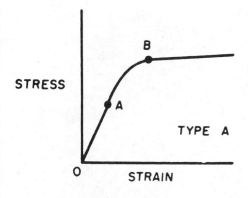

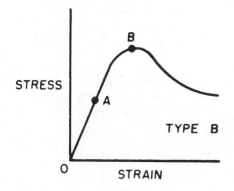

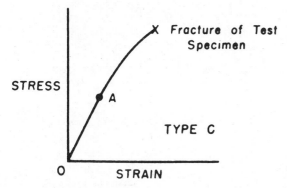

Fig. 3-11. Stress-strain curves. (A) Gradual yielding; (B) Abrupt yielding; (C) Fracture occurs at low strains before yielding (*All sketches on "Structural design considerations," courtesy of E. I. du Pont de Nemours & Co., Inc.*)

increasing strain, and a concept modulus accuracy is needed. Hence, we have chosen the term "modulus accuracy limit." As an arbitrary choice, an accuracy of 15% was chosen as an allowable error in most design calculations.

On a specific design problem, after the deflection is calculated, the strain or the stress

should be calculated. If the "modulus accuracy limit" is not exceeded, the calculation is sufficiently accurate. Otherwise, the geometry of the part must be changed to reduce the strain (or stress), or a correction must be applied to the calculation. Each designer and each application will have specific needs for accuracy, which are related to the concept and use of a safety factor. It is through an understanding and application of a modulus accuracy limit that these specific needs are met.

It is convenient that strain at the 15% modulus accuracy limit turns out to be essentially a material constant regardless of temperature or loading time.

It is possible that beyond a limiting strain value, marked inaccuracies in calculations for creep deflection and stress relaxation may occur. This possibility emphasizes the importance of using the modulus accuracy limit concept as a checkpoint until, or unless, future data show the limit is not necessary.

In order to define the yield point in plastics of Type A, as shown in Fig. 3-11, the true stress-logarithmic strain curve must first be calculated from the conventional stress-strain curve. In the conventional stress-strain curve, strain is calculated in terms of the original

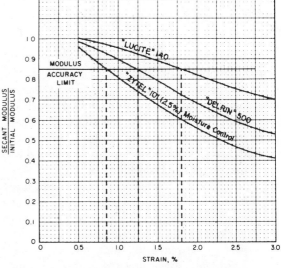

Fig. 3-12. Decrease of the ratio of secant modulus to initial modulus with increasing strain (73°F, 23°C) for three different plastics (top to bottom): acrylic, acetal, and nylon 6/6.

length of the test specimen and stress is calculated on the basis of the original cross-sectional area. Since the cross-sectional area of the test specimen changes during a tensile test, the true stress-logarithmic strain curve which is based on the instantaneous dimensional changes is more meaningful. The equations which convert conventional strain, ϵ, to logarithmic strain, $\bar{\epsilon}$, and conventional stress, S, to true stress, $\bar{S}$ are

$$\bar{S} = S(1 + \epsilon) \qquad (1)$$

and

$$\bar{\epsilon} = \ln(1 + \epsilon) \qquad (2)$$

The following examples illustrate how these equations may be used.

EXAMPLE 1. A specimen is stressed 5000 psi at a strain of 1%. What is the corresponding true stress and logarithmic strain?

True stress = $\bar{S}$ = 5000 (1 + 0.01) = 5050 psi

Logarithmic strain = $\bar{\epsilon}$ = 2.303 $\log_{10}$ (1 + 0.01)
= 0.0099 or 0.99%

EXAMPLE 2. Based on the original dimensions, a specimen is stressed to 10,000 psi at a strain of 10%. What is the corresponding true stress and logarithmic strain?

True stress = $\bar{S}$ = 10,000 (1 + 0.1) = 11,000 psi

Logarithmic strain = $\bar{\epsilon}$ = 2.303 $\log_{10}$ (1 + 0.1)
= 0.095 or 9.5%

By applying Eqs. (1) and (2), the true stress-logarithmic strain curves are constructed from the conventional stress-strain curve. Examples 1 and 2 show that the differences between the conventional and the true-stress logarithmic strain curves are substantial for relatively large strains and negligible for small strains within the modulus accuracy limit.

In design terminology, there are many different definitions of the yield point, and therefore the danger of confusion always exists. For materials which yield gradually, the yield point is determined from the true stress-logarithmic strain curve as is shown in Figs. 3-13, 3-14, and 3-15. The straight line BC is the work-hardening region of the tensile stress-strain curve, and point B, which marks the

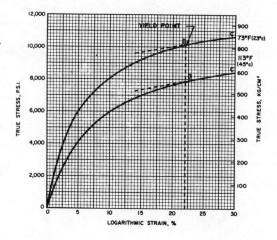

Fig. 3-13. Tensile stress-strain curves for nylon 6/6 (2.5% moisture content).

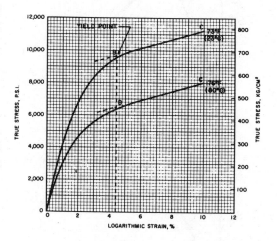

Fig. 3-14. Tensile stress-strain curves for acetal.

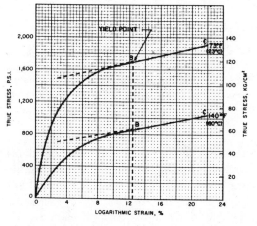

Fig. 3-15. Tensile stress-strain curves for low-density polyethylene.

beginning of work hardening, is defined as the yield point. For materials which yield abruptly, the yield point is defined as the maximum or hump of the conventional stress-strain curve, as shown in Fig. 3-11(b). The stress corresponding to the yield point is called the *yield-stress*. While this definition for materials which yield gradually differs from the ASTM definition, it is used in most texts on the theory of plasticity.

Figures 3-13, 3-14, and 3-15 are the true stress-logarithmic strain curves for nylon 6/6 (2.5% moisture content), acetal, and low-density polyethylene, measured at standard ASTM loading rates. These curves illustrate how the yield point may be determined. The yield strains are shown to be 22, 4.5, and 12.5% logarithmic strain, respectively. If a test specimen of a ductile plastic material is extended to relatively large strains, the test specimen will neck down and the cross-sectional area of the specimen becomes uncertain. The stress-strain curve beyond the yield point is of little value to the design engineer except that it adds confidence that stress concentrations can be distributed without failure by fracture.

Effect of Temperature, Loading Time, and Environment. Tensile testing of materials at ordinary temperatures has shown that most metals are characterized by one type of stress-strain curve. This is not the case for plastics which are, in general, more sensitive to temperature, rate of testing, and environmental conditions.

Nylon 6/6 is a good example. Testing rate and moisture content can change the type of stress-strain diagram, thereby changing the basic design properties such as yield stress. It is essential that the designer of plastic parts know under what conditions changes in material behavior may occur and adapt the design procedure accordingly.

It is an established fact that the yield stress increases with the rate of testing. In many practical applications the load is applied at speeds which are considerably faster than the ASTM recommended testing rates and therefore the material properties which are obtained from standard tests are sometimes misleading.

In some calculations, it is important for the designer to know the stress-strain curves in tension and compression. For example, in flexural design problems both types of stress-strain curves are needed because tensile and compressive stresses are present in the structure. Furthermore, various materials are considerably stronger in compression than in tension (for example, concrete). It has been shown that for large strains the compressive strength properties of various plastics are substantially higher than the corresponding tensile properties.

The tensile and compressive stress-strain curves for several plastics are given in Figs. 3-16, 3-17, and 3-18. From these curves, which

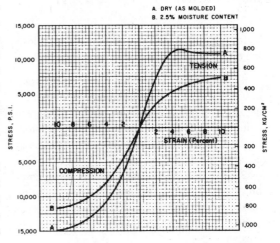

Fig. 3-16. Conventional stress-strain curves in tension and compression of nylon 6/6 (73°F, 23°C).

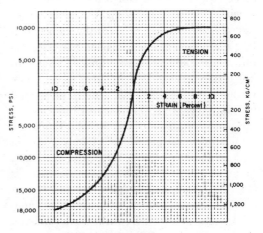

Fig. 3-17. Conventional stress-strain curves in tension and compression of acetal (73°F, 23°C).

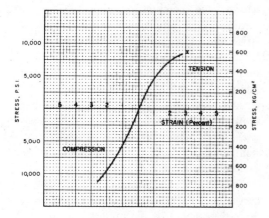

Fig. 3-18. Stress-strain curves in tension and compression of acrylic (73° F).

were obtained at ASTM strain rates, the following conclusions can be made:

1. The tensile and compressive stress-strain curves are, for all practical purposes, identical at small strains. Therefore the modulus in compression is equal to the modulus in tension.

2. It follows that the flexural modulus is equal to the tensile modulus.

3. For relatively large strains, the compressive stress is higher than the corresponding tensile stress. This means that the yield stress in compression is greater than the yield stress in tension.

Creep and Relaxation. The phenomena of creep and relaxation in plastic materials are of prime concern to the designer of plastic parts which, in use, can be only slightly deformed yet carry a load for long periods of time. Creep will take place even at low loads and low temperature and the designer must estimate its amount. For metals, analogous behavior is usually only encountered at elevated temperatures and therefore these phemonena are of less concern. Creep and relaxation can be illustrated by considering a test bar which is simply loaded in tension. In Fig. 3-19a, a force is applied to a specimen and an initial deformation is observed. With time, the test bar will permanently elongate. Creep is defined as non-recoverable deformation, with time, under a constant load. Consider a similar specimen stressed by the same force, but in this case the

specimen is confined to the initial deformation. As is shown in Fig. 3-19b, the force decreases with time. Relaxation is defined as a decrease of force, with time, required to produce a constant strain.

Basic creep and relaxation behavior are measured by a simple uniaxial tension test. The

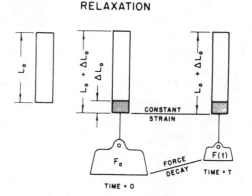

Fig. 3-19a. Creep and relaxation in tension.

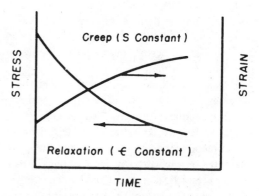

Fig. 3-19b. Graphical representation of creep and relaxation data.

data are usually given graphically. From the individual curves, the time-dependent moduli are calculated. The time-dependent creep modulus is defined by Eq. (3).

$$E_{t\ creep} = \frac{Stress}{Strain\ at\ time\ (t)} = \frac{F_o/A_o}{\Delta L_t/L_o} \quad (3)$$

where

F_o = Constant force, lb
A_o = Original cross-sectional area, in.
L_o = Original length, in.
ΔL_t = Increase in length at time (t), in.

The time dependent relaxation modulus is similarly defined as

$$E_{t\ relax} = \frac{Stress\ at\ time\ (t)}{Strain} = \frac{F_t/A_o}{\Delta L_o/L_o} \quad (4)$$

where

A_o and L_o = Are as before
ΔL_o = Initial deformation, in.
F_t = Force at time (t), lb

Tests so far have confirmed that Eqs. (3) and (4) are applicable when the initial strain is within the modulus accuracy limit. The modulus accuracy limit is used as a checkpoint when making creep (deformation) and relaxation (force decay) calculation.

It has been shown experimentally that the creep modulus and relaxation modulus are similar in magnitude and for design purposes may be assumed to be the same. That is $E_{t\ creep} = E_{t\ relax}$. Therefore, only one time-dependent modulus called the *apparent modulus* is necessary. The following examples illustrate the similarity between creep and relaxation moduli.

EXAMPLE 1. Determination of the apparent modulus from a creep experiment would be accomplished by placing a test bar in tension under a load of 95 lb. The bar has an original gauge length of 5.0 in. and a cross-sectional area of 0.063 sq in. After 1 year, the increase in length is 0.025 in.

$$Stress\ (constant) = \frac{95}{0.063} = 1500\ psi$$

$$Strain\ after\ one\ year = \frac{0.025}{5.0} = 0.005\ or\ 0.5\%$$

$$E(1\ year)_{creep} = \frac{Stress}{Strain\ after\ 1\ year} = \frac{1,500}{0.005}$$

$$= 300,000\ psi$$

EXAMPLE 2. Similar information would be obtained by elongating a specimen of the same dimensions to 0.5% strain, maintaining a constant strain and recording the force decay. After 1 year the force decreases to 95 lb.

$$Stress\ after\ 1\ year = \frac{95}{0.063} = 1500\ psi$$

$$Strain\ (constant) = 0.005$$

$$E(1\ year)_{relax} = \frac{Stress\ after\ 1\ year}{Strain} = \frac{1,500}{0.005}$$

$$= 300,000\ psi$$

In using apparent modulus in design procedure, in general, the standard strength of materials design formulas are used. However, the modulus which is chosen is an apparent modulus selected on the basis of the time the part is under load.

Design Methods. The major difference between metals and plastics design lies in the choice of structural properties to be used in standard design theory and formulas. For metals, these properties are relatively constant over wide ranges of temperature, time, and other measures of environment. For plastics, structural properties are more sensitive to comparable changes in environment.

The well-known stress distribution of a simple beam in flexure is the background for the recommended methods for using the concepts discussed above in plastics design. When a beam is flexed, two distinct regions of stresses are obtained which are separated by a neutral plane. On one side of the neutral plane, the fibers are in tension, while on the other side the fibers are in compression. Since an idealized stress-strain curve in tension and compression is symmetrical and linear at small deformations, the stress distribution in the beam in flexure also is symmetrical about the neutral plane as is shown in Fig. 3-20(a). This type of linear distribution, which exists only at relatively small deflections, is assumed in the derivation of the standard elastic design equations.

When a beam is flexed to carry maximum load, the stress distribution in the beam approaches the rectangular shape shown in Fig. 3-20(b). This can only occur if the material is

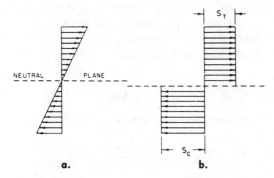

Fig. 3-20. Stress distribution in flexure: a. linear (elastic) stress distribution; b. stress distribution maximum load.

ductile and does not fracture before this distribution is reached. It is important to note that the neutral plane is no longer in the center of the beam since the yield stress in compression is greater than the yield stress in tension. Yield design equations, derived on the basis of this type of behavior for metals, can be modified for plastics. The basic assumption is that a structure will fail when all the fibers have yielded. In both those sections, illustrations with simple design calculations are intended to familiarize the reader with the general procedure.

Designing within the Modulus Accuracy Limit. Elastic design equations which were developed for metals can be applied to plastics. The deflection equations are expressed in terms of two material variables, the elastic (Young's) modulus and Poisson's ratio; while the stress equations are only dependent on load and geometry. These formulas can be converted to the appropriate time-dependent equations by replacing the elastic modulus with the apparent time-dependent modulus and assuming Poisson's ratio to be a constant.

For practical purposes, Poisson's ratio is constant with time. The other major assumption is that the stress distribution is initially linear and remains linear with time.

The following examples illustrate this method of designing within the modulus accuracy limit; terms in the equations are defined in Fig. 3-21.

EXAMPLE 1. Deformation of aerosol bottle base.

In the design of plastic aerosol bottles, the deformation of the base with time must be known. The slightest bulge of an initially flat bottom will cause the bottle to rock on its base. Hence, a skirt around the base is necessary if a flat bottom is intended for the container. The height of the skirt must be greater than the base deformation. To estimate the amount of deflection, the base is considered as a circular flat plate under a uniform load. The walls of a bottle add stiffness to the base. If we assume simple vertical support for the walls, we will calculate a height for the skirt that will assure that the bottle will stand erect. For this condition the equation for maximum deflection at the center of the plate is

$$Y_m = \frac{3pr^4(5 - 4\mu - \mu^2)}{16E_t T^3} \qquad (5)$$

and the correspondence maximum stress is

$$S_m = \frac{3pr^2(3 + \mu)}{8t^2} \qquad (6)$$

To show the complete calculation, consider a bottle made of acetal resin with internal pressure of 100 psi for 1 year at room temperature.

Geometry

Radius of plate = r = 0.75 in.
Thickness of plate = t = 0.20 in.

Physical constants at $73°F$ after 1 year

Apparent modulus
= E_t (1 year) = 175,000 psi
Poisson's Ratio = μ = 0.35

$$Y = \frac{3 \times 100 \times 0.75^4(5 - 4 \times 0.35 - 0.35^2)}{16 \times 175,000 \times 0.20^3}$$

$$= 0.015 \text{ in.}$$

$$S = \frac{3 \times 100 \times 0.75^2(3 + 0.35)}{8 \times 0.20^2} = 1750 \text{ psi}$$

$$= \frac{1750}{410,000} = 0.0043 \text{ in./in. or } 0.43\%$$

Geometry	Deflection Equation	Stress Equation	Nomenclature
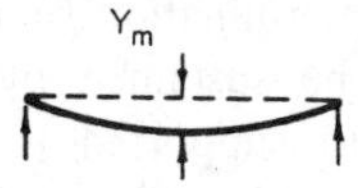 Centrally Loaded Beam	$Y_m = \dfrac{PL^3}{48\,E_t\,I}$	$S = \dfrac{3\,P\,L}{2\,b\,d^2}$	P = Load L = Span d = Depth b = Width I = Moment of Inertia E_t = Apparent Modulus S = Outer Fiber Stress Y_m = Maximum Deflection
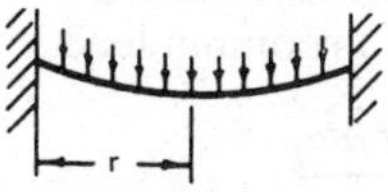 Uniform Load on Circular Plate (Edges Fixed)	$Y_m = \dfrac{3p\,r^4\,(1-\mu^2)}{16\,E_t\,t^3}$	$S = \dfrac{3p\,r^2\,(1+\mu)}{8\,t^2}$	p = Load per Unit Area t = Thickness r = Radius μ = Poisson's Ratio S = Maximum Stress
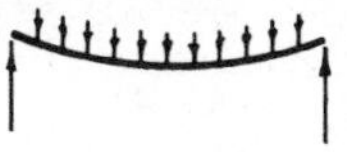 Uniform Load on Circular Plate (Edges Supported)	$Y_m = \dfrac{3p\,r^4\,(5-4\mu-\mu^2)}{16\,E_t\,t^3}$	$S = \dfrac{3p\,r^2\,(3+\mu)}{8\,t^2}$	p = Load per Unit Area t = Thickness r = Radius μ = Poisson's Ratio S = Maximum Stress
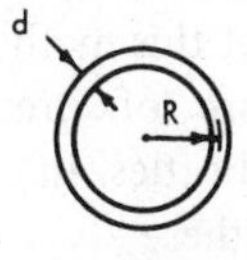 Thin Closed-End Pipe Under Internal Pressure	$Y = \dfrac{R}{E_t}\,\left(1 - \dfrac{\mu}{2}\right)\dfrac{P_i\,R}{t}$	$S_h = \dfrac{P_i\,R}{t}$	t = Wall Thickness R = Mean Radius P_i = Internal Pressure Y = Radial Displacement S_h = Hoop Stress

Fig. 3-21. Typical formulas for designing within the modulus accuracy limit.

Since the calculated maximum initial stress and strain are within the modulus accuracy limit (5300 psi, 1.25%), the calculated deformation of 15 mils is reliable. Thus, the height of the skirt which elevates the base should be at least 15 mils, so that the bottle will stand erect and firm even if slight deformation with time is encountered.

EXAMPLE 2. Radial displacement of a pipe under internal pressure.

In the construction of plastic piping networks, it is sometimes important to know the radial displacement of the wall of a pipe that is under internal pressure for a period of time. The radial displacement and the mean hoop stress of a thin closed-end pipe under internal pressure may be calculated using the following equations:

Radial displacement $= Y = \dfrac{R}{E_t}\left(1 - \dfrac{\mu}{2}\right)\dfrac{P_i R}{t}$ (7)

Mean hoop stress $= S_b = \dfrac{P_i R}{t}$ (8)

Consider, for example, a pipe made of nylon 6/6 conditioned to 2.5% moisture content. The pipe contains compressed air at 73°F and 50% relative humidity and is under a pressure of 150 psi for 5 years.

Geometry

Mean radius of pipe = R = 0.50 in.
Wall thickness = t = 0.10 in.

Physical constants at 73°F after 5 years

Apparent modulus

$$= E_t(5\text{ years}) = 100,000 \text{ psi}$$

Poisson's ratio = 0.40

$$Y = \frac{0.50}{10 \times 10^4}\left[(1 - 0.20)\frac{150 \times 0.50}{0.10}\right]$$

$$= 0.0029 \text{ in.}$$

$$S_h = \frac{150 \times 0.50}{0.10} = 750 \text{ psi}$$

$$\epsilon = \frac{S_h}{E_o} = \frac{750}{175,000} = 0.0043 \text{ in./in. or } 0.43\%$$

The calculation shows that the radius of the pipe would increase by 2.9 mils in 5 years. The radial displacement is a realistic estimate since both the initial stress and strain did not exceed the modulus accuracy limit for this specific nylon (1400 psi, 0.85%).

The two preceding examples apply only to calculating deformation. The biaxial stress systems, such as pipe and aerosol bottles, fracture due to excessive stress is another factor which must be considered in the overall design. Examples that illustrate the calculation of burst stress are given in the following discussion on yield design.

Yield Design. The previous discussion considered the deformation of structures under load. In ultimate load determinations, the designer is primarily concerned with the stress required for failure. In both cases, accurate design can be accomplished only if the effect of time and temperature on the mechanical properties is known.

Standard formulas used in limit design are found in various textbooks dealing with the theory of plasticity. The equations can be used in designing structures made of plastics if the yield stress at time of failure is known. Experimental evidence justifying this procedure has been obtained in a number of cases. The simplest example is the determination of the maximum load that can be sustained by a centrally loaded beam with supported ends. Failure occurs when all the fibers in tension reach the yield stress in tension and all the fibers in compression reach the yield stress in compression (Fig. 3-22). The following equation is used to calculate the maximum load:

$$P_{max} = K \frac{bh^2}{L}$$

where

$$K = \frac{2S_t - S_c}{S_t + S_c} \quad (9)$$

where b, h, L are the width, depth, and span, respectively, S_t is the yield stress in tension and S_c is the yield stress in compression.

At present, the greatest use of this method is made in predicting the failure of pressure vessels such as pipe and aerosol bottles. In order to obtain the solutions to these types of multiaxial stress problems, the von Mises-Hencky yield condition is applied to plastics.

$$(S_1 - S_2)^2 + (S_2 - S_3)^2 + (S_3 - S_1)^2 = 2S_t^2(t)$$

where S_1, S_2, and S_3 are the three principal stresses and $S_t(t)$ is the yield stress in tension at the time of failure.

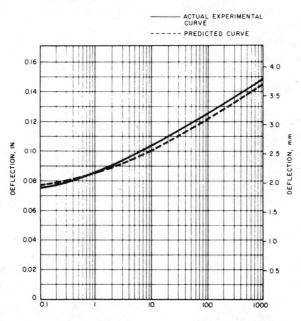

Fig. 3-22. Deflection of a centrally loaded beam with time (acetal, 73°C, end supports).

4

INJECTION MOLDING OF THERMOPLASTICS

Irvin I. Rubin [1]

A BASIC INTRODUCTION[2]

In the pages that follow, the author presents a complete analysis of injection molding machines, the injection molding operation, and molding theory and principles. This basic review and introduction has been prepared for readers unfamiliar with the process.

The basic concept of injection molding revolves around the ability of a thermoplastic material to be softened by heat and to harden when cooled. In most operations, granular material (the plastic resin) is fed into one end of the cylinder (usually through a feeding device known as a hopper), heated, and softened (plasticized or plasticated), forced out the other end of the cyclinder (while it is still in the form of a melt) through a nozzle into a relatively cool mold held closed under pressure. Here, the melt cools and hardens (cures) until fully set-up. The mold then opens and the molded part is removed (see Fig. 4-1).

Thus, the significant elements of an injection molding machine become: (1) the way in which the melt is plasticized (softened) and forced into the mold (called the injection unit); (2) the system for opening the mold and closing it

[1] The author gratefully acknowledges John Wiley & Sons, Inc., New York, N.Y., for permission to use material from his book on *Injection Molding of Plastics* (1973). Mr. Rubin is associated with Robinson Plastics Corp.
[2] Prepared by the editors.

under pressure (called the clamping unit); (3) the type of mold used; and (4) the machine controls.

(1) Types of Injection Units. Injection molding machines fall into categories delineated by the type of injection unit used. The oldest type, known as a single-stage plunger (Fig. 4-2), operates as follows:

The feed hopper feeds granular molding material into the injection cylinder using some kind of a measuring device to meter the amount fed. The forward motion of the injection plunger forces the granules along the cylinder and around a spreader. The section of the injection cylinder around the spreader is heated by electrical heaters and the material is spread so thin in passing between the spreader and the cylinder that it is melted to the point where it can be forced through the nozzle, a tapered sprue hole, and runners into the cavities of the mold. The mold is held tightly closed by the clamping action of the press platen. As a result, the plastic material is forced into all parts of the mold cavities, giving a perfect reproduction of the mold. When reverse action of the clamp opens the mold, releasing the formed part, ejector pins or strippers are used to assist in removing the part.

As the plastics industry developed, a second type of plunger machine appeared, known as a two-stage plunger (Fig. 4-3). This type of equipment involved two plunger units set one

Fig. 4-1. Injection molded high-density polyethylene pallet is removed from machine capable of molding up to 70 lb in a single shot. (*Courtesy Arco/Polymers*)

on top of the other—the one to plasticize the material and feed it to the other cylinder where the second plunger operates as a shooting plunger and pushes the plasticized material into the mold.

Later, still another variation appeared in which the first plunger stage (also known as a pre-plasticizer) was replaced by a rotating screw in a cylinder. In this case, the action of the screw serves to work and melt (plasticize) the resin and feed it into the second plunger unit where the injection ram forces it forward into the mold (Fig. 4-4).

The newest injection unit is the in-line reciprocating screw injection machine (Fig. 4-5), which has supplanted the plunger-type as the most popular machine being sold today. In a screw-type machine, the plunger and spreader that are key to plunger-type machines are replaced by a rotating screw which moves back and forth within the heating cylinder. As the screw rotates, the flights pick up the feed of granular material from the hopper and force it along the hot cylinder barrel. As the material comes off the end of the screw, the screw

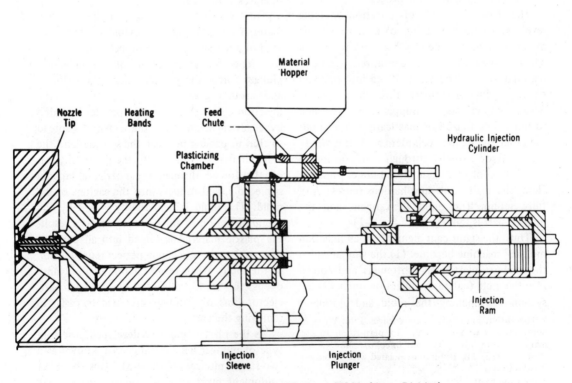

Fig. 4-2. Single-stage plunger unit. (*Courtesy SPI Machinery Division*)

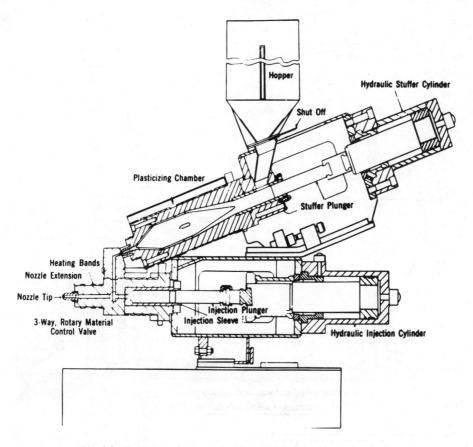

Fig. 4-3. Two-stage plunger unit. (*Courtesy SPI Machinery Division*)

moves back to permit the plastic material to accumulate. At the proper time, the screw is forced forward, acting as a plunger and propelling the softened material through the nozzle and sprue into the mold cavities. The size of the charge is regulated by measuring the back travel of the screw. In this case, the regulation of the plastics material condition is also accomplished by controlled screw rotation speed, as well as temperatures.

(2) **Clamping Units.** There are several techniques available for opening the molds and for generating enough force to hold the mold halves closed during injection. These include: (1) hydraulic clamps, in which an hydraulic cylinder operates directly on the movable parts of the mold to open and close it (Fig. 4-6); (2) toggle (or mechanical) clamps, in which the hydraulic cylinder operates through a toggle linkage to open and close the mold halves (Fig.

4-7); and (3) various mechanical hydraulic clamps that combine features of both.

Most clamps in use today are horizontal, with injection taking place through the center of the stationary platen. But for special jobs, vertical clamp presses are available, in which the injection mechanism is horizontal and injection takes place at the die parting line.

(3) **Molds.** The heart of the molding process is the mold. While its primary purpose is to determine the shape of the part, it performs other jobs as well. It must conduct the material from the heating cylinder to the cavity, vent the entrapped air, cool the part, and eject or strip the parts without marks or damage. The surface finish of the mold will determine the finish of the part. A detailed description of injection molds can be found in Chapter 6.

(4) **Machine Controls.** Perhaps the most active area of development in injection molding

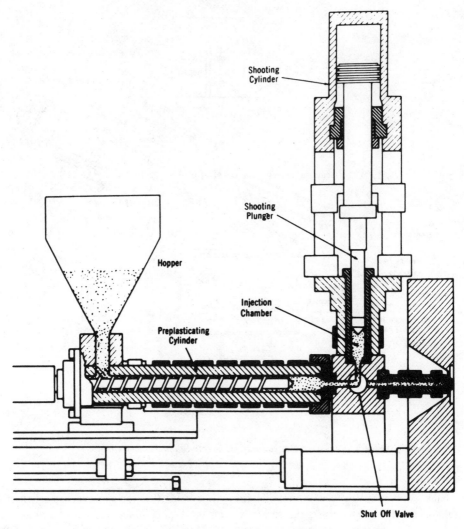

Fig. 4-4. Two-stage screw unit. (*Courtesy SPI Machinery Division*)

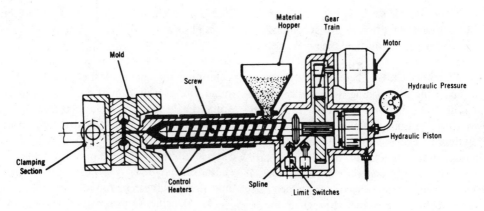

Fig. 4-5. In-line reciprocating screw unit. (*Courtesy SPI Machinery Division*)

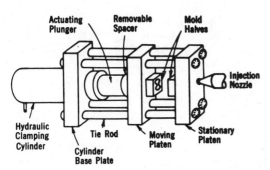

Fig. 4-6. Hydraulic clamp. (*Courtesy SPI Machinery Division*)

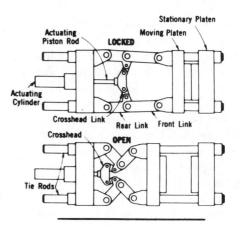

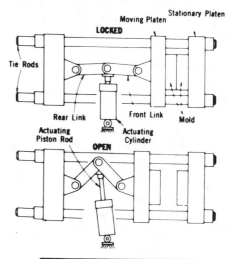

machine design in the late sixties and early seventies was machine controls (see Chapter 5 on Controls for Injection Molding). By the mid-seventies, solid state controls had virtually found universal acceptance among machinery manufacturers. Another significant area of approach was the introduction of monitoring devices to maintain a constant check on a number of different processing variables (temperature, pressure, etc.). These devices are designed to react to deviation by providing a warning signal (i.e., lights or a buzzer) or by shutting the machine off.

More sophisticated, however, has been the appearance of adaptive or feedback controls. Such controls are also designed to keep a continuous check on various processing variables. They differ from monitoring, however, in that they have the capability of energizing mechanisms that will change the variables (i.e., feed information back to the machine to raise or lower temperatures, increase or decrease pressures, etc.). Using these devices will automatically maintain processing operations within a fixed level of variance.

The ultimate system, however, involves connecting the various controls to a computer that will use feedback loops, monitoring devices, etc., for instantaneous self-correction of all processing variables. Machine cycles and processing parameters for a given part can thus be stored on tape or other electronic data storage device and simply plugged into the control cabinet of a computer-run machine whenever that particular part is to be molded. In essence, the computer will automatically direct the setting-up of the machine.

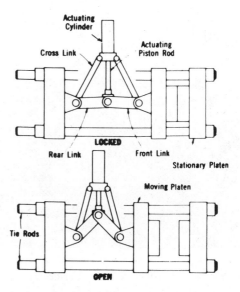

Fig. 4-7. Types of toggle clamps. (*Courtesy SPI Machinery Division*)

Specifying a Machine. In addition to considering the type of injection unit, the clamping system, and available controls, buyers and users of injection molding machines must also take the following press specifications into account:

(1) Clamping force, usually expressed in tons (see Fig. 4-8 for typical 2500-ton machine, among the larger units commercially available; a 5000-ton unit is already commercial).

(2) Injection capacity, usually stated in maximum shot size for general-purpose polystyrene. Note that there need not be any direct relationship between injection shot capacity and clamp tonnage. For example, a 450-ton clamp machine could have shot capacities from 24 to 68 oz; while 200 to 225-ton machines can offer 5 to 25-oz shot capacities.

(3) Plasticating rate, usually given in lb/hr or oz/sec for polystyrene with the screw running as an extruder. It must be remembered that the screw has only a fraction of the cycle available for plasticating and that the rating does not necessarily give the oz/sec of moldable material.

(4) Injection rate is the maximum speed at which general-purpose polystyrene can be injected through the nozzle and is usually given in cu in./min or sec at a stated pressure.

(5) Injection pressure. For plunger machines, the injection pressure is the pressure in psi on the injection plunger. For the reciprocating screw, it is the pressure on the material ahead of the screw.

The Versatile Injection Molding Machine. While the discussion that follows treats generally with the injection molding of thermoplastics and with conventional plunger and screw machines, it is well to point out that the versatility of the injection molding machine does go beyond these materials and machines.

In addition to molding conventional thermoplastics, injection molding machines can be used (as well as specially modified machines) to mold structural thermoplastic foams (see Chapter 20). Variations of reciprocating screw molding machines are also available to handle thermosets (see Chapter 9) and, more recently, rubber. Special injection molding grades are now available in such rubbers as polyurethane, ethylene-propylene, epichlorohydrin, nitriles, fluoroelastomers, etc. Finally, there are systems being evolved in which continuous metering-type feeders, a continuous polyester mixer into which chopped glass is also fed, a premix transfer device, and a reciprocating screw injection molding machine are combined together for molding glass-reinforced polyesters.

By sequencing various shots in order, it is also feasible to use an injection molding machine to produce two-color or three-color moldings. In another commercial application, a directional signal for tractors, a zinc die casting is placed in the mold and polypropylene is injection molded around the die casting to form a housing for the lens (Fig. 4-9).

Fig. 4-8. Size of 2500-ton injection molding machine is indicated by automobile shown between the platens. Machine is 52$\frac{1}{4}$ ft long, 20$\frac{1}{2}$ ft wide, and 17 ft high. With a 6-in. diameter screw, the plasticizing capacity is 1500 lb/hr (GP polystyrene). Injection capacity is 339 oz (GP polystyrene). (*Courtesy Ingersoll-Rand Co., Plastics Machinery Div.*)

Fig. 4-9. Directional signal (A) uses injection molding as an assembly technique. To produce it, a small zinc die casting (B) is placed in an injection mold and polypropylene is molded around it to form the sides of the signal light (C). On removal from the mold and while the plastic is still warm, acrylic lenses (D) are pressed into position. (*Courtesy American Die Casting Institute*)

Also available in injection molding equipment are multi-station rotary turntable injection molding machines. The advantages of a rotary are that several operations can be performed simultaneously at the different stations. These include loading, injection, stripping of parts, and the placing of inserts. Another advantage is that the part can cool as the table rotates. Rotary turntable machines can have from 2 to 32 stations and single injection heads of different sizes. Dual injection heads can be used on some machines to inject two different materials or colors.

INJECTION MOLDING OF THERMOPLASTICS[3]

Injection molding is a major processing technique for converting thermoplastic materials. About 4 billion pounds were molded in the United States in 1972.

[3] By Irvin I. Rubin.

The process was patented by John and Isaiah Hyatt in 1872 to mold camphor-plasticized cellulose nitrate (celluloid). The first multi-cavity mold was introduced by John Hyatt in 1878. Modern technology began to develop in the late 1930's and was accelerated by the demands of World War II. A similar surge in the technology of materials and equipment took place in the late sixties and early seventies.

This section will discuss the machine and the theory and practice of injection molding.

Injection Molding Machines

Figure 4-10 shows a 300-ton, hydraulic clamp, injection molding machine with a 2½ in. reciprocating screw hydraulically operated as the injection end. It will mold approximately 28 oz of polystyrene per shot. The clamping and injection ends of a molding machine are described and rated separately. All clamp ends use hydraulic force. Machines can be completely hydraulic, a combination of an hydraulic cylinder and a toggle mechanism, or other mechanical devices. It is rated by the number of tons of locking force it provides. The relationship is

$$F = \frac{P \times A}{2000} \qquad (1)$$

where

F = force (tons)
P = pressure $(lb/in.^2)(psi)$
A = sq. in.

As a rule of thumb, 2½ tons of force are required for each square inch of projected area of the molded parts. The projected area is the area parallel to the clamping force (i.e., the platens). For example, the 350-ton clamp machine could mold a polystyrene box 10 x 14 in. or 140 sq. in. (350 ÷ 2.5 = 140 sq. in.). The depth of the box would not be relevant in determining the clamp force requirements because the sides are not perpendicular to the clamping force.

Injection ends are classified by the method of plasticizing, the capacity per shot, and the amount of material that can be plasticized in 1 hour.

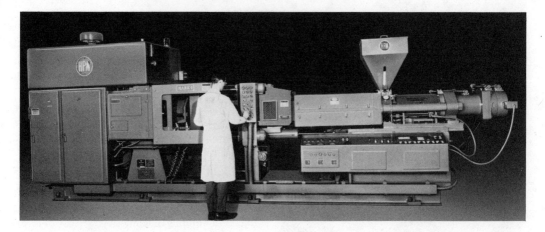

Fig. 4-10. Reciprocating screw injection molding machine with a 300-ton hydraulic clamp and a 28 oz injection shot capacity. (*Courtesy HPM Div., Koehring Co.*)

Hydraulic Clamping System

Figure 4-6 shows a schematic drawing of an hydraulic clamping system. To the stationary platen (which is attached to the molding machine) are attached four tie rods which go through and support the moving platen and then through the hydraulic clamping cylinder mounting plate. At each end of the tie rods are tie rod nuts. The hydraulic clamp ram is attached to the moving platen. The stationary side of the mold is attached to the stationary platen. The moving or knockout part of the mold is attached to the moving platen (the knockout or ejection mechanisms of the molding machine are not shown). The air space in the mold is filled with the molten plastic, and after solidification by cooling it is ejected therefrom.

It can be seen from the illustration that when oil is used in the forward port the clamp ram will move the moving platen until the mold parts make contact. As the pressure builds up, the force behind the clamp ram is transmitted through it, the moving platen, the mold, and the stationary platen to the tie rod nuts. This force stretches the tie bar which provides the clamping action. When the mold is to be opened, oil is sent to the return port and the forward port is vented to a tank. This retracts the clamp ram, moving platen, and moving part of the mold. The plastic part normally remains on the moving part of the mold and is ejected

or knocked out of the mold by the ejection mechanism (see Chapter 6 on Injection Molds). The ejection mechanism can be operated hydraulically, which is preferable because it allows the operator to control the timing, direction, and force of the stroke, or mechanically, wherein the knockout plate is stopped by the knockout bars of the machine.

Reciprocating Screw Injection End

Figure 4-11 shows a schematic drawing of the injection end of a reciprocating screw machine. The extruder screw is contained in the extruder barrel. It is turned in this instance by an hydraulic motor (as contrasted to an electric motor) through a screw drive system consisting of gears.

As the screw turns it picks up material from the hopper. As it progresses down the screw it is compacted, melted, and pumped past the non-return flow valve assembly. This, in essence, is a check valve which allows material to flow from right to left but not from left to right. As the material is pumped in front of the screw it forces back the screw, hydraulic motor, and screw drive system. In so doing, it also moves the piston and rod of the two hydraulic cylinders used for injection. Increasing this resistance requires higher pressures from the pumping section of the screw. This increased resistance from the hydraulic cylinders is called the *back pressure*.

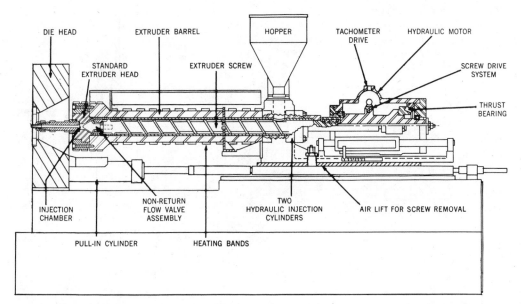

Fig. 4-11. Schematic drawing of injection end of reciprocating screw machine. (*Courtesy HPM Div., Koehring Co.*)

The screw will continue to turn, forcing the carriage back until a predetermined location is reached. At this time the rotation is stopped, and the amount of material in front of the screw will be injected into the mold at the appropriate time in the cycle. This is done by using the two hydraulic injection cylinders. It is evident that the first particles of material from the hopper will pass over the full flight of the screw. As the screw moves toward the right, succeeding particles will travel through a diminishing length of the screw. This action of the reciprocating screw limits the amount of material that can be plasticized by any given screw. It also results in a less homogeneous melt, though this is not important for most applications.

These limitations and others are overcome by using a two-stage screw-plunger machine shown in Fig. 4-12. Here, the material goes over the full length of the screw through an admission valve into the injection or shooting chamber. Here, too, the injection plunger is forced back until it reaches a predetermined point, at which time the screw stops. The rotary shutoff valve which had been closed is now opened so that when the injection plunger advances forward the material is injected into the mold.

Cycle of an Hydraulic Clamp/Reciprocating Screw Machine

Thus, the cycle of a reciprocating screw molding machine may be summarized as follows:

(1) Oil is sent behind the clamp ram closing the mold. Pressure builds up to develop enough force to keep the mold closed while the injection process occurs.

(2) Previously plasticized material in front of the reciprocating screw is forced into the mold by the hydraulic injection cylinders.

(3) Pressure is maintained on the material to mold a part free from sink marks, flow marks, welds, and other defects.

(4) At the end of this period, the screw starts to turn, plasticizing material for the next shot.

(5) While this is occurring, the plastic is cooling in the mold and solidifying to a point where it can be successfully ejected. This cooling is accomplished by circulating a cooling media, usually water, through drilled holes or channels in the mold base, cavities, and cores.

(6) Oil is sent to the return port of the clamping ram separating the mold halves.

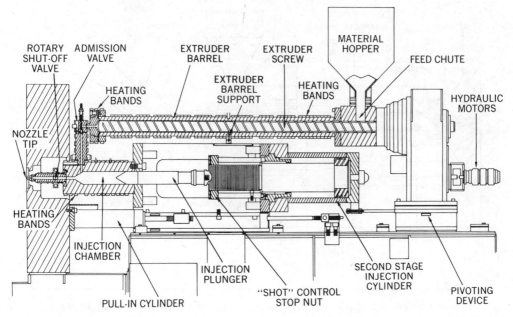

Fig. 4-12. Schematic drawing of injection end of two-stage screw plunger machine. (*Courtesy HPM Div., Koehring Co.*)

(7) As the moving platen returns, the knockout or ejection mechanism is activated, removing the pieces from the mold.

Plunger-type Injection Cylinder

Until the introduction of the reciprocating screw, the standard method of plasticizing was a plunger machine, Fig. 4-13. While relatively few are being built today, except in very small sizes, approximately one-third of the injection machines in use are of this type. Material from the hopper is fed into the feed chute to a volumetrically controlled chamber. The injection plunger comes forward and forces the cold granules into the plasticizing chamber over the torpedo and out through the nozzle tip into the mold. The heat is supplied by resistance heaters. The material is melted by conduction of the heat from the heating band through the cylinder wall into the plastic. Because there is relatively little convection in a plunger machine and the plastic has good insulating (or poor thermal conduction) properties, the temperatures of the material at the wall or the torpedo is significantly higher than the material in the center. This makes for poor homogeneity which will adversely affect the molded part.

The pressure is transmitted from the injection plunger through the cold granules and melted material into the mold. There is considerable pressure loss, averaging about 40%, but it can be as high as 80%. The smaller the clearance between the plunger and the torpedo, the higher the resistance to flow and the greater the pressure loss. Increasing this distance would decrease the pressure loss and also increase the capacity of the cylinder, but would lead to a progressively deteriorating condition of the melt because of the poor conduction of the plastic material. Therefore, the straight plunger is limited in its capacity, since these two design requirements oppose each other. One rarely finds a straight plunger machine larger than 16 oz.

The amount of heat that the plastic receives in a given plunger type cylinder is also a function of the residence time. Therefore, uneven cycles change the temperature of the melt and prevent good moldings. When the machine is off cycle even for a short period of time it is necessary to purge out the overheated material and continue with the molding. One of the principal advan-

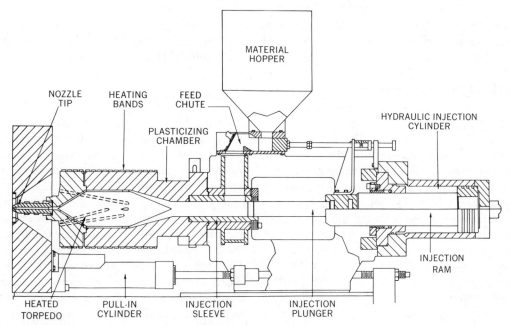

Fig. 4-13. Schematic drawing of an injection end of a single-stage plunger machine. (*Courtesy HPM Div., Koehring Co.*)

tages of the screw is that its heating is done by the shearing action caused by the screw rotation. This mechanical working puts the heat directly in the material. When the screw is not turning, the only heat the material gets is from the barrel heaters which is comparatively small.

Temperature Control

Most of the machines in service today use a bimetallic thermocouple as the sensing unit for the temperature of the heating cylinder. It feeds into a pyrometer which is an on-off type instrument. Since there is a large thermal over-ride from the heating band to the material, most pyrometers have compensating circuits. The pyrometers use relays to control the contactors for the heating bands. The heating bands are either fully on, or completely off.

Newer equipment today uses solid state controls in the pyrometer circuit. Thermistors can be used for temperature sensing. By means of SCR units, the amount of heat is regulated by the requirements of the heating cylinder; the heating bands will give a percentage of their heat output as required rather than the on-off action of older controls. This provides a much smoother temperature control system

with lower variations in material temperature. However, it should be noted that both of these systems measure the temperature of the heating barrel. The more desirable method is to measure the temperature of the plastic melt itself, which is now being done by means of thermocouples inserted in the melt stream at the machine nozzle.

Packing

Once the mold is initially filled, additional material is added to the mold by the injection pressure to compensate for shrinkage as the material cools. Adding too much material is called *packing*, and this results in highly stressed parts and may cause ejection problems. Insufficient material causes short shots, poor surface, sink marks, welds, and other defects. The material will continue to flow into the mold as long as there is injection pressure, provided the gate is not sealed. When no more material enters the mold, the contraction of the cooling material results in a rapid decrease in the pressure in the mold. The residual pressure is caused either by the return of the deformed parts of the mold to their original size or the parts not shrinking to smaller than the mold size. Because the pressure in a plunger system is transmitted inconsistently through cold granules, it is more difficult to control the amount of material forced into the mold than in other preplasticizing systems.

Preplasticizing

The obvious disadvantages of the single-stage plunger were initially overcome by putting one plunger on top of the other in a so-called *piggyback system*. The top plunger will plasticize the material and force the injection plunger back, in a similar fashion to the two-stage screw-plunger machine shown in Fig. 4-12. The second method of preplasticizing is in the reciprocating screw itself, where the shooting *pot* is the chamber which develops in front of the retracting screw. Preplasticizing gives rise to a number of significant advantages.

(1) The melt is more homogeneous because it mixes as it passes through the small opening connecting the two chambers.
(2) There is direct pressure on the material by the injection plunger.
(3) Faster injection is possible.
(4) Higher injection pressures and better injection control are possible.
(5) Better shot weight control is possible.
(6) The weight of a single shot can easily be increased by making the stroke of the shooting cylinder longer, or its diameter larger. This is not true in the reciprocating screw-type of preplasticizing.
(7) All the material passes over the full length of the screw.

Types of Machines

To summarize, the four types of injection ends are as follows:

(1) The straight plunger machine (Fig. 4-13).
(2) The two-stage plunger machine where one plunger plasticizes and the other shoots. (This machine is rarely used today.)
(3) The reciprocating screw (Fig. 4-11).
(4) The two-stage screw-plunger machine

where the screw plasticizes material which is forced into the shooting cylinder (Fig. 4-12).

Injection Screw

Since the screw is used to plasticize material in almost all new equipment, it is essential to understand its characteristics and how it works. Figure 4-14 shows a typical screw used in injection molding equipment. Its task is to convey the cold pellets at the hopper end, compact the material in the feed section, degas and plasticize it in the transition section, and pump it into the mold in the metering section. The feed section is approximately half the length of the screw, the melting or transition section one-quarter the length, and the metering or pumping section one-quarter the length. The outside diameter of the barrel (D_B) is determined by the pressure requirements of the cylinder. A two-stage, screw-pot machine requires a pressure that generally does not exceed 8000 psi. The barrel of a reciprocating screw must contain at least the 20,000 psi of injection pressure that is developed. The diameter of the screw (D_S) is the nominal diameter of the hole in the barrel. Experience has shown that a pitch of one turn per screw diameter works well for injection molding. This helix angle (θ) is 17.8°. The land width (S) is approximately 10% of the diameter. The radial flight clearance (δ) is specified considering the effects of leakage flow over the flights, the temperature rise in the plastics caused by this clearance, the scraping ability of the flights in cleaning the barrel, the eccentricity of the screw and barrel, and the manufacturing costs. The depth of the flight (h_F) is constant in the feed and metering sections. It is reduced from the deeper feed section to the shallower metering section in the transition section wherein the material is melted. The compression ratio of a screw is, in effect, the ratio of the flight depth in the feed section over that of the metering section. The length of the screw (L) is the axial length of the flighted section. The ratio of the length over the diameter (L/D) is a very important specification in a screw. High (L/D) ratios give a more uniform melt,

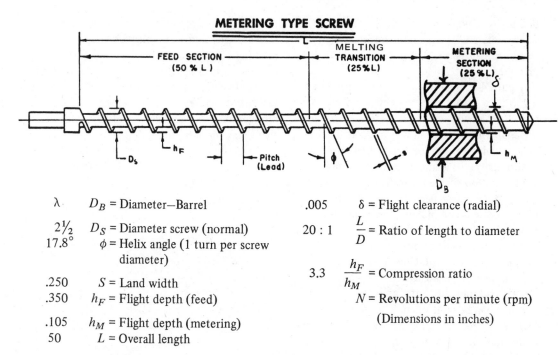

METERING TYPE SCREW

λ	D_B = Diameter—Barrel	.005	δ = Flight clearance (radial)
$2\frac{1}{2}$	D_S = Diameter screw (normal)	20 : 1	$\dfrac{L}{D}$ = Ratio of length to diameter
17.8°	ϕ = Helix angle (1 turn per screw diameter)		
.250	S = Land width	3.3	$\dfrac{h_F}{h_M}$ = Compression ratio
.350	h_F = Flight depth (feed)		N = Revolutions per minute (rpm)
.105	h_M = Flight depth (metering)		(Dimensions in inches)
50	L = Overall length		

Fig. 4-14. Typical metering screw used for reciprocating-screw type injection molding machines. (*Courtesy Robinson Plastics Corp.*)

better mixing, and higher output pressures. An L/D of 20 : 1 should be the minimum for molding machines, with higher ratios such as 24 : 1 more desirable.

Melting a Plastic in a Screw

To move forward, the plastic must stick more to the barrel than to the screw. If the coefficient of friction between the plastic and the screw and the plastic and the barrel were identical, there would be no flow of material. It would just rotate as a plug within the flight of the screw. Because the polymer molecule is anchored more to one side of the barrel than the screw, the polymer molecules slide over one another. This is called a shearing action. In effect, molecular forces are being broken and turned into heat. This is what is meant when we say we are converting the mechanical action of the screw into heat.

The solid pellets are picked up from the hopper (Figure 4-15A shows a schematic drawing of how plastic melts in a screw). They touch the barrel forming a thin film of melted plastic on the barrel surface. This melting is

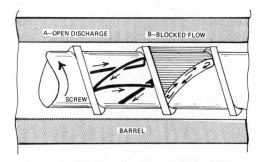

Fig. 4-15. Flow of plastic in a screw. A-open discharge. B-blocked flow. (*Courtesy Robinson Plastics Corp.*)

caused by conduction from the heater bands. The relative motion of the barrel and screw drag this melt, which is picked up by the leading edge of the advancing flight of the screw. This flushes the polymer down in front of it forming a pool, which circulates. Heat now enters the plastic by the shearing action whose energy is derived from the turning of the screw. Unmelted material is drawn into the melted area in increasing amounts down the flights of the barrel. The melting is complete at the point where the solid bed of unmelted plastic is completely melted. With a given screw, material, and temperature, the flow pattern in the screw flights varies with the back pressure.

Figure 4-15B shows a simplified view of the flight of a particle in a screw. If there is no resistance at all (open discharge), a particle of material will move as shown in A. It is moving in a circulatory motion from flight to flight. Additionally, it is moving forward along the axial direction of the barrel toward the nozzle. At no time does a particle ever flow rearward. If the nozzle is completely closed and the screw rotates, the particle will circulate within the flight. There can be no forward motion because the open end is closed. Obviously, by regulating the amount of resistance to the return of the screw (back pressure), the amount of mixing in the melt can be controlled. This is an important feature in reciprocating screw molding. The more homogeneous melt from the higher back pressures will improve color and additive dispersion, and often eliminate warpage and shrinkage problems.

Reciprocating Screw Tips

The reciprocating screw machine uses the screw as a plunger. As the plunger comes forward, the material can flow back into the flights of the screw. For heat-sensitive materials such as PVC, or viscous materials, a plain screw tip is used. For other materials this is not adequate and a number of different check valves have been designed. Figure 4-16 shows a sliding ring-type

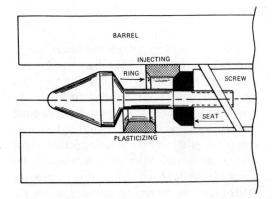

Fig. 4-16. Ring-type non-return valve for tip of screw. (*Courtesy Robinson Plastics Corp.*)

non-return valve where the ring is in the forward position while plasticizing, so that the material can flow past the seat and through its hollow portion in front of the screw. When the material injects, the ring slides back on to the seat and the screw acts as a plunger. This has a number of disadvantages, primarily relating to the wear caused by the sliding ring.

A better method is to use a ball-type non-return valve, Fig. 4-17, where a ball moves

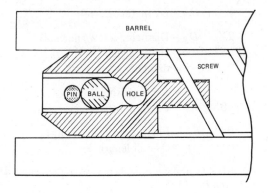

Fig. 4-17. Ball-type non-return valve in plasticizing position. (*Courtesy Robinson Plastics Corp.*)

back and forth, opening or sealing a hole. Other types of non-return valves have been developed and are in use.

Theory of Screw Plasticizing

To further understand the action of the screw a number of equations will be given which qualitatively describe what happens. The shear rate is defined as the surface velocity at the barrel wall divided by the channel depth.

$$\gamma = \frac{D \times N_x}{h} \tag{2}$$

where

γ = Shear rate
D = Diameter of screw
N = Rate of screw rotation
h = Channel depth

All materials have a maximum limiting shear rate, beyond which they degrade. For most plastics, it is 150 ft/min. For heat-sensitive plastics, it is closer to 100 ft/min. Therefore, the speed of a screw of given diameter is limited. For example, for 150 ft/min the maximum speed for a 2 in. screw is 230 rpm; a $4\frac{1}{2}$ in. screw is 130 rpm. As we shall see, the output of a screw depends, among other things, upon its speed of rotation, which is limited by the maximum allowable shear rate.

The work done by a screw in melting the plastic is measured by the rotational force, called *torque*. It is the product of the tangential force and the distance from the center of the rotating member. For example, if a 1 lb weight were placed at the end of a 1 ft bar attached to the center of the screw, the torque would be 1 ft x 1 lb or 1 ft-lb. Torque is related to horsepower as in:

$$hp = \frac{\text{Torque (ft/lb)} \times \text{rpm}}{5252} \quad (3)$$

It is clear that the torque output of an electric motor of given horsepower will depend on its speed. A 30-hp motor at 1800 rpm has a torque of $87\frac{1}{2}$ ft-lb. At 900 rpm it is 175 ft-lb. The strength of the screw varies with the cube of the root diameter. This strength limits the input horsepower. Overpowering the screw will cause it to break. For example, the maximum practical horsepower for a 2 in. diameter screw is 15 hp and a $4\frac{1}{2}$ in.-hp screw is 150 hp. This is the second limiting factor, namely, that the amount of horsepower input is limited by the diameter of the screw.

Obviously the maximum output of a system is controlled by the maximum input. The power required to raise the plastic material to a given temperature and pump it is:

$$hp = C \cdot (T_p - T_f) \cdot Q + \Delta P \cdot Q \quad (4)$$

where

    hp = Power
    C = Average specific heat
    T_p = Temperature of plasticized material
    T_f = Temperature of the feed material
    Q = Through-put
    ΔP = Back pressure.

The power required for pumping, $\Delta P \cdot Q$, is relatively small. The output of a screw, there-

fore, depends entirely on the horsepower input. There is no mention of the size of the screw, as it is not a factor in this equation; theoretically, a $\frac{1}{2}$ in. screw could put out as much as a 6 in. screw if you could put in enough horsepower. Molding material outputs range from 6 to 14 lb/hr for each horsepower input.

In considering the pumping or metering section of the screw the output is the forward component of flow (the drag flow) minus the pressure flow (the resistance to flow) and minus the slippage of the material over the flight which can be disregarded for our applications.

The drag flow is:

$$Q_d = \frac{\pi^2}{2} (D^2 \cdot h \cdot \sin\theta \cdot \cos\theta)N \quad (5)$$

where

    Q_d = Drag flow
    D = Barrel diameter
    N = Screw speed
    h = Channel depth
    θ = Helix angle

The drag flow is a product of the geometry of the empty area of the screw ($\pi^2/2 \cdot D^2 \cdot h \cdot \sin\theta \cdot \cos\theta$) and the screw speed, N. It is, in effect, describing a positive displacement pump. Obviously the density of the material conveyed will be a factor in the output. In converting Eq. (5) to consistent units the density appears.

Equation (6) describes the pressure flow which is the resistance to flow in the screw.

$$Q_p = \frac{\pi}{12} \cdot \frac{D \cdot h^3 \cdot \Delta P \cdot \sin^2\theta}{\eta \cdot L} \quad (6)$$

where

    Q_p = Output (m³/sec)
    D = Diameter of the screw (in)
    h = Channel depth (in)
    ΔP = Increase in pressure (psi)
    θ = Helix angle (17.8°)
    η = Viscosity (lb-sec/in²)
    L = Length of the metering section (in)

It must be emphasized that Equation (6) is resistance to flow and does not signify a physical rearward motion. In a given extruder, the pressure is highly sensitive to the viscosity. Pressure losses vary from 5 to 10%

From these equations can be seen the interrelation of the geometry of the screw, the allowable power input, and the speed of rotation. The manufacturer combines all of these to give a screw whose major specification is the screw recovery rate. This tells how many ounces per second are plasticized while the screw is running. It is tested in a standard jointly developed by the Society of Plastic Engineers (SPE) and the Society of The Plastic Industry (SPI) and adopted as a standard by SPI.

Screw Drives

One method of applying the driving force to turn the screw is to attach it to an electric motor, through a speed reducing gear train with different speed ranges. A second method is to connect the screw to a nonvariable speed-reducing coupling which is driven by an hydraulic motor. The third method is to attach a special low torque hydraulic motor directly to the screw. This can be done internally in the injection housing. This type drive has much to recommend it in that it has no gear speed reducers and developes minimum noise and vibration. No lubrication is required and maintenance is at a minimum.

The highest torque requirements are at the start of the rotation. This is primarily due to the inertia of the plastic material, i.e. viscosity. In addition, there is the inertial mass of the screw, coupling grears, and drive system. As the screw turns, the material is heated and the viscosity decreases, reducing the torque requirements. The available torque of an electric motor follows this pattern, with a 100% starting overload usually available. Safety devices of varying degrees of effectiveness are used to prevent the overload from snapping the screw.

The hydraulic motor on the other hand can never develop more than its rated torque. For this reason a larger hydraulic motor is required. The safety protection for an hydraulically operated screw is a simple relief valve, as the oil pressure to the hydraulic motor controls the torque. Fixed displacement hydraulic motors are used for screw drive and have a constant torque output at a given pressure. Therefore, their horsepower varies with the speed. The maximum torque depends upon the pressure setting of the controlling relief valve. The speed of the hydraulic motor is controlled by an hydraulic flow control valve. The outstanding characteristic of the hydraulic drive is its stepless control of torque and speed which permits the selection of the optimum conditions for molding. Additionally, the ease of making these changes encourages production people to do so. Electric drives do not have independent speed and torque control. The speed is changed by gear trains. Since the input and output power are constant, the change in either speed or torque will inversely affect the other. This restricts finding the optimum molding conditions. The electrical system has a high efficiency of approximately 95% compared to the 60-75% of the hydraulic system. This represents a significant dollar saving during the life of the machine. The hydraulic motor is smaller and lighter in weight than the equivalent electric system, thus permitting a lower initial back pressure. This gives better shot weight control.

In summation, the hydraulic drive is less efficient, has a higher margin of screw safety, is smaller and lighter in weight, provides stepless torque and speed control and gives the better melt quality. The electrical drive is more efficient, has ease of installation, particularly in converting machines from plunger to screw operation, and provides an acceptable melt quality.

Figure 4.18 shows a cut-away view of a 300-ton toggle machine with a $2\frac{1}{2}$-in. reciprocating screw. The screw is driven by an hydraulic motor (labeled No. 27 in the diagram), which is internally mounted in the injection housing.

Advantages of Screw Plasticizing

From the preceding discussion, we can see the major advantages of the screw as a plasticizing method. They develop because the melting is caused by the shearing action of the screw. As the polymer molecules slide over each other, the mechanical energy of the screw drive is

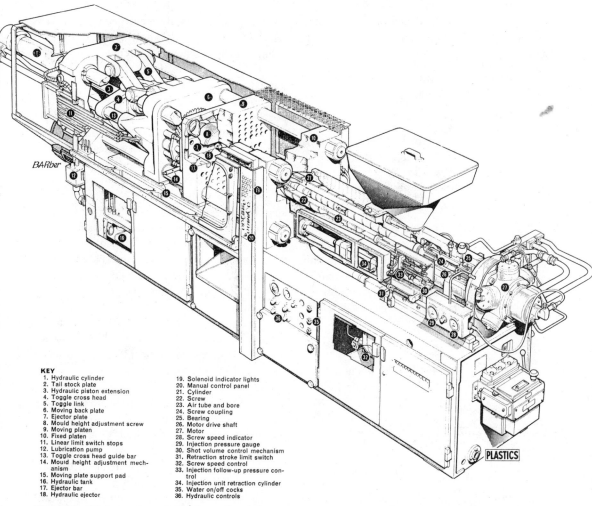

KEY

1. Hydraulic cylinder
2. Tail stock plate
3. Hydraulic piston extension
4. Toggle cross head
5. Toggle link
6. Moving back plate
7. Ejector plate
8. Mould height adjustment screw
9. Moving platen
10. Fixed platen
11. Linear limit switch stops
12. Lubrication pump
13. Toggle cross head guide bar
14. Mould height adjustment mech-anism
15. Moving plate support pad
16. Hydraulic tank
17. Ejector bar
18. Hydraulic ejector

19. Solenoid indicator lights
20. Manual control panel
21. Cylinder
22. Screw
23. Air tube and bore
24. Screw coupling
25. Bearing
26. Motor drive shaft
27. Motor
28. Screw speed indicator
29. Injection pressure gauge
30. Shot volume control mechanism
31. Retraction stroke limit switch
32. Screw speed control
33. Injection follow-up pressure con-trol
34. Injection unit retraction cylinder
35. Water on/off cocks
36. Hydraulic controls

Fig. 4-18. Cut-away view of $2\frac{1}{2}$-in. reciprocating screw, 300-ton toggle clamp machine. (A PECO injection molding machine, as originally published in British Plastics magazine)

converted into heat energy. The heat is applied directly to the material. This, plus the mixing action of the screw, contribute to its major advantages as a plasticizing method:

(1) High shearing rate. This lowers the viscosity, making the material flow easier.
(2) Good mixing results in a homogeneous melt.
(3) The flow is non-laminar.
(4) The residence time in the cylinder is approximately three shots compared to the 8 to 10 shots of a plunger machine.
(5) Most of the heat is supplied directly to the material.

(6) Since little heat is supplied from the heating bands, the cycle can be delayed by a longer period before purging.
(7) Can be used with heat-sensitive materials, such as PVC.
(8) The action of the screw reduces the chances of material holdup and subsequent degradation.
(9) The screw is easier to purge and clean than a plunger machine.

Nozzles

The nozzle is a tube-like device, the function of which is to provide a leakproof connection from the cylinder to the injection mold with a minimum pressure and thermal loss. There are three types of nozzles:

(1) An open channel with no mechanical valve between the cylinder and mold, Fig. 4-19.
(2) An internal check valve held closed either by an internal or external spring and opened by the injection pressure of the plastic, Fig. 4-20.
(3) A cutoff valve operated by an external source such as a cylinder.

It is impossible to mold correctly without controlling the nozzle temperature. With a very short nozzle, the heat conduction from the cylinder will usually be adequate to maintain it at the proper temperature. Usually nozzles are long enough to require external heat which

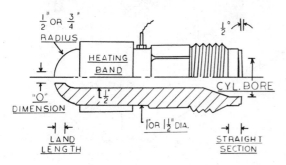

Fig. 4-19. Standard nozzle. (*Courtesy Robinson Plastics Corp.*)

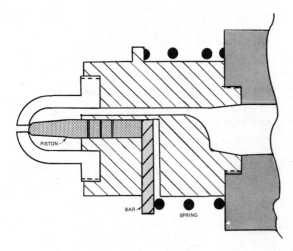

Fig. 4-20. Needle-type shut-off nozzle.. (*Courtesy Robinson Plastics Corp.*)

must be independently controlled and never attached to the front heating band of the cylinder. The cylinder requirements are completely different than those of the nozzle, and burned and degraded plastic may result as well as a cold nozzle plugging up the cylinder. While nozzles can be controlled by auto transformers or proportional timers, the best way is to use a thermocouple and a pyrometer. Heating bands can be used, although tubular heaters are preferable because they do not burn out when contacted by molten plastic. Nozzles are available with heating cartridges inserted therein which produce up to four times the amount of heat. Nozzles are kept streamlined to prevent pressure loss and hangup of the material with subsequent degradation. The land length is kept to a minimum consistent with the strength requirements of the nozzle. The nozzle need not be in one piece. Very often, the tip is replaceable since it is screwed into the nozzle body. This makes replacement and repair considerably less expensive.

The nozzle in Fig. 4-20 is sealed off internally by the spring acting through the bar on to the piston. When there is enough injection pressure to overcome the force of the spring, the piston is pushed back and the material flows. When the injection pressure drops to that predetermined point, the piston closes again, sealing off the injection cylinder from the mold. Nozzles are available with thermocouples that extend into the melt stream to monitor the temperature of the plastic rather than that of the nozzle.

Toggle Systems

The hydraulic clamp has been discussed previously (p. 90). Another type of clamp is the toggle. A toggle is a mechanical device to amplify force. In a molding machine, it consists of two bars joined together end-to-end with a pivot. The end of one bar is attached to a stationary platen and the other end of a second bar is attached to the moveable platen. When the mold is open the toggle is in the shape of a V (Fig. 4-18, No. 5). When pressure is applied to the pivot the two bars form a straight line. Mechanical advantage can be as high as 50 : 1.

The force to straighten the toggle is applied by an hydraulic cylinder. A 350-ton machine with a 50 : 1 toggle would require only 7 tons of force from the hydraulic cylinder.

Figure 4-21 shows a double toggle clamping mechanism. In the mold open position, the hydraulic cylinder has retracted, pulling the crosshead close to the tail stock platen. This pulls the moving platen away from the stationary platen and opens the mold. It is difficult to stop the moving platen before completion of the full stroke. Where this is important to achieve, nylon buffers can be used as a mechanical stop. To close the mold, the hydraulic locking cylinder is extended. The moving plate moves rapidly at first and automatically decelerates as the crosshead extends and straightens out the links. A small motion of the crosshead develops a large mechanical advantage causing the locking.

The two main advantages of a toggle system are the economy of running a much smaller hydraulic cylinder than a comparable fully hydraulic machine and the inherent speed of the design. Fully hydraulic clamps are capable of moving as fast as toggles, but the cost to achieve this is much higher than in an equivalent toggle system. Another advantage of the toggle system is that it is self locking. Once the links have reached their extended position they will remain there until retracted. The hydraulic system requires maintenance of line pressure all the time.

The toggle systems on molding machines have several disadvantages. The primary one is that there is no indication of the clamping force, and it is therefore difficult to adjust and monitor. The clamping force in an hydraulic system is read immediately by a pressure gauge and can be controlled in stepless increments. It is difficult to control the speed and force of the toggle mechanism, as well as starting and stopping at different points. A major disadvantage of the toggle system is that it requires significantly more maintenance than an hydraulic one. It is susceptible to much more wear.

In order to clamp properly, the toggles must be fully extended. Therefore, the distance of the tail stock platen has to be changed to accommodate different molds. One way to do this is to have a chain which simultaneously

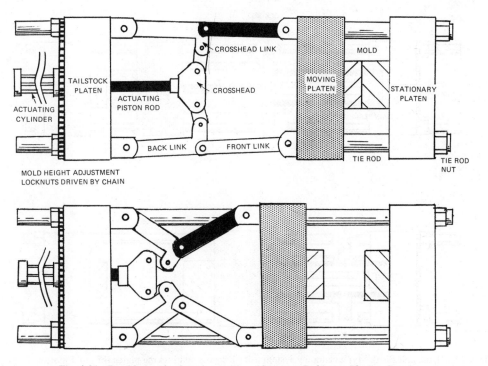

Fig. 4-21. Double toggle clamping system. (*Courtesy Robinson Plastics Corp.*)

moves the four locking nuts on the tail stock platen. This can be turned mechanically, electrically, or hydraulically.

Clamping for Large Machines

As machines became large, larger clamp capacities and longer strokes were required. This became unwieldly and costly for all hydraulic clamps. Many systems have been designed to overcome this, such as the one shown in Fig. 4-22; it is commonly called a *lock and block*. A small diameter high-speed rapid traverse cylinder is used to move the moveable platen. Spacers, which may be hollow tubes, are attached to the moveable platen and main clamping cylinder. At the end of the stroke of the rapid traverse cylinder, a locking mechanism, hydraulically operated, is inserted between the spacers. The large diameter short-stroke hydraulic cylinder moves forward approximately an inch to provide the locking force. This type of "lock and block" mechanism requires a mold height adjustment. Systems of this design, while losing a slight speed advantage because of the three motions, gain in economy because of smaller

hydraulic cylinder size, lower power requirements, and no need for huge toggle links.

One of the disadvantages of this system is its long unadjustable stroke, with a corresponding waste of floor space and long length of stressed tie bars. This is overcome in the design shown in Fig. 4-23. The rapid traverse cylinder moves the moving tail stock platen and clamping plate until the mold closes. The half nuts are closed over the tie bars anchoring the moving tail stock platen. The large diameter, limited stroke main cylinder extends, providing the clamping force. To open the mold the clamping cylinder is retracted, the half nuts are extended, and the rapid traverse cylinder is retracted. The tie bars are only stressed between the point where the half nuts are locked and the tie bars nuts behind the stationary platen. The advantages of this are adjustable stroke, minimum stress on the tie bars, increased rigidity, and smaller floor space.

Other Specifications

Aside from the clamping force, there are a number of other important specifications. The

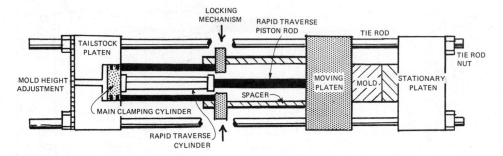

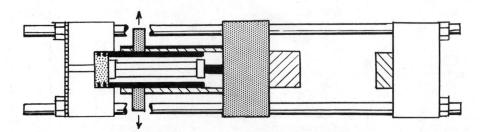

Fig. 4-22. "Lock and Block" type clamping system using a spacer and locking plates for rapid motion. (*Courtesy Robinsion Plastics Corp.*)

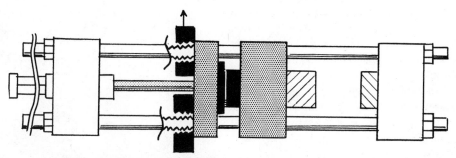

Fig. 4-23. Long-stroke, rapid-acting, clamping system, using 1/2 nuts which anchor on the tie bars to lock the moving platens. (*Courtesy Robinson Plastics Corp.*)

clamp stroke is the maximum distance the moving platen will move. The maximum daylight is the maximum distance between the stationary platen and the fully retracted moving platen. These two specifications will determine the height of the mold that can fit in the machine.

The clearance between the tie rods is the determining factor whether a mold of a given length or width will fit into the press. For example, a press with a 20-in. clearance vertically and 18 in. horizontally will take a mold up to 20 in. wide, and over 20 in. long vertically, and 18 in. or less high but over 20 in. long horizontally. The length and width dimensions of the molds are often determined by the side which is parallel to the knockout plate.

The clamp speed is an important specification. Losing $\frac{1}{2}$ sec per shot on a machine producing 100 shots per hour will waste 110 productive hours a year.

The knockout stroke determines the maximum knockout movement available. There are two types of knockout systems—a mechanical one which stops the knockout plate before the mold stops, and an hydraulic one which is independent of the machine action. The hydraulic knockout system is preferable.

Circuits

Space does not permit discussing electrical and hydraulic circuits. Older machines used electro-mechanical relays and timers. The newer machines are using solid state devices which are more reliable. Monitoring injection speed, melt temperature, pressure, and pressure of the material in the mold form the basis for the automatic control of the injection molding process. This is at its earliest stages of development and will eventually lead to completely automatic processing of quality parts with exceptional tolerance.

Automation

Many times one hears the expression "automatic machines" when referring to automatic molding. This is a misnomer, since all machines today are automatic. What makes automatic molding is the mold. There are a number of requirements for automatic molding—the machine must be capable of constant repetitive action, the mold must clear itself automatically, low pressure closing must be available so that any stuck parts will not damage the mold.

Automatic molding does not necessarily eliminate the operator. Many times an operator is present to pack the parts and perform secondary operations. Automatic molding gives a better quality piece at a more rapid cycle. Usually on automatic molding machines an experienced person attends several machines. Unless the powder, feed, and part removal are automated he will take care of them.

Automation is expensive to obtain. It requires excellent machinery, molds, trained employees and managerial skills. When the quantity of parts permit, it is a satisfactory and economical method of operation.

Materials Handling

Plastic material is shipped in 25 lb tins, 50 lb bags, 300 lb drums, 1000 lb Gaylords, tank trucks, and tank cars. In the latter two instances they are pneumatically removed into silos for storage. From there they are usually conveyed automatically to the molding machine.

The selection of the proper material handling technique can greatly reduce wastage caused by contamination and spillage. Regardless of the size of the plant, the material handling procedure should be constantly reviewed and updated. The subject is covered in Chapter 29.

Sprues, runners, and rejected parts can be reground and reused. A typical grinder is shown in Fig. 4-24. It consists of an electric motor which turns a shaft on which are the blades that cut the plastic. The reground material is usually blended in with virgin material. It can be done automatically by conveyor systems.

Certain materials such as nylon, polycarbonate, and acetate are hygroscopic—absorbing moisture. They require heating to dry them. This can be done in ovens which are thermostatically controlled and circulate heated air.

Fig. 4-24. Grinder for thermoplastics, showing flywheel and electric motor. (*Courtesy Injection Molders Supply Co., Inc.*)

The material is spread out in trays until is has been dried. Alternatively, hopper dryers fit on top of the molding machine and send filtered, heated, dried air through the material to dry it. Wet materials cause surface defects and seriously degrade the material.

Secondary Operations

Injection molded parts are fabricated and decorated with the same techniques used on plastics manufactured by other methods. Milling off gates and drilling, shaping and routing are very commonly done on molded parts. Frequently the cycle time allows these operations to be done at the press. Cementing, riveting, eyeletting, tapping, adding metallic threaded inserts and nuts and bolts are common methods of joining plastic to itself or to other materials.

Injection molded parts can be painted, silk screened, hot stamped, electrolytically plated, and vacuum metallized. Figure 4-25 shows some vacuum metallized lamp parts which have completely eliminated brass for this application.

The full scope of these operations is covered in the chapters on Finishing and Machining Plastics, Joining and Assembling Plastics, and Decorating Plastics.

Safety

The Occupational Safety Health Act of 1970 (OSHA) has declared the congressional purpose to "assure as far as possible every working man and woman in the nation safe and healthful working conditions and to preserve our human resources." It is the employer's duty to furnish his employees a place of employment free from recognized hazards causing or likely to cause death or serious physical harm. The SPI has many resources to help injection molders comply with this law.

Coloring

There are three methods of obtaining colored material—milled in, dry colored, and color concentrates.

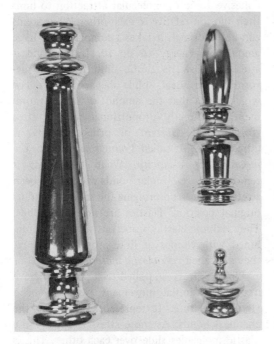

Fig. 4-25. Vacuum metalized plastics parts. (*Courtesy Robinson Plastics Corp.*)

Milled in color has the advantage of the best dispersion and the ability to match colors. It is the most costly method of coloring, requiring a large inventory compared with other methods and a significantly longer lead time for production. The base material is re-extruded with colorants, which under some conditions may lower its physical properties.

Dry coloring consists of tumbling the basic resin with a colorant. Sometimes wetting agents are used. Certain automatic material handling systems meter in the colorant during feeding. This is the most economical way to color. It requires a minimum inventory of the uncolored material. Colorants are usually packed in bags for mixing with 50, 100 or 200 lb of the resin. It has a number of disadvantages. The wrong colorant might be used or the colorant might not be completely emptied into the tumbling equipment. There is a lot of material handling and a chance for loss. It is basically a messy procedure and because of the dusting, may cause contamination problems. It is more difficult to clean the feed system. Colors which require several pigments, particularly in small amounts, are very difficult if not impossible to color evenly. Dispersion can be significantly poorer than in milled material, and in many instances is unacceptable. Dry coloring is much less successful in plunger machines than in screws. More recently, liquid colors have been developed for in-plant coloring at the machine.

A compromise between milled in material and dry coloring are color concentrates or master batches. High concentration of colorant are extruded into the same resin from which the parts will be molded. The concentrate is usually prepared so that 5 lb of concentrate are required to color 100 lb of base resin. While this is more expensive than dry coloring, it overcomes a number of its disadvantages. Colors difficult to match in dry coloring will usually work with color concentrates. Contamination dusting and cleaning problems are the same as for milled in materials. Compared to milled in materials, they are usually less expensive, require lower material inventory, and occupy less space.

Coloring methods are covered in more detail

in the chapter on Compounding and Materials Handling.

THEORY OF INJECTION MOLDING

This section will develop a qualitative theory of injection molding based on rheological data, concepts of energy levels, molecular structure and forces, and theories of heat transfer. This can be of great value in preventing problems, solving difficulties, and understanding the literature. Of necessity, coverage of this kind will contain certain generalizations, which while not mathematically exact, adequately serve the purpose.

The reasons that prevent a quantitative statement of injection molding become clear when we analyze the machinery, materials, and process. The factors which affect the process—material temperature, temperature profile, pressure, material velocities in the cylinder and mold, mold temperature and flow patterns are not measured at all or intermittently at isolated points. Until recently there has been no feed back of the processing variables to the molding machine to automatically compensate for changing conditions.

The material is never the same, having different heat histories, molecular weights, molecular weight distributions, degrees of polymerization, and impurities. It is exposed to moist air and compressed with it in the heating cylinder, with varying amounts of oxidation. It is heated by convection, conduction, and shearing. Pressure changes from 0 to possibly 30,000 psi.

The material is not linear in most of its functions. It is compressible, stretchable, elastic, and subject to changing properties after removal from the mold. They may differ depending on the direction of the material flow in the mold. It has time dependent properties which are strongly altered by its environment. In some materials there are varying degrees of crystallinity which are not reproducible or predictable.

Material exists in three forms—solid, liquid and gas. In plastics we are primarily concerned with the liquid and solid phases. In crystalline materials the change from solid to liquid is abrupt and easily discernible. In an amorphous (non-crystalline) polymer the change is not abrupt or readily apparent. The material softens over a wide temperature range and there is no dramatic visible change in its properties at any given point, such as would be found in the conversion of ice to water or water to ice. If we plot certain properties of an amorphous plastic such as the specific volume or heat capacity against the temperature, we notice that at a certain point there is an abrupt change of the slope of the line. In the thermodynamic sense it is called a *second order transition*. In polymer science it is called the *glass transition point* (T_g). Below the glass transition point the polymer is stiff and dimensionally stable, behaving like a solid. It has brittle characteristics with little elasticity and its properties are relatively time independent. Above T_g, the polymer will behave as a viscous liquid, being elastomeric and having highly time dependent properties.

The difference between the three forms of matter can be explained in terms of molecular attraction. In a solid, the closeness of the molecules to each other permits the strong cohesive force of molecular attraction to limit their motion relative to each other. While solids can be deformed, it takes a comparatively large amount of energy to do so. If the solid is stressed below its elastic limit it will be deformed. An ideal solid obeys Hooke's law which states that the amount of strain (movement) is directly proportional to the stress (force). The constant of proportionality, E (stress/strain), is called *Young's modulus* or the *modulus of elasticity*. When the stress is removed, the molecular bonds which have been stretched contract, bringing the solid back to its original position. Plastics are not ideal solids. They exhibit elastic properties which are combined with viscous or flow properties. Hence, they are called *viscoelastic materials*.

When a stress is applied to a polymer above its T_g, the initial movement is elastic of a Hookean nature. It represents stretching the chemical bonds. As more stress is applied the plastic molecules slide over each other. This is viscous flow which is not recoverable. There is a third type of flow which is a retarded type of elasticity. When the stress is removed the initial elastic deformation is instantaneously recovered. This is Hookean in nature. The retarded elasticity will eventually recover, but not instantaneously. The amount of time will depend on the temperature (viscosity) and the nature of the material.

There are two energy systems within a material. One is the potential energy of a Newtonian gravitational energy and is a measure of the forces between the molecules. The other is kinetic energy which is the energy of motion and is related to the thermal or heat energy of the system. It is the random motion of the molecules, called *Brownian movement*. The higher the internal (heat) energy of the system the more the random motion.

In a solid, the potential energy (forces of attraction between the molecules) is larger than the kinetic energy (energy of movement tending to separate the molecules). Hence, a solid has an ordered structure, with a molecular attraction strong enough to limit their motion relative to each other.

As more energy is put into the system it turns into a liquid where the potential and kinetic energy are equal. The molecules can move relative to each other but the cohesive forces are large enough to maintain a contiguous medium.

It would be appropriate here to briefly review the nature of the bonding forces in the polymer. The two major bonding forces in plastics are the covalent bond and the secondary forces sometimes called *van der Waals forces*. The atom consists of a relatively small nucleus which contains most of the atom's mass and the positive charges. The negative electrons spin around the nucleus. The orbits of the electrons form outer concentric shells, each of which has certain stable configurations. A covalent bond exists when the electron in the outer shells of two atoms is shared between them. This is typical of carbon-carbon bonds and is the primary bond found in polymers used commercially. It has a disassociation energy of 83 Kcal/mole.

The van der Waals forces are electrostatic in nature and have a disassociation of 2 to 5

Kcal/mole. They are between molecules and molecular segments and vary as the sixth power of the distance. They are obviously much weaker than the primary covalent bond. They are part of the resistance to flow. The energy attracting molecules, as contrasted to the bonds holding atoms together, is sometimes called *cohesive energy,* and is that energy required to move a molecule a large distance from its neighbor.

If we take a cubic inch of plastic and raise its temperature, its volume will increase. Since we are not adding any molecules to the cube, it is reasonable to believe that the distance between the molecules has increased. Since the van der Waals forces decrease with the sixth power of the distance, the molecules and their segments become much more mobile. Since these forces are decreasing in an exponential manner there will be a relatively narrow range in which the polymer properties change from "solid" to a "liquid". This is called the *glass transition point* (T_g). Since these cohesive forces form a major portion of the strength of the polymer, we can expect polymer properties to be very temperature dependent. This is the case.

Having a physical concept of a plastic molecule will make it easier to understand its flow properties and characteristics. For example, let us consider the polyethylene polymer. It is made by linking ethylene molecules. The double bond between the carbons in ethylene are less stable than the single bond in polyethylene. We can, with appropriate temperatures, pressures, and catalysts, cause the ethylene molecules to react with each to form polyethylene.

Ethylene   Polyethylene

The molecular weight of an ethylene molecule is 32. A typical Ziegler type polyethylene polymer might have 7000 ethylene molecules with a molecular weight of approximately 200,000. The polymerization does not proceed as simply as indicated above. Each polymer molecule will not be the same length, nor will it polymerize in a straight line of carbon linkages. When we discuss the molecular weight of polymers, we mean the average molecular weight. The molecular weight distribution is an important characterization of the polymer. If it is spread over a wide range its properties will differ from those with a narrow distribution. For example, a wide spectrum material shows more elastic effects and extreme pressure sensitivity. Viscosity, solubility, and stress crack resistance are some of the other properties affected.

To get some idea of the size of a polyethylene molecule, imagine that the methyl group

$$-\overset{\displaystyle H}{\underset{\displaystyle H}{\vphantom{|}C}}-$$ is 0.25-in. in diameter. A typical poly-

ethylene molecule would be one city block long. A molecule of water would be about the size of the methyl group. When one considers the possibilities of entanglement, kinking, and partial crystallization of the huge polyethylene molecule, compared to the small size and simplicity of the water molecule, it would not be expected to find considerable differences in flow properties. Flow of water is relatively simple (Newtonian). Viscoelastic flow of polymers is much more complicated and has not yet been mathematically defined quantitatively.

When ethylene molecules polymerize, they could theoretically do so to produce a straight line of carbon linkages, as shown in the top of Fig. 4-26. This would be called a *linear type material*. It is also possible for polymerization to take place so that the carbon atoms attach to each other in a non-linear fashion, branching out to form chains, as shown in the bottom section of Fig. 4-26. The amount of branching depends on the method of manufacture. High pressure processes have more branching than the low pressure systems. To further understand the nature of the polymer molecule, it should be noted that the carbon atoms are free to rotate around their bonds and can bend at angles less than 180°. This swiveling and twisting permit the molecules and segments of the molecules to twist and entangle each with the other. These cohesive forces consist of van

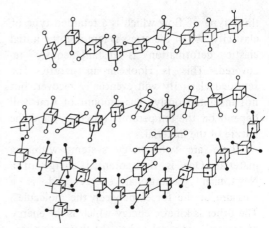

Fig. 4-26. Molecular structures of linear polyethylene (top) and branched polyethylene (bottom). Branching keeps the chains further apart, reducing density, rigidity, and tensile strength. (*Courtesy Robinson Plastics Corp.*)

der Waals type attraction. The other type of force in the polymer is, of course, the carbon-carbon (C–C) linkages.

With these simple concepts of molecular structure it should be possible to predict the different properties of the linear and branched materials.

Density. The linear structure of the polymer should permit the polymer molecules to come closer to each other, i.e., more dense. Obviously, the branched molecules prevent this so that they are less dense.

Yield. The higher the density, the fewer pieces of molded parts per pound of polyethylene can be produced. This is not an unimportant consideration in material selection.

Permeability to Gas and Solvents. Since the branched materials physically create larger voids in the polymer, it is more permeable to gases and solvents than the linear or high density material.

Tensile Strength. The linear materials, having the molecules closer together, should have stronger intermolecular (van der Waals) forces. Since tensile strength is a measure of strength of these molecular forces, it is higher in linear materials than in branched. For example, a .96 resin has a typical tensile strength of 4300 psi and a .915 resin a strength of 1400 psi.

Per Cent Elongation to Failure. Since the linear molecules can entwine and kink more than the branched molecules, one would expect that it would be more difficult to separate them, so that applying a strong tensile force would rupture the molecule rather than cause it to flow and elongate. The branched material having lower intermolecular strength would slide considerably more before rupturing. This is the case.

Stiffness. Linear polyethylene molecules being closer together than the branched has less room for segmental motion of the chains and bending of the backbone. Therefore, it is stiffer.

Heat Distortion. The heat deflection temperature under load is that temperature at which a specimen bar under given conditions of loading to produce an outer fiber stress of 66 or 264 psi, will deflect. At a given temperature the molecular forces of attraction of a high density material are greater than a low density material because the segments are closer together. Therefore, it will take a given amount of heat energy to separate the linear molecules to such a distance so that their attractive strength will be equal to that of the branched. This extra heat required means in effect that the linear material can absorb more heat for a given stiffness, so that its heat distortion temperature is higher. A low density polyethylene has a typical heat distortion temperature of 100°F, a medium of 130°F and a high density of 160°F.

Softening Temperature. Similarly the softening temperature of the high density material is higher than that of the low density material.

Hardness. Since linear molecules are closer together, the material should be harder. This is the case.

Resistance to Creep. Creep is the amount of flow (strain) caused by a given force (stress). As one expects, the higher intermolecular force of the linear material makes it more resistant to strain.

Flowability. Because of the stronger molecular attraction of the linear material, it should be more difficult to flow than the branched materials.

Compressibility. Since there is more open space in the branched material, it should compress more easily.

Impact Strength. Since the linear material has greater molecular attractive forces than the branched, one could normally expect that it would have a higher impact strength. This is not the case. Polyethylene crystallizes. It is well known that impact forces travel along the interfaces of a crystalline structure propagating breaks rapidly. Since the molecules in the linear material are closer together they will crystallize more readily. The higher crystallinity of the linear material is the reason for its lowered impact strength.

Increasing the density of polyethylene increases its tensile strength, percentage elongation to failure, stiffness, heat distortion temperature, softening temperature, and hardness. Increasing the density decreases the yield, permeability to gases and solvents, creep, flowability, compressibility, and impact strength.

Crystallinity

We have been able to predict and understand the differentiating properties of linear and branched polyethylene simply by having a conceptual catalog of their physical states and by understanding some very simple principles. The same technique can be successfully used regarding crystallinity. Crystalline materials consist of a combination of amorphous sections and crystalline sections. When a crystalline polymer is melted, the molecules are separated so that there is no longer an ordered structure. Large molecular segments vibrate and rotate to give a totally disordered structure. When the plastic cools, a point is reached where the forces of attraction are strong enough to prevent this free movement and lock the part of the polymer into an ordered position. The segments can now rotate and oscillate only in a small fixed location. Since the molecular configuration is the same throughout, the intermolecular distances should be the same and are controlled by the temperature. In a crystalline material the molecules are in an ordered structure, which takes up much less space than the amorphous state. Crystallization is indicated by a sharp decrease in volume. Therefore, crystalline polymers show greater shrinkage than amorphous ones. Since the amount of crystallization varies with the material and molding conditions, it is much more difficult to hold tolerances in crystalline materials than in amorphous ones.

The molecular segments are much closer together in the ordered crystalline lattices. To achieve this state, energy is required. An identical amount of energy is required to break up the crystals into an amorphous condition. The crystalline structure having the molecules closer together and the corresponding increase in intermolecular forces explains the properties of crystalline material.

Since crystallinity gives a more compact structure, the density increases with the crystallinity. The flexibility of a plastic depends on the ability of its segments to rotate. Therefore crystalline structures, which inhibit rotation, are stiffer. Since the crystalline structure has the molecules closer together, the tensile strength increases with crystallinity. The impact strength decreases with crystallinity primarily because of the propagation of faults along the crystalline structure. Shrinkage will increase with crystallinity. The heat properties will be improved because the crystalline material must absorb a significant amount of heat energy before the structure is analogous to the amorphous material. Increasing the crystallinity brings the molecules closer together and increases the resistance to permeability of gases and vapors. Increasing crystallinity lowers the resistance to stress cracking, probably following the same mechanism as the lowering of impact strength. Crystalline materials warp more than amorphous ones, probably because of the different densities within the same material. This sets up internal stresses.

FLOW PROPERTIES

We have been considering the static properties of polymers. Understanding flow properties is essential in polymer processing. These too are amenable to simple analysis.

Two men, Hagan and Poiseuille, independently derived the volumetric flow rate for a Newtonian liquid through a tube.

$$Q = \frac{\pi}{8} \cdot \frac{R^4}{L} \cdot \Delta P \cdot \frac{1}{\mu} \qquad (7)$$

where

$$Q = \text{Volumetric flow rate}$$
$$R = \text{Radius of tube}$$
$$L = \text{Length of tube}$$
$$\Delta P, = \text{Pressure drop}$$
$$\mu = \text{Viscosity}$$

Inspection shows the volumetric flow rate depends on three things. The first is the physical constants of the tube, R^4/L. A Newtonian liquid is extremely sensitive to the radius. This is less so with plastic as we shall shortly see where viscosity varies with the shear rate (velocity). Secondly, the greater the pressure (ΔP), the higher the flow rate. Finally, the more viscous (μ) the material, the lower the flow rate. The length of the tube (L) corresponds to the land length of a gate. When maximum flow into a cavity is desired, the land length is kept to a minimum. Varying the land length is used to help balance the flow of material into multi-cavity molds.

In rheology (the study of flow), the word stress is not used in the sense of a force acting on a body. It is a measure of the internal resistance of a body to an applied force. This resistance is the attraction of molecular bonds and forces. When we say that we increase the shear stress to increase the shear rate we really mean that we have to overcome increasing molecular resistance to achieve a faster flow rate. Shearing stress is the measure of the resistance to flow of molecules sliding over each other. It is reported in pounds per square inch (psi). Force which is measured in the same unit is different in two respects. Force acts perpendicular to the body while the shear stress acts parallel to the containing surface. Pressure is force per unit area while shearing stress is resistance to force. Newton developed this concept of viscosity, using concentric cylinders. It can be explained more easily by imagining a stationary plate over which there is a movable plate with the area of A moving at a velocity U with a force F, stationed at a distance X above the stationary plate. Neglecting the slip of the molecules on the stationary plate, we assume the velocity of the liquid at the stationary plate as zero and the maximum velocity U at the moving plate. The rate of change of velocity is the slope of the line connecting the velocity vectors, or du/dx. The force is therefore proportional to the area and velocity.

$$F = (f) \cdot A \, (du/dx)$$

The proportionality constant (f) is called the viscosity and designated μ (mu) for Newtonian liquids or η (eta) for non-Newtonian liquids. Shear force or stress is represented by the Greek letter τ (tau) and shear rate by γ (gamma). Rearranging the terms we have the following classical definition of viscosity.

$$\mu = \frac{\dfrac{F}{A} \text{ (shear stress)}}{\dfrac{du}{dx} \text{ (shear rate)}} = \frac{\tau}{\gamma} \qquad (8)$$

In a Newtonian liquid, therefore, the shear force is directly proportional to the shear rate. Doubling the unit force doubles the unit rate. In thermoplastic materials this is not the case. In the processing range a unit increase in the shear force may quadruple the shear rate. The viscosity is dependent on the shear rate and drops exponentially with increasing shear rate. This is shown in Fig. 4-27 which has an arithmetic plot of the viscosity and shear rate and of the shear stress and shear rate. It can be noticed that in the latter graph, there is a Newtonian portion at the beginning and end of the curve. It is more practical to plot such data on log-log plots. They will characterize the flow properties of a material. The information is obtained by using rheometers. The most com-mon one extrudes the polymer through a capillary tube while measuring the force and speed of the plunger. Viscosity is then a simple calculation. Such an instrument is called a *capillary rheometer.*

These flow properties can be easily under-stood with a conceptual picture of how the molecules move. Figure 4-28 shows a represen-tation of a number of different polymer mole-cules of the same kind. They are in a random pattern, their vibrations or movement being determined by their heat energy. This Brownian movement, named after its postulator, tends to locate the polymer sections in a random position, this being the lowest energy level. The plastic molecule is too large to move as a unit. Brownian motion occurs in segmental units of the polymer.

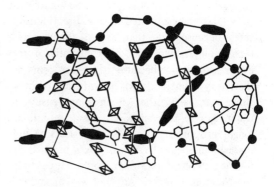

Fig. 4-28. Schematic representation of segments of polymer chains in their random position. This is a result of local vibration, thermally controlled, called "Brownian" movement. (*Courtesy Robinson Plastics Corp.*)

If a force is applied in one direction to a polymer above its glass transition point it will begin to move in a direction away from the force. As it starts to move, the carbon-carbon chains of the molecule will tend to orient in the direction of flow (Fig. 4-29). If the force is applied very slowly, so that the Brownian motion overcomes the orienting caused by flow, the mass of the polymer will move with a rate proportional to the applied stress. This is *Newtonian flow* and is the corresponding straight section at the beginning of the left curve on Fig. 4-27.

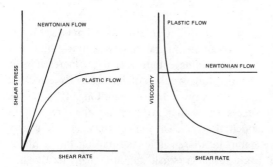

Fig. 4-27. Arithmetic plots of shear stress vs. shear rate and viscosity vs. shear rate for Newtonian and plastics materials. (*Courtesy Robinson Plastics Corp.*)

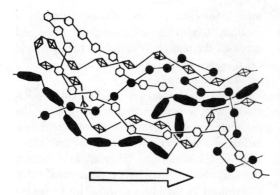

Fig. 4-29. Schematic representation showing the effect of a stress on the random structure of Fig. 4-28. Molecular segments tend to orient in the direction of flow. (*Courtesy Robinson Plastics Corp.*)

As we increase the flow rate, two things happen. The chains move so rapidly that there is not sufficient time for the Brownian motion to have an appreciable effect. Secondly, the molecules being oriented in one direction separate because of their side chains. This separation reduces the intermolecular forces exponentionally (because they are van der Waal's type attractions), permitting them to slide over each other much more easily. In other words, the increased shear rate is no longer proportional to the shear stress. A unit increase in shear stress will give a much larger increase in shear rate than would happen with the Newtonian liquid. Another way of saying it is that the unit force does two things accelerate the mass and separate the molecular segments. The proportion of each changes exponentially. This is the central portion of the curve in Fig. 4-27, and is a characteristic of plastic polymer flow; shear rate is no longer linearly proportional to shear stress. As the flow rate increases it reaches the final stage where all the polymer molecules have oriented to their maximum level. There is no further untangling. Therefore, any increase in the shearing stress in this range will give a proportional increase in the shearing rate and the material acts as a Newtonian fluid, which is indicated in the top portion of the curve in Fig. 4-27. It is this central portion of the curve which is met in injection molding.

The basic difference between a Newtonian and non-Newtonian fluid is the length of the molecule. Newtonian liquids such as water, toluol, and glycerine all have very short molecular lengths. It is evident from the previous discussion that the rheological or flow properties of a polymer depend in some measure on its molecular structure. It is also evident that the flow properties are highly temperature dependent, as the temperature is an indication of the molecular motion and the intermolecular distances. The relation is exponential and a plot of the log of the viscosity versus the temperature at a given shear rate is a straight line over narrow temperature ranges, and describes viscous flow fairly accurately (Fig. 4-30). This

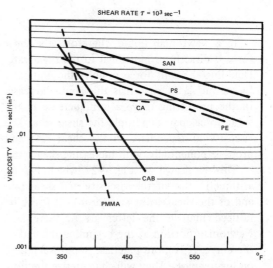

Fig. 4-30. Effect of temperature upon polymer viscosity. SAN-styrene-acrylonitrile; PS-polystyrene; PE-polyethylene; CA-cellulose acetate; CAB-cellulose acetate butyrate; PMMA-polymethyl-methacrylate (acrylic). (*Courtesy Robinson Plastics Corp.*)

type of information also has its practical value. For example, if cavities were not filling out when molding cellulose acetate, increasing the temperature would not have a great effect on the viscosity or hence the filling. It would be necessary either to increase the pressure or open the gate. On the contrary, acrylic is very temperature-sensitive and raising the temperature even a small amount will result in a considerable decrease in viscosity. This also means that temperature control of the material is more important for a viscosity/temperature-sensitive material like acrylic than for cellulose acetate or polystyrene.

From this discussion and Equation (7) the controlling parameters for consistent injection molding become evident. Equation (7) states that the viscosity and pressure control the output in a given geometric system. We have seen that viscosity in thermoplastic materials depends on two conditions—temperature and rate of flow. The temperature of the material is controlled by the conditions of the heating cylinder, nozzle, and mold. Therefore, in order to have consistent molding, the temperature, pressure, and rate of flow must be controlled. It is not possible to have successful automatic molding, or consistent molding without having a control of the flow rate. When the machine is instrumented to measure, compare the measurements to a preset standard, and change the conditions during the molding to meet this standard, it is possible to have consistent automatic molding. The technology is readily available to accomplish this.

Orientation

Orientation effects are very important. By orientation, we mean the alignment of the molecule and molecular segments in the direction of flow. The strength in that direction is that of the carbon-carbon linkage whose disassociation energy of 83 Kcal/mole is much greater than the 2-5 Kcal/mole van der Waal type forces holding the polymer together perpendicular to the line of flow. This means that the plastic that is oriented will be stronger in the direction of flow than perpendicular to it. The ratio will not be 83/5, as no material orients completely. The second major implication of this concept is that the oriented plastic will shrink more in the direction of flow than perpendicular to it. Shrinkage is a result of two things—the normal decrease in volume due to temperature change and the relaxation of the stretching caused by carbon-carbon linkages. Since there are more carbon-carbon linkages in the direction of the oriented flow than perpendicular to it, this phenomenon occurs. All plastics do not exhibit orientation to the same degree. Consider molding a rectangular plaque

of clear polystyrene 2 in. wide and 6 in. long, 0.090 in. thick and gated on the 2-in. end. If the molding was held between crossed polaroid filters a colored pattern could be seen. This is called *birefringence* and is used to measure orientation. The material that flows past the gate is randomized, and freezes as such on the walls of the cavity. This section is totally unoriented. However, one end of the molecule is anchored to the wall and a flow of the material pulls the other end of the molecule in its direction, giving a maximum amount of orientation. As the part cools, the orientation is frozen at the walls. The center of the section remains warmest the longest and allows the Brownian motion to disorient the molecules. Therefore the center section is least oriented. This is shown by birefringent patterns.

It can be easily demonstrated by milling off 0.030 in. of the thickness. In effect, one section is highly oriented and the center section which has been exposed by the milling is less oriented. If the milled piece is heated, the stretched carbon-carbon linkages should return to their normal position. Since the oriented section has the carbon-carbon linkages lined up more in one direction than in the less oriented sections, that part should shrink more. In effect, then, it would be acting as a bimetallic unit, one side shrinking more than the other, and the piece should bend over. This is what happens.

Since the amount of orientation depends on the flow and on the forces which aid or prevent the motion of the molecular segments, it is easy to see the conditions which can affect orientation. Anything that increases the mobility of the segments decreases orientation. Therefore higher material temperatures, higher mold temperatures, and slower cooling would decrease orientation. Pressure on the material would limit mobility. Therefore, low injection pressures and short ram forward time decrease orientation. A thicker part would decrease orientation because it takes a longer time for the center portion to cool.

We shall now examine some practical applications of orientation.

Consider molding a lid or cover 6 in. in diameter in a polyolefin (Fig. 4-31). The shrinkage in the direction of flow is

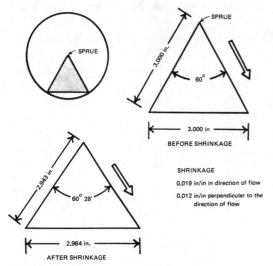

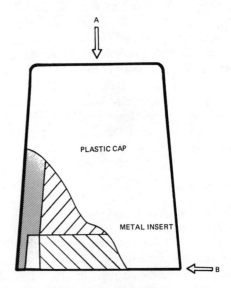

Fig. 4-31. Warping of center-gated polypropylene cover caused by the different shrinkages perpendicular and parallel to the direction of flow. (*Courtesy Robinson Plastics Corp.*)

0.019 in./in. The shrinkage perpendicular to flow is 0.012 in./in. This difference is caused by the differing number of carbon-carbon linkages in the direction of and perpendicular to the flow.

Consider a 60° segment of the cover immediately upon molding. Each side will be 3.000 in. Upon cooling, the two sides in the direction of flow will have shrunk to 2.943 in. and the segments perpendicular to flow will now be 2.964 in. A simple trigonometric calculation shows the central angle is now 60° 28'. The full 360° circle is now 362° 48'. Obviously this extra material has to go somewhere. If it cannot lie in a flat plane, it will warp. If the thickness of the material and the ribbing provide enough strength, the part might not visibly warp, but would be highly stressed. The way to minimize such warp or stress is to mold under those conditions which give the least orientation. Multiple gating is also effective, as is redesigning the cover.

Gate location affects the amount and direction of orientation. Figure 4-32 shows a cap with a metal insert which was used as a protective guard over the fuse mechanism of a shell. The dimensions were controlled by the brass cap which it replaced. The plastic was

molded over a threaded metal insert originally gated at point A. After some time in the field, cracking developed around the metal insert. The main strength was in the direction of flow rather than in the hoop or circumferential direction. Since the thickness of the material could not be increased, the effects of orientation were used by changing the mold and regating at point B. The material flowed in the hoop direction and gave the maximum strength there. This slight difference was enough to prevent failure in the field.

Fig. 4-32. Effect of orientation on a plastic cap molded with a metal insert. Gating at point A will give a cap strength along the walls. Gating at point B will give a cap strength in the hoop direction. (*Courtesy Robinson Plastics Corp.*)

Consider gating a deep polyolefin box (Fig. 4-33). Gating the box in the center (A) would give severe radial distortion for the same reasons illustrated in Fig. 4-31. It would be further complicated by the difference in flow length from the gate to point X and the gate to point Y. The wall would have to be heavy enough to overcome this stress. Gating it diagonally with two gates (B) would give a radial twist, for the same reasons. It would be much less than the center gate and require thinner walls for a stable part. It would also require a three-plate mold for the gating.

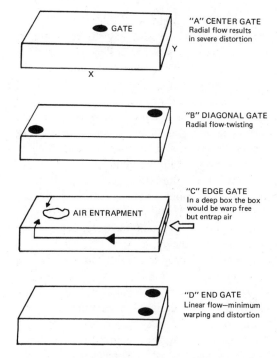

"A" CENTER GATE
Radial flow results in severe distortion

"B" DIAGONAL GATE
Radial flow-twisting

"C" EDGE GATE
In a deep box the box would be warp free but entrap air

"D" END GATE
Linear flow—minimum warping and distortion

Fig. 4-33. Effect of gate location on a deep molded polyethylene box. (*Courtesy Robinson Plastics Corp.*)

It would seem logical to gate on the edge of the Y portion, as shown in (C). This would be true for a relatively shallow box. With a deep box, the material flows around the sides faster than over the top and air is entrapped somewhere on the top. This is virtually impossible to eliminate by venting. Sometimes it can be overcome by increasing the thickness of the top. However, it is not the best method of gating. The preferred method is shown in (D) where there are two gates on the top end of the box. This gives maximum linear flow without air entrapment and produces a part with the least amount of warp. In most instances, a satisfactory part could be molded with one gate located on the top end. Another possibility is two submarine gates near the top. For large parts, it is sometimes necessary to multiple gate to insure relatively even orientation patterns and flow lengths. The main problems that can be encountered are air entrapment and weld lines.

Warpage is the result of unequal stress in the molded part when the stress is strong enough to strain or distort the piece. Warping can be caused by the nature of the material, poor part design, poor mold design, and incorrect molding conditions. (See Chapter 23 on Designing Molded Products for additional data).

Shrinkage

When a plastic material is heated, it expands. Upon cooling to the same temperature, it will contract to the same volume, neglecting the effects of crystallinity. This is not the only situation. During injection molding the additional factor of pressure is introduced. The material basically follows the equation of state, Pressure x Volume equals the Gas Constant x the Absolute Temperature, $PV = RT$. Mold shrinkage should not be confused with tolerance. Tolerance is the variation in mold shrinkage rather than the shrinkage itself. A brief discussion of what happens during injection molding will be helpful. The hot material is injected into the cold cavity, initially under low pressure. Cooling starts immediately as the parts in contact with the wall solidify. Since this specific volume (the volume of a unit weight of plastic) decreases with the temperature, the solid will occupy less room than the molten polymer. The material fills the cavity and the pressure builds up rapidly. The pressure does two things. It adds more material to the cavity to make up for the decrease in volume of the material that is already solidified and it adds more material to compensate for the decrease in volume that will occur when the rest of the material solidifies. If not enough material is put in, there will be excess shrinkage. If too much material is put in, there will be a highly stressed area at the gate. This is called *packing*. The effect of the machine pressure ceases when the injection pressure is stopped or the gate seals. The availability of a second injection pressure is helpful in controlling packing.

Overstressing the gate section of an opaque part is impossible to detect during molding. Highly stressed parts are much more likely to fail. This is one of the drawbacks of molding wherein quality cannot be immediately determined. This is also a good reason why economic conditions should not force improper molding.

To decrease shrinkage, reduce wall thickness, increase injection pressure and injection forward time, increase injection speed, increase the overall cycle, raise the material temperature slightly, lower the mold temperature, decrease the molecular weight distribution, and usually increase the gate size.

Tolerance in molding is beyond the scope of this section. Suggested tolerances are found on pages 668 through 687. Controlling molding tolerance requires a good mold, a good machine, good management, and the proper calculation of the price of the item. Though obvious, it is necessary to repeat that *unneeded tolerances are costly*. Good engineering specifies the minimum tolerances required for the application.

CORRECTING MOLDING FAULTS

Injection molding faults may appear when starting a new mold, changing to a new material, or during the regular operation of a mold. The causes of these faults can be as follows: machine, mold, operating conditions (time, temperature, pressure, injection speed) material, parts design, and management. Of these, the latter is the most important.

Before correcting a fault, it must be found. The purpose of quality control is to find the fault while it is being molded, rather than to discover the error some hours, shifts or days later when it becomes history rather than a force for increasing productivity and profit. It should be a continuing ongoing process starting with the ordering of the raw materials and molds and ending with the shipping of the finished part.

It is obvious that the condition of the machines, molds, auxiliary equipment, and work area contribute to preventing and correcting molding faults.

Poor communications and record keeping are other obvious sources of molding problems.

Before attempting to correct faults, the machine must be operating consistently, the temperature control must be consistent, the mold operation must be consistent, and the cycle time must be consistent.

A single-cavity mold should fill evenly. A multiple-cavity mold must fill evenly. If one cavity fills first, the gate may seal off so that it will not be fully packed. The material destined for this cavity will be forced into other cavities, overpacking them. This causes shrinkages, sticking, and other problems.

The machine should have a large enough clamping and plasticizing capacity. The plasticizing system should be appropriate for the job. The ejection system should be effective.

Most time functions are controlled by timers. If the mold is not run automatically, the major variable is the operator. The amount of time required to open the gate, remove the part, inspect the mold, and close the gate is variable. This often is a major cause of molding problems.

Temperature is normally controlled automatically. Poor design and malfunction of the temperature control system is not uncommon. In some areas of the country, voltage is not constant because of the inadequacies of the electrical power. This can be disastrous in the molding operation. The best way to recognize it is to use a recording voltmeter.

Pressure is also automatically controlled. Sometimes, overheating will change the viscosity of the oil in the machine, thereby changing the operating conditions. This is difficult to detect. An overheated machine is always a suspect. A malfunctioning or worn screw tip will effect the injection pressure on the material.

One of the major variables in the molding process is the material. This is inherent in the manufacturing process. Normally, there is little the molder can do other than recognize it. Changing materials will usually pinpoint this area.

Specific molding problems can be arbitrarily divided into the following categories:

Short shots are usually caused by the material solidifying before it completely fills the cavity. It is usually due to insufficient effective pressure on the material in the cavity. It can also be due to lack of material feeding to the machine. It may require increasing the temperature and pressure on the material, increasing the nozzle-screw-runner gate system, improving the mold design, and redesigning the part.

Parts flashing is usually caused by a mold deficiency. Other causes are the injection force being greater than the clamping force, overheated material, insufficient venting, excess feed, and foreign matter left on the mold.

Sink marks are usually caused by insufficient pressure on the parts, insufficient material packed into the cavity, and piece part design. They occur when there are heavy sections, particularly adjacent to thinner ones. These defects are predictable and the end user should be cautioned before the mold is built. Many times the solution lies in altering the finished part to make the sink mark acceptable. This might include a design over the sink mark or hiding it from view. (See Chapter 23 on Designing Molded Products.)

Voids are caused by insufficient plastic in the cavity. The external walls are frozen solid and the shrinkage occurs internally, creating a vacuum. Inadequate drying of hygroscopic materials and residual monomer and other chemicals in the original powder may cause the trouble. When voids appear immediately upon opening the mold. it is probably a material problem. When the voids form after cooling it is usually a mold or molding problem.

Poor welds (flow marks) are caused by insufficient temperature and pressure at the weld location. Venting at the weld where possible and adding runoffs from the weld section may help. Increasing the thickness of the part at the weld is also useful. The material may be lubricated or the mold locally heated at the weld mark. Poor welds have the dismaying proclivity of showing up as broken parts in the field. Flow marks are the result of welding a cooler material around a projection such as a pin. Their visibility depends on the material, color, and surface. They rarely present mechanical problems. Flow marks are inherent in the design of some parts and cannot be eliminated. This should be thoroughly explored before the part design is finalized.

Brittleness is caused by degradation of the material during molding or contamination. It may be accentuated by a part which is designed at the low limits of its mechanical strength.

Material discoloration results from burning and degradation. It is caused by excessive temperatures, material hanging up somewhere in the system (usually in the cylinder), and flowing over sharp projections such as a nick in the nozzle. The other major cause for discoloration is contamination, which can come from the material itself, poor housekeeping, poor handling, and colorants floating in the air.

Splays, mica, flow marks, and surface disturbances at the gate are probably the most difficult molding faults to overcome. If molding conditions do not help, it is usually necessary to change the gating system and the mold temperature control. Sometimes localized heating at the gate will solve the problem. Splay marks or blushing at the gate is usually caused by melt fracture as the material expands entering the mold. It is usually corrected by changing the gate design and localized gate heating.

Warpage and excessive shrinkage are usually caused by the design of the part, gate location, and the molding conditions. Orientation and high stress levels are also factors.

Dimensional variations are caused by inconsistent machine controls, incorrect molding conditions, poor part design, and variations in materials. Once the part has been molded and the machine conditions set, dimension variations should maintain themselves within a small given limit. The problem is usually detecting the dimensional variances during the molding operation.

Parts sticking in the mold results primarily from mold defects, molding defects, insufficient knockout, packing of material in the mold, and incorrect mold design. If parts stick in the mold, it is impossible to mold correctly. Difficult ejection problems are usually the result of insufficient consideration of ejection problems during the design of the part and the mold.

Sprue sticking is caused by improperly fitted sprue-nozzle interface, pitted surfaces, inadequate pull-back, and packing. Occasionally the sprue diameter will be so large that it will not solidify enough for ejection at the same time as the molded part.

Nozzle drooling is caused by overheated material. For a material with a sharp viscosity change at molding temperature, such as nylon, the use of a reverse taper nozzle or a positive seal type of nozzle is recommended.

Excessive cycles are usually caused by poor management. Proper records are not kept, standards not established, and constant monitoring of output not established. Other causes of excessive cycles are insufficient plasticizing capacity of the machine, inadequate cooling channels in the mold, insufficient cooling fluid, and erratic cycles.

Multi-component Molding*

Multi-component (or 'sandwich') molding is a new injection molding technique designed to produce a range of composite plastic products. Typical of these are sandwich structures with, for example, a solid outer skin and a central core of foamed plastic (Figs. 4-34 and 4-35). These composite moldings are made in one operation in an injection molding machine and take no longer to produce than conventional solid moldings.

Sandwich molding machines up to 1500-ton clamp pressure have been built and parts up to 4-ft square have been made (see Fig. 4-36). The combinations possible span a wide range: Polypropylene/low-density PE, solid polypropylene/foamed polypropylene, acrylic/ABS, polycarbonate/butyrate, EVA/polypropylene, and a number of others.

The principle is simple. Two polymer formulations from separate injection units are injected one after the other, into a mold through the same sprue. If one of the components of the structure is to be a foamed core, the mold is opened for a short time after filling to allow the foamable polymer to expand.

In producing a solid skin/foamed core sandwich, the operation proceeds as follows: The two injection units are charged with polymer, the screws retracted, and the valve to the closed mold is shut. The valve is then opened to the first injection unit and a partial charge of plastic A (which will become the solid skin) is injected in the mold. The sequencing valve then opens to the second injection unit and plastic B (which will become the foamed core) is injected into the mold, forcing plastic A

* Prepared by ICI America, Inc., Stamford, Conn.

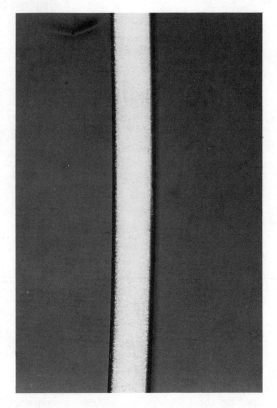

Fig. 4-34. Enlarged photo of section cut from "sandwich molded" product, showing solid skin and foamed core. (*Courtesy ICI America, Inc.*)

Fig. 4-35. Experimental concrete shutter trough, with solid polypropylene skin and foamed polypropylene core. (*Courtesy ICI America, Inc.*)

Fig. 4-36. A 1500-ton clamp machine for sandwich molding, using standard injection and press units. (*Courtesy ICI Plastics Division*).

to the edges of the mold without bursting through. The mechanism is analagous to blowing up a balloon. The air inside the balloon stays there provided that the rate of filling and amount of air is not excessive. The same idea is applicable to two polymer charges, provided the conditions of polymer temperature, mold temperature, injection rate, and other variables are controlled. Instead of air forcing the balloon out to the sides, however, it is polymer B that forces polymer A to the edges of the mold, laying down a uniform layer of polymer A as it does.

In the next step, a sufficient amount of plastic A in injected into the cavity to clear the sprue of foamable polymer B. Unless this additional charge were injected, a foam structure would be revealed on cutting the sprue from the molding and residual foamable polymer in the channel would appear in the skin of the next molding and cause poor surface finish on part of it.

Having filled the mold, closed the valve, and held full clamping pressure for a few seconds, the mold is then opened a small distance. Polymer B now foams to give a uniform foam structure enclosed or encapsulated within a thin skin of polymer A.

5

CONTROLS FOR INJECTION
MOLDING MACHINES

Glenn A. Tanner*

As will be evident in this new edition of the Handbook, refinements in controls for processing have been a dominant theme throughout the sixties and seventies and progress is continuing toward even greater sophistication.

This chapter will review the various controls for injection molding machines that have made the greatest impact in recent years—from solid state controls that have become fairly universally accepted to the more complex adaptive, feed-back controls that can correct deviations in a machine operation, to the eventuality of computer controls that can analyze and correct all possible variations. Table 5-1 lists these in order of increasing sophistication and control. Table 5-2 lists those systems commercially available today which are indicative of the industry's approach to various methods of control.

The inclusion of this single chapter on controls for injection molding is not to imply that there has not been equal activity in controls for other processing techniques, including an increased reliance on solid state controls and the introduction of monitoring, adaptive, or computer systems. Computerized controls, in fact, have advanced further in extrusion than they have in injection molding. Rather, this chapter is intended to provide a more basic insight into the entire subject of controls as used by the plastics processing industry today. The chapters that follow on individual processing methods will carry references to the specific work being done in each particular field.

In recent years, we have seen quite a metamorphosis take place in the control and operation of injection molding machines, with the introduction and increased sophistication of solid state controls.

As molding processes became more complex, molders required greater accuracy and increased variations in the types of cycles that could be adapted to their machines. These include variations in core pull sequences, alterations in the sprue break sequences, different ejection sequences, and even changes in the timing of high pressure application after clamp close in combination with injection. Each variation in itself may require but a simple change, although the numerous combinations become a nightmare for the machine design engineers. Although all the requested sequences are possible, to supply them all on one machine would multiply selector switches to a point of ultimate confusion, and the relays required would fill panels larger than the machine itself.

It is fortunate that as these new requirements for machines materialized, the development and manufacture of solid state components were reaching advanced stages of reliability and prices were coming down commensurately. For, as it has turned out, solid

* 9 Cardinal Drive, Hudson, N.H. 03051.

Table 5-1. Classes of Solid-State Control Systems

Class 1: *Accessory Components*
The simplest class involves the incorporation into the machine of solid-state accessory components, such as timers or heat controls.

Class 2: *Direct Replacement of Relays with Solid-State Switching*
This type of control is accomplished by taking the old relay logic circuit and replacing the relays with solid-state switching mechanisms.

Class 3: *Solid State Logic in Sequence Controls*
The machines utilize new circuits with full benefits of solid-state logic. Generally, this second generation includes integrated circuits offering greater versatility and complexity of cycles.

Class 4: *Replacement of Linear Controls*
This class includes machines which have gone beyond sequence controls and placed solid-state controls on the linear movements, either clamp, plasticizer, or both. This class could be coincidental to any other class.

Class 5: *Monitors*
These machines overlook the molding operation using watch-dog devices, recording devices, or even computers which accumulate statistics or data on production output or machine functions.

Class 6: *Process Controllers*
These units furnished with the machine or attached to it provide at least one closed-loop or feed-back to adjust a machine function when it leaves a pre-set limit. Double or multiple closed-loop circuits from any unit (even a computer) belong here.

Class 7: *Computer Operation*
Although computers can function on Class 5 or 6 machines, ultimate control is achieved where the computer is programmed to analyze many machine variables and actually operates in a decision-making capacity to determine the best of alternative adjustments to make on the machine.

Table 5-2. Some Typical Control Systems Commercially Available for Injection Molding.

(A) Monitoring systems (alarm buzzes or lights flash on deviation)
  1. System monitors high and/or low set points of specific conditions relative to pressure, temperature, displacement, and force.
  2. System uses time and distance relationships to produce an indicative reaction from the monitor. If reaction is consistent, parts are assumed to be the same.

(B) Feedback systems (deviation sets up corrective action)
  1. System measures what is called a viscosity index. If deviation is noted, compensation is provided by increasing or decreasing the primary injection pressure time in the same cycle in which the variation occurs.
  2. System uses closed-loop control to monitor cavity pressure (using an ejector pin and strain gauge). Control is through electro-mechanical relief valve on injection pressures and holding pressures.
  3. System monitors injection velocity, injection pressure, and melt temperature. Pressure is controlled by shifting from high-volume pumps to zero volume. Melt temperature is controlled by regulating screw speed, back pressure, and barrel temperature.

(C) Computer-controlled systems
  1. Mini-computer-based system uses programmed logic to control all machine functions and key process variables. The logic is entered into the computer memory from programmed cassettes, enabling different control parameters to be entered by merely changing cassettes.
  2. Mini-computer-based system controls up to 8 injection molding machines fitted with on-machine adaptive controls. The computer functions as a supervisory control and logs such data as production quantities, rejects, reject causes, machine status, etc.

state components have proved to be critical in achieving the complex circuits needed on today's molding machines.

These solid-state controlled machines offer many advantages to molders. New controls should not deter a purchaser because of fear of the unknown. The theory of solid-state components need not be completely understood to operate or indeed troubleshoot the new machines. For years, electrical maintenance men have run faultfinding procedures on relay control systems, including solenoid-operated air and hydraulic valves, without being able to explain hysteresis, inductance, or electromotive force. Indeed, they work daily with electric motors and few understand power factor correction.

These technical points are left to the trained engineer, and this will be the case with solid-state controls. Ordinary lay people can operate these machines as easily, if not more so, than the older systems. Maintenance men learn readily to use the new test equipment and where to look for trouble.

Solid-State Components

Although the theories and technical limitations can and should be left to the electrical design engineer, it is advantageous to have some understanding of the terms that have been introduced with this new technology.

Solid State. This is a relatively new term, but

it applies to an old phenomena. It applies to semiconductors that are solid in nature. A semiconductor is simply a material that has properties somewhere between a conductor and an insulator. It is really a non-conductor with some measurable or controllable level of conductance.

For years the predominant type of semiconductor in use was the vacuum tube, which could continuously, or on command, release certain electrical impulses from one direction at a controlled rate. Even older, though, were the natural crystals. There were crystal set radios using natural crystals before the vast radio networks were established, or television was a reality. Why certain crystals found in nature had these properties was a mystery. The crystals were studied to learn how they functioned and from these, man learned to make his own in various configurations.

These man-made solid crystals were called *solid semiconductors* or *semiconductors in a solid state* and hence the new term was born.

Diode. The simplest of these new components is the diode (Fig. 5-1). It is comprised

Fig. 5-1. Diode.

of two small bits of special metals, fastened back to back in intimate contact with each other so that the molecules of the two metals will interrelate and thereby exchange electrons under certain conditions. Actually, the condition in the case of the diode is an electrical current (a positive charge) applied to one side; in such an instance, the current will pass through the diode from that direction almost like a short circuit. When a similar charge is applied to the opposite side, however, the diode is incapable of passing the current and it will therefore act as an electrical rectifier, or as a check valve might act in a fluid circuit.

Diodes of various types are most useful. They are in common use in even the most advanced circuits today.

Transistor. Another solid-state component is the transistor (Fig. 5-2). The transistor is made

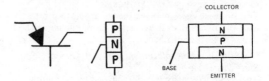

Fig. 5-2. Transistor.

up of three layers of metal, the outer two of which are the same. It is sometimes referred to as a double-based diode or a diode with two bases. You cannot connect two diodes by wire to accomplish this purpose because there must be that very intimate contact of the metals for the electron exchange. The transistor has three leads, one leading out from each layer. By applying the correct charge onto the lead to the center layer (base), the transistor is actuated so that a large current will flow between the two outer layers (collector and emitter). Such an action is the same as we would expect from a low voltage relay.

Because variations in the current flow applied to the center layer will permit proportionate variations in the larger current between the two outer layers, a transistor can also work effectively as an amplifier. This application is primarily seen in radios and television sets.

SCR. The next component in complexity is the "silicon controlled rectifier" (SCR), Fig. 5-3.

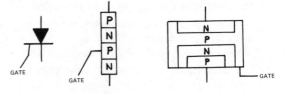

Fig. 5-3. SCR.

It might be classified as a four-layer diode in which alternate layers of these same thin special metals are used. A lead to one of the center layers is termed the *gate*. A small current applied to this gate activates the central layers, permitting a large current to pass through from outer layer to outer layer. Like the diode, it has only the ability to carry current in one direction. Once the gate is fired, the SCR will continue to conduct current as long as the anode terminal remains positive. Such an action might be compared to a motor-control relay with a holding contact that remains engaged until the current flow is cut by a motor-stop button.

Triac. A more complex and new unit is the triac (Fig. 5-4), which will conduct current in

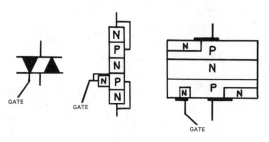

Fig. 5-4. Triac.

either direction. As such, it is similar to two diodes mounted in reverse directions across parallel leads, with a gate lead on one of the diodes. Once the gate is triggered, the triac will continue to conduct even alternating current until the voltage across the two main terminals drops to zero. In order for this unit to conduct ac, the gate must actually be fired every half wave of the cycle. A triac then, like the SCR, is an electronic switching mechanism.

There are also variations of these basic components, such as the Zener diode, the unijunction transistor (UJT), the quadrack, and others. In each of these, there is the same system of basic layers of special metals with leads fastened to them or other tiny bits of metal as spots on a layer. In fact, by placing the material in the right location on these sandwiches and adding more leads, they can, within one small unit, serve as many functions as if 40 transistors were wired in a circuit.

As these units became more complex with more layers and spots, it became extremely difficult to identify each unit by an individual name. Thus, the units, with their many variations, were called *integrated circuits.* These components are the basis of most of the latest solid state control circuits.

Advantages

Solid-state components, in general, offer a number of advantages to a machine owner, as follows.

(1) *Longer Life.* They are very durable. Since there is no mechanical movement, wear is no problem and their life is measured in the range of billions of cycles.

(2) *Faster Response.* These components require no initial warm-up as did vacuum tubes. But of more importance in the industrial field of 24-hr operation, is the fact that reaction time to a direct impulse is almost instantaneous. The advantage is obvious compared with the measurable reaction time for relays to trip and make contact.

(3) *Greater Flexibility.* This feature is most responsible for the growth of solid-state controls in plastic processing machines. The units can be used or combined in a number of ways, or special units selected that will accomplish a wide range of interlocking circuits without bulky and complex protection against impulses feeding back down the control wiring to undesirable points.

(4) *Greater Reliability.* In addition to longer life, the solid-state components are more uniform in their performance. Relays, even from the same production lot, often vary greatly in their response time.

(5) *No Contacts or Coils.* All motor controls and relays have contact points. After a time these points become pitted and worn. This not only alters the reaction time, but the maximum current that they will carry efficiently. In fact, the arc across these points can weld them together. Stuck points fault-out the circuit and require replacement. Weakening coils will cause delay in relay reaction time, and even worse, allow the contacts to bounce. Making contact and then dropping out often permits sufficient current passage to give odd and often damaging action to a machine. These problems are not encountered with solid-state controls, for they have neither contacts nor coils to wear and weaken.

(6) *Compactness.* The large amount of work one small component will do permits reasonably sized control panels. A typical plug-in relay with socket occupies at least 5 cu in.,

whereas a triac of similar value with its soldered mounting would occupy about 0.03 cu in. In fact, an integrated circuit of less than $\frac{1}{2}$ cu in. could easily replace a 12 in. x 12 in. relay panel (see Fig. 5-5)

Usage

Almost without exception, these small solid-state components are assembled (soldered) onto printed circuit boards. These boards, or cards, as they are often called (Fig. 5-6), are designed in a systematic way to group components of a like nature or more often to group components affecting one function of the machine. These cards are in turn plugged into a metal frame that contains sockets, guides, and catches designed to hold them. This compact unit is called a *card rack* (Fig. 5-7). The advantage, obviously, is quick removal and replacement of the solid-state cards.

Solid-state components have also been introduced into the accessory equipment in the control panel. In fact, many of these applications predate their use in the sequence logic. Solid-state temperature controls have been available for some years. Their small size initially made them suspect in the molding

community. However, they have proved themselves and designs have continued to improve. They are accepted now as reliable units with longer life and greater accuracy than the older designs.

Much the same can be said for the development of timing controls using solid-state systems. Some of the newer ones use digital read-out units with extremely sensitive settings. The many variations available permit selection of a specific type to match the needs for which it is required.

Since digital read-out units can accurately detect time intervals as low as $\frac{1}{10}$th of a second, they are most useful. If the expense does not seem warranted on each machine purchased, you might consider less expensive standard timers with a portable digital read-out timer that will plug into the machines. This digital timer accurately checks time settings as well as machine functions not covered by timing controls, such as clamp-close time, screw-back time, total injection time, or time of progress of injection at various stages.

Solid-state systems are now being used more frequently to detect and control linear movements on the machines. These include movement of the main ram to activate slow close,

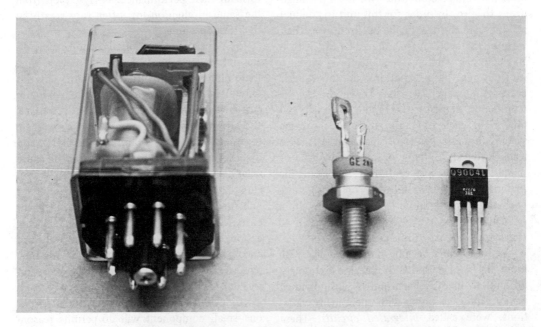

Fig. 5-5. Size comparison between various controls. From left to right: conventional plug-in, low-voltage relay; SCR (silicon controlled rectifier); plastic-encased Triac.

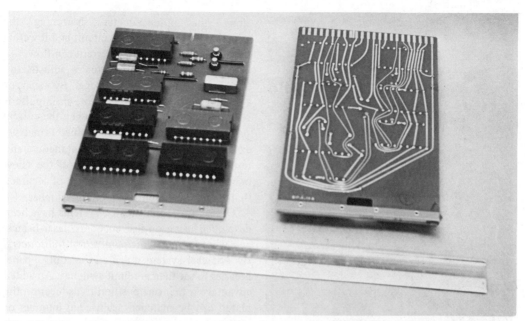

Fig. 5-6. Solid state card. All solid state components are mounted on a printed circuit board in accordance with the system used in the design. Shown here is a combination of resistors, capacitors, transistors, and integrated circuits.

Fig. 5-7. Solid state card rack. Cards holding solid state components are plugged into a rack. Each card differs from the others, depending upon system of organization. Cards are readily removable for maintenance, checking, or replacement.

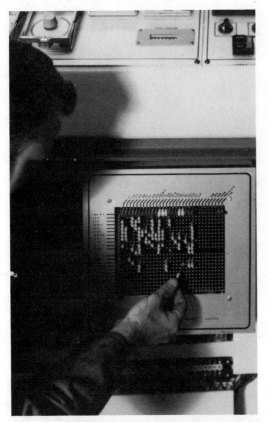

Fig. 5-8. Logic sequence control. This system uses diode plug-in connections at selected points in the logic board to determine desired variations of sequences in seconds. (*Courtesy Ingersoll Rand Co., Plastics Machinery Div.*)

low-pressure close, double-shot protection, slow open, fast travel, full open cushion, ejector operation, core actuation and sweep operation. On the plasticizer, they can be useful to detect screw travel for measurement of shot size, activate injection pressure variations at adjustable points in screw travel, packing pressure, measurement of screw full forward or cushion of melted material left in the barrel at the end of each shot.

The first of the new linear controls was accomplished by replacing the limit switches with a linear or rotary potentiometer. These controls work as position sensors of great accuracy. A *potentiometer* is a wire of uniform resistance wound around an insulation core. As the movement occurs, a contact finger moves along the potentiometer shortening the effective wire length in the circuit and lowering the resistance. Thus, more current can flow.

Another dial potentiometer is adjusted by the operator on the control panel. By means of a solid-state current comparator, an impulse is sent out to the next function when the current passing through the two potentiometers matches exactly. The reaction is instantaneous and results in sensing the movements of the screw or moving platen to a higher degree of accuracy than can be accomplished by limit switches.

Other more sophisticated systems have been developed using optical encoders, laser beams, and what is termed *digital position transducers*. By whatever system, a dial on the control panel is easier to set than are limit switches. An added advantage is that one position transducer on the platen can be matched against any number of settings on the control panel, causing each function to occur at the proper time. In effect, this eliminates not only problems of setting a number of cams and limit switches, but eliminates the complicated and often cumbersome mechanism from the machine. Elimination of maintenance and replacement of faulty limit switches is an added advantage.

With the advent of these accurate position transducers, considerably more positions can be sensed than was practical with limit switches. Thus, increased features can be added to process controllers and increased steps incorporated in the injection programming.

Earlier, we explained that one of the main advantages of the solid-state control systems was the increased variations available in machine sequences. Programmable sequence settings are now available in a number of schemes. Probably the first was the punch card type where individual cards have the machine sequence fixed by holes strategically placed in the card (Fig. 5-9). A separate card is furnished for each type of cycle. Another type is the matrix programmer where horizontal and vertical lines form a checkerboard pattern on a printed circuit board. At any juncture of these lines a contact can be made by inserting a diode (to prevent back feeds). Some of the boards are fixed in the machine and the diodes placed into a pattern for the desired sequence by the set-up

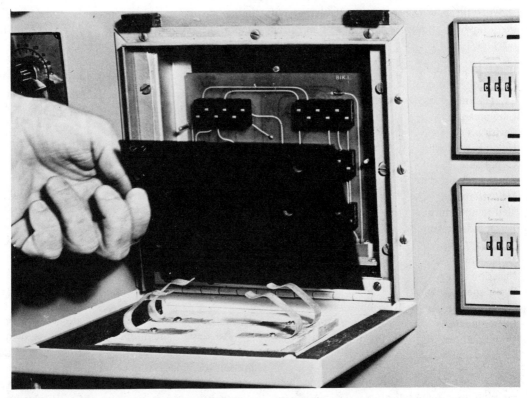

Fig. 5-9. Punched card controller. Logic sequence controller using miniature limit switches mounted on printed circuit board. Inexpensive punched card determines switches contacted and thereby establishes fast change to desired sequences. (*Courtesy BIPEL International, Inc.*)

man. In others, the boards themselves are replaceable and have diodes soldered in for fixed cycles on that one board.

All of these units are of particular value to the molding plant that runs a wide variety of molds. They offer a simple way to obtain complex cycles on the machine. Most of these units can be calculated to offer as many as two thousand possible cycle variations. It would be advisable not to add diode or holes in any cards that have not been recommended by the machine manufacturer. With all of the combinations available, only a fraction of these can have been checked for positive and safe operation, and false circuits could quite easily damage the mold or machine. They are a great convenience feature for the plastics industry, but they still need perfection for trouble-free operation and the relentless checking of all possible combinations.

Perhaps these sequence programmers will soon be outdated by the use of an ultra-small computer on each machine. These are being introduced with the capability of accepting, holding, and operating the machine sequence program directly.

Monitoring

For years, strip or rotary recording charts have been available as laboratory tools. In fact, they have been a common sight in generating plants and boiler rooms for years. Their adaptation to the production equipment for plastics awaited not only the responsiveness of solid-state control systems, but also the development of reasonably priced and accurate transducers.

Although a transducer is most commonly thought to be a pressure-sensing device, the term actually includes any unit which measures variables and converts that variable to a correspondingly variable electric impulse. The well-known thermocouple is a *temperature transducer.*

With the new transducers available, any of the following can now be measured on injection machines:

Injection pressure	Ejector pressure
Nozzle or melt pressure	Fill rate
	Time to fill
Gate pressure	Time to plasticize
Cavity pressure	Overall cycle time
Holding or packing pressure	Clamp force
	Hydraulic pressure-system
Screw back pressure	
Melt viscosity	Hydraulic pressure-clamp
Melt temperature	
Barrel temperature	Hydraulic pressure-injection
Nozzle temperature	
Mold temperature	Hydraulic system temperature
Injection speed	
Screw speed (rpm)	Hopper level
Plasticizing stroke	Ambient temperature
Injection stroke	Ambient humidity
Cushion at end of injection stroke	Decompression stroke

The new transducers introduced a wide range of monitor applications. The units commercially available fall into three classes: the recording monitor, the non-recording alarm type, and units serving both functions (Fig. 5-10). Although some units are designed to "watch-dog" one or two specific functions, many can be adapted to the requirements of the user. It would be wise before purchasing to determine how easily you can change to adapt or handle the particular parameters which are important in your operation.

The alarm-type unit will emit an audible or visual signal whenever a function being monitored exceeds the upper or lower limit set on the unit. It is then the responsibility of the operator to find the cause of the problem and to correct it manually. His job is simplified if this is one of the units that stores or records the readings of important variables during that faulty cycle. Some of the units have this information stored for "call-back" by the operator and others will have that cycle already printed out when he arrives at the machine. Units also can be furnished to reject the parts produced in the faulty cycle.

Process Controller

The next logical step in solid-state controlling was to design the monitor to make its own correction when it sensed a variable leaving a pre-set tolerance. This could only be possible with the development of servo-mechanisms. A *servo-mechanism* is a device that converts an electrical impulse into a mechanical action. An example is the electrical servo relief valve.

By selecting the correct servo-mechanism, it is now possible to have feed-back or closed-loop circuits. These are control systems whereby a variable is monitored by a simple unit (often referred to as a *black box*), a large and complex strip recording monitor, or even a computer.

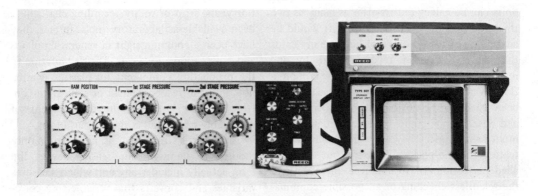

Fig. 5-10. Monitor with visual read-out. Continually watches three important functions of machine and will signal alarm when functions reach overly-high or overly-low settings. Visual display has storage capabilities for instant read-out of any of these functions to determine when and how they varied. (*Courtesy Reed-Prentice*)

Fig. 5-11. Process monitor. Strip chart unit has eight channel pens which continuously record graph of variables from one or more machines. The features monitored depend on transducers and amplifier input modules. (*Courtesy Control Process Inc.*)

Whichever it is, the unit is designed to sense when a variable is exceeding the limits dialed into it for that job. When this occurs, it feeds back to the servo-mechanism an impulse necessary to correct the out-of-line variation. The correction may take place on the same cycle or on the next cycle, depending upon its capabilities.

The name *process controller* appears to have been accepted by most people for these closed-loop controllers. However, some refer to them as *adaptive controls* since they adapt the functions of the machine to variations which occur. Although many purist control technicians argue that this latter term is misused, it still finds use in the industry today.

Closed-loop controls are not new. The industry has been familiar with them for years in temperature control systems. The heat would be cut off or turned on to hold the temperature, as sensed by a thermocouple, at the desired setting. Today, we go much further, however. For example, if the melt viscosity leaves the desired tolerance, the unit can alter the barrel temperature or the back pressure to correct it. Some units will even alter the injection pressure on the same cycle to compensate.

A process controller can be simple, controlling only one function on the machine, or it can be complex, where a number of variables of the molding operation are monitored and each one individually adjusted when required. The sophistication required is based on the difficulty of the molding and the length of the runs. Before purchasing, it is important to investigate the process controller thoroughly or even test it on the job.

Computer

The use of the computer in injection molding has already been demonstrated in production. Small computers can be used for each individual machine; larger ones can be connected to a battery of machines or for that matter, all machines in a plant. Recent thinking is more toward a small dedicated computer on each machine, with a large master computer overlooking all of the machines.

Before any of these functions are possible, it is necessary to develop a suitable interface. An *interface* is a unit which converts the electrical impulses generated by the transducers on the machine to the type of electrical impulse used by the computer. This can often be at different voltage, but it must be in an electrical impulse rate and tempo (called *language*) that can be understood by the computer. The interface can serve the reverse purpose as well, by taking a direction from the computer and converting it into electrical impulses capable of reacting on the servo-mechanism of the machine. Each interface must be tailored to a specific machine with specific options and, of course, to the specific computer.

In many cases (indeed, in most of the first applications), the computer was used as a production control tool. Here, it monitored cycles and outputs for production statistics. Later, it monitored machine functions for instant retrieval and accumulation of data for management's benefit.

More recently, in addition to the above uses, languages and interfaces were improved to allow the computer to work feed-back circuits and make machine corrections, thereby serving as a process controller. To date, the majority of such units have only had the capability of correcting one or two machine variables on individual closed loops. These techniques, programs, or instructions to the computer are termed *software*.

Computers have tremendous and varied capabilities to store and put out reactive (adaptive, if you prefer) information. Our human capabilities are the limits of the possibilities in these interrelated systems, for man himself must first learn and understand both the complexities of the molding operation and the limits of the computer system. Too often, molding is done by trial and error. To develop the software for computers, we must learn the interrelated actions of every aspect of the machine for the part being molded. We must be able to set the tolerances for every machine variable and know the best of the alternative corrections when it strays from those set tolerances. Such data are critical to exploiting the capabilities of the computer for molding operations.

It is anticipated that computer control may someday monitor 5 to 12 variables of the molding operation. If one or more of the variables fall or rise to limits that will result in reject parts or inefficient production, the computer will analyze all the machine functions and decide which one has to be adjusted. The decision aspect is the key to such an operation.

The early application for such computer-run jobs will necessarily involve extremely long runs, probably ones that leave the machine only when the parts are obsoleted by design change. As programs for one part are developed, they will assist in developing the next. Eventually, the system will reach sufficient maturity so that molds can be changed daily. Management need only advise the computer of the job number and batch number of material. The computer will automatically reset the machine to the settings on the last run, made weeks or years previously; at the same time, it will automatically compensate for changes in ambient temperatures, humidity, and material specifications.

This is, of course, the ultimate (and still somewhat in the future) control system. Engineers are moving quickly toward it. In the meantime, however, there is sufficient flexibility in commercially available equipment (see Table 5-2) to allow the molder to select an existing system best suited to his requirements.

6
DESIGNING MOLDS FOR INJECTION MOLDING

Irvin I. Rubin *

The injection mold incorporates the cavity into which the hot plasticized material is injected under pressure. Heat is removed from the material in the mold until it is rigid enough to be ejected so that the final part will conform to all its specifications. The quality of the part and its cost of manufacture are largely dependent on the mold design, construction, and excellence of workmanship.

Over the years, as the size of the molding machine increased, so did the cost of molds. Molds costing upward of $50,000 are no longer unusual.

Obviously the two critical areas in insuring the right mold are the design of the piece and the design of the mold. An acceptable product is not possible with an incorrect part design. An acceptable part may be possible with an incorrect mold design, but always under extremely difficult and uneconomical conditions. While the design of the part is not within the scope of this section, it should nevertheless be thoroughly reviewed by those people directly involved in the design of the mold and in the molding operation (see Chapter 23 on Designing

Molded Products). The last opportunity to change the design of the part is when the mold is being designed.

It is imperative that the user, moldmaker, molder, mold maintenance personnel, quality control department, and packaging people be consulted in the design of the mold. The selection of the method of gating, location of the gate, venting, parting line, method of ejection, and type of ejection affect the appearance and function of the part.

Moldmakers vary from the individual owner with several helpers to plants with hundreds of employees. The quality of the mold bears no relation to the size of the shop. In selecting a moldmaker, it is important to use one who has the best type of equipment for the job and who has had experience in building similar molds.

The overriding consideration in mold design is to be completely confident that the mold will work. If there are questions, they should be resolved before the mold is built, not afterward, as it is much easier to change a design on a piece of paper than to change a completed die. If, during the course of mold construction, questions of this kind arise, the work should be stopped and the problem resolved.

The injection mold is normally described by a variety of different criteria, including the following:

* The author gratefully acknowledges the permission of John Wiley & Sons, Inc., N.Y. to use material from his book on *Injection Molding of Plastics* (1973). Mr. Rubin is associated with Robinson Plastics Corp.

[1] Readers are also referred to Chapter 22 on Mold Making and Materials for additional information on the subject.

Number of cavities
Material
  Steel
  Stainless steel
  Prehardened steel
  Hardened steel
  Beryllium copper
  Chrome plated
  Aluminum
  Epoxy-steel

Parting line
  Regular
  Irregular
  Two-plate mold
  Three-plate mold

Method of
  manufacture
  Machined
  Hobbed
  Cast
  Pressure cast
  Electroplated
  EDM
    (spark erosion)

Runner system
  Hot runner
  Insulated runner

Gating
  Edge
  Restricted
    (pin point)
  Submarine
  Sprue
  Ring
  Diaphragm
  Tab
  Flash
  Fan
  Multiple

Ejection
  Knockout pins
  Stripper ring
  Stripper plate
  Unscrewing
  Cam
  Removable insert
  Hydraulic core pull
  Pneumatic core pull

Following these criteria, a typical mold might be described as follows: A four-cavity, machined steel, chrome-plated, hot-runner, stripper-plate tumbler mold.

Mold Bases

In the early years of injection molding, mold-makers had to build their own mold bases. Standard mold bases, which were pioneered by the D-M-E Co., provide the moldmaker with a precision base at a cost that is usually lower than if the moldmaker were to manufacture the base himself. The molder can buy replacement parts, such as sprue bushings, leader pins and bushings, push-back pins and knockout pins at a lower cost and with much higher quality than if he were forced to manufacture them himself. Additionally, they are stock parts immediately available. Inasmuch as they are common for many molds they can be stocked by the molder in his plant and thus minimize down time. A detailed discussion of standard mold bases appears on p. 148.

Two-Plate Mold

Figure 6-1 shows a so-called *two-plate mold* that certainly has more than two plates. It is the common name for a mold with a single parting line. The parting line of a mold can best be defined as that surface where a mold separates to permit the ejection of plastic (either the molded parts or gates and runners). The part being made in the mold shown in Fig. 6-1 is a shallow dish, edge-gated. The temperature control channels are in both the cavity and core. This is preferable to just cooling the plate as the boundary between the cavity or core and the plate has high thermal resistance and greatly lowers the rate of heat transfer. Note also the support pillars that prevent the support plate from buckling under the pressure of the injection material and the knockout bar that is attached to the machine and is the actuator for the knockout plates. A sprue puller of the "Z" type is used.

Suppose one wanted to mold a drinking tumbler in plastic. A design for a single-cavity mold is shown in Fig. 6-2. The material is fed directly from the nozzle through its sprue gate into the tumbler. The tumbler is ejected by a stripper ring. Note the excellent cooling of the cavity which is made in two pieces. The inside part is turned with a spiral on its outer wall. It is fitted into the shell and made leakproof by brazing, or by the use of O rings. The core could be similarly made. In this illustration, the core is cooled by a baffle system which forces the water to run up and down in the center of the core. Instead of gating through the sprue bushing, a hot nozzle could be used, gating directly into the tumbler and leaving no sprue to be cut off and reground.

Three-plate Mold

Suppose for example, one had to mold several cups or similar shaped items in one mold. It would not be possible to gate them except at the top. Otherwise, venting problems would make it impossible to mold. One way to produce such parts is to use a three-plate mold.

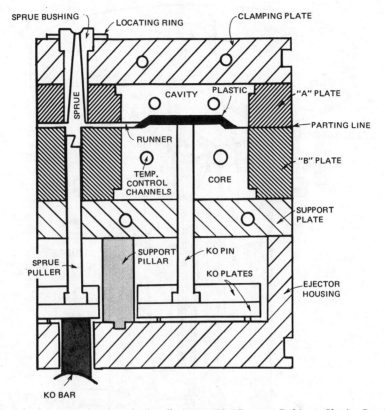

Fig. 6-1. Schematic drawing of a "two"-plate mold. (*Courtesy Robinson Plastics Corp.*)

A dome- shaped cup, as shown in Fig. 6-3, would be made in the mold shown in Fig. 6-4. This mold operates as follows. After the material has been injected and cooled, the machine opens. The latching mechanism keeps the A and B plates attached so that they move together. The initial opening of the mold is on parting line 1. The trapezoidal runner has been cut into the A plate. As the mold opens, the runner is held flush to the pin, plate C, because of the undercut pin, A. At this time, the gate is also broken and the free piece is contained between the A and B plates. This opening must be large enough to allow the operator to remove the runner system manually, if desired.

As the mold continues to open, the latch is activated, separating the mold at parting line 2. The amount of travel is controlled by the stripper bolt on the outside. The mold continues to move backward and the parts are ejected by the knockout pins. Concurrent with this, the pin plate is forced forward (parting line 3) stripping the runner system off the undercut pin. Since the runner is trapezoidal with its flat back on pin plate C, it should be free to fall by gravity. Often a pneumatically operated sweep or an air blast is used to help. Such molds can be easily run automatically.

Hot Runner Mold

In a three-plate mold the runner system must be reground and the material reused. It is possible to eliminate the runner system entirely by keeping it fluid. This is called a *hot runner mold* (Fig. 6-5). The material is kept plasticized by the hot runner manifold which is heated with electric cartridges. The block is usually thermostatically controlled. Heater bands which are individually controlled can be mounted around the nozzle. This is desirable, but not always necessary. The plastic is kept fluid and the injection pressure is transmitted through the hot runner manifold.

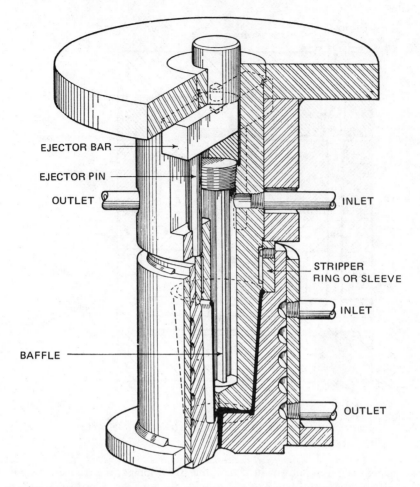

EJECTOR BAR

EJECTOR PIN

OUTLET

INLET

STRIPPER
RING OR SLEEVE

INLET

BAFFLE

OUTLET

Fig. 6-2. A single-cavity, sprue-gated mold. (*Courtesy E. I. du Pont de Nemours & Co., Inc.*)

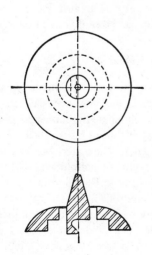

Fig. 6-3. Part to be produced in mold shown in Fig. 6-4. (*Courtesy D-M-E Co.*)

A hot runner mold presents certain difficulties. It takes considerably longer to become operational. In multi-cavity molds, balancing the gate and the flow and preventing drooling are difficult. A keen balance between non-drooling and freezing up of the nozzle must be achieved. The hot runner mold is highly susceptible to tramp metal, wood, paper, and other contaminants which quickly clog up the nozzle. Various screen systems have been devised to try to prevent this. Cleaning out a plugged hot runner mold is a long process. These molds are comparatively more expensive. The advantage is that in a long running job, it is the most economical way of molding. There is no regrinding with its cost of handling and loss of material. The mold runs automatically, eliminating the variations caused by operators.

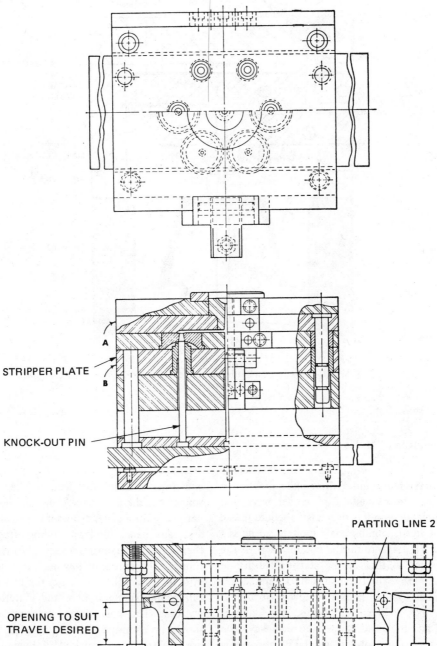

STRIPPER PLATE

KNOCK-OUT PIN

A

B

PARTING LINE 2

OPENING TO SUIT
TRAVEL DESIRED

STRIPPER BOLT

LATCH

Fig. 6-4. Design for center-gating multiple-cavity molds. (*Courtesy D-M-E Co.*)

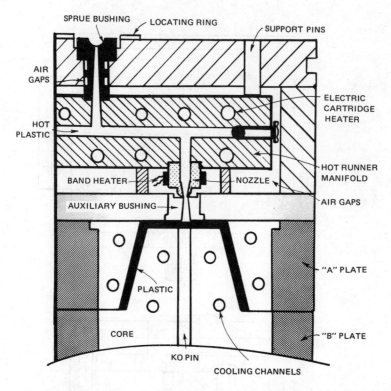

Fig. 6-5. Schematic drawing of a hot runner mold. (*Courtesy Robinson Plastics Corp.*)

Insulated Runner Molds

An insulated runner mold is a cross between a hot runner mold and a three-plate mold. If a large diameter runner, $\frac{3}{4}$ in. or over, were used in a three-plate mold, the outside would freeze and the inside would remain molten, acting as a runner. Half the runner is cut on each of the two plates so that during start-up, or if the runner freezes, the two plates are separated and the runner system quickly removed. As soon as the runner reaches equilibrium, the latch is closed and the mold is operated that way. The nozzles present problems similar to those of a hot runner mold. Usually a heated torpedo individually controlled is inserted into each gate area and kept on continually. The wattage is controlled and set pragmatically. Insulated runner molds are more difficult to start and operate than three-plate molds, but are considerably easier than a hot runner mold. There is, of course, no runner system to regrind.

Runners

There are two types of runners used: full round and trapezoidal in cross section. The full round runner has the advantage of having the smallest periphery for a given diameter or cross-sectional area, and hence the least chilling effect on the thermoplastic material as it passes through the runner. The material flowing from the runner into the gate will come from the center of the full round runner where the material is the warmest. The only time a trapezoidal runner should be employed is when it is impossible to use a full round runner, or in a three-plate mold where the flat back is essential for automatic ejection. The trapezoidal runners have a taper of about $5°$ per side. Standard cutters are available for them. Under no condition should a half round runner be considered, as a trapezoidal shape is much superior. (See also Chapter 22).

Gates

The gate is the connection between the runner system and the molded part. It must permit enough material to flow into the mold to fill out the part plus such additional material as is required to overcome the normal thermal shrinkage of the part. The location of the gate, its type and size, strongly affect the molding process and the physical properties of the molded part.

There are two types of gates—large and restricted (pin pointed). Restricted gates are usually circular in cross section and for most materials do not exceed 0.060 in. in diameter. The restricted gate is successful because the apparent viscosity of the plastic is a function of the shear rate. The faster it moves, the less viscous it becomes. As the material is forced through the small opening, its velocity increases. The shear rate is directly related to the velocity. In addition, some of the kinetic energy is transformed into heat, raising the local gate temperature and thus reducing its viscosity. A second effect is high mixing through a restricted area. It is virtually impossible to mold a good variegated pattern without going through a large gate. Dispersion or mixing nozzles on molding machines use the same principle.

The size of the restricted gate is so small that when the flow ceases, the plastic solidifies, separating the molded part from the runner system. Unless the cavity is completely filled before this happens, the part will be a reject. It becomes evident that in multi-cavity molds all parts must fill equally. If not, some cavities will receive more plastic than the others. This results in sticking, packing, and different sized parts. The only way to overcome this is by short shooting the shot and opening individual gates carefully until the parts fill equally.

An advantage of the small gate is the ease of degating. In most instances, the parts are acceptable if they are cleanly broken from the runner. Secondly, the rapidity of gate sealing when the flow ceases, tends to make it more difficult to pack the molded part at the gate. This type of stress is a prime cause for failure. Unfortunately, this type of failure might not show up for a long time after molding.

Figure 6-6 shows various types of gating. A *sprue gate* feeds directly from the nozzle of the machine into the molded part. It has the advantage of direct flow with minimum pressure loss. The disadvantages include the lack of a cold slug, the possibility of sinking around the gate, the high stress concentration around the gate area, and the need for gate removal. Most single-cavity molds are gated this way.

A *diaphragm gate* has, in addition to the sprue, a circular area which leads from the sprue to the piece that is also removed. In gating hollow tubes, flow considerations suggest the use of this type of gate. The part could be gated by a single, double or quadruple runner coming from the sprue. Four runners, 90° apart, would give four flow or weld lines on the molded part. This is often objectionable. Additionally, the diaphragm gate gets rid of stress concentration around the gate because the whole area is removed. The cleaning of a diaphragm gate is more difficult and time consuming than a sprue gate. There are instances when gating internally in a hollow tube is not practical. *Ring gates* accomplish the same purpose from the outside.

The most common gate is the *edge gate* where the part is gated either as a restricted or larger gate on some point on the edge. This is an easy method from the point of view of mold construction. For many parts it is the only practical way to gate. When parts are large, or there is a special technical reason, the edge gate can be fanned out. This is called a *fan gate*. If the flow pattern is required to orient in one direction, a *flash gate* may be used. This involves extending the fan gate the full length of the piece, but keeping it very thin, comparable to a piece of flash.

When the gate goes directly into the part, there may be jetting or other surface imperfection. To overcome this the gate might be directed against a wall or a pin. If this is not possible, it can be done artificially by extending a tab from the part into which the gate is cut. The tab will have to be removed as a secondary operation. This is called *tab gating*.

A *submarine gate* is one that goes through the steel of the cavity. When the mold opens the part sticks to the core and shears the gate at

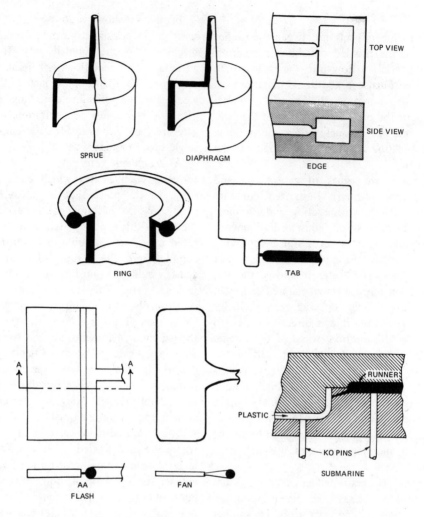

Fig. 6-6. Gating designs. (*Courtesy Robinson Plastics Corp.*)

the piece. The submarine gate works because of the flexibility of the runner and the judicious placement of the knockout pin. This is a highly desirable way of gating and is very often used in automatic molds.

See also Chapter 22.

Venting

When the plastic fills the mold, it displaces the air. Unless the air is removed quickly, several things may happen. It may ignite, causing the characteristic burn. It may compress enough to restrict the flow to the point where the cavity cannot fill or, to a lesser extent, where the rate of fill is extremely slow. To remove the air

from the cavity, slots can be milled, usually opposite the gate. These slots normally vary from 0.001 to 0.002 in. deep and from $\frac{3}{8}$ to 1 in. wide. After the vents have been extended for $\frac{1}{2}$ in. the depth is increased usually to about 0.005 in. Clearance between knockout pins and their holes also provide venting, though they usually cannot be relied upon. If more venting is needed a flat can be ground on the knockout pin. The gate location is directly related to venting. Often the gate cannot be located in certain places because it would be impossible to vent the mold or prevent air entrapment. A typical example is the impossibility of gating a tumbler on the rim. The plastic would flow around the rim preventing the air from being removed in the upper part of the tumbler. No amount of vent pins or venting devices would successfully overcome this.

Parting Line

When a mold closes, the core and cavity, or two cavities meet, producing an air space into which the plastic is injected. If one were inside this air space and looking out, the mating junction would appear as a line. It so appears on the molded piece and is called the *parting line*. A piece may have several parting lines. The selection of the parting line is largely influenced by the shape of the piece, method of fabrication, tapers, method of ejection, type of mold, esthetic considerations, post-molding operations, inserts, venting, wall thickness, number of cavities, and the location and type of gating.

A simple example follows:

Figure 6-7A shows a slab of plastic $\frac{3}{16}$ x $\frac{1}{2}$ x 1 in. With the parting line on the $\frac{1}{2}$ x 1 in. side (Fig. 6-7B), the projected area would be 0.5 sq in. Assuming a 1° per side taper the difference between the maximum and minimum dimension would be 0.006 in. Rounded edges could be as shown. This latter point could be very important, for example, if the part were used as a handle or attached to a flat surface. The cavity is relatively easy to make.

If it were parted on the $\frac{3}{16}$ x $\frac{1}{4}$ in. side (Fig. 6-7C), the projected area would be 0.093 sq in., the difference between the maximum and minimum dimensions of the sides, 0.035 in., and the cavity would be a very difficult one to polish. If the mold were parted on the 1 x $\frac{3}{16}$ in. side (Fig. 6-7D), the projected area would be 0.187 in²., and the difference in dimensions .017 in. Each piece would look different. If the 1 in. dimension was the important one, using the parting line shown in "C", would be least effective. When the material is injected into the mold, the mold has a tendency to open. Since the 1 in. dimension would be perpendicular to the parting line that tolerance would be the most difficult one to maintain. If the projected area was the limitation to the number of cavities on the mold, configuration "C" would result in more than five times the number of cavities than configuration "B". The

parting line on each one could be split down the middle, such as MM in Fig. 6-7B. This would result in three more different pieces. It can be seen, then, that by changing the parting line it is possible to obtain six different pieces, six different appearances, six different sizes, and possibly six different functions.

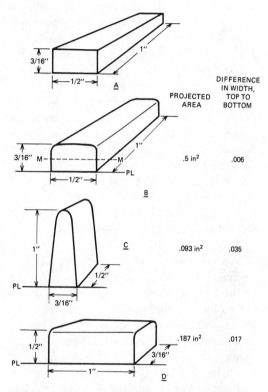

Fig. 6-7. Effect of parting line on shape of a molded part

Ejection

After a part has been molded, it must be removed from the mold. It is mechanically ejected by KO (knockout pins), KO sleeves, stripper plates, stripper rings or air, either singly or in combination. The selection of the type and location of the ejection mechanisms affect the molded part in ways similar to the selection of the parting line. The most frequent problem in new molds is with the ejection of the part, as there is no mathematical way to determine the amount of ejecting force needed. It is entirely a matter of experience. It is poor practice to build a mold unless the part will eject properly.

If enough knockouts cannot be provided on paper it is highly doubtful that it can be successfully done later when the mold is finished. Unless a mold ejects consistently, it is impossible to produce parts on an economical production basis.

The geometry of the part and the plastic material are the two most important parameters in ejection. It is desirable to have a minimum draft of 1° per side, though many parts will work with less draft. The quality of polishing is important. Many times, polishing and stoning only in the direction of ejection will solve a difficult problem. Since ejection involves overcoming the forces of adhesion between the plastic and the mold, the area of the knockout mechanism becomes important. If the area is too small, the force will be concentrated developing severe stresses on the part. In materials such as ABS and high-impact polystyrene, these forces can discolor the plastic. In other instances, it may so stress the part that it will fail immediately or in later service.

A common cause for sticking in a mold is related to the elasticity of steel and is called *packing*. When the injection pressure is applied to the molten plastic, the steel deforms. When the pressure is relieved, the steel will return to its original position, acting as a clamp on the plastic. Additionally, packing causes sticking by deeper penetration of the plastic into the pores of the mold. Very often, reducing the injection pressure, and/or the injection forward time, will eliminate the problem. Packing is also common in multi-cavity molds where the individual cavities do not fill equally. One cavity will seal off first and the material intended for that cavity will be forced into other cavities causing overfilling.

Ejector Mechanism

Ejector pins are made either from an H-11 or a nitriding steel. They have a surface hardness of 70 to 80 R_c to a depth of 0.004 to 0.007 in. The inside core is tough. The heads are forged and annealed for maximum strength. They are honed to a fine finish. They come in fractional and letter size diameters, each being available in a 0.005 oversized pin. These are used when the knockout holes in the cavity or core are worn and flash occurs around the pins. Figures 6-1, 6-4, 6-5, 6-8, and 6-9 show knockout pins in a mold. The right side of Fig. 6-8 shows the way a knockout pin (L) is mounted. The ejector plate is drilled and countersunk. The pins are held in by screwing the ejector retainer plate (J) to the ejector plate (G). All ejector pins, ejector sleeves, and return pins are located in this plate. This construction makes assembling of the mold easy since the pins can be entered one by one into the cavity plate. It is often difficult to assemble large molds with a great number of ejector pins, if the construction does not allow the pins to be inserted individually. It also makes it possible to remove one or two knockout pins without removing all of them.

There is nearly always a slight misalignment between the holes and the cavity plate and the ejector plate. Therefore, it is important to leave a clearance of from $\frac{1}{64}$ to $\frac{1}{32}$ of an inch around the heads of the pins and at least 0.002 in. clearance at K. This will permit the pins to find their proper location when the mold is assembled. Return pins should be installed rigidly if no other guide for the ejector plate is provided. In many molds, this is not necessary. If there are numerous pins or if there is a large ejector plate, it is best to use small leader pins and bushings to support the plates. Chamfers at point I are helpful for easy insertion of the pins. The holes for the ejector pins should be relieved to within a fraction of an inch of the face of the cavity or core to facilitate the alignment and operation of the pins. The area of the pins on the molded part should be as large as possible to reduce stress concentration. The top of the knockout pins will leave a circle on the molded parts.

Ejector Sleeve

Ejector sleeves (Fig. 6-8, part E) are preferred when molded parts have to be stripped off round cores. They are subjected to severe stress and wear. The inside and outside surfaces must be hard and finely polished. If they are not sufficiently hard, scoring of both cavity and core may take place. Additionally, both cavity

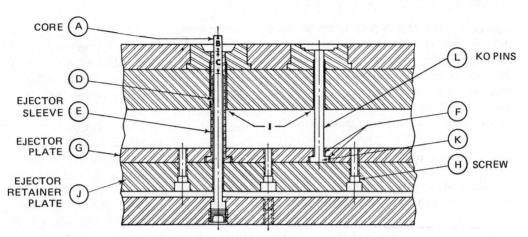

Fig. 6-8. Ejector (KO) pins, ejector plates, and stripper pin and sleeve. (*Courtesy D-M-E Co.*)

and core must be of a different hardness compared to the ejector sleeve. Two parts of equal hardness, regardless of how hard they are, will scour. The lower portion of the sleeve should be drawn to obtain maximum toughness, while the upper part should be left hard for the full length of the ejector movement.

The outside diameter of the sleeve should be about 0.001 to 0.002 in. smaller than the hole in the cavity. An equal clearance between the core and the sleeve should be maintained for a distance C. The inside diameter of the sleeve should be about $\frac{1}{64}$ to $\frac{1}{32}$ in. larger than the core, leaving a clearance as shown in D. The core (A) should be dimensioned so that the portion that extends into the molded article (distance B) is at least $\frac{1}{64}$ in. smaller in diameter than the lower part. If this is not done, the reciprocating movement of the sleeve will damage the fine finish of the core. Distance C should be at least $\frac{3}{8}$ in. longer than the entire movement of the ejector plate. If the clearance extends too far, the shoulder and end of the core may be damaged when the sleeve is retracted. It is also important to leave a clearance of $\frac{1}{64}$ of an inch around the outside of the sleeve. This clearance, however, should not extend too far since it is necessary to have a bearing at least $\frac{1}{2}$ in. long at the cavity.

Stripper Ring

The ejector sleeve requires a core being attached beyond the ejector plates, usually to the

back-up plate. In many instances a stripper ring can be used (Fig. 6-9). The core can then be attached to the core retaining plate. A ring is inserted which is directly attached by stripper bolts to the knockout plates. Another example of a stripper ring is shown in Fig. 6-2.

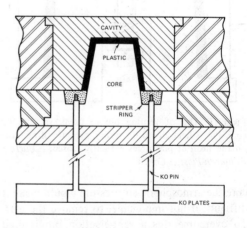

Fig. 6-9. Stripper ring type of ejection

Stripper Plates

If instead of mounting the stripper rings directly to the knockout plate, the stripper rings were mounted on their own plate and that plate mounted to the knockout plate, a stripper plate mold (Fig. 6-4) would result. Hardened inserts are usually put around the core and inserted into the stripper plate to prevent wear and for ease of maintenance. Stripper plate molds give a larger surface area for ejection than knockout pins. Heretofore, they have primarily been used for round cores. With the use of EDM equipment, irregularly shaped stripper plate molds are economically feasible.

Venting-Ejection Pins

Consider molding a deep container, as shown in Fig. 6-10. After the cavity is separated from the

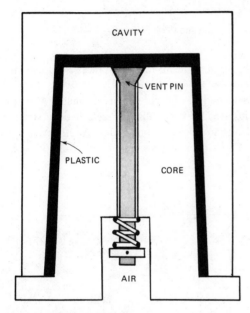

Fig. 6-10. Use of venting pin to break vacuum on core. Pin is held closed by spring; pressure of material on the head of pin forces it tightly closed. (*Courtesy Robinson Plastics Corp.*)

core the atmospheric air pressure would make it difficult, if not impossible, to remove the part. To overcome this a vent-ejection pin is used which is held in its normally closed position by the spring. When the material is injected the pressure of the material on the head of the pin forces it tightly closed. When the part is to be ejected the pin will move up when the knock-out system is activated, venting the interface between the core and the plastic. Additionally, air can be used to help blast the part off. In this instance, the combination of the force and plastic part act like an air cylinder. Often this provides enough ejection force.

Cam Actions

A cam acting mold is used to overcome the effect of an undercut, which is an interference by the mold to delay or prevent mechanical ejection of the part. There are two types of cam action—one moves the cams independent of the machine action, and the other moves the cam by the molding machine action.

Figure 6-11 shows a schematic section of a cam for making a hole in the side of a plastic box. The cam pin is attached to the cam slide which moves on wear plates. The slide is activated by either an air or hydraulic cylinder. The locking device is not shown, though it may not be required in this type of cam. This cylinder can be activated at any time in the cycle. If the plastic part is to stick to the core, the cam pin must be retracted before the mold opens. This would be impossible in a mold where the cam action is worked by the machine opening. Figure 6-12 shows the standard cam action most commonly used. The cam slide is attached to Plate A, which is the moving side of the mold. The cam pin and the cam lock are attached to the stationary side of the mold. In the mold closed position (A) the cam pin is held in place by the cam lock. As the mold opens (B) the cam slides up on the cam pin moving toward the right. The amount of the cam slide movement will depend on the cam angle and how far the slide moves on the cam pins. By the time the mold is opened the cam slide has moved enough so that the cam pin has been fully withdrawn from the plastic part. Following that, the part is ready for normal ejection. There are other ways of camming mechanically, but they all work on the same principle as that of an angle pin.

External threads can be molded readily by using cam pins and slides to separate the cavities themselves. Such movement of the cavity block is a common form of cammed mold.

Molding internal threads or external threads when cams are not used requires unscrewing devices. The usual methods include rack and pinion where the rack is moved by the machine action, an hydraulic cylinder, a pneumatic cylinder, or an electric motor. Unscrewing can also be obtained by the use of an individual

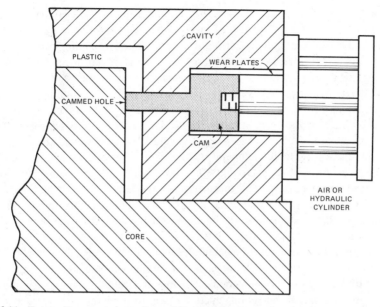

Fig. 6-11. Externally operated cam for molding hole in plastic. (Locking device not shown)

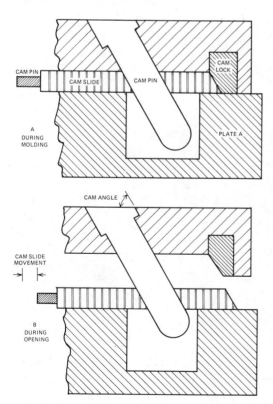

Fig. 6-12. Mechanical cam action using cam pins

hydraulic motor for each core, or by a gear and chain mechanism, motor driven. Another way to mold internal threads is to use a collapsible core.[2]

Before deciding on cammed or unscrewing molds one might ask if they are essential. Sometimes a hole can be drilled or a slot machined after molding. Often it can be done at the machine at no additional cost. Even if the work has to be done away from the machine it might be more economical and better to do it on short or even long running jobs. Sometimes assembling two parts can avoid cam action. Other times threaded inserts or tapping will eliminate automatic unscrewing molds. The part should be carefully reviewed at this stage of development.

Mold Cooling

The mold, in effect, is the containment for the molten plastic and the vehicle through which heat is removed from the plastic to solidify it to the point of ejection. It is obvious that proper selection of mold cooling channels is essential for its economic operation. Mold temperature control consists of regulating the temperature and flow rate of the cooling fluid. This is done

[2] Patented by D-M-E Co.

externally and is not discussed here. The functioning of the mold and the quality of the part depend in large measure on the location of the cooling channel. It is difficult to have "too much" cooling, because the amount is always regulated by the temperature and velocity of the circulating cooling fluid. If there are "too many" cooling channels they can always be blocked off and not used.

The second factor in mold temperature control is the selection of the material itself. Beryllium copper molds have a high thermal conductivity, approximately two times that of steel and four times that of stainless steel. Since the time required for cooling plastic in the mold is a function of heat removal, a beryllium copper cavity should cool about four times as rapidly as that of a comparable one in stainless steel. This does not mean that the cooling cycle for beryllium will be one-fourth of that of stainless steel. However, all other things being equal, a beryllium mold will run significantly faster than one in stainless steel.

Designing of cooling channels in molds will now be discussed. Cooling channels should be in the cavities and cores themselves, rather than only in the retaining plate. The rate of heat transfer is lowered tremendously by the interface of the two pieces of metals, no matter how well they fit. The cooling channels should be as close as possible to the surface of the plastic (taking into account the strength of the mold material), and should be evenly spaced to prevent uneven temperature on the surface of the mold. For practical purposes it can be assumed that the heat is removed in a linear fashion, i.e., equal distances from a cooling channel will be the same temperature.

The simplest type of cooling is to drill channels through the cavity or core. The minimum size should be $\frac{1}{4}$ in. pipe, with $\frac{3}{8}$ in. pipe preferable; $\frac{1}{8}$ in. pipe should never be used unless its use is dictated by the size limitations of the mold.

Figure 6-13 shows one of the simplest types of coring commonly used in injection molds for rectangular shapes, such as boxes and radio cabinets. It consists of a number of channels

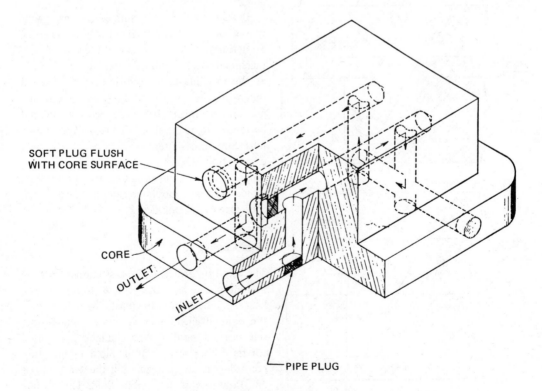

Fig. 6-13. Simple type of coring. (*Courtesy Consolidated Molded Products Corp.*)

drilled near the top of the core (the number depending on the size of the core), joined by several risers to circulate the cooling medium. The horizontal channels are drilled from one end of the core, and tapped, and counterbored to accommodate a soft plug, which should be a light drive fit, machined flush with the core surface, and then polished. Usually a con-

tinuous flow is used through the core, and should it be necessary to cross over channels already drilled, the flow of the fluid can be directed by using blank plugs which are a drive fit in the channels.

Figure 6-14 shows the construction of a more elaborate cooling system for use where uniformity of cooling is important. It involves

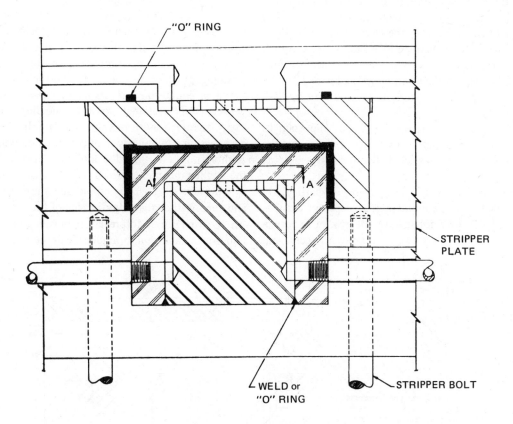

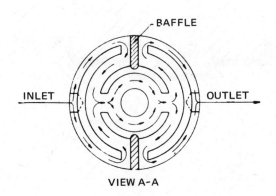

Fig. 6-14. Mold construction showing a more elaborate cooling system. (*Courtesy Consolidated Molded Products Corp.*)

the use of channeled inserts within the cavity block and core block.

The cavity block and core block are machined out to hold the inserts, circular (as shown) or rectangular, before the cavity and core are finished, hardened, and polished. The inserts are made to fit the spaces thus provided.

Grooves for the channels are machined in the face of the inserts in circular or rectangular pattern, and are then interconnected and baffled to provide the desired route of flow. The baffles are dimensioned to make a drive fit.

The inserts, not hardened, are fastened into place within the cavity block and core block by brazing or welding along the joint, or by "O" rings.

In designing the grooves, it is important to leave sufficient area unmachined to support the cavity or core against distortion by molding pressure.

This method of channeling is particularly adapted to the molding of transparent plastics, since it does not involve the risk of imperfections from plugs in the molding surfaces, and also since the uniformity of cooling that it provides tends to minimize optical distortion in the molded article.

When a mold is channeled in this way, there may be difficulty in finding place for ejector pins, and hence a stripper plate may be required.

Figure 6-15 shows a method of channeling a cavity or core, using "O" rings to seal off the cooling medium or confine it to each individual cavity or core in the chase. The underside of the cavity or core may have a pattern of grooves similar to that of the channeled insert in Fig. 6-14, or be simply as shown. The retaining shoulder on the cavity and the corresponding counterbore recess in the chase must be accurately matched, to insure holding the cavity tightly against the bottom plate. In the core section a bubbler type of cooling is shown. A short nipple is threaded into the chase and the cooling medium is forced through a channel in the frame through the nipple into the force, and then is discharged out the side of the force. An "O" ring is used to prevent the cooling medium from escaping on the underside of the cavity.

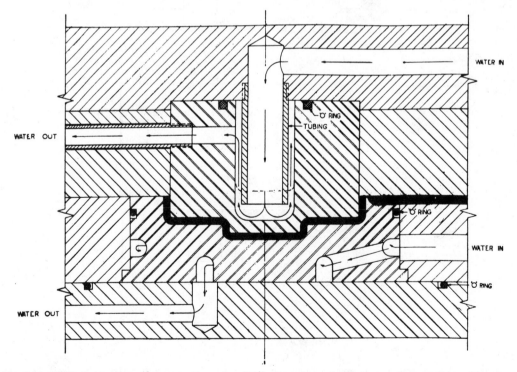

Fig. 6-15. Method of channeling a cavity using "O" rings to seal off the cooling medium. (*Courtesy Consolidated Molded Products Corp.*)

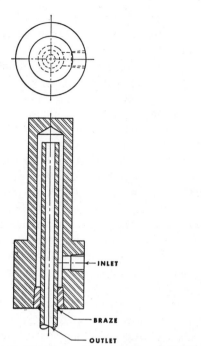

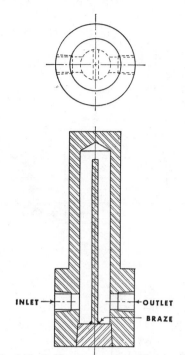

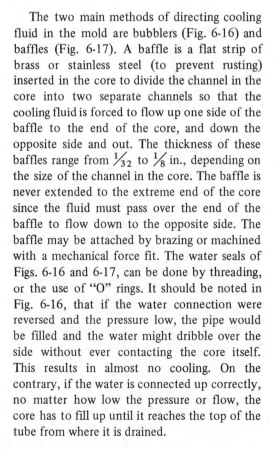

Fig. 6-16. Bubbler-type cooling. (*Courtesy D-M-E Co.*)

Fig. 6-17. Baffler-type cooling. (*Courtesy D-M-E Co.*)

The two main methods of directing cooling fluid in the mold are bubblers (Fig. 6-16) and baffles (Fig. 6-17). A baffle is a flat strip of brass or stainless steel (to prevent rusting) inserted in the core to divide the channel in the core into two separate channels so that the cooling fluid is forced to flow up one side of the baffle to the end of the core, and down the opposite side and out. The thickness of these baffles range from $\frac{1}{32}$ to $\frac{1}{8}$ in., depending on the size of the channel in the core. The baffle is never extended to the extreme end of the core since the fluid must pass over the end of the baffle to flow down to the opposite side. The baffle may be attached by brazing or machined with a mechanical force fit. The water seals of Figs. 6-16 and 6-17, can be done by threading, or the use of "O" rings. It should be noted in Fig. 6-16, that if the water connection were reversed and the pressure low, the pipe would be filled and the water might dribble over the side without ever contacting the core itself. This results in almost no cooling. On the contrary, if the water is connected up correctly, no matter how low the pressure or flow, the core has to fill up until it reaches the top of the tube from where it is drained.

Cooling with bubblers is excellent if they are not hooked up in series. Consider ten pins with baffles hooked up in series (Fig. 6-18). The first pin will be cooler than the last pin. The resistance to flow of each pin is in series, thus reducing the amount and velocity of the cooling fluid flowing through the system. This will markedly decrease the cooling capacity. Additionally, if there is any blockage because of dirt, or for any other reason in one pin, the whole unit is inoperative.

Figure 6-19 shows the correct way to cool, using parallel cooling. This can be done with

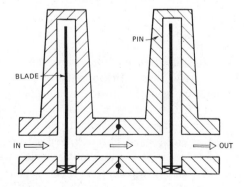

Fig. 6-18. Series cooling of pins, poor design. (*Courtesy Robinson Plastics Corp.*)

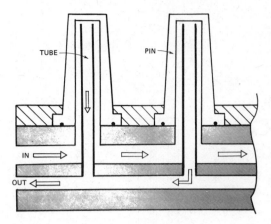

TUBE

PIN

IN

OUT

Fig. 6-19. Parallel cooling of pins, correct design. (*Courtesy Robinson Plastics Corp.*)

pins, or any other series of bubbler action in a core or cavity. This type of cooling will give significantly faster molding compared to that shown in Fig. 6-18.

The excellence of the cooling system depends on the ingenuity and ability of the mold designer and the moldmaker. There is no area in moldmaking that affects the cycle and quality of the molded part more than the selection and location of the cooling channel.

Moldmaking

The various techniques for manufacturing molds, as well as a basic guide in ordering molds, and a discussion of materials for molds can be found in Chapter 29.

STANDARD MOLD BASES*

History

The standardization of mold bases, unknown prior to 1940, was an important factor in the development of more efficient moldmaking. In 1940, I. T. Quarnstrom, a custom moldmaker in Detroit, recognized the lack of, and the need for, commercial standardization, and set upon an investigation of his own to determine the potential, and to what extent, molds could be

* John Andras, Product Engineering Manager, D-M-E Co., Madison Heights, Mich.

standardized. The results of this research was the introduction of standard mold bases for injection molds in 1942.

By standardizing on the mold assembly and the component parts, especially the ones that come in contact with the press platens such as the locating ring, sprue bushing and ejector plates and housing, the mold can be made interchangeable between various makes of injection molding machines. All of these component parts, together with various other parts such as leader pins and bushings, return pins, sprue puller pins, and ejector pins were made available as stock items for replacement to the moldmaker or molder. This standard mold base assembly for injection molds could also be rearranged for use in compression or transfer molds.

From a modest beginning of two standard mold base sizes being available in 1942, the program of standardization has progressed so that today it includes 54 standard mold bases in a range from $3\frac{1}{2} \times 3\frac{3}{4}$ in. to $23\frac{3}{4} \times 35\frac{1}{2}$ in. In many of these sizes there is a selection of cavity plates in thicknesses from $\frac{7}{8}$ to $5\frac{7}{8}$ in., in increments of $\frac{1}{2}$ in., with a choice of three steels, giving a possibility of more than 10,000 combinations. A choice is also available in ejector housing heights from $2\frac{1}{2}$ to 5 in. to allow proper ejector plate travel for various knockout requirements of molded parts. The more commonly required complete standard mold base assemblies are carried in stock for immediate delivery to the moldmaker.

All mold components incorporated in the mold base assemblies are also available as separate items so that the moldmaker can make up his own special base if the molded part and mold requirements indicate a deviation from the standard. These standard components include all plate items in addition to the leader pins, bushings, locating rings, sprue bushings, sprue puller pins and return pins. The plates are pre-finished, i.e., they are finish ground on all sides to extremely close tolerances, allowing the mold to be built to exacting requirements. The plates are available in medium carbon steel, in AISI 4130 type holder block quality steel, pre-heat treated to approximately 300 Bhn, and in AISI 4135 (modified), P20 type cavity steel, also pre-heat treated to approximately 300 Bhn.

Standard Designs

In general, a standard injection mold base consists of a stationary or upper half, and an ejector or movable lower half of the mold assembly. The upper half is made up of a top clamp plate and the upper cavity plate. This upper half contains a locating ring for positioning to the press platen, and the sprue bushing, which has an orifice through which the plastic material enters the mold under heat and pressure. The locating rings and sprue bushings are supplied in a variety of types and sizes to suit various presses. Leader pins are also contained in the upper cavity plate.

The ejector or lower half of the mold assembly is composed first of all of the lower cavity plate, then a support plate, which, as its name implies, "supports" or backs up the lower cavity plate, and an ejector housing onto which have been mounted the lower cavity and support plates. The ejector housing contains the lower clamp slots. The lower cavity plate carries hardened bushings in alignment with the upper cavity plate leader pins to give proper register when the mold opens and closes. The ejector housing incorporates within it the ejector plate and ejector retainer plate between which are held the ejector pins which eject the molded part from the mold. The return pins and sprue puller pin, standard parts of the assembly, are also held between these two plates. As their names imply, their function is to return the ejector plate as the mold closes, and to pull or "break" the sprue at mold opening. The sprue puller pin is usually located in the center of the lower half of the mold, and the return pins are located near the corners of the ejector plate to allow as much area as possible for the placement of cavities, cores, and ejector pins. Ejector plates are available in different lengths to accommodate knockout patterns of various presses.

An exploded view of the components of a standard injection mold base assembly is shown in Fig. 6-20. Improvements, incorporated subsequent to the original design and shown in Fig.

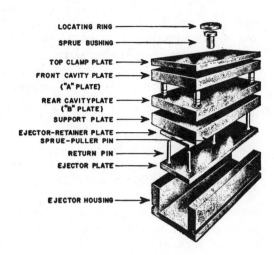

Fig. 6-20. Exploded view of a standard mold base, showing the component parts. (*Courtesy D-M-E Co.*)

6-21, include the use of tubular dowels (2) which carry the assembly screws, thus minimizing interference with the placement of water lines or core slides; clamp slots (1) incorporated within the top clamp plate, and within the ejector housing, eliminating the plate "extensions" which enabled presses of a more limited platen space to accommodate molds of larger cavity area; welded stop pins (3) on the underside of the ejector plate, replacing the "pilot" type which was subject to possible loosening and interference with ejection; and a one-piece ejector housing (4) which simplifies and strengthens the assembly.

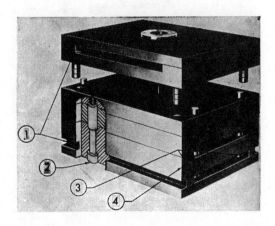

Fig. 6-21. Improvements in design of standard mold bases. (*Courtesy D-M-E Co.*)

Cavities generally can be mounted in any one of four ways: (1) surface mounted on the cavity plates (or cavity retainer plates as they are often called); (2) inserted in blind pockets milled part way through the cavity retainer plates; (3) inserted in holes cut completely through the cavity retainer plates; (4) cut directly into the cavity plates. In this latter case, it is important for the moldmaker to use plates made of a cavity steel, such as the P-20 pre-hardened type indicated earlier.

Another standard type mold base assembly, specifically designed for compression molding, is shown in Fig. 6-22. This assembly includes upper and lower cavity plates, to both of which are attached support plates and spacer clamping blocks, with ejector plates installed on both halves to allow ejection from either side. It is also possible, with minor changes, to sometimes adapt a standard injection mold base for use as a plunger-type transfer mold (Fig. 6-23). A hardened bushing of proper size is substituted for the sprue bushing, a hardened runner plate is added, together with a large-diameter support pillar located on center, and the "standard" part of the transfer mold base assembly is then complete.

Standard Specific Applications

In the general area of standard injection mold bases, since many new injection molding machines are continually appearing on the market,

Fig. 6-22. Standard mold-base assembly designed specifically for compression molding. (*Courtesy D-M-E Co.*)

especially appropriate mold bases have been designed for some of these specific presses and are available as standard mold bases. Much has been done to make these bases applicable to several makes of injection molding machines, each mold base containing, in composite, all of the special features required for each press, thus giving extreme versatility.

This concept of versatility has recently been broadened into what is known as an *adapter plate, universal mold base system*. This system consists of two specially designed plates which are fastened with screws to the platens of a selected group of various makes and sizes of presses (from the rear of the platens). Onto these adapter plates are then mounted the specially designed but standardized universal mold bases. Each mold half easily slips over dowel pins provided in the adapter plates, giving immediate, accurate alignment, and requiring only fastening by screws from the parting line, a very simple and quick operation. This arrangement is shown in Fig. 6-24. Removing a mold and installing another usually takes only 5 to 10 min.

This idea of quick installation and removal of a mold from the press was incorporated several years ago in another standardized unit called a *unit die*. This is a mold base designed so that the upper and lower cavity plates, support plates and ejector plate assembly including the ejector pins, can be quickly removed from the press, leaving the main mold or holder still clamped in position. A replacement unit assembly for molding a different part can then be mounted back into the holder, being fastened down with only two screws from each end for each mold half. Rapid and accurate alignment is obtained by vertical and horizontal keys in the holder assembly registering with corresponding keyways in the replaceable units, in addition to the conventional leader pins and bushings, both in the holder and in the replacement units. This unit die holder assembly with two replaceable units is shown in Fig. 6-25.

Standard Large Plate Components

While the present range of complete, standard mold bases now available (from $3\frac{1}{2} \times 3\frac{3}{4}$ in.

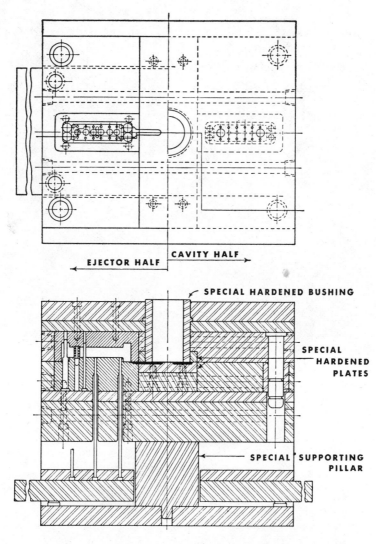

Fig. 6-23. Conventional standard mold-base assembly used in plunger-type transfer molding. (*Courtesy D-M-E Co.*)

to $23\frac{3}{4}$ x $35\frac{1}{2}$ in.) covers most molding requirements, a sizable segment of molding operations, especially in the furniture, automotive, sports equipment, and shipping container fields, requires much larger molds. To meet this need, an entirely new and much larger range of standard, pre-finished mold plates and mold build-up components has been introduced to the tooling industry. These plates are available in widths of $26\frac{3}{4}$, and then in 4 in. increments from $29\frac{3}{4}$ to $45\frac{3}{4}$ in., the lengths being 30 to 66 in. by 6 in. increments. Thick-

nesses are $1\frac{3}{8}$, $1\frac{7}{8}$, $2\frac{7}{8}$, $3\frac{7}{8}$, $4\frac{7}{8}$ and $5\frac{7}{8}$ in. They are available in the same three types of steels as standard mold bases.

Since in these larger size molds, the cavity and core (lower cavity side) are often machined in the solid, the plates have not been made into mold bases; the mold designer and moldmaker select from the wide variety the size combinations that fit the various individual requirements, and make up the assembly accordingly. To enable the moldmaker to start immediately on his machining or duplicating work, the

Fig. 6-24. Standard universal adapter plate mold base system shown mounted in press. (*Courtesy D-M-E Co.*)

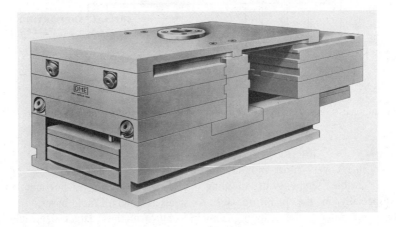

Fig. 6-25. Standard two-cavity unit die showing a standard replacement unit in a semi-removed position

plates are furnished with a ground finish on the two faces, with the four sides milled. To complete the plate line, a wide selection of ejector plates, ejector retainer plates, clamp plates and spacer blocks (or risers) are also available. Within the range covered, virtually any mold size or combination requirement can be met from the standard sizes available. The complete mold can be built up from the sizes listed.

Standard Cavity Inserts

As a further aid to mold building, standard, rectangular cavity insert blocks are also available as stock items. These come in sizes from 3 x 3 in. to 6 x 8 in., and in thicknesses from $\frac{7}{8}$ to $4\frac{7}{8}$ in. In addition, round cavity inserts are also available from stock in 1 to 4 in. diameters in $\frac{1}{2}$ in. increments, and in thicknesses from $\frac{7}{8}$ to $3\frac{7}{8}$ in. in $\frac{1}{2}$ in. increments. All of the cavity inserts have a ground finish on all sides and surfaces, ready for immediate cavity cutting. The rectangular inserts come in P-20 type steel, H-13 type steel and T-420 type stainless steel. The rounds come in P-20 and H-13 type steels. These rectangular and round inserts are shown in Fig. 6-26 and Fig. 6-27.

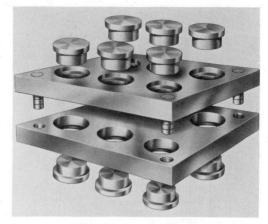

Fig. 6-27. Standard cavity insert rounds showing bored holes in upper and lower cavity plates to receive inserts

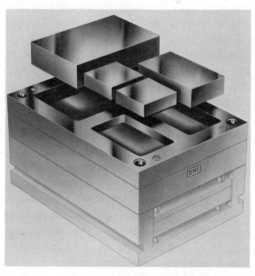

Fig. 6-26. Standard rectangular cavity insert blocks shown with pockets machined in standard mold base

Standard Collapsible Cores

A relatively recent tooling innovation for the injection molding of internal threads and undercuts was the introduction of a standard collapsible core. The unit consists of a circular, segmented core or sleeve into which the internal threads or configuration is ground. When a center core pin, which backs up and holds the segments in molding position, is retracted, with normal ejector plate travel, the segments collapse inwardly, due to inherent flexibility, pulling themselves away from the

molded undercuts and allowing the part to be ejected. The segments are alternately wide and narrow, and the narrow segments collapse farther inwardly, thus "making room" for the wider segments to collapse also, freeing themselves from the molded undercuts. The unit can be seen in Fig. 6-28. A third member, a positive collapse bushing, is not shown to allow greater clarity of collapsible core detail to be seen.

Fig. 6-28. Standard collapsible core in collapsed position

This unit, besides molding internal threads, can be used to mold parts which, until recently, had been considered impossible to mold. Some of these are internal protrusions, dimples, grooves for cap liners, interrupted threads, and window cut-outs completely around the part. These units eliminate complex unscrewing devices, simplify mold construction, and often reduce molding cycles.

A schematic, partial assembly is presented in Fig. 6-29, showing the collapsible core molding an internally threaded cap. The threaded configuration ground into the top of the segments

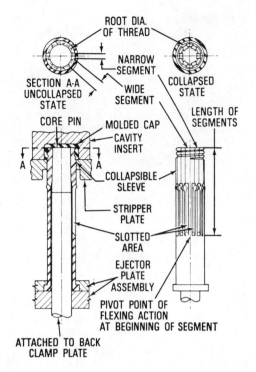

Fig. 6-29. Standard collapsible core shown in schematic partial mold assembly

can be noted in the open view at the right. The root diameter of the threads can be seen as lying within the collapsible core sleeve (indicated by dotted circle) in the uncollapsed state at upper left. However, in the collapsed state at upper right, after ejector plate travel "retracts" the center core pin, the root diameter of the threads is seen as lying outside (indicated by solid circle) the collapsed segments of the core, the "dove-tailing" of the wide and narrow segments at collapse being visible. The part can now be ejected.

As can be noted, the collapsible core in the mold is installed in the ejector plate assembly, and comes through the support plates (not shown) and the stripper plate, into the cavity insert area to mold the inside of the cap. The cener core pin is attached to the back clamp plate. In molding sequence, after the mold opens, the ejector plate and stripper plate come forward together, during which time the segments are collapsing away from the internal threads (since the center core pin retracts), and at the end of the ejector plate travel, the

stripper plate is actuated to lift the molded part off of the collapsed core.

These units can be used singly or in multi-cavity molds, and are available in six sizes, from 1 in. diameter approximately to $3\frac{1}{2}$ in. diameter, with corresponding collapses per side of approximately 0.043 in. for the smallest and 0.140 in. for the largest, at the top of the core. These collapses reduce by 0.020 in./in. along segment length within usable molding length.

Standard Early Ejector Return Unit

Another standard mold base accessory which has found wide application in recent years is a standard early ejector return unit. In mold construction, whenever a cam slide passes over an ejector pin, the ejector plate must be returned "early," or before the mold is closed; otherwise there will be interference between the slide and ejector pin with mold damage probably resulting. To preclude this possibility, the standard early ejector return unit was developed, as shown in Fig. 6-30. The unit consists of a bushing, a post with slidable cam fingers, and a cam actuating pin. The bushing is installed in the lower cavity plate, the post is attached to the ejector plate, and the cam actuating pin is installed in the upper cavity plate.

In operation, the early returning of the ejector plate is accomplished while the press is closing by the cam actuating pin pushing against the cams on the post, (and thus returning the ejector plate), until the cams release into a matching countersink, when the ejector plate is fully back. The cam pin then passes on through, and the mold continues to close. Timing can be arranged to meet the requirements of the mold merely by changing the length of pin. This type of standard early ejector return unit is specificially designed for standard injection mold bases. Installation is entirely in-board of the mold, eliminating possible interference with waterline connections. For very small molds, two units may be adequate if space is limited.

On most medium-sized standard molds four units are generally used. For the larger size standard mold bases, six or eight units may be advisable. These standard units are available as stock items.

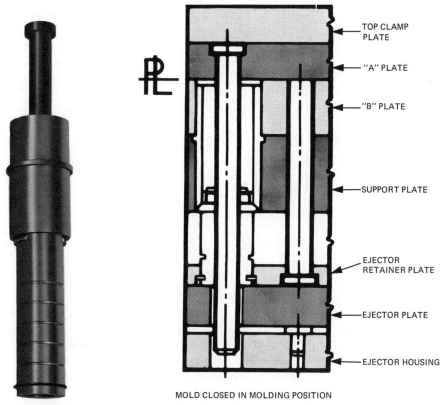

Fig. 6-30. Standard early ejector return units showing schematic assembly in mold

Standard Ejector Pin Locking System

Another aid in the standardization of mold construction is an easily applied standard system of locking ejector pins. When molded parts are contoured, ejector pins must be locked to prevent rotation. This often was done by milling slots adjacent to the ejector pin head, grinding or milling a flat on the ejector pin head, and installing a rectangular key. While this is an effective method, time and expense can be saved by using the standard system shown in Fig. 6-31. Instead of milling a slot, two shallow holes are drilled near the head of the pin located by means of a simple $1\frac{1}{4}$ in. x 3 in. available locating fixture. A right-angle key pin is then dropped into the slotted head of the pin and into the smaller of the drilled holes. This system is also used for core pins which mold non-symmetrical holes or openings. These standard slotted-head pins and accessories are available from stock in all standard sizes of ejector pins and core pins.

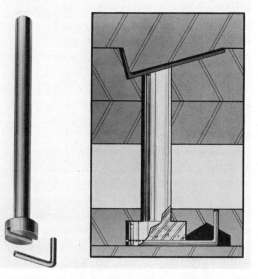

Fig. 6-31. Standard ejector pin locking system showing slotted-head pin and schematic partial mold assembly

7
EXTRUSION

As one of the basic processing techniques, extrusion—with its relatively low cost of melting, pressuring, delivering, and forming resins—dominates in the manufacture of continuous shapes, such as rods, tubing, film, sheet, filaments, and pipe (Fig. 7-1).

An over-simplified approach to extrusion might be to compare the technique to a household meat grinder or to squeezing toothpaste out of a tube. In the single screw machine, for example, which is the most common and most popular type of extruder (see Fig. 7-2), the machine takes the plastics material (in granular or powdered form), conveys it by the action of the screw, and squeezes and heats it to form a molten stream. As it does this, it develops pressure on the material, like a pump, and forces it through a die in the shape of the die opening (i.e., a round die opening produces pipe, a square die opening produces a square profile, etc.).

In addition to the production of continuous shapes (like the film, sheet, rods, tubing, pipe, and filaments described above), extruders are also used to apply insulation and jacketing to wire and cable or to coat substrates such as paper, foil, or cloth. In the blow molding process (see Chapter 13), an extruder can be used to form a molten tube (i.e., a parison) that is subsequently pinched between the two halves of the blow mold and expanded by air to form the finished product. The extruder is also integral to converting plastics into the pellet shape most commonly used in processing. In this type of operation, the die is replaced with specialized equipment such as a die plate/cutter assembly. As such, the extruder also functions as a device for compounding plastics (i.e., adding ingredients such as colorants to a resin mix). This is discussed in greater detail in Chapter 29 on Compounding and Materials Handling.

And finally, as is evident from Chapter 4 on Injection Molding, an extrusion-type screw is used to provide the melt for various injection molding processes.

EXTRUDER TYPES AND DESIGN

As already indicated, the single screw machine dominates the extrusion field and will be the focus of discussion in this chapter. It should be noted, however, that other types of extruders are available and in use. Instead of a single screw, for example, it is possible to use two or four screws. These so-called multi-screw extruders are discussed on page 199. They are often used for achieving improved dispersing and mixing as in the compounding of additives. Other claims for multi-screw extruders, most often of twin screw design, include high conveying capacity at low screw speed, positive and controlled pumping rate over a wide range of temperatures and coefficient of frictions, low

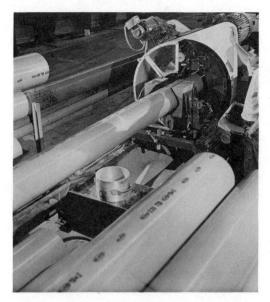

Fig. 7-1. Extruding vinyl pipe. (*Courtesy Goodyear*)

Ram extruders, in which a ram is used (instead of a screw) to build up pressure and to force the molten plastics into or through the desired shaping device are still in use today, but to a limited extent. Typical applications for ram extruders are in handling some of the extremely tough thermoplastics like ultra-high molecular weight polyethylene.

Ram extrusion techniques are also used to extrude some of the hard-to-process fluoroplastics; when so used, the technique has been described as "continuous compression molding and sintering within a tube." Lower-molecular weight fluoroplastics (like some grades of TFE, for example), however, can be mixed with a solvent and more conventionally extruded into thin-wall tubes or used to coat wire.

Ram-type continuous flow extruders have also found some use in extruding thermoset materials (e.g., phenolics, melamines, ureas, etc.), although generally, one does not consider the thermosets for extrusion. The discussion in this chapter applies principally to extruding thermoplastics.

A newer type of extruder is the so-called *elastic melt* or *screwless extruder.* This type utilizes the elastic energy storage characteristics of polymer melts to achieve mixing and extrusion. In essence, extruder pumping and extrusion results from imparting elastic (or recoverable) deformation to the plastic melt and then

frictional heat generation which permits low temperature operation, low contact time in the extruder, relatively low motor power requirements, and the ability to feed normally difficult feeding materials like powders. The twin screw extruders have, due to their low temperature extrusion characteristics, found increasingly wide usage in PVC extrusion. One such application is the extrusion of large diameter PVC pipe.

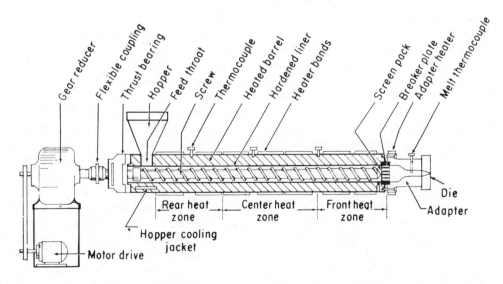

Fig. 7-2. Basic elements of a single screw extruder. (*Courtesy E. I. du Pont de Nemours & Co., Inc.*)

using this recoverable deformation to extrude the melt. In operation, this means shearing the plastic between two rotary discs to change it from a solid to a semi-liquid melt. To begin, plastic pellets or flake are introduced into a gap between a stationary plate and a rotating disc. As a result of a centripetal pumping action, the plastic is conveyed inward toward the axis of rotation and extruded out of the die opening.

However, even though these other types of machines are in use, it is the single screw extruder that is still the heart of today's extrusion industry. Basically, the operation of the screw extruder as shown in Fig. 7-2 is as follows.

Single Screw Extruder Operation

The plastics pellets (or powder) are fed from a feed hopper through the feed throat into the channel of a screw. The screw rotates in a barrel (which is heated or cooled to adjust or hold a particular melt temperature and serves almost as a pressure vessel to contain the operation) and, in rotating, conveys the plastic forward for melting and delivery. A drive motor provides the locomotion for the controlled rotation of the screw. The thrust of the screw is taken up by a thrust bearing.

Heat is generally applied to the barrel by electrical heaters, whose temperature is measured by thermocouples. As the material moves along the screw, it is melted and forced through a breakerplate which often carries a screen pack. The function of the plate and screen is to reduce rotary motion of the melt and remove large particles. In some extruders, an extrusion valve is used to regulate the operating pressure of the extruders.

Finally, the melt passes through an adaptor and into the die which dictates the shape of the final extrudate.

Usually, the extruder must be followed by some kind of a cooling system to remove heat sufficiently to solidify the final product. This can be as simple as using air to cool the part or it can involve water or cooled roll contact to accelerate the cooling process.

A typical technique is to draw the extrusion directly into a long water bath, keeping it submerged so that the water can cool the section. Small plates having the shape of the finished extrusion are often used to hold the plastic as it passes through the water. These are called *sizing plates*. Depending on the shape of the part, other means may be used, such as rollers, adjustable fingers, or blocks of metal which stand against the section. The water bath temperature may vary from cold to boiling hot, depending on material and section. Also, the bath may be made hotter nearer the die where the plastic enters and colder where the plastic leaves the bath.

Air cooling follows much the same procedure as water cooling, except that streams of air are blown against the extruded section. Again, sizing plates, fingers, rollers or metal are used to hold the section while it is moving and cooling.

Another method used to make hollow sections, such as tubes and pipe, is the vacuum box. In this system the extrudate is strung through a large metal box partly filled with water. At the entrance end there will be sizing rings or shaped fixtures. When the lid of the box is closed, a vacuum is induced inside the box. The air inside the section can now push walls of the section out against the sizing rings and thus hold the section firmly while it is cooled and set.

Near the end of the extrusion line, there usually is some sort of a powered device, called a *take-off*, which functions in gripping the extrusion and pulling it through the cooling phase. A number of different devices are available. The simplest might be a pair of pinch rolls much like a wringer on a washing machine. If the pinch rolls are replaced by a pair of opposed belts, we have a *caterpillar take-off*. Sometimes a single long belt or conveyor is used with counter-weighted rolls above to exert a gripping action. The whole trick is to grip the piece hard enough to pull it through without crushing or distortion. Of course, the take-off has a variable speed drive to permit a wide range of speed for different products. Also, the speed control is used to vary the draw down for precise control of section size.

Following extrusion, the extruded product or extrudate must be reduced to its most useful form. It may be spooled in long lengths or coiled. It can be sawed, sheared, punched, trimmed, or machined in many ways. Various post-forming techniques, including the thermoforming of extruded sheet or film (see Chapter 12) are also available. Some complex contours can be more easily made by extruding a simple shape—tube or sleeves—and continuously post-forming after exit from the die with specially designed sizing plates, sleeves, or simple brass fingers (such as is shown in Fig. 7-3).

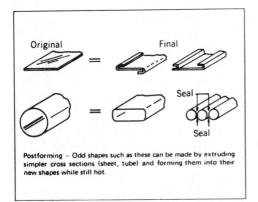

Postforming — Odd shapes such as these can be made by extruding simpler cross sections (sheet, tube) and forming them into their new shapes while still hot.

Fig. 7-3. Simple shapes can be extruded and post-formed into new shapes while still hot. (*Courtesy Celanses Corp.*)

Extruder Nomenclature

For the official SPI nomenclature as it applies to extrusion, readers are referred to the Glossary (p. 10). However, the discussion that follows will be useful in putting the various extrusion terms into perspective.

In delineating the size and output of extruders, a common expression is the L/D ratio, which is defined as the length of the flighted part of the screw divided by the inside diameter (ID) of the barrel. The L/D ratio is important in that it defines the available surface area of the barrel which can be used for heat transfer and the length of the screw which will be available for the various functions of feeding, metering, devolatilization, etc. This ratio has tended to increase over the years as a means of achieving higher outputs and better uniformity of melt. In the 1950's, ratios were often under 15:1; by the early 1970's, ratios of 28:1 were common

and some sheet extrusion lines were running as high as 30:1 and 32:1 (particularly advantageous where volatiles have to be removed).

However, the normal size specification used for extruders is based on the inside diameter expressed in inches (i.e., a 2½ in. extruder, a 6 in. extruder, etc.). Diameters vary from about 1 to 20 in. or more, though most of the sizes above 8 in. are considered as special orders. The output for each rpm increases rapidly for increases in extruder diameter, assuming the general screw dimensions are scaled in a consistent manner. In some applications, for example, a 4½ in. extruder may deliver about 4½ lb per hour per rpm (pph/rpm), a 6 in. extruder about 9 pph/rpm, and a 9 in. extruder about 18 pph/rpm. These outputs would shift considerably for different screw designs, operating temperatures, and resins. However, it is possible to calculate extruder output and a simplified method for accomplishing this is described on p. 170.

Maximum operating extruder screw speeds vary, but speeds to about 300 rpm are not too uncommon for some applications involving relatively large extruders. Some European extruders of small diameter have been capable of screw speeds above 700 rpm. Work has also been done domestically with these ultra-high screw speeds, but lack of temperature control generally produces higher temperature levels than desired and the short residence time tends to give non-uniform mixing.

Extruder Construction

In the design of the extruder, the screw (Fig. 7-4) obviously plays a basic role. In a typical extruder, the screw consists of a steel cylinder which may be either solid or cored. The coring

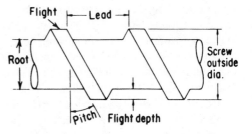

Fig. 7-4. Detail of screw. (*Courtesy Davis-Standard*)

is used for heat addition, or more often, heat removal. Coring can be for a portion or the entire length of the screw, depending on the particular application. Full length coring is generally preferred where large amounts of heat are to be removed and a drop in melt temperature is desired. Coring into only the early portions of the screw just after the feed hopper is usually desired when the objective is simply to help keep the feed area resin cooler; this can facilitate feeding of those resins which tend to soften easily.

The steel shaft or screw has a helical channel cut into it and the unit is rotated within the barrel. The helical ridge formed by the machining of the channel is called the *flights* (another way of describing it would be as a metal ridge, i.e., flight, wrapped like a screw thread around the cylinder). The distance between the flights (i.e., between two successive wraps of the ridge), or the lead, is usually constant for the length of the screw in single screw machines. The diameter of the screw from flight to flight is generally just a few thousandths of an inch (often 0.00075 to 0.002 in. radial clearance per inch of screw diameter) below that of the barrel inside diameter (*ID*). The exact clearance between screw and the *ID* of the barrel is usually the shortest distance that can be achieved within machining tolerances and without causing excessive wear. The close tolerance tends to prevent excessive flow of molten resin over the flights, it provides good wiping of resin from the wall to maximize heat transfer, and it prevents excessive amounts of material from holding up on the inside barrel wall.

The simplest extruder and screw combination would be the single screw with one continuous flight of constant pitch in a simple non-vented extruder.

When large quantities of objectionable volatiles are present in the feed or generated in the extrusion, a vented extruder design is sometimes used to prevent undesirable voids in the final product. The vents are basically open areas located along the barrel and designed so that volatiles can escape from the melted plastic to atmospheric or below atmospheric pressure. Usually, the sections are operated under vacuum to improve the devolatilizing effi-

ciency. Design and operation must be suitably controlled to minimize plugging of the vent, the possibility of resin escaping from this area, or loss of conveying in the extruder.

Obviously, there are many variations possible in screw design to accommodate a wide range of resins or applications. In fact, so many parameters are involved that the industry has moved to computerized screw design in which such variables of the plastication process as screw geometry, materials characteristics, operating conditions, etc., can be analyzed via mathematical models. The information thus derived permits optimum design of a screw for a given application by providing data on melt temperatures, pressures, flow, horsepower required, etc. This approach to screw design can, of course, be extended to the rest of the machine, thereby permitting computerized design of the entire extruder.

In practice, screw design can vary from the continuous flight screw with constant pitch described above to more recent sophisticated designs that include flow disrupters or mixing sections (Fig. 7-5). These mixer screws were developed to produce more thorough melt mixing and to balance heat distribution in the metering section of the screw prior to the time the melt enters the die. Mechanical means are used to break up and rearrange the laminar flow of the melt within the flight channel. They have also found use in mixing dissimilar materials (i.e., whether additive/resin combinations or simply dissimilar resins) and in improving extruder output uniformity at higher screw speeds (over 100 rpm). Some typical examples of mixing section designs are shown in Fig. 7-6. The fluted mixing section has proved especially applicable for extrusion of the polyolefins.

The use of a secondary undercut barrier flight for several turns is favored in many rigid vinyl applications (typically utilizing only moderate screw rpm). For some special dispersion problems, such as pigment mixing during extrusion, it can be advantageous to use rings or mixing pins or sometimes many parallel interrupted flights at a wide pitch angle.

Screw Zones. One of the basic parameters in screw design involves the ratio of lengths between the feed, transition, and metering

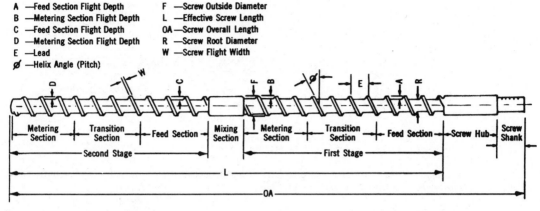

A —Feed Section Flight Depth
B —Metering Section Flight Depth
C —Feed Section Flight Depth
D —Metering Section Flight Depth
E —Lead
Ø —Helix Angle (Pitch)

F —Screw Outside Diameter
L —Effective Screw Length
OA—Screw Overall Length
R —Screw Root Diameter
W —Screw Flight Width

Fig. 7-5. Single flight, two-stage extrusion screw with mixing section. (*Courtesy SPI Machinery Div.*)

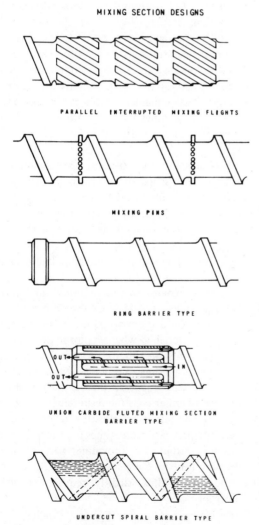

MIXING SECTION DESIGNS

PARALLEL INTERRUPTED MIXING FLIGHTS

MIXING PINS

RING BARRIER TYPE

UNION CARBIDE FLUTED MIXING SECTION BARRIER TYPE

UNDERCUT SPIRAL BARRIER TYPE

Fig. 7-6. Mixing section designs. (*Courtesy HPM, Div. Koehring*)

zones of the screw (Fig. 7-7). Each, of course, has its own special role. The feed section picks up the pellets, powder or beads and conveys them forward in the solid state. Generally, several flights of constant depth are provided to level out irregularities of feed. Because of the reduction in bulk effected as the material is moved forward and plasticated, a compression is built into the screw. This also prevents occluded air from being carried forward with the resin.

The transition zone, between the deep flighted feed section and the shallower metering zone, builds up pressure and starts fluxing the plastic. The transition may be either gradual or abrupt, depending on the material extruded.

On most conventional screws, the feed section usually has a higher capacity than the subsequent transition and metering zones. Thus, it functions in supplying enough material to prevent starving the forward zones. For many screw designs, feed depth is three to five times as deep as the metering zone (sometimes as high as six). This ratio of feed to metering section flight depth is known as the *depth* or *compression ratio.*

The resin should be fully melted into a reasonably uniform melt by the time it gets to the metering zone. The latter, as its name implies, should control uniformity of output rate. The importance of a good uniform feed rate, however, cannot be ignored to achieve uniformity of operation. A gross example of feed section problems is that caused by pellets

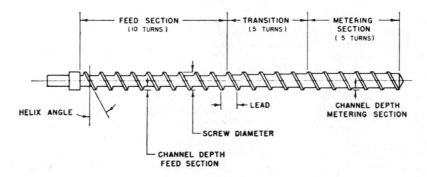

Fig. 7-7. Basic parameters in screw design involve ratio of lengths between feed, transition, and metering zones of screw. (*Courtesy E. I. du Pont de Nemours & Co., Inc.*)

becoming prematurely tacky with resultant "bridging" and loss of feeding characteristics. Suitable cooling of the feed hopper section is one normal prerequisite to reduce bridging problems. Bridging tends to become more severe with large extruder sizes, high screw speeds, low temperature softening resins, and hopper cooling with ineffective cooling capabilities.

Characteristically, higher feed efficiencies are achieved when friction between resin and barrel is maximum and friction between resin and screw is minimum. Helix angles of the screw are generally chosen to optimize the feeding characteristics with about 17.5° often typical. (It can be varied between 12 and 20°, however).

Screws are often fabricated from 4140 alloy steel, but other materials are also used. Generally, the screw flights are hardened by flame-hardening techniques or inset with a wear resisting alloy (e.g., Stellite #6) to resist wear. The degree of wear resistance will often depend on the abrasiveness of the resin system. Some filler materials, for example, are often quite abrasive.

Many screws are chrome plated to improve corrosion resistance and facilitate cleaning.

The Feed Zone. The feed section design, where granular resins enter the extruder, is quite critical. Its relationship to good feeding and prevention of bridging has already been mentioned. To help assure good cooling, the zone should have an efficient water cooling jacket. The inside throat area should be chrome plated to avoid rusting, especially since high

humidity sometimes causes water condensation on the cooled section. A relieved design to achieve an insulating air layer between the cooled hopper and heater barrel is desirable to minimize heat transfer between the two areas.

Feed openings are usually centered over the barrel and are often round, with diameter equal to barrier inside diameter. Alternately, oval and rectangular openings with a length about 1½ to 2 diameters and a width of 1 diameter are also used.

Specialized feed openings are sometimes used to feed strips of materials or materials which are difficult to feed (as distinct from the granular form generally used). These may include tangential hoppers and batch gate-type openings, with or without rams, screw feeders, or flanged openings for pumping liquid and melt.

Special batch or continuous metering assemblies may also be incorporated at or just before the feed hopper to permit feeding more than one material, as when a color master batch is mixed with some other resin. Another special technique sometimes combined with or just before the feed hopper is the use of heated or dried air to either raise the resin temperature or assist in prevention of moisture related problems. The preheat is sometimes used to improve extruder output by providing part of the necessary heat to the resin before it gets to the extruder and, thus, slightly reduce extruder power load.

Designing Screws for Specific Plastics. To provide some insight into the variations possible in screw designs, following is a representative range of supplier-recommended screws for extruding various plastics.

Acetal copolymer (recommended by Celanese Corp.): A metering type screw with at least a 4 or 5 flight metering section is strongly recommended. Metering sections as long as 11 flights have been used successfully in some applications. The channel in the metering section should be of medium depth. If too shallow, output would be reduced and overheating might occur. If too deep, the resin would not completely melt. Suggested screw dimensions for common extruders are shown in Table 7-1.

Channel depth ratio (compression ratio-feed/metering) should be at least 3:1, with 4:1 preferred. When ratios as low as 2:1 have been successful, other factors such as a high-resistance die or a valve in the extruder head provided the necessary back pressure for complete plastification. In general, screw channel depth, length of metering section and L/D barrel ratio are more important than channel depth ratio (feed/metering ratio).

Acrylonitrile-butadiene-styrene (recommended by Borg-Warner): In extruding ABS sheet, low-compression PVC or shallow-flighted polystyrene screws can possibly be used, but the production rates achieved may be relatively low. Output with either of these screws cannot be increased simply by increasing screw revolutions per minute because overheating of the plasticized material will result. On the other hand, a screw with an excessive flight depth may produce surge marks on the sheet at high production rates.

For this reason, a single-lead, full-flighted, constant-pitch screw with a progressively increasing root diameter and a compression ratio of 2:1 to 2.5:1 (see Fig. 7-8) is recommended for most ABS grades. In addition, when extruding ABS/PVC alloys, it is further recommended that the screw be chrome-plated (for corrosion-resistance).

Two-stage (vented) extrusion equipment offers an advantage in that venting assists the drying system and aids in removing volatiles that may be released by the compounded pellets. If the drying system is inadequate, two-stage extrusion can produce greater output rates than single-stage extrusion under the same conditions, primarily by venting off the moisture not removed in the dryer and improving the quality of the extruded sheet.

Polycarbonate (recommended by General Electric): Screws having L/D ratios of 20:1 or greater are preferred. A full-flighted, constant-pitch screw with a progressively increasing root diameter and a compression ratio between 2:1 and 3:1 gives a smooth, homogenous extrudate. This type of screw is normally used to process rigid vinyl. The screw contains no sharp transitions from feed to compression to metering sections. Generally, high-shear mixing screws should be avoided since they develop excess frictional heat at high temperatures, making extrudate control difficult.

Polyethylene (recommended by U.S. Industrial Chemicals): For polyethylene work, the screw (or the barrel) should be long, with an L/D of at least 16:1 to 30:1 to provide a large area for heat transfer, compounding, and homogenizing. The screw should be of the constant-pitch, decreasing-channel-depth, metering type (polyethylene screw) or constant-pitch, constant-channel-depth, metering type (nylon screw), with a compression ratio between 3 to 1 and 4 to 1 (see Fig. 7-9). A polyethylene screw is preferable for film extrusion and extrusion coating.

Table 7-1. Suggested Screw Dimensions for Acetal Copolymer.

Screw Diameter, In.	Pitch, In.	Depth of Feed Section, In.	Depth of Metering Section, In.	Width of Land, In.
1½	1½	0.260	0.065	0.150
2½	2½	0.440	0.110	0.250
3½	3½	0.500	0.125	0.350
4½	4½	0.560	0.140	0.450

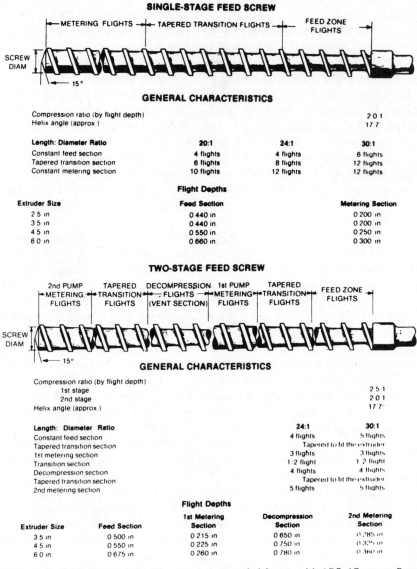

SINGLE-STAGE FEED SCREW

METERING FLIGHTS — TAPERED TRANSITION FLIGHTS — FEED ZONE FLIGHTS

SCREW DIAM

15°

GENERAL CHARACTERISTICS

Compression ratio (by flight depth)   2 0 1
Helix angle (approx)   17 7

Length: Diameter Ratio	20:1	24:1	30:1
Constant feed section	4 flights	4 flights	6 flights
Tapered transition section	6 flights	8 flights	12 flights
Constant metering section	10 flights	12 flights	12 flights

Flight Depths

Extruder Size	Feed Section	Metering Section
2 5 in	0.440 in	0 200 in
3 5 in	0.440 in	0 200 in
4 5 in	0.550 in	0 250 in
6 0 in	0.660 in	0 300 in

TWO-STAGE FEED SCREW

2nd PUMP METERING FLIGHTS — TAPERED TRANSITION FLIGHTS — DECOMPRESSION FLIGHTS (VENT SECTION) — 1st PUMP METERING FLIGHTS — TAPERED TRANSITION FLIGHTS — FEED ZONE FLIGHTS

SCREW DIAM

15°

GENERAL CHARACTERISTICS

Compression ratio (by flight depth)
  1st stage   2 5 1
  2nd stage   2 0 1
Helix angle (approx)   17 7

Length: Diameter Ratio	24:1	30:1
Constant feed section	4 flights	5 flights
Tapered transition section	Tapered to fit the extruder	
1st metering section	3 flights	3 flights
Transition section	1 2 flight	1 2 flight
Decompression section	4 flights	4 flights
Tapered transition section	Tapered to fit the extruder	
2nd metering section	5 flights	5 flights

Flight Depths

Extruder Size	Feed Section	1st Metering Section	Decompression Section	2nd Metering Section
3 5 in	0 500 in	0 215 in	0 650 in	0 285 in
4 5 in	0 550 in	0 225 in	0 750 in	0 325 in
6 0 in	0 675 in	0 260 in	0 780 in	0 360 in

Fig. 7-8. Design and general characteristics of screws recommended for use with ABS. (*Courtesy Borg-Warner*)

Nylon 6/6 (recommended by Celanese): Nylon 6/6 melts at approximately 500°F. For this reason, an extruder with at least an L/D ratio of 16 : 1 is necessary. A screw with a 4 to 1 channel depth ratio is recommended. The screws for nylon (see Fig. 7-10) are similar in geometry to those used for high density polyethylene.

Dry blend extrusion of rigid PVC: A large number of rigid vinyl products are being extruded and fabricated directly from dry blended compounds, rather than the pelletized compounds used with most other thermoplastics. Several extrusion systems are available:

(1) Single screw, single-stage non-vented extruder. The 20 : 1 L/D, non-vented extruder is but one type of system presently used for dry blend extrusion. This type can be used provided some means of air removal, such as a vacuum hopper, is available. The disadvantage of this system is the maintenance of vacuum under production conditions. Powder compounds are prone to bridge under vacuum and the extruder screw must be sealed at the rear of

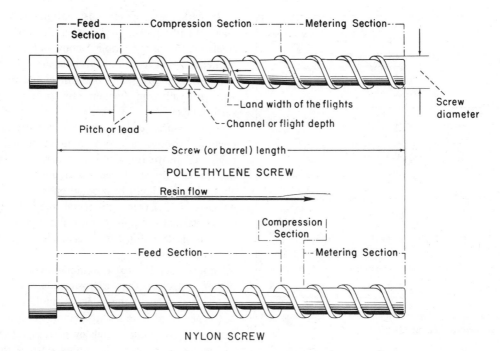

Fig. 7-9. Constant pitch, decreasing-channel-depth metering type "polyethylene screw" and constant pitch, constant-channel-depth metering type "nylon screw." To emphasize differences, screws are drawn out of scale. (*Courtesy U.S. Industrial Chemicals Co.*)

the feed section to prevent vacuum loss. The 20 : 1 *L/D* extruder with neither vacuum hopper nor vent port has proved to be of limited use with dry blend. Air pickup by the powder feed is a constant problem and generally limits extrusion rates. Various screw designs to eliminate air pickup have been tried with limited success.

In a single screw machine the material must not rotate with the screw or at least rotate at a slower rate than the screw. The force that keeps the material from turning with the screw, thus making it advance, is friction against the inside surface of the barrel. To increase the frictional surface, the length of the barrel has been increased. In most cases a 24 : 1 *L/D* has proven sufficient. The friction against the barrel varies considerably with temperature and different formulations.

A single-stage screw with a compression ratio of 3.0 : 1 has proven efficient in dry blend production provided a means of removing air and volatiles has been provided.

(2) Single screw, two-stage vented extruder.

This type of system normally consists of a 24 : 1 *L/D* two-stage screw and a vented barrel. The system provides a simple means of air and volatile removal, separately controlled fluxing and metering functions, and permits visual observation of the first stage melt conditions.

A two-stage screw with a compression ratio of 3.0-3.5 : 1 and a pump ratio of 1.7-2.2 : 1 is regarded optimum for rigid PVC dry blend processing. Output rates of 200-300 lb/hr are possible with this screw design on a 2½ in. lab extruder at high rpm. This design provides sufficient mixing for most types of rigid products tested at screw speeds cf 30-100 rpm on the lab extruder. Barrel temperatures and screw temperature greatly effect performance of the dry blend compound using two-stage screws.

All screws for single screw extrusion of PVC dry blends should be cored to accept oil cooling and the coring should extend to within ½ to ⅝ in. of the screw tip. Screw tips should be removable to allow the use of either a blunt nose tip and breaker plate or a conical tip.

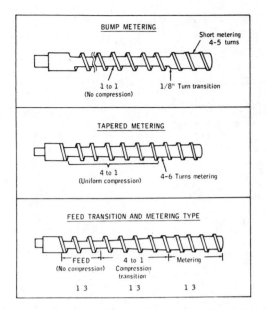

Fig. 7-10. Screws recommended for extruding nylon. (*Courtesy Celanese Corp.*)

Water or air has proved inefficient as a cooling medium for the screw. Heated oil cooling is generally used. Improper screw temperatures can result in burning on the screw tip or cold material accumulating near the screw tip.

(3) Twin screw and four screw extruders. These types of extrusion systems are now gaining wide acceptance for rigid vinyl production. They offer advantages of lower formulation costs and higher output rates than single screw extruders. In a twin screw extruder with intermeshing screws, the relative motion of the flight of one screw inside the channel of the other acts as a wedge pushing the material from the back to the front of the channel. In twin screw extruders as well as four screw extruders the lubricant system seems to have more overall importance than any other single additive. Since more physical mixing and shearing action take place in the barrels, less heat is necessary for proper fusion of the compound, thereby reducing level of heat stabilizer necessary. See also p. 199 on multi-screw extruders for additional information.

Extruder Barrels. The function of the barrel, of course, is both to house the screw and provide a transfer medium for heating and cooling.

In earlier extruders, most barrels were nitrided to provide improved wear. In nitriding, steels incorporating various amounts of alloying elements are heated in nitrogenous atmosphere. The nitrogen diffuses into the steel and reacts with the alloying elements to form a hard surface layer.

As pressures encountered in extrusion increased, however, a strong trend developed in using longer-wearing bimetallic barrels. These barrels use linings based on a durable alloy about $\frac{1}{16}$ in. thick. In production, the alloy is charged into a seamless metal tube, heated above its melting point, and centrifugally cast (i.e., using centrifugal force generated by rotation) onto the tube.

More recently, to accommodate the even higher pressures that extrusion may involve (i.e., when used in the injection molding process), newer techniques are being employed to make the bimetallic barrels even stronger. One technique is to cast the lining onto a thin-walled steel tube and then shrink fit the tube into an outer casing with a thicker wall. Another is to machine the end of the barrel and shrink-fit a reinforcing metal sleeve over it.

Many standard extruder barrels are rated for 10,000 psi maximum operating pressure to provide an adequate range of operating pressure.

Though it often is taken for granted, because it is considered the concern of the extruder manufacturer, it is important that the extruder barrel be properly supported to sustain the weight of barrel and auxiliary attached equipment, to resist vibration, to provide the longitudinal expansion and contraction with temperature changes in the barrel, and to maintain proper alignment of drive portions as well as the barrel itself.

The need for heating and cooling the extruder barrels is inherent in most extrusion operations. Temperature adjustments to the operating system and the need for heatup following shutdown almost universally require the installation of good heaters and controllers. Normally, the heat is applied by electrical heating which can be either of the conventional resistance heater type or by induction heated means. In the case of the resistance heaters, several heater bands or elements are usually wired together as zones or groups and controlled from the same temperature controller. Usually, a thermocouple or resistance unit advises the controller of the temperature for appropriate corrective action. Where resistance heaters are used, it is important that they have good uniform contact with the barrel to provide constant heat transfer and to prevent premature burnout of the heater element.

Sufficient heater capacity is required for the specific processing desired. Often, a watt density of 25 to 50 watts/sq. in., based on the inside diameter, is desirable. Inadequate levels can result in heatup delays or inability to reach prescribed temperature levels. Excessive watt densities can cause operating upsets due to large quantities of heat whenever the heater comes on, unless suitable controlling systems are employed to control current to the heaters instead of just controlling time. The aluminum or alloy cast heater represent a relatively recent type of resistance heater which tends to give good fit to the heated part and uniform element temperature as a result of the high thermal conductivity. Another added advantage of these units is that water cooling tubes can also be embedded in them for cooling, or they can be provided with fins for air cooling.

Though some extruders are not equipped with barrel cooling capabilities, it is at least desirable, if not mandatory, on many machines. It can help accelerate cooling on shutdown or when a drastic drop in operating temperature is necessary. Cooling of at least some portions of the extruder is sometimes employed to remove the excess heat which may have resulted from the screw working. Selective cooling can also result in better operating uniformity in some processes.

Cooling is commonly accomplished by either water or forced air. Water is the more effective, but, unless properly controlled, can cause too severe a change when it cycles on, with resulting upsets in uniformity of melt temperature, melt pressure, and/or output. Typical water-cooling systems include tubes within the cast heater units or cooling coils embedded in the barrel and jacket.

Air cooling is a more gentle change, but cannot remove the large quantities of heat of a liquid system. Typical air cooling is done with a series of blowers corresponding to the heating zones.

On many machines, the more efficient water cooling system is employed with a suitable proportioning controller to feed in only small quantities of water where only limited cooling is needed.

Some specialized extrusion applications utilize closed liquid or vapor systems. Refluxing pressurized or steam jacketed systems are also available to control the temperature.

Drives. The power for extruder screw rotation is supplied by a variable speed motor drive system. It is transmitted through a gear reduction unit, a coupling, and the thrust bearing. Gear reducers impart final speed and torque to the extruder screw. Most gear reducers use double-reduction helical gears for ruggedness, and to hold noise levels within acceptable limits. Worm and pinion gear combinations have been used on smaller extruders.

Thrust bearings absorb the thrust force exerted by the screw as it turns against the material in the barrel. The size or rating of the thrust bearing is chosen to give an anticipated number of operating hours under the operating conditions anticipated for the particular application. The operating pressure, size of the extruder, and the operating speeds are important factors in the thrust bearing specifications.

The motor size for the extruder must be sufficient to allow for the energy required to work, melt, and deliver the resin at the desired output condition and rate. More power is usually required, for example, if it is desired to deliver the same resin at $600°F$ than at $400°F$, if both are designed to operate without cooling of barrel or screw. Likewise, higher outputs of resin at higher screw speeds require more power than low outputs. The specific heat of the resin will also be a determining factor in the power requirements. As rules of thumb, many simple extrusions require 1 hp for every 10 to 15 lb/hr of product. For high energy mixing applications or where high temperatures at high screw speeds are required, 1 hp may be required for every 3 to 5 lb/hr. If the extruder operates at very high throughputs, it may be necessary to size the drive to supply 100% of the energy, or even more if cooling is involved.

Extruder types of drives cover a wide range, although advances in variable speed direct current drives are resulting in their use on increasingly larger numbers of extruders.

The DC drives are efficient over a wide range of speeds, compact, relatively quiet, and permit good speed regulation or reproducibility. Speed regulation is often 1% on these systems.

Mechanical drives are used on extruders with advantages including low cost, high efficiency, ease of maintenance, and simple operation. They can be as simple as a twin set of variable diameter pulleys to permit speed changes by altering the relative diameters of the two pulleys. Some disadvantages for the mechanical drive include bulkiness, poor speed regulation at high loadings, noise, and relative low upper torque limitations.

Many eddy current couplings are used in series with the drive motor to permit variable speed operation of the extruder. In the small and medium size units, these are often air cooled to remove heat. In the larger units, these are water cooled. Advantages for these systems include compactness, relatively low initial cost, and a wide range of speed control. Disadvantages can include noise level, inefficiency of power transmission at low speed, and the need for a satisfactory quantity of cooling water in the larger sizes.

A newer drive system consists of a hydraulic unit powered by an electric motor. Costs for these are close to the DC drive. The hydraulic units reportedly provide high available torque even at low speeds, accurate speed control, a high constant horsepower over much of the operating speed range, and the ability to set maximum torque to prevent shearing of a screw. Operating noise, however, can be relatively high on some of these units.

Although there is no single solution to satisfying every parameter involved in selecting the proper drive, a basic comparison between three adjustable speed drive systems as shown in Table 7-2 might make a good starting point.

Screenpacks and Valves. Before extruder valves were widely included on the end of the barrel, screenpacks (which also function in removing contamination or coarse particles from the melt stream) were used to permit developing the pressure for providing better mixing or higher melt temperature by increased working.

In operation, the screenpack is backed up by a breaker plate that has a number of passages, usually many round holes ranging from $\frac{1}{8}$ to $\frac{3}{16}$ in. in diameter. One side of the breaker plate is a recess that accommodates the round discs of wire screen cloth which go to make up the screenpack. By adding more discs of finer screening, the resistance to flow increases and the head pressure can be increased, resulting in harder working of the material.

More recently, external screen changers have been developed for use when screenpacks must be frequently changed. The packs are mounted outside the extruder between head clamp and dies and are changed via mechanical or hydraulic operation.

However, although the breaker plate-screenpack arrangement did permit higher pressure, it was without proper control and would often increase out of control as the screens plugged. The modern valve permits control of pressure at higher levels without the concern of the gradually increasing pressure or the breakage of screens. The valve, to do its job properly, must be streamlined to prevent holdup and degradation. It also must be adjustable over a fair range of movement to permit good control of the system, rather than like some of the older units which failed to cause much change in pressure until almost closed.

Extruder valves vary in detail considerably, but usually fall into two general types. The streamlined manually adjusted plug which permits the melt to pass between it and a stationary seat represent the one class of stationary valves, while the valves with one portion rotating represent the general class of dynamic valves. The latter usually have one portion of the valve attached to the rotating screw and the other portion stationary so that there is mechanical working in addition to the simple pressure working occurring.

Extruder Controls. There are a number of controls generally in use with extruders that function in maintaining quality and a smooth

Table 7-2. Comparison Characteristics of Three Adjustable Speed Drives*

Features and Characteristics	Direct Current Motor with Solid State Power Supply	Eddy-Current Drive with Solid State Controller	Adjustable Frequency W/ AC Motor and Solid State Power Supply
Cost	Medium	Low	High
Speed range	30 : 1 typical. Blower may be required. Can be extended to 100 : 1	30 : typical. Water cooling can be used on large drives. Can be extended to 100 : 1	10 : 1 typical max.
Constant torque capacity	Yes	Yes	Yes
Constant horsepower capacity	Yes	Yes	Yes
Rapid reversing	Yes	No	Yes
Dynamic braking	Yes	Simulated, with eddy-current brake added	Yes
Regenerative braking	Yes, with 12 thyristors required	Simulated, with eddy-current brake added	Yes, regenerative power can be exchanged between inverters on the same DC bus.
Totally enclosed Construction	Yes	With water cooled unit	Yes
WK²	Low	Medium	Lowest
Speed regulation without feedback	3%-4%	No inherent speed regulation. Feedback must be used.	0% frequency regulation if synchronous induction motors are used.
Speed regulation with feedback	Can be held to 0.1% after transients.	Can be held to 0.1% after transients.	Better than 0.1% with synchronous induction motors.
Speed drift	With armature voltage control, drift could vary 10% for the initial 30 minute warmup period. After the warmup period, drift would be 1%. With tachometer feedback and non-regulated field, drift would be approximately 1%. With tachometer feedback and regulated field, drift would be approximately $\frac{1}{4}$%.	.02%/° F.	.05% or better
Power supply size	Large	Small	Largest
Drive efficiency	Approximately 86%	Proportional to set speed	Approximately 90%

* Courtesy, F. E. Simo, Eaton Yale & Towne, Inc.

running production line. For example, to record melt temperature and pressure, an instrument as simple as a thermocouple or a high-temperature grease-filled pressure gage can be used. More sophisticated recording equipment would involve the use of electrical and pneumatic pressure transducers that would replace the gage and permit more accurate recordings as well as the possibility of controlling the pressure. Also available today are infrared thermometers that can measure average melt temperature without contact with the melt after it has exited from the die. These units can also measure temperature profile across a die, cooling behavior, etc.

To measure screw speed, a tachometer is generally used. Extruder power consumption is monitored by either an ammeter or wattmeter and is regarded as a useful indicator of the uniformity of the extrusion operation. An increase in power, for example, can mean excessive material working, as when cold spots develop; a decrease can signify a loss in the extrusion operation (i.e., an empty hopper or a bridged hopper). Screw temperature, when cooling is applied, is measured by a thermometer on the outlet water.

Barrel heating and cooling controls for most extruders are electrical controllers which sense the metal zone temperature and try to match it to a desired setpoint. Some units only supply heat, while others can control both heating and forced cooling units. The simplest controllers are on-off, with heat on until setpoint is reached and then off until the actual temperature drops below setpoint. The time proportioning controller is a modification of the on-off which cycles the length of time heat is on, depending on the degree of heat correction required. The current proportioning unit regulates the amount of current to the heaters, depending on the need. The time and current proportioning units tend to provide more uniform temperature control than the on-off units, and are generally preferred for most extruders.

The use of advanced concepts such as digital readout of various variables (e.g., screw rpm) or hooking computers up to monitor and control extruder operation is becoming increasingly popular. A typical example would be a computer set-up to control a sheet line. In such an operation, the variable speed motor driving the screw, the various temperature zones on the barrel, die, and chill rolls, the beta-gage system for measuring sheet thickness are all interfaced with a computer. Should the sheet thickness go out of the set specifications, the computer would bring the operation back into the tolerances required by either speeding up the lines or slowing screw speed.

CALCULATING EXTRUDER OUTPUT*

Of the three sections of a well designed extruder screw, output is dependent on only the metering section. If the geometry of this section is known, and if actual operating conditions can be assumed as known, it is possible to calculate extruder output.

In the metering section of the screw, three kinds of melt flow may exist:

(1) *Drag Flow.* This is the forward motion of the melt stemming from the conveying or pumping action of the screw with no flow restrictions.

(2) *Pressure Flow.* This means a theoretical back flow (loss of output) due to restrictions in the system such as the die or any pressure-control devices. In other words, pressure flow is the flow that one would obtain if screw rotation were stopped and molten resin were forced under pressure from the die down the screw channel which would act as a long rectangular orifice.

(3) *Leakage Flow.* This is defined as flow over the top of a screw flight into the preceding screw channel because of the presence of melt under pressure and clearance between the screw flight and the barrel wall.

The net output of the extruder screw is equal to the sum of these three flow components:

Output = drag flow-pressure flow-leakage flow.

Notice that pressure flow and leakage flow are output loss terms and are shown as subtractive quantities. (Mathematical expressions for these flow terms have been derived and the results are shown in Table 7-3, p. 171.)

Leakage flow is dependent on the third power of the clearance between the top of the screw flight and the barrel wall. This dimension is on the order of 0.010 in. on most equipment (except for badly worn screws), which lead to very small output losses. It is common to ignore leakage flow in extruder calculations, and no serious error results. With this assumption,

* By William G. Frizelle, 35D Joyce Ellen, St. Louis, Mo. 63135 (formerly with Emerson Electric); reprinted with permission from the Sept. 1970 issue of *Plastics Technology* magazine.

Table 7-3. Sample Calculation: How to Determine Extruder Output

Let's assume we're dealing with an extruder screw of the following dimensions:

Diameter: 1.75 in.
Metering zone length: 7 turns = 7(1.75) = 12.25 in.
Helix angle: 17.7°
Metering zone channel depth: 0.071 in.
Measured axial channel width = 1.59 in.
Single flighted screw

Other pertinent data:
Resin to be run: Polystyrene
Melt temp: 390 F
Screw speed: 75rpm
Operating pressure: 1000 psi

Channel depth-to-width ratio: $\dfrac{0.071}{1.59} = 0.044$

Therefore, F_d, F_p = 1 (Ref. 1.)

The symbols in our example are defined as follows:

F_d = Drag flow shape factor (which is 1 if the ratio of channel depth to width is less than 0.1).
F_p = Pressure flow shape factor
D = Nominal screw dia, in.
h = Metering zone channel depth, in.
W = Channel width perpendicular to flight face, in.
N = Screw speed, rps
n = No. of parallel threads (1 for conventional screw)
Φ = Helix angle (usually 17.7°).
μ = Melt viscosity, #-sec./in.^{2}
P = Change in pressure through metering zone, #/in.^{2}
L = Axial length of metering zone.
e = Axial flight width, in.
t = Screw thread lead (D for conventional screw)
b = Length of one flight on a screw
δ = Clearance between screw flight and barrel, in.

Using these dimensions, output Q is given in in.^{3}/sec.

Now to the calculations, plugging in to the appropriate formulas the numerical values listed above:

$$\alpha = \frac{n\pi Dbh \cos^2 \Phi}{2} F_d =$$

$$\frac{(1)(3.14)(1.75)(1.59)(0.071)(0.953)^2(1)}{2}$$

$$= 0.282 \frac{in^3}{rev.}$$

$$\beta = \frac{bh^3 \sin\Phi \cos\Phi}{12L} F_p =$$

$$\frac{(1.59)(0.071)^3(0.304)(0.953)(1)}{(12)(12.25)}$$

$$= 1.12 \times 10^{-6} \, in.^3$$

$$Q = \alpha N - \frac{B\Delta P}{\mu} = (0.282)\left(\frac{75}{60}\right) - \frac{(1.12 \times 10^{-6})(1000)}{\mu}$$

$$= 0.353 \frac{in.^3}{sec.} - \frac{1.12 \times 10^{-3}}{\mu} \frac{in.^3}{sec.}$$

Viscosity μ is a function of shear rate and temperature. Shear rate may be estimated by the expression

$$Y = \frac{\pi DN}{h} = \frac{(3.14)(1.75)}{0.071}\left(\frac{75}{60}\right) = 97 \, sec.^{-1}$$

The reader is referred to Bernhardt (1), p. 628, where viscosity at 390 F is estimated to be 0.14 #sec/in.^{2}

Thus,

$$Q = .353 \frac{in.^3}{sec.} - \frac{1.12 \times 10^{-3}}{0.14} \frac{in.^3}{sec.}$$

$$= 0.353 - 0.00802$$

$$= 0.345 \frac{in.^3}{sec.}$$

If the density of this material is considered to be 60.8#/ft.^{3} at 390 F atmospheric pressure, then,

$$Q = \frac{(0.345 \, in.^3)}{sec.} \frac{(3600 \, sec.)}{hr.} \frac{(1 \, ft.^3)}{1728 \, in.^3} (60.8 lb/ft.^3)$$

$$= 43.5 lb/hr.$$

Ref. 1: *Processing of Thermoplastic Materials,* E. C. Bernhardt, Reinhold Publishing Corp., (Van Nostrand Reinhold) New York, 1959.

then, the output of the extruder is only a function of drag flow and pressure flow:

Output = drag flow-pressure flow.

Inspection of the relationships illustrated in the calculations above will show that drag flow is dependent only on the geometry of the metering section and the speed with which it rotates. At the same time, pressure flow is dependent on the metering section geometry, the pressure drop through this section, and the melt viscosity. Thus, the output relationship described above can be presented in a more quantitative form:

$$\text{Output, } Q = \alpha N - \frac{\beta \Delta \rho}{\mu}$$

where α and β are functions of the metering section geometry, $\Delta \rho$ and μ represent the pressure drop in the metering section and the viscosity in the metering section. This situation can be described graphically with a "screw characteristic curve" (Fig. 7-11).

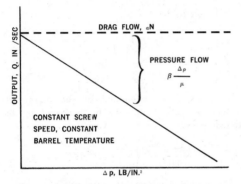

Fig. 7-11. Typical screw characteristic curve. Output decreases as pressure drop increases. (*Courtesy Emerson Electric*)

With a particular metering section geometry operating at constant speed and temperature, drag flow is shown at zero pressure drop (open discharge) and output decreases as pressure-drop increases.

The effect of operating changes can easily be seen (Fig. 7-12); for example, if screw speed is

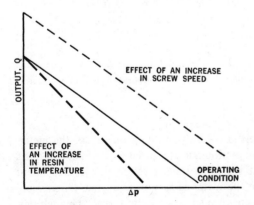

Fig. 7-12. Effects of operating condition changes on the typical screw characteristic curve. (*Courtesy Emerson Electric*)

increased, drag flow increases while pressure flow remains fixed If, instead, barrel temperature is increased, drag flow would remain unchanged, but pressure flow would increase because resin viscosity would decrease.

The general slope of the curves shown in Fig. 7-12 is a function of the metering section geometry (see Fig. 7-13), a variable that is fixed unless you are willing to replace the extruder screw.

Looking at Fig. 7-13, one should be aware that a deep-channeled metering section is unsuitable for high-pressure extrusion because the pressure-flow losses have too great an effect on screw output.

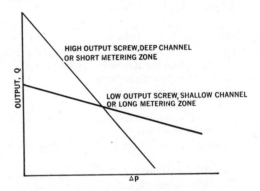

Fig. 7-13. Screw geometry affects the screw characteristic. Deep-channeled metering section is unsuitable for high pressure. (*Courtesy Emerson Electric*)

Our discussion thus far has centered only on the extruder screw. While there are devices (restrictor valves) to control pressure loss in the metering section, the main source of pressure loss is due to the extruder die. The die is a restriction to flow and thus, by definition, should be recognized as a pressure-flow device. Die output may be presented in a form similar to the pressure-flow portion of the extruder-output relationship:

$$Q \text{ die} = \frac{k \Delta P}{\mu}$$

where k is a function of die geometry and $\Delta \rho$ and ρ represent die-section pressure drop and die-section viscosity, respectively. Since die and extruder will always operate together, a "die characteristic curve" can be combined with the previously shown screw characteristic curve (Fig. 7-11). The interaction of these two curves represents the operating point for the system under that particular combination of conditions.

If the die temperature is increased, the extruder operates at a lower pressure with a higher output (Fig. 7-14). (Note: An increase in temperature produces a decrease in melt viscosity in the die.) If operating temperature were changed by the same amount in both barrel and die, the net result would be no change in output though the extruder would operate at lower pressure (Fig. 7-15).

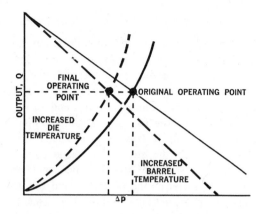

Fig. 7-15. Effect of temperature rise on operating point. When only die temperature is increased, extruder operates at a lower pressure with a higher output. (*Courtesy Emerson Electric*)

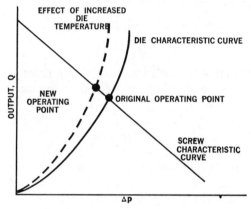

Fig. 7-14. Interaction of screw characteristic curve (Fig. 7-11) and a die characteristic curve represents operating point for the system under that particular combination of conditions. (*Courtesy Emerson Electric*)

In general, calculating extruder output is not as simple as the foregoing presentation might indicate. Unfortunately, in the extrusion process it is not possible to deal with independent variables; for example, these calculations indicate that extruder output depends on operating pressure and that a change in operating pressure leads to a new point on the old curve. In fact, change in pressure also leads to a temperature change (and thus a viscosity change as well), yielding a new point on a new curve. And to complicate matters even further, the magnitude of the dependent change is difficult to measure or control.

Nonetheless, the fundamental concepts presented here are still useful in an understand-

ing of the basic extrusion problems and process. The curves presented are technically correct and will be found to hold true in actual practice. There are three things to be kept in mind in making this calculation:

(1) As shown, the calculation assumes that pressure increases as material travels down the length of the metering section. In extruders using low-resistance dies (profile extrusion, for example), the pressure may decrease through the metering section. In this circumstance, Δp is negative and pressure flow is additive. Failure to account for the pressure profile may lead to errors of about 10%.

(2) The most troublesome part of the calculation is in determining a proper viscosity value to use in the pressure-flow term. Viscosity is a function of both melt temperature and shear rate. Check the resin suppliers to obtain this data. Reference should also be made to Bernhardt, E. C., *Processing of Thermoplastic Materials*, pp. 120 and 628, Reinhold Publishing Corp. (Van Nostrand Reinhold) New York, 1959.

(3) To be most meaningful, output should be presented on a mass basis, which requires that you know the density of the melt at the temperature under which it leaves the extruder or die. Use of room-temperature density values may lead to gross errors, particularly in the case of crystalline polymers.

DIE DESIGN AND PRODUCTION SET-UPS

Thus far, we have talked primarily about the extruder and its role in the extrusion operation. However, as is readily apparent, it is the die that really determines the form of the extrudate that will be coming out the end of the machine. While the die may be quite complex in construction, it is simply a shaped hole through which the stream of plastic melt is pushed. The shape of the die opening determines the shape of the product (even though subsequent shaping may be done downstream). The plastic melt always tries to take the easy way out and will try to crowd through the larger part of the opening. For this reason, the exact shape of the opening may be a little different than the exact shape of the extruded section.

Another difference will be in size. Almost always, the size of the opening will be larger than the finished section desired. The plastic is pulled away faster than it is extruded and will draw down or get smaller as it leaves the die. This is called *drawdown* and it helps in keeping the extrusion straight as well as in permitting fine size adjustment.

In essence, when a resin flows through a die or orifice there is a resistance to flow which is dependent upon the pressure differential in the die, the viscosity of the resin, the length of the orifice, and the configuration of the orifice. The relationship of these variables is helpful in designing dies.

Assuming that the flow through the die is laminar, the rate of flow, Q, through orifices of several simple shapes may be calculated from the equations in Fig. 7-16, in which:

Q = rate of flow (volumetric), in cu in./sec

μ = viscosity, in lb-sec/sq in.

P = pressure, in lb/sq in.

L_o = length of land of orifice, in inches

R_o = radius of orifice, in inches

D_o = mean diameter of orifice, in inches

W_o = width of orifice, in inches

H_o = height of orifice, in inches

S_o = annular opening of orifice, in inches

Plastics deviate from true Newtonian flow, but compensation for this deviation is made by

NEWTONIAN FLOW EQUATIONS

FOR ROUND ORIFICE:

$$Q = \frac{\pi R_o^4 P}{8 \mu L_o}$$

FOR RECTANGULAR ORIFICE:

$$Q = \frac{W_o H_o^3 P}{12 \mu L_o}$$

FOR ANNULAR ORIFICE:

$$Q = \frac{\pi D_o S_o^3 P}{12 \mu L_o}$$

PRESSURE (P) IS USED THROUGHOUT INSTEAD OF PRESSURE DROP (ΔP) ACROSS THE ORIFICE.

Fig. 7-16. Newtonian flow equations

measuring the melt viscosities of the resins at shear stresses which are typical of extrusion conditions.

Basically, however, the equations in Fig. 7-16 can be used to determine output and to estimate proper die design.

In the following pages, die design and production set-ups will be considered for each of the basic types of extrusion operations.

Blown Film

The blown film technique involves extrusion of the plastic through a circular die, followed by expansion (by the pressure of internal air admitted through the center of the mandrel), cooling, and collapsing of the bubble. In operation, the blown film is extruded upward (vertical extrusion is most common, although horizontal techniques have been successfully used) through guiding devices into a set of pinch rolls which flatten it. It can be slit, gussetted and surface-treated in-line.

A typical set-up for blown film extrusion is shown in Fig. 7-17. Arrangement can be vertical or horizontal (Fig. 7-17A).

The technique is widely used in the manufacture of polyethylene and other plastics films. Advantages of blown film include: the system's adaptability to producing tubing, both flat and gussetted, in a single operation; the relative ease

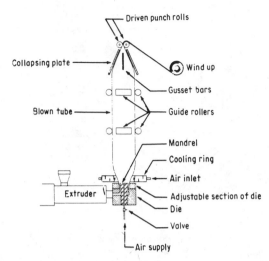

Fig. 7-17. Typical blown film set-up

of changing film width and caliber by controlling the volume of air in the bubble and the rpm of the screw; the elimination of the end-effects that result from flat die film extrusion (e.g., edge bead trim and non-uniform temperature); and the capability of biaxial orientation (as the film is stretched during the bubble blowing step). Film produced by this method generally is hazier than films made by other techniques that involve quicker cooling, although the industry is now working on concepts to overcome this limitation.

Screw Design.   This depends in large part on the particular resins being processed. However, the standard single-stage polyethylene screw does predominate (since polyethylene is the largest volume resin extruded by blown film techniques). Although most early film screws were equally divided in length between feed, transition, and metering zones, a noticeable trend has been evident toward longer metering sections (up to 65% of screw length) and consequent reductions in feed and transition sections. The deeper metering section provides higher output per screw revolution at lower melt temperature, with no sacrifice in melt homogeneity. The lower melt temperature is especially important in blown film techniques in that cooling of the bubble is the primary limiting step in the rate at which the process can run. Toward this end, a number of techniques (many of them proprietary and covered by patents) have been developed for faster cooling and higher production rates through the use of refrigerated air, secondary air rings, internal cooling mandrels, or combinations

Fig. 7-17A. High-density polyethylene blown film line. (*Courtesy NRM Corp., Subs. Condec Corp.*)

therefore. Typical of these is the patented tubular water bath process that reportedly combines the advantages of a blown film technique (gage randomization, increased output, no edge bead) with the advantages of a chill roll or conventional water-bath process. In operation, a bubble of molten film is drawn over a large-diameter mandrel and then collapsed down into a pit. Airstreams blown through the mandrel extend the film to the desired diameter. When that is reached, the film is plunged into a cold water bath. Another stream of cold water, within the mandrel, cools the metal skin of the mandrel. Thus, the cold water on one side of the extruded film and the water-cooled mandrel on the other can speed up the cooling process.

Another recent trend in screw design for blown film has been interest in the use of mixing screws similar to those used for extrusion coating, with appropriately greater depths to avoid overheating the resin.

Die Design. A blown film die has an annular orifice between a conical center mandrel and an outer ring which is adjustable to extrude films of various widths and thicknesses. The center core and outside ring must be perfectly concentric insofar as possible. Typical blown dies are shown in Fig. 7-18, 7-19, and 7-20. Film dies are either side-fed or bottom-fed.

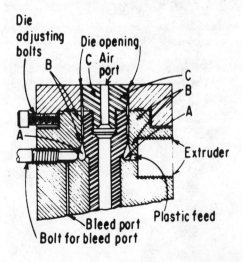

Fig. 7-18. Side fed blown film die. A-manifold; B-chokes; C-land (usually 1/2 to 1 in. long with 0.020 to 0.030 in. tolerance). (*Courtesy E. I. du Pont de Nemours & Co., Inc.*)

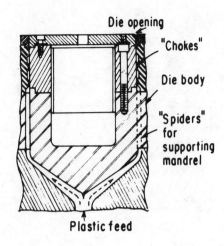

Fig. 7-19. Bottom fed blown film die with spider-supported mandrel. (*Courtesy E. I. du Pont de Nemours & Co., Inc.*)

Fig. 7-20. Bottom fed blown film die with spiral mandrel. (*Courtesy E. I. du Pont de Nemours & Co., Inc.*)

Adjustments to the center core and ring spacing along their common axis change the land length slightly. To produce 1-mil film, the die opening should be about 15 to 30 mils with $\frac{1}{4}$ to $\frac{1}{2}$-in. land.

At least one restriction should be provided within the die, followed by an expansion area. This prevents prominent weld lines, helps maintain the back pressure necessary for suitable working of the material within the barrel, and reduces pressure fluctuations.

To hold the conical center mandrel in position, spider arms are usually used (as in Fig. 7-19). However, they leave small weld lines on the film. To minimize these weld lines and to balance out small differences in film gage that result in annular lip adjustment, the die can be continuously rotated. This helps to prevent gage bands from building up in one spot on the wind-up roll and stretching the film. Rotation can only be used on bottom-fed dies, not on side-fed dies.

Special internal die configurations have also been evolved that eliminate the need for spiders. Typical is a spiral mandrel bottom-fed die (Fig. 7-20) that feeds the material through feed port and upward through a spiral flow channel, where blending and overlapping take place to reduce the weld lines.

Another configuration uses a spreader plate to split the extrudate flow into two separate streams. One stream feeds six passages on top of the plate; the other, six bottom passages. As the streams flow radially outward through the passages, they broaden and thin, finally welding at the outer rim of the passages. Passages are offset so each weld in one sheet of melt falls exactly between the weld lines on the other sheet of melt. Melt sheets are joined in the flow channel before extrusion. The result is a greatly reduced visual pattern of weld lines that extend only half way through the film thickness.

Bubble Blow-up. Once the tubular shape is being satisfactorily extruded and has been threaded through the collapsing tower and takeup, sufficient air is introduced through a port in the mandrel into the area between the die and the pinch rolls to form a bubble of the volume necessary to produce film of desired width and thickness. This air remains within the bubble surrounded by passing film and requires little augmentation. Expanding the bubble also imparts some transverse orientation to the film for the added strength.

Blow-up ratio of blown film is the ratio between the diameter of the final film tube and the diameter of the die orifice. Usually, ratios between 2 : 1 and 3 : 1 are used. Some of the effects of blow-up ratio on final film properties are indicated in Table 7-4.

As it leaves the die, the tube is transparent. Where the polyethylene crystallizes (solidifies), it becomes translucent or "hazy." this level is called the *frost line*. Blown film is cooled by air

Table 7-4. Effects of Variables on Blown Film[1] Properties.*

Property	Raising Melt Temperature	Increasing Blow-Up Ratio	Faster Cooling[2]
Clarity	Up	Up (to 4 : 1 ratio)	Up
Sparkle	Up slightly	Up (to 4 : 1 ratio)	(3)
Coefficient of friction	(3)	(3)	(3)
Max. sealing temperature	(3)	(3)	(3)
Tear strength			
Machine direction	Down	(3)	Down
Transverse direction	Down	Up	Up
Dart drop impact strength	Up	Up	(Varies)
Bag drop strength			
Machine direction	Up	(3)	Down
Transverse direction	Up	Up	Down
Blocking	Up	(3)	Less

1. Contained slip agent.
2. I.e., lower frost line.
3. No appreciable effect.
* Courtesy U.S. Industrial Chemicals, Div. National Distillers

blown against the surface. The level of the frost line depends upon several variables: thickness and temperature of the extrudate, rate of extrusion, and volume and temperature of cooling air. Cooling faster, of course, lowers the frost line. Raising the frost line reduces tear strength in the machine direction and may increase haze slightly. How frost line affects impact strength varies with the type of test.

Figure 7-21 is a schematic drawing of the blown film bubble. Given with the diagram are four simple equations which govern the relationships of blow-up ratio, die diameter, bubble (or tube) diameter, and lay-flat width. The equations can be used to calculate either the blow-up ratio if the lay-flat width and the die diameter are known, or the lay-flat width if the blow-up ratio and the die diameter are known.

Garbage Bags and Trash Liners. This new market for blown polyethylene film has led to the evolvement of several new production concepts. One is the so-called *double-bubble approach*, in which two blown films are extruded simultaneously using an extruder and a dual die. Proponents of this approach claim elimination of the need for side seals, since only the bottom of the bag has to be sealed. Balanced flow through the dies is achieved by using a valve in the manifold supporting and feeding each die. Some systems also use independent nips to give the processor the capability of controlling both bubble thickness and sizes independent of each other.

Another approach is a single-die extrusion to produce a wide lay-flat tube which is slit-sealed, printed on (if desired) and post-gussetted before entering the bag machine downstream (see Fig. 7-22). This approach requires a lower capital investment since only a single die is used.

Plastic Paper. Blown film techniques are also being used to produce high-density polyethylene papers, particularly for grocery bags, but also in thinner gages for tissue wrapping. Many of the resins involved do have a higher molecular weight so that screw and die design and even tower design are critical. In some instances, mixing screws are being used to generate efficient working of the material.

Tower design is also quite critical, since

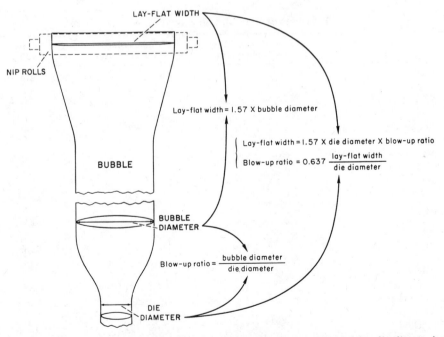

Fig. 7-21. Schematic drawing of a blow-up ratio of 2.5 (bubble diameter divided by die diameter) and the lay-flat width after the nip rolls have flattened the bubble to a double layer of film. The four equations describe the relationships between blow-up ratio, lay-flat width, and the bubble and die diameters. (*Courtesy U.S. Industrial Chemicals*)

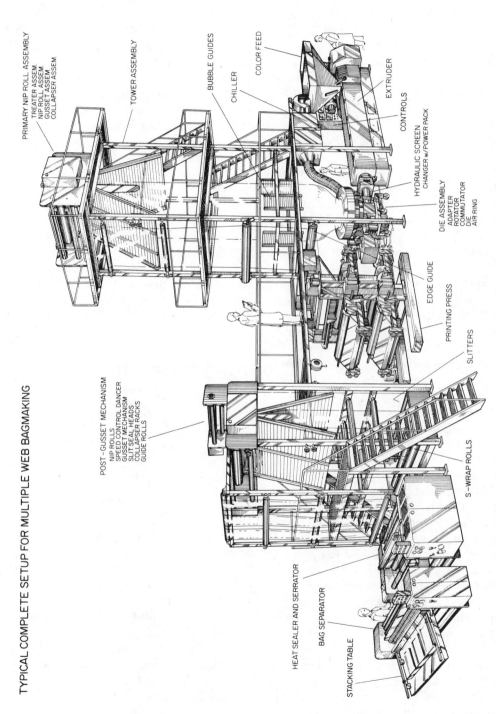

TYPICAL COMPLETE SETUP FOR MULTIPLE WEB BAGMAKING

PRIMARY NIP ROLL ASSEMBLY
TREATER ASSEM.
NIP ROLL ASSEM.
GUSSET ASSEM.
COLLAPSER ASSEM.

TOWER ASSEMBLY

BUBBLE GUIDES

CHILLER

COLOR FEED

EXTRUDER

CONTROLS

HYDRAULIC SCREEN
CHANGER w/POWER PACK

DIE ASSEMBLY
ADAPTER
ROTATOR
COMMUTATOR
DIE
AIR RING

EDGE GUIDE

PRINTING PRESS

SLITTERS

POST-GUSSET MECHANISM

NIP ROLLS
SPEED CONTROL DANCER
GUSSET MECHANISM
SLIT SEAL HEADS
COLLAPSER RACKS
GUIDE ROLLS

S-WRAP ROLLS

HEAT SEALER AND SERRATOR

BAG SEPARATOR

STACKING TABLE

Fig. 7-22. Blown film extruder set-up for trash-liner and garbage-bag production. (*Courtesy Gloucester Engineering Co.*)

high-density polyethylene, after chilling, will not stretch like low-density. If the film is collapsed too cold, it will wrinkle; if it is collapsed too warm, it will stick. You must find the optimum film temperature at which to work. Other factors that are critical include extrusion rate, melt temperature, film gage and ambient air temperature. So the tower itself must allow for adjustment in distance between die and nip rolls.

An adjustable vertical or even horizontal tower is used. The collapsing surface is fibrous to allow an air cushion to exist between the material and the collapsing surface for minimizing drag and sticking.

As for winding the thin tissue-like films, the ideal is automatic operation at 400 ft/min. line speeds.

At times you may have to anneal this tissue-like film over heated rolls (200 to 300°F), followed by cooling, depending on the resin you are running and what you want to do with the product.

Also in use for producing plastic paper for bags is a patented cross-laminated method. The material is produced from two high-density polyethylene oriented layers which are blown and slit at an angle.

Another technique for plastic paper is based on cast film uniaxially or biaxially oriented. This is discussed in further detail under cast film.

In industry parlance, plastic paper is sometimes distinguished from synthetic paper, in the sense that the latter product is aimed more at the graphics business (e.g., books, magazines, posters, etc.). Synthetic paper can be made either by paperizing plastic film (e.g., giving it paper-like qualities) or by using synthetic fibers or synthetic pulp made directly from polymers.

Paperizing a plastic film generally means you have to impart cavitation to the film to give it paper-like characteristics. This can be done internally by foaming, stretching, leaching, etc., or externally by coating, particle imbedding, abrading, laminating, etc. Typical of the internal methods used are (1) filling a polyolefin or polystyrene with clay, calcium carbonate, etc., and then stretching it so that the fillers become the nucleus for cavitation; (2) using two incom-

patible polymers (e.g., medium density polyethylene and polyisobutylene) so that, assuming one polymer forms a continuous phase while the other disperses, the dispersed polymer will function like a filler; (3) using a crystalline polymer (like polypropylene) so that the crystallites act as sites for cavitation if the film is stretched beyond normal orientation limits at temperatures below the crystalline melting point; and (4) using a composition of high-density polyethylene, mineral filler, and oil and then selectively leaching out the non-polymeric components.

Flat Film

The flat film extrusion process utilizes a die similar to extrusion coating in design (i.e., a "T" or "coathanger" die), but with somewhat thicker wall sections on the final lands to minimize deflection of the lips from internal melt pressure. The die lands should be comparatively long—a minimum of $\frac{5}{8}$ in. Typical dies are shown in Fig. 7-23.

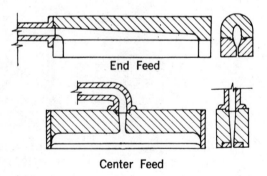

Fig. 7.23. Extruding film through a water bath requires a flat die of the type shown here. (*Courtesy Soltex Polymer Corp.*)

Following extrusion, the flat film can be quenched (i.e., cooled) either by being passed through a water bath or over a chromed chill roll. In the water bath technique (see Fig. 7-24), the film is simply extruded straight down into the water. The variables that must be adjusted to achieve the type of film wanted include: extrusion rate, material temperature, drawdown, distance from die exit to water surface, and water temperature. Since excessive

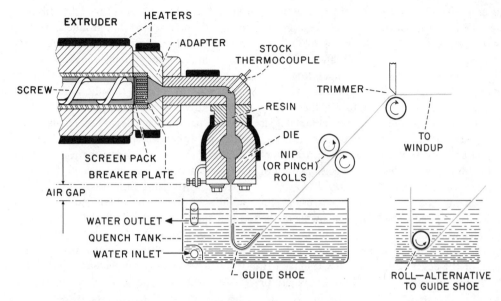

Fig. 7-24. Schematic cross-section of the front part of a flat film extruder, the quench tank, and the take-off equipment. (*Courtesy U.S. Industrial Chemicals Co.*)

water carry-over on the film may limit production rates, air jets or an air knife may be used to "wipe" the film surfaces dry. Wringer rolls and radiant heat may also be used.

In chill roll or cast-film extrusion, the hot melt extruded through the die slot is cooled by the surface of two or more water-cooled chill or casting rolls (see Fig. 7-25). The principle advantages of the film-casting technique as opposed to the quench-tank method are: better opticals; potentially increased output; and production of stiffer films.

In cast film extrusion, a die which extrudes vertically downward is generally preferred. However, dies which extrude horizontally or at any angle between vertical and horizontal can be used.

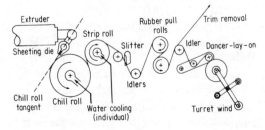

Fig. 7-25. Chill roll extrusion set-up. (*Courtesy E. I. du Pont de Nemours & Co., Inc.*)

In cast-film extrusion, the hot plastic web drops onto the first chill roll which it contacts tangentially. The alignment of this roll is critical in relation to the falling film. Wherever wrinkling of thin film occurs on the casting roll surface, the first roll must be carefully repositioned in relation to the die lands.

Temperature of the casting rolls is important. The optimum roll temperature for a given resin should be established. It is advisable to control the temperature gradient across each roll as closely as possible, preferably at 5°F (3°C) maximum variation.

The chrome-plated surface of the first roll must be highly polished because of its effect on film surface characteristics such as gloss and smoothness.

Most extruders use two, some three or more, chill rolls. The use of more than one roll has advantages in setting and cooling the film. The first roll takes some of the heat out of the web. This roll is generally kept at a temperature in excess of 100°F (40°C). The heat removal by the second, third, etc., rolls depends upon take-off speed, melt temperature, and the temperature of the first roll.

The film "S"-wraps around all chill rolls. It is desirable to maintain the shortest possible distances between the chill rolls and between the last chill roll and the windup equipment. Short distances minimize whipping of the film which is one cause of the wrinkling on the roll. The film should be at storage temperature when wound.

The use of an air knife to pin the molten web to the chill roll helps to eliminate air entrapment between the resin and the chill roll as well as speed heat removal. The final caliper of the film is determined by the combination of screw output and takeoff speed.

As in the case of any flat die extrusion, surface tension effects result in a narrowing of the extruded web in the air gap between the die and the chill roll. This causes a corresponding thickening of the edges which must be trimmed off.

"Paper" Based on Cast Sheet Uniaxially or Biaxially Oriented. This process involves extruding and casting a film or sheet through a water bath or over cooling rolls. Takeoff equipment consists of a series of rollers for machine direction orientation and a lateral stretching machine to obtain cross-direction orientation. Following is a description of both uniaxial and biaxial techniques.* In both instances, the extruded film uses the chill roll system rather than the water quench (which produces thick edges due to draw down for the die and may show corrugations because of water turbulence and water carry over into the orientation stage).

In a system for producing uniaxially oriented plastic paper, the system would consist of extruder, chill rolls, a longitudinal stretching arrangement, and a winder.

After extrusion and casting in the normal fashion, the film passes through the longitudinal stretching system. Generally, the system consists of the necessary rollers to preheat the extruded and cast sheet. Upon achieving the proper orientation temperature for a given polymer, the sheet is then passed through an adjustable short gap. This gap can be set to a given sheet thickness. Through the short gap method there is very little loss of width and molecular orientation is achieved at a loss in thickness only. As an example, a sheet may

*Based on a paper by Theodore R. Coburn, Film Div., Marshall and Williams Co., Providence, R.I.

enter the gap at a thickness of 18 mils and 62 in. wide, the oriented film produced could be 3 mils and 60 in. wide, depending on the polymer being used. The stretching ratio can easily be adjusted over a 10 : 1 range by adjusting the roller speeds one to another. The rolls are usually driven by a D.C., regenerative-type drive system with a high degree of regulation. The rolls are either water or oil heated and a close temperature control across the roll is absolutely necessary for gage and orientation control. After orienting, the film passes through several additional rollers to obtain some relaxation, annealing, and cooling of the finished film. An edge trim at this point is necessary on each side; the trim can be recycled, when possible, through a chopping and reblending system.

In making biaxially oriented plastic paper, the necessary equipment consists of an extruder, casting, and longitudinal stretching as outlined for uniaxially oriented material. However, the width is set to suit the lateral stretching ratio as it relates to the polymer. Annealing after longitudinal stretching is undesirable at this stage as it would inhibit the lateral stretching ability of the polymer.

The sheet is passed through a nip to apply back tension and then fed to the lateral stretcher clamps under tension. Once the sheet is clamped on both sides, it is conveyed into a preheat oven. The function of this oven is to raise the sheet to the required temperature for orientation. Once achieved, the rails are set on a diverging path to crosswise stretch the sheet. Lateral stretching ratios have been accomplished up to 10 : 1, here again depending on the polymer and the desired balance or imbalance of tensile properties required. The ratio of stretch is easily and infinitely adjustable over a given range and generally the equipment can be set to a 1 : 1 to 10 : 1 condition.

Lateral stretchers presently in operation are running at speeds in excess of 1000 fpm. Widths of 240 in. are commercial and oriented films and sheeting are being produced in 0.25 to 30 mils finished thicknesses.

The film is then stripped from the clamps by an additional nipping station. An edge trim is necessary and depending on the polymer the edge trim is recycled or used elsewhere.

Sheet Extrusion

Heavy-gauge flat-film material more than 10 mils ($\frac{1}{4}$ mm) thick is generally called *sheeting*. The upper limit for extruded sheet has usually been thought of as $\frac{1}{2}$ in. thick; however, more recently, dies for sheet that can run more than 10 ft wide and can produce thicknesses from $\frac{1}{2}$ to 1 in. are becoming more common, especially in the manufacture of sheet for thermoforming.

Extruders used for sheeting are of conventional design, resembling those used for other types of extrusion. There has been a tendency, however, to go to extruders with longer L/D ratios—30 : 1 up to 32 : 1 and higher. These longer L/Ds are particularly advantageous where volatiles have to be removed.

Sheeting is extruded horizontally from a die similar to a flat-film die (see Fig. 7-26). There are two basic differences, however—the use of an adjustable choke, or restrictor, bar, and longer die lands. In addition, sheeting dies are usually more heavily constructed to reduce warping of the die lands.

The choke bar is an adjustable bar located between the manifold and the die lands to distribute the molten polymer uniformly. The bar is used to smooth out gross variations in gage. It is usually set at the beginning of an extrusion run and not changed during the run.

The length of the die lands may vary with sheet thickness. However, since die pressure increases with die land length, the lands should not be longer than necessary to ensure high sheet quality. The surface of the lands must have a high polish for high-quality sheeting. Average die land lengths for extruding polyethylene sheeting are shown in Table 7-5.

One of the die jaws is adjustable to permit minor changes in die openings to maintain close control on sheet gage uniformity. The die opening should be set about 10% wider than the desired thickness of the finished sheet.

The take-off unit for sheeting usually consists of a vertical stack of three driven, highly polished, chrome-plated rolls and a pair of driven, rubber-covered pull-off rolls (see Fig. 7-27). The chrome-plated rolls serve three purposes—cooling, gage control, and imparting a desired finish to the sheet (such as a high polish or some embossed pattern). The cooling rolls are generally of twin-shell construction to permit close temperature control of the surface. The coolant is usually water of closely controlled temperature. The diameter of the rolls is generally in the range of 8 to 12 in. (20 to 30 cm), depending on the linear speed of the sheet and the amount of cooling required. The rubber rolls serve for tension control and as a pull-off device.

The spacing between the top and middle roles should be set 5% less than the desired thickness of the finished sheet. The spacing between the middle and bottom rolls should be 5% more than this thickness. The spacing between the rubber rolls should be a minimum, usually about 5% less than the sheet thickness, to maintain a uniform pull on the sheet.

The variable-speed drive for the take-off unit must have very uniform speed. It must be easy to adjust to permit making small changes in speed.

The framework for the take-off unit must be relatively free of vibration. Even slight vibrations may cause a wavy appearance of the sheet and pulsating variations in the gauge.

Sheet up to about 0.030 in (3/4 mm) can be rolled up like film. Sheet thicker than that should be cut to the desired length and stacked as flat sheet. A traveling power shear usually cuts the sheet automatically to the desired lengths.

For some applications that require sheet with special characteristics, it is possible to cap the extruded sheet with a plastic film. For example, extruded ABS sheet can be surfaced with such films as acrylic, polyvinyl chloride, polyvinyl fluoride, polyester, etc., primarily to impart additional weathering or scuff resistance, to upgrade chemical resistance, or to add special decorative effects. These films are laminated to the ABS sheet during the extrusion operation as shown in Fig. 7-28. Adequate bonds can be achieved with some films (like polyvinyl fluoride and polyesters) by using an adhesive that can be activated by heat and pressure. Acrylic, polyvinyl chloride, or polycarbonate are joined to the ABS sheet by heat and pressure alone.

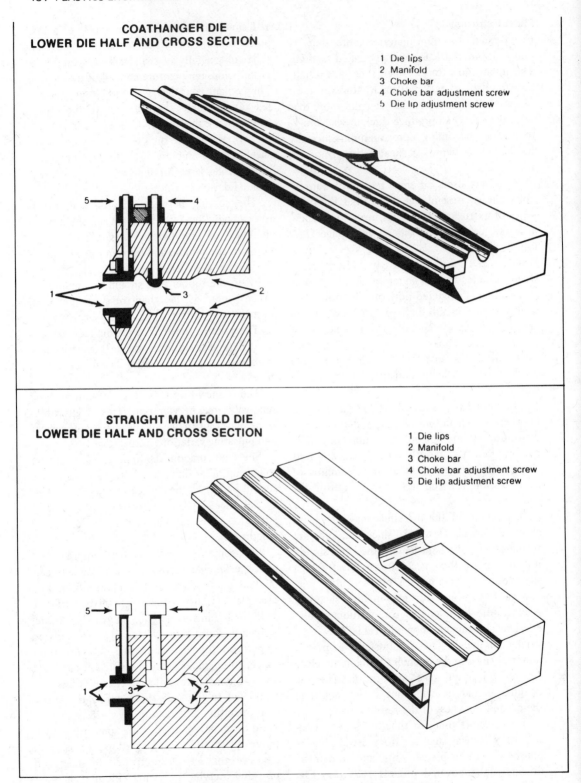

COATHANGER DIE
LOWER DIE HALF AND CROSS SECTION

1 Die lips
2 Manifold
3 Choke bar
4 Choke bar adjustment screw
5 Die lip adjustment screw

STRAIGHT MANIFOLD DIE
LOWER DIE HALF AND CROSS SECTION

1 Die lips
2 Manifold
3 Choke bar
4 Choke bar adjustment screw
5 Die lip adjustment screw

Fig. 7-26. Dies for extruding ABS sheet. (*Courtesy Borg-Warner*)

Table 7-5. Average Die Land Lengths for Polyethylene Sheeting.*

Sheet Gage		Die Land Length	
In.	mm	In.	mm
0.010-0.030	0.25-0.75	½	13
0.030-0.060	0.75-1.50	1	25
0.060-0.100	1.50-2.50	1½	38
0.100 or more	2.50 or more	2 or more	51 or more

* Courtesy U.S. Industrial Chemicals, Div. National Distillers

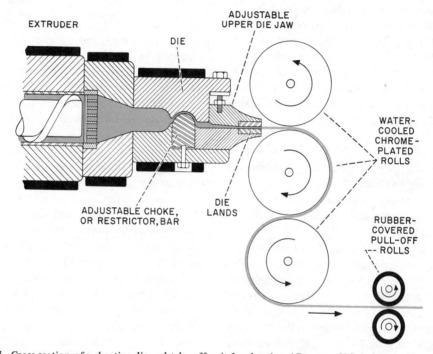

Fig. 7-27. Cross-section of a sheeting die and take-off unit for sheeting. (*Courtesy U.S. Industrial Chemicals Co.*)

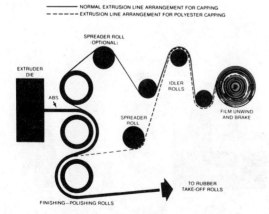

Fig. 7-28. Normal extrusion line arrangement for capping ABS sheet. (*Courtesy Borg-Warner*)

Extrusion Coating

The extrusion coating process combines a molten resin film emerging from a flat die with a moving solid web or substrate. The resins most commonly used are the polyolefins, such as polyethylene, ionomer, ethylene vinyl ace-tate copolymers, and polypropylene. Nylon, polyvinyl chloride, and polyester are used to a lesser extent. The substrates involved are com-monly paper, paperboard, metal foils, or trans-parent films, with woven and non-woven fabrics being used in special applications. Often, com-binations of the above resins and substrates are used to provide a multi-layer structure incorporating the particular characteristics of each layer in the final composite.

A related technique is called *extrusion laminating* and involves two or more substrates (e.g., paper and aluminum foil) combined by using a plastic film (e.g., polyethylene) as the adhesive and as a moisture barrier.

Coatings are applied in thicknesses of about 0.2 to 15 mils, although the common average is 0.5 to 2 mils. The substrates range in thickness from 0.5 to greater than 24 mils.

A Polyethylene Extrusion Coating Line. A basic line for polyethylene extrusion coating is shown in Fig. 7-29. In operation, the substrate leaves the unwind station and generally is subjected to one of various pretreatment techniques that enhance adhesion of the coating to the substrates. Factors affecting this adhesion include the chemical structure of the resin, the oxidation of the resin in the air gap between the extrusion die and the chill roll, melt temperature, line speed, coating thickness, and the pretreatment methods.

The pretreatment or priming can be accomplished by three basic methods: chemical, flame, and corona.

Chemical primers are usually of two general types, i.e., solvent base and water base. Usually, the solvent base primers are used on polymer films and foil while the water base primers are used on paper and board substrates. Both types of primers are normally applied to the substrate from a gravure system with a hot air dryer used to remove the diluent. In some cases, reverse roll coaters are used to apply the primer and

steam rolls are sometimes used (when line speeds are low) in place of dryers.

Flame treatment is accomplished by contacting the surface of the substrate (on the side to be coated) with one or more "oxidizing" flames. The flame exits from a ribbon-type burner, which is usually equipped with deckles to vary the flame width. Gas and combustion blowers supply the products for combustion to the burner. In most cases the substrate passes approximately 4 in. below the burner face.

Corona priming is accomplished in the same manner as treating a polyethylene surface for printability or glueability. The only difference is that the substrate surface is treated prior to coating instead of afterward. There are many types of generators available. In most cases, a grounded treater roll covered with a synthetic elastomer as the dielectric is used.

Following pretreatment, the molten resin emanating from the extrusion die is laid on the surface of the substrate. This point of combination is usually a nip between a chromed chill roll on the resin side and an elastomeric pressure roll on the substrate side. The two rolls press the resin and substrate together, leave the impression of the chill roll on the coating (i.e., the impression can be gloss, matté, or special design), and remove heat from the resin so that it can be stripped from the chill roll and handled without blocking.

After leaving the chill roll, the edge bead of resin can be trimmed off both sides and the final composite wound onto a finished roll. Vari-

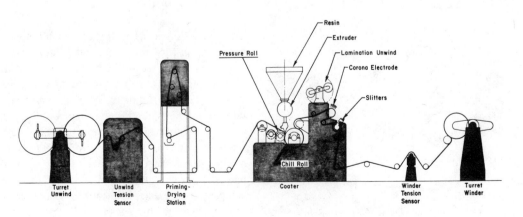

Fig. 7-29. Typical extrusion coating line. (*Courtesy E. I. du Pont de Nemours & Co., Inc.*)

ous auxiliary equipment can be added to the line to monitor and adjust tension, give surface treatment to the quenched resin (e.g., electrodes are placed between chill roll and winder to treat polyethylene surfaces to allow printability or to enhance adhesion), or provide a secondary unwind for a three-ply lamination. In addition, other extruders and chill rolls can be added in tandem to produce multi-layer structures in a single pass or to coat both sides of the substrate at the same time.

Die and Screw Design. The normal capabilities of an extrusion coating line generally permit a relatively high-speed operation in widths varying up to about 12 ft. Die design is similar to a flat film die. Two of the more popular designs for extrusion coating are: the "key hole" or uniform manifold cross-section type with long preland and the "coathanger" type where the preland is longest at the center and diminishes toward the ends to compensate for velocity and pressure changes along the manifold (see Fig. 7-30). Where the extrusion coating die does differ from the flat film die, however, is in the use of deckles (movable plug-off devices) that can be moved in or out to adjust the molten resin width to the width of

the substrate to be coated. Deckles are either internal rods or are devices clamped to the outside of the die.

Since the die controls the uniformity of the coating, it must be accurately machined of hard tool steel and must have highly polished manifold, slot, and lands. Its lower part has a "V"-shaped cross section so that it can be moved downward as close to the nip of the chill and pressure rolls as desired.

Coating dies must be wider than the substrate to be coated. Dies used in commercial coating are commonly 54 to 96 in. wide, but up to 12 ft is possible.

A uniform die opening across the entire width of the die is desirable. If this results in higher or lower output at the die edges, the melt temperature at these points can be adjusted to compensate for it.

The final lands on the die normally range between $\frac{1}{2}$ and 1 in. in length and the gap between the land is usually set at about 0.020 in. Adjustment of the gap can easily be accomplished with a series of adjusting bolts that help to equalize the uniformity of caliper across the die.

Temperature control is very important

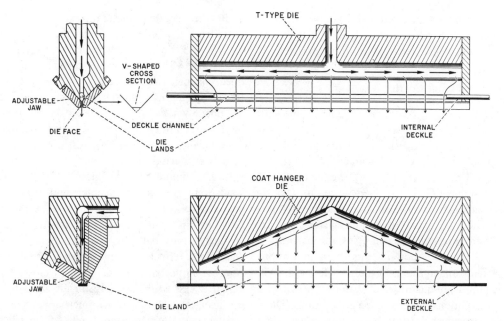

Fig. 7-30. Schematic cross-section of "T" and "coat-hanger" type extrusion dies. Locations of internal and external deckles are indicated. (*Courtesy U.S. Industrial Chemicals Co.*)

throughout the extruder and die since both caliper and adhesion to the substrate are affected. Melt temperature monitoring through both thermocouples and infrared devices are quite common to assist in achieving uniformity of temperature across the web. The melt temperature required depends on the type of resin and adhesion-related degree of oxidation. Thus, melt temperatures may range as high as 650°F to as low as 250°F for certain resin systems. However, in each case, the melt temperature uniformity should be held within a 10°F range. In order to achieve this, extruder screw designs have tended toward longer length-to-diameter ratios with long metering sections (also for higher output/spin at high screw speed). Most designs are of the long metering type with a feed section of constant depth, a transition section where depth shallows from that of feed to that of metering, and a shallow metering zone with a length of 50% or more of the screw.

Higher outputs or greater uniformity from a particular extruder diameter tax the capability of the long metering screw, and lead to special considerations rather than simply optimizing the length and depth of the three sections. The addition of mixing sections to the screw is one of the techniques available.

Valving of the extruder is quite common both to aid in homogenizing the melt and in attaining higher temperatures through shear.

Wire and Cable Coating

The covering or coating of wire and cable with insulating or sheathing plastics is an important application of extrusion and large quantities of various resins are used annually for this purpose. The requirements for power, control, and communication cable increase each year as increasingly complex and far-reaching systems are developed and the older systems are replaced or revised. Current plans, for instance, to install many power and communications systems underground in the next few years will require large quantities of various types of wire constructions used for these applications. These systems must perform under increasingly rigid requirements and must be engineered to exact-

ing specifications. This, in turn, requires continual improvement of wire coating equipment which must produce to more and more exacting electrical requirements.

The basic unit of the wire coating system, as before, is the extruder, which as already indicated, has been undergoing gradual change to improve outputs and temperature and pressure uniformity of the extruded melt. Temperature controller design, for example, has become increasingly sophisticated to avoid variations which may be acceptable for production of large diameter power cables, but unsatisfactory for manufacture of foamed insulation to the exacting capacitance requirements of television or telephone transmission lines. Desire for high resin outputs to reduce investment in new wire lines has resulted in drives with increasing horsepower. As an example of this, telephone insulation coating speeds over the past several years have increased to the point where it is not uncommon to manufacture at rates in the 3000 to 6000 ft/min. range. As outputs increase, screw design must be modified to hold melt temperatures at desired levels without sacrifice of melt uniformity. However, mixing action cannot be sacrificed, particularly where colorants are incorporated into the resin mix at the extruder.

In wire coating, molten plastic from the extruder (Fig. 7-31) passes into the crosshead die (Fig. 7-32) where the wire insulating process takes place. The melt passes through streamlined channels of ever decreasing cross section, traversing a 30 to 90 degree angle to effect alignment with wire flow. Since balanced flow in this area after rounding the turn is a prime requirement, a number of crosshead designs have been devised to accomplish this objective with a minimum of weld lines in the finished insulation. As with the extruder, uniformity of heat control is essential, since areas of excessive heat loss can result in rough coatings or changes in melt flow patterns that could have a deleterious effect on the finished product. Another important consideration in crosshead design is ease of disassembly for die and guide tip changes to reduce downtime between processing runs. Both adjustable and self-centering dies are used, depending on the wire con-

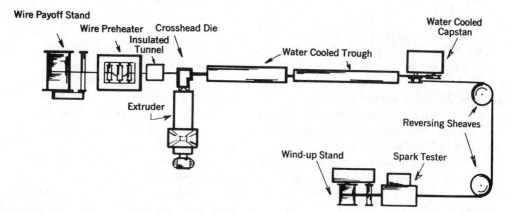

Fig. 7-31. Schematic of flow line for wire coating, showing preheater between pay-off and die. (*Courtesy Soltex Polymer Corp.*)

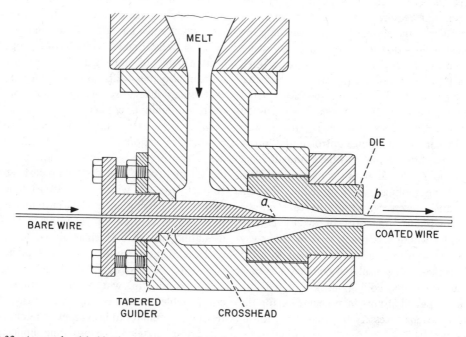

Fig. 7-32. A cross-head holds the wire-coating die and the tapered guider. (*Courtesy U.S. Industrial Chemicals Co.*)

struction that is being produced and the eccentricity and the wall thickness tolerances which are required.

The wire is positioned in relation to the die in the crosshead by the guide tip. Minimum contact and wear are desirable in this area and as a result, diamond inserts are often used in guide tips used in high-speed wire coating at rates of 3000 ft/min. and above.

Wire coating dies are of two general types.

With a pressure type die, the wire is coated within the body of the die while the plastic melt is still under pressure. With a tubing die, a tubular extrudate is drawn down upon the wire as the melt emerges from the die and a vacuum may be required to ensure tight contact of the coating to the wire.

The die design and type used depend on the particular plastic being processed, uniformity of the wire core being fed through the system, diameter of the insulated conductor, processing rates, and configuration of the finished product which may have widely varying shapes. Spacing between the guide tip and die in the angled section, called *gum space*, varies for the different types of plastic materials being applied to the wire. If clearances in this area are too small, melt fracture can lead to roughness on the surface of the finished product. If too large, in extreme cases, roughness can again result, along with the added possibility of stretching when insulation is being applied to small diameter conductors.

In pressure-type dies, the gum space may be varied by moving the guide tip back and forth within the crosshead. With tubing type dies, however, the required space must be carefully designed into the system before machining the finished guide tip and die. Since so many different materials are used in wire coating, contact with the resin manufacturer is desirable prior to designing the system to be used.

The spool of wire to be coated is carried from a payoff device, through various wire straighteners and tension control units, into a preheater, through the crosshead and die, and into a cooling trough, with the pull of a capstan beyond the cooling trough furnishing the uniform drive necessary. From the capstan, the wire flows into some sort of takeup unit or caterpillar whose sole function is to take away the insulated wire at a rate consistent with the capstan speed. The insulated conductor is then wound on spools, coiling blocks, or fed into various types of drums in preparation for the additional steps necessary to produce the finished product.

The wire and cable industry was one of the first to apply several coatings to the conductor in a single pass, and a number of applications of this technique are currently used. Single pass processing by this technique has a number of advantages over the old two- and three-pass techniques, including reduced investment in auxiliary equipment needed if several wire lines are used to produce the same finished construction. For a double coating, both tandem and common head extrusion are used and such equipment is available from machinery manufacturers of wire and cable equipment. For a triple extrusion, three extruders are mounted on the line and there is no reason that more could not be added, providing one of the extrusions is not rate limiting.

Payoffs of various types are used for high-speed wire coating applying thin insulation to small diameter conductors. These units may take the form of so-called flyer types. Dual units allow continuous operation by welding the ends of the wire together from one reel to the next. Very fine wire is often furnished on spools which may be mounted on controlled tension pay-offs, normally motor driven, with speed coordinated to that of the capstan at the end of the wire line.

For the large diameter conductors used in power cable construction, payoff is from large reels which must be well balanced and mounted on effectively braked units having supporting arms capable of moving in both vertical and horizontal planes to lift and position the often exceedingly heavy spools of conductor, normally aluminum or copper.

Wire preheaters, used to impart to the conductor various levels of temperature, depending on the wire construction, are generally of two types. The first and oldest type is a contact unit in which preheating is accomplished by passing an electric current through the wire just prior to its entrance into the insulating crosshead. Such a unit would be used on large diameter conductors where preheating is felt to be necessary, since the bronze or copper conducting wheels can sometimes offer considerable resistance to wire flow. Such resistance could result in excessive stretching and cause rejects in telephone insulating applications.

The second type of preheater in common use today is the non-contact design, which for many applications is used in preference to a contact unit. Little resistance to wire flow is encountered in the non-contact design and it is commonly used in high-speed insulating applications.

At one time, gas flames were used extensively for preheating. However, problems were occasionally encountered with small quantities of moisture on the conductor entering the insulating crosshead and causing undesirable voids in the finished construction.

The next unit in the line, and one which is quite often underrated in importance, is the quench trough. These troughs may vary in length from a few feet to several hundred feet, depending on wall thickness and coating speed. The longer designs are used in cooling heavy wall insulation, such as power cable, where slow cooling is necessary to prevent void formation in the finished product. High-speed insulating troughs may vary from 30 to 100 feet, but care must be used to prevent conductor stretching due to excessive water resistance. Temperature indicating devices are desirable where gradient cooling is used, since individual sections of the trough quite often come equipped with their own water heaters which may occasionally malfunction. In addition, where water contains dissolved solids, system plugging may occur unexpectedly. The water must completely cover the cooling insulation and must be introduced in such a way as to eliminate bubble formation which can cause pits in heavy wall insulation. Finally, care must be used to make certain that water droplets do not splash on the molten resin coming from the crosshead.

Capstan units of various types, depending on production rates desired and the wire constructions being produced, draw the conductor through the coating system at constant speed. If high tensions are required as, for instance, with power cables, both payoff and takeup capstans may be used for better cable support in the system.

Takeup units spool the finished product and are controlled in various ways to produce a level wind without introducing excessive vibration which can carry back through the system. These can vary from rather small dual units for continuous operation at high speed to much larger designs for power cable or communication cable jacketing operations.

With the current trends toward more automation and improved control, all types of electronic auxiliaries are installed in the modern wire line. These include capacitance monitors, diameter gages, electronic footage counters, spark testers, printing systems, and dancer units for improved tension control. In addition, since many of the current wire and cable resins contain varying quantities of carbon which can

pick up moisture at rapid rates, large capacity dehumidifying hopper dryers are quite often installed at or near the extruder. Since each manufacturing operation has its own requirements, engineering of the system includes careful analysis of equipment from various manufacturers to determine which best meets the needs of the particular type of wire coating under consideration.

It should also be noted that large quantities of insulation are cross-linked after extrusion to produce three-dimensional polymer networks with substantially higher melting points and, thus, improved temperature ratings in service. The polymer melt contains a heat activated cross-linking agent such as an organic peroxide. Cross-linking is accomplished by passing the insulated conductor through a high-pressure steam tube, with cooling in a high-pressure water section to prevent void formation. Irradiation techniques (on the extruded product) are also used.

Tubing or Pipe

Dies used for extruding hollow tubing or pipe do not differ markedly for the various thermoplastics that may be processed in this way. In general, flow passages should be streamlined and polished to eliminate stagnant areas and to assure uniform flow. In addition, of course, a mandrel or pin must be used to form the inside surface of the hollow tube or pipe (in contrast to the solid rods and profiles described below). Since melt pressures in the die may be high, the resin may swell more than predicted and die dimensions will have to be adjusted for this. On the other hand, if the melt cools or "sets up" slower than other materials and if cooling is not very effective, considerable "drawdown" may be experienced. In this case, an oversized die may have to be designed to achieve the final dimensions required. As with profile extrusions in general, the die design is largely a matter of experience and trial and error. One may expect to change or modify a die a number of times to achieve the desired part. If a die is available which is similar to the intended shape or designed for another material, it should be tried with the resin under test to aid in making a new die. It should be remembered however, that too much "drawdown" introduces stresses in the part which may ultimately cause undesirable distortions during use and should be minimized as much as possible.

Figure 7-33 illustrates a tooling arrangement for an extended mandrel pipe die. In this instance, the die is mounted in crosshead fashion on the end of the extruder so that the axis of the screw and the tubing take-off line are off-set. Straight-through dies are also used, but this design requires that the mandrel (which forms the inside of the pipe or tube) be supported by fins (i.e., a spider) or that a breaker plate plug holder be used.

Tubing or pipe can be cooled and sized by any of the standard techniques generally used. It is necessary, however, to have sufficient cooling available so that the melt will be quickly chilled to the point where it will not stick to the metal surfaces with which it comes in contact. It may be found necessary to lubricate the melt with water or a water-soluble oil to keep it sliding along a sizing ring edge or forming shoe.

In the production of plastic tubing or pipe, several methods are used to shape, cool, and control the size of the melt after it emerges from the extrusion die. Small tubing often can be made by "free extrusion" wherein the molten tube is simply flooded with cooling water by showers or immersing in a tank or trough filled with circulating water. The size in this case is determined by the size of the extrusion die and the drawdown. Tension is maintained on the cooling melt by pulling the tube at a faster rate than the melt is issuing from the die. The final tube or pipe is smaller than the die and quite often a very slight internal air pressure, introduced through the die mandrel, will keep the tube from collapsing and will maintain a circular cross section.

If the tube is about 0.25 in. in diameter or larger, it is difficult to maintain a constant shape or dimension by free extrusion. Guides and supports are needed as the material cools. A series of plates or sizing rings which will form and support the melt as it is drawn down through the water will enable exact sizes and shapes to be made. For tubing, the sizing is done by a series of circular holes, whose sizes decrease gradually to conform to the tube as it solidifies. Complex shapes may require rather elaborate guide plates which support or blend the melt as it cools.

Typical of external sizing methods are the

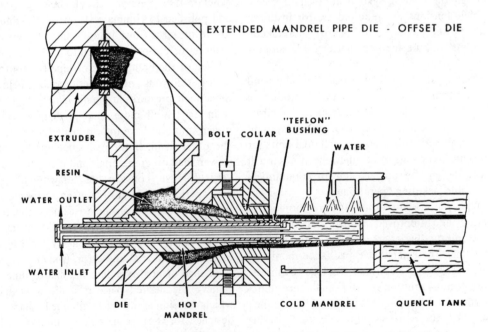

Fig. 7-33. Extended mandrel pipe die. (*Courtesy E. I. du Pont de Nemours & Co., Inc.*)

pressure-forming box technique, in which air or inert gas is passed through a hollow mandrel to force the molten plastic against the wall of the forming box; the vacuum box technique (Fig. 7-34), where vacuum is applied to the jacket

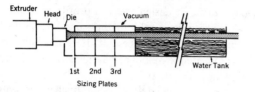

Fig. 7-34. Pipe sizing by vacuum box technique. (*Courtesy Celanese Corp.*)

surrounding the forming tube and atmospheric pressure forces the tube to conform to the forming box; and sizing plate calibration, in which the tube is drawn out of the die and through thin sizing plates. Internal sizing can also be used (Fig. 7-35), involving the pulling of the tube over a tapered, cooled mandrel as it exits the die, then conveying the tube into a conventional water quench bath.

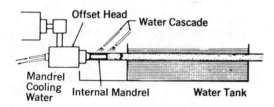

Fig. 7-35. Internal mandrel sizing of pipe. (*Courtesy Celanese Corp.*)

In using sizing rings or a vacuum sizing device, the extrusion die, of course, is larger than the finished shape and the melt is drawn down as it cools. Square tubing has been prepared using an internal forming mandrel combined with external cooling. Cold water, preferably 50°F or less, is circulated through the brass or aluminum forming mandrel. The external surface of this mandrel over which the plastic slides as it cools is roughened with a grit blast or with medium sandpaper and is tapered to allow for shrinkage as the tube cools. Slight variations in internal diameter of pipe or tubing can be made by changing the position at which the external cooling water first hits the two. A

number of sizes and shapes can be made with a single extrusion die, simply by changing the forming mandrel.

More recently, a number of new techniques have been evolved for the manufacture of different variations of plastic pipe and tubing. Corrugated tubing is typical. This design offers a number of advantages, including the fact that ordinarily stiff plastic pipe in thin-wall can now be processed into coils where long lengths are needed. In addition, the design offers added resistance to external pressure.

Thus far, the corrugated extrusions show potential for pipes, tubes, or conduits that have to be buried underground; for protective coverings, such as wiring harnesses for automotive use; for insulation for metal pipes and electrical cables; for protection of underground burial of house service lines; for vacuum sweeper hoses and pool hoses; etc.

Much of the detail of the manufacturing process is still shrouded in secrecy (because of the patent situation), but some basic facts are known. For short runs and for very large sizes, blow molding is used. However, it is difficult to produce continuous lengths in this way and the pieces are generally limited to less than 10 ft. Drainage tubing and most of the smaller tubing is made by continuous traveling chains of mold blocks. A long snouted die from an extruder extends into the junction point of the blocks and deposits the smooth tube into the corrugations of the blocks as they pass by. Expansion into the tube can be done by either pressure or vacuum through the blocks. Production speeds vary from 5 to 50 ft/min.

Another take-off on the manufacture of pipe involves the spiral winding of extruded sheets of plastic around a mandrel. The spiral winding technique is especially applicable for large diameter pipe.

Rods and Irregular Profiles

Round rod stock is the least complicated of extruded profile cross sections, but maintaining roundness and preventing shrinkage voids at commercial rates is often difficult.

When practical, rod is freely extruded into a water or air-quench system. Resins of higher melting point or higher crystallinity and large diameter rods may require a cooling sleeve, or, more likely, forming box methods. In this, the rod is shaped in a forming box while under constant melt pressure from the extruder; constant take-off is controlled by the haul-off. Forming boxes or sleeves may vary in length, depending on the cooling characteristics of the resin. The slow cooling rate, combined with a low shrinkage during solidification, however, facilitates the extrusion of heavy rod which is free of internal voids.

Irregular shapes are the most difficult type of profile to extrude. Quite often the melt flow equations for plastic melts have been applied in the design of some dies for various shapes. However, most dies are still made by the "cut and try" method. Some shape or profile extruders have stated that on an average this type of die makes three trips between the plant of the extruder and the diemaker shop before it is satisfactory for a production run. A completely correct design can be worked out in advance only when the variables to be encountered can be accurately calculated. The land length of a die is also important in controlling the shape of an extruded article.

For instance, a die with a rectangular opening will not give a rectangular cross section, but a pillow-like one, to the finished piece. If the length of the land is reduced at the sides, thus reducing resistance to flow of material at these points, the extruded part will be more nearly rectangular. The same result of course can be achieved by modifying the die opening and making it somewhat concave around the center area. Some drawdown is normally required to maintain control of the extruded shape (to prevent sagging, distortion, and so forth). Drawdown should normally be kept to a minimum so that internal stresses are minimized and a drawdown ratio of about 2 : 1 is usually adequate. Melt flow tends to increase linearly with a decrease in die land length and directly with the cube of the gap opening. In other words, shortening the land by a factor of 2 would increase the flow by a factor of 2. Increasing the gap by a factor of 2 would increase the flow by a factor of 2^3 (two cubed) or 8. For this reason, minor changes in flow are normally adjusted by changing land length.

Perhaps the best indication of the complexities of die design for extruded profiles is the following recommendations (from Celanese) on extruding polyethylene profiles:

Die design is one of the most important yet most difficult design areas in profile extrusion. The extrudate is seldom the same size and/or shape as the die orifice exit. Here is where art and knowledge of a particular plastic's rheology becomes important.

A well designed die is streamlined to prevent obstructions to flow (hang up or stagnation causes degradation and possible defects in the extrudate). The internal entrance or angle of approach to the die land section must not be excessive or extrudate surface appearance can be affected. Included angles 30-45° are considered maximum. This is another reason why small extruders are used; large barrel diameters would require large entry angles. For polycrystalline materials (like polyethylene) which are susceptible to melt fracture and very thin die openings, the included approach angle should be about 20°.

Designing the die orifice oversized is necessary because tension is required to pull the extrudate away from the die and through subsequent sizing equipment. The amount of the oversize depends on melt elasticity and strength, with less needed for high melt elasticity (high extrudate swell) materials. The effects of tension are greater on the width than on the thickness.

The unequal flow and cooling of complex shapes must be considered in developing the die orifice shape. All sharp projections and edges will cool faster and shrink more. The resultant changes must be compensated for in defining the orifice shape. In practice this means distorting the die profile as illustrated in Fig. 7-36.

Another important consideration in die design is the use of lands (length of constant cross section just prior to the die orifice exit) to control extrudate swell, and exit velocity. The land provides back pressure in the head and extruder barrel promoting mixing and a more homogeneous extrudate. It reknits weld lines due to mix and longer residence time under heat and results in a smooth, lustrous surface.

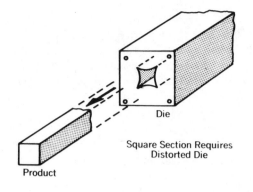

Square Section Requires
Distorted Die

Product

Die

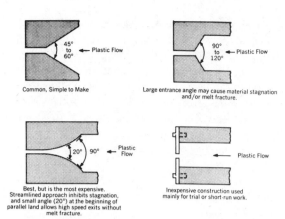

45°
to
60°
← Plastic Flow

Common, Simple to Make

90°
to
120°
← Plastic Flow

Large entrance angle may cause material stagnation
and/or melt fracture.

20° 90°
← Plastic
Flow

Best, but is the most expensive.
Streamlined approach inhibits stagnation,
and small angle (20°) at the beginning of
parallel land allows high speed exits without
melt fracture.

← Plastic Flow

Inexpensive construction used
mainly for trial or short-run work.

Fig. 7-37. Die entrances for profiles. (*Courtesy Celanese Corp.*)

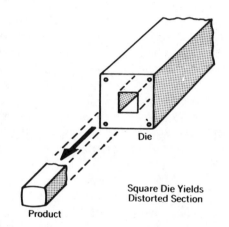

Square Die Yields
Distorted Section

Product

Die

Fig. 7-36. Compensating for die distortion. (*Courtesy Celanese Corp.*)

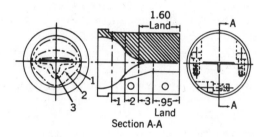

1.60
←Land→

1
2
3

├1├2├3├.95→
Land

Section A-A

Fig. 7-38. Die design for profiles. (*Courtesy Celanese Corp.*)

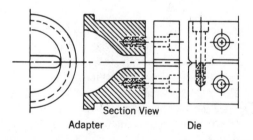

Section View
Adapter   Die

Fig. 7-39. Die design for profiles. (*Courtesy Celanese Corp.*)

It controls the velocity of the moving melt by offering resistance to flow. Because profiles are normally non-uniform in wall thickness, a velocity distribution occurs across the die opening. Thin sections are starved, while heavy sections pillow. Adjusting the land length is a means of controlling the extrudate shape and dimensions. Short lands result in higher swell and greater velocity. If too short, they cause poor surface finish and greater gage variation. If too long, they result in a reduced output. For any shape and material a constant land length-to-thickness ratio should be maintained. Profiled lands give optimum results and greatest dimensional control. For polyethylene, the land length should be 10 to 20 times the thickness of the extruded section.

Figures 7-37, 7-38, and 7-39 represent typical die designs. Figure 7-37 depicts a poor design—a flat plate die with a 180° entry angle. This should be used only with thermally stable resins and then only for prototyping or short runs. Figure 7-38 shows an ideal design—streamlined flow and long profiled land. This type of die is expensive and is used only where long continuous runs are required. Figure 7-39 depicts a compromise design. Here a common streamlined die body is used and various shaped dies are bolted on. This design is economical and versatile and is most common. It can be used for short and medium production runs.

Steel is the common material for die construction. Internal surfaces are chrome plated only for long runs. Lips can be hardened for stock dies after dimensions and shape have been developed. This will prevent damage. Brass and beryllium copper are often used because of ease of machining. They are used primarily for short run custom jobs.

Both straight-through and crosshead die bodies may be employed for solid plastic extrusions. A crosshead is necessary when coating a substrate such as metal foil. In both cases a low volume head should be employed.

Special Profiles. A number of techniques are available for producing profiles for special jobs. For example, dual extrusion is the combination of two different thermoplastics (although they may both be members of the same family; i.e., flexible vinyl and rigid vinyl) into a single homogeneous profile. The technique can also be applied to combinations of two different colored materials. In both cases, the materials are bonded together in a single die and extruded as a one-piece unit.

Another technique in use with extruded profiles involves metal embedment, with continuous lengths of wire or flat metal strips (in continuous coils) being embedded completely or partially within the extrudate. In operation, the metal insert is simply fed through the die while the plastic flows around it.

This type of technique can also be used to produce metallic-like trim or moldings. In such applications, an aluminum foil or a metallized polyester film is bonded to the surface of an extruded plastic profile. This operation is generally accomplished in a one-step extrusion process, using a complex die that combines the two materials before they exit as an integral extrudate. It is also possible to incorporate a thin aluminum foil inside an extruded profile based on a clear rigid plastic (e.g., butyrate) so that the plastic appears to have a metallic finish.

Foam Extrusion

Extruding foam materials became a major business for the plastics industry in the late 1960's, particularly the extrusion of foamed polystyrene sheet for egg cartons, meat trays, and the like (see also Chapter 20 on Cellular Plastics) and the extrusion of structural foam profiles (i.e., profiles with a solid exterior skin and cellular core).

Several techniques are currently in use for extruding polystyrene foam. In one, precharged pellets (incorporating a blowing agent) are extruded on conventionally designed equipment. In a second, a blowing agent, like dichlorodifluoromethane, is directly injected into the polystyrene melt so that it expands as it exits the die; standard extruders with long L/D in the 32 : 1 to 50 : 1 range are used. In a third technique, twin-screw extruders are used with polystyrene compounds containing a blowing agent (see also p. 199 on Multiple Screw Extruders). And in a fourth, a tandem extrusion set-up is employed in which one extruder (with a two-stage screw) melts and mixes blowing agents and polystyrene resin and feeds the melt to a second larger-diameter extruder which cools and extrudes the melt (Fig. 7-40).

Die designs for foamed polystyrene sheet vary, depending on the system used and include spider plates and crosshead types. Most common is the two-legged spider which is useful because of the short polymer flow through the die and the fact that the sheet can be slit on the spider lines. The key to optimum die design involves balancing the flow restrictors downstream of the spiders for gage uniformity and minimizing die-lip pressure drop without developing pre-expansion of the gas in the die.

In a typical operation, a water-cooled sizing mandrel is used, with the sheet being slit on both sides as it leaves the mandrel. The two sheets are then directed to a dual S-wrap draw-roll unit, which consists of four rubber-coated rolls. The two middle rolls are designed to form a nip for startup. After the sheet is threaded to the respective winders, the two middle draw rolls are moved apart and set adjacent to the top and bottom rubber rolls to form the S-wrap for normal production. Surface winding is generally recommended to produce a tight, evenly wound roll with build-up ratios in excess of 10 : 1. Center winding transmits torque through the inner layers,

Fig. 7-40. A 4½/6 in. tandem foam sheet extrusion line. (*Courtesy NRM Corp., Subs. Condec Corp.*)

causing them to slip, stretching the sheet, and compressing the inner layers.

Foam extrusion is also involved in coating wire with cellular polyethylene and in extruding polyolefin film or sheet for subsequent cross-linking (either by irradiation or by incorporating organic peroxides in the resin mix) These are discussed in greater detail in Chapter 20 on Cellular Plastics.

More recently, extrusion has come into consideration for producing structural foam profiles (e.g., vinyl, polystyrene, etc.) that can compete with wood in construction trim and moldings, picture frames, window frames, furniture, etc. Again, a number of techniques are available.

One is a patented system (the Celuka process from Ugine Kuhlmann, Paris, France), which is available for license. For more complete details, see the section on Structural Foams, Chapter 20. Basically, however, this process is distinguished from others in that die and sizer have identical internal dimensions and are situated immediately continuous to each other. A hollow profile is extruded from the die into the sizer where no change in external dimensions occur; rather, the hollow profile forces inward toward the center, creating a thicker skin.

Another method in use for structural foam extrusion today involves a conventional single-screw machine using plastics incorporating chemical blowing agents (although direct gas injection and tandem extrusion set-ups are also used) and fairly standard extrusion techniques.

A typical foam profile screw has a diameter of 2 to $2\frac{1}{2}$ in. and an L/D ratio of 24 : 1 or 30 : 1. Little has been revealed to date on screw design. It is known, however, that the process requires particularly careful mixing and temperature control to maintain uniformity and prevent premature blowing. One technique for more accurate temperature control, for example, uses aluminum fin shrouds between heater bands and barrels.

Dies are similar to other profile dies, though some have more extended lands and may have differential heating on each plane of the profile shape. The sizer is situated 6 to 8 in. beyond the die. It is about 4 to 6 in. long, is water-cooled, and may have vacuum rings applied. Further downstream, most set-ups use one or two cooling troughs, an endless-belt-type puller, a saw for cutting, and an automatic dump table.

COEXTRUSION

A relatively new variation of extrusion involves the simultaneous, or coextrusion, of multiple molten layers from a single extrusion system (in some instances, involving as many as four different extruders). In other words, an integral construction of two or more different plastics films or sheets can be produced on a single die. Thus far, the technique has been used for film, sheet, tubing, and extrusion coating, although profiles, wire coating, and even parisons for blow molding show potential.

The coextrusion technique has aroused the interest of a number of markets. By using melt temperatures to bond the various polymers together (or, in some instances, by using the center layer of the coextrusion as an adhesive for joining the outer layers), coextrusion becomes an economical competitor to conventional laminating techniques by virtue of the reduced materials-handling costs, raw materials costs, and machine-time costs resulting from producing the structures in one pass. Pinholeing is also reduced with co-extrusion, along with the problems of delamination and entrapment of air between the layers of plastic.

Typical coextruded combinations include a variety of packaging films that balance different plastics offering varying degrees of moisture resistance, gas barrier properties, economics, etc. For medical and sterile-packaged disposables, for example, a polyethylene/nylon/polyethylene combination has found use. For shrink film for pallet wrap and for shopping bags, coextruded low density polyethylene/high density polyethylene offers a balance of rigidity and low cost. Other uses: polystyrene/polystyrene foam coextrusions for egg cartons and meat trays (the polystyrene provides a glossy, attractive surface); polypropylene/saran/polyethylene coextrusions laminated to XT polymer, vinyl, or styrene for food containers; and PS/barrier plastics coextrusions for transparent "cans." In the sheet area, an immediate target of interest is ABS (for chemical resistance)/polystyrene (for economy) coextrusions for refrigerator door liners and margarine tubs; acrylic (for weather resistance)/polystyrene (for economy) coextrusions for outdoor building applications; polystyrene/polysulfone coextrusions for electrical and appliance applications where heat resistance is necessary; and ABS/rigid PVC coextrusions for shower stalls. The possible combinations are limitless.

A number of techniques for coextrusions are currently available to industry (some are patented and available only under license). Basically, however, there are three general concepts in use:

(1) The various melts are combined just upstream of the die (that is, before they enter the die) via a special adapter. Laminar flow keeps the layers from mixing together, so that the coextruded structure exits as an integral construction from the die. Most such systems are used with flat coating or film dies, with the adapter normally located at the feed port so that the layers can be maintained throughout the single manifold die.

Advantages of this system include: simplicity, lower cost (since combining is done in the external adapter, rather than in an expensive die), versatility (fewer limitations on thickness), and the potential for better bonding since the layers are in contact for longer periods within the die, as opposed to the other systems. However, polymers should reasonably match each other in viscosity.

(2) A second technique, known as multiple manifold coextrusion, involves the combination of the melts within the die (ess Fig. 7-41). Each inlet port leads to a separate manifold for the individual layers involved. The layers are combined at or close to the final land of the die and emerge as an integral construction through a

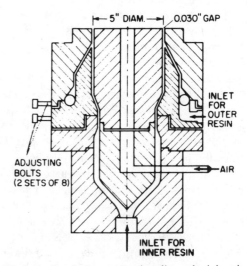

Fig. 7-41. In tubular coextrusion die, melts join prior to leaving the die and blowing into bubble. (*Courtesy E. I. du Pont de Nemours & Co., Inc.*)

single lip. This technique can be used with flat or blown film dies (Fig. 7-42). On the flat die, layer gage is controlled by restrictor bars for each manifold; with the blown film dies, the extruder speed is the controlling factor.

Figure 7-43 indicates this type of system involving the use of four extruders. The extruders are connected to a distribution block, on which the multi-layer die is secured with the interposition of a distributor adaptor. By adjusting this adaptor, it becomes possible to vary the layer sequence of the composite structure. The layout of the channel inside the distributor determines the number of possible variations.

Although multiple manifold coextrusion can be costly, it does have the advantage of more precise control over individual layer thickness.

(3) The third technique, sometimes known as multiple lip coextrusion, is designed to keep

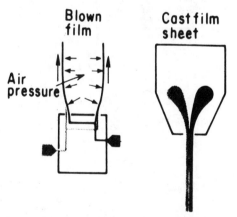

Fig. 7-42. Multiple manifold coextrusion dies.. (*Courtesy Davis-Standard*)

the individual layers of plastic isolated from each other until they exit from the die (Fig. 7-44). The layers are subsequently combined (after exiting) while still molten and just downstream of the die. Depending on the type of extrusion—flat die, blown film, or coating—various techniques are used to combine the layers. For flat film dies, pressure rolls are used; for blown film dies, air pressure inside the expanding bubble provides the necessary pressure for combining the layers.

This technique is also more costly than the laminar flow adapter system, but again, gage control of individual layers is more accurate,

pinholes are eliminated, and the system is easier to start up.

Regardless of which of the three methods are used, it is important in co-extrusion to maintain melt temperature of each layer above the freezing temperature of both layers. Otherwise, poor adhesion may result because of a freeze line between the two layers. In operation, the blown film processing temperature is calculated for each resin and both resins are run at the higher of these two values.

It is also desirable to be able to vary and control the thickness of the plies. Since there are two extruders feeding the separate inlet channels of the die, the first step is to calibrate them in terms of output rate vs. screw speed.

The next step is to establish the desirable width, thickness, and take-off speed to produce the film, and convert this into an output rate (pph) for each resin. From these two output rates and the calibration curves, screw speeds are established for each extruder.

MULTIPLE SCREW EXTRUDERS*

In the early days of the plastics industry, multiple screw extruders (that is, extruders with more than a single screw) found their major outlet in the United States as a compounding device for effectively and uniformly blending plasticizers, fillers, pigments, stabilizers, etc., into the basic polymer. This is still a major outlet for the extruders and is discussed in greater detail in Chapter 29, which covers Compounding.

More recently, following a trend begun in Europe, the multiple screw extruders have started to find use in the processing of plastics. Currently, a major use has been in the handling of materials that are heat-sensitive (like polyvinyl chloride) or in the production of high-quality rigid PVC pipe of large diameter. Multiple screw extruders have also found application in the extrusion of polystyrene foam sheet and are now under consideration for the extrusion of tubing, solid sheet, blown film, and parisons for blow molding.

Several types of multiple screw machines are on the market today, including intermeshing

*Portions of this section based on data supplied by Dr. Wolfgang A. Mack, Werner & Pfleiderer Corp.

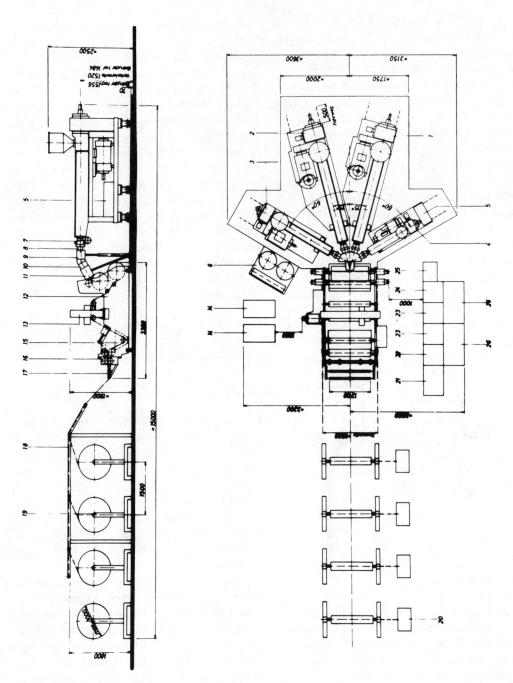

Fig. 7-43. Four extruders set up for a multiple manifold coextrusion (*Courtesy Reifenhauser*)

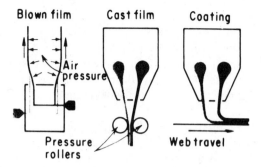

Blown film   **Cast film**   **Coating**

Air pressure

Pressure rollers   Web travel

Fig. 7-44. Multiple-lip coextrusion dies. (*Courtesy Davis-Standard*)

counter-rotating screws (in which the screws rotate in opposite directions and intermesh in that the flight of one screw moves inside the channel of the other), intermeshing co-rotating screws (in which the screws rotate in the same direction), and counter-rotating screws that do not mesh (i.e., non-intermeshing). Most of the commercial machines on the market today are intermeshing. One interesting feature of non-intermeshing twin-screw systems is the possibility of running each screw at different speeds, thus creating frictional relationships between the screws, which in some instances can be exploited for the rapid melting of polymer powders. However, it is the intermeshing screw design that does dominate today's market.

On some compounding machines, it is also possible to have a twin-screw system where one screw is significantly shorter than the other. This technique can be used for handling polymers that may be adequately conveyed by a single-screw mechanism once they are in the form of a melt, but which, because of low bulk density or because of very low coefficients of friction of the solid against the surrounding walls, are difficult to feed into the screw flights. In such a design, the twin-screws are, in essence, used only in the feed section of the extruder, conveying the material forward past the twin-screw stage to where the longer of the two screws functions as a single-screw.

Screws can either be constant in diameter, as in single screw designs, or they can be conical. The conical twin screw extruder is designed with a large diameter rear feed zone having a volume capacity greater than the compression

and metering zone. The idea in this design is to allow a greater amount of powder to be transported to the feed zone per rpm and to use the minimal clearance between flights to achieve a higher output rate with minimal shear and frictional heat accumulation.

Multiple screw machines can involve either two screws (the more common twin-screw design) or four screws. A typical four-screw machine would be a two-stage operation, in which a twin-screw plasticating section feeds into a twin-screw discharge section directly below it. This design is intended to up output rates.

Because different variables are involved in multiple screw machines than in single screw units, the multiple screws are generally sized on the basis of 1b/hr output, rather than on L/D ratios or barrel diameters.

Single-Screw vs. Multiple Screw. As described earlier, the single screw machines use friction between the resin and the rotating screw to make the resin rotate with the screw. The friction between the plastic and the barrel then functions to push the material forward.

In contrast, in twin-screw extruders with intermeshing screws, the relative motion of the flight of one screw inside the channel of the other acts as a wedge that pushes the material from the back to the front of the channel and from screw to screw. This pattern keeps the material moving forward, almost as if the machine were a positive displacement gear pump which conveys material at fairly low rpm with low compression and very low friction (no shear).

In other words, since the friction that causes material to move forward in a single-screw machine also generates heat, to increase the output rate in a single-screw by increasing the screw speed and/or screw diameter will naturally result in increased frictional heat buildup and higher temperatures. Twin-screw extruders, on the other hand, do not have to worry about frictional heat buildup since heat is controlled independently from an outside source and is not influenced by friction. This becomes especially critical when you're working with a heat-sensitive plastic, like polyvinyl chloride, and which is why multiple screws are finding use in processing the vinyls. By the same token, the multiple screws offer advantages in achieving higher output rates on products like large diameter pipe.

Although multiple screws are more expensive than single-screw machines, there are some economic advantages to be gained when using the multiple screws for vinyl. As described earlier (p. 166), twin-screws have been used effectively in handling PVC dry blends that can be compounded in-plant—representing a significant materials costs savings as compared to buying and using compounded PVC. In addition, since less heat is involved in running PVC on multiple screws than on single screws, the PVC compounds used in multiple screws generally require a lower level of heat stabilizer, representing another area of potential savings in materials costs.

In using twin-screw extruders for extruding expanded polystyrene sheet, the economic advantages are felt to be in using one machine to replace a tandem single-screw set-up such as is described on p. 196 (in which one machine mixes while the other extrudes). Since twin-screws do not generate as much frictional heat as single-screws, the second extruder for cooling is not needed. As used for foam extrusion, twin-screws offer positive conveyance and an ability to control melt temperature more accurately.

Another potential area for use of the twin-screw is in the processing of very high-molecular weight products like high-molecular weight polyolefins or some of the TFE-fluoroplastics. The ability of the twin-screw mechanism to gently melt these high viscosity products and convey them without pulsation through the die are the major advantages in this area.

Co-rotating vs. Counter-rotating Screws. One of the more common ways to classify multiple screw extruders is by the direction of screw rotation (see Fig. 7-45).

In a counter-rotating system, in which the screws rotate in opposite directions, the plastic is carried by the screw flights in such a way that all the material is forced to the center where the two screws meet, forming a "bank" or build-up of material similar to that found in a two-roll mill. There is a certain amount of material that will pass through the gap between the two screws and will rotate out with the parting screw flights, but not all the material can pass through here, because there is an obvious restriction in terms of the two screws intermeshing.

A basic advantage of the counter-rotating system is that the material that does pass through the nip of the two screws is subjected to an extremely high degree of shear, just as if it were passing through the nip between the two rolls of a two-roll mill. By varying this clearance—that is, the free space that is left after the two screws have intermeshed—it is possible to vary the portion of material that is

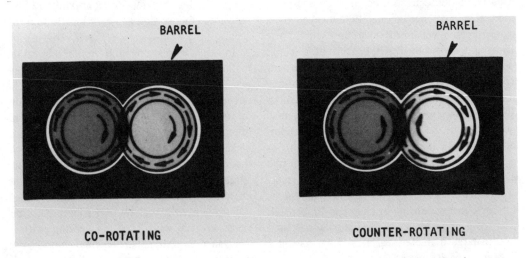

Fig. 7-45. Types of rotation in twin-screw extruders. (*Courtesy Werner & Pfleiderer Corp.*)

carried through in this direction vs. the portion of the material that is accumulated in the "bank" and just moves axially. The narrower the clearance between screws, the greater the shear force that will be exerted and the larger the portion of material that stays in the bank. Thus, it becomes easy to adjust the amount of shear to be applied and the frequency of shear that the operator might want to subject the material to.

In a co-rotating system, the idea is to let one screw transport the material around up to the point where the screws intermesh. Since at that point, we have two opposing and equal velocity gradients, nearly everything that one screw has carried with it is taken over by the other screw. There is little incentive for the plastic to pass through the gap between the two screws (only as much as the inevitable clearance between the two screws allows) because of the opposing velocity components. Thus, most of the plastic in this system will follow an 8-shaped path along the entire barrel length.

Advantages of the co-rotating system include: (1) statistically, chances are better that all the polymer particles will be subjected to the same shear; (2) with the relatively long 8-shaped path, the chances of influencing the polymer in terms of stock temperature control

are good; (3) at the point of deflection, the shear energy introduced can be regulated within very wide limits by adjusting the depths of the screw flight; and (4) co-rotating systems allow for a much higher degree of self-cleaning or self-wiping (one screw completely wipes the other screw), as in Fig. 7-46. This latter

Fig. 7-46. Intermeshing and self-wiping twin-screws. (*Courtesy Werner & Pfleiderer*)

characteristic is important not as much in terms of keeping screws clean, but in permitting greater control over residence time distribution (of special importance when working with heat-sensitive polymers).

8

COMPRESSION AND TRANSFER MOLDING

COMPRESSION MOLDING

History has failed to establish definitely the date of origin of the art of molding. It might be said that the art of molding originated with prehistoric man, when he learned how to make pottery from clay, using the pressure of his hands to form the shape and the heat of the sun to harden the clay.

The earliest application of compression molding as a manufacturing process was early in the nineteenth century, when Thomas Hancock perfected a process for molding rubber. The first patent on a process of molding in the United States was issued in 1870 to John Wesley Hyatt, Jr. and Isaiah S. Hyatt.

Dr. Leo H. Baekeland's development of phenol-formaldehyde resins in 1908 gave the industry its first synthetic molding material, which is even today one of the principal materials used in the compression molding process.

Technique and Materials

The process of compression molding may be simply described by reference to Fig. 8-1. A

* Reviewed and revised by E. J. Fitzpatrick, Stokes Division, Pennwalt Corporation, Philadelphia, Pennsylvania; John Hull, Hull Corporation, Hatboro, Pennsylvania; and F. W. Ducca, Union Carbide Corporation, Bound Brook, N. J.
See also Chapter 9 on Injection Molding Thermosets and Chapter 10 on Designing Molds for Thermoset Molding.

two-piece mold provides a cavity having the shape of the desired molded article. The mold is heated. An appropriate amount of molding material is loaded into the lower half of the mold. The two parts of the mold are brought together under pressure. The compound, softened by heat, is thereby welded into a continuous mass having the shape of the cavity. This mass must then be hardened, so that it can be removed without distortion when the mold is opened.

If the plastic is a thermosetting one, the hardening is effected by further heating, under pressure, in the mold. If it is a thermoplastic, the hardening is effected by chilling, under pressure, in the mold. The procedure is described in greater detail later in this chapter.

Compression molding is used principally for thermosetting plastics; much less commonly for thermoplastics (for which injection is the preferred method of molding).

Thermosetting Materials. Thermosetting materials are chemical compounds made by processing a mixture of heat-reactive resin with fillers, pigments, dyestuffs, lubricants, etc., in preparation for the final molding operation. These materials or molding compounds are, in most cases, in powder, granulated or nodular form, having bulk factors ranging from 1.2 to 10. Some are used in the form of rope, putty, or slabs.

The materials of lower bulk factor are

usually those having woodflour or mineral compounds as fillers, while those of higher bulk factor have as fillers cotton or nylon flock, rag fibers, pieces of macerated rag, tire cord, sisal and, for very high impact strengths, glass rovings.

Phenol-formaldehyde is the single most common resin and catalyst combination, generally called phenolic molding compound. If the filler is mineral, such as mica, the molded part will have good electrical properties. If the filler is glass fibers, say one-quarter inch long, the molded part will have good impact strength. Small hollow glass micro-balloons have been used as fillers to make low-density parts.

Other resin systems include melamine-formaldehyde (often used in plastic dinnerware), urea-formaldehyde (common in white or pastel heat-resistant handles for kitchenware, or outlet sockets for household use), alkyds and polyesters (often used in high voltage insulators in TV sets, or for arc resistance and insulation in circuit breakers and switch gear), diallyl phthalate (electrical connectors in computers), epoxy (housings for electronic components), and silicone (high temperature requirements to $600°F$ or more). Common fillers include silica, glass, wood flour, natural or synthetic fibers, and combinations of these.

Although most thermosetting formulations are dry and granular at room temperature, some are putty-like, some in the form of dry or moist matted fibers, and some a fine powder.

When subjected to heat, thermosetting formulations first become liquid, and then undergo an irreversible chemical reaction called *cure* or *polymerization*. If polymerization occurs under mechanical pressure, as in a closed mold, the resulting material is a dense solid. Polymerization is generally a time-temperature relationship, with shorter cure times when higher temperatures are used. Typical pressure, temperature, and time values, for a phenolic wall socket, in semi-automatic compression molding, might be 3000 psi, $300°F$, and $1\frac{1}{2}$ min.

Thermoplastics. Practically all compression molding uses thermosetting plastics. But in certain specialized applications, thermoplastic materials may be processed by compression molding.

These materials, when compression molded, become plastic under pressure and heat in a heated mold and flow out to the contour of the cavity. Molds must be arranged for rapid heating and cooling, since the molded articles cannot be removed from the mold until the material has been sufficiently cooled to harden. This process of softening the plastic by heating and hardening it by chilling can be repeated indefinitely.

Large plastic optical lenses, for example, may be compression molded from methyl methacrylate (acrylic). In this particular instance, using compression molding rather than the normal injection molding helps eliminate flow marks, warpage, and shrink marks.

For molding articles of heavy cross-section from thermoplastics a combination of injection and compression is sometimes advantageous. The mold is filled with hot softened material by injection, and is then subjected to pressure by means of a compression force plug. This positive application of pressure during cooling minimizes the development of voids and shrink marks.

Molding of Phenol-Formaldehyde Compounds

The details of the procedure of molding thermosetting materials can conveniently be covered by a description of the molding of phenol-formaldehyde compounds. The minor differences in procedure required by other thermosetting materials will be discussed in later sections.

A typical mold is made in two parts which, when brought together, enclose a cavity representing the article to be molded. The two parts are mounted in register, in a hydraulic, pneumatic, or mechanical press which serves to open and close the mold and to apply pressure to its contents.

The most common method of heating molds for compression molding today is with electrical heating cartridges or strips. Electrical heating gives good results with molds which have a uniform distribution of heating elements with adequate capacity. It has the advantages of high efficiency, is clean and simple to hook up, is easily adjustable to $400°F$, and, with modern temperature controllers, is accurate and reliable.

Steam molding, popular in the past, has declined in recent years, partly because of the difficulty of producing the higher pressures needed to reach the not uncommon 350°F and higher mold temperatures. With steam, the mold must be coned for circulation of steam under pressure. It is, however, useful in large, complicated molds because it automatically replaces itself as it condenses and holds the mold at an even temperature.

Molds are occasionally heated by other media such as oil, hot water, and gas flame. However, such media are limited in their usefulness since they do not have the latitude and adaptability of steam and electricity.

Phenolic molding is carried out by inserting a predetermined amount of material into the lower half of the mold (Fig. 8-1). The mold is closed under pressure. With heat and pressure, the material softens, and fills out the contours of the cavity created by the two halves of the mold.

The molding powder may be volumetrically fed, weighed out, or shaped into preforms. The charge may be fed as cold powder, or preheated. It can be preheated by radio frequency, an oven, or other methods. Preheating shortens cure time, reduces molding pressure required, and improves electrical properties.

Range of Compression Molding Conditions		Typical
Temperature	300-400°F	340°F
Pressure	2,000-10,000 psi on part	3,000 psi
Cure time	30-300 sec.	90 sec.

For most molded articles, the plastic material must be confined during molding by telescoping one half of the mold into the other half. A slight clearance is allowed between the halves in order to permit a slight excess of molding compound to escape; usually a clearance of 0.001 to 0.005 in. is sufficient. The material filling the cavity is held under heat and pressure to harden it, and then the mold is opened and the molded article removed. Thermosetting materials, having been hardened by a chemical change caused by the heat, can be ejected after the proper curing cycle without cooling the mold.

Usually, articles requiring special dimensional control can be removed hot from the heated mold and placed on a suitable fixture which holds them during cooling and prevents distortion.

Gassing or breathing the mold may be necessary in compression molding of phenolic. The procedure is to release the pressure on the mold either just before or after it has closed on the material charge. The mold should be opened just far enough to allow entrapped air and gas to escape from each cavity of the mold. Certain molding compounds and/or molded articles require a timed "dwell" in this open position before the mold is closed again. And, on rare occasions, two complete breathe cycles are required for optimum results from a given mold. Gassing or breathing the mold will result in denser moldings, reduce the chances of internal voids or blisters, and shorten the molding cycle.

The proper molding procedure produces

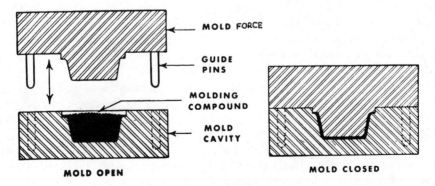

Fig. 8-1. Basics of a two-piece compression mold. (From "Plastics Mold Engineering")

cured articles of sound, uniform structure. Before these can be used, they must be roto-tumbled, blasted with a mild abrasive, or finished as described in Chapter 25.

General-purpose phenolic molding compounds are supplied in granular form, and can be loaded into the mold in this form in weighed or measured charges. But in commerical semi-automatic operations there is economy in pre-forming such material into tablets of the correct size and weight by means of automatic preforming equipment. This is less expensive than weighing out the individual charges; the tablets are easily handled and can be loaded into the cavity conveniently, either manually or by means of a loading board. Frequently, through the use of preforms the cavity loading chamber of the lower half of the mold can be made less deep than it would have to be in order to hold the charge in the bulkier form of looser granules.

A few high-impact phenolic molding materials, which contain fabric, glass, or asbestos fibers as reinforcement for the molded article, cannot be handled by automatic preforming equipment, and must be loaded as weighed charges or preformed by hand.

All plastics require heat in order to be molded. Because of their inherently poor thermal conductivity, the penetration of heat from the hot mold into the cold material is slow, and may not be uniform. Time is saved, and uniformity is promoted, by preheating the plastic before it is put into the mold. This converts the plastic into a uniformly softened mass, ready to flow cleanly and evenly as the mold closes. The practical benefits include better surface finish, freedom from flow marks, better uniformity of cure and less difficulty in produc-tion of articles of thick section, as well as the economy of shorter cure cycles.

Metal parts, or inserts, placed in the mold and held firmly in position, can be molded into the article. (See Chapter 28 on Design Stan-dards for Inserts.)

Molding of Urea-Formaldehyde and Melamine-Formaldehyde Compounds

The techniques employed in handling and molding urea-formaldehyde and melamine-formaldehyde are, in general, similar to those used for phenol-formaldehyde, but some dif-ferences in practice are frequently required.

Since these are usually light-colored materials, attention must be directed to pre-venting any contamination that will show up in the molded article. Dust from adjacent presses, the soiling of preforms, and incomplete removal of flash from the mold are frequent sources of contamination. By making the necessary pro-visions to prevent contamination, urea and melamine can be run with a low percentage of rejection for dirt.

In molding urea and melamine, the design of the article and of the mold is particularly important, because the translucency and light color of these plastics fail to conceal flow marks and gas pockets such as may be present but undetected in the dark, opaque phenolic materials. Hence it is desirable, whenever pos-sible, to design both the article and the mold to minimize such defects in appearance. Ribs, variations in thickness, louvers, and molded holes can be molded satisfactorily in urea when the article and mold are properly designed.

In molding these materials, it is frequently necessary to open the mold slightly and briefly, after it has been initially closed, to allow the escape of gas formed in the reaction of curing. Melamine and urea materials generate moisture under the heat of molding. Also, they entrap and effectively seal air during their compression from an initially high bulk down to one half or less volume in molded form.

Both urea and melamine may sometimes be preheated advantageously before being put into the mold. Melamine-formaldehyde may be elec-tronically preheated successfully, but urea does not react as well to electronic methods. For urea, heating by conduction, such as in a rotary-canister preheater or, for automatic presses, in an oil-bath heater, has gained more favor. Infrared lamps also have been used with some success. The mold temperature for urea materials is usually 325°F or less.

Molding of General-Purpose Polyester Molding Compounds

A line of general-purpose polyester molding compounds is available in pastel shades having excellent high stability. These materials have an outstanding characteristic in that there is no "after-mold shrinkage." This makes them desirable for appliance applications, especially when used in connection with metal inserts.

These polyester materials can be readily molded in either transfer or compression molds. The design of the molded article is facilitated by the fact that variations in wall thickness cause no trouble. Materials can be preheated to 250°F and molded at regular temperatures of 315°F without any danger of discoloration. The duration of cure is equal to or less than that of phenolic compounds.

Low pressure, glass-filled polyester molding compounds can be molded satisfactorily in regular compression equipment, if the molds have adequate restriction to insure filling out. Also, adequate vents are needed, since the polyesters effectively seal air in blind holes in molds, with resulting porosity in the molded article.

In most cases, the actual molding cycle involves simply loading the mold, closing the press, degassing, and completing the cure. However, there are applications which require more elaborate techniques, such as a slow close or a dwell period, to give the best results. In such cases, no over-all rule may be prescribed; the cycle must be worked out for each job.

Molding of Alkyd Materials

The techniques of molding alkyd thermosetting materials are similar, except that the chemical reaction is usually much faster, and therefore, faster-closing and -opening presses are required. In molding these materials, it is frequently necessary to breathe or degas the mold slightly and briefly, after it has been initially closed, to allow the escape of gas formed in the reaction of curing.

Molding of Thermoplastic Materials

The general procedure is the same as for thermosetting materials, but the molded article is hardened under pressure by cooling in the mold, i.e., by shutting off the steam and circulating cold water through the coring of the mold. Some experimentation will be necessary to establish at what point in the cycle the cooling should begin and how long it must be continued in order to harden the article sufficiently.

Thermoplastic molding compounds cannot usually be preformed.

Fig. 8-2. A self-contained straight ram hydraulic combination compression/transfer molding press, 200-ton capacity, with electro-mechanical controls for semi-automatic operation (operator initiates each cycle, but cycle automatically controlled). (*Courtesy Hull Corp.*)

Molding of Fluorocarbons

These are classed as thermoplastics, but are not fully amenable to the conventional technique of compression molding.

Chlorotrifluoroethylene (CTFE-fluorocarbon resin) requires mold temperatures in the range of 450 to 600°F. The necessity of alternately heating to so high a temperature, and cooling, makes for long cycles and large power consumption in conventional integrally heated compression molds. Hand- or bench-type molds are usually preferable. Separate platen presses are used, one electrically heated and one water-cooled. The mold is heated, and the shaping of the article is accomplished in the hot press; then the mold is quickly transferred to the cold press, to chill the articles, under pressure, so that it can be removed.

Tetrafluoroethylene resin (TFE-fluoro-carbon resin) does not soften and flow in the manner of conventional thermoplastics, and it is molded by a special technique as follows:

1. The granular material is preformed at room temperature by a pressure of 2,000 to 10,000 psi.

2. The preform is sintered into a continuous gel by exposure to a temperature of about 700 to 740°F in an oven or fluid bath.

3. The piece is then given its final shape by a pressure of 1,000 to 20,000 psi in a confining die cavity, in which it is cooled. This final step of which there are several variants ("cold-coining," "hobbing," "hot-coining"), depending on the shape of the article and the dimensional accuracy required, may be omitted with articles of thin section for which accuracy is not important.

Fluorinated ethylene-propylene resin (FEP-fluorocarbon resin) can be molded into simple shapes by an essentially conventional compression-molding technique, with gradual application of pressure up to about 1500 psi. For rapid molding, a temperature of 650 to 700°F is required, but a 3-mil sheet of aluminum must be interposed to prevent sticking to the surface of the mold. When this is undesirable, the molding can be done at 550 to 600°F, but more time will be required.

Cold Molding

Some materials are formed in presses with unheated molds, giving a very fragile molding. When this is sintered or baked in ovens at appropriate temperatures, the material hardens or fuses. After cooling, the part is relatively strong, and ready for use. The process is called *cold molding.*

This technique is used to make certain types of ceramic insulators. The molding compound may be ceramic powder with perhaps 15%

phenolic resin by weight, homogeneously mixed. The cold pressing compacts the mixture. Subsequent heating cures the phenolic which binds the ceramic particles together.

A more detailed discussion of cold molding can be found in Chapter 11.

Advantages of Compression Molding

Thermosetting and thermoplastic materials can be compression molded. Quantity production is possible through the use of multiple-cavity molds. Large housings, e.g., for adding machines, switch bases, furniture drawers, etc., are molded by this method. The size of the article which can be molded is limited only by the tonnage and size of available press equipment.

Compression molding of thermosetting materials has certain advantages as follows:

1. Waste of material in the form of sprue runners and transfer-culls is avoided and there is no problem of gate erosion.

2. Internal stress in the molded article is minimized by the shorter and multi-directional flow of the material under pressure in the mold cavity. In the case of high impact types with reinforcing fibers, maximum impact strength is gained. This results because reinforcing fibers are not broken up as is the case when forced through runners and gates in injection molding.

3. A maximum number of cavities can be used in a given mold base without regard to demands of a sprue and runner system.

4. Compression molding is readily adaptable to automatic loading of material and automatic removal of molded articles. Automatic molding is widely used for small items such as wiring device parts and closures.

5. This technique is useful for thin wall parts that must not warp and must retain dimensions. Parts with wall thicknesses as thin as 0.025 in. are molded; however, a minimum wall thickness of 0.060 in. is usually recommended.

6. For parts weighing more than 3 lb, compression molding is recommended since transfer or screw injection equipment would be very expensive for larger parts.
7. For high impact, fluffy materials, compression molding is normally recommended because of the difficulty in feeding the molding compound from a hopper to the press or preformer.
8. In general, compression molds are usually less expensive to build than transfer or injection types.

Limitations of Compression Molding

In the case of very intricately designed articles containing undercuts, side draws, and small holes, the compression method may not be practicable, because of the necessity of complicated molds and the possibility of distorting or breaking mold pins during the flow of the material under high pressure. Articles of 0.35 in. thickness or more may be more advantageously made by transfer molding, particularly a thick article of small area, in which there is little flow. Thus, for a heavy handle, compression molding would be slower than transfer or injection, since in transfer the plastic is thoroughly heated and is precompressed almost to its final density prior to entering the mold.

Frequently, insufficient consideration is given to the physical condition of thermosetting plastics at various stages of molding. The complete filling of a compression mold cavity is spoken of as resulting from the flow of the plastic, but because of its extremely high viscosity, the plastic must be mechanically forced to fill all parts of the cavity. To insure complete filling, most articles to be molded require that two parts of the mold fit telescopically into each other to prevent escape of plastic prior to the final closing of the mold. Also, in order to insure complete filling out of the mold, it may be necessary to place the charge of plastic into an optimum position in the mold, and in some cases to use preforms of special shape. This is particularly important if the mold does not provide a means of confining the charge. Polyester and alkyd compounds are particularly troublesome, and require positive means of confinement in order to fill the cavity completely.

All thermosets, during their period of flow in a mold, have an apparent surface viscosity which is so low that clearances between mold parts, even when held to less than one thousandth of an inch, become filled with plastic. This often results in damage to the mold if adequate escape is not provided for this leakage. It also requires that mating parts for the molds be cleaned between successive shots. Also slight fins or "flash" are to be expected on molded articles where the mold sections meet.

Another important consideration is their degree of rigidity at that point of final cure when ejection is to take place. Melamines are very hard and rigid, phenolics more flexible, unreinforced polyesters quite weak. Thus, a compression mold for phenolic may work with undrafted or even moderately undercut cores. With melamines, the same mold would require enormous pressure to open, and would probably crack the molded articles at the undercut. Articles of polyesters require very careful adherence to all rules for draft; they also require generous ejector areas to avoid fracture on release from the molds.

In some cases, compression molding of thermosetting material may be unsatisfactory for production of articles having extremely close dimensional tolerances, especially in multiple-cavity molds, particularly in relation to non-uniformity of thickness at the parting line of the molded article. In such cases, transfer or injection molding is recommended. For a further discussion of this method of molding see Transfer Molding, p. 220, and Chapter 9 on Injection Molding Thermosets.

Procedure for Compression Molding

The sequence of operations constituting the molding cycles is as follows:

1. open the mold;
2. eject the molded article(s);
3. place article in shrink or cooling fixtures when necessary to maintain close dimensional tolerances (if necessary);

4. remove all foreign matter and flash from the mold, usually by air blast;

5. place inserts or other loose mold parts, if any;

6. load molding compound (powder or preforms, cold or preheated);

7. close the heated mold (breathe if necessary);

8a. for thermosetting materials, hold under heat and pressure until cure is completed. Certain materials require cooling under pressure for best control of dimensions;

8b. for thermoplastic materials, hold under pressure and cool to harden the article.

The temperature of the mold and the pressure applied are extremely important, and it is advisable to follow the recommendations of the manufacturer for each grade of material used.

Thermosetting materials used in compression molding can be classified as conventional and low-pressure materials (the latter should not be confused with materials used in low-pressure molding of impregnated laminates).

There are five very important variables in the compression molding of thermosetting materials which determine the pressure required to produce the best molding in the shortest length of time. These are as follows:

1. Design of the article to be produced:
   (a) projected area and depth
   (b) wall thickness
   (c) obstruction to vertical flow (such as pins, louvers, and sharp corners)
2. Speed of press in closing:
   (a) use of slow- or fast-acting self-contained press
   (b) use of fast-acting press served by hydraulic line accumulator system
   (c) capacity of accumulator to maintain constant follow-up of pressure on material
3. Plasticity of material:
   (a) degree and type of preheating
   (b) density of charge (preform or powder)
   (c) position of charge in cavity
   (d) mobility of resin under pressure

   (e) type of filler (woodflour, cotton flock, macerated fabric, asbestos, glass or mica)
4. Over-all temperature of mold:
   (a) temperature variations within cavity and force of mold
5. Surface condition of mold cavity and force:
   (a) highly polished chrome-plated surface
   (b) polished steel
   (c) poor polish (chromium plating worn; pits, gouges, and nicks).

Molding pressures required for most thermosetting materials follow the pattern established for phenolic materials.

Conventional phenolic materials loaded at room temperature, i.e., without preheating require a minimum pressure of 3000 psi on the projected land area for the first inch of depth of the molded article, plus 700 psi for each additional inch of depth. Efficient high-frequency preheating, however, may reduce the required pressure to as low as 1000 psi on the projected land area, plus 250 psi for each additional inch of depth. The pressure required on high-impact materials may reach 10,000 to 12,000 psi. These recommendations of pressure are predicated on minimum press-closing speeds of 1 in./sec. The flow characteristics of thermosetting molding materials are changing continually during the molding, and the effect of this is particularly noticeable in slow-closing presses.

Low-pressure phenolic molding materials, efficiently preheated by high frequency, require a minimum of 350 psi on the projected mold area, plus about 100 psi for each additional inch of depth.

The table on p. 212 may be used as a guide for the pressure required for the depth of a given molded article.

The wide range of pressures given in the second column is necessary to cover the variety of molded pieces and the types of press equipment used. For example, an article of comparatively small area and great depth requires pressures at the lower end of the range given; an article of large area and great depth

Pressure Table
Pressure, psi of Projected Land Area

Depth of Molding (in.)	Conventional Phenolic		Low-Pressure Phenolic	
	Preheated by high frequency	Not preheated	Preheated by high frequency	Not preheated
0-¾	1000-2000	3000	350	1000
¾-1½	1250-2500	3700	450	1250
2	1500-3000	4400	550	1500
3	1750-3500	5100	650	1750
4	2000-4000	5800	750	2000
5	2250-4500	*	850	**
6	2500-5000	*	950	**
7	2750-5500	*	1050	**
8	3000-6000	*	1150	**
9	3250-6500	*	1250	**
10	3500-7000	*	1350	**
12	4000-8000	*	1450	**
14	4500-9000	*	1550	**
16	5000-10000	*	1650	**

  * Add 700 psi for each additional inch of depth; but beyond 4 in. in depth it is desirable (and beyond 12 in. essential) to preheat.
** Add 250 psi for each additional inch of depth; but beyond 4 in. in depth it is desirable (and beyond 12 in. essential) to preheat

may require pressure at the upper end of the range. Also, the thickness of wall of the article influences the pressure required; thin sections require more pressure than heavier sections. For articles involving a deep draw, fast-closing presses with speeds of over 20 in./min. for full depth of draw will make it possible to use lower pressures.

The time required to harden thermosetting materials is commonly referred to as the cure time. Depending upon the type of material, preheat temperature and thickness of the molded articles, the time may range from seconds to several minutes.

Types of Compression Molds

For most economical production, molds are made from high-grade steels so that they can be hardened and polished.

A hand mold is so constructed that it must be removed from the press manually, taken apart to remove the molded article, and assembled again for the next molding cycle. These molds are used primarily for experimental or for small production runs, or for molding articles which by reason of complexity require dismantling of mold sections in order to release them. These molds are usually small and light and contain not more than a few mold cavities. They are usually heated by means of electrically heated or steam-heated platens attached to the press. (See Fig. 8-3).

Semi-automatic molds are self-contained units which are firmly mounted on the top and bottom platens of the press. The operation of the press opens and closes the mold and also operates the ejector mechanism provided for the removal of the molded piece from the mold. This type is employed particularly for multiple-cavity work and for articles too large or too deep in draw for hand molding. The use of ejector pins make it a necessity to provide steam or oil channeling or electric heaters as part of the mold design.

Fully automatic molds are of special design and adapted to a completely automatic press. The complete cycle of operation, including the loading and unloading of the mold, is carried out automatically. A multiple-cavity mold may be used, and usually the molded article contains no insert or metal part.

Fig. 8-3. Small transfer press used for hand molding; mold is removed from press each cycle, opened, parts removed, mold closed, then placed in press again between electrically-heated plates for next cycle. Used for low-volume parts or complex cavities that may need considerable dis-assembly for parts removal. (*Courtesy Hull Corp.*)

Machinery and Equipment

Presses. Presses for hand molds range from small laboratory presses to production equipment of capacities from 5 to 100 tons. The heating plates are fastened directly to the top and bottom press platens. The mold is placed between these plates for the transfer of heat and pressure to it during the molding operation. Presses for semi-automatic molds range in size from 10 to 4000 tons and up. These presses have top and bottom platens with grids and parallels so that the molds can be readily mounted. The presses are provided with either mechanical or hydraulic ejecting apparatus to remove molded articles from the molds.

In compression molding articles with high vertical walls, it is important that the pressure be applied sufficiently fast to maintain maximum pressure on the material as it becomes soft and flows within the mold cavity. Hydraulic presses are usually considered best for articles with high vertical walls.

Self-contained hydraulic press units must have a pump capacity large enough and must be fast enough to take full advantage of the short period of maximum plasticity.

Hydraulic System. Hydraulic presses may be provided with individual sources of hydraulic power, or a group of presses may be operated from a single line. A pump driven by steam or oil provides the necessary volume of oil or water at the required pressure, and an accumulator maintains a reserve sufficient to serve all presses on the line without fluctuation of pressure.

In some cases, a high-and-low pressure accumulator system is used for economy. The low-pressure system is especially advantageous for operating presses with a long stroke; the low pressure system can advance the moving part of the mold up to the point where the material takes pressure, and then the high pressure is thrown on to complete the molding cycle.

Heating and Cooling of Molds. A hand mold is heated and cooled by contact with cored platens mounted in the press. Heat for automatic and semi-automatic molds may be supplied by steam flowing through channeled mold sections or by electric heaters installed in the platens of the mold or press.

For chilling thermoplastics, a supply of cold water is required, which flows through cooling channels or the hot oil circulating system is arranged for chilling as well as heating.

Automatic Compression Molding

Description. Fully automatic compression molding involves the automatic sequencing of necessary functions performed in making finished molded articles from granulated or nodular thermosetting molding compounds. Essentially speaking, it involves a compression-molding press equipped with a mold and additional equipment designed (1) to store a quantity of molding compound; (2) to meter, volumetrically or gravimetrically, exact charges of compound; (3) to deposit these charges into appropriate mold cavities; (4) to remove the finished article from the mold area following each cycle; and (5) to remove any flash or granular molding material not ejected with the finished article. Once an automatic press has been placed into operation, it can run unattended except for periodic recharging of the storage container (hopper) and removal of finished moldings, plus an occasional adjustment in the control system to accommodate minor variations in compound, ambient temperature, and humidity.

Automatic compression molding has been done for many years. Many basic improvements in some of the earlier types have led to faster operations, better molded products, and more economical production. Molders generally believe that automatic molding will expand many times over in years to come. Although screw-injection molding of thermosetting compounds (see Chapter on Injection Molding of Thermosets) is gaining in usage and popularity, automatic compression molding continues to prove ideal for many applications, and to play a major role in thermoset molding.

Applicability. Probably the main reason for selecting automatic compression molding is the need to produce a large quantity of the same article. An automatic press involves heavier financial investment than a manual or semi-automatic unit and the full set-up, including metering adjustment and alignment of the comb which removes articles from the mold, may take longer for automatic molding than for semi-automatic molding. These extra costs may be offset in a long production run by a high rate of production, a small number of rejects, and a sharp reduction in labor costs. Contrary to general belief, a mold for an automatic press is often no more expensive than a good semi-automatic mold, because both must be precisely machined and have essentially the same arrangement of knockout pin, hold-down pin, etc. In fact, an automatic mold usually requires fewer cavities than a semi-automatic mold to achieve a given production rate, and may therefore be less expensive.

For preliminary analysis of a particular molding requirement, the difference in cost between semi-automatic and fully automatic molding may be calculated on the basis of labor. For example, the semi-automatic operation may cost from $2.00 to $5.00 per hour more than the fully automatic molding. Other factors may include a higher production rate, less wastage of raw material, fewer rejects, and less down-time. Not to be overlooked is the fact that production rates in automatic molding can often be geared to the demand requirements of the molded article; inventories of the molded article may therefore be restricted, with additional savings.

An intelligent analysis of the various cost factors given above enable the molder to determine when fully automatic molding is feasible.

On the other hand, automatic molding may not be feasible if the article to be molded is subject to one or more design changes early in the life of the mold. Even this consideration is altered if the fully automatic mold has, say, four cavities and the semi-automatic mold has eight cavities.

Automatic molding is generally not feasible if the molded article must be made from molding compounds of bulk factor in excess of 3 (except in the case of certain nodular molding compounds) or of poor dry flowing properties in feeding devices.

Basic Requirements for Automatic Compression Molding

Design of Molded Article. For automatic compression molding, considerable attention must be given to the design of the molded article. Articles 3 in. or more in height may require daylight, stroke, or travel or ejector pin which exceeds the specifications of automatic presses now available. Keeping such deep cavities free from flash or impurities may also present obstacles to trouble-free operation.

Articles of particularly fragile sections, or requiring fragile mold pins or mold sections, are generally not practicable for trouble-free automatic molding. The molded article should be free of sharp changes in section, and should have adequate draft (1 degree or more) to facilitate ready removal from cavity or force. Such draft simplifies the manufacture of molds and helps to minimize their wear.

Articles normally requiring inserts molded in should be studied to determine the possibility of molding them without inserts, and of putting the inserts in as a secondary operation. The shorter cycles possible with this method generally result in an over-all economic saving. Some articles which normally require side cores or a split mold should be evaluated to determine whether a redesign is possible, or whether two plastic parts, both molded automatically, might be substituted for the one article and still prove more economical in the long run. The best results are obtained when the designer is thoroughly familiar with the techniques possible in automatic molding, with the techniques used by the moldmaker, and with the functional requirements of the final application of the article.

Mold Design. While the design of a mold for automatic molding incorporates most of the feature used in manual or semi-automatic molding, there are, however, two major features which require special attention. First, the molded article must be made to withdraw with the same half of the mold, either top half or bottom half, on every opening stroke. If, for example, bottom ejection is required, every precaution must be taken that the molded piece will not stick to the top half of the mold on the opening stroke. Undercuts, or ridges to grip the article, are frequently required. Top hold-down pins are often used. These are essentially the same as knockout pins, but are designed to push the molded piece away from the top half of the mold as it is opened, to ensure that the piece remains in the bottom half of the mold. Similarly, of course, bottom hold-up pins are used where top ejection is required. The design of the article indicates whether top ejection or bottom ejection is the better. In automatic operation, the molded article must be under positive control at all times. Automatic takeoff devices and scanning or checking devices are set to conform with a situation which must remain essentially unchanged from cycle to cycle.

The second is to insure that any flash will cling to the molded piece, and not to the mold face or cavity or force. Automatic presses generally utilize an air blast as a cleaning device to remove flash or excess powder from a mold just prior to loading the new charge of material. But such an airblast mechanism is never so thorough as an operator in ensuring that the mold is clean. Thus, every effort must be made to keep flash to a minimum, and to be sure that any flash which does develop is automatically removed with the molded article. In the mold, well-polished surfaces are necessary, chrome-plating of the actual cavities and forces is generally recommended, and all sharp edges or corners should be rounded wherever possible, to minimize accumulation of flash.

Today, with minaturization, small molded articles are frequently necessary. In automatic molds for small articles, sub-cavity design is frequently the most satisfactory solution. A number of the small cavities are arranged in a sub-cavity, which receives one charge of molding compound in each cycle. The complete cluster of molded pieces remains intact during ejection, being held together by flash or controlled thickness. Articles smaller than about $\frac{1}{2}$ in. diameter should usually be molded in this way.

Selection of Molding Compound. In addition to the above requisities, some consideration must be given to the molding compound. Generally speaking, automatic molding requires materials which flow easily in the dry state, and which do not tend to cake or bridge when in storage or when required to flow through feeder tubes. Bulk factors must generally be under 3.0, and sometimes the finer powders and coarser granules must be removed. The limitations on molding compounds develop not so much because of molding operation itself, but because of the problems involved in automatic feeding. Manufacturers of molding compounds have recognized this problem and have produced standard materials which provide excellent pourability for volumetric feeding of automatic compression presses. They also are able to give specific advice on materials which can be used and which will produce the best quality in the finished molded article.

Some automatic molding presses utilize vibrators or agitators in the hopper to minimize caking or bridging.

Equipment Available. Several different types of automatic molding presses are available, some of which have been in use for as long as twenty-five years or more. The conventional equipment is of the vertical type, with the platen moving either up or down during the compression stroke. As the press opens following each cycle, the molded pieces are stripped from the knockout pins by means of a comb or air blast; following this operation, the loading board moves over the mold to drop charges into the cavities. The loading board then leaves the mold area and the press closes for its next cycle. Operation is generally of a sequential nature, wherein each individual operation, when completed, initiates the next step. Controls are generally electrically actuated, and use pneumatic, hydraulic, or mechanical means to position the take off comb and loading board, and other moving parts. Some smaller presses depart from the sequential-type operation and remove the molded articles and load the new charge of powder simultaneously.

Another type of automatic press is the rotary press. As the name implies, these presses have a rotary movement of the main moving parts. In one type, a number of single-cavity molds are arranged in a large circle, the lower halves affixed to one platen and the upper halves to another platen. Generally speaking, these platens do not move in relation to one another. A traveling mechanism moves continuously around these two platens at a controlled speed. This mechanism performs, in the appropriate sequence, the functions of causing two individual mold halves to come together, and to remain together while the mechanism completes essentially 300 degrees of travel around the platen. The moving mechanism causes the mold to open, the molded pieces to be physically ejected and the force and cavities cleaned by an air blast, a new metered charge to be dropped into the cavity, and the mold halves closed again. Each mold is thus actuated individually, and at a uniform sequence after each preceeding mold. There may be ten stations, or even thirty stations, on a rotary press of this type. The cavities may all be alike, or they may be different, but the cure times of each must be compatible, as the over-all cycle is dependent on the time required for one complete rotation of the moving mechanism.

Also, rotary presses are made in which the "mechanism" remains stationary and the round platens with the molds rotate about a central axis. In operation, however, the principle is the same as that described above, except that all molded pieces must be of the same size.

Some of the advantages of rotary presses are listed below:

1. Individual cavities are easy to change, thus reducing loss of production time.
2. Some presses can incorporate molds of different volumes, since the material feed can be varied for individual cavities.
3. High productivity can be gained with minimum amount of labor.
4. Rotaries are especially adaptable to high volume, small parts such as wiring devices and closures.
5. When different size cavities are used in one press, molding cycles are determined by the part with the heaviest section.
6. Variable speed of rotation controls length of cure.
7. Unloading mechanisms can be used (e.g., wheel to unscrew molded threaded caps, air jet, stripper plates, knockout pins, twist ram).
8. Cavities do not have to be precisely matched for applied pressure.

Several limitations to the rotary press are as follows:

1. Cannot easily load inserts.
2. Part size is usually limited to small parts and is governed by the pressure available to each station. The mold load is usually in a range of 2 to 5 tons, with some machines as high as 15 tons.
3. Molds must be of the flash or semi-positive type. Complicated molds with features such as split cavities, cores or side draws cannot be readily adapted.

Types of Machines

Several commercially available automatic molding presses are shown in Figs. 8-4, 8-5, 8-6, and 8-7. In selecting the equipment for a specific job, the different characteristics of the various presses should be studied to ensure that

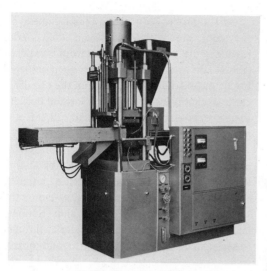

Fig. 8-4. A 75-ton hydraulically-actuated fully automatic compression molding press. (*Courtesy Stokes Div., Pennwalt Corp.*

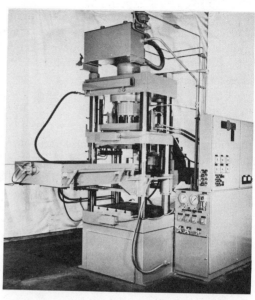

Fig. 8-6. A 225-ton hydraulically operated compression molding press. (*Courtesy Reed-Prentice*)

Fig. 8-5. A 200-ton fully automatic compression press, hydraulic straight ram downward moving clamp, arranged for non-preheated molding compound. (*Courtesy Hull Corp.*)

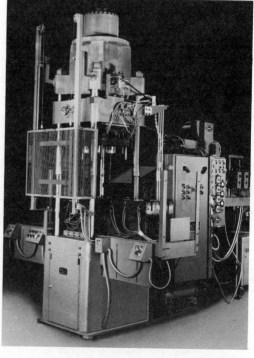

Fig. 8-7. A 200-ton shuttle press for compression molding with inserts, using automatic feeder for electronic preheating and loading molding compound. Two bottom halves of molds, arranged on shuttle table, and simple top half of mold, enable essentially continuous molding while operator loads inserts in bottom half for next cycle. (*Courtesy Hull Corp.*)

the job in hand can be done safely, effectively, and economically. Various safety features are available on such presses to ensure protection from double shots, from malfunction of accessories, etc.

Automatic Screw Compression

Recently, new processing machines were developed to combine the advantages of compression molding and the shorter cycle times available to screw transfer and injection molders of thermoset materials. This is accomplished by means of screws to preheat the material, prepare a measured charge, and put it into the compression mold. A complete machine is shown in Fig. 8-8.

Machines are arranged with one, two or three separate screws, each with its own speed control, water-jacket and shot measuring device (Fig. 8-9). Cups at the open end of the barrel nozzle accept, and condense the material controlling the weight of the preheated charge.

Fig. 8-8. A 25-ton air-toggle automatic compression press with three screws. (*Courtesy Stokes Div., Pennwalt Corp.*)

At the proper sequence in the machine cycle the cups are withdrawn from the nozzles leaving the preheated preform exposed (Fig. 8-10). A shearing device is then actuated to cut the extruded material from the nozzle. A chute assembly or a feed board located beneath the nozzle moves forward to direct the preforms into their respective cavities.

Preheating process: The temperature of the preheated charge can be as high as 280°F depending upon the material. Heat input is accomplished through convection from the water jackets (110 to 130°F), and frictional heat generated by forcing the material through the clearance between screw tip and nozzle wall.

Work in-put at the nozzle results in a minimum lapsed time between obtaining a high material temperature, and depositing the charge into the mold.

The preform cups are designed with minimum and maximum material weight charges. Each is adjustable to make preforms to any weight increment within its capacity. When the cup has been fully charged to its present measurement, a limits switch stops the screw rotation.

Cure time comparisons, shown in Fig. 8-11, reflect the extremes of part thickness. A part with a cross section thickness of 0.070 in. cures on an average of 24 sec. when using cold powder of an "eight" plasticity, and almost 45 sec. using an "eighteen" flow. The same part using screw compression cures in 12 and 15 sec., respectively, for the same materials.

A part with a cross section thickness of .370 in. cures on an average of 150 sec. as compared with 30 sec. with screw compression. With cold powder, even a 150 sec. cure does not ensure the part to be free from internal voids.

Automatic Electronic Preheating

Automatic compression molding of articles of relatively thick sections is generally more feasible when automatic preheating units are used. Preheating is generally done by high-frequency induction. These units are furnished for coupling to existing automatic presses. They

Fig. 8-9. A tri-screw unit after preheating a charge of thermoset material. (*Courtesy Stokes Div., Pennwalt Corp.*)

Fig. 8-10. A tri-screw assembly with measuring cups withdrawn. (*Courtesy Stokes Div., Pennwalt Corp.*)

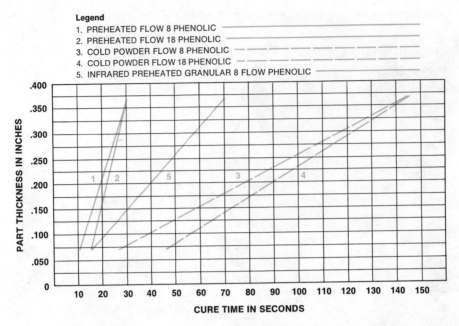

Legend
1. PREHEATED FLOW 8 PHENOLIC
2. PREHEATED FLOW 18 PHENOLIC
3. COLD POWDER FLOW 8 PHENOLIC
4. COLD POWDER FLOW 18 PHENOLIC
5. INFRARED PREHEATED GRANULAR 8 FLOW PHENOLIC

Fig. 8-11. Cure time comparison. (*Courtesy Stokes Div., Pennwalt Corp.*)

consist essentially of the automatic metering and feeding mechanism for handling the molding powder plus the unit in which preheating takes place. The charge is passed through a high-frequency unit in a timed operation prior to being dropped into the mold cavities. Wherever the thickness of the article exceeds $\frac{1}{8}$ in., careful consideration should be given to the possibilities of automatic preheating equipment in conjunction with an automatic press, since the increase in rate of production generally offsets the cost of operating and amortizing the additional equipment.

TRANSFER MOLDING

The term *transfer molding* is now generally applied to the process of forming articles, in a closed mold, from a thermosetting material that is conveyed under pressure, in a hot, plastic state, from an auxiliary chamber.

History

Prior to the development of transfer molding, thermosetting materials were handled only in compression molds, by methods adapted from the prior art used in forming articles from rubber, shellac and cold-molded compositions. The compression method is still widely used, and is entirely suitable for the production of a great variety of small and large articles of relatively simple outline and plain cross-section.

However, compression molding does not readily permit the forming of articles having intricate sections, thin walls and fragile inserts, and those on which close tolerances on "build-up" dimensions are desired. When compression molding is applied to such articles, the usual results are high costs of molds and maintenance, due to excessive wear and breakage.

To overcome these difficulties, transfer molding was introduced by Shaw Insulator Company, Irvington, New Jersey, in 1926. At that time, the process was carried out in conventional single-ram compression presses, with three-piece molds, as described later. To promote flow, it was frequently necessary to employ special, long-flow, premium molding compounds, and to preheat them in an oven. This type of transfer molding is now usually referred to as *pot type*.

Later, presses equipped with two or more rams were developed for transfer molding. In these, the main ram is attached to the lower or upper press platen in the usual manner, and holds the mold closed. An auxiliary ram is used to force the material into the closed mold. This form of transfer molding is frequently called *plunger molding*.

In the most commonly used American molding presses, the main ram is attached to the lower press platen and operates upward. When these presses are equipped for transfer molding, the auxiliary rams may be mounted on the side columns, operating at right angles to the main columns, operating at right angles to the main ram, or on the press head, operating downward in a direction opposite to that of the main ram, or on the press bed or within the main ram, operating upward and in the same direction as the main ram.

The introduction of high-frequency dielectric preheating during World War II, greatly accelerated the growth of transfer molding, particularly with auxiliary-ram presses. Speeds of transfer and of cure were increased, and lower pressures could be used for transfer and for clamping the mold.

During 1956, two new automatic transfer-molding machines were developed. One is a small horizontal machine, using loose powder and comparatively inexpensive molds. These and various other working parts of the machine are readily accessible and easily maintained. The other machine is much larger, with a 400-ton clamping capacity, and is extremely fast. It is designed to use preheated preforms.

In the mid-60's, a high speed automatic transfer press, with integral preforming and high-frequency preheating, entered the market, and yielded fast cycles as short as 25 sec. for multi-pin connectors and similar parts.

This progress in molding technology and in equipment has been matched by the development of improved materials, particularly the phenolics, which are better suited to the large multiple-cavity transfer molds now in use than were the special transfer materials previously employed. These new materials have more rapid rates of flow and shorter cure times than those previously available for transfer molding. With the aid of dielectric preheating, they exhibit the same total flow as did the special transfer materials formerly required, but accomplish this flow in a much shorter time. Rigidity on discharge from the mold has also been greatly improved. This all makes for faster production and lower molding costs, and further enhances the usefulness of transfer molding.

Utility of Process

As a molding technique, transfer molding offers the molder certain advantages. Some typical applications where this process is useful are as follows:

1. For parts requiring side draw core pins which must be readily withdrawn before discharging a part from the mold.
2. For applications where inserts must be molded into intricate parts.
3. For complicated parts where tolerance are very close in three dimensions.
4. For production of various small molded parts that are assembled together, a "family" transfer or plunger mold is economical. It also permits substitution of different cavities in one mold depending upon production requirements.
5. Finishing costs are reduced because of less flash on the transfer molded parts.

Description of Process

Essentially, this type of molding requires the transfer of material under pressure from a "pot" or "well" through runners and gates into cavities retained in a closed heated mold (Figs. 8-12 and 8-13). Usually, the charge has been preheated before placing it in the pot. By preheating, less pressure is required for transfer and in addition it reduces the mold cycle time. Basically, there are three variations of this technique:

1. *Transfer Mold in Compression Press.* With this type, a single, hydraulic ram is used. The plunger for the pot is clamped to the upper platen of the press. Pressure is developed in the pot by the action of the main hydraulic ram of the press. The area of the pot should exceed the area of the cavities by a minimum of 10%.

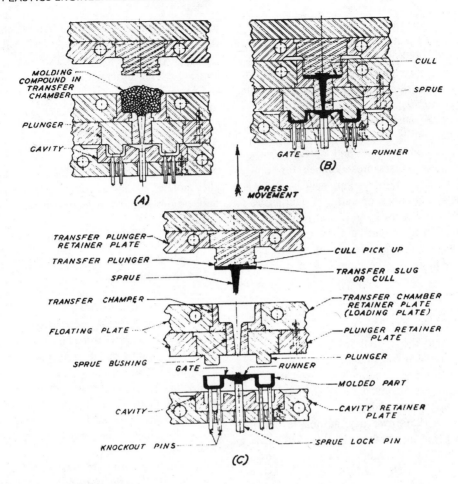

Fig. 8-12. True transfer or pot-type transfer molding, the forerunner of today's common "plunger" or conventional transfer molding. (From "Plastics Mold Engineering")

Thus, the wedging action of the material will not force the mold cavities to open and flash. Pot-type molding is often a manual operation.

This method is usually faster than compression molding, since cure time, particularly for thick sections, often is shorter because the hot material is injected into the closed cavity under high pressure. This results in molded parts with good dimensional control and uniform density. Mold cost for transfer molding is usually higher than for compression molding, however, and the process is not well suited to automatic operation.

A three-plate mold is used with the ram or plunger in the top plate (Fig. 8-14). Because the material enters the cavity at a single point, an orientation of any fibrous filler is produced in a direction parallel to the flow. The shrinkage of the molded part parallel to the line of flow, and shrinkage at right angles to the line of flow, may thus be different and rather difficult to predict, depending upon the geometry of the molded part and the position of the gate.

When a transfer mold is opened, the residual disc of material left in the pot, known as the cull, and the sprue (or runner from the pot into the cavities) are removed as a unit. To remove the molded part from the bottom section of the mold, the center must be raised.

In pot-type transfer, the taper of the sprue is the reverse of that used in injection molding since it is desired to keep the sprue attached to the cull so that it will pull away from the part. A detailed pot-type is shown in Fig. 8-12.

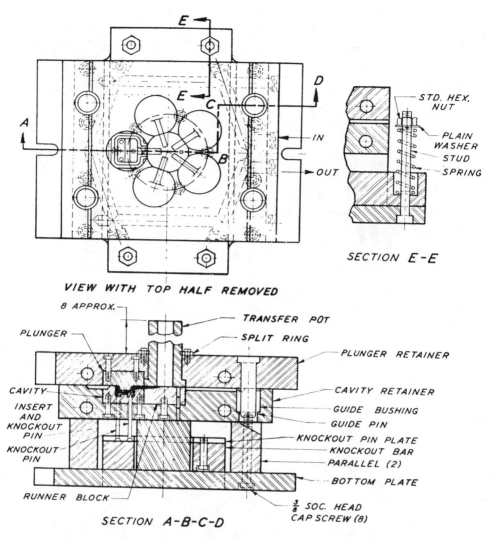

Fig. 8-13. Plunger-type transfer mold.

2. *Plunger Molding* (Sometimes called auxiliary ram transfer). Plunger molding is similar to transfer molding except that an auxiliary ram is used to exert pressure on the material in the pot. This forces the preheated material through runners into the cavities of the closed mold. A two-plate mold is used for this type. Most of this type molding is a semi-automatic operation with self-contained presses.

Basic steps of this process are similar to those of the pot or transfer molding method. When the plunger is withdrawn and the mold opened, the molded part may be removed from the cavity with the runners and cull still attached as a unit. The over-all molding cycle in plunger molding is usually shorter than for transfer molding, since removal of the sprue, runners, or cull does not require a separate operation. In plunger molding, it is essential that radio frequency preheated preforms be used to take maximum advantage of the fast cures obtainable. Figure 8-15 illustrates this process.

The mold temperature and pressure required to obtain a satisfactory molded piece must be predetermined for each type of thermosetting material used. A detailed plunger mold is shown in Fig. 8-13.

3. *Screw Transfer Molding* In this process,

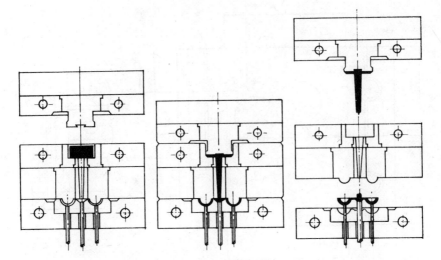

Fig 8-14. Molding cycle of a transfer mold. Material is placed in the transfer pot (left), then forced through an orifice into the closed mold (center). When the mold opens (right), the cull and sprue are removed an a unit, and the part is lifted out of the cavity by ejector pins. (*Courtesy Union Carbide Corp.*)

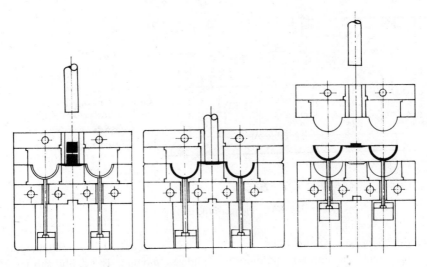

Fig. 8-15. Plunger molding. An auxiliary ram exerts pressure on the material in the pot (left) and forces it into the mold (center). When the plunger mold is opened (right), the cull and sprue remain with the molded piece. (*Courtesy Union Carbide Corp.*)

the material is preheated by preplasticizing in a screw and is dropped into the pot of an inverted plunger mold. The preheated material is transferred into the mold by the same method as shown in Fig. 8-15. The screw transfer process and the sequence of operation are shown in Figs. 8-16 and 8-16A. This molding technique lends itself particularly well to fully automatic operation. The optimum

temperature of a phenolic mold charge is $240° ± 20°F$, the same as that when RF preheating for transfer and plunger molding.

Screw-transfer presses are equipped with screw units capable of preparing from 3 oz to 5 lb shots. To provide adequate control of preheating the material in the screw units, certain requirements are necessary. The compression ratio (ratio of shallowest to deepest

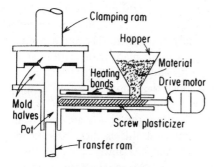

Fig. 8-16. Screw-transfer process.

depth of flight) for a thermoset screw transfer is much less than the ratio for screws available for thermoplastic materials. The length to diameter screw ratio, in most cases, is considerably shorter. Because of exothermic action of some thermoset resins, a means to carry away heat generated in the barrel is also provided. Aluminum water jackets surround the barrel permitting the circulation of water. The temperature of this water is regulated by separate control units. Normally, one water control unit is provided for each temperature zone of the barrel. Some machines are even provided with plasticizing screws which are channeled to permit the circulation of controlled temperature water. The object of the plasticizing unit is to prepare material as hot as possible without precuring the resin. Variables, such as the water temperature of the barrel jackets, screw rotational speed, and back pressure applied to the screw as it prepares the material, serve to control the amount of heat input to the material.

In the screw-transfer process (Fig. 8-16A), thermosetting material is gravity fed from a larger storage hopper through a hole in the barrel to the reciprocating screw preplasticizer. As the screw rotates, material travels forward along the flights and is thoroughly preheated by mechanical shearing action. Material flows off the end of the screw and begins to accumulate. This build-up of material pushes the screw back along its axis (away from the transfer pot) to a predetermined point which can be set by a limit switch. The amount of reverse travel of the screw establishes the volume of the charge. While the shot is being preplasticized, the

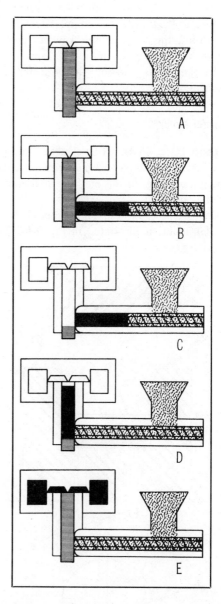

Fig. 8-16A. Screw-transfer sequence. In A, the material is preplasticized as it travels along the flights of the screw. In B, material builds up to the end of the screw and forces it to move backward a predetermined distance. In C, after the shot is formed, the transfer ram lowers to open the transfer pot. In D, the screw moves forward to push the material into the transfer pot. In E, the transfer ram then advances through a standard bottom-transfer molding operation to inject the material into the mold. (*Courtesy Stokes Div., Pennwalt Corp.*)

transfer ram is in the raised position, blocking the opening into the transfer pot.

After the shot is formed, the transfer ram returns to its lower position, leaving the opening to the transfer pot completely clear. At this point, the screw moves forward, pushing the preheated material into the transfer pot. With the press closed, the transfer ram advances.

Transfer Molds

Loose-Plate Molds. This classification may be subdivided into manual and semi-automatic types, depending on the method of mounting and the operation.

One of the earliest and simplest hand-transfer molds is illustrated in Fig. 8-17. This

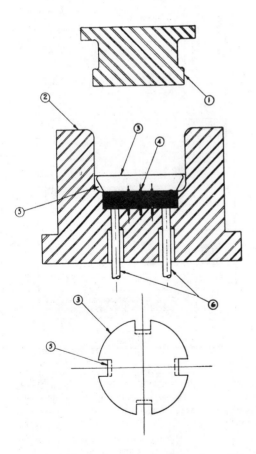

Fig. 8-17. Loose-plate or hand-transfer mold.

1. Plunger	4. Inserts
2. Cavity	5. Gate
3. Loose plate	6. Knock-out pin

mold is especially useful where the molded piece contains a group of fragile inserts extending completely through it. The mold consists of a plunger, a loose plate with orifices around its perimeter, and the cavity. The space in the cavity above the loose plate serves as the pot or transfer chamber.

In operation, the inserts are loaded into the loose plate and this is then inserted into the cavity so that the lower ends of the inserts enter the proper holes. The compound is loaded into the mold above the loose plate and is transferred by the plunger through the orifices into the closed mold. After curing, the molded piece, loose plate and cull are ejected by knock-out pins, and disassembled at the bench. Two plates may be used alternately to speed up production by keeping the press continually in operation.

These molds are most useful when the cost of the mold must be as low as possible, and when the volume of production is small.

Figure 8-18 shows a semi-automatic mold assembly of the loose- or floating-plate type. In this case, the floating plate is carried more or less permanently in the press and has a central opening that will accommodate stock-transfer pots and plungers of various sizes, as required by different molds. In this way the design of the mold itself is simplified. Presses equipped with these floating plates have been called *transfer presses*, but are not to be confused with auxiliary-ram transfer presses. Movement of these loose plates may be accomplished either by latches or bolts built into the mold, or by auxiliary pneumatic or hydraulic cylinders attached to the press.

Integral Molds. As the name implies, these molds are self-contained; each one has its own pot and plunger. This frequently increases the efficiency of the mold, since the transfer pot can be designed for best results with a specific cavity. A simple type is illustrated in Fig. 8-19.

Molds of this type can be designed for either manual or semi-automatic operation. The transfer chamber may be located above or below the mold cavity, and the material may flow through a sprue, runner and gate to the cavity, or the sprue may enter the cavity directly, as in the illustration.

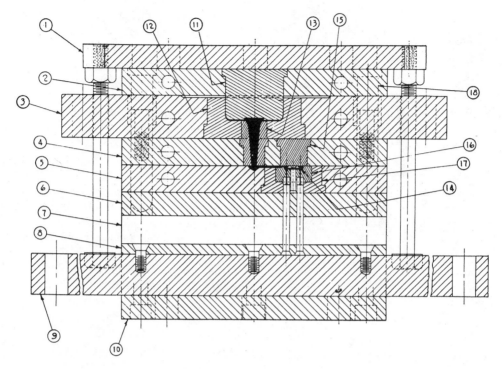

Fig. 8-18. Design of transfer mold for use in a transfer press. The press has a floating platen that will receive several standard sizes of transfer chamber.

1. Top plate	5. Cavity plate	10. Bottom plate	15. Force
2. Upper force plate	6. Backing-up plate	11. Upper force	16. Cavity
3. Floating platen	7. Parallel	12. Loading chamber	17. Chase
4. Force plate	8. Pin plate	13. Sprue plug	18. Guide pin
	9. Knock-out bar	14. Molded article	

Auxiliary-Ram Molds. Molds of this type are illustrated in Figs. 8-20 and 8-21. These are integral molds, in which the transfer plunger is operated by a separate double-acting cylinder mounted on the press head. Figure 8-22 shows a molded shot from the mold shown in Fig. 8-21.

As mentioned previously, this transfer cylinder and plunger may be mounted on the bottom press platen, within the main ram, or even on the tie rods or side columns of the press. These various methods of mounting are usually described as top-ram, bottom-ram, or side-ram mounting, and the molds are designated accordingly.

In comparing the advantages and disadvantages of the various designs of auxiliary-ram methods, it is difficult to generalize, because of limiting factors in each design and the method of operation. The statements that follow are therefore made with the realization that special conditions in individual plants may alter considerably the conclusions given.

Top-ram molds are frequently awkward to load because of the restriction of space between the bolsters and press head at the upper opening of the transfer tube. Aside from this, top-ram molds will normally permit a faster cycle, after loading, since the mold is already closed at the start of the cycle. There are certain exceptions to this statement, however, as will be noted later.

Another factor to be considered in any appraisal of top-ram molds is the difficulty of using top ejection mechanisms with this design, because of the space required for the transfer cylinder and ram.

Bottom-ram molds are easier to load, since they are open, and the transfer well is readily accessible, at the beginning of the cycle.

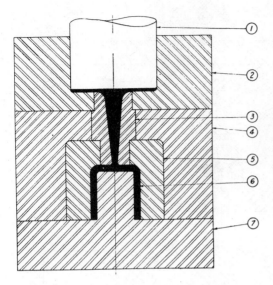

Fig. 8-19. The "integral" or "accordion" type of transfer mold.

1. Plunger	3. Sprue	6. Molded
2. Loading	4. Cavity plate	article
chamber	5. Cavity	7. Force

However, the molding cycle may be longer, because the mold must be closed, after loading, before the transfer ram is actuated.

When a bottom-ram mold is mounted in a conventional upstroke compression press, some of the stroke and opening must be sacrificed to provide mounting space on the lower press platen for the transfer cylinder. In some cases this can be offset by the use of longer side columns or strain rods.

In bottom-ram transfer and screw-transfer presses, the transfer cylinder and ram are contained within the main ram. This saves opening or daylight in the press.

Side-ram molds are less common, and are used primarily where the design of the molded article requires injection of material at the parting line. Theoretically, this design permits use of the full clamping area of the mold, but in actual practice, unless the molding material is extremely soft and fluid and the transfer pressure low, there is a distinct tendency for

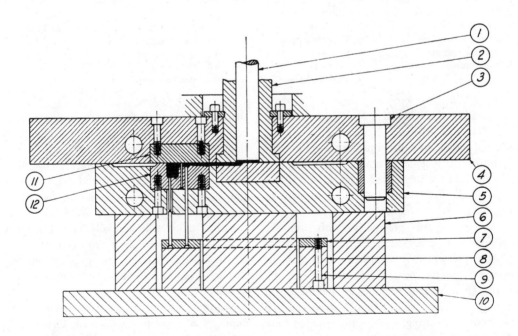

Fig. 8-20. Design of a pressure-type of transfer mold which makes use of the conventional press for clamping pressure only. Transfer is effected by means of an auxiliary ram.

1. Plunger	5. Cavity plate	9. Pin-plate screw
2. Loading chamber	6. Parallel	10. Bottom plate
3. Guide pin	7. Pin plate	11. Force
4. Force plate	8. Knock-out bar	12. Cavity

Fig. 8-21. Four-cavity auxiliary-ram transfer mold. (*Courtesy Cutler-Hammer, Inc.*)

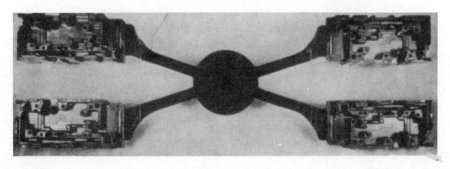

Fig. 8-22. Shot from plunger transfer mold illustrated in Figure 8-21. (*Courtesy Cutler-Hammer, Inc.*)

the mold to spring and flash as a result of localized, unbalanced transfer pressure near the gate.

Whatever method of mounting is used, it is now generally conceded that the auxiliary-ram method of transfer permits the highest production with the lowest mold cost and minimum loss of material in sprues, runners and culls. Other advantages, common to this and other types of transfer molding, are discussed below.

Mold design is covered in more detail in Chapter 10.

Advantages of Transfer Molding

Molding Cycles. Loading time is usually shorter in transfer molding than in compression molding, since fewer and larger preforms are used. These can be preheated most rapidly and effectively in dielectric equipment.

In entering the mold, the material flows in thin streams through small runners and gates. This promotes heat transfer, and may also momentarily add some heat to the material through friction and mechanical work. Most phenolic materials are also homogenized thereby, and volatile matter, which might otherwise remain in the piece and necessitate a longer cure to avoid blistering, is reduced in passing through these small channels and by escape through vents and clearance spaces around movable pins. All of these factors contribute to substantially shorter molding cycles than are possible in compression molds, without preheating.

Tool and Maintenance Costs. Deep loading wells are not necessary with transfer molds, and mold sections can be thinner than in compression molds, since they are not required to withstand the higher stresses involved during closing of the latter. This obviously permits an initial savings in tool costs.

There is less wear on the mold and much less tendency toward breakage of pins. In transfer molds, core pins can readily be piloted in the opposite half of the mold. When so designed, core pins having a length-to-diameter ratio of 8 : 1 can safely be used; in compression molds the largest ratio that can ordinarily be used is 4 : 1 using dielectrically preheated preforms and about 2.5 : 1 using cold powder. Transfer molds retain their original accuracy and finish considerably longer than do compression molds.

Molding Tolerances. Since the articles are produced in closed molds which are subjected to less mechanical wear and erosion by the molding material than are compression molds, closer tolerances on all molded dimensions should be possible in transfer molding. This is particularly true of dimensions perpendicular to parting lines, because of the very small amount of flash in properly designed and operated molds of this type.

Also, it should be possible to hold closer tolerances on diameters of holes and dimensions between holes, since forces in the mold which tend to distort or displace pins or inserts are much less.

This is the theoretical ideal which could be attained, if no other factors than those mentioned above were at work in the process. Actually, warpage and dimensional variations have occurred in numerous instances in plunger-molded articles, and it is not always possible in these cases to achieve the close dimensional tolerances hoped for.

Careful study has indicated that these difficulties are caused, in most cases, by abnormal shrinkage and internal strains set up in the molded articles by improper gating and excessive transfer pressure. In some instances, the article, although complicated in outline and section, may not lend itself to transfer molding because of peculiarities of its design.

Those who have studied the auxiliary-ram transfer process most thoroughly are convinced that in most instances close tolerances can be held, and optimum dimensional stability realized if, and only if, the preheating, the plasticity of the material and the transfer pressure are accurately controlled, and if the runners and gates are so located, and of such size and shape, as to permit rapid, free flow of the material into the closed mold.

Finishing Costs. Transfer molding reduces the costs of finishing all thermosetting materials, and especially those having cotton flock and chopped fabric as filler. In an article molded from these latter in a compression mold, the flash is frequently heavy, tough, and expensive to remove; this condition becomes even worse as the mold becomes worn. In a transfer-molded article, assuming a properly designed mold with adequate clamping pressure, the flash is quite thin or altogether absent.

Gates, except in certain cases with fabric-filled phenolics, can usually be made sufficiently thin, and can be so located, that their removal is easy and inexpensive. It may be possible to gate into a hole in the molded piece and to remove the gate by drilling.

Limitations of the Process

As might be expected, there are certain inherent limitations in transfer molding. These are discussed below:

Mold Costs. The statements concerning lower mold costs apply particularly to auxiliary-ram transfer molds. Pot-type transfer molds may, in some cases, be more expensive than equivalent compression molds, without off-setting advantages.

Loss of Material. The material left in the pot or well (the cull), and also in the sprue and runners, is completely polymerized and must be discarded. This loss of material is unavoidable, and for small articles it can represent a sizeable percentage of the weight of the pieces molded. In the auxiliary-ram molds, the cull is reduced to a minimum, and the main sprue is eliminated.

Until recently, transfer molding has not been well suited to the production of small molded items because of this loss of material. However, to meet the increasing trend toward miniaturization and increased demands for close tolerances and thin wall sections in molded electronics parts, small, fully automatic transfer molding presses permit economical production of such small items. While loss of material in cull and runners is still a factor in molding costs for these small parts, the savings elsewhere, including mold costs and finishing costs, usually offset the loss of material.

On the other hand, where large-volume production (in excess of 500,000 pieces) is required, and where cross-sections are simple and uniform, as in buttons and bottle caps, production by compression molding, in multi-cavity semi-automatic or automatic molds, with electronic preheating, may be cheaper.

Effect on Mechanical Strength. With wood-flour-filled phenolic materials, no significant loss of mechanical strength attributable to the transfer-molding process has been observed in laboratory tests on standard ASTM specimens. This conclusion has correlated reasonably well with general production molding experience.

Some instances of cracking around inserts and at weld lines have been noted in production molded articles, especially with mineral-filled phenolics and melamine materials, which have inherently lower mechanical strength. Changing the design and location of gates, altering the cycle of preheating, or reducing the volatiles content of the molding material, when tried separately or in combination, have overcome these difficulties in many cases; in some severe cases, however, no practicable remedy could be found within the framework of the transfer molding process, and it was necessary to mold the articles by compression in order to overcome the trouble.

With the improved impact materials containing fibrous fillers, a definite and sometimes marked decrease from values obtained in compression molding has been noted by a number of investigators. This effect on strength can be minimized by the use of lower transfer pressures and larger gates.

Comparison between Compression and Transfer Molding

A comparison of the two molding techniques is provided in Table 8-1, since it is recognized that many parts may be molded by both of these methods. The molding technique used frequently depends upon the molder's own economics and capabilities.

Materials Used

Phenol-formaldehyde and melamine-formaldehyde molding compounds are the materials most widely used in transfer molding. General results with these are excellent.

Urea-formaldehyde materials have been transfer-molded quite successfully in some cases, principally for small articles, but because of their reactivity and critical behavior in dielectric preheating, the process has not been found as generally satisfactory with them as with the phenolics and melamines.

The alkyd or polyester molding materials have been tried in transfer molds. In certain instances, they have proved suitable for smaller articles, but their extremely fast rate of reaction has prevented their use in larger molds and in articles where considerable plastic flow is required.

Table 8-1. Compression vs. transfer molding

Characteristic	Compression	Transfer
Loading the Mold	1. Powder or preforms 2. Mold open at time of loading 3. Material positioned for optimum flow.	1. Mold closed at time of loading. 2. RF heated preforms placed in plunger well.
Material temperature before molding	1. Cold powder or preforms 2. RF heated preforms to 220-280°F	RF heated preforms to 220-280°F
Molding temperature	1. One step closures— 350-450°F 2. Others—290-390°F	290-360°F
Pressures	1. 2,000-10,000 psi (3000 optimum on part) 2. Add 700 psi for each inch of part depth.	1. Plunger ram—6000-10,000 psi 2. Clamping ram–Minimum tonnage should be 75% of load applied by plunger ram on mold
Breathing the mold	Frequently used to eliminate gas and reduce cure time.	1. Neither practical nor necessary. 2. Accomplished by proper venting.
Cure time (time pressure is being applied on mold)	30-300 sec—Will vary with mass of of material; thickness of part and preheating.	45-90 sec—Will vary with part geometry.
Size of pieces moldable	Limited only by press capacity.	About 1 lb maximum
Use of inserts	Limited—inserts apt to be lifted out of position or deformed by closing.	Unlimited—complicated inserts readily accommodated.
Tolerances on finished products	1. Fair to good—depends on mold construction and direction of molding. 2. Flash—poorest Positive—best Semi-positive—intermediate	Good—close tolerances easier to hold
Shrinkage	Least	1. Greater than compression. 2. Shrinkage across line of flow is less than with line of flow.

Phenol-formaldehyde. Phenolic materials of all of the standard types, containing any of the usual fillers, can be successfully molded by transfer. The principal types used are listed below.

Type	Filler
general-purpose	woodflour
heat-resistant	asbestos
medium-impact	cotton flock or fabric
high-impact	cotton fabric
highest-impact	chopped cotton cord
low-electrical-loss	mica

For satisfactory use in transfer molding, a phenolic material must be extremely fluid at molding temperature in order to flow readily through runners and gates and form dense homogeneous molded articles. For most economical production, this flow should be accomplished in the shortest possible time, i.e., the material should have a rapid rate of flow. Phenolic materials possess these properties and, with the aid of dielectric preheating, give excellent results in all types of transfer molds.

Melamine - Formaldehyde. Mineral - filled, cellulose-filled and fabric-filled melamine materials have been employed in transfer molding with good results. Certain melamine-urea formulations also have shown good characteristics in transfer molding.

Melamine materials may require some modification of basic properties in order to produce best flow in transfer. These compounds as a class show greater after-shrinkage than do phenolics. This tendency exists more or less independently of the method of molding; it can be considerably reduced by proper control of the resin and its compounding, and by suitable preheating before molding.

Silicones. Glass-filled silicone materials are molded by transfer methods for special high-impact heat-resistant applications. Dielectric preheating is recommended to improve the flow. As with other fiber-filled types. precautions must be taken to use gates of adequate cross-section to maintain proper strength. After-baking at temperatures up to $200°C$ is usually practiced to obtain optimum strength and heat resistance.

Soft flowing epoxy and silicone molding compounds, molding at pressures of several hundred psi, are extensively used in transfer molding to "encapsulate" or mold plastic around electrical or electronic components, such as diodes, transistors, integrated circuits, resistors, capacitors, solenoid coils, and modules containing several such components. In such applications, the materials provide electrical insulation, moisture protection, and mechanical protection (see Fig. 8-23).

Theoretical and Design Considerations

There are many and diverse opinions, within the industry, concerning design of transfer molds, and molding pressures and molding techniques. Because of the many factors involved, no hard-and-fast rules can be given to serve as infallible guides in the design and operation of these molds. The principal factors are listed below, and recommendations based on experience and successful current practice are given. See also Chapter 10 on Designing Molds for Thermoset Processing.

Mold Design. It is considered good practice in multiple-cavity plunger-type transfer molds to mount the mold cavities, forces, and runners as separate hardened-steel inserts in the mold chases or retainer plates. This permits removal of individual mold cavities and runner blocks for repair or replacement when wear or breakage occurs. Where not precluded by other features of design, such as location of electric heating cartridges and ejector pins, a circular layout of mold cavities with short radial runners from the central well is preferred. However, there are numerous examples of successful transfer molds having various ladder and multiple-"T" arrangements.

The center pad beneath the pressure well should be hardened and well supported to prevent deflection and consequent flashing of the mold. For the same reasons, all mating mold surfaces should be ground smooth and perfectly flat to provide uniform contact. It is not imperative that cavities and runners be chrome-plated, but this is usually advisable to improve release from the mold and to reduce wear, particularly in the runners.

Fig. 8-23. Typical mold and loading frame, after molding, for transfer encapsulation of electronic integrated circuits, with a soft-flowing epoxy or silicone molding compound. (*Courtesy Hull Corp.*)

Runners, Gates and Vents. The design of runners and the size and location of gates can be a controversial subject in this field. The recommendations given below are in accord with current practice, but are not intended to be infallible pronouncements.

Main runners are usually $\frac{1}{4}$ to $\frac{5}{16}$ in. wide and $\frac{1}{8}$ in. deep, and semi-circular in section; branch runners, where used, should be about $\frac{1}{8}$ in. wide and $\frac{3}{32}$ in. deep. These runners are located in the half of the mold containing the ejector pins; this is usually the plate opposite the pressure well. The runners should be kept to minimum length and should extend into the pad at the bottom of the pressure well at least $\frac{1}{2}$ in.

Location, size, and shape of gates are quite important to proper operation of the mold. While there are many conflicting opinions concerning location, and while striking individual exceptions may be found on particular jobs, it is considered good practice to place the gate at the thickest section of the molded article, and whenever possible, at a readily accessible point on the article, so that it can be removed by simple sanding or filling.

The size of the gate will vary with the type of material molded, the size of the piece and the molding pressure available. For small pieces, and where general-purpose wood-flour-filled phenolics are used, it is recommended that the gate be 0.80 to 0.100 in. wide and 0.015 to 0.020 in. deep. With mineral-filled materials, a gate 0.125 in. wide and 0.030 in. deep is usually necessary.

The shape of gates for transfer molds has received considerable study. Depending on the size and shape of the molded part, gates of circular or rectangular cross section may be used. While the circular gate generally permits most rapid filling with smallest gate area, it may leave an unsightly scar on the finished part. Often fan-gating or edge gating is used, with gates from 0.010 to 0.020 in. thick, and as wide as several inches (perhaps as wide as the part). Such gates, if shaped such that the runner approaching the gate is always of thicker and larger cross-section than the gate itself, will permit clean break-off of the part from the runner with minimum scar and often no further finishing costs.

Because of the coarseness of the fillers used in high-impact materials, larger gates are usually necessary, both to lessen the transfer pressure required and to prevent impairment of the mechanical strength of the molded articles. In these cases, a width of 0.500 in. and a depth of 0.125 in. may be needed.

These are general recommendations; it may well be necessary, in individual cases, to enlarge gates beyond these dimensions, depending on circumstances. It is always easier, in any case, to remove metal from a mold than to add it.

Vents are almost equally important to permit escape of air, moisture and other volatiles, as the materials fill the mold cavity. It is frequently found that without proper vents the mold cavities will not fill properly even under high transfer pressures. Vents should be located on the same half of the mold as the runners, and opposite the gate. They are usually 0.003 to 0.005 in. deep, and about $\frac{1}{8}$ in. wide, and are extended to the outer surface of the mold block. With epoxies and silicones, vent depths are only 0.001 to 0.002 in. deep.

Vacuum Venting. To overcome the trapping of air or gas in a cavity, in locations difficult to vent effectively, molds may be designed such that all cavity vents feed into a space which is sealed from the outside of the mold (when closed) by an "O"-ring seal, and which is connected to a vacuum reservoir through a vacuum line containing a solenoid operated valve. In operation, as soon as the mold is closed, and the transfer plunger enters the pot, the aforementioned solenoid valve is automatically opened, causing the cavities to vent rapidly into the vacuum reservoir, before the molding compound has entered or filled the cavities. Two advantages result. First, the material, finding it unnecessary to "push" the air from the cavity through the vents, enters with a minimum of back pressure and thus fills the cavity more rapidly, leading to faster cures. Second, not only is there essentially no entrapped air, and therefore no voids in the

part, but such minute quantities of air that may be present are readily absorbed "into solution" in the molding compound, due to the molding pressure.

Vacuum venting has been used successfully in both semi-automatic and fully automatic transfer molding, and also in automatic in-line screw injection molding. (See Figs. 8-24 and 8-25.)

Selection of Material. The problem here is three-fold—to choose the type of material that will yield the desired physical properties in the molded article, e.g., impact strength, heat resistance, or low electrical loss; to choose from among a number of formulations of this type the one having the proper preheating and molding characteristics for the mold in question; finally, to choose the proper plasticity, which will provide the fastest transfer and cure. The last cannot be decided in advance; it must be determined by experiment.

Choice of the proper plasticity is very important. A material which is unduly stiff will be extremely critical in behavior during preheating, and may not fill the mold; a material that is too soft may flash the mold or require excessive preheating to attain a satisfactorily rapid cure.

A maximum transfer time of 10 to 15 sec. at a mold temperature of 320 to 350°F, with a preform temperature of 220 to 260°F, is usually found satisfactory. It will be realized that individual conditions may necessitate some departure from these values, but these represent the average or general experience.

Molding Pressure. It would be extremely difficult, if not impossible, to measure accurately, or to calculate theoretically, the magnitude and distribution of pressure within thermosetting material flowing in a transfer mold. If all of the factors involved were in perfect dynamic balance, there would be no unbalanced fluid pressure in the mold cavity, at the moment of complete filling.

Initially, with either pot or plunger molds, fluid pressure in the hydraulic line to the press is converted to mechanical pressure on the molding material, as the latter is compressed and forced into the closed mold. As the mold fills, and polymerization of the material

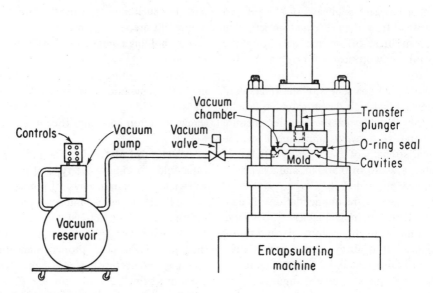

Fig. 8-24. Schematic diagram showing principle of vacuum venting with transfer molding. Principle also used with automatic transfer and screw injection molding where cavity configuration precludes adequate parting line venting. (*Courtesy Hull Corp.*)

Fig. 8-25. Multi-cavity transfer mold arranged for vacuum venting. Note rectangular O-ring seal surrounding the six bottom chases. (*Courtesy Hull Corp.*)

continues, this pressure builds up to a maximum. Ideally, at the moment when the mold is filled completely, all of the material will have hardened to infusibility, and the transfer pressure, transmitted through the hardened material, will be balanced exactly by the clamping pressure applied to the closed mold by the main ram.

Actually, no such equilibrium is attained. Usually, even after the mold is filled, the material is still in a semi-plastic state, and for a few seconds, at least, exerts an indeterminate fluid pressure on the interior surfaces of the mold. If this were not counterbalanced by external clamping pressure, the mold would open and the material would flash excessively at the moment of filling. Thus it is necessary to provide excess clamping area or pressure greater than that theoretically necessary.

Pot-type transfer molds are usually provided with only one source of molding pressure, the main ram; plunger molds, on the other hand, usually have two, more or less independent, sources of pressure for the main and transfer ram. It is usually, therefore, in designing transfer molds, to equate transfer area to total mold-clamping area for the pot type, and transfer pressure to clamping pressure for the plunger type, and then to allow 10 to 15% excess clamping area, or pressure, to prevent opening of the mold, and flashing. Usually, if the mold is properly supported in the press, this margin will be sufficient.

Because of the many variables involved in the design of an article and of the mold, and in the molding conditions, it is practically impossible to calculate the minimum transfer pressure required to mold a given article. Only empirical assumptions can be made, based on limited experimental data and on previous production-molding experience.

For soft-plasticity, general-purpose phenolic materials, electronically preheated, minimum transfer pressures of 8000 psi for plunger-type and of 12,000 psi for pot-type transfer molding are generally recommended. This assumes a total runner area of about 0.05 sq in., and is valid primarily for quantities of material of the order of 100 grams. For quantities of material greater than 100 grams, enlarge the runner and gate areas by 50 to 100%.

For fabric-filled phenolic materials also, a 50% increase in transfer pressure will usually be required to keep transfer time to an economical minimum. Here, too, larger gates, as recommended previously, should be used to avoid the secondary impregnation and consequent embrittlement of the material.

Design Calculations. The various factors included in design of plunger transfer molds, in so far as pressure is concerned, can be defined and illustrated as follows:

1. *Line Pressure.* The pressure in psi in the hydraulic line to the press, supplied either by a self-contained pump or from a central pump and accumulator system.
2. *Clamping-Ram Pressure.* The total pressure delivered by the main ram, determined by multiplying the ram area by the line pressure.
3. *Transfer-Ram Pressure.* The total force applied by the transfer ram, expressed in pounds, determined by multiplying the hydraulic ram area by the line pressure.
4. *Plunger Pressure.* The pressure exerted on the material in the transfer chamber by the plunger. It is equal to the injection-ram pressure divided by the area of the plunger.
5. *Mold-Clamping Pressure.* The effective pressure which holds the mold closed against the pressure exerted by the material within the mold cavities and runners. It is expressed in psi and is determined by dividing the clamping-ram pressure by the total projected area of cavities and runners in square inches.

These factors can be shown symbolically as follows:

L_1 = line pressure, psi, clamp
L_2 = line pressure, psi, transfer
A_m = projected area of mold cavities, lands and runners, sq in.
A_r = area of clamping ram, sq in.
A_i = area of injection ram, sq in.
A_p = area of plunger, sq in.
CRP = clamping-ram pressure $(A_r \times L_1)$, lb
IRP = transfer-ram pressure $(A_i \times L_2)$, lb
PP = plunger pressure, psi $\dfrac{(A_i \times L_2)}{A_p}$
MCP = mold-clamping pressure $\dfrac{(A_r \times L_1)}{A_m}$, psi

By definition, to insure safe operation and avoid flashing, mold-clamping pressure = plunger pressure +15%, or:

$$MCP = 1.15 \, PP$$

Practical application of this simple equation is demonstrated in the following example.

Example. How many cavities can be placed in a mold which is to operate under the following conditions:

projected area of molded article = 3 sq in.
estimated runner area per cavity = 1 sq in.
total area per cavity = 4 sq in.
diameter of main ram = 14 in.
diameter of injection ram = 6 in.
plunger diameter = 3 in.
line pressure = 2000 psi, both clamp and transfer

Let x = the number of cavities
L = 2,000 psi
A_m = 4x
$A_r = \dfrac{3.1416 \times 14^2}{4} = 154$ sq in.

$$A_i = \frac{3.1416 \times 6^2}{4} = 28.3 \text{ sq in.}$$

$$A_p = \frac{3.1416 \times 3^2}{4} = 7.0 \text{ sq in.}$$

$$CRP = A_r \times L = 154 \times 2{,}000 = 308{,}000 \text{ lb}$$

$$IRP = A_i \times L = 28.3 \times 2{,}000 = 56{,}600 \text{ lb}$$

$$PP = \frac{A_i \times L}{A_p} = \frac{56{,}600}{7} = 8{,}085 \text{ psi}$$

$$MCP = \frac{A_r \times L}{A_m} = \frac{308{,}000}{4x} = \frac{77{,}000}{x} \text{ psi}$$

$$MCP = 1.15 \, PP$$

$$\frac{77{,}000}{x} = 1.15 \times 8{,}085$$

$$x = \frac{77{,}000}{1.15 \times 8{,}085} = 8.3 \text{ or } 8 \text{ cavities}$$

In designing pot-type transfer molds, where the only source of external pressure is the main ram, the pot area is equated to the total mold area and, as mentioned previously, the latter is increased by 15% to prevent flashing.

Let L = line pressure, psi

A_r = area of main ram, sq in.

A_p = projected area of pot or material chamber, sq in.

A_m = projected area of mold cavities, lands and runners, sq in.

CF = clamping force $(A_r \times L)$, lb

MCP = mold-clamping pressure

$$\left\{ \frac{A_r \times L}{A_m} \right\}, \text{psi}$$

Then for equilibrium, with the factor of safety:

(1) $\quad A_m = 1.15 \, Ap$

(2) $\quad A_m = \dfrac{A_r \times L}{MCP}$

(3) $\quad \dfrac{A_r \times L}{MCP} = 1.15 \, A_p$

Other Considerations. It has been assumed in the foregoing discussion of transfer-molding pressure that flashing will be prevented if sufficient clamping pressure or area is provided. This is generally true if the mold is properly supported, particularly under the transfer chamber. However, if insufficient bolstering is used at this point, or if the mold is not properly aligned or hardened, deflection can occur and permit flashing, in spite of the excess clamping pressure provided in the design.

Transfer pressure should be sufficient to fill the mold, under ordinary circumstances, in 10 to 15 sec. Excessive transfer pressure should be avoided—it may cause flashing, it will probably cause undue wear of gates and runners, and it may decrease normal molding-shrinkage so much that molded articles will be outside normal tolerances.

The rate at which a phenolic material will be preheated in a dielectric preheater depends, among other factors, on the high-frequency output of the machine and on the electrical-loss characteristics of the material involved. Wood-flour-filled compounds having a higher electrical-loss factor, will preheat more rapidly than mica-filled materials. Modern dielectric preheaters of 2.5 KW output and with frequency of 100 mega Hertz are designed to heat approximately 1 lb of wood-flour-filled material from room temperature to 300°F in 30 sec.

Where a number of preforms are being preheated at one time, the uniformity of preheating of the material is quite important. This is related directly to the hardness and density, their temperatures will differ considerably and their behavior in transfer will be erratic. Once established, optimum preheating conditions should be maintained as nearly constant as possible for a given batch of material. To facilitate uniform heating of cylindrical preforms, preheaters with automatic roller electrodes cause the preform to rotate under the top electrode during the actual heating.

Numerous other factors affect transfer molding, but these are already familiar to molders employing this method and need not be described in detail. It has been the purpose of this chapter, rather to point out some of the important features of this method and the advantages in increased production and lowered production costs that may be achieved with it, if certain fundamental considerations are known and observed.

Some automatic transfer machines are shown in Figs. 8-26 and 8-27.

Liquid Resin Molding. A recent development, called *liquid resin molding* (LRM) is a

Fig. 8-26. Automatic transfer machine. (*Courtesy Stokes Div., Pennwalt Corp.*)

combination of liquid resin mixing and dispensing, and transfer molding. The process equipment includes a machine to proportion, mix and dispense a resin and catalyst directly into a transfer mold, generally through a parting line sprue feeding into the runner system. Another version, suitable for long pot-life liquid resin systems, dispenses the resin mix from a storage chamber in which the catalyst has previously been added. (See Fig. 8-28.)

Because the pressures of injection are approximately 25 to 50 psi, very fragile inserts can be molded, and mold wear is at a minimum. Some formulations for LRM also may be molded at temperatures as low as $200°F$, which permits encapsulation of some heat-sensitive electronic components which do not lend themselves to encapsulation at conventional transfer molding temperatures of $300°F$ or higher.

Liquid resin systems for LRM may be filled

Fig. 8-27. High-speed, fully-automatic transfer molding machine with integral preformer and electronic preheater. (*Courtesy Hull Corp.*)

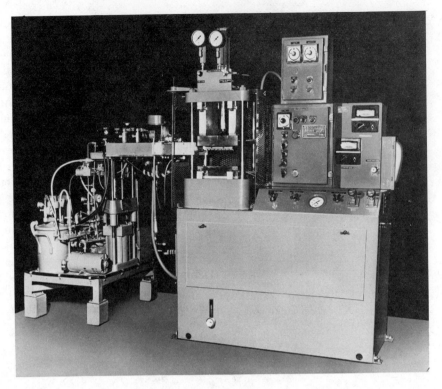

Fig. 8-28. Complete installation for liquid resin molding process. Apparatus at left proportions, mixes, and dispenses liquid resin and liquid curing agent, which is then injected into closed molds at parting line, at pressures as low as 25 psi. (*Courtesy Hull Corp.*)

or unfilled. They generally have a pot life of at least two hours at room temperature, but a "snap cure" of often less than one minute at mold temperature. Unlike liquid casting resins, such resin systems must include a mold release agent to enable easy parts removal from cavities.

See also Chapter 18 on Casting.

9

INJECTION MOLDING OF THERMOSETS

Around 1940 three processes were developed for the injection molding of thermosetting materials. Thermoplastic plunger injection molding machines were modified for these processes, which molded parts with uniform density throughout varying thicknesses of section and developed maximum physical characteristics of the molding compound.

The jet process. The attached sketch (Fig. 9-1) shows the principal features employed in jet molding. Careful control of temperatures along the heating cylinder was necessary, as was high pressure at injection. The material was held at lower than polymerization temperature in the cylinder and additional heat was added by electrodes at the nozzle, frictional heat through the nozzle, and heat of compression.

Front and rear sections of the nozzle had separate heat controls. Once the shot had been injected, the nozzle was immediately cooled to prevent polymerization of the material remaining in the nozzle.

Free flowing materials were hopper fed but fabric-, glass-, and asbestos-filled materials required hand feeding.

Molding cycles were dramatically reduced and, because of the nature of heating the material, thick sections cured as quickly as thin sections. This process tended to break the fibers

* Reviewed and revised by E. J. Fitzpatrick, Stokes Division, Pennwalt Corp. Additional material by S. Bodner, Livingston, N.J., and E. Vaill, Chatham, N.J.

of fibrous fillers, and in the case of urea, discoloration was possible with pastel colors.

The jet process could be used for natural rubber compounds at faster cycles than possible by compression molding.

The flow process was similar to the jet process except that nozzle temperatures were controlled by circulating hot oil, steam or high pressure hot water in place of the electrodes.

The offset process was so named because the injection plunger displaced the material into passages offset along its path. A heated preform was dropped into a yoke positioned in front of the stationary platen, immediately ahead of the plunger. With the mold closed, the plunger passed through the stationary half of the mold and forced the preheated preform into runners and cavities. The advantages claimed for the offset process were fast cycles, large capacities, and low plunger pressure requirements.

In all three processes the parts were ejected by the opening of the press actuating an ejector bar.

None of these techniques, however, ever really achieved significant commercial penetration.

In-line Screw Injection Molding

But in the late sixties, about two years after the introduction of screw transfer machines, the

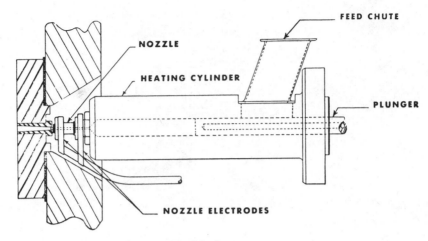

Fig. 9-1. Jet process.

concept of in-line screw injection molding of thermosets (also known as direct screw transfer or DST) was developed. In a short time, this techinque has had a significant influence on the thermoset molding business by virtue of reduced molding cycle time and the potential it offers for low-cost, high-volume production of molded thermoset parts.

Today, machines are available in all clamp tonnages up to 1200 tons and shot sizes up to 10 lb (see Fig. 9-2, 9-3, and 9-4).

Thermoset in-line screw assemblies are fitted to horizontal or vertical clamp machines. Most horizontal clamp thermoplastic injection machines can be converted to injection of thermoset by changing the screw barrel and nozzle. Because most thermoset barrels are shorter than their thermoplastic counterpart, it may be necessary to re-position the injection assembly closer to the clamp fixed platen.

During the past years, there has been a new series of thermoset molding materials developed specifically for injection molding. These materials have long life in the barrel at moderate temperatures (approximately 200°F), and react very rapidly when the temperature is brought up to 350 to 400°F. This unique development in materials has helped result in acceptance of injection molding as a production process.

How the Process Operates. A typical arrangement for in-line screw injection molding of thermosets is shown in Fig. 9-5.

Fig. 9-2. 100-ton parting line screw injection machine for thermoset molding with inserts. (*Courtesy Hull Corp.*)

The machine consists of two sections mounted on a common base. One section clamps and holds the mold halves together under pressure during the injection of material into the mold. The other section—the plasticizing and injection unit—includes the feed

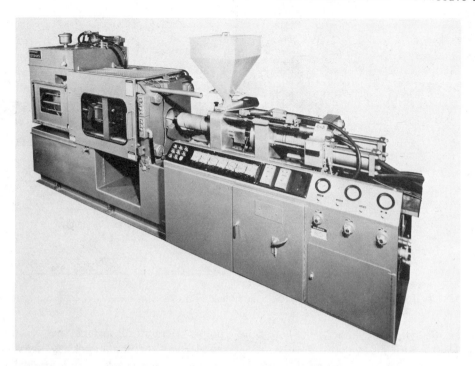

Fig. 9-3. Horizontal thermoset injection machine, 200-ton clamp with 28-oz. shot capacity. (*Courtesy Stokes Div., Pennwalt Corp.*)

Fig. 9-4. 275-ton fully automatic, in-line screw injection molding machine for thermosets. Features digital control system that controls each step in the molding cycle to 0.1-sec. accuracy. (*Courtesy Hull Corp.*)

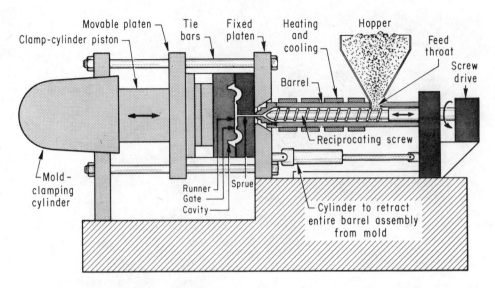

Fig. 9-5. Typical arrangement of direct screw transfer molding machine for thermosets. (*Courtesy S. Bodner*)

hopper, the hydraulic cylinder which forces the screw forward to inject the material into the mold, a motor to rotate the screw, and the heated barrel that encloses the screw.

Basically, the injection-molding press, whether for thermosets or thermoplastics, is the same, utilizing the reciprocating screw to plasticate the material charge. In most injection-molding of thermosets, the material, in granular or pellet form, is fed from the hopper by gravity into the feed throat of the barrel. It is then moved forward by the action of the flights of the screw. As it passes through the barrel, the plastic picks up conductive heat from the heating element on the barrel and frictional heat from the rotation of the screw.

For thermoset materials, the depth of flights of the screw at the feed-zone end is normally the same as the depth of flight at the nozzle end; this is a zero compression ratio screw. This screw configuration is the major difference between thermoset and thermoplastic molding machines—the latter have compression ratios such that the depth of flight at the feed end is $1\frac{1}{2}$ to 5 times that at the nozzle end.

As the material moves forward in the barrel it changes from a granular to a semiviscous consistency, and it forces the screw backward in the barrel against a preset hydraulic pressure. This back pressure is an important processing variable. The screw stops turning when the

proper amount of material has reached the nozzle end of the barrel, as sensed by a vernier-set limit switch. This material at the nozzle end—the charge—is the exact volume of material required to fill the sprue, runners, and cavities of the mold.

The screw is then moved forward by hydraulic pressure (up to 20,000 psi) on the plastic. The hot plastic melt is forced through the nozzle of the barrel, through the sprue of the mold, and into the runner system, gates, and mold cavities.

The granular plastic is fed directly into the barrel of the press, and no external auxiliary preforming or preheating equipment is required. Despite this, the temperature of the material entering the mold is higher and more uniform than that in other thermoset molding techniques because of the homogenizing effect of the screw upon the plastic.

Most barrels are covered with a metal jacket designed to permit heated water to flow across and around the barrel (Fig. 9-6). The water or temperature controlled fluid not only assists in heating the thermoset material as it moves along the flight of the screw, but helps prevent over-heating because the fluid will withdraw internal heat should this occur because of excessive screw rpm or back pressure.

Screw check valves or sleeves which are standard for thermoplastics (except for heat-sensitive materials like PVC) are not used with

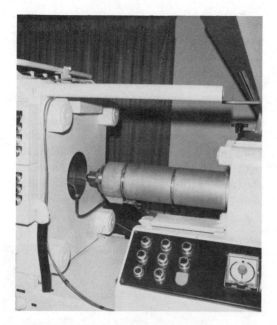

Fig. 9-6. Water-jacketed barrel. (*Courtesy Stokes Div., Pennwalt Corp.*)

the in-cavity cycle time. The direct screw transfer (DST) process substantially reduces cure time from that of other thermoset processes, particularly for parts having $\frac{1}{8}$ in. wall section or greater, Fig. 9-7. Cycle-time reductions on the order of 20 to 30% are common; even greater reductions have been reported.

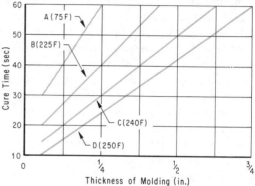

Fig. 9-7. Cure time vs. molding thickness for thermoset molding process and charge temperatures for general-purpose phenolic. (*Courtesy S. Bodner*)

thermosets. They provide restrictions to even flow of high viscosity materials, and eventually cause the material to set up in this area. Some material slippage or back flow along the flights of the screw does occur when the injection pressure approaches maximum. Because of this, one should not attempt to use maximum rated shot capacity. A rule of thumb for most materials and machines is to use up to 80% of rated capacity.

Nozzles at the end of the barrel are usually water cooled or controlled. It is used to maintain a proper balance between a hot mold (350-400°F), and a relatively cool barrel (150-200°F).

Process and Design Considerations for DST*

Thermoset molding resins require curing by a chemical reaction, or polymerization. Thus, the closer to the curing temperature the material can be placed into the mold, the shorter will be

* Section on process and design considerations prepared by Seymour S. Bodner, Consulting Engineer, Livington, N.J., and E. W. Vaill, Consultant, Chatham, N.J. Adapted from an article, "Injection molded thermoset parts," as it appeared in the Feb. 10, 1972 issue of *Machine Design*, and reprinted here with the permission of the publishers.

Four methods of processing thermosets are compared. Line A represents the original method of molding, wherein cold material is fed directly into the heated cavities. In this method, the heat required to polymerize the material is supplied by conduction from the mold. Since plastics are thermal insulators, the heat-transfer rate is low, and the cold-powder method is very inefficient, particularly for thick sections. A $\frac{1}{4}$-in. wall section requires a cure time of 60 sec. Line B shows that if the loose powder is preheated to 225°F prior to loading the mold, cure time for the same $\frac{1}{4}$-wall section is reduced by approximately one-third.

Line C shows cure times obtainable by the plunger, or transfer method, using high-frequency dielectric preheating of preforms (cold-sintered pellets) to a temperature of approximately 240°F. By this improvement, the $\frac{1}{4}$ in. section is now molded in about 30 sec. This represented the highest state of the art before injection molding was developed.

Line D demonstrates the still greater reduction in cure time obtainable by DST, for this same $\frac{1}{4}$-in. wall section is cured in less than

25 sec. Moreover, the injection process is automatic, so no operator time is required to charge the transfer pot or remove the molded parts and flash. This can cut overall cycle time by an additional 5 to 10 sec.

No direct labor or equipment is required in DST molding to preform or preheat the molding compound. Furthermore, certain high-bulk materials which do not lend themselves to transfer molding due to their inability to be preformed and preheated are adaptable to DST molding.

Molding a thermoset material on an injection-molding machine involves approximately the same cost factors as molding a thermoplastic material. Machinery costs, plant-space requirements, and labor costs are about the same. Therefore, the costs that makes the difference are basic material price and cycle time required.

Materials. Most materials used in DST are slightly modified from conventional thermoset compounds. These modifications are required to provide the working time/temperature relationship needed for screw plasticating. Additionally, material formulations may be altered according to the geometry of the part being molded, as is done for the other processes. The new compounds are priced approximately the same as conventional materials.

While all thermosetting materials have been molded by this technique, the most commonly used DST-molding materials are the phenolics. These materials were the first to be injection molded, and it is estimated that about 20 million pounds were processed by DST in 1970. Other thermoset materials being molded by the DST process include melamine, urea, polyester, alkyd, diallyl phthalate (DAP), and alloys such as phenolic melamine and phenolic epoxy.

A wide variety of fillers can be used to achieve the properties required in the finished product. Except in a few special cases, the choice of filler is of greater significance in end-product properties than in the method by which the product is molded.

The recent introduction of dust-free phenolic in pellet form brings additional advantages to DST molding. The dust-free material eliminates the former need for separate molding areas and equipment for phenolics. Now these

materials can be molded side-by-side with a thermoplastic operation, with no danger of contamination. In fact, with a change of the barrel, the same machine can be used for both materials.

Applications particularly suited for DST materials are automotive power brake, transmission, and electrical parts; wet/dry applications such as in steam irons, washer pumps, and moisture vaporizers; and communications and electrical distribution parts that require dimensional stability and electrical insulation. (See Figs. 9-8 and 9-9.)

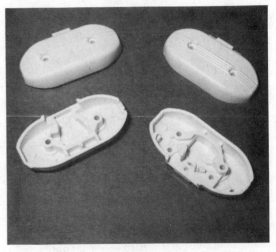

Fig. 9-8. Switch body molded of alpha-cellulose-filled urea, with flow characteristics adjusted for screw injection molding. (*Courtesy American Cyanamid*)

Fig. 9-9. Steam iron handle screw molded from phenolic. (*Courtesy Westinghouse Electric*)

Process Considerations in Design. Proper design of DST molded parts requires an understanding of the flow characteristics of material within the mold. These characteristics affect some properties of the finished parts, and they vary somewhat with the different materials. In that they have a flow pattern, DST parts are more similar to transfer-molded parts than to compression-molded parts.

Shrinkage, in most parts, is slightly more in the direction of flow than in the transverse direction. This difference is caused by the orientation of the filler (almost all thermoset compounds contain a filler) as the compound flows through the gate and into the cavity. Such shrinkage is particularly evident in edge-gated parts. To minimize the flow effect, center gating, which distributes the material more evenly is recommended where optimum shrinkage control is required, such as in molding a round piece within dimensional tolerance. Technology is now well advanced in the use of warm-runner molds (equivalent to hot-runner molds as used with thermoplastics) which, when used with center gating, provide substantial reduction in material consumed in sprue and runner—a cost saving over conventional thermoset molding. (See pp. 248 and 264.)

Flatness required in the finished molded part is another design parameter that is affected by process capabilities. Long, narrow parts that are required to be flat can be difficult to mold because of the variation in shrinkage. Similarly, a non-uniform wall section may warp as a result of non-uniform shrinkage. Choice of gate location can sometimes compensate for such conditions; close cooperation with the mold designer is recommended at this stage.

Surface finish on DST-molded parts is also different from that of traditionally molded thermosets. The greater the distance of material flow required, the more likely that flow marks will occur. Avoiding the large, flat, polished areas helps reduce rejects and keeps molding costs down. Here, too, gate location is a factor. If large, flat areas are necessary, such areas can be textured or patterned to mask slight irregularities. A rough surface texture is not desirable, however, because it may cause sticking in the mold cavity.

Gates used for DST moldings are usually smaller than those needed for transfer molding. This provides several advantages. A smaller gate gives a cleaner gate break that requires little or no hand finishing. A gate can be placed in an area where previously the appearance of a larger gate mark would have been objectionable. A small gate also contributes to faster cure cycles for DST parts since it increases frictional heat within the molding compound during filling. However, impact properties of compounds containing long-fiber reinforcement degrade when the material passes through a small gate. When such materials are specified, close cooperation between the material supplier and the molder is necessary. Other physical properties such as tensile and flexural strength can also be controlled by gate geometry.

Optimum gate cross sectional area as a function of part weight is shown in Fig. 9-10.

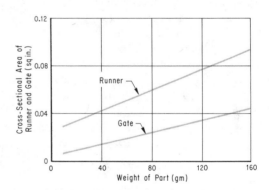

Fig. 9-10. Recommended areas for runners and gates for various weights of general-purpose phenolic parts. (*Courtesy S. Bodner*)

Modifications to the gate size indicated on the chart are made according to the number of cavities, runner length, wall thickness, and molding conditions. This chart can be used in deciding whether to allow or restrict a gate on a given surface as appearance requirements dictate.

Part-Design Considerations. As in most molding processes, wall-section uniformity is important. Molding cycles, and therefore costs, depend upon the cure time of the thickest section. Therefore, cross sections should be as uniform as design parameters allow, with a minimum wall thickness of $\frac{1}{16}$-in. A good working average for wall thickness is $\frac{1}{8}$ to $\frac{3}{16}$ in. Nevertheless, as part requirements dictate, heavy walls favor the use of thermosets because the cure rate of thermosetting materials is considerably faster than the cooling rate of thermoplastics. A rule of thumb for estimating cycle times for a $\frac{1}{4}$-in. wall section is 45 sec for thermoplastics and 30 sec for thermosets, as molded by DST.

Generous radii and fillets are recommended, as in other plastic processes, for maximum strength. Most plastic materials are somewhat notch-sensitive, so avoiding sharp corners is important. Generous fillets are not as likely to cause sink marks in the low-shrinkage thermoset materials as they do in some of the thermoplastics, so more freedom is available in the use of reinforcing fillets.

Draft to allow the release of the molded part from the cavity and force plug (core) should be at least $\frac{1}{4}°$ per side; greater draft is preferred if possible. Since the DST method is automatic, ease of mold release is essential for rapid ejection with minimum distortion of the part. Provision should be made in the part for the placement of ejector pins for this purpose. The use of as large as possible (at least $\frac{1}{8}$ in.) knockout pins consistent with part design and aesthetic considerations will provide trouble-free processing which is reflected ultimately in cost.

Venting of gases from the cavity must be done in the short time available during the filling of the cavity. The most effective vent is the parting line of the mold, so try to visualize the material flow from the gate to the most distant point of the part where the gases will collect. If this point is on a mold parting line, then the part is well designed for venting. Avoid long, dead-end corridors of flow and trapped wall sections that prevent venting.

Molded-in inserts are commonly used with thermosetting materials. However, since the DST process is automatic, the use of post-assembled inserts rather than molded-in inserts is recommended. Molded-in inserts require holding the mold open each cycle for placing the inserts. A delay in the manual placement destroys the advantage of uniformity and consistency of the automatic cycle, affecting both production and quality.

Tolerances of parts molded by the DST method are comparable to those produced by compression and transfer methods. Tolerances have been held as low as ±0.001 in./in., but ordinarily, tolerances of ±0.003 to 0.005 in./in. are economically practical for production.

All mold techniques and variations such as cams and movable die sections used in other forms of molding are adaptable to DST molding. Therefore, parts can include such features as molded-in side holes, threads, and undercuts. These refinements increase mold cost, of course, so the decision of how many refinements to build into a mold is usually an economics problem rather than a technical one.

Cold Runner Molding

Standard injection molds for thermoset materials are very similar to standard mold designs for thermoplastic. Because one cannot grind and reuse the thermoset runner system, there is always an interest in reducing the amount of material in a runner system required to produce parts on a production basis.

During 1969, a mold designed to maintain thermoset materials in a plastic state for the runner without ejecting it from the mold with each molding was introduced. In theory, it was not too different from "hot runner" thermoplastic mold designs, except that heated water was used to maintain a "runner" temperature of between 150-210°F, and cartridge heaters were used to maintain proper cavity and force temperatures for curing thermoset materials. This type of mold is known as a "cold runner" thermoset mold, or sometimes "warm runner" mold (see Fig. 9-11). In most cases only short sub-runners are rejected with the parts, resulting in very little waste material.

A multi-cavity standard injection mold for small thermoset parts may have 50 to 150% runner or waste material. Using a cold runner mold, this can be reduced to 10% waste matter.

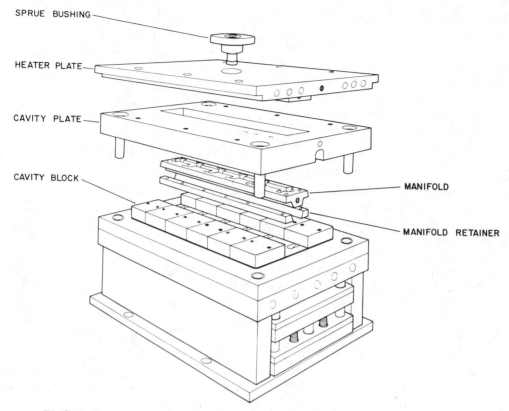

SPRUE BUSHING

HEATER PLATE

CAVITY PLATE

CAVITY BLOCK

MANIFOLD

MANIFOLD RETAINER

Fig. 9-11. Exploded view of a cold runner manifold. (*Courtesy Stokes Div., Pennwalt Corp.*)

Injection Molding Reinforced Polyesters*

In the mid-seventies, the injection molding of glass-reinforced thermoset polyesters began to take on more importance as better materials (e.g., low-shrink versions of polyester resins, pelletized forms of polyester/glass, etc.). and more efficient equipment and tooling became available. As such, injection molded reinforced thermoset plastics parts began to compete in such markets as distributor caps, fuse blocks, switch housings, power tool housings, office machines, and garbage disposal units. Other possibilities for the injection molding of bulk molding compounds (BMC) would be in competition with various metal die castings.

A Systems Approach. Injection molding FRP (fibrous glass-reinforced plastic) should be

* Adapted from "Injection molding FRP", by John F. Comey, Technical Service Mgr., Glastic Corp., Cleveland, Ohio, as originally appearing in the August 18, 1975 issue of *Plastics World,* published by Cahners Publishing Co.

thought of as a system operation, made up of material, press and mold. Each element must be compatible with the others to get the best results.

The first element is the material. It should have some specific characteristics. For example: it must flow easily at lower-than-mold temperatures without curing although traveling through long, complex patterns. It's important that resin, glass and filler components maintain integrity and not separate during the flow. Finally, when in place at mold temperature it should cure rapidly.

Shrinkage has always been a particular problem when working with any reinforced materials. Variations in shrinkage result because the rate of shrinkage between fibers is greater than it is along fiber lengths. Also, after traveling through runners, gates, and the workpiece, fiber distribution is relatively uneven. These factors combine to produce uneven stress that may lead to warpage and cracks. The problem is solved by eliminating most of the shrinkage.

This problem underscores the importance of the new developments in low-shrink FRP materials, particularly to injection molding, where longer material flows than either compression or transfer molding are required.

A traditional FRP material shrinks about 0.003 ipi during molding. Low-shrink versions shrink as little as 0.000 to 0.0005 ipi. Combined with proper tooling, low-shrink FRP permits production of pieces with dimensional tolerances of ±0.0005 ipi. In addition, machining is often made unnecessary and flat surface warping and characteristic FRP surface patterns are eliminated.

Formulating for specific properties enhances the material's capabilities to meet end-use requirements. Low shrinkage, flame resistance, improved wear resistance and even color are properties that may be included in specific compounds. The material has other qualities that are only moderately affected by formulation. Among these are electrical and thermal properties, dimensional stability and corrosion resistance. Strength properties, of course, are affected by the type and amount of glass used. The way the material is processed in the mold also has an effect on the achievement of potential strength.

Injection Molding Presses. There are a number of injection molding presses specifically designed to handle thermoset polyester operations.

All require stuffing cylinders because of the physical characteristics of the material. Most FRP is putty-like (BMC) or a fiber-like coated material, neither of which flow freely through normal hopper systems (Fig. 9-12).

On some machines the material or compound is forced from the stuffer cylinder from the top or side into the rear of a conventional screw injection cylinder. The screw acts only as conveyor that moves the material to the front of the cylinder rather than to provide a plasticating function. Then, the screw acts as a plunger. It does not turn as it pushes the material into the mold.

On other machines a plunger instead of a screw pushes the material into the mold. This

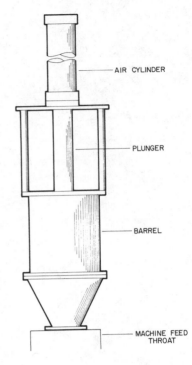

Fig. 9-12. Sketch of material stuffer for "gunk" type polyesters. The stuffer usually consists of a cylinder with a pneumatic operated plunger which maintains pressure against the polyester material while the screw is rotating. When the stuffer cylinder is nearly empty, the plunger is withdrawn and additional globs are deposited into the stuffer. (*Courtesy Stokes Div., Pennwalt Corp.*)

type densifies material as it pushes it into mold cavities.

Screws and plungers can be interchanged within the same machine frame requiring only changes in electricals.

Another type is the coaxial plunger machine. Material is dropped into a stuffer cylinder in-line with a smaller cylinder that pushes the material in the mold. Advantages claimed by this structure include the short distance the material has to flow. Angles and corners around which the material has to move have been eliminated.

Regardless of basic design, machines should have a capability for accurately controlling temperatures and pressures throughout the molding cycling with as little glass degradation as possible.

Fig. 9-13. Log of polyester material being fed into a reciprocating screw injection molding machine. (*Courtesy Stokes Div., Pennwalt Corp.*)

The object of controlling temperatures is to warm the material in the barrel enough to permit free flow to preheat it as it goes through the nozzle and cold runners, and to cure it as fast as possible after it has stopped flowing in the mold. Typical settings are: rear zone of barrel, 125°F; front zone, 150°F; cold runner, 175°F and both halves of the mold, 335°F. As a practical limit, material coming out of the nozzle should not be more than 200°F. Mold temperatures for most materials are between 300 and 350°F.

Experience shows that cure time of a $\frac{1}{8}$-in. section should be within 20 to 45 sec depending on part size.

Other typical press conditions are screw rpm, 20-75; back pressure up to 100 psi; injection time 1-5 sec; injection pressure, 5000 to 12,000 psi on material; and clamp pressure, 3 tons/sq in. max (based on cavity projected area).

Besides presses specifically designed for thermoset molding, conversion kits are also available to change over machines originally designed for thermoplastic operations.

Tooling requirements are similar to those for thermoplastic injection molding. The major difference is that adequate provision must be made for heating the molds. Molds should be fully hardened, polished and chrome plated. Tool steels such as AISI H13 are recommended.

The low-shrinkage characteristic of FRP makes ejection of pieces more difficult than with thermoplastics. Liberal draft, where possible, and provisions for positive ejection on both sides of the mold are generally required.

Properly hardened molds should wear well, but gate areas are susceptible to wear. Designing them as replaceable inserts make their repair and replacement less costly.

Gates should be located to minimize effects of knit lines that form after material has flowed around an obstruction in the mold. Also, multiple gates should be avoided to reduce the number of knit lines that form where material from gates join. Eliminating knit lines reduces chances of rejects. Large and short gates and runners give the strongest parts.

The mold should permit the material to flow so that it pushes air ahead of it into places where it can be vented out at parting lines or knockout pins. Vacuum extraction of air is another useful technique for air removal.

Sharp corners, restricted orifices and gates

that cause glass degradation should be avoided, if possible.

Because low shrink FRP can stick in conventional sprue bushings after curing, cold runners and sprues can be used. These elements can be oriented in different ways, but in all cases they are water cooled to somewhat below mold temperatures. A typical temperature for them is about 175°F. Thus the material actually gets a useful amount of preheating in the "cold runner." Runners should have relatively large diameters to provide for easier material flow and less degradation. Center-gated, three-plate molds should be considered for parts requiring maximum impact strength if cold runners are not used.

Problems of flashing common to compression molding do not occur with properly designed injection molds. What little flash that occurs is paper thin and easily removed.

10

DESIGNING MOLDS FOR THERMOSET PROCESSING

A successful and profitable thermoset molding operation is dependent on the selection of the proper type of molding equipment, the plant layout, the method of molding, and molds that have been engineered properly and that are of superior construction. A poor mold design or an inferior mold will produce a part that is costly in that it may require subsequent finishing operations before it can be shipped. The processor of thermoset raw materials must constantly keep abreast of new molding methods and new and improved mold steels or nonferrous alloys. He must also make sure his mold suppliers are aware of the newer, more advanced techniques for making molds (computer controlled machining equipment, spark or chemical erosion processes, etc.). This chapter will cover the methods of molding and mold design for thermosets.

Method of Molding

The trend is toward automation wherever possible. Therefore, the processor has an opportunity to select one of five methods of molding (taking into account the geometric design of the part, the finished part tolerances, the operating conditions, and environmental surroundings): (1) compression, (2) transfer, (3)

* Robert W. Bainbridge, Consulting Plastics Engineer, Hooker Chemicals & Plastics Corp. Durez Div., North Tonawanda, N.Y.

two-stage reciprocal screw transfer, (4) injection, and (5) extrusion (extrusion is a highly specialized, licensed process; production is quite limited).

Compression Molds

Compression molds may be designed for hand molds, semi-automatic, or automatic operations. Hand molds in the early stages of molding were quite popular but in recent years have been used only for prototype parts or parts with limited production. The custom molder does a large amount of molding using the semi-automatic method. The compression molds may be hand loaded with loose powder or preforms. The preforms are generally preheated to 200-260°F in high frequency preheaters. The trend being to automate, both custom and captive molders are automatically compression molding a wide range of thermoset parts. This is done on a variety of press equipment. Conventional single ram hydraulic or toggle presses may range from 15 to 300 ton capacity.

Rotary presses designed for compression molding are used to mold closures, wiring device parts, timer cams, outlet boxes, etc. The size of the part is limited to the clamp capacity of each individual hydraulic cylinder.

The automatic molding presses for compression molding are so designed that various methods of loading the cavities may be used. Material may be loaded with cold granular, high frequency preheated or infrared preheated granular, preheated preforms, or a charge which is preheated by a heated barrel and reciprocating screw.

A compression mold consists of heated platens onto which are mounted the cavity and force block or blocks. A knockout system, either top, bottom, or both, is incorporated to remove the parts from the mold. A typical cycle is: mold open after cure period, remove parts, clean mold, load material, close press. On closing the mold, the excess material is forced out through the flash escapement area.

In general, when parts are molded by compression, the part density is equal in all directions. It is for this reason that parts which have very close tolerances over the face of the part are recommended to be molded by compression. There is much less chance of uneven density causing shrinkage problems of out-of-round conditions or differences in shrinkage from width to length.

As explained previously, in a compression mold the material is loaded into an open mold and the excess material required to fill the part is forced out over the flash line. The least expensive compression mold is a "flash mold," as illustrated by Fig. 10-1. The greatest disadvantage to a flash mold is that there is no restriction to the flow of material and no back pressure is built up in the molded part. The flash is free to flow horizontally with no restriction. The parts will lack complete

density; therefore, the parts may be inferior as regards strength and molded finish. It is most difficult to maintain dimensional tolerances due to lack of density. Although the mold cost may be less expensive than other types of compression molds, the fact that we have no loading space except for the cavity of the part itself makes the material charge excessive. The percentage of flash loss may be extremely high, thus costly on a piece part basis. It is extremely difficult to maintain centerline relationship between force and cavity, and if guide pins wear, the force and cavity become mismatched. Very few semi-automatic or automatic molds are considered as "flash molds."

The second type of compression mold, the "semi-positive horizontal flash", is illustrated by Fig. 10-2. The majority of the semi-automatic and automatic compression molds are of

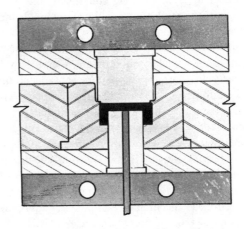

Fig. 10-2. Semi-positive mold.

this design. Ample loading space may be provided so that the percentage of flash loss may be held to a minimum. If the clearance between the force and cavity is held to 0.003-0.004 in., all the material is trapped in the cavity so that the part is of maximum density. Since there may be slight differences in the volume contained within the force and cavity in a multi-cavity mold, it is well to provide flash escapements in the form of flats or grooves 0.015-0.020 in. in depth. These should always be put on the force since they can be cleaned off each shot more easily than when put into the cavity.

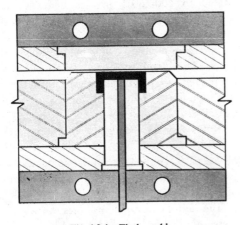

Fig. 10-1. Flash mold.

This type mold produces parts having maximum physical strength. If molds are "breathed" just before final close, porosity may be eliminated by releasing the trapped volatiles. Dimensional tolerance may be closely held, especially over the face of the part.

The third type of compression mold, the "semi-positive vertical flash," is illustrated by Fig. 10-3. As you will note, it is quite similar to

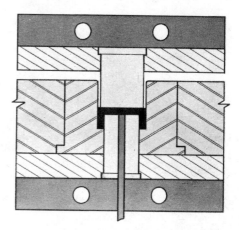

Fig. 10-4. Positive mold.

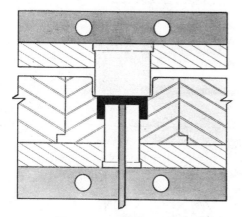

Fig. 10-3. Semi-positive mold.

the semi-positive horizontal flash, except that the force enters directly into the geometric design of the part itself for a distance of $\frac{1}{16}$-$\frac{1}{8}$ in. The clearance between the force and cavity at this point should be 0.005-0.007 in. to prevent scoring of the cavity side wall. If not done and scoring takes place, the molded article will have a scratched appearance. Again, flash escapements should be provided on the force. A part molded in such a mold will be of good density. The flash is vertical, which may be more desirable, and can be removed easily. The molded article, since it has good density, should have a good finish and maximum physical strength. This type mold is quite expensive since there are two areas between the force and cavity that must fit perfectly. There is absolutely no chance for mismatch between force and cavity, so concentricity of the article from cavity to force is as close as possible. Again, good dimensional tolerances are achieved over the face of the part.

The fourth type of compression mold, "full positive," is illustrated by Fig. 10-4. This type

of mold is generally used if the materials are of high bulk, or if the filler fibers are quite long as in glass roving. It becomes difficult to cut off such fibers to a thin flash. The full-positive molds give a vertical flash which may be easier to remove. It is difficult to produce parts of equal density if a multi-cavity mold is used. The clearance between force and cavity must be excessive or generous flash escapements should be provided. It may be desirable to also provide a slight taper on the force to help in flash escapement; in addition, this can prevent scoring of the cavity wall which would mar the molded part. Pressure pads, away from the loading area, must be provided to ensure proper control of thickness of part. Material or flash must be kept free from surface of landing pads.

If parts are extremely small so that it becomes difficult to charge the individual cavities, a sub-cavity mold can be used as illustrated in Fig. 10-5. The mold shown is a "positive" mold; however, a "semi-positive" design may be used.

In the design of the mold, variations of the basic design may be made to provide an improved tool design. The ejection of the part in an automatic or semi-automatic molding operation must be considered in the final mold design. The part design may make it almost impossible to provide knock-out pins. In this case a "stripper plate" knock-out system must be used. Figure 10-6 illustrates such a mold design. In the part illustrated, we shall assume that no undercut or special taper can be

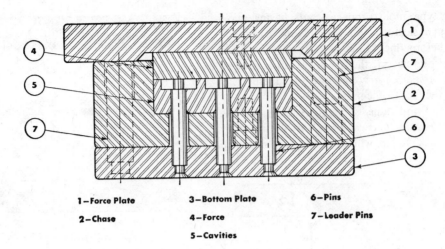

1—Force Plate 3—Bottom Plate 6—Pins

2—Chase   4—Force   7—Leader Pins

5—Cavities

Fig. 10-5. Sub-cavity mold.

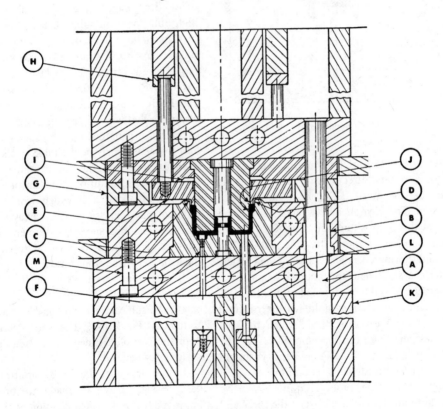

Fig. 10-6. "Stripper plate" knockout system.

allowed, to ensure that the molded part will stay either on the force or in the cavity when the mold is opened. In this case, both top and bottom ejection are provided, the top being the "stripper plate" and the bottom a normal knock-out system. When using the stripper plate design, care must be taken to provide for adequate flash removal. Stripper plate molds are used to mold parts such as pin connector boards containing large number of holes.

It may be advisable to call attention to certain details of the mold design here illustrated:

A. Clearance under the dowels or leader pins.

B. The use of a hardened dowel or leader-pin bushing in a soft plate.

C. Slight positive entry action of the force into the cavity, to ensure alignment of the mold and full density of the molded articles.

D. Flash grooves in the force.

E. Clearance between underside of stripper plate and top of cavity to allow escape of material from flash grooves.

F. Clearance holes under inserts to allow any flash to fall completely through the mold.

G. Pressure pads or blocks to control the thickness.

H. Retainer plate for top ejection pins.

I. Clearance space above stripper plate so that flash working up between the top force and the stripper plate will not be confined and can easily be blown out.

J. Inside diameter or stripper plate made slightly greater than outside diameter of the force which shapes the inside of the molded article, in order to avoid scoring of the force, which would mar the inside surface of the molded article.

K. Placing of lower supporting parallels to avoid blocking off the clearance holes under the dowel bushings.

L. Counterbore in underside of ejection pin holes to reduce friction.

M. Screws for holding mold parts together put in from undersides of plates, remote from the surfaces where molding material might get into the counterbored recesses for screw heads and cause trouble in cleaning.

Pressure pads, as illustrated as Item G in Fig. 10-6, are desirable on all compression molds. It is a safety measure that assures that the land area of the mold will not be damaged when mounting the mold in the press. The landing pads (or pressure pads) should be designed to be in contact when there is approximately 0.002 in. clearance between the force and cavity at the land area or other surfaces such as pins, blades, etc.

In the design of compression molds, especially "flash" and/or "semi-positive" types, it is important that the land area at cutoff in the outer landed area, or in the cavity itself, be carefully designed in relation to the molding pressure. If, for instance, a meter housing or an instrument door should have one or more rather large openings in the face surface it is imperative that the fin thickness at these areas be held to 0.005-0.010 in. in thickness for ease in deflashing. If the area of these openings exceeds 1-2 sq in., the thermosetting material will advance in cure to the point we are endeavoring to flow a set-up material across this land area, thus preventing the mold from closing unless excessive pressure is used. In order to eliminate this condition it is well to provide a relief area in the center portion which will inhibit the cure effect. Provide a land of $\frac{1}{8}$-$\frac{3}{16}$ in. in width inside of the opening and in this area relieve force or cavity to $\frac{1}{16}$-$\frac{3}{64}$ in. in thickness. In the case of a rather large round window or opening, it is desirable to insert a pin in the cavity and permit it to enter the top force. Inserting the pin will reduce the projected area of the piece by the area of the pin diameter, thus requiring less pressure to mold the part. Figure 10-7 illustrates these two part designs.

Semi-automatic or automatic molds may be designed for parts having undercut side surfaces such as found in bobbins or spools, pump housings, automotive brake parts, etc. The mold in Fig. 10-8 is used for such a part. The undercut areas may be molded by the use of split molds, removable pins, or sections from molds. Side cores activated by cams or side acting hydraulic cylinders are commonly used. In the latter case, locking devices (such as wedges) of sufficient pressure must be provided to keep the cores from being forced out of position by the internal force of the material.

One of the critical areas in this design is the flash removal. The flash must be completely removed from around the movable side cores. It is for this reason that the majority of this type of mold is operated semi-automatically. In

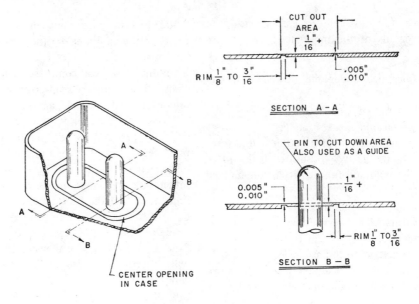

Fig. 10-7. Part designs using pins.

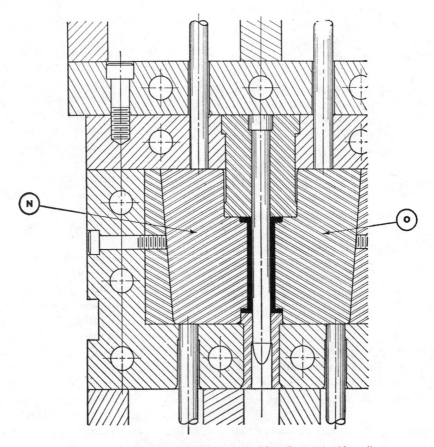

Fig. 10-8. Semi-automatic mold for part with undercuts in side walls.

order to provide economical operations, shuttle presses or reciprocating table presses are used. This design incorporates two or more bottom halves to one upper half of the mold. While the one shot is being cured the operator is removing the part and removable sections of the mold, replacing the removable section and preparing the lower half for the next cure cycle.

Raw material suppliers have formulated compounds that respond quickly to heat and pressure, and have thus helped to reduce cure cycles and make it possible for processors to adopt automatic, economical operations. Manufacturers of preheat equipment utilizing high frequency electrical current have provided equipment requiring less maintenance and more efficient preheat which has further reduced cures. Manufacturers specializing in compression molding equipment now offer solid state controls to provide trouble-free automatic or semi-automatic operations. Reduced time for dry cycles has permitted shorter over-all cycles. Also, reciprocal screws used in connection with heated barrels mounted on the molding press can provide an extruded charge of preheated material that will further reduce cure cycles. The charge can be deposited directly into the cavity or stored in a loading tray and several cavities fed at one time.

All of the above factors have contributed to the continuing use of compression molding for making a wide range of plastic parts.

Transfer Molds

Transfer molds will vary in design depending on the method of transfer molding. The basic methods are: (1) the original method—pot and plunger—utilizing a three-plate mold with the plunger-retaining plate mounted to the head of the press, a floating plate containing the pot and sprue bushing and top cavity or cavities, and the third retainer plate (mounted on grids) containing the bottom cavity. Figures 10-9 and 10-10 illustrate the above method. (2) The second method is known as the plunger transfer method. In this case, the plunger or plungers are activated by an independent hydraulic cylinder. The transfer cylinder may be mounted at the top of the press, at the bottom of the

press, or may incorporate a ram within a ram if the main clamp cylinder is located beneath the lower bed of the press. Figure 10-11 illustrates top plunger transfer. (3) The third is the two-stage, reciprocal screw transfer method. In this case a conventional designed hydraulic or toggle, top clamp press, with a bottom ram and

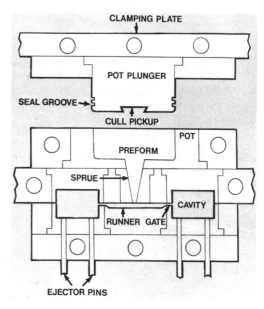

Fig. 10-9. Pot transfer mold.

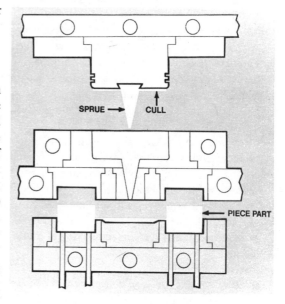

Fig. 10-10. Pot transfer mold.

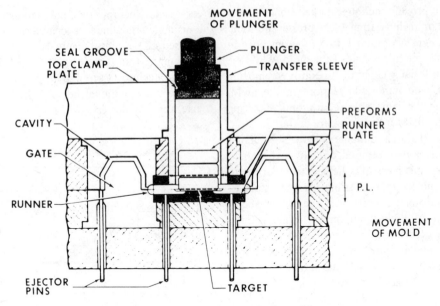

Fig. 10-11. Top plunger transfer mold.

transfer cylinder is used. The reciprocal screw and barrel is mounted horizontally and material is discharged into the transfer chamber through an opening in the transfer sleeve or cylinder.

There are certain design features that must be considered for each method of transfer molding and each method will be discussed separately.

The *pot and plunger method* of transfer molding is generally done in a conventional bottom ram clamp molding press. The majority of the molds are three-plate molds with the center section floating. There are instances, however, where the plunger is activated separately by an independent hydraulic cylinder into the pot, and in these cases the mold proper consists of a conventional two-plate mold.

Generally, the material is fed directly into the mold cavity from the pot through the sprue bushing. If multiple cavities are used when feeding directly into the part, the sprues may come from a common pot, or multiple pots may be used. The principle may also employ a sprue into a runner system feeding one or more cavities.

After the mold layout of the individual cavity or cavities has been determined, the next step is to accurately figure the projected area of the part or parts. If a runner system is used this

must be added. The square inches of area in the pot or pots required to clamp the mold is open for discussion.

It was considered good practice to make the area equal to or slightly greater than the total projected of all articles to be molded, including the sprue and runners. However, since the material is advancing, it need not be treated as a hydraulic liquid. Standard practice is to reduce the area in the pot to less than the above total area.

The next step is to determine the volume of charge to be transferred and make the pot of sufficient depth to accept the preheated charge. This may be granular, preforms, or extruded material.

The mold design should provide for the pot or pots to be centrally located so that the pressure is exerted uniformly in the center of the press.

The design of the steel retainer or mounting plate and section of steel containing the pot should be sufficiently strong to withstand the higher transfer pressures used to transfer material from pot to mold cavity. In order to prevent escapement of material around the plunger the clearance between the plunger and pot should be held to 0.002-0.003 in. on smaller diameter plungers and 0.004-0.005 in.

on larger diameter or rectangular plungers. To ensure no escapement of material it is well to incorporate one or two sealing grooves. These may be $\frac{5}{32}$-$\frac{3}{16}$ in. in width and $\frac{1}{16}$-$\frac{3}{32}$ in. in depth. A small radius is desirable at the bottom of the pot to provide extra strength to pot and to assure cull remains on force.

The pot may have a sprue cut into the solid piece of steel; however, in most cases it is desirable to insert sprue bushings. The larger diameter of the sprue should be at the face of the bottom of the pot. The length of the sprue bushing should be kept to a minimum. Figure 10-12 illustrates standard sprue bushings.

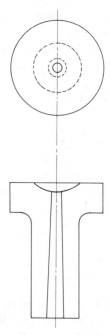

Fig. 10-12. Standard sprue bushings.

An undercut on the face of the plunger is provided to remove the sprue or cull from the sprue bushing. Figure 10-13 illustrates various types of cull-removing grooves and sealing grooves on the plunger.

Since the pot and plunger method of transfer does not normally incorporate runners and gates, they will be discussed later in this section.

The second and more popular method of transfer molding is the *plunger transfer method.*

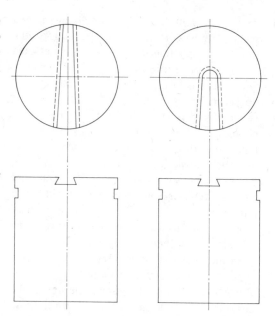

Fig. 10-13. Types of cull-removing grooves and sealing grooves on the plunger.

In this case, the press may be a toggle clamp or a full hydraulic. The clamping pressure may be applied either from top or bottom of the molding press. These may vary from 25 to 500 tons. Individual or multiple hydraulic cylinders are mounted in opposite direction to the clamping pressure to provide the transfer pressure. Limited presses are available that have the transfer ram within the clamp ram. The clamp ram in relation to transfer ram is generally 3 or 4 to one (25-35 ton transfer-100 ton clamp). The majority of presses are constructed with a horizontal parting line; however, presses are available with vertical parting lines. The advantage of the latter is that no unloading mechanism is necessary to remove parts from the mold.

The plunger transfer method of transfer molding can easily be designed for automatic molding, especially if the transfer ram is in the lower half of the mold. High frequency preheated pills or extruded stock may be charged automatically (by the use of a loading tray) into the transfer cylinder. In case of a horizontal transfer cylinder, the preheated material is dropped into an open section of the transfer cylinder.

The clearance between plunger and transfer

cylinder should be held to a minimum, and sealing grooves provided on the end of the plunger. Some transfer plungers are designed with removable tips to reduce maintenance. Other designs incorporate a bronze sleeve with no sealing grooves that permit clearances between plunger and cylinder to be held very close. As the bronze sleeve wears, it can be replaced with a new sleeve at limited cost and minimum downtime for change over. The cull developed between the face of the mold and face of transfer ram should be held to a minimum. The face of the mold or plunger should be designed so as not to cut off the flow of material into the runner system.

In the design of plunger transfer molds, it is important that ample clamping pressure is provided to keep from flashing the mold. The surface of the mold should be relieved in those areas where a sealoff of part, runner, gate, and plunger is not necessary. This surface should be relieved $\frac{1}{32}$-$\frac{1}{16}$ in. in depth. A sealoff surface of $\frac{3}{16}$ in. in width is sufficient, providing landing pads are located in the area of the guide pins. The molds should be well supported with pillar supports to keep the center section of the mold from bowing or sinking when under pressure of material. Any flashing of the mold will result in uneven pressure in the mold cavities.

The third method of transfer molding is known as the *two-stage, reciprocal screw method.* The method of transfer and mold layout is similar to the plunger method of bottom ram transfer. There is only one press manufacturer producing this type of equipment in ranges of 150 to 300 ton. It has a specially designed toggle clamp press. The reciprocating screw mechanism is mounted at right angles to the press. The reciprocating screw is used as a means of preheating the charge of material, thus eliminating need of preform equipment and preheater. The material is preheated to 220-260°F and fed into the transfer cylinder during the curing cycle. The operation is fully automatic with parts being removed from the knock-out pins by means of a comb. The parts may be degated from the cull and runner system, and then separated in one operation. If inserts are involved, a press can be provided

with a rotating table to permit load and unload of one section of the mold while the other section is curing.

Injection Molding of Thermosets

Injection molding of thermosets has become popular during the last few years and a substantial growth is forecast for the future.

The mold design for an injection mold is similar to a plunger transfer or two-stage reciprocal screw transfer method. The plunger transfer cylinder is replaced by a sprue bushing.

Since most injection presses have a horizontal action and vertical cutoff area, knock-out systems are somewhat different in that the parts and runner system fall by gravity force from the mold area. It is not uncommon to have combs installed to assure that all molded material is removed from mold area before the next cycle.

Faster molding cycles are obtained and higher mold temperatures are common. It is essential that more heating elements be provided to keep molds at the desired mold temperature. This is especially true with alkyd materials since their high specific gravity and mineral filler system tend to pull considerable heat from the mold.

Venting is extremely important since injection times are faster than in transfer molding.

Runner and Gate Design

Molds for all types of transfer and injection molding should be designed for the ultimate in economical operations; use of reduced transfer and injection pressure; minimum waste material in cull, runners, and gates; maximum heat buildup to obtain short cure cycles; and ease of removing gates from the article.

The runners should be circular in cross section since they offer minimum resistance to flow, and the most effective insulation of the center of the stock against the heat of the mold block which tends to advance the material. Advanced material can cause uneven density, warpage, and poor surface appearance. If a full round runner cannot be incorporated, a trapezoidal cross section may be used. Avoid a half

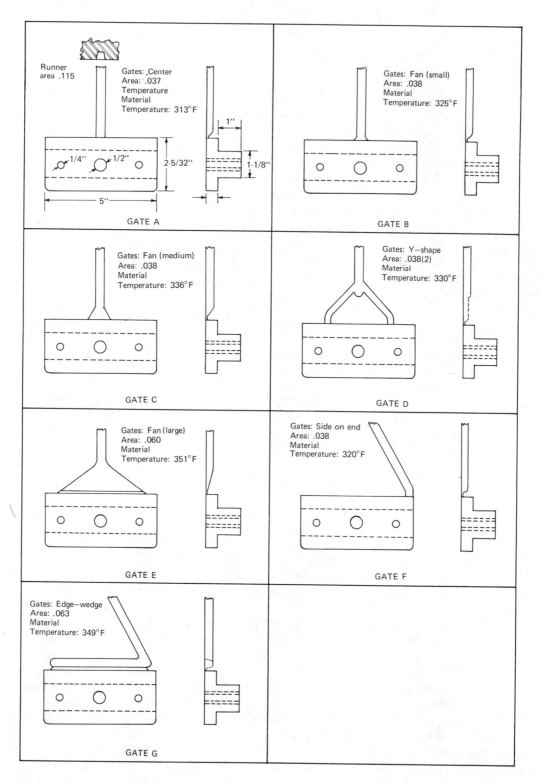

Fig. 10-14. Seven gate designs: (A) center; (B) small fan; (C) medium fan; (D) "Y" with two gates; (E) large fan; (F) side on end; and (G) edge.

round runner since it offers the least effective insulation of the center of the stock and precuring will certainly develop. Runners should be as short as possible to prevent waste and precure. Design to avoid sudden changes in direction and with generous radii when changing direction of flow. The type of material, number of cavities, and part design may have an effect on the size of the runners. In the design of the mold it is recommended that the runner section be a separate section of the mold, hardened to 60-65 Rockwell C Scale, and well polished. It may be desirable to chrome plate this section of the mold to facilitate better flow and release. Undercut wells may be provided with knock-out pins below the wells. This will provide anchorage in case the runner system is to be held on one side of the mold.

Gate design is extremely important when consideration is given to: part density, warpage, shrinkage, weld lines, and over-all physical strength. Gate size, shape, and location are responsible for material orientation, the pattern of flow, time and pressure to fill the cavity, and temperature of the material as it enters the cavity. Figure 10-14 illustrates a test specimen fed from a uniform runner cross section, gate sizes of approximately the same square inches of cross section, and the increase in temperature from a uniform stock temperature of 270°F and a mold temperature of 350°F. Also illustrated are gate designs—center, small fan, medium fan, "Y" with two gates, large fan, side on end, and edge. The large fan and edge gate on part of this design give the most uniform flow. The "Y" gate design is not recommended since it produces parts filled with ball joints and weld lines.

Figure 10-15 shows a ring gate used to avoid weld lines and uneven density within the molded part which causes warpage.

Figure 10-16 illustrates submarine or pin-point gates that are becoming quite popular in transfer and injection molds. The main advantage of the pin-point gate is the automatic degate principle. When pin-point gate design is used the gate area must be a removable section of the mold. The pin-point gate will erode quite rapidly and hardened gate sections may be replaced in a short period of time. Erosion of

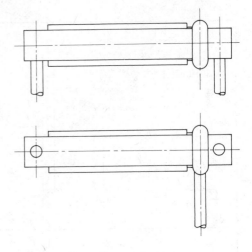

Fig. 10-15. Ring gate.

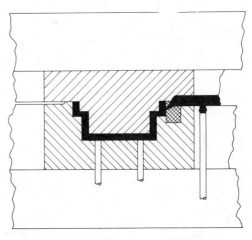

Fig. 10-16. Submarine or pin-point gate.

the gate will depend on the type of filler used in the molding compound.

It is a good practice to gate into the heaviest section of the molded part, if possible. Gates located below the surface of the part will eliminate erosion of the top surface area of the mold. If the part design permits, removable gate sections should be incorporated in gate layout. They should be very hard to minimize gate wear. Frequently, a special steel is used in the gate section. With removable gate sections provided in the mold, they may be replaced as needed to maintain a minimum gate size.

Runnerless molding, or cold or warm runner systems, have been developed to eliminate the

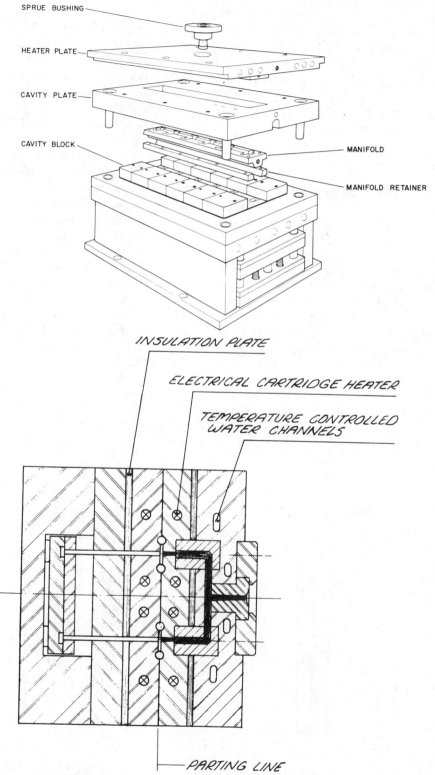

SPRUE BUSHING

HEATER PLATE

CAVITY PLATE

CAVITY BLOCK

MANIFOLD

MANIFOLD RETAINER

INSULATION PLATE

ELECTRICAL CARTRIDGE HEATER

TEMPERATURE CONTROLLED WATER CHANNELS

PARTING LINE

Fig. 10-17. Cold runner molds. Top: *Courtesy Stokes Div., Pennwalt Corp.*; bottom: *Courtesy New Britain Machine Co.*

large percentage of scrap material lost in sprues, culls, runners, and gates. These apply to thermoset injection molds. More experience and possibly additional development work in runner design and molding compounds are needed for completely trouble-free molding cycles. Figure 10-17 illustrates the two concepts that are now used.

Vents

Air or volatiles entrapped within a mold are often the cause of many molding problems and may be present in both compression, transfer, and injection molds. In the case of compression molding, the location of vented knock-out pins, proper venting at activated mold sections, and placement of vented activated pins at dead end sections of the mold, will provide adequate venting. Figure 10-18 illustrates vented pins.

As stated, in transfer and injection molding the location of the gate is important; likewise, the location of vents is equally important. If entrapped air and volatiles are not permitted to escape they will prevent the filling of the cavity, produce porosity in the part, which may cause warpage and dimensional problems, and may contain weld lines causing weak parts. The trapped air being heated by compression causes a burning of the material. Vents a minimum of $\frac{3}{16}$ in. in width and 0.003-0.005 in. in depth should be ground in the face of the sealing area of the cavity. Unless it is obvious where the knit area will be, it is best to make sample shots to determine where vents should be placed and then grind vents on the land area.

On particularly troublesome venting problems, it may be necessary to apply vacuum to the mold cavity. This requires: all vents from cavities discharge into an air groove and all cavities enclosed within the confines of a Teflon or silicone rubber gasket, located between two or more portions of the mold. An ample vacuum reservoir tank must be provided to pull 15 to 25 in. of vacuum immediately as the material enters the cavity.

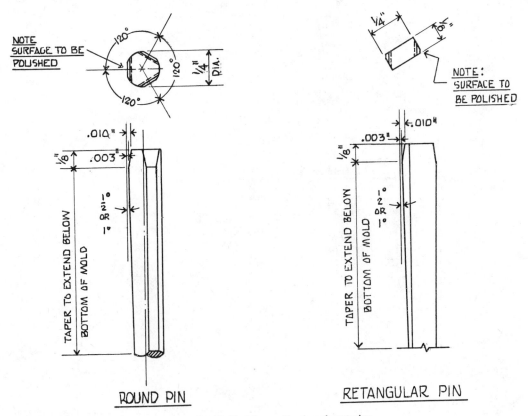

Fig. 10-18. Method for venting knock-out pins.

11
COLD MOLDING

History and General Considerations

Cold-molded plastics were introduced in this country in 1908 by Emile Hemming, Sr., just about the same time phenolic molding compounds came into being, and both were recognized as better electrical insulation than any other products available at that time. The materials and the methods of molding them, however, were quite different and each soon found its separate field of importance and activity. The term "cold-molding" has persisted through the intervening years because it describes a *method* rather than a *type* of plastics, and it must not be confused with current *cold-setting* resins and cements which have come into use recently for casting and laminating. The ingredients are utterly different and so are the results.

In fact, cold-molded plastics are entirely different from those produced by conventional hot-molding methods, both in the mix and in the technique employed to form and cure them. Therefore, this discussion will be confined strictly to the refractory and nonrefractory types of cold-molded plastics, which find their greatest service in the electrical insulating field, and in handles for pots and pans, where greater heat resistance is required than other plastics can provide.

The definite need for an efficient commer-

* Reviewed by J. Harry DuBois, J. Harry DuBois Co., Morris Plains, N.J.

cial heat-resisting molded insulating material for the electrical and automotive industries inspired the research work which resulted in the development of the cold-molding process. For many years porcelain was popular for such purposes because it was the only fireproof, heatproof and waterproof material available to electrical engineers, and its low cost was a strong recommendation for its use. Unfortunately it was brittle, as are all ceramics, and could not be used in many places where its insulating properties would otherwise prove valuable.

Cold-molded materials are inexpensive as compared with hot-molded insulation, and certain types are capable of a much higher rate of production because they are not cured in the press.

Contrary to conventional molding, where molding compounds are manufactured by materials suppliers and shipped in drums to molders, cold-molding compounds are mixed on the spot by the companies which mold them. The cold-molding compounds contain much higher moisture content when they go into the press and each batch is mixed and used within a period of hours, before it dries out. The very nature of the mix prohibits its preparation very far in advance, and thus it cannot be shipped and stored as other compounds are.

Methods of preparation are relatively simple when compared to hot-molding compounds. There is no need for polymerization kettles or stills. The ingredients, consisting variously of bitumens, synthetic resins, drying oils, fillers and solvents, are combined in conventional mixers and sometimes rolled to get thorough homogenization. Binders are often heated before mixing to make them workable. After these simple mechanical operations, the compounds are ready to use. Other compounds consist largely of asbestos, cement and water, and these will be explained later on.

Molding is done in conventional punch presses (Fig. 11-1), except that no provision has to be made to heat or cool the molds. The prepared compound is accurately weighed or measured in the exact amount required for the finished piece, placed in a cold mold and subjected to pressures from 2,000 to 12,000 psi. The pressing operation is continued just long enough to compress the cold-molding material thoroughly into the desired shape. Then is is removed from the mold and transferred to a heating oven and baked until it becomes hard. Finishing operations consist of removing fins or excess compound by sanding, filing, etc., after the piece is baked.

Cold-molded articles do not have the luster usually associated with hot-molded or cast plastics, nor is this necessary, because the prime requirement of insulating material is to resist arcing, moisture and heat.

Those cold-molded products in which phenolic resin is used as a binder have a somewhat smoother surface, but, generally speaking, cold-molded products are more or less dull and unattractive in appearance. Colors are usually dark brown, black or gray. Color in cold-molding is secondary to function, and the finish or surface of cold-molded parts is satisfactory for all practical purposes unless the mold is old or worn.

Cold-Molding Materials and Their Preparation

Cold-molding compounds ordinarily are divided into two distinct groups according to the binder and filler used. These are *nonrefractory (organic)* and *refractory (inorganic)*.

(1) **Nonrefractory (Organic).** In general, the binders suitable for nonrefractory cold-molding compounds are asphalts, oil deriva-

Fig. 11-1. Punch presses are widely used for cold molding.

tives, coal-tar pitches, stearine pitches, residue from distillation of oils and fats, various solutions of gums or resins, oxidized oils, wax compounds, etc. Fillers consist for the most part of asbestos fibers, silica and magnesia compounds. To transform these materials into molding compound, the solid binders must first be dissolved into a liquid or viscous state. This is done by adding solvents such as coal-tar oils, petroleum oils, drying oils, naphtha, benzene, turpentine, etc., and cooking the mixture in a large kettle until the proper viscosity is reached. The solution is poured into a mixer, where it blends with the filler. If coloring pigments are to be included, they are added to the mix. The resulting compound is then ground and screened to get uniform particles for molding, after which it is allowed to "age" for a brief period, during which time the solvents evaporate and the material is conditioned for molding.

The ageing process is critical and often presents problems because the molding mixture must be conditioned to the consistency best suited for molding, so that the articles will come from the press neither too soft nor too hard to prevent their safe removal from the mold. It should be remembered that cold-molded articles are uncured when they come from the mold; therefore they may be easily damaged unless the molding compound is of proper formulation and age to prevent sticking. They must be handled with considerable care until they have been baked.

This problem is not as acute in conventional hot-molding because dry powder is fully cured by heat and pressure in the mold before being removed. But cold-molding compounds are already plastic when placed in the mold. Their moisture content must be carefully controlled in order that the pressure applied in the press will bring them into intimate contact with all portions of the mold to give them shape. Since no chemical change takes place in the mold, and since no heat is applied to cause the compound to dry out, the exact amount of compound must enter the mold each time if duplicated products are to be obtained. The compound must not be allowed to dry out beyond the point where it will flow readily

under pressure. Since it is also affected by atmospheric conditions it must not be exposed too long before molding, as it is likely to harden to some extent and fail to mold properly.

Experience alone is the guide used by cold-molders, and they vary their compounds from time to time to achieve the best results in molding articles of various designs. A composition of binder in which phenolic or melamine resin is used comes under the heading of nonrefractory compounds and the procedure for mixing and preparing the material is much the same as with natural binders. Melamine cold mold compounds have done much to open new markets to the technique.

(2) Refractory (Inorganic). Refractory cold-molding compounds basically are mixtures of cement, lime or silica, as the binder, with a filler of asbestos, blended together with water. This mixture, like other cold-molding compounds, depends upon pressing to shape, then baking to cure or dry. The presence of cement in the mixture, however, requires somewhat different treatment in the drying process. The mixture itself is blended by any suitable mechanical process, but the resulting compound is difficult to weigh or measure accurately because the long fibrous fillers have a tendency to hang together when wet. It is also very abrasive on the mold when the tremendous pressures required to mold it are applied.

It is even more critical during the curing period because, if dried too quickly, cement has a tendency to crack. Therefore, molded articles as they come from the press are stacked and covered with canvas or burlap and kept moist for a considerable time before they are placed in the oven to cure. Even in the oven they are dried during the first period in a moist humid atmosphere provided by permitting steam to circulate throughout the oven. This steam bath is gradually diminished as time goes on, and the final cure, requiring several days, is accomplished in dry hot air. Once cured, refractory cold-molded articles are exceedingly dense and hard, and will stand considerable abuse.

(3) Glass Bonded Mica. Glass bonded mica bridges the gap between the organic plastics and the ceramics. It has the moldability and machinability of the plastics plus the total dimensional stability, arc and nuclear radiation resistance, freedom from outgassing in vacuum, high thermal endurance (1800°F), nil moisture absorption and superior electrical properties of the ceramics. Glass bonded mica is a 1500 degree thermoplastic with a thermal expansion rate that matches stainless steel. It is molded into sheets for fabricated products and molded with or without inserts for those products that require properties above those available in the organic plastics and the ceramics. It is a costly material and should not be considered unless the properties of the plastics cannot meet the demands of the application.

Glass-bonded mica is composed of a mixture of natural or synthetic mica blended with the ingredients that form a glass when raised to the melting point of the glass. Sheet materials are compression molded. Molded products are injection molded in transfer type molds. Preheat temperatures range from 1600 to 2000°F. Mold temperatures run between 700 and 950°F. Molding pressures up to 40,000 psi are used depending on the flow problems. Compounds for these materials are developed by their molders in a variety of formulations as needed to meet the application requirements.

Cold-Molded Insulation

Among the properties which recommend cold-molding materials to the electrical field, for which they were primarily developed, are good arc-resistance and superior resistance to heat. Another advantage lies in the fact that the raw materials for cold-molding compounds are plentiful and cheap and, once the formula is established, may be mixed without the necessity of excessively costly equipment or highly skilled labor. Fewer molds are required also, and this reduces mold cost to a minimum right at the start. Production runs can be made economically and rapidly in a single-cavity mold, because the molding cycle is short. It consists simply of filling the mold, closing the press, and removing the article. No time is required to cure the article in the press, and since no gases develop, which would have to be allowed to escape during the molding opera-

tion, the press may be closed rapidly and full pressure applied at once. The operation may be likened to preforming more than to conventional molding, and may be done as rapidly as the operator can remove pressed articles and refill the mold.

Larger articles frequently require more time in loading the mold because the material has to be distributed largely by hand, since it does not flow. It cannot be bunched in the center of the mold with the hope that it will distribute itself evenly throughout the cavity when pressure is applied. This is especially true of refractory compounds in which asbestos and cement are used. On the other hand, some cold-molding compounds are capable of more rapid production than in any other form of molding. Ink-bottle tops, buttons, and similar small items have been molded in a conventional preforming press, where a continuous flow of nonrefractory organic compound is fed into the cavity from a hopper and some six hundred molded articles discharged from the press every minute during the operating period. These are discharged upon metal trays as they come from the press, then placed in an oven to cure. Much more oven capacity is required than press capacity, and a prodigious number of pieces are produced in a single press.

Limitations of Cold-Molded Articles

Detracting somewhat from the desirable qualities of cold-molded plastics are certain limitations which must be taken into account whenever use of the material is planned. Unless molds are watched carefully, and repaired or renewed when they become worn as a result of the heavy pressures required and the abrasive action of the materials, molded articles may have uneven surfaces and burrs which must be removed after the articles have been baked. This entails more expensive finishing operations than otherwise would be necessary, and runs up the cost per article. Unavoidable shrinkage is another factor which may cause trouble. The evaporation of solvents during the baking and curing is responsible for shrinkage anywhere from 0.002 to 0.020 in. per in., and sometimes the finished product is slightly warped. Taking shrinkage into consideration in the design of the cold-molded article, however, reduces this hazard to a minimum. Wall thickness must be more carefully planned and specified than for hot-molding, because the shrinkage takes place after the piece has been removed from the mold and at a time when no follow-up pressure may be applied. If wall thickness is not planned to avoid more rapid drying in one part of the piece than in other parts, trouble may result. Positioning of inserts and holes for attachment is likewise critical in design and should be carefully planned before the molds are made.

Other possible drawbacks which distinguish cold-molded articles include dull finish, lack of brilliant color, low tensile strength, and the restriction that cold-molding compounds are not available unless mixed on the spot and used almost as soon as they are made.

Molds and Fixtures

It has been pointed out that cold-molded articles are still soft when they come from the mold. They are merely compressed to shape and have very little strength; therefore, the mold must be designed and made to permit easy removal of the delicate pieces without damage. In general, exactly the right amount of compound is placed in a positive mold. The upper plunger descending upon the material forces it into all parts of the mold and high pressure is exerted to squeeze and compress the compound into the exact shape of the mold. After the piece has been formed, the lower plunger, which forms the base of the mold cavity, rises to bring the piece up out of the mold, and it may be safely removed. Conventional knockout pins used in hot-molding would likely destroy the article if they were used in cold-molding operations; thus the mold is constructed so that the lower plunger acts in their place. The walls of the mold are tapered slightly to facilitate removal of the articles.

Special steels having high carbon and high chromium content are usually used in making molds, because of their greater resistance to the abrasive action of the raw materials that go into the mix. The greatest single factor, perhaps, in assuring the successful operation of a mold is to have it built by the molder who is to produce the articles. In this way it can be tailored to accommodate his particular brand of mix, and his experience with his own compounds will prompt its proper design to handle the mix expediently and turn out evenly dense parts. It will also be built to accommodate the pressures he has available without distorting, and, as a result, the mold will have a longer life. Molds can be built by any engineer, of course, but he must have all the facts before him when the design is laid out and he must understand the special requirements of the cold-molding compounds to be pressed and have full knowledge of the presses on which they will be used.

Fillets or corners should never be sharp, because they may be too easily damaged when the comparatively soft material is removed from the press or during curing processes. Walls should be sufficiently thick to stand expected strains, and a generous bevel, or draw, should be provided for easy removal of the pieces. In general, if these principles are followed, quite intricate cold-molded articles can be successfully made. Obviously, there is not space in this brief chapter to detail the many variations of mold-construction, but it is safe to leave them in the hands of an experienced cold-molder who is familiar with the requirements and will be able to provide the proper mold to do any cold-molding job.

Finishing Operations

As in all molded plastics, parting lines on cold-molded pieces are often removed to improve the appearance of the article and facilitate assembling. Parting lines are caused by excess material which squeezes out where the plunger enters the mold, and as the mold wears this becomes more acute. Such flash is removed by sanding, or buffing with a wire brush wheel, and sometimes it is removed with a file. Surfaces may be ground flat to improve assembly, but drilling and threading cold-molded articles is not easy. They are sometimes tumbled if better appearance is desired. This removes small burrs and other defects from the surface, and a final tumbling with waxed shoe-pegs gives a smooth polished appearance.

Glass, diamonds, metallic carbides, etc., are required for cutting tools in finishing operations because ordinary steel, or chromium steel, soon dulls, and becomes useless as a result of the hard abrasive character of the surface of cold-molded pieces.

Markets for Cold-Molded Materials. Although the volume of cold-molded materials is relatively small when compared to other plastics materials, it still finds application where the following characteristics may be desired: elevated temperature resistance, 500 to 1300°F; exceptional arc resistance; nontracking under repeated arc; low material cost; high yield per cavity and low tool investment; and low cost of prototype.

12

THERMOFORMING PLASTIC FILM AND SHEET

The shaping of plastic film and sheet has been known by many names over the years. Originally, shaping was considered one of a variety of fabrication techniques available to transform so-called plastic secondary goods (e.g., film, sheet, profiles, tubing, etc.) into finished products. As the technique grew in importance, however, and took on its own identity, the ambiguous term *forming* came into general use. Finally, the industry settled on the term *thermoforming* (a contraction of *thermoplastics* and *forming*) as the more apt description, and this is now the accepted terminology.

This chapter will discuss the various aspects of the thermoforming process. It should be noted, however, that while the expression *plastic sheet* will be used throughout this chapter, in many instances it is used to include both film (10 mils and under) as well as sheet (over 10 mils). While the general tendency is to think of film as a soft, flexible material and sheet as rigid extrusions, the availability in recent years of tough new plastics has made it possible to feed thinner-gauge materials (e.g., biaxially oriented styrene) into a thermoforming press as easily as the thicker-gauge sheets.

Basic Concepts

In its simplest explanation, the thermoforming process consists of heating thermoplastic sheet to its softening temperature and forcing the hot

and flexible material against the contours of a mold by mechanical means (e.g., tools, plugs, solid molds, etc.) or by pneumatic means (e.g., differentials in air pressure created by pulling a vacuum or using the pressures of compressed air). When held to the shape of the mold and allowed to cool, the plastic retains the shape and detail of the mold. Obviously, since softening by heat and curing by the removal of heat is involved, the technique is applicable only to thermoplastics materials and not to thermosets.

In the early days of thermoforming, the process involved heating of sheet in a separate oven and then transferring the hot sheet to a forming tool. As the industry became more sophisticated, however, automatic machinery combining heating and forming in a single unit (later, trimming as well) became available. At the same time, the industry moved from the original concept of sheet-fed machines (i.e., feeding a single die-cut sheet into the press) to continuous operations, feeding off a roll of plastic or directly from the mouth of an extruder.

More recently, the thermoforming industry has begun to evaluate the idea of reducing the amount of heat applied to a thermoplastic sheet, or even eliminating it entirely, and then using the higher pressures of stamping presses to form a part. These techniques, variously known as *cold forming*, *forging*, *solid-phase forming*, etc., are discussed in detail on p. 318.

Materials

Virtually every thermoplastic sheet can be, and has been, thermoformed (Figs. 12-1a through 12-1d). In fact, as the thermoforming process has become more and more important, the emphasis of extruders has turned to producing sheet specifically to meet the demands of thermoforming markets (i.e., larger sheet for the production of products like boats, pallets, signs, siding for houses, etc.). Current maximum dimensions for thick sheet are limited by the presently available manufacturing facilities (10 ft in width and 14 ft in length). Thinner sheeting (produced by extrusion, calendering, or continuous casting) is available in rolls of several hundred feet in length.

The section on p. 295 discusses some of the specific processing variables to be considered when thermoforming individual plastics. It may be well at this point, however, to note some of the unique characteristics and properties of thermoplastic sheets that must be taken into account when considering thermoforming.

Plastic Memory. Most thermoplastics possess enough elasticity when hot so that when stretched, by mechanical or pneumatic means, they tend to draw tight against the force which stretches them, and also to stretch as evenly as possible. These characteristics permit forming against a single mold, which is a common practice in many thermoforming methods (as distinguished from matching male and female molds used in most other plastic processing techniques). Reheating a formed part to the original forming temperature will activate the so-called plastic *memory* and cause it to relax back to its original shape (i.e., a flat blank). Hence, errors in the contours of formed parts can often be corrected by reforming. It is also possible to decorate a formed part, then relax it back to the original flat, and use the "distorted" pattern to print subsequent sheets "in distortion" and form them into products in which the decorative elements will be in perfect register on the three-dimensional surface.

Hot Elongation. All thermoplastic sheet materials can be stretched when hot, but this property varies greatly with different materials, and under different conditions, and is measurable. It is intimately related to temperature and to speed of elongation and, in many of the methods to be described later, it is of critical importance. Because of its dependence on correct temperature, methods of heating, methods of stretching or forming, choice of material for molds and the related methods of cooling on the mold, it will be referred to frequently.

Some commercial sheet stocks can be stretched as much as 500 or 600% over their original area; others as little as 15 to 20%.

Fig. 12-1a. Range of modern thermoformed products includes (left to right): acrylic sink rigidized with reinforced plastics backing; a thermoformed impact styrene drawer; cabinet door based on vacuum formed wood-grained plastic skin laminated to flat wood or chip wood base; and impact styrene decorative panels. (*Courtesy AAA Plastic Equipment, Inc.*)

Fig. 12-1b. Sailboat hull, 11 ft long, is formed of ABS sheet, filled with expanded polystyrene for buoyancy. (*Courtesy Borg-Warner Chemicals*)

Fig. 12-1d. Removing ABS boathull with PS foam flotation from forming press. (*Courtesy Arco/Polymers*)

Fig. 12-1c. Tool and tote boxes are vacuum formed of polyethylene. (*Courtesy U.S.I. Chemicals*)

Naturally this has great influence on what shapes can be produced and the quality of what is produced.

Hot Strength. Some materials at forming temperatures become almost putty-like and respond to a minimum of pressure, either pneumatic or mechanical, in such a way as to pick up every detail of the mold. Others exhibit strong resistance, and thus require heavier equipment and tools. The limited differential of pressure available in vacuum methods may not suffice to provide small details in some formed articles.

This property is somewhat related to ability to be stretched while hot, but does not run parallel to it.

Temperature Range for Forming. The thermoplastics considered in this chapter do not have sharp melting points. Their softening with increase of temperature is gradual, and each material has its own range of temperature, wide or narrow, within which it can effectively be formed. Thus one may have a forming range of 275 to 420°F; another may become soft enough for forming at 360°F, but melt at 400°F. Also, a material may stretch well at a given temperature but tear easily if heated a few degrees higher or dropped a few degrees colder. This single factor is probably the most important of all in the forming of thermoplastic sheet.

Redistribution of Material. This is a matter of importance to the design engineer as well as to the fabricator. The design engineer must detail what he wants, based on performance standards and on what the basic material will do. Since it is obvious that formed pieces made from sheet cannot be made with thickened sections such as ribs or bosses, or may even, by nature of the shape or the method, have areas which may be thinner than desired, it is important that the engineer understand basic methods, and how to predict results from those methods. Unlike the drawing of metal to shape, in which a developed pattern of stock is drawn into a new position, most formed articles of plastic are produced from a sheet which is restrained around its edges. Thus, when additional area is created by stretching the stock into a new shape, the average thickness is reduced. Here is the place for the skill and judgment of the fabricator, but the engineer must realize that wall thickness in the finished article will be reduced from the original sheet thickness in reverse proportion to the increase in area of the formed piece over the area of the original sheet.

Methods of Forming

Because there are so many different techniques for thermoforming plastic sheet, the basic forming methods will be reviewed in this section. Most of the illustrations (Figs. 12-2 through 12-15, inclusive) will be based on equipment incorporating parallel-acting frames, heaters above and below the sheet stock (called *sandwich* heaters and capable of heating both sides of the sheet), male molds and plug assists above the stock (except in the case of drape and slip forming where the male mold is below the stock), and female molds below the sheet. There are, of course, many variations in thermoforming techniques that go beyond this simplified approach, such as folding book-action frames, single heaters above or below, alternating molds, etc.

It should also be noted that, according to general terminology, the use of the words "male molds" connotes a primarily convex mold; "female mold" is a primarily concave mold; the "ring" is the initial plane from which the sheet is drawn to shape it; a "plug" (which may or may not be the male mold) is the component of the mold which moves through the ring and carries the sheet of plastic with it.

Vacuum Forming Into a Female Mold. To vacuum form a thermoplastic sheet into a female mold without prestretching, the ratio of depth to minor dimension of a given section should not be greater than 1 : 1 and no sharp inside radii required. The sheet stock is locked in a frame around its periphery only, is heated to a predetermined temperature or for a predetermined time, and then brought into contact with the edge of the mold. This contact should create a seal so that it is possible to remove the air between the hot plastic and the mold, allowing atmospheric pressure (about 14 psi) to force the hot plastic against the mold.

Most thermoplastics sheets (with the exception of cast acrylic) can be easily formed with vacuum. The reasons for using a female mold: greater details can be achieved on the outer surface of the part (the side against the mold); multiple cavities can be placed closer together; and it's easier to work with when close tolerances are needed on the outside of the part.

Vacuum forming is illustrated in Fig. 12-2.

When this method is used to make an article having an irregular ring line or periphery, the stock is not held by a frame, but is draped manually to the required contour in such a way as to make a seal possible. After that, the same procedure is followed.

Pressure Forming into Female Cavity. Instead of relying on atmospheric pressure against a vacuum (as in vacuum forming above), forming by compressed air at up to 500 psi is now also common. Positive air pressure is applied against the top of the sheet to force it into a female mold. As contrasted to vacuum forming, pressure forming offers a faster production cycle (the sheet can be formed at a lower sheet

THERMOPLASTIC SHEETS

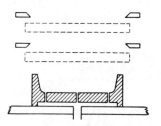

Female mold on platen—frames open—heaters idle.

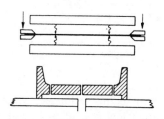

Stock in place—frames closed—heaters active.

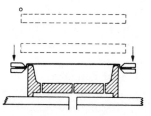

Heaters idle—frames lowered, drawing stock
into contact with mold.

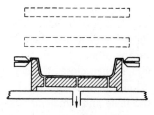

Vacuum applied—stock cooling.

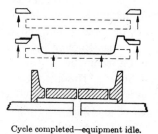

Cycle completed—equipment idle.

Fig. 12-2. Vacuum forming.

temperature), greater part definition, and greater dimensional control. (Fig. 12-3)

Free Forming. This variation, also called *free blowing* has been used with acrylic sheeting to produce parts that require superior optical quality. The periphery is defined mechanically by clamping, but no mold is used, and the depth of draw is governed only by the vacuum or compressed air applied. Visual control or an "electric eye" is used to cut off the pressure when the required depth or height is reached.

Plug-Assist Forming. Straight cavity forming is not well adapted to forming a cup or box shape. The sheet, drawn down by vacuum, touches first along the side walls and then at the center of the bottom of the box-shaped mold and starts to cool there, with its position and its thickness becoming fixed. As the sheet continues to fill out the mold, solidification continues in such a way as to use up most of the stock before it reaches the periphery of the base, and hence this part of the article will be relatively thin and weak. This is undesirable in a cup shape, and usually unacceptable in a box shape, in which this thinness will be most marked at the corners of the base.

To promote uniformity of distribution in such shapes, the plug-assist is used. This is any type of mechanical helper which carries extra stock toward an area which would otherwise be too thin. Usually the plug is made of metal, and heated to a temperature slightly below that of the hot plastic, so as not to cool the stock before it can reach its final shape. Instead of metal, a smooth-grained wood can be used, or a thermoset plastic such as phenolic or epoxy; these materials are poor conductors of heat and hence do not withdraw much heat from the sheet stock.

The plug-assist technique is illustrated in Fig. 12-4. Note that it is also possible to reverse mold and plug assist from the bottom platen to give better material utilization around edges.

Both cavity and plug-assist forming make possible the production of shapes having protuberances on their inner surfaces, formed in contact with plugs projecting from the interior of the female mold. In the sides of an article, such protuberances constitute undercuts, and hence ordinarily the plugs which form them

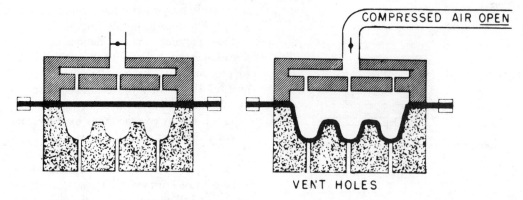

Fig. 12-3. Pressure forming into female cavity. Heated sheet is clamped over cavity and compressed air pressure forces the sheet into the mold. (*Courtesy Dow*)

must be withdrawn in order to release the article from the mold. But some undercut articles made from tough flexible sheeting can be sprung out of the mold.

Plug-assist techniques are adaptable both to vacuum forming and pressure forming techniques. The system shown in Fig. 12-4, is known as plug-assist vacuum forming in that a vacuum is drawn after the plug has reached its closed position to complete formation of the sheet. In plug-assist pressure forming, the process differs in that after the plug enters the sheet, the air under the sheet is vented to the atmosphere. When the plug completes its stroke and seals the mold, air pressure is applied from the plug side (through the plug or from behind the plug). As opposed to plug-assist vacuum forming, pressure forming offers more uniform material distribution over the entire formed part (see Fig. 12-5).

Drape Forming. This method is also adaptable to either machine or manual operation. After framing and heating, the stock is mechanically stretched over a male mold to allow the framed edge to make a seal with the periphery of the mold. This stretching serves to redistribute or preform the sheet preliminary to application of the vacuum. It has the disadvantage of allowing the stock to touch prominent projections, to freeze there, and perhaps rob other areas of sufficient mass to make acceptable articles. Careful control of the temperature of the mold, plus selective heating of the sheets, can alleviate some difficulties. Many articles are well adapted to this technique, which has the advantages of low cost of mold and machine, and rapidity of operation. This technique is illustrated in Fig. 12-6.

The term *drape forming* is sometimes applied to forming a hot sheet by laying it over a mold (either male or female) and allowing it to conform by gravity.

Matched Mold Forming. In recent years, a number of mechanical techniques that use neither air pressure nor vacuum have been evolved. Typical of these is matched mold forming (Fig. 12-7). In this operation, the plastic sheet is locked into the clamping frame and heated to the proper forming temperature. A male mold is positioned on the top or bottom platen with a matched female mold mounted on the other.

The mold is then closed, forcing the plastic to the contours of both molds. The clearance between the male and female molds determines the wall thickness. Trapped air is allowed to escape through both mold faces. Molds are held in place until the plastic cools and cures.

Matched mold forming offers excellent reproduction of mold detail and dimensional accuracy. Internal cooling of the mold is desirable in this technique.

Trapped Sheet. This thermoforming technique involves the use of both contact heat and air pressure (Fig. 12-8). The plastic sheet is inserted between the mold cavity and a hot blow plate (which is flat and porous to allow air to be blown through its face). The mold cavity

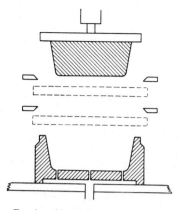

Female mold and plug assist mounted—
frames open —heaters idle.

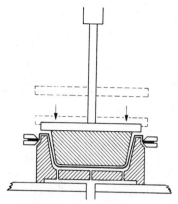

Plug assist lowered—prestretching the stock.

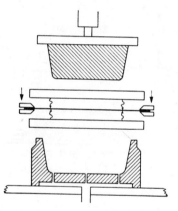

Stock in place—frames closed—heaters active.

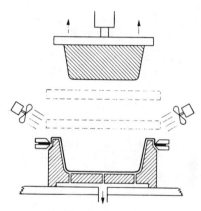

Vacuum applied—plug assist retracted—fans operating.

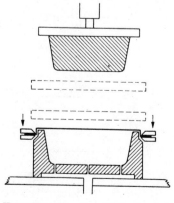

Heaters idle—frames lowered, drawing stock
into contact with mold.

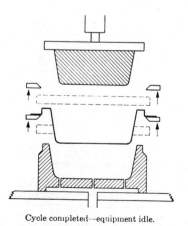

Cycle completed—equipment idle.

Fig. 12-4. Plug-assist forming, using vacuum.

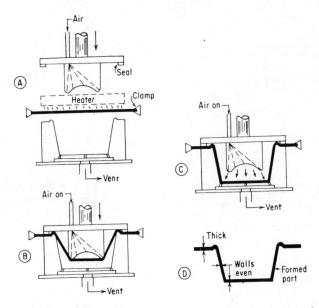

Fig. 12-5. Plug-assist pressure forming. (*Courtesy Modern Plastics Encyclopedia*)

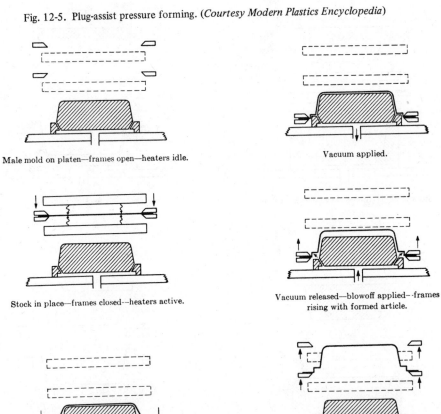

Male mold on platen—frames open—heaters idle.

Vacuum applied.

Stock in place—frames closed—heaters active.

Vacuum released—blowoff applied—frames rising with formed article.

Frames lowered to stretch stock and to make a seal at edge of mold.

Cycle completed—equipment idle.

Fig. 12-6. Drape forming.

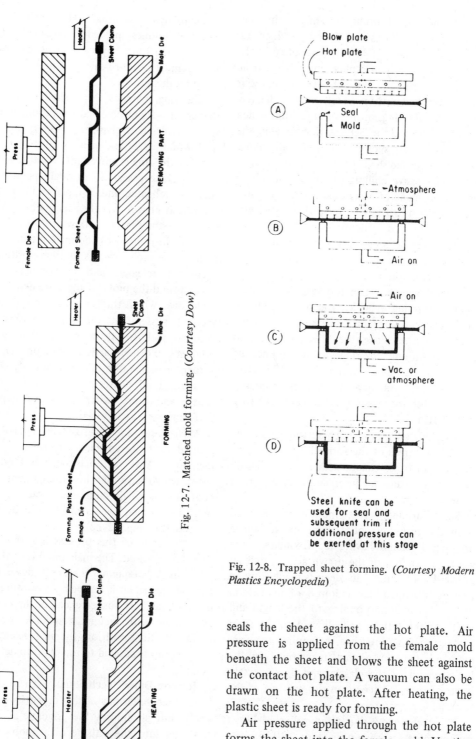

Fig. 12-7. Matched mold forming. (*Courtesy Dow*)

Fig. 12-8. Trapped sheet forming. (*Courtesy Modern Plastics Encyclopedia*)

seals the sheet against the hot plate. Air pressure is applied from the female mold beneath the sheet and blows the sheet against the contact hot plate. A vacuum can also be drawn on the hot plate. After heating, the plastic sheet is ready for forming.

Air pressure applied through the hot plate forms the sheet into the female mold. Venting can be used on the opposite side and steel knives can be inserted into the molds for sealing. After forming, additional closing pressure can be exerted.

Plug-and-Ring Forming. This is the simplest type of mechanical forming which involves more than a fold into two planes. The equipment consists of a plug which is the male mold, and a ring matching the outside contour of the finished article. Stock may be heated away from the ring, or on the ring, after which the plug is forced through the ring, drawing the plastic with it in such a way as to redistribute the stock over the shape of the plug. In order to prevent excessive chilling of the plastic, either the plug is made of a material of slow thermal conductivity or, if made of metal, its temperature is controlled. This method, illustrated in Fig. 12-9, has been used for the manufacture of single and multi-cavity trays, with excellent yields in large volumes.

Ridge Forming. This is a variation of plug-and-ring forming in which the plug is reduced to a skeleton frame which determines the shape of the article. Since the sheet comes into contact with only the ridges of this frame, the intervening flat areas are free from mold marks or "mark-off" and have better surface quality than if formed against a solid plug. (See Fig. 12-10.)

If a skeleton frame which surrounds a plane is used, the areas of the formed piece are plane surfaces. In other shapes with ridges which do not fall in a plane, the intervening surfaces tend to be concave.

Slip Forming. This method is adapted from the technique of drawing metal from a pre-shaped blank having approximately the same area as that of the drawn article; thus an article 4 x 4 in. and 2 in. deep is made from an irregularly shaped blank about 8 x 8 in., held in pressure pads corresponding to the length and width of the article, and allowed to slip from these as drawing proceeds. As applied to forming of plastics, this method is restricted by the hot strength of the plastic, and by the likelihood of its being scored in slipping out of the restraining pressure pads.

This technique is illustrated in Fig. 12-11a; a variant of it is shown in Fig. 12-11b. The springs shown in these drawings can be replaced by air cylinders with relief valves, to improve the control. Automotive carpeting is formed in this way.

Snap-Back Forming. The vacuum snap-back techniques preceded so-called vacuum forming machines by several years. In the 1940's most point-of-sale plastic displays were made by this process.

The ring matches the desired periphery of the article. In mechanical operation, the cold sheet of plastic is clamped to the ring close to its forming edge, and heated; the vacuum box beneath the ring is moved into contact with the ring to make a seal. In manual operation, the sheet is preheated and clamped to the ring, and the ring is placed over the vacuum box. A small bubble of the plastic is drawn into the box, by vacuum, just ahead of the male mold. As soon as the male mold reaches its final position, the vacuum is released, and the hot plastic snaps back around the mold. If there are details in the mold over which the hot plastic may bridge, a vacuum system in the male mold can be used to pick up those details. This method produces excellent articles, except that sharp corners may be quite thin. It is entirely possible to control all phases of this procedure with timers and limit switches, so that repetitive performance is guaranteed.

Most cases and luggage shells made from grained or textured sheet are formed by this technique. Also, materials, such as acrylic cast sheet, which have high hot strength and do not conform to small radii under vacuum alone, must be formed over a male mold by such a process. (See Fig. 12-12).

A variation of this process is known as the "air-slip" method. The male mold is confined in a box in such a manner that, as it moves toward the hot plastic, air is trapped. As pressure builds up, the plastic is pushed ahead of the mold just as is done with vacuum snap-back. When the mold is in final position, the air is released and the vacuum is applied to the mold to define the details.

Reverse-Draw Techniques. There are at least three variations of this process worthy of study, but all employ the principle of pushing the plug into the outside of a bubble of hot plastic, thus accomplishing a folding operation which permits deeper draws than are possible by any other common practice. The final shape may be determined by a male mold or a female

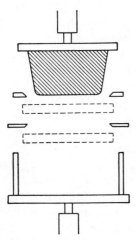

Male mold mounted—risers on platen—frame
open above forming ring—heaters idle.

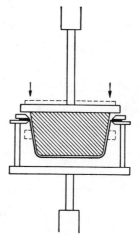

Plug lowered—stretching stock to shape.

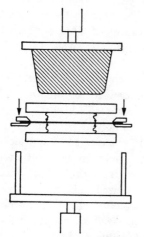

Stock on ring—frame closed—heaters active.

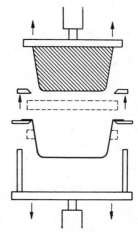

Cycle completed—equipment idle.

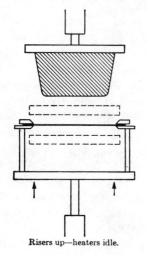

Risers up—heaters idle.

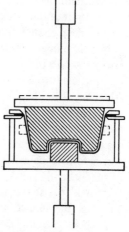

Variation: with female plug added.

Fig. 12-9. Plug-and-ring forming.

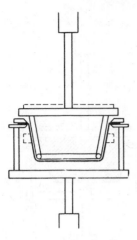

Fig. 12-10. Ridge forming.

the height of the bubble, and the temperature and speed of travel of the plug assist, as well as the release of air pressure and application of vacuum, it is possible to closely control the thickness of any given section of the article. One exception is that corners will never be thin, since the folding or rolling action as the plug moves downward gathers extra stock around the corners. Since articles formed into a cavity shrink away from the mold, little draft is required in the mold.

By this process articles may be drawn with uniform wall thicknesses less than 25% of the original stock thickness, with repetitive quality, and usually without apparent impairment of the physical properties of the original material. But in some cases a uniaxial orientation of stress may develop, so that the finished article is crack-sensitive parallel to the direction of stretch. This becomes more noticeable if the flanges are trimmed off.

Reverse Draw with Air Cushion. A variant of this method, known as reverse draw with air cushion, promises to be important in the forming of materials having limited hot strength, such as polyethylene and polypropylene. The bubble is created by hot air, and the plug assist has holes in the bottom through which hot air is forced. Thus mechanical contact is minimized, and the plastic is kept heated and is pushed ahead of the plug until ready to be finalized against the walls of the female mold. (See Fig. 12-14.)

Reverse Draw on a Mold or Billow-Up Vacuum Snap-Back. The third version of this method is used with grained or polished stock on a male mold, to preserve the finish. The blank must be oversize to provide a bubble of sufficient size and shape so that when the mold comes into contact with the outside of the bubble, the stock which touches the bottom of the mold freezes there and need not be reduced further in thickness, nor change its shape. Corners should be "cut" to avoid "webbing." The forming ring matches the finished article and assists in "wiping" the stock around the mold as the mold moves into final position. (See Fig. 12-15).

The mold should be run as warm as is compatible with efficient cooling of the article.

mold. The latter is recommended except when a grained stock is used, and the grain is on the outside of the article, or when the original surface gloss of the sheet might be impaired by contact with a mold.

The material is thinned at the center of the bubble before contact with the plug. A free-blown hemisphere has at its center a thickness about one-third that of the original sheet. The thinned area, however, remains in contact with the plug (or male mold), and is carried by it. Further stretching occurs in the thicker side area, and thus the finished article has approximately uniform thickness.

Reverse Draw with Plug Assist or Billow-Up Plug Assist. (Fig. 12-13). The stock is placed in the frame and heated. Then the female mold is raised up into the stock close to the edge of the frame, deep enough to maintain a seal while a bubble is blown upward. Air is introduced into the mold to a pressure of about 1 to 3 psi, or more, sufficient to create a bubble upward toward the plug. The plug, in this case only an assist, is heated to a temperature only slightly less hot than the stock itself. As it is lowered, it pushes the redistributed stock down into the cavity in a folding or rolling action. The plug is designed slightly undersize, but should stretch the stock to a shape quite close to its final position in the female mold. The final shaping is then accomplished with vacuum being applied to the female mold.

By regulating the temperature of the stock,

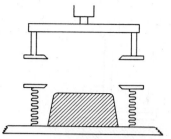

Male mold on platen—spring-loaded pads open—
forming ring with risers above.

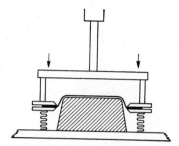

Stretching in progress—stock slipping inward.

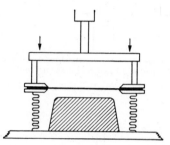

Stock in place—ring moving downward.

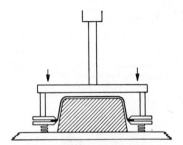

Forming completed—stock slipped to optimum.

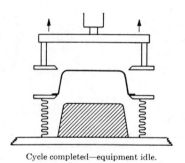

Cycle completed—equipment idle.

Fig. 12-11a. Slip forming.

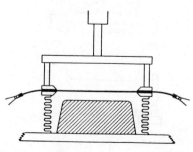

Variation: narrow ring- -clips to restrict slippage.

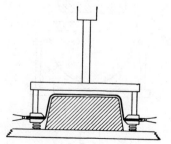

Forming completed—clips against ring.

Fig. 12-11b. Slip forming.

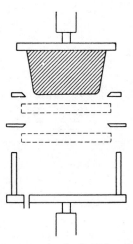

Male mold above—vacuum chamber below—frame open—
forming ring and heaters idle.

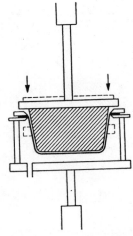

Forming completed—vacuum released for snap-back.
(Vacuum should not be released until male mold is in place.)

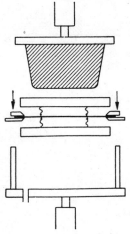

Stock on ring—frame closed—heaters active.

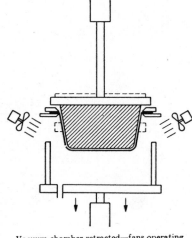

Vacuum chamber retracted—fans operating.

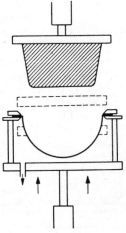

Vacuum chamber up—heaters idle—vacuum applied.

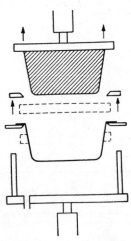

Cycle completed—equipment idle.

Fig. 12-12. Snap-back forming.

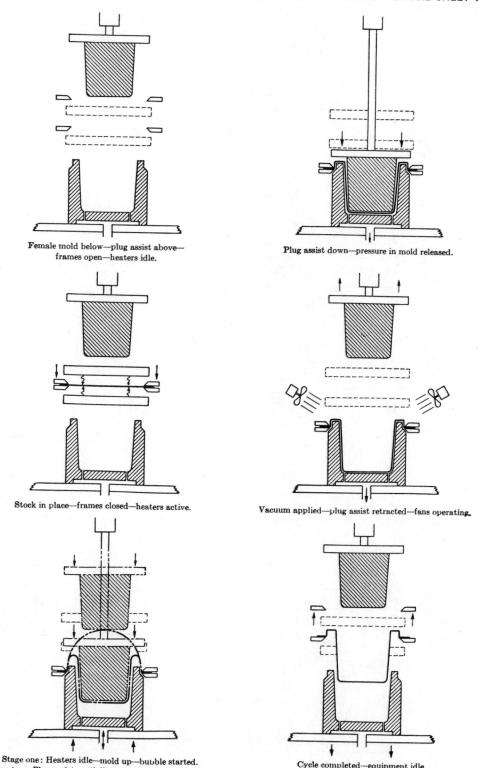

Female mold below—plug assist above—
frames open—heaters idle.

Plug assist down—pressure in mold released.

Stock in place—frames closed—heaters active.

Vacuum applied—plug assist retracted—fans operating.

Stage one: Heaters idle—mold up—bubble started.
Stage two: Plug assist partially down—bubble maintained
around perimeter.

Cycle completed—equipment idle.

Fig. 12-13. Reverse draw with plug assist.

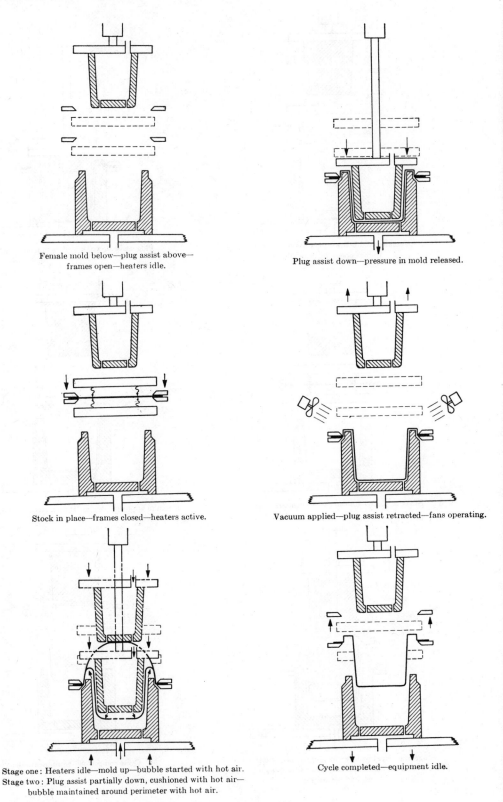

Female mold below—plug assist above—
frames open—heaters idle.

Plug assist down—pressure in mold released.

Stock in place—frames closed—heaters active.

Vacuum applied—plug assist retracted—fans operating.

Stage one: Heaters idle—mold up—bubble started with hot air.
Stage two: Plug assist partially down, cushioned with hot air—
bubble maintained around perimeter with hot air.

Cycle completed—equipment idle.

Fig. 12-14. Reverse draw with air cushion.

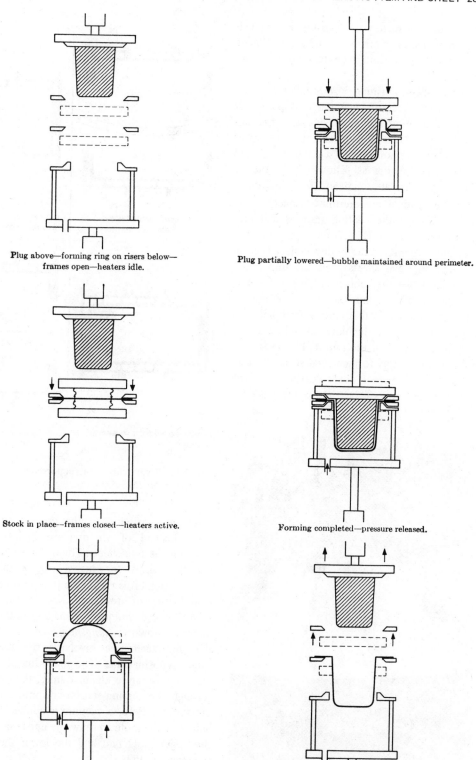

Plug above—forming ring on risers below—
frames open—heaters idle.

Plug partially lowered—bubble maintained around perimeter.

Stock in place—frames closed—heaters active.

Forming completed—pressure released.

Heaters idle—ring raised—bubble blown to meet plug.

Cycle completed—equipment idle.

Fig. 12-15. Reverse draw on a plug.

It should be made of a material which is thermally a slow-to-medium conductor; if made of aluminum, it should be temperature controlled.

Twin Sheet Forming. A number of techniques have recently been made available for the production of hollow products by thermoforming. (Fig. 12-16). Typical is the concept of twin sheet thermoforming (see Fig. 12-17 and 12-17a). It operates as follows: Two rolls of plastic sheet are automatically fed, one above the other with a predetermined space in between, through the heating stations and into the forming station. Here, a blow pin enters at the central point of the hollow object (i.e., in between the two sheets) and the upper and lower halves of the tool close onto the sheets and pinch off around the entire perimeter. High-pressure air is then introduced between the two sheets from the blow pin and a vacuum is applied to each of the two mold halves. The hollow object thus formed (and sealed around the periphery) then indexes forward and the next two segments of sheet move into place for forming.

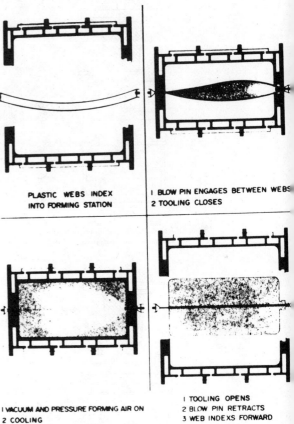

PLASTIC WEBS INDEX INTO FORMING STATION

1 BLOW PIN ENGAGES BETWEEN WEBS
2 TOOLING CLOSES

1 VACUUM AND PRESSURE FORMING AIR ON
2 COOLING

1 TOOLING OPENS
2 BLOW PIN RETRACTS
3 WEB INDEXS FORWARD

Fig. 12-17. Twin sheet thermoforming. (*Courtesy Brown Machine Div., Leesona*)

In one variation of the process, urethane foam instead of air pressure is introduced between the two sheets. The urethane bonds to the two skins forming a rugged sandwich construction. This technique is applicable to the manufacture of urethane foam-filled boat hulls.

Another technique for making hollow products is known as *clam-shell forming*. This is a sheet-fed technique involving the use of a rotary-type machine. The individual sheets are placed in separate clamp frames, then indexed through the heating station and into the forming station. Here, a vacuum is applied to both halves of the mold to draw the upper sheet into the upper mold half and the lower sheet into the lower mold half.

Still another method, known as *twin-shell forming*, involves the use of a series of continuously moving molds (traveling on belts)

Fig. 12-16. Fourteen ft.-long high-density polyethylene ski launch is based on a twin sheet thermoforming technique. (*Courtesy Phillips Petroleum*)

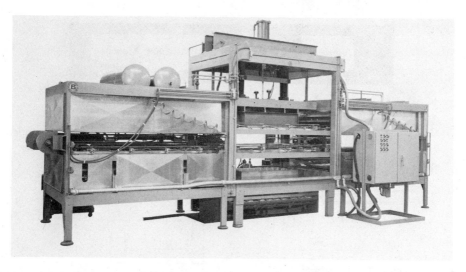

Fig. 12-17a. Equipment for twin sheet thermoforming. The two thermoplastics sheets are held in separate independent clamp frames and indexed into ovens. After the desired temperature has been reached, clamp frames move the material interdependently into the forming area. (*Courtesy Brown Machine Div., Leesona*)

that clamp onto the sheets (feeding off rolls) and travel with them as the vacuum is pulled and the sheet halves formed.

Thermoforming Machines

Industrial thermoforming machines are as varied as the products they were designed to produce. They range from relatively simple, shallow drawn, manually controlled equipment to the complex, computer-controlled rotary and in-line machines designed for production and efficiency.

Any thermoforming machine must provide the following: (1) A method for heating sheet to the pliable, plastic state called the *forming temperature*; (2) A clamping arrangement to hold the plastic sheet for heating and positioning for forming; (3) A device to raise or lower the mold into the plastic sheet or to move the clamped sheet over the mold; (4) A vacuum system; (5) An air-pressure system; (6) Controls for the various operations; and (7) Safety devices.

Machines are often classified according to the number of operations they perform, as follows:

Single-Stage, Sheet-Fed Machines. A single-stage machine is one which can perform only one operation at a time and its total cycle will be the sum of the times required for loading, heating, forming, cooling, and unloading. In a typical operation, the sheet is clamped in a frame, the frame is moved between the heaters (or under a single heater), and back to the forming station for thermoforming. (See Fig. 12-18.)

Multiple-Stage, Sheet-Fed Machines. A two-stage machine is one which can perform two operations simultaneously. It usually consists of two forming stations and a bank of heaters which move from one to the other. Machines with three stages or more are usually built on a horizontal circular frame and are called *rotaries* (Fig. 12-19). The rotary operates like a merry-go-round, indexing through the various stations. A three-stage machine would have a loading and unloading, a heating, and a forming and cooling station, and would index 120 degrees after each operation; a four-stage machine would have a loading and unloading, a preheating, a heating, and a forming and cooling station, and would index 90 degrees after each operation). Since there is always a sheet in each of the stations, it provides considerably higher output than a single-station machine.

In-Line, Sheet-Fed Machines. Here, the sheet follows the same pattern as a caterpillar track. The sheet is clamped in a frame which travels into the heating station, then indexes

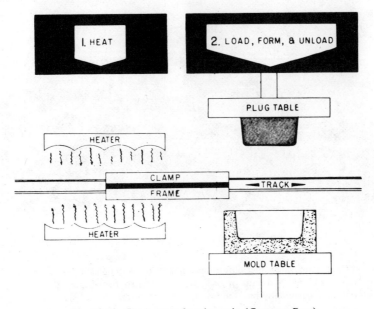

Fig. 12-18. Single-stage forming unit. (*Courtesy Dow*)

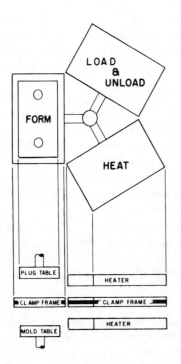

Fig. 12-19. Three-stage rotary unit. (*Courtesy Dow*)

through the forming station and on to the unloading station where the part is removed by the operator. This type of machine has a total of five clamp frames in use at all times.

Continuous Machines. In the early days of thermoforming, continuous forming machines that fed off a roll of plastic or directly from an extruder used either a rotating cylinder as the mold (drum former) or used conventional molds that traveled horizontally at the same speed as the sheet (the upper and lower platens moved with the sheet but had a reciprocating motion so that they could index to the next unformed section of sheet, once the forming cycle was complete).

The workhorse of the industry today, however, is the intermittently fed continuous forming machine (used for thin wall containers, disposable cups, lids, etc.). This machine is fed either from a roll or directly from an extruder. As opposed to the systems described above, however, the sheet is indexed through the machine intermittently. It is considerably faster than the other techniques where the sheet is continuously moving through the machine and the molds have to move back and forth.

In a typical operation, the sheet feeds off a roll at the rear of the thermoformer into a set of conveying chains that indexes the sheet intermittently forward through heating, forming, and trimming. Once the roll of material is threaded through the system, it functions completely automatically and can cycle as fast as two seconds.

Packaging Machines. Although most of the equipment already discussed can be used to form plastic packages, there are several machines which have been adapted to combine the forming operation with other packaging functions. Such machines are used chiefly for skin or contour packaging and blister packaging. Both processes involve the sealing of paperboard or plastic to the formed plastic sheet, with the product to be packaged enclosed. Most conventional machines can be equipped to handle skin packaging, since the loading operation is usually manual. These machines may operate from roll or sheet stock and some are equipped to trim the packages as they leave the machine.

Rather elaborate machines are now available for complete blister packaging. These machines will form the blisters, provide an area for manual or automatic loading, seal paperboard to the back side and die-cut the finished packages. Generally, they operate with intermittent or continuous feed of roll stock, and the various functions are performed simultaneously at a sequence of stations in line.

More recent has been the appearance of form/fill/seal equipment, in which a package is formed, immediately filled with the product to be packaged (in some cases, a liquid product can also be used as a pressure media to assist in forming), then sealed. Because aseptic features are required for machines of this type, all operations are generally enclosed in pressurized, sterilized air chambers.

Trimming. Since most forming operations require a surplus of material for clamping, and since this flange area is usually warped by variance in the heat applied, most fabrication requires a final trimming operation. Numerous methods are known and proven; the proper choice depends upon the material to be cut, the shape of the trim line, and the number of articles to be trimmed.

Many jobs are trimmed by saws or routers. The most extensively used trimming method,

except in packaging, is by router. These systems are operated by hand, using a trimming fixture mounted underneath tables or overhead. Various types of router bits and circular saws are used in these routers. See Fig. 12-20a.

Certain operations with small articles provide for trimming off the flange while the article is still in the mold. This is usually done by the hot-knife method, which requires only limited power and yields a clean cut, partially "healed" by the heat applied. This method is practicable only in relatively large-volume operations. It can be used in a multiple setup. It is most efficient with materials most sensitive to heat and in the thinner gages.

Another and similar method re-registers the formed articles in a subsequent stage of the operation, permitting in-line forming and trim-

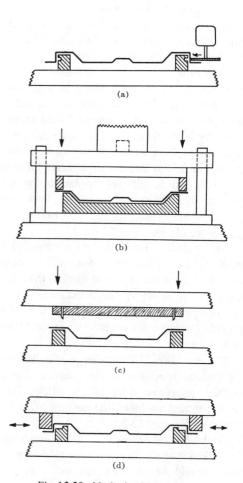

Fig. 12-20. Methods of trimming.

ming. Accurate registry is provided by dimples formed into the periphery, to register and then be trimmed away with the scrap.

The method most often used in trimming (and punching) large, high-production, relatively flat articles, such as refrigerator door liners, requires handling by the operator, who places the untrimmed piece into a press, activates it through the cycle, and then removes the article and the trim scrap. This will be described more fully under equipment and tools.

Tools for trimming will be described below.

Shear Dies. These are the obvious tools for trimming large production runs of large articles, if they are trimmed on a single level. These dies are built the same as if they were to trim metal articles, are mounted on die shoes, and operated in metal-working presses. Some plastics require quite close tolerances between die members. The power required is about one-third of that required to shear mild steel of equal thickness. (See Fig. 12-20b.)

Steel-Rule Dies. These are made of strip steel about $\frac{3}{32}$ in. thick and 1 in. wide, with one sharpened edge. The strips are formed to the shape of the trim line, and held to that shape by birch die stock. They are practical in small to medium runs, and for most thermoplastics in thin to medium thicknesses. Some of the more brittle plastics can be cut by this process only when warm, before they have cooled after forming, or by post-heating the part or the die. (See Fig. 12-20c.)

Walker Dies. These dies are known also as envelope dies or high dies. They are a heavy-duty version of the steel-rule dies, in that they are forged to about $\frac{5}{16}$ in. thickness, and they are available up to 4 in. high, and thus may provide clearance for projections in the article to be trimmed.

Planetary Dies. These dies provide for side motion and progressively shear vertical flanges on the respective sides of an article. They require special machines that are now becoming available. (See Fig. 12-20d.)

Machinery for Trimming. It is obvious that tools similar or identical to those used in metal work should be used on the same type of equipment regardless of material. Therefore, trimming of plastics is often done on punch presses, press brakes and other toggle-action machines. Hydraulic presses are entirely satisfactory, and are frequently moved into the forming area so that the operator who runs the forming machine is able to trim articles just formed, within the duration of the next forming cycle.

The special contour dies such as the steel-rule and envelope types are used on clicker and dinker machines such as those used in the leather industry, on continuous or clutch-type toggle presses, and on hydraulic presses.

Saws are used for many operations where a vertical flange is to be trimmed. A radial-arm type is readily adjustable, and direct drive yields appropriate blade speed. If articles are registered from the table area, clearance beneath the saw blade can be provided. Bandsaws in both vertical and horizontal positions are efficient in trimming "hat" shaped parts. Wood-shaper machines are also used, but are not as adaptable.

Problems of Shrinkage. As stated earlier in this chapter, mold shrinkage occurs not only when the article is removed from the mold, but during a considerable period after removal. Therefore it is important, especially with large items that go into precise assemblies, that all trimming be done on schedule. If it is most practicable to trim as the article is removed from the mold, a shrinkage factor must be built into the trim die. If it is desired to "build a bank" of parts, a different factor must be experimentally determined. A "no-shrink" die for trimming large articles is probably practicable only after an overnight cooling period.

Trimming in the Mold. This innovation makes it possible to trim formed articles at the forming station. One method of trimming-in-place incorporates a knife edge around the periphery of the forming die and a movable heated ring directly over the die. After the object is formed in the conventional manner, but before it is cool, the heated ring drops down and presses against the knife edge in the mold and pinches the part from the web. The process may also work in reverse by placing a heated knife edge on the upper platen and bringing it in contact with the sheet against a flat mold surface. The problem with this method is that plastic may build up on the trimmer and reduce its effectiveness.

Another method uses shearing dies which either punch a section of sheet that is then carried into the mold and formed, or which trim the finished part before it is removed from the mold. These techniques are more involved and require expensive tooling.

THERMOFORMING VARIABLES*

Today, thermoforming is recognized as one of the fastest growing segments of the plastic industry. This has been brought about by the development of new materials and techniques, coupled with the production of specialized equipment capable of providing the production efficiency necessary for modern industry.

Thermoforming is basically a method of thermoforming or shaping sheets of thermoplastic materials. The plastic sheet is heated until it becomes soft and pliable. This hot, now flexible material, is forced or drawn against the contours of a mold and held until it cools. When cooled, the plastic retains the shape and detail of the mold.

Based on this very simple procedure, many "thermoforming techniques" have been developed in order to provide accurate control of wall thickness and to increase the depth of draw to which the material may be formed.

History

Thermoforming dates back to the days of the Pharoas of Egypt. Ancient craftsmen found that the hard shells of the tortise could be softened and fashioned into a variety of shapes and for useful and decorative vessels.

Commercial thermoforming had its start in the 19th century with the Swiss development of nitrocellulose. While this material originated in Europe, early progress came through the development of cellulose nitrate by an American, John Wesley Hyatt.

In the early nineteen fifties, the thermoforming technique was a relatively minor

* By William K. McConnell, Jr., McConnell Co., Inc., Ft. Worth, Texas, 76110.

plastics processing method. During the mid and late fifties, it began to make its mark, however, with such applications as thermoformed acrylic windshields and "bubbles" for use in the aircraft industry, thermoformed internally lighted outdoor signs as well as decorated point-of-purchase displays, thermoformed packages, skin packs, and blisters, and thermoformed styrene inner door liners for refrigerators.

During the sixties, the process experienced exceptional growth and by the end of the decade, it had become one of the major plastics fabricating techniques. Today, thermoformed applications run the gamut from 20-ft boats to luggage, from business machine housings to inner cabinets for refrigerators, from automotive interior panels, shields and decks to one-piece drawers, kitchen cabinet shells, and even chairs and tables rigidized with foam. Future applications for thermoforming are expected to be even more dramatic in terms both of size (e.g., full-size pallets, boats, house sidings can now be formed), of product versatility (e.g., hollow products like bottles can now be made and glass-reinforced thermoplastics sheets, heat-resistant thermoplastics sheets, and cross-linked thermoplastics sheets could open new industrial and consumer markets), and of economics (new grades of sheet are available that can be worked in existing stamping and metalworking equipment).

The advantages of thermoforming relative to most other methods of processing plastics lie in much lower tooling and machinery costs. The disadvantages are having to trim the parts and having to start with sheet material instead of the less-costly resin.

Modern Production Equipment

Many modern machines are now roll-fed, providing "continuous forming" operations. These machines are designed primarily for producing thin-wall containers, disposable cups, lids and similar items. Sheet-fed single station, in-line and rotary machines are used for heavier materials (40 to 500 mils) in making large parts such as refrigerator case liners, machinery housings, cases, boxes, chests, boats, campers, luggage, chairs, drawers, shutters, and walls.

To meet the needs of production speed, thermoforming equipment is available to encompass multiple stages with loading and unloading stations, one or more heating stations, and a forming station. Variations of production equipment range from simple, single-stage units to more complex machines having four or more stations. A range of commercial equipment is shown in Figs. 12-21 to 12-26.

Most high production equipment is automatically operated, cycling through pre-set elapsed times at each station along the production line.

Heating

One of the real keys to efficient thermoforming is the ability to bring the plastic sheet to a uniform and even forming temperature. The sheet of plastic that is "hotter" or "cooler" in

Fig. 12-21. Low-cost, manual control thermoformers for short-run production, research and development, and employee training. (*Courtesy AAA Plastic Equipment, Inc.*)

spots, will have a different stretch strength at these spots. Thus, if a vacuum is drawn (or pressure applied), the plastic will stretch more (or less) at these points of temperature variation. Results could range from a break or tear in the plastic sheet to "lumps" where the plastic did not stretch evenly, to heavier or thinner sections in the formed part.

The type of material, its thickness, and its forming temperature are factors which must be considered. Far infrared gas-fired and electric heaters are made with accurate controls to produce even, over-all heat. Most heating is now done by far infrared radiant heat rather than air re-circulating ovens as in the past.

Control systems for heaters include timers designed to cycle heating elements, allowing the heat to "soak in" or saturate the plastic so that the center-line of thick gage materials may be brought to the same temperatures as the outer surfaces. For most efficient heating throughout the sheet, the far infrared emitting surfaces should be at the proper temperature for the particular type of plastic sheet. Most heating sections for industrial equipment include "sandwich heaters" so that both sides of the plastic sheet are heated simultaneously. This helps to avoid overheating one side and underheating the other, causing strains and stresses, excessive sag, and degradation of material toward the heater.

Heating the sheet is critical. Fast and uniform heating methods are desirable. Radiant heat is by far the fastest, cheapest, and most popular. The most common method of radiant heat is far infrared tubular heaters. These are economical and long lasting, but relatively inefficient.

Because proper, fast heating of the sheet is the real key to thermoforming, industry is turning to more efficient, controllable techniques. Flat, wide area emitting surface heater banks are filling the bill. The large, dense, radiant heat emitting surfaces give much faster and more uniform heat. Heating cycles are cut 50 to 75%. Both electric and gas are used, with electric usually being preferred.

Screening the rays by layers of shielding material such as screens and flat metal has been common for some time. However, a more

Fig. 12-22. Thermoforming machine for industrial production has a 60 by 60 in. mold capacity and is fully automated. (*Courtesy AAA Plastic Equipment, Inc.*)

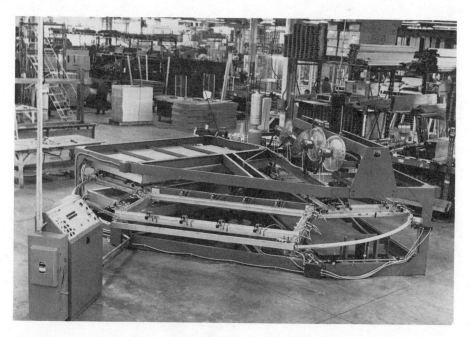

Fig. 12-23. High-speed production is achieved through rotary or in-line equipment, which includes multiple stations for the various aspects of the thermoforming cycle. This particular model is a three-station rotary (loading/unloading; heating; forming) that accommodates a molding area of 48 by 96 inches. (*Courtesy AAA Plastic Equipment Co.*)

Fig. 12-24. In high-speed continuous in-line operation, sheet is fed from a roll (far right) into the forming press (both compressed air and vacuum are available so that a variety of forming techniques can be used). From there, the formed web feeds into a variable speed trim press (far left). (*Courtesy Brown Machine Div., Leesona*)

Fig. 12-25. Another version of a rotary thermoforming machine, in which clamping frames rotate through various stations in the work area. (*Courtesy Brown Machine Div., Leesona*)

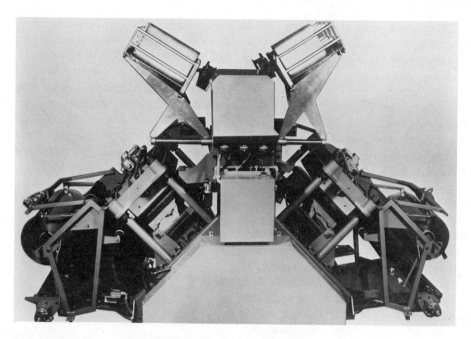

Fig. 12-26. Twin-feed thermoformers consists of dual feed sections located at opposite sides of the machine and twin thermoforming sections that alternately rotate (around a shaft) to a central trim station. (*Courtesy Packaging Industries*)

accurate method is by individually controlling the temperature of small modular flat elements. Profile heating or pattern heating has become important in obtaining uniform wall thickness on deep-draw heavy-wall parts. Each area of the sheet will receive the exact amount of heat required using individual heater modules that operate according to a carefully calculated grid pattern.

Plastics Flow

An understanding of the movement or flow of plastic in the thermoforming process is also important. When heated, the plastic nears a fluid state. It is rubbery, pliable, and elastic. With air pressure or vacuum, the plastic may be free blown or drawn into a bubble. So long as the material has been heated evenly, the plastic flows to provide a uniform thickness throughout the bubble.

When forming against a mold, however, the hot plastic sheet comes in direct contact with the mold itself. As the mold temperature must be at the "set" temperature of the plastic or lower, there is a wide difference in tempera-

tures. Because of the relatively cold mold, drag or chill lines will often appear on the side of the part during forming, particularly over a male mold. Chill lines are a rapid change in material thickness. This occurs because the area first contacting the mold becomes chilled and stretches less than the surrounding hot sheet. The cooler area of the sheet will, consequently, stretch less than the warmer portion as the forming process is finished. Basically, when any part of the hot sheet touches the cooler mold, that part of the sheet will do very little stretching or sliding over the mold. The main reason for this is the fact that chilling gives more hot strength to those parts of the sheet in contact with the mold (as opposed to the hotter portion which has not made contact).

Straight drape forming over a male or into a female mold will, therefore, leave thick, unstretched areas where mold contact is first made. Deeper drawn areas are stretched thin.

To overcome this, the deep-draw thermoforming machines provide mechanized top and bottom air pressure/vacuum platens (i.e., platens are moveable platforms which hold the molds). These platens, which operate independently, allow the operator to use a plug-assist or to pre-stretch the plastic in a manner similar to blowing up a balloon. This pre-stretching provides uniform wall thickness before the mold is brought into place.

To explain what actually happens, let us examine one technique, "vacuum snap-back" forming. Using this technique, the bottom platen is fitted with a vacuum, pre-stretch box which is slightly larger than the male mold which is positioned on the top platen.

The heated plastic sheet, held in the clamping frame, is positioned in the molding area. The bottom platen is raised until the pre-stretch box contacts the plastic. On sealing against the hot plastic sheet, a vacuum is drawn, pulling the plastic into a bowl shape within the vacuum box. The top platen then lowers the mold into the pre-stretched bubble with the mold sealing around the perimeter of the box. Vacuum is drawn through the mold and released in the vacuum box.

Thus, the pre-stretched plastic bubble is pulled (snapped-back) against the contours of the mold and held by the vacuum until cooled. On withdrawal of the top mold, air pressure is sometimes used through the mold to break the formed part away.

Using this vacuum snap-back technique, the plastic maintains near uniform wall thickness throughout the bottom and sides of the product since it was pre-stretched by a vacuum bubble prior to the introduction of the mold itself.

A variety of thermoforming techniques has evolved to meet the production needs of the industry (see Figs. 12-2 through 12-15). An analysis of the job to be done, the tooling, the material, and the finished product specifications will determine the selection of the most advantageous technique.

Manual Control Equipment

While complex, automated equipment is used for production forming, the need for economical prototype and short-run production of plastic parts has created markets for manual control, thermoforming machines. These machines are also often used to augment production equipment through providing research and development of molds and techniques for use in high-speed production equipment.

With these machines, techniques for proper forming are developed to test tooling and various types of materials considered for use in the final production item. The relative cost of such machines enables the manufacturer to complete testing without interruption of normal production of his assembly line.

This type of equipment is also adaptable for educational purposes. There are presently hundreds of schools teaching plastics with R&D type thermoformers (Fig. 12-21).

Automation

Manufacturers of thermoforming machines have utilized technology to produce machines for efficient performance with accurate control of each phase of the thermoforming process. While automation can mean a simple temperature controlling device which maintains the sheet temperature at a given setting, it can also mean a sophisticated system which controls all functions of extruding, thermoforming, and finishing.

Most thermoformers draw a median line between these extremes to create a balance of initial cost, man hours, the need for technical skills, material costs, and maintenance along with production output and production expense. These factors determine the equipment needed to enter a competitive arena of mass production.

Most automated thermoforming equipment is designed to automatically control all thermoforming functions in a timed-sequence, except for sheet loading and unloading. (There are now, however, more and more automatic sheet load-unload systems on the market, and, of course, many that are automatically roll fed). Thus, after the operator loads the raw plastic sheet into the thermoformer and initiates the cycle, the machine functions on pre-set programming.

This programming generally includes:

1. Actuation of a pneumatic clamping frame to hold the sheet during forming.

2. Movement of the clamped plastic sheet to the heater section.
3. Application of controlled far infrared radiant heat in a timed cycle.
4. Movement of the heated sheet to the forming section.
5. The forming sequence. This portion of the forming cycle includes several sequenced and timed operations depending on the thermoforming technique utilized. Automated controls include:
   (a) Platen movement to position. Top and bottom platens may move independently or in unison or by sequence stages.
   (b) Air and/or vacuum pressure applied from either or both platens independently, in unison or in sequence.
   (c) Cooling of the formed part (through the mold and/or by convection).
   (d) Breaking the seal between the mold and the formed part.
   (e) Removal of the mold from the formed part.
6. With rotary or multiple in-line machines, movement of the formed part to the unloading-load station.
7. Releasing the clamping frame for unloading the part.

Control Systems

With this over-all view of the job to be accomplished, various systems are available to provide automated control.

Electro-Mechanical. The oldest and most common method of automatic control in thermoforming equipment is the electro-mechanical system. Multiple relays, individual timers, and limit switches initiate successive steps in the sequence. The sequence operation must be selected by an arrangement of toggle switches or multi-plexers. Limit switches and operating cams are adjusted to provide interlocking of various functions such as platen advance and speed, vacuum bleed, and air blow. Safeguard features such as fully retracting platens and material clamp frame positions can

also be included. This system can use timers to monitor the duration of each stage of the thermoforming cycle. With these adjustments, the machine is now ready to operate in the production of the desired article; however, should a change in forming technique be required, the entire procedure of switching and adjusting must be repeated.

The end result is that this method of control is somewhat cumbersome, slow, and requires a proficient, experienced operator to set up the equipment. Its advantages are a relatively low, initial cost.

Mod-Logic. Pneumatic controls, known as "fluidics" or "mod-logic," have been used in recent years. As with the electro-mechanical system, air limit switches are used to interlock the machine functions. Individual timers are employed for each function. Each of the standard thermoforming techniques must be considered in designing this system so that each operating sequence is a part of the device arrangement. Function cannot be altered without circuit changes. Each sequence is set by manipulation of selected switches.

The pneumatic system is highly reliable. Trouble-shooting can be provided by average maintenance men with the use of a simple air pressure gauge to check the various circuits. The major disadvantages of this system is a high initial cost, lack of flexibility, and the need for a clean, oil and dirt-free air supply. Mod-logic is also somewhat cumbersome and requires a proficient, experienced operator to set up.

Computer Control. Solid-state control systems have appeared recently and do an excellent job in automating thermoforming equipment. Sophisticated circuitry is necessary to utilize the solid-state controls to provide a fully flexible system. Here again, as in the electro-mechanical system, changes from one mold to another require switch manipulation.

The solid-state system is extremely reliable and capable of operating millions of cycles virtually trouble-free. It can be quite compact and is absolutely silent in operation.

The principal disadvantage of the solid state computer system is a relatively high initial cost.

Electro-Mechanical Solid State. This system provides precise control of all machine functions employed in thermoforming (Fig. 12-27). The system employs a solid-state control coupled with a drum-type programmer which eliminates multiple relays, individual timers, and limit switches. The control circuitry is simplified and the original equipment investment is considerably below computer type systems.

The standard system controls up to 17 electrically isolated load circuits through 12 sequential steps. The duration of each step is determined by the setting of a calibrated dial on the front panel. The time control for each step may be supplemented by external controls from limit switches, photo-cells or other sensors mounted on the thermoformer.

This combination of time-base and in-point control provides control flexibility with complete equipment and personnel protection. Actuation of each of the load switches is independently controlled by plug-in pins in the program drum. Position of the pins determines the machine function which can be actuated

sequentially, simultaneously or overlapping. The time period for each step can be adjusted for efficient operation with resulting cycle timing for efficient production.

Programming to change the function of the machine for another forming task is relatively simple and fast. The program drum is easily removed and extra drums with pre-set programs may be stored for quick interchangeability. Program changes on a single drum are made by the re-positioning of program pins. Program sheets covering frequently used forming techniques can be used to record specific operating settings for each project.

Forming Temperatures

All thermoplastic materials have specific processing temperatures. Basically, these ranges apply regardless of how the material is being processed. Table 12-1 shows the processing temperature ranges for thermoforming of some popular thermoplastic materials.

Fig. 12-27. Coupling solid state controls with a pin-drum programmer is one form of automation used in modern thermoforming machines. The system provides timed and sequenced control for the complete thermoforming cycle. (*Courtesy AAA Plastic Equipment, Inc.*)

Table 12-1. Thermoforming Processing Temperature Ranges

Material	Set Temperature (°F)	Lower Processing Limit (°F)	Normal Forming Temperature (°F)	Upper Limit (°F)
HDPE	180	260	295	360
ABS	185	260	295	360
Acetate	160	260	310	360
Acrylic	185	300	350	380
Acrylic/PVC	175	325	370	400
Styrene	185	260	295	360
Polycarbonate	280	335	375	400
PVC	150	200	245	300

1. *Set Temperature.* The temperature at which the part may be removed from the mold without warpage. Sometimes parts can be removed at higher temperatures if cooling fixtures are used.
2. *Lower Processing Limit.* This represents the lowest temperature that the material can be formed without creating undue stresses. This means that the sheet material should touch every corner of the mold before it reaches this lower limit. Material processed below the lower limit will have stresses and strains that later could cause warpage, brittleness or other physical changes in the finished item.
3. *Normal Forming Temperature.* Temperature at which the sheet should be formed under normal operation. This temperature should be reached *throughout* the sheet. Shallow draw projects with fast vacuum and/or pressure forming will allow somewhat lower sheet temperature and thus a faster cycle. Higher temperatures are required for deep draws, prestretching operations, detailed molds, etc.
4. *Upper Limit.* The temperature point where the thermoplastic sheet begins to degrade or becomes too fluid and pliable to thermoform. These temperatures normally can be exceeded only with an impairment of the material's physical properties. Injection molding and extruding use much higher temperatures but only for short durations.

Thermoforming Molds

One of the most attractive features of the thermoforming process is the lower cost of producing tooling for the process because of the low pressures involved, relative to other methods of processing plastics. Tooling can range from hardwood molds used for short runs to epoxy, urethane, and cast aluminum for production runs.

The most common mold material is cast aluminum which provides a good combination of durability, light weight, and thermal conductivity, as well as ease of manufacture.

Commonly used materials include gypsum, softwood, hardwood, pressed wood, cast phenolic resins, filled or unfilled high-temperature polyester or epoxy resins, cast solid urethane, sprayed metal, cast aluminum, cast porous aluminum, and machined steels.

In tooling design, a male primary mold will allow deeper draw than a female mold since the plastic can be draped or prestretched over it. However, when a male plug assist is used to pre-stretch the material for a primary female mold, the advantage is nullified.

In general, female molds provide easier release, are less likely to get scratched or damaged, produce thicker and stronger rims in containers, can use smaller sheet blanks, and provide sharpest definition on the outside of objects. Ordinarily, female molds have the disadvantage of producing products with thin bottoms; however, good plug assist design and operation can largely eliminate this problem. Male molds are generally cheaper to produce.

All thermoforming molds used with vacuum or air pressure techniques require holes, channels or ducts for the evacuation of air or the build-up of pressure. The holes, or slits, must be kept small, usually 0.010 to 0.025 in. to avoid visible marks on the surfaces of parts.

Vacuum holes are usually kept to the minimum needed for full, quick and uniform air removal. Careful placement of the holes will be helpful in providing fast, efficient air flow in forming. A slot instead of a hole should be used where possible for fast air passage.

In cast resin molds, vacuum holes can be provided by including greased wires in the casting for later removal.

For greater detail such as graining, stitching, and relief work, cast porous aluminum molds should be used.

Undercuts in thermoforming molds can be handled by the use of split molds. Some molds use removable parts that pull out of the mold after forming. Both methods allow release of the formed object. The first system is used with containers, the latter is more common in the manufacture of complex parts. With some materials, the natural elasticity of the plastic allows the part to be pulled forcefully from the undercut.

In the design of molds, it is common practice to include a 2 to 3° draft angle for female molds and a 5 to 7° angle for male molds. A straight side angle in the direction of the draw makes the parts difficult to release. This is especially true with male molds, where the natural shrink of the plastic material is toward the mold. With advanced forming techniques, parts with up to zero degrees of draft and a one and one-half to one forming ratio have been successfully produced. In female molds, the tendency is to shrink away from the mold.

Various sheet materials have different mold shrinkage factors ranging from almost no shrinkage up to as much as 4%. The percent of shrinkage is not as important as the consistency of the factor. Molds may be designed to allow for shrinkage to the proper size. For precision work, careful pre-testing is required. Cooling conditions will also affect the rate of shrinkage of the material. Restraining the part, either before or after release, will tend to limit total shrinkage. Mold temperature, cooling speed, and cooling fixtures are factors that should remain constant in order to maintain uniformity of specification.

Finishing Thermoformed Parts

In finishing any plastic, primary requirements are sharp tools and clean surfaces. Thermoplastics may be sawed, filed, wet or dry sanded, lapped, buffed, and polished. Generally, cuts should be as coarse as possible to achieve the desired finish. Where extreme smoothness is desired, start with a coarse cut and finish with finer cuts.

In the thermoformed part, the first step is trimming the part from the formed sheet. This may be accomplished with hand or power tools such as routers, band saws, tablesaws, arbor presses, punch presses, etc., or in production runs with special steel rule or punch press dies and heated platens. Matched metal dies are utilized for trimming where volume will allow amortization of die costs.

While one of the values afforded by thermoplastics is the molded-in color, surface marking or coating may be desirable. Materials for painting or printing plastics are readily available which are suitable for silk screening, spraying, dipping, vacuum depositions or electrostatic means. Most difficulties accrue from lack of cleanliness of the article or items used to decorate it.

Other problems are created by improper selection of solvent materials. Rapid drying solvents that chill the surface can cause blushing. Uniform, even drying and controlled humidity are important control factors in this finishing area.

Crazing is caused by solvents cutting along lines of strain in the formed article. These strains come from stresses set up in the forming process and are difficult to control. Forming at higher temperatures, slower cooling, or annealing will often prevent this difficulty. But crazing may call for design changes.

Process Variables

The thermoforming process has many variables which will affect the appearance, quality, size and material distribution in the formed part. An understanding of these variables can often solve difficult production problems encountered in the thermoforming process. They will also give an indication of the forming technique best suited to the specific project at hand.

Variation in Sheet Plastic:

Sheet Thickness. Allowable variation in the thickness of plastic sheet should be held to not more than 4 to 8%. Where greater variation is found, a slower heating cycle should be considered.

When variations are found within a blank, a slower and longer heating cycle is suggested, allowing heat to soak through the thick areas to provide uniform temperature at both thick and thin spots. The probability of a good finished part is improved with uniform temperatures throughout the sheet.

Where tolerances are too great, "blow-outs" at excessively thin spots will result.

Sheet Viscosity or Melt Index. A wide change in viscosity from one batch to the next can create forming problems. This is especially true of polyolefins.

If the melt viscosity (lower "melt index") is high (requires higher temperature to reach forming state), impact strength increases, hot strength improves, elongation improves, and mold shrinkage increases.

Regrinding and re-extruding tend to reduce melt index (raise the viscosity). For this reason, reground trim should be mixed with virgin material when possible to stabilize the over-all properties.

Sheet Density. Increased density of a specific material increases shrinkage, tensile strength, hardness, and stiffness.

Sheet Orientation. The "Stretch or Strain" effect caused when the sheet is extruded or calendered affects the forming operation. The condition is created when a stretching is set up between the polishing rolls and the take-off rollers of the extruding equipment. Orientation in the cross direction of extrusion as well as in the machine direction can be caused by calen-dering (controlling gage with the cooling-polishing rolls).

Generally, a minimum of orientation is desirable. A simple orientation test may be conducted heating 1 x 10 in. strips of the sheet in both extrusion and cross machine directions to forming temperatures in a free, unrestrained state. The amount of shrinkage or expansion gives a good measure of orientation within the sheet.

Orientation normally runs from 12 to 15% in the extrusion direction and from 2 to 5% in the cross direction in a free state. Very low orientation sheet can be obtained with 5% or less shrinkage for pre-screening work.

With orientation, close tolerance parts can be produced so long as the strain remains consistent from batch to batch.

Sheet Temperature. As a general rule, to achieve the fastest cycle, the sheet is formed on the lower side of the forming range of the particular material.

For example, ABS forms at sheet temperatures as low as 260°F and as high as 360°F. Simple draws can be made at temperatures of 285-298°F *where fast vacuum is used*; normal to deep draws at about 300°F and complex parts at 330°F. The lower the temperature within this range (providing the desired detail), the more preferable it is considered. Again, the sheet should definitely reach all portions of the mold before the sheet temperature drops to the lower forming limit

With many types of thick sheet, best results are obtained when a delay of a number of seconds between heating and forming can be introduced. This allows the sheet to "soak," giving a more uniformly heated sheet and relieving of strains.

Pre-Stretch Variables:

On deep draw ratios, pre-stretching through drawing or blowing a uniformly shaped bubble is necessary to obtain even material distribution in the finished article. The bubble size should stretch the plastic to the approximate thickness desired in the bottom of the formed part. Once the mold touches the plastic, little stretching takes place at those points of contact.

Unevenly blown bubbles, caused by uneven

heating, variations in sheet thickness, or a non-uniform raw material mixture, are an immediate cause of trouble.

Normally, the maximum, practical height of a bubble is about one-half to three-quarters the shorter dimension of the clamped sheet. A delay of several seconds after the desired bubble size is reached will allow the bubble to complete its shape. This permits the warmer (thicker) portions of the sheet to continue stretching while the cooler (thinner) portions contract.

In drawing a bubble into a vacuum box, the sides of the box should be 3 to 5 in. larger than the desired bubble to avoid chilling the outer areas of the sheet from the cold side-walls of the box.

Air Temperature. In some parts, a chilling problem may develop from using room temperature blowing air. With pre-heated air, temperature should remain about 10° below the sheet temperature to avoid heating the plastic and thus affecting its shape. A diffuser or deflector over the air entry hole prevents spot chilling.

Plugging Variables:

Plugging involves the use of a "plug-assist" to carry sheet material into a mold cavity and/or to stretch the portions of the sheet which will form the sides of the finished part.

Plug Shape. The plug should conform closely to the shape of the mold cavity but should be smaller in size. With molds 5 in. or more, the plug should be from 10 to 20% smaller in length and width. A $\frac{1}{2}$ in. clearance between plug and mold cavity when the plug bottoms is desirable. Smaller plug-assist molds should allow at least $\frac{1}{4}$ in. clearance between the finished part and the plug to avoid non-uniform thinning of the material. As the plug enters the mold area, pressure is built up, keeping the hot sheet material away from the mold and against the plug until vacuum is applied. As clearance decreases, pressure in the mold cavity increases and the plug mechanically forces the sheet against the mold instead of allowing the vacuum to shape the plastic.

The plug should follow the mold contour with ridges following the contours of grooves in the mold. Such ridges will carry extra material into the grooves for added thickness. For pockets in the mold, there should be corresponding projections in the plug. In the case of deep recesses in mold sidewalls, it may be desirable to incorporate cam-actuated plugs designed to carry more material into these areas. Normally corners of the plug should have generous radii.

Plug Material. Strip heated aluminum is one of the best materials; however, aluminum-filled epoxy with cartridge heaters or even hardwood plugs covered with flannel or flocked suede rubber work well. If the plug is not heated, a non-heat-conducting material should be used. Wood plugs, not covered with flannel, should be greased or Teflon (du Pont) sprayed frequently to prevent drying out and splitting and to give better slippage.

Plug Temperatures. Surface temperatures should be maintained at or just below the temperature of the sheet being formed. While temperatures are not as critical in the plug as in the mold, temperatures too low will cause "chill" marks and thick sections on the finished part. A Teflon spray coating on the plug will avoid sticking and assist slippage. Cloth covered plugs are usually not heated.

Plug Height. Height should be greater than the cavity depth of the mold to provide adequate adjustment. The best ratio of bottom and wall thickness is usually achieved with plug penetration of 70 to 85% of the mold cavity.

Plug Speed and Amount of Vacuum. Normally, vacuum bleed in the mold should begin when the plug first touches the bubble. Plug speed should balance the rate of vacuum bleed. If the plug speed exceeds the vacuum bleed, air compression in the mold cavity is increased. A "roll of material" should follow the plug into the mold cavity, thus keeping the sheet from touching the relatively cold mold until the last second.

Sheet Materials in Plug-Assist Forming. High hot-strength materials such as acrylic and ABS require relatively high pressures (15 to 50 psi) in the mold cavity to maintain a proper "roll of material" around the plug assist. Lower hot-strength materials will require only 2 to 6 psi in forming.

Mold Variables:

Vacuum Holes. Normally the faster a vacuum is drawn, the better the part. Slow vacuum causes the sheet first touching the mold to chill faster than the balance of the sheet, usually producing thin sections or incompletely formed parts. For this reason, instantaneous air evacuation would be considered ideal.

In mold construction, the vacuum is drawn through holes or slots in the surface of the mold. Hole diameter should be small enough not to show on the finished part. When forming thin gage materials such as polyolefins, small diameter holes (0.021 to 0.0135 in.) should be used. Higher hot strength materials and thicker materials can permit larger vacuum holes (0.035 to 0.020 in.). Where fine detail is needed, vacuum holes should be spaced as close as $\frac{1}{4}$ in. apart or cast aluminum porous molds used. On large, flat surfaces, 1 to 3 in. spacing is practical.

To increase speed of draw, tiny vacuum holes may be countersunk on the reverse side to within $\frac{1}{4}$ in. of the mold surface. Where possible, long slots improve air evacuation speed. In forming door liners for refrigerators, vacuum draw time has been reduced to $\frac{1}{2}$ sec using slots as compared with 2 to 5 sec with normal vacuum holes.

Mold Surface. In thermoforming, the finished part will acquire the appearance of the surface of the mold. A matté-finished mold will yield a dull finish to the part. A highly polished mold will yield a glossy part.

When forming low hot strength materials, such as the polyolefins, a smooth mold is difficult to use. Air trapped between the hot-fluid-like sheet and the mold results in "pock-marked" surfaces on the finished part. Roughened, sand-blasted cast aluminum provides an excellent mold surface. Large numbers of closely spaced small vacuum holes will help on a smooth surface.

Mold Temperatures. The temperature of mold surfaces play an important role in thermoforming. Normally, mold temperatures should be just below the "set" temperature or the heat distortion point of the material (e.g., ABS: 160-185°F; high density polyethylene: 165-190°F; styrene: 150-185°F; vinyl: 120-145°F).

Mold temperatures affect the appearance of the formed part, length of the forming cycle, and size of the finished part. The ratio of final shrinkage to the mold temperature is approximately the same as the thermal coefficient of expansion of the material. For example, a 25°F mold temperature change while forming 30 in. long high-density polyethylene trays will alter its length by $\frac{1}{16}$ in. Increases in mold temperature increases part shrinkage and cycle time of the forming technique.

Trouble-Shooting Guide

The thermoforming trouble-shooting guide presented on p. 308 attempts to pull all these (and other) variables into perspective.

The problems covered include:

1. Blisters or bubbles.
2. Incomplete forming or poor detail.
3. Sheet scorched.
4. Blushing or change in color intensity.
5. Whitening of sheet.
6. Webbing, bridging, or wrinkling.
7. Nipples on mold side of formed part.
8. Too much sag.
9. Sag variation between sheet blanks.
10. Chill marks or "mark-off" lines on part.
11. Bad surface markings.
12. Shiny streaks on part.
13. Excessive post shrinkage or distortion of part after removing from mold.
14. Part warpage.
15. Poor wall thickness distribution and excessive thinning in some areas.
16. Non-uniform pre-stretch bubble.
17. Shrink marks on part, especially in corner areas (inside radius of molds).
18. Too thin corners in deep draws.
19. Part sticking to mold.
20. Sheet sticking to plug assist.
21. Tearing of part when forming.
22. Cracking in corners during service.

New Developments

While the thermoforming process gained popularity during the fifties and early sixties, the late sixties and seventies has been a period of

Trouble-Shooting Guide for Thermoforming*

Problem	Probable Cause	Suggested Course of Action
1. Blisters or bubbles	A. Heating too rapidly	1. Lower heater temperature 2. Use slower heating 3. Increase distance between heater(s) and sheet
	B. Excess moisture	1. Predry 2. Preheat 3. Heat from both sides 4. Do not remove material from moisture-proof wrap until ready to use 5. Obtain dry material from supplier
	C. Uneven heating	1. Screen by attaching baffles, masks, or screen 2. Check for heaters or screens out
	D. Wrong sheet type or formulation	1. Obtain correct formulation
2. Incomplete forming, poor detail	A. Sheet too cold	1. Heat sheet longer 2. Raise temperature of heaters 3. Use more heaters 4. If problem occurs repeatedly in same area, check for lack of uniformity of heat
	B. Clamping frame not hot before inserting sheet	1. Preheat clamping frame before inserting sheet
	C. Insufficient vacuum	1. Check vacuum holes for clogging 2. Increase number of vacuum holes 3. Increase size of vacuum holes
	D. Vacuum not drawn fast enough	1. Use vacuum slots instead of holes where possible 2. Add vacuum surge and/or pump capacity 3. Enlarge vacuum line and valves avoiding sharp bends at tee and elbow connections 4. Check for vacuum leaks 5. Check vacuum system
	E. Additional pressure needed	1. Use 20-30 psi air pressure on part opposite mold surface if mold will withstand this pressure 2. Use frame assist 3. Use plug, silicone slab rubber, or other pressure assist
3. Sheet scorched	A. Outer surface of sheet too hot	1. Shorten heat cycle 2. Use slower, soaking heat
4. Blushing or change in color intensity	A. Insufficient heating	1. Lengthen heating cycle 2. Raise temperature of heaters
	B. Excess heating	1. Reduce heater temperature 2. Shorten heater cycle 3. If in same spot on sheet, check heaters
	C. Mold is too cold	1. Warm mold
	D. Assist is too cold	1. Warm assist
	E. Sheet is being stretched too far	1. Use heavier gauge sheet or more elastic, deep draw formulation 2. Change mold design

* Prepared by Wm. K. McConnell, Jr., President, McConnell Co., Inc., Ft. Worth, Texas, 76110.

Trouble-Shooting Guide for Thermoforming (*Continued*)

Problem	*Probable Cause*	*Suggested Course of Action*
	F. Sheet cools before it is completely formed	1. Move mold into sheet faster 2. Increase rate of vacuum withdrawal 3. Be sure molds and plugs are hot
	G. Poor mold design	1. Reduce depth of draw 2. Increase draft (taper) of mold 3. Enlarge radii
	H. Sheet material not suitable for job	1. Try different sheet formulation or a different plastic material
	I. Uncontrolled use of regrind	1. Control percentage and quality of regrind
5. Whitening of sheet	A. Cold sheet stretching beyond its temperature yield point	1. Increase heat of sheet; increase speed of drape and vacuum
	B. Sheet material dry colored	1. If above action won't correct, check with sheet supplier for availability of other types of coloring. Some colors do not lend themselves to dry or concentrate coloring. 2. A hot air gun can be used to diminish or eliminate whitened surfaces on formed part.
6. Webbing, bridging or wrinkling	A. Sheet is too hot causing too much material in forming area	1. Shorten heating cycle 2. Increase heater distance 3. Lower heater temperature
	B. Melt strength of resin is too low (sheet sag too great)	1. Change to lower melt index resin 2. Ask sheet supplier for more orientation in sheet 3. Use minimum sheet temperature possible
	C. Too much or too little sheet orientation	1. Have sheet supplier reduce or increase orientation
	D. Insufficient vacuum	1. Check vacuum system 2. Add more vacuum holes or slots
	E. Extrusion direction of sheet parallel to space between molds	1. Move sheet 90° in relation to space between molds
	F. Draw ratio too great in area of mold or poor mold design or layout	1. Redesign mold 2. Use plug or ring mechanical assist 3. Use female mold instead of male 4. Add take-up blocks to pull out wrinkles 5. Increase draft and radii where possible 6. If more than one article being formed, move them farther apart 7. Speed up assist and/or mold travel 8. Redesign grid, plug or ring assists
7. Nipples on mold side of formed part	A. Sheet too hot	1. Reduce heating cycle 2. Reduce heater temperature 3. Plug holes and redrill with smaller bit
	B. Vacuum holes too large	
8. Too much sag	A. Sheet is too hot	1. Reduce heating cycle 2. Reduce heater temperature
	B. Melt index is too high	1. Use lower melt index resin or different resin

(*Continued*)

Trouble-Shooting Guide for Thermoforming (*Continued*)

Problem	Probable Cause	Suggested Course of Action
		2. Have sheet supplier put more orientation in sheet
	C. Sheet area is too large	1. Use screening or other means of shading or give preferential heat to sheet, thus reducing relative temperature of center of sheet
9. Sag variation between sheet blanks	A. Variation in sheet temperature	1. Check for air drafts through oven using solid screens around heater section to eliminate
	B. Sheet made from different resins	1. Control regrind percentage and quality 2. Avoid resin mix-ups
10. Chill marks or "Mark-off" lines	A. Plug assist temperature too low	1. Increase plug assist temperature 2. Use wood plug assist 3. Cover plug with cotton flannel or felt
	B. Mold temperature too low—stretching stops when sheet meets cold mold (or plug)	1. Increase mold temperature, not exceeding "set temperature" for particular resin 2. Relieve molds in critical areas
	C. Inadequate mold temperature control	1. Increase number of water cooling tubes or channels 2. Check for plugged water flow
	D. Sheet is too hot	1. Reduce heat 2. Heat more slowly 3. Slightly chill surface of hot sheet with forced air before forming
	E. Wrong technique	1. Use different forming technique
11. Bad surface markings	A. Pock marks due to air entrapment over smooth mold surface	1. Grit blast mold surface 2. Add many additional vacuum holes.
	B. Poor vacuum	1. Add vacuum holes 2. If pock marks are in isolated area, add vacuum holes to this area or check for plugged vacuum holes
	C. Mark off due to accumulation of plasticizer on mold when using sheet with plasticizers	1. Use temperature controlled mold 2. Have mold as far away from sheet as possible during heating cycle 3. If too long, shorten heating cycle 4. Clean mold
	D. Mold is too hot	1. Reduce mold temperature
	E. Mold is too cold	1. Increase mold temperature
	F. Improper mold composition	1. Avoid phenolic molds with clear transparent sheet 2. Use aluminum molds where possible
	G. Mold surface is too rough	1. Smooth surface 2. Change mold material
	H. Dirt on sheet	1. Clean sheet
	I. Dirt on mold	1. Clean mold
	J. Dust in atmosphere	1. Clean thermoforming area; isolate area if necessary and supply filtered air
	K. Contaminated sheet materials	1. If regrind is used be sure to *keep* clean and different materials stored separately

Trouble-Shooting Guide for Thermoforming (*Continued*)

Problem	Probable Cause	Suggested Course of Action
	L. Scratched sheet	2. Check, supplier of sheet 1. Separate sheets with paper in storage 2. Polish sheet
12. Shiny Streaks on Part	A. Sheet overheated in this area	1. Lower heater temperature in scorched area 2. Shield heater with screen wire to reduce overheating 3. Slow heating cycle 4. Increase heater to sheet distance
13. Excessive shrinkage or distortion of part after removing from mold	A. Removed part from mold too soon	1. Increase cooling cycle 2. Use cooling fixtures 3. Use fan or vapor spray mist to cool part faster on mold
	B. Mold too hot	1. Lower mold temperature (the hotter the mold, the more the final part shrinkage)
14. Part warpage	A. Uneven part cooling	1. Add more water channels or tubing to mold 2. Check for plugged water flow
	B. Poor wall distribution in part	1. Improve pre-stretching or plugging techniques 2. Use plug assist 3. Check for non-uniformity of sheet heating 4. Check sheet gauge
	C. Poor mold design	1. Add vacuum holes 2. Add moat to mold at trim line 3. Check for plugged vacuum holes
	D. Poor part design	1. Break-up large flat surfaces with ribs where practical
	E. Mold temperature too low	1. Raise mold temperature to just below "set-temperature" of sheet material
	F. Part removed before cooled to set-temperature	1. Cool to below set-temperature
	G. Part cold formed	1. Heat sheet to proper temperature and apply fast vacuum
15. Poor wall thickness distribution and excessive thinning in some areas	A. Improper sheet sag	1. Use different forming technique such as mounting mold on top platen 2. Use vacuum snap-back technique 3. Use reverse vacuum snap back 4. Use billow-up plug assist or vacuum snap-back 5. Use different melt index resin 6. Try more orientation in sheet
	B. Variations in sheet gauge	1. Consult supplier regarding his commercial tolerances and improve quality of sheet
	C. Hot or cold spots in sheet	1. Improve heating technique to achieve uniform heat distribution; screen or shade as necessary 2. Check to see if all heating elements are functioning
	D. Stray drafts and air currents around machine	1. Enclose heating and forming areas

(*Continued*)

Trouble-Shooting Guide for Thermoforming (*Continued*)

Problem	Probable Cause	Suggested Course of Action
	E. Too much sag	1. Use screening or other temperature control of center areas of heater banks 2. Use lower melt index resin 3. Use more orientation in sheet
	F. Mold is too cold	1. Provide uniform heating of mold to bring to proper temperature 2. Check temperature control system for scale or plugging
	G. Sheet slipping out of frame	1. Adjust clamping frame to provide uniform pressure 2. Check for variation in sheet gauge 3. Heat frames to proper temperature before inserting sheet 4. Check for non-uniformity of heat giving cold areas around clamp frame
16. Non-uniform pre-stretch bubble	A. Uneven sheet gauge	1. Consult sheet supplier 2. Heat sheet slowly in a "soak" type heat
	B. Uneven heating of sheet	1. Check heater section for heaters out 2. Check heater section for missing screens 3. Screen heater section as necessary
	C. Stray air drafts	1. Enclose or otherwise shield or screen machine
	D. Non-uniform air blow	1. Baffle air inlet in pre-stretch box
17. Shrink marks on part, especially in corner areas (inside radius of molds)	A. Inadequate vacuum	1. Check for vacuum leaks 2. Add vacuum surge and/or pump capacity 3. Check for plugged vacuum holes 4. Add vacuum holes
	B. Mold surface is too smooth	1. Grit blast mold surface
	C. Part shrinking away	1. May be impossible to eliminate on thick sheet with vacuum only; use 20-30 psi air pressure on part opposite mold surface if mold will withstand this pressure
18. Too thin corners in deep draws	A. Improper forming technique	1. Check other techniques such as billow-up plug assist, etc.
	B. Sheet is too thin	1. Use heavier gauge
	C. Variation in sheet temperature	1. Adjust heating as needed by adding screens to portion of sheet going into corners 2. Cross hatch sheet with markings prior to forming so movement of material can be accurately checked
	D. Variation in mold temperature	1. Adjust temperature control system for uniformity
	E. Improper material selection	1. Consult sheet supplier or raw material supplier
19. Part sticking to mold	A. Part temperature too high	1. Increase cooling cycle 2. Slightly lower mold temperature, not much less than recommended by resin manufacturer

Trouble-Shooting Guide for Thermoforming (*Continued*)

Problem	Probable Cause	Suggested Course of Action
	B. Not enough draft in mold	1. Increase taper 2. Use female mold 3. Remove part from mold as early as possible; if above "set temperature," use cooling jigs
	C. Mold undercuts	1. Use stripping frame 2. Increase Air-eject air pressure 3. Remove part from mold as early as possible; if above "set temperature," use cooling jigs
	D. Wooden mold	1. Grease with vasoline 2. Use Teflon (du Pont) spray
	E. Rough mold surface	1. Polish corners of mold 2. Use mold release 3. Use Teflon (du Pont) spray
20. Sheet sticking to plug assist	A. Improper metal plug assist temperature	1. Reduce plug temperature 2. Use mold release 3. Teflon (du Pont) coat 4. Cover plug with felt cloth or cotton flannel
	B. Wooden plug assist	1. Cover plug with felt cloth or cotton flannel 2. Grease with vasoline 3. Use mold release compound 4. Use Teflon (du Pont) spray
21. Tearing of sheet when forming	A. Mold design	1. Increase radius of corner
	B. Sheet is too hot	1. Decrease heating time or temperature 2. Check for uniform heat 3. Preheat sheet
	C. Sheet is too cold (usually thinner gauges)	1. Increase heating time or temperature 2. Check for uniform heat 3. Preheat sheet
22. Cracking in corners during service	A. Stress concentration	1. Increase fillets 2. In transparencies check with polarized light 3. Increase temperature of sheet 4. Be sure part is completely formed before some sections are too cool for proper forming; thus setting up undue stresses in these areas 5. Change to a stress crack-resistant resin
	B. "Under-designed"	1. Re-evaluate design

near revolution. The packaging field, long one of industry's most ardent practitioners, has developed new techniques and applications (see p. 314). But, the real growth developed as other major industries found the process uniquely fitted to their needs. The automotive, recreation vehicles, construction, appliance, and furniture manufacturing industries have all moved to thermoformed components.

The dawning of the seventies saw developments that promise even more growth. Industry is currently producing plastic boat hulls up to 20 ft in length with a single thermoforming operation. The formed hull is rigidized using urethane foam.

Another relatively new process is one which combines thermoformed parts with other reinforcing materials. In this process, the thermoformed plastic part receives a backing of glass-reinforced polyester or rigid urethane foam.

The biggest application of this process now is in the plumbing industry. Marbelized and solid colored cross-linked cast acrylic sheets are vacuum formed into sinks, tubs, and shower enclosures. Afterward the thinly formed acrylic parts are rigidized by spraying or press laminating the reversed side. Chopped fibrous glass and polyester resin are used.

The process is also used in the manufacture of large in-the-ground swimming pools. In operation, sections of the pool are thermoformed at a central factory. These sections nest for shipment to installation plants around the nation. Glass-reinforced polyester resins are sprayed on the formed part for rigidizing prior to installation. The process is reported to reduce labor time as much as 95% as compared with conventionally reinforced plastic lay-up practices.

Other industries utilizing this new process include manufacturers of snowmobiles, cattle feeders, boats, golf carts, and large housings. Studies are underway on producing complete modular homes constructed of thermoformed, sprayed-up panels.

Other techniques have emerged to heighten the versatility of specific plastics. Thermoforming of polycarbonate sheet, for example, normally requires oven drying for a number of hours prior to forming. A technique of forming undried polycarbonate sheet for sign faces has recently been developed. This method involves using a combination of radiant and oven heat. Sandwich banks of far infrared heater banks are enclosed completely. The heaters are tuned to an emitting temperature of approximately 450 to 500° F. While the sheet is being heated the oven section is completely closed in. This slowly drives the moisture out and brings the sheet to forming temperature. The heating cycle is longer, but eliminates the pre-drying operation.

Other new techniques, such as twin-sheet forming (p. 290), solid-phase and cold forming (p. 318), and various packaging innovations (following), are discussed elsewhere in this chapter.

SPECIAL THERMOFORMING TECHNIQUES

Packaging*

A wide variety of new thermoforming materials, processing techniques, and specialized equipment permit thermoformers to produce packages that are tailored to the specific applications at hand. Virtually any combination of requirements—barrier properties, structural strength, optical properties, resistance to heat and cold, and seemingly impossible configurations—can be built into a single package at a reasonable cost. No longer limited to relatively simple trays and blisters, thermoformers are reaching into new markets. Plastic jars, cans, and bottles are now being produced on thermoforming lines; thermoformed set-up boxes with integral hinges are challenging for a major paperboard market; and the use of attractive, textured surface finishing techniques is creating the increased market acceptance required to move into new product areas.

The most significant factors in expanding the uses of thermoforming in packaging are the variations of the basic process which allow the processor to overcome previous limitations on thermoforming. A number of these involve the combination of thermoforming with other processing techniques in-line. In one such process, a makeup compact (see Fig. 12-28) is produced in an operation combining thermoforming with dielectric heat sealing. Rigid vinyl is first thermoformed to produce the inner frame for the pan and mirror sections, and an outer section which forms the external shape of the compact.

An outer covering, consisting of a single piece of flexible vinyl, is then dielectrically sealed over both frames. At the same time, a strap is formed which carries a button type snap. The metal snaps are applied automatically in a secondary operation.

The outer covering serves both a decorative and a functional purpose; functional in that the covering acts as an integral, flexible hinge for the compact; and decorative in that it allows·

* By Gerald J. Ardito, President, D.P.I. Deena Packaging, Inc., Long Island City, N.Y.

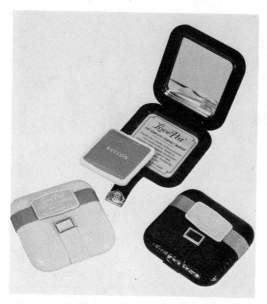

Fig. 12-28. Thermoformed compact has soft feel and realistic surfaces, such as alligator skin, tortoise shell, etc. (*Courtesy D.P.I. Deena Packaging*)

Fig. 12-29. Thermoformed and spin-welded containers. (*Courtesy Brown Machine Div., Leesona*)

the use of a number of highly attractive surface textures and patterns. The processor can duplicate the rich look and feel of alligator, leather, woodgrain, or create a wide variety of other effects.

The tooling costs required for thermoforming and dielectric sealing are considerably less than for injection molding and less than other processes used to fabricate compacts. The molds can be produced in as little as four to eight weeks.

Another processing technique which promises to broaden the markets available to thermoformers is thermoform/spin-welding. The process produces closed containers, such as bottles and jars, which are virtually impossible on standard thermoforming equipment (see Fig. 12-29).

The sequence of operations is as follows: the top and bottom halves of the containers are thermoformed and trimmed from the web; the bottom half of each container is then placed in a holding fixture, open end up, while the top half is placed, open end down, in a rotating mechanism directly above. As the top half is spun at high speed, they are brought together resulting in frictional heat which melt bonds

them. Due to the double thickness of material around the weld line a strong container can be made with less material than a blow-molded container of comparable strength.

The potential uses of thermoform/spin-welding are being increased by the use of a new technique that consists of spraying a coating of liquid saran on the inside of each container half prior to spin-welding. The increased barrier properties obtained in this manner are necessary in many beverage packaging applications.

While the processes thus far discussed involve the addition of secondary fabricating and finishing operations to the thermoforming line, at least two other new techniques combine thermoforming with other primary plastics forming procedures to give the processor added capabilities.

A highly specialized system (developed by Thompson International Corp., Phoenix, Arizona) produces printed container lids at very high speed—as many as 8000/hr. In this process, a 2-in. extruder produces a tube which is collapsed to form a two-layer laminate. The laminate is then thermoformed while still hot. Thus, a key economy is gained by eliminating the need for web heating and temperature control equipment at the forming station.

Another system (from Hoffco) shown in Fig. 12-30, produces thermoformed containers from compression molded preforms. The preform has the outer dimensions of the finished container;

Fig. 12-30. Variation in thermoforming involves the use of compression molded preforms that are then shaped into finished containers. (*Courtesy Hoffco*)

Fig. 12-31. Types of containers produced by system shown in Fig. 12-30. (*Courtesy Hoffco*)

therefore, the process results in a minimal amount of scrap. The machinery required, including completely automated handling of the preforms, is now commercially available in the U.S. for producing a wide variety of containers (Fig. 12-31).

A more recent technique available to formers is a combination injection molding/compression molding/stamping unit. In operation, a small injection molding machine injects into a mold in the first station of an indexing rotary table to form a disc. The table then indexes to the next station where the disc (still hot) is compression molded in matched metal dies to the specific contours and wall thickness desired in the finished part. Next, the disc indexes to the third station where a forming plug comes down and punches it into shape. Final station: eject. The technique has thus far been used for making wide-mouth styrene containers for dairy products, but other types of materials and other types of packages are being evaluated.

Variations of the basic thermoforming process itself are few, but there are several which could have a significant effect on expanding markets. A three-sheet thermoforming process should allow semi-rigid containers to gain a larger share of the market for vacuum and gas-flush packaging of granular products. Traditionally, problems have been encountered in keeping the materials in place during air evacuation or flushing resulting in slow packaging speeds. The new system uses a web of porous film between the two formed container halves to hold the granulate in place. Air can be withdrawn through the porous film without dislodging any of the contents.

Another thermoforming variation which could have a major impact on the packaging market is the development of cold forming techniques which eliminate web heating. The method is currently employed to produce ABS margarine tubs. Because the material is not heated during the forming process, the sheets can be preprinted. The entire container surface can easily be decorated with no subsequent distortion during the forming cycle. (See also p. 318.)

Concurrent with these developments in processing technology have been equally impressive advances in the area of resin development. The list of material properties available to thermoformers is considerably longer than it was just a year or two ago. Leading the excitement in new materials are coextruded sheets. These allow the processor to combine the best properties of several single-resin materials to meet the demands of almost any packaging use. Coextrusions are expected to play a major role in the development of thermoformed food and drug packaging where barrier resistance to moisture and vapors is critical. (See also p. 197).

The so-called barrier plastics are also being extruded into sheet and thermoform/spin-welded to produce bottles. The bottles have a potentially huge application in the carbonated beverage market where the material's high resistance to vapor permeation is a must.

The area of thermoforming technology which is undergoing the least apparent change is that of thermoforming machine design. However, the changes that have occurred have been of prime importance to the production of new shapes and types of packages. Advances in tooling design have been the most significant. Special tooling arrangements can now be used to form parts with severe undercuts, reverse tapers, and threaded necks.

To produce parts with reverse tapers and undercuts, molds using two or three piece female cavities have been developed. As the mold halves are separated vertically, the female mold sections move laterally to allow clearance of the part as it is removed.

Screw caps, a product previously unattainable by thermoforming, can now be thermoformed at rates exceeding 700/min, thanks to the development of a male mold half which rotates to "unscrew" itself from the part after the forming operation.

The other major area of machine improvement has been in temperature control systems. Exact control over sheet temperatures is required to form parts with unusual configurations or deep draws. Control systems have included infrared temperature detectors, photoelectric cells which detect sheet sag, and ovens which establish a temperature profile across the sheet by selectively heating preprogrammed areas of the sheet at different temperatures.

Taken as a whole, the technological developments in processing techniques, materials, and machinery represent formidable weapons with which to attack thermoforming markets. New concepts in package design are also promoting an increased use of thermoforming. For example, a thermoformed set-up box is already beginning to replace the paperboard type which has been the standard of the industry for years. (See Fig. 12-32). The box consists of a formed platform with individual cavities to hold and protect the products during shipping and handling; a polystyrene sheet which is attached to the bottom of the platform to provide structural support; and a thermoformed cover which pivots easily on an integral hinge. Although made from thin gauge sheet, the boxes are strong enough to be shipped cross country in an inexpensive chipboard sleeve

Fig. 12-32. Thermoformed set-up boxes offer equal or superior strength to boxboard containers. (*Courtesy D.P.I. Deena Packaging*)

without danger of damage to either products or package.

The kind of creative package designing evidenced in these applications, coupled with technical advances in processing, materials and machinery, will almost certainly open vast new markets for thermoformed packaging.

Industrial*

One refrigerator manufacturer is using a set of articulated molds to form, in one shot, a double-cavity liner (refrigerator and freezer compartments). In operation, large-size ABS sheets (47 by 71 in.) are indexed to a thermoforming station. A plug-assist pressure bubble method is used, in which the plug assist descends while air pressure forms two bubbles side by side. A hinged mold (with two halves) in spread position then rises inside the ABS bubble as the plug assist is removed. The molds pivot together and a vacuum is drawn.

Another manufacturer also uses a pressure bubble, plug-assist method to produce a steel shell liner (for refrigerators) with ABS formed over the shell. The key is a heat-reactive adhesive sprayed over the interior of the steel shell (which is held in the female mold). The adhesive is activated as the hot ABS sheet is drawn into the shell.

From a boat manufacturer comes this innovation in thermoforming: First, the deck and hull skins are thermoformed of ABS sheet (the boat is a 12-ft catamaran). The formed skins are then placed in a specially designed press with a 5 x 14 ft platen and with highly polished female molds conforming to the shape of the finished hull and deck. The presses are closed and polystyrene beads are forced into the cavity between the skin and press. When steam is applied, the beads expand, pressing the skin against the female molds to smooth out any irregularities. Hull and deck are later glued together. (See Fig. 12-1d).

Perhaps the most important of the new concepts to involve thermoforming has been the so-called *form-and-spray* method. In this technique, sheet is thermoformed into such products as boats, sanitaryware (sinks, vanities,

* Prepared by the Editors

tubs), recreational vehicle parts, etc., then rigidized by spraying reinforced plastics onto the back of the formed piece. Acrylic sheet is most commonly used. As this market has expanded, a number of commercial machines expressly intended to service form-and-spray has been introduced. One is an hydraulic rotary machine, with solid state and electro-mechanical controls, that can turn out acrylic shells for bathtub and shower enclosures (in sizes up to 10 x 20 ft) on a 2-min. cycle. Another is aimed at overcoming the difficulty of removing very large thin-gage acrylic parts (e.g., 40 mils thick and 15 x 20 ft) from the mold prior to rigidizing. In the new machine, the molds are mounted on carts and can be rolled out of the machine, the part rigidized while still on the mold, and then removed.

In other applications, urethane foam is used as the rigidizing element, particularly in boat manufacturing.

SOLID-PHASE FORMING*

In recent years, a considerable amount of research in the thermoforming area has been devoted to modifying conventional metalworking techniques as a means of fabricating thermoplastics materials. Basically, these techniques can be classified as: cold forming (performed at room temperatures with unheated materials and tooling); solid-phase forming (in which the material is heated below melt temperature and formed while in this heated, solid state); and compression molding (where glass-reinforced material is heated above its melt temperature but has the consistency of wet cardboard due to its reinforcement). Within this framework, the following forming methods are available: Forging, which includes closed-die forming, open-die forging, cold heading, etc., stamping, rubber pad or diaphragm forming, coining, brake press bending, rolling, spinning, explosive forming, and drawing.

Some materials that have been successfully solid-phase formed with one or more of these techniques include: ABS, ABS/polycarbonate,

* This review adapted from PLASTEC Report R42, "Solid-Phase Forming (Cold Forming) of Plastics," by J. B. Titus, PLASTEC, Picatinny Arsenal, Dover, N.J.

butyrate, polypropylene, and high molecular weight-high density polyethylene.

By using solid-phase forming, manufacturers can make more efficient use of ultra-high molecular weight, high density plastics and can achieve heavier section designs. These techniques also lend themselves to the production of parts with high rigidity, high strength, and higher impact strength. In terms of economic advantages, they can use existing metalworking equipment with minor modifications; tooling is inexpensive and production rates can be high; also, flash, trim, or weld lines are eliminated and preprinting and decorating can be achieved much easier with some techniques.

The principal disadvantage of these processes is that the advantages must economically outweigh the costs of preparing the billet (e.g., by compression molding, casting, extrusion, etc.) or sheet (e.g., by extrusion, casting, compression molding, etc.) needed for feeding into these systems. Another problem is that many of these parts have to be used at reduced service temperatures to prevent springback or excessive recovery of strains imposed during forming.

The following discussion covers some of the more important of the solid-phase forming and compression molding techniques.

Forging

1. *Closed-Die Forming.* Two opposingly shaped punches mate in a common floating ring to form a closed die. With the upper punch raised, a preheated billet of specific weight and shape is placed on the lower punch in the die ring. The upper punch closes and pressure is then applied, forming the billet to the desired shape. After a short dwell time, the upper punch is raised, the die ring is pulled down over the lower punch, and the lower punch is raised to eject the forging.*

Most of the work has been done with polypropylene, although there are indications that high-density PE, high molecular weight PE, nylon, acetal, ABS, polycarbonate and PVC can also be forged.

2. *Cold Heading.* Continuous rod or bar feed stock is cut to the required length, rough formed in one cavity, and finished formed in a second cavity. It has been used with several thermoplastics to make rivets, bolts, screws, etc.

Sheet Forming

1. *Stamping* employs a rapid application of force and is limited to shallow depth, normally $\frac{1}{4}$ in. or less. Mechanical presses are usually used in this process.*

Glass-reinforced thermoplastic sheet such as nylon, polypropylene and styrene-acrylonitrile has been stamped in conventional mechanical stamping presses into such products as: automotive light housings, outer housings for emission improvement assemblies in automobiles, batter trays, fender liners, luggage, tote boxes, etc.

In a typical compression molding stamping operation, the reinforced sheet is sheared or die cut into blanks somewhat smaller and thicker than the desired part. The blanks are then heated above their melt temperature on a wire or screen carrier in an infrared oven. The hot sheet is then fed to the press and stamped between cooled matched metal dies. Cycles can be between 15 and 20 sec. Dwell time depends upon part thickness and is adjusted to the part design so that it will not distort upon removal from the die. Cooling is achieved by stopping the press at the bottom of its stroke in water-cooled dies.

2. *Rubber Pad Forming.* This is similar to matched-metal stamping except that one of the metal dies is replaced by a block of solid rubber. A heated blank is placed on the rubber block in the lower die and the upper die, containing the mold, presses the blank into the rubber. The rubber is compressed, wiping the sheet into the mold intricacies. Material cannot be flowed to the extent that it can with matched-metal die stamping. However, more uniform pressure is exerted on the blank. A variety of non-reinforced thermoplastics sheet, about 0.040 to 1-in. thick, has been rubber pad formed.

3. In *diaphragm forming*, the rubber pad of the previous process is replaced by a fluid pressurized diaphragm which exerts pressure on the sheet stock. Forming may be adapted to

shallow draws as in stamping or to deeper drawn items. The process involves cheaper tooling cost, relatively high-priced machinery, and moderate pressures. It is advantageous for forming complex parts or large parts requiring only one finished side. Various thermoplastics (e.g., polycarbonate, ABS, cellulosic) have been formed in this way.

4. *Coining* is a mechanical reshaping of part surfaces. It is usually a secondary operation used in conjunction with other solid-phase forming techniques, such as deep drawing. It is accomplished by placing a part over a holding jig and bringing a male die to bear on the surface.

5. *Brake press bending* forms a flat sheet into an angle section by the application of force. The material is placed on the female die of a brake press and formed into shape by the downward movement of the punch or male dies. It has been used to form polyvinyl chloride.

Drawing

In the drawing and deep-drawing process for thermoplastic sheet, the material is converted into shapes with little if any change in sheet thickness. These processes are in contrast to the conventional thermoforming techniques described earlier in which the material is stretched into the desired shape with simultaneous decreases in wall thickness. Because of this advantage, and because they are amenable to continuous, automated or semi-automated operation, the drawing techniques have received considerable attention in the plastics industry.

One of the more successful candidates in this area has been ABS, which is already being cold formed commercially into packages. The cold forming of ABS will be discussed in the following pages.

COLD FORMING ABS*

Cold forming grades of ABS have been developed that can be formed at room temperature, using standard metal working techniques and equipment.

* Prepared by Borg-Warner Chemicals, Parkersburg, W.Va.

In comparison with metal working, cold forming of plastics usually requires lower tonnage. Common die materials and refined diemaking techniques are used. Cold forming grades of ABS can be preprinted in flat sheet. Generally, the as-formed finish is the final finish. Trim can be reprocessed and reused.

Cold drawing, stamping and fluid forming are cold forming techniques applicable with these lightweight ABS resins, offering the metal fabricator a competitive plastic material for many applications, at equal or better production rates than metal.

Cold Drawing

Cold forming grades of ABS can be cold drawn into most shapes made by metal forming methods. A few other thermoplastics can be cold drawn but most are too brittle or lack the dimensional stability to be processed in this manner.

The ductility characteristics of metal and plastic are measureably different (Fig. 12-33) and while ductility at room temperature is not generally associated with plastics, cold forming grades of ABS are ductile and tough enough to withstand the forces necessary to produce up to a 45% diameter reduction by cold drawing.

Often this initial drawing operation is the only reduction necessary to achieve a desired depth, but in cases that may require an additional drawing operation using a multiple die arrangement, sequencing or progressive press operations are used. Cold forming grades of ABS can be redrawn up to 35% reduction of the diameter resulting from the first draw. Further re-drawing operations, up to the 35% reduction in diameter may be performed to achieve final part diameter.

The punch nose radius of the first draw operation must be developed to suit the conditions required (for closed end features) of the second draw operation. With the proper first draw radius or configuration, the finished part of the second draw will not have a distorted wall. The part size from the first draw will be larger in diameter at the open end than the tool punch and die members used to form the part.

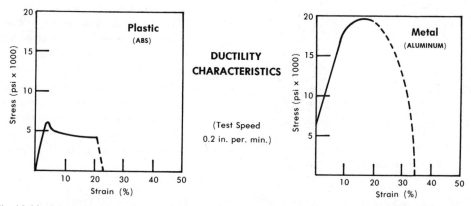

Fig. 12-33. Ductility characteristics of plastic (ABS) vs. metal. (*Diagrams, Courtesy Borg-Warner Chemicals*)

This must be considered when designing the redraw operation.

Figures 12-34 and 12-35 show the chronological steps in each of two drawing operations using ABS sheet. Figure 12-36 illustrates the restrike and trim operation.

Figure 12-37 represents the sequential steps in a two-draw operation beginning with cutting of the drawing blank and concluding with a cutting and stacking operation. The first and second draw in Fig. 12-37 can be related to Figs. 12-34 and 12-35, respectively.

Blank Development. The blank is determined by the design and number of operations required to produce a cold drawn part. One method of blank development is to calculate the surface area of the finished part and use this area to establish the approximate diameter of

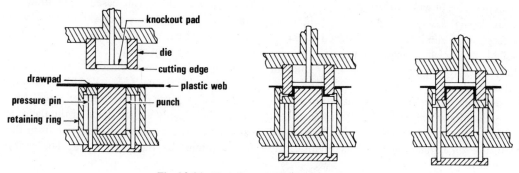

Fig. 12-34. First draw shown at three stages.

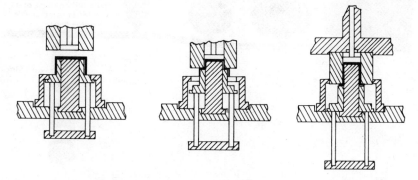

Fig. 12-35. Second draw shown at three stages.

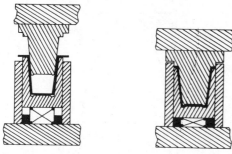

Fig. 12-36. Restrike and trim operation.

the blank, i.e., a finished part having a total surface area of 15 sq in. will have a blank area of 15 sq in., or equivalent to 4.361 in. blank diameter for a round drawn part.

A general rule, when using cold forming grades of ABS, is that a 45% diameter reduction is maximum for the first draw, and up to 35% diameter reduction is possible for subsequent drawing operations.

Stress Whitening. Stress whitening is the whitening effect that frequently appears on highly stressed areas of parts formed from ABS sheet. It generally occurs with all colored materials except white or off-white.

In some cases, however, such whitening effect can be eliminated or minimized either by using a special pigmented material or by keeping part configurations to the minimum stress level.

Reclaim Material. Reclamation of defective cold drawn parts is entirely feasible. If kept clean, they may be reground and reprocessed into the same grade sheet without significantly affecting its physical properties. A blend ratio up to 20% reclaim material mixed with 80% virgin pellets is recommended. Printed areas, of course, cannot be reground and reused.

Lubrication. For successful drawing of ABS plastics, lubrication is essential. Generally, lubricants fall into two categories—aqueous and nonaqueous. The nonaqueous type is more apt to be used as a pre-coat prior to coiling and the aqueous is preferred on at-the-press applications.

Helpful Hints for Drawing ABS Plastics

Surface Condition Control. The sheet surface to be used in drawing must be completely free of flaws such as score marks, spills, blemishes, or other surface defects. Surface irregularities impair control of the draw and result in an uneven flow and deformation of the material. Tearing follows.

Flaws in the Blank Periphery. The presence of burrs, nicks, or any raggedness on the outer edges of blanks will cause an effect much the same as that of earing metal when drawn. These flaws will be increased from draw to draw. The end result is a cracked uneven wall and a distorted rim atop the part. Blanking tools should be kept sharp.

Earing Tendencies in Plastics. An irregular, square-shaped flange (Fig. 12-38) is characteristic of a part drawn from oriented ABS sheet. The roughly square shape of the flange is caused by excess pulling of material from the flange where the flat sides occur. The earing

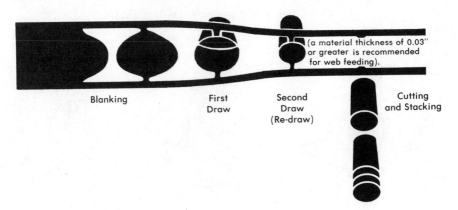

Blanking   First Draw   Second Draw (Re-draw)   (a material thickness of 0.03" or greater is recommended for web feeding).   Cutting and Stacking

Fig. 12-37. Representative cold drawing operation.

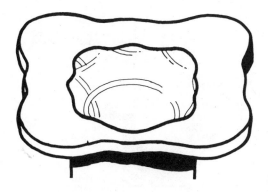

Fig. 12-38. Irregular, square shaped flange is characteristic of part drawn from oriented ABS sheet.

pinching of the wall area and crowd excess plastic into the flange (Fig. 12-39).

Punch and Die Radii

The radii of the punch and the mouth of the forming die should be of sufficient size and be smoothly blended to permit full freedom of material flow. Too sharp a radius on either the punch or the die opening will hinder the normal flow of material, causing uneven thinning of the part walls and eventually material failure. The drawing die, radii, blank holder and drawpad

effect becomes more pronounced with each redraw.

Blank Diameter Reduction Ratio. The reduction in diameter from blank to part should not be too great in a single drawing operation. Excessive reduction is accompanied by extreme thinning of material to the point where the bottom tears away as the punch extends. The area across the bottom of the punch is insufficient to overcome resistance to material flow at the die mouth. Reduction should rarely exceed 45% in the initial drawing operation and 35% in subsequent redraws.

In drawing metal parts, the reduction percentage is generally governed by a T/D factor (material thickness over the blank diameter). As an example, a T/D factor of 0.04 can be drawn to a 45% reduction, a T/D factor of 0.01 can be drawn to a 35% blank reduction.

With cold forming grades of ABS, however, a sheet having 0.015 in. thickness and a T/D factor of 0.023 can be drawn up to a 45% reduction. In terms of the number of operations involved, such blank reduction can prove to be significant in total production cost.

Gauge of Material

The use of uniform sheet stock that is within specified gauge tolerance limits is vital to the success of any drawing operations.

Deviations above or below these limits result in deformations of the part which is progressively magnified in redrawing. Improper punch and die clearance will cause thinning or

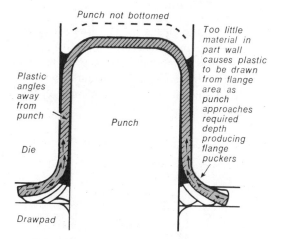

Drawing operations are adversely affected by plastic below gage limits.

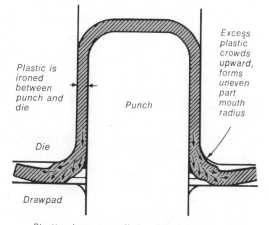

Plastic above gage limits distorts part mouth.

Fig. 12-39. Effects of material gage on drawing operations.

should be smooth and highly polished with no radial rings or lines, scratches or galling.

Alignment of Punch and Die

The punch must be centered and correctly aligned with the die for even distribution of material flow around the part being formed. Any degree of misalignment prevents material flow, distorts the flange, and causes non-uniform wall thickness and part geometry (Fig. 12-40). Parts thus formed cannot be redrawn successfully.

Drawpad Pressure

The pressure setting on the drawpad must be sufficient to maintain drawing control, yet not so great as to prevent plastic from flowing into the die. This setting is critical and only with the correct amount of pressure can even, controlled drawing be accomplished.

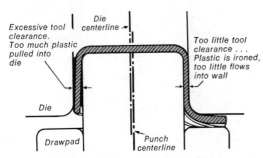

Fig. 12-40. Alignment of punch and die.

When too little pressure is used, loss of control permits an uneven flow of material into the part wall resulting in a non-uniform flange. Wrinkles will form which extend down along the part wall (Fig. 12-41).

Excess pressure prevents the flow of material around the die radius and into the part wall. The plastic fails as forming begins and the blank area beneath the punch will tear away.

Blank Diameter

The blank should not exceed the diameter of die and draw pad. Material overhang will

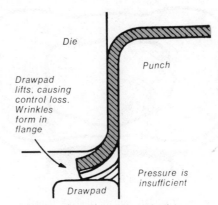

Flange wrinkles when part is drawn with too little pressure between drawpad and pressure face of die.

Fig. 12-41. Effect of drawpad pressure.

enlarge as it is pinched between the drawpad and die resisting flow causing the material to stretch and subsequent product failure.

Shallow Depressions

On shallow depressions (Fig. 12-42) with a sharp cornered configuration, it is recommended that the punch and die clearance be less than the material thickness. An example would be to have an 0.03 in. draw radius and 0.03 in. punch nose radius using 0.040 in. material with only 0.020 in. clearance per side. As much as 0.125 in. depth can be attained without fracturing the part.

Redraw Punch Radii

The end radii of redraw punches must be carefully blended for normal unimpeded flow

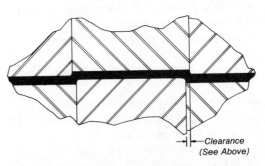

Clearance
(See Above)

Fig. 12-42. On shallow depressions with sharp-cornered configuration, punch and die clearance should be less than material thickness.

of material. Punch radii that are too small will produce rings or waves near the part bottom which result in wall thinning. This distortion becomes more pronounced with each redraw. Abnormally sharp radii restrict material flow, causing straining or stretching.

Shearing and Trimming

Punch and die clearance for shearing should be held to a minimum. The general accepted rule as applied to metals, cannot be applied to plastic. In practice, it has been found that the cold forming ABS materials will shear best with minimum clearance. Recommended punch and die clearance for material 0.015 in. thick up to 0.125 in. thickness should be between 0.0005 in. and 0.0015 in. per side.

Stamping

Essentially stamping is the same as cold drawing, differentiated by part depth ($\frac{1}{4}$ in. or less). Matched male-female dies should be used.

Typical Properties

Cold forming grades of ABS are less than one-third the weight of aluminum, one seventh the weight of steel, and when properly processed have a high modulus of elasticity, good rigidity, and will maintain their properties over a wide range of temperatures. The lighter weight of plastic makes it possible to produce more parts per pound of material than with metals. In addition, parts formed from these ABS materials have a hard surface that resists stains, rust and corrosion, and is further characterized by a high gloss appearance that requires no painting, buffing, or other finishing operations. Since color is integral, product surfaces never flake, peel or wear away. Generally speaking, the chemical resistance and physical properties remain virtually unchanged after cold forming.

Decorating

Cold forming grades of ABS can be pre-decorated using the distortion printing techniques already in use with metals; or, if preferred, they will accept a variety of post-decoration treatments.

Pre-decoration is more economical than post-decoration because of a combination of high speed, reduced reject rate and lower manpower requirements. Flat sheet can be printed much faster and far more efficiently than any formed object. Distortion printing with close control of registration of the blank allows decoration of recessed or irregular surfaces on a finished part. Both sides of the ABS sheet may be printed to decorate the inside and outside of the part. Etching of the surface to be printed is not required.

The flexography, rotogravure, or lithography printing processes, using very flexible inks and a high slip over-print varnish for the face forming the outside wall of the part, are recommended when decorating flat ABS sheet.

Laminates and embossed grains may also be used for decorative patterns.

13

BLOW MOLDING

Historically, the blow molding of thermoplastics materials began during World War II. While polystyrene was the first material used with the newly developed blow molding machines, polyethylene was used in the first large commercial application, which was a squeeze bottle for deodorant. Until 1957 virtually all blow molding was done with low density polyethylene to make squeeze bottles and other assorted containers up to about 15 gallons in size. At this time, two developments took place which resulted in the rapid growth and subsequent maturity of the blow molding industry: both blow molding machines and high density polyethylene became available to the plastics fabricator.

Heretofore all blow molding in the United States was done by a few companies operating with their own proprietary blow molding equipment. Today, a wide range of equipment (such as the two typical machines shown in Figs. 13-1 and 13-2) is commercially available from a number of different manufacturers. Also contributing to the phenomenal growth rate of blow molding in the 1960's was the commercial acceptance of high density polyethylene for packaging household chemicals such as detergents and bleaches and for toys and industrial applications.

Basically, blow molding is intended for use

* By B. T. Morgan, Technical Services, Plastics Div., Chemicals Group, Phillips Petroleum Co.

in manufacturing hollow plastic products (in some applications, like housings for electric tools, it is also adaptable to the production of double-walled products). Although there are considerable differences in the processes available, as described below, all have in common the production of a parison (a tube-like plastic shape), the insertion of the parison into a closed mold, and the injection of air into the parison to blow it out against the sides of the mold where it sets up into the finished product.

Where differences do exist is in the way the parison is made (i.e., by extrusion or injection molding), whether it is to be used hot as it comes from the extruder or injection molding machine (as in conventional blow molding), or stored cold and then reheated (as in cold preform molding), and the manner in which the parison is transferred to the blow mold or the blow mold is moved to the parison.

The basic process remains the same, however:

1. Melt the material.
2. Form the molten resin into a tube or parison.
3. Seal the ends of the parison, except for an area in which the blowing air can enter.
4. Inflate the parison inside of the mold.
5. Cool the blow molded part.
6. Eject.
7. Trim flash, if needed.

Fig. 13-1. Extrusion blow molding machine (with alternate presses) handles bottles and containers up to 1-gal. size. (*Courtesy Kautex Machines, Inc.*)

Fig. 13-2. A 400-ton injection blow molding machine that can produce finished bottles, jars, or other containers up to 48 fluid ounces. (*Courtesy Reed-Prentice Div., Package Machinery Co.*)

Processes

The two basic processes are extrusion blow molding and injection blow molding. The extrusion process utilizes an unsupported parison; the injection process utilizes a parison supported on a metal core. The extrusion process is by far the more widely used; the injection process is, however, gaining acceptance.

Before reviewing these methods, it is suggested that the reader refer to the chapter on Extrusion for additional information on single-screw and twin-screw extruders. While single-screw extruders have traditionally been hooked up to blow molding machines to provide the parison for blowing, there has been interest in using multiple or twin-screw extruders for processing heat-sensitive materials like PVC. At least two such units are commercially available.

Extrusion Blow Molding:

Continuous Extrusion. In continuous extrusion a molten parison is produced from the die without interruption. There are two basic concepts used in continuous extrusion blow molding machines. These deal with the manner in which the molds are mounted and moved. In one instance, molds are moved on rotating vertical wheels; in another, on a rotating horizontal table. Two types of vertical wheel arrangements are shown in Figs. 13-3 and

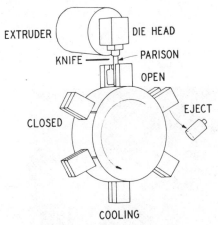

Fig. 13-3. Continuous vertical rotation of a wheel carrying mold sets on the periphery. (*All sketches and charts on blow molding, Courtesy Phillips Petroleum Co.*)

13-4. Machines of this type with up to nine one-gallon molds are in use. Figure 13-5 shows a schematic arrangement for mounting four molds on a rotating horizontal table. Due to the mold cost and down time required to change molds, rotary machines are best suited for long runs and large volume applications.

The second basic type of continuous extrusion blow molding machine moves the mold with a reciprocating action. The mold is reciprocated between a molding position and a position under the die head. When the parison is extruded, the mold moves under the die

head, closes on the parison, and moves to a blowing station. Typical positions of mold to die head are illustrated in Figs. 13-6 and 13-7. In some cases the parison is moved to the mold with a transport system as shown in Fig. 13-8. Because controlling the position of the parison in the mold presents a problem, machines of this type have been superseded by the reciprocating mold concept. A popular feature of the reciprocating mold machines is that finished containers can be produced directly off the machine. Neck finishing and detabbing are done either in the mold or on the machine. The manner in which the bottles are finished in the mold is shown in Fig. 13-9.

Multiple die heads can be used with resins which are heat stable. Such resins include the polyethylenes, polypropylenes, polystyrene, etc. When heat-sensitive resins such as polyvinyl chloride are to be blow molded, the number of die heads is generally limited to two. While one-gallon containers can be molded, the reciprocating mold machines are used primarily to mold containers of less than one quart.

Intermittent Extrusion. The intermittent extrusion of parisons is achieved in two ways. They are: (1) a reciprocating screw extruder and (2) ram accumulator. The function of the accumulator, of course, is to accumulate and hold in readiness the volume of melted plastic needed to make the next part or parts.

A reciprocating screw blow molder is shown in Fig. 13-10. As the screw rotates, melted plastic is pushed forward to the end of the extruder barrel and the screw is pushed in the opposite direction. When the amount of plastic melt to make the next extrusion has collected, a hydraulic cylinder moves the screw forward, thus forcing plastic out the die head to make the parison. Typical die head and mold positions for machines commonly used to mold containers are shown in Fig. 13-11. The molds are all mounted on common platens. Depending on the machine size and the size of the container to be molded, the number of molds will vary from one to as many as eight. With all molds mounted on one press, the removal of containers to the subsequent trimming operation is greatly facilitated. Since the screw diameter and backward travel of the screw

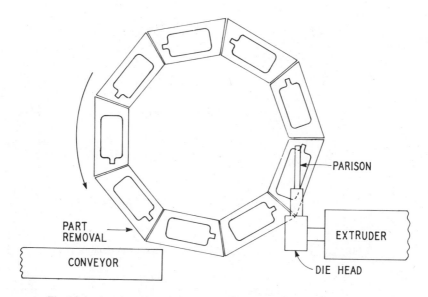

Fig. 13-4. Another type of set-up for a vertical wheel blow molding machine.

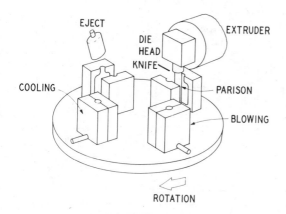

Fig. 13-5. Rotating horizontal table for continuous extrusion blow molding.

determine the total amount of plastic melt that can be extruded in a single shot, the use of reciprocating screw machines is limited to making parts that require less than a five-pound extrusion or shot.

The ram accumulator blow molding machine is essential for blow molding large items weighing over five pounds. Since accumulator machines were in use before the reciprocating screw machines, many of them were used in the same manner and the same applications as the reciprocating screw machine. A schematic of a typical ram accumulator blow molder is shown

in Fig. 13-12. The size of the extruder is completely independent of the capacity of the accumulator. For example, a 50-lb accumulator can be filled with any size extruder, and in some cases more than one extruder is used to fill the accumulator. In actuality, the expected production rate determines the extruder size. The production rate is easily estimated from the shot size for each part, cycle time, and number of molds. The actual shot size needed to make a part or parts, if more than one mold is involved per shot, will determine the capacity requirement of the accumulator.

Ram accumulator blow molding machines are widely used in molding toys, industrial parts, and large shipping containers. The machines are available with both single press and twin press arrangements. Several sizes of die heads are available so that one large die head per press can be used to make larger items or two or more smaller heads per press can be used to produce smaller items. With the variety of options available, the ram accumulator blow molding machines are very versatile and well suited to meet the needs of both the custom and proprietary blow molder.

Injection Blow Molding. Injection blow molding differs from extrusion blow molding primarily in the manner in which the parison is formed (e.g., injection molded rather than

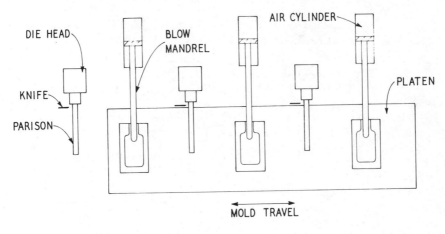

Fig. 13-6. Horizontal movement of mold between die head and molding position.

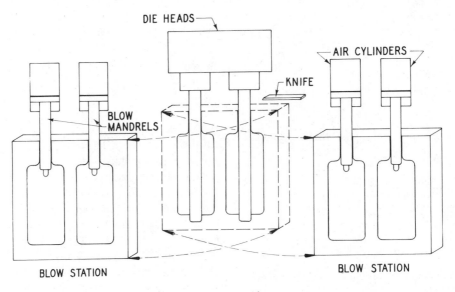

Fig. 13-7. Oblique mold movement.

extruded) and handled (e.g., in the way in which it is transferred to the blow mold). In one system, for example, the parison is injection molded at the top half of the mold; the parison arm (with the molded parison on it) is then rotated 180 degrees to the cavities at the bottom half of the mold where the parison is blown into the finished product. Another system is a three-station (or four-station) rotary indexing unit in which the parison is injected around a core rod at one station, indexed to a second station where it is blown, and then indexed to a third station for ejection (see Fig. 13-13). Three-position rotary machines are fast gaining acceptance in the non-handle ware container field. While limited to very small containers in the past, bottles up to 48 oz are now being made.

The main advantages of injection blow molding machines are: (1) no post molding or secondary operations are required as a finished container is produced; (2) dimensional control of neck finishes and bottle wall thickness is better than with the pinched parison technique; and (3) there is no pinchoff on the bottom of a bottle.

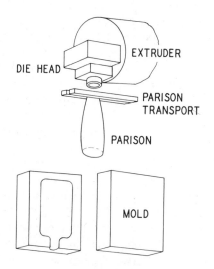

Fig. 13-8. Transport system is used to carry parison to the mold.

Wall thickness control of the blown part is established by the dimensions of the core rod and injection cavity. Production rates are determined by the size of part to be molded and size of the machine. Small, 4-oz bottles can be produced with ten core rods per station while only three 48-oz bottles can be made on the same sized machine.

The cost factor alone indicates injection blow molding is primarily best suited for large volume applications. Mold cost will be considerably higher for the injection blow molding system as the production of a single container requires one injection mold, three core rods, and one blow mold.

A wide variety of thermoplastics have been processed successfully on injection blow molding machines. In addition to high and low density polyethylene and polystyrene, other resins such as acrylic multipolymer, nylon, "barrier" plastics, and acetal copolymer, have been successfully processed.

Other Blow Molding Systems. A system now becoming popular in the U.S. uses a preformed cold parison. In this process, tubing with closely controlled tolerances is extruded and cut to length. After the cold tube is reheated on both ends a special mandrel which forms the neck is inserted and the bottom of the tube is closed and welded. The closed tube, still on a mandrel, is again reheated and positioned in a blowing mold. The heating is such that both orientation and wall thickness control are achieved during the blowing operation. The process offers the following advan-

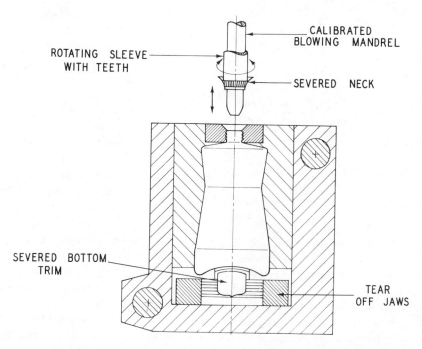

Fig. 13-9. In-mold finishing (i.e., neck finishing and detabbing).

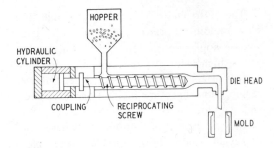

Fig. 13-10. Reciprocating screw blow molder.

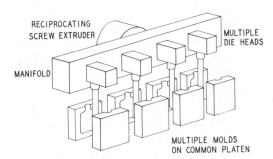

Fig. 13-11. Multiple molds mounted on a common platen.

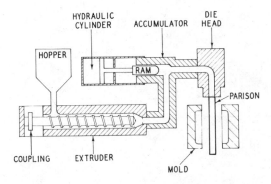

Fig. 13-12. Typical ram accumulator blow molder.

tages: (1) conversion of the pipe into the preform is scrap-free; and (2) final blowing of the preform into the finished bottle takes place in the thermoelastic rather than thermoplastic temperature range which results in a true orientation of the plastic. The system is designed to produce 10,000 twelve ounce bottles per hour or 5000 one quart bottles per hour.

In the United States, another system in commercial production also uses orientation (i.e., stretching the plastic to align molecules) to produce a clear polypropylene bottle (although the system is also adaptable to bottles based on other plastics, such as high-nitrile barrier resins, thermoplastic polyesters, etc.). In addition to clarity, physical properties of the polypropylene bottle are greatly enhanced. In both processes, bottles of equal or better toughness are produced at a lower weight as compared to unoriented bottles made with the same material.

Here is how one typical system for producing oriented bottles would operate. It is based on a two-step operation (or "cold preform" or "parison reheat," as it is also known) that involves parisons which are extruded or injection molded separately, then stored for later use. The parisons, which are generally open at both ends, are fed via an automatic floormounted loader onto a series of pins mounted on a continuously moving conveyor and delivered into a controlled temperature oven to be brought up to orientation/forming temperature. The pins rotate the parisons to insure uniform heating and at one point, the parisons pass by a series of 11 horizontally mounted individually controlled quartz heaters that provide the equivalent of parison programming (by varying the heat applied). When the parisons have reached forming temperature, two pickers descend to grasp the top of the parisons (which will be the bottom of the bottle; they're blown upside down) and shuttle them over to the blow mold. In the mold clamp, neck forming dies close around the bottom of the parison (i.e., the top of the bottle) and blow plugs move into position to form the neck finish of the bottle. With the parison thus clamped top and bottom, the picker assembly moves upward to give the parison the desired degree of stretch or vertical orientation (from 100 to 125%). The molds then close on the stretched parisons and the containers are blown to the mold configuration, providing orientation in the horizontal direction. As the molds close, the pickers return to the stand-by position, ready to start the next cycle.

(EDITOR'S NOTE: An innovative departure in blow molding techniques—and still too new at this writing to properly evaluate—is the so-called *dipping mandrel process*. Here is how it operates: Two pairs of dipping mandrels, equipped with neck-forming tools, operate in tandem. One pair is dipped simultaneously into two separate extruder-fed ports containing molten material (which can be any moldable thermoplastic, except PVC). Each pair of mandrels is coated with melt. At this stage, the neck finish is formed under low pressure, and a rotating potentiometer assures uniformity of the plastic coating on the mandrels. The coated mandrels, with the preformed necks, are then shuttled left or right (depending on the stage of the sequence) and descend into a twin-cavity mold for blowing. Finished bottles, which show no weld or flow marks, are then deposited upright onto take away conveyors. The advantages offered are: virtually no limit on the ratio between neck diameter and bottle height; high production rates; close control of neck finishes and wall thicknesses; and elimination of need for deflashing and trimming.)

Materials

While most thermoplastic resins can be blow molded, the three most widely used are the polyethylenes, polypropylenes and polyvinyl chlorides. Consumption in 1974 was:

	Million Pounds
Polyethylene, high density	957
Polyethylene, low density	64
Polypropylenes	37
Polyvinyl chloride	75

As will be noted, high density polyethylene is by far the most popular. The excellent balance of physical and processing properties, in addition to low cost, makes high density polyethylene well suited for blow molding containers, toys, and industrial parts. An ideal resin for the extrusion blow molding process should be heat stable and have good melt strength. Heat stability is necessary to minimize polymer degradation during processing and to permit reuse of all scrap generated. Melt strength is needed to prevent the parison from stretching or necking down. The larger the parison, the more melt strength required. These same properties apply to resins used for injection blow molding except for melt strength which is not as important because the core rod supports the parison.

Polyethylenes are widely used in all types and sizes of containers, in various industrial and consumer applications (including such products as picnic coolers, Fig. 13-14, automotive gasoline tanks, tool boxes, etc.), and in toys. Polypropylene and polypropylene copolymers are used mostly in containers such as those used in cosmetics, drug and some food containers (e.g., for syrup), especially where hot filling is required, for medical-surgical fluids, etc. Polyvinyl chloride has found acceptance in cosmetics and some household chemical con-

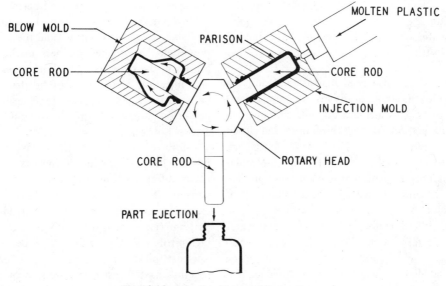

Fig. 13-13. Injection blow molding system.

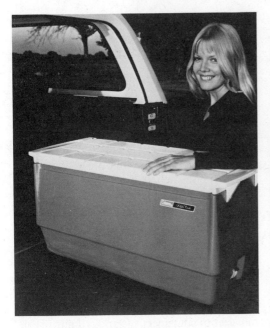

Fig. 13-14. Blow molded high-density polyethylene picnic cooler. (*Courtesy Phillips Petroleum Co.*)

Fig. 13-15. Blow molded 50-gal. high-density PE shipping drum. (*Courtesy Chemplex*)

tainers. Although the use of many thermoplastic resins in blow molding is limited due to their costs, they are used when their unique physical properties are an advantage. The barrier plastics and thermoplastic polyesters, for example, are under serious consideration for bottles for carbonated beverages and beer (and, in fact, may even prove out more economical than glass). Similarly, premium engineering plastics, like polycarbonate, are now being used for blow molded milk bottles and 5-gal. containers for bottled water.

Interest has also been increasing steadily in using blow molding for the manufacture of very large parts. This includes shipping containers in 15-, 30, and 55-gal. sizes, both open and closed head types (Fig. 13-15). These containers, molded mostly of high-density, high molecular weight polyethylene, are expected to compete with steel drums especially in the shipping of corrosive products. The automotive gasoline tanks, pallets, picnic and ice chests, and water and beverage coolers mentioned earlier also represent large-size industrial parts that can be blow molded of high-density polyethylene.

Although polyethylene (both high and low),

polypropylene, polyvinyl chloride, and polystyrene dominate the blow molding field, other resins have been successfully used. Acetals, for example, have been blow molded into aerosol bottles (Fig. 13-16). And as indicated above, the availability of the so-called barrier plastics or high-nitrile plastics (which show improved resistance to gas permeation) and thermoplastic polyesters has opened up new market possibilities in the blow molding (both by extrusion and injection techniques) of bottles for carbonated beverages and beer. The thermoplastic polyesters are also being blow molded into aerosol bottles.

Another innovation that has attracted interest in blow molding is the use of two or more different layers of plastic in a single parison. This can be accomplished in the extrusion blow process by co-extruding the parison (e.g., two different plastics, say a high-nitrile resin for its barrier properties and a low-cost plastic for economy, are extruded together as the inside and outside of a single parison; see chapter on Extrusion for further

Fig. 13-16. Blow molded 3-oz acetal copolymer aerosol container. (*Courtesy Celanese Plastics Co.*)

details on co-extrusion). The co-extruded parison is then blown into a container. In the injection blow process, two different plastics can be combined into a single parison by first molding the parison in one material, then using this parison as a core insert around which a second plastic is molded; the double-layer parison is then transferred to a blow mold for blowing.

Molds

Most molds for blow molding in the United States are made with either aluminum, beryllium copper, or a zinc alloy. All of these metals are good thermal conductors of heat and can be cast. Nearly all of the larger molds are cast with aluminum to minimize weight. With many small molds, there is a choice between a cast mold or a machined mold. Well constructed molds made by either casting or

machining will give good performance. Many of the small molds are now being machined from forged aluminum with outstanding results. Although zinc alloys are always cast, steel or beryllium copper inserts are added where the mold must pinch the parison.

Well designed molds are vented, as entrapped air in the mold prevents good contact between the parison and mold cavity surface. When air entrapment occurs, the surface of the blown part is rough and pitted in appearance. A rough surface on a shampoo bottle, for example, is undesirable because it can interfere with the quality of decoration and can detract from the over-all appearance. Molds are easily vented by means of their parting line, with core vents and with small holes. A typical mold parting line venting system is shown in Fig. 13-17. The venting is only incorporated in one mold half. This type of venting can be used on all sizes of molds. When certain areas of the mold cavity are prone to trap air, core vents as shown in Fig. 13-18 can be used. Venting in the mold cavity should be anticipated in the mold design and layout of the cooling channels so provisions can be made for their locations. For cast molds the cooling channel baffles can be located over areas to be vented, as shown in Fig. 13-19. The vent opening will pass through a boss in the baffle to the back or outside of the mold. In machined molds, care must be taken so that vents miss the drilled cooling channels. When core vents cannot be used because the slots mark off on the blown part and show, small drilled holes can be used. The effect of the size of hole on the surface of the part is shown in Fig. 13-20. If the hole is too large, a protrusion will be formed; if too small, a dimple will be formed on the part. Venting can also be incorporated in molds that are made in sections. A 3 to 10 mil gap between the two sections with venting to the outside of the mold is a very effective vent. For small containers a 2 to 3 mil opening is used while up to a 10 mil opening has been used on large parts such as a 20-gal. garbage container.

The mold cavity surface has an important bearing on mold venting and on the surface of the molded part. With polyethylenes and polypropylenes a roughened mold cavity

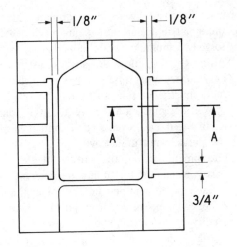

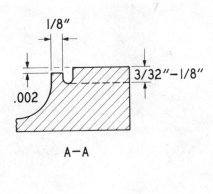

Fig. 13-17. Parting line venting.

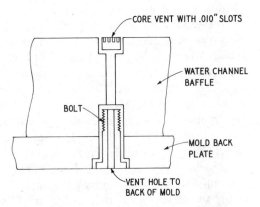

Fig. 13-18. Core venting.

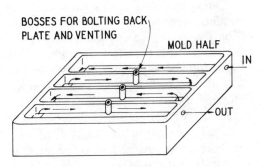

Fig. 13-19. Location of vents in baffles.

reproduce on the clear plastics and this is not normally desirable.

Programming

Parison programming consists of varying the wall thickness of the parison. Two methods of parison programming are in use. One method is to vary the orifice size of the extrusion die and the second method is to vary the extrusion rate of the parison. A variable orifice programming

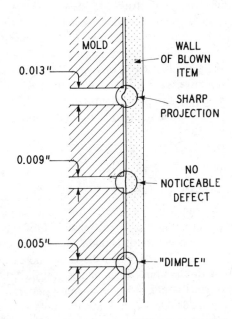

Fig. 13-20. Effect of vent hole size on part surface.

surface is necessary for the smoothest surface. Grit blasting with 60 to 80 mesh grit for bottle molds and 30 to 40 mesh grit for larger molds is a common practice. The clear plastics such as PVC and styrene require a polished mold cavity for the best surface. A grit blasted surface will system can be used on either continuous or intermittent extrusion machines while the variable extrusion rate system can be used only with intermittent extrusion blow molding machines.

In Fig. 13-21, the effect of extrusion rate on part weight is shown. Variable parison extrusion rate is achieved by controlling either the pressure or the flow of hydraulic oil to the hydraulic piston on the ram or on the reciprocating screw. Various timers or cams, flow control valves, or pressure regulators are combined to vary the extrusion rate of the parison. While the variable orifice programming system is used on all extrusion blow molding machines, the variable extrusion rate system is usually used only on the large ram accumulator blow molding machines.

The variable orifice programming system varies the orifice opening by moving either the mandrel or the bushing. A converging or diverging type extrusion die, as shown in Fig. 13-22, must be used. Movement is obtained by hydraulic actuators which act directly or through a mechanical linkage on the moving part of the die head. A typical variable orifice programmed die head is shown in Fig. 13-23.

As mentioned previously, this system is in wide use for both large and small blow molding machines.

The value of parison programming for part-weight production and control of wall thickness has become fairly universally accepted in modern blow molding machine design. Electronic systems are available that can fix from 5 to 50 points on the parison. In addition to programming wall thickness, systems are also available to control parison length, a factor of special importance in blow molding large industrial pieces.

Die Shaping

While programming varies the entire wall thickness of the parison, die shaping introduces variations in the cross-sectional area of a parison. A well designed die head extrudes a parison that is round and that has a uniform wall thickness. Whenever a round uniform walled parison is blown to form a square-shaped item, the wall thickness of the blown part will be less in the edges and corners than in the flat side surfaces. This is so because the parison must stretch farther to reach the edges and corners.

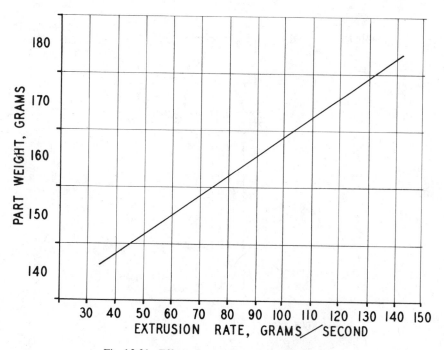

Fig. 13-21. Effect of extrusion rate on part weight.

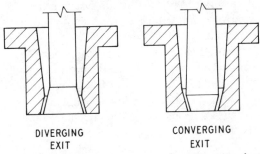

DIVERGING
EXIT

CONVERGING
EXIT

Fig. 13-22. Diverging or converging-type extrusion dies.

To overcome this problem the parison is made thicker in that section which stretches the farthest so that the wall thickness in the edges and corners of the molded part is increased. Thick areas in the parison are made by removing metal from the corresponding section of the die. The metal can be removed from either the mandrel or the bushing. A square-shaped item with and without die shaping is shown in Fig. 13-24. A typical die shaped for an oval bottle is shown in Fig. 13-25. The die mandrel or bushing can be easily shaped by machining on a lathe or on a milling machine.

Auxiliary Equipment and Options

All blow molding machines require certain auxiliary equipment. A complete blow molding system could include: (1) bulk storage for the resin; (2) hopper loader which proportions virgin resin, regrind and color concentrate; (3) granulator for grinding scrap and rejected parts; (4) trimming equipment; (5) air compressor; (6) water chillers; (7) surface treating equipment for parts to be decorated; (8) decorating equipment; and (9) packers. If the part is a container to be filled, filling and capping equipment would be needed. Pneumatic conveyors to move the resin from storage and surge tanks and conveyors to move parts through the operation will be required. Depending on the size of operation, the cost could vary from about $40,000 to $400,000 for a single blow molding machine and auxiliary equipment ready to operate.

Most suppliers now offer some type of in-mold trimming system for non-handle items. This is usually accomplished by transferring the parison from the extrusion station to a separate blowing station where calibration of the neck and trimming takes place.

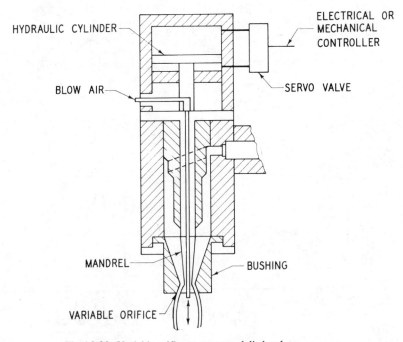

Fig. 13-23. Variable orifice programmed die head.

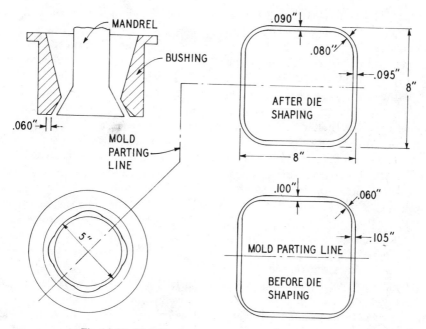

Fig. 13-24. Wall distribution with and without a shaped die.

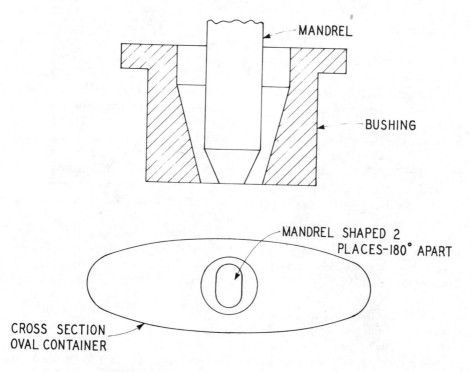

Fig. 13-25. Typical die shaped for an oval bottle.

There is an even more dominant trend in designing equipment to deflash products (even handleware) automatically within the machine. After top and bottom flash is trimmed from the parts, they are transferred to an integral station where the handle slugs are punched out.

In addition to the auxiliary equipment needed to operate a blow molding line there are a variety of options that can be obtained on a molding machine. The blow molder who is faced with the task of molding a variety of sizes will have a wide choice of options for his machine such as:

1. Single or double press. This is determined by the volume of business expected.
2. Press size. This is determined by the length and width of the molds to be run.
3. Press stroke and maximum daylight. This is determined by the maximum diameter part expected to be made.
4. Shot size. This is determined by the maximum amount of plastic that is expected to be needed for a single extrusion. (One or more die heads can make up a single extrusion; also more than one mold per head can be involved.)
5. Number and size of die heads per press. This determines the versatility and efficiency of the machine. For example, one large head or two or more smaller heads can be used with one press, depending on the size of the items to be made.
6. Extruder size. This is determined by the maximum amount of plastic melt required per hour for the most demanding operation anticipated.
7. Type of programming system. This is a judgment factor depending upon the design of the parts to be made and the actual need. Variable orifice programming is considered the most versatile and also the most expensive.
8. Timers. Additional timers will be required if the following special operations are to be used:
   a. Internal cooling with CO_2. This can reduce cycle time 20-40% but is not good for all applications.
   b. Parison prepinchers. This is used in conjunction with preblow air to increase the diameter of the parison and it is often required when blowing a part with a needle.
   c. Needle blow. A needle is used when a minimum sized hole is required in the part such as a wheel.
   d. Source of variable hydraulic pressure. When molds with moving sections or when blow pins are used to compression mold a segment of the part hydraulic cylinders are often needed. The hydraulic pressure can be provided from the blow molding machine or from a separate power source.

Selection of a blow molding machine is relatively simple if the items to be made are precisely known. For example, the selection of equipment to blow mold one-gallon dairy containers is relatively easy. The quantity of containers required will dictate the amount and size of equipment needed. The blow molding machine will be equipped with only those features necessary to make a finished one-gallon milk bottle. The selection of blow molding equipment to make both a one-gallon and one-quart container could present more of a problem. The total number for each container as well as the total number for each run for each container needs to be known. The choice of blow molding equipment could now be between one large machine that can be converted to mold both gallons and quarts or two machines probably smaller in capacity, one for each size of container. The length of run for each container on the larger machine is important because the time to convert the machine from one container to the other is non-productive time. A six-hour change over time on a blow molding operation, if done once per week, will result in a loss of 5% of the operating time. If done once per month, the loss is 1.5%.

Future Trends

Blow molding is generally recognized to be one of the fastest growing areas in the plastics industry today. Further, the in-plant blow molding of containers is increasing rapidly. While the greatest number of in-plant operations are now in the dairy and household chemical markets, penetration into the pharmaceutical, toiletry and cosmetic markets is taking place. Factors influencing the trend toward in-plant blow molding are: (1) reduced costs, (2) difficulty in obtaining small volume custom bottles, and (3) increased reliability of blow molding machines and systems now available.

At the same time that in-plant operations are expanding in the container field, proprietory blow molding continues to expand in the non-container industrial market. Here again, the reasons are the same. However, the custom molder will continue to fill an important need in this fast growing fabrication area just as they have in injection molding and other segments of the plastics industry.

PROCESS VARIABLES*

Obviously, the process parameters to be considered in blow molding will be conditioned by the type of resin used (e.g., making an acetal product would involve higher blow pressures than would be required for polyethylene), the type of blow molding unit used, and the product being made.

The discussion below deals primarily with the extrusion blow molding of high-density polyethylene bottles—the technique, material, and application in most common use today. The process variables discussed cover the extruder die (for making the parison) and the blow mold. Details on the extruder itself can be found in the chapter on Extrusion.

Die. In a sense, the parison die has become the key element in blow molding because it controls material distribution in the finished item, and in turn, the economics of the final product. Therefore, increasing attention has been devoted to making the programming die work in favor of improved economics as well as improved properties. The main control factor in parison programming is the core pin. This pin

* Adapted from Bulletin P3C, "Fortiflex® Polyethylene – Blow Molding," issued by Soltex Polymer Corp., a subsidiary of Solvay & Cie S.A. (Belgium).

can be given greater latitude by providing a taper at the die face and providing for movement of the pin so the opening at the face of the die can be made larger or smaller as required to deliver parisons with thicker or thinner walls. Such a movable core pin is schematically diagrammed in Fig. 13-26.

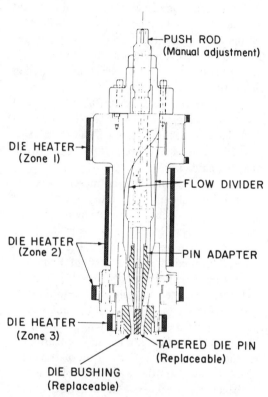

Fig. 13-26. Manually variable die. (*All illustrations on Processing Variables, Courtesy Soltex Polymer Corp.*)

Die Dimension Calculations. In selecting the die bushing and mandrel dimensions to be used for the production of a blow molded polyethylene product, several features must be considered.

For bottles, the weight, minimum allowable wall thickness, and minimum diameter are important considerations. Likewise, the need, if any, to use a parison within the neck area and whether there may be adjacent pinch-offs.

The type and melt index of the resin used are factors because of swell and elasticity characteristics.

Die land length and cross sectional area must be considered.

Part of the die dimensions will also depend partly on processing stock temperature and extrusion rate anticipated for production.

Mathematical formulas have been developed to permit the selection of die dimensions. Although these calculated dimensions are intended as approximations or starting points in die selection, they have been found to yield products, in the majority of cases, within ±5% of the design weight. In some cases, only slight changes in mandrel size or stock temperature and/or extrusion rate are necessary to obtain the desired weight.

Formulas for Calculating Die Dimensions. The formulas presented here are for use with long land dies, those having 20-30/1 ratio of mandrel land length to clearance between mandrel and bushing.

In their use, consideration must be given as to the anticipated blow ratio, the ratio of maximum product outside diameter to the parison diameter. Normally, ratios in the range of 2-3 : 1 are recommended. The practical upper limit is considered to be about 4 : 1.

For large bottles with small necks, this ratio has been extended as high as 7 : 1 so that the parison fits within the neck. In such a case, a heavier bottom and pinch-off results from the thicker parison. Also, less material is distributed in the bottle walls 90° from the parting line than in similar bottles with lower blow ratios.

When the neck size of a bottle or the smallest diameter of the item is the controlling feature (as when the parison must be contained within the smallest diameter), the following approximations may be used to calculate die dimensions:

Free Falling Parison

$$D_d \cong 0.5N_d$$
$$P_d \cong D_d^2 - 2B_d t + 2t^2$$

where

D_d = Diameter of die bushing, in.
N_d = Minimum neck diameter, in.
P_d = Mandrel diameter, in.
B_d = Bottle diameter, in.
t = Bottle thickness at B_d, in.

This relationship is useful with most polyethylene blow molding resins, and is employed when bottle dimensions are known and a minimum wall thickness is specified. It is particularly useful for round cross sections.

The 0.5 figure presented for selecting the diameter of the die bushing may change slightly depending on processing conditions employed (stock temperature, extrusion rate, etc.), resin melt index, and die cross sectional areas available for flow. It may be slightly lower for a very thin die opening (small cross section) and higher for large openings.

If product weight is specified rather than wall thickness for a process employing "inside-the-neck" blowing, the following approximation may be employed.

$$P_d = D_d^2 - 2W/T^2 Ld$$

where

W = Weight of object, grams
L = Length of object, in.
d = Density of the resin, g/cc
T = Wall thickness, in.

This system is applicable to most shapes and is of particular advantage for irregularly shaped objects.

A controlled parison is one in which the dimensions are partially controlled through tension; i.e., the rotary wheel, the falling neck ring, etc.

Because of this, the following relationships are employed:

$$D_d \cong 0.9N_d$$
$$P_d \cong \sqrt{D_d^2 - 3.6B_d t + 3.6t^2}$$
$$P_d \cong \sqrt{D_d^2 - 3.6W/T^2 Ld}$$

Derivation of Formulas. Blowing-inside-the-neck system.

When a polymer is forced through a die, the molecules tend to orient in the direction of the flow. As the extrudate leaves the die, the molecules tend to relax to their original random order. Parison drawdown, the stress exerted by the parison's own weight, tends to prevent complete relaxation. This results in longitudinal shrinkage and some swelling in diameter and wall thickness.

Through laboratory and field experience it has been found for most high-density polyethylene blow molding resins that

$$D_a \cong 0.5 N_a$$
$$A_a \cong 0.5 A_b$$

where

D_a = Die diameter
N_a = Minimum neck diameter
A_a = Cross-sectional area of the die
A_b = Cross-sectional area of the bottle

$$A_D = \frac{\pi}{4}(D_2{}^2 - P_a{}^2)$$

$$A_B = \frac{\pi}{4}\left[B_a{}^2 - (B_a - 2t)^2 \right]$$

where

P_a = Mandrel diameter, in.
B_a = Product diameter, in.
t = Product thickness at B_a, in.

$$\therefore A_D = 0.5\, A_B = 0.5\,\frac{\pi}{4}(B_a{}^2 - B_a{}^2 + 4B_a t - 4t^2)$$

$$\frac{\pi}{4}(D_a{}^2 - P_a{}^2) = 0.5\,\frac{\pi}{4}(-4t^2 + 4B_a t^2)$$

Dividing through by $\dfrac{\pi}{4}$ and rearranging terms

$$P_a{}^2 = D_a{}^2 - 2B_a t + 2t^2 \quad \text{or}$$

$$P_a = \sqrt{D_a{}^2 - 2B_a t + 2t^2}$$

Also $A_B = \dfrac{W}{Ld}$

where

W = Object weight, grams
L = Object length, in.
d = Resin density, g/cc

$\therefore$ since $A_D = 0.5\, A_B$

$$\frac{\pi}{4}(D_a{}^2 - P_a{}^2)\,4 = 0.5\,\frac{W}{Ld}$$

$$D_a{}^2 - P_a{}^2 = \frac{4}{\pi}\,0.5\,\frac{W}{Ld}$$

$$P_a{}^2 = D_a{}^2 - 2\,\frac{W}{Ld}$$

$$Pd = \sqrt{D_a{}^2 - 2W/\pi\,Ld}$$

The same derivation is employed for controlled parisons except that

$$D_a = 0.9 N_a$$
$$A_D = 0.9 A_B$$

As shown, the size selected for die bushing and mandrel depend on wall thickness of the finished blow molded part, the blow ratio, and certain resin qualities included in the above formulas for various polyethylene blow molding resins. These qualities are parison swell (increase in wall thickness as the parison exits the die) and parison flare (ballooning or puffing out of the parison as it exits the die). Both depend on processing conditions. It has been shown that calculations can be made for the general die dimensions. The other dimensions of the die—approach angles and lengths— vary widely with machinery capabilities and manufacturer's experience. Calculations for these dimensions will therefore not be given here. Instead, a few rules of thumb: the land length of the die (see Fig. 13-27) is generally

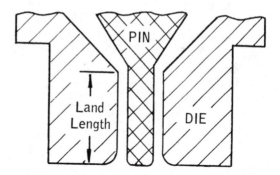

Fig. 13-27. Die and pin.

eight times the gap distance between the pin and the die. In simple tabular form, this works out to be:

Gap Size (in.)	Land Length (in.)
Above 0.100	1-2
0.030 - 0.100	$\frac{3}{4}$-1
Below 0.030	$\frac{1}{4}$-$\frac{3}{8}$

Notice that the land length is at least $\frac{1}{4}$ inch, regardless of gap size. This land length is necessary to get the desired parison flare.

The die should be streamlined to avoid abrupt changes in flow which could cause polymer melt fracture. When no further changes are expected in die dimensions, the die mandrel and bushing should be highly polished and chrome plated. This keeps the surface cleaner and eliminates possible areas of resin hangup. Finally, the edges of the pin (mandrel) and die should have slight radii to minimize hangup within or at the exit of the die area. The face of the mandrel should extend 0.010″-0.020″ below the face of the die to avoid a doughnut occurring at parison exit.

Air Entrance. In blow molding, air is forced into the parison expanding it against the walls of the mold with such pressure that the expanded parison picks up the surface detail of the mold. Air is a fluid, just as is molten polyethylene, and as such it is limited in its ability to flow through an orifice. If the air entrance channel is too small, the required blow time will be excessively long or the pressure exerted on the parison will not be adequate to reproduce the surface details of the mold. General rules of thumb to be used in determining the optimum air entrance orifice size when blowing via a needle are summarized below:

Orifice Diameter (in.)	Part Size (vol.)
$\frac{1}{16}$	Up to one quart
$\frac{1}{4}$	1 quart - 1 gallon
$\frac{1}{2}$	1 gallon - 55 gallons

Normally, gauge pressure of the air used to inflate parisons is between 40 and 80 psig. Often, too high a blow pressure will "blow-out" the parison. Too little, on the other hand, will yield end products lacking adequate surface detail. As high a blowing air pressure as possible is desirable to give both minimum blow time (resulting in higher production rates) and finished parts that faithfully reproduce the mold surface. The optimum blowing pressure is generally found by experimentation on the machinery with the part being produced. The blow pin should not be so long that the air is blown against the hot plastic. Air blowing against the hot plastic can result in freeze off and stresses in the bottle at that point.

Moisture in the blowing air can cause pock marks on the inside product surface. This defective appearance is particularly objectionable in thin-walled items such as milk bottles. A system of separators and traps to dry the air taken into the air compressor can prevent this problem.

Molds. Compared with injection molds, those for blow molding can be considerably less rugged in construction. Clamping pressures applied to blow molds range from 100 to 300 psi and blowing pressures from 5 to 100 psi. This contrasts with the pressures reaching 50 tons used in some injection molding applications. A typical bottle mold is schematically drawn in Fig. 13-28.

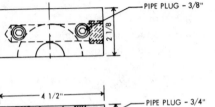

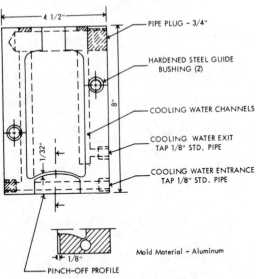

Fig. 13-28. Typical 12-oz container mold (half).

On many blow molding machines, the mold closes in two steps in conjunction with expanding the parison by the blowing air. On these systems, the mold closes rapidly until $\frac{1}{3}$ to $\frac{1}{2}$ in. of daylight remains. The final clamping is done at reduced speed, but with increased clamping force. This two-speed operation protects the mold surface from being marred by objects which may fall between the mold halves, and reduces mold surface wear.

The mold does not have to be positioned vertically between the platen area. It also does not have to be positioned so that there is equal resin all around the blow pin. This "off-setting" of the molds is commonly done when containers with large handles on one side (such as jugs for milk, bleach, or detergent) are blow molded from polyethylene. This off-setting of the molds makes it especially easy to "catch" the hollow handle of such containers.

There are three main ways of making molds: machining, casting, and hobbing. Cast molds are most common because they require little machining or tooling work. Hobbed molds exactly duplicate the hob in the same way that cast molds duplicate the original pattern. These two types of construction are the most useful where several identical molds are required. The main areas of attention to a mold designer are the "pinch-off" areas at the top and bottom of the mold and the provisions for cooling.

The "pinch-off" areas pinch the ends of the plastic parison and seal the edges together when the mold closes. These surfaces are subject to more wear than any other part of the mold. The high heat conductive metals preferred for blowing molds, like aluminum and copper alloys, are generally less wear resistant than steel. Steel inserts are often used for the pinch-off areas of molds of the softer metals. An additional advantage of pinch-off inserts is that they can be made replaceable in the event of wear or damage. A neck pinch-off insert is sketched in Fig. 13-29.

Also important is the insert constructed at the "tail" section of the untrimmed plastic part. If improperly constructed, the end products may break along the bottom pinch-off weld line on drop impacting. Generally, the pinch-off insert is made of hard steel while the rest of the mold is made of a non-ferrous metal. This pinch-off insert should not have knife edges as these will tend to yield a grooved weld as shown in Fig. 13-30b. A good pinch-off insert, as shown in Fig. 13-30a, has a 0.012 to 0.125 in. land before it flares outward in a relief angle of approximately 15°. Figure 13-31 illustrates the way a blowing mold with these features looks. The weld line should be smooth on the outside and form an elevated bead on the inside.

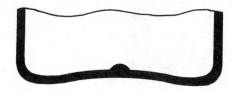

A – Good Weld Line

B – Poor Weld Line (Weak) Pinch-off had Knife Edge; Relief Angle was Either too Large or too Small.

Fig. 13-30. Weld lines.

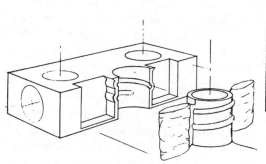

Fig. 13-29. Replaceable neck insert.

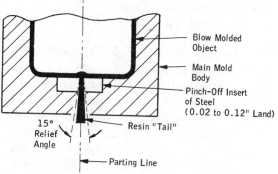

Blow Molded Object

Main Mold Body

Pinch-Off Insert of Steel (0.02 to 0.12" Land)

15° Relief Angle

Resin "Tail"

Parting Line

Fig. 13-31. Recommended pinch-off and insert.

Other uses for inserts include replaceable neck finishes to allow making bottles with different neck sizes from a single mold, and also to incorporate in products textured areas or lettered information or trademarks which can be changed, making some molds more versatile.

Molds must have pins which perfectly align the mold when closed. These guide pins and bushings can be arranged to suit the opening and closing operation required for a given set of equipment.

Cooling. Cooling is particularly important in mold design because it consumes much of the cycle time and therefore bears on product economics. Cooling can take two-thirds or more of the entire mold-closed time in a cycle. Best economics require cooling as quickly as possible. Mold cooling is a function of three factors: the coolant circulating rate, coolant temperature, and efficiency of the overall heat transfer system.

Coolants can be circulated through the mold in several ways. One of the simplest is to "hog" out the back of each mold half, leaving an open area, and enclose this cavity with a gasketed back plate. Then with simple entrance and exit openings, the coolant can be circulated in fairly large volume. Another method is to drill holes through the mold block from top to bottom and from side to side so that they intersect. Excess openings are plugged and hose connections inserted for the exit and entrance lines. This type of cooling arrangement can provide extra cooling where needed in the mold, such as at the neck finish and bottom pinch off areas where the plastic material is thicker and requires more cooling.

Results are best when uniform temperatures are maintained throughout the mold.

The biggest problem frequently encountered is inadequate mold cooling. It can be improved by increasing the rate of coolant flow through the mold or by making the mold of material with better heat transfer. Flow of coolant can also be improved by increasing the size of cooling channels (including the adjacent piping) and by increasing the coolant flow rate. Heat transfer can be enhanced by lowering the temperature of the coolant or by using a mold material with better heat transfer properties.

Beryllium-copper is the best mold material. Kirksite, aluminum, and steel are the next best, in order, of commonly used materials.

Mold materials like aluminum and beryllium-copper are often porous in some areas, and even with careful mold making, coolant can penetrate through to the inner mold surface and cause problems. However, when this happens, leakage can be stopped by applying sealants.

The most common mold coolant is tap water. Hard water should be softened before use. The overall efficiency of a mold cooling system would be greatly reduced if water scale were allowed to plate out on, or partially block, cooling channels. Generally, cooling water circulates first through a cooling system which lowers its temperature to between 40 and 70°F. If the temperature of the molding shop is high, cooling water can cause moisture to condense on mold surfaces. This will cause defects in finished parts, small blotches or pock marks on the surface. Condensation problems can be eliminated by raising the mold coolant temperature, or by air conditioning the molding shop. This solution merits careful economic consideration: overall production increases resulting from shorter cycles can pay for the initial capital investment in air conditioning.

It is often useful to cool the neck rings and the blow pin independently: here, as with mold cooling, basic rules of heat transfer optimization must be applied.

In addition to those external systems which circulate liquid coolants through the molds to cool both mold and part simultaneously, there has been recent interest in internal cooling systems. These use circulating air, a mixture of air and water, or carbon dioxide injections to cool the inside of the parts while they are in the mold. Typical of the commercial methods currently in use: (1) liquid carbon dioxide is injected into the blown part, followed by vaporization, superheating, and exhausting of the coolant as hot gas through the blow pin exhaust; (2) highly pressurized moist air is injected into the part where it expands to normal blow pressures, producing a cooling effect by lowering the air temperature and freezing the moisture present into ice crystals

(crystals and cold air strike the hot walls of the part where they melt and vaporize); (3) air is passed through a refrigeration system and into the hot parison; and (4) normal plant air is cycled into and out of the blown parts by a series of timers and valves.

Mold Materials. Although beryllium-copper and Kirksite are better conductors of heat, aluminum is by far the most popular material for blow molds. This is due to the high cost of beryllium-copper and the short life of the Kirksite (soft zinc alloy) molds. Aluminum is light in weight, a relatively good conductor of heat, very easy to machine (but also easy to damage—if the mold is abused) and low in cost. Its potential porosity may easily be eliminated by coating the inside of the mold halves with a sealer—such as automotive radiator sealant.

Aluminum is used for single molds, molds for prototypes, and large numbers of identical molds, as might be used on wheel type blowing equipment or equipment with multiple die arrangements. Aluminum may tend to distort somewhat after prolonged use. Thin areas, as at pinch-off regions, can wear in aluminum and become damaged relatively easily. However, if carefully handled and used, these regions can be as durable as the rest of the mold. Steel pinch-off inserts, of course, could be replaced when damaged or worn. Alcoa 319 and 356 alloys are extensively used for cast aluminum molds.

Steel can be used for extremely long production runs where utmost durability is necessary. Cooling difficulties and the poor heat transfer characteristics of steel discourage its use in any great volume. It is not possible to cast steel molds, but multiple duplicates can be hobbed.

Surface Finish. Interior surface finish of blow molds should be sand blasted, but it is *not* recommended that they be highly polished and chromed. The natural finish of high density polyethylene blown items is a smooth matte surface which does not improve with mold polishing beyond the normal limits obtainable with abrasives. In addition, this rough surface of the mold accomplishes a venting function by providing small pockets where the air can go during blowing of the parison.

14
ROTATIONAL MOLDING

In rotational molding, the product is formed inside a closed mold or cavity while the mold is rotating biaxially in a heating chamber. To obtain this mold rotation in two planes perpendicular to each other, the spindle is turned on a primary axis, while the molds are rotated on a secondary axis (Fig. 14-1).

Rotational molding (also popularly known as *roto-molding*) is best suited for large, hollow products requiring stress-free strength, complicated curves, a good finish, a variety of colors, a comparatively short (or very long) production run, and uniform wall thickness. It has been used for products such as car and truck body components (including an entire car body), tilt trucks, industrial containers, portable outhouses, modular bathrooms, telephone booths, boat hulls, garbage cans, light globes, ice buckets, appliance housings (Fig. 14-2), toys (Fig. 14-3), and boat hulls (Fig. 14-4). The technique is applicable to most thermoplastics and has recently been adapted for possible use with the thermosets.

How it works

There are essentially four basic steps in rotational molding: loading, molding or curing, cooling, and unloading.

In the loading stage, either liquid or powdered

* By Dario J. Ramazzoti, Mgr. of Eng., McNeil Femco, Div. of McNeil Corporation.

plastic is charged into a hollow mold. The mold halves are then clamped shut and moved into an oven where the loaded mold spins biaxially. Rotation speeds should be infinitely variable at the heating station, ranging up to 40 rpm on the minor axes and 12 rpm on the major axes. A 4 : 1 rotation ratio is generally used for symmetrically shaped objects, but a wide variability of ratios is necessary for molding unusual configurations.

In the oven, the heat penetrates the mold, causing the plastic, if it is in powder form, to become tacky and stick to the mold surface, or if it is in liquid form, to start to gel. On most units, the heating is done either by air (as in a gas-fired hot-air oven) or by a liquid of high specific heat, such as molten salt; where jacketed molds are used (see below), heating is done with a hot liquid medium, such as oil.

Since the molds continue to rotate while the heating is going on, the plastics will gradually become distributed evenly on the mold cavity walls through gravitational force (centrifugal force is not a factor). As the cycle continues, the polymer melts completely, forming a homogeneous layer of molten plastic.

When the parts have been formed, the molds move to a cooling chamber where cooling is accomplished by either a cold water spray and/or forced cold air and/or a cool liquid circulating inside the mold. The mold continues to rotate during the cooling cycle so as to

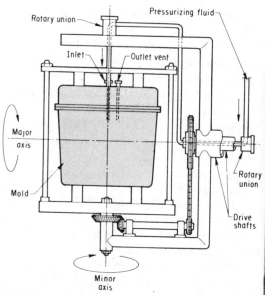

Fig. 14-1. In rotational molding, the product is formed inside a closed mold while the mold is rotated about two axes and heat is applied. The spindle is turned on a primary axis, while the molds are rotated on a secondary axis. (*Courtesy McNeil Femco, Div. of McNeil Corp.*)

Fig. 14-2. Rotationally molded polyethylene appliance housing. (*Courtesy U.S.I. Chemicals*)

Fig. 14-3. Rotationally molded polyethylene hobby horse. (*Courtesy U.S.I. Chemicals*)

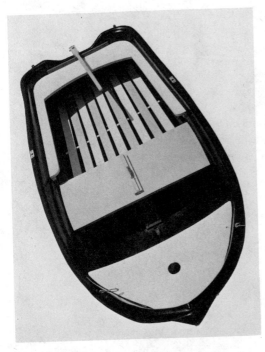

Fig. 14-4. Rotationally molded polyethylene boat hull. (*Courtesy U.S.I. Chemicals*)

insure that the part does not sag away from the mold surface, causing distortion.

Finally, the molds are opened and the parts removed. This can be either manually or by the use of forced air or mechanical means to eject the part.

Typical cycle times range from 7 to 15 min., but can be as short as 5 min. or as long as 30 min. for very large parts. Wall thickness of the parts affects cycle time, but not in a direct ratio. Normally, on a plastic like polyethylene, the cycle times increase by 30 sec. for each 25 mils of added thickness up to $\frac{1}{4}$ in. Beyond $\frac{1}{4}$ in. the heat-insulating effect of the walls increases cycle time disproportionately for any further increase in thickness, and cycle times usually have to be determined experimentally.

Machinery

Machines for rotational molding are generally characterized by the weight in pounds of the maximum load supported by the arms, including mold and weight of charge; and the spherical diameter, in inches, of the rotation

possible by the mold extremities in the chambers.

Modern-day machines feature up to 5000 lb capacities, with some carousel-type machines sweeping out a 140-in. spherical radius. Large parts now being molded include 500-gal. industrial containers, a 200-lb refuse bin, and as indicated above, a 500-lb car body. For producing small parts, an arm may hold as many as 96 cavities on each arm.

Following are the major types of machinery in use today:

Batch-Type Machines. This is the least expensive of the rotational molding machines because it is the least sophisticated and requires the most manual labor. In a typical batch operation, the charged mold is rolled into the oven for rotation and heating. At the completion of the cycle, the mold is removed and a newly charged one inserted in its place. The completed mold is then transferred manually on rollers to a cooling station for cure before removal of the parts.

Carousel-Type Machines. The most common rotational molding machine in use today is the carousel unit which is essentially a three-station rotary indexing type with a central turret and three cantilevered mold arms (Figs. 14-5 and 14-6).

In operation, individual arms are involved in different phases simultaneously, so that no arms are idle at any time. The arms or mold spindles extend from a rotating hub in the center of the unit. Thus, while one arm is in the loading/unloading station, another is rotating within the oven, while the third is rotating within the cooling station. All operations are automated and at the end of each cycle, the turret is rotated 120 degrees, thereby moving each mold arm to its next station.

While most carousel-type machines have only three arms—one for each of the three stations—some of the newer units use four or more arms to save operating time. With multiple-arm machines, more than one mold can be in the oven, with each arm indexing out at a predetermined time. A typical unit uses four rotating arms, each powered separately. Each arm can be indexed forward or backward independently. Thus, a mold may be indexed into the load/unload station immediately upon cooling, without waiting for the oven cycle required when a fixed turret carousel (where arms are fixed relative to each other) is used. This particular machine can handle molds weighing up to 3000 lb and can produce products up to 9 ft in diameter and 14 ft long.

Another variation involves a special unit designed to handle up to 5000 lb (mold plus part) per arm, in sizes from rectangular shapes

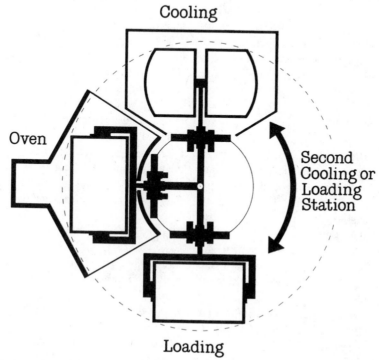

Fig. 14-5. Basics of carousel-type machine, indicating the three stations involved. (*Courtesy McNeil Femco*)

Fig. 14-6. Typical continuous-type three-arm machine. (*Courtesy McNeil Femco*)

$5\frac{1}{2}$ ft sq. by 12 ft high up to cylindrical 15-ft diameter tanks with 8 ft straight walls topped with a 4 ft high dome, or boat hulls up to 14 ft long. As distinguished from the normal carousel machine, the various stations in this unit (heating, cooling, loading/unloading) are laid out in a circle around a central turntable, with each mold arm mounted on a self-propelled carriage actuated by its own motor to drive the unit shaft. Each carriage rolls on tracks from the turntable into the various stations. The mold arms can thus handle tremendous weights since they are supported on each end, not cantilevered.

Straight-Line Machines. Used primarily for molding large parts, this is a shuttle carriage machine that is generally a straight-line operation with the oven on one side, the loading/unloading station in the middle, and the cooling station on the opposite end. The carriage is guided on parallel tracks that insure positive placement in the stations. Variations of this particular design, involving the placement and relationship of the oven, cooling stations, and loading/unloading stations, are also available for handling particular jobs.

Jacketed-Mold Machines. These units offer precise temperature control up to 300°C heat-transfer media temperature. This type of machine finds its biggest market in molding heat-sensitive polymers due to its accurate temperature control.

Key to the machines are double-walled jacketed molds that are charged with a hot liquid, usually oil, to attain temperature control. When the heating period is over, the oil is drained and a coolant is introduced.

The molds rotate biaxially through all production operations. Significant savings are effected because the heating media retains its heat and the heat transfer is more rapid, resulting in faster thermal cycling at lower temperatures.

Thus, while the initial cost of machinery and molds is higher for hot oil machines, the more efficient thermal cycle and potentially lower operating cost could be an advantage. Since the media temperature can be accurately controlled, this type of machine has been used to polymerize caprolactam directly in the mold.

Molds

Molds for convection units are generally inexpensive; however, this is entirely dependent upon the quality level of the end product required, the type of plastic being molded, and the operating temperature to be used in the process. Additionally, except for very large parts multiple cavities are used (Fig. 14-7).

Fig. 14-7. Multiple-cavity rotational mold. (*Courtesy U.S.I. Chemicals*)

Cast aluminum molds are probably the most widely used, and are the most practical for small to medium parts requiring a number of cavities. Wall thicknesses vary from $\frac{1}{4}$- to $\frac{3}{16}$-in. for use in hot-air machines, and up to $\frac{1}{2}$-in. for molten salt machines. Initial cavity cost may be relatively high, since a model and/or pattern are required. However, subsequent cavities are moderate in price, and a fair reproduction of the mold surface on the finished part may be expected. The process used for casting aluminum molds is somewhat specialized; an experienced rotational casting moldmaker is required.

Electroformed nickel molds are best where precise detail is required on the finished part, or where no parting line can be tolerated. This type of mold is not as durable as cast aluminum or sheet metal; however, it is widely used by molders of such objects as automotive headrests, armrests, etc., and by plastisol molders.

Sheet metal. For extremely large parts or single cavities requiring inexpensive tools, a simple sheet metal mold is generally adequate. Prototype molds are often fabricated in this method for reasons of cost, though eventual production molds are usually made of cast aluminum.

Open areas may be molded by simply insulating the mold in the area where plastic is not desired. Some flash may result, but it can be easily trimmed. Inserts can be molded in place by locating the inserts inside cavities while loading the mold.

When it is essential to maintain atmospheric pressure inside the mold during casting, a tube may be inserted through the mold. In effect, this vents the mold and prevents a vacuum on the mold interior as the part is cooled. This vacuum would cause flat parts to warp or cause blow holes at the parting line of the mold.

Molds may be attached directly to the extremity of the major axis of the machine arm, or may be mounted on a "spider" or holding platform built onto the extended arm.

Materials

Most thermoplastics can be used in rotational molding. The most popular, however, continues

to be the polyethylenes, the first powdered plastics to be used in this field. Both low-density and high-density polyethylene are being rotationally molded today.

Also used are: butyrate, propionate, polybutylene, nylon, polyacetal, polycarbonate, polyurethane, polyester, polypropylene, polystyrene, PVC plastisols, and PVC powders.

The physical properties of the molded parts usually vary considerably from injection molded parts. This is due to the long thermal cycles.

Some thermosets are being roto-molded on a limited scale, but large-volume use of these resins awaits solutions to a number of technical problems. A future breakthrough in this area could well revolutionize the industry, since thermosets do not require heat input to set up. This would eliminate the need for a high temperature oven (and the related problems in an era of energy shortages).

Composite rotational molding is a recent innovation in which two resins with different melting points are combined in the mold. One resin melts first, coating the mold and forming the outer layer of the product; then, the second melts and fuses to the first. The following products are typical of those made by this process: a lamp globe with a tough outer skin of butyrate, combined with a polystyrene interior to produce a diffused light, and an ice bucket with a polystyrene core and a polyethylene skin.

Also newly available for rotational molding is a thermally crosslinkable high-density polyethylene that offers outstanding environmental stress cracking resistance, impact strength, and over-all toughness. (Figs. 14-8 and 14-8a.)

For other special applications, it is possible to add fillers to low- and medium-density polyethylene to roto-mold a product with increased stiffness. For example, chopped strand fiber glass, about $\frac{1}{8}$ in. in length, can be added to higher melt index, low-density polyethylene resins. Satisfactory rotational moldings can be produced at about 8% glass in complex molds and up to 15% glass in simple molds. These reinforced articles approach the rigidity of high-density polyethylene (Fig. 14-9).

Most resins used in roto-molding are in the powdered form, but PVC and polyurethane are

Fig. 14-8. Auto bucket seat has rotationally molded crosslinkable high-density polyethylene structural shell. (*Courtesy Phillips Petroleum Co.*)

Fig. 14-9. Planters rotationally molded from fibrous glass-reinforced low-density polyethylene. (*Courtesy Chemplex*)

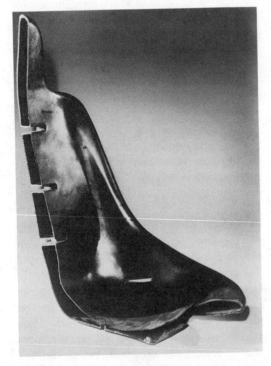

Fig. 14-8a. Cross-section of seat shell (see Fig. 14-8) shows double-wall construction. (*Courtesy Phillips Petroleum Co.*)

available both as liquids and as powders. Liquid PVC (plastisol) cycles in a shorter time, but the properties of both are about equal. See Chapter 16 on Processing Vinyl Dispersions.

Some recent developments have taken place in the rotational casting of liquid nylon caprolactam. During heating, the anionic polymerization of the liquid monomer takes place simultaneously with the molding, so that products like nylon tanks can be produced at a cost far below that of steel tanks, due to the relatively low cost of the materials. This process does, however, require expert technical assistance and should not be attempted without this aid.

Part Design

Dimensional tolerances of ±5% are the present general limits of roto-molding, both in lineal dimension and wall thickness. Tolerances are a function of shrinkage in the mold. Low-density PE shrinks less than high-density PE; shrinkage rates for other plastics vary with no particular relationship to their density, melt index, or any other characteristic.

Final size of the part is affected by both the

Design Tips

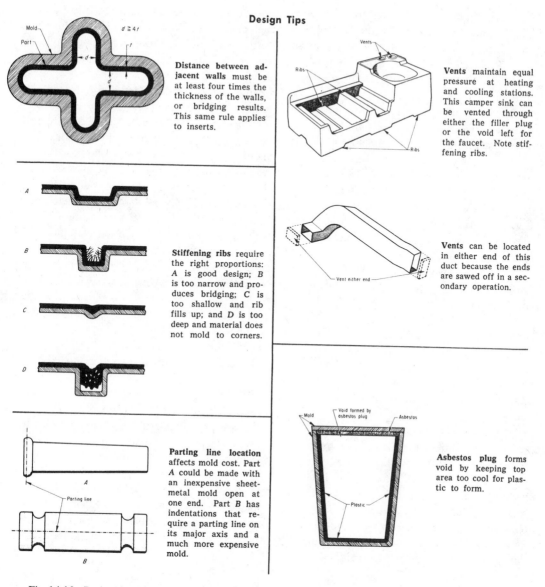

Distance between adjacent walls must be at least four times the thickness of the walls, or bridging results. This same rule applies to inserts.

Vents maintain equal pressure at heating and cooling stations. This camper sink can be vented through either the filler plug or the void left for the faucet. Note stiffening ribs.

Stiffening ribs require the right proportions: A is good design; B is too narrow and produces bridging; C is too shallow and rib fills up; and D is too deep and material does not mold to corners.

Vents can be located in either end of this duct because the ends are sawed off in a secondary operation.

Parting line location affects mold cost. Part A could be made with an inexpensive sheet-metal mold open at one end. Part B has indentations that require a parting line on its major axis and a much more expensive mold.

Asbestos plug forms void by keeping top area too cool for plastic to form.

Fig. 14-10. Design tips for rotationally molded products. (*Courtesy McNeil Femco, Div. of McNeil Corp.*)

rate and amount of shrinkage during cooling and part configuration. The degree of shrinking also depends on the adhesion of the plastic to the mold during cooling and how this condition changes as the part cools. Machines that can apply internal air pressure to hold the part against the mold help minimize shrinking. Different shrinkage rates and wall thicknesses sometimes occur within a given part. (Flat sections may vary as much as ±5%.) Like warping, this problem can also be minimized by adding reinforcing ribs or frames behind the flat sections.

Wall thickness can be decreased or increased to the limits of roto-molding (0.030 to 0.5 in.) by simply adjusting the amount of the charge and the cycle time. Heat-insulating plugs can be used to reduce wall thickness or to eliminate the wall entirely in a given area. Wastebaskets, for example, are made by using an asbestos disc at the open end to prevent plastic formation. Wall thickness can be maintained to a lesser degree by controlling the ratio of rotation so that the area that requires a thicker wall is at the bottom more often than the rest of the part. This technique might be used in making a tapered hopper, for example, with an opening at both ends.

Surface detail can be controlled only on the outside surface since the molds are female, and the inside of the part is in contact only with the air inside the mold. Cycle time and mold temperature determine the inside finish. Too little time or heat leaves part of the powder unfused, causing a rough texture on both surfaces. Too much time or heat causes degradation of the plastic, discoloration, or scorching.

Generally, roto-molded parts have a good outer surface appearance without sink marks. Special finishes, particularly matté or any type of grain are easily obtained. Uniform, high gloss surfaces are more difficult. Surface gloss is simply a reflection of the glossiness of the mold, and practically any plastic will produce a glossy surface if the mold is smooth and polished. However, a glossy finish on the mold increases its price, but still much less than a comparable injection mold.

Injection molding is usually superior to roto-molding where surface detail includes sharp edges. The high pressure of injection molding forces the plastic into these mold cavities to produce sharper edges, whereas edges tend to become rounded off in roto-molding. Blow molded parts, because of flash, trim, and thin spots at the corners, are usually inferior in appearance to roto-molded parts. Easy flowing, high-melt-index plastics generally produce the best inside surface. Low-density polyethylenes give a better inside surface than the high-density versions.

Inserts are as easily added in roto-molding as in injection molding. They may be of any size, but they must be positioned so that they will be completely surrounded by plastic, and anchored firmly. The insert should also be spaced away from any wall by at least four times the thickness of the wall to avoid bridging. Practically all inserts are metal, but higher melting plastics can also be used. Nylon is the most common plastic insert material.

In some cases, it may be more economical to leave a void in the part and make the insert a secondary operation. One way to create a void is to use an asbestos insulating plug to keep the void area cool during the heating cycle. Another approach is to use a fluorocarbon plug to prevent the plastic from adhering to a specific area. In either case, there may be a small amount of flash to be trimmed.

Part strength. Because only atmospheric pressure is used in the process, roto-molded parts are virtually stress free. Material tends to build up in the corners where more strength is needed, as opposed to blow molded products which tend to thin out at the corners. The naturally uniform wall thickness of roto-molded hollow containers provides structural strength that is generally superior to that obtained in blow-molded products. (See also Fig. 14-10.)

15

CALENDERING

The calendering process as used in the production of plastic webs is a continuous, synchronized method for converting raw materials into a flux and then handling the paste-like mass through the nips of a series of cooperating rolls into a sheet of specified thickness and width. This web may be polished or embossed, as required, and may be of either a rigid or flexible character with the "hand" controlled by the formulation and process.

The calendering of plastics involves heat as a catalytic agent and, therefore, differs from the paper process which it resembles to a great extent. Calendering is applicable to a wide variety of plastic resins, although it is best known in the formation of vinyl and ABS sheets. Some high-density polyethylene, some polypropylene, some styrene, and a variety of lesser known materials are calendered. The basic limitation of the calendering operation is the necessity to have a sufficiently broad melt index to allow a temperature range to the process. This permits the material to have relatively high viscosity in the banks of the calender. As a result of the viscosity, a shear effect can be developed throughout the process and especially between the calender rolls. Therefore, the calender forms the web as a continuous extrusion between the rolls. Unlike the process in an extruder or injection press,

the plastic mass cannot be confined when being calendered. Thus the shear effect and broad melt band are essential for calendering.

A typical calender plant (Fig. 15-1) starts with the inventorying of raw materials in either silos for powders, or tanks for liquids. The raw ingredients are delivered by tank cars or tank trucks and are then either pumped (if liquids), or airveyed (if dry) into the storage media. The only major ingredient not so handled is titanium dioxide which has a tendency to cake if stored in bulk.

Minor ingredients such as expensive stabilizers, pigments, and metallic or pearlescent flakes are manually handled. Then, based on the compounder's recipe, various proportions of the materials are delivered to blenders where a powdery premix is produced. This mix allows for the distribution of all ingredients and, in the case of plasticizers, permits the absorption of these liquids into the dry powdery resin.

The blending of the recipe is a most critical part of the entire process. Blending must produce a uniformly colored and stabilized product in powder form. After blending, the rate of consumption dictates the temperature of the process. Since the plastic is processed between the process temperature and its critical temperature for degradation, time becomes an important part of the process. The residence time of the plastic flux at high temperatures must be limited.

*By S. Everett Perlberg, S. Everett Perlberg, Inc., Hackensack, N.J.

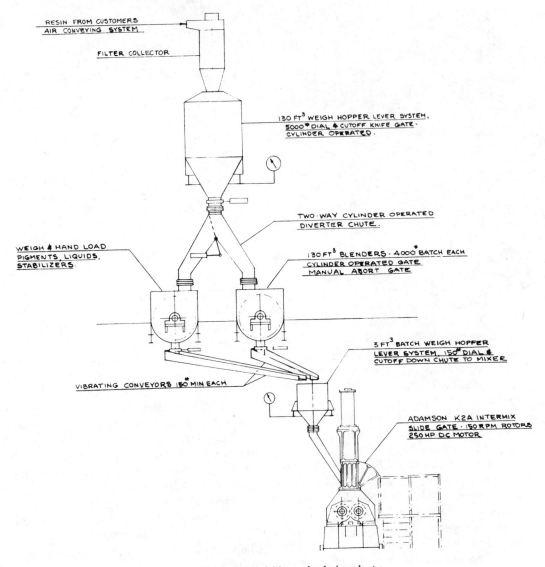

Fig. 15-1. Flow of material in a calendering plant.

The feed to the blenders may be either manual or automatic, and in some cases is even computerized and controlled by punch cards or tapes. Weighments are frequently made automatically and in sequence. Manually fed additives such as pigments or stabilizers must be acknowledged in an automatic system to be sure of the fact that they have been added.

Types of Blenders

Several types of blenders have been used including high speed, high intensity types, exotic types of various configurations, and ribbon blade types. While all produce the same result, the ease of cleaning and quality of product from the blender should determine the type used. Heat can either be added to the blender by means of jackets, or can be developed in the blender by conversion of mechanical energy. Residence time in the blender is determined by the processing character of the blend and the blending temperature. The blender cycle and flux rate dictate the blender size. For example, a blender cycle of 60 min. and a flux rate requirement of

5000 lb/hr would necessitate a blender capacity of 10,000 pounds usually divided into two 5000 lb. units. A smaller unit might be desired for sampling or short runs. Similarly, a blender cycle of 2 min. and an output rate of 5000 lb/hr would necessitate a blender capacity of about 165 lb per batch. A second blender would still be needed to permit time for color change clean-up. A small blender would not be required for sampling or short runs since the batch size is not large.

From a capital investment standpoint, the high speed, high intensity blender might seem to have an advantage. This is not as true as seems evident at first glance. For instance, scrap handling is not easily accomplished through the high intensity blender. Secondly, no opportunity exists for establishing the quality of the blend. Unless some other control can be developed, the accuracy of small frequent weighments is not as good as that of large weighments. Thus, the majority of calendering plants utilize ribbon blenders with a relatively slow cycle, and more uniform weight and temperature control.

After blending of the compound in accordance with the requirements of the chemical formulation, the material is fed to a fluxing unit where the addition of heat converts the product from a powder to a hot paste called a *flux*. The transformation requires the addition of a great deal of mechanical energy and may be accomplished in a continuous or batch-type intensive mixer. Continuous mixers may be of either the screw type such as a single-screw extruder, dual or multiple screw extruder, extruder with high compression ratio and a number of shear planes, or a combination screw and rotor unit.

Batch-type mixers are composed of a pair of rotors encased in a chamber wherein the work is done either between the rotor tips and the body as in the sigma blade mixer, or between the rotors as in the interlocking rotor type mixer (Fig. 15-2). Dependent on clearances, rotor speeds, and pressure on the batch, the connected horsepower can be determined, and the conversion of the mechanical energy of the motor establishes the throughput rate, or batch cycle time. For high shear materials, such as

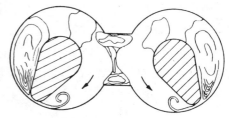

SIGMA BLADE MIXERS

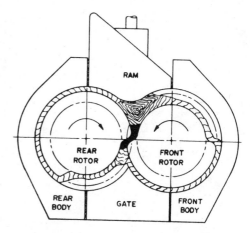

Fig. 15-2. Batch-type mixers.

rigid plastics, cycle times as short as 60 sec. are possible. For low shear materials such as highly plasticized products, cycle times may increase 50%. Batch sizes are established by the mixer capacity based on chamber volume.

In continuous type fluxing units, the throughput rates are limited by virtue of the fixed residence work versus time relationship in the machine. Thus, even though the time in the fluxing unit is shorter, the energy per pound input remains constant, and the ability of the machine to extract heat is the determinant in establishing the range of throughputs. A graph of this effect is shown in Fig. 15-3. Another difficulty encountered in the use of continuous type mixers is that of scrap feed. The hoppers of these units are relatively small to prevent plastic regurgitation, and the scrap particles have a tendency to bridge or clog up the hopper.

Scrap and Cold Trim

Scrap and cold trim handling from the product line pose one of the most difficult

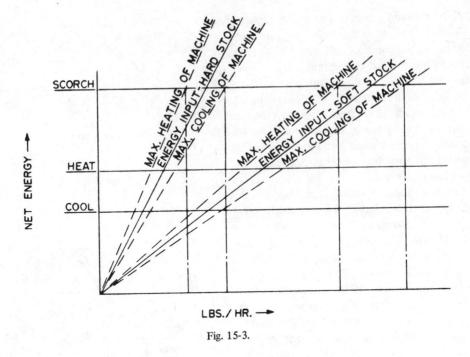

Fig. 15-3.

problems. Scrap and trim represent from 10 to 40% of the product mix, dependent on the width of the calender in relation to the sheet width. The flux rate and energy required for remelting the scrap are considerably less than that required to flux the virgin resin. The result is that a danger of decomposition of material exists whenever scrap is handled, and yet for optimum uniformity, it is necessary to prepare the compound so that a blending of both new and old material allows for a standardized product. The reprocessed material is therefore best added to the blender where the standard can be established, although in many plants it is fed directly into the fluxing unit. Careful control of the scrap percentage to the total mix is essential to obtain a good quality product.

In turn, the converse of material proportioning poses problems in quality control. For a standardized quality, the scrap level generated in the process line may be more or less than required at any given time, so a reservoir or storage of scrap is mandatory. Since the cost of reprocessing material is high in relation to the profit margin, greater economic gain can be made by keeping the rework to the minimum. In addition, color degradation and contamination frequently result from varying percentages of scrap in a given run.

The power requirements needed to directly convert scrap to flux are a small part of the total cost of the rework. They amount to only 37.5 KW per 1000 lb of material. To this, however, must be added the cost of granulating, which can approximate the same power cost, the cost of milling, if any, straining, calendering, and cooling. The total power consumed per 1000 lb of reworked material will be in the neighborhood of 350 hp or 270 KW. Assuming a power cost of 1¢ per KWH, the electrical cost alone represents approximately $\frac{1}{4}$¢ per pound. To this must be added handling costs, labor costs, additional plasticizer costs and, if necessary, additional stabilizers. These additives are needed because some of the liquids are volatilized during the original processing. Pigments may also have to be introduced to keep color matching control. Then, the cost per square foot of floor space for the scrap storage must be considered.

Mills and Strainers

The discharge from the intensive mixer may be fed to a mill or directly to a strainer. The mill, if used, has the disadvantage of being manually controlled and is, therefore, less uniform in the processing of vinyls.

The size of the mill is determined by the throughput rates. Yields of up to 3000 lb/hr can be handled on a 22 x 22 x 60 in. mill, while above 3000 lb and below 6000 lb, a 26 x 26 x 84 in. mill should be used. If more than 6000 lb/hr is to be processed, additional mill capacity will be required in the form of a second or third unit.

The mill can strip feed the strainer, or the strainer can be fed by a batch from an intensive mixer.

The size of the strainer should be determined not by the theoretical throughput rate, but rather by the temperature buildup that can be tolerated in the machine. A small strainer must have the screw running at a higher speed and, therefore, the frictional heat generated becomes greater than a slow screw speed on a larger diameter. If all of the material to be processed is highly plasticized, a strainer size could be one size smaller than if hard materials have to be handled. Generally, strainer temperature rise should be limited to $15°F/lb$. To achieve this result, it is necessary to determine the maximum flux capacity of the line and establish a diameter for the screw which will permit the discharge rate at the limited temperature rise prescribed.

Calender Design

The material is fed from the strainer to the nip of the calender where, for the first time, the material assumes the form of a sheet. This sheet is then progressively pulled through two subsequent banks in order to resurface each of the two sides. Plastic calenders are generally made in four basic configurations:

1. The "L" calender, wherein the offset roll is on the bottom, and the take-off is from the top roll (Fig. 15-4).
2. The Inverted "L" calender, wherein the offset roll is on the top and take-off is from the bottom roll (Fig. 15-5).
3. The "F" calender wherein the offset roll is on the top, and the take-off is from the

middle roll, on the offset roll side (Fig. 15-6).

4. The "Z" calender, either flat or inclined. This calender has two offset rolls, and the take-off may be made either from the top or bottom offset roll, or from the back side of the stack roll (Fig. 15-7).

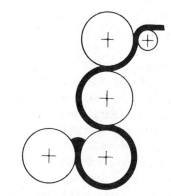

Fig. 15-4. "L" calender.

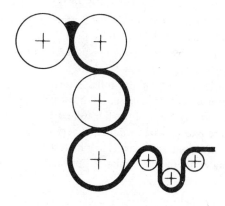

Fig. 15-5. Inverted "L" calender.

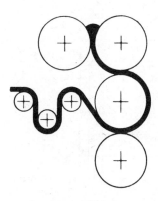

Fig. 15-6. "F" calender.

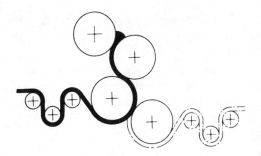

Fig. 15-7. "Z" calender.

The temperature of the plastic material begins to increase sharply as work is done between the several calender rolls. The bank goes into high velocity rotation and develops a vortexing action as it enters the nip. This causes high frictional heats which are of short duration and, therefore, tolerable. Should the temperature rise too high, degradation in the form of yellowing or burning begins to take place.

As the web moves through the calender nips, it is widened out. The width of the sheet on the calender rolls is established by the final product trimmed width (Fig. 15-8). The sheet on the number two roll should be at least the width of the trimmed product in order to reflect the surface finish of the rolls. The fluid temperature in the drill passes of the calender rolls is invariably lower than the roll surface temperature except when running at very low speeds of 10-15 ypm on plasticized materials. On rigid materials the heat balance between roll surface

and product occurs at much lower speeds than on a plasticized material.

Since plastic webs release better from cold rolls than from hot rolls, it is frequently desirable to cool the surface of the roll from which the web is "taken off." This is difficult to accomplish when taking off the number three roll.

The take-off or stripper section of the calender usually consists of a series of rolls of suitable diameters, such as 8 in., arranged to form an "S" wrap (Fig. 15-9). These rolls run at a higher surface speed than the calender rolls in order to put a positive draw into the sheet. This draw is desirable from the standpoint of yield since the hot web is stretched by the percentage of speed increase over the finishing roll of the calender. Little heat buildup is encountered at the stripper rolls. The amount of stretch is determined by the orientation desired in the plastic product. If low residual shrinkage is required, this stretch must be limited. For some products, however, residual shrinkage is not a criteria of physical properties.

After leaving the stripping section, the web is fed into an embossing unit which can be arranged either vertically or horizontally. The embosser consists of an engraved steel roll pressing against a rubber covered roll. The hardness and thickness of the rubber covering together with the pressure applied by the steel roll determines the depth of the indentations or embossing in the plastic sheet. In order to remove plasticizers and to keep the rubber cover cool, cooling water is circulated inside the steel arbor and on the outside of the rubber

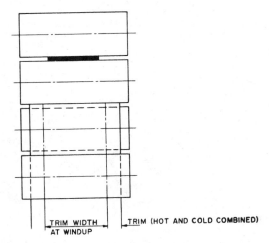

Fig. 15-8.

Fig. 15-9. "S" wrap, 8 in. diameter stripper rolls.

covering. The water on the rubber surface must be squeegeed off before contacting the plastic to prevent water marks on the film.

The wrap of the film around the embossing roll is controlled by a stripping roll. This roll should be adjustable in position to provide from 0 to 180° of contact. The embossing roll is cooled to a temperature well below the film surface. This chilling effect tends to set the film surface. Frequently, an embossing pass is used for no other reason than to reduce the temperature of the plastic.

From the embossing roll section, the web is carried through one or more series of cooling drums and the beta ray gauge is usually placed between these drums to determine the basis weight. Basis weight can then be correlated to thickness, depending on the formulation used. The beta ray gauge can be used to control product thickness, either automatically by adjusting the bottom roll position, or manually by showing the operator an out-of-tolerance condition.

Cooling is a very inefficient operation and since some of the heat remains in the center of the plastic sheet, cooling is inversely proportional to the sheet thickness. It is necessary to have more cooling cans for heavy gauge products than for thin products. Actually, at the same speeds, a sheet of double the thickness takes six times longer to cool. If the film is wound up too hot, it has a tendency to shrink during the cooling process and undesirable results such as wrinkles, reduction in embossed pattern depth, and exceptionally tight rolls develop in the product.

Between the cooling drums and the windup is an edge slitting or trimming unit which removes the excess sheet width. The slitters are adjustable in position so that varying widths can be run on the same calendering line. The cold trim should be at least 2 in. per side, more if required.

Careful market analysis should be made to establish calender width. When vinyl films were first produced commercially, it was necessary to provide a wide range of widths from a single calender. With the number of calenders existing in each plant, it is usually sufficient to have only one for wide films. The other calenders should be tailored to suit the product widths based on marketing requirements.

Calendering lines designed for rigid processing are essentially the same as those for plasticized materials except that the temperature ranges must be higher. Since plasticized processing temperatures require better heat extraction from the roll surfaces, it is desirable to use high pressure water circulating at high velocities for good heat transfer. For processing rigid products, higher roll surface temperatures are required, and rather than use extremely high pressure water systems, the tendency has been to use low pressure circulating oil heating

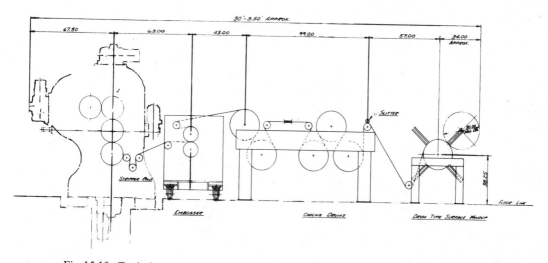

Fig. 15-10. Typical set-up for inverted "L" calender and train. (*Courtesy Adamson United*)

systems. The thermal conductivity of oil, based on its specific heat, is poorer than that of water. Therefore, it is essential to determine the type of product to be produced before the calender heating system is selected.

The windups will be different for heavy gauge rigid than for light gauge plasticized materials and the core sizes on which the rolls are wound are of major importance.

There has been a tendency to use automatic "cut and start" center wind units for many applications where it would have been desirable to use surface type windups. Windups are still one of the major problems in many processing plants simply because they have become too sophisticated for the range of applications currently being made.

Much study is presently taking place on efficient methods of calendering of plastic products, and it is believed that a substantial and drastic revision of process methods will develop during the seventies.

16

VINYL DISPERSIONS

Vinyl dispersions are fluid suspensions of special fine particle-size polyvinyl chloride resins in plasticizing liquids (Fig. 16-1). The plasticizers, e.g., esters of phthalic anhydride and 6 to 13 carbon aliphatic alcohols, have little effect on the resin at room temperature. When the system is heated to about 148 to 177°C (300 to 350°F), fusion (mutual solubilization of resin and plasticizer) takes place. The dispersion turns into a homogeneous hot melt. When the melt is cooled below 50 to 60°C (122 to 140°F), it becomes a tough vinyl product.

With vinyl dispersions, the processor can use convenient liquid handling techniques—spraying, pouring, spread coating, dipping—to make products which would otherwise require heavy melt processing equipment.

The term *plastisol* is used to describe a vinyl dispersion which contains no volatile thinners or diluents. Plastisols often contain stabilizers, fillers, and pigments along with the essentials, dispersion resin and liquid plasticizer, but all ingredients have very low volatility under the processing and use conditions. Plastisols can be made into thick fused sections with no concern for solvent or water blistering as with solution or latex systems; hence they are described as being "100% solids" materials.

* By C. A. Clark, Senior Development Scientist, Avon Lake Technical Center, B. F. Goodrich Chemical Co.; T. J. Kraus, Sales Product Engineer, B. F. Goodrich Chemical Co.; D. W. Ward, Development Technical Leader, Avon Lake Technical Center, B. F. Goodrich Chemical Co.

It is convenient in some instances to extend the liquid phase of a dispersion with organic volatiles which are removed during fusion. The term *organosol* applies to these dispersions.

About 10% of the polyvinyl chloride resin used in the United States is dispersion grade resin. Thin and heavy gauge, solid and cellular coatings are continuously applied to moving substrates of cloth, metal, and paper to make upholstery (Fig. 16-2), floor coverings, wall coverings, prefinished metal building panels, and integral carpet cushioning. A typical modern automobile incorporates a number of different vinyl dispersion applications (Fig. 16-3). Molded and dipped articles made from fluid dispersions compete with high-pressure-molded products, particularly in the production of hollow items from low cost molds and machinery (Fig. 16-4).

Historical

The development of flexible vinyl compounds in fluid dispersion form was a goal of the early researchers in the 1930-1940 period. In Waldo L. Semon's patent disclosure[1] of the plasticization of poly(vinyl chloride) in 1933, he mentions application of highly plasticized combinations by the spread coating of hot melts. In 1939 and 1940, patents were issued in Great Britain[2] and the United States[3] which fully

[1] Superscript numbers refer to References on p. 421.

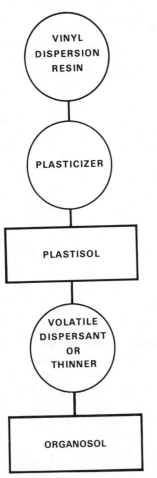

Fig. 16-1. Types of vinyl dispersions. (*All photographs and illustrations in this section (unless otherwise indicated) are courtesy of B. F. Goodrich Chemical Co.*)

Fig. 16-2. Upholstery based on vinyl dispersions.

describe vinyl dispersions as we know them today.

The discovery of the plastisol method for processing polyvinyl chloride was one of the citations included in the 1968 presentation of the Morley Medal to Dr. Semon of the B. F. Goodrich Company by the Cleveland Section of the American Chemical Society.

The earliest commercial resins were a product of the Union Carbide Chemicals Co. fellowship at Mellon Institute. These emulsion-polymerized resins could be dispersed by grinding in an organic media to make stable dispersions, provided the suspending phase contained enough polar liquids to have some solvating action on the resin but not enough to dissolve

it. This technique was introduced to the cloth-coating industry in 1943, and the resin was introduced commercially in 1944. The significance of this development was recognized by the plastics industry with the award of the John Wesley Hyatt Medal in 1950 to G. M. Powell of Union Carbide.

Notable progress in the field of vinyl dispersions occurred in 1947 with commercial introduction, by the B. F. Goodrich Chemical Company, of a plastisol-type resin which simplified the preparation of dispersions. This resin could be mixed with relatively low levels of plasticizer by a simple stir-in technique and volatile thinners were no longer required.

Resin developments over the last twenty-five years have been primarily improvements in the stir-in type aimed at special properties such as fusion below 148°C (300°F), high gloss, predictable viscosity behavior, and reactive functionality.

In the mid-70s, there were significant changes in government regulations relating to vinyl chloride and occupational health. These are discussed at the end of the chapter.

Resins

Resins for vinyl dispersions are unique from those designed for calendering, extrusion, or

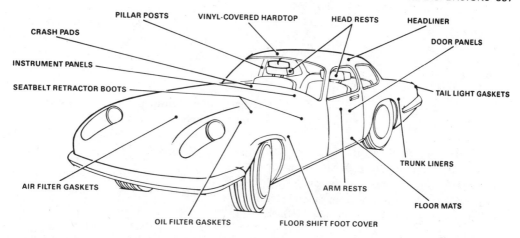

Fig. 16-3. Automotive applications.

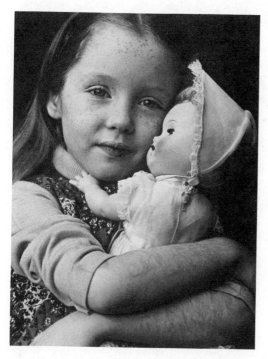

Fig. 16-4. Dolls made from vinyl dispersions have the "real-live" touch.

injection molding. The latter make up the bulk of the resins consumed in the vinyl industry and are mostly the result of suspension polymerization. The most common suspension resins are coarse, porous blends of particles in the range of 75 to 200 microns.

Vinyl dispersion resins are polyvinyl chloride homopolymers and copolymers with vinyl acetate. They are polymerized as emulsions of very fine (0.2 to 15.0 microns) particles in water. The water is removed, usually by spray drying, and the resultant loose agglomerates are reduced by dry grinding. The product is a mixture of the ultimate individual particles and particle clusters no larger than about 50 microns.

For good flow characteristics, vinyl dispersions require that at least 50% of the polymer be dispersion grade resin. In some applications the remaining polymer can be a blending resin selected from specially prepared suspension and bulk polymerized PVC.

In 1974, the suppliers of resins for vinyl dispersions were as follows:

Suppliers of Resins for Vinyl Dispersions

Company	Trade Name
Borden Chemical	Borden VC
Diamond Shamrock	Diamond PVC
Ethyl	Ethyl EH
Firestone	FPC
B. F. Goodrich Chemical Co.	Geon
Goodyear	Pliovic
Stauffer Chemical	Stauffer SCC
Tenneco	Tenneco
Union Carbide	Bakelite
Uniroyal	Marvinol

A wide variety of resins and technical assistance is available from these suppliers. One supplier provides sixteen separate resins for vinyl dispersion manufacture.

Rheology

Viscosity (η), the resistance to liquid flow, can be more precisely defined as the ratio of shear stress (τ) to shear rate ($\dot{\gamma}$) in laminar flow.

$$\eta = \frac{\tau}{\dot{\gamma}} = \frac{\text{shear stress}}{\text{shear rate}} \qquad (1)$$

Shear stress is the tangential force per unit area applied to a liquid layer. Shear rate is the ratio of the resulting velocity of the layer No. 1 in Fig. 16-5 to the distance from a reference layer No. 2. Shear rate is more precisely defined as the rate of change of velocity with distance in the system of laminae, thus dv/dr.

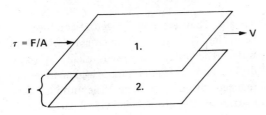

Fig. 16-5. Schematic illustrating laminar flow.

In simple systems, viscosity is independent of shear rate (Newtonian example in Fig. 16-6). Few plastisols exhibit this type of flow behavior and it is necessary to consider application of shear rate when compounding for flow properties and to approximate it in some way when measuring the viscosity of the compounded dispersion. Most dispersions show the shear rate thinning, thickening, or mixed dependencies of Fig. 16-6.

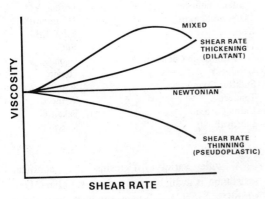

Fig. 16-6. Types of flow, illustrating shear rate dependence.

The particles in vinyl dispersions interact to form temporary structures and the dispersion viscosity reacts to the time necessary to break down and rebuild this structure. Time dependency, as well as shear rate dependency, must therefore be considered in selecting instruments and test methods for viscosity measurement. Common terms for shear rate and time dependencies are:

Shear Rate Thinning (pseudoplastic)—decreasing viscosity at increasing shear rate
Shear Rate Thickening (dilatant)—increasing viscosity at increasing shear rate
Mixed—shear rate thinning and thickening at different shear rate ranges
Rheopectic—increasing viscosity with time of agitation at constant shear rate
Thixotropic—decreasing viscosity with time of agitation at constant shear rate

Plastisols may also exhibit "yield value," a limiting stress required to initiate *any* flow. Methods of measuring yield value have been described.[4] They all require extrapolation of shear stress/shear rate data to an intercept with the stress axis and, thus, can be influenced by thixotropy.

Our discussion of time dependent phenomena up to this point has dealt with short-term, reversible viscosity changes caused by structures that will rebuild after being broken. Irreversible viscosity changes also occur over relatively longer time-aging periods of the order of days or weeks.

Although plasticizers are very weak solvents for PVC, some absorption into the particles occurs and some deagglomeration of clusters takes place with aging. Dispersion viscosity rises fairly rapidly during the initial hours of aging, tapers off in days and, after about a month, no more aging can be measured (Fig. 16-7). Evidently, plasticizer absorption is limited by the internal crystallite structures in the particles, and deagglomeration finds a limiting particle-particle bond strength which cannot be broken in the clusters. In any event, plastisols have been stored as long as 24 years and were still pourable.

Because of the shear rate and time dependence of dispersion viscosity, the compounder

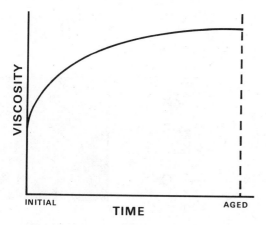

Fig. 16-7. Viscosity aging of vinyl dispersions.

must consider application shear rate and expected plastisol age when designing and testing formulations.

Accurate determination of application shear rate in an actual production situation is difficult. Approximations are possible. For example, the shear rate of a knife over roll coating going on a web (see Fig. 16-26) can be calculated by dividing web speed by coating thickness.

By using standardized measuring techniques and by making comparison studies in the appropriate shear rate *range*, the compounder can select materials and predict production performance quite accurately. Test methods, the rheological effects of compounding ingredients, and specific application requirements are discussed later in this chapter.

Fusion

After the liquid dispersion has been applied in its final coated or molded shape, it must be fused into a homogeneous solid. This conversion is a true fusion (melting) of the crystallite structures in the polymer particles followed by solution of the molten polymer in the plasticizing vehicle. Since no chemical "cure" takes place, the time to fuse is negligible once the stock reaches the fusion temperature.

The fusion temperature of plasticized vinyl is a complex function of molecular weight, amount of comonomer, plasticizer level, and polymer-plasticizer interaction parameter.[5]

Since dispersions are fused with heat alone, without mechanical working, the particle size, particle size distribution, and interference from non-vinyl constituents at particle surfaces also affect fusion temperature.

Since the crystallites in PVC melt over a fairly wide temperature range, some strength is developed at temperatures below the ultimate fusion point. This cannot be increased by increasing the time of heating (Fig. 16-8).

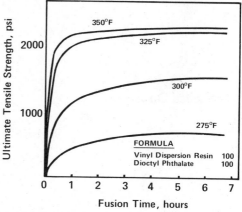

Fig. 16-8. Effect of time on plastisol fusion-temperature-tensile strength relationship.

As the plastisol is heated to the fusion point, it changes in several important ways (Fig. 16-9). In the early stages of heating the viscosity is affected by two competing phenomena. Plasticizer viscosity drops with increasing temperature up to about 38°C (100°F). This causes a decrease in viscosity of the dispersion (A). As the temperature continues to rise, plasticizer begins to permeate the particle, the particles swell (B), and viscosity increases at an accelerating rate. Further heating and more plasticizer absorption lead to contact between swollen particles and a weak bond (C). If the material were cooled to room temperature at this point, it would have a weak, cheesy structure and it would be opaque. Additional heat would produce a clear matrix (D) and a strong product when cooled. Transparency, however, is not as good a criterion of complete fusion as is strength development.

In practice, the fusion temperature of any dispersion system is usually determined em-

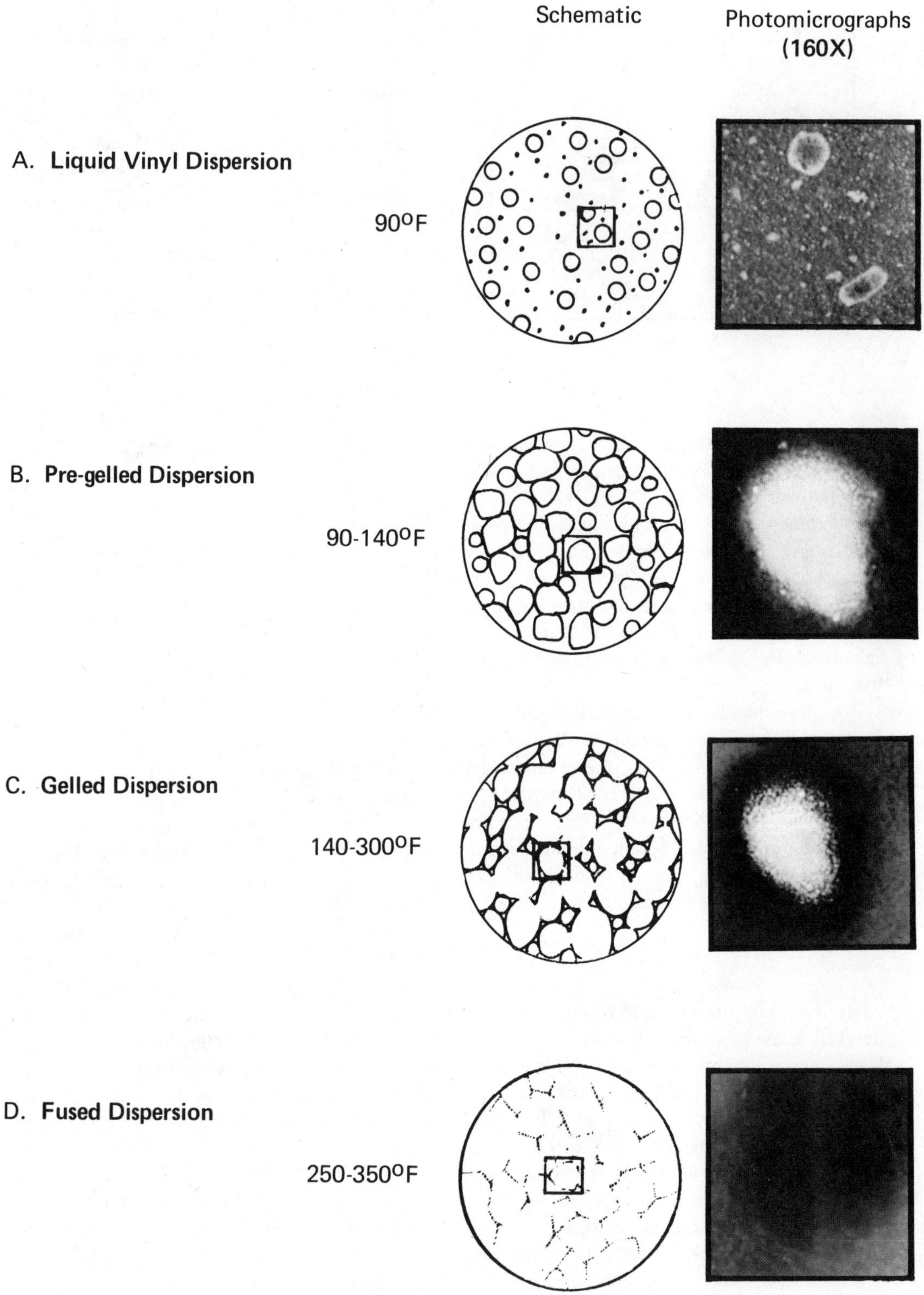

Schematic

Photomicrographs
(160X)

A. **Liquid Vinyl Dispersion**

90°F

B. **Pre-gelled Dispersion**

90-140°F

C. **Gelled Dispersion**

140-300°F

D. **Fused Dispersion**

250-350°F

Fig. 16-9. Sequential effects of heating plastisols.

pirically from a curve of physical properties (usually the ultimate tensile strength and ultimate elongation at break) of cast film heated at various temperatures (Fig. 16-8). The "true" fusion temperature is considered to be the temperature which produces maximum properties.

Dispersion Preparation

Vinyl dispersion resins are supplied in multi-wall bags containing 50 lb. The very fine nature of these resins has presented problems in bulk handling, but systems are near commercialization for the convenience of the large volume consumer. Suspension resins used as blending resins and fillers can be handled in bulk, but their low volume consumption at any one location precludes bulk handling.

Plasticizers, the second largest constituent of vinyl dispersions, are available in drums, tank trucks, or rail cars.

The resins can be dispersed in plasticizer with relatively simple mixers. Only enough shear is needed to break up the loose agglomerates. It is important that the temperature be kept low, approximately 20 to 40°C (70 to 105°F), depending on the sensitivity of the plasticizer and resin to mutual solvation. Low-speed, high-shear equipment is considered best for dispersion. It is often water jacketed for temperature control.

The planetary mixer is typical of this equipment. Two examples of production mixers are shown in Figs. 16-10 and 16-11. Laboratory size models are available.

Internal mixers such as that shown in Fig. 16-12 are also quite versatile. In addition to provision for excellent shear mixing, these units are most frequently provided with water jackets and tight fitting covers for mixing under vacuum conditions. The latter is important for many applications where occluded air on particles and mixed-in air can cause bubbles and blisters on fusion.

Recent developments in internal mixers are shown in Figs. 16-13, 16-14a, and 16-14b. These combine the basic principles of vinyl dispersion mixing with provision for temperature control and vacuum on the mixing chamber.

Fig. 16-10. Planetary mixer with timed mixing control provides thorough blending and mixing. (*Courtesy Hobart Mfg. Co.*)

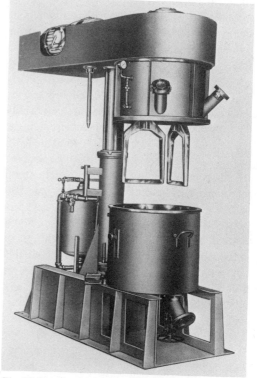

Fig. 16-11. Planetary mixer for low-speed, high shear mixing. (*Courtesy Charles Ross & Co.*)

Fig. 16-12. Universal mixer in discharge position. (*Courtesy Baker-Perkins, Inc.*)

High-speed, intensive mixers such as shown in Fig. 16-15 do an excellent job on special-purpose, low-viscosity mixes such as organosols for metal coil coating. This latter subject is covered in greater detail in the section on applications, below.

Some ingredients, such as pigments, require more intensive mixing than a vinyl dispersion resin. These are ordinarily pre-dispersed on three-roll mills using a viscous plasticizer as a grinding vehicle.

The order of addition of ingredients and the sequence of mixer operations are necessarily determined by the type of mixer and the ingredients which are to be combined. In low-speed high-shear mixers, a plastisol is mixed by charging all the dry ingredients and around 50-55 parts by weight of plasticizer per 100 parts by weight of resin along with any pre-dispersed ingredients. The low level of liquid content results in high viscosity; thus, high shear and the dispersion is usually mixed this way for about two-thirds of the total cycle. At this point the remaining plasticizer and any volatile thinner is charged and the cycle is completed. The total cycle ranges from 10 to about 40 minutes.

In high-speed, high-shear mixing equipment, there is no need to limit the liquid content to get high shear. All—or nearly all—of the liquid can be weighed into the mix pot. Pigments can be added and the mixer can be run to disperse these before resin addition. Running time after resin addition is short (on the order of 3 to 5 minutes). High speed leads to rapid shear heating and these mixes must be carefully time and temperature controlled. It has been demonstrated[6] that a final temperature "quench," by adding withheld plasticizer and thinner just before mix completion, will cool the mix and help reduce viscosity aging.

The prepared vinyl dispersion will contain dispersed air unless the mixing is performed in a vacuum chamber. The air is introduced (1) as a result of the turbulence and (2) from occluded air on any dry powders entering the mix. Ideally, where air in the system can cause bubbles and blisters on fusion, the dispersing takes place in a vacuum. Where this type of mixer is not available, a vacuum is applied to the prepared mix. Two equipment approaches to this are shown in Figs. 16-16 and 16-17. Sometimes, if the application allows and the dispersion easily releases air, simple aging in drums will release most of the mixed-in air but not that occluded on particles. Some compositions are passed through a three-roll or five-roll mill for the same objective.

In the laboratory, a desiccator is frequently used for vacuum deaeration. Some type of mechanical agitation helps this operation by bringing entrapped air to the surface.

After preparation of the vinyl dispersion it is important that the homogeneity of the mix be protected during storage and handling to the point of application. Although there is generally no interaction between carbon steel and vinyl dispersions, coated steel or stainless steel is frequently used. This is done primarily to prevent the inadvertant introduction of rust to the dispersion. Not only would this be a color contaminant, but such iron oxides may cause undesirable resin decomposition in the fusion operation.

The prepared vinyl dispersion needs additional protection prior to application. Pumps, if used, are normally of the positive displacement type and operated in a manner to minimize heat buildup. Air pressured vessels or gravity feed often supplies material to the point of use.

To minimize viscosity increase, storage

Fig. 16-13. Mixer equipped with plow type blade and high-speed choppers which cause the dispersing, mixing action. (*Courtesy Littleford Brothers, Inc.*)

Fig. 16-14a. Production installation with two Conical Nauta mixers. (*Courtesy J. H. Day Co.*)

should be under 38°C (100°F) and as near 24°C (75°F) as possible (Fig. 16-18).

Formulating the Vinyl Dispersion

Much information is available on the general properties of vinyl compounding ingredients.[7,8] This discussion will emphasize the ingredient characteristics which specifically affect vinyl dispersions. In dispersions, the rheological contributions of compounding ingredients are of great concern as is the effect of the ingredients on transitional characteristics such as rate of gelation and fusion point.

Many manufacturers, wishing to utilize the unique adaptability of vinyl dispersions for their operations, may not wish to establish a formulating unit within their organization. Fortunately, there is a large number of companies specializing in this area who can develop formulations for specific needs and supply the vinyl dispersions from drum lots to tank truck or tank car. These specialists are widely distributed throughout the United States and may

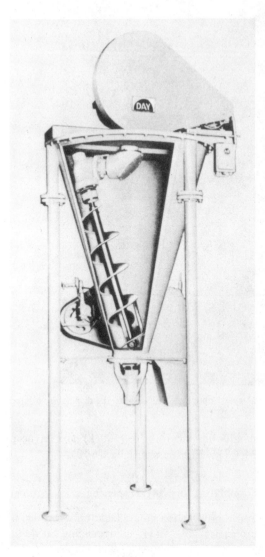

Fig. 16-15. Typical high-speed, high-shear mixer used for organosols. (*Courtesy Morehouse-Cowles, Inc.*)

Fig. 16-14b. Schematic of the mixer shown in photograph, Fig. 16-14a. (*Courtesy J. H. Day Co.*)

be easily contacted through the aid of materials' suppliers and the Society of the Plastics Industry (SPI) National Headquarters. These vinyl dispersion manufacturers are organized as a special Vinyl Dispersion Division of SPI.

Resins There are more than fifty commercial vinyl dispersion resins. Most are homopolymers of polyvinyl chloride. These differ in: molecular weight, particle size, type of emulsifier, and level of emulsifier remaining on the product. Inherent viscosity, or I.V. (see ASTM D1243-66) is the most common measure of

Fig. 16-16. Typical vacuum deaerator system for batch deaeration. (*Courtesy Teknika, Inc.*)

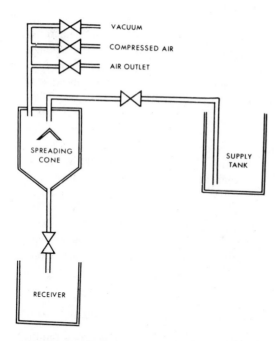

Fig. 16-17. Schematic of a deaeration system for continuous or batch type operation. (*All photographs and illustrations in this section (unless otherwise indicated) are courtesy of B. F. Goodrich Chemical Co.*)

molecular weight. Dispersion resins range in I.V. from about 0.7 to 1.35. As molecular weight increases, many properties are affected. Increasing molecular weight leads to increased fused strength, but slower gelation. It increases the viscosity of the vinyl melt which, in turn, has implications for reduced flow-out on moldings and coatings and requires compound adjustment to give smooth expansion of chemically blown foam. Changes in molecular weight, though they affect fused product strength, have little influence on fusion temperature per se. Fusion temperature is much more radically affected by inclusion of co-monomers like vinyl acetate which break up the crystalline structure in the polymer.

Copolymers are available with about 5 to 10% vinyl acetate content. Increasing the vinyl acetate content lowers the fusion point, but at the expense of the ultimately available strength (Fig. 16-19). Other effects of increasing vinyl acetate are higher initial viscosity, more rapid viscosity aging, softer fused products, more rapid gelation and poorer heat stability at any given processing temperature. Since processing temperature drops with increasing co-monomer content, this should be taken into consideration when making heat stability comparisons.

Residual non-vinyl content of dispersion resins can affect a number of characteristics such as clarity, heat stability, taste, odor, acceptability for food uses, and windshield fogging in autos. Specialized dispersion resins

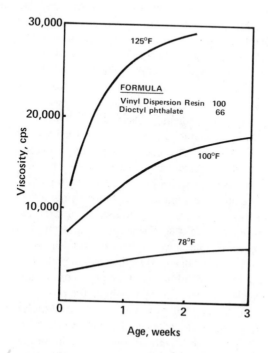

Fig. 16-18. Effect of temperature and storage time on plastisol viscosity.

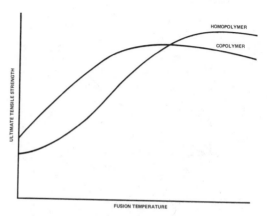

Fig. 16-19. Tensile strength development during fusion, homopolymer vs. copolymer dispersion resin.

which vary in residual emulsifier level and type are available for critical applications. When these specialty resins are low in surface-active materials they can be expected to be sensitive to surface-active compounding ingredients.

Functionally active resins are being developed. A recent publication describes the properties attributable to carboxyl groups on the vinyl chain.[9] The carboxyl functionality increases adhesion to a number of metal and fabric substrates. The effect on nylon is shown in Fig. 16-20.

Resin blending is common practice in the formulating of vinyl dispersions. The objectives are property adjustment and, frequently, lower cost. In addition to the obvious blending of various dispersion resins, extensive use is made of selected, large particle, essentially nonporous vinyl resins made by the suspension and bulk polymerization processes. Upon substitution of the latter for a portion of dispersion resin up to about 45% replacement, the viscosity of the dispersion is lowered because of more efficient particle packing.

Blending resins vary widely in particle size and in inherent viscosity. Particle size is usually stated in terms of the range in which the middle 80% of the particles fall. For typical large particle resins, this is about 50 to 125 microns; for typical fine particle resins, the size range drops to 25 to 50 microns. Coarser resins fuse more slowly, give a matte surface appearance, and may settle in low viscosity plastisols. They decrease viscosity significantly and make formulation of very low plasticizer level, high-hardness compounds possible.

Blending resins are also available in special forms such as copolymer types to hold fusion temperature requirements low, especially when blending is needed in compounds based on copolymer dispersion resins. Other special blending resins are designed to produce high viscosity and flow properties to minimize fabric penetration of coating.

Plasticizers

In dispersions, plasticizers provide the fluidity and also affect the finished properties. Plasticizer type and level have an important influence on viscosity, viscosity aging, gelation and fusion characteristics, and on the end product properties. The reader is directed to encyclopedia issues of industry trade journals[7,8] for over-all descriptions of plasticizers available and to suppliers' literature. Most plasticizers utilized in vinyl dispersions are esters. They may be esters of long-chain alcohols with aromatics such as phthalic anhydride; with straight-chain dibasic acids such as sebacic or adipic; with phosphoric acids; or polyesters such as those based on propylene glycol and sebacic acid.[10-12]

As a general rule, the initial viscosity of a vinyl dispersion immediately after mixing directly correlates with the plasticizer viscosity as shown in Fig. 16-21. On aging, viscosity will generally rise somewhat as noted in the above section on storage. Of course, the higher the proportion of the plasticizer component of a dispersion, the lower is the dispersion viscosity.

A theory that works quite well in predicting the effects of plasticizer type on viscosity aging involves the percent by weight of the plasticizer molecule present as aromatic ring structure(s). Higher aromaticity means higher relative ability of a plasticizer to interact with the resin. In Fig. 16-22 the viscosity changes with rising temperature within the dispersion are traced for a

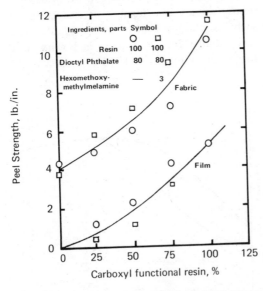

Fig. 16-20. Vinyl-to-nylon adhesion performance of simple laminating adhesives with various ratios of carboxyl-active and low fusion copolymer resins in a blend.

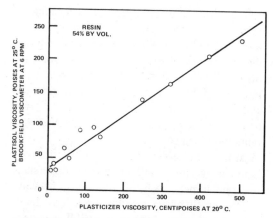

Fig. 16-21. Effect of viscosity of plasticizer on initial viscosity of a plastisol.

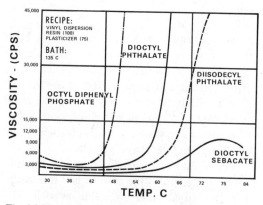

Fig. 16-22. Plastisol viscosity and gelation vs. temperature using different plasticizer systems.

variety of phthalate plasticizers as well as one based on sebacic acid.[13] As either the alcohol component becomes a larger weight fraction of the ester or as the aromatic component disappears, it can be noted that the temperature at which a rapid viscosity rise takes place, moves to a higher range. This type of measurement is an SPI Vinyl Division Standard Test Procedure (SPI-VD-T15).

Phthalate esters are used more often than any other plasticizer in the formulation of vinyl dispersions and, most frequently, this is di-2-ethylhexyl phthalate (DOP). DOP combines most of the desirable properties of a plasticizer such as minimal interaction with the resin at room temperature, good fusion properties, low volatility, acceptable low-temperature flexibility in the product, and low cost.

Shorter chain alcohol phthalate esters such as dihexyl and dibutyl phthalates are more volatile but fuse faster in dispersions because of their higher percent aromaticity. Longer chain alcohol phthalates are also used, primarily for their greater permanence in the product because of lower volatility. Linear alcohol phthalates, based on 7 to 11 carbon alcohols, have enjoyed success in recent years since they provide lower dispersion viscosities and somewhat better low-temperature properties than DOP at similar cost.

Phosphate esters, such as tricresyl phosphate (TCP), and octyldiphenyl phosphate, are highly solvating and find special use in vinyl dispersions. These, except for trioctyl phosphate (TOF), furnish rather high initial and aged viscosities and lower fusion temperature when compared with DOP. Phosphates contribute a degree of flame resistance through increasing char formation during combustion.

Other highly solvating plasticizers designed to furnish good strength properties when the fusion temperature allowed by an application is necessarily low are butyl benzyl phthalate and esters of benzoic acid. These are frequently used with the copolymer dispersion and blending resins (see application section, Froth Casting on Carpet) to attain usable physical properties at fusion temperature below 138°C (280°F).

Long-chain alcohol esters of straight-chain dibasic acids, such as dioctyl adipate, sebacate, and dioctyl azelate are frequently used in blends for their low initial and aged viscosity and excellent low-temperature flexibility in fused compounds. They are more volatile than DOP and have higher gelation and fusion temperatures.

Specialized monomeric esters that are gaining some favor in formulating are the trimellitates. Some of these are trioctyl trimellitate (TOTM), triisooctyl trimellitate (TIOTM), triisononyl trimellitate, and similar esters of linear alcohols. Their contribution is primarily very low volatility and resistance to extraction by solvents and oils.

Polymeric plasticizers, such as the polyester of propylene glycol and sebacic acid, find utility in vinyl dispersions where they contribute to very low volatility and excellent resistance to extraction from the fused product by solvents. These plasticizers, however, are generally very viscous and are usually blended with lower viscosity plasticizers.

Other specialized materials also find some utility as vinyl dispersion plasticizers. Liquid nitrile rubber, though of high viscosity, contributes oil resistance and can be vulcanized. Low-volatility acrylate monomers such as triethylene glycol dimethacrylate contribute fluidity to the liquid dispersion and cure during fusion to a hard polymer.

Another class of plasticizers is differentiated from the foregoing by their limited compatibility, color, or odor. These are mainly used for cost reduction in blends and are hydrocarbons of both highly aromatic and highly aliphatic types. The aromatic types provide moderate to high viscosity and fairly good compatibility with relatively poor heat and light stability. The aliphatic types provide low viscosity but may exude from aged fused products. An example of this group is chlorinated hydrocarbon. The latter contributes some flame resistance and is sometimes used for this reason.

Epoxidized soya and linseed oils are also used in small amounts because of their heat and light stabilization, as well as for their plasticization action.

One can thus see that the selection of a plasticizer or, more frequently, a blend is crucial to the overall performance of the vinyl dispersion. The primary goal remains—a practical balance between liquid and fused properties.

Stabilization

The word *stabilizer* in the plastics industry is a term applied to chemicals added to polymer systems to keep them intact during processing into products and during the service life of the product. In the case of flexible vinyl plastics, the main concern is to retard the heat or light influenced decomposition of PVC and copolymers. When decomposition of PVC occurs, the major results are the evolution of hydrogen chloride as a gas and the development of color in the remaining polymer.[7,8,12,14,15]

Stabilizers most used in vinyl dispersions are barium, cadmium, and zinc salts of long-chain fatty acids such as stearic, as well as lead-based materials (e.g., dibasic lead phosphite). Also, tin organics such as dibutyl tin dilaurate and the more recently developed alkyl tin mercaptide and alkyl tin derivativies of thio acids[14] are used when high stabilizing effectiveness warrants their high cost.

Most stabilizers used in vinyl dispersions are liquid in nature or ink-mill ground dispersions of solid types, such as the leads, in vinyl plasticizer. Stabilizer manufacturers have developed many proprietary types to fill the needs of the vinyl dispersion industry.[7,8]

Fillers

Fillers are usually added to reduce costs. Most fillers used in vinyl dispersions are selected fine-particle calcium carbonate or clays. A prime selection critera is the effect of the filler addition on the viscosity. A guide is oil absorption.[10,14] High oil-absorption fillers raise dispersion viscosity markedly, while low oil-absorption fillers have less effect.

Fillers in PVC generally result in lowered physical strength of the product and this is true with the fused vinyl dispersion as well. The formulator must balance this loss in strength, the viscosity effects, and the effects of filler addition on specific gravity (fillers have about twice the specific gravity of the base compounds they modify) against cost and performance.

Pigments

Vinyl dispersions and general-purpose extrusion, molding, and calendering vinyl compounds use the same organic and inorganic pigments. Some typical colors are carbon black, phthalocyanine blues and greens, chrome yellows and titanium dioxide white. Pigments are incorporated as ink-milled grinds in part of the plasticizer system when the vinyl dispersion itself is to be prepared by intensive low-speed mixing. They can be incorporated as dry powders when high-speed, high-shear mixers are used.

In production, where many different colors must be made, for *example* rotationally molded toys or press-molded auto carpet heel pads, it is most practical to master-mix unpigmented plastisol in large-scale intensive mixers and blend in the pigment grind(s) as needed. This can be done easily in drum lots with propeller mixers.

Pigment effects on dispersion rheology must be considered. High surface area blacks act as yield-building agents, for example. If this property is undesirable, it can be minimized by addition of surfactants to reduce particle-particle interaction in the pigment. Since dispersions undergo little or no shear during fusion, leafing pigments (aluminum flake or opalescent pigments) do not align as well as they do in extrusion, and it is more difficult to get bright metallic hues. On the other hand, heat-sensitive materials such as fluorescent pigments have an advantage since heat history in vinyl dispersions is usually much less than for melt-processed compound.

Thinners

Volatile organic materials such as VM&P naphtha are mostly aliphatic in nature and, when added to a vinyl dispersion, will markedly reduce the viscosity. However, they do not contribute to resin solvation and must be, in most cases, evaporated at lower temperatures prior to fusion, or bubbles and blisters occur. These volatile organics are frequently called *diluents*.

Another group, called *dispersants*, are volatile organics, such as diisobutyl ketone or isophorone, which do have some solvating effects on the resin. They are particularly helpful in developing good fusion in compounds containing plasticizer levels below about 35 parts of plasticizer per 100 parts of resin. A plasticizer is sometimes called a *nonvolatile dispersant*.

Frequently, in compounds containing low plasticizer levels, a balance of diluent-dispersant blends is found most effective against the objectives of viscosity reduction and good fusion properties.[14]

Specialty Ingredients

Much can be done through judicious plasticizer, resin, and filler choice to meet dispersion compounding requirements, but these are major constituents and a compound change to enhance one property usually will have adverse effects on others. Many proprietary and generic chemical materials are available which modify viscosity, processing, or property characteristics of dispersions when used at low concentration.

Viscosity modifiers are one class of specialty ingredients. They usually are designed to reduce viscosity. They are liquid surface-active agents which act by reducing the inter-particle structure in the dispersion. Nonionic surfactants, polyethylene glycol and its fatty acid esters, and low-molecular-weight silicone oils are typical. There are several proprietary types available as such, or in stabilizer package blends, from stabilizer manufacturers. Viscosity depressants act in conjunction with surfactants already present in the resin and stabilizer system, hence must be tested in each formulation. Even the mode of addition can be important. Addition at the beginning of a mix before the resin is wetted with plasticizer usually is most effective. The advantages of viscosity depression via modifiers must be weighed against possible loss in film clarity and heat stability.

Techniques for increasing viscosity with low level ingredients are seldom needed. Addition of part of the vinyl resin as a plasticizer solution or use of low-molecular-weight polystyrene liquids are two relatively inexpensive approaches. More often, the compounder needs to increase only plastisol yield value. Formation of oil-soluble soap micelles with aluminum distearate is a classical approach but it is plasticizer-sensitive and not reproducible. Silica aerogels are excellent controllable thickeners if they are well dispersed before compounding. A $12\frac{1}{2}\%$ ink-mill dispersion in plasticizer makes a heavy "grease" which can be sheared into a plastisol. More convenient thickeners are available in proprietary liquid calcium organic complexes which can be stirred into finished plastisol.

Silicone additives at higher molecular weight

than the viscosity-depressant types assist air release. Above about 10,000 centistokes, polymethoxysiloxane is insoluble in plasticizer. Tiny dispersed droplets of this low-surface-tension material make "flaws" in bubble films and break them. These same high-molecular-weight silicones are incompatible in fused stocks. They bleed to the air surface and provide a low-tack, slippery surface.

A number of non-vinyl polymeric modifiers for dispersion systems have been mentioned—liquid nitrile, liquid polystyrene and, of course, the polyester plasticizers. In situ polymerized epoxy resin, B-stage phenolics, and derivatives of natural rosin are examples of materials that have been used as adhesion promoters for plastisols. Powdered nitrile rubber has been used as a compounding adjunct to promote specific surface effects in thin films and it adds a degree of impact-cut resistance to coal mine conveyor belting.

Reduction of flammability has been mentioned with respect to phosphate plasticizers. Antimony oxide at up to 6 phr remains the most effective flame-quenching material available. If this white pigment cannot be used, chlorinated or brominated phosphate esters are useful, as are hydrated metal borates.[16]

Plastisols incorporate many types of chemical blowing agents for foam and surfactant profoamants for mechanical frothing. (see Cellular Vinyls section in Chapter 20).

Property Determination

Many tests have been devised by raw material suppliers, users, and industry trade groups for characterizing vinyl dispersions. Tests are available for the properties of the liquid dispersions, for the transitional properties as the dispersions are heated, and for the physical properties of the finished, fused stock. Many tests are run to ensure the quality of the raw materials used in the dispersions and the uniformity of mixing the dispersion themselves. Table 16-1 lists some standard dispersion test procedures.

The complex flow behavior, the sensitivity to mixing technique, and the importance of temperature history during mixing and storage, complicate the problem of measuring dispersion flow properties. If the above viscosity tests are to be used for characterization of raw materials, it is important that the standardized mixing and storage methods included in these tests be adhered to. Since plastisols and organosols vary in viscosity depending on the shear rate, instruments are needed which give an indication of the total behavior of the dispersion. There are several versatile but expensive wide-range cone and plate or capillary-type instruments which will measure viscosity across several decades of shear rate. In practice, the Brookfield viscometer for low shear rate measurement (SPI-VD-T1)-1959 and the Severs Extrusion Rheometer for high shear measurement (SPI-VD-T2)-1959 represent a reasonable compromise in shear rate range, cost and ease of operation.

In the Brookfield viscometer, the principle involved is to rotate a suspended disc or cylinder at a predetermined rate in the liquid and measure the torque required. Depending on the model, the rotational speed can vary from as little as 0.5 to as much as 100 rpm. Regardless, the shear rate is low, relative to high-speed strand coating or spread coating, and other instrumentation must be found for characterization at high shear. A single-point determination (at one speed) with the Brookfield gives an apparent viscosity which is useful for inter-comparison of low shear viscosity. Methods have been devised for determining relative thixotropy and yield value of dispersion systems. These require changes in spindle rotational speed and in the time that the spindle is allowed to rotate before torque is read.[4]

The Severs Extrusion Rheometer is a simple, effective instrument for measuring high-shear characteristics of dispersions. The instrument consists of a precision capillary attached to a reservoir in which the dispersion can be subjected to pressures up to 100 psi. Fluids are assumed to pass through the capillary in laminar flow and thus obey Poiselle's law. The shear stress in this instrument can be controlled by varying the applied pressure. The shear rate is directly proportional to the volume of effluent liquid. Since several capillary orifices are provided, ranging from nominal $\frac{1}{16}$ to $\frac{1}{8}$ in. diameter, a wide range of efflux and, thus, shear rates, can be obtained. Viscosity measurement

Table 16-1. SPI Vinyl Dispersion Division Tests

SPI-VD-T1 (1959)	Procedure for measuring the viscosity of a vinyl dispersion at low shear rates by the Brookfield viscometer.
SPI-VD-T2 (1959)	Procedure for measuring the viscosity of a vinyl dispersion at high shear rates by the Severs Extrusion Rheometer.
SPI-VD-T3 (1959)	Procedure for measuring the weight per gallon and specific gravity of a vinyl dispersion.
SPI-VD-T4 (1959)	Procedure for measuring shore hardness of a vinyl dispersion.
SPI-VD-T5 (1959)	Tentative procedure for determining the heat-resistance of fused plastisol.
SPI-VD-T6 (1959)	Tentative procedure for the determination of the deaeration of vinyl dispersions.
SPI-VD-T7 (1959)	Tensile properties of thin plastic sheets and films.
SPI-VD-T8 (1959)	Volatile loss from plastic materials.
SPI-VD-T9 (1959)	Tear resistance of plastic film and sheeting.
SPI-VD-T10 (1962)	Procedure for the determination of the degree of dispersion of vinyl dispersions using the "precision vinyl dispersion gauge."
SPI-VD-T11 (1962)	Flammability of plastic foams and sheeting.
SPI-VD-T12 (1962)	Tentative procedure for measuring water extraction of plasticized sheeting—activated carbon technique.
SPI-VD-T13 (1962)	Tentative procedure for measuring loss by extraction of plasticized vinyl film, when immersed in mineral oil at 50.0°C.
SPI-VD-T14 (1962)	Statement on low-temperature flexibility methods.
SPI-VD-T15 (1962)	Method for the determination of the gelation characteristics of a plastisol using a Brookfield viscometer.
SPI-VD-T16 (1962)	Tentative migration test.
SPI-VD-T17 (1964)	Tentative procedure for the preparation of fused samples of vinyl dispersions for physical and chemical testing.
SPI-VD-T18 (1965)	Hot bench gelation temperature test.
SPI-VD-T19 (1968)	Procedure to determine a plastisol's apparent viscosity utilizing a torque rheometer (Brabender Plasticorder).
SPI-VD-T20 (1972)	Procedure for determination of air release.
SPI-VD-T21 (1972)	Procedure for determining non-volatile content of vinyl dispersions thin film technique.
SPI-VD-T22 (1972)	Procedure for measuring the force necessary to compress a fused plastisol 10% on the RPC test equipment.
SPI-VD-T23 (1975)	Procedure to determine a plastisol's or organosol's apparent gelation and fusion (hot melt) viscosity characteristics utilizing a torque rheometer with a programmed (increasing) mixer temperature and oil-heated mixer head.

SPI Cellular Vinyl Division Tests

SPI-CV-1 (1971-1)	Same as SPI-VD-T1 (1959)
SPI-CV-2 (1971-1)	Same as SPI-VD-T2 (1959)
SPI-CV-3 (1971-1)	Same as SPI-VD-T10 (1962)
SPI-CV-4 (1971-1)	Same as SPI-VD-T15 (1962)
SPI-CV-5 (1971-1)	Same as SPI-VD-T18 (1965)
SPI-CV-6 (1972-1)	Proposed test method for determining settling in plastisols.
SPI-CV-7 (1971-r)	Proposed test method for plastisol frothability on an Oakes Foamer.
SPI-CV-8 (1971-I)	Interim procedure for measuring froth density of mechanically frothed vinyl plastisol foams.
SPI-CV-9 (1971-I)	Interim procedure for measuring compression set of vinyl foam (Constant Deflection Method).

(Continued)

Table 16-1. SPI Vinyl Dispersion Division Tests (*Continued*)	
SPI-CV-10 (1971-I)	Interim procedure for measuring compression resistance of vinyl foam.
SPI-CV-11 (1971-I)	Interim Procedure for determining density of vinyl foam.
SPI-CV-12 (1971-I)	Same as SPI-VD-T3 (1959).
SPI-CV-13 (1971-I)	Interim test method for resistance to exudation in vinyl foam.
SPI-CV-14 (1971-I)	Proposed test method for determining fusion of a vinyl foam.
SPI-CV-15 (1971-I)	Proposed procedure for measuring low-temperature flexibility of vinyl foams.
SPI-CV-16 (1971-I)	Interim test method for determining ash content of a vinyl foam.
SPI-CV-18 (1971-I)	Interim test method for measuring flammability of a vinyl foam by the Methanamine Pill Test.
SPI-CV-19 (1972)	Interim test method for migration of ingredients in supported and unsupported vinyl foams.

ASTM Tests

D 1823-66	Test for apparent viscosity of plastisols and organosols at high shear rates by Castor-Severs Rheometer.
D 1824-66	Test for apparent viscosity of plastisols and organosols at low shear rate by Brookfield Viscometer.
D 1243-66	Test for dilute solution viscosity of vinyl chloride polymers.

at any one shear stress gives a single apparent viscosity reading. If two materials of different viscosity are to be compared at the same shear *rate*, it is necessary to make a series of measurements at various shear *stress* values and interpolate.

In many applications, it is important that a plastisol be readily able to release air trapped during pouring or pumping. Once air bubbles have coalesced and risen, for example, to the top of a dip tank, the tendency for them to break is a function of viscosity in the bubble film and surface tension of the plastisol. These are difficult to measure and the most satisfactory empirical test (SPI-VD-T20)-1969 seems to be film break time across holes drilled around the circumference of a stainless steel cylinder. When this cylinder is immersed in plastisol, removed and allowed to drain with the axis vertical, drainage along the wall creates dynamic flow conditions in the exposed films across the circumferencial holes so that the breaking time of these films correlate with air release ability in production.

The rate of gelation of a plastisol depends, as does fusion rate, on such things as particle size, resin composition, emulsifier coverage, and plasticizer solvating effectiveness. Gelation and fusion are not directly correlatible, however, and means of testing gelation independent of fusion are needed.

Three basically different methods of testing are used. They all are based on the principle that plastisol undergoes an extremely rapid rate of change of viscosity with temperature just prior to the point where it solidifies to a dry but very weak gel. They differ mainly in that they shear the plastisol at varying rates during the heat up. In the hot bench gelation test, (SPI-VD-T18)-1965, a 10-mil layer of plastisol is applied to a thermal gradient plate and allowed to heat for a specific period of time (usually 50 seconds). A strip of foil lightly applied along the gradient will adhere to the still wet plastisol. When the foil is removed, the temperature at the line of demarcation between the wet and dry material can be measured. In this test, plastisol is in a completely static, unsheared state during the heatup.

The method for determination of gel rate using the Brookfield viscometer (SPI-VD-T15)-1962 involves continuous measurement of plastisol viscosity while the material heats up in an oil bath. This apparatus (Fig. 16-23) provides a complete curve of viscosity versus temperature and is especially useful in characterizing materials that are in motion during fusion. But it is incapable of measuring consistency at, and just below, the point where solidification occurs. The highest shear method of measuring gelation (and fusion) is the Brabender Plasticorder (SPI-VD-T19-1968 and SPI-VD-T23-1975). This instrument shear-heats plastisol in a mixing head and records torque versus temperature. It can cover the torques involved in mixing liquid plastisol through mastication of gelled material to mixing molten stock. The shear rate is high and indeterminate and tests with this instrument must be considered to be characterization tests.

In theory, fusion in vinyl material is defined as the minimum temperature at which the most perfect crystallite structures melt and the matrix becomes amorphous. Practically, it is the temperature required to give maximum physical properties. With experimental compounds the fusion temperature is easily determined by casting plastisol on a convenient surface, casting paper, or metal sheet, and heating successive samples at increasing temperatures. Measurement of physical properties (tensile strength, tear strength, etc.) yields a curve of strength-versus-fusion temperature and the temperature at the maximum point can be easily deter-

mined. Tests of tensile strength require regular samples. When a test is needed to determine the degree of fusion of a finished molded part, the processor often turns to one of the solvent-resistance tests. When a vinyl piece is underfused, and then stressed while in contact with a mild solvent such as acetone or ethyl acetate, it will break apart rather than simply swell as would be the case with well-fused material. The exact effect is a function of the solvent, and common practice is to compare parts off the production line with parts that have been reheated in an oven to assure that they are fused.

Physical tests on fused stock follow closely the tests common to other thermoplastics and will not be covered in detail here. They are usually modified to meet the unique processing characteristics of vinyl dispersions, however. Sheets and thicker sections for testing are usually cast rather than processed in heavy equipment. Heat stability tests are regularly run with sections of cast sheet under static oven post-heating conditions. Many of the test requirements for dispersion systems parallel those used in the paint industry more closely than those of the plastics industry.

VINYL DISPERSIONS, PROCESSES AND APPLICATION

Plastisols and organosols can be applied by spread coating, molding, and other specialized processes. Spread coating processes include knife, roll, and curtain coatings and by saturating. Molding is done by dipping, rotocasting, slushing, and cavity or in-place molding. Specialized processes include strand and spray coating and extrusion. All these techniques take advantage of the ease of convertibility from a liquid to a thermoplastic solid and the ability to formulate without volatiles.

Spread Coating

Of the application methods, spread coating is the most important in terms of compound consumption. This process is widely used to make products as diverse as roll goods flooring,

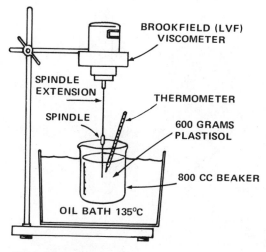

BROOKFIELD (LVF) VISCOMETER

SPINDLE EXTENSION

SPINDLE

THERMOMETER

600 GRAMS PLASTISOL

800 CC BEAKER

OIL BATH 135°C

Fig. 16-23. Typical equipment set-up for viscosity-time-temperature testing.

apparel fabric and automotive padding (Fig. 16-24). The relative processing ease makes plastisols quite competitive with calendered vinyls for short-to-medium runs of material, for thin films, which can be made strain-free, and for chemically blown foams. Plastisols are 100% solids with no solvent to be removed. This means that thick coatings can be applied in one pass and that the cost of solvents and of the heat for solvent vaporization is eliminated. On the other hand, plastisols depend on the plasticizer vehicle for their fluidity. Proper formulation and careful selection of materials are most important for successful use in coating.

Knife Coating. Fabric, paper, and metal substrates are coated by a process which, in its simplest form, consists of let-off and take-up equipment, a coating head, a fusion zone, an optional embosser and cooling zone (Fig. 16-25). There are at least twenty generic types

Fig. 16-24. Apparel and accessory fabrics are produced from vinyl dispersions using spread coating techniques. (Handbag, courtesy Pandel-Bradford; boots. Bonan)

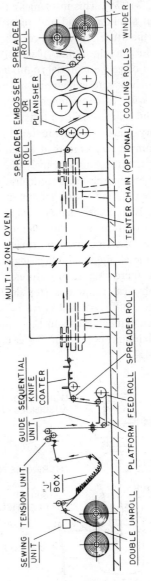

Fig. 16-25. A knife coating line. (Courtesy Machinery Div., Midland Ross Corp.)

of applicators used in applying coatings to webs.[17] Knife coaters, as in Fig. 16-25, and reverse roll coaters are the most important spreading mechanisms for applying dispersions. These coaters are more capable of handling the relatively viscous and somewhat viscoelastic dispersions than are, for example, wire-wound rod coaters, direct-roll coaters, and air knives.

Knife-coating heads are, in principle, the simplest applicators for applying dispersion coatings to flat webs. Figures 16-26, 16-28, and 16-29 illustrate the three most common types. These are differentiated by the method used for supporting the web. All depend on the clearance between a coating blade and the web for control of coating thickness.

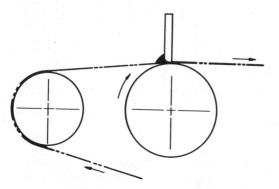

Fig. 16-26. Knife-over-roll coating. (*Courtesy Machinery Div., Midland Ross Corp.*)

Fig. 16-27. Sixty-in. knife/roll coater, web exit side. (*Courtesy Egan Machinery Corp.*)

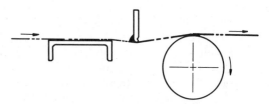

Fig. 16-28. Floating knife coating. (*Courtesy Machinery Div., Midland Ross Corp.*)

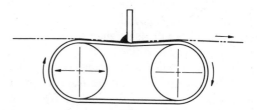

Fig. 16-29. Knife-on-blanket coating. (*Courtesy Machinery Div., Midland Ross Corp.*)

Knife over roll coaters (Figs. 16-26 and 16-27) apply from 3 to 125 mils of dispersion compound at the fixed gap between the knife and the backing roll. Elastomeric backing rolls are usually used in coaters intended for the heavier coating range. More precise control of thin coatings is afforded by accurately ground, steel backing rolls. Coating speeds on knife machines range up to 100 ft/min with rates of 60 ft/min common for coatings around 10 mils. Since there is a fixed gap between blade and roll, any variations in thickness of substrate show up as variations in coating gauge and the result is a uniform thickness composite which may vary in coating thickness. Knife angle has little effect on thickness, but the angle may be varied to change the quality of the coating.

Floating knife coaters, Fig. 16-28, carry the web, unbacked, between support channels or rolls. Coating weight is determined by knife configuration, knife angle, and web tension (that is, the pressure of knife against web). By controlling these variables, it is possible to apply uniform coatings to non-uniform substrates. Thick coatings are difficult to apply with floating knife coaters. Optimum results are obtained in the range from about 1 to 3 mils. In the coater shown in Fig. 16-25, a knife-over-roll and a floating knife coater are used in tandem. The knife-over-roll applies a heavy fill coat and levels the fabric and the floating knife applies a light "color coat."

The knife-over-blanket coater (Fig. 16-29) represents a compromise between knife-over-roll and floating knife. In this coater, the web is supported on a driven rubber belt rather than by its own tension, and weak webs can thus be processed without tearing.

In formulating for knife coating, it is necessary to consider both the high and low shear rheology of the compound.

Compounds with high yield value may be needed to prevent striking into the weave of open fabric before and immediately after the web has passed under the knife. True yield value can prevent flow of material from the puddle behind the knife into the interstices of the web. Often, transverse backing dams are set behind the knife to retain the coating bank in a small area and minimize bank-to-web contact time. True thixotropy (recoverable structure in the liquid) helps to keep the applied coating from striking into the web after coating and before fusion.

In many knife-coating processes, high rates of shear are induced into the dispersion as it passes under the knife. Thus, careful consideration of the high shear rheology is also important. The shear rate at the knife is directly proportional to the gap between knife and web; and it can be calculated roughly by dividing web speed in inches per second by the coating thickness in inches. The high shear rheology of dispersions has two parameters: viscosity per se at the coating knife and the extent to which the dispersion is shear-rate-thickening or shear-rate-thinning as it passes through the shear rate range just prior to attaining the maximum at the coating gap. It is difficult to differentiate between the effects of these two parameters, so the usual goal of the compounder is to minimize both shear thickening and high shear viscosity for optimum coatability. Compounds which are too shear-thickening or too viscous at high shear may actually act more like solids than liquids as they pass under the knife. They may actually shatter and produce skipped spots or droplets that deposit at random on the downstream side of the knife (termed *spitting*). Unlike skipping or spitting, streaking of coat-

ings is usually a simple matter of foreign material lodged under the coating knife. This can be prevented by care in mixing to ensure complete dispersion and by screening the material before it goes to the coating head.

The proper choice of coating knife configuration can help prevent skipping and spitting in the high shear range as well as striking through in the low shear range. There are roughly five different knife shapes used in applying dispersions. Fig. 16-30 shows four of these. The upstream or coating bank side of all of the knives in this figure is on the left.

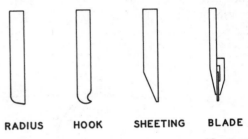

RADIUS   HOOK   SHEETING   BLADE

Fig. 16-30. Coating knife contours. (*Courtesy Machinery Div., Midland Ross Corp.*)

The radius knife is the one most commonly used for applying heavier coating weights and for smoothing coatings. Radius knives vary in blade thickness and in radius of the upstream side. An extreme example of a radiused knife (not shown) is the bull-nose knife with a $180°$ rounded surface in contact with the web. The bull-nose knife is used for smoothing very heavy coatings. The hook or "j" knife is a modification of the radius knife intended to prevent spitting from shear-thickening compounds. In practice, it provides a pocket to collect globules of material which separate from the matrix and it presents a very sharp trailing edge as the coating leaves the knife. The sharp trailing edge minimizes the forces which would otherwise tend to cause the departing coating to stick to the knife instead of adhering uniformly to the web.

For thin coatings and for coatings intended to fill surfaces without penetrating into the web, the sheeting or the blade knife is used. Visualize these knives as scraping material off the web rather than pushing it into the web.

The blade knife is most commonly used in *Spanishing*, a decorating technique which is described later. Knife angle and gap are usually micrometrically adjustable. The knife can be rotated around an axis at the coating edge to vary the quality of the coating being deposited. This knife angle usually ranges from 85 to 90° relative to the upstream side of the web.

Roll Coating

As has been indicated, coating weight is inherently difficult to control on knife coaters. Problems can also occur with strike-through and streaking because the coating is carried in excess on the web and is doctored off under a stationary knife. These problems can be minimized and faster coating rates can be obtained by applying the dispersion compound via roll coater.

The simplest roll coating devices are the direct or forward roll coaters such as are illustrated in Fig. 16-31. In direct roll coating, liquid is picked up or supplied to a roller which travels in the same direction as the web. The liquid film splits as it comes into contact with the web and the resulting deposit must level before drying or fusion can take place.

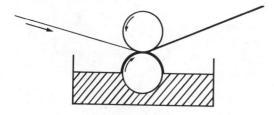

Fig. 16-31. Direct roll coater (useful only with very low viscosity dispersions). (*Courtesy B. F. Goodrich Chemical Co.*)

Vinyl dispersions are usually high enough in viscosity and viscoelastic enough to make this splitting and leveling very difficult to control. As a result, roll coating of dispersions is usually carried out by reverse roll coater in which the complete deposit of coating on an applicator roll is wiped off on a web running in opposition to the roll.

Figure 16-32 illustrates a typical three-roll nip-fed reverse roll coater. Dispersion com-

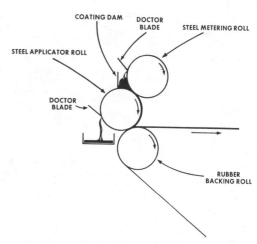

Fig. 16-32. Three-roll nip fed reverse roll coater. (*Courtesy B. F. Goodrich Chemical Co.*)

pound is metered through a precise nip between a slowly rotating metering roll and an applicator roll running in the opposite direction. The coating on the applicator roll is then wiped off onto a web which is traveling in the opposite direction. The applicator roll is run faster than the web travels, so the entire amount of coating is wiped off and no splits or kiss marks develop. The metering roll is usually rotated so that it constantly presents a fresh surface and foreign matter cannot clog the nip and streak the coating. The amount of coating deposited by a reverse roll coater is a function of the amount carried by the applicator roll and the relative speed of applicator and web. Shear rate is controllable by varying these two parameters, and liquids of widely different viscosities can be adequately coated. It has been reported that viscosities ranging from water-thin to 100,000 cps can be handled by this type of machine and that, with highly fluid systems, coatings can be applied at rates up to 1000 ft/min. Reverse roll coaters vary considerably in their design and cost depending on the precision with which the coating must be applied.

Figure 16-33 shows a coater designed to deposit thin coatings precisely. In this coater, fine adjustments of nip spacing are made by flexing the beams which carry the metering roll bearings.

There are several other configurations of

Fig. 16-33. Three-roll nip fed reverse roll coater. Detail shows precision adjustment for metering roll. (*Courtesy Egan Machinery Corp.*)

reverse roll coaters designed, for example, to eliminate abrasion of the metering roll against a metal web, to provide a bottom reservoir of recirculatable coating compound, or to eliminate splashing of high-speed applicator rolls running in pans of low viscosity fluid (Fig. 16-34). These differ in detail but not in principle from the nip-fed coater and are described fully elsewhere.[17,18,19]

Fusion

The applied coating must be fused. Temperatures around 148 to 177°C (300 to 350°F) are required. Time of temperature is not a consideration since fusion is a physical rather than a chemical phenomenon. The most commonly used oven-type for web coating is the straight-pass tunnel oven (Fig. 16-35). These vary in length from 15 to 250 feet depending on the coating rate and fusion requirements. Frequently they are divided into two or three stages so that temperatures can be varied to drive off thinners prior to fusion. Because of the temperature requirements, steam heating is impractical and oil-fired, gas-fired, electrical

heating, or hot oil heat exchangers are most commonly used.

Although dispersions rarely need to reach more than 190°C (375°F) to fuse, the plenum temperatures in forced hot-air ovens used in

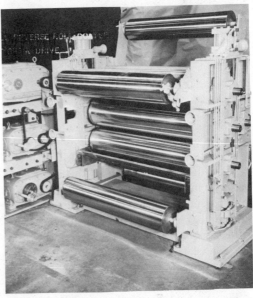

Fig. 16-34. Four-roll contracoater. (*Courtesy Black Clawson Co., Dilts Div.*)

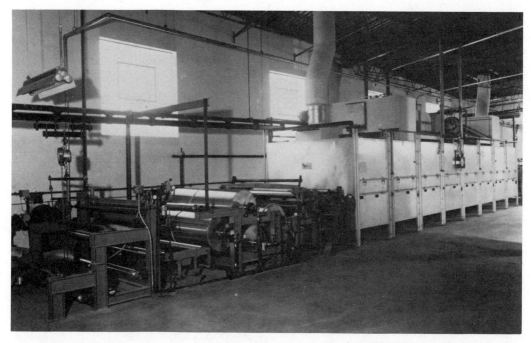

Fig. 16-35. Plastisol coating line; tunnel type fusion oven at right. (*Courtesy Dawson Engineering*)

spread coating lines usually range much higher—up to 427°C (800°F). Some method of aiming the hot blast directly at the surface to be fused and away from sensitive substrates is usually provided. Fast through-put time and minimum web damage are the result but some method of indexing the heat source away from the web is necessary to prevent overheating in case of line stoppage.

Sufficient exhaust must be provided for removal of flammable organosol vapors. In the case of both plastisols and organosols, volatile plasticizer and thinner vapors must either be condensed and removed from the exiting gas or they must be incinerated. Figure 16-36 is a schematic diagram of an installation for vapor incineration with the heat of vapor combustion reclaimed and used to preheat the makeup and recirculating oven air.

Other fusion devices used in plastisol coating applications include heated platens, heated rolls, and radiant resistance rods or quartz lamp heating. Radiant heat units are especially useful in modifying existing equipment for greater heat input. These units usually operate at high temperatures and must be installed so they can index away from the web in case of line stoppage.

Web Handling and Finishing

At the minimum, web handling equipment for plastisol coatings must include let-off and take-up equipment and accumulators for splicing rolls of material without stopping the line. "J" boxes or expandable dancer rolls (loopers) commonly perform this function. Cooling rolls, utilizing refrigerated water, bring the coating down to a temperature at which it will not block when rolled up.

Plastisol coating lines frequently provide for one or more finishing operations on the coated web. Embossing can be done in-line with an engraved steel roll applying pressure to the web as it passes over a rubber back-up roll immediately at the oven exit. The fused plastisol must be chilled to set the emboss and allow it to release cleanly from the embossing roll. Alternatively, coated unembossed fabric can be reheated by passing it over heated platens and embossed in a separate operation (Fig. 16-37). Separate finishing operations can allow econ-

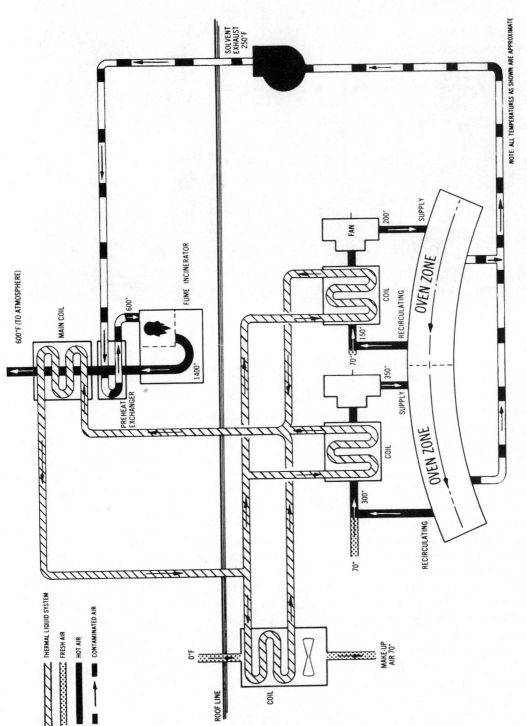

Fig. 16-36. Schematic of fume incinerator with heat reclaim package. (*Courtesy Ross Engineering, Machinery Div., Midland Ross Corp.*)

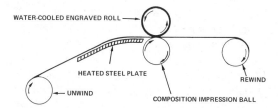

Fig. 16-37. Embosser for separate decoration of coated webs. (*Courtesy B. F. Goodrich Chemical Co.*)

omy by widening processing versatility and not tying line speed to a slow operation.

Spanishing—application of a contrasting color in the emboss valley by means of a tight knife pass—and top coating usually are performed with solvent-based materials. Top coating protects exposed surfaces from abrasion, dirt, and stain pickup. Typically, top coatings are formulated with high-molecular-weight homopolymers to minimize plasticizer migration from the substrate. Acrylic solution polymers and cellulose acetate butyrate blended with the vinyl add to the blocking resistance and the dryness of the coating. Other typical finishing operations are printing and flocking. In flocking, copolymer or carboxy functional plastisol resin dispersions are used as the adhesive. A complete discussion of coating applicators and their auxiliary equipment appears several places in the literature.[17,18,20]

Fabric Coating

Complete lines can vary widely in complexity and cost depending on the operation being performed and the substrate being coated. Perhaps the simplest operation is coating of fabric for such products as shade cloth, wall covering, and fabrics for shoes and handbags. Base fabrics range from lightweight cotton used in rainwear to heavy duck used in awnings. A specially designed striping box is used in awning coating. Multicolored stripes are laid down side by side simultaneously from separate slots in the applicator.

Vinyl dispersions have inherent adhesiveness to cotton and rayon; synthetics, like nylon and polyester fabrics, require the use of adhesive primers. Nitrile latexes modified with borated casein and nitrile/phenolic solvent systems have been used in the past. More recently, carboxyl-modified vinyl dispersion resin systems have been found to be useful primers[9]. Carboxyl resin primer systems increase the adhesion of subsequent top coatings by as much as three or four times that obtained with vinyl chloride vinyl acetate primers.

Most base fabrics are strong enough to be carried through the fusion oven with no support other than the oven rolls. Weak fabrics and non-heat-stable base materials are carried on tenters. One of the most useful fabrics for vinyl coating is knit cotton. Its stretch characteristics and hand make it ideal for upholstery, outerwear, and sundry applications such as handbags and boots. The open mesh and dimensional instability of knit backing make it necessary to use a radically different coating process for plastisol application. This process, commonly called *transfer* or *cast coating*, builds the coating in reverse (see Fig. 16-38). A casting substrate of thermoset-resin-treated paper, silicone-treated fabric, or metal is coated with the plastisol in multiple coats, starting with the eventual wear surface. Each coating is gelled before application of the next and the final composite is heated just enough to leave a viscous, partially gelled surface. The base fabric is laid into this coating with light pressure and the final assembly is fused. After cooling, the finished-coated knit fabric can be stripped from the carrier and rolled.

The type of carrier used is determined by the economics of the process and the finished product desired. Thermoset-resin-treated casting paper is the least expensive base but it has a limited use life. This material cannot be expected to withstand more than about six cycles. Primary hazard is from mechanical abuse. Casting papers are supplied in many finishes: high gloss patent leather, matte, and a number of emboss patterns. For long runs of transfer coating, stainless steel belts, resin-treated fiberglass, or resin-treated cotton have the longest life expectancy. If emboss finishes are desired, they must be made separately with these casting surfaces. The high cost of stainless steel belting and the time consumed in chilling coated material to permit removal from the substrate are limiting factors.

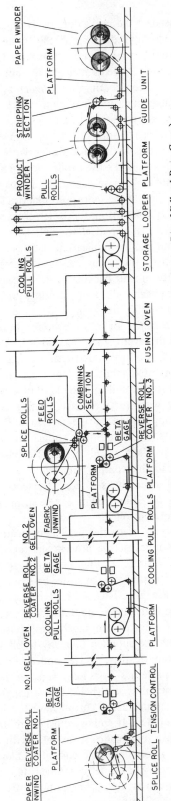

Fig. 16-38. Cast coating line for coating by transfer from paper carrier. (*Courtesy Machinery Div., Midland Ross Corp.*)

Much of the coated knit fabric made for upholstery and apparel applications contains an intermediate layer of vinyl foam about 20 mils thick topped with a 6 to 8 mil wear surface. This construction gives a very supple leathery hand which is generally quite superior in aesthetic properties to unexpanded coatings. Figure 16-38 illustrates a typical transfer (cast) coating line for foam fabric. This line has three coating stations. The wear surface is applied at Station No. 1. This coating is then gelled in Oven No. 1. After cooling, the foam innerlayer goes on at Station No. 2. This is gelled but not expanded on passing through Gell Oven No. 2. Since a third coating Station (No. 3) is available, the extent to which the foamable coat is gelled is not critical. A base coating of wet plastisol serves as the adhesive for vinyl-to-knit back.

The latter comes off the roll just above and to the left of the fusing oven. The coated fabric—with its fused-surface and expanded, fused inner layer—is separated from the casting paper in the stripping section and product and paper are wound up. Formulating dispersions for cast coated foam/fabric requires coordinating the melt flow characteristics of the foam to the temperature at which the blowing agent decomposes to produce gas for expansion. Formulating parameters to determine the coordination of these characteristics are discussed more fully elsewhere in this book.[21]

Another use of vinyl foam is as resilient backing for commercial carpet. Here a dispersion precoat of about $1\frac{1}{2}$ lb/sq yd is knife-applied to the back of carpet which is tufted through a polypropylene nonwoven fabric. A liquid froth of plastisol is then applied to this wet, or partially gelled, precoat at a rate of about 2 to $2\frac{1}{2}$ lb/sq yd using a stationary roller as the applicator. Foam coatings of this type must be fused at low temperatures in the order of 135°C (275°F) to prevent heat shrinkage and melting of the polypropylene. Hence, materials for this application are formulated with copolymer resins and plasticizers of as high solvating effectiveness as possible compatible with the low viscosity necessary for foaming in mechanical foamers.[21]

Screen printing and dot printing of plastisols

onto fabrics produce a number of very interesting products. Soles for children's sleepwear (Fig. 16-39) and abrasion-resistant surfaces for mail sacks and cotton picking bags have been made using coaters such as those shown in Fig. 16-40. Here, a rubber squeegee knife forces the compound through perforations in a cylinder rotating in the web direction. The plastisol is formulated with yield-building agents, usually specialty fine-particle dispersion resins, so that the dots deposited will not slump before fusion.

Shapes other than simple round perforations have been used to produce decorative coatings for upholstery. Discontinuous coatings of this

Fig. 16-39. Dot printing makes wear-resistant soles for children's sleepwear. (*Courtesy B. F. Goodrich Chemical Co.*)

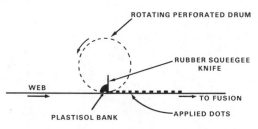

Fig. 16-40. Plastisol dot printer. (*Courtesy B. F. Goodrich Chemical Co.*)

sort provide maximum abrasion protection for minimum coating weight and, in addition, provide breathability and moisture vapor permeability, an added comfort factor.

Rotary wire screens replace the perforated cylinders of Fig. 16-40 when more intricate designs are needed. The screen can be blocked so as to produce a decorative pattern. Specialty floor and wall coverings have been made by screen printing plastisol flock adhesive onto continuous vinyl coated webs and then applying nylon or cotton flock via electrostatic or beater bar methods.

Some very soft fabric coatings are applied by a hot-melt coating technique. On fusion, vinyl dispersions pass from the fluidity of a two-phase dispersion through a weak cheesy gel to a homogeneous liquid hot melt. The viscosity of this melt depends on a number of factors but notably the level of plasticizer. Hence, with high plasticizer and the resulting very low dispersion viscosity, it is convenient to preheat and flux the dispersion and flow this melt from a hopper-type coater onto the substrate. Since the material must be held for a period at the melt temperature, excellent heat stabilization is a prime requisite in this type of application.

Coating Paper and Nonwoven Materials

Paper and nonwoven fabrics comprise another important family of casting substrates. In general, the paper adds dimensional stability and ease of application inexpensively. Products from coated paper include packaging board, shelf lining papers, shoe liners, auto door panel finishes, and masking tape. For tapes, the plastisol coating acts as a release surface for the adhesive mass. Formulating requirements depend on the substrate, processing characteristics, and finished properties of the coating. Substrates vary from box board to kraft to special latex saturated papers for flooring and automotive use. Adhesion of dispersion resin coatings to most paper substrates is good. In the case of super calendered kraft, however, it may be necessary to add modified rosin derivatives and other tackifiers to ensure adhesion. Papers treated with latex, or tapes utilizing rubber adhesive masses which are plasticizer-sensitive, require special formulating of the coating. Polyester polymeric plasticizers are used here to prevent migration into and softening of the continuous non-vinyl polymer.

Nonwoven fabrics, scrims, and felts are used in a number of instances as the base for vinyl dispersion coatings. One application, coating carpet tufted through nonwoven polypropylene has been mentioned. Another utilizes dots printed onto nonwoven polyester linings for apparel—as the melt adhesive system for attachment to the garment. Hot-iron pressing melts the dots of gelled plastisol adhesive which then flow into the outer garment fabric to make a permanent bond. Latex systems are used for this same function when the somewhat higher cost is offset by the need for dryclean-resistance.

By far the most important coating operation involved with felts or nonwovens—in fact, the most important application of dispersion systems—is the manufacture of roll goods flooring. This application probably consumes from 20 to 25% of the total dispersion resin produced in the United States. Roll floorings have passed through about three evolutionary stages since vinyl supplanted oleoresinous wear surfaces in the early 1950's.[22] Such products are backed with either asphalt or latex-saturated cellulosic or mineral fiber felt. One of the earliest versions consisted of printed paper, coated with 4 mils of vinyl organosol and laminated to asphalted felt. In later constructions, asphalted felt has been directly coated with barrier coats, fill coats, and print coats of latexes. After printing, a vinyl organosol or plastisol wear layer has been applied and fused. At the present time, one of the most popular types of flooring consists of an asbestos felt coated with about 20 mils of vinyl foam which is printed and then top coated with about 7 mils of hard, clear plastisol. A process of chemical embossing produces emboss valleys in perfect register with the printed pattern. One such technique[22] utilizes a chemical inhibitor that retards the gas evolution of the azodicarbonamide in the foamable layer during the final fusion stage.

The specific processes used in manufacturing resilient flooring are highly proprietary. Figure 16-41 is an idealized conception of a produc-

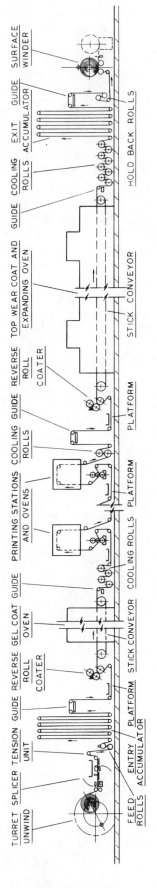

Fig. 16-41. Resilient vinyl floor covering line. (*Courtesy Machinery Div., Midland Ross Corp.*)

tion line. Two reverse roll coating stations are provided to apply the foamable resilient inner layer and the wear coating. The decoration is printed onto the gelled, cooled foamable layer at two printing/drying stations using presses similar to the 154 in. three-roll press shown in Fig. 16-42 to make seamless flooring up to 12 ft in width.

Formulation of plastisol systems for the wear and resilient layers of this type of flooring is a highly sophisticated technology. There is a high degree of interdependence between the foam compound, the embossing agent, and the wear layer compound. Requirements in the wear layer are stringent. The exact gloss needed must come from the compound itself, since post-planishing would destroy the emboss already extant. Crystal clarity is needed for maximum visual impact from the printed pattern beneath the wear layer; and this clarity must be maintained despite contact with aqueous solutions of typical floor cleaning detergents. Maximum wear life requires that the top coating be high in hardness and abrasion-resistance and that it resist indentation, marking, and staining from contact with shoe heels, furniture legs and common household chemicals. Several modifications of this resilient flooring concept provide a thick ($\frac{3}{16}$ to $\frac{1}{4}$ in.)

vinyl foam backing (Fig. 16-43). These give a highly resilient product which combines many of the best properties of soft carpeting with those of tough and easily cleaned vinyl flooring. The felt is retained or replaced with woven fiber glass or fiber glass scrim to eliminate cold-flow creep and to spread applied loads over the foam backing, thus reducing indentation problems.

Film Casting

Many spread coated products entail no permanently attached substrate. Decals and decorative roll goods films are manufactured by casting a vinyl dispersion on previously printed release paper, fusing, then casting a solvent-based adhesive mass on the back of the dispersion. The release paper is designed so that the printed pattern transfers from the paper to the dispersion. When unrolled, the adhesive mass is temporarily attached to (originally) the bottom surface of the casting paper and the product is right side up. Such contact adhesive films are convenient materials for household use and for automotive decorations (with more permanent adhesives) such as wood grain panels for station wagons.

Since dispersion systems can be cast on a

Fig. 16-42. Three-roll press for printing wide webs of resilient flooring. (*Courtesy Egan Machinery Corp.*)

Fig. 16-43. Interior foam layer in resilient vinyl floor is soft underfoot. (*Courtesy B. F. Goodrich Chemical Co.*)

heat from beneath and the surrounding oven heat. The fused, cooled matting is stripped and cut to shape. Since a moderate size walk-off mat can contain as much as six pounds of vinyl compound, compounding requires minimizing cost without jeopardizing floor surfaces from plasticizer migration or ruining the appearance of exposed borders. Heat for fusion is usually minimal and fast-fusing copolymer resin systems are often used. See Fig. 16-44.

Spread Coating Metal Substrates

Except for the newer carboxyl functional polymers, simple vinyl dispersion systems have no adhesion to metal substrates. Proprietary dispersion systems have been developed with self-adhesive characteristics and a number of adhesive adjunct plasticizers are available.[23] The most common expedient with metal is to use one of the commercially available solvent-based adhesive primers. These are usually nitrile/

substrate, fused, and cooled thoroughly before stripping, the likelihood of inducing strains by distorting warm or hot film is reduced to nearly zero. Thus, such film can withstand exposure under high ambient temperatures without experiencing shrinkage as a result of "plastic memory." Production of strain-free, thin calendered, or extruded film is very difficult and the use of more expensive dispersion systems is warranted.

Production of "walk-off" mats and runners is a related cast coating process. These mats contain carpet squares or continuous strips backed with and bordered by homogenous vinyl sheet about $3/16$ in. thick. The carpet serves as an excellent dirt pickup medium at entrances to buildings and the heavy vinyl back prevents slipping and bunching up. In this process, plastisol is cast onto a continuous, treated fiber glass belt, the carpet is laid in, leaving the desired borders uncovered, and the composite is fused by a combination of platen

Fig. 16.44. Carpet tiles made with cast coated plastisol backing have wide design potential. (*Courtesy Collins and Aikman*)

phenolic/epoxy combinations and require flash baking or partial curing before application of the dispersion.

One of the most interesting recent applications for dispersion spread coating is in the coil coating industry. Coil coating is a process through which metal is painted in flat form, coiled, and then, subsequently uncoiled and formed into finished articles. Despite the limitations which prepainting imposes on the subsequent shaping and fastening operations, it often is more economical than painting the finished article. In fact, The National Coil Coaters Association, has developed around this technology.[24] Exterior siding finishes, down spouts, siding for mobile homes and appliance covers represent markets where the flexibility of vinyl coatings is essential. Vinyl organosols provide a moderate cost alternative to vinyl or acrylic solutions. Some coil lines have tandem stations for priming and finishing operations[25] and standard primer systems can be used in conjunction with vinyl dispersions. For single-station, one-pass operation, self-adherent formulations are available. These formulations combine dissolved carboxyl-functional solution vinyl resins with dispersion resin. Solvents are selected to minimize swelling of the disperse phase. High gloss, one mil, opaque, weatherable films result from coatings fused under fast coil-line cycles (e.g., 1 min. at 260°C (500°F). See Fig. 16-45.

Fig. 16-45. Metals coil-coated with vinyl dispersions offer durable pre-painted material for making appliance housings, metal furniture, and other formed metal products. (*Courtesy B. F. Goodrich Chemical Co.*)

Miscellaneous Coating Applications

Vinyl dispersions are well suited for use as adhesives for vinyl calendered film to synthetic and natural fabric and for fabric-to-fabric laminations, especially when formulated with fast solvating plasticizers. Thus, they will bite into and adhere to plasticized vinyl film at temperatures low enough to preclude damage to the film. Laminators using dispersion adhesives usually have a configuration such as Fig. 16-46. Here, one web of the two being laminated is coated with adhesive by knife-over-roll. The two webs are brought together while the adhesive is still wet and the combination is passed around a steam or electrically heated drum with the least-heat-sensitive web in con-

tact with the drum. Electric resistance rod units are frequently positioned above the drum for additional heat, if needed.

Dip saturation (Fig. 16-47) is a technique used to obtain thorough saturation in products like conveyor belting. Passing the belt around rolls stacked, as in the illustration, opens up alternate surfaces and ensures good penetration into the interstices. Squeeze rolls positioned as shown drive out the air and yield a homogenous vinyl mass between the fibers. The excess compound is doctored off both surfaces of the belt before it enters the vertical fusion oven. Mine belting is wet-woven, then primed and top coated in two such coating stations (Fig. 16-48).

In curtain-coating, liquid dispersion flows from a slit in a hopper-shaped container onto parts passing through on a conveyor beneath. The material by-passing the parts enters a catch basin beneath the conveyor and is recirculated. Curtain-coating is infrequently used with plasti-

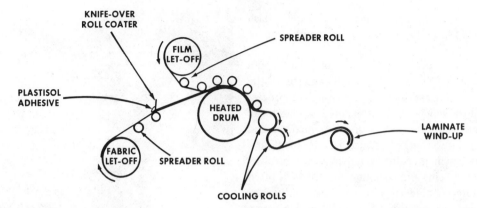

Fig. 16-46. Laminator for joining film to film or film to fabric with vinyl dispersion adhesives. (*Courtesy B. F. Goodrich Chemical Co.*)

sols since they are generally too viscous to handle conveniently by this method. The process is, however, a good way to coat irregularly shaped parts. Specialty packaging applications, such as strippable foam coatings on automotive fenders and bumpers, have dominated.

Spread coating technology has matured considerably since the late 1950's. Improvements in ingredients and technology have caused continued growth. Advances will probably come through improved resins of the reactive sort for special purpose coatings. Also, harder, tougher coatings from dispersions of resin in monomers which can be radiation cured in-line

are being developed. Certainly, higher-solids organosols and lower volatility plastisols will result from the continuing search for materials which are more compatible with the economic and environmental needs for reduced effluent vapors in processing.

Fig. 16-48. Conveyor belting made by plastisol dip saturation techniques offer excellent flame and abrasion resistance in mines. (*Courtesy B. F. Goodrich Chemical Co.*)

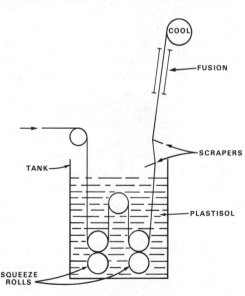

Fig. 16.47. Dip saturators are used for impregnation of heavy webs. (*Courtesy B. F. Goodrich Chemical Co.*)

MOLDING

Dip Coating and Dip Molding

Dip coated or molded products are readily produced by rapid, economical deposition of light or heavy layers of decorative, chemically resistant, tough plastisol coatings. Thickness of the coating can range from 1 to over 125 mils, or $\frac{1}{4}$ in.

Basically, this process consists of dipping a mold into a plastisol and then fusing the coating that remains on the mold. There are many variations of the process: hot or cold dipping, or both; more than one dip may be used for heavier coatings or special effects; the coating may be stripped from the mold and the mold re-used; or the coating may become a functional part of a finished product.

Generally, the term *dip molding* refers to the process of dipping, fusing, and stripping a mold of its plastisol coating to produce hollow products. Examples of plastisol dip-molded products are shown in Figs. 16-49 and 16-50. Figure 16-49 shows several designs of handlebar grips and Fig. 16-50 shows automotive seat belt retractor boot covers. When the plastisol coating is to become a functional part of the mold, it is called *dip coating*, and the mold may, or may not, have an adhesive primer. Examples of dip coatings are illustrated in Fig. 16-51, where several different designs of glass bottles have plastisol protective coatings.

If adhesion to the substrate is important and physical encirclement is not sufficient (e.g., wire coatings), an adhesive primer will be required. Primers can be applied by brushing, dipping, spraying, or flow coating. Where it is to be applied to a selective area, such as required on tool handles, the dipping method is most commonly used.

Dip molding and coating may be done with a batch process or by a highly automated continuous process. The batch process is most frequently used where only a few similar parts, or a large number of dissimilar parts, are to be

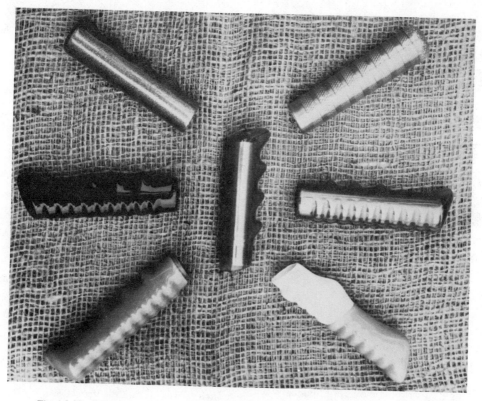

Fig. 16-49. Several designs of dip molded handle-bar grips. (*Courtesy Steere Enterprises Inc.*)

Fig. 16-50. Hot dip molded automotive seat belt retractor boot covers. (*Courtesy Steere Enterprises Inc.*)

Fig. 16-51. Hot dip coated bottles. Coating offers protection as well as aesthetic appeal. (*Courtesy Chemical Products Corp.*)

coated. The continuous dipping process is used for rapid production with minimum labor. Several molds mounted on a conveyor go through priming, pre-heating, coating, fusing and cooling operations, all automatically. Some systems even strip the molds automatically with compressed air; in this case, the only labor

required is for monitoring purposes. Figure 16-52 illustrates a highly automated, continuous dip coating line for producing surgical examination gloves.

Plastisol drips may occur at the lower edge or points of a mold after it has been withdrawn from the plastisol or during the gelation of the plastisol coating. For many applications, this is unimportant; for others, it is undesirable and should be avoided. Dripping can be controlled by proper compounding and by various processing techniques including: (1) the withdrawal rate of the mold from the wet plastisol; (2) inverting the mold just before the coating is gelled; and (3) jarring the mold to knock off the drips.

The molds used in dip molding are usually solid and are made from cast or machined aluminum, machined brass, steel, or ceramic.

Some examples of commercial dipping applications include plating racks, dish racks, electrical parts, spark plug covers, tool handles, traffic

Fig. 16-52. Automated continuous dip molding line for producing plastisol household and surgical examination gloves. (*Courtesy McNeil Femco, Div. of McNeil Corp.*)

safety cones, work gloves, and many household and surgical examination gloves. Protective coatings are applied to many types of glass bottles, automotive bumpers and hub caps, and to protect the threads of carefully machined nuts and bolts.

Hot Dipping Process

Hot dipping is the most widely used technique because it takes advantage of the inherent heat gelling characteristics of a plastisol, which cause it to form a gelled coating on a hot mold. This technique is useful in building multiple coats and for coating over cured adhesives. Figure 16-53 illustrates glove molds coming out of the pre-heat oven and being dipped into a plastisol tank.

The coating weight, or thickness of the coating, will depend upon the following factors:

(1) Mass and shape (ratio of area to volume), not only of the whole object, but also of sections of it.

(2) Temperature of the object at the instant of dipping.
(3) Heat capacity and conductivity of the mold.
(4) Dwell time of the mold in the plastisol.
(5) Gelation characteristics of the plastisol.
(6) Temperature of the plastisol at the time of dipping.
(7) Withdrawal rate of the object from the plastisol.

Uniformity of the coating is controlled by balancing the withdrawal rate and the dwell time. The dwell time should be adjusted so that the plastisol is almost completely gelled on the surface at the conclusion of withdrawal. The flow-off and dripping are controlled by the withdrawal rate, which must be reasonably slow and uniform. All other factors being equal, the pickup is primarily controlled by the preheat temperature of the mold which usually is in the range from 49°C (120°F) to 260°C (500°F). Slow-speed agitators and cooling coils are recommended for plastisol hot dip tanks to con-

Fig. 16-53. Automatic dip molding line for gloves. Preheated glove forms are shown going from oven to plastisol dip tank. (*Courtesy McNeil Femco, Div. of McNeil Corp.*)

trol the plastisol temperature. A significant rise in plastisol temperature, resulting from repeated hot dips, will increase plastisol viscosity and change the dipping requirements considerably.

Finally, the flow properties of the plastisol, or rather their change with change in temperature, may have a significant effect on the appearance and thickness of the fused coating. If the partially gelled film of plastisol undergoes a marked reduction in viscosity during fusion, sagging will result, with local thinning of the coating, or irregular surface appearance.

Cold Dipping Process

When a cold (or room temperature) mold is dipped into plastisol, removed from the plastisol, and then fused, the process is referred to as *cold dipping*. The thickness of the coating depends upon the low-shear-rate viscosity and the yield value of the plastisol, in contrast to gelling the plastisol with a preheated mold as is done in hot dipping. Cold dipping is used when: (1) there is low-heat capacity or conductivity in the mold to be dipped, and preheating is impractical, (2) the plastisol is extremely sensitive to heat and very high viscosities would result from dipping hot molds, or (3) the mold is irregular and/or has fine surface detail which would not be reproduced in hot dipping due to premature plastisol gelation.

Withdrawal rates of the mold from the plastisol have a significant effect on the plastisol pick-up and are usually slower than the withdrawal rates used in hot dipping. A careful adjustment must be made between the viscosity at low shear rates and the yield value of the plastisol so that it flows well enough to form a uniform coating and yet does not drip. Higher viscosity plastisols require less yield value, but the important factor is the ratio between viscosity and yield value. To be effective, this ratio must not increase from the time the dipping is completed until the coating has gelled. The desirable viscosity range of the plastisol for a particular dipping application is obtained primarily by the dispersion and blending resins used and the plasticizer levels and type. Small amounts of volatile thinners are sometimes added to a plastisol to lower the viscosity. High yield values are obtained with gelling additives like colloidal silicas, high oil absorbing fillers, and metallic-organic complexes.

Automatic Hot Dip System

One of the automated plastisol hot dip coating systems (manufactured by the W. S. Rockwell Co.) illustrates how one such system operates. This system is designed to coat a variety of tool handles, such as pliers and wire cutters, (Figs. 16-54 and 16-55) at the rate of 450 to 1000 pieces per hour. The handles to be coated are steel and are vapor degreased prior to entering the coating machine. Coating thickness is approximately $\frac{1}{8}$-in. thick. This machine, of the closed-loop design, requires a minimum of floor space and requires only one operator for loading and unloading. This system will prime, allow the primer to dry, preheat, dip coat, post-fuse, and cool the parts to a handling temperature.

The length, width, and height of the machine are determined by how many tool handles must be coated at one time and how long it will take to preheat, dip, and fuse the coatings. Sometimes parts are required to be vertically rotated 180° to allow drips to flow back. This will result in adding to the overall height. The indexing cycle required for this system is approximately 1 minute, as determined by adding up the time of all the motions necessary to accomplish the coating. For a particular tool handle on this indexing cycle, a preheat of 12 minutes at 191°C (375°F) is required. The conveyor system consists of two parallel strands of chain closed in a loop, with magnetic cross bars on 16-in. centers between the chains. The chains are supported by guides and sprockets and driven by a hydraulic motor. The tools are held magnetically to the cross bars.

The primer dip tank is located six stations ahead of the preheat oven to allow 6 minutes for the primer to air dry. The primer dip station and drain station are vapor areas containing quantities of flammable vapors from the primer, and must be enclosed and well venti-

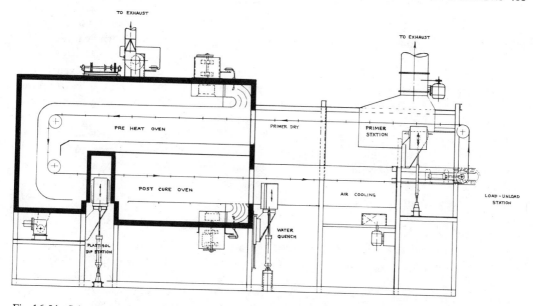

Fig. 16-54. Schematic drawing of a completely automated hot dip coating line capable of priming, preheating, dipping, postheating, and cooling of parts. (*Courtesy W. S. Rockwell Co.*)

Fig. 16-55. Photo of automated dip coating line similar to schematic in Fig. 16-54. (*Courtesy W. S. Rockwell Co.*)

lated to meet *all* safety standards. The primer tank moves up and down within the enclosure by a hydraulic system. A large exhaust fan pulls air in through the end openings and out through the top to an exhaust duct.

The preheat oven and conveyor are designed to give 13 minutes of heating time. The temperature range goes to a maximum of 260°C (500°F). Adjustable shields are provided in the oven to allow for easy balancing of air distribution.

Directly between the preheat and post-heat oven is the plastisol dip station. The dip elevator is adjustable between 2 and 12 in. The plastisol dip tank is a tank within a tank, with the plastisol recirculating up through the inner tank and flowing over into the outer tank to maintain the same plastisol level at all times. The outer tank drains back to the supply drum, and is jacketed with water-cooled coils to keep the plastisol at the optimum working temperature.

As the conveyor advances from the dip station to the post-fusion oven, the entire cross-flight bar is rotated 180° to allow the teardrops to flow back and blend into the coating. They are held in this position for one station, which is adequate to gel the plastisol and keep it from flowing. As the conveyor indexes to the next station, the cross-flight bar is returned to its original position for the remainder of the post-fusion cycle. This oven is

also capable of fusion temperatures up to 260°C (500°F) and is controlled separately from the preheat oven.

After fusion, the parts are indexed to the water quench station where an elevator lifts an overflowing water quench tank to submerge the plastisol coating. The tank is lowered and air is blown over the parts to remove water and cool.

The conveyor then indexes to the load-unload zone where the coated parts are removed and new, uncoated parts are loaded.

Rotational Molding

Rotational molding is a unique process, as compared to injection molding and blow molding, and it allows the producer greater flexibility in end-product design, especially for hollow parts, double-wall constructions and large sizes where conventional tooling would make end-product cost prohibitive. (See also Chapter 14 on Rotational Molding.) Products can be rotational molded, ranging from small syringe bulbs up to 17 ft x 17 ft x 8 ft modules, storage tanks, or shipping containers. Custom-designed machinery and molds, as illustrated in Fig. 16-56, are required for large objects.

The molding process involves four individual stages: loading the molds with raw materials, rotating and fusing the part, cooling the part, and unloading. In loading the molds, a predetermined amount of raw material is placed

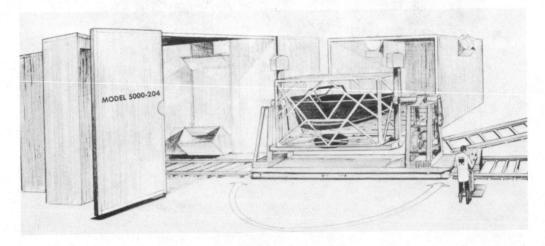

Fig. 16-56. Shuttle type of rotational molding machine for producing large objects such as storage tanks and boat hulls. (*Courtesy McNeil Femco, Div. of McNeil Corp.*)

within the molds and the mold halves, or sections, are closed and locked. In the casting and fusing stage, the mold is rotated in two planes perpendicular to one another while being subjected to heat (Fig. 16-57). As heat penetrates the mold, the raw material is gelled and builds up in an even distribution on the interior wall surfaces of the molds. While the mold is still rotating, it is exposed to cool air and water, after which rotation is stopped and the molded parts are removed.

Typical products rotationally molded from plastisols include boat bumpers, mannequins, dolls, squeeze toys, play balls, basketballs, footballs and automotive armrests and headrests. Soft, flexible products with much detail as in Fig. 16-58, or rigid functional products such as the drum table in Fig. 16-59, are being molded from plastisols.

Molds. Molds used in rotational molding may be constructed from one or more of several metals. Cast aluminum is often used for rapid heat transfer and lowest cost. Machined

aluminum will generally give parts which are free of surface porosity or voids. Electroformed nickel is best for fine surface detail reproduction. Sheet metal is used for large, simple shapes that do not have undercuts. To obtain good surface and contour reproduction, powdered metal sprays over a model can be used.

In general, regardless of the material used, the mold should meet the following requirements:

(1) It should have good thermal conductivity.
(2) Wall thickness should be the minimum necessary for strength and dimensional stability, and should be uniform.
(3) Provision for venting (Fig. 16-60) should be made to allow for pressure equalization and also to aid in producing stress-free parts.
(4) Provision must be made for clamping mold sections together firmly to prevent leakage.

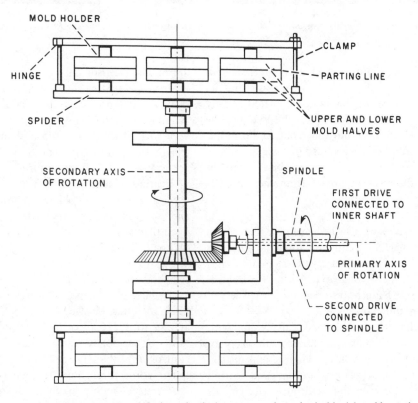

Fig. 16-57. Schematic presentation of a typical mechanical system used to obtain biaxial mold rotation in two perpendicular planes. (*Courtesy U.S. Industrial Chemicals Corp.*)

Fig. 16-58. Rotationally molded toy banks used for promotional items. (*Courtesy Royalty Industries, Inc.*)

(5) The mounting must allow for free circulation of heating medium so that all surfaces of the mold are heated equally to get uniform distribution of the plastisol within the mold. Uniform distribution, regardless of irregularities in configuration of the product, is perhaps the most important requirement in rotational molding.

Molds may be attached one at a time to the molding machine (common for large parts) or they may be mounted several at a time on a "spider" which is then attached to the machine. Figure 16-61 illustrates open and closed sets of molds mounted on spiders.

The loading or charging of the molds with plastisol is easily accomplished with many of the commercial metering, mixing, and dispensing units. Most systems are designed so that the operator can fill multiple-cavity molds easily and quickly without excess material being carried over or dripping onto the edges of the molds. Plastisol spills or drips on the mold exterior or parting lines will result in mold maintenance problems and poor moldings. Accurate metering is required and, since there is very little flash or waste in molding plastisols, exact part weight must be put into the molds to get accurate part-to-part uniformity.

Heating Systems. The heat sources most used in rotational molding are forced hot-air and sprayed molten salt. Other sources are open-flame, infrared, and recirculating hot oil.

Recirculating hot air may range from 177 to 482°C (350 to 900°F). The higher the oven temperature within this range, the shorter the cycle. Plastisol fusion cycles may range from 4 to 30 minutes. Most cycles are 5 to 8 minutes at 288 to 371°C (550 to 700°F). It is important to accurately direct the air flow and control the velocity and uniformity of the temperature throughout the oven. Maximum heat transfer with a minimum of localized hot spots in the oven can be obtained with baffles and high velocity air blowers. Excess heat

Fig. 16-59. Rigid functional drum table rotocast from acrylic-modified plastisol. (*Courtesy Whittaker Corp., Advanced Structures Div.*)

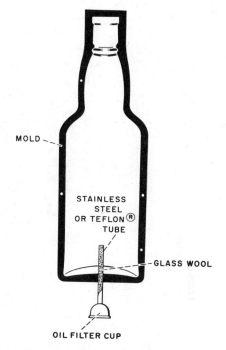

MOLD

STAINLESS
STEEL
OR TEFLON®
TUBE

GLASS WOOL

OIL FILTER CUP

Fig. 16-60. System of venting a rotational mold for pressure equalization during heating and cooling. (*Courtesy U.S. Industrial Chemical Co.*)

capacity of the oven is desirable so that the required operating temperature can be attained rapidly after insertion of the cool, charged molds. Gas-fired, oil-fired, or electrically heated hot air ovens are available.

Heating by means of sprayed molten salt is also used in rotational molding of plastisols. The molten salt is a melted standard mixture of low melting point heat transfer salts suitable for temperatures within the usual range of operations from 177 to 288°C (350 to 550°F). The salt mixture is brought up to operating temperature in a melting tank containing gas-fired tubes or electrical immersion heaters. The molten liquid is then pumped through a series of spray nozzles onto the rotating mold assembly as in Fig. 16-62. In this heating system, heat is transferred by conduction, rather than convection as with hot air. Major advantage of hot-liquid spraying is that heat transfer is very

rapid and plastisol fusion may be achieved faster with a lower temperature heat source. This method is especially suitable for molding pieces with heavy walls and for the production of complex shapes. It is essential that molds fit tight and form a good seal to keep the molten liquid from getting in and ruining the part. Molds are usually rinsed with water to remove salts. Molten salt equipment must have an efficient means to recover and reuse the heating medium because it can mean the difference between profit or loss.

Special-purpose or prototype rotational molding equipment may use infrared or open-flame heating. Radiation heaters may be electrical or gas-fired. Open flame may be gas or oil fed. Fusion of the part is accomplished by applying direct flame to the exterior of the mold. These methods are fast and efficient but only simple shapes and single molds can be used. Complex shapes or multiple molds will generally shield the heat from various sections of the molds, thus causing uneven heat transfer resulting in poor uniformity or unfused sections in the end product.

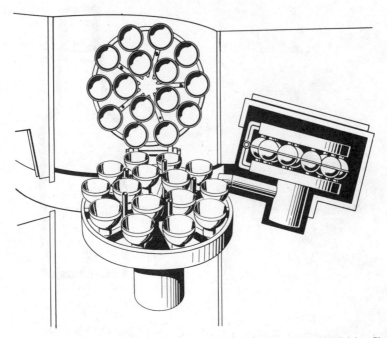

Fig. 16-61. Left: An open set of 15 molds mounted on a spider ready to be loaded. Right: Closed set of molds, containing molding material, entering fusion oven. (*Courtesy U.S. Industrial Chemicals Co.*)

Fig. 16-62. Mold carrier in oven chamber with spray nozzles mounted above for delivery of molten salt. (*Courtesy E. B. Blue Co.*)

Recirculating hot oil with jacketed molds is also used to some extent in rotational molding. This will be discussed under machinery types.

After completion of the fusion cycle, the molds may be subjected to a combination of cooling methods, including forced cold air, cold water spray or cool liquid circulating through jacketed molds. It is important that the molds continue to rotate between fusion and cooling, and during cooling to control shrinkage and to eliminate hot melt sag from occurring.

Machinery Set-ups. Rotational molding machines utilize different approaches to rotating, heating, and cooling of the molds. The molding facility can be set up to be a batch type, continuous, shuttle, or jacketed-mold system.

Key construction points to consider in deciding which rotational molding facility to set up include:[26]

(1) Four to five times the floor space of the rotomold machine selected should be available for material and product storage, part finishing operations, and mold storage.

(2) A concrete floor is desirable for support and safety since the equipment and charged molds can be very heavy and the ovens are capable of operating up to $482°C$ ($900°F$).

(3) Typical production machines require a minimum free ceiling height of 13 to 20 feet, which allows 3 to 5 feet above the exhaust fan housing on top of the machine.

(4) Equipment should be positioned for free vertical access to the roof for exhaust and cooling stacks. The most direct route of exhaust is desirable due to the high skin temperatures reached and also for cost reasons.

(5) Utility needs of the equipment deserves thorough consideration as to whether equipment should be gas, oil or electrically heated and what, if any, pollution control equipment may be necessary.

(6) Consideration needs to be given as to whether special mixing, blending and/or drying equipment will be needed and whether special handling equipment will be required such as bulk storage facilities or a crane for loading and unloading molds and parts.

Batch-type machines are normally used for small to medium size parts. They are largely manual-type machines involving considerable operating labor. There are two batch machines used commercially; the fixed-spindle type and the pivoting-arm type.

The fixed-spindle machine requires a roller-top table where the mold assembly is transferred between heating, cooling, and loading-unloading. Two drive systems, one in heating and one in cooling, are required to rotate the molds. With this system the mold or molds are loaded, rolled into the fusion oven, locked onto the spindle, and then fused. They are then manually rolled to the cooling chamber, locked in place, cooled, and returned to the load-unload area. Fig. 16-63 presents a schematic of this type of equipment and Fig. 16-64 shows a production machine of the fixed-spindle type.

In the pivoting-arm batch machine, the molds are attached to a spindle arm which swings in a 90° arc when moving from the oven to the cooling chamber (Fig. 16-65). A second pivot spindle and a cooling chamber may be added to this system, which will utilize the same oven and result in twice the production rate, with the same labor requirement.

Rotational molding machines utilizing the fixed-spindle or the pivoting-arm concept may be designed to have a double-center-line mounting (most common) or an offset-arm mounting (Fig. 16-66) on the molds. The double-center-line mounting attaches several molds to spiders that lie directly above and below the spindle. Easier mold loading and unloading in the production of numerous small items are possible. The offset arm mounting is primarily used for molding large items. It also minimizes variations in wall thickness by bringing the center of volume of the mold close to the intersection of the primary and secondary axes of rotation.

Continuous rotational molding set-ups are most commonly of the carousel-type, using multiple arms. This type of machine may have

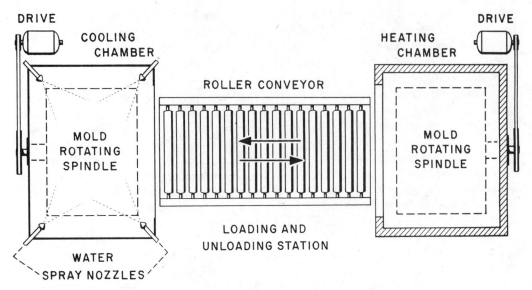

Fig. 16-63. Schematic of a fixed spindle batch type rotational molding machine. Separate drives are required for the oven and cooling chamber. (*Courtesy U.S. Industrial Chemicals Co.*)

Fig. 16-64. Production model fixed spindle batch type of rotational molding machine. (*Courtesy E. B. Blue Co.*)

3, 4, or 5 arms located in a circular pattern and equidistant from each other. Figure 16-67 is a schematic of a 4-arm machine. Figure 16-68 shows a production unit having 3 arms. One of the arms shown is for double-center-line mounting and the other arm shown is for offset mounting.

The arms, each carrying a group of molds or a single, large mold, are mounted on a common hub. The molds are indexed from station to

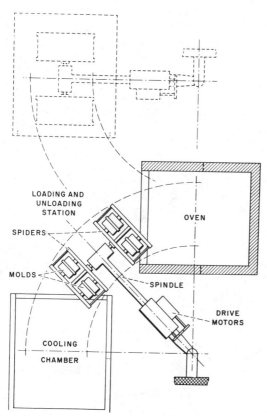

LOADING AND
UNLOADING
STATION

SPIDERS

MOLDS

OVEN

SPINDLE

DRIVE
MOTORS

COOLING
CHAMBER

Fig. 16-65. Schematic of a pivoting-arm rotocasting machine. Spindle and molds swing in a 90° arc between heating and cooling chambers. Addition of a second cooling chamber, pivot, and spindle (dotted lines) will utilize the same oven and result in twice the production with the same labor requirement. (*Courtesy U.S. Industrial Chemicals Co.*)

station. Each arm may function independently as to speeds and ratios of rotation. All portions of the cycle are automatic with the exception of loading and unloading finished parts.

Another type of continuous rotational molding uses a continuous conveyor system with individual rotomolding stations which travel on a conveyor. A complete cycle includes automatic filling of the molds, automatic closing, engagement of rotational gearing, fusion of the plastisol while rotating, cooling of the mold by water spray or cold air, automatic opening of the mold, and manual removal of the finished parts.[14] Mold rotation ratios are set by the crown and pinion gear assembly on each mold station and the rotation gear rack on the conveyor bed. The rotation ratio is there-

fore fixed and nonadjustable, and the product types that can be produced are limited. The equipment is operated with a minimum of labor and is capable of achieving a high production rate. Mold investment is high due to the large number of molds necessary for automatic cycling.

Shuttle-type rotational molding systems may be relatively simple as in Fig. 16-69, or quite complex as in Fig. 16-56. Shuttle machines contain features from both batch and continuous machines. This type of equipment is generally used for very large objects. Support of the mold may be on one or both ends. Two or more arms may be utilized in the same machine. The process is semi-automatic with the molds shuttling on a continuous track, moving from station to station horizontally.

The jacketed mold type of rotational molding is both expensive and sophisticated in relation to the molds and the machinery. This is a single-station operation where the mold is cored, or has channels in the mold walls, which will allow hot and cold liquids to circulate. After the mold is loaded and closed, it is rotated while a hot liquid, such as oil, is circulated through the mold walls and fuses the plastisol. Upon completion of fusion, a chilled liquid is circulated through the walls to cool the part. The mold does not change stations and there is no oven or cooling chamber, thus, space requirements for this system are minimized. An insulated heating tank and cooling oil storage are required along with a series of valves and pumps. Fast cycles for long production runs of symmetrical-shaped objects are the primary advantages of this system.

Formulating Considerations. Plastisols allow the rotational molder to explore new markets with relative ease (Fig. 16-70). They can be custom formulated not only to meet the end product requirements but also to meet processing requirements which may result from equipment limitations or difficult mold shapes.

The most important factors influencing successful rotational molding of plastisols are the flow properties at room and elevated temperatures. The plastisols should have low viscosity at low shear and very little or no yield value. Along with these, appropriate gelation charac-

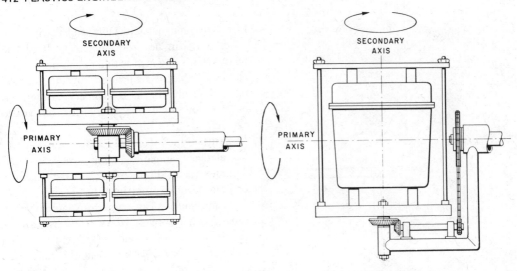

Fig. 16-66. Mold mounting systems. Left: Double-Center-Line mounting. Right: Offset-arm mounting. (*Courtesy U.S. Industrial Chemicals Co.*)

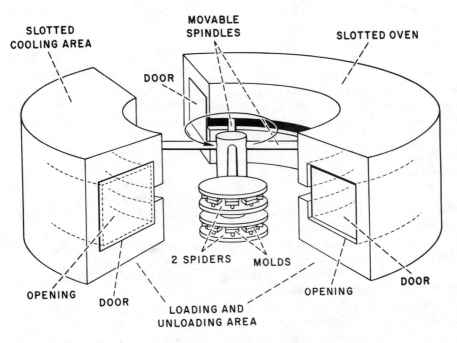

Fig. 16-67. Schematic of a typical rotary, multiple-spindle (4) rotocasting machine. Center spindle is in the loading-unloading area, 2 are in the oven, and 1 is in the cooling area. (*Courtesy U.S. Industrial Chemicals Co.*)

teristics are necessary to get an even distribution of material without bridging occurring in close tolerances. Optimum viscosity appears to be in the range of 1000 to 4000 centipoises.[27] This range provides the plastisol the greatest

opportunity to wet the mold prior to gelation. Plastisols lower than 1000 cps have a tendency to leak from the mold, while those greater than 4000 cps offer increasingly less control of the cycle during gelation.

Fig. 16-68. A multiple-spindle (3) production rotocasting machine. One of the arms shown is for double-center line mounting and the other is for offset mounting. (*Courtesy McNeil Femco, Div. of McNeil Corp.*)

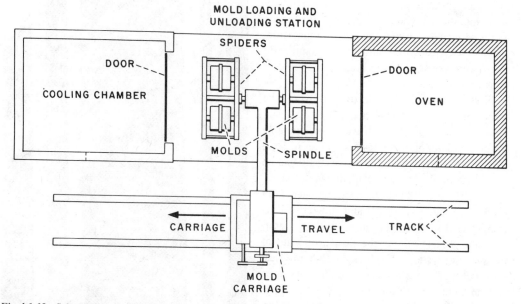

Fig. 16-69. Schematic of a shuttle type rotational molding machine. The mold carriage moves spiders and molds back and forth on a track through the mold loading and unloading station, the oven, and the cooling chamber. (*Courtesy U.S. Industrial Chemicals Co.*)

The gelation cycle is very critical in plastisol rotational molding. Gelation is usually completed within 2 minutes after the mold is placed in the oven. Therefore, distribution of the plastisol[28] must take place within that time period.

Even distribution and gelation of the plastisol can be accomplished by mechanical

Fig. 16-70. Large inflated vinyl balls with a variety of handle designs are rotationally molded from plastisols. (*Courtesy Sun Products Corp. Div. of Tally Industries*)

plastisol gelation. Stronger solvating plasticizers will increase plastisol viscosity more rapidly and gelation will be sooner, while lower solvating plasticizers will give slower gelling plastisols with good viscosity aging characteristics.

Additives used to adjust flow and physical properties include thinners and monomers. Volatile thinners can be used in limited quantities to adjust flow properties.[29] Generally, not more than 5 parts per 100 of resin should be used. Acrylic monomers are used in plastisol formulating to lower viscosity and then polymerize during fusion to produce rigid products. Rigid pipe fittings in Fig. 16-71, and the drum table in Fig. 16-59 were rotocast from acrylic modified plastisols.

In formulating plastisols for a rotocast product, aside from the type and amount of ingredients necessary for the final properties,

processing means and also, to some extent, by adjusting the plastisol formulation. Processing methods include varying the oven temperature, air velocity, heat input, and the rpm and ratio of the two axes of the arm holding the mold or molds.

Plastisol formulating adjustments would be made primarily through a change in resin, plasticizer or special additives. By changing from a homopolymer dispersion resin to a copolymer dispersion resin, the viscosity can be increased and the gelation can be made to occur sooner. The use of blending resins will generally result in lower viscosity and slower gelling characteristics. The plasticizer system determines to a large extent the flow properties of the plastisol system, as well as the final physical properties of the finished product. The type and amount of plasticizer affect flexibility, hardness, low-temperature properties, odor, migration characteristics, and cost. Both the viscosity and solvating power of plasticizers affect plastisol flow properties. The viscosity and level of the plasticizer will determine the plastisol viscosity initially; however, the solvating action of the plasticizer will have a marked effect on aged viscosity and also on the

Fig. 16-71. Rigid pipe fittings used to connect extruded vinyl pipe are rotocast from acrylic modified plastisols. (*Courtesy Whittaker Corp., Advanced Structures Div.*)

consideration must be given to the gelation characteristics best suited for the machine conditions. In other words, compound to meet the end product requirements and then make a rotocast trial. Two variations in formulation should be made if the resulting plastisol distribution is undesirable. First, change the compound so as to speed the gelation characteristics. Second, change the compound to slow the gelation characteristics. Both changes should be significant enough to noticeably change the gelation of the plastisol. Usually a 25% replacement of a high or low solvating plasticizer is sufficient. Rotocast these variations and note the effect on distribution. This will quickly indicate which formulating direction should be followed to obtain optimum gelation characteristics and distribution.

In general, the larger the product to be rotocast from plastisol, the faster the gelation characteristics should be. Whenever making a rotocasting trial on a new product, complete fusion to obtain the maximum physical properties is essential to the trial. Increasing the molding cycle time or temperature to the point of slight degradation, and then backing off, will generally give sufficient fusion for production development trials.

Slush Molding

Slush molding is another excellent method of producing hollow objects. A wide variety of products can be manufactured by this process, including rainboots, shoes, hollow toys and dolls and automotive products such as the protective skin coatings on arm rests, head rests, and crash pads.

The basic process of slush molding involves filling a hollow mold with plastisol, exposing the mold to heat, gelling an inner layer or wall of plastisol in the mold, inverting the mold to pour out the excess liquid plastisol and then heating the mold to fuse the plastisol. The mold is then cooled and the finished part removed. Slush molding can be a simple hand operation for limited production or an elaborate conveyorized system for long runs. This process can be a one-pour method, where finished or semi-finished products can be made by one slushing, or a multiple-pour method where two or more slushings are used.

In the one-pour method, molds are placed on a conveyor belt or system, having positions for as many molds as desired. Molds are filled to the top and carried through an oven where temperatures between 93 to 316°C (200 to 600°F) gel the plastisol next to the mold wall. The thickness of the gelled plastisol at a given oven temperature is determined by several factors: the thickness of the metal wall of the mold; the length of time the mold is in the oven; and the gelling characteristics of the plastisol. Next the mold is turned upside down and the remaining liquid plastisol is dumped out, while the gelled plastisol next to the wall remains in place. The mold is conveyed to a second oven where fusion of the plastisol is completed. Passage through a cooling chamber cools the mold sufficiently to permit removal of the finished piece by blowing it out with compressed air or collapsing it with vacuum. It is advisable to rotate the molds in the second oven to maintain an even temperature on all sides of the mold to evenly fuse the plastisol.

The multiple-pour method of slush molding is used when the mold has fine surface detail, a heavy wall thickness is desired, or when different types of plastisol are used in the same product. For example, boots, as in Fig. 16-72 and 16-73, are often produced by the multiple-pour method. A cold mold is filled with plastisol and then emptied. The mold, with a thin coating of plastisol, is then heated, refilled while hot with the same plastisol or another type, and emptied again. The mold with the multiple coating of plastisol is heated to fuse the plastisol. It is then cooled and the product is removed. It is possible to reproduce fine detail of the mold suface with this method because of the cold pour and thorough wetting of the mold prior to gelation of the plastisol. The second or third pour may consist of a cellular plastisol which, upon fusion, will provide a soft lining and insulation properties which are often necessary in cold weather boots.

Molds. Molds used in slush molding are produced from sprayed molten metal, spun aluminum, machined aluminum, or ceramics or

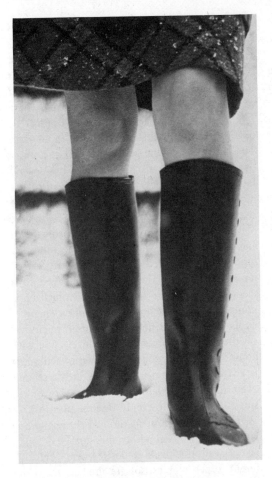

Fig. 16-72. Slush molded all-weather boots produced from plastisols. (*Courtesy Chemical Products Corp.*)

(1) The molder's technique and equipment.
(2) Variations in shape, size, and wall thickness of the article.
(3) Physical properties required in the finished article, such as color, flexibility at various temperatures, grease resistance, electrical characteristics, etc.

Plastisols used in slush molding are very often formulated similarly to those used in rotational molding. Brookfield viscosities may range from 1000 to over 10,000 cps. Usually, low Brookfield viscosity (under 5000 cps.) is preferred. Good viscosity stability is desired because of the constant heating and cooling the plastisol undergoes in slush molding.

Physical qualities of slush-molded plastisols allow exceptionally accurate work. Because the material is not fused under pressure, it exhibits very little shrinkage when molded. Colors are almost unlimited. Plastisol articles may be readily cemented and heat-sealed. When properly formulated, they resist acids, alkalies, and a wide range of solvents. They resist aging and sunlight. They can be made to have excellent electrical insulating properties and flame resistance.

by electroforming. An electroformed mold is relatively inexpensive, easily reproduced and very durable. Some molds of this type have lasted for many years. Molds should be cleaned occasionally with a liquid honing machine.

Regardless of the type of mold, the article molded from plastisol will faithfully reproduce the surface of the mold whether matte or glossy. Since molds occasionally have, or develop, surface porosity, the finish will be affected and will usually have low gloss. In cases where a highly glossy surface is desired, or where transparency is necessary, molds can be coated on the interior with epoxies, phenolics, and even porcelain.

Formulating Considerations. Plastisol used in a slush molding job will depend on three factors:

Fig. 16-73. Stylish ladies boots slush molded from plastisols with semi-rigid heels and soles and flexible uppers with expandable sides and foam liner. (*Courtesy Bonan Co.*)

Cavity, In-Place and Low-Pressure Molding

Plastisols are used in the manufacture of many different products that are solid moldings. Solid molded parts are usually made by cavity, in-place, and/or low-pressure molding.

The fusion time for solid molded parts is dependent upon the mold thickness and heat conductivity, the heat available and the cross-sectional thickness of the article to be molded. Generally, for every $\frac{1}{4}$ in. of cross section, 15 minutes at 177 to 191°C (350 to 375°F) is sufficient to complete fusion. Plastisols to be molded in thick sections ($\frac{1}{2}$ in. or more) may require special heat stabilization to protect the vinyl from degradation during the long fusion cycle.

Cavity Molding. Cavity molding is the simplest process for the manufacture of solid molded parts. It consists of pouring plastisol into a mold, heating the mold to fuse the plastisol, cooling, and stripping the molded part from the mold. Toy kits for making replicas of insects and for figurines use this process. Other products made by cavity molding include fishing lures and sink and disposal stoppers.

Plastisols used in cavity molding may be very fluid and pourable or they can be formulated to have a stiff consistency such as modeling clay. Wetting the mold surface with a thin film of plasticizer before pouring in plastisol will aid in reproducing surface detail.

In-Place Molding. The excellent moldability, low cost and excellent physical and chemical properties of plastisols have yielded several large volume applications for plastisols for gasketing. In-place molding is a process of molding directly in, or onto, a finished article.

Largest volume applications for in-place molding are jar lid and bottle cap liners as shown in Fig. 16-74. Automatic cap production lines operate at very high speeds and thousands of plastisol gaskets are applied to jar lids and bottle crown caps per hour. Plastisols must be formulated to have low viscosity at the very high shear rates encountered in high-speed production lines.

The automotive air filters shown in Fig. 16-75 constitute another large volume use for in-place molded plastisols. The filter paper and

support elements are placed in a ring mold which has been filled with plastisol. The mold is passed through an oven to fuse the plastisol. Once out of the oven, the filter is turned over and the other end is placed in the ring mold

Fig. 16-74. Jar lid and bottle cap liners are molded in place from plastisols. (*Courtesy White Cap Div. of Continental Can Co.*)

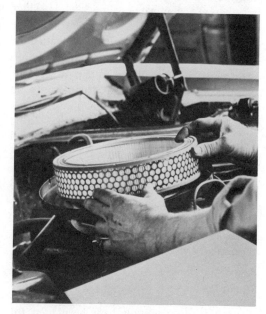

Fig. 16-75. The end caps for most automotive air filters are molded in-place from plastisols. (*Courtesy Chemical Products Corp.*)

with plastisol and this, in turn, is fused in the oven. The fused plastisol provides an air-tight gasket for assembly in the air cleaner of the car and it also bonds the ends of the filter paper into a sealed assembly.

Sewer pipe gaskets have been made from in-place molded plastisols for many years. Special molds are placed over the ends of the clay or ceramic sewer pipe. Plastisol is poured into the molds and fused around the pipe ends. The plastisol is formulated to a specific hardness range to permit the pipe ends to be joined together easily to form a tight seal. Solvents are used to lubricate the pipe ends during assembly and to cement the ends together upon drying. Sewer pipe can be laid in place very rapidly with this gasket system. Another advantage of the plastisol in this application is its ability to compensate for lack of roundness of the pipe.

Low-Pressure Molding. Plastisols can be injected into closed molds with low pressure positive displacement pumps. The molds are then placed in ovens (or they can be put into a heated press) to fuse the plastisol. Mold detail is accurately reproduced and products up to 2-in. thick may be produced.

The molds must be well matched to form a tight fit and they can be clamped or bolted together. To eliminate air pockets in filling closed molds, small bleeder holes must be correctly positioned in the mold so that all the air is ejected and the cavity completely fills with plastisol. After filling, certain of these air bleeder holes and the inlet may have to be closed off so that the plastisol will not escape. This is particularly important because during heating of the plastisol the viscosity will decrease and the plastisol may become quite fluid prior to gelation. Well deaerated plastisols should be used in low pressure molding to ensure void-free products.

Products made from low pressure molding of plastisol include shoe soles, printing plates, and encapsulated electronic parts.

Strand Coating

Plastisols and organosols have been used for many years as protective coatings on various filaments, wires, and woven cords. Fibrous glass yarns are strand-coated at speeds of 500 to 600 ft/min with the speed varying according to the thickness of the coating desired and the type of fusing heat used (radiant or convection). This coated fibrous glass is then woven into various types of screening. Certain types of "spaghetti" wire tubing commonly used in the electrical industry are also strand-coated with plastisol.

There are generally three methods of strand coating: the set-die method, the floating-die method, and a method using no die. In the set-die method, the die is rigidly set and centered around the strand to be coated. The strand must be held sufficiently taut so that all bends or imperfections are straightened out before it enters the die. Plastisol is poured or pumped over the face of the die, the strand passes through it, the die wipes the excess plastisol off (Fig. 16-76), and the strand then passes through a fusion tunnel. The excess plastisol that was wiped off is recirculated. Although this method will work for wide ranges of viscosities, it is particularly suitable for viscous organosols or plastisols and thus will provide a heavier coating per pass.

In the floating-die method, the die is loosely held in place and allows a certain freedom of movement in one plane. The theory is that the die will center itself around any slight bend or imperfection. If the viscosity of the plastisol or organosol is too high, the material will pile up under the die and restrict the freedom of movement of the die. The plastisol or organosol

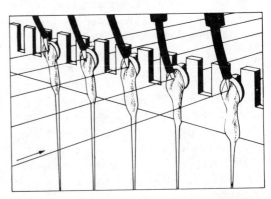

Fig. 16-76. Plastisol protective coatings on filaments, wires, and woven cords are applied at very high speeds via strand coating. (*Courtesy B. F. Goodrich Chemical Co.*)

must be low enough in viscosity to flow freely off the die, with only slight accumulation.

With the method using no die, the strand is passed vertically through the plastisol or organosol, the viscosity of which is adjusted to give a uniform pickup of desired thickness with no runbacks. A runback occurs if the wire or line is passing too fast through the coating material or if the viscosity of this material is too high. In either case there is too great an initial pickup of material which tends to flow back down the strand and cause an uneven coating. This method is generally suitable for dispersions of relatively low viscosity. Maximum pickup without runbacks is ordinarily small and several passes may be necessary to build thicker coatings.

Fusion of plastisol or organosol strand coatings is dependent upon speed of travel, length of oven or tower, fusion temperatures available and thickness of coating.

In formulating for strand coating, both the requirements of the end use and the processing conditions must be considered. A formulation for coating fibrous glass for screening will be quite different than a formulation for coating electrical wire. In coating rope, thread or woven cords, it must be determined if the coating should penetrate the strand or merely coat the surface. These examples show the importance of selecting the proper ingredients for the vinyl dispersion to meet the requirements of the job.

Spray Coating

Spraying of vinyl dispersions has grown with the expanding use of plastisols in specialty applications and with the increased interest in new protective and decorative coatings on metal furniture, appliances, and building products. Spraying is the only sure way of coating many intricate or irregular objects. Sprayed vinyl dispersions can lay down a coating from a few mils up to 60 mils.

Organosols are widely used for spray coating because of easier flow, but the extra body of plastisols permits heavy coats on vertical surfaces without sags in the coating. Standard spray guns can be used. By employing special spray heads, striking, decorative veil and spatter patterns can be applied.

Flow properties of vinyl dispersions for spraying are very important to successful application. Both low and high-shear conditions are encountered. Standard air spray gun orifices used in plastisol work range from about 0.07 to 0.1 in. in diameter. In passing through these orifices at pressures ranging from about 30 to 100 psi, the materials undergo a high degree of shear. Then, upon being deposited at the work surface, the low shear flow properties of the plastisol will dictate just how smooth a film will result and how much can be laid up without sags or runs. Balancing low-shear properties for production of thick coatings is an exacting procedure. Too little yield for the coating thickness will cause the film to sag, curtain, and run off. Too much yield will inhibit leveling and result in extreme orange peel surface.

Extrusion

Plastisols may be extruded as a means of preparing partially fused materials for subsequent molding or as a means of producing finished goods. Plastisol extrusion is an ideal way to produce vinyl formulations of low durometer reading. These compounds are very difficult to produce by the conventional two-roll mill because of adhesion to the rolls and the low viscosity of the hot melt.

Almost any extruder will extrude plastisols but the important factors are quantity of production and uniformity of output. The screw in an extrusion machine works as an inefficient high-pressure pump with plastisols, with the friction between the screw and the cylinder wall creating the forward driving force. Plastisols offer little resistance to the action of the screw forcing them forward. Therefore, it is possible to increase the screw speed two to three times its normal usage rate. If the extruder is being used as a step in the compounding of a plastisol, it should be extruded incompletely fused so that it may be readily ground up or cubed. Final fusion then will take place in an injection or compression molding machine.

The method of feeding the plastisol to the

extruder will depend upon its viscosity. It is desirable that the machine have a large opening or throat for high viscosity materials so that they may flow into the hopper at a rate which will maintain a reservoir of material and avoid the entrapment of air or surging of the material. If the material is of low viscosity, it is desirable to suspend the drum of plastisol over the machine and pipe it to the hopper, with suitable gaskets to prevent leaking. It is possible to attach a cover to the drum and inject a safe amount of air to increase the flow of the plastisol in the extruder. The screw should have shallow flights and, if possible, a heated screw should be used.

The advantage of plastisols over conventional molding compounds in extrusion lies in the production of low-durometer formulations. Extrusion of such formulations gives results more uniform than can be obtained from a mill. Plastisols are advantageous in extruding delicate shapes, because their fluidity permits the use of low pressures.

APPENDIX

Vinyl Chloride and Occupational Health

A new dimension in terms of environmental concerns relating to PVC was introduced in early 1974. At that time the B. F. Goodrich Chemical Company disclosed its concern, to appropriate federal and state agencies, about an occupational health problem involving exposure to vinyl chloride gas at levels which probably were 100 times higher than those existing in plants today.

The finding was reported after the discovery, from medical and hospital records, that several people who had worked at one plant for many years had a rare form of liver cancer (angiosarcoma) at the time of their deaths. These were the first known occurrences of angiosarcoma among vinyl chloride workers anywhere in the world. In cooperation with government agencies, an industry-wide investigation was begun immediately to determine the scope of the problem and to plan corrective action. Epidemiological studies have identified 16 such deaths in this country from among the many thousands of workers who have been employed in the manufacture of polyvinyl chloride over the past 35 years.

Between the Goodrich announcement in January 1974, and the effective date of the Occupational Safety and Health Administration (OSHA) vinyl chloride standard April 1, 1975, the industry has made significant progress in reduction of worker exposure to vinyl chloride. Manufacturers of both vinyl chloride monomer and polyvinyl chloride resins devoted thousands of man-hours to the identification and solution of worker health problems with OSHA, as well as with the Environmental Protection Agency (EPA) and the Food and Drug Administration (FDA).

It was essential that vinyl chloride emissions be controlled in three primary areas: (1) manufacturing work areas of monomer plants, resin manufacturers, processors, and fabricators, (2) residual monomer in resins and compounds shipped by resin manufacturers to their processor and fabricator customers, and (3) emissions to the air, in water and in solid waste. The industry's first and most immediate objective following discovery of the vinyl chloride hazard was reduction of ambient vinyl chloride in work areas.

The first step in the program was establishment of an intensive program designed to locate sources of vinyl chloride leakage. To accomplish this, a number of monitoring methods were used, including stationary probes in work areas, portable analyzers to pinpoint leaks, and personal systems worn by workers in polyvinyl chloride manufacturing areas. Simultaneously, engineering techniques were introduced to confine vinyl chloride gas or to recover it in closed systems to reduce its presence in the work atmosphere and the environment.

The PVC industry as a whole is running plants at levels somewhere between non-detectable vinyl chloride and 5 ppm. This marks a significant drop from the 50 to 100 ppm level the industry was operating at in the months just prior to announcement of the health hazard.

The OSHA standard calls for a vinyl chloride exposure level limit in specified plant areas of 1 part per million (ppm) for 8 hours and 5 ppm for 15 min., both on the basis of time-weighted averages. However, the 1975 standard also allows higher levels up to a maximum of 25 ppm, provided that respiratory protective equipment is available to employees who choose to wear these devices. Employees are required to use respirators whenever the

25 ppm level is exceeded. Beginning in 1976, employees must use respirators whenever vinyl chloride levels exceed the specified levels of 1 ppm and 5 ppm.

The industry has also made dramatic progress in the reduction of vinyl chloride monomer in finished vinyl chloride products—both resins and compounds. For example, one major producer has reported reductions ranging from 50 to 90% (1975 versus 1974), and another producer reports that it has virtually eliminated Vinyl Chloride monomer in products accounting for 90% of its business.

Resins and compounds with low levels of residual vinyl chloride monomer result in levels below 1/2 ppm of vinyl chloride in work areas in the plants of processors. This is important because these fabricators are exempt from the complex requirements of the OSHA standard when worker exposure in their plants does not exceed 1/2 ppm or below—the "action level" specified in the OSHA standard.

In the third problem area—that of plant emissions—the industry has been working closely with the Environmental Protection Agency. Continuing discussions covered technical exchanges and in-depth review of a wide range of processes and problems. Monitoring programs are being conducted at several plants throughout the United States, and the data developed from this activity will be used to assist EPA in developing and establishing effective and realistic emission standards. Good progress has been made in this area already and new engineering controls, scheduled by leading producers for installation in 1975, provide for more complete collection and recovery of vinyl chloride during manufacturing, which will result in still lower emissions.

The vinyl chloride health hazard has provided the PVC industry with perhaps its greatest problem and its greatest challenge. The progress made to date has been so encouraging that solid growth potential for polyvinyl chloride is foreseen by industry analysts.

References

1. Semon, W. L., U.S. Patent 1,929,453 (Oct. 10, 1933), assigned to B. F. Goodrich Company.
2. Johnson, G. W., British Patent 500,298 (Feb. 7, 1939), I. G. Farbenindustrie.
3. Semon, W. L., U.S. Patent 2,188,396 (Jan. 30, 1940), Assigned to B. F. Goodrich Company.
4. Bowles, R. L., Davie, R. P., and Todd, W. D., Modern Plastics, 33, 140 (Nov., 1955).
5. Nielsen, L. E., Mechanical Properties of Polymers, pp. 30-35, Van Nostrand Reinhold Co., New York, 1962.
6. Biteman, J. H., Formulating Parameters for Organosol Metal Finishes, J. Paint Technol., 39, No. 504, 51-55.
7. Modern Plastics Encyclopedia, 1972-1973.
8. Plastics World, Directory of the Plastics Industry, Vol. 30, No. 11, (Aug., 1972).
9. Ward, D. W., Reactive Functionality in Dispersion Resins, SPE Journal, 28, 44-50 (May 1972).
10. Ritchie, P. D., Plasticizers, Stabilizers, and Fillers, The Plastics Institute, Iliffe Books, Ltd., London, 1972.
11. Whittington, L. P., A Guide to the Literature and Patents Concerning Polyvinyl Chloride Technology, Society of Plastics Engineers, Stamford, Conn., 1963.
12. Darby, J. R. and Sears, J. K., Plasticizers, Encyclopedia Poly. Sci. and Technol., 10, 228-306, 1969.
13. Todd, W. D., Esarove, D. and Smith, W. M., Modern Plastics, 34, No. 1, p. 159 (1956).
14. Sarvetnik, M. A., Plastisols and Organosols, New York, Van Nostrand Reinhold Company, 1972.
15. Encyclopedia of Polymer Science & Technology, Vol. 14, New York, Interscience Publishers, Div. of John Wiley & Sons, Inc., 1971.
16. O'Mara, M. M., Ward, D. W., Knechtges, D., and Meyer, R. J., Fire Retardation of PVC and Related Polymers, Chap. 18 in Flame Retardant Science and Technology of Polymeric Materials, Marcel Dekker, New York. N.Y.
17. Higgins, D. G., "Coating Methods Survey", Encyclopedia of Polymer Science, Vol. 3, p. 766, New York, John Wiley & Sons, Inc., 1965.
18. Jacobs, R. J., "Fundamentals to Consider in Selecting Coating Methods", Reprint from Paper Film and Foil Converter.
19. Booth, George, L., Take Your Choice of Coating Methods, Modern Plastics Magazine (Sept. and Oct., 1958).
20. Booth, George L., Coating Equipment and Processes, Lockwood Publishing Co., Inc., 1970.

21. See Chapter 20 on Cellular Plastics in this book.

22. Conger, R. P., *SPE Journal,* **24** (3), 43 (Mar., 1968).

23. Eastman Chemicals Publications L163, L164, and L165 on Adhesion Promoting Plasticizers, Eastman Chemical Products Co., Kingsport, Tenn.

24. National Coil Coaters Assoc., 1900 Arch St., Philadelphia, Pa. 19103.

25. See for example, Poll, G. H., "Painting Aluminum at 200 fpm", *Products Finishing,* pp. 2-9. (July, 1969).

26. Ramazzotti, How to Plan a Rotational Molding Facility, *Plastics Technology,* p. 19. (Jan. 1972).

27. Bruins, P. F., *Basic Principles of Rotational Molding,* p. 276. Gordon and Breach, Science Publishers, Inc., 1971.

28. *Ibid,* p. 278.

29. *Ibid,* p. 262.

17

POWDER COATINGS

J. J. Sokol and R. C. Hendrickson
The Sherwin-Williams Co.
Chicago

Introduction

For at least the past thirty years, industry has strived toward developing a solventless coating: a coating that is not dependent upon a sacrificial media such as solvent, but is based on the performance constituents of the film. A powder coating is such a coating, based on a solid resin, thermoplastic or thermoset, pigments, fillers, and additives such as hardeners or flow agents. A powder coating is a homogeneous blend of these ingredients in the form of a dry, fine particle size compound similar to flour.

The research and development of powder coatings has been accelerated by such potential advantages as minimized air pollution and water contamination, increased performance from the coating, and improved economics to the user in application and handling. However, it should be kept in mind that a powder coating is basically a chemical coating and, therefore, has many of the same problems as solution paints. If not properly formulated, the powder coating may sag at high film thicknesses, show poor performance when undercured, show film imperfections such as craters and pinholes, and have poor hiding at low film thicknesses, etc.

The following sections discuss the composition, application, and manufacture of powder coatings. There are practically an infinite number of formulations in each resin class. This section centers on the most available generic resin types and their general performance and handling characteristics.

Polyvinyl Chloride (PVC)

Polyvinyl chloride resins are one of the most versatile groups of plastic materials in terms of formulation, physical properties, and processing. They are also among the least expensive of the fluid bed coating resins.

Vinyl powders show excellent exterior durability, coupled with chemical and high impact resistance. When used over a properly formulated primer, vinyl powders show outstanding resistance to salt spray, detergent, and humidity, along with high dielectric strength and good abrasion resistance.

Vinyl powders comprise several different groups, many of which are tailor-made to meet exacting specifications such as dishwasher rack coatings and electrical grade insulation. Even the so-called "general-purpose" type must meet the customer's individual requirements such as rate of fusion, film build, and initial heat stability.

One advantage of vinyls is their formulating latitude. The vinyl resin may be suspension or bulk polymerized; the main considerations are molecular weight, inherent viscosity, plasticizer absorption, and particle size. The plasticizer, stabilizer, pigments, and antioxidants used are determined by the degree of coating performance required, such as electrical, anti-rust, flexibility, or weathering properties. For example, specific stabilizers are needed in semi-rigid vinyl coatings due to the low plasticizer level. The drying agent is selected on the basis of its effect on the free flowing and fusion properties of the completed formulation.

Vinyl powders are normally processed by dry blending or melt mixing. The dry blend technique is the least expensive and the most popular of the two. Both processes are described in the section on Manufacturing.

The major drawback of vinyl powders is the fact that dry film thickness is normally in excess of 5 mils. Most vinyls are applied by fluidized bed techniques which result in heavy film. The technology for formulating electrostatic spray-grade vinyls is just developing; a few are already on the market. Laboratory work has also shown that the volume resistivity and the particle size of traditional vinyl powders can be adjusted to permit the application of one and two mil films.

Another drawback is that most vinyl powders have poor adhesion unless applied over a specially formulated primer. The liquid primer is applied by dip or spray techniques and cured in the preheat cycle prior to applying the vinyl powder.

Vinyl powders are used in the appliance industry because of their resistance to hot water, detergent, and heat. The high dielectric strength, durability, and corrosion resistance makes the coating acceptable for pole transformers, outdoor furniture, pipe, battery clamps, etc.

Polyethylene

Polyethylene is another low cost thermoplastic material. Polyethylene powders are relatively easy to formulate due to the wide ranges of density and melt index available. Most formulations use a low density resin to achieve maximum overall results.

Polyethylene coatings show good chemical and humidity resistance, high dielectric strength, and extreme toughness. Their major limitations are poor adhesion and abrasion resistance. As with vinyls, polyethylenes are also used at rather high film thicknesses. Polyethylene powders have mostly been used for carpet backing and rotational molding. However, new formulations are producing quality polyethylene powders for coating pipe, marine parts, and containers.

Polypropylene

Polypropylene powders have much the same advantage as the vinyls and polyethylene coatings. They have excellent chemical, humidity, and abrasion resistance. The principal limitations of polypropylene have been adhesion, UV resistance, and impact resistance. The impact resistance can be greatly improved by a water quench after the fusion bake. An adhesion-promoting agent is now available from a polypropylene supplier which practically eliminates the need for a primer.

Cellulose Acetate Butyrate (CAB)

As with the other thermoplastic resins, CAB requires a fusion temperature in excess of 400°F, usually 450 to 500°F. Normally, CAB coatings are high in film thickness and are applied by fluid-bed techniques. Recently, new grades of CAB powders have been formulated which can be applied at film thicknesses as low as 3 mils.

Over a solvent-based primer, the CAB coating has good performance characteristics, including exterior durability. However it has limited chemical resistance, poor heat stability for electrical applications, and a high fusion temperature. Its decorative coating properties are used primarily for metal furniture, fencing, bicycles, etc.

Epoxy

Epoxy powder coatings have outstanding adhesion to clean, bare steel. However, as with all coatings requiring maximum adhesion and resistance to corrosion, etc., pretreated substrates are necessary. Epoxies have excellent resistance to most chemicals, salt spray, and humidity. A pencil hardness from 3 to 6H is common, while retaining flexibility and resistance to failure from 160 in.-lb reverse impact. The coating also has abrasion resistance and high dielectric strength.

The success of a protective coating depends on proper formulation and application to a properly prepared substrate. Most epoxy powders are based on a Type 4 resin with a WPE (weight per epoxide) around 1000. The greater the WPE, the better the flexibility, edge coverage, and package stability. Decreasing the WPE results in decreasing the properties described above but in increasing the flow or film appearance and the chemical resistance. The lower limits for stability as a powder are a softening point of 70°C and WPE of about 600 to 700.

Since epoxies can be applied in thin films, 2 mils or less, pigment loadings from 3 to 100 phr (parts per hundred resin) are required; with the high film thickness thermoplastic formulations, the pigment loading varies from 1 to 30 phr. The phr is a function of the pigment's hiding power, the dry film thickness applied, and the contrast ratio desired. The same heat-resistant pigment grades found in conventional baking enamels or plastics are used in epoxy powders.

Since small quantities of additives are needed, they are melt mixed to a 5 to 20% concentration for accurate addition and a better mix. One additive is needed to adjust flow and minimize craters and pinholes. Another, usually a silica, is used to give better edge coverage.

The hardener level varies from 2 to 10 phr depending on the cure cycle. Dicyandiamides, amines, and anhydrides are normally used as hardeners.

The advantages of epoxy coatings have already been mentioned. The two drawbacks are early exterior chalking and a tendency to yellow during over-bake and when exposed to UV light. Decorative and protective epoxy powder coatings are used for the following applications: pipe, metal furniture, electrical insulation, interior automotive parts, sewing machines, appliance parts, etc.

Polyesters

Polyester powders are just coming into the market place with all the properties of epoxies plus increased exterior durability and resistance to yellowing. Both thermoplastic and thermoset polyester powders are available. Due to the high molecular weight of the thermoplastic resins, poor film appearance is obtained at film thicknesses less than 3 mils and a high fusion temperature is required (around 450°F). Thermosetting polyesters can be formulated to flow out and give good film appearance, even at 1 mil, while being cured at temperatures around 350°F.

The polyester formula is considerably more complex than an epoxy. The following polyester resin characteristics must be evaluated: melt and glass transition points, melt or reduced viscosity, and acid and OH number. The pigments must be heat resistant and have good exterior durability. The flow, thixotropic, and curing agents are selected based on the resin characteristics and the cured film appearance desired. As in the processing of epoxies, polyesters are processed by dry blend, melt, mix and solution techniques. These techniques will be discussed below.

The polyester powders are a little more costly than the epoxies and generally require a 25% increase in cure time at the same bake temperature. It is not unreasonable to expect that both of these disadvantages will be eliminated.

The potential uses are in the following industries: automobile, appliance, coil coating, aluminum extrusions, building products, fire extinguishers, metal furniture, etc.

One of the battles that has been going on in solvent-based paints also exists in powder coatings—that is, can a polyester weather as well as an acrylic coating.

Acrylics

Thermoplastic and thermoset acrylic powder coatings are in the development stage. A thermoplastic acrylic formulation consists of resins, pigments, plasticizer, flow and thixotropic additives and possibly a UV absorber.

A thermoplastic acrylic powder would be primarily designed to meet automotive specifications. Here, they have been successful in showing excellent durability, good adhesion to primers and resistance to humidity, cold check, salt spray, and chipping. The disadvantage is in mediocre film appearance (e.g., an orange peel film results with a fusion temperature of 350°F or less). To correct this, the molecular weight must be lowered or the monomers or plasticizer composition adjusted without detracting from the performance.

To achieve film appearance at lower molecular weights and maintain performance, thermoset acrylic powders are being developed. A thermoset possesses the ideal characteristic of being able to flow at low temperatures and then cross-link to develop performance. Thermoset acrylic powders have been formulated to show good resistance to detergent, humidity, salt spray, cold check, and heat. As in solvent-based paints, one grade of acrylic powder will probably be designed to meet automotive and general industrial specifications and another grade to meet appliance specifications. The main research efforts are aimed at reducing volatiles during cure, minimizing orange-peel, and increasing the rate of cure.

Manufacturing

Powder coatings are processed by one of three different methods: dry blend, melt mix, or solution.

Dry Blend. A dry blended powder is a relatively uniform blend of all the raw materials in the formulation. A dry blend is batch processed through a high intensity mixer or a ball mill.

In the first case, all the raw materials are added to the high intensity mixer (see Fig. 17-1). The mixer cycles for several minutes at 1800 rpm. The batch is then sieved through 100 mesh and packaged. This grade of powder is used only in fluid-bed applications where gloss, film thickness, and appearance are not critical.

An exception to this may be the vinyls. A high grade of vinyl powder can be produced by dry blending (or agglomerate mixing as it is also called). The process consists of loading a high intensity mixer with the vinyl resin, pigments, and other solid additives. The mixer is run at about 1800 rpm for a predetermined cycle with

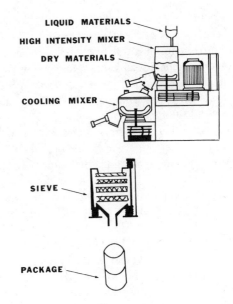

Fig. 17-1. Dry blend process.

or without heat added to the mixing jacket. External heat is sometimes added to shorten the cycle. The plasticizer and other liquid additives are added at a set rate. The batch is normally around 100°C for optimum plasticizer absorption. At the end of the cycle, the batch is dropped into a cooling mixer.

The cooling mixer is a low shear unit, about 100 rpm, such as a ribbon blender. When the batch is 35°C or less, the drying agent is added under low shear. On completing the cycle, the batch is sieved through at least 40 mesh and packaged.

The ball mill is also a batch process. The formula ingredients are added to a ball mill which tumbles for 10 to 20 hours. The cycle can be shortened by using a vibratory ball mill. Due to the temperatures developed during this technique, the hardener is sometimes added during the last 2 hours of processing to prevent partial curing. After the process, the batch is sieved and packaged.

Ball-milled powder is used in fluid bed and electrostatic application systems. The quality of the product is better than that from the high intensity mixer but more expensive on a volume per unit time basis. The quality of the product is inferior to that processed by melt mixing.

The first powders formulated for the pipe and electrical insulation industries were processed by the dry blend technique. The present decorative and protective coating market demands a quality for thermosetting powders that can only be produced by melt mix or solution technique.

Melt Mix. The melt mix process (see Fig. 17-2) is the most popular for manufacturing thermoset powder coatings. It consists of the following steps: premix, masticate, cool, flake, pulverize, sieve, and package.

In the melt mix process, all the formula ingredients are added to the premixer. The premixer may be a cone blender, ball mill, high intensity mixer, ribbon blender, or a similar piece of equipment. The particle size of the mix should be between 20 and 60 mesh for a good mix and feed to the masticator.

After a satisfactory blend is achieved, the batch is dumped into the feed hopper of the masticator. The process changes from a batch process to a continuous process.

The masticator may be a kneader, sigma-blade mixer, extruder, twin-screw mixer, etc. The blended batch is metered into the feed section of the masticator maintained at 30°C. Temperatures above 50°C will promote softening and bridging of the material in the feed section and reduce the throughput.

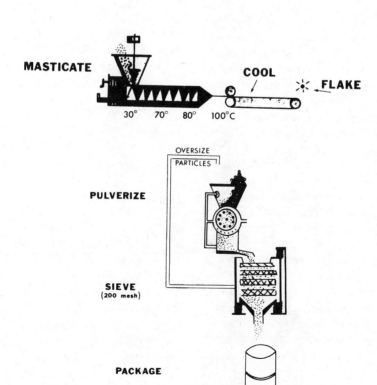

PRE - MIX

MASTICATE   COOL   FLAKE

30°   70°   80°   100°C

OVERSIZE PARTICLES

PULVERIZE

SIEVE (200 mesh)

PACKAGE

Fig. 17-2. Melt mix process.

The following temperature profiles of the masticator depend on the rheological properties of the powder; that is, glass transition temperature and melt temperature or melt viscosity and gel time. For example, for an epoxy with WPE of 1000 and a T_m of 105 to 120°C, the die head should be set at 100 to 110°C, the barrel next to the die at 80 to 100°C, the center section at 70 to 90°C, the feed section at 30°C, and the screw at 70 to 90°C. These temperatures represent a steady-state position; higher temperatures may be used at the start of the run.

From the die head, the molten mix is fed to a chill belt, cooled rolls, chilled air, or an equivalent system to drop the temperature of the material from 100 to 40°C for crushing. A coarse crusher prepares the material for pulverizing to a fine powder.

The most common pulverizers are hammer, hurricane, and fluid energy mills. The hammer mill transfers a large portion of its energy to heat. This problem is solved by a higher glass transition temperature, lower feed rates, or cryogenics. The hurricane mills run cooler due to the large volumes of air. The air volume makes the powder collection system bulky and difficult to clean. The fluid energy mills require large volumes of air at high pressures. The units usually have a low output and are bulky and difficult to clean from batch to batch.

The particle size distribution is a function of the type of pulverizer, the operation of the pulverizer, and the rheological properties of the powder. It is necessary to control the distribution since slight changes affect the package stability, the handling characteristics, the electrostatic charge acceptance, and the appearance of the fused film. The distribution can be determined by microscopic or automatic counters or by mechanical or sonic sieves.

The sieving step consists of a 140 or 200 mesh screen. This step may be used to alter the particle size distribution but normally it is just for scalping the oversize for further processing. The material passing through the screen is packaged.

Solution. The solution technique is most popular with current paint manufacturers because it is more in line with their paint equipment and technology. The process consists of manufacturing the coatings just like a conventional solvent-based coating. After the batch is standard for quality, color, etc., the solvent is removed by either spray drying, a devolatizer, or suspending in water.

The spray drying approach is dependent on the percentage of solvent, the type of solvent, and the viscosity of the paint solution. The solution is reduced to 10 to 20% solids with methylene chloride or acetone. Spray drying produces a narrow particle size distribution which can be adjusted by varying the above parameters. For example, the higher the solids, the coarser the particle size distribution. The spray dry technique also produces a spherical particle, whereas the other techniques yield jagged, irregular-shaped particles. The spherical particles can be advantageous in applying and handling powders. However, this technique is generally considered to be the most costly per pound output.

Devolatizers are of many designs, such as thin film evaporators. They are affected by the same parameters as the spray dryer. However, in this process, a film or flake is normally produced. This flake is then pulverized by a hammer, hurricane, or fluid energy mill.

By the proper selection of solvents, the paint can be dispersed in water to form a powder slurry. The water can be removed by spray drying or a devolatizer. However, the slurry can be also applied as a conventional solvent-based paint.

Application

The three basic methods for applying powder coatings are fluidized bed, electrostatic spray, and electrostatic fluidized bed.

Fluidized Bed. The earliest powders were applied by blowing or sprinkling powder onto a hot object or rolling a hot object in a container of powder. The invention of the fluidized bed was the birth of serious application methods for powder coatings.

Powder is placed in the bed (see Fig. 17-3)—basically a container, such as a fiber or metal drum, with a false bottom, The bottom is a porous membrane. The membrane does not

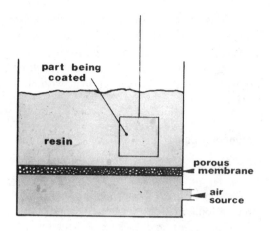

Fig. 17-3. Fluidized bed coating.

allow the powder to fall through but does allow, and evenly distributes, an upflow of dried air.

The air rises through the powder contained in the bed resulting in a fluid action of air and powder. This action is necessary to give an even distribution of powder and to allow an object to enter the bed.

The object to be coated is preheated to a temperature above the melting point of the powder coating. The preheated object is then immersed in the fluidized powder where the powder particles melt and fuse together to form a film on the object. A post-heat is sometimes required to give the film more flow or an improved appearance or complete cure.

The final film thickness is determined by the following parameters: preheat temperature, the object's mass and its ability to retain heat, immersion time, the object's movement in the bed, and the velocity of fluidizing air. The

thickness is in excess of 4 mils and generally 10 to 15 mils.

The particle size of the powder is also important. A particle size that is too fine (e.g., around 30 microns) will be difficult to fluidize in the bed and will cause dusting over the bed.

The basic advantages of the bed techniques are the simplicity of the equipment and the technique, the low cost, and the ease of applying heavy film thicknesses. Some of the disadvantages are: an inability to apply thin films (less than 5 mils); the object to be coated is limited by its ability to retain heat from the preheat oven to the point of application; the type of object is also limited by the size of the bed; and color changes involve considerable cleaning unless multiple beds are used.

Electrostatic Spraying. In the 1960's, electrostatic charging principles were applied to the application of powder coatings. Most powders are insulators with relatively high volume resistivity values. Therefore, they accept a charge (positive or negative polarity) and are attracted to a grounded or oppositely charged object.

An electrostatic spray system (Fig. 17-4) consists of: a powder reservoir, the powder feed mechanism, the gun design, a powder generator, the application booth, and powder recycling equipment. The powder reservoir is usually a fluidized bed or mechanically agitated hopper. From the reservoir, the powder is fed to the gun. This is accomplished by a Venturi air pump. Arriving at the gun, the powder/air mixture is charged. The corona charge occurs internally, at the tip of the gun or at an electrode close to the powder exit. The charge from the power pack varies from 60 to 120 Kv

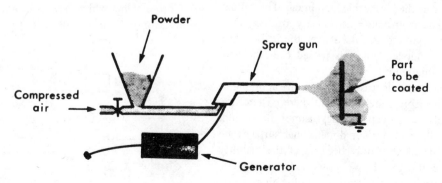

Fig. 17-4. Electrostatic spraying.

and 200 to 400 microamps. Also at the tip of the gun is a diffuser to direct the powder and shape the powder cloud.

The thickness of applied powder is dependent on the object's speed through the cloud, the cloud pattern, the powder feed rate, the air currents in the booth, the gun arrangement, and the resistivity of the powder.

The primary advantage of electrostatic powder spray is low film thickness coatings applied to a cold part. The system is also readily automated, even if there are various size parts. The initial transfer efficiency varies from 50 to 80%, but the system may achieve 98% powder usage with an efficient powder recycling system. The recycling system is generally a cyclone followed by a bag collector. Powders collected in the cyclone are sieved and sent back to the powder reservoir to be blended with fresh powder. The bag collector is used to catch the fine particles going past the cyclone. This quantity is normally small and therefore is just discarded. This complex system also presents a problem in changing the powder in the system. The cleaning process is long and difficult.

Another disadvantage of electrostatic spray is the field effect which prevents powder deposition in small openings and tight angles. Also, due to the resistivity of the powder, thick films in excess of 8 mils can be difficult to deposit.

Electrostatic Fluidized Bed. The process is a combination of the fluidized bed and electrostatic spray methods. A current of air passing upward through the bed fluidizes and suspends the powder. The powder is charged by electrodes in the permeable membrane. The voltage applied determines the density of the cloud when a ground is passed over the bed. Here a cold, grounded object is passed over the bed, not immersed into the bed.

The same electrostatic principles applying to powder spray are pertinent here. Therefore, objects with deep recesses cannot be coated. Also, long vertical or dimensional surfaces are not coated uniformly because of the inability of the unit to "throw" powders long distances. As with the other methods, color change is a problem.

Improvements in design and performance are made every day. Thin, uniform films can be applied at high rates for such industries as coil, webbing, fencing, etc. The powder handling system is considerably simpler than with electrostatic spray.

Production Line

The flow chart of a production line (see Fig. 17-5) is quite similar to a solvent coating system, involving the hanging of parts, surface preparation, preheat (optional), application, post-heat (cure), and package. No matter how well the application section is designed, if the whole line is not carefully considered, the final product is likely to be unacceptable. For example, the better the surface preparation (metal pretreatment), the better the coating performance. The post-heat cycle must match the cure cycle required by the coating—an uncured film just does not perform. Again, the line must be considered as one unit consisting of finely tuned complimentary sections for the optimum in product satisfaction.

Economics

Many items and facets contribute to the overall economics of powder coatings, but one of the most important is the fact that *the customer uses all of the products that are purchased.* When an individual or a company buys powder coating materials they use what they purchase and everything they purchase. This factor contributes most to the overall economics.

It is the purpose of this section to cover in brief those items which have a bearing on the overall economics of powder coating materials.

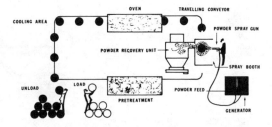

Fig. 17-5. Typical powder coating production line.

When the three types of manufacturing processes discussed above are considered, it is clear that the first two, the dry blending and the melt mix method, can contribute to a lowering of cost. The dry blending step, as explained previously, is relatively simple and fast and is, from a time standpoint, a very economical process. The melt mix process is also economical and, while longer and more complex than the dry blend method, does effect production economy. Both of these can easily be converted, through a series of metering and blending operations, into a continuous process. Once the continuous process level is achieved, large reductions in cost can be effected. The unit cost or manufacturing charge per pound is brought down and continues to be reduced throughout the extension of a long run.

While the design and formulation of powder coatings may be complex, the number of ingredients used is usually less than for some of the complicated liquid-based coatings. The many additives and the numerous items that are used in a liquid-based coating are frequently unnecessary in powder coatings.

The storage of raw materials in the manufacturer's plant and the storage of finished goods in the user's plant is another economy that is gained with the saving of floor space and the reduction in fire hazard. The space required for 10,000 lb of dry powder is much less than the space required for the 20,000 lb of liquid paint containing 10,000 lb of dry solids.

The handling of the powder coating raw materials and finished powder coating products is also more economical, since volume is lower. Less labor and manpower is required to handle fewer pounds of the coating required for the same number of objects. This savings in handling fewer pounds also converts itself into a savings in shipping. A full truckload of liquid paint, i.e., 30,000 to 40,000 lb represents, on the average, 15,000 to 20,000 lb of dry solids coating. However, freight is paid on the full 30,000 to 40,000 lb of liquid. The same amount of dry powder would weigh 15,000 to 20,000 lb and freight in this case would only be paid on the 15,000 to 20,000 lb shipped. The freight paid on the liquid is only one part of the overall expense. The liquid must also be purchased and disposed.

The overall quality of the current powder coating materials is exceptionally high. At the present time and state of the science, there are no powder coatings used or manufactured that would compare in quality with the very low cost and low physical property coatings used for many applications in American industry. The powder coatings as they are currently formulated and used would all fall in the better or best overall quality of finishes.

The application of powder coatings is totally different from the liquid type of paint and brings many different phases of economics into play, as follows:

1. The primary economy is achieved through the high efficiency of application deposition and recovery of the powder. When the powder is initially sprayed on to a part, the powder is electrodeposited onto the substrate and overspray is then collected. This collected overspray is sieved and cleaned and then reused after mixing with virgin powder. In this manner an overall deposition of 96 to 98% has been achieved on various production applications in practice. Various suppliers of liquid spray coating equipment, both electrostatic and nonelectrostatic, have indicated through practice that efficiencies of spraying liquid do not approach those of powder.

Quite obviously, the application methods of fluidized bed and electrostatic fluidized bed are also extremely efficient; in these cases, the only powder that is removed from the container is that powder which is applied to the parts. In this type of situation, efficiencies of 95 to 98% are also achieved.

2. Obvious economies are achieved through the application of powder and the deposition of heavy films of dry powder in one application step. In contrast with liquid paints where frequently multiple coats are required to build up 3, 4, or 5 mils of dry film thickness, as much as 4 to 6 mils of electrostatic spray powder can easily be deposited in one step while extremely heavy coatings of 15 mils can be deposited with the fluidized-bed techniques.

3. The baking requirements of powder coatings obviously require the use of high heat

ovens. These high heat ovens may require additional heating facilities as far as temperature is concerned, but do not require near the amount of Btu energy for heating the air that is exhausted from an oven used for powder coating as contrasted with an oven used for solvent reducible finishes. The air exchange and the requirements for air make up for safety purposes in an oven used for curing solvent paint finishes are far higher than for a powder coating. The Btu's which must be expended to heat this high air exchange are extremely high and reduction in this overall gas or fuel cost will certainly effect economies in the normal oven requirements for powder coatings.

4. Economies are also achieved in terms of floor space requirements and in the construction of a powder coatings system. As mentioned above, normally all of the required film thickness can be applied with one pass; as a result, this means only one spray booth and normally no flash-off time between coats. Once again, there is no time required as a flash-off period with the powder coatings between the application and the oven; this obviously saves space. There is no large installation of paint mixing stations, paint reducing stations, and paint circulation tanks required with powder as there is with liquid paint.

5. Powder coatings do not require expensive sludge removal systems for removing overspray and disposing of this wasted material. Again, *the customer uses everything he buys.* The powder coating material is caught with a cyclone, reused, and put on the work in this manner. This step also results in a savings to the user.

In terms of labor requirements, there are a number of savings which can be realized with powder coatings. As mentioned previously, they do not need mixing and reduction in the user's plant. Spray booth cleaning is reduced to a minimum and sludge removal is not necessary. Savings are also obtained by eliminating the need to buy solvent recovery or after-burning equipment to dispose of pollutant-type materials.

A comparison of the purchase price between a liquid coating and a powder coating is useful. However, liquid coatings are sold by the gallon and powder coatings are sold by the pound.

The following formula can be used to convert a

COST PER DRY POUND

For Liquid Systems:

Cost per Dry Pound =

$$\frac{\$ \text{ Cost per Gal.} + \$ \text{ Cost of Thinner}}{(\text{Weight per Gal.}) \times (\% \text{ N.V. by Weight})}$$

For Powder Systems: Quoted Price

liquid gallon price to a dry pound price. With this formula the cost of the two types of coatings can be compared. The second formula

ACTUAL APPLICATION COSTS

Liquid:

100% *Effective* Efficiency =

$$\frac{\$ \text{ Cost per Dry Lb. of Paint}}{\text{Efficiency of Liquid Spray}}$$

Powder:

100% *Effective* Efficiency =

$$\frac{\$ \text{ Cost per Lb. of Powder Paint}}{\text{Efficiency of Electrostatic Spray}}$$

shows the actual application costs between liquid and powder paint as effected by the efficiency of the type of application.

As already indicated, there are a number of areas for savings with powder coatings, such as less energy usage, less space required, no solvents, and higher efficiency of material usage. It is further imperative that the unit material and applications costs be determined. The following data detail unit cost calculations and the comparisons between powder coating and liquid system materials. These material unit costs along with the savings and applied unit costs are normally quite sufficient for the adoption of this technology.

UNIT COST CALCULATIONS

(a) The coverage of one pound of powder can be determined by the following equation:

$$\text{I.} \quad \frac{\text{Efficiency} \times 192.3}{\text{Specific Gravity}}$$

$$= \text{coverage sq ft/mil}$$

(b) The cost per square foot can be found by using the following equation:

II.
$$\frac{\text{Specific Gravity} \times \text{Film Thickness} \times \$/\text{Lb} \times 100}{\text{Efficiency} \times 192.3}$$
$$= \text{¢}/\text{sq ft}$$

To determine your unit or part cost, merely multiply the cost per square foot by the area in feet of your part.

$$\text{¢}/\text{sq ft} \times \text{sq ft}/\text{unit} = \text{¢}/\text{unit}$$

Example: Epoxy Powder Coating

> Film Thickness: 1.4 mils
> Specific Gravity: 1.2
> Cost: $1.50 per pound
> Efficiency: 98%

Using Equation I

$$\frac{.98 \times 192.3}{1.2} = 157 \text{ sq ft/mil}$$

or

$$\frac{157 \text{ sq ft/lb}}{1.4 \text{ mil}} = 112 \text{ sq ft at 1.4 mil}$$

Using Equation II

$$\frac{1.2 \times 1.4 \times 1.5 \times 100}{.98 \times 192.3} = 1.3\text{¢}/\text{sq ft}$$

(c) The coverage of one gallon of paint can be determined by:

III.   Efficiency × % Volume Solids × 1604 = coverage sq ft/mil

(d) The cost per square foot can be found by:

IV.
$$\frac{\$/\text{gallon} \times \text{Film Thickness} \times 100}{\text{Efficiency} \times \% \text{ Volume Solids} \times 1604}$$
$$= \text{¢}/\text{sq ft}$$

Example: Liquid Paint

> Film Thickness: 1.4 mils
> Volume Solids: 35%
> Cost: $3.85 per gallon
> Efficiency: 65%

Using Equation III

$$.65 \times .35 \times 1604 = 365 \text{ sq ft/mil}$$

or

$$\frac{365 \text{ sq ft/mil}}{1.4 \text{ mil}} = 261 \text{ sq ft at 1.4 mil}$$

Using Equation IV

$$\frac{3.85 \times 1.4 \times 100}{.65 \times .35 \times 1604} = 1.5\text{¢}/\text{sq ft}$$

As can be seen, the concept of powder coatings with plastics offers a number of economic and functional advantages. It is anticipated that it will continue to play an even more important role in the years to come.

MOLDING PARTS BY ELECTROSTATIC COATING*

Plastic parts are formed by this process without the use of pressure. The raw material used is powdered resin of a grade suitable for fluidizing. During the process this powder is projected, in an electrostatic field, against a grounded open metal mold to which it adheres. This mold is heated on the uncoated side to fuse the resin and allow it to flow out. The part is then water cooled by spraying the back of the mold and removed. The cycle is then repeated.

This cycle is achieved practically by mounting the molds in panels on the curved face of a cylindrical drum. Inside the drum at the axis are mounted two sets of spray nozzles. One set is supplied with liquid heat exchange salt at a controlled temperature, pumped into the system from an auxiliary heating tank, the other with cold water. The angles of the two sets of sprays do not overlap and baffles minimize overspray. The salt is returned via a sump to the tank, the water, most of which flashed off as steam, is run to waste. This drum is mounted above one, or two, fluidized beds and is driven peripherally so that molds pass over the bed in sequence and are then escaped to the salt and water sprays in turn. The finished parts are removed by a synchrononus pick-off mechanism. (See Fig. 17-6).

The materials best suited to the process are those which retain an electrostatic charge,

* Prepared by K. J. Chivers, Spraywell Ltd., Waterloo, Ontario, Canada.

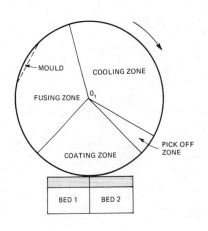

Fig. 17-6. Molding parts by electrostatic coating.

fluidize well, fuse rapidly with degradation, and flow out to form smooth surfaces. Each of these factors affects the rate of output of a machine. Most suitable powders are readily available without regrinding (e.g. polypropylene), some (e.g. polyethylene) are more often in pellet form. The particle size distribution has to be a balance between the cost of grinding to smaller particles and their effectiveness as a coating medium. The system lends itself best to high volume production on thin-walled parts, principally in the container industry. Since a single mold is used, no over-thickness is required to overcome the problem of thin spots in critical areas. This leads to a saving of material. The elimination of cut-off scrap also contributes to a cost savings.

18
CASTING

By definition, casting applies to the formation of an object by pouring a fluid monomer-polymer solution into an open mold where it finishes polymerizing (i.e., cures) or to the formation of film or sheet by pouring the liquid resin onto a moving belt or by precipitation in a chemical bath.

As such, casting differs from many of the other techniques described in this handbook in that pressure is generally not involved (although large-volume, complex parts may use pressure casting methods). In addition, the starting material is usually in liquid form, rather than in the granular or powdered forms that go into molding systems, and is generally a monomer rather than a polymer. In chemical parlance, a monomer is a relatively simple compound which reacts to form a polymer. Thus, the resins that are used for most of the other molding processes discussed in this book have already been polymerized and are polymers. However, the plastics industry being what it is, there are casting techniques which do not exactly adhere to these definitions, as in the solvent casting of thermoplastics films described below. Generally, however, most of the other methods discussed fit into this overall framework.

Some take-offs on the casting process, like rotational molding (which is also known as rotational casting) have developed into highly individual processing methods with their own unique technology (see p. 348). The chapter on vinyl dispersions (p. 365) also carries information on specialized casting techniques.

This chapter, will discuss: (1) the casting of various thermosets and thermoplastics; (2) two specialty market areas—tooling and encapsulation—that are based on casting technology and have grown large enough in their own rights to have evolved special casting procedures; and (3) liquid resin molding—a potting and encapsulation technique that bridges the gap between conventional methods of liquid potting and powder transfer molding (i.e., combining liquid casting and pressure molding).

It should be noted that elsewhere in this book, readers will find the expression "casting" used to apply to other processing techniques. For example, the "cast" polyethylene and polypropylene films described in the chapter on Extrusion, do not fall into the "casting" definition that is being used here. There, the name is derived from the technique of extruding the resin onto a highly polished water-cooled chill or "casting" roll which takes the heat out of the web.

The chapter on Reinforced Plastics also carries data on a technique known as centrifugal "casting". In centrifugal casting, chopped fibrous glass and resin are placed inside a hollow mandrel, the mandrel then is rotated inside an oven, uniformly distributing the reinforced plastics on the inner surface of the mandrel, where it cures into a finished hollow product.

History

Resins for casting first emerged a little over a half century ago, but formulations suitable for increasingly widespread use date back only about thirty years.

About 1906, Dr. Leo H. Baekeland announced a process of making a synthetic resin by condensing phenol with formaldehyde. This was a thermosetting resin in the form of a sirupy liquid. It could be cast in molds of lead or glass, and then cured by heat and thus transformed into a solid, infusible plastic. The use of early phenolic castings was restricted by limitations on size, and by their tendency to craze, crack, and discolor with age.

Another early casting resin was a urea-formaldehyde compound first marketed around 1928. Castings made from this thermosetting resin proved unsatisfactory because of cracking. Subsequent uses of urea-formaldehyde were largely confined to molding powders, such as are now commercially available.

Improved phenolic resins, specifically developed for casting, began to attract substantial attention during 1925-1930. These could be cast in large pieces, which were generally free from cracking or other deterioration in service. The cast phenolics gained popularity because of a number of factors, not the least of which is their eye-appeal. They became available in a wide range of stable colors, from pastels to dark tones, in various degrees of translucency from transparency to opacity, and with mottled effects to simulate marble or gems.

Plastics available for casting were augmented about 1934 by acrylic resins. These are readily cast in the form of sheets, rods, or tubes, with excellent finish, in molds of various types.

National defense needs during World War II not only spurred the uses of phenolic and acrylic resins, but also accelerated the development and utilization of other casting plastics. Thermosetting resins of unsaturated polyester, while principally used with glass fiber and other filling and reinforcing materials in laminated structures, were found to be adaptable to casting applications.

Other thermoset resins, like the epoxies and the urethanes, soon followed into the casting field. The epoxies, in particular, were quick to find acceptance in a wide diversity of uses.

Epoxy resins exhibit negligible shrinkage; a high degree of dimensional stability; good dielectric strength; exceptional bonding properties and adherence to a wide variety of materials; resilience; and notable resistance to wear and abrasion. They are resistant to acids and alkalies over a broad range of concentrations, and are inert to most chemicals and contaminants. Their impact strength encourages their use as facings for composite forms which may embody a core of foamed plastic or other relatively inexpensive material. They resist crazing, chipping, and cracking. They are available in special formulations when required to provide unusual flexibility and resilience, and are serviceable for use as an embedment material around large inserts, such as electronic circuits, and delicate inserts, such as glass structures and diodes.

In the field of thermoplastic casting resins, acrylic is the dominant factor. More recently, however, there has been interest in the casting of massive nylon parts (e.g., fuel tanks) via the anionic polymerization technique (Fig. 18-1).

Outstanding Characteristics

There are advantages which may be said to be common to all cast plastics. Among these are low mold cost, making small runs practicable; relatively inexpensive auxiliary processing equipment, such as baking ovens, when cures by heat are involved; feasibility of large pieces and thick sections; and minimum of post-casting machining and finishing, accomplished speedily and without special expensive equipment.

Apart from the general overall properties, each of the castable plastics has its own outstanding characteristics or combination of characteristics, which differentiates it from the others and serves as a guide in the selection of the most suitable material for any particular application (see the individual sections, below).

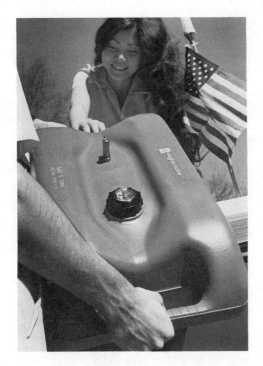

Fig. 18-1. Cast seamless nylon fuel tank for boats combines light weight, corrosion resistance, and a low permeation rate. (*Courtesy Goodyear*)

CASTING ACRYLICS*

In the technology of acrylic plastics, casting is employed mainly for the production of basic stock such as sheets, rods, and tubes, which can be used in those forms or thermoformed, machined, and cemented to produce a variety of useful shapes. Thick blocks and embedments are made on a smaller scale, and usually on a custom basis.

For the production of acrylic stock with the highest clarity and the best physical and optical properties, casting is superior to injection molding or extrusion. This is especially pertinent when large sheets are required for architectural glazing; aircraft windows, canopies, and windshields; boat windshields; and other applications with severe requirements for optical quality. Even in many less demanding applications, cast acrylics are used because the casting process can produce sheets with dimensions and properties not economically obtainable by other routes.

* By Dr. J. F. Woodman, Rohm & Haas Co.

Cast sheets, which make up the major portion of the output of cast acrylics, have been produced since the mid-1930's in individual glass molds or cells. There have been continual refinements in chemical composition, catalysts, purity of ingredients, additives, and mechanical details of operation, but the basic process has remained essentially unchanged. Recently, processes for production of acrylic sheet in a continuous ribbon have gained an important place in the market.

Both the cell-cast and the continuous process for sheet production involve the polymerization of acrylic monomers mixed with a catalyst and special additives or colorants, and exposed to a carefully controlled thermal cycle.

In the cell-cast process, the liquid mix is poured into molds, each consisting of two parallel sheets of glass of high surface quality, separated by a peripheral gasket of the correct size for the sheet thickness to be made. The surfaces of the glass may be plain or patterned to produce sheet with special surface contours. The mold and contents are heated in a circulating air oven, autoclave, or liquid bath in a carefully controlled temperature cycle ranging up to approximately 200°F, until polymerization is complete. After cooling, the sheet is removed as a finished piece, usually requiring only edge trimming.

In the continuous process, the prepared monomer mix is introduced continuously between moving stainless steel belts with edge gaskets, which carry it though several thermally controlled zones to emerge as a solid sheet. The continuous sheet is trimmed and cut into sheets of any desired length, or rolled onto cores which may accommodate several hundred feet in a single length as a standard package.

Cast acrylic sheet is now available in thicknesses from 0.030 to 4.25 in. Some thicknesses can be supplied in widths up to 120 in.; cell-cast sheets may be as long as 168 in., whereas the length of sheets made by the continuous process is limited only by problems of handling and shipping.

Rods are cast either in metal molds or in tubular bags of plastic film. Tubes are cast in rotating horizontally mounted aluminum tubes. Embedments and other thick castings usually start with a slurry of acrylic polymer in monomer, to minimize the shrinkage which accompanies polymerization.

CASTING PHENOLICS AND EPOXIES

Phenolic casting resins are available as sirupy liquids produced in huge kettles by the condensation of formaldehyde and phenol at high temperature in the presence of a catalyst, and removal of excess moisture by vacuum distillation. These resins, after having been blended with a chemically active hardener, are cast and solidified in molds of various materials. Polymerization in the mold may occur at room temperature, but is more usually attained by carefully controlled heating, ranging from a few hours to several days. This heat cure may take place in a closed vessel, an autoclave under pressure, or a standard baking oven. Oven-baking permits handling a large number and variety of molds at one time; it is less restricted by the size of the casting, and in general less expensive than other methods.

Epoxy materials range from liquids to semisolids. Some epoxy resins intended for special uses require heat for polymerization, but most of them can be set at room temperature with the aid of a noncorrosive chemical hardener. Curing is further simplified in some instances by the availability of epoxy resins prepackaged with the required amount of the proper catalyst, ready to be added.

Open molds are satisfactorily used in curing epoxy castings, since the process of curing is not adversely affected by exposure to air. Depending mainly on the material of the mold, and in some cases on the chemical nature of the plastic, the surfaces of the mold are generally treated with protective coatings and release agents.

Molds

The mold in which a liquid casting material is formed into a solid cast shape is actually a container shaped to the contour of the piece to be formed, and in the dimensions of the desired piece with shrinkage allowance as necessary. Considerations in selecting the type of mold to be used include the size and contour of the article to be cast and the chemical make-up of the casting plastic that is used. Among the major types in current use are the following:

Draw Molds. A tapered steel dipping arbor or mandrel is machined to the dimensions of the finished casting and attached to a funnel plate and handle. The mandrel is dipped into molten lead, which coats the steel. The lead solidifies on removal from the pot, and is stripped from the mandrel as an open lead mold.

Obviously, undercuts cannot be cast by this method. Flutes, beads, scallops, and other design features can be incorporated in the castings, if they run in the direction in which the castings are to be removed from the mold.

Phenolics are often cast in molds of this kind. After curing, the phenolic casting is forcibly removed from the lead mold, typically by tapping the end of the mold with a mallet. To facilitate removal, the mold is tapered in order to reduce friction and allow the vacuum to be broken.

The taper and the forcible method of removal dictate adequate thicknesses for the walls of castings made in draw molds. In general, castings intended for large radio cabinets, boxes, plates, long tubes, and similar items must have walls at least $\frac{3}{16}$ in. thick. Smaller tubes, boxes, and other articles which can be more readily removed from the molds require wall thickness of $\frac{5}{32}$ in. A wall $\frac{1}{8}$ in. thick will usually suffice for very small castings, such as finger rings and hollow bottle closures.

Split Molds. These are two-piece molds closed at all points except for the gate through which the liquid casting resin is poured.

Undercuts may be incorporated into articles produced from split molds, so long as they do not prevent release of the casting from the separate sections of the mold.

Die-casting machines are often employed in producing split molds. The machines make one half of the lead mold at a time, to interlock with the back of the opposite end of the adjacent mold when the successive sections are clamped together for the casting procedure. (It is also possible to make two half sections of lead draw molds to interlock by stapling the halves together.)

After the halves of a split mold are joined, the liquid resin is poured into the gate, and curing takes place in the manner prescribed for the casting material being used. Walls of the casting should not be less than $\frac{3}{16}$ in. thick. It is also found that castings smaller than ten to the pound cannot ordinarily be economically produced in split molds.

Cored Molds. These are employed to produce large castings with decorative faces, hollow castings of hemispherical shape, and pilasters with compound curves.

A dipping arbor is made, reproducing the outside dimensions and outer shape of the finished piece, and a lead mold is made on this arbor. A metal core which represents the inner shape and dimensions of the desired casting is then fitted against the lead mold to form a cell. The liquid resin is poured into this cell and cured. When curing is complete, the core is removed and the casting withdrawn from the lead mold.

Cored molds require no taper. They permit a wide latitude in design, and can be used with virtually no limit to the size of the casting. Examples are recorded of housings 20 in. square and 12 in. deep regularly cast in cored molds, and pilasters as long as 36 in. and with a periphery of 18 in. are likewise successfully produced in this way. Conversely, the operating costs would make it economically unattractive to use cored molds for casting units smaller than 4 in. in each dimension.

With a usual minimum of $\frac{3}{16}$ in., there is virtually no maximum limit upon the wall thickness of phenolic and epoxy castings made in cored molds. The thickness can be as great as required for strength, without danger of distortion, undercuring, or defective curing.

On exceptionally large castings produced by this method, allowance for mold shrinkage must obviously be carefully watched, if there is to be no sacrifice of dimensional accuracy.

When cast phenolic pieces are fastened to other materials needed to manufacture a finished product, oversized holes should be drilled to compensate for differences in shrinkage. Where pilasters or other extremely long pieces are involved, the castings should be fastened only at the top or bottom, not at both ends, and should ride in grooves along the sides to provide leeway for expansion and contraction with temperature changes, and for shrinkage with age. If changes in length are impeded by fastening at both ends, stresses are likely to develop which will cause the casting to crack.

Flexible Molds. This procedure begins with fabrication of a model of the article to be produced. The model may be made of porcelain, hard wood, metal, or other nonporous material.

The flexible mold is built up upon the model to a thickness of about $\frac{1}{8}$ in. by dipping or brushing successive coats of rubber latex, plastisol, rubber-like epoxy, or other synthetic rubber or elastomeric plastic.

The mold is cut in two along a convenient parting line. A shell of plaster of Paris is cast around it to hold it in shape. The shell is parted along the same line, and provided with dowels to align the two parts, and with an opening for the introduction of the liquid casting resin into the mold cavity. In substance, the unit is a plaster mold lined with the flexible material.

After the resin is poured and cured, the two halves of the mold are separated and the flexible mold is stripped from the casting. Castings formed this way usually have good surface finish, which can be buffed or polished to produce a high luster.

Plaster Molds. Any article, model, or pattern can be duplicated in a plaster mold. Plaster molds are most often used when no undercuts or backdrafts are required. However, a plaster mold can be made when there are undercuts or backdrafts, by using loose pieces at the places where they occur, with mechanical locks to locate them exactly in place. This method has been successfully used in plastic tooling applications, where extreme accuracy is vital, and only a single casting is required.

Fabrication of a plaster mold is relatively simple. Successive coatings of plaster of Paris are spread over a master pattern until sufficient thickness is attained to assure adequate strength. The inside of the plaster mold may be treated with a sealing agent, and waxed; a parting agent is applied before liquid phenolic or epoxy resin is poured into the cavity. The casting is cured according to the method appropriate to the plastic. Castings are easily removed from plaster molds, and they have a smooth surface which ordinarily requires minimum polishing or finishing.

The speed and simplicity of the plaster-mold technique have encouraged its adoption in applications where military needs or competitive conditions dictate frequent changes in design for modification and improvement—such as in plastic tooling for aircraft, automobiles, and appliances. Plaster molds are therefore extensively employed in fabricating drill jigs, stretch presses and other metal-forming tools cast from phenolics and epoxies.

Methods of Fabrication

Cast plastics are widely used in fabricating a diversity of consumer products such as costume jewelry, umbrella and cutlery handles, brush backs, radio cabinets, trophy bases, and jukebox components, as well as many industrial components, such as machine housings.

These articles frequently are cast to shape, and require little finishing. Finishing operations consist mainly of the removal of any flash that exists, the cutting of any required openings which cannot he readily formed by casting, and any necessary final ashing or polishing.

In other instances, it may be more feasible for a fabricator to use castings in the form of rods, sheets, tubes, and profile shapes, and machine these to provide the desired article.

Phenolic, epoxy, and acrylic castings in standard and special shapes are readily machined in much the same way as metals and hard woods, with standard tools and low setup costs. Band saws and abrasive cutoff wheels are commonly used for slicing plastic castings. Ashing to remove tool marks is done with a cloth wheel and wet pumice, and is followed by final polishing with a dry, untreated soft buffing wheel. Many small and simple pieces, such as dice and poker chips, are finished by tumbling. Other operations that may be involved in finishing cast shapes include wet and dry grinding, drilling, tapping, turning on lathes and spindles, sanding, lapping, rouging, and buffing.

Typical Uses

Phenolics recommend themselves by their low cost, versatility, and ease of handling. These resins can be treated and precolored to provide a wide diversity of visual effects in the finished castings. These castings are tough and resilient, with good compressive strength, and are not softened by high temperatures. When formulated for maximum chemical resistance, they are waterproof, and resistant to all common reagents except alkalies and strong acids.

Cast phenolics are used largely for their esthetic qualities in jewelry, ornaments, imitations of ivory, marble, wood, etc. Billiard balls can be cast of phenolic.

Epoxies owe their extraordinary popularity to their unique combination of superior properties, among which can be included lack of shrinkage, dimensional stability, volume resistivity, chemical resistance, color stability, and availability in all desired pigmentations except light pastel shades. Their durability has made epoxy castings successful as molds for thermoforming various plastics, and their abrasion-resistance has spurred their uses in metal-forming. They come in many specialized types, offering, e.g., a combination of resilience and rigidity; rubber-like flexibility; putty-like consistency with thixotropic behavior which permits spreading, even on vertical surfaces, without run-off. They bond successfully with wood, metal, glass, and ceramics. They can be formulated as low-viscosity compounds, and in special-purpose blends, such as with lead for use in radiation shielding.

Pressure Casting

As mentioned above, several of the newer casting techniques for plastics involve the use of some degree of pressure. Typical is the so-called *pressure gelation process* (patented by Ciba-Geigy) for producing large and small epoxy castings. In essence, pressure gelation bridges the gap between conventional casting (used for large components but subject to long mold cycles) and transfer molding (which is fast, but limited to small castings). See Fig. 18-2.

Three critical features provide necessary

Fig. 18-2. Switchgear component, weighing 37 lb, was demolded in 25 min. using pressure gelation process. (*Courtesy Ciba-Geigy*)

temperature control to the pressure gelation process:

1. Use of highly reactive epoxy resin casting systems that liberate heat quickly and are fast gelling.
2. Heated molds that are a minimum of 50° higher than the casting mix.
3. Application of moderate pressure to the mold in the mix until gelation is completed.

How it Works. Pressure gelation utilizes the heat of reaction given off during polymerization reaction to heat the resin mass. In conventional casting, the heat is allowed to dissipate in the mold. Molds are heated to temperatures close to those of the casting mix temperature. In pressure gelation, a relatively cool resin system (including premixed resin, filler and curing agent) is dispersed into a hot mold. Curing moves from the mold wall toward the core while the heat of reaction is being generated.

The continuous feeding of liquid resin into the mold during curing compensates for shrinkage. Moderate application of 15 to 35 psi pressure to the resin feed provides the flow required.

Setting up the Production Line. Only slight modifications are needed to convert equipment from conventional casting to pressure gelation with the subsequent improvement in output (see Fig. 18-3). Mold conversion normally requires improving the seal and adding some means of accommodating and maintaining pressure in the mold cavity during gelation. Molds should also be heated directly for uniform mold temperature.

Semi-automatic and automatic equipment for material delivery, metering, and mold opening and closing is suggested to eliminate the time-consuming manual operations that lower production rates. A fully automatic production line incorporating in-line mixers, automatic mold handling equipment, material delivery pumps, and control valves has an output rate approaching that of conventional transfer or compression molding techniques.

The operation sequence of a sample automatic pressure gelation set-up is as follows:

1. Premixed, vacuum degassed material, either from an automatic metering and mixing facility or a batch mixing facility, is fed under pressure via a solenoid operated valve into an electrically heated mold.
2. When the mold is filled, a bolt valve is activated to cut off the material feed and simultaneously apply continued pressure to the reservoir of mix maintained at the mold.
3. Upon completion of gelation, the core insert retracts, the mold automatically opens, and the part is removed.

DECORATIVE CAST POLYESTERS*

Polyester resin casting has been happening for a great many years and has found its way into a

* By V. L. Montani, Technical Services, Reichhold Chemicals, Tuxedo Park, N.Y.

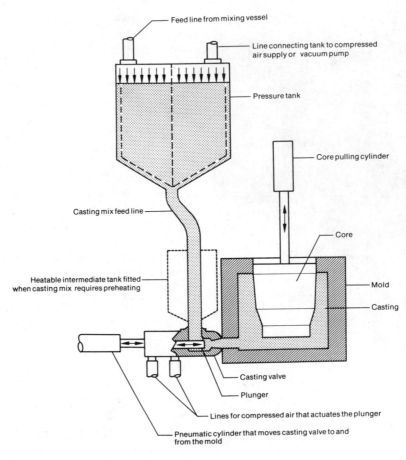

Fig. 18-3. Basic principles of pressure gelation process. (*Courtesy Ciba-Geigy*)

spectrum of diversified applications. This can be attributed to its low cost, relative simplicity, and excellent versatility. The size and shape can vary from small and smooth to large and intricate and be virtually cast with equal ease and fineness of detail, with the entire process performed at room temperature in inexpensive molds. See Fig. 18-4.

The basic method involves pouring liquid catalyzed polyester into a mold where it chemically polymerizes into a solid, taking on the shape of the mold. Polyesters are particularly well suited for this process because of their ability to polymerize in varying lengths of time and temperature, into objects possessing varying degrees of hardness.

Their properties can be further modified by introducing inert fillers or extenders into the liquid resin before it is catalyzed. The resulting casting would then tend to simulate the

qualities of the filler in both "looks" and "feel" (i.e., using crushed stone as the filler produces a casting that looks like crushed stone). See Fig. 18-5.

In some instances the casting may be required, in addition to the above properties, to serve a load-bearing or structural application. This is accomplished by molding the resin around a center core for reinforcement. The core material (which is usually wood, metal, or glass reinforced polyester), is designed to meet the structural requirements. The outer skin of cast polyester provides protection against surface abuse, as well as decoration.

Methods

The basic casting concept involves transferrence of the liquid polyester resin through the compounding stage, where the fillers or pigments

Fig. 18-4. Decorative base illustrates the precision that can be achieved with cast polyesters in duplicating ornament detail. (*Courtesy Flex Moulding, Inc.*)

Fig. 18-5. Sink bowl cast from polyester.

are added, into the catalyst step where the chemical hardner is added, and finally into the mold where polymerization or "gelling" takes place. This entire operation is often accomplished by what is commonly referred to as "the hand pouring technique." The method utilizes operators from beginning to end, without any specialized equipment. It is usually performed in three stages:

1. The first stage operator master-batches resin and the required fillers and pigments according to formulation.
2. The second stage operator draws off smaller portions of the master-batch, which he catalyzes and pours into the molds.
3. After the required curing time, or time needed to develop sufficient demolding strength, other operators remove the parts from the mold.

Variations of this method involve machines that serve to automate the process and increase production output. Other casting methods, such as centrifugal or rotational casting, also involve machines that permit the molding of highly specific shapes and configurations.

Equipment

Completely automating the cast polyester process is seldom contemplated and often impractical. This is because cast polyester has found most of its use in the field of "decorative reproduction"; which is subject to many variations of style and taste. These variations lead to low production requirements for individual pieces, making it difficult to design machinery that can handle the many different sizes, shapes, and forms it would have to encounter. Significant savings in time and labor can be achieved, however, with various types of automated and semi-automated equipment, when higher production rates are warranted.

The first piece of equipment, the automatic dispenser, is designed to airlessly mix, catalyze, and dispense the filled or unfilled polyester. Use of the dispenser enables fast, more bubble-free filling of the molds, and increases the mold cycle time.

To increase mold turnover still further, the installation of a radio frequency curing oven is highly effective. The device electronically stimulates the polyester molecules, causing friction and extremely rapid increases in temperature. The high temperature accelerates the action of the catalyst, resulting in mold cycle times of less than two minutes. Its primary advantage is in allowing a high production rate from relatively few molds.

The next two pieces of equipment—the centrifugal and rotational casting machines—are used for molding hollow shapes, rather than to automate the process as the dispenser and R.F. oven. The purpose of the centrifugal machine is to create hollow castings by spinning the mold on a horizontal axis at approximately 100 rpm. The rotational machine (of which there are several variations) is designed to spin the mold along two axes. It operates at less than 10 rpm on the principle of slushing rather than centrifugal force. This machine has the added attraction of being able to produce a hollow casting of any configuration. (See also Chapter 14 on Rotational Molding.)

Molds

Molds can be classified into two categories: rigid and flexible. Rigid molds are used wherever there is sufficient draft and no undercuts to interfere with demolding. The advantages of rigid molds are low tooling costs and long life. Several of the materials used in construction of rigid molds are: fibrous glass, cast epoxy, cast metal, cast urethane, and vacuum formed thermoplastic sheet.

Construction of rigid molds varies with the choice of material. Cast metal molds can be made by tooling shops. Cast epoxy and urethane molds can be made in-house by pouring a two-component resin system over a master, where it solidifies on curing. More difficult to make, but stronger and lighter, are the reinforced epoxy or polyester molds. They are made by first spraying a gel coat over the master and allowing it to gel or set. A piece of fibrous glass mat is then cut approximately to size and saturated with more resin (referred to as the tooling resin). Before it gels, it is laid over the master and "worked" with a stiff brush to conform to the shape of the master. When the laminate has gained sufficient strength, it is removed from the master and allowed to continue curing overnight. (See also p. 448 on Tooling).

Another type of rigid mold is the vacuum formed mold. Vacuum formed molds are made by vacuum drawing pre-warmed thermoplastic sheet over a master. They are fast and inexpensive to make, but have limitations in thickness and reproducibility of detail.

When undercuts, or inadequate drafts make rigid molds impractical, flexible molds are recommended. The most widely used flexible-mold material is silicone rubber. Silicone rubbers are extremely tough and resistant to chemical attack and high exothermic temperatures. In addition, they have unique self-releasing properties which eliminate the need for release agents and subsequent degreasing operations.

Urethane elastomers are also used to make flexible molds. They resemble silicone rubber in flexibility and toughness, but do not possess the built-in release characteristics.

A new and more recently used mold material is vinyl plastisol. Vinyls look and handle differently than silicones or urethanes. They do not gel and cure in the same manner and can be extremely flexible. They are supplied as liquids and must be heated slowly to an elevated temperature (usually 350°F) and then poured over the master, which must be preheated. When the casting cools, it can be stripped from the master and put into service. After the mold has deteriorated, it can be remelted and the process repeated. The high flexibility of vinyl plastisol lends itself well to molds with deep undercuts, but its low chemical resistance leads to dimensional instability.

Materials

Resins. Polyester casting resins vary from highly flexible to rigid, transluscent to water clear. Resin selection is made in accordance with the application and desired properties. Often two resins are employed, a flexible and a rigid, that are blended by the molder to his exact needs.

The flexible polyesters or blends containing large amounts of a flexible resin are generally considered to be better suited for casting because of their high impact strength and more workable properties. The flexible or slightly resilient resin can also be cast in massive cross-sectional thicknesses without fear of stress-cracking. When a more structural application has to be met, the same resin can be molded around a wood or steel core for added versatility.

The rigid casting resins are primarily used as blending resins, or in applications requiring high surface hardness. They are also less susceptible to the natural degradation process of air, light, heat, and chemical attack.

Clear casting polyester resins are water-white in appearance and highly transparent. They are used for small decorative items and encapsulations which can be molded at room temperature to produce clear, tough, mar-resistant objects. The clear casting resins receive special consideration over the previously mentioned resins because of the different conditions under which they are molded. Their application is almost always centered around their high quality of optical clarity. This eliminates the use of filler and the advantages offered in the form of reduced exotherm and shrinkage. Secondly, because the demolded article does not get painted, as most other articles, the surface characteristics are even more important. The resins usually range from resilient for castings of high cross-sectional thickness, to rigid for high-strength requirements.

Some activity has also progressed in extending unsaturated polyester resins by dispersing tiny droplets of water in the resin prior to gelation. Castings from this type of formulation show up as cellular, white materials resembling plaster of paris. As such, they have found use in replacing plaster of paris applications (e.g., lamp bases, wall plaques, and statuary).

Fillers. The properties of a polyester casting largely depend on the type and amount of the filler used. Its concentration usually varies between 20 and 75 weight percent of the total system. The primary properties affected are: cost, flexural strength, impact strength, density, and tooling characteristics.

In the field of decorative furniture parts and accessories, the best balance of properties appears to be achieved through the use of organic or hollow cellular fillers. One of the more popular is pecan shell nut flour. Accounting for its large usage are favorable handling properties and moderate density. It is readily dispersed in polyester with simple agitation. Because its density nearly matches polyester, the dispersion remains uniform and stable over long periods of time.

The two most popular "hollow" fillers are Perlite and glass micro-balloons. These fillers are used when trying to simulate the low density and nailing-stapling properties of wood. Because the parts produced with these fillers are actually cellular, they form what is termed *polyester syntactic foam.*

In the area of construction, where density is not a critical factor, the heavier inorganic calcium carbonate, silica, and glass fibers are widely used. The low oil absorption and inertness of the inorganic's result in high filler loadings, reduced costs, and increased flexural strength. This has led to their extensive use in the production of synthetic marble, slate and statuary. Glass fibers in either the finely chopped or milled form are frequently used in small amounts (between 1 and 10%) to give added strength to any casting. Their severe restriction of flow properties limits their use in casting formulations.

Barrier Coat. In the polyester casting operation, the term "barrier coat" refers to spraying the mold with a specially formulated lacquer that becomes transferred to the polyester casting during its exotherm and while it is still in the mold. The purpose of the barrier coat is to prevent resin cure inhibition on the face of the casting that is against the mold. The mold itself, depending on its composition, can act as a deterrent to prevent complete polymerization of the resin surface in contact, thereby resulting in "sticky" or "tacky" parts. Use of the barrier, through elimination of tack, also ensures faithful reproduction of detail and surface smoothness.

Release Agents. Release or parting agents are used to prevent adhesion of the castings to mold surfaces that do not have built-in release properties. Typical molds requiring release agents include: urethane elastomer, epoxy, and latex. In order for release agents to be totally effective, they should only be used against mold surfaces for which they were designed. They should also be capable of easy application in thin uniform thicknesses, so as not to interfere with mold definition. If the casting is to be finished, the release agent should be investigated for adhesive strength using the particular finishing system. Several common types of release agents used are: waxes, silicones, and fluorocarbons. The fluorocarbons appear to function best when applied to rigid molds, while the waxes and silicones serve mostly the flexible molds.

General Casting Procedure

Mixing. The resin is weighed into a container large enough to accommodate the resin and filler. If fibrous glass is used, it should be added first, followed by the remainder of the fillers. Care should be taken during the mixing operation to avoid air entrapment through controlled agitation. Once the fillers are dispersed, the mixing operation is ended. It is then advisable to allow any entrapped air a chance to escape. A quick way of doing this is to slow the speed of the agitator until it is barely turning. This helps break up and release entrapped air bubbles. If low density fillers are used, the compound cannot be left standing unagitated for more than a few hours, or the filler will begin to separate and float to the surface of the resin. If this does occur, the filler can easily be redispersed with further agitation.

Hand Pouring. The amount of polyester needed to fill the mold is determined by calculation or sample pouring. Once determined, this weight of resin is drawn from the master batch into a suitable container. Containers are selected to be as nearly full as possible in order to ensure good catalyst mixing and minimal air entrapment. Polyethylene buckets are normally used, since they are reusable. The appropriate amount of catalyst (usually 1% of the resin weight) is then weighed or volumetrically sighted, added, and thoroughly mixed. If the volume of compound is small, it is hand mixed with a spatula. Larger amounts require motorized mixing. Average mixing time in either case ranges between 30 sec and 1 min.

The compound is poured into the mold, making sure to pour resin on top of resin and create as little turbulence as possible, in order to avoid entrapping air on the surface of the mold. If a core is to be inserted in the mold, the mold is only partially poured, the core inserted, and more material added if needed.

Demolding. The castings can usually be demolded within 7 to 10 min. To determine this point, the thumbnail should be pushed into the casting. When sufficient resistance is met, without crumbling, there is usually sufficient strength to demold the part. When demolding strength has been reached, the mold is carefully separated from the part to avoid damaging delicate sections. To aid in demolding, the flexible molds are designed to be easily removed from the mold box in order to take full advantage of their elasticity. Whenever possible, open-faced molds are designed to support themselves without a mold box.

Typical Advantages and Uses

The casting method offers the designer and molder almost unlimited freedom in both design and product character. The equipment costs are minimal and a relatively small amount of technology is required. The simple hand-poured technique, though not suitable for high production, allows profitable short-run molding with inexpensive self-made tools. The faithful reproduction of detail is known to be the finest possible. Additional benefits include excellent mold life and low reject rates.

Cast polyester is currently being used in the production of furniture parts, lamp bases, wall plaques, boat boards, sign letters, wall panels, embedments, encapsulations, edge castings, sink bowls, wall tile, etc. (See Figs. 18-4 to 18-7).

CASTING URETHANES

Because most urethanes—foams as well as elastomers—start out with liquid formulations that are dispensed into a mold or onto a belt (as

Fig. 18-6. Cast polyester lamp bases. (*Courtesy Nova Mfg. Co.*)

Fig. 18-7. Cast polyester chair back. (*Courtesy Nova Mfg. Co.*)

in the production of flexible urethane foam sheeting), the processing techniques are generally take-offs on casting procedures. More recently, however, techniques have been developed (e.g., liquid injection molding) that compete directly with conventional molding systems. (See chapter on Cellular Plastics.)

Urethane elastomers for finished parts or coatings can be cast by several methods: hand batch and machine casting; centrifugal and rotational casting; solvent casting; and spray casting. Although one-shot systems have been used, the discussion below is based on two-component systems—a prepolymer mixed with a curative or curing agent.

In the hand-batch system, the procedure is as follows: the prepolymer is first degassed to remove entrapped air, then mixed with the curative. This mix, too, must be degassed to remove any air whipped in during mixing. The mix is then poured into a mold and cured by heat.

Although these operations were originally done by hand, mixing and metering equipment is available today to process the urethanes continuously and automatically. This equipment is similar to that used for urethane foams (see Chapter 20 on Cellular Plastics), but is modified to permit the degassing needed and the mixing and depositing of the system in a bubble-free and air-free manner.

In centrifugal casting, the polymer/curative mix is introduced into a rapidly spinning mold where centrifugal force pushes the plastic against the inner surface of the mold walls where it is cured into a finished part by the application of heat. In rotational molding (see Chapter 14), the mold rotates in two planes simultaneously until the mix cures. In solvent casting, the prepolymer and curative are dissolved in a common solvent, cast onto a surface, and allowed to dry (i.e., the solvent is driven off). In spray casting, the two components are both dissolved in a common solvent (as in solvent casting), and sprayed through a nozzle onto a surface, where it is allowed to dry. It should be noted that with both solvent and spray casting, the surface can either be a release surface (in which case, the urethane is stripped off as a solid film or sheet after the solvent has evaporated) or a substrate such as fabric (in which case, a urethane-coated fabric is produced).

CASTING NYLONS

Nylon 6 caprolactams can be cast directly into finished parts at low pressures by using in-situ polymerization techniques. In essence, a liquid nylon monomer, mixed in with a catalyst, can be poured directly into a mold (or rotationally cast), where it polymerizes anionically while still in the mold. At the conclusion of the polymerization process, the mold is opened and a finished casting is removed.

This technique is especially applicable to the production of large, complex shapes (as large as 1500 pounds per casting) that could not possibly be made by the more conventional plastics processing techniques. Key factors are the low heats and low pressures involved. Although the polymerization process is exothermic, the relatively low heat of polymerization of nylon caprolactam (coupled with its low melting point) makes the process easy to control and simplifies the heat transfer problem generally associated with the production of massive parts. In terms of pressure requirements, the low

viscosity of the molten monomer and its small shrinkage on polymerization, make pressure unnecessary and permit the use of low cost molds. Further economies are attributable to the low cost of monomer as compared to polymer.

Cast nylon parts made by anionic polymerization exhibit higher molecular weights and a highly crystalline structure, and are therefore slightly harder and stiffer than conventionally molded nylon 6.

Applications for cast nylon include: huge gears (e.g., a 300-lb nylon gear for driving a large steel drum drier) and bearings, gasoline and fuel tanks, building shutters, school furniture, buckets, and various components for paper production machinery and mining and construction equipment. Recently, activity has centered on adding reinforcing materials to the nylon monomer before polymerization to produce parts with higher heat distortion, impact strength, and tensile strength.

SOLVENT CASTING THERMOPLASTICS

Several thermoplastics films (e.g., cellulosics, vinyl, etc.) can be produced by a process known as *solvent casting*. It is similar to the system already described for casting urethanes. In essence, the polymer and/or other materials are dissolved in a solvent. This solution is then spread on a belt and passed through an oven. Under heat, the solvent evaporates and the remaining solid material is stripped off the belt in the form of film or sheet. There are, of course, many variations to this process, depending on the plastic involved and the thicknesses and quality desired in the film.

TOOLING WITH PLASTICS*

Since World War II there has been a steady increase in the utilization of plastics in the construction of tools, dies and fixtures, for use in fabrication of articles from metal, plastics, glass, wood, etc. Such tools can be produced

* Reviewed and revised by Vincent Sussman, Director of Research, Epoxy Products Co., Div. of Allied Products Corp., New Haven, Conn.

rapidly and economically. Today, tooling with plastics is an established and widely accepted technique.

It has been recognized that further development and growth of tooling with plastics must be based upon sound engineering standards. Test methods and standards are now being established through cooperative efforts of the Society of the Plastics Industry (Plastics for Tooling Division) and the American Society of Tool Engineers. The ASTE Research Fund, with the help of the plastics industry, has initiated a research program at Purdue University, and the industry members of SPI have begun to issue practical methods of test, which are being accepted and utilized extensively.

Through the proper use of plastics in tooling, several definite advantages can be realized over tools of metal and/or wood:

(1) tools can be cast or laid up to the desired final shape and dimensions in one operation;
(2) relatively inexpensive equipment and labor, and fewer man-hours, are required;
(3) delivery of tools is faster, reducing the required lead-time between design and production of the end-use article;
(4) more frequent changes in design are feasible;
(5) duplication is easier, where several tools are required;
(6) revisions and repairs are simpler, promoting efficient development of design and reducing down-time;
(7) the tools are relatively light in weight and easy to handle;
(8) they are resistant to corrosive atmospheres, lubricants, and weather, and thus can be stored outdoors.

However, careful evaluation of the end-use must dictate the selection of material for each specific application.

Recognition of the advantages offered by tooling with plastics, together with current efforts toward standardization, is resulting in a consolidation of this branch of the industry. Hence, the subject matter of this chapter has been limited to a brief description of materials

and methods, without detailed discussion of applications, testing, etc.

Materials

Tools utilizing plastics are usually made from viscous liquids referred to by the industry as "resins." These are convertible into solids by the action of a chemical (catalyst, hardener, accelerator, etc.), sometimes assisted by the application of heat. Pressure may or may not be required.

Only a few basic raw materials are finding application in tooling with plastics. The very strict requirements as to stability and dimensional tolerance have limited the choice of materials. Epoxy, phenolic, polyester resins, and urethane are the four major types used in tooling. The polyamide and polysulfide types are seldom used alone, but are used in combination with epoxy resins to provide several excellent flexible tooling materials. Vinyls and silicones find application as parting or release agents.

When properly formulated, the epoxy and polyester resins have good dimensional stability (Fig. 18-8).

Fig. 18-8. Epoxy tool is cast to tolerances of better than 0.003-in. maximum.

Phenolic resins are not satisfactory for close-tolerance work.

The need for careful and expert chemical control of tooling resins as used by the fabricator of the finished item has brought about the founding of special industrial firms known as "formulators," who are an essential link between the chemical concerns manufacturing the uncompounded base resin and the fabricators, who design and produce tools from the resins.

In most cases, so-called "fillers" constitute an important part of the finished tool. Glass fibers, metal, sand or gravel, etc., may be used. Sapphire and graphite fibers offer extremely high physical properties. Core structures of foamed plastics may be used to reduce weight where high strength is not required.

Phenolic Resins. Phenolic specialty resins were among the earliest raw materials used by the industry to make plastic tools. Chemically engineered to overcome most of their natural shrinkage on curing, the materials are cast to shape. Fairly strong acids are used as catalysts to cure the resin. Tools of large size may be cast and hardened in a short time, with little labor and at low material cost.

Polyester Resins. Although the term "polyester" is being applied to various plastics, certain very specific compounds under this designation are manufactured and formulated for tooling with plastics. Monomeric styrene is, as a rule, one of the reactants in these polyester tooling resins. Proper reinforcements, such as glass fibers, are used to impart strength and toughness, especially in thinner sections. These structures have good resistance to wear.

Epoxy Resins. The favorable shrinkage and stability characteristics of epoxy resins, when properly formulated and cured, are mainly responsible for the increasing use of these materials. Various fillers are used to improve impact strength and resistance to heat or abrasion. Development of heat by the exothermic reaction of cure must be restricted, in order to minimize shrinkage; therefore, recommended and established procedures must be strictly followed. Normal shrinkage can be further minimized by applying face-casts to rigid cores, and also by using special formulations containing large proportions of fillers. (See Fig. 18-9).

Urethane resins have been utilized to obtain tooling compounds with high impact properties. Hardness must be controlled from a rigid Shore D of 80 to a rubber-like Shore A of 50. These resins exhibit high abrasion resistance.

Methods

Individual skill has created various auxiliary techniques in the making and repairing of plastic tools, but there are only three basic methods in use today, singly or in combination, namely: casting, laminating, and troweling and splining.

Casting is most commonly employed. The molds used for casting plastic tools can be made from almost any material strong enough to hold the resin. Rigid or flexible materials may be employed. Casting to close tolerances requires costlier molds such as reinforced plastics, laminated, or steel molds.

The type of resin, the size, and contours of the casting determine the choice of the mold. It is common to use a wood master or a plaster splash taken from the master. Sometimes, an actual metal part may serve as a mold.

Since phenolic casting resins attack metals, metal molds should be protected with a thin coat of Tygon.

The strong adhesion of epoxies to many types of materials makes it necessary to apply parting agents to the mold surface. Among these agents are waxes, vinyl alcohols, silicones, polyethylene, and fluorocarbons. They may be applied by brushing, wiping, slushing or spraying.

Prior to casting, molds made of plaster or other porous materials have to be thoroughly dried or sealed with moisture-barrier sealants. Several coats of a sealant are usually applied, each being lightly sanded with 400 grit paper. The surface is waxed and buffed before applying a parting agent.

Since the cast resin is capable of reproducing minute imperfections that appear on the mold surface, great care must be taken in preparing the mold. Flaws have to be corrected, and mold preparation should include sanding off blemishes, filling in hollows, and smoothing down

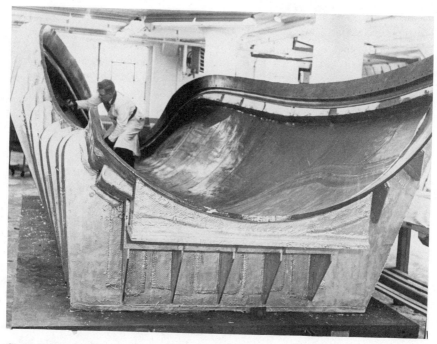

Fig. 18-9. Glass-reinforced epoxy mold is 25 ft long and weighs over 6000 pounds. It is based on an epoxy formulation that withstands prolonged exposure to temperatures in excess of 400° F.

any weaves or wave-like patterns which may have occurred on the surface of the mold.

The resin-hardener mix is poured into the prepared mold and cured at room or elevated temperatures, depending on the nature of the resin and the size of the casting.

Advantages of the casting process include low cost, speed, handling ease, as well as excellent pick-up of detail.

It is common to reinforce plastic tools to impart strength and to sustain the wear of longer runs. Glass and synthetic fibers are the most popular among the reinforcing materials used.

To prevent serious erosion of contour, metal inserts are employed at critical locations to strengthen plastic tools. The inserts can be set in place prior to casting, or they can be fit into cavities milled into the cured and hardened tool.

Laminating. Reinforcing the resin with glass fibers or other synthetic fiber cloth is termed *laminating*. Laminating is done by first applying a plastic gel coat to a prepared form and allowing the gel to become tack-free. The gel coat provides the tool with its surface properties like abrasion resistance and chip resistance. The coat is backed up to a predetermined thickness by successive layers of fibrous glass cloth or other fabric, impregnated with a laminating resin that may be brushed directly onto the consecutive layers of the reinforcing material.

Splining and Troweling are procedures accomplished with the use of paste-like resins. These are thixotropic compounds that allow work on vertical surfaces. They are applied over the outside of male forms or on the inside of female forms.

Applications

The applications of tooling may be divided into four types:

(1) Prototypes or Models. Prototypes are full-size, actual working items used to prove and evaluate finished products. Models are scale or full-size replicas of products and are generally used to illustrate arrangement.

(2) Transfer Tools. Transfer tools are used to transfer a surface, hole, point, or area from a prototype, model, or master tool to another tool which is not normally used as a master.

(3) Manufacturing Tools. These tools are used in production of products. They include:

    (a) dies, e.g., draw dies, bending dies, stretch-form dies, drop-hammer dies, vacuum forming, etc.;

    (b) jigs and fixtures, e.g., drill jigs, assembly fixtures, welding fixtures, scribing fixtures, etc.;

    (c) foundry patterns and core boxes;

    (d) molding tools for plastics.

(4) Inspection Tools. These tools are used to check surface, hole, point, or area during or after a manufacturing operation.

Metal Forming Tools. Metal sheets of varying thicknesses can be formed on plastic tools (including draw dies, stretch dies, drop hammer dies, hydroforming and rubber-pad forming dies).

Plastic draw dies can be constructed by several methods. They can be cast to exact contours from epoxy resins. They may have a cast plastic core and a laminate surface, or they may be made by casting plastic over a metal core.

Plastic draw dies are primarily used in developmental work and for drawing of prototype parts. They are sometimes employed for actual production runs, usually in drawing parts of light-gauge steel or aluminum.

Plastic stretch dies have been made with working face areas of well over 100 sq. ft. The most usual method of constructing them is to cast an epoxy resin over a cured core of foam (generally phenolic, although other resins are used). Inexpensive high strength cores are produced by mixing granite chips with a special core resin. (See Fig. 18-10)

Laminated shells, cross-ribbed for added strength, are occasionally utilized in making such tools as stretch dies.

The construction of plastic drop-hammer dies often involves combining various types of plas-

Fig. 18-10. Stretch press die for forming metal sheet.

tic resins with other materials. Plastic punches with metal bottom dies, punches made with a facing of laminated plastic over a body of cast plastic, and punches strengthened by steel weldments are commonly employed.

An interesting method of constructing drop-hammer dies involves casting a filled resin inside a laminate for the bottom half of the die, which is then placed in a frame where it serves as the mold for making the mating punch. A parting agent is applied, and the punch is cast from a resilient epoxy face over a lead or over a Kirksite core.

Easily adaptable to quick changes in design, plastic hand hammer forms are popular metal forming tools for production runs where a small number of metal parts are to be furnished.

Hand hammer forms are cast solid or by pouring a casting resin over a foamed plastic core.

Precision-cast inserts of brass and zinc alloy may be locked into place prior to casting by slots or bolt heads protruding into the plastic. The inserts can be set into the tool after the tool is cast, by chipping away a portion of the tool and casting the metal into position with additional resin.

The development of low-exotherm casting resins has overcome former limitations on the size of plastic dies for metal forming.

Dies weighing as much as six tons have been constructed in a continuous round-the-clock operation taking approximately one day for pouring and one night for setting. These large dies are often made with a core of a bulk-mass casting resin extended with gravel for cost reduction, and with an epoxy-glass cloth laminate for the face. Such tools, which have been made as long as 17 ft, are about half as heavy as comparable cored metal tools.

Fixtures, Jigs, and Gauges. Plastic fixtures to check contour flanges and hole locations of parts are accurately made from master models or duplicates of master models. Either the casting or laminating technique may be used. Fixtures and gauges for complicated shapes are often made by hand lay-up of epoxy resins and fibrous glass cloth against models. An actual part, as well as a model, can be used as a mold for casting a plastic holding form.

Plastic chucking fixtures have been found to hold die-cast parts more securely than machined steel jaws. The plastic fixtures retain their accuracy under consistent wear, and have recorded satisfactory service life of as long as a year or more.

Duplicates and Models. Plastic duplicates and models provide automobile manufacturers with a fast and inexpensive way of making copies of actual or projected parts to send to accessory suppliers and assembly plants for reference.

Such duplicates are often produced by applying a gel coat and a laminate over a wooden die model. A framework of tubing or sheet stock, bonded into place, serves as a base. After setting, the female form is removed from the model. Parting agents are applied to the female form and the laminating process is repeated to make as many duplicates as may be necessary.

Casting and laminating resins are used to make plastic study parts for planning of design, tooling and assembly operations.

Foundry Applications. Casting, laminating, and troweling epoxy resins are all widely used in the foundry field. They lend themselves to simple duplicating techniques for making patterns that require almost no machining, and that have smooth and pit-free surfaces for improved sand release.

Available formulations stand up well under slinger operations. High-strength and high-heat resistant resins are used to fabricate core driers for dielectric baking. Plastics are now utilized for coating wooden patterns and for patching worn spots in patterns and metal core boxes.

Carvable epoxy resins with dimensional stability greater than wood are often employed to fabricate sturdy and inexpensive master patterns.

Plastic-Forming Tools. Plastic tools and dies are being increasingly used for forming plastics as well as metals.

Epoxy dies are utilized in processes whereby thermoplastic films and sheets are shaped under heat and atmospheric pressure. Plastic dies for vacuum forming and drape forming are cast at a fraction of the time required for more costly, although no more accurate, machined metal dies.

One-piece bathtubs and huge **sectionalized**

prefabricated swimming pools are made on plastic molds. A great number of reinforced plastic products of various sizes are made by vacuum-bag molding on cast and laminated epoxy molds.

Plastic matched dies are being employed for low-pressure molding of plastic components. Plastic tools are used for forming crash helmets and foamed plastic parts such as automobile cushioning and crash pads.

Plastic pattern duplicates speed up mass production of products manufactured by latex dipping.

Using heat and pressure, sometimes employing vacuum, cast and laminated plastic dies are serving to shape precut and preheated acrylic sheets to produce aircraft windows and canopies, astrodomes, airfield landing light covers, and large advertising signs and displays.

Materials suppliers and molders are production-testing injection molds made from new epoxy resins with high heat transfer and high heat resistance.

ENCAPSULATION*

The term *encapsulation* covers a wide variety of processes used to apply solventless protective coverings to electrical and electronic hardware: (1) potting, (2) casting, (3) vacuum impregnation, (4) trickle impregnation, and (5) conformal coating, used singly or in combination. Terms such as *pressure stalling, liquid injection molding, hot melt coating*, and *thermoset injection molding* have gained some publicity recently and will be discussed as variants of the basic process mentioned above. While it can be argued, in a strict sense of definition, that encapsulation can be accomplished using solvent applied varnishes, this section will be confined to a discussion of solventless systems, the materials used, and the above-mentioned processes.

Applications

Applications include almost every known electrical or electronic device from stators and

* By J. E. Carey, Technical Service, Shell Development Co., Shell Chemical Co., Woodbury, N.J. 08096.

armatures of the largest power generating equipment through motors, transformers, solenoid coils, integrated circuits, printed circuits, ignition coils, electronic ignition systems, magnetic amplifiers, and the whole gamut of solid state devices (Figs. 18-11 through 18-13). Considerations involved in encapsulating include:

1. Electrical insulation
2. Mechanical stabilization
   a. vibration
   b. impact
   c. electrical impulse
   d. freeze-thaw
3. Environmental protection (Ref. 1, p. 458)
   a. atmospheric humidity
   b. corrosive atmospheres
4. Thermal shock.

Encapsulating generally induces sensitivity to thermal shock and therefore must be called a negative consideration which must be dealt with.

Formulation Considerations

Electrical components give off heat in service; they are also expected to operate over fairly wide temperature ranges because of fluctuations in ambient or storage conditions before start-up. For example, in automotive or aerospace applications, a part can be at rest in

Fig. 18-11. Commutator ring made with epoxy resin.

Thermal shock mechanical abuse arises primarily because the plastic material used in the insulating encapsulant is a poor conductor of both electricity and heat and has a thermal coefficient of expansion much higher than the electrical components.

To accommodate the real physical facts of linear thermal coefficient of expansion (TCE), differences of metals and plastics, and poor thermal conductivity of the plastic, one must choose a plastic compound which has either:

1. High elongation,
2. Sufficient strength to withstand the load, or
3. Internal thermal expansion compensation by use of fillers.

Fig. 18-12. High performance ignition coil encapsulated with epoxy resin.

High elongation of solventless thermoset plastics is generally accompanied by low mechanical strength at elevated temperatures. While high mechanical strength is not always necessary in an encapsulant, it is often desirable where this covering also acts as a mechanical attachment or is expected to resist mechanical displacement due to magnetic influence. Thus, obtaining an encapsulant with high elongation is not always a suitable choice.

To achieve high enough strength in the plastic to resist mechanical loads or thermal shock or both is a near-impossible design problem because of necessary variations in thickness imposed by the shape of the structure. Thus, choice number 2 (above) is a very limited solution.

The use of metal oxide fillers whose thermal coefficients of expansion are generally lower than metals offers a fairly effective route for development of a plastic encapsulating system whose average thermal coefficient of expansion is closer to that of the metals encapsulated. In general, silica sand fillers have been preferred in the electrical industry because they do not contain sodium, calcium, or magnesium ions which are known to migrate under electrical stimulus in wet or high humidity conditions. The normal crystalline products obtained from many commercial silica sand beds are used widely. Such silica sands show thermal coefficients of expansion (TCE) of from 5 to 14×10^{-6} cm/cm/°C depending on whether

Fig. 18-13. Epoxy resin-encapsulated radio circuit.

an environment of ice at −65°C and someone will come along and turn on a switch which generally initiates electrical service and heating of the device. It also initiates a severe cycle of thermal shock on the electrical insulation or encapsulant. Table 18-1 shows thermal expansion properties of some materials used in the electrical industry.

Table 18-1. Linear Thermal Coefficient of Expansion (TCE) of Some Materials used in the Electrical Industry

Material	TCE cm/cm/°F x 10^{-6}	TCE cm/cm/°C x 10^{-6}
Aluminum	13.	23
Iron	6.5	12
Copper	9.2	17
Silica / / to Axis	4.4	8
$\perp$ to Axis	7.4	13
Fused quartz	.2	.5
Epoxy resin		
1. Low temperature		75
2. High temperature		50
Urethane rubber		100
Glass		7-10

the TCE is measured parallel or normal to the axis of the crystal. However, when the silica sands are fused at high temperatures to what are called fused or vitreous quartz glasses, the resultant products show TCE in the range of 0.5×10^{-6} cm/cm/°C. Some handbooks report negative TCE (ca-0.25 x 10^{-6} cm/cm/°C) for fused quartz at temperatures below 16°C.

Choice, treatment, and handling of fillers are major considerations in designing and tooling for production of encapsulation systems.

When the low TCE fillers are blended with the higher TCE resin systems, composites whose average TCE is in the same range as the metals to be encapsulated can be developed.

Figure 18-14 shows the effect of weight percent silica on TCE of an epoxy resin system.

Materials and Formulations

Thermoset resins of the type which do not release water or other volatile products during cure are the major materials used for electrical encapsulation where severe service requirements exist. However, the heat fusible vinyl plastisols and various powdered thermoplastic resins are sometimes used in "conformal coating" processes, particularly as color or identification covers where mild service conditions are expected.

Epoxy, polyester, silicone, polyurethane, and polysulfides comprise the most commonly discussed chemical types of resins used. These resins are then "formulated" to accommodate the end-use requirements and the production

process to be used. Some factors which must be included in developing a satisfactory formulation include:

1. Basic resin classes.
2. Curing agent, which generally dictates:
    a. Cure conditions necessary.
    b. Service conditions acceptable.
3. Filler choice.
4. Flow control agents.
5. Anti-settling agents.

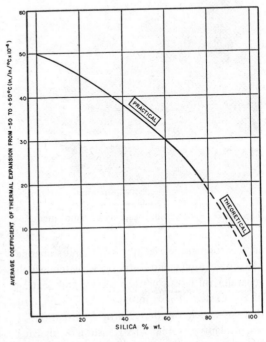

Fig. 18-14. Effect of weight percent silica on thermal coefficient of expansion of an epoxy resin casting.

6. Foam breaking agents (for vacuum processing).
7. Pigments or dyes.
8. Production process.

The knowledge of how each of these factors dovetail into formulation, process and end-use requirements is beyond the scope of this chapter. However, they must be considered and indeed occupy a massive part of the literature on electrical insulation.

Processes for Encapsulation

(1) Casting refers to the use of a permanent mold to receive the encapsulating compound which contains the electrical objects to be protected. The processing may be done with preheated molds, parts, or compounds or may be done in vacuum. Use of semi-permanent molds or even one-time-use molds which are stripped from the cured encapsulated product can also be considered in casting processes.

(2) Potting refers to processes which use a receptacle or jacket which remains as a permanent exterior skin on the final product. Either ambient temperature, elevated temperature, and/or vacuum may be used in such processes.

(3) Vacuum impregnation applies to any process in which vacuum is used to degas the electrical assembly before and during the encapsulation operation and implies that part of the formulated plastic compound penetrates into or through some very closely spaced components and surrounds them.

(4) "Trickle impregnation" is a fairly recent addition to the processing technology for protection of electrical elements. Generally the assembly of electrical components is preheated in a vacuum chamber on rotating fixtures and the liquid resin/curing agent combination is allowed to drip or run in a thin stream onto the assembly. Capillary action along with the absence of air will generally cause the liquid plastic to penetrate the fine structural spaces and develop a thin final covering.

In some instances, the objects are preheated in vacuum by electrically overloading the actual circuitry of the assembly to such an extent that heating results.

(5) Conformal coating with thermosetting systems differs from vacuum impregnation and trickle impregnation in that the electrical assembly is dipped into a plastic formulation which contains fillers and even fibers but is removed from the dipping bath before final cure. The use of fillers, fibers, and thixotroping agents assist in developing an encapsulating composition which will be left as a thick skin on the electrical components. Such compositions have very little capability to flow into voids and cannot be expected to penetrate between closely spaced structures, although if the dipping is done in vacuum, a limited amount of penetration will take place provided the fillers, etc., do not pack at the entry to void spaces preventing movement of material.

(6) "Pressure stalling" or "liquid injection molding" are terms used to define either casting or potting processes where the mold or pot is connected directly to a feed system where a liquid slurry of the desired encapsulating formulation is forced into the mold and components. A prime requirement of such processes is that the feed line (runner, sprue) is the last part of the plastic formulation to polymerize and harden. Thus, liquid material is always available to fill in behind the polymerizing material which is shrinking (see also p. 458 on Liquid Resin Molding).

Several novel procedures for maintaining pressure by use of spring or pneumatically loaded accumulator sections and check valves have been used. Molds which contain such liquid accumulators make it possible to feed and remove molds from the injection station before gelation of the product has occurred.

(7) Thermoset injection molding is somewhat similar to the above mentioned variations except that the formulated compound is received at the encapsulating station as a solid which must be melted in a reciprocating screw extruder before introduction to the mold. (See also Chapter 9 on Injection Molding Thermosets.)

(8) "Hot melt coating" or powder coating is a relatively recent development for placing conformal coatings on electrical components. A wide variety of processes are used ranging from dipping heated objects into powders which are

suspended as slurries in flowing gases (fluidized bed) to spraying positively charged plastic powders onto negatively charged substrates which are subsequently heated to fuse and form coatings (electrostatic deposition). See also Chapter 17 on Powder Coating.

Cleaning and Drying

Drying of electrical components before encapsulation is a well honored tradition in the industry. Even vapor phase degreasing of metallic elements which have been treated with oils is common. However, until recently (Ref. 2, below), no one seems to have proposed washing components with distilled or deionized water after degreasing and before drying to remove water-soluble salts which might be left by handling.

While this discussion is not complete in any detail it can be seen that the solventless encapsulation industry has made great strides over the last few years and that a wide variety of opportunities in both materials and processes are available from which to make engineering choices.

References

1. ASTM F 74-71 T "Determining Hydrolytic Stability of Plastic Encapsulatants for Electronic Devices."
2. Tautscher, C. J., *Insulation/Circuits,* p. 32 (June, 1972).

LIQUID RESIN MOLDING*

The need for low-pressure encapsulation of electrical and electronic devices has long been realized by customers and producers throughout the military, space, and commercial indusries.

This need has been met in part through the introduction of liquid resin molding. By bridging the gap between the conventional methods of liquid potting and powder transfer molding, liquid resin molding has been proven to be a practical process for producing reliable, high volume low-cost molded assemblies and devices.

Basic Process

Liquid resin molding (LRM) is a process involving an integrated system for proportioning, mixing, and dispensing two component liquid resin formulations and directly injecting the resultant mix into a mold which is clamped under pressure (see Fig. 18-15).

Two basic metering systems are currently

* By Hull Corp., Hatboro, Pa.

Fig. 18-15. Liquid resin molding system attached to 50-ton transfer press.

available. The first and simplest of these systems consists of a single pressurized reservoir into which the premixed resin and hardener are placed. This system has successfully met those applications involving long pot life materials. However, it is not suited for those applications requiring controlled material flow and consistent molding pressures. Caution must also be exercised when working with filled compounds; otherwise, short shots and rejected devices will result.

If the use of rapid pot life materials is envisioned, a positive displacement-type metering, mixing and dispensing machine must be utilized. This second system separates the resin and catalyst by use of individual pots; material is forced from the pots to a mixing chamber where the actual blending of materials takes place. A multitude of mixing ratios can be used as well as a wide range of viscosities ranging from less than 5,000 up to 750,000 centipoise. Facilities for mixing and heating of both components can also be provided when using viscous or heavily filled materials.

Positive displacement systems have successfully metered and mixed materials with filler contents of 70% and more. Occasionally, vacuum must be employed on these heavily filled resin systems to eliminate air pockets.

From either system, the premixed material is fed into a water-cooled injection nozzle which mates against the mold sprue bushing during injection.

Sophisticated clamping devices are generally not required (except for very high volume production). One reason is that mold injection pressures are in a range of only 5 to 12 psi. But because many of today's most popular dispensing systems rely upon press hydraulics for motive power, many manufacturers have reported maximum versatility when clamping devices used were conventional transfer presses. Quick disconnects are provided on various machine parts to easily convert from either transfer or liquid resin molding.

The mold is placed between heated platens of the press for use of hand molds; or, if a production mold containing its own integral heaters is used, it is bolted onto the press platen. Molded devices and runner systems can be transferred by use of work loading fixtures for production continuity.

Liquid resin molding systems adapt more easily to automation than do conventional powder systems. The main reason is elimination of material handling. The operator does not physically come in contact with LRM materials. Pushing buttons for cycle sequencing is all that is required. A typical sequence would include:

A. Clamp close.
B. Injection nozzle advance.
C. Injection of material.
D. Injection hold.
E. Injection nozzle retract.
F. Cure.
G. Clamp open.

By analyzing the above sequences, one can see the opportunity for automation. The operator has only to initiate sequence A (clamp close) and thereafter the equipment takes over sequencing automatically. Between cycles, the operator can be busy loading additional work loading fixtures, or performing secondary non-related functions.

As with any automated system, and particularly when handling liquid materials, machine functions should remain uncomplicated. Disassembly procedures of the entire injection cylinder and nozzle system must be fast and simple in case of inadvertent material cure. To prevent accidental material set-up, equipment manufacturers have options available ranging from audible alarms to automatic solvent purge systems.

Mold Design

Aluminum and soft steel molds have proven successful for developmental and short run programs. However, use of through hardened tool steel is still recommended for those applications involving leads or terminals. Several manufacturers report having successfully used aluminum mold bases with steel cavity inserts. In using this approach, a doweling arrangement must be considered to prevent unacceptable cavity mismatch.

One important design consideration is that of part ejection. Liquid resin materials, especially clear epoxy systems, have exhibited very long flow characteristics. Ejector pin tolerances must be held extremely close, otherwise flashing will occur down the pin, "freezing" the pin in location.

Cavity and cavity related areas must still be highly polished, preferably chrome-plated. This will allow easier part ejection and will sometimes eliminate the need for ejector pins altogether, allowing the operator to simply air blast the molded devices from the cavities.

Mold layout, although basically similar to transfer type molds, must make provisions for injection on the parting line.

Width and depth of runners, gates, and vents depends upon shot size, material viscosity, and size of filler particles. Half and full round runners have proven superior to other types. Runner size should be as small as possible to prevent turbulence and resultant air entrapment. (See Fig. 18-16)

Generally speaking, it has been observed that gates should be located as low as possible on the molded device. Centrally located gates in most cases are acceptable. Vents should be positioned on the highest point of the encapsulated device to prevent air entrapment.

Vertical gates and vents have proven successful in those troublesome cases where everything else failed to produce void free devices. A typical example of this problem is a simple 14 lead DIL (Dual In-Line) device. By utilizing conventional parting line gating techniques and by incorporating vertical vents, air voids were totally eliminated in the upper corner opposite the gate.

Conversely, experience has taught that radiused corners, rather than square corners are easier to fill and keep polished. This is particularly true on relatively small packages, including flat packs and integrated circuits.

On those circular strip-mounted devices that prevent adequate venting, vertical strip molding techniques can be employed. Unfortunately, mold design becomes much more complex, sometimes necessitating 3 or 4 plate mold construction and mechanical caming arrangements.

Advantages of Liquid Resin Molding (LRM)

The greatest contributions that liquid resin molding can offer over conventional molding techniques are lower molding pressures (5 to 12 psi) and lower mold temperatures (200°F and lower). Additionally, it is anticipated that more

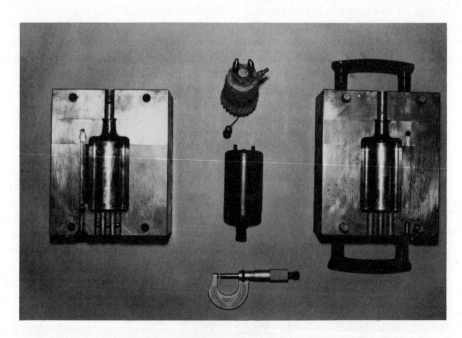

Fig. 18-16. Typical LRM mold, showing runner and sprue details. Device shown is ignition coil.

rapid cures can be expected on automatic systems as compared to transfer and potting techniques. Electrostatic and fluidized coating techniques should not be confused with LRM even though they permit extremely rapid cycles. LRM is concerned with encapsulating to a definite configuration other than simply coating in a protective envelope of epoxy.

Other advantages of liquid resin molding are found in the materials themselves as it appears there is more freedom to formulate for specific end properties with liquid epoxies than with powders.

To date, in addition to epoxies, liquid silicones, polyesters, and polyurethanes have been used experimentally. These new materials will eventually lend themselves to liquid resin molding design and application concepts.

Perhaps the greatest contribution of LRM has been involved with the production of light-emitting and light-seeking devices such as light-emitting diodes. By utilizing specially formulated clear epoxy materials, liquid resin molding equipment has attracted attention in this developing market.

Materials research is continuing to both lower LRM material costs and to seek the ideal blending of resins, hardeners, fillers, pigments and catalyst to give batch-to-batch consistency.

Applications

In addition to light-emitting and light-seeking devices, LRM has attracted great interest from coil and transformer manufacturers where low pressures and temperatures are important factors to eliminate distorted windings and bobbins. In many instances, the combination flow and pressure of conventional transfer molding crushes and rips out vital component sections. Utilizing the slower and softer flows of LRM materials has solved difficulties in such applications as flat packs, reed relays, pulse transformers, modules and MOS devices (Fig. 18-17).

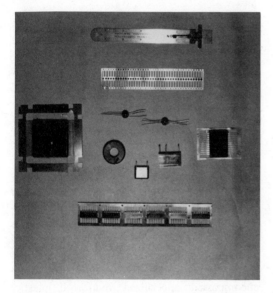

Fig. 18-17. Electrical and electronic molded devices, including pulse transformer, reed relay module, and coil.

19

REINFORCED PLASTICS

Reinforced plastics (RP) hold a special place in the industry. They are at one and the same time both unique materials unto themselves and part and parcel of virtually every other segment of the plastics industry.

In the early days of plastics, the so-called "reinforced plastics" (or "low-pressure laminates," as they were more commonly known) were easy to identify. The basic definition then, as it still is today, is simply a plastic reinforced with a fibrous or nonfibrous material. Thermoset resins, like polyesters and epoxies, and fibrous glass reinforcements dominated. Most processors did their own compounding of resin, catalyst, filler, pigment, etc.

What essentially characterized reinforced plastics in those days, however, was their ability to be molded into extremely large shapes (well beyond the capabilities of other processes available at that time) at little or no pressure and, in some instances, with much less heat than that required for other processes. Methods used included hand lay-up, bag molding, autoclave molding, and the like. Consequently, reinforced plastics went by the name of low-pressure laminates.

The term high-pressure laminates was reserved for melamine- or phenolic-impregnated

* Much of the data relating to fibrous glass reinforced plastics is adapted from information supplied by Owens-Corning Fiberglas. Data on design with glass reinforced plastics adapted from an Owens-Corning Fiberglas publication, "Fiberglas/plastic design guide."

papers or fabrics compressed under high heat and pressure to form either decorative laminates (under such trade names as Formica, Micarta, etc.) or industrial laminates for the electrical and other industries.

By the early sixties, however, the processing of reinforced plastics began to involve higher and higher pressures (e.g., matched metal molding). The name "low pressure laminates" was therefore dropped in favor of simply reinforced plastics. But even then, the name referred primarily to reinforced thermoset resins and encompassed specialized reinforced plastics molding methods.

By the mid-sixties, the industry started to notice a significant change. The volume of glass, asbestos, and other forms of reinforcements going into the thermoplastic resins (as opposed to the thermosets) began to increase. Today, the reinforced thermoplastics (RTP) have become an accepted part of the reinforced plastics business, although still much smaller than the reinforced thermosets. Their appearance, however, also opened the door to the use of many of the conventional processing techniques previously associated with the unreinforced thermoplastics—like injection molding, rotational molding, stamping, etc. More recently, some of these reinforced thermoplastic processing methods have been turned around once again and are now being considered for processing reinforced thermosets, as in the injection molding of polyester premixes or the rotational molding of glass-reinforced thermosets.

During the late sixties, the reinforced plastics industry discovered still another new branch—advanced composites. Although by definition, all reinforced plastics are composites (i.e., combinations of two different materials—resin and reinforcement—that act synergistically to form a new third material—reinforced plastics—with different properties than the original components), the term "advanced composites" or "high strength composites" or "structural composites" has taken on a special meaning. As generally used in the industry, the term is applied to stiffer, higher modulus combinations involving exotic reinforcements such as special glass, graphite, boron, or other high modulus fiber and resins like epoxy or some of the newer high heat-resistant plastics—polyimides, polyamideimide, polyquinoxalines, etc. Prime outlets for these materials so far have been in the aerospace and aircraft industries.

Reinforced Plastics

Reinforced plastics are composites in which a resin is combined with a reinforcing agent to improve one or more properties of the plastic matrix. The resin may be either thermosetting or thermoplastic.

Reinforced plastics provide the designer, fabricator, equipment manufacturer, and consumer with a flexibility in engineering properties to meet different environments and to create different shapes. They can be designed to provide practically any of a variety of characteristics. For this reason, they are used in practically all industries. Economical, efficient, and sophisticated parts are made, ranging from toys to re-entry insulation shields and miniature printed circuits.

Typical resins used in reinforced plastics include polyester, phenolic, epoxy, silicone, diallyl phthalate, alkyd, melamine-formaldehyde, polyamide, fluorocarbon, polycarbonate, acrylic, acetal, polypropylene, acrylonitrile-butadiene-styrene copolymer, and polyethylene. However, the thermosetting resins are predominantly used. The reinforcement is a strong inert material bound into the plastic to improve its strength, stiffness, or impact resistance. The reinforcing agent can be fibrous, powdered, spherical, crystalline, or whisker, and made of organic, inorganic, metallic, or ceramic material. Fibrous reinforcements are usually of glass, although asbestos, sisal, graphite, cotton, and other fibers are also used, in woven or nonwoven form. To be effective structurally, there must be a strong adhesive bond between the resin and the reinforcement.

Thermoset Resins, Catalyst, and Promoters

Polyester resins are by far the most widely used, largely because of their generally good properties, relatively easy handling, and relatively low cost. For special uses, however, other types are significant: epoxies for higher strength and greater chemical resistance, phenolics for higher strength and greater heat resistance, and silicones for their electrical properties and heat resistance. In addition, furanes, melamines, thermosetting acrylics, and diallyl phthalates (DAP) have some applications.

All these resins are thermosetting, and must be used in conjunction with a system of catalysts or curing agents. The type and amount strongly affect the properties, working life, and molding characteristics of the resin.

The polyesters and epoxies are most often mixed with a catalyst system just prior to molding; hence the choice of catalyst is particularly important with these resins. The most widely used catalyst for polyesters is benzoyl peroxide. Where heat is not available for curing, special catalyst-promoter systems can be used. With epoxies, an amine curing agent that reacts with the resin is most often used. However, there are literally dozens of types to choose from.

The polyester, epoxy, and thermosetting acrylic resins are usually thick liquids that become hard when cured. For this reason, they are most often combined with the reinforcement, by the molder, by dipping or pouring. There are available, however, preimpregnated reinforcements (i.e., prepregs) for the molder who wants to keep his operations as uncomplicated as possible.

By contrast, the phenolics, silicones, furanes, melamines, and DAP's are solid or semisolid before molding and hence must be used in the form of preimpregnated reinforcement. The catalyst systems are already incorporated into the material. Most preimpregnated reinforcements contain proprietary formulations of resin.

As the plastics industry has grown over the years, a number of specialty polyester resins has been made available for the variety of applications that plastics are now being used. In addition to the general-purpose polyesters, there are flexible and semi-rigid grades, weather-resistant grades, chemical-resistant grades, and high heat distortion varieties. Others can be selected for such manufacturing characteristics as hot strength, low exotherm, extended pot life, air dry and thixotropic.

Two of the newest variations in this field are the low shrink and low profile polyester systems. Low shrink and low profile systems are, in essence, combinations of thermoset resins (e.g., polyesters) and thermoplastic resins (e.g., acrylic). The low shrinks contain up to 30% thermoplastic polymers by weight of total resin (when in liquid form, e.g., styrene solutions; in powder form, the thermoplastics are normally added up to 15% by weight); low profile systems contain from 30 to 50%. Low shrink gives the molder minimum surface waviness in the molded part, as low as 1 mil/in. mold shrinkage, and is easily pigmented. Low profile offers no surface waviness and it is possible to achieve a 0.5 to 0 mil/in. mold shrinkage, depending upon the amount of thermoplastic additive. Although early low profiles could not be pigmented, recent developments have seen several pigmentable systems become commercially available for use in limited applications.

Thermoplastic Resins

In the field of reinforced thermoplastics, virtually every type of thermoplastic material can be, and has been, reinforced and commercially molded. The more popular grades on the market today include nylon, polystyrene, styrene-acrylonitrile, polycarbonate, poly-

propylene, acetal, polyethylene, ABS, polyvinyl chloride, polysulfone, polyphenylene sulfide, and thermoplastic polyesters. In most instances, the addition of a reinforcing material will improve the tensile and flexural strength, temperature resistance, and impact resistance of the thermoplastic (Fig. 19-1).

Reinforced thermoplastics are generally made available to the processor in the form of injection molding pellets (into which the glass has been compounded) or concentrates (also a pellet, but containing a much higher percentage of reinforcement, and designed to be mixed with non-reinforced pellets). In injection molding, it is also possible for the processor to do his own compounding of chopped glass and thermoplastic powder. In rotational molding and casting techniques, the processor usually adds his own reinforcement. For stamping, glass-reinforced thermoplastic laminates are sold commercially.

Most recently, research has been proceeding into the feasability of structural foam molding

Fig. 19-1. Some typical injection molded glass-reinforced thermoplastic automotive applications: glass/nylon radiator fan, oil suction tube, thrust washer, valve stem oil seal; glass/polysulfone switch housings; glass/polypropylene air conditioning squirrel cage fan, sleeve bearings, smog filter fan. (*Courtesy Owens-Corning Fiberglas*)

of glass-reinforced thermoplastics and of reinforcing molded urethane foams.

Reinforcing Materials and Fillers

Reinforcements are materials used to increase the strength of plastics several-fold. Fillers are intended primarily to reduce costs, although fillers often give a slight improvement in strength. Fillers are generally used only when the content of reinforcement is low (50% or less).

Although many types of reinforcements are used with plastics, the glass fibers predominate. Fibrous glass reinforcements are available in many forms: woven fabrics, nonwoven matting, bulk-chopped and milled fibers, and unidirectional rovings and yarns.

Rovings are rope-like bundles of continuous untwisted strands for use in such processes as preform press molding, filament winding, spray-up, pultrusion, and centrifugal casting. They can also be converted into chopped strand mats or cut into short fibers for molding compounds.

Chopped strands are fibers ranging from $\frac{1}{8}$ to 2 inches and are cut from continuous strands. They can be used in wet slurry preforming or in premix and injection molding compounds. Chopped strands can also be laid down in a random pattern to create a reinforcing mat designed to provide nondirectional reinforcement (i.e., strength in many directions as contrasted to unidirectional forms which are continuous fibers, like roving, that provide strength in one direction). These mats are available in a variety of thicknesses, usually expressed in weight per square foot. In order to hold the fibers together, a resin binder is generally used, the type depending on the resin and molding process. In some cases, the mats are stitched or needled, instead of using the resin binder.

Another type of reinforcing mat is known as a continuous filament mat and is made from swirled strands of filament. These various types of mats provide essentially the same degree of reinforcement. The difference is in the handling characteristics. Needled mats are softer and more easily draped; chemically bonded mats are stiffer but stronger; continuous mats can be molded into more complex shapes without tearing.

Twisted yarns are generally woven into fabrics of varying thickness and with tight or loose weaves, depending upon the application. Most are balanced weaves (equal amounts of yarn in each direction), although some are unidirectional (more fibers running in one direction). Although costly, they offer a high degree of strength. Rovings can also be woven into a fabric that is less costly than the woven yarn fabrics, coarser, heavier, and easier to drape.

Milled fibers are short nodules of filamented glass produced by hammer milling continuous strands. They are used in premix compounds and as an anti-crazing filler reinforcement in casting resins.

Another distinction of glass reinforcements is that they can be pre-impregnated with resin. When yarns, rovings, etc., are sold already impregnated with resin, they are known as prepregs. A special category is sheet molding compounds (SMC) in which glass, resin, catalyst, etc., are supplied in sheet form. These are discussed in greater detail below.

In addition to glass fibers, glass microballoons or spheres are also available commercially for reinforcing purposes.

It should be noted here that although the rest of this chapter will be devoted to glass-reinforced plastics (since glass continues to dominate the field and accounts for, by far, the largest volume), there are a number of other reinforcing materials in growing use today. These will be discussed briefly below. In many cases, the molding and fabrication processes discussed later on for glass-reinforced plastics are applicable (depending on application, resin, etc.) for some of the reinforcements covered here.

Asbestos is used in the form of loose fiber, paper, yarn, felt, and cloth. In some applications, combinations of asbestos and glass fibers are used. The two largest uses of asbestos in plastics are with polyvinyl chloride in vinyl-asbestos tile and with polyesters and polypropylene. Shorter grades of asbestos fiber are also used as a filler in resins to give flow control during processing and to improve modulus and heat-distortion temperatures.

High-modulus graphite and carbon yarns are playing a more and more important role in reinforced plastics, both by themselves and in combination with glass fibers. These yarns demonstrate high tensile strength and add stiffness and rigidity to parts. As price has come down, they have also moved into the reinforced thermoplastics field for injection molded products like carbon-reinforced nylon tennis racket frames (Fig. 19-1a). One of the first large commercial applications for carbon reinforcements was a filament-wound carbon/epoxy golf club shaft.

Most natural and synthetic fibers do not have the strength required for a reinforced plastics part. However, when intermediate strengths are satisfactory, they can be used. In this category are nylon, rayon, cotton fabrics, and paper. Sisal fibers have also found use as a low-cost reinforcing material in premix molding compounds.

Surfacing and overlay mats are a type of semi-reinforcement used primarily to give a smoother surface to molded articles of reinforced plastics. These are thin veils of fine glass, nylon, rayon, or other fibers that are placed over the regular fabric, mat, or preform reinforcement.

Boron filaments, with outstanding tensile strengths, are usually used in the form of prepreg tapes and have been primarily evaluated for the aerospace and aircraft industry. Several other high modulus fibers have also been put on the market, but suppliers have not revealed their make-up.

Outside of the fiber reinforcement field, considerable interest has been generated in the use of mica platelets as a reinforcing agent for such thermoplastics as nylon, ABS, and styrene-acrylonitrile. The mica serves to increase stiffness, reduce shrinkage, and increase heat distortion temperatures. The largest application to date is in mica-reinforced nylon fender extensions for the automotive industry.

Most reinforcements are treated with sizes or finishes to promote maximum adhesion by the resins. Many sizes and finishes are available, often tied in directly with a given resin or a given processing technique. It is advisable in these instances to check directly with the reinforcement supplier.

Glass Reinforced Plastics

The remainder of this chapter will be devoted primarily to the processing of glass-reinforced plastics materials. Before proceeding, however, it would be well to note the reasons why reinforced plastics have achieved such popularity in such a short time: design flexibility, corrosion resistance, high strength, dimensional stability, parts consolidation (e.g., one molded part can replace an assembly of several metal pieces), light weight, low tooling costs, and low finishing costs.

As indicated, the addition of reinforcement can significantly improve the properties of a thermoplastic or thermoset (see Table 19-1). Mechanical properties, of course, are also directly affected by the glass fiber content, usually exhibiting a straight line relationship. Typical curves for the strength-to-glass content relationships for thermosets and thermoplastics are shown in Figs. 19-2 and 19-3. Thermosetting compounds are subject to considerable variation in content, processing, method and

Fig. 19-1a. One-piece tennis rackets are injection molded of graphite-reinforced nylon structural foam. (*Courtesy Como, a subsidiary of PPG Industries*)

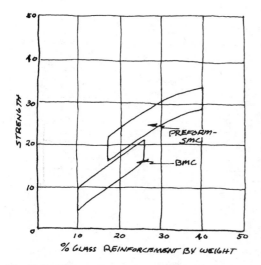

Fig. 19-2. Effect of glass content on mechanical properties of thermoset preform-SMC (sheet molding compound) and BMC (bulk molding compound). (*Courtesy Owens-Corning Fiberglas*)

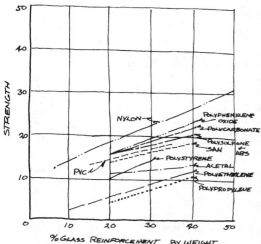

Fig. 19-3. Effect of glass content on mechanical properties of various thermoplastics. (*Courtesy Owens-Corning Fiberglas*)

molding technique. The strength-to-glass data are therefore expressed as a range of values for every glass loading.

As will be evident throughout this chapter, markets for reinforced plastics cut across the entire spectrum of American industry. Typical uses: transportation applications (including entire automobile bodies and truck cabs, as well as numerous interior and exterior components);

electrical applications (Fig. 19-4); marine applications—from small component parts up to 75-ft and larger yachts (Figs. 19-5a and 19-5b), even full-size minesweepers; construction applications (Fig. 19-6); sporting goods; luggage; business machine housings; pipe and storage; tanks; corrosion-resistant applications; aircraft and aerospace components; and a long list of other consumer and industrial goods.

MOLDING METHODS

As already indicated, reinforced plastics can be fabricated by a wide variety of techniques to accomplish many different purposes. Table 19-2 summarizes them briefly. More detailed descriptions, as well as design guidelines, will be found below.

Fig. 19-4. Non-conductive reinforced plastics ladders and platforms protect utility linemen constructing extra-high-voltage transmission lines. (*Courtesy Owens-Corning Fiberglass*)

Table 19-1. Reinforced plastics vs. un-reinforced plastics and metals*

Mechanical and Physical Properties

	Material	% Glass Fiber By Weight	Flexural Strength psi x 10³	Flexural Modulus psi x 10⁵	Tensile Strength at Yield psi x 10³	Tensile Modulus psi x 10⁵	Compressive Strength psi x 10³	Ultimate Tensile Elongation %	Impact Strength Izod Ft-lb/in of notch	Thermal Conductivity BTU/hr/Ft²/°F/in (K Value)	Specific Heat BTU/lb/°F
Glass Fiber Reinforced — Thermoset	Sheet Molding Compound (SMC)	15-30	18-30	14-18	8-18	16-25	8.0-16	.3-1.5	8.0-16	1.3-1.7	.30-.35
	Bulk Molding Compound (BMC)	15-35	10-20	14-18	4-10	16-25	3-6	.3-.5	3-6	1.3-1.7	.30-.35
	Preform/Mat (Compression Molded) Polyester	25-35	25-40	13-18	12-20	9-20	15-30	1-2	10-20	1.3-1.8	.30-.33
	Cold Press Molding-Polyester	20-30	22-37	13-19	12-20	—	—	1-2	9-12	1.3-1.8	.30-.33
	Spray-Up—Polyester	30-50	16-28	10-12	9-18	8-18	15-25	1.0-1.2	4-12	1.2-1.6	.31-.34
	Filament Wound—Epoxy	30-80	100-270	50-70	80-250	40-90	45-70	1.6-2.8	40-60	1.92-2.28	.23-.25
	Rod Stock—Polyester	40-80	100-180	40-60	60-180	40-60	30-70	1.6-2.5	45-60	1.92-2.28	.22-.25
	Molding Compound—Phenolic	5-25	18-24	30	7-17	26-29	14-35	0.25-0.6	1-6	1.1-2.0	.20-.30
Thermoplastic	Acetal	20-40	15-28	8-13	9-18	8-15	11-17	2	0.8-2.8	—	—
	Nylon	6-60	7-50	2-26	13-33	2-20	13-24	2-10	0.8-4.5	—	.30-.35
	Polycarbonate	20-40	17-30	7.5-25	12-25	7.5-17	14-24	2	1.5-3.5	—	—
	Polyethylene	10-40	7-12	2.1-6	6.5-11	4-9	4-8	1.5-3.5	1.2-4.0	—	—
	Polypropylene	20-40	7-11	3.5-8.2	5.5-10.5	4.5-9	6-8	1-3	1-4	—	—
	Polystyrene	20-35	10-17	8-12	10-15	8.4-12.1	13.5-19	1.0-1.4	0.4-2.5	—	.23-.35
	Polysulfone	20-40	21-27	8-15	13-20	15	21-26	2-3	1.3-2.5	—	—
	ABS (Acrylonitrile Butadiene Styrene)	20-40	23-26	9.2-18	11-16	8	15	3-3.4	1-2.4	—	—
	PVC (Polyvinyl Chloride)	15-35	20-25	9-16	14-18	10-18	13.4-16.8	2-4	0.8-1.6	—	—
	Polyphenylene Oxide (Modified)	20-40	17-31	8-15	15-22	9.5-15	18-20	1.7-5	1.6-2.2	—	—
	SAN (Styrene Acrylonitrile)	20-40	15-21	8.0-18	13-18	9-18.5	12-23	1.1-1.6	0.4-2.4	—	—
	Thermoplastic Polyester	20-35	19-29	8.7-15	14-19	13-15.5	16-18	1-5	1.0-2.7	—	—
Unreinforced Thermoplastics	Acetal		13-14	4	8-10	4-5	5	25-60	1.2-2.3	—	.35
	Nylon		5-18	2-4	9	2-5	7-10	29	1-4	—	.40
	Polycarbonate		13	3	9-11	3.5	12	100-130	16	—	.30
	Polyethylene (High Density)		—	0.7-2.6	4	0.6-1.5	2.7-3.6	30-900	0.6-2.0	—	—
	Polypropylene		5-8	1.2-2.7	3-5	1.2	3.7-8	200-700	0.5-20.0	—	—
	Polystyrene (High Impact)		3	4	3-5	3-4	4-9	15-30	0.7-3.6	—	—
	Polysulfone		1.5	4	10	3.6	14	50-100	1.3	—	—
	ABS (High Heat)		9	3	6.6	2.8-4.1	6.8-12.5	10-20	2.5	—	—
	PVC		13-16	4	6-7	4	—	—	2-20	—	—
	Polyphenylene Oxide (Modified)		15	4	10	3.7	15	50-100	1.5-1.9	—	—
	SAN		9.7-17.5	5	9-11	5	14-17	2.5-3.7	0.4	—	—
Metals	Gray Cast Iron		10	N.Av.	15-30	120	25	1	4-4.4	288-408	.13-.19
	Low Carbon Steel (Cold Rolled)		28	300	29-33	300	28	38-39	A.Ap.	260-460	.10-.11
	Stainless Steel		30-35	280	30-35	280	30	50-60	8.5-11.0	96-185	.12
	Aluminum, Wrought		20	100	6-27	100	N.Av.	30-40	N.Ap.	810-1620	.22-.23
	Aluminum, Die Cast		8-26	100	8-26	100	9	6-8	N.Ap.	610-1100	.22-.23
	Magnesium, Die Cast		14	65	8-30	65	10-14	4-6	3	288-960	.245-.25
	Zinc, Die Cast		N.Av.	N.Av.	10-25	N.Av.	N.Av.	10	4.3	764-792	.10
	Brass, Plain Yellow Wrought		14	150	14	150	N.Av.	60-65	N.Ap.	804	.09

S.E.—Self extinguishing   N.Av.—Not available   N.Ap.—Not applicable

* Prepared by Owens-Corning Fiberglas.

Hand Lay-up

The hand lay-up process is the oldest and simplest method for making glass fiber-reinforced parts. Male or female molds can be made of easily worked materials such as wood, plaster, or reinforced plastics.

In hand lay-ups, resin and glass fibers in the form of fabric, woven roving, or mat are simply placed in the mold manually. Entrapped air is then removed with squeegees or serated metal rollers. Successive layers of glass and resin can be added to build the part to the desired thickness (Fig. 19-7). If a smooth, colored surface is required, a pigmented material (called a *gel coat*) can be sprayed on the mold before lay-up. The mold side of the part upon completion thus becomes the finished outer surface.

Contact Molding

There are several variations of the hand lay-up method. In contact molding, the resin is in contact with the air and the wet lay-up normally hardens at room temperature. Heat may be used to reduce hardening time. A smoother exposed side may be achieved by wiping on a film such as cellophane or polyester, which can be removed after the hardening process.

Contact molding uses low-cost molds (Fig. 19-8), requires a minimum amount of equipment, and presents no size restriction. It offers maximum design flexibility and changes in part configuration can be readily made in the molds. This process is used to make boats, prototypes, pools, tanks, ducts, truck body components, housings, and corrugated and flat sheets.

Table 19-1. Reinforced plastics vs. un-reinforced plastics and metals (*Continued*)

Flammability in minutes or as noted	Rockwell Hardness	Dielectric Strength Volts/Mil	Specific Gravity	Density lb/in³	Heat Distortion °F at 264 psi	Continuous Heat Resistance °F	Thermal Coeff. of Expansion in/in/°Fx10⁻⁶	Chemical Resistance Key: Excellent: outstanding. Good: acceptable. Fair: test before using. Poor: not recommended.				
								Weak Acids	Strong Acids	Weak Alkalis	Strong Alkalis	Organic Solvents
Slow	H50-H112	300-450	1.7-2.1	.061-.075	400-500	300-400	8-12	G to E	F	F	P	G to E
to	H80-H112	300-450	1.8-2.1	.065-.075	400-450	300-400	8-12	G to E	F	F	P	G to E
Non-burning	H40-H105	300-600	1.5-1.7	.054-.061	350-400	150-400	10-18	G to E	F	F	P	G to E
Non-burning	H40-H105	300-600	1.5-1.7	.054-.061	350-400	150-400	10-18	G to E	F	F	P	G to E
Burn to S.E.	H40-H105	200-400	1.4-1.6	.050-.058	350-400	150-350	12-20	G to E	F	F	P	G to E
Slow to S.E.	M98-120	300-400	1.7-2.2	.061-.079	350-400	500	2-6	E	F	E	G	E
S.E.	H80-112	200-400	1.6-2.0	.058-.072	325-375	150-500	3-8	G to E	F	F	F	G to E
S.E.	M90-99	150-370	1.7-1.9	.061-.069	400-500	325-350	4.5-9	F	P	F	P	F
1.1-0	M78-M94	—	1.55-1.69	—	315-335	—	19-35	F	P	F	P	E
—	—	400-500	1.47-1.7	.049	300-500	300-400	11-21	G	P	E	F	G
Non-burn to S.E.	M75-M100	—	1.34-1.52	—	285-300	—	12-18	E	G[2]	G	F	P[4]
—	—	—	1.16-1.28	—	200-260	—	17-27	E	G[2]	E	E	G[7]
0.85-0.75	R95-R115	300-450	1.04-1.22	—	230-300	—	16-24	E	G[2]	E	E	G[7]
—	M70-M95	350-425	1.20-1.29	.045-.048	200-220	180-200	17-22	E	G[2]	G	G	P[4]
N.B. to S.E.	M85-M92	—	1.38-1.55	—	333-350	—	12-17	E	E	E	E	G
—	M75-M102	—	1.23-1.38	—	215-240	—	16-20	E	G[2]	E	E	P[5]
S.E.	M80-M88	—	1.45-1.62	—	155-165	—	12	E	G	E	E	P[5]
S.E.	M95	—	1.20-1.38	—	220-315	—	10-20	E	E	E	E	G[8]
—	M77-M103	—	1.22-1.40	—	210-230	—	16-21	G	G[3]	G	G	P[5]
Slow-burning	R118-M70	560-750	1.45-1.61	—	380-470	—	24-33	F	P	P	P	E
1.0	M78-M94	465-500	1.42	.052	230-255	185-250	45	F	P	F	P	E
Slow-burn to S.E.	R108-118	300-470	1.12-1.14	.039-.041	122-129	250-300	55-63	G	P	E	F	G
S.E.	M70	400-425	1.20	.043	265-290	275	39	E	G[2]	G	F	P[4]
—	—	—	0.95	—	93	—	6	E	G[2]	E	E	P[4]
—	R50-R110	—	0.9	—	125-140	—	38	E	G[2]	E	E	G[7]
1-2	M12-M45	500-700	1.05	.039	175-203	150-180	22-56	E	G[2]	G	G	P[4]
S.E.	M69	—	1.24	—	345	—	31	E	E	E	E	G
Slow-burn	R113	—	1.05	—	215-230	—	41-52	G	G[3]	G	G	P[5]
S.E.	D80	—	1.4	—	155-165	—	—	E	E	E	E	P[5]
5	M75	—	1.06	—	375	—	30	E	E	E	E	G[9]
0.8-1.2	M80	—	1.08	—	190-220	—	36	G	G[3]	G	G	P[5]
N.Ap.	B93	C	7.19	.26	N.Ap.	N.Av.	6	Rusted by water, oxygen, and salt solutions; poor acid resistance; good alkaline resistance.				
N.Ap.	B72	C	7.8	.28	N.Ap.	N.Av.	6-8	Rusted by water, oxygen, and salt solutions; poor acid resistance; good alkaline resistance.				
N.Ap.	B90	C	7.92	.29	N.Ap.	N.Av.	9-10	Poor acid resistance (especially hydrochloride and sulfuric); poor chloride solution resistance; good alkaline and organic resistance.				
N.Ap.	B1-B5	C	2.6-2.8	.10	N.Ap.	N.Av.	12-13	Poor acid resistance (especially hydrochloric and sulfuric); poor chloride salt resistance (must be chemically treated for appearance when exposed to weather).				
N.Ap.	E59	C	2.57-2.96	.09	N.Ap.	N.Av.	12-13	Poor acid resistance (especially hydrochloric and sulfuric); poor chloride salt resistance (must be chemically treated for appearance when exposed to weather).				
N.Ap.	E50-E59	C	1.81	.07	N.Ap.	N.Av.	14-16	Poor acid resistance (except hydrofluoric); good alkaline resistance, corrodes in presence of salt, salt spray, or industrial atmospheres.				
N.Ap.	B44	C	6.6	.24	N.Ap.	N.Av.	15-16	Poor resistance to strong acids and bases; poor resistance to steam; good resistance to atmosphere.				
N.Ap.	F58-F64	C	8.5	.31	N.Ap.	N.Av.	11-12	Good resistance to atmosphere; poor resistance to soft water and high-salinity water.				

Notes: [1]Acid resistance improves with use of corrosion resistance resins; [2]Attacked by oxidizing acid. [3]Disintegrates in sulfuric acid. [4]Soluble in aromatic and chlorinated hydrocarbons. [5]Soluble in ketones and esters, aromatic and chlorinated hydrocarbons. [6]Soluble in ketones and esters, softened in alcohol. [7]Below 176° F (80° C.) [8]Softens in some aromatic and chlorinated aliphatics. Resistance to alcohol. [9]Soluble or swells in some aromatic and chlorinated aliphatics. Resistance to alcohol.

Vacuum-bag Method

Hand lay-up can also be done with the vacuum-bag method in which cellophane or polyvinyl acetate film is placed over the lay-up, joints are sealed and a vacuum is created. Atmospheric pressure then eliminates voids and forces out entrapped air and excess resin (see Fig. 19-9).

Used to make aircraft radomes, electronic components, boats, boat components, and prototypes, the vacuum-bag method retains the advantages of contact molding while providing higher glass loading, a better unfinished side, and fewer voids. Though better adhesion in sandwich constructions is possible, it requires more labor, and quality often depends on the operator.

Pressure-bag Method

Another variation is the pressure-bag method in which a tailored bag, normally rubber sheeting, is placed against the lay-up (Fig. 19-9a). Air or steam pressure up to 50 psi is applied between a pressure plate and the bag. Cylindrical shapes can be made, undercuts are possible and cores and inserts can be used. This process is commonly used to make boats, safety helmets, containers, instrument cases, aircraft components, luggage, necked tanks, and large tubing.

Autoclave

Autoclave is a modification of the pressure-bag method. After lay-up the entire assembly is

Table 19-2. Processes for reinforced plastics*

Process	Reinforcement	General comments
Open molding	All types	Hand lay-up, spray-up, etc., where no pressure or relatively little pressure is applied during curing of resin.
Closed molding	All types	Matched die, etc., where pressure is applied during curing of the resin.
Hand lay-up or contact molding	Chopped strand, fabric, woven roving	Process requires little equipment. Can be extremely efficient. Large articles easily formed. Main disadvantage–only one good surface and variable thickness. Properties can be varied easily in different parts of molding. Unless good control is exercised, properties may be variable.
Spray-up	Chopped rovings	More complicated process utilizing spray guns that spray resin and chop rovings simultaneously. Consistency of product depends on operator skill. Large moldings easily formed. Only one good surface unless spray-up is put into matched die molds.
Vacuum bag	Mat, chopped strand, fabric, woven roving	Two good surfaces are obtained; quality of surface next to bag depends on how well bag fits mold. Maximum pressure 14 psi.
Pressure bag, autoclave, and hydroclave	Needled mat, chopped strand mat, chopped roving	As in vacuum bag except pressure is exerted on pressure bag up to 50 psi. Size limited by capacity of autoclaves.
Cold pressure molding	Needled mat, chopped strand mat, continuous strand mat	Rapid exotherm cures moldings on a relatively rapid cycle. Plastics or concrete tooling used.
Centrifugal casting	Needled mat, chopped strand mat, chopped rovings	Used for round objects, e.g., pipe. Reinforcement positioned in mold and is rotated; resin distributed through pipe impregnates reinforcement by centrifugal action.
Encapsulation	Milled fibers, chopped strands, etc.	Mixed compound is poured into open molds to surround and envelope components. Cure may be at room temperature with heated postcure.
Matched die molding	Mat, fabric, preform, chopped strand, etc.	Compound is put in two-part mold, and cured under pressure and heat. Process can be highly automated for mass production.
Preformed matched die molding	Chopped strand mat, needled mat, fabric	Chopped glass strands and resin binder are formed into a blanket on a rotating screen shaped like the part to be made. Binders are applied as the glass strands fall onto the screen. The resulting preform is then oven baked to make it easy to handle and prevent washout of fibers during molding. Final part is made by fitting the cured preform into matched metal dies, adding catalyzed resin, and press curing. Only the size of commercially available equipment limits size of the part.
Bulk molding	Roving, yarn, chopped strand	Resin, reinforcement, filler, and pigment are blended together in either a putty-like or granular mix. The compound is then fed in premeasured amounts into the cavity of matched metal die molds. Principal advantages are that the process produces complex parts with sections of varying wall thickness along with molded-in ribs, inserts, bosses, or threads. Also properties can be varied widely by varying the amount and type of material. Parts can be provided with molded-in color, and postmold shrinkage is low.

* Various combinations of these individual processes can also be used.

By D. V. Rosato. Reprinted from the 1972-1973 *Modern Plastics Encyclopedia*, with permission of the publishers, McGraw-Hill, Inc.

Table 19-2. Processes for reinforced plastics (*Continued*)

Process	Reinforcement	General comments
Sheet molding	Roving, yarn, chopped strand mat	The compound generally consists of a pre-impregnated glass mat or chopped strand glass in sheet form ready for matched die molding. The glass is impregnated with a resin, usually a low profile (smooth) polyester. Sheet molding using layers of glass fabric (swirl type) with thermoplastics is now also important. Because the materials are oriented and ready to use, the process is ideal for mechanized production and high outputs. The process combines the best advantages of mat molding and premix molding, and may eliminate the need for preforms in many cases.
Continuous pultrusion	Roving, needled mat, chopped strand mat, fabric, woven roving	Used for continuous production of shapes (rod, tube and angle, etc.). Generally gives high strength along length of product. High output. More recent developments use curing by infrared or irradiation rather than the usual heating elements.
Continuous laminating	Chopped strand mat, fabric, chopped roving	Used for production of construction panels, e.g., translucent corrugated sheet curtain walling. High production, uniform thickness.
Injection molding	Chopped strands	Parts can be made in standard plastics injection molding machines of the plunger or screw type. The process is associated with high production runs, low labor costs, high reproducibility molding of complex detail, and the ability to make delicate, precision parts.
Filament winding	Roving, yarn, tape	Process principally noted for production of high strength-to-weight ratio parts. However, parts are generally restricted to surfaces of revolution such as pipe, pressure bottles, rocket cases, etc.
Cold forming	Roving, yarn, chopped strand mat, spiral woven fabric	Fast cold-forming techniques such as are used for metal stamping can be used with reinforced thermosets or thermoplastics (SMC). Thermal curing is generally used after stamping thermosets. Infrared heating is generally used before stamping thermoplastics.

placed in a steam autoclave at 50 to 100 psi. The additional pressure achieves higher glass loadings and improved air removal required for certain types of boats, aircraft parts, and components for large electronic equipment. (See Fig. 19-10).

Spray-up

In the spray-up process, chopped glass fiber roving (chopped strand) and catalyzed resin are deposited simultaneously in a mold from special spraying equipment (Fig. 19-11). The roving is fed through a chopper and inpinges into a resin stream; the resin mix precoats the strands and the merged spray is directed onto the mold by the operator. The glass fiber-resin mix is rolled by hand to remove air, lay down the fibers, and smooth the surface. The part is usually air cured, but it may be heated to accelerate curing.

Like hand lay-up, spray-up uses low cost molds usually made of wood, plaster, reinforced plastics, or sheet metal. An important advantage of spray-up is that the equipment is portable, making possible on-site fabrication. Another benefit is that spray-up uses roving that is the lowest-cost form of glass fiber reinforcement. Also, complex shapes with reverse curves can be formed with minimum waste and less labor than with hand lay-up.

The spray-up process is widely used for

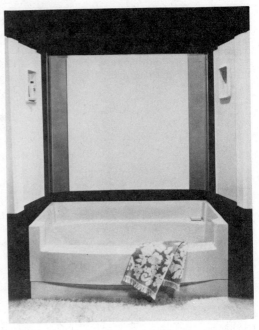

Fig. 19-5a. Mold for 75-ft reinforced plastics yacht shown in Fig. 19-5b.

Fig. 19-6. Modular tub-shower molded of reinforced plastics consists of four components: tub, two end-walls, and a back wall section. (*Courtesy Owens-Corning Fiberglas*)

Fig. 19-5b. Over the years, reinforced plastics have moved from small boats to the construction of large commercial and industrial (fishing boats, minesweepers, etc.) vessels. (*Courtesy Owens-Corning Fiberglas*)

Fig. 19-7. Applying resin to fibrous glass that has been manually laid up in the mold. (*Courtesy Owens-Corning Fiberglas*)

making boats, display signs, prototypes, tank linings, large integral moldings, truck roofs, and roofing.

The trend in spray-up has been toward more and more automation. It is possible, for example, to use contour-controlled programmable manipulators or robots for spray-up. In this type of set-up, an operator first guides the arm of the unit through the required motion patterns to spray up a particular product. These motions are recorded simultaneously on magnetic tapes in a control console. The unit will then repeat the program motions and sequences automatically over and over again (at faster or slower work speeds). The technique has been used for spraying up small boats which varied only 3% in weight from hull to hull.

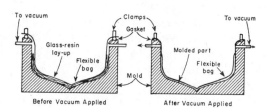

Fig. 19-9. Vacuum-bag molding.

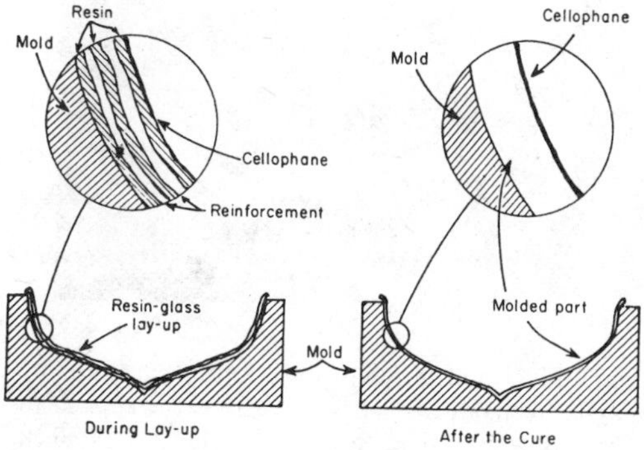

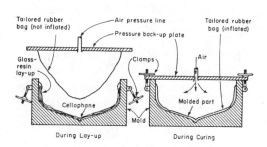

Fig. 19-9a. Pressure-bag molding.

Fig. 19-8. Contact molding.

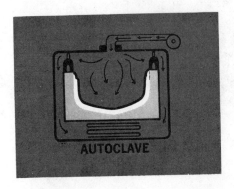

Fig. 19-10. Autoclave molding. (*Sketches and illustrations, courtesy Owens-Corning Fiberglas*)

Filament Winding

A processing technique that combines continuous filamental reinforcement with a thermosetting resin binder, filament winding produces structures having exceptional strength and high specific efficiency. Winding shapes are mostly confined to axial bodies of revolution, but 6-sided, box-like structures have been successfully manufactured. Simple cylinders are the optimum form for filament winding, but conical shapes, isotensoids, cups, spheres and ogives—some of which demand rather elaborate winding-pattern geometry—are regularly produced by this method to take advantage of the material's superior strength and, in most cases, reduce fabrication costs.

In its simplest explanation, filament winding operates as follows: Rovings or single strands of glass are fed through a bath of resin and wound onto a suitably designed mandrel. Preimpregnated rovings can also be used. Special lathes lay down the continuous glass fibers in a predetermined pattern to give maximum strength in the directions required. When the layers have been applied, the wound mandrel is cured at room temperature or in an oven. (See Fig. 19-12).

The superior strength of filament-wound products is based on the fact that the reinforcing filament is applied under carefully applied tension; thus, each strand of the reinforcing bears its equal share of stress under load.

Substantial gains have been made in recent years in the areas of filament-wound pipe,

storage tanks (Fig. 19-13), and pressure vessels. Filament-wound pipe installations made 12 to 14 years ago in corrosive environments show no signs of deterioration, and it appears that they will last at least another 12 or 14 years, or for even longer periods.

The aerospace industry continues to use filament winding wherever it is practical to do so. Large rocket motor cases, interstage shrouds, high-pressure gas bottles, etc., have

Fig. 19-11. Molding reinforced plastics furniture by spray-up technique. First, white gel coat is applied to the mold (top). Then, polyester resin and chopped fibrous glass roving are sprayed simultaneously into the mold (bottom).

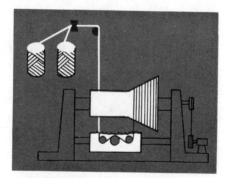

Fig. 19-12. Filament winding.

become standardized equipment for military rockets.

Matched Metal Die Molding

Over the years, this technique has developed into the primary method for manufacturing high-volume glass-reinforced plastic components and assemblies.

Basically, it is a compression molding process that uses a temperature-pressure-time cycle to polymerize a thermosetting resin combined with glass reinforcement. The matched metal dies are heated (250 to 325°F) to initiate the exothermic reaction. Hydraulic pressures

applied to the mold may vary from 250 to 1200 psi, depending on the viscosity of the resin, the glass content, and the complexity of the part being molded. Time is dependent on molded part thickness and the configuration of the part. (Fig. 19-14).

While the molding technique is common to the various systems in use today, there are major differences relating to the way in which the reinforced plastic material is fed into the molding press. The three forms available today include:

(1) *The Wet System.* This was one of the first techniques developed for reinforced plastics matched metal molding. It involves the use of a preform shaped from fibrous glass (with or without resin) to the approximate shape of the part to be made, and placed in the mold. Instead of a preform, it is also possible to cut the shape directly from reinforcing mat.

(2) *Bulk Molding Compounds or Premix.* This was the second system to be developed and involves the mixing of the glass (usually chopped strand), resin, filler, catalyst, etc., into a putty-like mixture that can then be placed in the mold. The mixture can be either extruded into a rope-like form for easy handling or used as is in bulk form.

Fig. 19-13. Filament wound underground storage tank for service stations is produced in 15,000 and 20,000-gal. capacities.

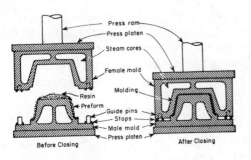

Fig. 19-14. Matched-die molding. (*Courtesy SPI Engineering Handbook*)

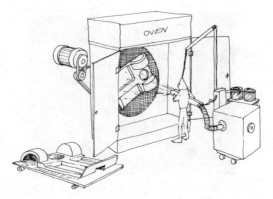

Fig. 19-15. Directed fiber process.

(3) *Sheet Molding Compound.* This is the newest of the available techniques and is fast developing into one of the more popular ones for large-volume production. In this system, glass, resin, and other ingredients are precombined into a sheet form that can easily be laid into the mold, either manually or by automatic mechanical handling techniques.

Each process is described in greater detail below; recommended design information on parts made by each of the techniques begins on p. 484.

Wet Process

Glass fiber reinforcement is prepared in the shape of the part to be molded. Liquid resin mix is then added to the reinforcement and the combination is molded in heated matched-metal molds under pressure.

The reinforcement can be shaped by "preforming" chopped fibers or by cutting reinforcing mat into the proper shape. Basically, however, there are two methods of making a preform; mat, however, can also be used for certain parts.

Directed Fiber. Directed fiber preforming is used for large complex shapes such as an automobile underbody or a fender.

The quality of the preform made by this technique depends on the skill of the machine operator.

The preform is made by depositing chopped glass fibers and resin binder onto a suction screen. The binder resin is cured by forcing hot air through the preform. The preform is then removed from the screen and placed in the mold. (See Fig. 19-15).

Plenum Chamber. Plenum chamber preforming is generally restricted to smaller parts of relatively simple shape. Operator skill is less involved in this preforming operation, so there is less chance of human error.

The preforming is accomplished by depositing chopped fibers and resin binders onto a rotating screen. The screen is the general shape of the part to be molded. The whole operation is done in a closed plenum chamber.

As in the directed fiber, the binder is cured to give the preform "handling body". The preform is then removed from the plenum chamber (Fig. 19-16) and placed in the mold.

Mat Molding. When a design calls for a simple shape of relatively constant thickness with only shallow draw, glass fiber reinforcing mat, cut to shape, can be used instead of a preform.

Fig. 19-16. Plenum chamber process.

Reinforcing mat is made of chopped glass strands or continuous strands held in mat form by a resin binder. The mat must be cut oversize to allow for draw and part contour. Resin can be added to the mat in the mold or before placing the mat in the mold. (See Fig. 19-17).

Bulk Molding Compound (BMC)

Bulk molding compound is a putty-like mixture of thermosetting resin, filler, and chopped glass reinforcements. After mixing, BMC is often extruded into "logs" or "rope" for easier handling and mold charge preparation.

BMC is commonly molded by three matched-metal mold processes:

(1) *Compression Molding.* Compression molding is the most popular molding process for BMC (Fig. 19-18). BMC is charged to a telescoping mold. Pressire is applied to the material by one mold half entering the other. The telescoping mold seals the edges before pressure is applied to the charge. The mold is sealed at the parting line by a "shear edge" with a small clearance (under 0.008 in.).

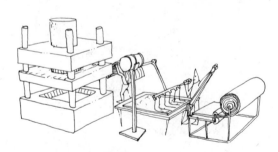

Fig. 19-17. Mat molding process.

(2) *Transfer Molding.* BMC is charged into a "pot" connected to the mold. The mold is closed before pressure is applied to the BMC by a ram, forcing the BMC from the "pot" to the mold cavity. Transfer molding is generally used when precise part thickness is required, or when vertical flash lines cannot be tolerated, or when complex molds require pressure to fit intricate detail.

(3) *Injection Molding.* BMC is charged into the cylinder of a screw-type injection molding machine, similar to the one used for thermoplastics (Fig. 19-19). See also Chapter 9 on Injection Molding Thermosets.

Sheet Molding Compound (SMC)

Sheet molding compound (SMC) is basically a polyester resin mixture reinforced with glass strands and formed into a sheet that can be handled easily, cut to shape, and charged into a compression mold. The mixing action requires that the viscosity of the resin be low, but the resin must reach a high viscosity to attain a tack-free, handleable form.

Although SMCs originally were introduced to the market as a prepreg-type of material sold by suppliers to processors, the stronger trend is now toward making SMC in-house by the processor himself. Some SMCs are still sold on a merchant basis.

In terms of an overall concept, a sheet molding compound is made by mixing resin and other ingredients into a uniform paste. The paste is doctored onto two carrier films (polyethylene), each of which will form the outer

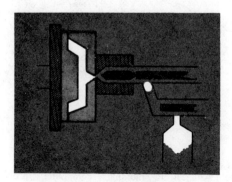

Fig. 19-18. Premix/molding compound.

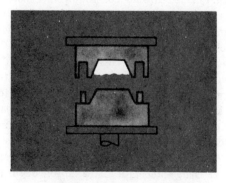

Fig. 19-19. Injection molding.

layers of the SMC (with the film as the exterior surface). Before the two layers come together, chopped glass is added in between. The assembly then is carried on a conveyor under a series of compacting rolls and onto a take-up roll. After winding, it is allowed to thicken, then blanked to shape.

In a step-by-step operation, SMC is made as follows (Fig. 19-20):

(1) Resin Mixing. Polyester resin is mixed with a thermoplastic resin for shrink control, inorganic fillers for modulus control, internal release agents for improved processing characteristics, and chemical thickeners.

(2) Resin Metering. The resin mixture is fed to two carrier films (upper and lower) in a uniform layer of desired thickness.

(3) Fiber glass roving is chopped ($\frac{1}{4}$ to 3 in. lengths possible) and gravity-fed onto the lower carrier sheet (already covered with resin).

(4) Upper carrier film (resin-covered) is applied to the lower film, placing the chopped strands between two layers of resins.

(5) Compaction rollers force complete wetting of glass, and removal of excess air from the resin-glass mixture.

(6) Heated compaction rolls aid the wetting operation by lowering the resin viscosity. Resin thickening process is also started by this heating action.

(7) Roll-up. Sheet is rolled and packaged for aging and delivery to press.

(8) Maturation. Final thickening is a time-temperature dependent reaction usually requiring several hours, or days, before the compound is ready to mold. Recent activity has been directed toward controlling this maturation period via the use of special additives that permit the processor to tailor the SMC chemical thickening performance to this requirements. In other words, the processor can achieve the viscosity levels he requires very quickly (in minutes versus hours) and maintain them for extended storage periods. The hoped-for-goal: an in-line operation from the SMC-making machine direct to the press.

(9) Molding. Matured SMC rolls are moved to the charge-preparation area, unrolled, slit to width, cut to length, stripped of carrier films, and fed to the press. High-tonnage plastics compression molding presses are most commonly used. Some work has also been done in converting mechanically driven metalworking presses to molding SMC (also for forming glass-reinforced thermoplastic laminate sheets). Alterations consist mainly in obtaining a controlled release of the press strain energy following the end of the cure cycle and having an adjustable slow speed full capacity drive.

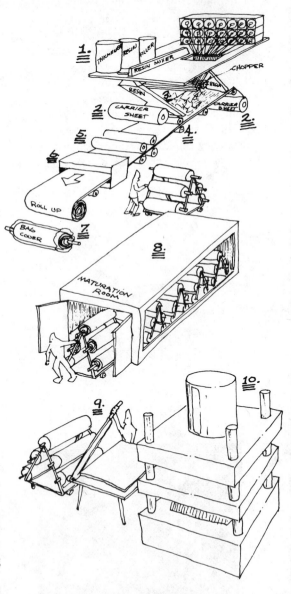

Fig. 19-20. Sheet molding compound process. Numbers relate to text description, at left.

Activity has also been noted in attempting to automate the SMC molding line. One idea is to use a series of presses fed by a number of molds on a programmed cycle. The molds are moved to the presses, loaded with SMC, molded, moved away for cure, then demolded on a continuous conveyor. Another variation uses a single press fed by a 12-mold carriage rack-track set-up. The molds are loaded, conveyed into the press, conveyed through the curing cycle, and into the unloading station. Thus, while parts are curing, other parts would be molded, to keep the press in continuous operation.

There are, of course, variations in production methods of making SMC, but the above is a typical one.

SMC markets already span a wide range—from air conditioner housings to furniture to construction applications. One of the major areas of use in in such automotive applications as front end and front valance panels, one-piece grille opening panel and fender extensions, hoods, and rear wheel opening covers (Fig. 19-21).

In the mid-70s, a second generation of SMCs were introduced requiring reduced molding pressures – 200 tons as compared to the conventional 700 tons of pressure. Also made commercial were new high-density SMCs with higher glass loadings (70% glass, 30% resin) for higher strength at lighter weight.

Pultrusion

Reinforced plastic profiles in continuous lengths are produced by pultrusion, a method which involves pulling the raw materials through shaping dies and curing operations. (See Fig. 19-22).

Some forms of pultrusion date back 20 years to the production of fishing-rod blanks. Today, pultrusion is an established world-wide operation and designers can specify off-the-shelf products ranging from $\frac{1}{32}$ in. rods to 8 in. I-beams.

Markets for pultruded products fall principally in the electrical and sporting-goods field and areas where corrosion resistance is a requisite.

Resins. Polyesters account for 90% of pultrusion binders and epoxies for the balance. Liquid-resin systems are formulated with low viscosity (around 2000 cps) for rapid fiber wet-out, catalyst balance to produce a strong gel which resists coating the die surface, internal release agents to promote mold release, and reasonable (8 hr) pot life.

Reinforcements. Except for flat laminates

Fig. 19-21. One-piece front end for truck is molded of SMC (sheet molding compound) and weighs only 70 lb.

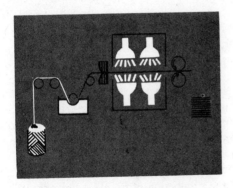

Fig. 19-22. Continuous pultrusion.

where paper is also used, fibrous glass in the form of roving, mat, and cloth is the principal reinforcement used in pultrusion. Parallel-oriented roving is the major ingredient of rod stock. It is also designed into high-stressed areas of many profiles and is braided, spirally wrapped, and hoop wound as further support for solid rods and hollow tubes.

Lateral strength is achieved at lowest cost with mats which are available as bonded surfacing, bonded chopped and combinations needed to parallel roving and cloth.

Equipment. Although at least one equipment manufacturer offers a complete pultrusion machine, most manufacturers build their own lines. These generally include:

Let off devices. Simple, lightweight guides and roll stands are adequate for pultrusion because of the low speeds involved.

Impregnation. Easily cleaned, open-top tanks from 3 to 6 ft long serve for wet-out of the reinforcement.

Resin Control. Orifices are most often used to meter off most of the resin and deliver only a slight excess to the die.

Preforming. Gradual shaping of the uncured wet-out material prior to entering the die is done with spiders, rings, and tubes to preserve optimum fiber orientation.

Dies. Steel dies 18 to 60 in. long are most common except for use of TFE-fluorocarbon linings where R.F. cure is employed. Cold junctions prevent cured resin build-up on metal dies which are, of course, equipped for heating. R.F. generators of 6 to 10 K.W. at 40 to 80 MHz will cure approximately 150 lb/hr of

glass reinforced polyester pultrusions. Epoxies require microwave frequencies.

Pulling. Although small sections can be pulled with simple belt or wheel pullers, large sechions require caterpillar types capable of exerting up to 1,000,000 lb pull.

Cut off. Diamond-faced cut-off saw systems are used.

Processes. Pultrusion processes can be intermittent (stop to cure) or continuous.

Intermittent methods include such techniques as indexing through a split matched-metal mold in a press, through a closed die as for continuous pull, and through ovens when film is wrapped and cut to length.

Some continuous methods pull through multiple rings (simple profiles), sectional dies (with interrupted cure), closed dies (preferred) and, in one process, a liquid melt (for heat and pressure). Other continuous methods rely on supporting the laminate with a plastic film (removable after cure) or an external wrap of braid or spiral fiber. Such supported material can be oven cured with simple shape assists or often just guides, as in the case when pultruding rods and tubes.

Cure rates are 1 to 5 ft/min. and are sensitive to gel control with a requirement to confine peak exotherm results in cracks and voids, whereas too low an exotherm produces an undercured product which often delaminates. Heated mold temperatures are in the range of 220 to 250°F.

Radio frequency cure heats the material and not the die, but exotherm control is still necessary. Oven cure rates are faster—more than 25 ft/min. for thin-wall profiles—and use wider temperature ranges of 180 to 350°F.

Hollow profiles produced with stationary mandrels supported at one end only are best produced vertically, although concentricity is still subject to some variation because of the lateral movement of materials.

Cold Press Molding

This technique is recommended for manufacturing intermediate volume products (200 to 8000 units), using low pressure (around 50 psi), room temperature cure, and fairly low-cost molds. In cold molding, fibrous glass and room-temperature curing polyester resin are placed between matching male and female plastic molds for forming and curing. The resulting part has two smooth, dimensionally accurate surfaces. Color may be molded-in by adding pigment to the resin system. (See Fig. 19-23).

Continuous Laminating

Fibrous glass fabric or mat is passed through a resin dip and brought together between cellophane covering sheets. The lay-up is then passed through heat and the resin is cured. Laminate thickness and resin content are controlled by squeeze rolls as the plies are brought together. This technique is used for making construction and glazing panels (Fig. 19-24 and Fig. 19-25).

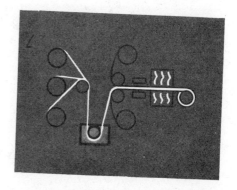

Fig. 19-24. Continuous laminating.

Fig. 19-25. Glass/polyester panels come off the production line at rates up to 50 ft. per minute. (*Courtesy Corrulux Corp.*)

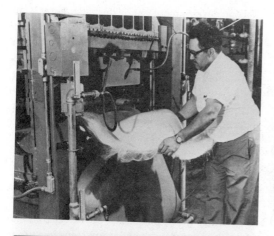

Fig. 19-23. Cold molding process (top) is used to produce reinforced plastics chair (bottom).

Centrifugal Casting

Round objects such as pipe and containers can be formed by centrifugal casting. Chopped strand mat is positioned inside a hollow mandrel. The assembly is then heated and rotated. Resin mix is distributed uniformly throughout the glass reinforcement. Centrifugal action forces glass and resin against the walls of the rotating mandrel prior to and during the cure. To accelerate cure, hot air is passed through the mandrel. (See Fig. 19-26).

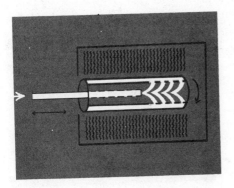

Fig. 19-26. Centrifugal casting.

A new technique for producing large-volume storage tanks (up to 3000 cu ft) combines centrifugal casting and spray-up. In operation, the centrifugal mold is fed simultaneously with resin and reinforcement (i.e., chopped glass fibers) through a spray-type depositor head fitted on a telescope jib and operating inside the rotating mold.

Injection Molding Reinforced Thermoplastics

The injection molding technique, as discussed in Chapter 4, is as applicable to reinforced thermoplastics as it is to the unreinforced grades. Basically, glass fibers and resin are introduced into a heating chamber where the mixture softens and is injected into a matched-metal mold cavity. The part then cools and solidifies to the desired shape. Finally, it is ejected from the mold and trimmed (Fig. 19-19).

At the present time, there are three ways in which reinforced thermoplastics can be fed into the machine:

Compounds. Commercially available reinforced thermoplastic compounds are most commonly used by injection molders. These compounds consist of pellets about 1/8 in. in diameter by 1/8 in. or more in length, most frequently containing between 20 and 40% glass reinforcement by weight. The pellet system may also contain appropriate additives such as stabilizers, pigments, and lubricants. A selected compound can be varied to optimize strength, impact resistance, surface finish, or the needed properties required for a specific application.

Concentrates (Compound). These are supplied to the injection molder in the form of pellets normally containing 60 to 80% glass combined with the selected thermoplastics. The concentrate is mixed with non-reinforced thermoplastic pellets to achieve the desired glass to resin ratio in the molded part.

In-Plant Compounding (Direct Molding). This process involves purchase of chopped strand and preblending of the glass with thermoplastic powder (in-plant). The blend is then fed into the screw injection machine. Special equipment is required because simple pellets are not being handled.

Injection Molding Reinforced Thermosets

Recent developments indicate much activity in the injection molding of reinforced thermosets (e.g., polyester bulk molding compounds). In this case, the injection screw or plunger is maintained at room temperature and the mold is heated to 300 to 325°F. The thermosetting material cures and hardens after about one minute under heat and pressure in the mold.

A typical system includes a continuous metering-type feeder for resin and chopped glass, a continuous mixer, a premix transfer device (to serve as a buffer tank or accumulator to permit the integration of a continuous mixing operation with an intermittent injection molding operation), and the reciprocating screw injection molding machine. (See also Chapter 9 on Injection Molding Thermosets.)

Rotational Molding

A thermoplastic powder is used in this method. Reinforcements can be added in the form of chopped strands at any time during the cycle. The thermoplastic powder and chopped strands are charged into a hollow heated mold, capable of being rotated in one or two planes. After the material is completely fused, the mold is chilled so that the product can be stripped out. (See also Chapter 14 on Rotational Molding.)

Cold Stamping

A reinforced thermoplastic sheet is pre-heated and placed into matching metal molds which are closed rapidly to form the part. The molds are typically kept at (or slightly above) room temperature. It takes 10 to 12 seconds to complete the molding cycle and 3 to 4 seconds dwell time to sufficiently chill the part prior to removal. (See also Chapter 12 on Thermoforming Plastic Film and Sheet.)

In a typical operation, the glass/thermoplastic laminate is heated above its flow temperature. This hot laminate is then transferred into a relatively cold mold, after which the material is molded to shape by a press wired to stop on bottom, time out, and then restart. The operation uses mechanical stamping presses, although hydraulic presses have been used satisfactorily.

The stroke rate of the press is in the same range as for metal stampings. Faster cycle times are possible, compared to injection molding, due to the contact of mold metal surface to plastic at all times, thereby allowing good heat transfer from the part. The fact that longer fiber glass reinforcements can be used also means higher strength/weight ratios compared to the shorter fibers in injection molding materials.

Form-and-Spray

A thermoplastic sheet (e.g., acrylic) is formed to shape in a regular thermoforming machine. It is then removed (it can also be left in the mold) and rigidized with a backing of reinforced plastics sprayed up on the inside surface of the part. It is also possible to combine molding and thermoforming to produce parts with excellent surface appearance, weatherability, and strengths. In this process, the skin is thermoformed, placed in a standard cold molding mold, and the resin/glass laminate placed on top of the skin. The cold molding then proceeds in the normal fashion.

Encapsulation

Milled glass fibers or short chopped strands are combined with catalyzed resin and poured into open molds. Fibers decrease shrinkage and crazing and increase useful temperature range. (See also Chapter 18 on Casting.)

Designing with Reinforced Plastics

The techniques for designing with reinforced plastics are many and varied—obviously depending on the type of resin and process involved. They differ in many ways from the design considerations when working with unreinforced materials. As a guide to the subject of design and an example of how the process can affect design concepts, you are referred to the series of illustrated charts beginning on p. 484. Design details are provided for reinforced plastics parts based on the wet process (Fig. 19-27), on bulk molding compounds (Fig. 19-28), on sheet molding compounds (Fig. 19-29), and on injection molding thermoplastics (Fig. 19-30). The guide was prepared by Owens-Corning Fiberglas.

Tooling

The uniqueness of reinforced plastics—in that, depending upon the resin used, forming can be done with heat and pressure, without heat and pressure, with heat alone, or with pressure alone—makes it impossible to generalize on the type of tooling to be used. For contact molding involving little or no pressure, plaster molds have been used. Reinforced plastics molds have also been used. On longer runs and as pressures increase, sprayed metal, aluminum, steel, etc., have been used. In the injection molding of reinforced thermoplastics, the same rules apply as in the molding of non-reinforced thermoplastics. The choice will depend on the job.

Finishing and Assembly

All normal cutting operations can be performed on reinforced plastics. Cutting is best done with an abrasive wheel. Where a bandsaw is necessary, a skiptooth blade is recommended. Lathe, shaper, and milling operations should be performed at the cutting speeds used for brass; carbide-tipped or diamond tools are necessary

WET PROCESS
Design Details

1. Corner Radii.

Radii should be generous to permit good glass-to-resin distribution in the part. 1/8" inside radius is generally considered to be the minimum.

2. Flash Lines or Shear.

Flash or shear should be vertical. Horizontal flash creates a thick edge that is difficult to deflash

If horizontal shear is necessary it is sometimes better to add material to be trimmed in secondary operations after molding.

3. Holes.

Holes and openings can be molded in if the hole area is greater than 5% of the total part area, or if the hole size is 4" x 4" or over. Molded-in holes must be located on a surface perpendicular to the press ram action. Spacing between holes should be 4" or more.

Holes should be drilled or punched when there are many small holes or holes are located in a section parallel to press action.

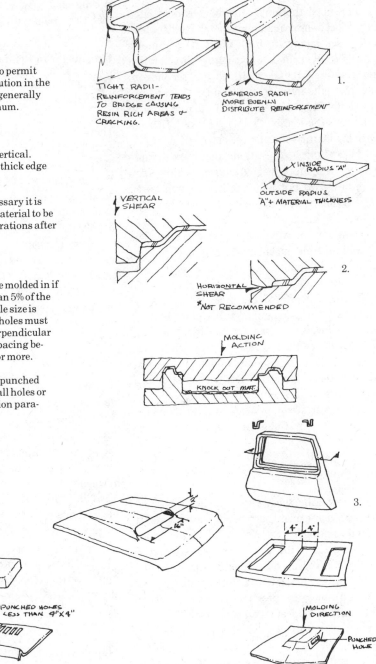

Fig. 19-27. Wet process design details. (*All illustrations for this RP design section, courtesy Owens-Corning Fiberglas*).

4. Joints.

When mechanical fasteners are used, the load applied by the fastener should be spread over as large an area as possible. Use of large washers is recommended. Screw heads or nuts should never be in direct contact with the FRP.

When using adhesive bonding the bond material should be used in .010" thicknesses if possible.

5. Bosses.

Bosses are not recommended in the wet process. However, it is possible to use BMC or SMC in conjunction with wet molding to form bosses.

6. Inserts.

Metal plate inserts are possible, but not recommended. Mold preparation is time consuming.

7. Wall Thickness Variation.

Variation in panel thickness should be kept to a minimum. A strip of mat can be added to the preform to achieve uniform glass distribution in thick areas, if part requires variation in thickness.

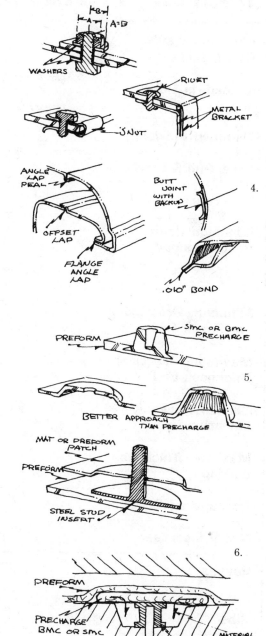

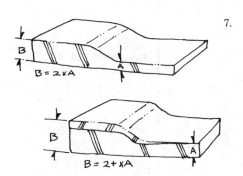

Fig. 19-27. Wet process design details. (*Continued*)

preform mat

·Minimum Inside Radius (in.)	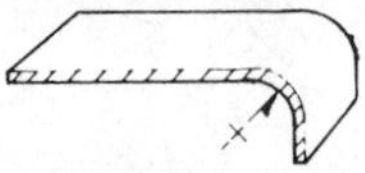	1/8"MINIMUM MORE RECOMMENDED
·Molded–In Holes	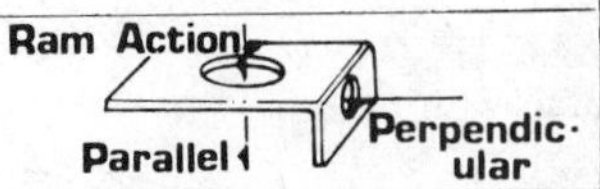	YES- PARALLEL TO RAM ACTION
·Trimmed In Mold		YES
·Core Pull & Slides	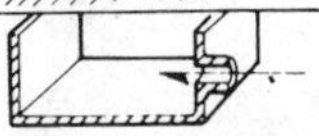	NO
·Undercuts	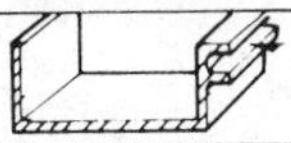	NO
·Minimum Draft Recommended	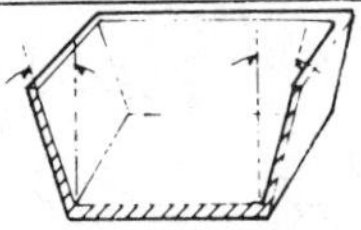	1/4"- 6" DEPTH – 1°-3° 6"+ DEPTH – 3°+ or as REQUIRED
·Minimum Practical Thickness (in.)	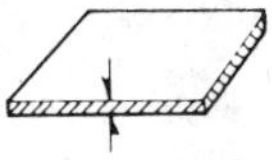	.030
·Maximum Practical Thickness (in.)	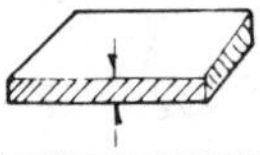	.250
·Normal Thickness Variation (in.)	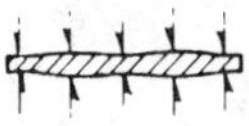	±.008
·Maximum Thickness Buildup - Heavy Buildup- Increased Cycle	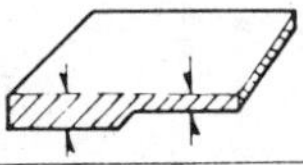	2 TO 1 MAXIMUM
·Corrugated Sections	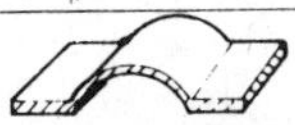	YES
·Metal Inserts		POSSIBLE BUT NOT RECOMMENDED
·Bosses	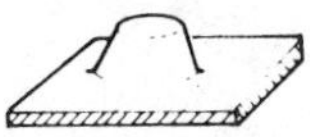	YES- PRECHARGE BMC or SMC
·Ribs	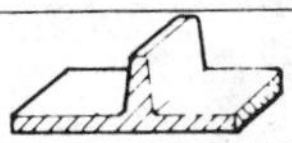	NOT RECOMMENDED
·Molded–In Labels		YES
·Raised Numbers		YES
·Finished Surfaces ·Reproduces Mold Surface	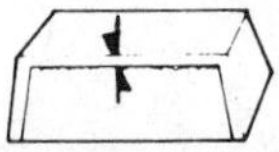	TWO

Fig. 19-27. Wet process design details. (*Continued*)

BULK MOLDING COMPOUND
Design Details

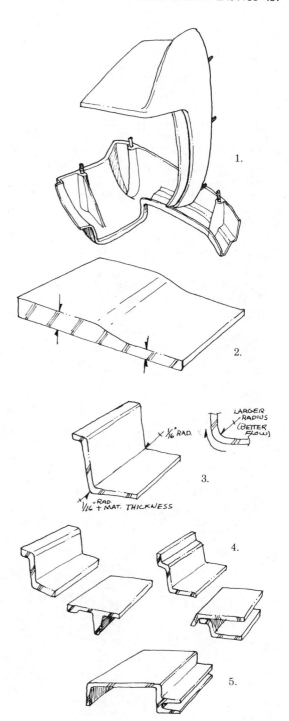

1. **Shape.**

   Virtually any shape that can be removed
   from the punch of a mold can molded in
   BMC.

2. **Wall Thickness.**

   It is generally felt that .050" is the
   minimum practical thickness and
   1.00" the maximum. Thin walls require
   shorter molding cycles.

3. **Corner Radii.**

   Minimum recommended inside radius is
   1/16" with outside maintained at uniform wall thickness. Rounded corners
   relieve stress concentrations.

4. **Flanges.**

   A variety of flange designs is possible.
   Offset flanges use less material and
   provide stiffness through design.

5. **Undercuts.**

   External undercuts are accomplished
   at the periphery of the part with
   slides or pulls.

Fig. 19-28. Designing with bulk molding compounds (BMC).

6. **Flash Lines.**

   Flash lines should be vertical. Horizontal flash lines are less effective in containing the material under pressure.

7. **Molded-In Holes.**

   Holes should be located 3x diameter from an edge of the part and 2x diameter from another hole.

   Blind hole design permits good material flow. Through holes can create knit lines when BMC flows around the core.

8. **Bosses.**

   Bosses are used to reinforce holes, to anchor inserts or studs, and to provide attachment points for assembly.

   Thickness should be limited to that of adjoining wall, or not more than 2x diameter of hole.

   Bosses should be located in corners where possible. Free standing bosses should have fillets for support and for good material flow.

9. **Inserts.**

   Inserts can be molded in place or driven in with a secondary operation.

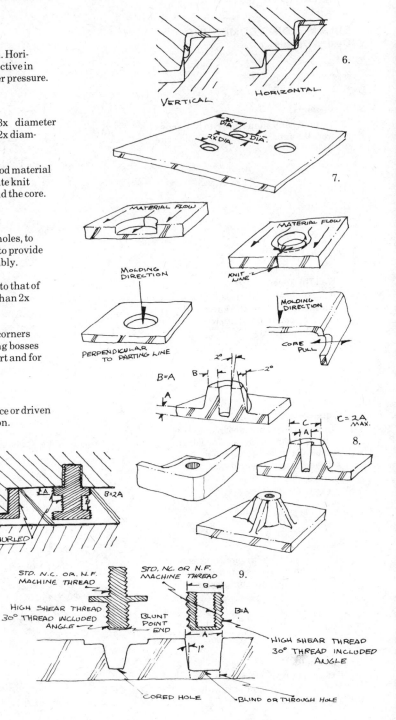

Fig. 19-28. Designing with bulk molding compounds. (*Continued*)

bulk molding compound

·Minimum Inside Radius (in.)	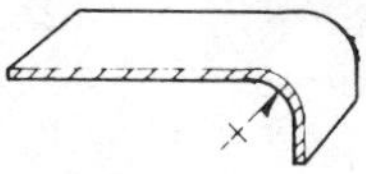	1/16" MINIMUM
·Molded–In Holes	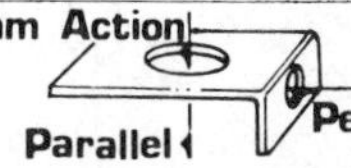 Ram Action / Parallel / Perpendicular	YES– PARALLEL OR PERPENDICULAR TO RAM ACTION
·Trimmed In Mold		YES
·Core Pull & Slides	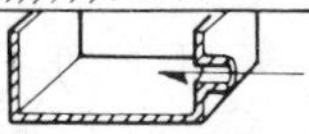	YES
·Undercuts	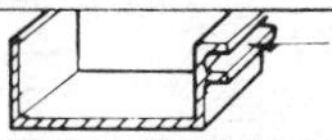	YES– WITH SLIDES
·Minimum Draft Recommended	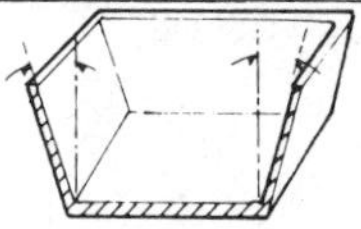	1/4"-6" DEPTH – 1°- 3° 6"+ DEPTH – 3°+ or as REQUIRED
·Minimum Practical Thickness (in.)	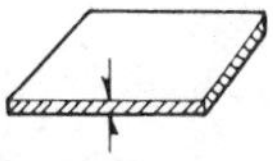	.060
·Maximum Practical Thickness (in.)	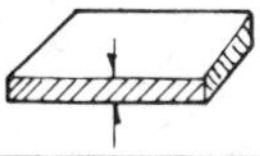	1
·Normal Thickness Variation (in.)	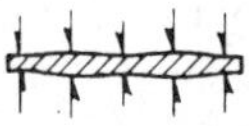	±.005
·Maximum Thickness Buildup - Heavy Buildup- Increased Cycle	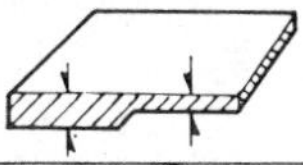	AS DESIRED
·Corrugated Sections	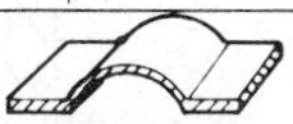	YES
·Metal Inserts		YES
·Bosses	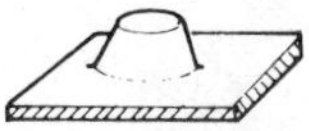	YES
·Ribs	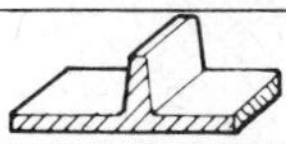	YES
·Molded–In Labels		YES
·Raised Numbers		YES
·Finished Surfaces ·Reproduces Mold Surface	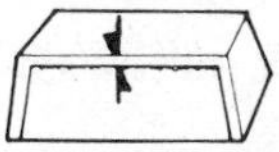	TWO

Fig. 19-28. Designing with bulk molding compounds. (*Continued*)

SHEET MOLDING COMPOUND
SPECIFIC DESIGN DETAILS:

1. **Draft.** Draft angles on all surfaces parallel to the direction of the platen movement must have a minimum draft angle of 1°. More is preferred. Draft angle should increase as depth of part increases.

2. **Pinch Off.** Pinch off design must be sufficent to insure a good seal and minimum flash. Two different designs are shown for basic resin systems.

3. **Knock-Outs.** Knock out areas must be designed to prevent material entrapment, which would create hydraulic action and keep the mold from closing.

4. **Slides.** Slides used to mold external undercuts should be located at the periphery of the part to be molded. Slides are mechanically or hydraulically actuated.

5. **Cores.** Cores can be readily molded into an SMC product. Cores should be located close to the periphery of the part to be molded. Cores are usually incorporated to remove excess material between bosses or to mold holes for mechanical fasteners. Core pulls are operated hydraulically or mechanically.

6. **Ejectors.** Ejector systems should be used whenever low-shrink SMC is being molded. This is particularily true of any part with molded-in ribs or bosses. This system may be actuated mechanically or hydraulically.

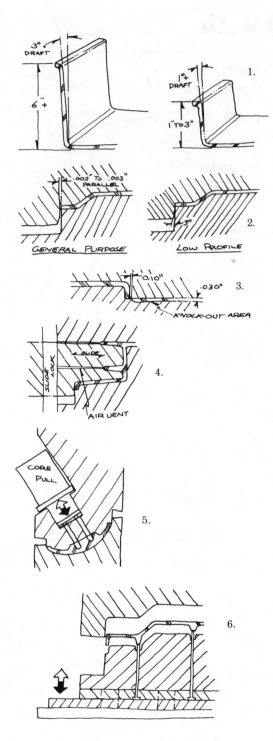

Fig. 19-29. Designing with sheet molding compounds (SMC).

7. **Ribs.** Ribs should have a 1° taper in each side and a lead-in radius of 1/8". When ribs must be placed in flat areas of the part they should be kept as thin as possible, but not less than 1/8" thick. If the rib cross section becomes too massive, the problem with "sink" marks on the opposite surface will occur. Visible sink marks can be eliminated by placing ribs behind textured areas or directly behind contour changes.

8. **Bosses.** Bosses follow the same general rule as ribs. Holes should be molded-in to reduce the mass of a boss. The shoulder width of the boss should be no larger than the diameter of the hole.

9. **Flanges.** Flanges provide rigidity and prevent warping of edges. Flange or edge thickening detail depends on part configuration.

10. **Joints.** There are many ways to mold SMC to accomodate the assembly of molded parts. Some of the techniques are shown. Parts can be molded to fit each other in such a way as to minimize assembly fixtures.

11. **Inserts.** A variety of inserts can be molded into the part, or mechanically driven into the part after molding.

12. **Thickness Variation.** Unlike conventional sheet materials, SMC can be molded with thickness variations. Thickest section should be kept under ½". Excessive thicknesses increase molding time.

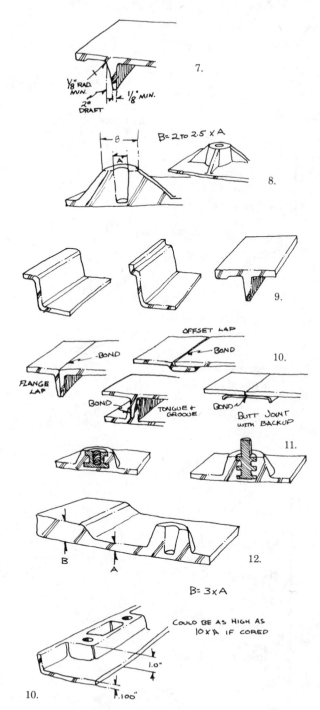

Fig. 19-29. Designing with sheet molding compounds. (*Continued*)

sheet molding compound

·Minimum Inside Radius (in.)	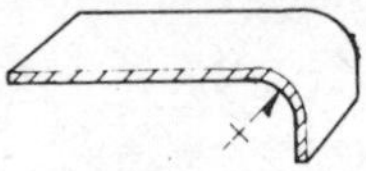	1/16" MINIMUM
·Molded–In Holes	 Ram Action Parallel · Perpendic-ular	YES - PARALLEL OR PERPENDICULAR TO RAM ACTION
·Trimmed In Mold	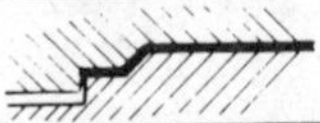	YES
·Core Pull & Slides	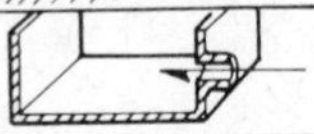	YES
·Undercuts	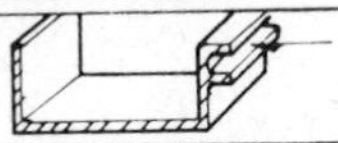	YES
·Minimum Draft Recommended	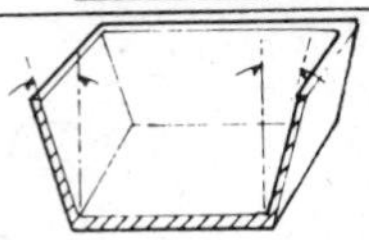	1/4"- 6" DEPTH – 1°- 3° 6"+ DEPTH – 3°+ or as REQUIRED
·Minimum Practical Thickness (in.)	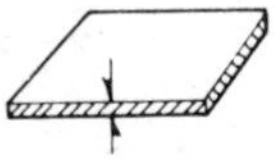	.050
·Maximum Practical Thickness (in.)	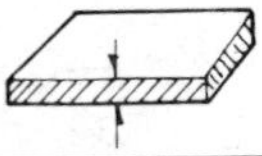	1
·Normal Thickness Variation (in.)	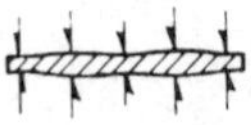	±.005
·Maximum Thickness Buildup - Heavy Buildup - Increased Cycle	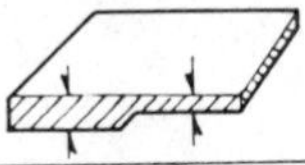	AS DESIRED
·Corrugated Sections	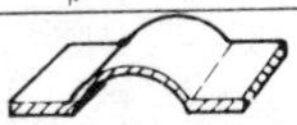	YES
·Metal Inserts		YES
·Bosses	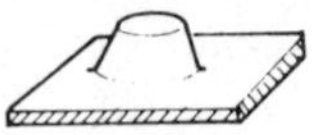	YES
·Ribs	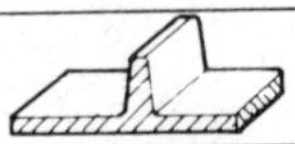	AS REQUIRED
·Molded–In Labels		YES
·Raised Numbers		YES
·Finished Surfaces ·Reproduces Mold Surface	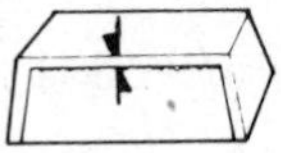	TWO

Fig. 19-29. Designing with sheet molding compounds. (*Continued*)

to avoid excessive wear. Surface and centerless grinding are easily performed and are preferable to cutting where practicable.

Coolants are desirable, but not necessary. If dry cutting is performed, an exhaust system should be provided to remove the rather obnoxious (but not dangerous) cutting dust.

Shearing, blanking, and punching are very practical. The spring stripper mechanism should be extra strong to overcome the tendency of the tool to seize the piece.

Pieces of reinforced plastics can be connected by all conventional systems, but adhesive bonding is preferred when a joint of high strength is required. Epoxy resin adhesives are, in general, the strongest and most reliable. Since reinforced plastics do not form strong threads, self-tapping screws are used only with light loads, and bolts should extend through the reinforced plastic. Large washers should be used on bolted connections, to prevent damage by the bolt head and nut during stress. For the same reason, washers should be used with rivets. Aluminum rivets should be used to prevent damage to the reinforced plastic during upsetting.

The bolt or rivet holes must be aligned exactly, to prevent local concentrations of

Design Details

1. Stress Concentrations.

Sharp radii produce areas of high stress concentration and should be avoided. By being generous with all radii, greater structural strengths can be achieved.

Stress concentration is reduced 50% by increasing ratio of radius to thickness from 0.1 to 0.6.

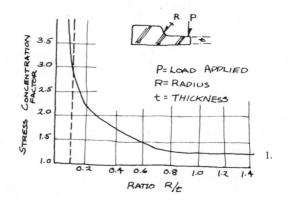

2. Section Thickness.

Part should be designed with minimum wall thickness that will provide required structure.

Wall thickness should be kept as uniform as possible.

Wall thickness variations, if required, should blend as gradually as possible. If it is not possible to blend thickness variations, it may be better to make an assembly of two or more parts.

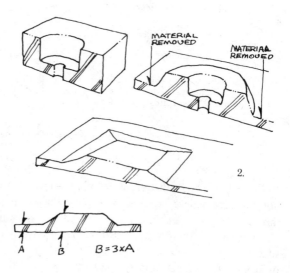

Fig. 19-30. Designing for injection molded thermoplastics.

3. Ribs.

Ribs, flanges and beads are used to increase strength without increasing overall thickness. They also improve material flow and warpage control.

Width of rib at juncture with main panel should be less than panel thickness. A ratio of .75 to .90 is optimum.

Ribs should taper 1° per side minimum.

Unsupported heights should be no greater than 10 times nominal wall thickness.

Ribs should be perpendicular to parting line, to allow ejection.

When ribs are behind a critical surface, visible sink marks can be kept to a minimum by texturing the surface or by designing rib to fall behind a contour change.

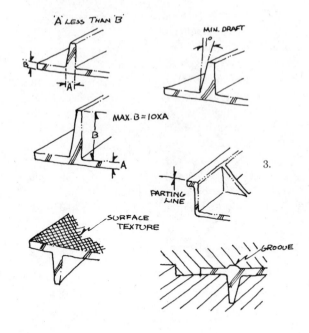

4. Bosses.

Bosses can be used to reinforce holes in the part and to provide mounting locations.

Bosses must be molded with rounded corners and adequate fillet radii.

Locating bosses in a corner or juncture of two or more surfaces helps to balance material flow.

Height of a boss should be limited to twice its diameter. Vertical surfaces must have at least 1° draft per side.

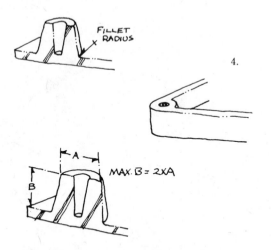

Fig. 19-30. Designing for injection molded thermoplastics. (*Continued*)

stress. Unlike metals, the reinforced plastic does not yield to relieve stresses caused by slight mis-matching of holes.

Painting of reinforced plastics has been successfully accomplished with a wide variety of paints. The preparation of the surface is of utmost importance. Most parting agents leave a residue that makes good adhesion of paint impossible. This residue must be removed by solvent washing or by sanding. Many moldings have minute cracks and voids that tend to collect moisture. Therefore, they must be thoroughly dried. These same tiny defects also are magnified in appearance when they are

5. Fillets.

Fillets should be as generous as
possible to facilitate material flow,
improve part strength, and permit
easy ejection of part from mold.

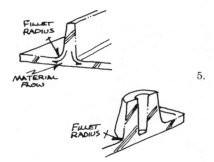

5.

6. Drafts.

Draft requirements vary according to
molding resin and wall thickness. Draft
should be as generous as possible, can
be as low as 1/4° for polyolefin materials.

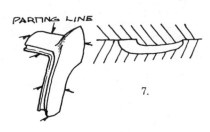

7. Parting Lines.

Parting lines should be designed into
an inconspicuous edge to improve
appearance and minimize finishing.

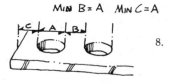

7.

8. Holes.

Hole-to-hole distance should be at least
one diameter of the hole.

Distance of hole from side or edge of
part should be one diameter of hole.

Through holes are easier to mold and
give support to core pins.

Blind holes should be limited in depth
to two times diameter of hole. A stepped
design in blind hole permits more
depth vs. hole diameter. This also
reduces wall thickness and die costs.

Holes parallel to parting lines require
core pulls.

Provide a 1/64" vertical step (minimum)
at open end of hole.

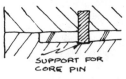

8.

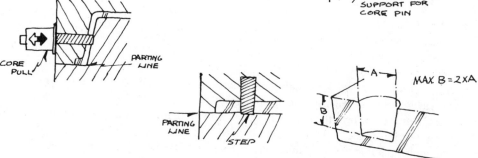

Fig. 19-30. Designing for injection molded thermoplastics. (*Continued*)

9. Inserts.

Inserts should be located perpendicular to parting lines, and secured so they will not be displaced during material flow under pressure.

Pin supported inserts require a hole diameter of 1/8".

Inserts less than 1/4" diameter require a boss diameter of at least two times the diameter of the insert.

Molded-in inserts must be deeply knurled to allow good material attachment and holding power. They must also be cylindrical in shape. Square corners on inserts tend to produce areas of high stress concentration.

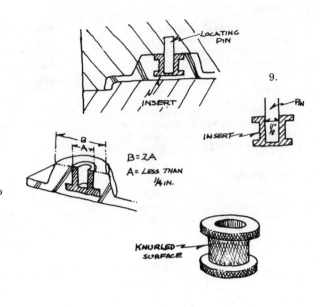

10. Undercuts.

External undercuts can be produced by use of slides or split molds.

If less than 1/16" in depth and well-rounded, undercuts may be molded without slides or split molds.

Extreme undercuts of narrow dimensions should be avoided.

Draft is required on all surfaces perpendicular to the parting line.

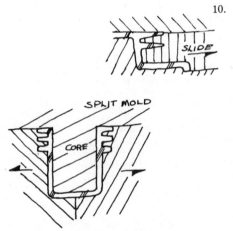

Fig. 19-30 Designing for injection molded thermoplastics. (*Continued*)

covered with a glossy paint; hence sanding of the surface yields a higher quality. To keep surface fibers from showing, the use of a surfacing mat is recommended.

Another popular method for finishing reinforced thermosets is with a gel coat (usually a polyester or a polyester-type coating). The gel coat is sprayed on a mold prior to molding; when the reinforced plastic part is then laid up in the mold, the gel coat becomes the exterior layer. Gel coats can be based on propylene glycol or neopentyl glycol (minimal water absorption, excellent weatherability, more resis-

tance to crazing and blistering). Gel coats are readily pigmentable.

Polyurethane coatings, which offer excellent gloss retention and color, are also available for finishing RP. Low bake vinyl organosols are another contender.

Plain colors or printed patterns can also be molded directly onto BMC or SMC parts in the press using a two-stage process involving a decorative paper or foil (usually limited to straight draw components). In stage one, the paper (printed side up) and the BMC are loaded into the mold and molded. The paper

injection molding

·Minimum Inside Radius (in.)	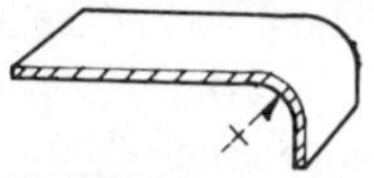	1/16" MINIMUM
·Molded–In Holes	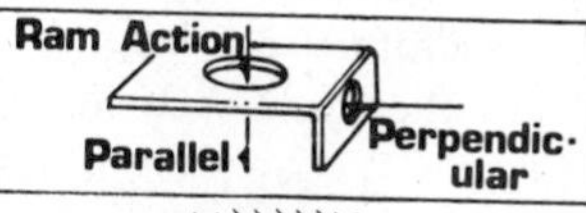 Ram Action Parallel · Perpendic-ular	YES-PARALLEL OR PERPENDICULAR TO RAM ACTION
·Trimmed In Mold	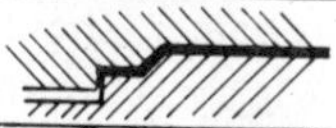	NO
·Core Pull & Slides	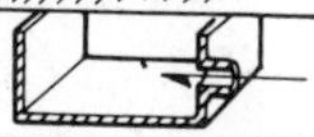	YES
·Undercuts	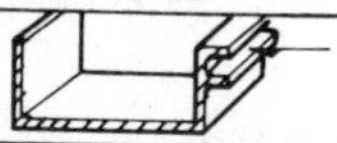	YES
·Minimum Draft Recommended	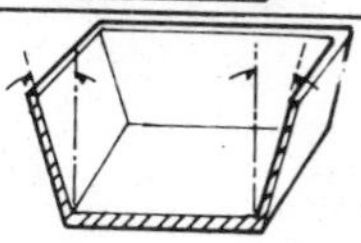	1/4"- 6" DEPTH – 1°-3° 6"+ DEPTH – 3°+ or as REQUIRED
·Minimum Practical Thickness (in.)	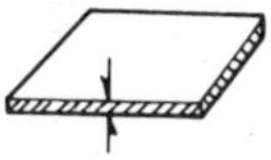	.035
·Maximum Practical Thickness (in.)	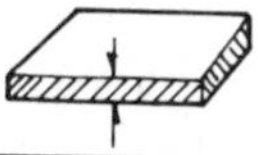	.500
·Normal Thickness Variation (in.)	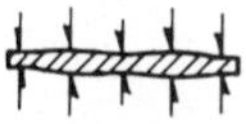	±.005
·Maximum Thickness Buildup – Heavy Buildup – Increased Cycle	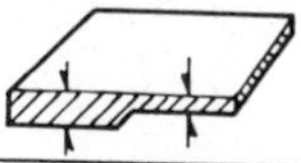	AS DESIRED
·Corrugated Sections	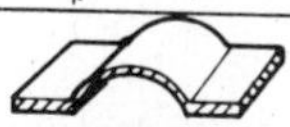	YES
·Metal Inserts		YES
·Bosses	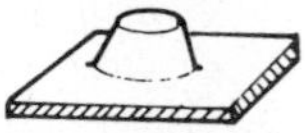	YES
·Ribs	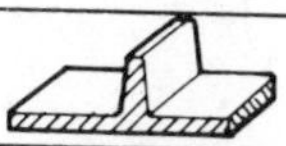	YES
·Molded–In Labels	 OCF	NO
·Raised Numbers		YES
·Finished Surfaces ·Reproduces Mold Surface	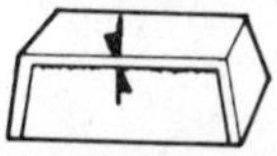	TWO

Fig. 19-30. Designing for injection molded thermoplastics. (*Continued*)

(unimpregnated until then) is partially impregnated during this first molding step from the resin in the BMC (or SMC). The paper also adheres to the plastic in this step. A glaze resin is then applied to the molding to complete impregnation (and provide a protective glaze) and the mold is closed and the cycle completed.

When molded reinforced thermosets are damaged, the breakage is usually localized. These damaged areas are easily repaired. The first step is to remove the broken portions and feather the broken edge by sanding. Then a patch of wet resin and reinforcement is laid up over the hole, with at least one ply of reinforcement lapping the hole on each side. The patch is then cured, and the repaired area sanded and painted, if desired. The materials used for the patch should be similar to those of the original construction.

20

CELLULAR PLASTICS

As with the reinforced plastic materials discussed in Chapter 19, the cellular plastics or foamed plastics have virtually become an industry unto themselves and at this writing, it appears as if they may even be ready to spawn an even newer, separate industry in the area of structural foam plastics.

Basically, cellular plastics are of two general types as regards structure: the closed-cell type (in which each individual cell, more or less spherical in shape, is completely closed in by a wall of plastic) and the open-cell type (in which individual cells are intercommunicating). They can be produced in a wide range of densities—from 0.1 lb/ft^3 to over 60 lb/ft^3. They can be rigid, semi-rigid, or flexible, colored or plain, and it is possible today to produce virtually every thermoplastic and thermoset material in cellular form. In general, the basic properties of the respective plastics are present in the foamed products, except of course, those which are changed by conversion to the cellular structure (Fig. 20-1).

In this chapter, the more common and popular foamed plastics in use by industry today will be discussed. Readers are cautioned, however, that there can be little in common between many of these materials, other than some degree of cellular structure. And even that aspect has changed over the years. As originally conceived, most foamed plastics were completely cellular in structure; today, it is possible

to isolate the cells so that a product may have a fairly solid skin surface and a cellular core underneath the skin.

Readers are also cautioned to be aware of the many ways in which cellular plastics can be used and processed. Cellular plastics can be produced in the form of slabs, blocks, boards, sheets, molded shapes, sprayed coatings, and extruded insulation. They can also be "foamed-in-place" in an existing cavity (i.e., poured into the cavity in liquid form and allowed to foam). More recently, it has become possible to process foamed plastics via conventional plastics processing machines like extruders and injection molding machines. It should be noted that the form used or the process employed can affect the properties of the end-product.

The ways in which a cellular structure can be effected in plastics also vary widely. The following are among those discussed in this chapter:

(1) air is whipped into a suspension or solution of the plastic, which is then hardened by heat or catalytic action or both;
(2) a gas is dissolved in the mix and expands when pressure is reduced;
(3) a liquid component of the mix is volatilized by heat;
(4) water produced in an exothermic chemi-

cal reaction is volatilized within the mass by the heat of reaction;

(5) carbon dioxide gas is produced within the mass by chemical reaction;

(6) a gas, such as nitrogen, is liberated within the mass by thermal decomposition of a chemical blowing agent;

(7) tiny beads of resin or even glass (e.g., micro-balloons) are incorporated in a plastic mix.

Finally, readers should be cautioned on the terminology currently in use in the cellular plastics industry. Unfortunately, it is loose and generally relates to the starting material (e.g., polyethylene) from which the foam is made. However, an expression like "polyethylene foam" can have many meanings today. It can refer to special low-density polyethylene foams and special high-density polyethylene foams that are quite different in character. It could include cross-linked polyethylene foams, which differ even more, or it can refer to polyethylene foam films made by extruding low-density polyethylene with a nitrogen blowing agent. Or, to complicate matters still further, it could also encompass low-density polyethylene structural foams or high-density polyethylene structural foams that bear no resemblance or relationship

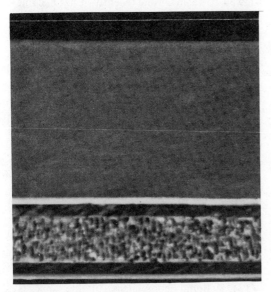

Fig. 20-1. Blowing agents expand plastics to give more product per pound. (*Courtesy Uniroyal*)

to any of the other foams that have been mentioned thus far. In this chapter, an attempt will be made to maintain these distinctions and to point up the differences between the various cellular plastics whenever possible.

History

In a sense, the first foamed plastic was what resulted from the reaction of phenol and formaldehyde before Dr. Leo H. Baekeland, seeking to develop a nonporous resin, found means of preventing the development of voids. Thus the recent development of foamed plastics has involved learning how to do, under controlled conditions, the very thing that Baekeland so profitably avoided doing.

The development of foamed phenolics dates from about 1945, and the use of phenolic "balloons" in the special "syntactic" type of foam (see below) dates from about 1953.

Epoxy foams were first developed, in 1949, to meet the need for a suitable material of light weight for the encapsulation of electronic components. Further study led, in 1953, to the development of epoxy foam of a density of 2 lb/ft^3 for the Bureau of Ships and for the aircraft industry.

At about this time there was also developed a special type of cellular plastic ("syntactic"), comprising tiny hollow spheres of phenolic resin embedded in epoxy resin. Later, similar hollow spheres of urea resin became available, and resins other than epoxy came to be used for foams of this type. Also, there was developed a further type in which tiny solid beads of polystyrene, containing dissolved gas, were mixed into epoxy resin and were expanded by the exothermic heat released during the cure of the epoxy.

Urea-formaldehyde foams were first produced in Germany, prior to World War II, in the form of slabs used as thermal insulation. During the war, such foam, manufactured in the United States, was used to float ammunition, food and other supplies to shore during amphibious operations.

The first published reports on foamed vinyls appeared during World War II in Germany. The toxicity of the blowing agents thus disclosed delayed commercial development of chemically expanded vinyls in the United States, but subsequently nontoxic blowing agents became available. A process of mechanical blowing, by absorption and expansion of an inert gas, was developed in the United States prior to 1950. The development of foamed vinyls in recent years has been greatly assisted by the availability of vinyl plastisols (see Chapter 16), which can be fused without application of pressure. Concurrently, pressure-molding techniques were developed for production of closed-cell vinyl foams. Extrusion methods have recently been developed for production of expanded shapes.

The technology of the urethanes was well developed in Germany by the mid-1930's, and in World War II urethane foams based upon alkyd-type polyesters came into wide use there—rigid foams as structural reinforcement in aircraft and ships; flexible foams in cushioning uses, replacing rubber foams for which latex was not obtainable.

Information brought from Germany by survey teams after the war, the licensing of German-owned patents, and service contracts to the aircraft industry, stimulated research in the United States which resulted in the development of practical techniques, improvements in production equipment, and formulations based upon starting materials other than alkyd-type polyesters.

The extrusion of foamed polystyrene was based upon developments in Sweden in the mid-1930's. Commercial production in the United States was begun in 1944.

Expandable polystyrene was first developed in Germany about 1952. Subsequently, a similar product was developed in the United States, and introduced to the market in early 1954.

Development work on foams of silicone resins was started in 1950, to meet the need for a material of low weight which would retain good mechanical strength after long exposure to temperatures of 400 to 700°F. The first commercial product, which appeared in 1952, required compounding by the user, and was superseded in 1953 by premixed powders re-

quiring merely to be melted, whereupon they foam and are cured under the influence of heat. Since for these powders the temperature required is high (sometimes too high to be tolerated by other materials in an assembly), and since the minimum density of the foam is high ($10 \, lb/ft^3$), attention has been given to a newer series of resins, which can be expanded and cured at room temperature, and yield rigid foams having a density of 4 to 5 lb/ft^3.

Cellular polyethylene, a closed-cell foam, was developed in 1944, specifically as a low-loss insulation for wire and cable. Extruded polyethylene foam then become available in the form of slabs and blocks.

Cellular cellulose acetate was developed, at the time of World War II, primarily to meet the requirements of the U.S. Air Force for reinforcement in a monocoque construction, to supersede the balsa wood used in the British "Mosquito" bomber.

In the mid- and late-sixties, plastic foams made some important strides forward in terms of technology. A number of commercial techniques for producing structural foam plastics (e.g., the Engelit process, the Union Carbide process, the USM process, etc.) was introduced to industry. Advances were also made in the direct injection extrusion of polystyrene foam as opposed to the more conventional bead molding. In the polyolefins area, there were such developments as chemically- and radiation-cross-linked polyethylene foams, low density closed-cell polypropylene foams (also chemically cross-linked) for automotive cushioning, and polyethylene foam film (for bath mats, table mats, embossed wall coverings, etc.) extruded from conventional low-density polyethylene and a nitrogen blowing agent. The sixties also saw the introduction of a rigid ABS foam, ionomer foams, and even acrylic foams.

More recently, there has been activity in making some of the high performance plastics (like polyimides) available as foams for use by the aircraft and other industries requiring outstanding heat resistance. The polyimide foams, for example, show good thermal and oxidative stability and are characterized by high glass-transition temperatures.

CHEMICAL BLOWING AGENTS*

Knowledge of the chemistry of blowing agents and an understanding of their decomposition products can be helpful in selecting blowing agents for particular applications and can be valuable in approaching problems associated with their use.

The development of organic blowing agents has been a major contribution to the field of cellular rubber and plastics. These organic materials show important advantages over inorganic gas-forming agents such as sodium bicarbonate which has long been used in the manufacture of open-cell expanded rubber. In contrast to "soda" the commercially available organic blowing agents are capable of producing a fine, closed-cell structure in rubber and plastics and their use has grown rapidly in the last decade, particularly in cellular plastics.

Generally speaking, commercial organic blowing agents are organic nitrogen compounds which are stable at normal storage and mixing temperatures but undergo controllable gas evolution at reasonably well-defined decomposition temperatures. Two comprehensive reviews of organic substances which have been suggested or used as blowing agents have been given by Reed.[1] A more recent summary has also appeared.[2]

Choice of Blowing Gas

Nitrogen is an inert, odorless, non-toxic gas. For this and other good reasons nitrogen-producing organic substances have been preferred as blowing agents. van Amerongen[3] measured the permeability of natural rubber to several gases and has given the following relative values (based on hydrogen as 100):

$$CO_2 = 260$$
$$H_2 = 100$$
$$O_2 = 46$$
$$N_2 = 17$$

Organic blowing agents which have been offered vary widely in their properties. Importantly, they vary in the temperature at which

* By Dr. Byron A. Hunter, Senior Research Associate, Uniroyal Chemical.
1 References on page 510.

they produce gas and in the nature of their decomposition products. Some produce odor while others may yield colored or toxic substances on decomposition. The agents vary considerably in their response to other materials present in the expandable polymer which may function as activators or retarders. These factors are important in the selection of a blowing agent for a particular use.

The present discussion is concerned with the chemistry of those blowing agents which have attained commercial importance as expanding agents for rubber and plastics in the United States. Reference will be made to a number of products which have not achieved commercial status. The discussion will emphasize the chemistry and decomposition mechanisms of chemical blowing agents insofar as this has been made known.

Diazoaminobenzene

One of the earliest organic blowing agents introduced to the rubber trade was diazoaminobenzene.[4] This compound is made by the partial diazotization of aniline:

The compound is quite soluble in rubber and is capable of producing an extremely fine unicellular structure. A disadvantage of the chemical lies in the fact that the decomposition products produce a dark color in the expanded product and may stain other materials with which it may come in contact. Diazoaminobenzene melts with decomposition near 100°C and produces substantially the theoretical amount of nitrogen gas [115 cc/gram (STP)]:

The diphenylamine residue is notorious as a staining chemical.

Diazoaminobenzene has been shown to be carcinogenic in mice[30] and is no longer recommended as a blowing agent.

Azobis(isobutyronitrile)

An interesting blowing agent came out of Germany during World War II.[5] This compound, azobis(isobutyronitrile):

$$(CH_3)_2C-N=N-C(CH_3)_2$$
$$\quad\ \ |\qquad\qquad\quad\ |$$
$$\quad\ CN\qquad\qquad CN$$

was introduced for the production of sponge rubber articles. Later the material was recommended for expanding polyvinyl chloride. The material is completely non-discoloring and non-staining and yields white PVC foam of fine uniform cell structure. The compound is manufactured from hydrazine, acetone, sodium cyanide and acid, using chlorine to oxidize the intermediate hydrazo bis(isobutyronitrile):

(3) $2(CH_3)_2C=O + H_2NNH_2 + 2\ NaCN + H_2SO_4 \longrightarrow$

$$\qquad\quad\ \ \ H\ \ H$$
$$(CH_3)_2C-N-N-C(CH_3)_2 + Na_2SO_4 + 2\ H_2O$$
$$\qquad\ \ |\qquad\qquad\ \ |$$
$$\qquad\ CN\qquad\qquad CN$$

$$\qquad\qquad\ \ H\ \ H$$
(4) $(CH_3)_2C-N-N-C(CH_3)_2 + Cl_2 \longrightarrow (CH_3)_2C-N=N-C(CH_3)_2$
$$\qquad\qquad\ |\qquad\qquad\ |\qquad\qquad\qquad\qquad\ |\qquad\qquad\ |$$
$$\qquad\qquad CN\qquad\ CN\qquad\qquad\qquad\qquad CN\qquad CN$$

$$\qquad\qquad\qquad\qquad\qquad\qquad\qquad + 2\ HCl$$

The decomposition of azobis(isobutyronitrile) proceeds as follows:

(5) $(CH_3)_2-C-N=N-C(CH_3)_2 \xrightarrow{\triangle} N_2 + (CH_3)_2C-C(CH_3)_2$
$$\qquad\qquad\ |\qquad\qquad\ |\qquad\qquad\qquad\qquad\qquad\quad |\ \ |$$
$$\qquad\qquad CN\qquad\ CN\qquad\qquad\qquad\qquad\qquad CN\ CN$$
$$\qquad\qquad\qquad\qquad\qquad\qquad\qquad\qquad\qquad\ (toxic)$$

The decomposition occurs rapidly at temperatures above $100°C$ and more slowly at lower temperatures. Complete decomposition yields 137 ml. of gas per gram (measured at standard conditions). The decomposition residue, tetramethyl succinonitrile, is a toxic substance and precautionary measures must be taken to eliminate this hazard [reference 5 (c)]. There has been reluctance in the United States to use the chemical as a blowing agent and the principle manufacturer here does not recommend the material for this use. It has been stated in Europe[1] that azobis(isobutyronitrile) can be successfully employed to expand PVC without accident if adequate ventilation is provided.

The decomposition of azobis(isobutyronitrile) yields free radicals and this substance has been employed quite widely as a polymerization initiator.

Dinitroso pentamethylene tetramine

$$\qquad\qquad CH_2-N\ \ \ CH_2$$
$$\qquad\qquad\ \ |\qquad\quad |$$
$$\qquad ON-N\quad CH_2\ N-NO$$
$$\qquad\qquad\ |\qquad\qquad |$$
$$\qquad\qquad CH_2-N\ \ \ CH_2$$

This compound is prepared by the nitrosation of hexamethylene tetramine (reaction product of formaldehyde and ammonia).

(6)

$+ 2\ NaNO_2 + H_2SO_4$

The chemical when heated alone or in the presence of inert diluents decomposes near $195°C$ ($383°F$) but when used in rubber or plastics in the presence of certain activators produces gas within a temperature range of $130\text{-}190°C$ ($266\text{-}374°F$).[2] The quantity of gas produced from the undiluted material is close to two moles per mole of blowing agent (near 265 ml/gram measured at standard conditions).[2]

The decomposition products of dinitroso pentamethylene tetramine have not been fully elucidated. It has been reported[1] that nitrogen and nitrous oxide* are formed as well as amines and water. The presence of formaldehyde can also be detected in the decomposition of the dry material. The amine residue gives rise to a characteristic "fishy odor" in the expanded product. The odor can be at least partially suppressed by the addition of urea, melamine and certain amino compounds.

* One manufacturer has found no evidence of nitrous oxide but reports ammonia as being present.

The temperature of decomposition of dinitroso pentamethylene tetramine can be substantially lowered and the decomposition accelerated by acidic substances such as salicylic acid and phthalic anhydride and these are commonly used as activators. Urea (BIK)† and hydroxy compounds (as ethylene glycol) have also been employed as activators.

Dinitroso pentamethylene tetramine is used widely as a blowing agent in the rubber industry but is of limited use in plastics because of the high decomposition exotherm and the unpleasant odor of the residue.[2] The chemical is non-discoloring and non-staining.

N,N′-Dinitroso-N,N′-dimethylterephthalamide

This substance in the undiluted state is described[7] as a "weak explosive which is sensitive to impact and friction". The material is successfully desensitized by treatment with mineral oil and the commercial product ("Nitrosan") is described as a relatively safe material and can be handled without difficulty if proper precautions are observed.[8] The product must be carefully protected from sparks and open flames.

When heated in small amounts this blowing agent melts with decomposition at 105°C (220°F). Prolonged heating at lower temperatures in the range of 80-90°C (175-195°F) will lead to sustained and rapid formation of gas. Practically the theoretical amount of nitrogen is evolved (125 cc/gram.).

(7)

N,N′-Dinitroso-N,N′-dimethylterephthalamide can be prepared from commercial dimethyl

†A surface treated urea supplied by Uniroyal Chemical.

terephthalate by reaction with mono methylamine and nitrosation of the resulting N,N′-dimethylterephthalamide with sodium nitrite and nitric acid:

(8)

This blowing agent is unique in the fact that the low decomposition temperature permits the expansion of a vinyl plastisol prior to gelation (at 200°F). Subsequent fusion of the fragile foam at 350°F produces open cell vinyl form. Closed cell vinyl can be made in a closed mold at 350°F and, after cooling the mold, releasing the pressure and subsequently heating the plastic at 100°C in a circulating air oven. The products are free of color.

Azodicarbonamide

First suggested as a potential blowing agent for plastics in Germany during World War Two,[5] azodicarbonamide was introduced in the United States in the early 1950's. Important today in plastics expansion, the chemical is prepared by reacting hydrazine with urea under controlled conditions to produce the intermediate hydrazodicarbonamide (9), which is oxidized to azodicarbonamide:

(9) $H_2NNH_2 + 2H_2N-\overset{O}{\underset{\parallel}{C}}-NH_2 \xrightarrow[H^+]{\Delta} H_2N-\overset{O}{\underset{\parallel}{C}}-\overset{H}{\underset{}{N}}-\overset{H}{\underset{}{N}}-\overset{O}{\underset{\parallel}{C}}-NH_2 + 2NH_3$

(10) $H_2N-\overset{O}{\underset{\parallel}{C}}-\overset{H}{\underset{}{N}}-\overset{H}{\underset{}{N}}-\overset{O}{\underset{\parallel}{C}}-NH_2 \xrightarrow{(O)} H_2N-\overset{O}{\underset{\parallel}{C}}-N=N-\overset{O}{\underset{\parallel}{C}}-NH_2 + H_2O$

Effective oxidants are dichromate, nitrates, nitrogen dioxide, chlorine, etc.

Azodicarbonamide (azobisformamide) is a yellow crystalline solid which decomposes to produce a high yield of gas (220-240 cc/gram,

STP). Unlike most other blowing agents, azodi-carbonamide does not support combustion and is self extinguishing. The white decomposition residue is odorless, non-toxic, non-discoloring and non-staining.

The dry decomposition temperature is high (195-216°C, depending on the mode of pre-paration). This range can be lowered by a variety of activators, e.g., lead, cadmium, zinc and barium salts, and others, thereby making azodicarbonamide successful in the expansion of rubber as well as plastics.

The decomposition gases have been analyz-ed. In diethylene glycol, the decomposition produces a gaseous mixture of nitrogen (62%), carbon monoxide (35%) and ammonia and carbon dioxide (total 3%). Reed[1] has reported a similar analysis of the gases from dry decom-position (190°C) of azodicarbonamide and has also analyzed the solid decomposition products. He found gases (noted above), 32% by weight; the weights of the solid products as percentages of azodicarbonamide were: solid residue, 41% (urazole, 39%, biurea, 2%); and sublimate 27% (cyanuric acid, 26%, cyamelide, 1%.)

However, Reed found the solid products of decomposition at 190°C under liquid paraffin were markedly different in proportion: (% by weight): gaseous products, 34%, solid residue, 61% (urazole, 27%, biurea 34%): and sublimate 5% (cyanuric acid 5%, cyamelide 0%).

Why this difference? Reed suggests that primary decomposition of azodicarbonamide can follow two courses:

(11) $H_2N-C-N=N-C-NH_2 \rightarrow NH_3 + CO + N_2 + HNCO$

(12) $2 H_2N-C-N=N-C-NH_2 \rightarrow H_2N-C-N-N-C-NH_2 + N_2 + 2HNCO$

Urazole can result from the elimination of a mole of ammonia from biurea:

(13) $H_2N-C-N-N-C-NH_2 \rightarrow O=C \quad C=O + NH_3$

An equilibrium between cyanic and iso-cyanic acid and trimerization of each can be written:

(14)

Cyanuric acid   Cyamelide

Moisture can affect the blowing character-istics of azodicarbonamide. It is known that hydrolysis of the chemical can occur at high temperatures in the presence of either acid or base to produce biurea, nitrogen and carbon-dioxide.[5]

Benzene Sulfonyl Hydrazide

Benzene sulfonyl hydrazide is the simplest aromatic compound in the class of sulfonyl hydrazides. It is prepared by treating benzene sulfonyl chloride with hydrazine in the pre-sence of a base (like ammonia):

(15) $SO_2Cl + H_2NNH_2 + NH_4OH \rightarrow$
$SO_2NHNH_2 + NH_4Cl + H_2O$

The product is a white crystalline solid that melts and begins to decompose near 105°C. It is capable of producing a white unicellular foam when incorporated into a PVC plastisol. Un-fortunately the odor is strong and unpleasant, remindful of thiophenol.

Interestingly enough, experiments to detect thiophenol in decomposition vapors have been negative. Instead, the odor may actually be due to diphenyl disulfide. Decomposition involves an internal oxidation-reduction of the sulfonyl hydrazide group. A possible mechanism of the gas forming reactions has been suggested:[11a]

(16) $SO_2NNH_2 \rightarrow N_2 + \left[SOH \right] + H_2O$

(17) $4\left[SOH \right] \rightarrow S-S$
$+ S-S + 2 H_2O$

The hypothetical intermediate benzene sulfenic acid (I) is apparently incapable of existence and immediately disproportionates to diphenyl disulfide (II) and phenyl benzene thiosulfonate (III) as shown in equation (17).

Diphenyl disulfide was identified in the decomposition products by early investigators.[12] A thiosulfonic ester was identified in a recent study in the similar decomposition of p-toluene sulfonyl hydrazide[13] (discussed next).

The strong odor where it is a blowing agent in PVC vanishes when the chemical is used to expand rubber. A likely explanation is that in rubber expansion the sulfur-containing residues (or their reactive intermediates) react with the rubber, leaving non-odorous combinations. Supporting this is the observation that certain difunctional sulfonyl hydrazides can function as curing agents in rubber compositions.[14]

p-Toluene Sulfonyl Hydrazide

The compound, p-toluene sulfonyl hydrazide, is similar to benzene sulfonyl hydrazide except that the melting point and decomposition temperatures are higher (120°C and up). Decomposition mechanisms and odor problems are similar. Deavin and Rees[13] isolated ditolyl disulfide and p-tolyl p-toluene thiosulfonate as

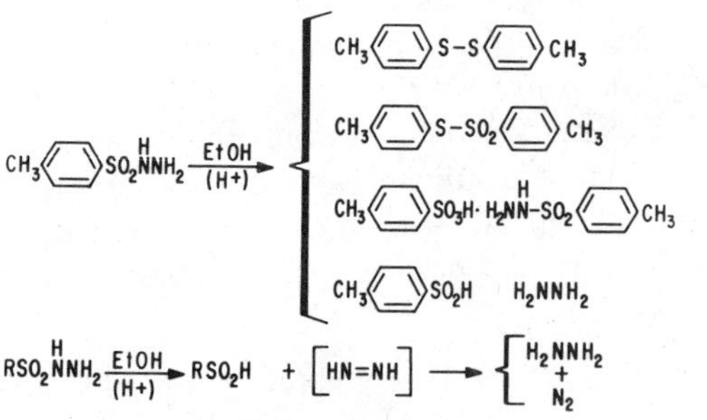

Fig. 20-1b. Decomposition of p-toluene sulfonyl hydrazide in boiling ethanol (acid present).

main decomposition products. Other decomposition products are p-toluene sulfinic acid, hydrazine and a small quantity of the p-toluene

sulfonyl hydrazide salt of p-toluene sulfinic acid (see Fig. 20-1b). The presence of p-toluene sulfinic acid and the salt of p-toluene sulfonic acid can be explained by the reaction shown in Fig. 20-1b and by the disproportionation reactions of sulfenic and sulfinic acids.[15]

p,p-Oxybis(Benzene Sulfonyl Hydrazide)

For all practical purposes this blowing agent eliminates odor in applications with cellular plastics. It is prepared by chlorosulfonation of diphenyl ether with chlorosulfonic acid and subsequent reaction with hydrazine in the presence of a base:[17]

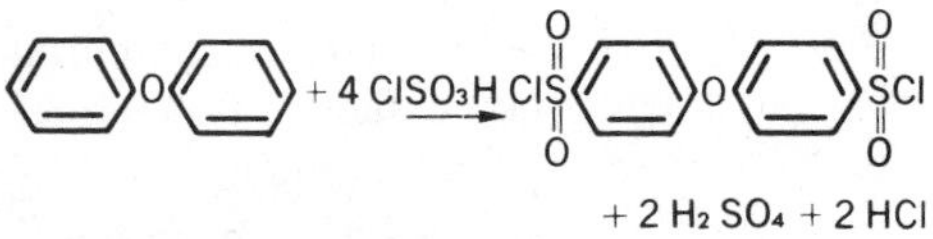

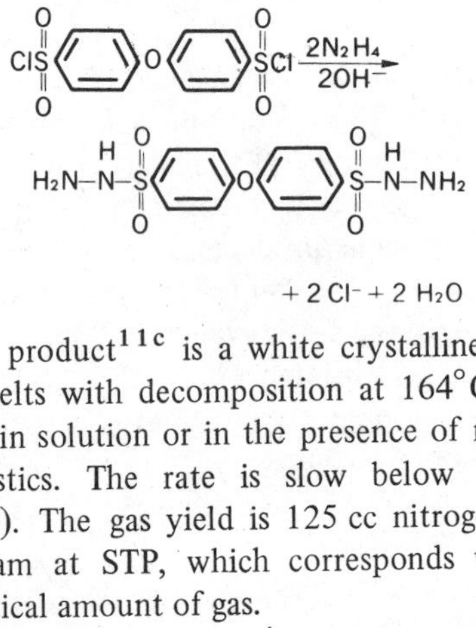

The product[11c] is a white crystalline solid that melts with decomposition at 164°C, and lower, in solution or in the presence of rubber or plastics. The rate is slow below 120°C (248°F). The gas yield is 125 cc nitrogen gas per gram at STP, which corresponds to the theoretical amount of gas.

Decomposition of p,p'-oxybis(benzene sulfonyl hydrazide) can be postulated by analogy with equations (16) and (17) for benzene sulfonyl hydrazide[11c]:

The polymeric residue (IV) exhibits practically no odor. That the residue is indeed polymeric can be shown by carefully igniting the blowing agent with a heated glass rod. Decomposition begins with a smooth gassing off at the point of ignition and spreads with moderate rapidity through the mass. No flame is seen as an insoluble expanded polymeric foam appears.

A high decomposition temperature $(235°C)$ makes this chemical useful in plastics that expand at high temperatures such as high density polyethylene, polypropylene, rigid polyvinyl chloride, ABS polymers, polycarbonates and nylon, among others.[22,23] A similar chemical is p,p'-oxybis(benzene sulfonyl semicarbazide) **(VI)** which decomposes at $215°C$.

$$(20) \quad H_2N-N-\overset{O}{\underset{O}{S}}-\text{⟨⟩}-O-\text{⟨⟩}-\overset{O}{\underset{O}{S}}-N-NH_2 \xrightarrow{\Delta}$$

$$\left[HOS-\text{⟨⟩}-O-\text{⟨⟩}-SOH \right]$$

$$+ 2N_2 + 2H_2O$$

$$(21) \quad n\left[HOS-\text{⟨⟩}-O-\text{⟨⟩}-SOH \right] \longrightarrow$$

$$\left[-S-\text{⟨⟩}-O-\text{⟨⟩}-S-+-S-\text{⟨⟩}-O-\text{⟨⟩}-\overset{O}{\underset{O}{S}}- \right]_{n/2}$$

IV $+ n H_2O$

$$H_2N-\overset{O}{\underset{}{C}}-N-N-\overset{O}{\underset{O}{S}}-\text{⟨⟩}-O-\text{⟨⟩}-\overset{O}{\underset{O}{S}}-N-N-\overset{O}{\underset{}{C}}-NH_2$$

VI

Upon cooling, the residue readily crumbles to an insoluble polymeric ash. Decomposition within a polymer is more controllable and the residue remains colorless.

This agent is widely used in rubber and plastics: in the extrusion of cellular polyethylene for wire insulation,[18] in the expansion of PVC plastisols,[16] in epoxy and phenolic resins,[19] in expanded rubbers[16] and in rubber-resin blends. It presents no problems with odor, toxicity or discoloration. It also has the unusual ability of simultaneously expanding and cross linking rubbery polymers in the absence of any other conventional curatives.[14]

This agent will decompose to produce gas in an alkaline latex system at steam temperature. Under these conditions the blowing agent forms acidic decomposition products (probably sulfinic or sulfonic acids) which coagulates or gels the latex in the foamed state and provides a unique method of producing an expanded rubber product.[20] Impregnation of paper or non-woven fabric with latex so treated can produce an interesting fiber-reinforced rubber article.

Sulfonyl Semicarbazides

$$CH_3-\text{⟨⟩}-\overset{O}{\underset{O}{S}}-N-N-\overset{O}{\underset{}{C}}-NH_2$$

V

One of the newest blowing agents in the U.S. is p-toluene sulfonyl semicarbazide **(V)**.[21,22,23]

Both can be used for rubber expansion since activators lower their decomposition temperatures. Both have relatively high gas yields (around 143-145 cc/gram, measured at STP). This corresponds to about 1.5 moles of gas per mole in the case of p-toluene sulfonyl semicarbazide. On the other hand, p,p'-oxybis(benzene sulfonyl semicarbazide, being a difunctional molecule, gives nearly 3.0 moles of gas per mole (1.5 moles gas per sulfonyl semicarbazide group). This strikingly high gas yield is due to a considerable volume of carbon dioxide as well as nitrogen in the decomposition gases.

Sulfonyl semicarbazides can be made by two methods:[21,24]

(1) The reaction of a sulfonyl chloride with semicarbazide in the presence of an acid sequestering agent:

$$(22) \quad RSO_2Cl + H_2N-N-\overset{O}{\underset{}{C}}-NH_2 \rightarrow RSO_2N-N-\overset{O}{\underset{}{C}}-NH_2 + HCl$$

and (2) the reaction of a sulfonyl hydrazide with a source of cyanic acid:

$$(23) \quad RSO_2 NNH_2 \xrightarrow{(HOCN)} RSO_2 N-N-\overset{O}{\underset{}{C}}-NH_2$$

The decomposition products from sulfonyl semicarbazides vary substantially from those formed from the parent sulfonyl hydrazides. The presence of considerable carbon dioxide in the decomposition gases has already been mentioned.

The non-gaseous products are also quite different, indicating a distinct difference in the decomposition mechanisms between the two classes of blowing agents. The greater stability (higher decomposition temperature) of the sulfonyl semicarbazides is significant in this connection.

The products of dry decomposition of p-toluene sulfonyl semicarbazide at 235°C are shown in Fig. 20-1c.

Decomposition Products of p-Toluene Sulfonyl Semicarbazide[23]

It would appear that the carbon dioxide and the ammonium bicarbonate (or carbamate) must result from hydrolysis of the carbonamide group. The necessary water could result from an oxidation-reduction of the sulfonyl hydrazide structure in the molecule. Mechanism postulated is indicated in equations (24) through (28). Brackets indicate speculative transitory intermediates.

(24) $R-\overset{O}{\underset{O}{S}}-\overset{H}{N}-\overset{H}{N}-\overset{O}{C}-NH_2 \rightarrow \left[R-\overset{O}{\underset{O}{S}}-N=N-\overset{O}{C}-NH_2 \right] + H_2O$

(25) $\left[R-\overset{O}{\underset{O}{S}}-N=N-\overset{O}{C}-NH_2 \right] + H_2O \rightarrow \left[R-\overset{O}{\underset{O}{S}}-N=N-\overset{O}{C}-OH \right] + NH_3$

(26) $\left[R-\overset{O}{\underset{O}{S}}-N=N-\overset{O}{C}-OH \right] \rightarrow [RSOH] + N_2 + CO_2$

(27) $5[RSOH] + NH_3 \rightarrow RS-SR + RSO_3NH_4 + 2H_2O$

(28) $3NH_3 + 2CO_2 + H_2O \rightarrow (NH_4)HCO_3 + NH_2COONH_4$

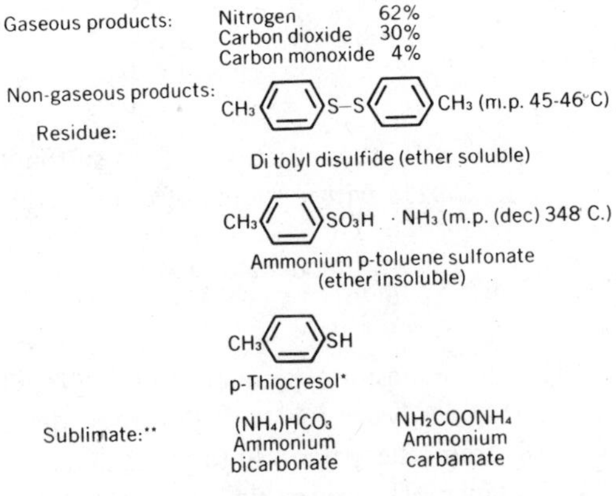

Gaseous products:
Nitrogen 62%
Carbon dioxide 30%
Carbon monoxide 4%

Non-gaseous products:
Residue:

CH_3⟨⟩$S-S$⟨⟩CH_3 (m.p. 45-46°C)

Di tolyl disulfide (ether soluble)

CH_3⟨⟩SO_3H · NH_3 (m.p. (dec) 348°C.)

Ammonium p-toluene sulfonate
(ether insoluble)

CH_3⟨⟩SH

p-Thiocresol*

Sublimate:** $(NH_4)HCO_3$ NH_2COONH_4
Ammonium   Ammonium
bicarbonate carbamate

Fig. 20-1c.

* Detected by titration with iodine. Known to oxidize readily to ditolyl disulfide in the presence of ammonia. ** Sublimate not rigorously identified.

Studies[23] directed at quantitative determination of the various decomposition products of p-toluene sulfonyl semicarbazide are in good agreement with the representation shown in equation (29):

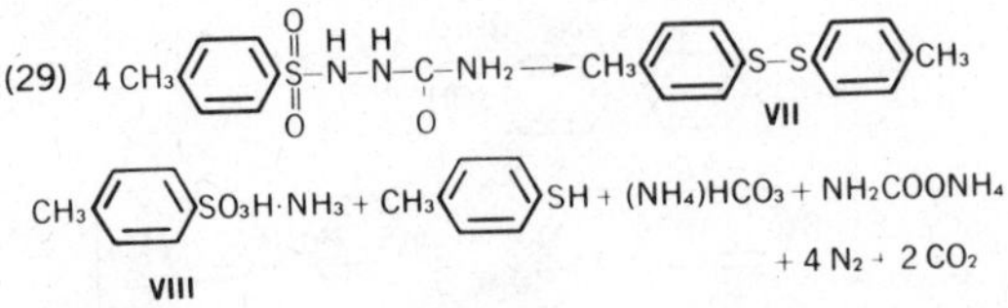

(29) $4CH_3$⟨⟩$\overset{O}{\underset{O}{S}}-\overset{H}{N}-\overset{H}{N}-\overset{O}{C}-NH_2 \rightarrow CH_3$⟨⟩$S-S$⟨⟩$CH_3$ **VII**

CH_3⟨⟩$SO_3H·NH_3 + CH_3$⟨⟩$SH + (NH_4)HCO_3 + NH_2COONH_4$

VIII $+ 4N_2 + 2CO_2$

The formation of six moles of gas from four moles of the blowing agent agrees with the observed evolution of 1.5 moles of gas per mole. The ratio of nitrogen to carbon dioxide (2 : 1) is in close agreement with the analysis (62% nitrogen, 30% carbon dioxide). The proportions of ditolyl disulfide (**VII**) and ammonium p-toluene sulfonate (**VIII**) isolated from the decomposition residue agree reasonably well with the values required by equation (29). p-Thiocresol was not isolated in the study but titration of the products of decomposition of p-toluene sulfonyl semicarbazide in dioctyl phthalate with standard iodine solution gave values in good agreement with that expected for the oxidation of one mole of p-thiocresol as indicated in equation (29). The sublimate, believed to consist of ammonium bicarbonate and ammonium carbamate, was not subjected to rigorous analysis.

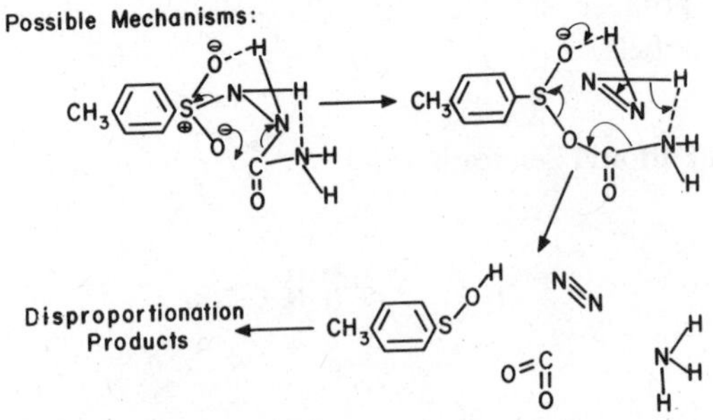

Possible Mechanisms:

Disproportionation Products ←

Fig. 20-1d. Spatial representation of the sulfonyl semicarbazide molecules indicates opportunities for strong hydrogen bonding.

Equations (24) through (29) are useful in representing the chemistry and stoichiometry of the decomposition of *p*-toluene sulfonyl semicarbazide but probably do not reflect the actual mechanism of the decomposition. The rapid decomposition at 235°C suggests that a concerted mechanism is involved. Spacial representation of the sulfonyl semicarbazide molecule as shown in Fig. 20-1d indicates opportunities for strong hydrogen bonding which may account for the relatively high heat stability. At the high decomposition temperature the excitation of the molecule is sufficient to produce electron shifts resulting in breaking and making of bonds in a concerted manner to produce the products indicated. Figure 20-1e indicates a similar proposed mechanism resulting in the formation of some carbon monoxide.

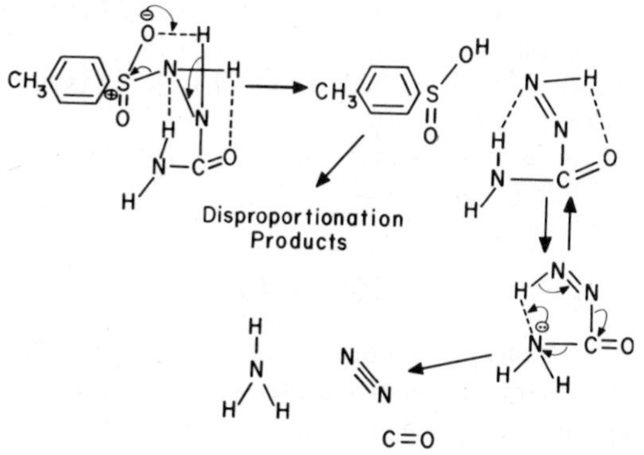

Fig. 20-1e. A proposed mechanism similar to Fig. 20-1d, resulting in the formation of some carbon monoxide.

Esters of Azodicarboxylic Acid

Related to azodicarbonamide are the alkyl esters of azodicarboxylic acid. These materials are quite soluble in the plasticizers normally employed in polyvinyl chloride plastisols and can be easily dispersed. Normally, these materials give little gas on heating, even to temperatures in excess of 240°C. However, in the presence of certain activating substances these azo esters decompose to produce a mixture of gases at temperatures as low as 100°C.[24]

It has been reported[24] that the gases produced from the decomposition of diisopropyl azodicarboxylate at 150°C (activated by lead stearate) consisted of nitrogen (53%), carbon monoxide (43.3%) and carbon dioxide (3.7%). The organic residue was found to consist mostly of diisopropyl carbonate (90-95%) along with diisopropyl ether (5-10%) and 2,3-dimethyl butane (1-3%). The following decomposition mechanism has been suggested:[24]

(30) $(CH_3)_2CHO{-}C(=O){-}N{=}N{-}C(=O){-}O{-}CH(CH_3)_2 \longrightarrow N_2 + 2\,(CH_3)_2CH{-}O{-}C(=O)\cdot$

(31) $(CH_3)_2CHOC(=O)\cdot \longrightarrow (CH_3)_2CH{-}O\cdot + CO$

(32) $(CH_3)_2CH{-}O{-}C(=O)\cdot + (CH_3)_2CH{-}O\cdot \longrightarrow (CH_3)_2CH{-}O{-}C(=O){-}OCH(CH_3)_2$

(33) $(CH_3)_2CH{-}OC(=O)\cdot \longrightarrow (CH_3)_2CH\cdot + CO_2$

(34) $(CH_3)_2CH{-}O\cdot + (CH_3)_2CH\cdot \longrightarrow (CH_3)_2CH{-}O{-}CH(CH_3)_2$

(35) $2\,(CH_3)_2CH\cdot \longrightarrow (CH_3)_2CH{-}CH(CH_3)_2$

The development of activators for azodicarboxylic acid esters has been a recent accomplishment and these materials are yet in the experimental stage.

Salts of Azodicarboxylic Acid

Only the barium salt of azodicarboxylic acid is available commercially. This material, prepared by the hydrolysis of azodicarbonamide in the presence of barium hydroxide[25] is a yellow crystalline solid which is reported to decompose at 240-250°C to produce nitrogen, carbon monoxide and barium carbonate:

(36) 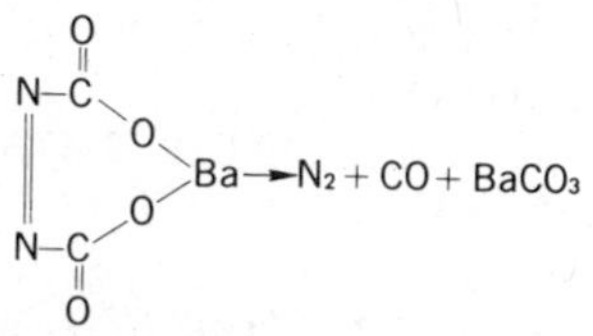

The sodium salt of azodicarboxylic acid is useful as a blowing agent for expanding liquid polysulfide rubbers.[26,27] This material decomposes quickly in the presence of water and must be stored under scrupulously anhydrous conditions. Certain metal hydrides can be employed in a similar manner.[28] Sulfonyl hydrazides and other hydrazine compounds can also be used in expanding liquid polysulfide rubbers.[29]

References

1. (a) Reed, R. A., "Plastics Progress" (1955), Iliffe and Sons, Ltd., London, pp. 51-80; (b) Reed, R. A., *Brit. Plastics,* **33**, (10), 469 (1969).
2. Lasman, H. R., Modern Plastics Encyclopedia (1966), New York, pp. 394-402.
3. van Amerongen, G. L., *Rubber Chem. & Tech.*, **20**, 503 (1947).
4. U.S. Patent 2,299,593 (duPont).
5. (a) B.I.O.S. Final Report 1150(PB79428), pp. 22-3
   (b) Lober, F., *Angew. Chem.*, **64**, 65-76 (1952).
   (c) Stevens and Emblem, *Industrial Chemist*, **27**, 391-4 (1951).
6. U.S. Patent 2,491,709 (ICI).
7. Fuller, M. F., *Ind. Eng. Chem.*, **49**, 722 (1957).
8. Bulletin, "Foaming Vinyl with duPont Nitrosan", E. I. duPont de Nemours & Co., Inc., Wilmington, Del.
9. Newby, T. H. and Allen, J. M., U.S. Patent 2,692,281 (Uniroyal).
10. Lober, F., Bogemann, F., and Wegler, R., U.S. Patent 2,626,933; British Patent 691,142 (Farbenfabriken Bayer).
11. (a) Hunter, B. A., and Schoene, D. L., *Ind. Eng. Chem.*, **44**, 119 (1952).
    (b) Hunter, B. A., and Stander, R. S., British Patent 686,814 and British Patent 693,954. (Uniroyal).
    (c) Schoene, D. L., U.S. Patent 2,552,065 (Uniroyal).
12. Curtius, T. and Lorenzen, F., *J. Prakt. Chem.* (ii) **58**, 160 (1898).
13. Deavin, A. and Rees, C. W., *J. Chem. Soc.* 4970 (1961).
14. U.S. Patents 2,849,028; 2,873,259 (Armstrong Cork Co.)
15. Kharasch, N., Potempa, S. J. and Wehrmeister, H. L., *Chem. Revs.* **39**, 269-332 (especially 276) (1946).
16. Bulletin, "Celogen OT", Uniroyal Chemical, Naugatuck, Conn.
17. (a) Sundholm, N. K., U.S. Patent 2,640,853 (Uniroyal).
    (b) Stemple, G. H., U.S. Patent 2,830,086 (General Tire Co.).
18. U.S. Patent 3,068,532 (Union Carbide Corp.).
19. Prel. Bulletin SC53-52-R (Dec., 1954) "Foamed Epon Resins", Shell Chemical Corp.
20. Fairclough, J. T., U.S. Patent 2,858,282 (Uniroyal).
21. Hunter, B. A., U.S. Patents 3,152,176; 3,235,519 (Uniroyal).
22. Bulletin IOC-122 "Celogen RA", Uniroyal Chemical, Naugatuck, Conn.
23. Hunter, B. A., Root, F. B., and Morrisey, G., *J. Cellular Plastics*, **3**, (6) 268 (1967).
24. Sheppard, C. S., Schack, H. N., and Mageli, O. L., *J. Cellular Plastics*, **2**, (2) 97 (1966).
25. U.S. Patent 3,141,002 (National Polychemicals, Inc.).
26. Hunter, B. A., and Kleinfeld, M. J., *Rubber World*, **153** (No. 3), 84 (1965); U.S. Patent 3,095,387 (Uniroyal Chemical).
27. "Findings" No. 8, (March, 1966) Uniroyal Chemical, Naugatuck, Conn.
28. Hunter, B. A., U.S. Patent 3,114,724 (Uniroyal Chemical).
29. Hunter, B. A., U.S. Patent 3,114,723 (Uniroyal Chemical).
30. Kirby, A. H. M., *Brit. J. Cancer*, **11**, (3) 290 (1948).

URETHANE FOAMS*

The family of urethane foams is one of the most versatile members of the cellular plastics group and growing more versatile each year. Depending on the starting ingredients, it is possible to produce a range of products from extremely soft flexible foams to tough rigid foams to integral-skinned foams.

Chemistry

Urethane foams are prepared by reacting hydroxyl-terminated compounds called *polyols* (usually of the polyester or polyether family) with an isocyanate.

The structural formula of a typical polyester polyol is as follows:

$$HO\text{---}(CH_2)_2\text{---}O\text{---}(CH_2)_2\text{---}O\text{---}\overset{\displaystyle O}{\overset{\displaystyle \|}{C}}\text{---}(CH_2)_2 \cdots \cdots$$
$$(CH_2)_2\text{---}OH$$

---

* Reviewed and revised by Mobay Chemical Co., Div. of Baychem Corp., and Upjohn Co.

The structural formula of a typical iso-cyanate used in the production of urethane foams—in this case, a mixture, 80/20 of the 2,4- and 2,6-isomers of toluene diisocyanate (TDI)—is as follows:

2,4-isomer   2,6-isomer

In the production of urethane foam, two reactions occur. The first is a reaction of the isocyanate, which is in excess, with the hydroxyl groups of the polyol. This lengthens the chain of the latter and ensures termination of the chain by —NCO groups:

polyurethane

The second reaction, which generates the gas and produces the expanded structure, may be chemical or physical in nature. In the chemical reaction, the polymeric structure of the foam is formed by the reaction of the isocyanate with the polyol. Simultaneously, the isocyanates react with water (if present) to form an intermediate product, carbamic acid, which decomposes to give a primary amine and the carbon dioxide gas that functions as the blowing agent to expand the structure.

The physical reaction involves the volatilization of a blowing agent (i.e., an inert chemical) to provide expansion. In a fluoro-carbon expanded system, for example, water is usually not used. Instead, the blowing agent (in this case, a fluorocarbon) is added to the formulation and the foam structure is formed by volatilization of the low-boiling fluoro-carbon under the exothermic heat produced by the polyol-isocyanate reaction. As the fluoro-carbon volatilizes, gas is released to create the cellular construction. Of special interest in some applications (e.g., insulation) is the fact that the gas formed by vaporization of the fluorocarbon is entrapped in the cells, imparting a very low thermal conductivity (K-factor) to the foam. Other low-boiling solvents, such as methylene chloride, can also be used in certain flexible urethane foams for density and hardness control.

Polyols and Isocyanates

It is possible to vary the two major ingredients of the urethanes—polyols and isocyanates—to create a wide range of products.

Isocyanates in use today include toluene diisocyanate, known as TDI, which has found wide use in flexible foams, crude methylene bis(4-phenyl isocyanate), known as MDI, various polymeric isocyanates (which have become important in one-shot processing, cold-cure foams, etc.), and different types of blends, such as TDI/crude MDI.

Polyols, the other major ingredient of a urethane foam, are active hydrogen-containing compounds, primarily variations of polyesters and polyethers. It is possible to prepare many different types of foams by simply changing the molecular weight of the polyol (since it is the molecular backbone that supplies the reactive sites for cross-linking, which in turn is the principal factor in determining whether a given foam will be flexible, semi-rigid, or rigid). For example, using a single isocyanate, it is possible to prepare soft, flexible foams by using polyols with an hydroxyl equivalent weight of about 1000 to 2000 or very dense foams suitable for rigid furniture parts by using polyols with functionalities of more than 2 and an hydroxyl equivalent weight of about 100 to 200.

As a rule of thumb: High molecular weight,

low functionability polyols produce molecules with a low amount of cross-linking and, consequently, a flexible foam. Conversely, low molecular weight polyols of high functionability produce a structure with a high degree of cross-linking—a rigid foam. Of course, it is possible to vary the formulation to produce any degree of flexibility or rigidity between the two extremes.

Modifiers and Additives

In addition to the basic polyol, isocyanate, and blowing agent (if used), urethane formulations may also include the following:

Catalysts (primarily organotin compounds) are added to control and accelerate the rate of reaction, so that maximum rise is synchronized with gelation to regulate cell size and prevent collapse of the foam.

Surfactants control surface tension and are used to develop a fine, uniform cell structure, or, in some cases, to produce large-celled foams. They are usually silicone copolymers.

Modifiers are inexpensive fillers that reduce the cost of the foam, or improve a specific physical characteristic such as processing or strength.

Modifiers like halogenated or phosphorous compounds can also be added to the urethane foam formulation.

Dyes and pigments can be added to color the urethane. Work is also continuing on the concept of using reinforcing materials like fibrous glass to improve the physical characteristics of the various urethane foams.

One-Shots and Prepolymers

The reactions by which urethane foams are produced can be carried out in a single stage or in a sequence of several. The two principal methods are one-shot and prepolymer.

In the one-shot method, all of the ingredients [polyol, isocyanate, blowing agent (if used), catalyst, etc.] are mixed simultaneously and the resulting mixture is allowed to foam.

In the prepolymer method, a portion of the polyol is pre-reacted with a large excess of isocyanate to yield a prepolymer. The prepoly-

mer is then subsequently mixed with additional polyol, catalyst, and other additives to effect foaming. In a quasi-prepolymer method, the polyol is reacted with isocyanate to form one component and the additional polyol, plus other additives, are mixed together to form the second component. The two components can then be mixed together to effect foaming. They are usually mixed in equal quantities.

Rigid Urethane Foam

The rigid forms of urethane foam are highly cross-linked products that differ from flexibles in that they do not recover when deformed. As indicated above, the rigid foams are generally produced by reacting low molecular weight polyols with a functionality of two or more with polymeric isocyanates in the presence of catalysts, surfactants, and blowing agents. Either polyether or polyester polyols can be used, but at the present time the polyethers dominate. Generally, they are formed by the reaction of ethylene or propylene oxide with sorbitol, trimethylolpropane, pentaerythritol glycerol, sucrose methyl ethyl glucoside, and aliphatic or aromatic amines.

The normal density range for the rigid urethanes is on the order of 1 to 3 lb/ft^3. However, some packaging applications can go down to 0.5 lb/ft^3, while furniture applications can use densities up to 20 to 60 lb/ft^3 (approaching solids). Many of these higher density foams belong to the category of integral skin foams (see below).

Rigid urethane foams have outstanding insulating values (their K values are twice as high as the next comparable material). They also have excellent compressive strength, good dimensional stability, and outstanding buoyancy characteristics. When used in the molding of furniture parts, rigid urethanes will faithfully replicate the surface of a mold to create a woodgrain or other decorative effect. Rigid urethane foam furniture will also accept nails or screws and can be finished at the same time as wood surfaces on the assembly line.

Because of these characteristics, rigid urethane foams have found application in such markets as construction, insulation, refrigeration, marine, furniture, packaging, and transportation. (See Fig. 20-2a through 20-2c).

In the construction industry, the high strength-to-weight ratio and excellent insulating

Fig. 20-2c. Roof of cold storage warehouse is insulated with double layer of rigid urethane foam boardstock. (*Courtesy Mobay Chemical Co.*)

Fig. 20-2a. Steel pipe for carrying heating oil is factory-insulated with rigid urethane foam to keep temperature loss to well under 1°F./mile at 750 gallons per minute. (*Courtesy Mobay Chemical Co.*)

Fig. 20-2b. Concrete wall slabs for school building are insulated with spray-up of rigid urethane foam. (*Courtesy Mobay Chemical Co.*)

ability of urethane foams has led to a number of significant uses, including cores for wall panels, insulation for siding, cores for extruded vinyl window frames, various sandwich constructions, etc. Builders have found that the strength and rigidity contributed by a urethane foam core in wall panels permit the use of thinner (by 60% or more), lighter weight, lower-cost surfacing materials with no sacrifice in structural strength. Urethane also makes an excellent core material since it adheres permanently to most surfaces, eliminates need for adhesives, resists water absorption, has good sound control properties, and is unaffected by high or low temperatures.

Another popular use for spray-up urethane foam is as insulation for fuel oil tanks, which must be kept at temperatures between 100 and 200°F so that the contents can maintain viscosity.

Rigid urethane foam is also under consideration as a bonding and insulating medium in road beds. Here, the foam serves to protect highways from frost damage, load stresses, and other destructive forces.

The outstanding insulating ability of rigid urethane has found ready acceptance in commercial refrigerators, refrigerated trucks and railcars, deep-freeze display units, and cold storage buildings. Since urethane foam is twice as efficient as the next best insulating material, it does the job at half the thickness of other materials. Thus, by using thin-wall urethane-insulated models, refrigerator manufacturers can increase capacity by 50% without increasing exterior dimensions.

In the marine industry, rigid urethane foam is used both for its insulating characteristics (e.g., to convert freighters into refrigerator ships, to insulate fishing boat freezer holds, etc.) and its flotation characteristics. At a standard commercial density of 2 lb/ft^3, a cubic foot of rigid urethane foam will float 60 lb of deadweight, which makes a boat virtually unsinkable. It is used both in smaller dinghies, run-abouts, catamarans, etc., as well as in larger vessels. An additional advantage of the rigid urethanes in this application is that the closed-cell construction of the rigid foams minimizes water absorption.

In the packaging industry, rigid foams complement flexible foams (see below) in the protective shipping of heavy, bulky items. Parts as large as a steam turbine generator have been successfully shipped across country using rigid urethane foam as a support. Sandwich panels based on wood or reinforced plastic skins and urethane foam cores are finding increasing use in containerization concepts.

Furniture manufacturers have adapted rigid urethane foams to various decorative and structural components for a number of reasons: low cost tooling (rubber, epoxy, or metal), low capital investment required, low molding pressures, excellent reproduction of mold detail (textured, wood-grained, or smooth), solid and wood-like feel, readily nailed, glued and screwed, and design freedom in contour shapes, cross sections, and surfaces. Currently, case-good components such as drawer and door fronts, decorative trim, legs, table tops, one-piece molded drawers and end panels, chair frames, and shells are being made from urethane foam. There is also extensive use of rigid foam in such decorative parts as wall decor, mirror frames and lamp bases and for structural components, as well as decorative components, in TV/stereo consoles.

Rigid urethane foam's ability to be foamed-in-place (see below) has also encouraged its application to composite structures for furniture parts. Thermoplastic table and chair bases are being filled with foam for rigidity and chair shells and frames involving combinations of vinyl, ABS, or reinforced plastic skins and urethane foam cores are becoming an accepted practice.

Typical rigid urethane foam properties are shown in Table 20-1.

Processing Rigid Urethane Foams

To meet the requirements of these many applications, a number of techniques are available to produce rigid urethane foam, including: foam-in-place (or pour-in-place), spraying, frothing, molding, slabs, and laminates (i.e., skins and foam cores produced as a single integral unit). These are discussed briefly below. A more extensive review of the equipment used in processing the rigid and flexible foams can be found in the section beginning on page 524.

In foam-in-place, a liquid urethane chemical mixture is simply poured into a cavity (or in more up-to-date production lines, metered in by machine). The reactive liquid flows to the bottom of the cavity and starts to foam up, filling all cracks and corners and forming a

Table 20-1. Typical Rigid Urethane Foam Properties.

Density	Compressive Strength	Compressive Modulus	Shear Strength	Shear Modulus
(pcf)	(psi)	(psi)	(psi)	(psi)
1.5-2.0	20-60	400-2000	20-50	250-550
2.1-3.0	35-95	800-3500	30-70	350-800
3.1-4.5	50-185	1500-6000	45-125	500-1300
4.6-7.0	100-350	3800-12,000	75-180	850-2000
7.1-10.0	200-600	5000-20,000	125-275	1300-3000

strong seamless core. The cavity, of course, can be anything from the space between two walls of a refrigerator or freezer or rail car to the space between the top and bottom hull of a boat. Where the cavity is the interior of a closed mold, the technique, of course, is then known as *molding*. The major difference in this instance is that release agents are used in the molding operation so that the shaped piece can be removed from the mold, whereas adhesion to the inside of the walls that form the cavity is a prime requirement in many foam-in-place applications.

Spraying of rigid urethane foam is accomplished with a two-component spray gun and a urethane system in which all reactants are incorporated either in the isocyanate or the polyol. It can be used for applying rigid foam to the inside of building panels, for insulating cold storage room, for rigidizing flexible structures (as in the military concept of erecting inflatable temporary shelters in the field), for insulating railroad cars, etc.

Frothing is a take-off on foam-in-place, in which the foam mixture is dispensed in a partially expanded state (like shaving cream), rather than as a viscous liquid. It is based on the use of a mixture of blowing agents that permit a two-step action. The first blowing agent is usually a low boiling agent that expands almost immediately to blow the mixture into a froth; the second blowing agent is a higher boiling point agent that requires the exothermic heat that evolves from the more advanced reaction of the ingredients. Thus, there is a slight delay between the application of the foam as a froth and the final blow. This allows narrow-walled cavities to be more easily and uniformly filled and is especially applicable for high-rise applications.

Rigid urethane foam is also available in the form of slab stock that can be cut to size and inserted into a cavity or shaped to any other component (e.g., corner supports for shipping heavy objects). In the slab process, the one-shot technique is used. The reactants are metered separately into a mixing head where they are mixed together and deposited evenly onto a moving conveyor. The head generally traverses back and forth to insure an even deposition. As it is laid on the conveyor, the foam begins to rise, forming a continuous bun or slab. Side plates control the foaming to the side of the bun, while top leveling or "loaf-leveling" techniques are available to reduce the amount of foam crowning. In a matter of minutes, the foam is firm enough to be cut into lengths and stored for curing and subsequent fabrication into board stock.

Since the rigid urethane foam, as it rises, can structurally bond itself to most substrates, it is also possible to substitute a surface skin for the conveyor belt. The liquid is then metered directly onto this skin and usually run through a double belt conveyor. It is also general practice to apply a second skin on top of the foaming mixture, using the upper belt of the conveyor to control the thickness of the foam. Thus, board stock with integral skins already attached to the surface of the foam can be produced. The technique has been used to make sandwich construction building panels.

Flexible Urethane Foams

Flexible foams are usually in the density range of 1 to 6 lb/ft^3 and differ from rigids in that they can be bent and depressed easily and will return to their original contours once the applied force is removed.

In flexible foam formulations, high molecular weight polyols are used, usually triols based on trimethylol propane and glycerol reacted with propylene and ethylene oxide. For textile and some packaging applications, polyesters are used. In textile applications, the polyester foam is bonded to a wide variety of fabrics to produce lightweight sportswear or top coats and is used for its resistance to dry-cleaning solvents, its ease of pigmentation, and its adaptability to flame lamination to fabrics (i.e., the foam is passed through a flame that melts the surface and permits it to act as an adhesive when applied to a substrate).

The most popular isocyanate in flexible foam formulations is toluene diisocyanate, although more recent innovations such as cold cure foams, semi-rigid foams, and integral skin foams, call for special isocyanate modifications and blends.

Also included in the flexible foam formulation are silicone surfactants, a tertiary amine, and an organotin catalyst. Expansion is usually through carbon dioxide evolved from the isocyanate-water reaction. But blowing agents like fluorocarbons can be used in conjunction with water to vary the softness of flexible foams.

Overall, flexible urethane foams offer outstanding cushioning characteristics, are easy to fabricate and produce, have excellent energy absorbing properties, and are long-wearing and low in cost. Principal market outlets include: furniture cushioning, carpet underlay, bedding, packaging, automotive seating, and safety padding. (See Figs. 20-2d and 20-2e).

As can be seen by the series of photographs, the urethane foams do lend themselves to adaptation to automated production, particularly where composite structures (e.g., producing shell, cushion, and frame in one piece) are involved.

Processing Flexible Urethanes

The two major production techniques in the flexible foam area are slab stock and molding.

Flexible slab is made in the same way as rigid foam slab or board stock, that is, the ingredients are fed into a mixing head, mixed, and deposited evenly onto a moving conveyor or substrate. Flexible slab, however, is usually wider and higher than rigid slab and can go up to 8 ft wide and 50 in. high. As for rigids, the one-shot technique is used. After the foam rises, it is cut into buns and generally stored for 12 to 24 hours before fabrication. A number of techniques are available today to contour flexible slab to most any shape desired or to simply peel off thin sheets of a given thickness.

Until recently, most molding techniques for the flexible foams involved metering the liquid ingredients into a mold cavity (with or without inserts such as springs, frames, etc.), then using heat for cure. In a typical application, molds would be filled and closed, then heated rapidly to 300 to 400°F to develop maximum properties, particularly compression set.

A good deal of recent activity, however, has centered on the so-called "cold cure" technique. As originally conceived, cold-cure foams,

involving special polyols and isocyanates, were formulated so that they could cure in a reasonable amount of time to their ultimate physical properties without the need for additional heat over and above that supplied by the exothermic reaction of the foaming process. As such, foams that are truly cold curing in nature offer a number of advantages, including: no need for curing ovens; short de-mold times and rapid production rates; adaptability to low-cost

Fig. 20-2d. Office chairs (bottom) are made by laying reinforced plastics chair shell over fabric cover in mold (top), then injecting the urethane foam cushioning into the space between shell and cover (center). (*Courtesy Mobay Chemical Co.*)

Fig. 20-2e. Sofa and chair ensemble is based on self-curing flexible urethane foams. (*Courtesy Burris Industries and Mobay Chemical Co.*)

molds (reinforced plastics or elastomeric); ability to produce a wide range of molded parts from 1 to 100 lb in weight; ability to produce highly contoured molded shapes which could not be produced by other techniques; good surface appearance; and the ability to foam the urethane in or behind potentially heat-sensitive skins. In the latter instance, this might apply to cold-cure foaming of urethane flexible cushions directly into a vinyl skin or wrapping a product in film and cold-cure foaming flexible urethane around the part as a protective cushion for shipping.

As currently used in the industry, however, the major outlet for cold-curing formulations has been in the production of what is known as *high-resilient foams*. In trade parlance, a high-resilient foam is one that has a sag factor (i.e., the ratio of the load needed to compress foam by 65% to the load needed to compress foam by 25%) of 2.7 and above. Sag factor, of course, is most important to cushioning characteristics. With high sag you get a better cushion that will give easily when you begin to sit on it, but will become firm when you put your whole weight on it.

True cold-cure foams will produce a sag factor of 3 to 3.2; hot-cured foams will run from 2 to 2.6.

However, truly cold-cure foams and high resilient foams are not mutually inclusive. There has been a trend in some industries, most notably automotive seating, to produce high resilient foams using heated molds and heat cures. Such foams have a generally higher level of mechanical properties, but their sag factor of 2.7 to 2.8 is not as high as true cold-cure foams. Optimum mold temperatures are in the range of 120 to 140°F, depending on the material of construction and an in-mold cure time of about 10 min. at 140 to 160°F is usually used before the part is stripped. Although these requirements are far less than that for conventional hot-molded flexible foams, these foams are not truly cold cure in the original sense. Since they are high-resilient foams, however, and since the same advantages of faster production rates and production economies apply, the industry trend has been to lump them all together under an overall heading of cold cure or high-resilience foams. The processor has a choice of matching formulation to the type of production line he wishes, from room temperature cure up to 140 to 160°F cure.

Typical properties of a cold-cure foam are shown in Table 20-2.

Table 20-2. Properties of cold-cure foams*

Density, pcf.	2.5–4.5+
Tensile strength, psi.	10-14
Elongation, %	90-110
Tear strength, pli. (core foam)	0.7-0.9
Tear strength, pli. (with surface skin outside)	0.9-1.1
90% compression set for 22 hr, %	less than 8
50% compression set for 22 hr, %	less than 5
Indention load deflection ILD, lb/50 in.^{2} at approx. 3″ thickness at:	
20% rest	7-50
65% rest	25-160
SAG factors	2.8-3.2
Compression load deflection CLD, psi. at	
25%	0.10-0.40
20% rest	0.15-0.35
50%	0.25-0.50
Rebound, %	50-60
Shrinkage, perpendicular to rise, %	2-3
Shrinkage, parallel to rise, %	0.3-0.5
Gas loss, %	4-5

* Mobay Chemical Co.

Semi-Rigid Foams

These foams are characterized by low resilience in that they recover very slowly from compression and by high energy absorbing characteristics. As such, they have found prime outlet in the automotive industry for applications like safety padding, arm rests, sun visors, horn buttons, etc. These foams are also cold curing in nature and usually involve special polymeric isocyanates. They are usually higher in density than regular flexible foam (up to 12 lb/ft^3) and are generally applied behind vinyl or ABS skins. The liquid ingredients are simply poured into a mold into which the vinyl or ABS skins and metal inserts for attachments have been laid. The foam fills the cavity, bonding to the skin and inserts.

The most recent advance has been the use of integral skin semi-rigid (or semi-flexible) foam in which the foam comes out of the mold with a continuous skin that can replace the separate vinyl or ABS surface (see below).

Microcellular Foams

These foams range in densities from about 30 to 60 lb/ft^3. The chemistry is essentially the same as that for solid urethanes, that is, a 2000 weight polyol and low molecular weight extenders reacted with methylene bis(phenylisocyanate). The microcellular urethanes offer the properties of solid urethanes, but are lighter in weight. They are used in shoe soles, automotive bumpers, seals, gaskets, and vibration pads.

Integral Skin Foams

Formulations and processing techniques are available to produce self-skinning semi-rigid and rigid urethane foams, usually fluorocarbon-blown one-shot systems based on polymeric isocyanates. Basically, in any of the self-skinning processes, urethane chemicals are poured into a mold, where they react to form a nonporous, relatively solid-surface skin at the mold face, while simultaneously developing a foam core on the inner sections of the part. This can be accomplished either chemically through formulation or by such processing techniques as using a heat-conducting mold surface that will remove heat and cause a skin to form just inside the surface of the mold (i.e., by suppressing vaporization of the blowing agent at the surface so that the foam does not expand at the same rate as in the center). No matter how it is made, however, a cross section of an integral skin foam will generally show a regular cellular core at the center that grows denser and less cellular in construction as it extends from the center to the outside surface, where the urethane is dense and almost solid.

Historically, the first integral skin foams were developed in the semi-rigid (or semi-flexible) area primarily for use in such parts as automotive safety padding, automotive bumpers, athletic padding, etc. Prior to its availability, most of these products were made by producing the skin in a separate operation (e.g., by vacuum forming a vinyl or ABS sheet) and then, in a second operation, pouring the foaming chemicals into the skin.

In a typical operation for producing a semi-rigid integral skin foam, a metal mold is coated with a mold release agent and heated to proper temperature. This is a very critical step since each integral skin foam has its own optimum mold temperature. If the mold temperature is too high, the foam skin will become porous and difficult to finish; if it is too low, the foam density will be excessively high, causing voids in the foam skin. After heating, the mold is filled with the urethane foam chemicals, using conventional metering/mixing equipment (accurate metering and efficient mixing are especially critical) and allowed to rise (provision is made for the air inside the mold cavity to escape during the foam rise). The part is then de-molded, ready for finishing.

More recently, self-skinning rigid urethane foams have moved into prominence for such applications as furniture, automotive components, construction parts, computer housings, etc. Parts over 100 lb can be made by these techniques and the resulting product has a sandwich-like structure that offers unusual stiffness at relatively low weights. High strength and good dimensional stability are other characteristics of the integral skin rigids. The methods used to produce these foams are also much faster cycling than conventional techniques for high-density foams and as such are adaptable to mass production techniques.

In industry terminology, these self-skinning rigids are sometimes included in the family of structural foams (see p. 580). Although characteristics are similar, we will limit the later discussion on structural foams to the thermoplastics and cover the urethanes at this time.

Molding Rigid Urethanes

One of the most active areas in R&D in the 1970s took place in the concept of molding rigid urethane foams. Today, the common terminology for the technique is reaction injection molding (RIM) or the closely related liquid injection molding (LIM). Originally, however, the self-skinning rigid urethane foams had their beginnings with such proprietary developments as Bayer's Duromer process (or Mobay's Baydur system, as it also known), Uniroyal's Rubicast process, and Upjohn's Isoderm process. As a starting point for understanding the historical

development of RIM and LIM, these three techniques are discussed in greater detail below:

Duromer. One fast-cycling method that is akin to injection molding can produce parts that range from 15 to 50 lb/ft^3 at the core up to 70 lb/ft^3 at the skin surface. It is known as the Duromer process (Mobay Chemical Co.) and is illustrated schematically in Fig. 20-3. Average values for a Duromer structural foam part are given in Table 20-3 and some typical applications are shown in Figs. 20-4a and 20-4b.

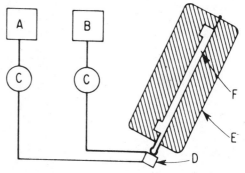

Fig. 20-3. In urethane structural foam process, chemical components are metered from tanks A and B, through pipe C to metering head D. Mix then enters mold E and foams in cavity F. (*Courtesy Mobay Chemical Co.*)

The starting materials are polyisocyanates and polyfunctional alcohols which are liquid at room temperature. These materials as well as stabilizers, activators and blowing agents can usually be preblended and used in the form of a two-component system that can be best handled through high pressure mixing equipment. Tertiary amines are most often used as catalysts for these formulations and by controlling the rate of chemical reaction affect the cycle times of the molding operations. Halogenated hydrocarbons, such as monofluorotrichloromethane, are the preferred blowing agents in these foaming systems.

The processing of the liquid starting raw materials into finished parts follows a very well defined pattern. The following sequence of events takes place in making a finished item.

1. Chemical reactants are prepared by preblending to a minimum of two components for ease of handling in foam

Table 20-3. Typical average properties of Duromer urethane foams.*

Skin thickness	$\frac{1}{64}$-$\frac{3}{32}$ in.
Overall density	25-75 pcf
Tensile strength	1400-2900 psi
Tensile modulus	
Skin	370,000 psi (approx.)
Composite	85,000-200,000 pri
Compression strength	2300-4300 psi
Flexural strength	5000-7000 psi
Flexural modulus	100,000-200,000 psi
Impact strength	3.3-6.9 ft-lb./in.^{2}
	(Charpy type)
Elongation	5-20%
Surface hardness	70-90 shore D
Temperature range	−60-+230°F (from DTA and Clash Berg tests)
Color stability	Will darken on exposure to light
Repairability	Epoxies, polyesters, etc.
Shrinkage	< 0.5%
Expansion coefficient	3-5 x 10^{-5} in./in./°F
K-Factor	0.4 at 25 pcf, 0.7 at 45 pef BTU−in./hr°F ft^2
Water absorption	< 1% (by weight)

* By Mobay Chemical Co.

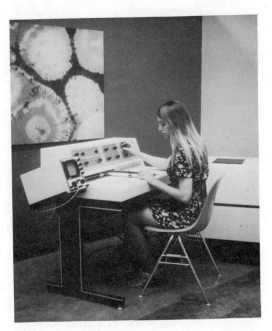

Fig. 20-4b. Control console and magnet housing for spectrometer use toughness, lightweight, and dimensional stability of structural urethane foam to advantage. (*Courtesy Varian Instrument and Mobay Chemical Co.*)

Fig. 20-4a. Sailboat hull is molded of self-skinning structural urethane foam. (*ourtesy Tedruth Plastics and Mobay Chemical Co.*)

machines. The usual procedure suggests that polyol, blowing agent and catalyst form one component and the polyisocyanate the other.

2. The components are accurately metered and thoroughly mixed using a proper foam machine. The materials are mixed in a fraction of a second.

3. Mixed materials are injected directly from the machine into a closed mold cavity only partially filling it. This operation takes from 2 to 15 sec.

4. Chemicals react, expanding the polymer mass to fill the mold. This takes 30 to 90 sec.

5. Foam then cures and partially cools to form a rigid item. Depending on many variables, this phase of production takes from 3 to 20 min.

6. Part is de-molded.

7. Part is held for continued cooling.

8. Part is prepared for finishing.

9. Part is finished in a conventional manner.

The use of a foam machine is absolutely

necessary in the production of parts because the raw material systems react extremely fast. Processing by hand is not recommended even for experimental purposes since wrong conclusions may be drawn from poor data. Generally, the raw materials can be processed using most of the known two component metering-mixing machines as long as the following conditions are fulfilled.

1. The mixing chamber volume should be kept as small as possible while still providing for good mixing.
2. The mixing chamber should not be sensitive to pressure variations.
3. The mixing head should be as light as possible and be connected to the metering machine by means of long flexible hoses for ease of handling and connecting to molds.

Easy attachment of additional pumps for the separate metering of blowing agents and catalyst into the resin component should also be possible. Machine metering of additives into the resin stream gives the option of controlling the foam formation within certain limits by variation of the blowing agent content and the activity of the mixture. The delivery rates of the metering pumps for the different components should be continuously variable so as to give a freedom of choice in chemical reactant.

The process requires only moderate temperature endurance and pressures, thus broadening the choice of mold material and allowing the use of relatively low-cost tooling.

The quality of molds is important, for the mold surface is reproduced in exact detail in the part. Scratches and pinholes on the mold surface will show up on each part made in that mold. The pressures developed during foaming cannot be predicted in every case, but usually fall between 35 and 45 psi. Therefore, all molds and clamps should have a designed strength of about 100 psi for safety and performance reasons. These pressures are much lower than injection molding pressures, thus allowing a lighter mold construction. Care should be taken, so that there is sufficient stiffness to prevent bulging of the surface of larger molds.

The formation of solid uniform skin is not only dependent on the chemical formulation, but also requires a specific temperature of the mold during foaming. Generally, the mold temperatures will be between 120 and 150°F, and should be controlled to +/-3°F. For medium and large production runs one must be able to control the mold temperature and this can be satisfactorily accomplished only with metal molds. The best temperature control can be obtained with a cooling-heating liquid being circulated through the mold. Other methods such as oven heating and water-spray cooling are also feasible. The type of metal molds that may be used are: cast aluminum, steel, electroformed nickel, fabricated aluminum or steel, and heavy-metal alloys. Aluminum filled epoxy resin tools can be used for prototype runs.

The correct mold position depends on the geometry of the part, the flow pattern and vent arrangement of the mold. The final positioning of the mold can be determined best by trial and error.

The geometry of the part also determines the general filling techniques. However, for the most trouble-free operation and the best-looking part, the sprue (injection part) should be positioned near the bottom of the mold. Any air bubbles present at the beginning of the filling cycle are on the surface of the reaction mixture, and can be removed at the highest point in the mold through venting. Again, venting arrangements can be established only by trial and error.

Rubicast. A second technique available for self-skinning rigid foams differs in that the outer skin is formed chemically at atmospheric pressure so that no special dispensing systems are needed and metal molds need not be used. It is known as the *Rubicast method* (Uniroyal) and produces parts ranging in density from 8.5 to 50 lb/ft^3.

As with conventional rigid urethane foam systems, this one is the product of two liquid chemicals: an isocyanate and a resin blend. When the components are mixed together and cast into a mold, the carefully controlled polymerization produces a weak polymer layer around the entire outer periphery of the rising foam. Because of this weak polymer layer, formation of a very high density, noncellular "skin" is readily formed with minimal molding pressure. Molding pressures for the self-skinning structural urethane foams are generally 10 to 15 psi for moldings up to 1 in. thick. Because of these low molding pressures, conventional mold materials such as GRP, silicone rubber, and urethane elastomers may be used for production. Also, the system may be processed on conventional rigid urethane foam casting equipment with only very minor adjustments necessary.

The processing parameters are simple but critical. Due to the relatively rapid reaction initiation of the system, the mixed liquid components are generally charged into closed molds, and for most applications the mold must be charged within 20 sec. This means that the output of one's casting machine will determine the largest possible part size which can be successfully molded. The casting machine's recirculation and dispensing pressures must be balanced, and there can be no excessive pressure in the mixing head. The range of temperatures of the system components in recirculation is as follows: resin, 60 to 75°F; isocyanate, less than 75°F. Depending upon the size and configuration of the article to be molded and the mold material, the mold temperature may vary from 70 to 130°F, although the vast majority of moldings is done from 85 to 110°F. Molded items may range in thickness from $\frac{3}{8}$ to 3 in., although maximum molding efficiency is generally experienced when the molded articles range in thickness from 0.5 to 1.0 in. Mold dwell times may run from five to 30 min depending upon the size and thickness of the molded article. By exact adherence to the correct processing conditions as well as careful control of the chemical technology, one is able to mold articles of diverse sizes and shapes; and those articles, upon de-molding, will possess an evenly distributed, distinct, hard skin around their entire outer surfaces, even when the foam rises 6 ft or more.

Isoderm. A third system known as the *Isoderm process* (Upjohn) is used for making thermally stable foams (can operate at service temperatures up to 150°C) and is based on the use of methylene bis-(phenylisocyanate) and a halocarbon blowing agent. As with the other techniques, two reactive components are mixed and poured into the mold cavity where they react and foam. Key to obtaining the dense skin is the use of a highly heat conductive mold. Since a halocarbon-blown system is used, the center of the part will blow while the surface nearest the mold will pack in advance of gelation. Because of this, a uniform hydrostatic pressure exists, resulting in a part with a uniform skin structure. The pressure collapses the cells at the mold surfaces and forces the halocarbon into the bulk of the foam. The free-rise density of parts made in this way can run from 3 to 5 lb/ft^3, while the packing at the skin surface can lead to an overall density of 15 to 50 lb/ft^3. Molding pressure is less than 100psi and parts up to 4 by 8 ft in size can be made in a single pour.

Reaction Injection Molding

More recently, as the technology for producing large polyurethane elastomer and rigid foam parts advanced, and as more machinery manufacturers and urethane raw materials suppliers entered the field, a common deniminator for classifying the various new methods for molding became evident in the fact that high-pressure impingement mixing (HPIM) was being used in contrast to the low-pressure, mechanical mixing systems used in standard liquid casting techniques. In HPIM, two or more pressurized reactive streams are impinged in the mixing chamber prior to injection into the mold. The reactants are delivered through nozzles to the chamber at 2000 to 3000 psi, though pressures in the chamber and in the mold remain quite low, around 15 to 75 psi. By comparison, the low-pressure equipment widely used for standard liquid casting of urethanes into molds involves feedline pressures of only about 40 to 200 psi and some means of mechanical mixing (with attendant disadvantages of periodic cleaning after each shot into a mold, except as described in some of the newer LIM techniques noted below).

Today, the various high-pressure impingement mixing systems are more widely classified as reaction injection molding (RIM). As contrasted to low-pressure molding and mechanical mixing, RIM offers the following advantages: (1) Since the mixing head is self-cleaning, the solvent flush, which is required in low-pressure molding, is eliminated; (2) since there is no mechanical mixing, outputs are higher and much faster reacting urethane mixes can be used; (3) since reaction time is faster, mold residence time is reduced because the material cures more rapidly; this, in turn, can reduce the mold cycle time by as much as 75%; and (4) high-pressure impingment mixing reduces air entrapment in the reaction mixture, improving part appearances, eliminating surface defects, etc.

RIM has also been compared to injection molding for the manufacture of such parts as automotive bumpers, etc. In this case, the advantage of RIM lies in the fact that mold pressures are very low—below 70 psi—so that a 100-ton press is adequate. In addition, it is possible to use less costly molds. On some runs, in fact, aluminum-filled epoxy tools have been used. On longer, quality runs, however, it is generally recommended that a nickel- or chrome-plated steel be used. Big savings, however, can be achieved in the machinery required. Injection molding, of course, requires high-torque extruders and very high tonnage presses to process high viscosity melts and mold large parts; RIM can achieve these goals with much less expensive equipment.

RIM is suitable for making flexible, semi-flexible, or rigid materials and has been used for producing solid, microcellular, or foamed parts. Its principal use thus far, however, has been with urethane elastomers and integral skin rigid foams.

The Process. The major components of a RIM system generally include: the storage tanks for the liquid components of the urethanes (tanks are heated and fitted for recirculation); high-pressure, high-volume pumps for metering the components to the mixing head; the mixing head itself, which must be designed to handle two or more streams; a mold (generally one that can be automatically opened and closed); and the press (300-ton capacity maximum is sufficient).

Precise control of material flow and tem-perature must be maintained for high through-put with recirculation. Heating of the delivery and storage equipment is required to maintain the low viscosities (1000 to 1500 cps) that are needed.

During operation, the urethane components, usually in two streams (the polyol and the isocyanate), continuously recirculate through the system. The pumps must be capable of filling the mold completely within 2 to 3 sec, which entails throughputs as high as 650 lb/min and pressure to 3000 psi.

The heart of the system is the mixing head. The impingement mixing chamber is very small—1 to 5 cu cm—and impingement pressures run between 1500 and 3000 psi. Most of the heads have a self-cleaning feature, accomplished through continuous recycling until the moment of mixing. No solvent flush is required.

The mixed components, upon being injected into the mold, usually at pressures below 60 psi, begin reacting immediately. Mold positioning is often important; it must be aligned in the best attitude to allow filling without turbulence and to provide for the fast venting of the gasses.

Liquid Injection Molding. There are at least two other techniques which come under the category of liquid injection molding (LIM) that are not quite the same as RIM, but accomplish essentially the same thing and that are used for much the same purpose. One (from USM) involves feeding the reactive components into the mixing chamber where they are acted upon by a mechanical mixer (instead of using high-pressure impingement mixing). However, unlike low-pressure mixing, the mixer does not normally need to be flushed, since a special feed system automatically dilutes the residue in the mixer with part of the polyol needed for the next shot, thereby keeping the ingredients from reacting. One advantage is that higher viscosity materials can be handled in this way.

Another departure (from Dow) involves still a second variation on LIM, plus a special urethane formulation. The formulation is different in that, in addition to the polyol and the isocyanate, it contains an organic modifier that functions as a heat sink to absorb the exothermic heat of reaction that usually results when an isocyanate and a polyol are combined. This means that very large, thick-sectioned parts (up to 6 in.) can be processed on very short cycles (1 min or less). The end-product is neither a foam or an elastomer, but a cross-linked, rigid solid urethane plastic. The LIM equipment used for the system also differs. In operation, the two liquid components (one of which also contains the organic modifier) is pumped into a mechanical mixer where it is mixed prior to injection into the mold (or molds). To overcome the need for flushing out the reacted material that might set up in the mixer, gates, runners, etc., the system uses two molds and a special valving device. By timing the catalyst kick-off in the formulation to coincide with the operation of the molds, it is possible to fill one mold in 20 to 30 sec (reaction takes place in the mold, not in the mixer, gates, or runners), close the valve, and divert the flow of un-reacted material to the second mold which fills while the material in the first mold is curing. The process then alternates, so that there is a continuous flow of unreacted material through the system. Shots of up to 40 lb and perhaps as high as 100 lb are considered feasible with the technique.

URETHANE PROCESSING EQUIPMENT*

Items which should be considered prior to purchasing urethane processing equipment are:

1. The most efficient method of manufacture in terms of:
    a. Product dollar output per dollar invested. A good example of this is the difference between molded items and slab products. For an equal investment cost, product dollar output of a molding line may be $\frac{1}{10}$th of that of a slab process. This must be balanced against a greater product loss from the slab process.
    b. Degree of mechanization or instrumentation required. There are two main facets: manual labor beyond a certain basic amount directly reduces productivity per unit of time; it is extremely difficult to reproduce

* By R. L. Bowyer, Martin Sweets Co., Louisville, Ky.

quality of product without a basic level of mechanization and instrumentation.
2. Versatility desired in the original equipment installation. This refers to the capability of the equipment to be adapted to changing production requirements or to take advantage of new or different technology.
3. Degree of durability required. This refers to the difference between intermittent production (light duty) rated machines and continuous production (heavy duty) rated machines.
4. Cost.

Following is a list of most of the types of equipment that are currently available to the industry together with very generalized comments on their type and capability. Most equipment manufacturers today offer a full line of standard equipment. Generally, however, the bulk of equipment sold falls into the semi-custom or custom-type category; other variations are possible.

The following information is divided into six main sub-groups based primarily on the type of application for which the equipment is used. These are:

1. Equipment for spray applications.
2. Equipment for pour-in-place applications.
3. Equipment for molding applications.
4. Equipment for the production of slab foam.
5. Equipment for frothing applications.
6. Equipment for cast elastomer applications.

It should be noted that Number 5, frothing equipment, and Number 6, elastomer equipment, have generally been, to a great degree, listed with Number 2, pour-in-place equipment. It is now felt, however, that broad growth and technological advances in these areas merit the listing and discussion of this type of equipment on an individual basis.

Equipment for Spray Applications

Spray machines are required to apply mixed foam chemicals in a largely atomized condition onto a vertical, horizontal, or other type of surface. The equipment available today includes machines designed primarily for field use, in-plant or factory use, laboratory use, or combinations of all. These units are generally designed for two components and are available with recirculating type systems when desirable and where applicable (Fig. 20-5). The type of metering systems available cover the entire range from pneumatic and hydraulic piston pumps through vane pumps and gear pumps, both internal and external. Mixing types include both external and internal systems with high and low pressure machines being used.

The production capacity of the usual spray unit usually lies between 2 and 10 lb of chemicals per minute. Some of the machines are capable of covering this entire output range with any formulation. Others are fixed at some point within this range and may be limited as to the type of formulation in regards to ratio and viscosity with which they can be used.

Accessories often required for successful application include temperature control devices for the chemical streams, instrumentation, automatic level control of the chemicals, solvent flush systems, air dryers and filters, additional hose length, and so forth.

It should be noted that some of the spray machines currently available, particularly those with positive internal mixing, are capable of being used for spray applications, pour-in-place applications, and froth applications merely by substituting tips and turning the atomizing air off.

It should also be noted that although the greatest interest in the spray field lies with the rigid and semi-rigid foams, equipment, techniques, and technology are available to spray flexible foams.

Equipment for Pour-In-Place Applications

Pour-in-place applications are those in which the liquid is poured into an existing cavity and allowed to bond itself intricately to the walls of the final product. The cavity may be brought to a fixed in-plant installation to be filled, or the

Fig. 20-5. Typical two-component, gear pump-type spray machine. (*Illustrations courtesy Martin Sweets Co.*)

equipment may be taken to the site of the cavity.

The equipment available includes those primarily designed for field use, in-plant or factory use, laboratory use, or combinations of all. Those machines for the production of rigid foams are primarily two component. Those for flexible may be three or more components. Full recirculating type equipment generally has the advantage here in terms of reproducibility and reliability since the product rarely can be removed for inspection without destroying it; thus, the urethane mix must be applied correctly each time.

Flowrates of equipment in this category can vary from as little as a few grams per minute up to several hundred pounds per minute with many installations being regarded as multiples having several machines pouring simultaneously. The bulk of the work, however, would be accomplished in the 20 to 100 lb/min range. It should be further noted that there has been a steady increase in the average flowrate of rigid processing machines as technological advances are made in the industry.

Metering pumps are generally gear pump systems, both internal and external (Fig. 20-6); however, we should not overlook the piston-type pumping units which utilize volumetric displacement to maintain metering accuracy. The piston-type pumping units are making advances in certain fields such as refrigeration since theoretically, once the system is set up for a given ratio and all air has been purged from the systems, there is virtually nothing that can change the ratio of the two materials being processed. Therefore, metering accuracy is guaranteed. There are, however, some limitations as far as the piston units are concerned. Ratio adjustment is limited to a small range without changing pistons or piston rods, temperature control of the materials in the pistons must be maintained in order to duplicate repeated pours, and shot size is limited to the size and number of pistons on the unit.

Available mixing impellers for the pour-in-place type equipment include low shear and high shear impellers, and generally are of the rotating type with mixing speeds generally in the 3000 to 6000 rpm range. The proper choice

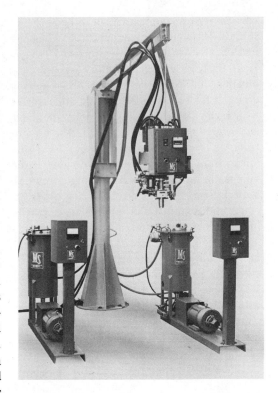

Fig. 20-6. Two-component, gear pump-type machine for processing urethanes.

depends partly on the duration and frequency of the pours with the high shear impellers having the advantage when self-cleaning and fast cycling are desired.

Production capacity of this type of machine ranges from the very smallest available, less than a quarter of a pound per minute total chemical flow rate, up to some of the largest rigid machines, more than 500 lb/min (Fig. 20-7). The majority of the applications seem to require units in the 20 to 100 lb/min range with capacity being determined both by the size of the cavity and the cream time of the formulation to be used since this determines the available time for pouring the proper quantity desired. Again, it should be noted, however, that the average flow rate of rigid foam machinery is on the increase.

Accessories often required for successful applications include temperature control devices, instrumentation, solvent flush systems, automatic level control of chemicals, bulk storage tank farms, filters, additional com-

Fig. 20-7. Comparison of small 1/4-lb/min. mixing head (right) and large 1000 lb/min. mixing head.

ponents, air bleed systems, special seals, special pumps, corrosion resistant fluid circuits, and so forth.

Equipment for Molding Applications

Molding applications are those in which the liquid mixture is poured into a cavity in an object from which the foamed article is later removed. The equipment is very similar in design and capacity to the pour-in-place type equipment and differs primarily in being somewhat more sophisticated with a greater degree of instrumentation and control. This type equipment is more apt to be associated with conveyor and oven equipment for greater production efficiency.

Equipment for the Production of Slab Foam

Foam slab producing equipment is designed to produce a continuous length of foam generally from 4 to 7 ft wide and from 2 to 3 ft high. As usually used, it consists of the chemical handling, metering, and mixing equipment, the spreading equipment for distributing the chemicals on the conveyor, paper handling equipment for lining the conveyor continuously with paper and removing it from the foam, ventilating hoods, various types of curing equipment, and various types of cutting and shaping equipment. (See Figs. 20-8 and 20-9)

The various segments can be bought separately or as a complete package. If purchased separately, however, they must be knowledgeably integrated with each other. Most reliable equipment manufacturers are able to offer turnkey plants equipped with a guarantee to turn out a previously agreed upon quality of product. This type of equipment is generally much more complex than the previous classifications and includes what is generally called "boardstock machines" of various types, as well as the open and closed top flexible and rigid slab complexes.

The flow rate of these machines vary from as little as 50 lb/min up to and somewhat above 1000 lb/min with the actual flow rate being determined more from the size of the product desired than from any economic considerations. The capacity, of course, is directly related to the height and width of the foam block to be produced. (Fig. 20-10)

Most of these installations receive their raw materials by tank car or truck; some utilize shipping containers directly for daily storage with in-plant storage of sufficient size to permit switching from one tank car to the next without interruption of flow and without mixing two tank car deliveries of material.

In the larger installations, preblending or mixing has been minimized through the use of multiple-component metering units able to handle each chemical separately.

Equipment for Frothing Applications

Frothing equipment, as mentioned earlier, is very similar to pour-in-place or molding equipment. However, there is a difference between the three, not so much in the type of equipment required, but mainly in the complexity of operation of the frothing equipment.

Frothing is the practice of pre-expanding the foam before polymerization begins so that it

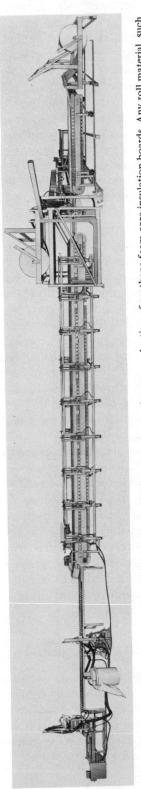

Fig. 20-8. Full length view of conveyor line (minus foam machine) for the continuous production of urethane foam core insulation boards. Any roll material, such as paper, foil, plastic, etc., can be used as the skin of the board.

does not tear itself apart traveling between two rather confining walls while increasing in viscosity.

Today's technology seems to indicate the limitations of vertical liquid pours with high surface to volume ratios. Using truck trailer bodies as an example, it has been extremely difficult to obtain good foam properties in the last two feet of an eight foot rise in a single pour. As another example, in the refrigeration industry, some manufacturers foam the refrigerator cabinet in the "throat down" or "breakers down" position, since a liquid unfrothed foam injected into the cabinet between the outer case shell and the inner case liner will be more susceptible to leakage than a froth foam would be. (Of course, the seal between the shell and liner is a consideration in the selection of liquid pour or froth.)

The solution that has been offered for these problems and for producing a very light density rigid foam in the 1.5 to 2 lb/ft^3 range is frothing. In addition, the frothing technique allows the user to "froth down" to a lower density than a pour-in-place system would permit, thereby making the frothing technique economically more attractive.

The complexity of operating a froth machine, as mentioned earlier, is due to the fact that Freon-12 must be introduced into the mixing chamber under pressure as a liquid and allowed to reach atmospheric pressure and turn to a gas at a controlled rate. This sounds as if it is a simple task; however, it is not quite as cut and dried as it may seem. Opening a let-down valve too quickly after a froth pour, or installing a wrong-size pour hose on the mixing chamber of a static helix mixer can create numerous problems.

Another reason the system is a bit more difficult to operate is that a thorough working knowledge of the fluid metering system must be had in order to understand what complications can arise when the metering systems have to deal with higher dispensed pressures created by injecting the Freon-12 into the mixing chamber during the mix cycle. It should be noted that generally speaking, there are two types of froth used on conventional frothing equipment. One is a static helix mixed froth

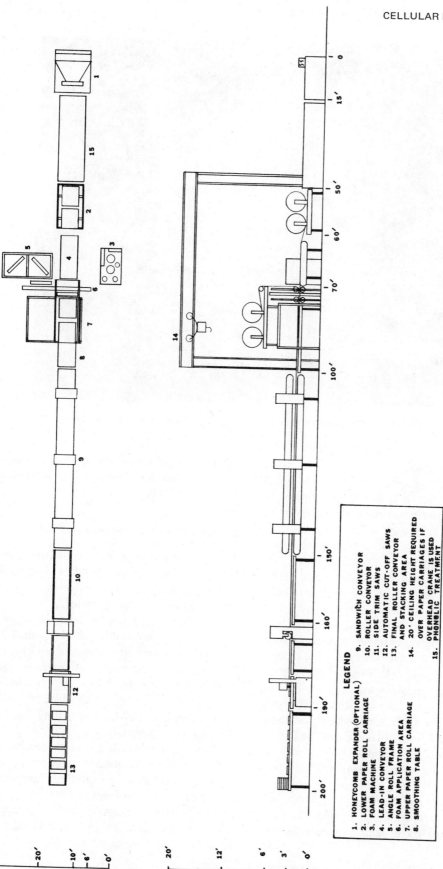

LEGEND

1. HONEYCOMB EXPANDER (OPTIONAL)
2. LOWER PAPER ROLL CARRIAGE
3. FOAM MACHINE
4. LEAD-IN CONVEYOR
5. ANGLE ROLL FRAME
6. FOAM APPLICATION AREA
7. UPPER PAPER ROLL CARRIAGE
8. SMOOTHING TABLE
9. SANDWICH CONVEYOR
10. ROLLER CONVEYOR
11. SIDE TRIM SAWS
12. AUTOMATIC CUT-OFF SAWS
13. FINAL ROLLER CONVEYOR AND STACKING AREA
14. 20' CEILING HEIGHT REQUIRED OVER PAPER CARRIAGES IF OVERHEAD CRANE IS USED
15. PHONBLIC TREATMENT

Fig. 20-9. Top and side view of typical panel production line.

Fig. 20-10. Multi-component machine for processing flexible slabstock or rigid boardstock.

whereby the actual mixing of the material is done by the turbulence created in the helical mixer by the Freon-12 injected into the system forcing the foam material to spiral down the nonrotating helix mixer. The second is a let-down valve frothing system whereby the urethane components as well as the Freon-12 are mechanically mixed by a driven mixer, but the let down of the Freon-12 from high pressure to atmospheric pressure is controlled by a valve with a variable orifice located at the discharge end of the mixing chamber. It is also claimed by some that the helix system is a less complex system to operate both from an operator and the machine standpoint, although the let-down system will give the operator much better control of the foam system in regard to the quality of the froth obtained, and the usage of Freon-12 will be considerably less with the let-down valve system than with the helix system.

Accessories required for successful operation of froth equipment would be identical to the pour-in-place and molding machines with the addition of a system for accurately and satisfactorily metering the Freon-12 into the mixing chamber as a liquid.

Equipment for Cast Elastomer Applications

Here again, elastomer equipment has been listed to a great degree as pour-in-place equipment. From a practical standpoint, however, there are differences which should be noted.

Elastomer casting can be defined as a technique whereby a liquid resin is poured or otherwise dispensed into an open mold where it cures without application of pressure, but possibly with the application or removal of heat. A special case of casting where the mold itself becomes part of finished casting is generally referred to as *potting*.

Mixing and metering machines capable of processing low performance elastomers are almost identical to the conventional equipment for handling urethane foam systems, but equipment capable of processing high performance elastomer systems (e.g., Dupont's Moca and

Adiprene systems) are considerably different inasmuch as these machines must be capable of thoroughly degassing the elastomer components to within 50 millimeters of mercury pressure, maintaining the components to a maximum of 285°F and also capable of mixing and depositing the system in a bubble or air free manner. There are different approaches to all three of these required capabilities.

Degassing is generally accomplished in much the same manner on all makes of equipment. However, there are considerably different approaches to the problem of maintaining the various components at the required elevated temperatures and to the requirement of pouring the elastomer system so that no air is entrapped in the liquid itself during the pour. Some of the recent technological advances surrounding the low performance or room temperature elastomer systems have made the machine processing of these systems less complex. However, elastomer equipment to process the high performance systems continues to remain as a somewhat sophisticated piece of equipment.

Accessories required for successful application and operation of high performance elastomer systems include much the same equipment as previously mentioned for pour-in-place and molding equipment, with the addition of degassing equipment, ovens, temperature-conditioned tanks, temperature-conditioned material lines and mixing head, specialized and temperature controlled automatic refill systems, temperature-conditioned pumps if not submerged, specialized equipment for eliminating all air from the liquid chemicals and also from the liquid pour, and so forth.

Accessory Items

The following accessory items may or may not be required for any or all of the above mentioned types of equipment.

1. *Heat Exchangers:* These are not always included in competitive quotations, but are almost universally mandatory for good quality temperature control of the urethane components and efficient production. These exchangers range in type from simple tube and shell heat ex-

changers requiring external sources of hot or cold water to highly efficient scraped surface one-pass types with self-contained heating or refrigeration units. It should be noted that tube and shell type exchangers are very efficient when utilized with a material having a viscosity of 2500 cps or less. When higher viscosity materials are being processed, it is advantageous to utilize the scraped surface type exchanger due to the fact that the higher viscosity material tends to build up a thick coating of material on the cooling surface; if not continually scraped clean as in the scraped surface type exchanger, the coating drops the heat transfer efficiency drastically. It should also be noted that the portion of the heat exchanger which comes in contact with the urethane component being processed should be of stainless steel construction. If an alternate material is used in the construction of the exchanger, care should be taken that it will not have any adverse effects on the liquid urethane component being processed (e.g., a brass exchanger in an amine metering system may result in an amine complex problem).

2. *Heating and Cooling Units:* There are many types of heating and cooling units provided with the mixing and metering equipment on the market today. All are generally claimed to maintain temperature control of the urethane components being processed to within ±1°F. In general, this equipment consists of some means of heating or cooling and pumping the heating or cooling media which is generally ethylene glycol and water to the above mentioned tube and shell heat exchanger. Also usually included in the package is a temperature controller for sensing the liquid urethane temperature and controlling what water, whether hot or cold, is required in the exchanger. These units will range from a totally manual unit whereby the hot and cold water must be manually valved into the tube and shell heat exchanger to

totally automatic units which will automatically sense the need for a change of temperature of the water in the tube and shell and make the change automatically. Even in the totally automatic units, one must be careful how this operation is accomplished since some units use manually adjustable flow control valves or solenoid valves to valve hot or cold water into the tube and shell heat exchanger which creates a bucking condition between the hot and cold water conditioning units. In contrast, there are totally automatic units with single dial control which utilize a common heating/cooling media (water-antifreeze mix) throughout the heating/cooling cycle with a null point controller to prevent sine-wave or over/under shoot controlling predominate in units which use flow control valves or solenoid valves with separate hot and cold water supplies. This arrangement completely eliminates the bucking condition between the heating and cooling units thereby preventing a sine wave type temperature control of the urethane chemical systems.

3. *Instrumentation:* Reproducibility of product is difficult if not impossible without sufficient production records to know how previous desirable products were made. Desirable information includes temperature, pressure, flowrates, process speed rates, and so forth, for all components. This instrumentation can consist of anything from a simple dial thermometer to monitor temperature and a pressure gauge at the mixing head to monitor material pump pressure to a complete solid state computerized metering and monitoring system. This system could be operated by one man, and it would be fully automated by control multiple streams to $\frac{1}{2}$% of total flow rate of each individual component. Systems such as this generally feature a digital read out in pounds per minute for each material stream and a master total in pounds per minute dispensed, an audible and visual alarm system to warn the operator of any irregularities in the system, a digital read out in material pressure on each component being pumped, and a digital read out of the material pump rpm of each component.

4. *Solvent Flush Systems:* Some sort of solvent flush is mandatory for efficient equipment utilization. The types available range from manual type squeeze bottle to the more efficient completely automatic push-button types wherein solvent and air are introduced alternately for complete purging. The completely automatic systems feature the flexibility of adjusting the delay timer from the end of the pour cycle to the start and end of the flush cycle. In addition, in keeping with some of the new OSHA Regulations, there are available from some manufacturers a completely closed solvent flush system to prevent any fumes from escaping into the operating area and to prevent any spillage or operators coming in direct contact with the solvent being used.

5. *Additional Components:* Most types of foam formulations can be handled with a two- or three-component system. For reasons of limited pot life or variable quality or even greater efficiency of operation due to less pre-mixing and handling involved, other components may be added. The most efficient and versatile rigid machines generally have three components or more. The most efficient flexible machines usually have five components or more. Most reliable equipment manufacturers today offer standard two-component urethane equipment with a multiple-component mixing head. This gives the customer the advantage of not having to replace the mixing head at a later date if an additional component is desired.

6. Other accessories that are sometimes necessary or desirable are liquid level controls and/or indicators for the chemical tanks, air dryers and filters, additional or special type hoses, corrosion resistance, all air operation, frothing attach-

ments, alternate type mixing impellers, air bleed systems, special pumps and seals, "know-how" books, and so forth.

The following accessory items are generally associated with urethane slab stock equipment.

1. *Traverse or Spreader Units.* These units are used to spread the mixed chemicals on the casting surface whatever it might be. Most usual types involve a reciprocating carriage on a frame above the casting surface which carries the mixing head back and forth. Those available range from simple and reliable pneumatic motor-driven types to elaborate mechanical, electrical, or hydraulically driven types. The pneumatic motor type is popular because of its capability of immediately discharging the energy absorbed in the reversal of the mass during the traversing cycle. However, the hydraulic system when installed as a two-speed system is also popular due to its capability of controlling the lay down pattern of the liquid mix to a much finer degree.

2. *Paper Handling Equipment:* This equipment is generally used to line the conveyor on slab production units. These generally take the paper from one or more rolls and shape it wrinkle free in the form of a wide "U" on which the mixed chemicals are spread. Multiple paper roll systems are available which enable the operator to cover both sides, top and bottom, of the conveyor with individual continuous sheets of paper.

3. *Curing Equipment:* These are usually of a conventional nature and their use is so diverse as to make generalizations impractical. Most any heat source from infrared lamps to a sophisticated steam, gas or electrical oven can be used.

4. *Cutting Equipment:* This is generally associated with foam slab production. Types available are widely diversified, and the product range is virtually unlimited. Usually, such equipment include: horizontal incline types (both automatic and manual adjusting in nature), vertical incline types (conveyorized or manual push type), convoluters, scooping cutters, peelers, borers, shoulder pad cutters, hot wire cutters, and so forth. (See Fig. 20-11).

Fig. 20-11. Accessory equipment—horizontal saw or slitter for cutting flexible foam bunstock.

LOW DENSITY POLYSTYRENE FOAMS*

Polystyrene, which is widely used in injection molding and extrusion, is also one of the most versatile raw materials available for the manufacture of plastic foams.

A number of factors determine the suitability of a polymer for producing an expanded plastic—low cost, availability, ability to be processed, and physical characteristics such as weight, thermal conductivity, water resistance and strength. Polystyrene is an inexpensive plastic commercially available in large quantities. It is easily processed at relatively low temperatures and pressures, and its solubility characteristics permit the use of many solvents as expanding agents. These factors result in a readily available inexpensive cellular material.

The suitability and acceptance of a foam plastic for an application is determined not only by its cost, but also by its physical properties. Many of these properties are controlled directly by the character of the base polymer. Polystyrene produces a light, rigid, closed-cell foam having low thermal conductivity and excellent water resistance, meeting the requirements for low temperature insulation and buoyancy mediums.

Two different types of low density cellular polystyrene are now available for use by the fabricator, molder or ultimate consumer.

Extruded Polystyrene Foam

This material has been produced for over thirty years, and is manufactured as billets and boards. The foam is made by extruding molten polystyrene containing a blowing agent, under elevated temperature and pressure, into the atmosphere where the mass expands. The billets and boards can be used directly, or can be cut into many different forms. Common tools for fabrication are bandsaws, hot-wire cutters, planers, and routers. Boards may have a cut or planed surface or may have an integrally extruded skin of extremely small cells.

Many sizes are available in extruded cellular polystyrene. Billets can be as large as 10 in. x 24 in. x 9 ft. long and boards are avail-

* By William H. Werst, Jr., Construction Materials TS & D, Dow Chemical U.S.A.

able ¾ in. to 4 in. thick by 24 in. by up to 9 ft. long.

Expandable Polystyrene for Molded Foam

Expandable polystyrene is produced in the form of free-flowing beads, symmetrical shapes, and strands containing an integral blowing agent. When exposed to heat, without restraint against expansion, these particles "puff" from a bulk density of about 35 lb/ft^3 to as low as 0.25 lb/ft^3. In the case of beads intended for molding, this very low density leads to difficulty with collapse during subsequent molding; hence the usual limit for bead preexpansion is 1.0 lb/ft^3. The shapes and strands which are used as loose fill cushion packaging are not processed beyond preexpansion.

Molding expandable polystyrene beads ordinarily comprises two separate steps:

1. Pre-expansion of the virgin beads by heat.
2. Further expansion and fusion of the pre-expanded beads by heat within the shaping confines of a mold.

In unconfined pre-expansion, the translucent beads grow larger and become white in color. Confined and subjected to heat, the pre-expanded beads can produce a smooth-skinned, closed-cell foam of controlled density, registering every detail of an intricate mold.

Pre-expansion

To minimize formation of a density gradient and to ensure uniform expansion throughout the molded piece, expandable polystyrene beads are pre-expanded to the approximate required density, by control of time and temperature, since the process of molding does not increase the density.

The continuous steam prefoamer is the most widely used method of expanding expandable polystyrene beads. In the continuous steam prefoamer, a steam Venturi or a screw auger injects the beads into the prefoamer. As the beads expand, a rotating agitator prevents them from fusing together as the lighter density, expanded beads are forced to the top of the drum and out the discharge chute. The expanded beads are then collected in storage bins or hoppers for aging prior to molding. A typical continuous steam pre-expander is seen in Fig. 20-12.

Fig. 20-12. Continuous expandable polystyrene pre-expander. Unfoamed beads enter, along with steam, at the bottom of the expander, are agitated by the rotating blade, and exit through the chute at the top of the unit. ((*Photos and illustrations, courtesy Dow Chemical U.S.A.*)

Strands of expandable polystyrene are usually pre-expanded on a wire mesh conveyor belt which passes through a steam chamber. The unfoamed strands are positioned on the belt by the tumbling action of a rotating wire mesh hopper. As the strands pass through the steaming chamber, the heat causes them to expand. After a short aging period, the particles become resilient and are ready for use.

Shaped particles of expandable polystyrene such as "s" shape or "doughnut" shape particles can be expanded in a rotating drum expander which takes on particles and steam at one end and expels the foamed particles at the other (Fig. 20-13).

Continuous Hot Air Prefoaming. Continuous heated air prefoaming is another bead expansion technique used by molders. Because of the dry atmosphere used for expansion, a dry prefoamed bead structure is formed. Because hot air has a lower heat content than an equal volume of steam, the very low densities and fast throughput rates obtainable with steam prefoaming are not usually achieved. Hot air prefoamers are used where higher densities are needed in applications such as the foam cup. Close control at high densities (5-10 lb/ft^3) can be maintained in a hot air pre-expander.

Batch Prefoaming. Batch prefoaming is used by molders primarily for producing small

Fig. 20-13. Expander for free-flowing loose-fill packaging material of expandable polystyrene. Unfoamed particles enter at the left and exit at the chute on right.

quantities of prefoamed beads, generally in a laboratory operation. As in continuous steam pre-expansion, density is controlled by varying the bead inventory time in the batch prefoamer as well as by controlling the prefoamer temperature. Any one of three types of heat source is used in batch prefoaming: hot air, hot water, and steam.

Steam at 212°F is the most desirable heating medium in batch prefoaming. Hot water heated to 212°F is used where a steam source may not be available and where slightly higher densities are desired. Oven batch prefoaming, utilizing air heated to between 220 to 240°F, is also used where a steam source is not available. In both steam and hot-air batch prefoaming, beads must first be coated with a liquid detergent in water to inhibit fusion. In hot water prefoaming, this is not necessary because of the suspension of the beads in water.

Molding

Molding of expandable polystyrene beads requires exposing the pre-expanded or virgin beads to heat in a confined space.

A number of conventional heating media are available for the fabricator to use in molding expandable polystyrene. The preferred medium is steam that is directly diffused through the pre-expanded beads in the mold cavity. Other techniques involve conductive heating through the mold wall with the heat being supplied by steam or by some other energy source.

Steam Chest Molding. The steam chest or jacketed mold (Fig. 20-14) is double-walled with a perforated inner wall and is mounted in a molding press. A typical press is shown in Fig. 20-15. During the heating cycle, steam enters the jacketed space and is distributed through the perforations into the cavity. Water is circulated in the same steam chamber during the cooling cycle. Introducing the cooling water through a system of spray heads inside the steam chest is more effective than flood cooling.

The steam chest method of molding is highly versatile. It is useful in molding sections smaller than $\frac{1}{8}$ in to more than 24 in. in thickness and is recommended for densities ranging from less than 1 to greater than 5 lb/ft^3.

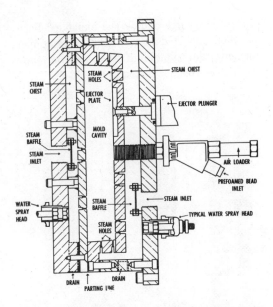

Fig. 20-14. Cross-section of typical steam chest mold. Steam is injected into the mold cavity through the steam holes in the mold plate.

Prefoamed beads of a bulk density equivalent to that desired in the finished part are normally blown into the mold by use of an air fill gun. A plunger moves forward and closes off the fill port. Then steam is vented into the mold cavity, causing the beads to expand, filling the mold and fusing together. The cooling cycle depends on the efficiency of the cooling water. The part must be cooled adequately so it can be removed without danger of post molding expansion.

Block Molding. A large, solid block of polystyrene foam can be made by a machine which uses a modified steam chest technique. The machine and the mold are one; spacers can be used to vary the size of the block to be produced. Steam is injected into the prefoamed beads from a steam chest. Cooling is commonly accomplished by spraying the outside of the mold with cooled water.

Block molding machines of this type can produce blocks of foam which are commonly between 12 and 24 in. thick, 4 ft wide and between 8 to 16 ft long. Block densities produced are normally 1 to 1$\frac{1}{2}$ lb/ft^3. The prefoamed beads are fed into the mold directly from the top, after which a typical molding sequence for steam chest molding is followed.

Fig. 20-15. Typical steam chest molding machine for producing small packaging or novelty parts of expandable polystyrene.

The termination of the cooling cycle is generally indicated by the reduction of pressure on the lugs holding the mold together. A pressure diaphragm can be mounted in the mold so it contacts the mold wall and thus determines the end of the cooling cycle.

Foam Cup Molding. Special machines have been designed to rapidly produce thin-walled parts such as the foam cup. An extremely small bead is used at a prefoamed density of approximately 4 to 5 lb/ft^3, allowing easy and rapid filling of the mold and producing a part having the proper stiffness for handling.

Cup machines all utilize air loaders for mold filling and incorporate some form of steam injection into the mold cavity to complete the fusion step. Conductive heating is used in some machines as an adjunct to the fusion cycle in order to smooth out the surfaces. The cooling cycle is much faster than that of other steam chest molding machines because of the very thin cup wall.

Steam Probe Molding. Another method of introducing steam into a mold cavity is by a probe which is inserted through a hole in the mold wall. When expansion is nearly complete, the probe is withdrawn. When the probe is withdrawn quickly after steaming, the hot foam continues to expand, closing off the probe hole. Multiple probes are used for large cross sections. The probe method is best for sections at least 1 in. thick and for densities of less than 2 lb/ft^3. It is the only practical method available for molding between structural facings.

Continuous Billet Molding. Boards or panels can be made continuously from expandable polystyrene by injecting steam from probes or through steam jets into the space between machine-mounted conveyor belts while beads are being blown by air fill guns into the back of the machine for filling. The block or laminate passes from the fusion zone into a cooling zone and ultimately out of the machine to be cut to the desired lengths.

Radio Frequency Molding. Another method of generating heat for expansion of expandable polystyrene in a closed mold is by means of radio frequency energy. Since formulations of expandable polystyrene have a low loss factor when high frequency voltage is applied, very little heat is generated when they are placed in a radio frequency field. However, when an electrolyte is added to the surface of expandable polystyrene, the heat generated from molecular excitation of the electrolyte provides heat for the expansion of the beads.

The mold itself must be made of a material such as polypropylene, epoxy, or polyester, which will not become heated when exposed to the radio frequency field. The advantages in this technique are reduced mold cost and a reduction in cooling time. Because the plastic mold has a reduced heat transfer rate, the overall cycle is approximately the same as in steam chest molding.

Properties

These two types of cellular polystyrene differ in technique of expansion and handling, but, as might be expected, most of their properties are quite similar. Significant differences in properties may be seen in Table 20-4.

In a number of applications, more than one type of polystyrene foam can be used. The choice will depend upon such factors as eco-

nomics, appearance, and the physical characteristics desired.

Extruded Polystyrene Foam Markets

One of the largest markets for extruded polystyrene in the form of boards is in low temperature insulation. Space insulation is used in freezers, coolers, and other types of refrigerated rooms. Truck bodies and railroad cars also are insulated with the board form. Ease of handling and lightness of weight add to its excellent low temperature properties. Refrigerated pipelines as well as the base of low temperature storage tanks for such things as liquefied natural gas are insulated with foam boards, or fabricated pipe covering (Fig. 20-16).

The board form is also useful as roof deck insulation. The foam is used to insulate the roof membrane on flat roofs. The foam is placed in the last hot bitumen layer of the roof then covered with gravel or stone to hold it in place. This system of insulating built up roofs is patented by the extruded foam supplier (Fig. 20-17).

A new and growing market for extruded polystyrene boards is in the insulation of residential housing by using the foam in place

Table 20-4. Properties of Polystyrene Foams

	Extruded	Molded
Density range, lb/ft^3 (D1622-59T)*	1.3-4.5	1.01-10.1
Tensile strength, psi (D1623-64)	40-250	20-220
Compressive strength, psi (D1623-64)	10-200	10-200
Flexural strength, psi (C203-58)	40-200	25-330
Shear modulus (C393-62)	800-1600	300-1000
Thermal conductivity, °K (Btu/in./ hr/ft^2/°F at 40°F mean temperature) (C518-70)	0.18-0.28	0.24-0.27
Water absorption (% by vol.) (C272-53)	0-1.0	0-2.0
Water-vapor-transmission rate (Perm-in) (grains/hr/ft^2/in./in. of Hg vapor pressure differential) (355-64)	0.1-1.1	0.5-3.5
Heat-distortion temperature, °F	165	180
Linear thermal coefficient, avg. (in./in./°F)	3.5×10^5	3.5×10^5
Dielectric constant, 10^2-10^3 cps	less than 1.05	less than 1.05

* Parentheses indicate ASTM Tests

Fig. 20-16. High-density polystyrene foam is used to insulate the floor of a liquefied natural gas tank.

Fig. 20-17. Two by four foot sections of extruded polystyrene foam are used as roof deck insulation.

of conventional sheathing. Thus, the insulation is placed over the studs and sill plates which are a source of thermal shorts in batt insulation buildings (Fig. 20-18).

In the agricultural area, extruded polystyrene boards are used to insulate livestock buildings, keeping the livestock cooler in the summer and warmer in the winter. It is also used to insulate low temperature produce storage buildings such as potato and onion storage. The insulation is generally applied on the outside of the studs or framing members of the building, then covered with metal siding. The board can also be applied to the inside of existing buildings by nailing to the studs and rafters (Fig. 20-19).

Extruded polystyrene foam boards are used as the core material for structural sandwich panels. This usage is particularly prominent in the growing recreational vehicle and motor home industry. The walls, floors, and roofs of many of these vehicles are constructed of polystyrene foam core sandwich panels.

White and green colored extruded polystyrene foams are used in the floral, novelty, hobbycraft, and display fields either as boards or billets or fabricated into shapes such as balls, cones, cylinders, rings, etc. Floating docks, marker buoys, and flotation for small boats are applications for large billets of this foam.

Extruded polystyrene foam sheet in thickness of $\frac{1}{16}$ to $\frac{1}{4}$ in. has found significant usage as a replacement for molded paper pulp board in meat and produce trays and egg cartons. The foam sheet has the advantages of a clean, bright appearance, excellent cushioning properties and is nonporous. The foam is extruded as a sheet and is subsequently vacuum formed into the desired shapes for packaging (Fig. 20-20).

Foamed polystyrene sheet and film are manufactured by a tubular film extrusion process using conventional methods with specially treated expandable polystyrene pellets, or by injecting a propellant directly into a section of the extruder barrel with standard polystyrene resins and additives. In both techniques the extrudate passes through an annular tubing die, and is expanded either by blowing air inside the tube or drawing the tube over an internal sizing mandrel. This enlargement of the tube (blow-up ratio of three to one or greater) is necessary to eliminate the normal tendency of the foam to "corrugate" as it emerges from the die.

Fig. 20-18. Another application for polystyrene board is insulation for residential housing. Boards are placed over the studs and sill plates of the house.

Fig. 20-19. Interior walls and ceiling of a dairy barn are insulated with expanded polystyrene boards.

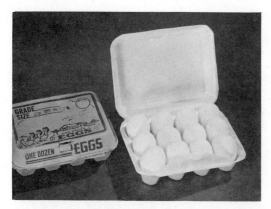

Fig. 20-20. Vacuum formed polystyrene foam sheet carton (right) replaced molded pulp board carton (left).

Molded Polystyrene Foam Markets

The expandable bead foam is utilized in many similar applications to those of extruded foam because of their similar properties. Board insulation is produced in 1 to 12 in. thicknesses, and up to 4 ft widths by 16 ft lengths. Greater lengths can be produced by a continuous molding process.

The ability of expandable polystyrene to duplicate accurately the details of molds led to its use in making toys, novelties, and displays. Silk screened, dry colored, and artfully berib-

boned articles in this category have found large markets (Figs. 20-21 and 20-22).

Its closed cell structure, low thermal conductivity (K factor), low water vapor permeability, low water absorption, and general adaptability at very low densities find molded polystyrene foam a ready market in low temperature insulation.

One of the largest markets for expandable polystyrene is the hot and cold drink cup market. These cups, because of the excellent insulating properties of expandable polystyrene, keep hot drinks hot while allowing the person drinking to hold the cut without burning his hand (Fig. 20-23).

Molded drip trays for refrigerators, freezer evaporators, evaporator supports and liner shields of expandable polystyrene foam are being extensively used in the domestic refrigeration industry.

Consumer, industrial, and military packaging problems have been successfully solved with easily molded, shock-resistant, low density expandable polystyrene. Its strength, water resistance, and insulating qualities have led to its use for shipping containers for whole blood, plasma, and perishable drugs. Delicate instruments and intricate parts are cushioned against shocks in transportation and air drops by

Fig. 20-21. Duck decoys are molded of expandable polystyrene beads. The two decoys in foreground use colored beads; the two in the background have been painted.

Fig. 20-22. Display items molded of beads and decorated by painting and silk screen printing.

Fig. 20-23. Hot and cold drink cups are major outlet for molded beads.

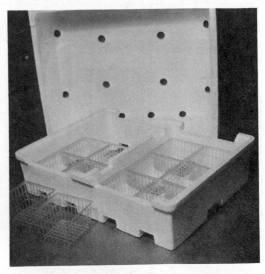

Fig. 20-24. Shock-absorbing lugs of expandable polystyrene facilitate the protective shipment of delicate fruits and vegetables.

dunnage readily molded from expandable polystyrene beads (Fig. 20-24).

Low density foamed expandable polystyrene has exceptional buoyancy characteristics, and in-place moldings are being used in boats, buoys, etc., to replace air tanks with buoyancy that functions virtually unimpaired if punctured.

Expandable polystyrene strands, stars, "s"-shapes and rings are used for loose fill cushion packaging. Their low density, high resiliency, cleanliness, and versatility result in vastly improved performance over more tradi-

tional loose fill packaging materials, such as excelsior and shredded paper (Fig. 20-25).

EXPANDABLE POLYSTYRENE MOLDING*

Expandable polystyrene molding is a two-step process. The beads that are supplied to the molder are solid and contain a blowing agent.

* By Walter E. Johnson, Mgr., Machine Design, Springfield Cast Products, Div., Hoover Ball and Bearing, Springfield, Mass., 01109.

Fig. 20-25. Delicate instruments are protected from damage in transit by the use of expandable polystyrene loose fill packaging materials.

In the first step, the beads are passed through a heating process to reduce the density to that desired in the final product. The mold is filled to capacity with the expanded beads. Heat is again applied to the mold and beads to cause them to expand enough to fill all the voids between the beads and to heat seal all of the beads together.

This process for molding polystyrene foam offers the molder an opportunity to make products with a wide range of characteristics. The density can be easily controlled. It ranges from below 1 lb/ft^3 to above 20 lb/ft^3. Molded products can have a wide range of wall thickness. The density of the part is uniform throughout. Most parts are molded with only a very thin skin but, with an additional step in the process, it is possible to produce a part with a hard skin. The molded product is stress free and quickly stabilizes. Molding pressures and temperatures are low.

In the low-density range, 0.5 to 1.0 lb/ft^3, expanded polystyrene is used in boats as flotation, in packaging as energy absorbers, in building as insulation and a moisture barrier.

In the middle-density range of from 0.1 to 4.0, the material is used in packaging as a structural support as well as energy absorber, in the construction field for such things as concrete forms, in the foundry industry as mold patterns, as insulated containers of all sizes and shapes, and in material-handling pallets.

Then, in the high-density range from 5.0 to 20.0, the material exhibits almost wood-like properties. Such products as thread spools, tape cores, and furniture parts have been made.

The typical molding operation is shown in Fig. 20-26. Raw material is fed into a continuous pre-expander. Low pressure steam is used to heat the material. After expansion, the expanded beads are screened, to remove any large clumps, and blown into a storage hopper where the beads are allowed to dry and stabilize. They are again moved by air to the machine hopper; steam is also used again to heat the beads and molds; water is used to remove the heat prior to opening of the press.

Storage of In-Coming Material

In the molding plant, raw material is received and stored in 200- to 1000-lb containers. The material is generally not transported in the usual bulk-handling equipment used for other types of plastic pellets because of the blowing agent used. The beads lose their blowing agent

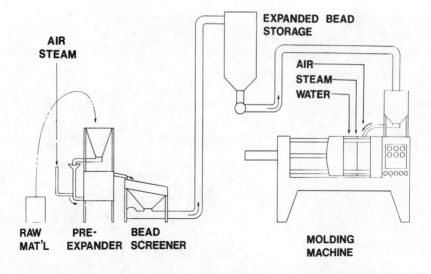

Fig. 20-26. Typical expandable polystyrene molding operation. (*Sketches courtesy Springfield Cast*)

rather quickly when stored. Care must be taken in setting up a molding operation because an accumulation of the blowing agent can create a fire hazard, particularly since movement of the beads generates static electricity.

Pre-Expansion (First Step)

There are several methods of heating the material during pre-expansion. Expandable polystyrene beads have been successfully expanded by immersing them in hot water, agitating them in hot air and/or steam, and exposing them to radiant heat. Pre-expansion equipment is available that operates either continuously or on a batch basis.

The continuous steam expander, the most widely used, consists of a mixing chamber in which air, steam, and the unexpanded beads are introduced at the bottom. As the density is reduced, the material is allowed to flow upward through the mixing chamber, and out the top. A close balance between heat input and material flow is required to assure final uniform results.

The success of a molding operation to a large extent hinges on the pre-expansion operation. Beads that are properly expanded contain a narrow range of densities. Molded parts exhibit properties similar to the lowest densities in the expanded mix. The narrower the density range in the bead, the lower the median can be and

still mold a product with the desired characteristics. Since the raw material makes up at least half the cost of the product, careful control at this step can offer real savings.

Storage of Expanded Material

If the material is not used within a few minutes of expansion it requires a period of time to stabilize. The air is allowed to migrate into the cells so that on re-heating, the air in the cell can aid in the expansion of the bead. In some molding operations, it is necessary to lower the blowing agent content to allow proper molding. Too high a blowing agent content can prevent even fusion throughout the molded piece. Consistent handling of expanded bead, prior to the molding step, can reduce scrap by making it unnecessary to adjust the molding cycle to the material condition.

Molding (Second Step)

On the surface, the molding of expandable polystyrene seems to be a rather simple operation. The temperatures and pressures are low, but when the range of products and the widely divergent physical properties that are available from this one material are considered, the complexities encountered become obvious.

Heating the beads during molding has been

confined to two general methods: passing steam through the beads, and generating steam with RF energy within the part to be molded. The two processes are quite different, and require an entirely different approach from the standpoint of equipment. The steam molding method is the one dealt with below.

Molding Cycle

A typical molding cycle consists of: preheating the molds to drive out water from the previous cycle, closing, complete filling of the cavity with the expanded beads, heating of mold and beads, cooling, opening and ejecting.

Molds

Metal molds are used for steam molding. Nonferrous materials are used for the molds because of the presence of steam. The molds are heated to approximately 230°F and cooled to approximately 90°F each cycle; it is therefore desirable to keep the mass of the mold as low as possible to conserve energy and time.

Figure 20-27 shows a typical mold.

The steam that is used to heat the mold is also used to heat the beads. Openings are made in the mold to allow the steam from the back side of the mold to pass through the beads. Several different types of openings are used; holes 0.020 to 0.050 in. are used as well as standard foundry core vents and machined slots 0.005 to 0.015 in. wide.

On multicavity molds, individual feeders are expensive, and add extra maintenance problems. The slide runner (Fig. 20-28) was developed so that a large number of parts could be filled without resorting to sprues and trimming after molding.

The range of size and density of the molded parts that can be made from expanded polystyrene has prompted the industry to "tailor" molding equipment to the product.

Expandable polystyrene cups are molded on small vertical machines, four at a time. A typical cycle time for this operation is 10 sec. The molds have thin walls, so they can be heated and cooled quickly. The machine feeds large amounts of steam and cool water quickly and accurately to each mold separately. The

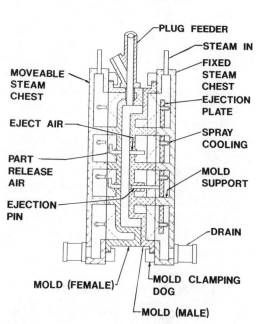
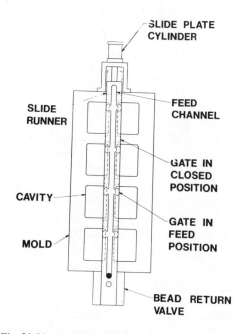

Fig. 20-27. Typical mold for expandable polystyrene molding.

Fig. 20-28. The slide runner was developed so that a large number of parts could be filled without resorting to sprues and trimming after molding.

operation is made completely automatic by the use of cup carry-away and stacking equipment.

At the other extreme, on limited runs of very large parts, molds are made strong enough to hold the internal pressure of molding, and are operated without a press by clamping the halves together.

Many boats, surfboards, chair frames, and couches have been made in this way. There seems no limit to the range of size of product that can be made from this process.

The custom molding industry uses, primarily, a horizontal single- or double-acting machine. Since runs are often short, and competing processes have short deliveries, molding equipment must operate multicavity molds automatically at a short cycle, and allow for quick tool changes. To do this over a wide range of products, the machines are equipped with several different methods of filling and ejecting, and are furnished with a steam chest that is, in effect, the back half of the mold. The chest contains the facilities for heating, cooling, and ejecting the molded parts. Molds in this case consist only of face plates. The cost of tooling is greatly reduced and mounting time is held to a minimum.

EXPANDED VINYL*

The production of expanded vinyl is about as old a process as the commercial utilization of plasticized PVC itself. In the early 1940's the search for effective chemical blowing agents for elastomeric products resulted in the discovery of a group of non-staining aliphatic compounds, the azonitriles.[1] One of these, azobisisobutyronitrile, was used during World War II for the production of expanded flexible and rigid PVC compounds. Incorporation of gas into PVC compounds can improve the economics while retaining the major aesthetic and physical property advantages of this polymeric material. Insulative, acoustical, density, resilience, and general appearance properties are improved.

Plastisol, calendering, extrusion, injection molding, and powder techniques are used to

* By D. W. Ward, Dev. Tech. Leader, Avon Lake Tech. Ctr., B. F. Goodrich Chemical Co.
[1] Superior numbers in text refer to References on p. 557.

process vinyls into expanded vinyl products. These technologies produce both rigid and flexible PVC foam products. The dominating products at this time are the flexible foams, but the utilization of cellular rigid vinyl extrusion products is growing rapidly.

Plastisol Technology

Plastisols are the most widely used route to flexible expanded vinyl products. Many types of upholstery, garment fabrics, and floor covering products are made from coatings with expandable plastisol compound. Molded products are dolls, gasketing, and resilient covers for tool handles. Proper utilization of plastisols requires knowledge of the compounding of the homogeneous stock as well as specific concern for formulating to get the desired expansion. Since plastisol compounding and processing is covered in depth elsewhere in this book,[2] this section deals mainly with the technology and changes in the technology required to make foam instead of solid products.

Definitions

Common usage has subverted the terminology originally borrowed from the rubber industry in which chemically blown plastisol compound was called *sponge*, and plastisol expanded from liquid was called *foam*. The term foam has become so common that some special definitions must be used in the discussion to follow. *Expanded* or *cellular* will be used as the most general term to describe froths, foams and pressure sponges. *Froths* are the result of incorporating gas into *liquid* plastisol. *Foams* result from expansion of thermoplastic plastisol *melts* at or near *atmospheric pressure*. *Pressure sponge* will be used to describe various products which must be held in a high-pressure mold during evolution or incorporation of the expanding gas.

Characterizing Expanded Vinyl Products

The Vinyl Dispersions Division of the Society of the Plastics Industry (SPI) and the SPI Cellular Vinyl Committee of the Cellular

Plastics Division have published test methods for characterizing both the general properties of plastisols and the specific properties that result from cellular technology. These are listed in Chapter 16 on Vinyl Dispersions.

Density and compressive properties are important measurements. Density is expressed in various ways for expanded products. Units used are grams/cc and lb/ft^3; for wet froths, it's grams/liter. Compression-set is a measure of permanent deformation resulting when either a unit load is applied to the cellular product under standardized time-temperature conditions or of the permanent deformation from compression to a standard percentage of the original product thickness, again under standard time-temperature conditions.[3] Compression-deflection resistance (often termed simply *compression resistance*) measures the resiliency of cellular flexible vinyl in terms of the pressure required to deform a standard thickness of foam to some set fraction of its original thickness, usually 75%.[4] While cell structure, cell structure uniformity, degree of closed cells, and physical strength of finished expanded vinyl products are also properties of importance, they are difficult to measure on a strictly quantitative basis.

General Principles

Plastisols are expanded by four general processes.

1. Physical incorporation of gas by mechanical agitation or by evaporation of dissolved liquids (physical frothing).
2. Chemical generation of gas to froth the liquid dispersion (chemical frothing).
3. Chemical generation of gas after the dispersion has fused and become a hot melt (chemical foaming).
4. Incorporation of gas while the plastic is held under high pressure with subsequent expansion of the high-pressure gas at optimum conditions for expansion of the matrix (pressure sponge).

Regardless of the condition of the plastisol when the gas enters, the material must be fused to make a useful product. Bubbles incorporated in room-temperature liquid, must therefore withstand heating and thermal expansion during the heat cycle. During this expansion, many bubble walls thin out and break and the cellular product resulting from physical or chemical frothing of plastisols is highly open in structure. Chemical *foaming*, on the other hand, takes place in the melt after fusion. The bubbles do not have to withstand much, if any, thermal expansion and the result is a highly unicellular or closed-cell structure. The unique process in pressure sponge allows for expansion under optimum conditions to produce nearly completely unicellular structure.

Physical Incorporation of Gas

The earliest practical use of plastisols in expanded products took place via physical gas incorporation. Processes for introducing nitrogen into fused, plasticized vinyl under high pressure and for solution, and subsequent release of CO_2 to form froths, have been used in the past but are only in limited use now. These are described fully in other references.[5,6]

Foamers. Mechanical frothing devices such as the Firestone, Texacote, Euromatic, and Oakes foamers were originated for use with rubber latexes. Development of surfactants, processing and formulating techniques has made their use with plastisols quite practical. Figure 20-29 is a general view of a laboratory scale foamer; Fig. 20-30 is a schematic of the same unit. Plastisol is fed through a positive displacement pump to a cross head where it meets a metered stream of incoming gas, air or nitrogen. The combined gross mixture passes into the foamer head where the shearing action of the intermeshing teeth on stator and rotor (Fig. 20-31) breaks up the gas into a fine dispersion of bubbles. The froth passes around the rotor and out the exit side through a delivery hose to the point of deposition. Gas and plastisol are quantitatively metered, so the density of the froth produced can be controlled. Pumping rate, hose length, and viscosity of the froth determine the back pressure which, in turn, acts to retain the froth in the foamer head until the proper degree of froth refinement is attained. With plastisols, the

Fig. 20-29. View of continuous mixer showing pump, foamer head, and foam delivery hose. (*Courtesy Oakes and B. F. Goodrich*)

temperature in the foaming head is important. Too high a temperature, especially at the rotor shaft bearing seals, can cause gelation and plugging. The usual procedure is to jacket the foamer head and cool it with circulating cold water. With the head jacketed, plastisol exit temperatures around 38°C (100°F) are quite satisfactory.

Profoamant Surfactants. Mixtures of anionic and nonionic surfactants have been used since the early 1960's to stabilize the froth from mechanically frothed plastisol. The Fomade® series of profoamant surfactants from R. T. Vanderbilt Co. is an example.[7] These were used for some early cushioning applications and to make self-inking stamp pads. The hygroscopic nature of soap surfactants makes them ideally suited for formulations in which affinity for polar stamping inks is needed, but moisture trapped in the liquid plastisol and moisture incorporated during surfactant manufacture makes fusion of thick foam sections difficult. Further, the finished, fused foam tends to pick up moisture readily. Development of silicone-type surfactants for mechanical foaming opened the market to commercial and indoor-outdoor carpeting.

Carpet Applications. Tufted carpeting consists of a synthetic yarn tufted through a primary backing fabric of either jute or a nonwoven synthetic, usually polypropylene. The tufts are loosely held by the primary back and some sort of polymeric binder must be spread on this surface to hold them in place. This binder serves an additional purpose of adhering a secondary backing web in place. In "contract" carpet, carpeting installed in commercial and public buildings (Fig. 20-32) and large residential developments, it has proved economical to replace the secondary backing web with a foamed-in-place pad, thus saving both the cost of the secondary backing web and

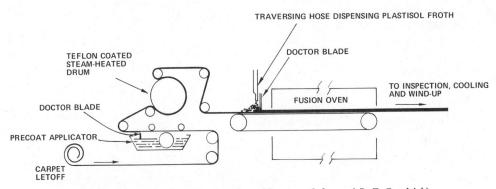

Fig. 20-30. Schematic of continuous mixer. (*Courtesy Oakes and B. F. Goodrich*)

Fig. 20-31. Rotor head of continuous mixer, showing mixing vanes. (*Courtesy Oakes and B. F. Goodrich*)

the dual installation costs of pad and carpet on the job. This led to the development of flowed-on high-density foamed-rubber-latex carpet cushioning. Table 20-5 lists the specifications for such cushioning as defined by the Carpet and Rug Institute.

Mechanically frothed vinyl plastisols represent a second generation of foamed-in-place

Fig. 20-32. Vinyl foam-backed carpet. (*Flooring courtesy Collins & Aikman*)

carpet cushioning with advantages in strength, durability, and aesthetics over early rubber latex types. Application can be via standard carpet industry coating equipment with slight modifications. Figure 20-33 is a schematic of a typical foam processing line for vinyl-backed carpet. It shows direct-roll application of a non-expandable plastisol precoat. A doctor knife removes the excess. The carpet is then passed over a heated drum to gel the precoat and subsequently is coated with froth carried to the doctor blade from a foamer such as is described above.

Formulating for Mechanical Froth. The substrate, often polypropylene, used in carpet

Table 20-5. Specification for High-Density Foam Rubber Carpet Cushion (Flowed On).

Weight	min. 38 oz/yd^2
Thickness	min. $\frac{1}{8}$ in.
Density	min. 17#/ft^3
Compression Set (24 hrs. 158° F)	max. 15%
Compression Resistance (25% deflection)	min. 5#/in^2
Delamination Resistance	min. 2#/in.
Accelerated Aging (24 hrs./275° F)	min. discoloration and degradation
Ash	max. 50%

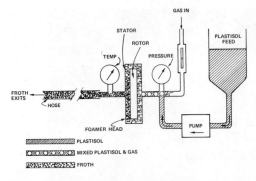

Fig. 20-33. In-line coater for applying plastisol pre-coats and mechanically frothed foam backings to carpets. (*Courtesy B. F. Goodrich*)

manufacture is heat sensitive and will melt above 135°C (270°F). This limits the temperature which can be used in the fusion oven of Fig. 20-33. Accordingly, plastisol foaming compounds for this application are made with low-temperature-fusing copolymer resins. The viscosity limits for mechanical frothing are such that plasticizer blends with moderately high solvating effectiveness but with low dispersion viscosity are needed. Silicone profoamants reach their optimum effectiveness when they have just marginal compatibility with the plasticizer system. About fifteen types are available, tailored for the most popular plasticizer blends.

Properties. Mechanically frothed plastisol is quite open in cell structure and can be made in very fine cell size. This results in a sheet material with a soft velvety hand and low compression resistance relative to other vinyl foam types. Densities range from about 15 lb/ft³ upward. Low density limits are set by the ability of the froth to retain gas without "blow by," the separation of large gas bubbles from froth in the delivery hose. Another limiting factor is the tendency for low-density froths to be viscous like whipped cream and difficult to cast out in a sheet without defects.

Some thin coating applications such as poromeric coatings for fabric, low cost, flame-retardant, crushed foam drapery backings and some molding applications are currently in developmental stages.

Chemical Blowing Agents

Many chemical blowing agents have been developed for elastomers and plastic materials. Of these, about five or six are commonly used in day-to-day compounding of expanded plastisols.[9] Three chemical types cover the range of plastisol processing as shown in Table 20-6.

R. A. Reed[10] has listed the requirements of an ideal blowing agent as:

1. Gas release over definite short temperature range.
2. Gas release at controllable, reasonably rapid rate.
3. The blowing agent and its residue must be noncorrosive.

Table 20-6. Important Chemical Blowing Agents.

Chemical Type	Abbrev.	Decomposition Point in Air (°C)	Decomposition Range in Vinyls (°C)	Gas Yield (ml/gm)	Trade Name	Manufacturer
NN'dimethyl-NN'-dinitrosoterephtha-lamide	(NTA)	118	80-105	180[1]	Nitrosan[2]	DuPont
4-4'-oxybisbenzene sulfonyl hydrozide	(OBSH)	164	130-160	125	Celogen OT	Uniroyal
1,1''-azobisformamide (azodicarbonamide)	(ABFA)	195-200	150-200	220	Kempore	National Polychemicals
					Celogen AZ	Uniroyal

[1] Based on 100% active material. Commercial grade material is 70% active.
[2] Sales of Nitrosan were discontinued in mid-1975. At the time of this writing, replacement products are believed to be in development. It is not known if a direct substitute product is commercially available.

4. Ready dispersibility.
5. Low cost.
6. Storage stability.
7. Low odor residue.
8. Colorless, non-staining residue.
9. Nontoxic residue.
10. Low decomposition exotherm.
11. No effect on rate of fusion.
12. Utility in closed molding.

Except for item 12, azodicarbonamide (ABFA) fulfills these requirements efficiently and thus is the most widely used blowing agent for vinyls. Evolution of cyanuric acid has, however, been said to stain rigid cellular vinyls and plate-out on molds.[11] ABFA decomposition can be adjusted through proper choice of metal organic activators so that the gas is evolved over a narrow range within the wide range given in Table 20-6. Typical activators are the common lead and zinc stabilizers used for protection of the vinyl against heat degradation. Specific information on activators can be obtained from the suppliers of vinyl stabilizers and manufacturers of ABFA.[11b] Decomposition of ABFA can be as low as 150°C (302°F). Thus, plastisols expanded with ABFA will be in a melt state during gas evolution and the ABFA must be activated so that it is decomposing fairly rapidly at the time when the vinyl matrix is at proper melt fluidity. The parameters of melt fluidity will be discussed in a later section.

Chemical Foaming. A few applications for vinyl dispersions require that the plastisol be completely fused at temperatures below 150°C (302°F). 4-4'-oxybisbenzene-sulfonylhydrazide (OBSH) has been used successfully to make good quality foam when these compounds are in the melt state. OBSH is a more expensive blowing agent than ABFA on a cost/efficiency basis. It does not respond to chemical activation as does ABFA. Another blowing agent, N,N' dimethyl-N,N'dinitrosoterephthalamide (NTA), decomposes at temperatures which are appreciably lower than those for either OBSH or ABFA. This material was removed from the market in mid-1975. Substitute products are in various stages of development at this time but no direct substitute seems to have reached wide commercial usage. The decomposition range for

NTA is 80 to 105°C (176 to 221°F). This is just about the maximum temperature at which normally compounded plastisols undergo the transition from fluid dispersion to gel. As a result, plastisols which utilized this blowing agent required special compounding to retain their fluidity during expansion. Agents which decompose in the range between those of NTA and OBSH are of limited use since the plastisol is in a viscous semi-fused state and will not expand smoothly. Satisfactory substitutes for NTA will therefore probably decompose at or below the 80 to 105°C (176 to 221°F) range characteristic of NTA.

The gas yield values listed in Table 20-6 and elsewhere in the literature can be used to calculate the theoretical density of cellular vinyl products if the density of the unexpanded matrix is known.

$$F = \frac{1}{fm\ vm + fb\ vb}$$

Here, fm and vm are the weight fraction and specific volume (ml/g) of the matrix vinyl compound and fb and vb are the weight fraction and gas yield in ml/g of the blowing agent. F is the theoretical specific gravity of the resulting foam.

Chemical Frothing

As mentioned, NTA was used rather extensively for chemical frothing of plastisols despite the fact that it decomposed in the temperature range of 80 to 105°C (176 to 221°F). By combining slow-gelling plasticizers, high-molecular-weight resins and special additives for gel retardation, the compounder was able to hold off the gelation of the matrix until the blowing agent had a chance to decompose. Thus the liquid (rather than melt) matrix expanded and the term chemical frothing applied.

The only other chemical frothing agent of real significance at this time is sodium borohydride. This chemical is especially interesting for two reasons. First, it makes a froth at room temperature which can subsequently be fused into a homogeneous plasticized matrix; and second, sodium borohydride is one of very few inorganic chemical blowing agents that have been found useful in vinyls expanded under atmospheric pressure conditions.[12] When sodium borohydride is dissolved in water, hydrolysis ensues, producing hydrogen and sodium metaborate:

$$NaBH_4 + 2H_2O \longrightarrow NaBO_2 + 4H_2 \uparrow$$

This reaction is rapid at low pH and essentially nil at a pH above about 11.5. One method of plastisol expansion with sodium borohydride utilizes a water solution of $NaBH_4$ buffered to a high pH with sodium borate. This is one-half of a two-component expandable plastisol system. The other component is acidified with a weak acid such as oleic acid or dispersed phthalic anhydride. When the components are combined in a mixer head, the pH of the disperse droplets of sodium borohydride solution drops rapidly and hydrogen gas is evolved to act as the blowing agent. This reaction takes place at room temperature. Plasticizer selection for the plastisol differs from that with NTA. The need is for rapid rather than retarded gelation so that the froth will quickly thicken on heating and not collapse.

Chemically frothed plastisols are used in molding applications predominately. The plastisol can be injected into a mold in the unfrothed state and then expanded with minimal heat. The expanding plastisol flows easily since it is in the fluid dispersion state and thus conforms well to embossed surfaces and intricate internal mold configurations. Molds must be open-faced or vented in the high points to allow the air displaced by the expanding froth to escape. Slipper soles are molded from chemically frothed plastisols. The soles are molded to the upper by injecting plastisol with frothing agent into a hot mold cavity which is covered by a last carrying the fabric upper. The wet plastisol expands to fill the cavity and is fused in place to adhere to the upper.

A few coating or casting applications have developed but only because the chemical froth provides some unique character such as hand or open-cell structure.

Chemical Foaming in the Melt

By far the most important technique for coating or casting vinyl foam is via blowing agents like ABFA which expand the vinyl in the melt after fusion. Coated fabrics for upholstery, handbags (Fig. 20-34), and outerwear, and coated felts for resilient roll flooring derive much of their appeal from relatively thin layers of chemically foamed plastisol beneath the wear surface. As mentioned earlier, ABFA is a nearly ideal blowing agent for plastisols. It has had much to do with the growth of these applications over the last 10 or 15 years.

The technology of formulating plastisols for ABFA centers on achieving a proper balance between melt flow and ABFA decomposition rate. When the melt is at correct viscosity during blowing, a fine, uniform, semi-closed cell structure results in the foam. With too viscous a melt, expansion is erratic if it occurs at all, and gas tends to channel and coalesce, forming coarse non-uniform cells and a warty, unpleasant surface. Conversely, if the melt becomes too fluid during or after blowing, the cell walls lose strength and the structure tends to coarsen and collapse. Although there are a number of

Fig. 20-34. Vinyl foam adds a leather-like "hand" to handbags and other accessories. (*Courtesy Pandel-Bradford and B. F. Goodrich*)

techniques for measuring melt viscosity of thermoplastic materials, none as yet is capable of yielding meaningful data in the actual shear and temperature range that exist when dispersion systems expand with ABFA.

The compounder is faced with experimenting empirically with the parameters known to affect melt rheology and with judging the effect from appearance and properties of the finished foam. Some of these parameters include: average molecular weight of the resin; plasticizer level; plasticizer solvating efficiency; stock temperature, per se; and rate of change of stock temperature during blowing agent decomposition. In practice, plasticizer level is fixed within narrow limits by the intended end use. The availability of a wide range of resin molecular weights and the ability to vary heating conditions still allow for excellent results in foams ranging from about 50 parts of plasticizer for resilient flooring to those with over 100 parts for supple-coated fabrics used in accessories. Densities available in melt-blown foams range from about 10 lb/ft^3 up to that of the matrix itself. Foam stabilizers are available which are said to stabilize foams, compounded to lower densities, against collapse and to protect higher density foams against collapse under higher-than-normal oven heat conditions.

Coating applications for chemical foams are mentioned elsewhere in this book.[2]

Conveyor-cast chemical foam is made in thicknesses up to about $\frac{1}{2}$ in. The material can be cast over a vinyl film to make a smooth-surfaced, easily cleaned product or the cast, expanded matting can be deep embossed with a cold roll as it leaves the fusion/expansion oven. Deep embossing produces a patterned surface with improved wear resistance because of collapse and coalescence of the cellular structure directly at the surface. Chemical foam cast on Teflon-treated fibrous glass belting supplements the mechanical frothing process of manufacturing vinyl foam-backed carpeting. Figure 20-35 is a schematic diagram of a typical tandem line for making chemical-foam-backed carpeting. The foam compound is applied to the belt and expanded in the first oven. Carpet precoated with a fast-fusing plastisol system is applied while this foam is still hot so that the retained heat will fuse the adhesive and bond the carpet to the backing. After a limited additional post-heating cycle, the finished foam-backed carpet is rolled up for shipping. This type of backing has advantages over mechanically frothed backing in that the partly closed cells are more resilient at a given density. Also, the cells tend to be more resistant to moisture pick-up from weather and spills. The strength of chemical foam is inherently higher than the more disconnected structure typical of mechanical froth. The closed structure backing, on the other hand, accentuates compression set. Once gas has diffused through the cell walls under the influence of a heavy point load on a carpet, there is too little restoring force—upon load removal—to cause gas to diffuse back and return the backing to its original thickness.

Molding applications for chemical foam are restricted to products which do not require filling mold cavities since the melts are viscous and do not flow well. Dip molding and slush

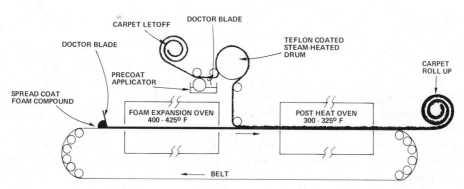

Fig. 20-35. Schematic of coating-lamination line for carpet backing with chemical foam. (*Courtesy B. F. Goodrich*)

molding of insulating layers in gloves and boots are quite feasible. Processing is analogous to molding non-expanded parts as described elsewhere in this book.[2]

Pressure Sponge

Flexible Closed-Cell Sponge. So far, the processes discussed take place at, or near, atmospheric pressure. This means that as gas is generated, the plastic is free to expand to the limit of its ability. In pressure sponge, however, plastisol or other expandable vinyl compound is confined in completely closed molds with little or no free space for expansion. The molds then are heated to decompose the blowing agent and fuse the plastisol while it is held in presses under 2,000 to 16,000 psi. Under these conditions, the gas evolved is forced to dissolve in the matrix. The gas remains dissolved or in the form of microscopic high-pressure bubbles when the parts are cooled under pressure and removed cold. Subsequent heating to the softening point of the vinyl, about 93 to 121°C (200 to 250°F), allows the part to expand. Material expanded in this way has optimum ability to hold gas without cell rupture and completely closed-cell parts can be made at densities as low as 2 lb/ft³. Low densities in this range require rather high levels of blowing agent. The decomposition exotherm of the blowing agent becomes an important consideration. Decomposition of high levels of ABFA at its high decomposition temperature and exotherm will produce enough extra heat to scorch the center of thick, molded pressure sponge. Aside from this limitation, the requirements for blowing agents for pressure sponge are much less stringent than for other processes because expansion is not tied to the decomposition temperature and range. Inexpensive inorganic blowing agents with wide temperature gas evolution ranges can be used quite satisfactorily. The processing requirements make this a slow, batch manufacturing technique. It is used for products which require the combination of low density and high resiliency such as athletic mats, or for products which must be completely unicellular, such as fish net floats, boat bumpers (Fig. 20-36),

Fig. 20-36. Pressure techniques are used to produce closed-cell marine flotation devices. (*Courtesy B. F. Goodrich*)

lobster pots, and life rings. The expansion of pressure sponge at a softening temperature which is below the fusion temperature leaves the expanded part with many internal strains which attempt to return it to the unexpanded shape. As a result, shrinkage can be excessive if the part is not heated enough during expansion.

To reduce shrinkage in pressure sponge, the molding is held at a high temperature of 121°C (250°F) for an extended tempering period. This allows for some diffusion and rearrangement so that subsequent shrinkage is minimized. By incorporating curable elastomeric material such as nitrile rubber in the formulation, the strength of the heat-softened matrix can be increased to prevent the expanding gas from tearing up the molding. These polyblend formulations resist higher expansion temperatures and, thus, end up with less inherent stresses and less tendency to shrink on standing.

Cellular Rigid Vinyl

One of the methods of making rigid (structural) PVC foam utilizes a process similar to the two-step method we have been describing. The process began with the concept of the Carpenter patent[18a] issued in this country in 1951. It was further refined through the efforts of Landler, Lebel and associates at Kleber-Colombes in France. Basically, a dispersion PVC system of the following recipe is fused in a press under high pressure (around 5000 psi).

Typical NCO-PVC Foam Recipe

PVC	100
Toluene diisocyanate (TDI)	53
Azobisisobutyro nitrile (AIBN)	10
Maleic anhydride	20
Styrene	10
Foam density (lb/ft^3)	1.9

In a cycle of about 10 min. at 175°C, the AIBN promotes copolymerization and grafting of the maleic anhydride and styrene. It also decomposes to produce minute, high-pressure bubble nuclei. The fused "prefoam" is removed from the press and permeated with water vapor, either in a steam chest or in boiling water. The water reaction with anhydride and isocyanate forms carbon dioxide for expansion of the slab, plus a complex rigid cross-linked macromolecular structure which has become rigid because the plasticizing action of TDI is no longer present. The result is a structural vinyl foam with better physical properties at any given density than a comparable rigid polyurethane. Greatest utility is in applications where both strength and insulating properties are important. The cost has been said to be equivalent to some modified urethanes.[17] Major uses are in insulating paneling in refrigerated transport trucks and rail cars, and as curtain wall paneling for commercial and industrial construction.

Calendering

Cellular vinyl sheeting and coated fabric are made by calendering as well as by plastisol processes. The process is relatively inflexible since long runs are required before the economics of the process became attractive. This is true of unexpanded goods as well as expanded products. Chemical expansion adds limitations to the process. Calendering is a high-shear melt processing technique. For chemical foam, the compounds must be processed at low temperatures to eliminate the danger of premature blowing agent decomposition. Then, a final expansion oven exactly the same as would be required for plastisol, must be added to the line. The lower cost of general-purpose resins and equipment availability still make calendering desirable.

The method consists of dispersing azobis-formamide (ABFA) in PVC sheeting compound on the mill or in a Banbury; calendering; laminating the calendered sheet to backing fabric and to an abrasion-resistant PVC surface layer; expansion of the composite in an oven; embossing; chilling and finishing.

The maximum acceptable processing temperature prior to expansion has been said to be 160°C (320°F)[13] and there is undoubtedly some premature blowing agent decomposition at this temperature since blowing agent efficiencies are reportedly lower with azodicarbonamide in calendering than with plastisol foam.[14] Compounding, then, requires that easy-processing, medium-molecular-weight resins be used along with high-solvating plasticizers. In order to minimize mixing time at processing temperature, pigments and the blowing agent should be predispersed in plasticizer much as would be done for plastisol.

The compounded, expandable stock is calendered and combined with base fabric and previously calendered surface skin by any of several methods. The unfoamed sheet can be cooled, rolled, and combined with skin and fabric in a three-web lamination process similar to that shown in Chapter 16. Or the expandable stock can be combined with the base fabric at the calender, then cooled, and, finally, laminated to the skin. In either case, the skin can carry a printed decoration.

The combined film/foam/fabric must be heated to expand the foam layer. Ovens used are similar to those pictured in Chapter 16 for plastisol coating. Pin or clip tenters are recommended for dimensional stability in the web. Expansion temperatures must be high to decompose the blowing agent. Final embossing in a separate embosser and top coating completes the typical process.

Extrusion

Processes have been described for extrusion of low-density profiles from plasticized PVC,[15,16] but these require post-expansion ovens or careful coordination of blowing agent decomposition with cure of nitrile elastomer and, so, have not become commercially important. Cellular flexible PVC extrusion products, at densities above about 30 lb/ft^3, are practical for uses in gasketing, for example. The compound can be made by powder-mixing all ingredients or it can be prepared by mill mixing and granulating all ingredients except the blowing agent which is then tumble-mixed with the granules.

The extruder should have a L/D of 15 to 20 : 1, a compression ratio of 1.25 to 2.5 : 1, and a decreasing flight depth and constant pitch. Plasticized cellular vinyl is processed at about 360°F.

The melt expands very shortly after leaving the die. The rate of expansion is rapid and high shearing forces are exerted on the die edge. To ensure smooth skins and no tearing, the extrudate should be removed rapidly from the die to prevent too much expansion close to the die. Expansion should be finished within several inches. The die should be cooler than the stock temperature and should be designed with a constant taper and no land. The cooler die will prevent bubble formation within the skin and increase its strength, reducing the chances of tearing.

The general procedure for producing either flexible or rigid cellular PVC extrusion products is the same: the PVC must be converted to a plastic mass containing blowing agent; this mass must be expanded during or after extrusion and, in the case of rigid materials, the expanded product must be deplasticized to rigidify the foam. Table 20-7 outlines the techniques which have been devised to prepare rigid cellular PVC products.

While all the combinations implied by this table have been demonstrated to be feasible, only two have had commercial significance: the process of using reactive monomers and isocyanates and subsequently polymerizing and reacting with water, and the process of simple heat plasticization of a rigid vinyl stock. The former process is basically a dispersion process and has been described earlier.

Processing cellular rigid PVC is analogous to the processing of cellular flexible stock as described above except that the exit end of the screw is modified to ensure streamline flow with no dead or hot spots and to shear the melt intensely to momentarily produce temperatures of around 221°C (430°F).[17]

Cellular rigid vinyl holds a unique place in the field of structural thermoplastic foams and in the wood-replacement market because it looks, feels, sounds, nails, saws, planes, and otherwise handles so remarkably like wood that it is difficult to tell them apart.

At the present time, the price of wood per unit is less than cellular rigid vinyl, but the overall cost of producing small, finished, complicated shapes in wood is usually more expensive. Also, it is possible to make many difficult shapes, by direct extrusion, which are impossible or impractical to make out of wood—shapes such as the trim around plywood paneling, windows, doors, floors, and ceilings.

Thin sheets and thick boards are not yet in regular production but applications are certain to develop. Thin sheets offer the potential to be vacuum formed into intricate patterns.

The outstanding chemical resistance properties of cellular rigid vinyl plastic will be important in some end uses, whereas its resistance to rot and weathering will be more

Table 20-7. Processes for Rigid Cellular PVC.

Plasticization with	Deplasticization by
1. *Heat*	
(a) high-temperature plasticizers	Cooling
(b) high-melting plasticizers	
2. *Temporary Plasticizers*	
(a) Volatile	Evaporation
(b) Reactive	
1. Monomers	Polymerization
2. Anhydrides	Reaction with H$_2$O
3. Isocyanates	Reaction with H$_2$O

Table 20-8. Properties of Cellular Rigid Vinyl.

Specific Gravity	Tensile Strength (psi)	Elongation (%)	Hardness Durometer "D"	Flexural Strength	Flexural Modulus
1.0	3600	112	71	7125	239,000
.9	3000	100	69	6300	197,000
.8	2800	77	64	5090	178,000
.7	2400	70	57	4220	156,000
.6	2000	61	53	3660	137,000
.5	1650	40	45	3020	119,000
.4	1250	25	40	2215	101,000

Coefficient of thermal expansion 5.5×10^{-5} /°C.
Thermal conductivity 2.101×10^{-4} cal/cm/sec/°C.

significant in certain building product applications.

Cellular rigid vinyls have been extruded and properly sized at specific gravities below 0.4; however, most production is run in the range of 0.4 to 0.6 specific gravity, with 0.5 being a good compromise between cost and quality. The gravity can be controlled by the extrusion operation and by adjusting the compound formulation. Physical properties vary with the specific gravity, as shown in Table 20-8.

References

1. French Patent 890,162 (1943) M. Borgmann, R. Schroetter, and P. Stoeklin to I. G. Farbenindustrie A. G.
2. Chapter 16—Vinyl Dispersions.
3. See, for example, ASTM D-1565, which is often run at room temperature for flexible vinyl, and Interim Procedure SPI-CV-9 (1970-1).
4. See ASTM D-1667-64 and SPI Interim Procedure SPI-CV-10 (1970-1).
5. Sarvetnik, H. A., *Plastisols and Organosols*, Chapter 6, "Foaming," by Henry Lasman. Van Nostrand Reinhold Co., New York, 1972.
6. Patents for CO_2 incorporation include U.S. Patent 2,666,036 to E. H. Schwenke (Jan. 12, 1954) and U.S. Patent 2,763,475 to I. Dennis (Sept. 18, 1956).
7. U.S. Patent 3,288,729 to R. T. Vanderbilt Co.
8. Wheeler, D., "Benefits of Vinyl Backed Carpeting," Proceedings of the Seminar of the Cellular Vinyl Committee, Cellular Plastics Division SPI, May 10, 1971.
9. See reference #5 above, pp. 86-91.
10. Reed, R. A., The Chemistry of Modern Blowing Agents, p. 51, "Plastics Progress Papers", British Plastics Convention, 1955.
11a. *Plastics Technology Magazine*, p. 36, (Sept. 1972).
11b. See for example, "Activating Effect of PVC Stabilizers on Kempore," National Polychemicals Inc., Technical Bulletin, No. OKE-44-0566.
12. Canadian Patent 565,916.
13. "Cellular Vinyl Coated Fabrics Using the Calendering Method", National Polychemicals Inc. Technical Bulletin, No. OKE-46-0665.
14. "Foamed Vinyl Plastics", by A. C. Werner and W. M. Smith, *Handbook of Foamed Plastics*, pp. 296-305, Lake Publishing Corp.
15. Esarove, D. and Meyer, R. J., "Low Density Cellular Vinyl by Post Expansion Methods", *Plastics Technology*.
16. Meyer, R. J., Trends in Foamed Vinyl Processing, *Chem. Eng. Progr.* 57, No. 11.
17. Wherley, F., Rigid Cellular PVC, Wayne State University Polymer Lecture Series, 1967.
18. (a) U.S. 2,576,749 (1951); (b) British 921,068 (1963); (c) French 1,256,549 (1961); (d) U.S. 3,200,089 (1965); (e) U.S. 3,256,217 (1966); (f) British 1,014,502 (1965); (g) British 993,763 (1965); (h) French 1,345,107 (1963); (i) French 1,366,979 (1964); (j) British 997,318 (1965).

CELLULAR POLYETHYLENE*

Production

The production of cellular polyethylene involves only one chemical reaction, i.e., the thermal decomposition of a blowing agent at a specific temperature, which liberates an inert gas.

For a product for electrical service, the correct choice of blowing agent is critical, in view of several unusual requirements. The blowing agent itself, the gas which it liberates and the residual by-product must not absorb moisture, which would impair the electrical properties of the product. In addition, it is desirable that the residue left by the blowing agent be nonpolar, to avoid losses at high frequencies.

The blowing agent used in producing cellular polyethylene must be made to liberate its gas under controlled conditions. This can most readily be achieved in standard extrusion equipment, which makes it possible to maintain continual pressure on the material up to the point of foaming. Only a few changes in normal extrusion procedure are necessary.

Heat in the extruder causes the blowing agent to liberate the inert gas, but pressure within the barrel, head and die prevents expansion of the gas before the material emerges from the die. When the extruder is performing properly, the finished cellular product has a smooth surface, and the cells are of a uniform size. Temperatures higher than necessary result in nonuniformity of size of cells, roughness of surface, and difficulties during cooling of the product. To establish the optimum operating temperatures for a given machine, the temperatures of barrel and head should be set first at 150°C, and the specific gravity of the product measured. Then the temperature of the barrel is raised, by small increments, until the usual specific gravity of 0.42 is achieved.

For best results, the ratio of length of screw to diameter should be at least 16 : 1. Screws with high compression ratio have been found somewhat more effective than those having a low ratio.

* Reviewed by R. L. Boysen, Union Carbide Corp.

To prevent premature expansion of the gas, sufficient pressure must be maintained in the barrel and head. A head temperature lower than the compound temperature also tends to prevent premature expansion. Pressure in the head can be maintained by making the die land sufficiently long, in proportion to the size of the annular opening between the die and the wire, or core. Premature expansion is recognized by "die plating" (a deposit of the compound on the face of the die), and roughness of surface of the extruded product.

Under proper operating conditions, the expansion (to sp gr 0.42) more than doubles the volume. Hence the annular area between the die and the wire, or core, must be one-half the desired cross-sectional area of the insulation. The relationship of the inside diameter of the die (D_1) and the outside diameter of the insulation (D) on a wire or core of diameter d is given by the formulas shown below. Practical experience has indicated that this formula often does not yield the optimum diameter since other variables such as adhesion and back pressure may make more or less drawdown desirable. The formula remains a good starting point, however.

$$D = \sqrt{2D_1{}^2 - d^2}$$

$$D_1 = \sqrt{\frac{D^2}{2} + \frac{d^2}{2}}$$

The temperature of the screw should be as low as needed to maintain uniform head pressure and temperature. Some cooling is always recommended for best output rates (at 150°C maximum) and product uniformity.

The conductor should be preheated to 100 to 150°C before it enters the die head, to promote a smooth surface on the insulation. Lower temperatures may reduce the foaming and conductor adhesion. Higher temperatures result in over-foaming at the conductor with possible voiding and flattening of the insulation.

Several seconds of air cooling before first water contact is recommended in order to allow full expansion of the polymer. Hot or cold water can be used after this.

Cellular polyethylene is normally made to have a specific gravity of 0.42. For uses requiring higher density, the increase is best achieved by blending granules of cellular stock with granules of straight polyethylene. This lowers the concentration of the blowing agent. Extruding at lower temperature to limit the degree of expansion cannot be controlled to yield constant results. Extrusion must always be conducted under conditions which will ensure essentially complete liberation of the gas available from the blowing agent.

Properties

Since cellular polyethylene comprises about equal volumes of polyethylene and gas, its properties differ from those of ordinary polyethylene. Comparisons of several properties are given in Table 20-9.

Cellular polyethylene offers the advantage of a much lower dielectric constant (hence lower electrical losses). The composition of polyethylene (dielectric constant 2.3) and an inert gas (dielectric constant 1.0) has a dielectric constant of 1.5. In terms of electrical insulation, this lower dielectric constant permits a reduction in space between inner and outer conductors without changing the characteristic impedance. Consequently the attenuation may

be reduced by increasing the size of the inner core without increasing the over-all diameter, or the weight may be reduced by decreasing the over-all diameter without decreasing the size of the inner conductor.

The lower density of the cellular material presents the advantages of lower cost and weight, per unit of volume.

Cellular polyethylene is of closed-cell type. Hence its permeability to moisture, while several times as large as that of solid polyethylene, is still desirably low.

Figure 20-37 shows percentage of moisture absorbed by solid and cellular polyethylene during immersion for 100 days in tap water at 20°C. Figures 20-38 and 20-39 illustrate the effect of the exposure on the power factors and dielectric constants. This test is severe, and the results indicate that the cellular material will give good performance in alternately wet and dry environments.

The moisture-resistance of cellular polyethylene is valuable in such uses as antenna lead-in wire for UHF television. This and other high-frequency applications require that the power factor and dielectric constant should not be affected by changes in frequency. Variations would result in power losses, e.g., a weakening

Table 20-9. Typical Properties of Cellular Polyethylene Insulation and Comparison with Solid Polyethylene

Property	Solid Polyethylene	Cellular Polyethylene
Tensile strength, psi at 23°C	2180	620
Elongation, %*	600	300
Dielectric strength, v/mil, ASTM D-149-55T		
short-time at 0.125 in.	550	150
Specific gravity, 23/23°C	0.92	0.42
Mandrel bend at −55°C and 2X†	no failure	no failure
Dissipation factor		
at 1 kc	0.00020	0.00025
10 kc	0.00020	0.00025
50 kc	0.00020	0.00025
Dielectric constant		
at 1kc	2.28	1.51
10 kc	2.28	1.51
50 kc	2.28	1.51

* #14 AWG wire + 93 mils.

† #14 AWG wire + 32 mils; coiled at twice the outside diameter of the insulation.

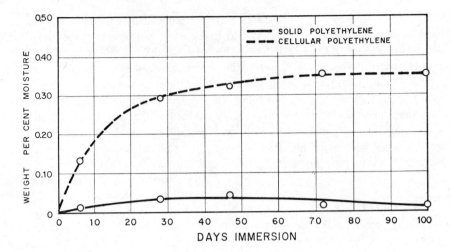

Fig. 20-37. Weight per cent moisture absorbed vs. days immersion in tap water at 23°C.

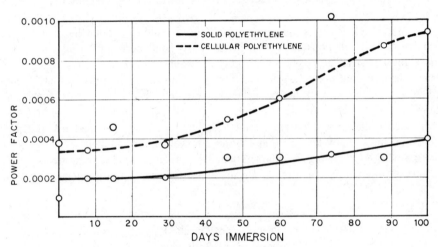

Fig. 20-38. Power factor at 1 megacycle vs. days immersion in tap water at 23°C.

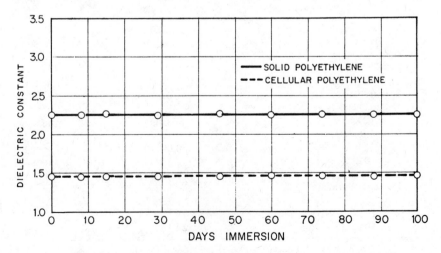

Fig. 20-39. Dielectric constant at 1 megacycle vs. days immersion in tap water at 23°C.

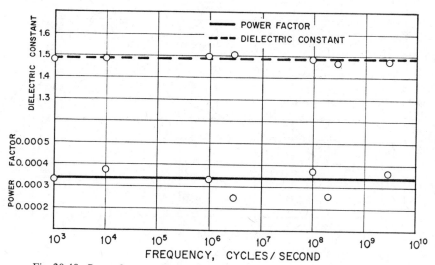

Fig. 20-40. Power factor and dielectric constant of cellular polyethylene vs. frequency.

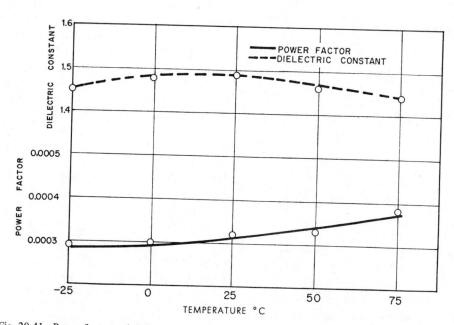

Fig. 20-41. Power factor and dielectric constant at 1 megacycle of cellular polyethylene vs. temperature.

of the signal brought in by an UHF antenna. It is evident from Fig. 20-40 that cellular polyethylene meets these requirements.

Figure 20-41 indicates that its dielectric constant and power factor are not materially affected by changes of temperature.

Figure 20-42 shows that the power factor of cellular polyethylene is independent of its specific gravity, while dielectric constant increases with increase in specific gravity.*

* The power factor in this figure is higher than those in the preceding figures because the material tested contained coloring ingredients which increased the power factor.

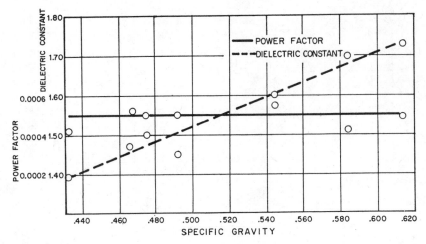

Fig. 20-42. Power factor and dielectric constant at 1 megacycle of cellular polyethylene vs. specific gravity.

Applications

The ease of handling of cellular polyethylene in modern wire-coating equipment, and the economies which it provides in size and weight of insulated conductors, indicate its utility in many electrical applications. Major applications are in coaxial cables (CATV, military, and other) and in twin leads.

LOW DENSITY POLYETHYLENE FOAMS*

Low density polyethylene foams have experienced high commercial interest because of their unique properties. They possess all the advantages of the base polymer, such as excellent resistance to most chemicals, both organic and inorganic. The low density polyethylene foams are generally closed cell and are classed as "semi-rigid" and "flexible" depending upon their densities and shapes, and can be made to feel as "soft" as low density flexible urethane foam and as "hard" as rigid polystyrene foam.

There are basically two types of low density polyethylene foam: extruded and cross-linked. The extruded foam is produced in a continuous process by first blending molten polyethylene polymer with a foaming or blowing agent (usually a halogenated hydrocarbon gas) under high pressure, conveying this mixture in a temperature controlled screw extruder through a die opening to a continuous conveyer exposed

* By L. Robert Schanhals, Construction Materials TS&D, Dow Chemical U.S.A.

to atmospheric pressure (Fig. 20-43). When the hot viscous liquid-gas solution is exposed to atmospheric pressure the gas expands to form individual cells. Simultaneously the mass is cooled to solidify the molten polyethylene, thus trapping the blowing agent in the intersticial cells. The degree of expansion, the cell size, and cell orientation can be controlled by varying flow rate, heating and cooling temperatures, gas-liquid ratio, and pressure drop through the die opening.

In this process, cross sections of up to around 50 in.^{2} are attainable with tolerances acceptable to most end uses. Presently low density polyethylene foam sheet having densities from 2.0 lb/ft^3 and sizes up to $\frac{1}{2}$ in. thick by 72 in. wide can be produced.

"Planks" of densities from 2/15 lb/ft^3 and cross sections of 1 x 14 in. to 4 x 12 in. are also attainable in this process. "Rounds" from $\frac{1}{4}$ to 8 in. diameter, ovals, and other non-standard cross sections are also products which have been

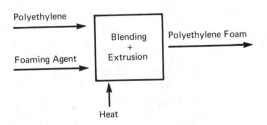

Fig. 20-43. Schematic of conventional extruded polyethylene foam.

successfully produced by the extruded process.

The other basic type of low density polyethylene foam is called "cross-linked" and can be produced by batch and continuous processes and cross-linking accomplished by chemical or irradiation methods.

The chemical cross-linked polyethylene foam is produced in a batch process (Fig. 20-44) and because of production economics is limited to producing "plank" products. Presently, plank sizes up to 48 x 48 x 3 in. are manufactured by this process. Solid polyethylene is blended with a chemical cross-linking agent, usually a solid which decomposes at a certain temperature T_{c1} and with a chemical foaming agent, usually a solid which decomposes at a temperature T_{f1} which is a higher temperature than T_{c1}. The blend is then subjected to a temperature T_{c2} higher than the decomposition temperature (T_{c1}) of the cross-linking agent to initiate cross-linking of the polymer. It is important that this temperature is maintained below the premature foaming. After the cross-linking has proceeded to desired levels the temperature is raised to T_{f2} (higher than T_{f1}) to initiate and maintain foaming. Generally these two unit operations are carried out in the same vessel (the mold) and under pressure to help control the rates of decomposition of the cross-linking and foaming agents. Presently only 2 lb/ft^3 density productions are made in this process.

The radiation cross-linking process (Fig. 20-45) allows the continuous production of cross-linking polyethylene foam but is limited to producing relatively thin cross sections (up to $\frac{3}{8}$ in.) or sheet products. The maximum width possible with this process is determined by practical considerations in the densified sheet extrusion and in the radiation operations. Generally, widths up to 48 in. are practical.

The chemical, mechanical, and thermal properties of the extruded and the cross-linked low density polyethylene foams are very similar. The largest difference between the two is that the cell size of the cross-linked foam is generally smaller and more uniform than the extruded products. Also the gauge tolerance or thickness dimension is more tight and uniform in cross-linked sheet as compared to extruded sheet. In addition, the cross-linked products possess a softer "feel" than the extruded products.

The properties of cross-linked plank (chemically cross-linked) generally are not as uniform throughout the plank and from plank to plank, whereas the extruded plank possesses more uniformity within a given plank, and from lot to lot.

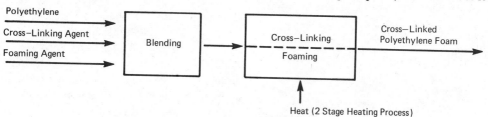

Fig. 20-44. Schematic of chemically cross-linked polyethylene foam process.

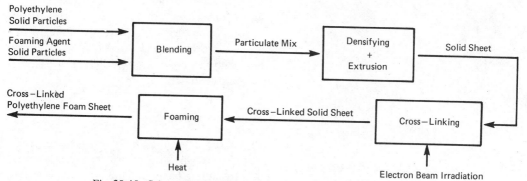

Fig. 20-45. Schematic of radiation cross-linked polyethylene foam process.

The mechanical properties (compressive, tensile, dynamic cushioning, etc.) of sheet products are usually much different than those of the plank.

In the interest of generalizing typical properties and characteristics of low density polyethylene foams, the following sections have been written by categorizing sheet versus plank. Where no significant differences exist, the category is "polyethylene foam."

To obtain exact values of properties for specification or design use, the reader is encouraged to contact suppliers of low density polyethylene foams.

Compression Characteristics

Compression Deflection. As mentioned previously, polyethylene foam exhibits properties similar to flexibles and semi-rigid foams as evidenced in compression deflection characteristics. A typical stress-strain relationship for plank is shown in Fig. 20-46. The compression or compressive strengths of polyethylene sheet are a bit lower than the respective density plank products and as one might expect the compressive strength increases as density increases both in the plank and sheet products.

Compressive Creep. When any material is loaded continuously over a period of time, it tends to creep or lose a portion of its original thickness. Generally, lower density products exhibit this more than high density products. The creep characteristics are important because they reflect long-term load carrying ability and affect the cushioning ability of the material. In Figs. 20-47 through 20-49 the compressive creep characteristics of various density polyethylene foam planks are presented, along with some data at elevated temperatures.

Compression Set Recovery. Compression set is the amount of thickness a material fails to recover after compressive creep has been experienced and the load is removed. Figures 20-50 and 20-51 show this property for both short-term loading and long-term loading for various densities of polyethylene foam plank.

Tensile and Tear Strength

The tensile and tear strengths for polyethylene foam are very high compared to similar properties of other plastic foams. These values for polyethylene foam, however, vary considerably over the different axis of the foam. This can be explained mainly by the cell orientation. Also, tensile elongation varies considerably from polyethylene foam to polyethylene foam and since there is such a wide variation, no attempt to characterize this has been made. Figure 20-52 shows the typical minimum tensile strength versus product density for sheet and plank. As might be expected, the tensile strength for sheet is greater than the tensile strength for the same density plank. This can be explained by the fact that most sheet has two skin surfaces which contribute significantly to the tensile properties. This is particularly true of extruded polyethylene foam. The same effect can be seen on typical tear strength of properties shown in Fig. 20-53. Again, these are minimum tear strengths regardless of the direction of orientation.

Dynamic Cushioning

One of the largest uses of polyethylene foams is in the packaging area to protect product from damage during handling, shipping, and storage. Polyethylene foams are excellent energy

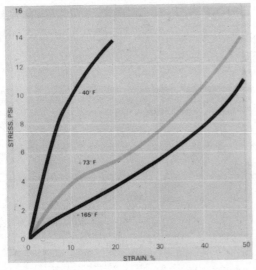

Fig. 20-46. Compressive stress-strain relationship for 2.2 lb/cu ft density polyethylene foam plank at −40°F, 73°F., and 165°F.

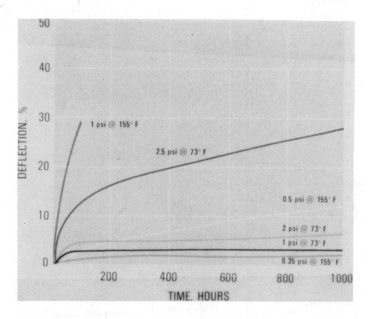

Fig. 20-47. Compressive creep of 2.2 lb/cu ft polyethylene foam plank under constant static load.

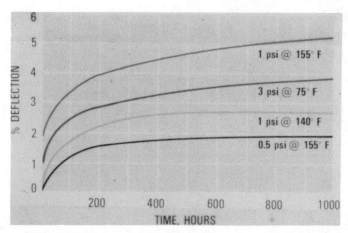

Fig. 20-48. Compressive creep of 4.0 lb/cu ft polyethylene foam plank under constant static load.

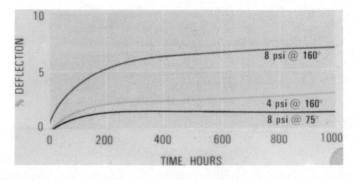

Fig. 20-49. Compressive creep of 9.0 lb/cu ft polyethylene foam plank under constant static load.

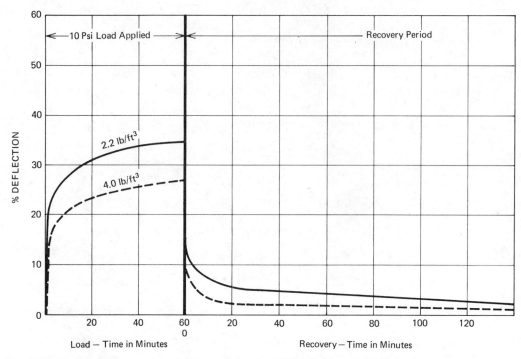

Fig. 20-50. Compressive set-recovery characteristics for 2.2 lb/cu ft and 4.0 lb/cu ft polyethylene foam plank—short-term loading.

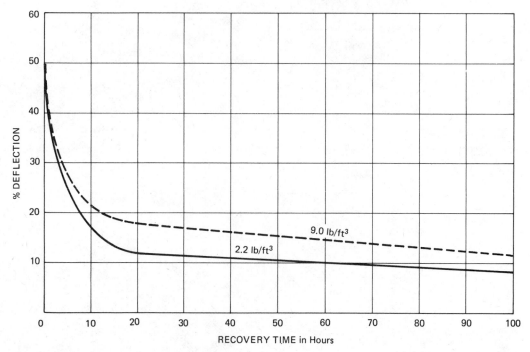

Fig. 20-51. Compressive recovery of 2.2 lb/cu ft and 9.0 lb/cu ft polyethylene foam plank after 22 hr at 50% deflection.

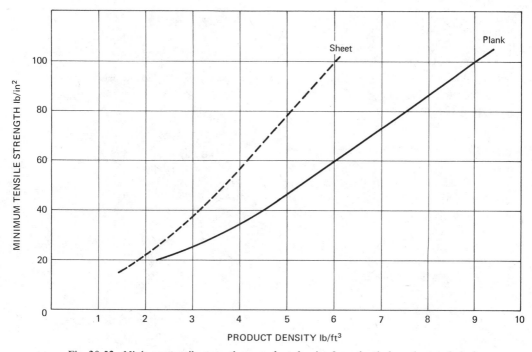

Fig. 20-52. Minimum tensile strength vs. product density for polyethylene sheet and plank.

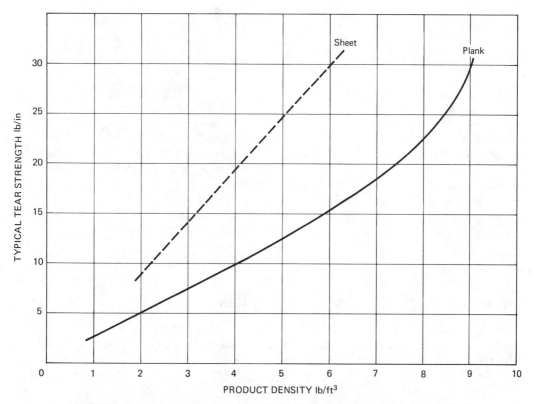

Fig. 20-53. Typical tear strength vs. product density for polyethylene foam sheet and plank.

absorbers and because of the wide range of dynamic cushioning properties afforded with different densities, a wide spectrum of packaging demands can be met with polyethylene foams.

Dynamic cushioning curves, which are the plot of peak deceleration values versus static stress, are generated as follows:

A platen of specific area A is weighted to a specific weight W which provides a static stress $P = W/A$ and is dropped from various heights onto various thicknesses of polyethylene foam. The platen is instrumented with an accelerometer which measures the peak deceleration value incurred by the platen during its travel into the foam. The platen is then weighted with additional weight W_2 providing various static stresses $P_{2-n} = W_2 - n \div A$ and dropped from various drop heights onto various thicknesses of foam and the peak deceleration values recorded.

Polyethylene foams provide optimum cushioning characteristics for objects exerting static stresses from around 0.4 to over 15 psi. There are no other flexible or semi-rigid foams commercially available which provide optimum cushioning characteristics in this static stress range.

Water Properties

Polyethylene foams are used extensively in buoyancy applications because of their excellent water-resistant properties. The polyethylene foams are basically closed-cell multicellular products and absorb less than 0.5% by volume of water after being immersed for 24 hr. (See Table 20-10.)

Because of its low density, the buoyancy properties are also excellent.

Thermal Properties

Thermal Conductivity. The measure of heat transmission through a foam material is determined by many factors; the base material, the cell size, degree of closed cell structure, and others. This property, called *thermal conductivity*, varies somewhat between the dif-

Table 20-10. Water Properties of Polyethylene Foams.

Water Absorption—ASTM C-272-53
Volume pick-up after 24 hr total immersion = <0.5%

Buoyancy

Maximum weight to be supported by 1 ft³ of:

2.2 lb/ft density product = 55 lb
4.0 lb/ft density product = 53 lb
6.0 lb/ft density product = 51 lb
9.0 lb/ft density product = 48 lb

ferent polyethylene foams; however, within the limits of a respectable insulation at a range of .28 to .40 Btu's/hr/ft²$-°$F/in. as indicated in Table 20-11.

Thermal Stability. The thermal stability of polyethylene foams is dependent upon the load imposed on the foam, the exposure time, and

Table 20-11. Thermal Conductivity of Polyethylene Foams.

Sheet—0.28–0.34 Btu-in. hr ft² $°$F
Plank—0.37–0.40 Btu-in. hr ft² $°$F

the temperature involved. Therefore, no specific maximum use temperature can be given for polyethylene foams without qualifying these requirements. Figure 20-54 shows a typical thermal stability time relationship for 2 to 9 lb/ft³ density polyethylene foams exposed to 180° for 96 hr. The dimensions do change over the heating period and the thickness change is usually more than the change in length and width. Again these are typical properties and should not be used for design purposes.

Electrical Properties

The excellent dielectric characteristics of polyethylene are retained when this plastic is expanded to make foam. Expanded polyethylene foam is a candidate for many electrical material uses requiring good properties of dielectric strength, dielectric constant, dissipa-

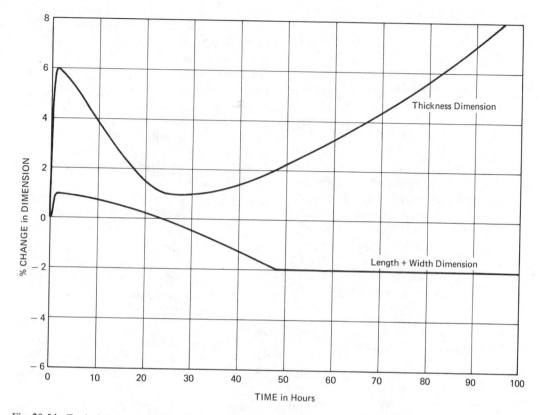

Fig. 20-54. Typical thermal stability time relationship for 2-9 lb/cu ft polyethylene foams exposed to 180°F for 96 hr.

tion factor and volume resistivity. Specific values are shown in Table 20-12.

Chemical Solvent Resistance

Polyethylene foam is chemically inert and contains no water-soluble constituents. It is resistant to most chemicals and solvents at room temperature. The material is unaffected by contact with fuel oil and heavier hydrocarbons but exhibits slight swelling when immersed for an extended period of time in gasoline. Acids and alkalis normally do not affect polyethylene foam but strong oxidizing agents may eventually cause degradation especially at higher temperatures. At temperatures

Table 20-12.   Typical Electrical Properties of Polyethylene Foams of Various Densities.

Density	Lb/Ft³			
	2.2	4.0	6.0	9.0
A. Dielectric strength (volts/mil ¼ in. thick)	52	59	86	133
B. Dielectric constant at 10^6 cps	1.05	1.06	1.07	1.15
C. Volume resistivity (ohm/cm)	10^{16}	10^{16}	10^{16}	10^{16}
D. Dissipation factor at 10^9 cps	$1\text{-}2 \times 10^{-4}$	$1\text{-}2 \times 10^{-4}$	$1\text{-}2 \times 10^{-4}$	$1\text{-}2 \times 10^{-4}$

above 130°F it becomes more susceptible to attack by certain solvents.

Light Stability

Over an extended period of exposure, ultraviolet rays in sunlight cause some degradation of polyethylene. This degradation is slower in higher density products. It is first noted in a yellowing of the foam surface; some degradation of physical properties will be experienced with longer periods of exposure. To give some idea of ultraviolet degradation, three months exposure in the Arizona summer sun results in some degradation of $2 \, lb/ft^3$ polyethylene foam; however, exposure for one year under Michigan solar conditions had very little effect. For applications where long-term performance is required under direct sunlight, a protective coating should be used.

Fabrication

Ease of fabrication is one of the many advantages of polyethylene foam. It can be skived to precise thickness, cut and shaped to form custom parts, and joined to itself or other material, without major investment in complex equipment. It also can be vacuum formed. Several of the common fabrication and finishing techniques used for polyethylene are described briefly below.

Methods and Equipment

Polyethylene foam can be fabricated with conventional power tools used for woodworking. Its flexible nature, however, makes the choice of equipment more critical than with rigid materials. For example, machines with blades or bits having a slicing-type action give best results. These include band saws with scalloped edge blades, slab splitters, and router bits with spiral cutting edges. Circular saws and dados, straight edge router bits, and band saws with toothed blades can be used successfully, but should be driven with as high a peripheral speed as is practical.

A typical splitter used to cut polyethylene might employ a scalloped blade that cuts with a slicing action and leaves no dust. Thicknesses down to $\frac{1}{16}$ in. can be obtained by this method. When cutting polyethylene into thin sheets, the product should be held securely to the table to obtain good tolerances. A vacuum table does an excellent job of clamping the material securely with a minimum of labor.

With a scalloped blade, cutting speeds of 50 ft/min are not uncommon. This speed, however, will vary with the thickness and width of the part and the required tolerances.

Both band saws with toothed blades and circular saws may be used to fabricate polyethylene, but they should be operated with as high a peripheral speed as is practical. These blades do not provide as smooth a cut as either the scalloped or the straight edge blades.

For parts requiring close dimensional tolerances, it may be necessary to age sliced stock for up to five days to allow it to become dimensionally stable before die cutting the parts. Stresses inherent in the product tend to be relieved by the slicing operation, resulting in a certain amount of dimensional instability in the sliced stock during the stress relaxation period.

The use of electrically heated resistance wires is another method of fabrication. A single wire can suffice, but to make multiple cuts simultaneously, several wires can be connected in parallel. For special shapes or contours, a heavier resistance wire can be bent into the desired form.

Longer runs of special shapes will often justify the use of contoured heated molds which form cavities by melting the foam. These molds are heated from 250 to 350°F for forming the part, and then cooled to stabilize the part and facilitate its removal from the cavity. In some cases, a hot mold and a chilled mold are used to achieve a smooth part with minimum cycle time. Hot molds must be covered with a release agent to prevent adhesion of the polyethylene to the hot metal surface.

Bonding

Expanded polyethylene will adhere to itself by the use of heat alone. Hot air or a plate heated to approximately 350°F can be used to simultaneously heat the surfaces of two sections of foam to be joined. When softening begins, the pieces are quickly joined with moderate pressure and an excellent bond is formed, with only a short cooling time necessary. A coating of fluorocarbon resin or silicone dispersion on the heating surface will aid in the release of the melted foam.

This foam may also be bonded to itself, and to other materials, by the use of commercially available adhesives. Adhesives with solvents can be used without fear of dissolving the foam. For best results, the adhesive should be applied to a cut surface. The exposed cellular structure will increase the contact between the adhesive and the foam resulting in a stronger bond.

Properly applied, many pressure-sensitive rubber-based adhesives give bonds stronger than the tensile strength of the foam. Such adhesives should always be used in accordance with the manufacturer's recommendations for open time, temperature range, and water resistance,

IONOMER FOAM*

Conventional processes for extrusion or foam injection molding of ionomers are used to produce a tough, closed-cell foamed structure. The low melting point, high melt strength of the resin, and compatibility with nucleating agents or fillers combine to provide systems that have utility in athletic products, footwear, and construction.

A low melting/freezing point and melt hot tack can provide product advantages. These same resin properties, however, have to be recognized in fabrication as limitations in certain processing equipment. In extrusion, it is necessary to consider air/water circulation for the inside cooling drum to prevent hot foam from sticking as it exits the die. In injection molding, the lower freezing point will result in a longer cycle time on single station equipment. On the other hand, these same melt characteristics also help to produce a tougher skin in the low density foamed sheet and a better surface finish in the higher density injection molded products. The higher tensile strength and low

* By R. H. Kinsey, E. I. duPont de Nemours & Co., Inc., Wilmington, Del.

melt point characteristics of an ionomer give stronger heat seal seams for fabricated sections used in packaging applications.

Products ranging in density from 3 lb/ft^3 are tougher and more solvent-resistant than equal density foams made from polyethylene or styrene. The ionomer foamed sheeting may be vacuum formed, laminated, stitched, glued, and modified for flame-retardant requirements.

The resilient nature of the polymer in medium density sheeting (6 to 9 lb/ft^3) has application in floor and carpet backing, swimming pool construction, and tennis court underliner insulation.

In foam injection molded parts, the smooth skinned, resilient foam produces a tough, energy-absorbing structure which is being used as a wood substitute in athletic products like lacrosse sticks (Fig. 20-56) and hockey stick blades. Protective structures for helmets and automotive parts are under development. All are being produced in densities ranging from 0.35 to 0.7 grams/cc. In those applications where added stiffness is desirable, the use of glass or titanate fiber reinforcement with the foamed structure is very effective.

Ionomer foams are competitive with polystyrene, polyethylene, and some urethane structures. Where the combination of lightweight,

Fig. 20-56. Ionomer foam is injection molded into a lacrosse stick head of exceptional toughness. (*Courtesy E. I. du Pont de Nemours & Co., Inc.*)

toughness, and/or direct paint adhesion are prerequisites, the foamed ionomer systems are more than competitive by reducing the overall cost through lower part weight and less involved finishing.

PHENOLIC FOAMS*

Phenolic foams are of two types—"reactive" and "syntactic." The reactive type involves chemical reaction between phenol, formaldehyde, and catalyst, and the liberation of gas by a blowing agent. The syntactic type is a mixture of microscopically small hollow beads of phenolic resin with a binder, such as phenolic, polyester, or epoxy resin.

Reactive-Type Foams

The starting material for a foam of this type is an incompletely condensed resin made by interaction of phenol and formaldehyde. When this is mixed with a catalyst and blowing agent, an exothermic reaction of further condensation causes liberation of gas by the blowing agent; the increasing viscosity of the mix prevents escape of this gas, and hence the mass expands; finally, the resin sets up, and the gas is thus trapped in a rigid foam. Lower density foams are essentially open-celled while higher density foam may contain up to 75% closed cells.

The density of the foam can be controlled by the amount and nature of the catalyst and the blowing agent, and also by the degree of condensation of the resin used, which influences the amount of exothermic heat developed in completing its condensation.

Resins. Resins of low reactivity are generally used for making foams of high density; resins of high reactivity for foams of low density. Blending of resins of the two types provides further means of controlling the density.

Blowing Agents. Fluorocarbons, such as trichlorofluoromethane, can be used to reduce the density of the phenolic foam.

Catalysts. A mineral acid catalyst is normally used. Concentrated hydrochloric, sul-

* Reviewed by Dr. Don Leis, Union Carbide Corp., Chemicals and Plastics.

furic, and various substituted sulfonic acids have been reported. Mixed boric/oxalic acids have been successfully used. The concentration of acid regulates the speed of the foaming reaction. The use of higher catalyst concentration can result in the foam setting before full expansion takes place, with a resultant increase in density.

Surfactants. Foam structure can be improved by the use of silicone surfactants, such as siloxane oxyalkylene copolymers, and wetting agents, such as polyoxyethylene sorbitan fatty acid esters. By varying the concentration of surfactant the cell structure can be varied from a coarse foam with large, thick-walled cells to a fine-celled, tough, resilient foam.

Mixing Procedures. The foam ingredients are mixed at or just below room temperature. The reaction is usually very rapid with cream times of 1 to 2 sec and rise times of 50 to 60 sec. Mechanical equipment has been developed for the metering and mixing of phenolic foam.

Typically, the blowing agent and the surfactant are thoroughly mixed with the phenolic resin. After addition of the catalyst, the mixture is rapidly agitated by a propeller-type mixer and quickly poured into the mold or the container in which the final foam is to be formed.

The interval between the addition of catalyst and the foaming (pot life) is short—from 20 sec to 3 min depending on the reactivity of the resin and the catalyst concentration. (See Fig. 20-57.)

During the foaming reaction, the expansion takes place by vaporization of the blowing agent and water as a result of the exothermic heat of condensation. It is normally desirable to preheat the container or mold to minimize heat loss and achieve maximum blowing.

The foam can be made with a tough protective skin by lowering the temperature of the metal mold. The thickness of the skin can be controlled by the temperature of the mold.

Suitable protection should be taken as the reaction can be nearly explosive under certain conditions.

Mechanical Properties. The prime charac-

Fig. 20-57. Technique for insulating tall narrow cavities with phenolic foam. Foaming begins almost as soon as liquid ingredients are poured into cavity (left photograph). The complete reaction is completed in about 60 seconds. (*Courtesy Union Carbide Corp.*)

teristics of reactive-type phenolic foam are light weight, stability, heat resistance (up to 250°F), and low cost. Phenolic foam has outstanding thermal insulation properties and offers several possibilities in very promising acoustical properties.

Figure 20-58 shows the performance of phenolic foams of various densities in terms of flexural, compressive, tensile, and shear

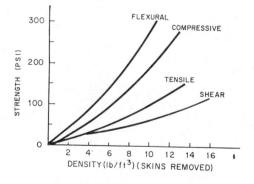

Fig. 20-58. Mechanical properties of phenolic foam.

strengths. The phenolic foams act like brittle solids, failing sharply at strains of 2 to 5%.

Other properties of the reactive-type phenolic foams that should be taken into consideration in deciding whether they are suitable for your needs include: the service temperature involved; the effects of water and humidity; transmission of water vapor; thermal conductivity; the coefficient of linear thermal expansion; toughness; adhesion to surface; corrosiveness; and biological resistance. These are discussed below.

Service Temperature. The top service temperature depends upon the catalyst used in foaming. For foams catalyzed with sulfuric and phosphoric acid, it is 250°F, regardless of density. For foams catalyzed with hydrochloric acid, it is 250°F in the lower densities, but in densities higher than about 10 lb/ft^3 decreases with increases in density, to about 200°F, being influenced by thermal conductivity and by the presence of a phosphoric acid.

Effects of Water and Humidity. When exposed to 100% RH, phenolic foam picks up

Table 20-13.   Service Temperature of Phenolic Foams

For foams catalyzed with sulfuric and phosphoric acid	250°F (regardless of density)
For foams catalyzed with hydrochloric acid	250°F (in the lower densities) 200°F (in densities higher than about 10 lb/cu ft)

weight at a fairly constant rate. When totally immersed in water, it gains weight until all of the cells are filled with water.

Virgin foam exposed to 50% RH loses weight, and approaches equilibrium in about 500 hr. The loss in weight is due to escape of water vapor and other gases.

Rates of change are reduced by the presence of a surface skin, in accordance with its thickness, or by an appropriate coating or barrier.

The dimensional effects of moisture content, for the core of a foam of density 2 lb/ft^3, at room temperature, are shown in Table 20-14, together with effects on weight.

Transmission of Water Vapor.   The presence of a skin on phenolic foam reduces greatly the rate of transmission of water vapor, as shown in Table 20-15.

Since the thickness of the skin usually increases with the density of the foam, the transmission usually decreases as density increases.

Thermal Conductivity.   Figure 20-59 shows that the thermal conductivity, measured at a mean temperature of 95 to 105°F, is minimum at densities 2 to 3 lb/ft^3, with a value 0.20 Btu/ft^2/hr/°F/in.

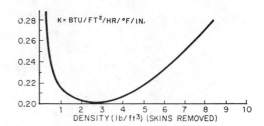

Fig. 20-59.   Thermal conductivity vs. density for cellular phenolics.

Coefficient of Linear Thermal Expansion. Measurements on a foam of a density of 2 lb/ft^3, over a range 27 to 93°C, show little difference in values for thermal expansion parallel and perpendicular to the direction of foaming. The values ranged from 1 to 3 x 10^{-5} in./in./C.

Toughness.   Phenolic foams of low density are relatively friable. At densities 5 lb/ft^3 and higher they become materially tougher.

Adhesion to Surfaces.   Phenolic foams made in contact with brick, wood, plasterboard, sheet asbestos, asphalt-coated surfaces, kraft paper, and asbestos-cement shapes adhere firmly to these materials. They do not adhere well to steel, copper, brass, galvanized iron,

Table 20-14.

Initial Condition	Final Condition	Change in Weight (%)	Coefficient of Linear Change, (in./in.)
50% RH	dryness	−2.2	−4 x 10^{-3}
50% RH	95% RH	+23	+3 x 10^{-3}
50% RH	total immersion		
	perpendicular to direction of foaming	+1000	+0.7 x 10^{-3}
	parallel to direction of foaming	+1000	+4.2 x 10^{-3}

Table 20-15. Transmission of Water Vapor, in Perms* by Phenolic Foams 0.3 in. thick

Density, (lb/ft³)	Sample with Skin Exposed to Wet Side	Sample with Skin Removed
2.1	1.4	71.0
5.5	0	63.0

* Perms: grains per hour per square foot under a partial water-vapor differential of 1 in. of mercury.

stainless steel, aluminum alloys, chromium plate, polyethylene, asphalt roofing felt, and waxed plywood; thus some of these latter may be used as forms from which the foam may be readily released.

Corrosiveness. On extended exposure at high relative humidity, acid-catalyzed phenolic foam shows corrosion of all common metals. In a series of experiments where metals were embedded in phenolic foams and aged 1200 hr at 50 to 55°C and 100% relative humidity, the following results were obtained:

Metal	Hydrochloric Acid Catalyst Corrosion Rate mils/year	Boric/Oxalic Acid Catalyst Corrosion Rate mils/year
Steel	30	2
Galvanized	1	0.2
Aluminum 3003	3	0.1
Copper	2	0.4
304 Stainless	1	0.07

Biological Resistance. Phenolic foams are only weakly toxic to insects, and thus probably cannot serve as a mechanical barrier to most insects. Nor do they inhibit the growth of fungi in the presence of adequate nutrient. On the other hand, some limited burial tests have resulted in no severe deterioration by contact with moist nonsterile soil.

Applications. Low density phenolic foam has found wide spread use as supports for floral arrangements.

The extensive use of phenolic foam has been hindered by the presence of the corrosive mineral acid catalyst residues in the foam.

One successful application is glass-reinforced phenolic foams as insulation in steel shipping containers for uranium hexafluoride cylinders. The boric/oxalic acid catalyzed phenolic foam containing 10% by weight of chopped glass

rovings is used at a density of 6 lb/ft³ and a thickness of 5 in.

Reactive vs. Syntactic. As indicated above, the reactive phenolic foams do differ from the syntactic-type of phenolic foams in terms of properties and performance. The syntactics will be discussed further below.

Syntactic Foams

These foams, consisting of commercially available tiny hollow spheres of phenolic resin held together by a binder, are of the closed-cell type.

Compositions. The binders may be phenolic, polyester or epoxy resins. The uncured mixtures have the consistency of putty. They are cured, or hardened, into foams having a density of 10 to 40 lb/ft³, according to the proportions of spheres and binder.

Typical formulations are given in Table 20-14.

Table 20-14. Polyester Syntactic Foams

Composition	No. 1 (parts by wt)	No. 2 (parts by wt)
Polyester resin	50.0	49.0
Peroxide catalyst	0.5	0.5
Cobalt naphthenate accelerator	0.5	0.5
Phenolic spheres	50.0	25.0
Foam density, lb/ft³	10.5	21.0

Epoxy Syntactic Foams

Composition	No. 1 (parts by wt)	No. 2 (parts by wt)
Epoxy resin	30.0	40.0
Epoxy hardener	9.0	10.0
Phenolic spheres	120.0	25.0
Foam density, lb/ft³	8.6	21.0

Technique. A propeller-type mixer is satisfactory for experimental batches, but on a commercial scale a kneader-type should be used. With the machine in operation, the hardener is added to the epoxy resin, or the catalyst and accelerator to the polyester resin. Peroxide and naphthenate must be added separately, since together they react explosively. Next, the phenolic spheres are thoroughly mixed in. The putty-like mass is transferred to a suitable mold, troweled onto a surface, pressed into cavities, or placed as a core in a sandwich structure. Curing will take place at room temperature, but can be hastened by heating, up to 100°C.

Properties. Typical properties of syntactic foams are shown in Table 20-15.

Table 20-15

Binder	Density, (lb/ft³)	Compressive Strength (psi)	Specific Strength (psi/sp gr)
Polyester	10.5	52	306
Polyester	12.5	145	725
Polyester	18.7	205	683
Epoxy	8.6	86	623
Epoxy	10.7	159	930
Epoxy	14.0	218	973

Applications. Syntactic foams are gaining in use as core materials in sandwich structures used in construction of aircraft and of hulls and decks of boats, and are being investigated for use in roofing structures. Experimental sandwiches of fibrous-glass mats with cores of polyester syntactic foams containing loose fibers of glass have shown flexural strengths of 1700 to 4500 psi, tensile strengths of 1600 to 2900 psi, and compressive strength as high as 12,000 psi.

The ease of preparation of syntactic foams and the ease with which they can be troweled, or otherwise worked by hand, into inaccessible spaces, have led to their use in filling out or repairing various structures, such as wooden structures damaged by rot.

CELLULAR CELLULOSE ACETATE

Manufacture

Cellular cellulose acetate (CCA) is produced by the controlled release of pressure from superheated solvent within a dough of cellulose acetate and solvent. The nature of the cellular structure is governed (1) by incorporating into the dough a finely-divided inert material, which provides nuclei for volatilization and thus influences the number and uniformity of the cells; (2) by controlling the degree of expansion, and thus the density of the product.

A further ingredient is desirable, namely, chopped glass fiber, which contributes to the strength of the product.

The hot dough is metered under pressure to the orifice of an extruder, where the expansion takes place. The orifice is fitted with a sizing box which governs the cross-section of the expanding product.

A screw extruder has been used, but a four-cylinder piston-type extruder gives better results.

CCA cannot be foamed in place.

Properties

CCA is a white rigid closed-cell foam, commercially available as boards (0.5 × 8 in., 0.75 × 6 in., 1 × 4 in., 1.25 × 4 in.) and as rods (diameter 2.25 in.). The product, continuously extruded, is usually cut to a length of 6 ft.

As extruded, CCA has a dense smooth skin, which is usually removed in machining to the standard commercial dimensions.

CCA does not tend to become brittle or frangible. Its mechanical properties are substantially constant between 77°F and −70°F. It will tolerate temperatures as high as 350°F, and even higher temperatures for short periods.

Fabrication

Since CCA cannot be foamed in molds, end-use shapes must be made by machining. Objects of larger than the standard cross sections available are made from blocks prepared by cementing together pieces of the CCA board. For cementing, almost any adhesive is suitable if it contains no ketone or ester solvent.

Sandwich constructions with facings of, e.g., sheet aluminum and polyester-impregnated glass cloth or glass mat, are made, with the aid of suitable adhesive, in a conventional platen press.

CCA is satisfactorily machined within close tolerances by ordinary shop equipment.

Applications

The use for which CCA was first developed was as core in monocoque construction. It has found use in control surfaces of aircraft; radome housings; filler blocks under fuel cells; structural members in reinforced-plastic boats; geodesic domes; panels used in house construction, containers, trailer bodies; flotation devices; x-ray and electronic equipment; etc.

EPOXY FOAMS

General

Products currently made by foaming of epoxy resins are of rigid type; some are preponderantly of closed-cell structure, some of open-cell structure. These commercial products include (1) prefoamed boards, sheets and rods, (2) formulated "pack-in-place" systems and (3) formulated "foam-in-place" systems. Current efforts may lead to development of special shock-resistant and semirigid types, or even flexible types, and of a type applicable to surfaces by spraying. In addition, there are now available certain foams in which the epoxy resin is not itself foamed, but serves as binder for microscopic hollow gas-filled beads of phenolic resins, urea resins and polystyrene; these are discussed elsewhere in this chapter.

Prefoamed Sheets, etc. Boards and planks are available in various types of formulations. They are free from directional grain and do not warp. They can be postformed at temperatures of 250 to 300°C.

Pack-in-Place Systems. These are mixtures, supplied as such or compounded from separate ingredients. They are pushed or tamped into the cavity to be filled (in the manner of molding sand in a foundry). During cure, at room temperature or with the aid of heat, a continuous porous mass is formed, which is completely or preponderantly of closed-cell type.

Pack-in-place systems are used for embedment of electronic components, and in microwave lenses, radome cores and antennae, in light sandwich structures, in thermal insulation, and as adhesive and caulking compounds, e.g., for filling voids and for reinforcing honeycomb structures.

Foam-in-Place Systems. In this case, an appropriate liquid mixture of epoxy resin with blowing agent and other ingredients is poured into the bottom of the cavity to be occupied by the foam, and foaming is induced *in situ*, at room temperature or with the aid of heat. Foams of either closed-cell or open-cell type are made by this method.

Foam-in-place systems are used in encapsulation of electrical equipment, in thermal insulation, in building materials and structural members, and in tooling.

Manufacture

The process of foaming the epoxies involves proper balancing of the formulation and proper control of the exothermic reaction to yield the desired uniformity of structure. The hardening of the resin by cure should take place at the moment of maximum expansion; in this connection, control of temperature is important.

The presence of solvent-diluents (which modify the viscosity of the mass, and may function as auxiliary blowing agents) and of other additives influences the thermal behavior of the mixture, as do also the initial temperature, and the size and shape of the batch.

Properties of Epoxy Foams

Certain characteristics of the epoxy resins are advantageous in their use in foamed form.

Versatility of Formulation. Epoxies can be cured, without formation of by-products and hence without large shrinkage, by use of a large variety of curing agents and hardeners, and are compatible with many modifiers. Thus they offer a wide choice of formulations and the opportunity to achieve a desired combination of properties.

Handling Properties. Epoxies have indefinitely long shelf life. Most systems can be worked at room temperature, or with only moderate heating.

When feasible, the more volatile amines used as accelerators should be replaced by longer-chained amines, in order to avoid dermatitis in manufacture and use.

Dimensional Stability. Being thermoset resins, the cured epoxies are dimensionally stable up to moderately high temperatures.

Their dimensions are very little affected by humidity.

Mechanical Properties. The foams are tough and strong, and can be shaped with wood-working tools.

Adhesion. These resins adhere strongly to most metals, ceramics, fabrics, plastics and concrete. Thus the foams will bond themselves firmly in cavities in which they are produced, or to inserts. (If adhesion must be avoided, a suitable release agent must be used.) Preformed foams may be readily bonded to themselves or to other materials.

Resistance to Chemicals, Solvents and Water. Epoxy foams are resistant to alkalies, most acids, solvents and water.

Electrical Properties. The electrical properties of epoxies make the foams useful in electrical applications where lightness of weight is essential. Foams can be made with dielectric constants ranging from 2 to 7.

SILICONE FOAMS

Three types of silicone foams are discussed in the paragraphs which follow.

Premixed Powders

Silicone foaming powders of standard type are expanded by means of a blowing agent which decomposes into nitrogen gas and an alkaline by-product at about 300°F. The silicone resins used are solventless polysiloxanes with a melting point of 120 to 140°F. In the presence of an appropriate catalyst, they become thermoset through the condensation of hydroxyl groups.

To make a foam, the powder containing the resin, blowing agent and fillers is simply heated above 320°F. After the resin liquefies, the blowing agent decomposes. Nitrogen gas expands the resin, while the amines given off act as catalysts for the condensation of the resin. Expansion and gelation are thus synchronized, so that the resin gels at maximum expansion.

Of the three types of powder, Type A can be foamed to densities ranging from 10 to 14 lb/ft^3; Type B from 12 to 16 lb/ft^3; Type C from 14 to 18 lb/ft^3. The first two may be

foamed in place; Type C can be foamed satisfactorily only as a block or sheet, but it is stronger than the others, and retains more compressive strength at high temperature.

Foamed structures produced from Type A are the most resistant to thermal shock. Samples have been cycled repeatedly between room temperature and 600°F without cracking, and have withstood 700°F for 72 hr with only slight dimensional change. The total loss of weight shown by the sample in this case is less than 10 per cent.

Similar to Type A in most respects, Type B retains a considerable amount of its compressive strength at elevated temperatures, especially if postcured for 48 hr at 480°F. In many applications, this foam will cure further, and become stronger, with use.

Molds for these silicone foams may be made of metal, wood, glass, etc. The molds do not require preheating. The powder is poured into the cavity, and heat is applied by circulating-air oven, strip heaters, heat lamps or similar equipment. To minimize shrinkage, the structure should be exposed to the expansion temperature for at least 4 hr. If postcuring is not required, the foam is now removed from the mold. If postcuring is required, the temperature is raised by increments of 50°F to the required temperature.

Room-temperature-curing Resins

The newer type, which can be expanded and cured at room temperature, is based upon chemical reaction between two silicone components, in the presence of catalyst. The reaction is slightly exothermic, but temperatures seldom exceed 150°F, even in very large pours. Hydrogen gas is liberated as the expanding agent, but the quantity is small and has not presented any explosive hazards. However, the usual precautions should be observed when large quantities of foam are processed. The reaction is complete in 15 min., and maximum strength is developed in 24 hr.

These materials are supplied in the form of two liquid components. These are blended in a high-speed mixer for 30 sec and poured. Expansion to 7 to 10 times the initial volume is complete within 15 min., but the foam remains soft and tender for about 2 hr. After 10 hr, the foam is hard enough to be cut and handled. Its density is usually between 3.5 and 4.5 lb/ft^3, depending upon the geometry of the cavity.

Finished foams based on these materials can be used continuously at 600°F. They have low thermal conductivity, good electrical insulation properties, and low water absorption. The heat of reaction and the expansion pressure are so low that molds of cardboard or heavy paper can be used.

Elastomeric Foams

These lightweight rubbery foams are made by mixing two components. The mixing requires only 30 sec, and the currently available materials must be poured immediately, because expansion begins promptly upon blending. Negligible pressure is generated, and the articles can be removed from the molds within 5 min., at which time the foam will have developed about 80 per cent of its ultimate strength. Maximum strength is developed after 24 hr. When cast against glass cloth or asbestos paper, the foam adheres strongly. Finished pieces can be easily bonded to each other, or to metal, by appropriate silicone adhesives.

Applications

The original interest in silicone foams centered on their possible use as core materials in high-temperature sandwich structures, but most of the foams thus far developed have proved too brittle to withstand the vibration encountered in such service. However, foams based on the powdered premixes have proved useful as molded components for aircraft, and as insulating materials for instruments.

UREA-FORMALDEHYDE FOAMS

Urea-formaldehyde foams are of two types, open-cell and closed-cell.

Open-cell Foams

These are prepared by whipping air into a specially prepared resin, and then setting the resin.

Formaldehyde (30%) is condensed with urea. The resin is cooled and brought to pH 8.0. An internal plasticizer is added, and more urea, to a urea/formaldehyde ratio 1.0/1.7.

A second solution is prepared, containing phosphoric acid and soap. The two solutions are mixed, in a beater which has air blown into it, and the result is a texture like that of whipped cream. This is fed upon a moving belt, which is passed through a heated zone to effect an initial cure. The foam is then rough-cut and further cured in ovens, and finally cut to size for shipment. Present maximum size is 24 x 24 x 6 in. Friability limits the minimum dimension to 1 in.

The product in this form of a block usually has a density of 0.8 lb/ft^3, but it can be made over a range 0.5 to 1.5 lb/ft^3. Its thermal conductivity at a mean of 10°F is 0.21 and at 70°F, 0.22 Btu/hr/ft^2/in./°F. Exposed to 100% RH for a long period, it will absorb up to 0.15 lb/ft^3 of moisture. It will support a compression load of 1.5 psi with negligible distortion.

The block material is often shredded into particles about $\frac{1}{32}$ to $\frac{1}{8}$ in. in size. In shredded form, the density is 0.8 to 1.0 lb/ft^3, and thermal conductivity at a mean of 75°F is 0.21 to 0.23 Btu/hr/ft^2/in./°F.

The open-cell urea-formaldehyde foams are used principally to support floral displays, and in thermal insulation.

For insulation, both the block and the shredded material can be used. However, its open-cell structure makes its insulating character vulnerable to entrance of water under wet conditions, unless suitably enclosed.

The various urea-formaldehyde foams are noncorrosive and nontoxic, and will not support growth of mold or bacteria.

It cannot be used for insulation at high temperatures. At 130°F, prolonged exposure results in shrinkage. At 300°F, the shrinkage may be as much as 15% by volume.

Closed-cell Foams

Microscopically small hollow spheres are made from liquid urea-formaldehyde resin by a special technique of spray drying, followed by a post-treatment to improve water resistance. The spheres range in diameter from 2 to 60 microns. The particle density is about 18 lb/ft^3; the bulk density of a mass of spheres is about 3.7 lb/ft^3. The compressive strength is 30 to 50 psi, determined by subjecting the spheres to pressure of nitrogen until rupture occurs. Decomposition of the spheres begins at 350°F.

This powdery-appearing material (spheres) was designed to be used as a floating blanket or roof on crude oil in storage, to minimize evaporation of volatile constituents. Such a blanket, 0.5 in. thick, decreases the rate of evaporation by 80 to 90%.

The material is used also in making low-density aggregates, produced by mixing the tiny spheres with an appropriate binder, e.g., liquid resins, such as polyester and epoxy (and then curing the resin), or thermoplastic materials such as polyethylene of low molecular weight, and paraffin. Such mixtures are useful in light-weight structural members, thermal insulation, potting compounds, caulking compounds, etc.

A family of proprietary urea-formaldehyde foams has been developed (U.F. Chemical Corp.) and is finding uses in the following applications: In the construction industry, the foam-in-place material results in a very useful thermal and acoustical insulation.

In agricultural use, the foam acts as a reservoir for water and nutrients increasing the rate of growth and reducing water requirements. It also permits the growth of vegetation on otherwise unplantable substrates (such as slag heaps).

The urea-formaldehyde foam will absorb oil and other hydrocarbons from polluted water.

Other applications for the foams that look promising include several in such fields as medical and cosmetic, ship-raising, solvent regeneration, rat and insect protection, paper, fertilizer and mine safety.

STRUCTURAL FOAMS*

To define structural foam is in itself a major undertaking.

* Portions of this section adapted from the brochure "Structural Foam", issued by The SPI's Cellular Plastics Division.

The SPI defines *structural foam* as a plastic product having integral skins, a cellular core and having a high enough strength-to-weight ratio to be classed as *structural*. What minimum value this strength-to-weight ratio must be to be classed as *structural* is dependent entirely upon the application.

This definition allows for many variations of the general "Structural Foam" concept. It includes both thermoplastic and thermoset polymers and covers a wide range of density.

The thermoplastics that can be produced as structural foams already run a wide gamut—ABS, acetals, acrylics, styrenes, polyethylenes, nylon, PVC, polycarbonate, modified polyphenylene oxide, polypropylene, polysulfone, thermoplastic polyesters, and various glass-reinforced nylons, polyethylenes, polypropylenes, and other thermoplastics. Among the thermosets, urethane structural foams are most in use and are covered in greater detail on p. 519. This section will treat primarily with thermoplastic structural foams and the wide range of markets into which they are currently going (Figs. 20-60 through 20-63).

Physical Tests and Standards

With any non-uniform material, physical testing becomes more difficult. The added variable of skin formation around a cellular core makes it almost impossible to get consistent values.

Up until recently, the only test standards available for testing structural foam parts have been the standard ASTM tests used for solid materials. Much work is being done to come up with tests specifically designed for structural foams.

However, until a great deal more experience is obtained using these new tests, the standard ASTM tests will continue to be the ones used for relative testing of structural foams.

Physical test data on several thermoplastics structural foams are shown in Tables 20-15 through 20-17. Data on urethane structural foams can be found on page 520.

Properties of Structural Foams

Basically, structural foam molding is intended

Fig. 20-60. Random group of structural foam molded products and components indicates the versatility of the process.

to produce tough, rigid, lightweight thermoplastic products with solid skins and a cellular core. The solid skin gives the structural foam its strength and rigidity; the cellular core helps structural foam to resist impact.

The importance of structural foam can best be understood by reviewing the general practices used in designing a solid molded part that is considered to have insufficient stiffness. This inadequacy is generally overcome by in-

Fig. 20-61. High-density polyethylene structural foam multi-duct conduit weighs only 76 lb, replaces 1600 lb of concrete ducting. (*Courtesy Union Carbide Corp.*)

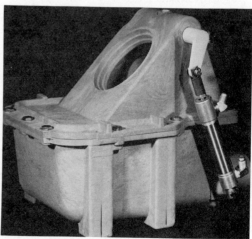

Fig. 20-62. Injection molded hopper for low-flush toilets is based on Noryl structural foam; replaces aluminum model.

Fig. 20-63. Lightweight high-density polyethylene structural foam enclosures are potential successors to concrete vaults for underground transformers. (*Courtesy Arco/Polymers*)

foam structure is employed. This is an excellent method of increasing rigidity, since stiffness is a function of thickness to the third power. The integral skins also improve the stiffness of the foam cross section. A linear decrease in the modulus of elasticity is assumed.

Making Structural Foams

Although there are many different systems—and an even more diverse range of machines—available for producing structural foams, most molding techniques (extrusion techniques are covered on p. 591) fall into one of two categories.

One category includes low-pressure systems that involve the use of nitrogen gas or chemical blowing agents as expansion devices. In a low-pressure system (so-called because the molds are under very low pressure), the molds are only partially filled (i.e., a "short shot") with the melt; the melt is then expanded by the nitrogen gas or by the gases released by the decomposing agents to fill the mold. In the processes involving nitrogen gas, the resin in pellet form is fed into an extruder where it is plasticated and mixed with the gas before injection into the mold. When chemical blowing agents are used, they are either already incor-

creasing the thickness of the part or adding ribs. Either solution adds material and cost to the product. The second approach sometimes significantly detracts from its appearance. However, the thickness of the part can be increased without a corresponding increase in weight if a

Table 20-15. Physical Properties of High Impact Polystyrene Structural Foam.				
Sample Thickness	¼"	⅜"	½"	ASTM Test No.
Density g/cc	.7	.7	.7	
Flexural modulus, psi	210,000	190,000	195,000	D-790-66
Flexural strength, psi	4,500	4,000	4,000	D-790-66
Tensile strength, psi	1,800	1,500	1,200	D-638-64T
Compressive strength				
Perpendicular to skin	3,300	2,300	2,000	D-695-63T
psi—10% Deflection				
Parallel to skin	5,200	3,700	3,600	D-695-63T
Coefficient linear exp.				
$\times 10^{-5}$ in./in. °C	9.0	8.4	8.6	D-696-44 (61)
Heat Distortion temp.				
°F @ 66 psi	189	187	193	D-648-56
°F @ 264 psi	176	161	165	D-648-56
Vicat softening point, °F	204	204	204	D-1525-65T
Durometer Hardness				
"D" Scale	74	70	70	D-2240-64T
Notched Izod impact				
ft-lb/in. of notch	1.05	1.10	1.00	D-256-56
Un-notched Izod, ft-lb/in.	2.3	1.25	1.14	
Rod drop impact, ft-lb	1.25	2.25	6.75	

Table 20-16. Physical Properties of Polypropylene Structural Foam.

Sample Thickness	$\frac{1}{4}$"	$\frac{3}{8}$"	$\frac{1}{2}$"	ASTM test No.
Density, g/cc	.6	.6	.6	
Flexural modulus, psi	120,000	110,000	105,000	D-790-66
Flexural strength, psi	3,200	3,000	2,500	D-790-66
Tensile strength, psi	2,000	2,300	1,500	D-638-65T
Compressive strength				
Perpendicular to skin	1,600	2,300	3,200	D-695-63T
psi–10% Deflection				
Parallel to skin	3,400	2,800	4,000	D-695-63T
Coefficient linear exp.				
$\times 10^{-5}$ in./in. °C	9.4	9.2	9.3	D-696-44 (61)
Heat distortion temp.				
°F @ 66 psi	167	162	184	D-648-56
°F @ 264 psi	132	115	118	D-648-56
Vicat softening point, °F	301	286	302	D-1525-65T
Durometer Hardness				
"D" Scale	63	54	64	D-2240-64T
Notched Izod impact				
ft-lb/in. of notch	*	*	*	D-256-56
Un-notched Izod, ft-lb	1.26	.99	.95	
Rod drop impact, ft-lb	.75	2.25	3.75	

* Values obtained were too low to be meaningful in this test.

Table 20-17. Physical Properties of High-Density Polyethylene Structural Foam.

Resin Density–0.962 grams/cm^3
Melt Index–8.0 grams/10 min.

Sample Thickness	$\frac{1}{4}$"	$\frac{3}{8}$"	$\frac{1}{2}$"	ASTM Test No.
Density, grams/cc	0.6	0.6	0.6	
Flexural modulus,	120,000	112,000	118,000	
psi in 25°C air				D-790-66
Flexural strength	2,730	2,450	2,510	
psi–25°C–5% Strain				D-790-66
Tensile str/elongation, psi–25°C/%	1310	1260	1390	D-638-64T
Compressive strength				
psi–10% Deflection				
Perpendicular to Skin	1,090	978	1,090	D-695-63T
Parallel to Skin	1,840	1,640	1,860	
Coefficient linear exp.				
$\times 10^{-5}$ in./in. °C	12.0	11.9	11.8	D-696-44 (61)
Heat distortion temp.				
°F @ 66 psi	129.6	138.2	131.8	
°F @ 264 psi	93.5°F	99°F	103°F	
Vicat softening point, °F	254°F	244°F	244°F	D-1525-65T
Dielectric strength volts/mil.	309	217	194	D-149-64
Durometer Hardness				
"D" Scale	54	51	55	D-2240-64T
Un-notched Izod, ft-lb	2.50	2.23	2.31	
Rod drop impact, ft-lb	3.25	3.5	7.75	
Shrinkage average–0.0175 in./in.				

porated into the molding pellet or dry-mixed into the resin by the processor or supplied in the form of an agent/resin concentrate.

The machinery involved in low-pressure foam molding systems also varies widely. In the nitrogen process, special equipment is required. With chemical blowing agents, it is possible to use conventional injection molding machines with minor modifications or machines modified especially for structural foam molding with such equipment as oversize injection pumps, large platen areas, multiple-mold stations, etc. (see discussion below)

The second major category in structural foam molding is the high-pressure molding system. As distinguished from the low-pressure system, it involves injecting the polymer melt and the blowing agents under higher pressures into the mold cavity to completely fill the mold; the mold then expands or mold inserts are withdrawn to accommodate the foaming action. Since tooling is the critical element in this system, conventional machines can be adapted; special machines with accumulators are also available.

Specialty techniques that can also be applied to structural foam molding include the expansion casting of ABS and other thermoplastics foams (see p. 593), rotational molding, and multi-polymer molding.

Low-pressure/Nitrogen

In the nitrogen process, the equipment consists of an extruder, a nozzle, a hydraulic accumulator, and a mold. Resin in pellet form is fed into the extruder. There it is melted and mixed with nitrogen gas which is injected through the extruder barrel. This material is pumped by the extruder into the accumulator and held there under pressure. When the accumulator has enough material to fill the mold the nozzle is opened and the plastic and gas mixture is forced out of the accumulator and into the mold. The nozzle is then closed causing the accumulator to refill, while the part is cooling in the mold. At the end of the cycle the part is cool and ready for ejection. When the mold is empty and reclosed the accumu-

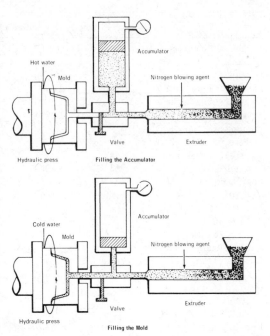

Fig. 20-64. Low-pressure structural foam molding system, using nitrogen gas as the blowing agent.

lator is full and the valve is opened causing the mold to be filled again. (See Fig. 20-64).

Accumulator pressures are generally between 2000 and 3000 psi while the mold pressure is about 300 psi. Only enough plastic to fill about one half of the mold is delivered by the accumulator, but since it contains gas it expands and fills the mold with foam. As the foam flows through the mold the surface cells collapse and solid skins are formed. These skins are beneficial since the maximum tensile and compressive stresses occur on the surface when a member is subjected to bending.

Because the mold is never packed with solid material, the pressures are low, and lower-cost molds can be used (see section on "Tooling," below).

Low-pressure Pros and Cons

Advantages. Parts molded by the low-pressure foam process are essentially stress free and exhibit only minor warping. Once the measured charge of polymer is delivered to the mold, the filling of the cavity is accomplished by the expanding gas. This greatly reduces the level of molded-in stresses.

Low spots or "sink marks" are not found in structural foam parts. Since the gas exerts an internal pressure the parts tend to swell or "puff" where there is a thick section adjacent to a thin section. This can be controlled by proper cooling. These parts generally have perfectly flat surfaces. "Sink marks" at a "T"-shaped intersection have been eliminated.

The integral solid skins on the surface of structural foam parts are formed in the molding cycle, requiring no separate operations. As the foam flows through the mold the cells that rub the mold surface rupture and densify, forming a solid skin. Pressure exerted on this skin by the cellular core also helps to solidify it. Melt temperature, mold temperature, the amount of gas in the melt, and the speed of melt injection all contribute to the formation of the skins. These conditions can be varied to produce the desired skin thicknesses. Parts with very thin skins have low stiffness while parts with thick skins are stiffer but become quite heavy. The preferred skin thickness on a $\frac{1}{4}$ in. part ranges from 15 to 50 thousandths of an inch. This material does not leak acids or alkalies and the permeation of gasoline satisfies the automotive requirements.

Structural foam parts can be molded to have the same weight as injection molded parts but they will be 3 to 4 times as stiff or rigid. However, they may also be molded significantly lighter than injection molded parts and still be $1\frac{1}{2}$ to 3 times as rigid. This provides a great economic advantage. Structural foam has been used successfully in some commercial items which previously required wood or metal for high rigidity. Injection molded plastic could not compete in these areas.

Limitations. Low-pressure foam parts generally have a swirl surface pattern. This texture is preferred for some applications. For furniture the part can be molded with a wood grained surface and painted as is done with injection molded parts. Because of the inherent ruggedness of this material it has many industry uses where a glossy smooth surface appearance is not necessary. Some pigments hide the swirl pattern more than others and molding conditions can be adjusted to vary the appearance. By increasing the temperature of the mold the

surface can be greatly improved but the cooling cycle is then lengthened. It is also possible to hide the swirl by special finishing techniques.

Another limitation of this process is minimum part thickness. Thicknesses under $\frac{1}{4}$ of an inch are generally avoided except in very small areas, because the skins constitute a larger portion of the cross section and the advantage of a foam is lost. The resistance to flow is also increased and parts with thin sections are difficult to fill.

This process is especially well suited for large parts. Items weighing less than $\frac{1}{4}$ lb are usually avoided.

Properties. The specific gravity of molded parts can be varied from 0.4 to that of the solid resin but the best balance of toughness and stiffness is generally found in parts that have a specific gravity range of .55 to .65. With the thickness increase of foam improving stiffness by a cubic function and the modulus decrease affecting stiffness less than 50% the result is a foam beam that is 3.4 times as stiff as a solid beam of the same weight. This is the reason for using structural foam in applications where the requirement is for stiffness and not tensile strength.

Notched izod tests on structural foam are not indicative of molded part performance since the notch exposes the cellular core and a tear propagates easily through this structure. Drop and tumble tests have been conducted on molded foam articles and the results have been extremely favorable when compared to their injection molded counterparts. Impact testing has been conducted at temperatures as low as $-85°F$ and the foam was still tough at this point. The cellular structure is apparently an excellent energy absorber contributing heavily to the toughness of these parts.

Tooling. The following materials are listed in order of present acceptability: aluminum, Kirksite, beryllium-copper, steel, epoxy, and wood.

Aluminum is most widely used because of its cost, weight, and ability to be cast and machined. It also has excellent thermal conductivity which is necessary for good skin formation and fast cooling rates. Aluminum's only disadvantage is that it does not reproduce detail quite as well as Kirksite or beryllium-copper. However, it is adequate for many applications.

Kirksite is being used for many wood grain furniture molds since it reproduces detail very well. It also has good thermal conductivity and machinability. The cost of Kirksite is quite low but it has a high density and Kirksite molds are very heavy.

Beryllium-copper is used where the highest quality mold is needed and cost can be justified. The raw material is expensive and machining costs are high. Beryllium-copper casts well and reproduces mold detail better than any other material. It has a high density, but since it also has very high strength, beryllium-copper molds are not as heavy as ones in Kirksite. The thermal conductivity is about the same as Kirksite but not as good as aluminum.

Steel is used in injection molding tooling because the strength is needed to withstand the great molding pressures. Structural foam pressures are low and the properties of steel are not what this process needs. The thermal conductivity is poor and the weight is high. It takes nearly twice as long to cool a piece of foam in a steel mold as in an aluminum mold.

Epoxy and epoxy-aluminum alloy molds have been used with limited success. While molds with fine detail can be cast easily at relatively low cost, they are not suitable for production tooling in the structural foam process. The thermal conductivity of these materials is too low to permit reasonable cooling cycles. This low thermal conductivity also allows the mold surface to get hot which causes very thin or non-existent skins on the surface of the molded part. This is undesirable since the parts would not be representative of what would be produced in a metal mold.

Wood has also been investigated. A mold can be shaped very inexpensively but the heat transfer problems in wood are even more severe than in epoxy alloys.

Molds for structural foam are constructed differently than those for blow molding or injection molding. There are some similarities to blow molding tools because of the low molding pressure and the rapid cooling require-

ment. The walls of the molds are usually about one inch thick with ribs on the back side. These ribs stiffen with walls and also serve as baffles for the cooling jackets. These ribs are then covered with a plate which seals the cooling chamber and strengthens the mold since it completes a boss structure. Whenever a plate is used on the top or bottom of a mold where press pressure is applied, the plate must be recessed below the pressure surface. This is done to protect the gasket around the water jacket from the extreme pressure changes as the molding press opens and closes. If this practice is not followed, leaks are likely to occur.

In some situations battery cases have been molded where the internal cell dividers had zero degrees draft.

Since the stress level in structural foam parts is low, the shrinkage is slightly less than in solid plastic parts. The following figures give average shrinkage:

High density polyethylene	1.8%
Polypropylene	1.5%
Polystyrene	.5%

These figures may vary slightly as part design and molding conditions change.

Drilled cooling holes have been used and are satisfactory when a surface is regular and easy to follow. However, for complex contours, drilled holes are not practical. Remember to get as much cooling in these molds as possible since the foam behaves like insulation.

Standard knock-off practice can be applied to these molds. Mechanical stripping by pins, rings, or plates is most common and an air assist is often used on deep parts. The air assist prevents collapsing of the sides or bottom as shallow draft parts are removed.

Structural foam parts seem to strip a little easier than injection molded parts and the following draft angles have been found to be more than adequate for these materials:

High density polyethylene	$\frac{1}{4}°$
Polypropylene	$\frac{1}{2}°$
Polystyrene	$1°$

Low Pressure/Chemical Blowing Agent*

In this low pressure system, several techniques are available for combining resin and blowing agent. The processor, for example, can simply tumble resin pellets and dry blowing agents together; more sophisticated are continuous on-stream mixers for proportioning exact amounts of the two materials in powdered form. In addition, suppliers do make available concentrates of resin and blowing agent to simplify incorporation into the resin pellets at the molding machine. Most recently, pre-compounded materials for structural foam molding have begun to appear on the market.

Normally, 0.2 to 1.5% of chemical blowing agent by weight of resin will produce parts with a good foamed structure. The recommended amount for each blowing agent may vary slightly. Use of excessive amounts is not recommended since it can increase cycle time. Basically, when molding with chemical blowing agents, the wall thickness of the part determines the percentage of blowing agent needed to produce the desired density of foamed resin.

No matter how the blowing agent is added to the resin, however, the molding process proceeds in the same way. A metered short shot is injected into the mold cavity as fast as possible, only partially filling the mold. During processing, the blowing agent decomposes with heat and releases gas through the melt, expanding the plastic to fill the mold completely.

Theoretically, standard screw injection molding machines can be used with this technique. However, a degree of modification is necessary as follows:

In the low-pressure system, the injection rate into the mold cavity is the most important factor in influencing the cell structure of the molded foam. And it is especially important to introduce the foamable melt into the cavity as fast as possible. The decomposition of the chemical blowing agent is spontaneous upon entering the mold and if injection speed is slow,

* Portions of this section adapted from "Injection Molding of Thermoplastic Rigid Foams", by N. E. Frailey, Shell Chemical Co., Woodbury, N.J.

the foam formed initially during the shot will be collapsed by the subsequent high-pressure melt.

Therefore, to make injection speed as fast as possible, machines must be modified via such approaches as: using booster circuits or oil diverted from the clamp; fitting the machine with oversize pumps; or using a pressure switch to delay forward movement of the screw and produce a forceful "explosive" filling of the mold.

In addition, it is necessary to prevent drooling of the expandable melt from the nozzle during the plastication phase of the molding cycle. This can be done by using a nozzle shut-off valve with an opening pressure of 500 psi. By this means, the mold can be pressurized at a pre-set level prior to the injection stroke. This shut-off nozzle (minimum $\frac{3}{16}$ in. diameter orifice) can either be mechanically or hydraulically actuated with the latter preferred on long production runs. The ideal nozzle shuts off at the very tip. If a shut-off nozzle is not used or if a shut-off is located at the rear of the nozzle or behind it, the nozzle may drool between the part ejection and the injection of the next shot. In addition, nozzle extensions should be avoided since these will impede the injection speed capabilities of the machine.

Although modified standard machines can therefore be used for low-pressure/chemical blowing agent systems, there has also been a trend toward the design of special molding machines expressly to take advantage of the unique molding characteristics of structural foam. Generally, these involve larger-capacity injection ends, larger platen area clamps; and lower-tonnage clamping capacity. With standard machines, the processor is generally wasting large clamp capacity on relatively small (for foam) platens and shot sizes. A 1000-ton standard machine, for example, will run only 120 oz or 7.5 lb capacity.

Foam molding also involves relatively long cooling cycles and this has increased interest in the use of rotary presses for structural foam molding. By using six or more molds set up on an indexing turntable, it becomes possible to meter the resin/blowing agent into the mold at one station, then index it around the turntable while it cools.

Typical of the new types of equipment developed for this market is a two-stage machine in which a plasticizing screw feeds a melt accumulator. A non-return check valve permits the material to pass from the screw into the accumulator, but guards against backflow. In the second stage, an hydraulically operated shut-off nozzle (between melt accumulator and mold) receives the charge from the transfer cylinder. When the material is ready to be injected (i.e., before premature expansion occurs), the nozzle opens and a reciprocating injection piston discharges the desired amount of melt.

Finishing Low-Pressure Structural Foam Parts*

Systems representing the current technology in finishing structural foam are discussed below. These systems are basic and have the flexibility to adapt to end-product requirements and finishing operations limitations. Since the parameters for every application are different, these systems may be varied in regard to the number of steps and in the paints and thinners employed in order to achieve the required quality. However, once the system is finalized and the procedure set, the capability to produce consistently finished parts is excellent, thus giving a low reject rate.

1. Smooth, high gloss finish for styrene, ABS, and polycarbonate. For the smooth gloss finish, a multi-step system up to 7 or 8 steps, including three primers and one color coat is required. Sanding is needed before priming and between the prime coats. The quality of the topcoat, although a lacquer, approaches the thermoset or two-component systems in finish quality. The lacquer systems offer both a single component finish plus low bake and cure requirements without demeaning the quality of the finished product.

2. Smooth, high gloss, two-component finishing systems for PPO and polycarbonate. Again, due to the smooth high gloss requirement, this system also requires some sanding. However, due to the high solids capabilities of two component polyester and urethane coat-

* Prepared by Bee Chemical Company

ings, only four or five processing steps are required including one or two prime coats and one color coat. This system is primarily employed in the business machine market and appliance area where the highest quality system is needed.

3. Two-component texture finish for business machines of foamed PPO and polycarbonate. Due to the hiding capabilities of a rough finish, sanding is not required and only one or two painting steps are needed. Again, this system finds use primarily in high quality markets.

4. Flat or semi-gloss one coat lacquer systems of interior and exterior quality for styrene and ABS. The coatings utilized in this system are one package acrylic type lacquers which give easy handling, quick dry and packaging properties. This system is exemplified in the building products market by structural foam styrene shutters and in the appliance market by structural foam styrene television bezels.

5. Woodgraining finish for polystyrene, ABS and polypropylene. The crux of this finishing operation lies in the barrier coat application. The barrier coat provides barrier protection to the substrate, hides the swirl pattern and gives a uniform appearance. Its application also allows for proper handling of the subsequent finishing steps, such as staining, toning and topcoating.

High-Pressure Systems

A high-pressure foam process is characterized by injecting a "full" shot of foamable melt into a closed mold at near normal injection molding pressures. Foaming takes place either by allowing the escape of excess material back into a runner system or by moving a plate or plates to open the mold cavity itself.

Perhaps the most significant advantage of the high-pressure injection process with expansion molds is the savings of 50% or more on weight over solid plastics. Resultant material savings and weight advantages translate directly into cost reduction. Since skin and foamed interior are integral, the process eliminates the necessity of filling cored-out parts with wood or other plastic foam for cabinet doors, drawer fronts, dividers and other parts demanding finished appearance on two opposing sides.

Density of a molded part produced by this process can be accurately and consistently controlled by establishing the ratio of starting volume to end product volume. To a degree, the density range depends on the resin being processed, but most unmodified resins can be molded as low as .25 specific gravity. Experience to date on commercial parts indicates density ranging from 0.5 to 0.6 specific gravity.

In general, high-pressure foam molding has one major advantage over low-pressure foam molding, i.e., the high filling pressures employed should allow for better reproduction of a mold surface. This advantage has to be weighed against a major disadvantage compared to low-pressure foam molding, i.e., the tooling for high-pressure foam molding is considerably more expensive.

Solid Skin. Because the process completely fills the mold with plastic before beginning the foaming action, a solid skin is formed which is free from surface gas splay marks. The solid skin promotes fine reproduction of whatever detail is required—wood grain, miter joints, tooled leather, etc.

Even the farthest corner of the mold is filled by a known level of injection pressure, ensuring excellent surface definition on all portions of the part.

Concentration of mass in the solid skin reduces sound transmission, hence the possibility of sound barrier applications.

Foamed Interior. Complete mold fill prior to the onset of foaming has the effect of closely controlling the foaming action. Foaming is carried out in-place, without any large-scale material migration required. Voids and open cells are thus avoided.

Closed cells are impermeable and will absorb no moisture. Uniform cell construction is important for impact properties in cushion applications, and essential if fasteners are to be attached to rigid structural members. Foamed parts will have uniformly low thermal conductivity, slightly lower than can be predicted by weight difference from solid to foam.

Parts produced with the process exhibit excellent flexural strength and stiffness-to-weight ratio. The mass of the solid skin separated by a low-density foam core, simulates an I-beam and provides the most efficient structure for strength per unit of weight. Other specific physical properties are shown in Table 20-18 which compares engineering constants for identical solid and foam sections. While all the basic properties listed are lower for foam than solid, this is to be expected because there is simply less mass to the foam. However, the structure of the foam is not taken into account by these basic properties and must be calculated for each application. When this is done, considerable strength-to-weight advantages are seen.

For the sections being compared, the solid is .150 in. thick, and the foam is .300 in. thick. At .5 specific gravity, the foam has essentially the same overall weight as the solid part.

The moment of inertia is thus calculated as .00337 for solid, and .027 for foam.

While Table 20-18 indicates an almost 3 : 1 relationship in flexural modulus between solid and foam, the real measure of load carrying capability is EI (foam) = 106,300 x .027 = 2870. EI (solid) = 295,700 x .00337 = 996.5.

The resultant foam/solid stiffness ratio is 2870/996.5, or 2.9 : 1. A foam part with stiffness equal to a solid part must always be about 38% thicker, which represents a final

Property	ASTM Test	Solid	Foam (.5 sp gr)
Hardness, R	D785	80	39
Tensile modulus, psi	D638	285,000	104,000
Flexural modulus, psi	D790	295,700	106,300
Shear modulus, psi	D2236-64T	93,400	29,600

Table 20-18. Engineering constants for a medium impact polystyrene 0.5 specific gravity.

weight savings of about 70% of the comparable solid part.

Design Features. This foam molding process opens many opportunities to the designer. The ability of the foam to simulate highly crafted wood products, both in surface embellishments and part weight, plus the inherent capability of plastics to form a complete component in a single shot, enables the designer to think in terms of a totally integrated product such as a chair, bed headboard, or complete drawer.

Foam core densities can vary from .25 to solid, an advantage for functional component design calling for acoustical, impact, and strength properties. Solid sections can be selectively designed into a part to meet functional requirements, such as for faster placement, or to provide maximum shear or impact properties.

A major advantage of this process is the uniformity, consistency and size of core foam cells. Structural integrity is significantly improved by eliminating voids in the foam core, and design requirements are eased for fastener placement, since foam exhibits strong holding power for staples, screws, nails, or other fasteners.

In addition, the advantage of using multi-cavity molds and short cycle times broadens the economic impact of structural foam molding. It is now feasible to foam mold very large parts as well as smaller, high-production parts.

Where good surface detail is not a design requirement, as with interior structural parts of a sofa, etc., low pressure molds without expanding members can be used successfully with the process.

How the Process Works. The process consists of plasticizing a polymer foaming agent blended in a specially modified reciprocating screw injection molding machine, then injecting it into a specially designed expandable mold.

Unlike other processes, the molding machine is dual purpose. For use in conventional molding, simply turning a switch in the main control panel selects the proper hydraulic and electrical operating sequence. Modifications and improved technology necessary for foam molding also considerably enhance solid-part molding.

Foaming agents may be added to the polymer in hopper-blenders at the machine or in tumble blending units prior to arrival at the machine. Preblended resin is also available commercially, which offers the choice of wider ranges of foaming agents not easily compounded by the processor. A third method is the use of foam concentrates now available for many polymers.

After plastication, the melt is injected into the mold. Controlled injection causes a solid skin to form at the surface while balanced mold expansion allows uniform cell growth in the interior core.

Tooling. The mold expansion cycle is the point where the machine more noticeably departs from conventional practice. The platen motion begins at the completion of the mold fill, and expands the cavity volume at a controlled rate.

Varying degrees of mold complexity can be used to meet different requirements for the expansion step. The simplest is shown in Fig. 20-65 as Technique 1. This is a two-piece mold: a cavity and force plug. The relative motion between force and cavity (upon expansion) occurs at the outer edge of the part. This means that a crease will be seen at the edge of the finished part. Two good sides, top and bottom, result.

Technique II, as shown in Fig. 20-65, moves the characteristic crease to the bottom or back side of the part. This is accomplished by adding a third plate in the mold which stays closed during the expansion step. It is not always necessary to hide the crease; often it can be integrated into the aesthetic lines of the part. Such would be the requirement for chair arms, legs or backs which might be viewed from all sides.

A third technique, essentially an extension of Technique II, is shown in Fig. 20-65. This shows that the expansion methods are not limited to the plane of the platen, but can be built into the mold and controlled independently.

Although most high-pressure systems involve some means of expanding the mold or using removable mold inserts, etc., to accommodate the foaming of material, at least one commer-

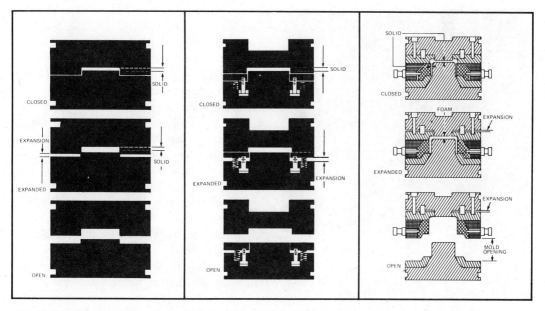

Fig. 20-65. Three techniques for mold expansion as used in high-pressure structural foam molding. Technique 1, as referred to in text, is on the left, Technique 2 in the center, and Technique 3 on the right.

cial technique appears to be a cross between high-pressure and low-pressure molding. This system involves a two-stage injection molding machine (specially designed), special molds, and specially compounded blowing agents. Polymers are first plasticized in a screw-type unit, then transferred to a plunger unit where melts can be maintained under pressures up to 30,000 psi for high-speed injection into specially heated and cooled molds. Venting systems in the mold remove moisture and gases.

Other Molding Systems

In addition to the high-pressure and low-pressure systems, a technique, known as *sandwich molding*, is also available to produce a structural foam part (i.e., a solid skin and cellular core). This system involves an injection molding machine with two injection units. Resin from one is partially injected into a mold cavity; then, resin (with blowing agents incorporated) is injected from the second unit into the same mold. This second shot forces the first polymer to the edges of the mold cavity where it sets up to form the solid skin. The second shot (with the blowing agent) then expands to form the cellular core beneath the skin. For further details see Chapter 4.

Still another approach to producing structural foam parts is through rotational molding. Using solid chemical blowing agents, it is possible to rotationally mold a multi-layer system of solid skin and cellular core. By formulating to proper decomposition temperatures, it is possible to lay down a layer of solid polymer near the mold surface to function as the skin. As heating continues, expansion then takes place in the rest of the polymer toward the end of the cycle and simultaneously with complete fusion.

Extruding Structural Foams

Structural foams with solid integral skins and cellular cores can also be extruded in the form of profiles, pipe, tubing, sheet, etc. Methods in use today involve conventional single-screw machines handling plastics that incorporate chemical blowing agents (although direct gas injection and tandem extrusion set-ups have also been used). These techniques have been used to extrude polystyrene and vinyl profiles for use as building trim and moldings, picture frames, etc. (Fig. 20-66). See also the chapter

Fig. 20-66. Extruded styrene foam molding (center) competes with decorative pine (left). Section on right is styrene foam as extruded and before decorative pattern is added. (*Courtesy Crane Plastics*)

on Extrusion, p. 156, and the section on Expanded Vinyl in this chapter. The following section on ABS foam also carries data on structural foam extrusion.

In addition, there is a proprietary extrusion system that is available for licensing (the Celuka process, Ugine Kuhlman). Briefly, the process consists of extruding a special thermoplastic formulation through a die having about the same outer dimensions as those desired in the finished profile (Fig. 20-67). In addition, there is a mandrel in the center of the die so that the extruded profile comes out in the shape of a formed "tube." This extrudate immediately goes into a shaper which also has the same dimensions as the die. The surface of the extrusion is quickly cooled to form a solid, unexpanded skin; at the same time, expansion of the thermoplastic takes place internally, thereby forming a uniform cellular core. Upon leaving the shaper, the profile goes through the normal downstream equipment for cooling and take-up. Standard extrusion lines can be used. Only specially designed dies and shapers and a slight downstream alteration are necessary to make the process work.

Almost any thermoplastic can be handled by this process, although most of the current work has concentrated on PVC, polystyrene, and the polyolefins. Products made by this technique include: profiles to replace pre-finished wood moldings, hollow wall-panel systems, door frames, window frames, siding, and picket fence posts. It is also possible to use this process with a cross-head on the extruder. This enables an expanded thermoplastic to be coated over a metal core, such as copper wire or tubing.

A variation of this system is also commercially available (Gatto Machinery). It is based on the idea that decomposition of the blowing agent is not complete within the extrudate, but continues at a reasonable rate when the extrudate emerges from the die. The system begins with a screw extension into the die head to delay the foaming action and allow the extrudate to continue foaming as it leaves the head of the die (which is about 80% of the finished profile dimensions) and enters the sizing fixture (which is the same dimension as the finished profile). Foaming is outward so

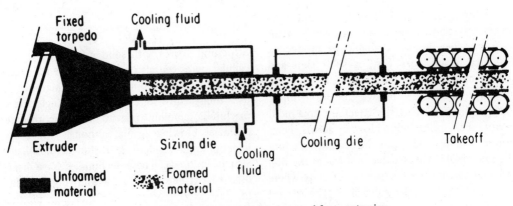

Fig. 20-67. Celuka process for structural foam extrusion.

you form a tough skin. Idea of bringing the extrudate out of the die and into the sizer while it is still foaming is to allow precise sizing, good cell structure, and good skin configuration.

ABS STRUCTURAL FOAM*

ABS structural foam can be processed by injection molding, through conventional or low-pressure injection machines; by expansion casting in rotational molding machines, or conveyorized oven systems; or it can be extruded into profiles through conventional extruders. For injection molding and expansion casting, the ABS contains a dispersed chemical blowing agent which, when heat is applied, decomposes into an inert gas forming a closed cellular structure. With the extrusion grade, the ABS pellets are dusted with a blowing agent prior to extrusion. Various ABS grades are also available that meet the approvals needed for business machines, TV, appliance, and other markets.

Expandable ABS is supplied in the form of pellets (0.093 in. cubes) having a specific gravity of 1.04. Depending on type of expansion process, the finished part may have a specific gravity from as low as 0.30 (20 lb/ft^3) up to the gravity of the solid material (64.9 lb/ft^3). Specific conversion techniques and part size determine the density achievable in a part. Injection molded parts are normally produced with specific gravities in the range of 0.7 to 0.9; expansion casting produces parts ranging from 0.4 to 0.6 and profile extrusion normally produces parts ranging from 0.3 to 0.7. Structural foam is produced in natural (cream-colored) and in a wide variety of other colors.

Because of its excellent balance of properties at reasonable cost, ABS structural foam is finding applications in the automotive, furniture, construction and materials handling industries.

Part Surface

The appearance of an expanded ABS part is dependent upon the mold surface and the conversion technique. All processes produce a

* By Borg-Warner Chemical Co., Parkersburg, W. Va.

"closed-cell" impervious surface. Extruded parts have a dense, smooth (non-glossy) surface. Expanded cast parts produced in rotational molding or other ovens have a uniformly textured surface. Injection molding produces a swirl pattern and striated surface.

Advantages

Many unique properties are combined in this type of product since ABS structural foam parts have no grain, which is inherent in wood. In expansion cast parts, density and physical properties are virtually uniform throughout. In extruded or injection molded parts, the dense, rigid skin on the surface provides extra rigidity due to its similarity to "sandwich core" construction. In certain applications, it may be desirable to use a combination of materials such as metal inserts. This combination of materials feature allows the engineer additional latitude in designing for lightweight, rigidized construction. The low specific gravity also offers buoyancy, desirable for marine applications such as buoys or hatch covers for boats. In addition, expanded ABS also exhibits an excellent stiffness-to-weight ratio. As can be seen from Table 20-19, it is possible to replace other materials of construction to produce parts with an equivalent stiffness and yet of much lighter weight.

Table 20-19. Stiffness to Weight Ratio.

	Weight for Equal Stiffness
Steel	7.95
Linear polyethylene	6.40
Polypropylene	4.38
ABS	4.29
Polystyrene	3.90
Aluminum	3.66
ABS structural foam (0.5 density)	3.06
Wood (Maple)	1.62
(White pine)	1.00

Figure 20-68 illustrates the dimensional stability of ABS structural foam after water immersion. ABS structural foam is substantially better than wood; indicating, for example, that

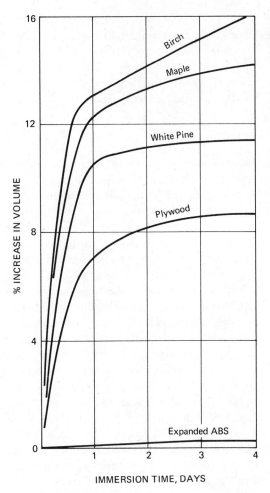

IMMERSION TIME, DAYS

Fig. 20-68. Volume change/water immersion at 73° F.

an expanded ABS drawer will have less tendency to swell and stick than a wood drawer. Since the wall thickness is 0.250 in. or more, the expanded ABS drawer would have the mass, sound, and feel of wood, and would offer greater toughness since it has no joints to come apart.

The screw and staple holding properties of ABS structural foam are shown in Table 20-20. These tests were performed in accordance with ASTM D-1761-64, Standard Methods of Testing Metal Fasteners in Wood. The pilot holes for screw withdrawal tests were 0.110 in. diameter. Reduction of the pilot hole size or increase in the length of the screw can increase the screw holding value by more than 30%.

Physical Properties

The physical properties of expanded ABS are shown in Table 20-21. The modulus is lower than with solid ABS; however this is overcome in structural foam applications by the increased thickness. Since part "stiffness" varies with thickness, a 50% increase in thickness increases part stiffness by a factor of more than 3. The notched Izod test for impact is not a very meaningful guide to the performance capabilities of expanded ABS. Performance is more reliably indicated by falling dart impact. In most expanded ABS applications the superior weight-to-stiffness performance, abuse resistance, resistance to creep under load, and lightweight characteristics of the material have proved to be most important.

Table 20-20. Screw and Staple Pull-Out Strengths.*

| | Screw | | Staple | |
Material	1″ Long	³⁄₄″ Long	³⁄₁₆″ Long #903	¹⁄₂″ Long #908
Expanded ABS	270 lb	193 lb	11.3 lb	38.0 lb
Birch	642 lb	442 lb	13.5 lb	46.3 lb
Maple	668 lb	463 lb	22.8 lb	—**
White Pine	239 lb	168 lb	3.1 lb	17.0 lb

* ASTM D 1761-64 "Standard Methods of Testing Metal Fasteners in Wood"
1. Removal speed: 0.1 in./min.
2. Number 10 gauge steel flathead wood screws embedded to depth of threads.
3. Pilot hole for screws .110 in. diameter.

** Unable to staple for test purposes.

Table 20-21. Properties of ABS Structural Foam.

Mechanical		density (grams/cc)		
		.35	.50	.75
Tensile properties				
Tensile strength @ 160°F	psi		800	
@ 73°F			1,400	
@ -40°F			2,200	
Flexural properties				
Flexural strength @ 160°F	psi		900	
@ 73°F		1800	2,500	4000
@ -40°F			3,600	
Flexural modulus @ 160°F	psi		60,000	
@ 73°F		78,000	100,000	162,000
@ -40°F			140,000	
Impact properties				
Notched Izod impact @ 73°F	ft-lb/in.		1.2	
@ -40°F			0.6	
Dart impact, (0.5 in. thick) @ 73°F	ft-lb		8	
@ -40°F			4	
Hardness	shore D		60	
Thermal				
Coefficient of linear thermal expansion	in./in./°C		9.7×10^5	
Deflection temperature (unannealed, 264 psi)	°F	130	150	
Thermal conductivity	Btu/hr/ft² /°F/in.		0.58	
Analytical				
Specific gravity	lb/ft³	22	32	47
Water absorption	%		0.60	
Mold shrinkage	in./in.		0.008	
Electrical				
Dielectric constant 50 cps			1.63	
10³ cps			1.56	
10⁶ cps			1.59	
Power factor 50 cps			0.002	
10³ cps			0.0035	
10⁶ cps			0.007	
Arc resistance	sec		66	
Dielectric strength	V/mil		—	

ABS structural foam has demonstrated creep resistance superior to that of high impact polystyrene and polyethylene foams (Fig. 20-69). This is particularly important in load bearing applications such as pallets, tote boxes, furniture, and parts buried under earth loads.

Expansion Casting

The expansion casting process includes rota-tional molding machines and conveyorized oven systems. The heating medium may be hot air or molten salt. Production molds constructed of cast or fabricated aluminum are relatively in-expensive. As a result of this low mold cost, the casting process is ideal for extremely large parts in all quantities, from full production to a few prototypes.

The expansion casting process is the sequen-tial heating and cooling of matched metal molds.

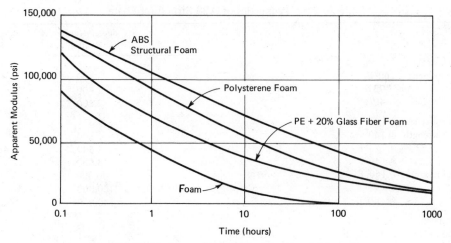

Fig. 20-69. Creep of foam materials (all at 0.55 sp gr).

The closed mold is filled pneumatically or by gravity with expandable ABS pellets, through an orifice or "filling port" in the mold. Once the mold is filled, the port is sealed and heating is begun. During the heating cycle (450-500°F) the material expands and fuses. There is very little flow within the mold; therefore, part density and properties are consistent throughout. The mold is then cooled, the part removed, and the process is ready to begin again.

Although the overall cycle for the mold may be on the order of 20 to 45 min, production economy is accomplished by using multiple molds and the fact that several molds are being processed at various stages at the same time. Thus, the effective "mold time" or net cycle time may be 2 to 10 min.

In the expansion casting process, it is possible to have a number of different parts, colors, or sizes in process at the same time provided the wall thicknesses are similar. Therefore, several sizes of furniture drawers can be molded as a package while simultaneously making furniture frames, pallets or other parts that can be molded in the same cycle. With injection molding, this could only be accomplished with a number of costly machines and with more expensive tooling. Also, an injection molded foam part will weigh substantially more than an expansion cast foam part, due to the lower density that can be achieved with the casting process.

The most effective application areas for expansion cast parts are in large parts (10-100 lb) where the strength of ribs or variability of wall thickness are needed. In smaller (1-20 lb) parts where the volume is not large enough for most plastics processing methods, expansion casting is also of value. The relatively low tooling cost for expansion casting, as compared to that required for injection molding and the fast cycle, show to advantage here and gives the process a unique place in the plastics industry.

Some typical applications:

(1) A saddle tree made by a midwestern saddle manufacturing company. This part is processed in cast aluminum molds in a series of single station ovens. Wood was formerly used for this application until production volume and labor cost reached a point to justify plastics. Production volume does not justify the tooling cost of injection molding. The thermally conductive stems of the cast iron saddle horn help to heat the thick body sections in the same time as the thin sections.

(2) A man-hole cover and locking ring (Fig. 20-70), weighs 25 lb and is intended for access in low traffic areas. The assembly will support 97,500 lb as tested by a Pittsburgh testing Lab.

(3) A 4 x 10 ft hydroponic tray used to grow vegetables in a water environment. Formed in a mold fabricated of $\frac{1}{2}$ in. aluminum plate strengthened with steel I-beams,

71 lb of material is used in each part. Expansion casting offers an advantage in this 10-ft part since there is no variation in density from gate-to-gate, as might be the case in injection molded foam. This, coupled with the absence of molded or formed in stresses, helps produce a part free of warpage.

(4) Specialty pallets, those that do a particular job or carry a particular part, are excellent applications for expansion casting. The relatively large part size and the fact that weight savings are possible over injection molded parts are two main advantages. ABS and most other engineering-type plastics are limited to 0.7 to 0.8 specific gravity by the injection process. Structural foam is consistently and uniformly in the 0.5 to 0.55 density range.

Injection Molding

Expandable ABS grades with an internal blowing agent can be processed on standard injection molding equipment, while some standard ABS grades are easily processed on the larger, low-pressure foam machines using nitrogen as the blowing agent.

The same processing concept is used on both conventional equipment and foam machines. The blowing agent is decomposed and the gas generated is contained under pressure to prevent expansion until the material enters the mold. Then, the gas is released to force the material under low pressure to fill the mold cavity.

Single-cavity molds are usually recommended because expansion in the runner system makes it difficult to balance multiple-cavity molds. Family molds should never be considered if a lower density is the prime consideration. One cavity will fill and pack before the other cavity, causing one part to have a higher density than desired.

The main reason for using a foam material is the ability afforded by low mold pressures to produce large, integrated parts with excellent rigidity because of thickness. There are various conditions that control the quality of a foam part. These are stock temperature, mold temperature, injection speed, part thickness, type

Fig. 20-70. Manhole rings and locking covers are expansion cast of an expandable grade of ABS. (*Courtesy Borg-Warner Chemicals*)

of molds, runners, and gates, and the overall machine cycle.

The minimum drying recommendation for expandable ABS is 1 to 2 hr at 170 to 180°F. Overdrying should be avoided or a loss of blowing agent will result.

For the best results, a 430 to 510°F stock will usually produce the lowest density part. Residence time should be minimized to prevent material degradation at high temperatures.

On conventional equipment using 100 psi back pressure, the injection screw should recover just prior to the mold opening. In low pressure systems, the accumulator should be full just before the mold opens.

Mold temperature recommendations range from 70 to 150°F. Higher temperatures within these limits generally produce a smoother surface. However, in some instances, such as a very short molding cycle, a cold mold may be necessary for maximum cooling of the part before ejection from the cavity.

A refrigeration unit coupled to the mold will optimize the cooling cycle.

Submersion of molded parts in a cold (70°) water bath located near the molding machine is another cooling technique commonly used. However, when cooling in water, care should be taken to prevent any breaks in the skin surface. Hot cellular material in the center has a tendency to absorb water. After cooling, this is not a problem. Cooling lines should enter the center of the mold, across the direction of material flow in the cavity. This allows the temperature controlled water to pick up heat as the water passes through the gate area and carry the heat to the outer edges of the cavity thus allowing the overall mold temperature to be better balanced.

The injection speed should be as fast as possible to reduce the packing effect caused by the expansion of the material leaving the nozzle. A slow speed usually increases part weight.

Part thicknesses should be as consistent as possible to allow uniform expansion, minimum density and optimum part appearance. If the part wall thickness is less than 0.187 in., a higher density part will result.

Under normal conditions, expandable ABS materials can be molded in inexpensive molds; materials such as aluminum or kirksite, either cast or machined. However when long-term production or cam-operated molds are needed, then steel molds should be considered.

In order for ABS to expand in the mold, the sprues, runners, and gates should be a minimum of 1 to 2 times larger in diameter than those recommended for solid ABS. Larger passage ways will better facilitate movement of the material at a fast ram speed.

More generous mold venting is necessary for expanded ABS than for solid ABS materials. Improper or inadequate venting will restrict material expansion and cause incomplete filling of the part. Common venting techniques use undersized knock-out pins or preferably milled grooves in the parting line from 0.005 to 0.010 in. deep and from 0.020 to 0.500 in. wide. Use of vents this size is possible because of the low cavity pressure.

The overall length of the cycle will depend on the length of time it takes to cool the part. The thicker the part, the longer the cycle. Cycles will usually be 20 to 50% longer for foamed ABS than for standard ABS.

Expanded ABS regrind will not expand substantially upon remolding. Therefore, to consistently mold the lowest density parts, it is recommended that no more than 10% of clean, non-degraded regrind be processed with virgin material. Low-pressure machines will process higher percentages of regrind but 20% is the maximum amount of regrind recommended in this case.

To purge expanded ABS material from the conventional injection molding machine, general-purpose styrene or acrylic at 400°F is recommended for best results.

Finishing of expanded ABS parts, using standard techniques for thermoplastic materials, can be easily accomplished. Expanded ABS can be cemented, welded, sawed, drilled, painted, and lacquered.

Profile Extrusion

Recent developments in the area of ABS structural foam are materials designed and tailored exclusively for profile and sheet extrusion. This unique material can be converted into expanded profile shapes using conventional extrusion equipment. Because of self-skinning characteristics, the end-product is smooth and retains the balance of properties typical of the ABS materials (Table 20-22). Co-extrusion can be used to obtain a solid skin-foam core construction.

The extraordinary hot strength of ABS permits expanded profile to be post-formed to different shapes and sizes from a single die of the general intended cross section. This is a very real processing advantage in terms of production set-up and turn-around time, as well as being distinctly economical for short runs.

The blowing agent for profile and sheet extrusion is not an integral part of the pellets fed to the extruder. Instead, the blowing agent is dusted on the ABS pellets by the processor in a twin-cone type blender after the ABS pellets have been dried for two to four hours at 180°F in a dehumidifying dryer. In order to promote

Table 20-22.   ABS Structural Foam for Profile Extrusion

	Unexpanded	Expanded	ASTM Test Method
Heat distortion	188°F	163°F	D648
Tensile strength	6850 psi	1800 psi	D638
Tensile modulus	320,000 psi	100,000 psi	D638
Tensile elongation	35%	20%	D638
Flexural strength	10,900 psi	3600 psi	D790
Flexural modulus	3.3×10^5 psi	1.4×10^5 psi	D790
Hardness	106R	68 Shore D	D785
Coefficient of thermal expansion	9.2×10^5 in./in./°C	9.8×10^5 in./in./°C	D696
Specific gravity	1.03	0.55	D792

adhesion of the blowing agent to the pellets, 0.5 parts/100 lb of an inert liquid such as mineral oil is added and blended for five minutes, after which approximately 0.4 parts/100 lb of an azodicarbonamide-type blowing agent is added. Blending is continued for an additional ten minutes.

Successful profile extrusions have been performed on $1\frac{3}{4}$ and $2\frac{1}{2}$ in. extruders with many different die shapes and types of sizing equipment. A single-stage typical ABS screw with a compression ratio of 1.8 to 1 or 2 to 1 is recommended. Screws with compression ratios approaching 3 to 1 will develop excessive heat in the metering zone, resulting in premature decomposition of the blowing agent. For this reason, screws with a substantially lower compression ratio are highly desirable.

Barrel temperature controllers are adjusted to provide an extruded stock temperature in the range of 390-410°F. Minor adjustments are made to yield the planned degree of expansion and proper external skin formation on the extrudate. The size of the die opening is appreciably smaller than the finishing sizer in the cooling trough. A size differential ratio of 1 : 2 will result in a profile specific gravity nominally 0.5, presuming there is no draw-

down. Thus, starting with a 1.03 specific gravity ABS pellet, the profile can be controlled to a specific gravity of 0.3 to 0.6, depending on configuration.

After the extruder conditions have become stabilized, rates can be improved by further adjustments. As an example, a $2\frac{1}{2}$ in. extruder can produce cove shapes with a specific gravity of 0.5 at a rate of 32 ft/min, using 160 lb/hr of material. The output rate is, of course, influenced by profile size, shape, and specific gravity requirements as well as equipment thermal limitations.

The extruder may be purged following the normal practice of removing the die and purging with a high nitrile rubber.

Finishing

Many finishing techniques are available for expanded ABS parts produced by these processes. Parts can be post-finished by laminating with a cap sheet or wood veneer, by flocking, painting, or a variety of other coatings. Like wood, parts can be sawed, nailed, stapled, glued, screwed or machined; however, unlike wood, the expanded ABS resists splitting and is dimensionally stable.

21

RADIATION PROCESSING

In many of the chapters preceding this one, the reader will note continuous reference to the use of radiation techniques for curing plastics, cross-linking plastics, and so on. Because of the growing importance of these techniques in plastics processing, this section is intended as a brief introduction to the subject. It will cover both high-energy ionizing radiation (e.g., via radioisotopes or particle accelerators) and the so-called "non-ionizing" systems like ultraviolet, infrared, induction, dielectric, and microwave.

IONIZING RADIATION*

In the mid-forties, interest in applied radiation chemistry was stimulated by the availability of high activity radioisotopes. High costs, however, prevented any significant commercial application in the plastics industry. In the late fifties and early sixties, however, as prices started to come down, a number of plastics processing techniques involving the use of radiation evolved. These included: curing thermosets, cross-linking thermoplastics, making graft copolymers, producing wood/plastic and concrete/plastic composites, and curing plastics

* Portions of this chapter adapted from PLASTEC Report R41, "Applications of Ionizing Radiations in Plastics and Polymer Technology," published by Plastics Technical Evaluation Center, Picatinny Arsenal, Dover, N.J.

coatings. Radiation is also used to sterilize medical disposables; since it is a cold process, it is especially applicable to plastics disposables since it will not distort them.

In these applications, two types of radiation sources are used. These are either a radioisotope (e.g., cobalt as Co^{60} or cesium as Cs^{137}, both giving gamma radiation) or a particle accelerator (e.g., direct electron beam). The intensity of the latter is a function of the design.

Gamma radiation has the advantage, when compared with accelerators, of being directly available without the need of constant supervision, does not involve shutdowns, has greater penetration, and is currently suitable for high-capacity production where relatively long dwell times are allowable. Disadvantages are the high cost of installation, an unwieldy design resulting from thick shielding requirements, and slow transmission of energy.

Electron radiation produced by machine acceleration has an advantage, when compared with radioisotopes, in its capacity for being readily switched on or off. Also, shielding requirements are less than with gamma radiation in relation to radiation output. (Accelerator beams are directional, gamma radiations are spherical in geometry.) Electron radiation provides high dose rates and is well suited for on-line applications with high-speed processing. As a disadvantage, an accelerator requires careful handling and service and usually calls for the use of well-trained personnel. There are five available types of electron accelerators: Dynamitron (and Dynacote); Insulating core transformer; Linear accelerator; Resonant transformer; and Van de Graaff accelerator.

Cross-Linking Thermoplastics for Wire and Cable

An interesting phenomena that occurs when thermoplastics are subjected to high energy radiation is the formation of free radicals that then combine to cross-link the molecular chains. What this means in practice is that you change the characteristics of the base material (e.g., polyethylene) to improve its temperature resistance, boost tensile strength, improve chemical resistance and weatherability, and lower dielectric losses at high temperatures.

This has made radiation especially useful as applied to wire and cable coatings or to polyethylene film used as an insulating wrap. Polyethylene is the major material so far adapted to cross-linked wire coatings, but vinyl is being increasingly used.

In the form of insulating film, cross-linked polyethylene is especially advantageous in terms of high temperature resistance, resistance to stress cracking, and superior electricals (as compared to non-treated polyethylene). Irradiated polyethylene film tape also shows some interesting and useful shrink characteristics. When the tape is wound upon a conductor, coil, or other object, under moderate tension, and then heated above 110°C. (preferably to 135 to 150°C), the inherent orientation in the tape causes it to shrink in the lengthwise direction, exerting pressure and causing the layers underneath to bond together into a substantially uniform sheath of excellent electrical and physical properties.

It is also possible to produce irradiated polyethylene tubing. In this form, the elastic memory of irradiated polyethylene makes the tubing of value as an insulative protective sheath for soldered terminations, cable markers, capacitor covers, line splices, and insulated connectors.

Cross-Linking Thermoplastics for Packaging Film

By the same token, polyethylene film can also be irradiated by electrons to produce a heat-shrinkable packaging wrap. Irradiation cross-links the polyethylene to the point where the film is stretchable but does not become fluid at its original melting temperature of 105 to 110°C. This means that the film can be placed around the object to be wrapped and heated to 100°C so that it will shrink tightly around the object.

More recently, other types of applications using the elastic memory of the irradiated plastics have been made commercial. One such use is a series of fasteners, rivet devices, or other mechanical joining system. The fasteners are produced by irradiating polyethylene rods (by either electron or gamma irradiation) and stretching them under heat. The cooled rods, now longer and thinner, are cut to size and inserted into the appropriate openings (i.e., through the two parts to be joined). Upon the application of heat, the inserted rivets expand back to their original shape (i.e., thicker), filling the opening and providing an effective lock.

Cross-Linked Foamed Resins

A number of thermoplastic foams cross-linked by radiation are available. Cross-linked styrene foam, for example, offers superior mechanical and chemical properties. On a typical production system for producing such foams, a mixture of polystyrene, a functional monomer, and a blowing agent is irradiated with 1 to 10 Mrad of electron radiation, at temperatures of non-dissociation of the foaming agent, in order to cross-link the resin. Then, by further heating, the foaming agent is dissociated, to give the foamed product. Or, alternatively, the heating for dissociating of the foaming agent is done while the molding is irradiated.

The more popular base for cross-linked foam structures, however, is polyethylene. These cross-linked foams offer higher thermal stability and mechanical strength, better insulation characteristics, and improved energy absorbing characteristics (as compared to conventional polyethylene foams). Most can be thermo-formed, embossed, printed, laminated, or punched using conventional equipment.

There are a number of techniques for producing cross-linked thermoplastic foams. One technique proceeds as follows. A mixture is made that incorporates one or more of the monomers selected from the group consisting of acrylamide, methacrylamide, acrylic acid, and methacrylic acid; a foaming agent in solid or liquid state at room temperature; a cross-linking agent having more than two polymerizable double bonds in the molecule; and any monomer, if necessary, that is copolymerizable with the monomers mentioned above. The mixture is polymerized to a thermoplastic body containing the foaming agent homogeneously, and the plastic body is then heated to a temperature above the softening point of the plastic in order to give, simultaneously, the foaming and the cross-linking reaction between the polymer molecules. Radio-induced polymerization or catalytic polymerization followed by irradiation are reported to produce the best results. (See also Chapter 20 on Cellular Plastics).

Graft Copolymers

The use of radiation processing to manufacture graft copolymers is finding increasing use in the production of improved plastic materials and tailored products. These copolymers differ from other copolymers in that a main polymer backbone is actually attached to another homopolymer by grafting on the side chain.

Because of this, the properties of the homopolymers are retained to a great extent, but can be improved by the addition of the second homopolymer. One example would be grafting a small amount of hydrophillic material to a hydrophobic material—not to make it hydro-

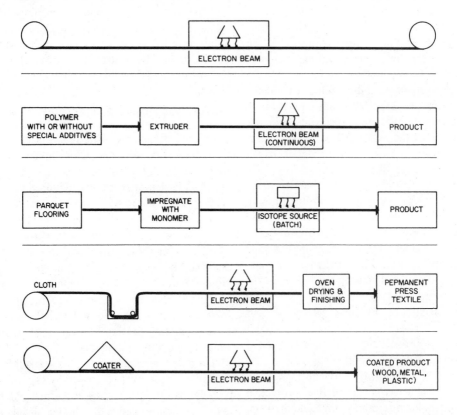

Fig. 21-1. Five applications for radiation processing (top to bottom): cross-linking film; cross-linking wire insulation; making wood-plastic composites by irradiation; finishing textiles (grafting); and curing of coatings (cross-linking). (*Courtesy Plastics Technology Magazine*)

phillic, but to give it a high tolerance for water (e.g., cellulose acetate grafted to styrene which has been used for desalinization of water to remove up to 99% of salt and other impurities in one pass). Another potential is combining materials for optimum UV resistance and gas permeability.

Grafting can be effected with powders or polymers in solution; it can also be applied to materials in fiber and film form. Films are reportedly particularly easy to graft by two techniques: (1) pre-irradiation, where the film is irradiated, then passed into a liquid or vapor monomer environment; or (2) mutual irradiation, where the film is irradiated directly in the presence of the monomer.

Radiation-Induced Curing of Coatings

Electron beams are beginning to find wide use in curing coatings on both metal and plastics. With irradiation, the need for elevated temperature cure is eliminated, production rates are increased, and the resulting coating has less porosity.

One of the more important new commercial techniques is the so-called *electrocure* process, originally developed for curing painted plastics in the automotive industry. (For a fuller description of the technique, see p. 733) In operation, the coatings, generally an acrylic-based material, are cured via electron beams, at a rate reportedly 750 times faster than conventional techniques. With this method, the manufacturers claim superior resistance to chipping and peeling, no need for antipollution equipment (since the uncured resin contains no conventional solvents), and a simplified production line. The electron beam curing system also allows paints to adhere to polypropylene, which is generally difficult with most curing systems. The system further permits greater use of large plastic body parts which would ordinarily distort under the heats involved in conventional thermal curing methods.

Curing Thermosets

By passing glass fiber-reinforced polyester prepreg materials under an electron accelerator, it was found possible to increase the rate of cure (i.e., shorten the curing cycle). In addition to the more rapid cycle, irradiation eliminates any need for heat or for the catalysts currently used in the curing of reinforced plastics (e.g., peroxides). Research is currently continuing in this area.

Work is also in progress on cross-linking thermoplastics sheets for cold-forming. Such sheets easily distort and revert to original flat geometries with increases in temperature. Irradiation processing can prevent this from happening. A glass fiber-reinforced vinyl sheet, for example, can be cold-formed using modified metalworking techniques and equipment, then cross-linked at ambient temperature with dosages in the 5 to 10 megarad range.

Irradiated Wood-Plastic Composites

Wood-plastic combinations are made by first removing the air and moisture from the wood by means of vacuum and heating techniques. Subsequently, the wood is impregnated with a liquid monomer. The impregnated wood is irradiated with gamma radiation, generally from a Co^{60} source, resulting in polymerization of the monomer and creating a unique type of composite material. Much of the initial work has been done with methyl methacrylate and vinyl acetate monomers, but styrene, ethyl acrylate, butyl acrylate, acrylonitrile, and 2-ethylhexyl acrylate have also been considered.

The major advantage of the combination is that while it retains the grain and appearance of wood, it can be two to three times harder than untreated wood. Other advantages include: higher mar, abrasion, and scratch resistance; improved resistance to warping and swelling; no finishes are required; and increased compression/static bending/shear strengths. The largest application of the material to date is in parquet flooring, but it is also being considered for door thresholds, garage doors, plywood paneling, wall tile, and outdoor furniture.

Irradiated Concrete-Plastic Composites

Similarly, it is also possible to impregnate concrete with plastic and polymerize the plastic by radiation to produce an unusual combination material. In application, preformed concrete is dried, evacuated, and monomer soaked or coated. The monomer is polymerized *in situ* either by Co^{60} gamma radiation or by thermal-catalytic initiation (or a combination of both methods). For premix concrete, part of the water is replaced with monomer or the monomer is added to the fresh concrete mix. Monomers tested include vinyl acetate, acrylonitrile, methyl methacrylate, styrene, styrene-acrylonitrile, polyester-styrene, and epoxy-styrene.

Improvements through the use of polymer loading are increased compressive strength (four times better than untreated concrete), tensile strength, modulus of elasticity, and modulus of rupture. Freezing and thawing resistance was also improved, while water permeability and water absorption were significantly decreased. Corrosion resistance was further improved.

Applications of high potential for concrete-polymer systems include: pipe (irrigation water, sewage, municipal/industrial water), housing (beams, wall/floor panels, load-bearing columns), structures resistant to chemical attack (desalination, corrosive wastes), underwater use, and prestressed pressure vessels. Colors can be incorporated to lend aesthetic and architectural advantages.

"NON-IONIZING" RADIATION*

These techniques go by such names as ultraviolet, infrared, induction (magnetic loss), dielectric, and microwave radiant energies. They are generally not considered radiation systems in the sense that the ionizing radiation methods described above are looked on. And unlike ionizing radiation, they have been in use by the plastics industry for many years. They are, however, playing an increasingly important role in the plastics industry and are included here simply to put the subject into perspective.

The approximate order of wavelengths and

* Portions of this chapter adapted from PLASTEC Report R43, "Plastics Fabrication by Ultraviolet, Infrared, Induction, Dielectric, and Microwave Radiation Methods," published by Plastics Technical Evaluation Center, Picatinny Arsenal, Dover, N.J.

frequencies of the various radiations discussed in this section are as follows:

Radiation	Wavelength (millimicrons)		Frequency, cycle/sec (Hertz)
Ultraviolet	10^1	to 10^2	$(>10^{16})$ to $(>10^{15})$
Infrared	10^3	to 10^5	$(>10^{14})$ to $(>10^{11})$
Induction	10^{11}	to 10^{12}	$(>10^6)$ to $(>10^5)$
Dielectric	10^9	to 10^{10}	$(<10^8)$ to $(>10^5)$
Microwave	10^7	to 10^8	$(<10^{10})$ to $(<10^9)$

Each will be covered briefly below. But the reader should note that most are discussed in greater detail elsewhere in this handbook.

Ultraviolet Radiation

Ultraviolet radiation has found application in the plastics industry in the curing of polyesters and other polymers. Usually, the light sources used in such reactions are mercury lamps. More recently, glass-polyester prepregs have been made available that can cure simply by direct exposure to sunlight.

Work has also been done in systems for rigidizing expandable resin-impregnated structures in space. It is claimed that space ultraviolet (also in conjunction with natural infrared radiation) would be suitable for cross-linking diallyl phthalate polyester resin containing benzoin (as a photosensitizer) and benzoyl peroxide. In operation, a typical structure (e.g., a communications satellite) could consist of a sandwich of flexible polyester film for the skins and a lightweight fabric (impregnated with an ultraviolet-activated polyester resin) as the core. Once in space, the polyester would cure on exposure to UV, stiffening and rigidizing the structure.

Ultraviolet energy has also been used in curing coatings (based on polyester and other polymers) and in curing orthopedic casts based on plastic-impregnated bandages.

Infrared Radiation

Typical plastics fabrication or finishing techniques that use infrared energy include: thermoforming, film extrusion, orientation, embossing, coating, laminating, ink drying (on printed plastics), and fusing.

One of the major outlets for infrared is in the heating of plastic sheets for thermoforming (see Chapter 12 on Thermoforming).

It has also found use in curing filament wound tubular structures (using glass roving impregnated with such resins as epoxy, polyester, and nylon-phenolic) and in the manufacture of the so-called "split or slit" polypropylene yarns. In this latter application, polypropylene film is slit into strips which are subsequently heated for stretching and orientation into yarns and fibers. Take-up machines twist and wind the fibers onto bobbins. A radiant heater tunnel is used in many installations for heating such fibers or yarns.

Paint drying and baking, however, is still the largest single use for infrared energy systems.

Inductive (Magnetic) Energy

The prime use for inductive energy in the plastics industry is in various induction bonding techniques. These are discussed in Chapter 27 on Assembly.

Dielectric Energy

Dielectric (or radio-frequency) energy has found a number of uses in the plastics industry, including the following:

(1) Radio-frequency preheating of molding materials (e.g., phenolic) prior to compression or transfer molding. See Chapter 8 on Compression Molding.

(2) Radio-frequency curing of reinforced plastics pipe (e.g., glass/epoxy-phenolic systems).

(3) Dielectric preheating of cast epoxy resins prior to pouring the resin into the mold (i.e., to decrease gel time and increase production rates).

(4) Dielectric heat sealing of thermoplastics film and sheeting. See Chapter 27 on Assembly.

(5) Radio-frequency curing of cast furniture and other decorative parts based on filled polyester and high density urethane foam. Electronic heating is used in this instance to shorten the cure.

(6) Dielectric heating and drying of inks and coatings on web substrates.

(7) Radio-frequency energy for molding expandable polystyrene beads. See Chapter 20 on Cellular Plastics.

(8) Flow molding. A die-cut plastic blank (either solid or expanded vinyl or vinyl-coated substrate) is placed in a mold (usually silicone rubber) and power is applied via a high-frequency RF generator to melt the plastic so that it flows into the mold to the desired shape and with the desired texture.

Microwave Energy

Microwaves can be used to create heat in, and thus accelerate the curing of, such materials as polyesters, epoxies, urethanes, etc. Among the possibilities that have been investigated are the microwave curing of cast epoxies and polyesters and of urethane elastomers.

Work has also progressed in using microwaves for cast rigid urethane foam furniture parts. Microwaves have proved especially applicable in this instance since most materials used in mold construction for rigid foams are generally poor conductors of heat. Since heating with microwaves is independent of thermal conductivity, the temperatures at which the urethane foam is poured into the molds can be more carefully controlled with this type of system.

Other potential uses for microwaves are in the molding of reinforced plastics (it may be possible to improve the resin penetration and viscosity by maintaining fine temperature control through the use of a microwave applicator); in the faster and more accurate curing of molded flexible urethane foam parts (like automotive seat backs); and in the bonding of untreated polyolefin surfaces. In the latter area, microwave radiation units are used as an energy source to melt a filled polyolefin layer between two untreated polyolefin surfaces. This melted layer provides a structural bond between the surfaces.

22

MOLDMAKING AND MATERIALS

As is evident by skimming through the various chapters on processing, the mold (or tool, as it is also called, or die, as it is generally referred to when applied to the shaping device used in the extrusion of plastics) is the heart of the process. It determines the size, shape, dimensions, finish, and often the physical properties of the finished product.

You will therefore note that considerable space is devoted in each chapter on processing to the specific design and construction of the molds and dies that are used with a particular process (e.g., injection molding, blow molding, extrusion, thermoforming, rotational molding, etc.). This section is included as an overall basic guide to the making of the molds, the specifying of molds, and the materials generally used in manufacturing molds. Emphasis will be primarily on molds for injection molding (since much of this data are applicable to other molds for plastics processing) and, to a lesser degree, casting, blow molding, structural foam molding, rotational molding, and thermoforming. Since extrusion dies are covered in extensive detail in the chapter on Extrusion, they will not be discussed in this section.

MANUFACTURING METHODS FOR INJECTION MOLDS*

Before manufacturing cavities and cores it is necessary to have an approved blueprint. The blueprint from which the moldmaker works should be signed, if possible, by the end-user and the engineer in charge. Very often the blueprint from the purchasing department is significantly different than the one finally approved for production. Among other things, the blueprint should provide for the taper, surface, finish and shrinkage. All plastic materials shrink when cooled from their molding temperature to room temperature. The amount of shrinkage varies with the material and the thickness. In $\frac{1}{8}$ in. sections, for example, polystyrene shrinks 0.04 to 0.05 in./in., and polycarbonate shrinks 0.006 in./inch. The crystalline materials have a broader shrinkage range; nylon for example, shrinks 0.014 to 0.022 in./in. This is accounted for by the

* By Irvin Rubin, Robinson Plastics Co., 965 Garfield Ave., Jersey City, N. J. The author gratefully acknowledges the permission of John Wiley & Sons, Inc., New York, N.Y., to use material from his book on "Injection Molding of Plastics" (1973).

varying amounts of crystallinity in the molded products.

Fabrication of molds is done in a tool room. A well-equipped tool room will have drill presses, milling machines, lathes, surface and cylindrical grinders, band saws and cutoff saws, a jig borer, two- and three-dimensional duplicating equipment, EDM machine, welding equipment, hardening and annealing ovens, and a hardness tester.

Machining. Machining is the most common way of manufacturing cavities and cores. One should be very careful about specifying tolerances as they are costly. For example, if rough turning a steel casting to a tolerance of ±.015 costs $1.00, then grinding to a tolerance of ±.0005 would cost $10.00. It is well to review mold making tolerances and open them up to the maximum.

Duplicating. A form of machining used to make cavities and cores is called duplicating. This is a mechanical reproduction of a master which is proportional in size to the finished part. Duplicators are horizontal or vertical millers with hydraulically controlled feed. The stylus passes over the master, transmitting its location to the cutting head. Most duplicators can adjust the ratio of the movements of the stylus head to the cutting head. Many of these machines are tape controlled. Large duplicators are used for making parts which cannot be readily turned. Small duplicators are used for making hobs or engraving small designs and letters and numerals on cavities. A major use is to make the carbon masters for EDM equipment. A major disadvantage of duplicating is its comparatively poor surface finish as duplicated.

Hobbing. Hobbing is the cold forming of metal. In tool making, it is used to describe the process analogous to forcing a coin into a piece of clay to obtain a reverse image. The steel hob is hydraulically forced into a hobbing iron. It is often necessary to anneal the hobbing between the pushings. The hobbing blank is carefully polished and fitted into the hobbing ring. This ring is a large piece of steel to hold the hobbing blank and contain the pressure of the hobbing press. The iron of the hobbing blank yields slowly. Figure 22-1 shows a hardened metal hob and the hobbed blank as it comes from the hobbing press. After the hobbing is complete, the hobbing must be cut to size, hardened, and polished. It is case hardened, giving a hard outside and a ductile core. Hobbing is a fast and economical way to produce multiple cavities that are the same size compared to each other and the hob. If the hob is highly polished, the cavities will be highly polished, saving considerable time. Figure 22-1 also shows the hardened polished cavity machined to size, and the molded part.

Pressure Casting Beryllium Copper. To pressure cast beryllium copper, the hob is put face upward in a container. The molten beryllium is poured on top of it and pressure is applied to the top of the molten beryllium to obtain a good surface area and a dense structure in the beryllium casting.

Since the beryllium contracts on cooling, the hob for the cavity must take into account not only the shrinkage of the molded plastic, but the shrinkage of the beryllium itself. The nature of the process permits hot hobbing of thin sections and projections, intricate details and delicate shapes not hobbable in steel. The steel would not be strong enough to prevent the thin sections of the hob from breaking. Pressure cast beryllium cavities are usually supplied to the mold maker annealed and hardened, ready for machining. They are approximately $R_c 35$ to $R_c 40$. Chrome plating adds to their life, which is about the same as steel. The cost of beryllium copper cavities and steel cavities are usually similar. The selection depends upon the characteristics desired in the mold. Cast beryllium cavities are sold on a poundage basis. It has a specific gravity of about 8.09, or about .292 lb/in.^{3}

Casting. Casting techniques have improved to the point where any metal can be cast suitably for injection molding. Typical is a patented technique known as the *Shaw process.* A sample or a plaster reversal of the part in plastic, wood, metal, or other material is cast against a ceramic slurry which is fired. It results in a reversed ceramic reproduction with a micro-grained structure filled with small air gaps. These gaps act as vents so that the molten metal can achieve a good reproduction of the surface. The resultant cavity is not as dense as

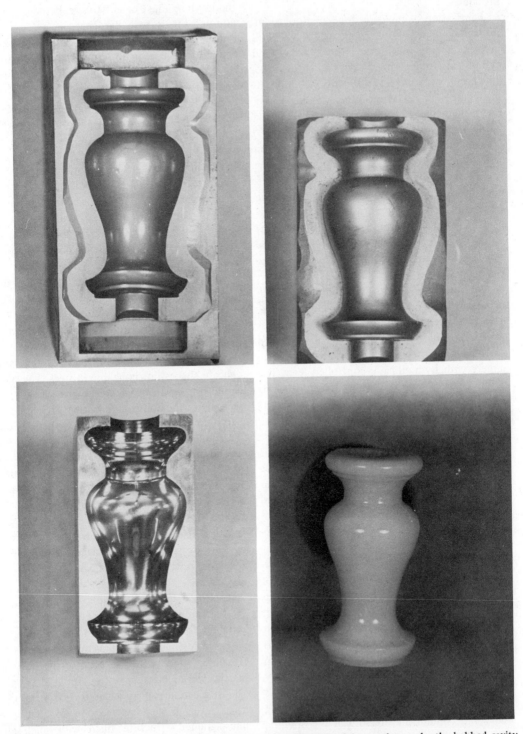

Fig. 22-1. Hobbing. Operation starts with hardened metal hob, upper right, used to make the hobbed cavity (upper left). Cavity is then machined to size, hardened, and polished (lower left). Molded part produced in the cavity is shown at lower right. (*Courtesy Robinson Plastics*)

those produced by other metals, and there is the possibility of small pits. These can usually be corrected. The major advantages are speed and cost. A cavity can be made in less than a week.

Electroplating. Electroplating can be used for cavities and cores. It offers excellent accuracy and has been adapted to the molding of such products as slide rules, electrical connectors, gear cavities, and externally threaded parts. A mandrel, which is the exact reverse of the cavity, is used as the cathode. A nickel cobalt compound is deposited at approximately 0.0005 in./hr to a total depth of 0.15 in. A plating of pure electrolytic copper goes behind this. Backing materials are added as required. The cavities are not hardened and are used as plated, giving no heat distortion. The mold surface resists corrosion and the material has exceptionally high thermal conductivity. The process is used when the application requires such qualities as reproduction of fine surfaces, very close tolerances, and shapes difficult to fabricate by other means.

Electrical Discharge Machining (EDM). A relatively new addition to the tool room is an EDM spark erosion machine. An electrode, usually of carbon, is made in the reverse of the shape to be reproduced. The metal of the cavity and the carbon are the electrodes which are immersed in a circulating solution. This flushes away the eroded material and cools the work. DC current is discharged between the carbon and the steel, creating a spark which erodes the steel. The steel is eroded about eight times as fast as the carbon. Roughing electrodes are used to bring the cavity to its approximate shape and a finishing electrode brings it to size. One of its main advantages is that it can cut hardened steel so that no heat distortion takes place. It can also cut very thin slots. By eroding a number of cavities on a hardened plate, the size of the mold may be reduced. Irregular parting lines can be made to fit perfectly by using the two parts as the electrodes. EDM is widely used in changing and correcting hardened steel cavities. A consideration of this method of operation will reveal many advantages not available in other techniques, including an ability to run unattended.

Optimum Number of Cavities. The opti-
mum number of cavities will usually be selected by the molder and end-user. The cost of the molded part is usually computed as the sum of the cost of material plus waste, packing material, secondary operations, and the hourly rate divided by the number of parts produced per hour. Obviously, the larger the number of cavities, the more parts per hour, and the lower the cost. This has to be balanced by the additional expense of building a larger mold. In this respect, it is important to take into consideration the intended use of the part—whether it is a repetitive item or planned for one-time use. Sometimes, adding more cavities will require a machine with a larger injection or clamping capacity. This, in turn, might increase the hourly rate.

There are also non-economic reasons for limiting the number of cavities. These usually relate to dimensional control and aesthetic appearance. If there are secondary operations to be performed at the machine, it is sometimes more economical to make a mold with fewer cavities, so that the operator can keep up with machine production.

SPECIFYING AN INJECTION MOLD

For those processors and end-users who purchase molds from outside moldmakers (and even those with in-house mold shops may buy some of their requirements on the outside), buying an injection mold involves more than simply handing over a rough sketch of the desired part to the moldmaker. To ensure that the mold conforms exactly to the buyer's needs, it is essential that the buyer communicate to the moldmaker such basic data as: critical dimensions, exact details and sizes, the type of plastic to be used, and any special requirements or limitations (e.g., will the part be post-decorated, will subsequent machining or assembly be involved, what are the gating requirements, when is delivery expected, etc.).

It is also important that the buyer select the moldmaker who can answer his specific needs. Moldmaking capabilities can vary widely. Some shops specialize in small molds, some in large ones, some in completely automated molds, some in inexpensive molds, etc.

Further, a buyer should be aware of the type of equipment available to the moldmaker. There have been major developments during the sixties in all phases of cutting or forming steel. Most significant has been the activity in three-dimensional contour milling equipment, either using computerized equipment or in combination with electrical discharge equipment. Using a modern measuring machine feeding into a computer, a tape can be produced directly from the model which will automatically run a contour milling machine for cutting cavities and cores. Also available today are automated or N/C lathes, drills and mills, jig borers, boring mills, machining centers for doing multiple surfaces, bridgeports, and duplicators of all descriptions.

It should be noted too that quality can be controlled at many stages of mold construction, but the periods before hardening and after hardening are the critical times for inspection. Incorrect machining can be corrected or devi-ated without costly repair work before harden-ing. Inspection after hardening can set guide-lines for work to be done during final grinding and assembly.

There are many techniques available for inspection: comparators, measuring devices that indicate and record dimensions, microscopes, and, of course, all the inspection tools for measuring that have been used over the years. Many moldmakers will run the mold in their own shop to establish the parameters for operating and obtaining acceptable parts. A more recent trend has been for moldmakers to combine with machine builders to supply a complete package—machine and mold—and eliminate any need for adjustments.

Molds can be very simple or as complex and highly engineered as the three types shown in Figs. 22-2 through 22-4. Special mold devices for special jobs are also available (Fig. 22-5). But although all molds are uniquely individual, there are many standards in designing and

Fig. 22-2. Six-cavity cold runner mold. Complex parts with fine tolerances were produced from this family mold. The molded halves had to assemble interchangeably with the mating halves of this mold and two other molds involved in this program. Gate marks and sinks were also highly critical. The mold was tunnel gated to provide automatic separation of parts from runner. Interchangeable inserts were supplied to produce a variety of configurations from the same basic molded part. (*Courtesy Husky of America Inc., Bolton, Ontario, Canada*)

Fig. 22-3. Six-cavity cold runner mold. Because of a complicated design with large variations in wall thickness, there was difficulty in getting adequate cooling to the thick sections. The mold was built using a special high heat-conducting material and great care was taken in providing cooling channels as close as possible to the molding surfaces. The part design necessitated side core action and two stage ejection due to the double-wall construction. Gate marks were highly critical. Accuracy in molded dimensions was extremely important. Runners were separated from the parts by tunnel gating the mold. (*Courtesy Husky of America Inc., Bolton, Ontario, Canada*)

building molds that processors and end-users should understand. The section below* details some of the basic parameters.

Injection Mold Design

Figure 22-6 shows how numbers are arranged in the cavity layout. Multiple cavities are arranged so that all are alike and can be manufactured as duplicates.

* Adapted from a paper, "Quality in the Injection Mold," by Robert G. McGee, currently with Gerard Design, Leominster, Mass., delivered at the 1971 SPE Antec. Reprinted by courtesy of the Society of the Plastics Engineers.

Figures 1 through 6 in Fig. 22-6 illustrate cavity layouts for single-cavity molds through thirty-two cavity molds.

Figure 7 in Fig. 22-6 shows circular layout for small parts generally used in very small machines or hot or insulated runner designs.

Figure 8 shows a layout of cavities that causes the runner to get larger as it moves away from the center on the left side or the gates to get larger as shown on the right side.

There are, of course, variations on all of these layouts, but many companies stay with Figures 1 through 6 in Fig. 22-6 for quotation and construction purposes.

Figure 22-7. Molds shown are generally in

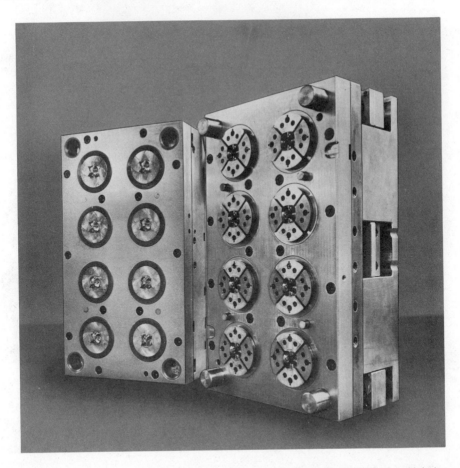

Fig. 22-4. Eight-cavity hot runner reflector mold. A good example of the use of standard mold design concepts applied to very complex parts. The 8-cavity hot runner valve gate system is completely standard. The mold is shown in the open position with the ejection system forward. The tulips are actuated by mechanical ejectors and also provide the side core action necessary to remove the parts. Surface finish of the molded part is extremely critical because the part is metallized afterwards. There is no gate protrusion allowed, which is the reason for using a valve gate hot runner system. (*Courtesy Husky of America Inc., Bolton, Ontario, Canada*)

the low-quality group. Figure 1 in Fig. 22-7 shows a purchased mold base with the cavity and core cut directly into the "A" and "B" plates. This usually is the fastest way to construct a mold provided that the quality of the steel available is acceptable.

Figure 2 shows a purchased ejector box and support plates with the cavity and core set mounted directly to the purchased parts. The cavity and core can be made in almost any stock suitable to the product or production required.

Figure 3 in Fig. 22-7 shows a purchased mold base with pockets cut away to receive a cavity and core. This is done to reduce the size

of the required tool steels, beryllium copper, or aluminum, etc., in making the cavity and core. This type of design is the basis or beginning for action molds of all types.

Figure 22-8 shows the different methods used to strip parts from molds.

Figure 1 in Fig. 22-8 shows a molded part with a ring or rectangle insert in the stripper plate interlocked with the cavity half when the mold is closed. This type is operated in a conventional manner from the ejector plate. Note the guide bushings around the push pins.

Figure 2 shows a molded part with irregular parting line and with a ring or rectangle made solid or in sections embedded in the core. These

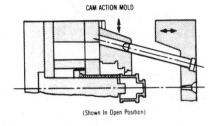

CAM ACTION MOLD

(Shown In Open Position)

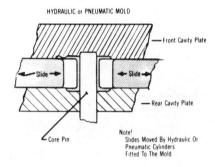

HYDRAULIC or PNEUMATIC MOLD

— Front Cavity Plate

← Slide →   ← Slide →

— Rear Cavity Plate

Core Pin

Note!
Slides Moved By Hydraulic Or
Pneumatic Cylinders
Fitted To The Mold.

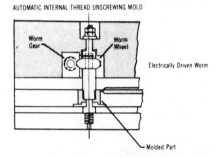

AUTOMATIC INTERNAL THREAD UNSCREWING MOLD

Worm Gear   Worm Wheel

Electrically Driven Worm

Molded Part

Fig. 22-5. Special mold devices such as cam action mechanisms, hydraulic or pneumatic core pullers, and automatic thread unscrewing mechanisms may be required to automatically produce parts of complex geometry. These devices are essential when large undercuts or threads have to be produced in the molding operation. (*Courtesy Celanese Plastics Co.*)

are attached to the ejector plate and operated in a conventional manner. Most moldmakers also bush these push pins.

Both mold types in Fig. 1 and 2 in Fig. 22-8 are readily manufactured using commercial mold bases.

Figure 3 in Fig. 22-8 shows a simpler mold design using three-piece construction. The stripper plate is actuated by machine knock-out rods and the travel limited by stripper bolts.

Figure 4 shows the same type of mold construction as Fig. 3, but the stripper is actuated with chains or pull bars attached to the fill half. This saves machine knock-out assembly and simplifies installation.

Figure 5 shows the clearance on moving parts and has no interlock, because the interlock is placed on a simpler surface or the mold.

Interlocking this type of mold is done with extra leader pins and bushings tighter than those between moving parts. Cone type and round pieces of steel are laid half and half on the parting line.

Figure 6 in Fig. 22-8 shows the fitted areas and clearances for all interlocked stripper molds.

When designing with Figs. 1 or 2, double ejection can be built-in with this type of design. When designing with Fig. 4, fill and ejector halves can be readily reversed.

Figure 22-9 shows various types of side action. The mold in Fig. 1, for example, uses an angular pin for a slide to pull a core pin only. This type of construction can be complicated by water requirements through multiple core pins. The slide is then widened and water spouts put through the core pins. All parts used in this action are hardened tool steel to prevent galling.

Figure 2 in Fig. 22-9 is an end view of the same action to show that water is generally placed outside the angle pin itself so that locks, etc. will not be weakened by cut-outs. Wear plates are installed to prevent galling and grease grooves are used on all molds having over 1 in. travel.

Figure 3 in Fig. 22-9 shows a typical side action that forms part of the article. The cavity insert is attached to the slide and either part can be water channeled. This slide is locked on both fill and eject halves to prevent movement in molding.

Figure 4 in Fig. 22-9 shows an alternate method of Gibway construction for very accurate guiding of slides. Rods can also be used when cylinders are the motion force.

Figure 5 shows a method for holding slides in the "out" position when used horizontally or vertically to prevent mold damage. The spring is on the outside of the mold and is readily seen if broken.

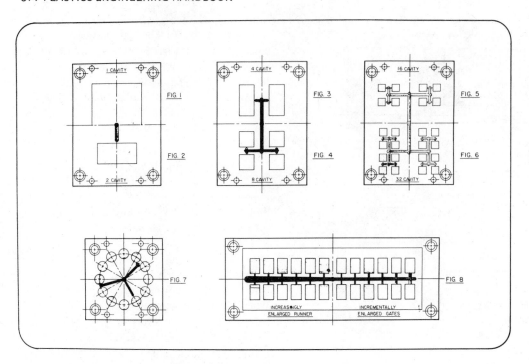

Fig. 22-6

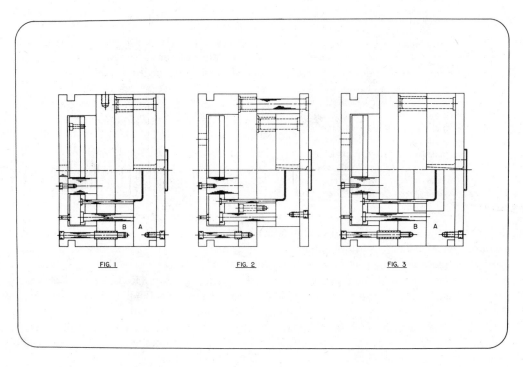

Fig. 22-7

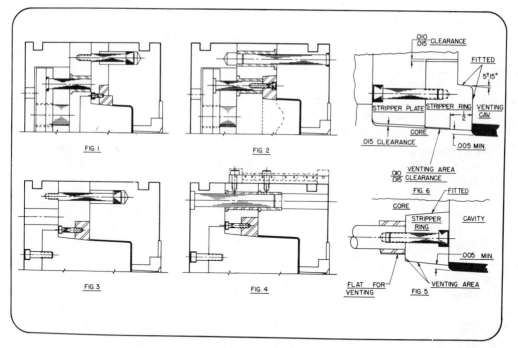

Fig. 22-8

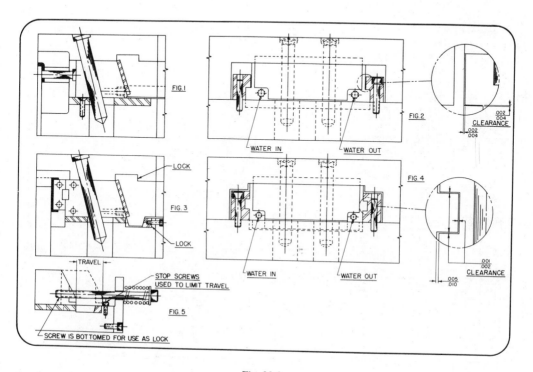

Fig. 22-9

The dimensions shown are critical for the proper clearances used in these actions.

Figure 22-10 shows molds for making threaded parts. In Fig. 1, we see a section through an internally threaded part and the bearings along the shaft required for unscrewing. The stripper plate is spring loaded to follow the thread lead so that the part will smoothly fall from the mold.

Figure 2 in Fig. 22-10 shows a section through the same internally threaded part but a lead screw retracts the core.

These designs are generally actuated with a hydraulic cylinder pulling and pushing on the racks with sufficient travel, as shown in Fig. 5, and is used when a cap has a thread that must start at a given point corresponding to decoration or assembly with a mating part.

Figure 3 in Fig. 22-10 shows a section through the same internally threaded part but this particular shaft is driven with a worm and worm wheel through a gear chain and hydraulic motor mounted on top of the mold, as shown in Fig. 6 in Fig. 22-10

Figure 4 illustrates the shaft rotated by either method above which rotates the cavity allowing core gating and interchangeable cavities. The last quarter turn pulls the part from the cavity for removal of the part from the mold.

Figure 4 in Fig. 22-10 also shows a method for cooling this type of cavity.

Figure 22-11 illustrates various types of gating. Area 1 shows most all of the typical gating methods used when parting line runners are required.

Runner size and gate shape are generally dictated by the article size and material to be molded in conjunction with the number of cavities in the mold.

The gates shown in Area 1 are generally self-breaking, although some parts would require a gate-cutting fixture for appearance. Figure 4, of course, cuts itself on mold opening.

Area 2 shows the remainder of gating methods for three-plate molding and molding of round parts. Figure 11 shows a gating method now being used to hide the gate and still allow design with the more economical standard mold.

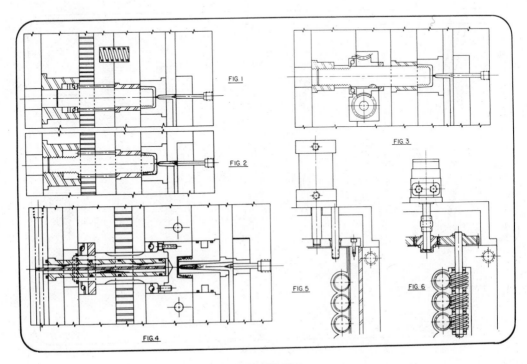

FIG 1

FIG. 2

FIG 3

FIG. 4

FIG. 5

FIG 6

Fig. 22-10

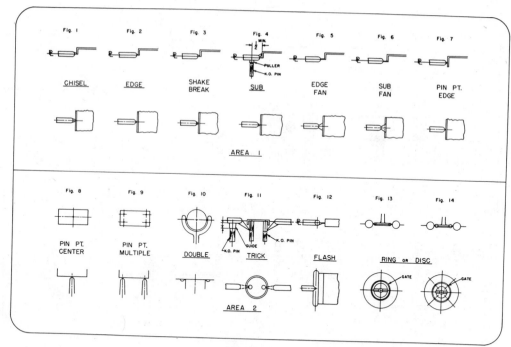

Fig. 22-11

No attempt is made to establish standard gate sizes because of the wide variety of plastics now available to injection molders.

Figure 22-12 illustrates runner systems. Area 1 shows the typical shapes of parting line or three-plate runners.

Figures 2, 3, and 4 in Fig. 22-12 have sharp corners, or as in Fig. 4, thin sections to set up the runner for faster cycles.

Figures 2 and 3 in Fig. 22-12 also have variations on shape and are used mostly when the runner is cut in one plate.

Area 2 in Fig. 22-12 shows insulated runner types. The first two are the simplest.

Figure 5 is used in running polyethylene where no heat is required.

Figure 6 in Fig. 22-12 is used when closer control is required at the gate.

Figure 7 is used when heat is required to a lesser degree but the runner must stay open because of length and only short risers are used.

Figure 8 shows auxiliary heaters in the plates which are more costly, but allow start ups and closer control of melt at all areas of the insulated runner. This approaches hot runner molding but does not require a manifold block and generally does not get as hot.

Figure 9 in Fig. 22-12 shows a typical three-plate layout with runner cut in Plate "A". Pullers are attached in Plate "C" and pull the riser and break the gate. "B" Plate is then pulled by stripper bolts or chains and the runner is stripped off the pullers. The center sprue is pulled by the puller and is stripped off the puller when the parting line opens. This causes the runner to collapse or break and fall out of the mold, in many cases without auxiliary sweepers.

Figure 10 in Fig. 22-12 shows a typical layout used for the risers in Area 2.

Many designers fail to enlarge the sprue area, causing stoppages. This is very important for start-ups on an insulated runner mold.

The enlarged gate view shows two tricks used in this type of molding. "A" shows a raised portion generally between 0.015 and 0.030 so that the gate break off will not interfere with the part sitting flat. "B" shows a small puddle of thicker plastic which stays hotter longer than the base, allowing the excess of the gate break off to shrink back into it, thereby eliminating the gate protrusion.

Figure 11 in Fig. 22-12 shows a typical heated hot runner manifold with bands around

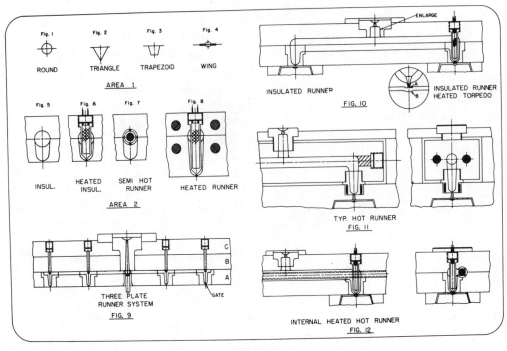

Fig. 22-12

the nozzles. These nozzles are designed in various ways depending on the shape of the article being molded. Cartridges are used internally as shown and when these manifolds get excessively long, strip heaters are attached to the outside of the manifold. Of course, the air gap is always used to insulate the manifold from the mold, but it is well supported with buttons.

Figure 12 in Fig. 22-12 shows a hot runner method with controlled heat in the melt at all times. This is used when melt temperatures of materials are very critical. Calrod is used as the heating element in this manifold.

Today there are available a number of manifolds and risers from moldmaker suppliers to meet most needs. The type and design of the runner system is also part of the quality built into a mold and must be thoroughly engineered to mate with the material and molded part.

Figure 22-13 shows water channeling on all types of mold construction. It is not intended to show a specific wall section, but merely the water channeling design of 95% of the molds built today.

Computers are now available to mold designers (through chiller manufacturers) to ascertain during mold design whether the water channeling will remove sufficient heat to mold the part at the planned cycle.

Figure 1 in Fig. 22-13 shows a conventional part. The dimensions shown are average. If channeling is put closer to the article, cold spots could show on the molded part.

Item A, Fig. 1, shows a method of putting a channel through a surface forming part of the article. A hole is drilled, counter bored and undercut, as shown, and a plug of the same material as the core is driven in and blended to the core.

Figure 2 in Fig. 22-13 illustrates the channeling generally used when designing molds for square or rectangular articles and some of the dimensions considered for optimum mold construction.

Item A, Fig. 2, would be used for warm or cold water and cannot be overlooked as a separate circuit. Dimensions X, Y, and Z will also vary with the size of the part or channels used.

The article shown in Fig. 3 in Fig. 22-13 would have a round cavity set in a retainer plate. Note that all of the water movement is circular and should flow in opposite directions

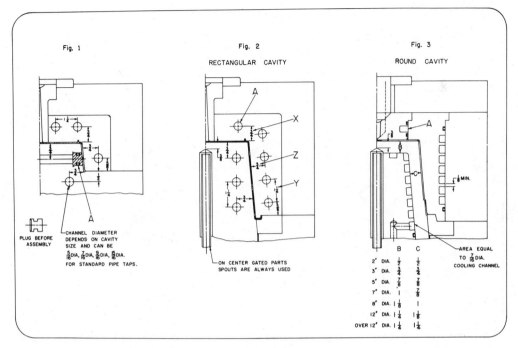

Fig. 22-13

for optimum cooling. Item A, Fig. 3, would also be used for warm or cold water.

Figure 22-14 is indicative of the various venting designs in injection molds.

Figure 1 shows the flow of material on a square or rectangular part center gated and the areas which would receive the initial pressure from gassing caused by the material. Most molds are only vented at the last point of pressure-caused burning. This is for economy in mold construction. Since quality must begin in engineering, planned venting must be a part of design.

Figure 2 in Fig. 22-14 shows the same part edge gated. It would require essentially the same venting, even though the flow is different.

Figure 3 shows a round, center gated item with the flow equal in all directions.

If the mold is built properly, equal venting should be designed around the cavities every 15° and carried down through the stripper ring, in many cases with venting completely around the item as shown in Fig. 4.

Following the material flow from the gate, one can readily design venting into almost any shaped item, always remembering that gas also

traps in blind ribs and vented pins must be installed as shown in Fig. 5 in Fig. 22-14.

Figure 6 shows vent pins which are installed when heavy sections cause the material to return on itself and cause blind burn spots on the tops of parts.

Figure 7 in Fig. 22-14 shows a valve-type vent pin which is also used to break the vacuum caused by the ejection of the molded part.

Venting of molds at the initial test can save many hours of repolishing due to sticking and cracking caused by high pressures used to fill an area not properly vented.

Figure 22-15 is intended to explain the surfaces of steels used in mold construction and is a guide to their use.

Figure 1 in Fig. 22-15 shows prehardened steel and the hardness expected. It is used for cavity and cores and can be highly polished. Unless care is used, it is easily orange peeled at the finish. This steel is also post-treated as is the steel in Figs. 2 and 4.

Figure 2 shows steel which is soft but when case hardened, takes on a hard outer surface and a tough inner core. It must be remembered that this steel should only be hardened as

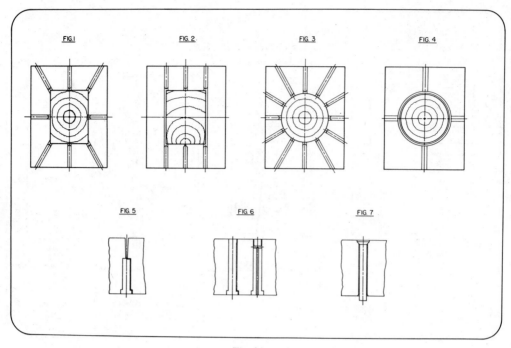

Fig. 22-14

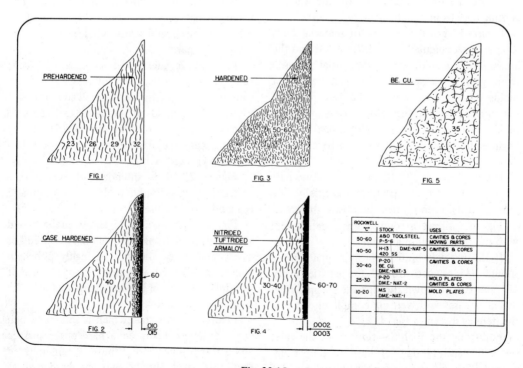

Fig. 22-15

dictated by the thickness of the part. Breakage will result if parts are too thin.

Figure 3 in Fig. 22-15 shows tool steel hardenable through. This steel is generally more useful in designing molds, cavities and cores, and also for moving parts. Since it is difficult to alter by welding, alterations are generally best made by inserting specific areas.

Figure 4 shows post-treatment of all steels which will give a very high Rockwell skin hardness and which is used in sliding assemblies.

Figure 5 illustrates beryllium copper for cavities and cores. It can be gravity cast and used either "as cast," such as for furniture, or heat treated for longer life. It is also cast under pressure and hardened for finer detailed items that cannot be readily cut or hobbed.

STEEL MOLDS FOR PLASTICS*

The steels most commonly used for molds for plastics are identified in the accompanying tables. They are described by the characteristics that influence their selection for specific molding requirements. Also there are certain general factors to be considered with respect to over-all performance of mold steels.

Inherent Quality. Steels are frequently referred to by their chemical compositions, which denotes nothing more than the quantity of various elements used by the melter. The chemical composition alone in no way reflects the general quality of the material.

A steel of given chemical analysis may be melted by any one of several methods, each with its own characteristics and influences on the finished product. Therefore, for proper evaluation of mold steel, the method of melting may be just as important as chemical composition.

The type of melting employed does not in itself regulate the soundness of the finished product; it only influences it. Molten steel solidifies after it has been poured into an ingot mold. As such, it is a cast product and of little value to the mold-maker. The degree of soundness, relative porosity, and uniformity of struc-

* Reviewed by W. Young, Technical Consultant, Colt Industries, Crucible Inc., Specialty Metals Division, Syracuse, New York.

ture are all dependent upon the methods of processing used to reduce the ingot to a finished bar or forging. The fastest or least expensive way to accomplish this would be in one heating, with maximum reduction on the largest and fastest mill available. This is commonly referred to as "tonnage mill" practice. It is a highly desirable method but only when the primary concern is the production of maximum tons at minimum cost, rather than quality.

To achieve "mold-steel quality" it is necessary to give consideration to many factors which influence the quality of the steel from the time it is poured into the ingot mold. The first factors are the rates of cooling from pouring temperature to solidification, and from solidification to room temperature, which often requires several days. Of equal importance are the percentage of reduction from ingot to finished bar, the relationship of size and shape of ingot to size and shape of finished product, and the method of reduction—whether by hammer, press, or rolling. Other considerations are the temperature at which reduction is started and stopped, and the annealing and heat-treating procedures. These are what are generally described as "tool-steel practices." Steel which is melted in an electric-arc or induction furnace, and processed by tool-steel practices, gives mold-builders the best assurance of consistently obtaining a steel suitable for molds for plastics.

Size Factor. Though we recognize the importance of techniques of processing in creating steels suitable for molds for plastics, we must recognize also certain physical limitations of the metal. An ingot of molten steel solidifies first at the bottom, then the sides, and finally at the top and center. Because certain gases continue to be evolved during the solidification, the center comes to be the most porous part of the ingot. Subsequent working will tend to reduce this condition. But it is impossible to work the very center of an ingot into equivalence with the outer portions. Every bar of steel therefore has a center portion which is more porous than the rest. In bars of small cross-section, this zone becomes so small and elongated as to be of negligible consequence. But in larger cross-sections this "center" condition increases, not only in size but also in severity. It is, therefore, unrealistic to expect the same internal soundness in a bar of mold steel 15 to 20 in. thick as would be found in a bar only 4 to 6 in. thick.

However, by using special forging practices and large forging presses, it is now possible to produce very large forgings, 50 in. wide by 25 in. thick, of P20 steel which are just as sound as 4 to 6 in.-thick sections.

Inspection. Various tools are commonly available for non-destructive inspection of mold steels. Macro-etch has been used for many years by makers of tool steel. By this method, a disc is cut from the end of a bar or forging and etched, usually in hot hydrochloric acid. The acid differentially attacks voids, porosity, inclusions, and areas rich or lean in alloy, which become prominently visible to the naked eye.* A more intensive investigation can be made by the use of the micro-etch. Here the mold steel to be examined is polished, lightly etched and examined under the microscope. The observations made by either or both of these methods are meaningful to the trained metallurgist. But it is evident that these observations are confined to the specific portion from which the discs were cut. Metallurgists must assume that the balance of the bar or forging is of similar quality. Because processing techniques are uniform, this usually is a fair and accurate assumption. However, the demands for high internal quality of mold steel have become so strict that mold-builders are not always willing to rely wholly upon this metallurgical assumption. For this reason, normal inspection methods have been supplemented by the use of ultrasonics.

Ultrasonic inspection of mold steels has become a standard practice in recent years. This method of inspection resembles radar in principle. An ultrasonic sound wave is transmitted through the steel. When it reaches the opposite side, it is reflected back to the point of transmission. The time required is recorded on an electronic cathode screen by the position of a sharp point of light. Because of the brevity of the time, the recording on the screen creates

* See standards set forth in AISI Steel Products Manual—"Tool Steels."

the illusion of a straight line. But any interruption of the sonic wave will shorten the time required to complete the reflection, and any irregularity in this is reproduced on the screen as irregularity in the straight line. The magnitude of deviation of the line can be measured on the screen grid. This expression in itself is not meaningful, nor necessarily indicative of the quality of the mold steel. Internal pits, voids and other undesirable discontinuities will cause an interruption of the sonic wave, but so also will segregations, and irregularities in size and orientation of grains, and other factors which are in no way objectionable or detrimental to the quality of the steel.

Sonic inspection cannot be considered a "go" or "no go" tool, such as a plug or ring gauge. It merely reflects a message, which must be interpreted by an experienced technician. Even a highly skilled sonic operator usually coordinates the sonic indications with macro-etch findings to establish the internal quality of a steel.

Hardness. The recommended hardnesses of steels for molds for plastics are based upon the practical effects of hardness, as follows.

Polishability. All things being equal, a steel will polish in direct proportion to its hardness. A minimum of Rockwell C30 is generally considered essential for a good mold finish. In instances where extremely high luster or optical finish is required, a hardness of at least Rockwell C54 is mandatory. This is easily attained by quenching and tempering the high-carbon steels, or by carburizing the low- and medium-carbon steels. It is recommended that soft cavities never be prepolished before heat treatment.

A chrome flash will not increase the ability of a steel to be polished, or rectify defects in finshing. Chrome flashing serves only to accentuate the existing finish.

Wear-Resistance. Resistance to abrasive wear is of primary importance in compression and transfer molds. During molding, thermosetting resins are generally more abrasive than the thermoplastics. This is particularly true of the phenolics. For all such applications, a hardness of 54 to 60 Rockwell C should be required. Carburized medium-carbon alloy tool steels lend themselves readily to such applications. The carburized case provides the necessary wear-resistance, and the tough core supports the case, to absorb the pressures of compression or transfer molding.

Abrasion-resistance is of primary importance in molds for cold molding. These molds should be considered as though they must be wear-resisting tools, rather than merely molds for plastics. Modifications of the D2 or D7 types with higher contents of carbon and vanadium are usually recommended, at a hardness of 62 to 64 Rockwell C.

Peening. Other than its influence on ability to be polished, the hardness of injection molds needs be considered only in relation to their resistance to peening or mechanical deformation. The physical properties of any commercially marketed steel in the annealed condition are adequate to withstand the low pressures of injection and clamping in injection molding. But more than a very low order of hardness is needed to resist the peening action of flash, drool, or other debris not removed from the mold face between cycles, and to retain positive contact in shut-off areas. A minimum hardness of Rockwell C30 is adequate to resist deformation of this type.

Greater hardnesses are often justified to resist mutilation resulting from bad shop practices, such as failure to use proper tools, or proper techniques, for removing molded items that have not been properly ejected. If local shop conditions require a mold to withstand major abuses, a minimum hardness of Rockwell C48-50 should be used.

Because nylon has a greater tendency to drool between cycles, and has a high compressive strength, it must be considered as an exception with respect to mold hardnesses. Adhering nylon drool will peen mold faces of all except deephardening tool steels heat-treated to 60-62 Rockwell C.

Flashing. Molds will sometimes flash under the load of injection pressure. This is usually the result of flexing below the elastic limits of the steel. It is a popular misconception that flexing can be reduced, or stiffness improved, by increasing the hardness or changing the composition of the mold steel. This is not true.

Greater rigidity can be achieved only by increasing the cross-sectional area or thickness of the mold.

Polishability. The influence of hardness on polishability has been discussed under Hardness, above. There are also other inherent qualities in a steel which influence its ability to be polished. Some steels lend themselves to processing with greater freedom than others from internal porosity, pits, flake, and other discontinuities. This is especially true of the higher-alloy types in large cross-sections.

Heat treatment also can influence the ability of a steel to be polished. Steels that are capable of being hardened by quenching in air have a sluggish response to hardening. If such steels are quenched from too high a temperature, or are held at temperature too long, they become so sluggish or retarded in their response to quenching that they do not fully harden. This results in a condition which metallurgists call "retained austenite." To the mold polisher, this is a soft constituent existing at random throughout the steel. It does not take a polish. It has a tendency to drag, tear, or pull out. Because of the resulting differential in hardness, it promotes a random pattern of hills and valleys frequently described as "orange peel." This condition can be alleviated and often corrected by subsequent retempering at 900 to 1000° or freezing to −100 to 150°F, followed by a conventional temper.

The polishing characteristics of carburized mold steels can be influenced by the temperature and cycle of carburizing. It is highly desirable to create fine, evenly distributed carbides in the primary case during carburizing. This presents the most uniformly hard and homogeneous surface to the polisher. It reduces the tendency for pullouts, comet tails, and shadow which are associated with large globular carbides and carbide networks. Better polishing surfaces, therefore, can usually be obtained by carburizing at the lower end of the temperature range and holding for a longer period to achieve the desired depth of case.

Another cause of orange peel and pitting is overpolishing. When the surface of the mold is stressed during polishing beyond the yield point of the steel, permanent plastic deformation occurs. If still greater polishing pressure is applied, the stress may even exceed the ultimate or breaking strength of the steel and the surface will disintegrate. It usually shows a uniform pitted condition. Mechanical polishing, when high pressures are applied, can cause orange peel and pitting especially on softer steels.

Chromium Plating. Chromium plating gives an extremely hard surface to a mold which can be finished to a very high luster. However, chromium plating will not fill pits or polishing defects. In fact, chromium plating magnifies defects on the mold surface about ten times. Chromium plating may also cause hydrogen embrittlement with subsequent pitting and flaking. To prevent this, the mold should be "baked" at 400°F for about 4 hr to remove the hydrogen.

EDM. Electric-discharge machining (EDM) is being used for sinking mold cavities. It is very useful in cutting deep and narrow slots.

When the metal is sparked away from the surface, a layer of remelted or hardened material will be produced on the surface of the mold. This is usually called the *white layer* and may vary from 0.001 to 0.005 in. deep. It is extremely brittle and, therefore, has been the cause of pitting during polishing or afterward when the mold was in service. It is good practice to remove the white layer by grinding or stoning. Should this be impractical, the mold should at least be tempered. The tempering treatment, which should be low enough so it will not soften the entire mold, will remove the stresses on the surface of the EDM and soften and toughen the hard and brittle surface layer. It has been found beneficial to temper the mold even though the white layer has been removed.

Most heat-treating and tempering salts have a tendency to attack or etch the steel. Although the reaction is usually so slight as not to be visible to the naked eye, it is important to remove all adhering salts immediately after use, for contact for even only a few hours is likely to cause minor random pits in a highly polished surface.

Heat Treatment. Linear movement of steel is a normal result of heat treatment. The extent and direction of his movement are dependent upon so many variables that they cannot be predicted with any degree of accuracy. Composition, mass, geometry, and heating and quenching techniques all are contributing influences. Also there are factors other than linear movement that cause change in size, or distortion, during heat treatment. These can be classed as mechanical, and may be controlled to a large extent.

Stress-relieving is perhaps the cheapest insurance against excessive distortion. Steels subjected to severe hogging or cutting operations become highly stressed. At room temperature the steel usually is stable enough to resist change of dimension by these stresses, but at higher temperatures the stresses take effect. At temperatures of about 1000 to 1300°F the steel becomes able to yield to residual stresses. By thus heating the steel, the stresses are relieved and the steel will retain its newly acquired shape.

Heat treatment for purposes of hardening carries the machined steel inevitably through the zone of temperature in which residual stresses are relieved. Any distortion resulting from release of stresses then becomes permanent in the hardened mold, and this distortion may result in very serious difficulty and expense in subsequent matching or aligning of the components of the mold.

This trouble can be avoided by relieving the stresses prior to final machining. The mold, rough-machined to within $\frac{1}{8}$ to $\frac{1}{4}$ in. of finished dimensions, is heated throughout to a temperature of 1250 to 1300°F, held until the temperature has equalized throughout the entire mass, and cooled. Thus stresses are relieved and the steel is left in annealed condition for machining to final dimensions. Thereafter, the risks of distortion upon further heat treatment are minimized.

Extremely large molds may become distorted during heat treatment by other factors, purely mechanical. It must be remembered that steels at their hardening temperatures have relatively little strength. If a mold is not properly supported on the hearth of a furnace or in a salt bath, or during removal from the furnace or bath for quenching, it will be distorted by its own weight. Mold builders frequently can help the heat treaters in this respect by providing convenient eye bolts or certain harness arrangements.

Molds of more intricate or complex shape are more frequently the victims of distortion resulting from variations in thickness. Such distortion may be kept to a minimum by stress-relieving before final machining, by selecting steels having maximum stability in heat treatment, and by employing the best techniques of heat treating.

A mold-builder should not overlook the advantages to be gained by working closely with the heat transfer. By this cooperation many of the factors tending to cause distortion can be effectively controlled or eliminated.

Types of Steel for Molds

Tables 22-1, 22-2, 22-3 are compilations of information concerning steels most commonly used for the various types of molds and accessories used for plastics. Throughout these tables the AISI-SAE classification system for identification of steels has been used. The few steels frequently used by the plastics industry and not covered by this system are identified by brief description.

The contents of the three tables are indicated by their titles:
Table 22-1 Recommended steels for molds, and their accessories, for plastics.
Table 22-2 Chemical compositions of recommended steels.
Table 22-3 Data on heat treating of recommended steels.

It should be noted that there has been some increasing interest recently in the use of stainless steel for making molds which will not readily rust or corrode. Refrigerated water used in cooling molds, for example, can bring about condensation of water on mold surfaces; this moisture, however, will not affect stainless steel. Rusting in water lines or in other internal cooling areas of molds also will not occur.

Molding of corrosive plastics such as PVC does not affect stainless steel molds. In contrast with chrome-plated molds, there isn't any possibility for pitting or flaking which might necessitate stripping and related operations.

Stainless steels for molds are available which can be heat treated to various hardnesses. Machinability varies from that of many of the conventional mold steels, but excellent operation in shops has occurred with present-day mold stainless steels. Welding can be successfully accomplished with the correct rod and proper temperature during welding.

Prehardened stainless steel is readily machinable, can be polished to a superior mirror finish, and can be textured. Thermal conductivity of stainless is lower than non-stainless, but this has a minor effect on productivity. It does not cause a problem on heavier cross sections and can be used to advantage in separating hot from cold zones.

OTHER MOLD MATERIALS

In recent years, especially with the development of plastics processes that require considerably lower pressures than those involved in injection molding (e.g., structural foam molding, thermoforming, rotational casting, etc.), materials other than steel have found growing markets in moldmaking for plastics. This group includes molds cast from beryllium copper, aluminum, or Kirksite; electroformed molds based on nickel-copper; and plastics molds —both flexible and rigid.

Beryllium-copper

An alloy of beryllium and copper is frequently used instead of steel for the making of molds, particularly for injection molding. Beryllium-copper is not to be considered as a universal replacement for steel, but it performs some tasks more competently than steel.

The advantages of beryllium-copper for certain applications result principally from three characteristics:

(1) its high thermal conductivity, which tends to shorten the cycle of injection molding, wherein the plastic in the mold must be hardened by removal of heat;
(2) its ability to absorb mechanical shock during the molding cycle;
(3) its ability to be hardened more than superficially by a simple heat treatment,

Table 22-1. Recommended Steels for Molds, and Their Accessories, for Plastics

AISI-SAE Steel Designation	General Characteristics and Uses	AISI-SAE Steel Designation	General Characteristics and Uses
	A. Injection Molds		*A. Injection Molds (continued)*
P1*	For short-run, small, inexpensive molds where exceptional ease of hobbing is desired and where maximum detail and sharpness of design are paramount.	O1 and O2	Can be heat-treated to high hardness for maximum strength. Usually used for small inserted molds where strength rather than toughness is an important consideration.
P2* and P3*	Relatively difficult to hob, but provide high-strength core for good resistance to sinking during operation.	H11, H12, and H13	Offer good combination of high strength, good toughness, and dimensional accuracy in hardening.
P4*	Air-hardening hobbing steel. Exhibits minimum distortion in heat treatment and develops a high core hardness. Suggested for relatively shallow hobbed impressions for molding close-tolerance articles, and for molding temperatures higher than normal.	A2 and A6	Popularly used for long-run injection molds requiring optical finishes.
		420 stainless	Suggested primarily for molds subjected to adverse atmospheric conditions and for molding PVC.
P5*	Offers very good compromise of ease of hobbing and core strength.		*B. Compression Molds*
P6*	Quite difficult to hob, but provides exceedingly strong core. Used primarily for machine-cut cavities, but shallow impressions can be hobbed satisfactorily.	P1*	Suggested for maximum ease of hobbing, if mold pressures are extremely low.
		P2* and P3*	More difficult to hob than P1 but suitable for medium-high molding pressures.
P20	Usually supplied prehardened at Brinell 300. Can also be carburized to increase surface hardness where change in size in heat treatment is not critical. Suitable for all types of machine-cut injection molds. Also popular for holding-blocks.	P4*	Used for shallow hobbed impressions for maximum resistance to high molding pressures. Particularly applicable where minimum change of size in heat treatment is of paramount importance.
		P5*	Offers very good compromise of ease of hobbing and resistance to high molding pressures.
Precipitation-hardening	Supplied prehardened at a hardness of approximately Brinell 340. Suitable for all types of machine-cut injection molds and high-strength holding-blocks.	P6*	Popularly used for large compression molds requiring maximum strength characteristics. Used both for shallow hobbed impressions and for machine-cut cavities.
L2	Available both prehardened at approximately Brinell 300 and also annealed. Suitable for all types of injection molds. Popularly used for holding-blocks as well as for cavities.	P20	When carburized, suitable for all types of compression molds. Very popular for large molds.
6115	Carburizing type of alloy steel suitable for all types of injection molds where change of size in heat treatment is not critical. Not a tool steel.	Precipitation-hardening	Since this material cannot be carburized, it should be used, in compression molding, only with nonabrasive plastics and under low molding pressures.

* These steels require a carburizing treatment to develop the required properties on the surface of the mold.

(continued)

Table 22-1.   Recommended Steels for Molds, and Their Accessories, for Plastics *(continued)*.

AISI-SAE Steel Designation	General Characteristics and Uses	AISI-SAE Steel Designation	General Characteristics and Uses
	B. Compression Molds (continued)		*D. Accessories (continued)*
6115	When carburized, suitable for all types of compression molds. Popularly used for large molds. Not a tool steel.		S1 or Nickel Shock Steel categories. Heavier pins are made in O1, O2, A2, or A6. Nitrided H13 is also popularly employed for all types of core pins, particularly when resistance to surface damage is important.
O1 and O2	Suggested for small or inserted molds for highly abrasive plastics.		
H11, H12, and H13	High hardenability and minimum change of size in heat treatment make these steels popular for relatively large compression molds requiring intermediate hardness characteristics.	3. Holding-blocks	Inexpensive holding-blocks are made from AISI-SAE 1020 steel, although this has a very limited strength. More popular choices are prehardened P20 or prehardened AISI SAE 4145. Heat-treated holding-blocks for maximum strength and durability are usually made from H11, H12, or H13.
A2 and A6	Suggested for small, long-run molds for highly abrasive plastics or for super-critical finish requirements.		
D2	Used for relatively small molds demanding very great hardness and the ultimate in abrasion-resistance.	4. Slides	It is usually desirable that slides possess a hard, wear-resisting surface to minimize galling and seizing. Carburized P20 or L2, and nitrided precipitation-hardening-steel are popular choices. Heat-treated O1, O2, A2, and A6 are also widely employed.

C. Transfer Molds

The intricate nature of transfer molds and the multitude of properties required necessarily make it exceedingly difficult to recommend specific types of steel. In general, however, the following steels are the most popularly used:

P20 and 6115	Suitable only if carburized. Not recommended for complex molds or where change of size in heat treatment is objectionable.	5. Sprue bushings	L2, O1, O2, A2, A6, or D2 are suggested.
			E. Master Hobs
O1 and O2	Popular for general-purpose use, since they provide an optimum balance of toughness, hardness, and minimum change of size in heat treatment.	S1	Usually requires carburizing to develop sufficient surface hardness. Popularly used for intricate hobs requiring good toughness.
A2 and A6	Suggested for long-run molds requiring maximum abrasion-resistance and minimum change of size.	S4	Suggested for maximum toughness where change of size in heat treatment is not critical.
		Ni Shock	Recommended for maximum toughness and minimum change of size in heat treatment.
	D. Accessories	O1 and O2	Possess optimum balance of toughness and wear-resistance for hobs of average shape and dimensions.
1. Ejector pins	Available in finished form with upset heads. Made from either H13 nitrided or a nitrided Nitralloy steel.	A2, A6 and D2	Recommended for high-strength hobs and maximum resistance to abrasion where toughness is not a major requirement.
2. Core pins	Long, slender core pins subjected to deflection loading are usually made from a shock-resisting steel in the	H11, H12, H13, and H23	Used for masters in hot hobbing of beryllium-copper molds.

Table 22-2. Chemical Compositions of Recommended Steels

AISI-SAE Type	C	Si	Mn	W	Cr	V	Ni	Mo	General Classification
				Typical Composition					
A2	1.00	–	–	–	5.00	–	–	1.00	air-hardening for cut molds and master hobs
A6	0.70	–	2.00	–	1.00	–	–	1.00	low-alloy air-hardening for cut molds and master hobs
D2	1.50	–	–	–	12.00	–	–	1.00	air-hardening high-carbon high-chromium for cut molds and master hobs
H11	0.35	–	–	–	5.00	0.40	–	1.50	5% Cr hot-work for cut molds and hot hobs
H12	0.35	–	–	1.50	5.00	0.40	–	1.50	5% Cr hot-work for cut molds and hot hobs
H13	0.40	–	–	–	5.00	1.00	–	1.50	5% Cr hot-work for cut molds and hot hobs
H23	0.30	–	–	12.00	12.00	–	–	–	heat-resisting hot-work for hot hobbing of beryllium-copper
O1	0.90	–	1.00	0.50	0.50	–	–	–	Mn oil-hardening for cut molds and master hobs
O2	0.90	–	1.60	–	–	–	–	–	Mn oil-hardening for cut molds and master hobs
L2	0.50	–	–	–	1.00	0.20	–	–	Cr-V low-alloy for cut molds, both prehardened and heat treated
P1	0.10	–	–	–	–	–	–	–	low-carbon ingot iron for hobbed molds
P2	0.07	–	–	–	2.00	–	0.50	0.20	Cr-Ni-Mo alloy for hobbed molds
P3	0.10	–	–	–	0.60	–	1.25	–	Cr-Ni-Mo alloy for hobbed molds
P4	0.07	–	–	–	5.00	–	–	–	air-hardening for hobbed molds
P5	0.10	–	–	–	2.25	–	–	–	medium-Cr alloy for hobbed molds
P6	0.10	–	–	–	1.50	–	3.50	–	Ni-Cr high-strength alloy for hobbed and cut molds
P20	0.30	–	–	–	0.75	–	–	0.25	Cr-Mo low-alloy for cut molds, both prehardened and heat treated
S1	0.50	–	–	2.50	1.50	–	–	–	W shock-resisting for master hobs
S4	0.55	2.00	0.80	–	–	–	–	–	Si-Mn shock-resisting for master hobs

(continued)

Table 22-2.   Chemical Compositions of Recommended Steels (continued).

Steels not classified by AISI-SAE but popularly used in plastics molding	Typical Composition								General Classification
	C	Si	Mn	W	Cr	V	Ni	Mo	
Ni-Al precipitation-hardening	0.20	–	–	–	–	–	4.10	1.20 Al	Ni-Al precipitation-hardening prehardened for cut molds
Ni shock steel	0.55	1.00	1.00	–	0.50	–	2.70	0.50	Ni shock-resisting for master hobs
420 stainless	0.30	–	–	–	13.00	–	–	–	stainless steel for cut molds
6115	0.18	–	0.80	–	1.00	0.20	–	–	low-carbon Cr-V for cut molds
1020	0.20	–	–	–	–	–	–	–	low-carbon steel for holding-blocks
Nitralloy	0.37	–	0.35	–	1.10	–	0.25	1.30 Al	nitriding steel for ejector pins

so that it reaches, throughout its body, a hardness of Rockwell C35 to C40.

In addition, it has good resistance to abrasion and to fatigue.

A choice between beryllium-copper and the conventional steels will be governed by several factors.

For molding an article of simple shape, for which the mold can be satisfactorily produced by cold hobbing of steel, beryllium-copper usually offers no advantage. But for making the molds for an article involving thin projections, intricate engraving, fluting, serrations or complicated parting lines, or, in general, for making molds which cannot be hobbed, or which present the risk of breakage of the hobbing tool, beryllium-copper offers a practical solution.

When a program of production requires the shortest possible molding cycle, the thermal conductivity of beryllium-copper may dictate the choice of it, rather than of steel.

Beryllium-copper can be cast, wrought and drawn. For making molds for plastics it is usually pressure-cast.

A master steel pattern is encased in a heavy retainer ring, and the molten alloy is cast around it. Hydraulic pressure is applied, and maintained until the alloy has solidified.

In this operation the steel pattern is subjected to repeated thermal shock, and to hydraulic pressure. Hence a pattern which is to be used repeatedly should be made from hot-worked steel of good grade, such as AISI-SAE H13 or H23, so that all the castings will have the same dimensions.

In making the steel pattern, allowance should be made for a differential shrinkage, usually 0.004 in./in.

In injection molding, pressure-cast molds of beryllium-copper have yielded successful runs of several million pieces, some in the tens of millions.

Beryllium-copper molds have become very popular in recent years for the injection molding of wood-like plastics parts, primarily for the furniture and appliance industries (e.g., cabinets, frames, radio housings, etc.). In addition to economy and durability, BeCu reproduces wood-grain patterns with excellent fidelity. The lead time for making BeCu molds is also often less than for molds machined in steel.

In compression molds, beryllium-copper has been less successful, and its use in this field has been limited principally to cases in which molds of steel are mechanically or economically impracticable, by reason of intricacy of shape or the risk of breakage of cold-hobbing tools.

Beryllium-copper has been used also for making trimming dies for removing flash from molded articles.

Table 22-3. Data on Heat-Treating of Recommended Steels

Item	AISI-SAE Designation	A2	A6	D2	H11	H12	H13	H23
	Forging							
1.	Start forging at, °F	1850-2000	1900-2050	1850-2000	1950-2100	1950-2100	1950-2100	1950-2100
2.	Do not forge below, °F	1650	1600	1700	1650	1650	1650	1800
3.	Normalizing, °F	(F)	(F)	(F)	(F)	(F)	(F)	(F)
	Annealing							
4.	Temperature, °F	1550-1600	1350-1375	1600-1650	1550-1650	1550-1650	1550-1650	1600-1650
5.	Rate of cooling, °F, maximum per hr	40	20	50	50	50	50	50
6.	Approximate hardness, Brinell	202-229	217-248	217-255	192-229	192-229	192-229	217-255
	Hardening							
7.	Carburizing temperature, °F	—	—	—	—	—	—	—
8.	Hardening temperature, °F	1700-1800	1525-1600	1800-1875	1825-1875	1825-1875	1825-1900	2200-2300
9.	Quenching medium	air	air	air	air	air	air	oil
10.	Tempering temperature, °F	350-1000	300-900	400-1000	1000-1200	1000-1200	1000-1200	1200-1300
11.	Approximate hardness, Rockwell C	62-57	60-50	61-54	54-38	55-38	53-38	45-40
12.	Depth of hardening	deep	deep	deep	deep	deep	deep	deep
13.	Non-deforming properties	best	best	best	very good	very good	very good	fair

Item	AISI-SAE Designation	O1	O2	L2	P1	P2	P3
1.	Forging Start forging at, °F	1800-1950	1800-1925	1800-2000	2200-2350 or 1450-1550 (A)	1850-2050	1850-2050
2.	Do not forge below, °F	1550	1550	1550	1900 or 1250 (A)	1550	—
3.	Normalizing, °F	1600	1550	1600-1650	(G)	(G)	(G)
4.	Annealing Temperature, °F	1400-1450	1375-1425	1400-1450	1350-1650	1350-1500	1350-1500
5.	Rate of cooling, °F, maximum per hr	50	50	50	50	50	slow, in furnace
6.	Approximate hardness, Brinell	183-212	183-212	163-196	81-100	103-123	109-137
7.	Hardening Carburizing temperature, °F	—	—	—	1650-1750	1650-1700	1650-1700
8.	Hardening temperature, °F	1450-1500	1400-1475	water: 1450-1550 oil:　1500-1700	1450-1475 (B)	1525-1550 (B)	1475-1525 (B)
9.	Quenching medium	oil	oil	oil or water	water or brine	oil	oil
10.	Tempering temperature, °F	300-500	300-500	300-1000	300-500	300-500	300-500
11.	Approximate hardness, Rockwell C	62-57	62-57	63-45	64-58 (C)	64-58 (C)	64-58 (C)
12.	Depth of hardening	medium	medium	medium	carburized	carburized	carburized
13.	Non-deforming properties	very good	very good	water: poor oil: fair	poor	good	good

A – Do not work between 1600 and 1900° F.
B – After carburizing.
C – Carburized case hardness.
D – After cooling from annealing temperature, reheat to the lower temperature and cool slowly.
E – Do not temper between 700 and 1100° F.
F – Do not normalize.
G – Not required.

(continued)

Table 22-3.   Data on Heat-Treating of Recommended Steels (*continued*)

Item	AISI-SAE Designation	P4	P5	P6	P20	S1	S4
	Forging						
1.	Start forging at, °F	1850-2050	1850-2050	1950-2150	1850-2050	1850-2050	1850-2050
2.	Do not forge below, °F	1600	1550	1700	1600	1600	1600
3.	Normalizing, °F	(F)	(G)	(G)	1650	(F)	(F)
	Annealing						
4.	Temperature, °F	1600-1650	1550-1600	1550	1400-1450	1450-1500	1400-1450
5.	Rate of cooling, °F, maximum per hr	25	40	30	50	50	50
6.	Approximate hardness, Brinell	116-128	105-110	207	150-180	183-229	192-229
	Hardening						
7.	Carburizing temperature, °F	1675-1700	1650-1700	1650-1700	1600-1650	1650-1700	—
8.	Hardening temperature, °F	1725-1750 (B)	1550-1600 (B)	1450-1500 (B)	1500-1600 (B)	1650-1800	water: 1600-1700 oil: 1650-1750
9.	Quenching medium	air	oil or water	oil	oil	oil	oil or water
10.	Tempering temperature, °F	300-500	300-500	300-450	300-500	400-1200	350-800
11.	Approximate hardness, Rockwell C	64-58 (C)	64-50 (C)	58-61 (C)	64-58 (C)	58-40	60-50
12.	Depth of hardening	carburized	carburized	carburized	medium	medium	medium
13.	Non-deforming properties	very good	good	good	good	fair	fair

Item	AISI-SAE Designation	Ni-Al precipitation hardening	Ni shock steel	420 stainless	6115	1020	"Nitralloy"
	Forging						
1.	Start forging at, °F		1875-1975	1950-2050	1800-2000		
2.	Do not forge below, °F		1700	1750	1550		
3.	Normalizing, °F		(F)	(F)	(G)		
	Annealing						
4.	Temperature, °F	Not applicable; always furnished in pre-hardened condition	1550-1600 1100-1125 (D)	1550-1650	1400-1450	Not applicable; not usually heat treated	Not applicable
5.	Rate of cooling, °F, maximum per hr		25	50	50		
6.	Approximate hardness, Brinell		255-285	179-240	150-180		
	Hardening						
7.	Carburizing temperature, °F		–	–	1650-1700		
8.	Hardening temperature, °F		1600	1800-1900	1500-1550 (B)		
9.	Quenching medium		air or oil	oil or air	oil		
10.	Tempering temperature, °F		350-850	300-700 1100-1200 (E)	300-500		
11.	Approximate hardness, Rockwell C		60-50	52-30	64-58 (C)		
12.	Depth of hardening		deep	deep	shallow		
13.	Non-deforming properties		very good	good	good		

A—Do not work between 1600 and 1900° F.
B—After carburizing.
C—Carburized case hardness.
D—After cooling from annealing temperature, reheat to the lower temperature and cool slowly.
E—Do not temper between 700 and 1100° F.
F—Do not normalize.
G—Not required.

Electroformed Molds*

The process of electroforming consists of electroplating against a conductive surface (called a pattern or mandrel) for a long period of time to reproduce a reverse of that surface. The electroform cavities or cores thus formed can be backed up as required.

Patterns, or mandrels, may be made of almost any material that does not absorb moisture, or does not expand excessively through a 100°F heat differential. Expansion during the first half-hour of plating will crack the very thin electroform. The adhesion of the plating to the mandrel is part of the state of the art since it must be strong enough to retain the surface detail and contour, but weak enough to be removed after plating. The adhesion is determined by the film deposited on the pattern or mandrel. Metal mandrels will usually be nickel chrome plated. In this case, the chrome acts as an electrical conductor as well as a release agent for the nickel since the electroform face will not permanently adhere to the chrome surface of the mandrel.

Mandrels with a non-conductive surface must be sprayed with reduced silver which is a liquid silver reduced to a metallic film by the addition of formaldehyde or with a highly conductive silver paint. The first method is only 3×10^{-6} thick, and therefore reproduces far more detail than the paint spray method.

The type of plating used in electroformed molds is normally nickel, copper, or a laminate of both. Other metals including iron, gold, silver, palladium, platinum, and rhodium have been used but the high cost and other factors limit their usefulness. Nickel copper molds, 50% nickel and 50% copper, are very effective for most liquid plastisols, polyethylene, and polypropylenes. Dry solid materials, however, etch the copper and a nickel copper nickel laminate is recommended.

For molds requiring machining, such as most compression molds or die inserts, a nickel face with copper back up is used due to the poor machinability of nickel. The copper back-up

* Adapted from a paper on Electroformed Molds, by J. L. Jansen, Electro Mold Corp., delivered at the SPE Retec on "Advanced Methods in Tooling for Molding," Buffalo, New York, Feb. 23, 1972.

can then be easily machined using the nickel surface as a reference. On molds which require a large number of holes, welding, soldering or brazing, solid nickel is used to prevent the possibility of delamination which exists above 1000°F. This would pertain to thermoformed molds where the holes would be either conventionally drilled or laser drilled, and molds where inserts, flanges, or a large number of mounting brackets may have to be brazed to the mold.

The plating rate is normally 0.012-0.015 in./24 hr day. This rate may be increased when plating copper, but stress and brittleness also increase. Plating does go on 24 hr/7 days a week, requiring 9 to 10 days to produce a mold 0.125 in. thick.

Molds are separated from mandrels in three basic ways. One-piece molds with undercuts require a disposable mandrel which is usually melted or etched out of the mold. A mold with sufficient draft is usually pried loose from the mandrel mechanically at one or two points, then, by the use of air pressure, removed completely. An electroformed mold with little or no draft can normally be released by cooling the pattern while heating the electroform. The expansion of the electroform will usually be sufficient to remove it from the mandrel.

Advantages of Electroformed Molds. Electroformed molds have many advantages and some disadvantages when compared with standard tooling methods. The major advantages are extremely accurate reproduction of detail in each cavity when compared with steel for a given price, zero porosity compared with aluminum castings, and zero shrink when compared with beryllium copper castings. In most cases, the electroformed mold is far less costly than steel, comparable in price to beryllium copper, and more costly than aluminum. As will be seen later, in some cases electroforming is the only way some molds can be made.

Disadvantages of Electroformed Molds. The disadvantages are that the electroformed mold is relatively soft, except in certain cases, compared to steel. It usually tests at the low end of the Rockwell C scale or high end of the B scale. Delivery of multiple cavities is slow without multiple mandrels or when design changes are to be made, compared to aluminum or beryllium castings. Limited configuration is also a disadvantage. Configuration plays an important part in whether or not an electroformed mold can be used, because of the difficulty in electroforming to a uniform thickness over the complete surface of the mold. At present, principles of electromagnetic force dictate that recesses deeper than they are wide should be avoided. Sharp corners and most internal bosses or projections should also be avoided, although some bosses may be inserted into the mandrel as "grow ins" and the plating will securely attach itself to the bottom of the boss and the "grow in" becomes an integral part of the mold. Suggestions for matrix manufacture are indicated in Fig. 22-16.

Applications for Electroforming. Today, electroformed molds are used for virtually every type of plastics processing, as follows:

Rotational Molding. Typical of the products roto-molded in an electroformed mold would be a meat tray. The advantage of electroforming in this application is a highly polished surface finish. This design also lends itself to electroforming because two trays are roto-molded at the same time; therefore, if the two mold halves have a slight mis-match or excessive flash, these defects can be cut out of the part during trimming. The meat tray mold would be all nickel, 0.125-in. thick.

Another typical application would be the roto-molded skin for an automobile head rest. An electroformed nickel-copper mold, 0.110-in. thick, would offer excellent grain reproduction at low cost for this application.

Compression Molding. Most compression molds used are relatively thin and are mounted to a heavy platen. Typical, would be a phonograph record die, which uses electroformed molds to achieve more accurate sound wave reproduction.

Thermoforming. Automobile door panels, for example, are formed on an electroformed mold shell. Two major advantages of using electroforms in this application are the ability to produce a multi-grained part in one operation and, with the addition of cooling coils, a reduction in cycle time. Molds for a part of this type would generally be all-nickel, 0.080-in. thick, with 0.007-in. diameter laser-drilled holes over the entire face.

Injection Molding. Cavities where detail or tolerance make hobbing or machining extremely expensive or virtually impossible can often be made by electroforming over a reverse of the cavity cut into a mandrel. Electroformed molds also can be plated to 45-50 Rockwell C, which is considered satisfactory for many injection molding compounds.

Tooling Repair. Tool steel molds can be repaired or reworked by the use of the electroforming process. Nickel with 45-50 Rockwell C can be added to the spot to be repaired, then machined to a finished size.

Other. In addition, electroformed shells are particularly adaptable to reproducing wood grains for urethane foam parts or any surface finish on integral skin foam parts, blow-molded parts, and structural foam parts. When used for structural foam molding, electroformed molds offer outstanding pattern reproduction and the ability to place cooling lines in the most desirable locations to provide very efficient cooling systems. However, since the plated shell is thin and the cooling lines are near the surface, it is extremely difficult to modify the mold should a change in design be required.

Aluminum Molds

As new plastics processes requiring lower molding pressures (e.g., structural foam molding) have evolved, materials like aluminum have become more and more important in moldmaking (not only for foam molding, but for thermoforming, rotational molding, etc.). Where the process is such that aluminum can be used, it offers: fast heat transfer (resulting in shorter cooling times and faster production cycles), lower cost (as compared to mold steels), lightweight and ease of handling. The metal does not reproduce surface detail, however, as well as Kirksite or beryllium copper.

Aluminum molds can be machined from solid stock where size and shape permit (i.e., simple shapes) or be built up from blocks, bolts, and weldments. They can also be pressure cast, although in some applications, like structural foam molding, it is especially important that dense castings be used so that water does

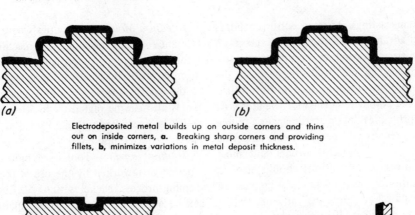

Electrodeposited metal builds up on outside corners and thins out on inside corners, **a.** Breaking sharp corners and providing fillets, **b,** minimizes variations in metal deposit thickness.

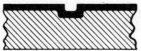

Recesses should be wider than deep. It is difficult and sometimes impossible to electrodeposit into deep, narrow recesses.

Use fillets at least equal to metal deposit thickness for strong inside corners.

Holes can be "spotted" for subsequent drilling by providing depressions in pattern.

When feasible provide slight taper (0.001 in. per ft) to aid mandrel removal.

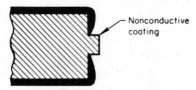

Nonconductive coating

Eliminate drilling and reaming operations by providing masked or nonconductive studs on pattern. Hole diameters can be held to ±0.0002 in. and have excellent surface finish.

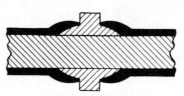

Flanges and bosses should fit tightly and be flared or tapered to mandrel diameter.

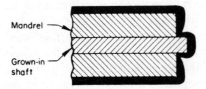

Mandrel

Grown-in shaft

Extend internal piece beyond end of surrounding part to assure deposition on sides as well as end of internal "grow-in" piece. This provision assures bonding of internal shaft to outside cylin-

Fig. 22-16. Matrix manufacture for electroformed molds. (*Courtesy Electro Mold Corp.*)

not pass from the cooling channels through the mold walls and into the cavity.

Kirksite

Again, in processes where pressures are relatively low and short runs are anticipated (usually in the thousands), Kirksite molds can be used. Because the material pours so well, it is generally cast and Type A Kirksite is usually used. Since the pouring temperatures are low (800°F for Kirksite as compared to 3000°F for steel and 2000°F for beryllium copper castings), it is possible to cast copper tubing cooling lines directly into place in Kirksite. More important, the low casting temperatures (and the retention of fluidity for a relatively long period of time) enables Kirksite to pick up fine detail from the pattern over a very large casting area. This means that Kirksite molds will reproduce pattern detail in the molded parts (it falls somewhere in between aluminum and beryllium copper in this regard) and, as such, has found application where fine patterns like wood grains are required (e.g., furniture parts). Shrinkage is about 0.008/in.

Kirksite is lower in cost than most other metals and machines well. But it is non-magnetic and therefore may need clamping for grinding. It has a tendency to load grinding wheels badly. Kirksite is also heavier than aluminum and only slightly lower in weight than steel or beryllium copper. It is not as strong as either of these other metals and therefore will require heavier wall sections, making it more difficult to handle. Cycles with Kirksite molds are usually shorter than steel, but longer than aluminum molds.

Plastics Molds

While plastics tooling has long found application in the metal-working industries (see Cast Plastics for Tooling, p. 448), it has only recently begun to really find its way into plastics processing per se. Today, both flexible and rigid molds are being used.

Flexible Molds. Molds of this type received their big impetus in the late sixties when furniture manufacturers turned increasingly toward their use in the casting of polyurethane and polyester parts (especially in applications involving undercuts, such as in furniture pieces simulating intricately carved styles, as in Fig. 22-17).

Strictly speaking, most of the materials used for this type of flexible mold are in the family of elastomers (including depolymerized natural rubber, as well as silicone, urethane, and polysulfide). Flexible molds cast from silicone rubber are probably the best known in this area. Overall, they offer durability, good stability, and resistance to cold flow. Many of these rubbers also show relatively poor tear strength, but a number of improved formulations with fairly good tear resistance are now commercially available.

Polysulfide flexible molds are less costly than the silicones, cure relatively fast, and reproduce details with great fidelity, but they

Fig. 22-17. Typical complex furniture part cast in a flexible mold. (*Courtesy Flex Moulding, Inc.*)

have problems of cold flow and warp and must be stored very carefully. Depolymerized natural rubber which costs even less than polysulfides, also reproduces detail well, is dimensionally stable, has a fairly good life in the cured state, and is resistant to cold flow. Its major disadvantage is the length of cure time required and difficult in handling. Polyurethane is extremely tough and durable, but is very susceptible to moisture.

Silicone Rubbers.* A full range of rubbers has been developed to provide the industry with various desirable properties for improved service life of the completed molds, as well as to simplify moldmaking techniques. Flexible molds are produced from a two-component system (base and catalyst) and cure at room temperature, or at elevated temperatures with a heat accelerated catalyst if quicker cure cycles are desired. You can choose the strength, hardness, work and cure times to best suit your job requirements and select from a range of viscosities—from crevice-penetrating free flow to a non-slump, butter-on paste.

Selection of the right grade of base rubber and catalyst depends upon the end properties desired in the finished mold and the type of pattern employed in making the mold. The finer detail the pattern has, the lower the viscosity of the uncured rubber material that will be required to duplicate the surface. If extreme distortion and stretching of the mold are required to extract the part, a high tear-strength and high-elongation material should be selected. Pot life, or working time, usually two to four hours for normal mixes, may also be a factor, although most base rubbers can be used with several catalysts to vary this parameter.

If the silicone molds are to be used to cast polyurethane or polyester furniture parts, be sure to select a rubber that is resistant to attack from these materials. Rigid polyurethane foam contains an amine which eventually attacks silicone rubber. Suitable barrier coatings applied to the silicone mold before each usage that also transfer to the molded parts as an integral prime coat increase the mold life in this

application. Polyester resins eventually stick to the mold after a number of castings. Remove the cast part from the mold as soon as it is set up to reduce contact time between the rubber and the polyester. Also, baking the rubber mold for about one hour at temperatures in the range of 500 to 550°F on a regular basis appears to help prolong mold life when used with polyester.

Two basic types of molds are made with silicone rubber. The *skin mold* is a thin covering of rubber over the pattern and is made by techniques such as brushing or troweling. Multiple coats of rubber are used and a reinforcing fabric is used between coats, although no fabric should be exposed on the molding surface. After this rubber shell has cured, with the pattern still in it, a rigid back-up shell such as plaster, polyester or urethane foam is cast around the rubber. If the pattern has a front and back, as opposed to an item such as a wall plaque that has only one detailed side, the rubber skin and backing must be made with a parting line.

The second type of mold, the *cast-in-place mold*, may also be either one piece or multipart to fit together to form an enclosed mold as the pattern demands. As the name implies, the cast mold is poured into place over the pattern which is secured in a frame to form the rubber perimeter. When using a wood pattern, be sure to nail or secure the pattern in place since it will float in the liquid rubber. Parting lines are determined from the pattern geometry much in the same manner as for any other molding technique. Cast wall sections of $\frac{3}{16}$ to $\frac{1}{4}$ in. are sufficient in the rubber, and as the silicone is relatively expensive, the framing should be made to core out blank corners and other areas to conserve material.

One of the major problems encountered in the early days of silicone moldmaking was the incomplete curing of portions of the mold at the contact surface with the pattern. This is called *inhibition* and is easily recognized since the surface is gummy, reproduces no pattern detail and sometimes sticks to the pattern. The problem has not been solved, but at least we know some ways to avoid it. Sulfur-containing materials such as natural rubbers, some plastics

*Adapted from an article "Flexible molds cast from silicone rubbers," by Seymour S. Bodner, appearing in *Plastics Machinery & Equipment*, October 1972, and reprinted by courtesy of the publishers.

and some modeling clays have been known to cause inhibition. In addition, certain silicone rubbers and other flexible mold materials will produce inhibition. Generally, wood, metal, plastic, plaster and stone are compatible with RTV silicone rubber and can be used without inhibition. For other materials, a small patch test is advisable. Also, to be sure that the rubber mold releases from the pattern, use a mold release as specified for the type of formulation being molded.

Rigid Molds. In the casting of plastics parts, epoxy, phenolic, and polyester molds have been used (see also Cast Plastics for Tooling, p. 448). Cast epoxy molds have also been developed which have proved suitable for prototype tooling in low-pressure processes like structural foam molding and for moderate production runs in injection molding, vacuum forming, etc. In most instances, special epoxy formulations with high heat distortion temperatures are used, and it is most common to use aluminum fillers to improve impact strength and resistance to heat. The major drawback to the use of plastic molds is low thermal conductivity and the longer cooling cycles required.

FINISHING METAL MOLDS

Chrome-plating. This technique, as applied to molds, is common practice in the industry and offers molders the following advantages:

(1) Chrome plate has a low coefficient of friction, and thus facilitates the flow of plastic in the mold, and also the removal of the molded article.

(2) Chrome plate has excellent resistance to abrasion, and thus tends to protect the mold from wear.

(3) Chrome plate, on a previously well-polished surface, provides a bright smooth finish which is duplicated on the molded article.

(4) Chrome plate protects the surface of the steel mold against rusting during shut-downs and periods of storage.

A simple method of checking lack of coverage of chrome on steel in remote areas, or on worn molds, is to apply a solution of the following composition, at room temperature:

$$CuSO_4(5H_2O) \quad 2 \text{ oz/gal}$$
$$H_2SO_4 \quad 1 \text{ oz/gal}$$

If the copper plates out on the surface, this is evidence that there is no chrome present.

Thickness of chrome plate can be checked, on many items, by means of micrometers, etc. For more difficult shapes, many types of magnetic devices are available, some utilizing an AC connection and others relying on a calibrated magnet only. The plater and his customer should agree on a suitable instrument for checking the thickness.

Deviations from the thicknesses listed above may be desirable in certain cases. If the mold is made of prehardened steel, the thickness of the chrome plate should be greater; e.g., 0.001 in. instead of 0.0005 in. Cast steels should have a minimum of 0.001 in.

On beryllium-copper pressure castings, the deposit should be heavier on large pieces than on small.

On nickel electroforms used for injection molding, a thickness of 0.0003 to 0.0005 in. is ample, but for compression molding this should be as much as 0.002 in.

Nickel Plating. In molds for some processes requiring plated surfaces to facilitate mold release, as in sheet molding compound processing using low-shrink resins, evaluation is going on in the concept of nickel plating. Costs are comparable to chrome and the process is more flexible than chrome plating (i.e., minor alterations can be made and the tool nickel plated while still in the press). Nickel plating, however, unlike chrome plating, is not applicable to large parts. Nickel usually needs more time in the moldmaking process, but in parts involving a deep draw, it can do a better job than chrome.

Polishing of Molds. The making of a mold should be planned, from the start, with a view to the ultimate finish required.

Coarse tool-marks will lead to expensive bench work. The last step in each machining operation should be a shallow cut with a sharp tool, or grinding against a dressed wheel. Hand work should be done prior to hardening, while the metal is still relatively soft and workable. But at this point no effort should be made to achieve high luster.

If the mold is to be made by hobbing, both the hob and the blank should have a high polish, in order to minimize the need of polishing the resulting mold.

Also proper procedures in hardening can save many hours of labor in polishing.

For removal of scale resulting from heat treatment, polishing by hand may be effective, but it is unduly costly. Other methods are available which are effective without leading to embrittlement of the mold by hydrogen:

(1) use of solutions of proprietary alkaline chemicals, with periodic reverse of electric current;

(2) vapor blasting;

(3) immersion in ultrasonic pickling solution.

Vapor blasting is entirely satisfactory if the operator is trained to recognize the uniform, "shadow"-free appearance which indicates that the job has been completed. Complete removal of scale is essential if the surface is to be chrome-plated.

Equipment and materials for polishing should include: polishing stones from #100 grit to #800 grit, in a variety of sizes and shapes to fit any contour; sticks; felt bobs; bristle brushes; hard sewn buffs; soft muslin buffs; polishing cloths; polishing compounds and pastes; flexible-shaft grinding machines; and hand grinders.

A typical sequence of operations on steel, to produce a polish required on a mold for a lens or for a melamine dish, is as follows:

(1) Starting with the least coarse stone required by the initial condition of the surface, stone the surface with progressively finer stones, ending with #800.

(2) Polish with a bristle brush dressed with a lanolin-based paste and #900 grit, until all stoning lines have been eradicated.

(3) Polish with a hard-sewn wheel dressed with fast-cutting compounds for stainless steel.

(4) Polish with a soft wheel dressed with a final polishing abrasive such as chrome-oxide green rouge.

The foregoing procedure produces a mirror finish, with no pitting or dragging of the surface. If less than this is sufficient, the procedure may be cut short before completion.

Some mold-makers prefer diamond compounds as being faster. Here again the sequence follows from the coarse to the finest abrasive.

Molds of beryllium-copper are usually supplied with a vapor-blasted finish which is in many cases good enough. If a higher polish is required, a stoning and a polishing with a felt bob and tripoli will usually suffice. For aluminum, the same procedure is effective, followed by buffing with a soft muslin wheel and a lime compound.

Texturing Molds. Texturing, the opposite of polishing, involves the formation of a three-dimensional effect on the surface of a tool. It is designed to add a surface pattern to a molded part to improve its aesthetics, hide unsightly sink marks and flow lines, etc. Techniques for texturing include hand engraving, electroforming, casting, machine engraving, photochemical etching, and electrical discharge machining.

DESIGNING MOLDED PRODUCTS

Considering the enormous diversity of plastics materials and the processes by which they are made into products, it becomes a formidable task to try to summarize in a single chapter any basic rules of product design. While this chapter concentrates on products made by the injection or compression molding process, readers should note that different parameters apply when the part is blow molded or thermoformed or made by any of the reinforced plastics molding processes. Wherever possible, this type of information has been included in the specific chapter pertaining to an individual process (readers are especially recommended to the excellent design charts in Chapter 19 on Reinforced Plastics).

However, this guide to the fundamental principles of molded product design is useful as a starting point in understanding how plastics differ from other materials and how they can best be approached in achieving an economical and serviceable design. Again, however, the reader should be cautioned to be aware of the physical differences in the many plastics materials and the influences these will have on design criteria. Obviously, for example, a flexible plastic like polyethylene represents less of a problem in the design of a piece with undercuts (since it can be snapped out of the mold) than a rigid plastic like styrene. Similarly, the molding of structural foam plastics can call different criteria into play than the molding of solid plastic. In determining the minimum cross-sectional thickness for a structural foam part, for example, the designer must be aware that in thin sections (e.g., less than $\frac{3}{16}$-in.), there is a chance that there will be a relatively small percentage of cellular construction and such a section may well approach the density and mechanical properties of the solid material. Similarly, with foam, designers will find that thickness variations are less troublesome to obtain than when working with solid material. The low residual stress and the low residual density of the foam reduce the chance of warpage, shrinkage, and sink marks during cool-down, even in 1- to 2-in. sections.

Bearing all these variations in mind, however, this chapter will attempt to formulate the basic ground rules in designing a molded product. For a quick finger-tip review, readers are recommended to the two charts, Figs. 23-1 and 23-2, prepared by the Custom Molders Division of the Society of the Plastics Industry, Inc., as a key to successful design. In the pages that follow, each of these design criteria (e.g., bosses, ribs, fillets, tapers, etc.) is covered in more extensive detail.

The last third of the chapter is a discussion on designing plastics to avoid warpage. It is included so that readers may better understand the relationship between part design, mold design, and molding conditions in terms of their influence on the finished product. In this

STANDARDS AND PRACTICES OF PLASTICS CUSTOM MOLDERS	Engineering and Technical Standards

Keys to Successful Design of Plastics Molded Parts

Threads

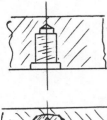

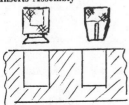

Molded threads should be of the larger sizes and coarse class. For good molding design, care should be taken to chamfer, counterbore, or recess at least the depth of one thread. Minimum size should be #6-32, class 1 or 2.

Cut threads, if possible, should be of the larger sizes coarse pitch and thread classes. Minimum size should be #6-32, class 1 or 2 and 60 per cent thread depth.

Inserts preferred for threads (brass or aluminum) when possible, especially where fine pitch, small diameter and tight tolerance classes are required. Thread should be reamed after tapping to facilitate molding.

Inserts Molded In

Where inserts are required to be molded in place, problems could develop due to plastic flashing in hole, or up-side of metal. Secondary operations may be necessary to remove flash.

Inserts Assembly

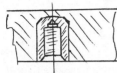

Inserts and holes in plastic can be designed in some cases to assemble insert at press while plastic is hot. Also, there are many screw machine type inserts that are assembled by expanding metal into plastic when parts are cold. Sonic pressing of inserts in some material is now possible.

Fillets and Radii

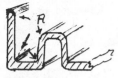

Avoid sharp corners where possible, and add maximum radii to strengthen part, to minimize strains, to assist flow of plastic in mold and to strengthen mold members.

Tapers

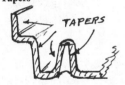

Tapers are necessary in plastic parts for good molding practice. Pins, projections, holes and cavities, where possible, should be tapered. Under 1″ length — 1° per side, over 1″ length — ½° per side.

Fig. 23-1.

Concentricity

The concentricity of a part must allow for maximum T.I.R. tolerance. The geometry of the part governs tolerances. However, minimum .005" for 1" diameter, others in proportion.

Out-of-Roundness

Many designs of shells, cups and rings will distort and go out-of-round; however, the geometry of the part and the gating methods are controlling factors. Design for approximately .007" on 1" diameter, other sizes in proportion.

Gates

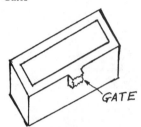

Gates are necessary for most plastic moldings. However, some type gates through necessity are in an objectionable position and must be removed. Secondary operations can become costly. We suggest the following allowance (recessed or protruding gate scar).

Break gate ± .025"
Clip gate ± .010"
Machine gate ± .005"

Mismatch Designs

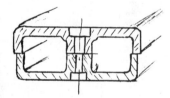

Design mating parts with a planned mismatch. It will prevent many problems. Also plan mismatch for hole alignment.

Wall Sections

Design uniform sections and walls. Heavy and non-uniform sections are subject to sinks in thermoplastic material. Minimum and maximum wall sections depend upon the geometry of parts.

Ribs and adjoining wall should be approximately 60 per cent of the normal wall to prevent sink.

Flatness and Warpage

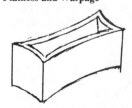

Design to allow for warpage in your plastic parts where possible. Warpage or distortion will vary with the type plastic. Allow .010" for part 3" long. Other lengths in proportion.

Fig. 23-2.

instance, the phenomena involved is known as material shrinkage—usually defined in plastics as the difference between corresponding linear dimensions of the mold and the molded piece.

Designing the part*

To insure proper design, close cooperation is required between the industrial designer, the engineer, the draftsman, the tool-builder, the molder, and the raw material supplier. Preferably, each must become involved to some extent at the very beginning of the design process. A general step-wise procedure for the development of a commercial part would be about as follow:*

Define the end-use requirements: As an initial step, the designer should list the anticipated conditions of use and the performance requirements of the article to be designed. He may then determine the limiting design factors and, by doing so realistically, avoid the various pitfalls which can cause loss of time and expense at a later stage of development. Use of the check list in Table 23-1 will be helpful in defining the various design factors.

Drafting the preliminary design: With the end-use requirements meaningfully defined, the designer is ready to start developing designs using the properties of the plastic being considered.

Prototyping the design: This gives the designer an opportunity to see his product as a three-dimensional object, and his first practical check of the engineering design. Perhaps the most widely used technique for making prototypes is to machine them from rod or bar stock. In some cases, it may be preferable to build a sample cavity of a multi-cavity mold. The sample cavity technique not only allows economical production of a quantity of prototype parts but also serves to develop mold design data for building the production mold. Another method of prototype fabrication is to injection mold parts in existing diecast, compression or transfer tools. However, the surface finish and dimensions of parts thus produced often must be modified by machining.

* Adapted from a brochure on "Designing with Delrin acetal resins." published by E. I. du Pont de Nemours & Co., Inc.

Table 23-1. Check list for product design

General Information
 What is the function of the part?
 How does the assembly operate?
 Can the assembly be simplified by using plastics?
 Could it be made and assembled more economically?
 How will it be made and assembled?
 What tolerances are necessary?
 What space limitations exist?
 What service life is required?
 Is wear resistance required?
 Is light weight desirable?
 Are there acceptance codes and specifications such as SAE, UL?
 Do analogous applications exist?

Structural Considerations
 How is it loaded in service?
 Magnitude of loads?
 For how long will it remain in service?
 How much deflection can be tolerated in service?

Environment
 Operating Temperature?
 Chemicals, solvents?
 Humidity?
 Service life in the environment?

Appearance
 Style?
 Shape?
 Color?
 Surface finish?
 Decoration?

Economic Factors
 Cost of present part?
 Cost estimate of part in plastic?
 Will redesign of the part simplify the assembled product and thus give rise to savings in installed cost?
 Are faster assemblies and elimination of finishing operations possible?

Testing the design: Every design should be subjected to some form of testing while in the prototype stage to check the accuracy of calculations and basic assumptions.

Actual end-use-testing of a part in service is the most meaningful kind of prototype testing. Here, all of the performance requirements are encountered, and a complete assessment of the design can be made.

- Simulated service tests are often conducted with prototype parts. The value of this type of testing depends on how closely the end-use conditions are duplicated. An automobile engine part might be given temperature, vibration and hydrocarbon resistance tests; a luggage fixture could be subjected to impact and abrasion tests, and a radio component might undergo tests for electrical and thermal insulation.
- Accelerated tests of a mechanical or chemical nature often are used as a basis for prototype evaluation. Such procedures, when used by an experienced and qualified person, can be very meaningful.
- Standard test procedures such as those developed by the ASTM generally are useful as a design guide but, normally, cannot be drawn upon to predict accurately the performance of a part in service. Again, representative field testing may be indispensable.

Taking a second look: A second look at the design helps to answer the basic question: "Will the product do the right job at the right price?" Even at this point, most products can be improved by redesigning for production economies or for important functional and aesthetic changes; weak sections can be strengthened, new features added, and colors changed. Substantial and vital changes in design may necessitate complete evaluation of the new design. If the design has held up under this close scrutiny, specifications and details of production can be established.

Writing meaningful specifications: The purpose of a specification is to eliminate any variations in the product that would prevent it from satisfying the functional, aesthetic or economic requirements. The specification is a complete set of written requirements which the part must meet. It should include such things as: method of fabrication, dimensions, tolerances, surface finish, parting line location, flash, gating, locations where voids are intolerable, warpage, color, decorating and performance specifications.

Setting up production: Once the specifications have been carefully and realistically written, molds can be designed and built to fit the processing equipment. Tool design for injection molding should be left to a specialist or able consultant in the field, because inefficient and unnecessarily expensive production can result from improper design of tools or selection of manufacturing equipment.

Controlling the quality: It is good inspection practice to schedule regular checking of production parts against a given standard. An inspection check list should include all the items which are pertinent to satisfactory performance of the part in actual service to its assembled cost. The end-user and molder should jointly establish the quality control procedures that will facilitate production of parts within specifications.

BASIC DESIGN THEORY

In the sections that follow, the individual design criteria relating to the placement of ribs, bosses, radii, etc., are discussed in detail.

Designers should be aware, however, of the uniqueness of plastic materials as used in molding processes. Chemically, the molecules of thermoplastics consist of long chains of repeating units. When melted and injected under high pressures into a closed mold, the polymer withstands forces and undergoes changes. Injection molding has been compared by some to stuffing coil springs into a cavity. If the cavity has generally rounded and uniform contours, it is relatively easy to fill. If, however, it has sharp corners and thick-and-thin areas, it will not only be more difficult to fill but, when filling is completed, the springs will be more compressed, stretched, and distorted. When the analogue of this happens in molding plastics, a part is said to contain *molding strains*. Molding strains, though undesirable, are present to some degree in every plastic part. When a molding is heated, it warps because it is pulled in different directions by the strains within it and is less able, because of softening (when the temperature is raised sufficiently), to maintain its shape.

When the polymer is injected into every portion of the mold cavity with about equal force and uniformly cooled, the distribution of internal stresses will tend to balance each other and yield a part with less tendency to warp. High levels of mold strain, on the other hand, are especially detrimental when the part is subjected to further external strain or stress, whether it be physical force, heat, or a stress crack agent (i.e., certain materials which disturb the intermolecular bonds of strained molecules and provide an 'opening wedge' for a crack to develop).

Wall Thickness

Under favorable conditions, the design of wall thickness normally depends upon the selection of the material. Occasionally, however, limitation of space precludes this, and the selection of material becomes predicated partly upon the wall thickness available. Whichever the path of approach, the determination of wall thickness should be the result of an analysis of the following requirements:

Requirements of Use	Manufacturing Requirements
1. Structure	1. Molding
2. Weight	a. Flow
3. Strength	b. Setting
4. Insulation	c. Ejection
5. Dimensional stability	2. Assembly
	a. Strength
	b. Precision

The foregoing requirements are intimately related. Purely from an economic standpoint, a wall thickness which is too great or too small can affect the design adversely.

Plastic parts should be designed with the minimum wall thickness that will provide the specific structural requirement. This thickness results in an economy of material and high production levels due to the rapid transfer of heat from the molten polymer to the cooler mold surfaces. Occasionally, the use of the article may involve little need of strength, but still adequate strength must be provided to withstand ejection from the mold and to facilitate assembly operations.

Tables 23-2 and 23-3 give preferred minimum, average, and maximum wall thicknesses for the various thermoplastics and thermosetting plastics.

It should also be remembered that molding phenomena, such as flow and cure, can influence the choice of wall thickness. Basically, wall thickness should be made as uniform as possible to eliminate part distortion, internal stresses, and cracking. Figure 23-3 illustrates the incorporation of uniform section thickness in part design. If different wall thicknesses must be used in a part, blend wall intersections gradually, such as shown in Fig. 23-4. When possible, the material should flow from thick to thin sections (filling thick sections from thin ones can result in poorly molded parts). Also, under such circumstances, consideration should be given to the use of assembly techniques for assembling two or more molded components to make the desired part.

As indicated earlier, it is important to understand flow in determining wall thickness. If consideration is given to flow within the mold, one can see that it is possible for material to flow past a pocket or depression leaving this area to be filled later as pressure builds up. It is also possible, due to restrictions of flow, for the material to momentarily hesitate until sufficient pressure has been developed to overcome resistance to flow, and then spurt forward causing part of the material to be chilled. Re-mixing with the oncoming melt can result in surface defects in both cases.

The formation of welds relates directly to wall thickness and is an important controlling factor. In molding terminology, a weld is created when two fronts of molten material meet, such as when the flow is interrupted by a pin. If the interruption is long or if the wall is thin, poor flow may cause a weak weld.

Flow within the mold could be severely restricted in walls thinner than the minimums recommended in Tables 23-2 and 23-3, and the mold might not fill properly. At the other extreme, voids or sink marks could develop in very thick cross sections.

Another factor to be taken into consideration in determining wall thickness is the process of curing—a function of heat-transfer, from or to the mold. Obviously,

Table 23-2. Suggested wall thicknesses for thermoplastic molding materials*

Thermoplastic Materials	Minimum Inches	Average Inches	Maximum Inches
Acetal	.015	.062	.125
ABS	.030	.090	.125
Acrylic	.025	.093	.250
Cellulosics	.025	.075	.187
FEP fluoroplastic	.010	.035	.500
Nylon	.015	.062	.125
Polycarbonate	.040	.093	.375
Polyethylene (L. D.)	.020	.062	.250
Polyethylene (H. D.)	.035	.062	.250
Ethylene vinyl acetate	.020	.062	.125
Polypropylene	.025	.080	.300
Polysulfone	.040	.100	.375
Noryl (modified PPO)	.030	.080	.375
Polystyrene	.030	.062	.250
SAN	.030	.062	.250
PVC–rigid	.040	.093	.375
Polyurethane	.025	.500	1.50
Surlyn (ionomer)	.025	.062	.75

* Reprinted from "Plastics Product Design," by Ronald D. Beck, with permission of the publishers, Van Nostrand Reinhold Co.

Table 23-3. Suggested wall thicknesses for thermosetting molding materials.*

Thermosetting Materials	Minimum Thickness	Average Thickness	Maximum Thickness
Alkyd–glass filled	.040	.125	.500
Alkyd–mineral filled	.040	.187	.375
Diallyl phthalate	.040	.187	.375
Epoxy glass	.030	.125	1.00
Melamine–cellulose filled	.035	.100	.187
Urea–cellulose filled	.035	.100	.187
Phenolic–general purpose	.050	.125	1.00
Phenolic–flock filled	.050	.125	1.00
Phenolic–glass filled	.030	.093	.750
Phenolic–fabric filled	.062	.187	.375
Phenolic–mineral filled	.125	.187	1.00
Silicone glass	.050	.125	.250
Polyester premix	.040	.070	1.00

* Reprinted from "Plastics Product Design," by Ronald D. Beck, with permission of the publishers, Van Nostrand Reinhold Co.

thin sections set more rapidly than thick ones and, since contraction or shrinkage occurs simultaneously with setting, irregularity in thickness causes irregularity in contraction and creates internal stresses. These will tend to relieve themselves either by forming concave depressions, known as "sink marks," on the thick sections, or by causing warping. This difficulty can frequently be eliminated by coring a thick section so as to divide it into two thin sections, when that is feasible.

Core removal should be designed parallel to the motion of the mold platen. Cores at right angles to the platen movement require cam or hydraulic actions which increase mold cost. Movable and loose-piece cores are expensive, but they can be used to mold internal undercuts and threads.

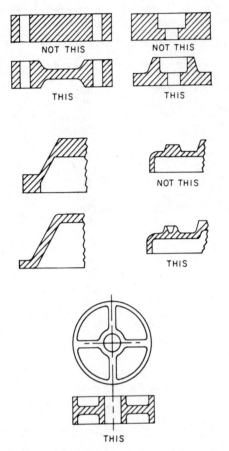

Fig. 23-3. Typical methods for maintaining uniform wall thickness.

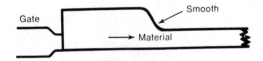

Fig. 23-4. Gradual blending between different wall thicknesses. (*Courtesy General Electric*)

Core size is generally determined by part design. As an example, the following are recommendations on core design in polycarbonate materials. The recommended core length for blind cores larger than 0.187 in. in diameter is 2.5 times the diameter. For blind cores smaller than 0.187 in. in diameter, the suggested core length is twice the diameter. Coring recommendations are summarized in Fig. 23-5. The length of cored-through holes should not exceed six times the diameter for core

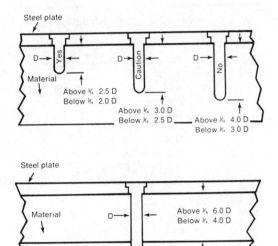

Fig. 23-5. Coring recommendations for polycarbonate. (*Courtesy General Electric*)

diameters greater than 0.187 in. and four times the diameter for core diameters smaller than 0.187 in. Draft should be added to all cores. If the polycarbonate parts require deep holes of small diameter, it may be necessary to drill the holes rather than mold them. Drilling usually proves more satisfactory than trying to maintain alignment of fragile core pins. Also, if threads are required, the holes can be tapped during the drilling set-up. See Fig. 23-6.

Similar recommended practices for other plastics are available from appropriate suppliers. Figures 23-7 and 23-8 are included to indicate how these wall thickness selection criteria are put to successful commercial use in the design of ABS chairs and grilles.

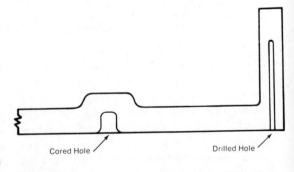

Fig. 23-6. Coring-out thick wall sections. (*Courtesy General Electric*)

tration at these points and produces a molding with greater structural strength. Fillets provide streamlined flow paths for the molten polymer in filling the mold and permit easier ejection of the part from the mold. Also, the use of radii and fillets produces economies in the production of the mold, since they are easier to machine and less subject to damage. All inside and outside rounded corners should have as large a radius as possible to reduce stress concentrations. The recommended minimum radius of 0.020 to 0.030 in. is usually permissible even where a sharp edge is required, and the radius ensures more economical, longer-life molds. A larger radius should be specified wherever possible.

Figure 23-9 illustrates the effect of a fillet radius on stress concentration. Assume a force "P" is exerted on the cantilever section shown. As the radius "R" is increased, with all other dimensions remaining constant, R/T increases proportionally, and the stress-concentration factor decreases as shown by the curve. The stress-concentration factor has been reduced by 50% (3.0 to 1.5) by increasing the ratio of fillet radius to thickness six-fold (from 0.1 to 0.6). The figure illustrates how readily the stress-concentration factor can be reduced by using a larger fillet radius. A fillet of optimum design is obtained with an R/T of 0.6, and a further

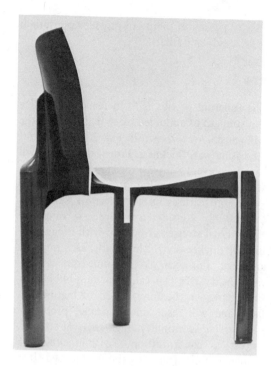

Fig. 23-7. Cross-section of wall design in molded ABS chair. (*Courtesy Borg-Warner Chemicals*)

Fillets and Radii

Sharp corners in plastic parts are perhaps the greatest contributors to part failure. Elimination of sharp corners reduces the stress concen-

Fig. 23-8. Different wall thicknesses in a molded ABS grille. (*Courtesy Borg-Warner Chemicals*)

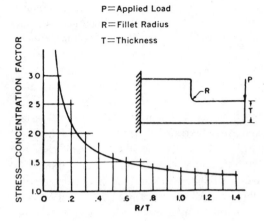

Fig. 23-9. Effect of a fillet radius on stress concentration. (*Courtesy E. I. du Pont de Nemours & Co., Inc.*)

increase in radius reduces the stress concentration only a marginal amount. This is true in general for most shapes; however, other ratios may have to be used on specific parts because of other functional needs.

The overall advantages of radii and fillets in the design of a molded article are several:

(1) The molded article is stronger and more nearly free from stress.
(2) Elimination of the sharp corner automatically reduces the hazard of cracking as a result of notch-sensitivity, and increases the over-all resistance to sudden shock or impact. Elimination of all internal sharp corners by using a radius of 0.015 in. or $\frac{1}{64}$ in. will greatly improve the strength.
(3) The flow behavior of the plastic will be greatly improved. Rounded corners permit uniform, unretarded and less stressed flow of plastic into all sections of the mold, and will improve the uniformity of density of the molded sections.
(4) Mold members, force plugs and cavities will be stronger because there will be less tendency for these parts of the mold to develop internal stresses due to notch-sensitivity created by concentration of stress at sharp corners. Mold parts frequently crack in hardening as

a result of failure to provide appropriate radii and fillets.

Ribs

The function of ribs is to increase the rigidity and strength of a molded piece (e.g., such as the ABS postal tray shown in Fig. 23-10) without increasing wall thickness. Proper use of ribs will usually prevent warpage during cooling and, in some cases, they facilitate flow during molding.

Several features in the design of a rib must be carefully considered in order to minimize the internal stresses associated with irregularity in wall thickness. Width, length, etc., must be analyzed. For example, in some applications, thick heavy ribs can cause vacuum bubbles or sink marks at the intersection of mating surfaces and will result in appearance problems, structural discontinuity, high thermal stresses, and stress concentration. To eliminate these problems, long, thin ribs should be used. It is also possible to core ribs from the underside.

In general, the width of the base of a rib would be less than the thickness of the wall to which it is attached. This can be demonstrated by examination of Fig. 23-11. Ribs of the proportions shown are frequently used in molded articles. However, when the circle R_1 is placed at the junction of the rib and wall, it will be seen that the thickness is increased by about 50%, and this may produce sink marks on the surface under normal molding conditions. By reducing the width of the base of the rib one-half the wall thickness, as shown in Fig. 23-12, the increase in thickness at the junction

Fig. 23-10. Rib design in a molded ABS postal tray. (*Courtesy Borg-Warner Chemicals*)

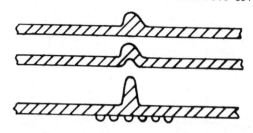

Fig. 23-13. Disguise for sinks. Section design (top), sink on molded part (center), and disguise (bottom). Disguise can consist of engraving emblems or corrugated patterns into the cavity. (*Courtesy E. I. du Pont de Nemours & Co., Inc.*)

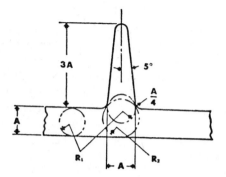

Fig. 23-11.

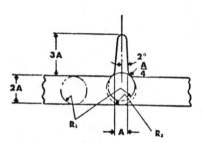

Fig. 23-12.

is made less than 20%. Sink marks are improbable when these proportions are used. The use of two or more ribs is better than to increase the height of a single rib more than is shown. When two or more ribs are required, the distance between them should be greater than the thickness of the wall to which they are attached.

The base of a rib may be tapered in cross section for easy ejection from the mold. Ribs or beads on side walls may be perpendicular to the parting line to ensure easy ejection. Spaced ribs or a groove should be added to the wall surface behind a bead of any size, if appearance is important, because large beads may cause sink marks on this surface during cooling. A fillet should be used where the rib joins the wall to minimize stress concentration and provide additional strength.

Figure 23-13 illustrates the effect of heavy beads on part surface and shows one solution (the bottom cross section) to the problem of surface imperfections (i.e., sink marks may be disguised by engraving emblems or corrugated patterns into the cavity so that these will be reproduced in the molded part).

Where wall thicknesses are increased to provide added strength, as for bosses surrounding holes, or where machining is to be done, varying thicknesses should be blended gradually into each other.

In some cases, it may be desirable to corrugate a surface instead of using ribs to increase stiffness and support. Another way to stiffen a panel is to give it a slight dome shape. A fairly flat conical shape may serve the same purpose more economically if the area is circular.

Undercuts

Modern molding practice dictates certain principles of design which should be observed if molded articles are to be produced successfully. Most elementary is the fact that the piece must be easily removed from the mold after it is formed. The point is frequently overlooked and many products are designed with undercuts which makes it impossible to eject them directly from the mold cavity. If undercuts are essential, then split molds or removable mold sections are required, and these increase the cost of molds and of the molded articles. Typical devices are the side pull core used for external undercut molding (see Fig. 23-14) and split-pin molding for internal undercuts (as shown in Fig. 23-15, in which a part of the pin is located in each mold platen).

Basically, undercuts can be classified as external or internal. As the names imply, undercuts located in the outside contours of the piece are called *external undercuts*; located on the inside contours, they are called *internal undercuts*.

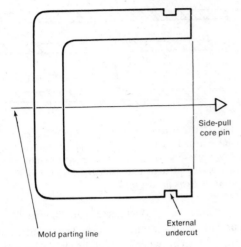

Fig. 23-14. Side pull core for external undercut molding. (*Courtesy General Electric*)

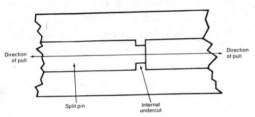

Fig. 23-15. Split-pin molding for internal undercuts. (*Courtesy General Electric*)

An example of external undercuts is shown in Fig. 23-16. This relatively simple shape must be produced in a split-cavity mold to permit ejection of the part. When undercuts such as these are necessary, strict attention must be given to mold parting lines if appearance is a

major consideration. It is possible to produce such parts with almost invisible parting lines through the use of molds that are precisely fitted.

When external undercuts are very shallow it may be possible to snap or strip the part out of the cavity. When contemplating this technique, the mold must be designed to operate so that the core shown in Fig. 23-17 has completely cleared the part before ejection occurs. This is necessary since the part is placed in compression and the inside surface must be free of restraint. To minimize part distortion, it may be desirable to provide ring or plate ejection, rather than ejector pins.

Some of the important variables to be considered when stripping a part are the allowable part strain, mold temperature, molding cycle, forces required for ejection, and undercut geometry. Since the definition of these parameters varies with resin and part geometry, it is suggested that experienced processors be consulted before undercut specifications are finalized.

Internal undercuts as illustrated in Fig. 23-18 can be provided in injection molding. However, as with external undercuts, tool and

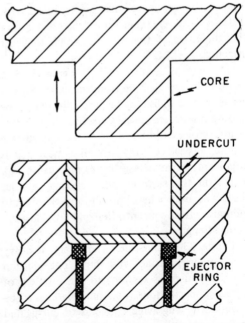

Fig. 23-17. Mold for external undercuts. (*Courtesy E. I. du Pont de Nemours & Co., Inc.*)

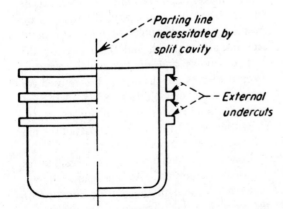

Fig. 23-16. External undercuts. (*Courtesy E. I. du Pont de Nemours & Co., Inc.*)

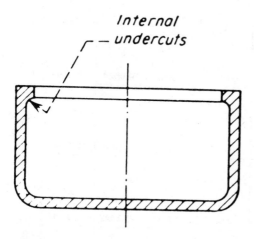

Fig. 23-18. Internal undercuts. (*Courtesy E. I. du Pont de Nemours & Co., Inc.*)

processing costs may be increased when internal undercuts are specified, particularly if collapsible cores are required.

As with external undercuts, it may be possible to snap or strip internal undercuts from the mold. With internal undercuts the part must be free to stretch around core projections during ejection. The basic concepts presented for stripping external undercuts also apply to internal undercuts. Again, suggestions and recommendations from experienced processors and suppliers are highly desirable.

Taper or Draft

In the design of articles produced from moldable rigid and elastomeric plastics, it is important that consideration be given to the easy removal of the piece from the mold cavity. Draft or taper should be provided, both inside and outside. Also, it is important that the surface of the molded piece, and hence that of the mold, particularly on the vertical or tapered walls, be polished by a high luster finish. If the molded piece is straight-sided or vertical, it is necessary to exert a strong pull in order to open the mold, but by the use of taper and a highly polished surface the article is easily released.

In deep-drawn articles, converging tapers assist by creating a wedging or compressing action as the mold is closed, and, in compression molding of thermosetting materials, tapers

of this type increase the density of the plastic in the upper sections.

There are no precise calculations or formulas for taper. The amount of taper required will vary with the depth of draw, and common sense will dictate how much. When the draw is relatively deep and the part shape complex, for example, one would generally allow more draft on internal walls. Draft will also vary according to the molding process, wall thickness, and the particular plastic being molded.

Since a designer should be on the alert to avoid details in design which will obstruct free ejection from the mold, he should provide the most liberal taper which the design will tolerate.

Figure 23-19 is a table of relation of degree of taper per side to the dimension, in in./in. It illustrates the effect of taper for various depths of piece (as shown in Fig. 23-20). A piece 4 in. in depth can carry a 4-degree taper, or 0.2796 in. per side. A 4-degree taper may be more than can be used in a deep article. Thus, in a 10-in. piece, it will amount to 0.6990 in. per side.

Bosses

Bosses are protruding studs or pads and are used in design for the reinforcement of holes or for mounting an assembly. The same general precautions to be considered in the design and use of ribs may apply to the utilization and design of bosses (Fig. 23-21). Wherever practicable, they should be located at the apex of angles where the surface contours of the part changes abruptly (e.g., corners) to get a balanced flow of material. If possible, use beads as material runners.

It is considered good design practice to limit the height of the boss to twice its diameter to obtain the required structural strength. But higher bosses can be provided in an injection molded part at the discretion of the designer.

In compression molding, the use of high bosses tend to trap gas, which decreases both the density and the strength of this molded section. This is particularly true in straight compression molding and where the boss is formed by the upper half or plunger of the mold.

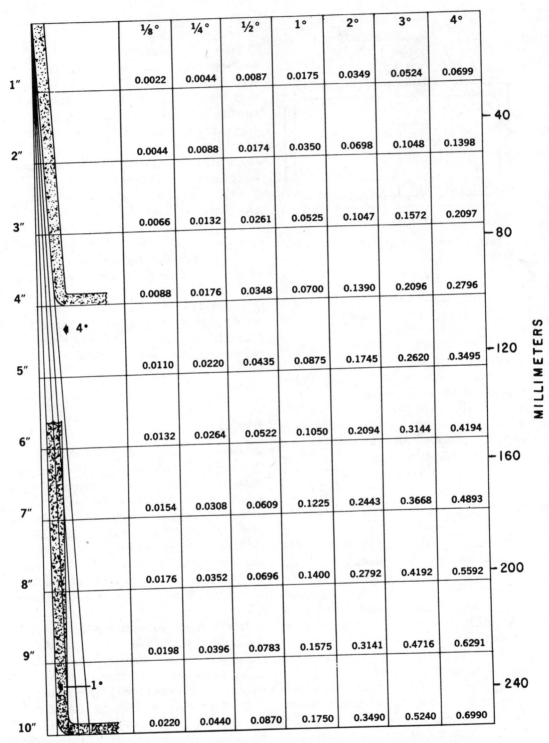

	⅛°	¼°	½°	1°	2°	3°	4°
1"	0.0022	0.0044	0.0087	0.0175	0.0349	0.0524	0.0699
2"	0.0044	0.0088	0.0174	0.0350	0.0698	0.1048	0.1398
3"	0.0066	0.0132	0.0261	0.0525	0.1047	0.1572	0.2097
4"	0.0088	0.0176	0.0348	0.0700	0.1390	0.2096	0.2796
5"	0.0110	0.0220	0.0435	0.0875	0.1745	0.2620	0.3495
6"	0.0132	0.0264	0.0522	0.1050	0.2094	0.3144	0.4194
7"	0.0154	0.0308	0.0609	0.1225	0.2443	0.3668	0.4893
8"	0.0176	0.0352	0.0696	0.1400	0.2792	0.4192	0.5592
9"	0.0198	0.0396	0.0783	0.1575	0.3141	0.4716	0.6291
10"	0.0220	0.0440	0.0870	0.1750	0.3490	0.5240	0.6990

MILLIMETERS

Fig. 23-19. Relation of degree of taper per side to the dimension in inch/inch.

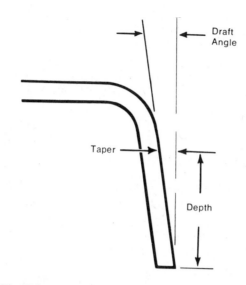

Fig. 23-20. Relationship of draft angle, depth, and taper. (*Courtesy General Electric*)

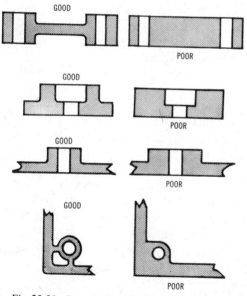

Fig. 23-21. Boss design. If possible, bosses should be located at the junction of two or more surfaces to favor balanced flow of material into the mold cavity. (*Courtesy E. I. du Pont de Nemours & Co., Inc.*)

Ribs may be employed on the sides of the boss to assist the flow of the material. In many cases, bosses are interlocked with other bosses or to side walls to provide structural stability. When the boss is in contact with external surfaces, proper rib design is essential to avoid sink marks on appearance surfaces. Fillets at

the junction of the boss to the wall section are important. In some instances, where possible, a bubbler should be used on the cavity side of the mold opposite the boss to minimize sink on the outside wall.

Data are available from various suppliers (as in Fig. 23-22 for polycarbonate) that provides specific information on boss design for individual plastics.

Holes

For a variety of reasons, holes are often required in a molded piece. They should be designed and located so as to introduce a minimum of weakness and to avoid complication in production. This requires the careful consideration of several factors.

The distance between successive holes or between a hole and adjacent side wall, should be at least equal to the diameter of the hole as shown in Fig. 23-23. It is always wise to provide as thick a wall section as is practicable, since cracking around holes in assembly is

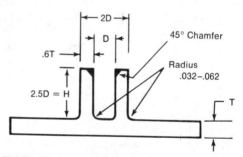

Fig. 23-22. Designing bosses for a polycarbonate part. (*Courtesy General Electric*)

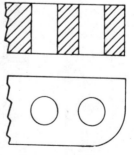

Fig. 23-23. Spacing recommendations for holes. (*Courtesy E. I. du Pont de Nemours & Co., Inc.*)

generally traceable to a disregard of this fundamental consideration.

The problem of design is always more complicated when a threaded hole must be used, because of the concentration of stress which causes notch-sensitivity in the region immediately surrounding the hole. Laboratory determinations provide a wealth of evidence indicating that the linear distance between the edge of the hole and the edge of the piece should be three times the diameter of the hole, if the stress at the edge of the hole is to be reduced to a safe working figure. As shown in the section on Bosses, increased boss areas are often used to provide additional strength.

Through Holes. Through holes are usually more useful for assembly than blind holes, and also they are easier to produce. Through holes should be preferred whenever possible, because the mold pins which form the holes can then be supported in both parts of the mold. Through holes may be produced by either a single pin supported at each end or by two pins butted together. The first method is generally considered the better. Where the two-pin method is used, one pin should be slightly larger in diameter than the other to compensate for any misalignment (see Fig. 23-24). The butting ends of the pins should be ground flat, and provisions must be made in the mold assembly for sufficient clearance to avoid upsetting the ends of the pins. While opinion differs regarding the amount required, a clearance of 0.005 in. appears to be considered good practice.

Blind Holes. The designer must remember that a blind hole is formed by a core pin, which through necessity can be supported only at one end. Thus this pin can be distorted, bent or sheared off during the molding operation as a result of the unbalanced pressure exerted by the flow of the plastic. The depth of blind holes, i.e., the length of the core pin, should be limited to twice the diameter of the hole, unless the diameter is $\frac{1}{16}$ in. or less, in which case the length should not exceed its diameter (see Fig. 23-25).

Drilled Holes. Often it will be found less expensive to drill holes after the molding, rather than attempt to mold them, particularly when they must be deep in proportion to their diameter. Broken and bent core pins are an expensive item.

Good manufacturing practice calls for provision of drill jigs for the accurate drilling of molded pieces. Drill jigs will do much to reduce the cost of broken drills and too-frequent resharpening. However, in some instances it is possible to provide spot points on the molded part which can be used to locate the holes, so that jigs are not needed.

Threaded Holes. See discussion of threads, below.

Side Holes. Side holes are difficult to produce and present problems which are not easily solved, because they create undercuts in molded pieces. Holes which must be molded at right angles to each other necessitate split molds or core pins and therefore are more costly, particularly in compression molding. The core pins, being not parallel to the applied pressure, may be distorted or sheared off, particularly if of small diameter. Also, the

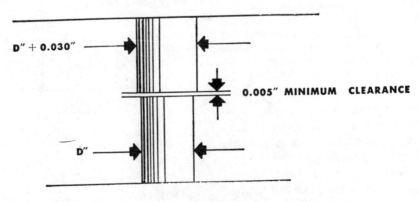

Fig. 23-24.

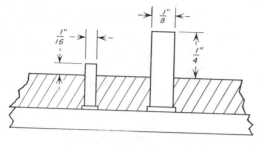

Fig. 23-25.

molding of such holes adds to the time required for molding, in that it is necessary to withdraw these core pins from the molded piece before it can be removed from the cavity. These problems are less serious when transfer or injection methods are used, but there remains the necessity of withdrawing the core pins from the molded piece prior to its removal.

Another problem brought about by the presence of core pins is imperfect welding of the plastic, in the area back of or adjacent to the pin and on the side opposite to the direction of flow. This condition will be encountered more frequently in thermoplastic materials, but exists also in pieces made from thermosetting materials in a compression mold.

Other Design Considerations. A stepped hole will often permit more depth than can be obtained with a single-diameter hole. Notching of a side wall may also be utilized to reduce the depth of a hole. These two techniques are illustrated in Fig. 23-26.

Holes which run parallel to the parting line may increase the cost of the mold because of the necessity of retractable core pins or split dies. Frequently, this problem is overcome by placing holes in walls which are perpendicular

to the parting line, using steps or extreme taper on the wall. Several of these possibilities are shown in Fig. 23-27.

Allowance should be made for a minimum $\frac{1}{64}$ in. (0.4 mm) vertical step at the open end of holes. A perfect chamfer or radius at the open end of a hole demands precision in the mold which may be unattainable from a practical or economic standpoint. This is illustrated in Fig. 23-28.

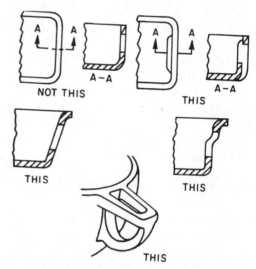

Fig. 23-27. Recommendations for placing holes in walls. Bottom sketch shows hole perpendicular to parting line. (*Courtesy E. I. du Pont de Nemours & Co., Inc.*)

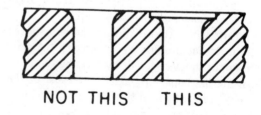

NOT THIS   THIS

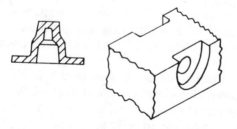

Fig. 23-26. Stepped hole (left); counterboring (right). (*Courtesy E. I. du Pont de Nemours & Co., Inc.*)

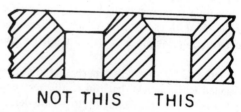

NOT THIS   THIS

Fig. 23-28. Chamfer on open holes. (*Courtesy E. I. du Pont de Nemours & Co., Inc.*)

External and Internal Threads

External and internal threads can be produced economically in plastic parts with modern molding techniques (or by machining or self-tapping). Screw threads, produced by the mold itself (using rotating core pins or loose-piece inserts) eliminate expensive post-molding threading operations. Although almost any thread profile can be produced, the Unified Thread Standard is the most suitable for molded parts. This form eliminates the feather edge at both the root and tip of the thread. The Unified Thread Standards are divided into the following classes:

Class 1A, 1B*—Most threaded nuts and bolts fall in this class and are used for most assembly work.

Class 2A, 2B*—A tighter fit than Class 1 with no looseness and widely used.

Class 3A, 3B*—Used in precision work and obtainable under exacting molding conditions with constant dimensional checking.

Coarse threads can be molded easier than fine threads, and threads less than 32 pitch should be avoided. Generally, threads of Class 1 or 2 are adequate for most applications. In many cases it is desirable to provide an interference fit between two threaded parts. This is easily accomplished with plastic materials and helps prevent a nut, bolt, or screw from loosening under mechanical vibration.

Parts with external threads can be removed from the mold either by unscrewing the parts from the mold cavity or by locating the mold parting line along the axial center-line of the screw profile. Internal threads can be produced by a threaded core which is unscrewed from the part or, if a part has internal threads of short engagement (few threads), it may be possible to strip the part over the threaded core, thus eliminating the unscrewing mechanism. An example of a threaded part which utilizes this economical principle is a cap for bottles and jars.

Figure 23-29 shows that parts should not be designed with threads that run to the very end of the part, nor completely up to a shoulder. A

* The letter A refers to the external thread and B to the internal thread.

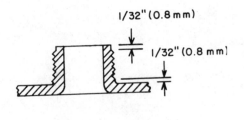

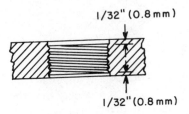

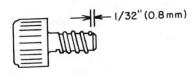

Fig. 23-29. Thread design. (*Courtesy E. I. du Pont de Nemours & Co., Inc.*)

clearance of at least $\frac{1}{32}$ in. (0.8 mm) should be provided at each end of the thread. The wall thickness supporting either an external or internal thread should be at least equal to the depth of the thread.

Inserts

Inserts in plastic parts can act as fasteners, load supports, or may simplify handling or facilitate assembly. Inserts may be functional or purely decorative, but they should be used sparingly since they increase costs.

Inserts derive a good deal of their holding power from the fact that plastic materials, when cooling, shrink around a metal insert.

Design of inserts must assure a secure anchorage to the plastic part to avoid rotation as well as pull-out. Sharp corners on inserts may result in areas of stress concentration and should be avoided. For techniques for securing inserts, see Fig. 23-30.

Knurling should be deep enough to permit material flow into the depressions and should be provided with a smooth surface where an insert protrudes from the molded part.

For ease of molding, inserts should be located perpendicular to the parting line. In-

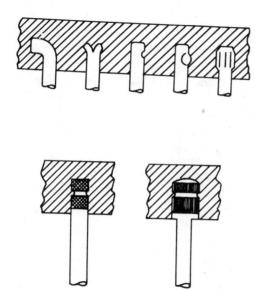

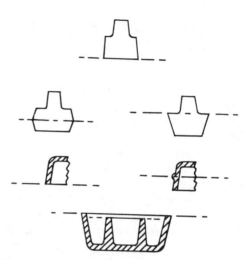

Fig. 23-31. Parting line locations. (*Courtesy E. I. du Pont de Nemours & Co., Inc.*)

Fig. 23-30. Techniques for securing inserts. (*Courtesy E. I. du Pont de Nemours & Co., Inc.*)

serts should be located so that they are not displaced during the injection portion of the molding cycle. Pin-supported inserts require a minimum hole diameter of $\frac{1}{8}$ in. (3.17 mm) to avoid damage to pins under molding pressures. For applications where the insert O.D. is less than $\frac{1}{4}$ in. (6.35 cm) the boss outside diameter should be at least twice that of the insert O.D. Radial ribs may be used to strengthen the boss and insert.

Parting Lines

Parting lines on the surface of a molded object, produced by the parting line of the mold, can often be concealed on a thin inconspicuous edge of the part. This preserves the good appearance of the molding and in most cases, eliminates the need of any finishing. Figure 23-31 shows parting line locations on various part configurations. Close coordination between the designer and molder in the early stages of design will usually resolve the best location of parting line.

In compression molds, the closing or telescoping of the two parts of the mold results in flow of material into the clearance between these parts. This material is known as flash and occurs at the parting line of the molds. The removal of flash from the article leaves a flash line, which is unavoidable and generally unsightly. The problem of flash lines is one that requires careful consideration by the designer, as the attractiveness of the product may depend to a large degree on a careful location of the flash line where it will not be seen. When a piece is molded in a transfer or injection mold, the problem of flash is greatly simplified, because the mold is already locked in closed position before the plastic enters it.

When flash lines are located improperly, the molded article may be marred either during its removal from the mold or later at the finishing bench, because the design does not permit easy removal of the flash. Articles of irregular shape must, as a rule, be smoothed along the flash line with a hand file, and then polished on a buffing wheel. A design which necessitates such work may impair the appearance of the article and will certainly increase its cost.

Surface Treatment

The designer should make full use of the many different types of surface treatment to which plastics so readily lend themselves. The possibilities are limited only by the ingenuity of the designer plus the fundamental requirement that the piece must be removed from the mold. They range from mirror-like surfaces to a dull finish (created by sandblasting the molded surface, as one technique). Surface treatments such as fluting, reeding, stippling, fine straight lines, diamond-knurl cut, leather grain—to mention only a few—may be used to advantage. They tend to conceal surface blemishes which might prove objectionable on a highly lustrous surface.

Various designs can be created in the mold and transferred to the part either by debossing or embossing the mold surface. It is usually less expensive to use raised letters on a part as these may be readily engraved in the mold. On the other hand, the converse means removing all metal but the letters, a more costly procedure. Either may be done, however, depending on the effect desired.

Knockout sleeves or pins are generally used to remove a part from the mold. Some consideration as to the placement of such knockout mechanisms should be given, since they do leave a visible mark on the part. Gate location should be considered carefully for the same reason. Consultation with the mold designer is recommended when it is thought that functional and aesthetic considerations may clash.

The appearance of a product is greatly influenced by the surface treatment employed.

The designer should be familiar with the problems encountered by the molder. This is particularly true when large flat surfaces with a high lustrous finish are called for. Such surfaces are difficult to produce. The molder must invariably resort to special molding techniques which are both time-consuming and costly. The designer, understanding these problems, can do much to simplify the molder's difficulties, through modification of design.

Wide, sweeping curves and domed rather than flat surfaces should be employed. Improved flow and distribution of material, during the molding operation, are thus achieved. In addition the tendency to warp is greatly reduced. As a result, the appearance of the molded piece is improved.

Unfortunately there are no simple rules or formulas for solving the problems of surface treatment. What may prove to be a sound solution in one case may not work in another. Each new piece presents its own individual problems. The designer must draw freely upon his ingenuity, but always remember that the molded article must be easily removed from the mold after it has been formed.

The telephones shown in Fig. 23-32 are typical of the pleasing appearances that can be

Fig. 23.32. ABS telephones typify high quality of surface finish obtainable with proper mold surfacing. (*Courtesy Borg-Warner Chemicals*)

gained by combining wide sweeping curves, domed surfaces, and a lustrous finish.

Molded Lettering

Names, monograms, dial numbers, instructional information, and the like, are frequently required on molded articles. The lettering must be applied in such a manner as not to complicate the removal of the article from the mold. This is accomplished by locating it perpendicular to the parting line and providing adequate draft.

While both raised and depressed letters are possible, the method to be used in constructing the mold will dictate which is the more economical.

When the mold is to be made by machining, raised letters on the molded piece will be less costly. A raised letter on the molded piece is formed by a corresponding depression in the mold, and it is far less costly to engrave or machine the letters into the face of the mold than it is to form a raised letter on the surface of the mold by cutting away the surrounding metal. On the other hand, if the mold is to be formed by hobbing, then the letters on the molded piece should be depressed, since it is the hob which must be machined. In making the hob, the letters are engraved into its surface. As the hob is sunk into the steel blank to produce the mold cavity, the letters are raised on the surface of the latter. These raised letters on the mold in turn produce depressed letters on the molded piece.

To improve legibility, depressed lettering, filled in with paint is sometimes required. When hobbing of the whole mold is not practicable, the desired result can generally be accomplished by setting in a hobbed block carrying the lettering. When this insert is treated as a panel and the fin line is concealed by fluting, the appearance is not unpleasant.

The two applications shown in Figs. 23-33 and 23-34 are indicative of the type of precision lettering that can be accomplished in plastics products today with modern techniques.

References

Considerable data are available today to help designers in working with plastics. It is also suggested that suppliers be contacted for available literature on the subject. For example, a good deal of the foregoing section was adapted from design brochures which are obtainable from the following companies: E. I. du Pont de Nemours, Inc. ("Designing with Delrin" and "Designing with Nylon"), Celanese Corp. ("Designing with Celcon"), Soltex Corp. ("Designing with Fortiflex"), Borg-Warner ("Designing with ABS"), Exxon ("Designing with Polypropylene"), and General Electric ("Designing with Lexan" and "Designing with Noryl"). Most other suppliers in the industry have similar data available on request.

AVOIDING WARPAGE PROBLEMS*

Warpage is defined as "dimensional distortion in a plastic object after molding." It is directly related to material shrinkage; i.e., as shrinkage increases the tendency for warpage to occur increases.

Anisotropic properties, such as are found in some filled materials, can also contribute to warpage.

The purpose of this discussion is to offer guidelines to the designer and the processor on the factors which can cause warpage through

Fig. 23-33. Molded-in lettering in ABS typewriter head. (*Courtesy Borg-Warner Chemicals*)

* By Nelson C. Baldwin, Technical Service Engineer, Celanese Plastics Co.

Fig. 23-34. Molded-in lettering in phenolic fuse (right). (*Courtesy General Electric*)

uneven shrinkage. They are as follows: Part design, mold design and molding conditions.

Part Design

Warpage caused by part design is the worst type, being nearly impossible to correct by molding conditions. For this reason, it is imperative that the part be designed to *prevent* objectionable warpage. Since shrinkage is directly proportional to wall thickness, then wall thickness is directly related to warpage. This means that wall thickness must be uniform to provide uniform shrinkage. Different wall thicknesses in the same part *must* result in either warpage, through stress-relief, or molded-in stress.

The following examples illustrate warpage due to non-uniform wall thicknesses.

EXAMPLE 1:

Results in

EXAMPLE 2:

Results in

This varying wall thickness condition is probably the single largest cause of warpage.

Another type of part design warpage concerns ribs and bosses. Indiscriminate location of ribs and improper selection of rib thickness can result in shrinkage patterns which will alter the shape of the entire molding.

Ribs should be no more than 50% of the adjacent wall thickness of the part to avoid sinks and possible distortion. However, ribs which are very thin compared to the main body can also cause distortion due to different degrees of shrinkage.

Bosses can affect the shape of the molded part if they are of a different wall thickness than the base to which they are attached, or if they are connected to a side wall of different thickness. Initial wall thickness for bosses should be the same as ribs.

Example 3 depicts distortion from a rib tied to a side wall.

EXAMPLE 3:

Ribs and bosses can also affect part geometry due to changes in heat transfer from the mold. This will be covered later.

Mold Design

One of the most critical aspects of mold design, in unfilled and glass-filled crystalline polymers is gate location. This is due to many factors, including the inherent high shrinkage of the material and the aniostropic behavior it may exhibit. Aniostrophy refers to a shrinkage differential between the flow direction and the direction perpendicular (transverse) to flow.

With unfilled materials, the greater shrinkage is usually encountered in the flow direction. Shrinkage in the transverse direction, conversely, usually ranges from 70 to 98% of the longitudinal shrinkage (direction of flow) depending on gate size and part thickness. Thinner parts do not exhibit the degree of aniostrophy shown in thicker parts.

On the other hand, polymers reinforced with glass fibers show the opposite condition. Shrinkage in the flow direction is *less* than transverse shrinkage, because of orientation of the fibers to the flow direction. The percentage difference between the shrinkage in each direction is dependent on the wall thickness, gate size, fiber length, etc., and is therefore difficult to pinpoint. However, the average difference is approximately 50%, with the shrinkage in the flow direction being the lesser. It is recommended that material molding manuals or preferably a technical service representative be consulted prior to any actual mold fabrication to assist in determining shrinkage values.

It can be seen from the above discussion on anisotrophy, that orientation by gate location can have a significant effect on warpage even in an ideally designed part.

Examples 4 and 5 illustrate warpage of this type with unfilled material. Glass reinforcement would have an opposite effect.

The correct way to minimize warpage from orientation is to provide a longitudinal flow

EXAMPLE 4:

EXAMPLE 5(a):

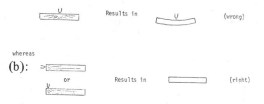

EXAMPLE 6:

For circular parts which are cored in the center other types of gating can be used to achieve uniform flow. Examples 7 and 8 depict gating systems of this type.

EXAMPLE 7:

EXAMPLE 8:

path for rectangular parts such as shown in Example 5, or a radial flow path for circular items as shown in the example below:

Due to the large degree of aniostrophy in glass-filled material, gating systems such as shown in Example 8 above, will sometimes still give enough orientation to cause slight out-of-roundness or warpage. For this reason, it is recommended that wherever possible, full ring gating (or if on the I.D., full disc gating) be used for these materials.

Example 9 illustrates the orientation patterns developed by four-point gating systems with glass-reinforced material. For purposes of illustration, the resultant shapes are exagerated.

EXAMPLE 9(a):   (b):

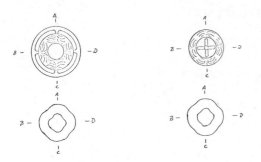

Example 10 also shows flow orientation for a cylindrical shape utilizing four-point gating in a two-plate mold.

EXAMPLE 10:

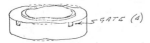

For this reason, full ring or disc gating is recommended for critical applications where TIR and flatness are of paramount importance, especially for glass-reinforced materials.

Another form of warpage associated with gating involves pressure distribution within the cavity. Refer back to Example 6 and assume that the disc has a uniform wall thickness.

Bear in mind the following three things: (1) warpage is the result of shrinkage differences in the same part, (2) shrinkage is affected by pressure (high pressure → low shrinkage) and (3) pressure decreases as flow increases.

Using these principles, it can be said that the pressure at the outside diameter of the part is less than the pressure at the gate. Therefore, shrinkage is greater at the O.D. than the gate area even though the wall is constant. On this basis, warpage *must* result if stress relaxation is allowed to take place. The only way this part can be held flat through molding conditions is by using techniques which give the lowest shrinkage (i.e. decreased mold and melt temperature, fast fill, increased injection pressure, plunger forward time, and overall cycle time, large gate, etc.). This procedure actually locks in the stresses so that the part retains its shape.

Using the three principles mentioned above,

a suggested remedy for a part of this shape is to taper the thickness from the center to the O.D., the thicker section being in the center. A gradual taper would reduce the pressure drop allowing more effective pressure at the outer periphery. This would reduce shrinkage in this area, and at the same time increase shrinkage at the center sections because of the thicker wall.

This leads into the second consideration for minimizing pressure differences, and this is to locate the gate at the thickest section of the part. Again referring to the factors governing shrinkage, thicker sections shrink more than thin ones and high pressure areas shrink less than low pressure areas. Therefore, if a gate is located in a thin section which feeds a thicker one, the highest pressure (developed near the gate) is generated in the section which will shrink the least due to its wall. As you can see, this is a direct invitation to warpage. The gate should be located in the thicker wall so that the highest pressures are developed in this area to help minimize shrinkage from wall thickness and therefore minimize shrinkage differentials in the part.

The last comment regarding gating involves gate size. The size of the gate regulates not only the volume of material allowed to enter the cavity, but also the effective pressure transmitted for packing out this material.

If the gate is too small, there is the possibility that it will prematurely freeze-off before the part is adequately filled. This causes low effective pressure and inadequate packing in the mold cavity with resultant increased shrinkage. In addition to causing undersize dimensions, the high shrinkage will magnify any trends toward warpage which may be present.

The product bulletins and preferably a technical service representative should be consulted prior to actual mold construction for gating recommendations.

Another condition encountered in mold design which can lead to warpage is the use of dissimilar metals in cavity construction.

Thermal conductivity varies with different metals. If two metals are used in fabricating the cavity and core, the one with the lower thermal conductivity will retain heat longer and thus create a differential mold temperature.

Since shrinkage increases as mold temperature increases, the plastic part will, in this case, bow toward the hotter side. Result? Warpage.

Example 11 shows this condition when using tool steel and beryllium copper for cavity construction.

EXAMPLE 11:

Since beryllium copper dissipates heat much more rapidly than tool steel, it runs cooler even though the same external heat is supplied to both mold halves. The solution for overcoming this condition is to use separate mold controllers for each half of the mold and adjust until both halves are the same temperature. It may also be necessary to include additional coolant channels in the tool steel half to achieve balance.

The final comment on mold design is concerned with non-uniform heat dissipation from the mold. This condition is perhaps the second largest contributor of warped parts and is usually not taken into account during the design stages of mold fabrication.

Non-uniform heat dissipation occurs when one side of the mold must dissipate more heat than the other. This is caused by the shape of the part and the area of the steel in contact with the hot resin. Ribs and bosses are prime causes of non-uniform temperatures in the mold as shown in Examples 12 and 13.

EXAMPLE 12:

EXAMPLE 13:

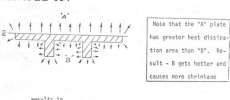

Non-uniform dissipation causes one side to run hotter than the other, and since thermoplastic resin shrinkage increases with increased mold temperature, more shrinkage occurs at the hotter face of the mold with warpage inevitable.

Several things can be done to offset this condition and they include the following:

1. Provide additional cooling channels, or bubblers, in the areas which must dissipate the most heat.
2. Use differential mold temperatures to provide extra heat in the areas which do not have much initial heat to remove. (This method can also be used to purposely warp parts the opposite way so that post-molding shrinkage will result in a flat part.)
3. Take advantage of thermal conductivity differences in dissimilar metals by using metals with high conductivity in areas requiring greater heat dissipation. An illustration would be to make the core pin in Example 12 in beryllium copper.

Evidence has recently come to light that suggests that in the case of glass-filled resins there may be a random "bunching" of fibers at sharp corners. These fibers are disoriented from the normal flow and may, because of anisotropic shrinkage, cause a pulling in of adjacent walls.

Example 14 shows this possibility and Examples 15 and 16 show possible solutions to off-set the condition.

EXAMPLE 14:

EXAMPLE 15: EXAMPLE 16:

Additional studies will be conducted to determine whether, in fact, this "bunching" actually causes warpage in glass-filled materials.

In any case, however, all sharp corners should be radiused 25-75% of the wall thickness to reduce stress concentrations which may lead to premature failure.

Improper ejection of molded parts may also cause apparent warpage, although it is actually deformation. This can be caused by undercuts in the cavity or an inadequate ejection system and should be checked along with the other design considerations.

Molding Conditions

Molding conditions can either contribute to warpage or aid in minimizing it. Thus, it is important to know how the various machine functions affect the molding material and its resultant post-molding behavior. We will review the pertinent functions one by one, and offer comments from a shrinkage, or warpage point of view.

A. *Filling Speed*. Filling speed is based to a large extent on gate size. However, assuming that gate size is correct, it is recommended that the cavity be filled as rapidly as possible. A fast fill ensures that the resin temperature will be constant throughout the cavity with a resultant uniform shrinkage.

B. *Mold Temperature*. A high mold temperature is preferable for optimum properties and surface finish. The molded parts cool more slowly thus relieving molded-in stress. However, this stress-relieving action will cause a tendency to warp, if the configuration of the part is such that uneven shrinkages will take place. If this is the case, then mold temperatures must be lowered to retard shrinkage.

Mold temperatures can be as low as necessary providing that the parts do not stress-relieve to a warped condition during post-molding shrinkage or in end-use operation.

If parts are to be used in a high temperature environment, it is suggested that the following test be performed on parts molded with a cold mold:

Condition a small quantity of molded parts for 10-12 hr at 180-190°F. Follow this with a slow cool to room temperature. If the parts are warped after cooling, the mold temperature is too cold. Resample with a warmer mold and repeat the test until the parts remain flat after testing.

Note that this test should only be used to compliment actual end-use testing. The values obtained must be correlated with conditions known to give acceptable parts, before the test can be of benefit.

C. *Material Temperature*. Temperature settings must be high enough to give a homogeneous melt. If settings are too low, incomplete melting will occur and varying degrees of shrinkage will take place with resultant warpage. The shrinkage differential is caused by varying pressure.

This same condition can occur if there are localized hot spots in the machine heating cylinder, or if too much of the machine capacity is being taken out each shot. The melt, in these cases, will have a varying temperature and therefore a varying shrinkage rate. Once again, warpage may result.

To avoid this, the cylinder should be checked for hot spots or areas where the material can hang up. The maximum shot size should be no more than 75% of the rated machine capacity.

D. *Cycle Time*. Too short a cycle will result in ejection of parts which have not cooled sufficiently to maintain structural stability. Post-molding shrinkage, higher because of decreased cycle, will in this case follow a stress-relieving pattern. As stresses relieve, any uneven shrinkage will result in warpage. Also, the parts are subject to deformation during ejection from the short cycle.

Too short an injection forward time can also lead to warpage problems. If the ram is retracted prior to gate seal, material will backflow from the cavity. This causes low and non-uniform pressure in the cavity resulting in increased and non-uniform shrinkage.

E. *Injection Pressure*. A low injection pressure allows increased material shrinkage. In addition to dimensions being undersize and erratic, the increased shrinkage promotes warpage if the configuration of the part is prone to differential shrinkage.

Too high an injection pressure, on the other hand, can lead to localized over-packing in the cavity. This causes differential shrinkage and possible warpage.

F. *Lost Contact With Cavity Surface.* This condition is usually due to part configuration and/or inadequate venting, rate of fill, injection pressure, etc. The condition appears as a pulling away of the plastic from the cavity, leaving an air gap. This air gap acts as an insulator causing a slower cooling rate. The result is the same as if this area was hotter than the rest of the mold. A higher shrinkage takes place and warpage is the result.

Change in mold temperature, or the use of differential mold temperature, usually corrects this condition, assuming that other corrections such as venting, faster fill speed or increased pressure have been tried first.

Trouble-shooting Guide

Thus, it is evident that warpage is the result of nothing more than distortion from shrinkage differences in a molded part. By recognizing this, and understanding the various conditions which affect shrinkage, it is possible to properly design and mold parts which are essentially warp-free.

The key, of course, is the initial design of the part. If this is not done properly, it is almost impossible to make corrections through changes in molding conditions.

Warpage caused by mold design or molding conditions can sometimes be corrected by annealing or fixturing. Warpage caused by part design, however, is usually made worse by annealing. Therefore, it should be emphasized that warpage must be *prevented* through proper part design so that it need not have to be corrected in molding.

Table 23-4 is a trouble-shooting guide for warpage. It is meant to assist those who have already reached the point where warpage is a problem, as well as those who are just beginning to design a part.

Table 23-4. Trouble Shooting Guide—Warpage

Cause	Solution
1. *Part Design*	
A. Wall thickness variation*	Core heavy sections
B. Ribs and bosses	Follow recommendations for wall thickness, especially when connecting to main body walls.
2. *Mold Design*	
A. Gate location	
1. Flow orientation*	Locate gate, or gates, for uniform orientation
2. Pressure distribution	Gate into heavy section
B. Gate size	Enlarge or reduce gate. Too small a gate causes premature freeze-off. Too large permits localized over-packing. In general, however, a large gate is preferable to a small one.
C. Materials of construction	Use materials to provide uniform dissipation. See also "D" below
D. Non-uniform dissipation*	Provide extra coolant passages where required. Also use differential mold temperature or dissimilar metals.
E. Ejection	Provide adequate, uniform ejection. Check for undercuts.
3. *Molding Conditions*	
A. Fill speed	Use fast fill whenever possible.
B. Mold temperature	Alter to suit. Hot mold increases shrinkage, stress relieving and warpage. Cold mold results in molded-in stress,
C. Melt temperature	Ensure homogenous melt for particular molding conditions. Non-uniform melt gives non-uniform shrinkage. Melt too cold gives inadequate pressure and/or non-uniform shrinkage.
D. Cycle	Keep injection forward time long enough for gate seal. Overall cycle should be long enough for part to maintain structural stability after ejection.
E. Pressure	Use minimum pressure consistent with good fill. Too low a pressure results in excessive shrinkage. Too high a pressure results in localized over-packing.

* Key items in design of part and mold.

24

STANDARDS FOR TOLERANCES ON MOLDED ARTICLES

The purpose of this chapter is to indicate the magnitude of practical tolerances on the dimensions of articles molded from a variety of thermosetting and thermoplastic materials, other than laminates.

The materials that are covered in this chapter include the following:

This information is given in the form of tables based upon data obtained from representative suppliers and molders of plastics by means of a questionnaire. It must be stressed that these tables are not to be construed as offering hard-and-fast rules applicable to all conditions. They can best be used as a basis for establishing standards for individual molded articles by agreement between the purchaser, or his design engineer, and the molder.

The tables were all developed by the Society of the Plastics Industry, Inc.

The questionnaire was based upon a hypothetical molded article, of which a cross section is shown in the tables.

It should be recognized that extreme accuracy of dimensions in molded articles is expensive to achieve. The closer the tolerances demanded, the greater will be the cost of the molds, because of the precision required, and also the greater will be the operational costs of molding, because of the greater care required to maintain uniformity of conditions. In some cases a further expense arises from the need of using cooling fixtures after the molding.

Dimensional tolerances in a molded article are the allowable variations, plus and minus, from a nominal or mean dimension.

Fine tolerance represents the narrowest possible limits of variation obtainable under close supervision and control of production.

Standard tolerance is that which can be held under average conditions of manufacture.

STANDARDS AND PRACTICES OF PLASTICS CUSTOM MOLDERS

Engineering and Technical Standards
ABS

NOTE: The Commercial values shown below represent common production tolerances at the most economical level. The Fine values represent closer tolerances that can be held but at a greater cost.

Drawing Code	Dimensions (Inches)		Plus or Minus in Thousands of an Inch
A = Diameter (see Note #1)	0.000 — 0.500 — 1.000 — 2.000		
B = Depth (see Note #3)	3.000 — 4.000		Fine / Commercial
C = Height (see Note #3)	5.000 — 6.000		

Drawing Code	Dimensions (Inches)	Comm. ±	Fine ±
	6.000 to 12.000 for each additional inch add (inches)	.003	.002
D = Bottom Wall (see Note #3)		.004	.002
E = Side Wall (see Note #4)		.003	.002
F = Hole Size Diameter (see Note #1)	0.000 to 0.125	.002	.001
	0.125 to 0.250	.002	.001
	0.250 to 0.500	.003	.002
	0.500 & Over	.004	.002
G = Hole Size Depth (see Note #5)	0.000 to 0.250	.003	.002
	0.250 to 0.500	.004	.002
	0.500 to 1.000	.005	.003
Draft Allowance per side (see Note #5)		2°	1°
Flatness (see Note #4)	0.000 to 3.000	.015	.010
	3.000 to 6.000	.030	.020
Thread Size (class)	Internal	1	2
	External	1	2
Concentricity (see Note #4)	(T.I.R.)	.009	.005
Fillets, Ribs, Corners (see Note #6)		.025	.015
Surface Finish	(see Note #7)		
Color Stability	(see Note #7)		

REFERENCE NOTES

1 — These tolerances do not include allowance for aging characteristics of material.

2 — Tolerances based on ⅛″ wall section.

3 — Parting line must be taken into consideration.

4 — Part design should maintain a wall thickness as nearly constant as possible. Complete uniformity in this dimension is impossible to achieve.

5 — Care must be taken that the ratio of the depth of a cored hole to its diameter does not reach a point that will result in excessive pin damage.

6 — These values should be increased whenever compatible with desired design and good molding technique.

7 — Customer-Molder understanding necessary prior to tooling.

STANDARDS AND PRACTICES OF PLASTICS CUSTOM MOLDERS	Engineering and Technical Standards ACETAL

NOTE: The Commercial values shown below represent common production tolerances at the most economical level. The Fine values represent closer tolerances that can be held but at a greater cost.

Drawing Code	Dimensions (Inches)	Plus or Minus in Thousands of an Inch 1 2 3 4 5 6 7 8 9 10 11 12 13 14 15 16 17 18 19 20 21 22 23 24 25 26 27 28
A = Diameter (see Note #1)	0.000 0.500 1.000 2.000	
B = Depth (see Note #3)	3.000 4.000	Fine Commercial
C = Height (see Note #3)	5.000 6.000	

		Comm. ±	Fine ±
	6.000 to 12.000 for each additional inch add (inches)	.004	.002
D=Bottom Wall (see Note #3)		.004	.002
E = Side Wall (see Note #4)		.004	.002
F = Hole Size Diameter (see Note #1)	0.000 to 0.125	.002	.001
	0.125 to 0.250	.003	.002
	0.250 to 0.500	.004	.002
	0.500 & Over	.006	.003
G = Hole Size Depth (see Note#5)	0.000 to 0.250	.004	.002
	0.250 to 0.500	.005	.003
	0.500 to 1.000	.006	.004
Draft Allowance per side (see Note #5)		1°	½°
Flatness (see Note #4)	0.000 to 3.000	.011	.006
	3.000 to 6.000	.020	.010
Thread Size (class)	Internal	1B	2B
	External	1A	2A
Concentricity (see Note #4)	(T.I.R.)	.010	.006
Fillets, Ribs, Corners (see Note #6)		.030	.015
Surface Finish	(see Note #7)		
Color Stability	(see Note #7)		

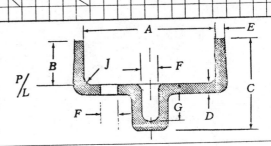

REFERENCE NOTES

1 — These tolerances do not include allowance for aging characteristics of material.

2 — Tolerances based on ⅛″ wall section.

3 — Parting line must be taken into consideration.

4 — Part design should maintain a wall thickness as nearly constant as possible. Complete uniformity in this dimension is impossible to achieve.

5 — Care must be taken that the ratio of the depth of a cored hole to its diameter does not reach a point that will result in excessive pin damage.

6 — These values should be increased whenever compatible with desired design and good molding technique.

7 — Customer-Molder understanding necessary prior to tooling.

STANDARDS AND PRACTICES OF PLASTICS CUSTOM MOLDERS

Engineering and Technical Standards
ACRYLIC

NOTE: The Commercial values shown below represent common production tolerances at the most economical level. The Fine values represent closer tolerances that can be held but at a greater cost.

Drawing Code	Dimensions (Inches)	Plus or Minus in Thousands of an Inch
A = Diameter (see Note #1)	0.000 – 0.500 – 1.000 – 2.000	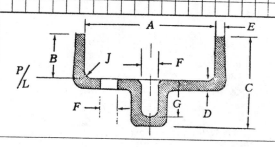
B = Depth (see Note #3)	3.000 – 4.000	
C = Height (see Note #3)	5.000 – 6.000	

Drawing Code	Dimensions (Inches)	Comm. ±	Fine ±
	6.000 to 12.000 for each additional inch add (inches)	.003	.001
D=Bottom Wall (see Note #3)		.005	.003
E = Side Wall (see Note #4)		.005	.003
F = Hole Size Diameter (see Note #1)	0.000 to 0.125	.003	.001
	0.125 to 0.250	.003	.002
	0.250 to 0.500	.004	.002
	0.500 & Over	.005	.003
G = Hole Size Depth (see Note#5)	0.000 to 0.250	.004	.002
	0.250 to 0.500	.004	.002
	0.500 to 1.000	.006	.003
Draft Allowance per side (see Note #5)		1°	½°
Flatness (see Note #4)	0.000 to 3.000	.010	.007
	3.000 to 6.000	.015	.010
Thread Size (class)	Internal	1	2
	External	1	2
Concentricity (see Note #4)	(T.I.R.)	.010	.006
Fillets, Ribs, Corners (see Note #6)		.030	.015
Surface Finish	(see Note #7)		
Color Stability	(see Note #7)		

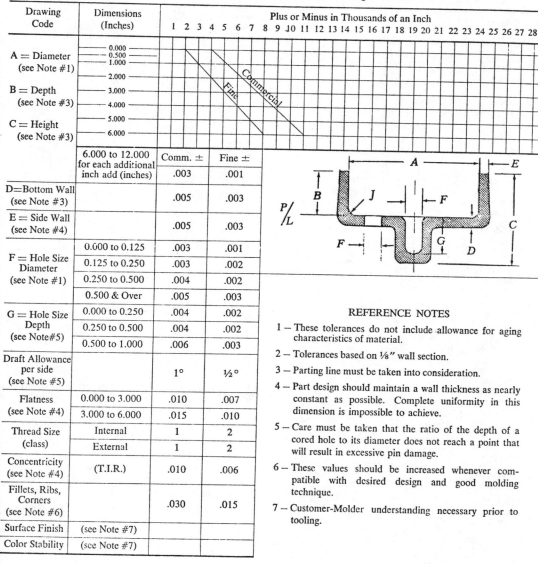

REFERENCE NOTES

1 – These tolerances do not include allowance for aging characteristics of material.

2 – Tolerances based on ⅛" wall section.

3 – Parting line must be taken into consideration.

4 – Part design should maintain a wall thickness as nearly constant as possible. Complete uniformity in this dimension is impossible to achieve.

5 – Care must be taken that the ratio of the depth of a cored hole to its diameter does not reach a point that will result in excessive pin damage.

6 – These values should be increased whenever compatible with desired design and good molding technique.

7 – Customer-Molder understanding necessary prior to tooling.

STANDARDS AND PRACTICES OF PLASTICS CUSTOM MOLDERS

Engineering and Technical Standards
ALKYD

NOTE: The Commercial values shown below represent common production tolerances at the most economical level. The Fine values represent closer tolerances that can be held but at a greater cost.

Drawing Code	Dimensions (Inches)	Plus or Minus in Thousands of an Inch 1 2 3 4 5 6 7 8 9 10 11 12 13 14 15 16 17 18 19 20 21 22 23 24 25 26 27 28
A = Diameter (see Note #1) B = Depth (see Note #3) C = Height (see Note #3)	0.000 — 0.500 — 1.000 — 2.000 — 3.000 — 4.000 — 5.000 — 6.000 —	Commercial / Fine

	6.000 to 12.000 for each additional inch add (inches)	Comm. ± .0015	Fine ± .001
D=Bottom Wall (see Note #3)		.002	.001
E = Side Wall (see Note #4)		.002	.001
F = Hole Size Diameter (see Note #1)	0.000 to 0.125	.002	.001
	0.125 to 0.250	.002	.002
	0.250 to 0.500	.002	.002
	0.500 & Over	.004	.003
G = Hole Size Depth (see Note#5)	0.000 to 0.250	.002	.002
	0.250 to 0.500	.002	.002
	0.500 to 1.000	.002	.002
Draft Allowance per side (see Note #5)		1°	½°
Flatness (see Note #4)	0.000 to 3.000	.010	.010
	3.000 to 6.000	.005	.005
Thread Size (class)	Internal	1	2
	External	1	2
Concentricity (see Note #4)	(T.I.R.)	.005	.005
Fillets, Ribs, Corners (see Note #6)		.062	.031
Surface Finish	(see Note #7)		
Color Stability	(see Note #7)		

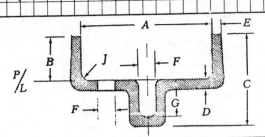

REFERENCE NOTES

1 — These tolerances do not include allowance for aging characteristics of material.

2 — Tolerances based on ⅛″ wall section.

3 — Parting line must be taken into consideration.

4 — Part design should maintain a wall thickness as nearly constant as possible. Complete uniformity in this dimension is impossible to achieve.

5 — Care must be taken that the ratio of the depth of a cored hole to its diameter does not reach a point that will result in excessive pin damage.

6 — These values should be increased whenever compatible with desired design and good molding technique.

7 — Customer-Molder understanding necessary prior to tooling.

STANDARDS AND PRACTICES OF PLASTICS CUSTOM MOLDERS

Engineering and Technical Standards
CELLULOSICS

NOTE: The Commercial values shown below represent common production tolerances at the most economical level. The Fine values represent closer tolerances that can be held but at a greater cost.

Drawing Code	Dimensions (Inches)	Plus or Minus in Thousands of an Inch 1 2 3 4 5 6 7 8 9 10 11 12 13 14 15 16 17 18 19 20 21 22 23 24 25 26 27 28	
A = Diameter (see Note #1)	0.000 / 0.500 / 1.000 / 2.000		
B = Depth (see Note #3)	3.000 / 4.000	Commercial / Fine	
C = Height (see Note #3)	5.000 / 6.000		
	6.000 to 12.000 for each additional inch add (inches)	Comm. ± .003	Fine ± .002
D=Bottom Wall (see Note #3)		.003	.002
E = Side Wall (see Note #4)		.003	.002
F = Hole Size Diameter (see Note #1)	0.000 to 0.125	.002	.001
	0.125 to 0.250	.003	.002
	0.250 to 0.500	.004	.002
	0.500 & Over	.005	.003
G = Hole Size Depth (see Note #5)	0.000 to 0.250	.003	.002
	0.250 to 0.500	.004	.002
	0.500 to 1.000	.005	.003
Draft Allowance per side (see Note #5)		1°	½°
Flatness (see Note #4)	0.000 to 3.000	.007	.005
	3.000 to 6.000	.010	.007
Thread Size (class)	Internal	1B	2B
	External	1A	2A
Concentricity (see Note #4)	(T.I.R.)	.005	.004
Fillets, Ribs, Corners (see Note #6)		.005	.005
Surface Finish	(see Note #7)		
Color Stability	(see Note #7)		

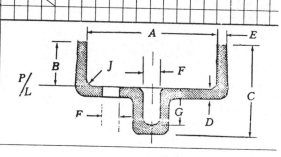

REFERENCE NOTES

1 – These tolerances do not include allowance for aging characteristics of material.

2 – Tolerances based on ⅛" wall section.

3 – Parting line must be taken into consideration.

4 – Part design should maintain a wall thickness as nearly constant as possible. Complete uniformity in this dimension is impossible to achieve.

5 – Care must be taken that the ratio of the depth of a cored hole to its diameter does not reach a point that will result in excessive pin damage.

6 – These values should be increased whenever compatible with desired design and good molding technique.

7 – Customer-Molder understanding necessary prior to tooling.

STANDARDS AND PRACTICES OF PLASTICS CUSTOM MOLDERS

Engineering and Technical Standards

DIALLYL PHTHALATE

NOTE: The Commercial values shown below represent common production tolerances at the most economical level. The Fine values represent closer tolerances that can be held but at a greater cost.

Drawing Code	Dimensions (Inches)	Plus or Minus in Thousands of an Inch
A = Diameter (see Note #1)	0.000 / 0.500 / 1.000 / 2.000	
B = Depth (see Note #3)	3.000 / 4.000	
C = Height (see Note #3)	5.000 / 6.000	

		Comm. ±	Fine ±
	6.000 to 12.000 for each additional inch add (inches)	.002	.001
D=Bottom Wall (see Note #3)		.005	.003
E = Side Wall (see Note #4)		.003	.002
F = Hole Size Diameter (see Note #1)	0.000 to 0.125	.002	.001
	0.125 to 0.250	.002	.001
	0.250 to 0.500	.002	.001
	0.500 & Over	.0025	.002
G = Hole Size Depth (see Note#5)	0.000 to 0.250	.002	.001
	0.250 to 0.500	.003	.002
	0.500 to 1.000	.005	.003
Draft Allowance per side (see Note #5)		1°	½°
Flatness (see Note #4)	0.000 to 3.000	.010	.005
	3.000 to 6.000	.012	.008
Thread Size (class)	Internal	1	2
	External	1	2
Concentricity (see Note #4)	(T.I.R.)	.005	.003
Fillets, Ribs, Corners (see Note #6)		.062	.031
Surface Finish	(see Note #7)		
Color Stability	(see Note #7)		

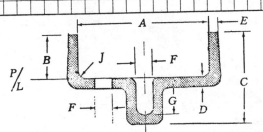

REFERENCE NOTES

1 — These tolerances do not include allowance for aging characteristics of material.

2 — Tolerances based on ⅛" wall section.

3 — Parting line must be taken into consideration.

4 — Part design should maintain a wall thickness as nearly constant as possible. Complete uniformity in this dimension is impossible to achieve.

5 — Care must be taken that the ratio of the depth of a cored hole to its diameter does not reach a point that will result in excessive pin damage.

6 — These values should be increased whenever compatible with desired design and good molding technique.

7 — Customer-Molder understanding necessary prior to tooling.

STANDARDS AND PRACTICES OF PLASTICS CUSTOM MOLDERS

Engineering and Technical Standards
EPOXY

NOTE: The Commercial values shown below represent common production tolerances at the most economical level. The Fine values represent closer tolerances that can be held but at a greater cost.

Drawing Code	Dimensions (Inches)	Plus or Minus in Thousands of an Inch 1 2 3 4 5 6 7 8 9 10 11 12 13 14 15 16 17 18 19 20 21 22 23 24 25 26 27 28
A = Diameter (see Note #1)	0.000 / 0.500 / 1.000 / 2.000	
B = Depth (see Note #3)	3.000 / 4.000	
C = Height (see Note #3)	5.000 / 6.000	

	6.000 to 12.000 for each additional inch add (inches)	Comm. ±	Fine ±
		.0015	.001
D=Bottom Wall (see Note #3)		.002	.001
E = Side Wall (see Note #4)		.002	.001
F = Hole Size Diameter (see Note #1)	0.000 to 0.125	.002	.001
	0.125 to 0.250	.002	.002
	0.250 to 0.500	.002	.002
	0.500 & Over	.004	.003
G = Hole Size Depth (see Note#5)	0.000 to 0.250	.002	.002
	0.250 to 0.500	.002	.002
	0.500 to 1.000	.002	.002
Draft Allowance per side (see Note #5)		1°	½°
Flatness (see Note #4)	0.000 to 3.000	.010	.010
	3.000 to 6.000	.010	.015
Thread Size (class)	Internal	1	2
	External	1	2
Concentricity (see Note #4)	(T.I.R.)	.005	.005
Fillets, Ribs, Corners (see Note #6)		.062	.031
Surface Finish	(see Note #7)		
Color Stability	(see Note #7)		

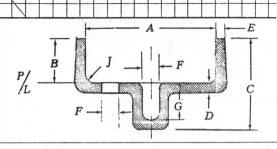

REFERENCE NOTES

1 – These tolerances do not include allowance for aging characteristics of material.

2 – Tolerances based on ⅛″ wall section.

3 – Parting line must be taken into consideration.

4 – Part design should maintain a wall thickness as nearly constant as possible. Complete uniformity in this dimension is impossible to achieve.

5 – Care must be taken that the ratio of the depth of a cored hole to its diameter does not reach a point that will result in excessive pin damage.

6 – These values should be increased whenever compatible with desired design and good molding technique.

7 – Customer-Molder understanding necessary prior to tooling.

STANDARDS AND PRACTICES OF PLASTICS CUSTOM MOLDERS

Engineering and Technical Standards
MELAMINE-UREA

NOTE: The Commercial values shown below represent common production tolerances at the most economical level. The Fine values represent closer tolerances that can be held but at a greater cost.

Drawing Code	Dimensions (Inches)		Plus or Minus in Thousands of an Inch
A = Diameter (see Note #1) B = Depth (see Note #3) C = Height (see Note #3)	0.000 0.500 1.000 2.000 3.000 4.000 5.000 6.000		1 2 3 4 5 6 7 8 9 10 11 12 13 14 15 16 17 18 19 20 21 22 23 24 25 26 27 28

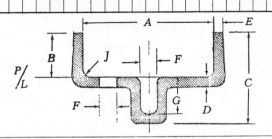

Drawing Code	Dimensions (Inches)	Comm. ±	Fine ±
	6.000 to 12.000 for each additional inch add (inches)	.003	.002
D=Bottom Wall (see Note #3)		.005	.003
E = Side Wall (see Note #4)		.004	.002
F = Hole Size Diameter (see Note #1)	0.000 to 0.125	.003	.002
	0.125 to 0.250	.003	.002
	0.250 to 0.500	.004	.003
	0.500 & Over	.005	.004
G = Hole Size Depth (see Note#5)	0.000 to 0.250	.003	.002
	0.250 to 0.500	.004	.002
	0.500 to 1.000	.005	.002
Draft Allowance per side (see Note #5)		1°	½°
Flatness (see Note #4)	0.000 to 3.000	.012	.008
	3.000 to 6.000	.018	.013
Thread Size (class)	Internal	1	2
	External	1	2
Concentricity (see Note #4)	(T.I.R.)	.007	.005
Fillets, Ribs, Corners (see Note #6)		.030	.005
Surface Finish	(see Note #7)		
Color Stability	(see Note #7)		

REFERENCE NOTES

1 – These tolerances do not include allowance for aging characteristics of material.

2 – Tolerances based on ⅛" wall section.

3 – Parting line must be taken into consideration.

4 – Part design should maintain a wall thickness as nearly constant as possible. Complete uniformity in this dimension is impossible to achieve.

5 – Care must be taken that the ratio of the depth of a cored hole to its diameter does not reach a point that will result in excessive pin damage.

6 – These values should be increased whenever compatible with desired design and good molding technique.

7 – Customer-Molder understanding necessary prior to tooling.

STANDARDS AND PRACTICES OF PLASTICS CUSTOM MOLDERS

Engineering and Technical Standards
NYLON

NOTE: The Commercial values shown below represent common production tolerances at the most economical level. The Fine values represent closer tolerances that can be held but at a greater cost.

Drawing Code	Dimensions (Inches)	Plus or Minus in Thousands of an Inch
A = Diameter (see Note #1)	0.000 / 0.500 / 1.000 / 2.000	
B = Depth (see Note #3)	3.000 / 4.000	
C = Height (see Note #3)	5.000 / 6.000	

Drawing Code	Dimensions (Inches)	Comm. ±	Fine ±
	6.000 to 12.000 for each additional inch add (inches)	.003	.002
D = Bottom Wall (see Note #3)		.004	.003
E = Side Wall (see Note #4)		.005	.003
F = Hole Size Diameter (see Note #1)	0.000 to 0.125	.002	.001
	0.125 to 0.250	.003	.002
	0.250 to 0.500	.003	.002
	0.500 & Over	.005	.003
G = Hole Size Depth (see Note #5)	0.000 to 0.250	.004	.002
	0.250 to 0.500	.004	.003
	0.500 to 1.000	.005	.004
Draft Allowance per side (see Note #5)		1½°	½°
Flatness (see Note #4)	0.000 to 3.000	.010	.004
	3.000 to 6.000	.015	.007
Thread Size (class)	Internal	1	2
	External	1	2
Concentricity (see Note #4)	(T.I.R.)	.010	.006
Fillets, Ribs, Corners (see Note #6)		.020	.012
Surface Finish	(see Note #7)		
Color Stability	(see Note #7)		

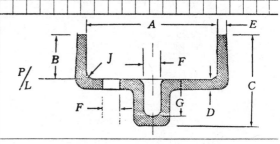

REFERENCE NOTES

1 — These tolerances do not include allowance for aging characteristics of material.

2 — Tolerances based on ⅛″ wall section.

3 — Parting line must be taken into consideration.

4 — Part design should maintain a wall thickness as nearly constant as possible. Complete uniformity in this dimension is impossible to achieve.

5 — Care must be taken that the ratio of the depth of a cored hole to its diameter does not reach a point that will result in excessive pin damage.

6 — These values should be increased whenever compatible with desired design and good molding technique.

7 — Customer-Molder understanding necessary prior to tooling.

STANDARDS AND PRACTICES OF PLASTICS CUSTOM MOLDERS

Engineering and Technical Standards

FIBRE-FILLED PHENOLIC

NOTE: The Commercial values shown below represent common production tolerances at the most economical level. The Fine values represent closer tolerances that can be held but at a greater cost.

Drawing Code	Dimensions (Inches)	Plus or Minus in Thousands of an Inch 1 2 3 4 5 6 7 8 9 10 11 12 13 14 15 16 17 18 19 20 21 22 23 24 25 26 27 28
A = Diameter (see Note #1) B = Depth (see Note #3) C = Height (see Note #3)	0.000 0.500 1.000 2.000 3.000 4.000 5.000 6.000	Fine Commercial

	Dimensions (Inches)	Comm. ±	Fine ±
	6.000 to 12.000 for each additional inch add (inches)	.003	.002
D=Bottom Wall (see Note #3)		.006	.004
E = Side Wall (see Note #4)		.004	.003
F = Hole Size Diameter (see Note #1)	0.000 to 0.125	.002	.001
	0.125 to 0.250	.003	.002
	0.250 to 0.500	.004	.003
	0.500 & Over	.005	.003
G = Hole Size Depth (see Note #5)	0.000 to 0.250	.004	.002
	0.250 to 0.500	.005	.003
	0.500 to 1.000	.007	.004
Draft Allowance per side (see Note #5)		1°	½°
Flatness (see Note #4)	0.000 to 3.000	.014	.008
	3.000 to 6.000	.021	.014
Thread Size (class)	Internal	1	2
	External	1	2
Concentricity (see Note #4)	(T.I.R.)	.007	.004
Fillets, Ribs, Corners (see Note #6)		.030	.005
Surface Finish	(see Note #7)		
Color Stability	(see Note #7)		

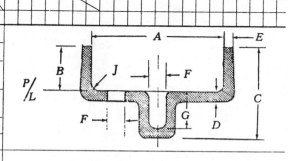

REFERENCE NOTES

1 — These tolerances do not include allowance for aging characteristics of material.

2 — Tolerances based on ⅛″ wall section.

3 — Parting line must be taken into consideration.

4 — Part design should maintain a wall thickness as nearly constant as possible. Complete uniformity in this dimension is impossible to achieve.

5 — Care must be taken that the ratio of the depth of a cored hole to its diameter does not reach a point that will result in excessive pin damage.

6 — These values should be increased whenever compatible with desired design and good molding technique.

7 — Customer-Molder understanding necessary prior to tooling.

STANDARDS AND PRACTICES OF PLASTICS CUSTOM MOLDERS

Engineering and Technical Standards

GENERAL PURPOSE PHENOLIC

NOTE: The Commercial values shown below represent common production tolerances at the most economical level. The Fine values represent closer tolerances that can be held but at a greater cost.

Drawing Code	Dimensions (Inches)	Plus or Minus in Thousands of an Inch
A = Diameter (see Note #1) B = Depth (see Note #3) C = Height (see Note #3)	0.000 0.500 1.000 2.000 3.000 4.000 5.000 6.000	1 2 3 4 5 6 7 8 9 10 11 12 13 14 15 16 17 18 19 20 21 22 23 24 25 26 27 28 Fine Commercial

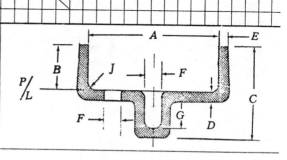

	Comm. ±	Fine ±
6.000 to 12.000 for each additional inch add (inches)	.002	.001
D=Bottom Wall (see Note #3)	.008	.005
E = Side Wall (see Note #4)	.005	.003

F = Hole Size Diameter (see Note #1)		Comm.	Fine
	0.000 to 0.125	.002	.001
	0.125 to 0.250	.002	.001
	0.250 to 0.500	.003	.002
	0.500 & Over	.003	.002
G = Hole Size Depth (see Note#5)	0.000 to 0.250	.004	.002
	0.250 to 0.500	.004	.002
	0.500 to 1.000	.005	.003
Draft Allowance per side (see Note #5)		1°	½°
Flatness (see Note #4)	0.000 to 3.000	.010	.005
	3.000 to 6.000	.012	.010
Thread Size (class)	Internal	1	2
	External	1	2
Concentricity (see Note #4)	(T.I.R.)	.005	.003
Fillets, Ribs, Corners (see Note #6)		.062	.031
Surface Finish	(see Note #7)		
Color Stability	(see Note #7)		

REFERENCE NOTES

1 — These tolerances do not include allowance for aging characteristics of material.

2 — Tolerances based on ⅛" wall section.

3 — Parting line must be taken into consideration.

4 — Part design should maintain a wall thickness as nearly constant as possible. Complete uniformity in this dimension is impossible to achieve.

5 — Care must be taken that the ratio of the depth of a cored hole to its diameter does not reach a point that will result in excessive pin damage.

6 — These values should be increased whenever compatible with desired design and good molding technique.

7 — Customer-Molder understanding necessary prior to tooling.

STANDARDS AND PRACTICES OF PLASTICS CUSTOM MOLDERS

Engineering and Technical Standards
POLYCARBONATE

NOTE: The Commercial values shown below represent common production tolerances at the most economical level. The Fine values represent closer tolerances that can be held but at a greater cost.

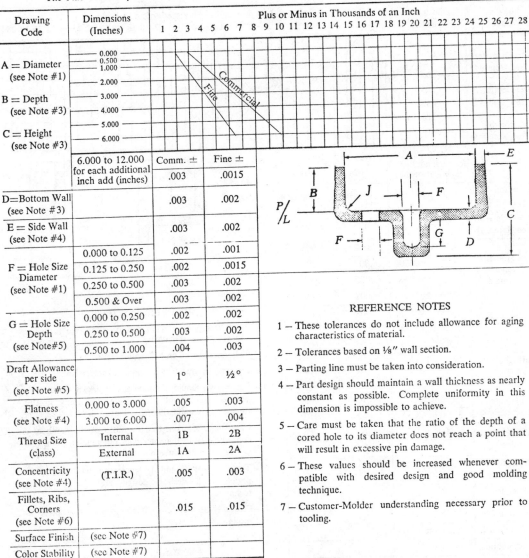

Drawing Code	Dimensions (Inches)		
A = Diameter (see Note #1) B = Depth (see Note #3) C = Height (see Note #3)	0.000 / 0.500 / 1.000 / 2.000 / 3.000 / 4.000 / 5.000 / 6.000		
	6.000 to 12.000 for each additional inch add (inches)	Comm. ±	Fine ±
		.003	.0015
D = Bottom Wall (see Note #3)		.003	.002
E = Side Wall (see Note #4)		.003	.002
F = Hole Size Diameter (see Note #1)	0.000 to 0.125	.002	.001
	0.125 to 0.250	.002	.0015
	0.250 to 0.500	.003	.002
	0.500 & Over	.003	.002
G = Hole Size Depth (see Note #5)	0.000 to 0.250	.002	.002
	0.250 to 0.500	.003	.002
	0.500 to 1.000	.004	.003
Draft Allowance per side (see Note #5)		1°	½°
Flatness (see Note #4)	0.000 to 3.000	.005	.003
	3.000 to 6.000	.007	.004
Thread Size (class)	Internal	1B	2B
	External	1A	2A
Concentricity (see Note #4)	(T.I.R.)	.005	.003
Fillets, Ribs, Corners (see Note #6)		.015	.015
Surface Finish	(see Note #7)		
Color Stability	(see Note #7)		

Plus or Minus in Thousands of an Inch
1 2 3 4 5 6 7 8 9 10 11 12 13 14 15 16 17 18 19 20 21 22 23 24 25 26 27 28

REFERENCE NOTES

1 — These tolerances do not include allowance for aging characteristics of material.

2 — Tolerances based on ⅛″ wall section.

3 — Parting line must be taken into consideration.

4 — Part design should maintain a wall thickness as nearly constant as possible. Complete uniformity in this dimension is impossible to achieve.

5 — Care must be taken that the ratio of the depth of a cored hole to its diameter does not reach a point that will result in excessive pin damage.

6 — These values should be increased whenever compatible with desired design and good molding technique.

7 — Customer-Molder understanding necessary prior to tooling.

STANDARDS AND PRACTICES OF PLASTICS CUSTOM MOLDERS

Engineering and Technical Standards

HIGH DENSITY POLYETHYLENE

NOTE: The Commercial values shown below represent common production tolerances at the most economical level. The Fine values represent closer tolerances that can be held but at a greater cost.

Drawing Code	Dimensions (Inches)	Plus or Minus in Thousands of an Inch
A = Diameter (see Note #1)	0.000 / 0.500 / 1.000 / 2.000	
B = Depth (see Note #3)	3.000 / 4.000	
C = Height (see Note #3)	5.000 / 6.000	

1 2 3 4 5 6 7 8 9 10 11 12 13 14 15 16 17 18 19 20 21 22 23 24 25 26 27 28

Commercial / Fine

Drawing Code	Dimensions (Inches)	Comm. ±	Fine ±
	6.000 to 12.000 for each additional inch add (inches)	.006	.003
D=Bottom Wall (see Note #3)		.006	.004
E = Side Wall (see Note #4)		.006	.004
F = Hole Size Diameter (see Note #1)	0.000 to 0.125	.003	.002
	0.125 to 0.250	.005	.003
	0.250 to 0.500	.006	.004
	0.500 & Over	.008	.005
G = Hole Size Depth (see Note #5)	0.000 to 0.250	.005	.003
	0.250 to 0.500	.007	.004
	0.500 to 1.000	.009	.006
Draft Allowance per side (see Note #5)		2°	¾°
Flatness (see Note #4)	0.000 to 3.000	.023	.015
	3.000 to 6.000	.037	.022
Thread Size (class)	Internal	1	2
	External	1	2
Concentricity (see Note #4)	(T.I.R.)	.027	.010
Fillets, Ribs, Corners (see Note #6)		.025	.010
Surface Finish	(see Note #7)		
Color Stability	(see Note #7)		

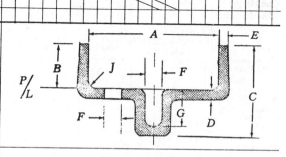

REFERENCE NOTES

1 — These tolerances do not include allowance for aging characteristics of material.

2 — Tolerances based on ⅛″ wall section.

3 — Parting line must be taken into consideration.

4 — Part design should maintain a wall thickness as nearly constant as possible. Complete uniformity in this dimension is impossible to achieve.

5 — Care must be taken that the ratio of the depth of a cored hole to its diameter does not reach a point that will result in excessive pin damage.

6 — These values should be increased whenever compatible with desired design and good molding technique.

7 — Customer-Molder understanding necessary prior to tooling.

STANDARDS AND PRACTICES OF PLASTICS CUSTOM MOLDERS

Engineering and Technical Standards
LOW DENSITY POLYETHYLENE

NOTE: The Commercial values shown below represent common production tolerances at the most economical level. The Fine values represent closer tolerances that can be held but at a greater cost.

Drawing Code	Dimensions (Inches)		
A = Diameter (see Note #1) B = Depth (see Note #3) C = Height (see Note #3)	Plus or Minus in Thousands of an Inch		
	6.000 to 12.000 for each additional inch add (inches)	Comm. ±	Fine ±
		.005	.004
D=Bottom Wall (see Note #3)		.005	.004
E = Side Wall (see Note #4)		.005	.004
F = Hole Size Diameter (see Note #1)	0.000 to 0.125	.003	.002
	0.125 to 0.250	.004	.003
	0.250 to 0.500	.005	.004
	0.500 & Over	.006	.005
G = Hole Size Depth (see Note#5)	0.000 to 0.250	.003	.003
	0.250 to 0.500	.004	.004
	0.500 to 1.000	.006	.005
Draft Allowance per side (see Note #5)		2°	1°
Flatness (see Note #4)	0.000 to 3.000	.020	.015
	3.000 to 6.000	.030	.020
Thread Size (class)	Internal	1	2
	External	1	2
Concentricity (see Note #4)	(T.I.R.)	.010	.008
Fillets, Ribs, Corners (see Note #6)		.025	.010
Surface Finish	(see Note #7)		
Color Stability	(see Note #7)		

Dimensions scale: 0.000, 0.500, 1.000, 2.000, 3.000, 4.000, 5.000, 6.000

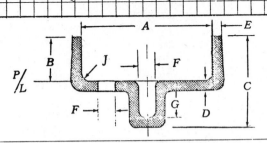

REFERENCE NOTES

1 — These tolerances do not include allowance for aging characteristics of material.

2 — Tolerances based on ⅛″ wall section.

3 — Parting line must be taken into consideration.

4 — Part design should maintain a wall thickness as nearly constant as possible. Complete uniformity in this dimension is impossible to achieve.

5 — Care must be taken that the ratio of the depth of a cored hole to its diameter does not reach a point that will result in excessive pin damage.

6 — These values should be increased whenever compatible with desired design and good molding technique.

7 — Customer-Molder understanding necessary prior to tooling.

STANDARDS AND PRACTICES OF PLASTICS CUSTOM MOLDERS

Engineering and Technical Standards

POLYPROPYLENE

NOTE: The Commercial values shown below represent common production tolerances at the most economical level. The Fine values represent closer tolerances that can be held but at a greater cost.

Drawing Code	Dimensions (Inches)	Plus or Minus in Thousands of an Inch
A = Diameter (see Note #1)	0.000 / 0.500 / 1.000 / 2.000	
B = Depth (see Note #3)	3.000 / 4.000	
C = Height (see Note #3)	5.000 / 6.000	

	Dimensions	Comm. ±	Fine ±
	6.000 to 12.000 for each additional inch add (inches)	.005	.003
D=Bottom Wall (see Note #3)		.006	.003
E = Side Wall (see Note #4)		.006	.003
F = Hole Size Diameter (see Note #1)	0.000 to 0.125	.003	.002
	0.125 to 0.250	.004	.003
	0.250 to 0.500	.005	.004
	0.500 & Over	.008	.006
G = Hole Size Depth (see Note#5)	0.000 to 0.250	.005	.003
	0.250 to 0.500	.006	.004
	0.500 to 1.000	.009	.006
Draft Allowance per side (see Note #5)		1½°	½°
Flatness (see Note #4)	0.000 to 3.000	.021	.014
	3.000 to 6.000	.035	.021
Thread Size (class)	Internal	1	2
	External	1	2
Concentricity (see Note #4)	(T.I.R.)	.016	.013
Fillets, Ribs, Corners (see Note #6)		.028	.015
Surface Finish	(see Note #7)		
Color Stability	(see Note #7)		

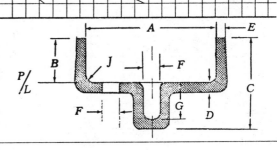

REFERENCE NOTES

1 — These tolerances do not include allowance for aging characteristics of material.

2 — Tolerances based on ⅛″ wall section.

3 — Parting line must be taken into consideration.

4 — Part design should maintain a wall thickness as nearly constant as possible. Complete uniformity in this dimension is impossible to achieve.

5 — Care must be taken that the ratio of the depth of a cored hole to its diameter does not reach a point that will result in excessive pin damage.

6 — These values should be increased whenever compatible with desired design and good molding technique.

7 — Customer-Molder understanding necessary prior to tooling.

STANDARDS AND PRACTICES OF PLASTICS CUSTOM MOLDERS

Engineering and Technical Standards

POLYSTYRENE

NOTE: The Commercial values shown below represent common production tolerances at the most economical level. The Fine values represent closer tolerances that can be held but at a greater cost.

Drawing Code	Dimensions (Inches)	Plus or Minus in Thousands of an Inch
A = Diameter (see Note #1) **B = Depth** (see Note #3) **C = Height** (see Note #3)	0.000 / 0.500 / 1.000 / 2.000 / 3.000 / 4.000 / 5.000 / 6.000	Fine / Commercial

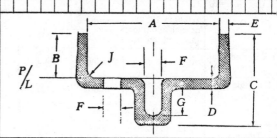

Drawing Code	Dimensions (Inches)	Comm. ±	Fine ±
	6.000 to 12.000 for each additional inch add (inches)	.004	.002
D=Bottom Wall (see Note #3)		.0055	.003
E = Side Wall (see Note #4)		.007	.0035
F = Hole Size Diameter (see Note #1)	0.000 to 0.125	.002	.001
	0.125 to 0.250	.002	.001
	0.250 to 0.500	.002	.0015
	0.500 & Over	.0035	.002
G = Hole Size Depth (see Note#5)	0.000 to 0.250	.0035	.002
	0.250 to 0.500	.004	.002
	0.500 to 1.000	.005	.003
Draft Allowance per side (see Note #5)		1½°	½°
Flatness (see Note #4)	0.000 to 3.000	.007	.004
	3.000 to 6.000	.013	.005
Thread Size (class)	Internal	1	2
	External	1	2
Concentricity (see Note #4)	(T.I.R.)	.010	.008
Fillets, Ribs, Corners (see Note #6)		.015	.010
Surface Finish	(see Note #7)		
Color Stability	(see Note #7)		

REFERENCE NOTES

1 – These tolerances do not include allowance for aging characteristics of material.

2 – Tolerances based on ⅛" wall section.

3 – Parting line must be taken into consideration.

4 – Part design should maintain a wall thickness as nearly constant as possible. Complete uniformity in this dimension is impossible to achieve.

5 – Care must be taken that the ratio of the depth of a cored hole to its diameter does not reach a point that will result in excessive pin damage.

6 – These values should be increased whenever compatible with desired design and good molding technique.

7 – Customer-Molder understanding necessary prior to tooling.

STANDARDS AND PRACTICES OF PLASTICS CUSTOM MOLDERS

Engineering and Technical Standards
FLEXIBLE VINYL

NOTE: The Commercial values shown below represent common production tolerances at the most economical level. The Fine values represent closer tolerances that can be held but at a greater cost.

Drawing Code	Dimensions (Inches)			
		Plus or Minus in Thousands of an Inch

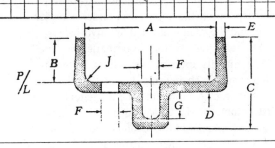

Drawing Code	Dimensions (Inches)	Comm. ±	Fine ±
A = Diameter (see Note #1) B = Depth (see Note #3) C = Height (see Note #3)	6.000 to 12.000 for each additional inch add (inches)	.005	.003
D = Bottom Wall (see Note #3)		.007	.003
E = Side Wall (see Note #4)		.007	.003
F = Hole Size Diameter (see Note #1)	0.000 to 0.125	.004	.003
	0.125 to 0.250	.005	.004
	0.250 to 0.500	.006	.005
	0.500 & Over	.008	.006
G = Hole Size Depth (see Note#5)	0.000 to 0.250	.004	.003
	0.250 to 0.500	.005	.004
	0.500 to 1.000	.006	.005
Draft Allowance per side (see Note #5)		1½°	1°
Flatness (see Note #4)	0.000 to 3.000	.010	.007
	3.000 to 6.000	.020	.015
Thread Size (class)	Internal	(Not Recommended)	
	External	(Not Recommended)	
Concentricity (see Note #4)	(T.I.R.)	.015	.010
Fillets, Ribs, Corners (see Note #6)		.030	.010
Surface Finish	(see Note #7)		
Color Stability	(see Note #7)		

REFERENCE NOTES

1 — These tolerances do not include allowance for aging characteristics of material.

2 — Tolerances based on ⅛″ wall section.

3 — Parting line must be taken into consideration.

4 — Part design should maintain a wall thickness as nearly constant as possible. Complete uniformity in this dimension is impossible to achieve.

5 — Care must be taken that the ratio of the depth of a cored hole to its diameter does not reach a point that will result in excessive pin damage.

6 — These values should be increased whenever compatible with desired design and good molding technique.

7 — Customer-Molder understanding necessary prior to tooling.

Coarse tolerance is acceptable when accurate dimensions are not important.

How To Use the Tables

The basic format of the tables is similar for each of the materials that are covered — whether for thermoplastics or thermosets.

Users will find that two separate sets of values are represented. The commercial values represent common production tolerances that can be achieved at the most economical level. The fine values represent closer tolerances that can be held, but at a greater cost. The selection will depend on the application under consideration and the economics that are involved.

By referring to the hypothetical molded article and its cross-section illustrated in the table, and by then using the applicable code number (e.g., A represents the diameter) in the first column of the table and the exact dimensions as indicated in the second column, readers can find the recommended tolerances either in the chart at the top of the table or in the two columns underneath. (Note that the typical article shown in cross-section in the tables may be of round or rectangular or other shapes. Thus, dimensions A and B may be diameters or lengths.)

For example, an ABS part with a diameter (A) of 2 inches would show tolerances of plus-or-minus 0.004 inches (fine value) and 0.007 inches (commercial value). If the dimensions go up to 6 inches, the tolerances would change to plus-or-minus 0.008 inches (fine) and 0.013 inches (commercial). If the dimension is greater than 6 inches, however, the tolerance for 6 inches is increased by the amount indicated on the lines directly below the chart in the table. Thus, if dimension A is 10 inches, the tolerance (commercial) is plus-or-minus 0.025 inches (i.e., 0.013 inches for a 6-inch dimension plus 0.003 for each of the additional four inches).

Other tolerances for the various dimensions shown in the typical cross-section are as indicated in the two columns running under the chart.

Special notes for users of the charts:

(1) The tolerances indicated in the tables for diameter and hole size do not include allowance for aging characteristics of the particular plastics material under consideration.

(2) Tolerances are based on 1/4-inch wall sections.

(3) For depth, height, and bottom wall dimensions, parting line must be taken into consideration.

(4) In terms of side wall dimensions, flatness, and concentricity, part design should maintain a wall thickness as nearly constant as possible. Complete uniformity in this dimension is impossible to achieve.

(5) In determining hole size depth and draft allowance per side, care must be taken that the ratio of the depth of a cored hole to the diameter does not reach a point that will result in excessive pin damage.

(6) Values for fillets, ribs, and corners should be increased whenever compatible with desired design and good molding technique.

(7) Where surface finish and color stability are concerned, customer and molder must come to an understanding as to what is necessary for the particular job under consideration prior to the decision to begin molding.

Design Considerations

"Do I need, and can I afford, the tolerances I am specifying?" If the designer will also ask himself these two questions before specifying tolerances, excessive mold and processing costs can often be avoided with no penalty in part performance. Regardless of the economics, it may be unreasonable to specify close production tolerances on a part when it is designed to operate through a wide range of environmental conditions. Dimension changes due to temperature variations alone can be three to four times as great as the specified tolerances. Also, in many applications, close tolerances with plastics are not as critical as with metals because of the resiliency of plastics.

Many factors control the production of precision parts and it is beyond the scope of this handbook to completely review this complex subject. However, realizing that the proper specification of tolerance is important to the designer, the following suggestions are offered:

(1) A design for a plastic part should indicate the conditions under which the dimensions shown must be held. For example, a drawing should state that dimensions shown are to be as specified after molding, after annealing, after moisture conditioning, etc.

(2) Overall tolerances for a part should be shown in inches per inch, not in fixed values. As an example, a title block might read, "All decimal dimensions ±0.00X in. per in. (±0.00X mm/mm), unless otherwise specified."

(3) Only those tight tolerances required for specific dimensions should be labeled as such. Less important dimensions can be controlled by overall tolerances.

(4) Generous molding tolerances should be allowed in any areas which will be machined after molding. Production variables such as the number and size of cavities in a mold also affect tolerances. Where compromises in tolerances may be acceptable from a performance standpoint, discussion of such tolerances with the molder may result in economies.

For example, the use of a multi-cavity mold is usually an economical production method. But, as the number of cavities per mold increases, so must the tolerances on critical dimensions. An increase of 1-5% per cavity is about average. For example: dimensions of a part produced in a single-cavity mold may be held to ±0.002 in. per in. (±0.002 mm/mm). When the number of cavities is increased to 20, the closest tolerances obtainable may be ±0.004 in per in. (±0.004 mm/mm), an increase of 5% per cavity, or a total increase of 100%.

25
FINISHING AND MACHINING PLASTICS

Many plastic products will require some degree of machining and finishing after they have been processed and before they are put on the market place.

A basic purpose of many finishing and machining operations is to remove such imperfections as scratches, dents, dull spots, marks of misalignment, slight ridges where the mold sections joined, or distinct flash. In transfer or injection moldings, rough spots may be left by breaking off the gates. Extruded pieces may exhibit longitudinal scratches from the edge of the extrusion die. Thermoformed pieces may show surface defects which were present in the original sheet. Cast plastic formed in a rigid mold will have a slight mark where the halves of the mold are joined; even where flexible molds are used, a gate mark may be left where the fluid plastic entered the mold.

Many finishing operations on molded articles can, of course, be avoided by careful design of the product or the mold (e.g., placing flash lines and gates in accessible positions). But where finishing or machining is necessary, a wide range of techniques are available for use with plastics, including tumbling, filing, sanding, buffing, tapping, turning, sawing, piercing, trimming, routing, grinding, etc.

Obviously, repair work and defect removal are not the only reasons for finishing and machining. In some applications, it may be necessary to saw, drill, tap, etc., simply to

prepare the plastic product for assembly or mounting. In other instances, machining and finishing may be used to meet very narrow tolerances by molding a critical dimension oversize, then machining it down to the desired tolerances.

Machining can also be useful in the early stages of design engineering for the creation of prototypes or in low-volume production. Prototypes can provide material and performance data to aid in material selection, or in successful final part production if the material has already been decided upon. Short production runs can be entirely machined, practically and economically, rather than commit relatively large sums of money for complex molds or dies. High-speed automatic screw machines, for example, are increasingly being used for low-volume runs.

Another aspect of the finishing operation is the decorating of the plastic part and this will be discussed in the section beginning on page 724. The discussion immediately following will concern itself with the various aspects of finishing and machining, other than decorating.

Finishing

In the general nomenclature, finishing operations are sometimes distinguished from machining operations in that they include such techniques as filing, grinding, sanding, buffing, and polishing, as used to finish-off, smooth-out, or polish-up the plastic part. Of course, such a distinction is rather vague in that some of the so-called finishing operations, like grinding, can also be used to fashion a part (i.e., machine it to a different shape), just as machining operations, like cutting, can be used to trim off flash (i.e., finish the product). For purposes of this discussion, however, we will follow this arbitrary distinction, assuming that the reader will be able to apply any of the techniques covered for his own particular needs.

Filing

Files are used for finishing molded articles, and for beveling, smoothing, burring, and fitting the edges and corners of sheets of plastic. For removal of flash, tumbling is to be preferred if feasible, but the shape, size, or contours of the article may require filing to remove both heavy and thin flash as well as gates of heavy section, or burrs left by machining operations such as drilling and tapping. Filing by hand is more costly than using special machine set-ups, but may be more economical on short runs.

Selection of the proper file is of great importance, as the shape, cut, size, and pattern of the file will determine the ease and speed of removal of stock, and also the appearance of the filed surface.

The type of file selected must be adapted to the plastic—its hardness, brittleness, flexibility, and heat-resistance. The size and shape of the file are determined by the size, shape, and contour of the article to be filed. For removal of flash, files should have very sharp, thin-topped teeth which will hold their edge, well-rounded gullets to minimize the tendency to clog, and the proper rake for clearing the chips.

Machine Filing. Removal of flash from circular or cylindrical articles in large production runs is sometimes accomplished by semi-automatic machines. A circular revolving table carries from 6 to 10 work-holding stations past a series of fixed stations equipped to remove flash, to polish, and to buff. The work-holding stations revolve on their individual axes as they pass the work stations. If the articles are perfectly round, the work stations may be equipped with tool-holders to accommodate files, etc. If the articles are slightly out of round, or vary in size, as may be the case with pieces produced in multi-cavity molds, then motor-driven sanding belts, spring-loaded files, or buffing wheels are used. The efficiency of these machines is limited by the quality of the molding. Excessively heavy flash or even a slight mismatch will seriously reduce production or make hand finishing necessary.

The edges of round articles can be finished in a simple button machine, comprising a drive chuck which rotates at high speed and can be moved axially by means of a pedal. The work is put against a stationary center, the drive chuck is brought forward to engage and rotate the work, and then the operator applies a file, usually about 70 degrees from the flash plane, to remove the fin. Frequently as many as three different flash edges may be smoothed in one handling. When the pedal is released, the work falls into a tray and the operator inserts another piece.

An excellent way to remove material rapidly is by using a rotary file or burr. Standard medium-cut, high-speed steel burrs operated at 800 to 1000 surface feet per minute are quite effective. Ground burrs provide better chip clearance than hand-cut rotary files and are, therefore, preferred. Carbide burrs offer no advantage over high-speed steel burrs. Because of their tooth geometry, carbide burrs remove less material than the high-speed steel burrs at the same speed, and when operated at higher speeds, tend to cause excessive frictional heat build-up.

Filing Thermoplastics. With thermoplastics, filing has become less popular than it was in the early days, since flash is more effectively removed by the three-square scrapers described below.

For the soft thermoplastics, fine files should be avoided, because they become clogged. Coarse, single-cut, shear-tooth files should be used, with teeth cut on a 45-degree angle, and in flat or half-round shape. The combination of coarse teeth and long angle promotes self-cleaning. Shear-tooth files are used with long sweeping light strokes, to avoid running off of the work.

Clogging is not encountered with the harder thermoplastics and fine files may be used (although the shear type, coarser files are still generally recommended). It is also well to note that some plastics with unusual toughness and adbrasion resistance, like nylon, are not easily filed.

For filing edges of sheet stock, milled-tooth files are recommended. The file should be held at approximately a 20-degree angle with the edge.

In general, when filing thermoplastics, it would be wise to check with suppliers on handling the various types, since some distinctions are evident. For ABS, for example, files are usually of the coarse texture type. For polycarbonate, a single-hatched file is better than a cross-hatched file, since the latter has a tendency to clog under heavy pressure. A No. 2 pillar file has no tendency to clog when used with light pressure. For acetal, best results are obtained by using the coarsest file consistent with the size of the surface and the finish required. Milled curved-tooth files with coarse, single-cut, shear-type teeth are particularly suitable.

Scraping Thermoplastics. Scraping involves smoothing a plastic surface by the use of a sharp-bladed tool, which is held so that the blade rides over the surface and has no tendency to dig under the surface.

A standard scraper is a three-edged tool of triangular cross section, the sides making up a scraping edge (at an angle of 60 degrees). The blade is tapered so that the scraping edge runs from a fairly long straight section at the handle end to a curved section neat the point. This variation in shape allows the careful scraping of a great variety of straight, concave, and convex surfaces.

Some scrapers are made in the form of an oblong metal plate, sized to be held between the thumb and three fingers. The ends of the plate are straight, while the side edges are slightly convex to allow scraping along narrow bands. The four-bladed scraper is a common variation on the plate scraper.

As with the other tools, the edges of the scraper must be free from burrs, dents, or scratches, which might mar the surface of the plastic.

Filing Thermosetting Plastics. Articles molded from thermosetting plastics always require some finishing operation for removal of flash from parting lines (or the removal of gates).

Flash should be filed off in such a way as to break it toward a solid portion rather than away from the main body, in order to prevent chipping. The file is pushed with a firm stroke to break off the flash close to the body, and then filing is continued to smooth the surface.

Since selection is made by trial and error, it is well to have a variety of files on hand.

Mill files in bastard and western cuts are used extensively to remove the flash from flat or convex surfaces of molded articles and on the corners and edges of sheets, to remove the burrs from sawing, or to bevel the corners. The western cut is slightly the coarser.

Milled-tooth files are recommended for large areas or for beveling the corners and edges of large sheets. They are relatively coarse files with curved teeth, and are available in both flexible and rigid types.

Various shapes of Swiss-pattern files are used, coarse enough to effect rapid removal of material, and fine enough to leave a good finish. These small files of various shapes, in cuts from No. 00 to No. 4, are used for small, intricate moldings, in which the surfaces to be filed are hard to reach. Round and half-round files are used for cleaning out holes or slots with rounded surfaces; knife and warding files are used to reach down between flutes and into narrow slots and grooves.

Files are designed to cut in one direction only; pressure should be applied in that direction, and relieved on the return stroke. Experience has shown that on some fine, intricate articles a steady pressure on both strokes may result in a better finish, but it is damaging to the file. As much as possible of the filing surface, on both length and width, should be used, so that the file will wear uniformly.

The life of the file is greatly shortened by improper selection, improper use and improper care. Since many resins and fillers cause rapid wear of cutting tools, it is imperative that the files be given proper care to retain their sharpness. Proper care must be given in storage.

Files should never be thrown into a drawer with other tools, or stacked on top of each other, since such treatment ruins the cutting edge of the teeth. They should be stored standing with the tangs in a row of holes, or hanging on racks by their handles, and in a dry place so that they will not rust. Files should be kept clean of filings or chips, which collect between the teeth during use, by tapping the end after every few strokes. A file card or brush should be used to remove the chips before storage. Oil or grease is removed by applying chalk and then brushing. When a file becomes dull, but is otherwise in good condition, it should be sharpened. Files can be resharpened as often as four times, even though they do not then do as good a job as when new.

Tumbling

Thermoplastics. Tumbling is used to round corners, to remove stumps of gates, and to apply finish to surfaces. It is the cheapest way of doing these things, for the equipment is not expensive and the only labor involved is in the loading and unloading of barrels. It is applicable chiefly to small objects which do not have projections that are easily broken off. Tumbling does not produce as high a finish as that obtainable by ashing and polishing, but for many articles a very high polish is not necessary and is not worth the higher labor cost involved.

The articles to be tumbled are placed in octagonal wooden barrels, 20 to 30 in. in diameter, which may be divided into two or more pockets to permit the handling of two or more colors or shapes simultaneously. The barrels are mounted on a horizontal axis and driven through pulleys and gears, and are usually run at a speed of 15 to 30 rpm.

Abrasives and hardwood pegs are put in the pockets of the barrels with the articles to be tumbled. The pegs serve to rub the abrasive against the articles during the tumbling. Sawdust and pumice are used in the first stage because of their rapid abrasive action; finer abrasives and polish are used later to give better finishes.

Tumbling of thermoplastics can be done wet, as well as dry. For wet polishing of a thermoplastic, like acetal, for example, a system might consist of aluminum oxide chips with a high-sudsing, burnishing compound (i.e., containing soap, detergent, alkaline cleaning agents, etc).

Automatic vibratory finishing machines are also available.

Thermosetting Plastics. Tumbling is used on all kinds of thermosetting materials, for removing flash. It is done in barrels of various types, with several different materials as filing agents.

A cylindrical ("cement-mixer") barrel, running at speeds of 15 to 25 rpm, is used for light articles which have little or very thin flash. The articles are allowed to roll by themselves, from 2 to 5 min as necessary.

An octagonal barrel with alternate closed and open sides, running horizontally, is used for heavier articles and where more positive action is needed. The open sections are covered by screens of suitable coarseness to let the fragments of flash fall out. Lignum vitae balls from $\frac{1}{2}$ to 1 in. in diameter are used to give a rolling action to the articles and to prevent chipping. It has been found that a mixture of two parts of balls to one part of moldings, by volume, generally gives good results. Hardwood blocks or scrap molded parts also are used on some jobs. The speed of this barrel should be variable from 5 to about 30 rpm. The time required varies with the work, some jobs running as long as 2 to 3 hr.

Tumbling may be used to reduce size of molded articles. This is done in a closed barrel running at a fairly high speed, and employes a cutting agent. Rubber impregnated with an abrasive is cut into various shapes and sizes and tumbled with the articles to give a satiny finish without a harsh cutting action. For fast removal of stock, strips of abrasive cloth or abrasive paper, mixed with the articles, are used. The barrel may be lined with abrasive cloth. For very light cuts, especially before polishing, pumice and small pegs are used. Octagonal barrels running at 20 to 35 rpm have proved best for this work.

With articles having grooves or projections, it is more difficult to achieve a uniform complete polishing, and the polishing agent tends to accumulate in grooves or other pockets on the surface. Such articles are best polished by being tumbled in barrels with string mops which have been impregnated and coated with specially compounded wax by being tumbled in clean barrels with balls of the wax. The tumbling together of articles and mop is done in a closed octagonal barrel having no screens, and no metal on the inside, which would scratch the surface of the articles. The barrel, on a horizontal axis, is rotated at speeds from 15 rpm (for large articles) up to 30 rpm (for small articles).

Outside of plain tumbling, most thermoset deflashing is accomplished in machines using the impellor wheel design. Using non-abrasive media such as vegetable grain (walnut shell, corn cob, apricot pit, etc.) or plastic pellets (like polycarbonate), these machines effectively handle large loads in a matter of minutes. In essence, these units are based on the use of a conveyor that carries the articles, in tumbling motion, through the stream of pellets projected at high velocity by a rotating bladed wheel (Fig. 25-1). A modified version of this type of equipment utilizes a rotary table or platform arrangement to pass parts under the impellor wheel without tumbling (Fig. 25-2). Production rates are slower with this type of equipment,

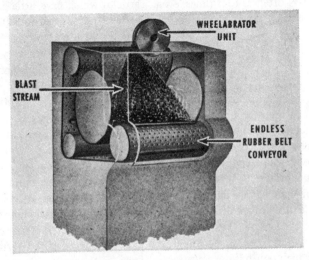

Fig. 25-1. Articles are carried by conveyor through stream of pellets.

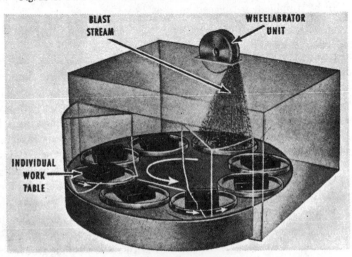

Fig. 25-2. Parts on a rotary table pass under an impellor wheel without tumbling.

but damage to delicate or fragile parts is avoided by eliminating the tumbling operation.

More recently, pressure blast type systems using individual blasting guns have been made available. A number of configurations can be used for handling parts, from simple hand machines through batch-type tumbling systems to the use of conveyors, rotary tables, etc.

Still another recent variation on technique has been the charging of small blocks of dry ice into the tumbling barrel along with molded parts and wooden pegs. In this case, the flash on the molded articles becomes cold and a cleaner flash break results. If dry ice is used, the barrel must be closed and have only small vents.

In some newer equipment, liquid nitrogen is used to freeze flash on the molded parts. Parts are then vibrated alone or with any standard deflashing media to break off the brittle frozen flash.

Grinding and Sanding

These operations, if conducted dry, require an exhaust system to dispose of dust.

Thermoplastics. Standard sanding machines of belt and disc types, run wet or dry, are used for form sanding, or for long production runs. A variable speed-control provides the proper speed in accordance with the amount of material to be removed.

Belts carrying coarse abrasive, and run dry, may be used for fast, rough cutting, if the speed is kept down to avoid excessive heating. In most cases, however, particularly in fine sanding, belts are run wet, as a precaution against overheating.

On *methyl methacrylate* resin (acrylic), the finest sandpaper that will remove the scratch or other defects (no coarser than grade 320) is used first. The paper should be of a waterproof type. The paper is wrapped around a soft block of felt or rubber, and the area is rubbed lightly with a circular motion, using water, or soap and water, as a lubricant. An area having a diameter two or three times the length of the defect should be followed by similar treatment with progressively finer grades of sandpaper (grade 360A, 400A and 500 or 600A), each of which removes the deeper scratches left by the pre-

ceding. The plastic must be washed after each sanding operation. Where a large amount of polishing is to be done, ashing compounds may be used in place of sandpaper.

When sanding *styrene*, the heat developed should be kept to a minimum to prevent gumming and loading of the abrasive belt or disc. For this reason, wet sanding is the best type of operation. Care should be exercised in filtering the coolant to prevent abrasive grit contamination of the surface of the molded article.

A vertical belt sander is most frequently used, with belts having grits to suit the operation. For rough work, 80 grit is suggested. For medium and fine work, 220-280 grit and 400-600 grit belts are best.

ABS can be centerless, surface, and spindle ground or sanded. Many abrasives can be used to obtain the desired stock removal or surface finish. Coolants and lubricants used to reduce frictional heat include water, soap, detergents, or soluble oil solutions. A variety of industrial abrasives is available for grinding parts to desired flatness or other preferred shapes. Wet sanding is preferable, since this method dissipates heat and retards loading of the abrasive belt. Silicon carbide belts or pads of No. 82 to 130 grit are ideal. Light pressure should be used to avoid over-heating. Surface speeds of grinding should be determined by starting at speeds of 4500 sfpm and adjusting the surface speed, feed rate, pressure, grit size and coolant to obtain desired finish or stock removal rates. Avoid excessive heat generation, or melting may occur.

Acetals can be wet-sanded using conventional belt and disc sanding equipment. Moderate feeds should be used in sanding to prevent overheating. After sanding to a smooth finish, acetal can be buffed to a high surface luster.

When working with *polycarbonate*, sandpaper or silicon carbide (emery) abrasive-coated belts may be used to remove sprue projections, toolmarks, or for flat surfacing sheet stock. Coarse abrasives will produce scratches difficult to remove, while too fine grades will cause the belts to fill. For a scratch-free, highly polished surface, a grit size not over No. 180 is suggested. Hence, it may be desirable to use a coarse grit sandpaper first and follow with a finer one. Belt speeds of approximately 3000 surface-feet-per-minute (sfm) are adequate, but somewhat faster speeds (4000 sfm) can be used so long as the pressure applied is not excessive.

Hand sanding must be done with light pressure in order to avoid clogging, especially with extremely fine grades. For example, 100 grit carborundum paper works well with light pressure, but clogs with heavy pressure. A 240 grit paper exhibits slight clogging even with very light pressure, but continues to produce a fine satin finish. Crocus cloth also shows a tendency to clog, but the use of kerosene lubricant prevents clogging and produces a fine satin finish.

Thermosetting Plastics. Belt sanding is commonly used for the removal of heavy flash and sprue projections, and for flat surfacing and beveling. The belts should carry silicon-carbide abrasive bonded by waterproof synthetic resin. Popular grit sizes are 50, 120, 180, 220 and 400—the coarse for heavy flash and sprues, the fine for lighter operations, Speeds recommended are from 2000 to 5000 linear fpm, but 4000 is most commonly used.

With a good exhaust system to remove dust and do some cooling by pulling air past the edge being sanded, thermosetting plastics can be successfully sanded at medium speeds on dry belts, thus avoiding the washing and drying required after wet sanding. But wet sanding offers the advantages of freedom from dust and overheating, longer life of belts, freedom from clogging of the belt, and the finer surface produced because of the lubrication.

On cold-molded articles, both disc and belt grinders are used. Disc grinders are run at 1750 to 2000 rpm; the grit ranges from 40 to 60 to 120. Belt grinders are operated at a surface speed of 3000 to 4000 fpm. All work is done dry.

It is seldom necessary to resort to sanding the surfaces of well-machined pieces, but sanding can be done with any good abrasive cloth or abrasive paper which can be used in belt form or in rotating discs. The finer the paper and the slower the cutting, the less pronounced are the marks left by the abrasive. Facets and other surfaces can be cut in this way, with care to avoid overheating and discoloring the work. Surface speed for sanding cast phenolics is about the same as, or slightly less than, that for sanding wood.

In the finishing of rough-cast phenolic castings it is frequently necessary to produce radii, bevels, etc., which cannot be formed in the usual straight-draw lead molds. The necessary shaping is done with tools, usually of bronze, honed in much the same manner as are tools for shaping wood, but having a larger number of teeth, and operated at a lower speed. These can be used to put a radius on the square end of a radio cabinet and to obtain bevels, beads, and other decorative lines.

Ashing, Buffing and Polishing

The finishing department requires polishing lathes, buffing wheels, and suitable compositions for the ashing, polishing, cutdown buffing and luster buffing which may be required.

Lathes for ashing, buffing, and polishing are available in types ranging from low-powered bench models, which are essentially converted bench grinders, to 50-hp floor models.

For finishing plastics a popular machine is a 2- or 3-hp floor-type lathe, with motor in base, and V-belt drive. This is preferred over the motor-on-spindle lathe because its speed can be changed by merely changing the diameters of the pulleys. Where speeds must be changed frequently, variable-speed lathes are available, in 3-, 5-, and $7\frac{1}{2}$-hp sizes, which permit changes of speed between 1500 and 3000 rpm generally without stopping the machine. With these machines, it is possible to operate at the most efficient peripheral speed, regardless of the diameter of the buff.

For production buffing of articles of simple contours, automatic buffing machines are available, which must, however, be engineered for the particular job.

All dry buffing and polishing operations require an efficient exhaust system. Suitable sheet metal hoods should enclose as much of the wheel as is practicable, and be connected to the exhaust piping. The exhaust should pass through a dust-collector rather than to open air. Since many of the dusts are combustible, care should be taken so that, if steel inserts are ground, incandescent metal particles are not drawn into the exhaust system.

Buffing wheels for finishing plastics are generally made up of sections of muslin discs, either with or without sewing, depending on the flexibility required. The cloth should be a high-count sheeting, such as 84 x 92, for the faster cutting, and a lower count, such as 64 x 68, for buffing and polishing. For waxing, canton flannel and 48-48 muslin are both very popular.

Where the contours of the article are regular, and fast cutting is desired, the buffing wheel is composed of sewed buffing sections with stitching spaced $\frac{1}{4}$ or $\frac{3}{8}$ in. Wider spacings, and narrower, down to $\frac{1}{8}$ in., are available. The wider spacings give the softer wheels. The next softer medium is the pocketed or folded buff, which presents pockets of cloth to the work and which makes for greater cooling of the surface, a faster cut than that given by a loose full-disc buff, and greater flexibility than that of conventional sewed buffs. Loose buffs made of full discs of muslin, while not cutting as fast as these, have a greater flexibility and give a smoother, more even, intermediate finish than do the harder wheels, and are generally to be preferred, especially for articles with curved or irregular contours.

For special cases, extremely soft wheels are needed, such as the packed buff or the string brush. The packed buff is made up of large discs of cloth alternated with smaller discs in the proportion of 1 : 1, 2 : 1, or 1 : 2, depending on the degree of hardness needed. The string brush is like a bristle brush, with cotton string substituted for the bristles. The wheel offers the maximum in flexibility, and in conjunction with greaseless compound is recommended for smoothing the edges of intricate articles.

Thermoplastics. In the ashing and polishing of thermoplastics, overheating must be carefully avoided, since it may soften and distort the surface into ripples. Hence, it is necessary to avoid excessively hard buffing wheels, excessive speed of wheels, and excessive pressure of the work against the wheel.

Ashing is frequently required for the re-moval of "cold spots," teardrops, deep scratches, parting lines, etc., from irregular surfaces which cannot be smoothed by wet sanding. A satin-smooth but dull surface is usually produced, requiring further finishing.

For ashing, wet pumice, grade No. 00 to No. 1, is used, on a loose muslin buff running at about 4000 linear fpm. The buff is often packed to increase its flexibility. The buff must be well hooded, since the wet pumice does not adhere to it, and tends to be thrown off. After wet ashing the articles must be washed and dried before being polished.

As wet ashing is essentially a messy operation, attempts are continually being made to get more cut in a buffing operation than is usual in cut-down buffing, so as to obviate the need for wet ashing. Certain commercial compounds are available which perform a fast cutting job, that can be called semi-ashing, on cellulose acetate, cellulose acetate butyrate, and acrylics.

For the buffing of thermoplastics, fine silica powders in special grease binders, differing considerably from mixtures that have been developed for finishing metals, are generally most successful. Compositions must be formulated to give sufficient lubrication to prevent excessive heating. Pocketed and ventilated buffs are an aid in prevention of overheating, although soft packed buffs are preferred for final luster buffing. Speeds generally run from 3000 to 4000 surface fpm.

Thermoplastics can sometimes be polished with greaseless compounds such as are used with thermosetting resins, but only with very soft wheels, and with special caution against overheating.

ABS. In buffing ABS, a two-step method is the general practice. First, the part is cut down or polished, using commercially available wheels made of unbleached cotton discs. The second step involves wiping or coloring to increase the luster, using the same kind of wheels. Both the polishing and wiping wheel should be about 6 in. thick and run at 1200-3000 rpm. A final softer wheel, kept free of polishing compound, is used for finishing off or wiping.

For ashing, a high-speed wheel (400 sfpm)

built up of muslin discs which have been sewn together, is employed. Water and pumice are used for the cutting action and heat production.

In polishing ABS, an exceptionally high gloss finish may be obtained by the use of muslin discs, felt, or sheepskin, with various polishing compounds. Detergents and soap solutions are used to reduce frictional heat and static electricity build-up.

Acetal. The ashing operation removes deep scratches, parting lines, or harsh surface imperfections. In many instances, the ashing operation in buffing can be by-passed if the surface imperfections on the part are not severe. Ashing is done on a wheel made up of alternating 12 and 6 in. diameter muslin discs wetted with a slurry of pumice and water. Wheel speed during the operation should be kept at approximately 1000 rpm. The acetal part should be held lightly against the wheel and kept in constant motion to prevent burning and irregular ashing.

The polishing and wiping operations are carried out on the same type of buffing wheel as is used for ashing. Instead of using a wet wheel, however, one-half of the wheel should be operated dry and the other half lubricated with a polishing compound, such as jeweler's rouge. The acetal part should first be held against the lubricated half for polishing, and then wiped clean of polishing compound with the dry half. Wheel speeds of 1000 to 2000 rpm are suggested.

Styrene. Ashing wheels are built up, alternating spaces and discs from muslin. A suitable mixture of #00 pumice mud is used as an abrasive and may be applied to the wheel.

In buffing, two operations are used. The first consists of polishing the piece, or "cutting it down." The second involves wiping or "coloring" to increase the luster. Wheels for the first operation usually consist of unbleached cotton discs laid up alternately with two layers of 5 in. and 2 layers of 12 in. discs. The wiping wheel is composed of 2 layers of 12 in. and 4 layers of 5 in. unbleached cotton discs laid up alternately. Both wheels should be approximately 6 in. thick and run at about 1200-3000 rpm.

After polishing the piece with a suitable compound, such as a rouge or greasy tripoli, a clean wheel which is made up softer than the polishing wheel is used for finishing off or wiping. Keep this wheel clean or it will tend to accumulate compound. Line or chalk "coloring" compound can be applied to the wiping wheel to remove any grease and bring out a lustrous surface. The pieces need only be wiped on this wheel.

In polishing styrene, care should be exercised to prevent a temperature rise over 175°F, or else crazing and gumming will occur. As a precaution, avoid polishing under too great a pressure, on too stiff a wheel, or with insufficient cooling.

Thermosetting Plastics. For thermosetting plastics, a polishing operation with greaseless compositions on muslin buffing wheels is recommended for the removal of surface defects, light or residual flash, and marks from machining operations, and for the smoothing of irregularities left by the belt sander. The cutting face is formed and maintained by periodic transfer of greaseless compound in bar form to the face of the revolving wheel. A wheel coated in such a manner presents a fast-cutting face similar to the surface of emery cloth, and because of its resilience has the ability to smooth irregularly shaped parts without distorting or gouging. Although conditions and materials differ somewhat with the individual job, it is common procedure to use a No. 220 greaseless compound on a full-disc loose muslin buff at 5000 linear fpm.

Cut-down buffing is a procedure that converts a dull sanded surface into a smooth semi-gloss, preliminary to final luster buffing. In cases where a very high luster is not needed, this becomes the final operation. For cut-down buffing of thermosetting plastics, compositions or "waxes," are used, composed essentially of a fast-cutting buffing powder in a grease binder. A dry, fast-cutting bar with no free grease is well adapted to this operation, and produces a minimum of buffing dirt. Sewed, pocketed or full-disc loose buffs are used, depending on the imperfections to be removed. Speeds range from 4000 to 6000 linear fpm.

Luster-buffing compositions or "waxes" are composed essentially of the finest abrasive buffing powders, such as levigated alumina, in a grease binder. The powders are finer and contain less grease than those used for cut-down buffing. Loose muslin buffs at 4000 to 5000 linear fpm are generally used.

Pigments can be added to white luster-buffing bars to match the shade of the plastic being buffed. This has the distinct advantage of coloring spots of the filler, such as wood flour, that may be exposed, and of not being notice-able if not thoroughly removed from the molded article. In cut-down bars, black pig-ments are often incorporated for the same reason, but light or bright-colored pigments are not effective, because of the dark color of the buffing powders used in fast-cutting bars.

If compositions with excess grease have been used, the soft residual film of grease can be removed by wiping with a clean, dry, soft buff to expose the lustrous surface.

MACHINING PLASTICS

Each type of plastic has unique properties, and therefore can be assumed to have different machining characteristics—far different from those of metallic materials familiar to many manufacturing engineers. The principal con-siderations are described in the Introduction. below.

Introduction*

Lower Modulus of Elasticity (Softness). Thermoplastics are relatively resilient when compared to metals; therefore, the forces in-volved in holding and cutting must be adjusted accordingly, and the material properly sup-ported to prevent distortion. Even within each family of plastics this characteristic will vary to some extent. For example, TFE-fluoroplastic

* The Introduction contributed by the Machinability Data Center, operated by Metcut Research Assoc., Inc., 3980 Rosslyn Dr., Cincinnati, Ohio 45209 for the U.S. Dept. of Defense, Defense Supply Agency. It originally appeared in the 1972-1973 Modern Plastics Encyclopedia Issue, published by McGraw-Hill, Inc., New York, N.Y., and is reprinted through the courtesy of the publishers. Tables 25-1 through 25-13 in this chapter are reprinted from "Machining Data Hand-book,' second edition, 1972, published by the Machinability Data Center.

and polyethylene have relatively low resistance to deformation, whereas polystyrene is harder and usually more brittle than either of the other two.

Plastic Memory. Elastic recovery occurs in plastic materials both during and after machin-ing, and provision must be made in the tool geometry for sufficient clearance to provide relief. This is because the expansion of com-pressed material, due to elastic recovery, causes increased friction between the recovered cut surface and the relief surface of the tool. In addition to generating heat, this abrasion causes tool wear. Elastic recovery that occurs after machining also explains why, if proper precau-tions are not used, drilled or tapped holes in plastics often are tapered or become smaller than the diameter of the drills that were used to make them and also why turned diameters often become larger than the dimensionals that were measured just after cutting.

Low Thermal Conductivity. Heat conduc-tivity of plastics is substantially less than that of metal. Essentially all of the heat generated by cutting friction between the plastics and metal cutting tool will be absorbed by the cutting tool. The small amount of heat con-ducted into the plastics cannot be transferred to the core of the shape, and the temperature of the surface layer will rise significantly. This heat must be kept minimal or be removed by a coolant to ensure a good job.

Coefficient of Thermal Expansion. The co-efficients of thermal expansion of plastics are greater, roughly by 10 times, than those of metals. Expansion of the plastics caused by the heat generated during machining increases fric-tion, and consequently, the amount of heat produced. Here again, adequate cutting-tool clearances are necessary to avoid rubbing.

Softening Point. The softening, deforma-tion, and degradation temperatures of plastics are relatively low. Gumming, discoloration, poor tolerance control, and poor finish are apt to occur if frictional heat is generated and allowed to build up. Thermoplastics having relatively high melting or softening points, such as nylon or TFE-fluoroplastic, have less ten-dency to become gummed, melted, or crazed in machining than do plastics with lower melting points. Heat build-up becomes more critical in plastics with lower melting points.

General Property Considerations. The modulus of elasticity for metals is 10 to 60 times greater than that of every plastic. Therefore, tool forces cause much greater deflection during the cutting operation for plastics. Tool forces increase considerably as tools become dull; therefore, it is essential that one use sharp tools. Another factor having considerable influence on tool force is the rake angle. Both the cutting force and the thrust force are higher for negative or zero rake angles than for the recommended positive rake angles.

The high coefficients of expansion of the plastics will cause problems in several ways. Dimensional control will be a problem because small variations in temperature will cause considerable dimensional change. The temperature rise during cutting, caused by heat generation from the shear zone and friction, will cause the workpiece to expand and rub, thus causing more heat.

Consideration of the properties of the work material is important in specifying the best speeds, feeds, depth of cuts, tool materials, tool geometries, and cutting fluids. The machining data given in Tables 25-1 through 25-13 represent starting recommendations provided by the *Machining Data Handbook*. It should be recognized that some of the materials may be cut at higher cutting speeds without resulting in loss of reasonable tool life. But higher speeds usually result in thermal problems with plastics.

Five general guidelines for tool geometry when machining plastics are:

1. To reduce frictional drag and temperature, it is desirable to have honed or polished surfaces on the tool where it comes into contact with work.
2. The geometries of tools should be such that they generate continuous-type chips. In general, large rake angles will serve this purpose because of the force directions resulting from these rake angles. Care must be exercised so that rake angles will not be so large that brittle fracture of workpieces result and chips become discontinuous.
3. Dr. A. Kabayashi has data* to indicate that there exists a critical rake angle in single point turning of plastics. He has found that the critical rake angle is dependent on depth of cut, cutting speed, and type of plastics material.
4. Drill geometry for plastics should differ from that for metals in two major respects: (a) wide polished flutes combined with low helix angles should be used to help eliminate packing of chips that causes overheatings and (b) the normal 118° point angle is generally modified to 70 to 120°. Table 25-1 gives drill-point geometry recommendations for several plastics materials.
5. Cutting-tool geometry for multiple tooth cutters can be arrived at by using the principles discussed above.

Drilling and Reaming

Tables 25-1 and 25-2 provide starting recommendations for drilling many thermosetting and thermoplastics materials. Drilling is a very severe operation because of restricted chip flow, inherent poor rake angles, and variable cutting speed across the cutting edge. Wide polished flutes will help provide chip clearance. The wide flutes also provide easy entrance for the cutting fluid. High clearance angles will help to prevent the drill flank from rubbing in the bottom of the hold. All tool surfaces in contact with the workpiece should be honed or polished to reduce frictional heat. Helix angle and drill point geometry are given in Table 25-1.

Drilling of Thermoplastics. In view of the variety of materials and operations involved, the following discussion is of a general nature. It is followed by some special instructions for specific thermoplastics.

Standard horizontal or vertical drill presses, single, gang, or multiple, can be used for drilling thermoplastics.

Drills are commercially available that are especially designed for plastics.

Drills having one or two wide and highly

* "Machining of Plastics," published by McGraw-Hill, Inc., New York.

Table 25-1. Drill geometry*

Material	Helix Angle	Point Angle	Clearance	Rake
Polyethylene	10°-20°	70°-90°	9°-15°	0°
Rigid polyvinyl chloride	27°	120°	9°-15°	–
Acrylic (polymethyl methacrylate)	27°	120°	12°-20°	–
Polystyrene	40°-50°	60°-90°	12°-15°	0°
Polyamide resin	17°	70°-90°	9°-15°	–
Polycarbonate	27°	80°	9°-15°	–
Acetal resin	10°-20°	60°-90°	10°-15°	–

* Reprinted from "Machining of Plastics" by Dr. A. Kabayashi, published by McGraw-Hill, Inc.

polished or chrome-plated flutes, narrow lands and large helix angles are the most desirable, since they expel chips with minimum friction, and hence with minimum overheating and gumming.

Points should have an included angle of 60 to 90 degrees, and a lip clearance of 12 to 18 degrees. A substantial clearance on the cutting edges makes for a smoother finish. Drill points must be sharpened frequently and carefully, with care to avoid loss of the desired point angles. Carbide-tipped points will hold cutting edges longest, may be used at high speeds, and in some applications do not require to be cooled by liquid.

The use of liquids as lubricants or cooling agents should be avoided if practicable, since it necessitates a subsequent cleaning of the articles. In the drilling of most holes, an air blast will suffice to assist in clearing chips, and to prevent overheating which would cause clogging.

The speeds used in drilling holes in thermoplastics will depend upon the type of material and size and depth of the holes. In general, speeds will be decreased with increase in the size of the hole, and increased with the hardness of the material.

Drilling equipment must be in good condition if accurate holes are to be obtained. Loose spindle bearings or bent or poorly sharpened drills will give inaccurate results with any material. The speed should be the greatest that will not cause burning or gumming, and the feed should be slow and uniform, to produce a smooth hole of uniform diameter. Chips should

be removed by frequently withdrawing the drill from the work to clear the hole. In deep holes, the application of cutting oil or other cooling agent will prevent sticking of the chips. Pressure should be relaxed near the termination of through holes, to prevent breakthrough. In deep holes, an intermittent relaxing of the drill pressure will reduce clogging and run-off.

Drilling can be expedited by specially designed drill jigs. It is also possible to automate machining operations via self-feed drills that are ganged up around a product and can simultaneously drill $\frac{5}{32}$ to $\frac{1}{4}$ in. diameter holes at the rate of 4 to 5 products per minute. The operator loads a shell into the machine, pushes the start button, and the units advance, drill, and retract.

(1) *Cellulose Acetate and Acetate Butyrate.* Standard twist drills developed for wood or metal are frequently used on acetate or butyrate, but the drills especially designed for plastics, mentioned above, are preferred.

Drilling with the conventional drills of the type used for metal requires much slower speeds and feeds to give a clean hole and to keep the material from gumming, and more frequent backing out of the drill to clear chips. Quality of work can best be controlled by air-operated feeds and mechanical feeding devices.

Reaming of holes drilled in acetate or butyrate is not recommended. Where accurate dimensions are required in thin sections, good results are obtained by drilling to within about 0.001 in. of size and then running a hardened polished rod through the hole to smooth it.

Table 25-2. Drilling*

Material	Hardness	Condition	Speed fpm	Feed, in./revolution Nominal hole diameter, in.								HSS Tool Material
				$1/16$	$1/8$	$1/4$	$1/2$	$3/4$	1	$1\frac{1}{2}$	2	
Thermoplastics												
Polyethylene Polypropylene TFE-fluorocarbon Butyrate	$31R_R$ to $116R_R$	Extruded, molded or cast	150–200	0.002	0.003	0.005	0.010	0.015	0.020	0.025	0.030	M10 M7 M1
High impact styrene Acrylonitrile-butadiene-styrene Modified acrylic	$83R_R$ to $107R_R$	Extruded, molded or cast	150–200	0.002	0.004	0.005	0.006	0.006	0.008	0.008	0.010	M10 M7 M1
Nylon Acetals Polycarbonate	$79R_M$ to $100R_M$	Molded	150–200	0.002	0.003	0.005	0.008	0.010	0.012	0.015	0.015	M10 M7 M1
Acrylics	$80R_M$ to $103R_M$	Extruded, molded or cast	150–200	0.001	0.002	0.004	0.008	0.010	0.012	0.015	0.015	M10 M7 M1
Polystyrenes	$70R_M$ to $95R_M$	Molded or extruded	150–200	0.001	0.002	0.003	0.004	0.005	0.006	0.007	0.008	M10 M7 M1
Thermosets												
Paper or cotton base	$50R_M$ to $125R_M$	Cast, molded or filled	200–400	0.002	0.003	0.005	0.006	0.010	0.012	0.015	0.015	M10 M7 M1
Fiber glass, graphitized, and asbestos base	$50R_M$ to $125R_M$	Cast, molded or filled	200–250	0.002	0.003	0.005	0.008	0.010	0.012	0.015	0.015	M10 M7 M1

* Reprinted from "Machining Data Handbook," 2nd ed. (1972), published by Machinability Data Center, Metcut Research Associates Inc.

(2) *Polystyrene:* The most important factor is the efficient removal of chips by the drill during operation. Chips often tend to pack in the flutes and fuse together because of the frictional heat developed and its effect upon styrene, which is thermoplastic. To minimize this fusing, highly polished flutes with a slow helix are used. Generous side relief will also help reduce friction, but the cutting speed and feed are the prime factors in drilling a clear, true hole.

The suggested drilling speed for styrene ranges between 75 to 150 ft/min for high-speed steel drills.

The most satisfactory type of drills used have been high speed steel with a thinned web and drill point angle of 90 degrees for small holes and increasing the angle to 118 degrees for large holes. The major manufacturers of twist drills supply special drills for use with plastic parts.

For accurate work and to minimize breakage of drills, guide bushings should be used (see Fig. 25-3). The design and heat treatment of these bushings are the same as used for drilling ferrous and nonferrous metals. Ample chip clearance is necessary if a through hole is drilled. Some provision in the jig is, of course, essential to ensure proper location of the work.

The use of coolants is suggested. These coolants have been successfully applied by

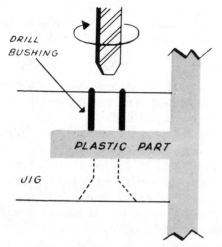

Fig. 25-3. When drilling polystyrene, guide bushings should be used to ensure accuracy and to minimize breakage of drills. (*Courtesy Monsanto*)

passing the drill through a felt wick wet with the coolant. This technique minimizes cleaning of the part.

Where a blind hole is drilled, a faster helix often facilitates chip removal.

(3) *Acrylics.* Because of the transparency of acrylics, it is usually desirable that the inside of a drilled hole have a high finish. Hence the drilling of acrylics requires extra care, and may call for special drills.

Though standard metal-type twist drills are satisfactory for the average drilling job in acrylics, they have a tendency to grab in large or deep holes. Better results are obtained with a sharp drill having a flute angle of 17 to 18 degrees, an included lip angle of 70 degrees, and a lip clearance angle of 4 to 8 degrees. The lands of the drill should be highly polished and about one-fourth the width of the heel. Special drills for acrylics, now commercially available, usually have a slow spiral with highly polished flutes. Outstanding results can be achieved with a jet drill, the point of which is cooled with a lubricant fed through a hole in the drill. Holes of large diameter can be cut with either hollow-end mills or fly cutters. It is important that tools for drilling acrylics be kept free of nicks and burrs. All types of tools can be used on standard vertical or horizontal presses.

Rates of feed can be determined only by experience. The proper feed will result in smooth, continuous spiral chips or ribbons. Feed should be slowed as the depth of the cut increases.

In drilling deep holes in acrylics, a lubricant is needed to prevent clogging and possible burning or scarring of the wall of the hole. An air blast to cool the drill will be found beneficial in all cases.

(4) *Nylon.* Nylon can be drilled satisfactorily with conventional twist drills, but more rapidly with special drills, designed for plastics, and having deep flutes, highly polished to facilitate removal of chips. In some, the flute leads are much longer than in conventional drills.

(5) *Polycarbonate.* Standard high-speed twist drills perform satisfactorily (Fig. 25-4). Drill life in polycarbonate is 5 to 6 times greater than in low carbon steel. For even

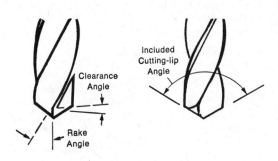

THESE ANGLES ARE:

Included angle	90° with coolant 60° without coolant
Clearance angle	15°
Rake angle	Depends on drill spiral. May be ground-to-zero or five negative for thin sheet drilling.

Fig. 25-4. Recommended drill design for polycarbonate. (*Courtesy General Electric*)

longer life and sharper cutting edges required for high-speed work, carbide-tipped drills are recommended. An added advantage of carbide-tipped drills is the absence of gumming, even without air or liquid coolants. There is no tendency for the drill to break out of the bottom of the piece or chip the edge of the hole, even when the drill is forced. Holes can be enlarged with larger drills without hogging or chipping.

Turning speeds for drilling will vary with the surface finish desired and the degree of induced surface strain acceptable. The best hole is obtained with surface speeds of 200 to 300 in./min for drills less than $\frac{1}{4}$ in. in diameter and speeds of 350 to 450 in./min for drill from $\frac{1}{4}$ to $\frac{1}{2}$ in. in diameter, when machined dry. A cooling medium should be used with speeds of 500 to 700 in./min for drills under $\frac{1}{4}$ in. in diameter, and 1500 to 1600 in. min for drills $\frac{1}{4}$ or $\frac{1}{2}$ in. in diameter. A feed rate of 0.001 to 0.0015 in. per revolution will give the best results.

No matter what type of drill is used, cutting edges must be kept sharp. Dull drills give poor surface finish and undersized holes. Burrs tend to heat up the work and induce machined-in stresses.

Use of a coolant is suggested to lower induced strain. Commonly used cooling media are: air, air-water spray mist, or very light machine oil. Avoid using standard cutting oils as they are not compatible with polycarbonate resin.

Annealing of drilled parts lowers strains and ensures optimum part performance.

(6) *Acetal.* Standard twist drills and special "plastic" drills are suitable.

Standard twist drills are normally supplied with drill point angles of 188 degrees, and lip-clearance angles of about 12 degrees. These drills work better on acetal if the included drill point angle is reduced to about 90 degrees, although the 118 degree angle can be used. The lip clearance angle should be maintained within the range of 10 to 15 degrees.

"Plastic" drills are furnished with included point angles of 60 and 90 degrees, and with lip-clearance angles of 10 to 15 degrees. The cutting edges of these drills are usually flattened slightly to provide zero or negative rake. They have extra-wide, highly polished flutes and a low helix angle to ease the removal of chips. For best performance with acetal the cutting edges of the drill should be sharpened in the usual manner (not flattened) so that a normal positive rake is obtained. A 60 degree included angle is of some advantage when drilling through thin sections (approximately $\frac{1}{16}$ in. thick), but the 90 degree point is better for thicker work.

During drilling, the work should be firmly supported and securely held. For deep holes, the drill should be raised frequently during drilling—about every $\frac{1}{4}$ in. of depth—to clear the drill and hole of chips. A jet of compressed air should be directed into the hole to disperse chips and cool the drill.

(7) *Polysulfone.* Normal steel-working tools work well with polysulfone. A configuration of 12 to 15 degrees clearance angle, 118 degree point angle, and 5 degree rake angle may be used for any drilling operation.

Small holes can be enlarged readily without chipping. When drilling completely through a piece of polysulfone, there is a tendency for the drill to break out of the bottom of the piece or chip the edge of the hole. This can be eliminated by backing up the piece and reducing the rate of feed.

(8) *PPO Resins.* Standard high-speed steel twist drills with a rake angle of 5 degrees and an included angle of 118 degrees perform well. It is important to use sharp drills to maintain the quality of machined parts. Recommended drilling speed are shown in Table 25-3.

Table 25-3. Recommended Drilling Speeds for Unfilled and Glass-Reinforced PPO Resins.

| Diameter (in.) | SPEED (IN./MIN.) | |
	Unfilled PPO	Glass-Reinforced PPO
1/8	150	700
1/4	300	750
3/8	450	750
1/2	500	750
5/8	650	875
3/4	775	800

(9) *ABS.* ABS may be drilled effectively because of its hardness and rigidity. The most important factor is the removal of chips by the drill. Chips will tend to pack in the flutes and, as a result of frictional heat, fuse together. To minimize fusion, highly polished flutes with a slow helix are recommended, with generous side relief, and the use of coolants. Cutting speeds and feed rates also influence the drilling of clear, true holes.

A standard drill press is adequate for ABS; however, the drill bit should be ground to scrape rather than cut. Customarily high-speed, double-fluted twist drills are satisfactory, but superior finish will result from bits having a point with a 90 degree included point angle, 300 degree helix angle, and wide polished flutes. Backing up the drilling surface is generally desirable, and is mandatory with thin stock.

A moderate feed rate of 0.001-0.005 in. per revolution will tend to avoid burring and overheating. A slow feed rate and consequent residence time may cause heat build-up and tool drift. Drill surface speed of 60-180 sfpm is recommended. Holes with a depth of up to five times diameter may be cooled with a jet of forced air; for deeper holes, cooling with water is recommended.

Single-fluted drills, with the required frictional characteristics for piloting and chip removal capacity, allow relatively good heat dissipation with surface speeds of 300 to 500 sfpm for small diameters, depth-to-diameter ratios of 25 : 1, and with plunging feed rates.

Burnishing double-fluted, small twist drills having 118 degree point angles and 0.003-0.005 in. off center, allow 65-100 sfpm with feed rates of 0.010-0.055 in. per revolution.

Hole saws with a skip tooth blade can be mounted in the drill press to cut larger holes.

Reaming Thermoplastics. Even though drilled holes are easy to make, they occasionally lack the precision or finish required for optimum performance and appearance. The answer is reaming. Like drilling, reaming is a simple process requiring only conventional tools.

Fluted reamers provide both accuracy and good finish. Properly sharpened, they assure shearing of material from side walls rather than trouble-making dislocation. Standard high speed steel units, straight or fluted, are usable without alteration of the cutting edge, chamfer or rake angles. The normal chamfer angle is 45 degrees; the normal rake angle is 5 degrees.

The helically fluted reamer is preferred over the straight-flute type because it provides a smoother cut, finer finish, and can be used in both through and blind holes. The straight flute reamer is limited to through holes.

While reaming can be done dry, the use of coolants will produce better finishes. Water is the preferred coolant, although light machine oils can be used. Standard cutting oils should be avoided.

Recommended conditions for reaming thermoplastics and thermosets are shown in Table 25-4.

Drilling of Thermosetting Plastics. The following table gives a general basis for selecting speeds of drills for conventionally molded thermosetting plastics.

Drill Size	rpm
No. 1 through No. 10	1700
No. 11 through No. 27	2500
No. 28 through No. 41	3000

Table 25-4. Reaming*

Material	Hardness BHN	Condition	High-Speed Steel Tool									Carbide Tool								
			Speed (fpm)	Feed Inches Per Revolution† Reamer Diameter Inches						Tool Mtl	Speed fpm	Feed Inches Per Revolution† Reamer Diameter Inches						Tool Mtl
				$1/8$	$1/4$	$1/2$	1	1.5	2			$1/8$	$1/4$	$1/2$	1	1.5	2	
Thermoplastics Polyethylene Polypropylene Fluorocarbons Butyrates	$31R_R$ to $116R_R$	Extruded, molded, or cast	250 to 300	.006	.008	.010	.010	.012	.015	M2 M7 M1	500 to 600	.006	.008	.010	.010	.012	.015	C-2
Nylon Acetals Polycarbonates	$79R_M$ to $100R_M$	Molded	250 to 300	.004	.006	.008	.010	.012	.015	M2 M7 M1	350 to 450	.004	.006	.008	.010	.012	.015	C-2
Acrylics	$80R_M$ to $103R_M$	Extruded, molded, or cast	200 to 300	.006	.008	.010	.010	.012	.015	M2 M7 M1	300 to 400	.006	.008	.010	.010	.012	.015	C-2
Thermosets Paper and cotton base reinforced	$50R_M$ to $125R_M$	Cast, molded or filled	200 to 250	.003	.003	.004	.005	.005	.005	M2 M7 M1	250 to 300	.003	.003	.004	.005	.005	.005	C-2
Fiber glass and graphitized base reinforced	$50R_M$ to $125R_M$	Cast, molded or filled	100 to 150	.002	.002	.002	.003	.005	.005	M2 M7 M1	150 to 200	.002	.002	.002	.003	.005	.005	C-2

* Reprinted from "Machining Data Handbook."
† Based on a 6-flute reamer.

Drill Size	rpm
No. 42 and up	5000
$\frac{1}{16}$ in.	5000
$\frac{1}{8}$ in.	3000
$\frac{3}{16}$ in.	2500
$\frac{1}{4}$ in.	1700
$\frac{5}{16}$ in.	1300
$\frac{3}{8}$ in.	1000
$\frac{7}{16}$ in.	600
$\frac{1}{2}$ in.	600
A and B	1700
C through O	1300
P through Z	1000

To drill holes in molded articles or to remove flash or fins in molded holes, it is best to use standard high-speed steel drills with deep flutes. Nitrided high-speed drills do not require frequent sharpening, and will last a long time. Drill points should be ground to an included angle of 70 to 90 degrees, and have a lip clearance or relief made by grinding the back away to $\frac{1}{16}$ in. wide, which reduces friction between the drill and the work and gives clearance for the chips. Backing off the cutting lip (rake angle) prevents the grabbing which occurs with a drill with a normal point, and will sometimes prevent chipping of the hole when the drill breaks through the under side of the work.

Most drilling is done without a lubricant, but a blast of air at the drill point will keep the drill and work cool, prolong the life of the point, and help clear away chips. Drill speeds should be from 100 to 150 fpm, or faster if proved by trial. Drills should be about 0.002 to 0.003 in. oversize. For drilling thin sections, the point of the drill can be ground with a sharper included angle to stop chipping around the hole.

Some manufacturers will make to order special drills for use in long-run or automatic drilling operations. These are made on slow-twist blanks and tipped with tungsten carbide. This is about the most economical drill for phenolics if the production will warrant the cost, and the best for very deep holes.

In drilling deep holes, good results are obtained, however, with steel drills having specially polished flutes and 0.0001 to 0.0002 in. of chronium plate.

For drilling through holes in canvas-filled materials, the drill may be specially ground. The end is ground like that of a wood drill. The outer edges of the drill are cut like circle-cutting tools, while the center acts as a pilot. Thus, at the breakthrough the cutting is done through a thin section supported on both sides by heavier areas, and the final chip is a disc with a hole.

Reducing the friction between the drill and the material by grinding the drill off-center (which results in a slightly larger hole) will often prolong the life of the drill.

Cast Phenolics. For small holes, drill speeds of about 2800 to 12,000 rpm are commonly used. Drills of diameter $\frac{1}{4}$ in. or more should have large flutes, for efficient removal of material, and the cutting edges should be ground with a negative rake. For the drilling of small holes for self-tapping screws, the hole is usually one drill size smaller than the screw.

Rapid production is obtained by a multiple drill assembly. Multiple drill heads are likewise effective. Where neither is available, drilling with a jig can be made both fast and accurate.

Tapping and Threading

General tapping recommendations for thermoplastics and thermosets are given in Table 25-5. Finish ground and polish flute taps are recommended because less frictional heat will be generated. It is recommended that oversized taps be used because of elastic recovery of plastics materials. Oversized taps are designated:

H1: Basic --- basic + 0.0005-in.
H2: Basic + 0.0005 basic + 0.0010-in.
H3: Basic + 0.0010 basic + 0.0015-in.
H4: Basic + 0.0015 basic + 0.0020-in.
H5: Basic + 0.0020 basic + 0.0025-in.
H6: Basic + 0.0025 basic + 0.0030-in.

Amount of oversize depends on elastic recovery properties of the material and sizes of holes. The number of flutes determines the chip space and the chip load per tooth; therefore, some compromise must be made. In general, the two-flute taps are preferred for holes that measure up to $\frac{1}{8}$ in.

Table 25.5. Tapping*

Material	Hardness	Condition	Speed (fpm)	HSS Tool Material
Thermoplastics	$31R_R$ to $125R_M$	Extruded, molded, or cast	50	M10, M7, M1
Thermosets	$50R_M$ to $125R_M$	Extruded, molded, or filled	50	M10, M7 M1
Reinforced plastics Silica fiber- reinforced phenolic resin (Refrasil)	55† to 75	Molded	25	M10, M7, M1

* Reprinted from "Machining Data Handbook."
† Barcol hardness.

Tapping Thermoplastics. Unless special accurate tapping machines with lead screws are available, it is unwise to attempt Class 2 or 3 fits. In any case, a higher percentage of rejects and higher costs may be expected, especially with nylon.

United States Standard (American Coarse Thread Series), Whitworth Standard (British Standard Series) and Acme are generally satisfactory. Sharp V-threads are to be avoided because the apex is easily broken. Coarse-pitch threads are preferred because they are stronger.

Bottom taps should be avoided whenever possible. If a bottom tap must be used, it should be used in a second operation done by hand, and only when a Class 2 or 3 fit is required. For maximum strength and dimensional stability, all tapped parts should be annealed to relieve the stresses set up by the tapping.

Before tapping, it is recommended that the hole be drilled to such size as to permit not more than 75% of a full thread to minimize difficulty in clearing the tap.

To obtain effective clearance of chips, with a minimum of friction, large, highly polished flutes are recommended. Taps should be nitrided or chrome-plated. All new taps should be stoned to remove burrs.

Taps for all thermoplastics should have maximum back clearance. In most cases, the pitch diameter should be 0.002 in. oversize. For tapping nylon, 0.005 in. oversize is recommended, unless a tight fit is desired.

Designs and speeds for several thermoplastics are as follows:

	Number of Flutes	Cutting Speed (fpm)
Cellulose acetate	2 or 3	50 to 100
Methyl methacrylate	4[a]	35 to 75
Nylon	3 or 4[b]	75 to 125
Polystyrene	3 or 4[c]	25 to 35

[a] Grind back rake angle to about 2° positive.
[b] Or use No. 2 flute spiral.
[c] Grind to zero rake.

The use of air or a lubricant is not essential in tapping, but it facilitates clearing the chips and permits faster tapping. The tap should be backed out before enough chips are formed to block the cooling agent.

In threading or tapping thermoplastics to fit a metal bolt or nut, allowance should be made for the difference in thermal expansion between the two materials. A slight increase over normal metal clearances is usually ample. But if variations in service temperature are to be extreme, dimensional changes will be too great to be accommodated in this way, and threading should be avoided.

Instead of being tapped, thermoplastics may be threaded on conventional lathes or screw machines. On automatic and semi-automatic machines, with self-opening dies, chasers should be ground to zero rake, highly polished, and chromium-plated. For nylon, a conventional rake, as for mild steel, is recommended. In most cases, two passes should be made, and the work flooded with water containing a high percentage of mild soap, or other cooling agent that will not attack the plastic. If a single-point tool is used in a lathe, the point should have a 2-degree side rake and a zero back rake (for nylon, a zero side rake and a 2 degree back rake). The best possible results will be given by diamond-pointed tools.

(1) *Polystyrene.* It is generally advisable to tap threads only in impact (modified) type styrene. As a general rule, National Coarse (NC) threads are preferred for the following reasons: Greater strength since more plastic area results due to the smaller minor diameter for the internal threads; and a slower helix results from the lesser number of threads per inch, making chip removal easier.

Spindle speeds of approximately $\frac{1}{2}$ drilling speed are suggested for tappings, and with the use of a coolant.

A 3-flute tap is self-centering and offers easier chip removal than a 4-flute type. Holes to be tapped should be slightly larger diameter than for metal in order to leave about 75% of a full thread. This will prevent the top of the thread from breaking or peeling off.

(2) *Polycarbonate.* Standard steel working taps are recommended. Taps that produce threads with slightly rounded root diameters are preferred. The use of a light machine oil during tapping is recommended to overcome resistance and reduce tap wear.

Self-tapping screws of the Parker-Kalon B-F National Screw—Type 25 thread cutting may be used with polycarbonate resin where environmental conditions permit. To ensure the best performance, it is important that the diameter of the hole is in proper relation to the diameter of the screw to be used. For example, the correct diameter of the hole for a No. 8 x $\frac{1}{2}$ in. screw is 0.147 in. and the ideal penetration is 0.75 in., or about four full threads on the screw. The wall thickness from the screw to an edge should be at least equal to the diameter of the screw.

The torque resistance of threads tapped in polycarbonate resin is high. A $\frac{1}{2}$ in. diameter bolt of 13 threads per inch, penetrating a block of resin to a depth of $\frac{9}{16}$ in. requires 48 ft lb of torque to strip the threads.

(3) *ABS.* Standard metal working tools are used. Taps should have a slight negative rake; and should produce threads with the root diameters slightly rounded in order to avoid any possibility of notch-effect weakening of the part. Taps and dies for copper and brass may be used effectively. Lubricants should always be used, and turning speeds should be very slow.

Thermosetting Plastics. Phenolics may be tapped with standard taps. The most durable are commercial ground taps with rather short chamfer and with 0.0001 to 0.0002 in. of chromium plate. If it is required to hold Class 2 or Class 3 threads, the taps should be oversize by 0.002 to 0.003 in. Holes should be chamfered to the maximum diameter of the thread. Here again is an instance where a planned mold design may help. Frequently a hole may be spotted or molded to a shallow depth, to be drilled and tapped to final depth later. In such cases, if the hole is to be tapped, the chamfer may be molded in at the same time as the hole, to eliminate one operation.

For long production runs a high-speed nitrided and chronium-plated tap, having three flutes rather than four, is recommended. Solid carbide taps will pay for themselves if used in a machine equipped with torque-control. A negative rake of about 5 degrees on the front face of the land, and ample clearance, are necessary to ensure accurate cutting and to prevent binding and chipping during backing out. For tapping mineral-filled material, which tends to dull the tap very rapidly, sometimes a carbide tap can be used, provided that the work is clamped tightly under a torque-controlled tapping spindle so as to prevent breakage of the tap.

Flutes of taps can be opened by grinding, to make room for clearance of chips. Most tapping is done dry, but oils can be applied as lubricants. Parrafin wax is sometimes applied to the point of the tap to help to prevent heating, but air blasts on the tap, operated by the stroke of the tapping head, will help to clear the chips and cool the tap and the work. This minimizes overheating, prolongs the life of the tap and promotes greater production per tap.

Peripheral speeds for tapping molded phenolics are from 50 to 80 fpm for taps up to $\frac{1}{4}$ in. diameter. Taps larger than this are impractical in phenolics.

A blind hole should not be machine-tapped unless there is plenty of clearance at the bottom for the tap.

Cast Phenolics. Cast phenolics may be easily tapped on vertical or horizontal tapping machines with standard taps. To provide strength, fairly coarse threads should be used. Tapped holes should be checked with plug gauges, as the abrasive action of the material causes wear. Standard machine screws can be used for assembly.

Cutting of threads of large diameter or coarse pitch, such as on bottle caps or jar covers machined from solid material, is done on a thread-milling machine, with a small milling cutter of the proper shape.

Turning and Milling

Tables 25-6 and 25-7 give the starting recommendations for turning many thermosets and thermoplastics. Table 25-6 covers turning, single-point, and box tools; Table 25-7 covers turning, cutoff and form tools. The definitions are as follows: (1) Single-point turning: Using a tool with one cutting edge; (2) Box tool turning: Turning the end of a workpiece with one or more cutters mounted in a box-like frame, primarily for finish cuts; (3) Turning cut off: Severing the workpiece with a special lathe tool; and (4) Form turning: Using a tool with a special shape.

Chatter, with resulting problems in tool life, finish, and tolerance, can be encountered in turning operations due to workpiece flexibility. The low modulus of elasticity of plastics makes it desirable to support the work to prevent deflection of stock away from the cutting tool due to cutting forces. Close chucking on short parts and follow rests on long parts are beneficial. Box tools are designed to support long turning operations, and should be used where possible. Water-soluble cutting fluids should be used, unless they react adversely with the work material, to reduce the surface temperature

generated at the shear zone and tool-chip interface. To reduce frictional drag and temperature, tools should be honed or polished at work contact points.

Face Milling. Table 25-8 gives starting recommendations for face milling plastic materials. Use fixtures that provide adequate support for workpieces. Milling cutters usually have multiple cutting edges. In Table 25-8, feed is given in in./per tooth, which must be converted to in./min of table travel. Too high a table feed will cause a rough surface; too low a feed travel will generate excessive heat that can cause melting, surface cracks due to high temperature, loss of dimensions, and poor surface finish. Use of mist-type water-soluble cutting fluids is recommended unless workpiece-cutting fluid compatibility problems are known to exist. To reduce frictional drag and temperature, tools should be honed or polished where they contact the workpiece.

Tables 25-9 and 25-10 give starting recommendations for end milling-slotting and end milling-peripheral. End milling is generally accomplished with a tool having cutting edges on its cylindrical surfaces as well as on its end. In end milling-peripheral, the peripheral cutting edges on the cylindrical surface are used; while in end milling-slotting, both end and peripheral cutting edges remove metal. Face milling, in contrast, applies to milling a surface perpendicular to the axis of the cutter.

Specifically, end mills are used on thermoplastics to finish off the surface when a part has been degated from the sprue. The end mill may be placed in a drill press and used as in a drilling operation. A jig should be used for safe practice and quality work.

In addition, a side mill or a reamer may be used to remove a heavy parting line from a molded thermoplastic piece. The removal of unsightly seams due to clamping and cementing may be removed in the same manner.

Finally, milling may be used to shape or fabricate a plastic rod or bar.

Thermoplastics. For turning and milling of thermoplastics, four cutting materials fill most needs: high-speed steel, high-speed steel chrome-plated, tungsten carbide, and diamond. They may be rated as follows (the range is arbitrary from 1 (best) to 10 (worst):

	Inter-rupted Cutting	Sur-face Finish	Wear-resistance	Uni-formity of Finish	Tool Cost per Cut	Accuracy
High-speed steel	1	5	10	10	10	10
Chrome-plated high-speed steel	1	4	8	8	8	8
Tungsten carbide	5	6	5	5	5	5
Diamond	10	1	1	1	1	1

Tools of standard carbon steel or high-speed steel may be used for short runs, if their cutting edges are kept very sharp, and their faces highly polished. But carbide-tipped or diamond tools are almost essential for long runs, because they hold a keener edge for a longer time. Such tools make possible a highly polished surface finish. To minimize surface friction at maximum cutting speed, the side and front clearances of the tools should be somewhat greater than those of standard turning tools. Both steel and diamond end-mills should have the same cutting angles and rakes as carbide tools.

The cutting edges of all tools should be kept honed very keen. Standard carbide-tipped tools are very satisfactory except that clearance angles should be ground to from 7 to 12 degrees. The top surface should be lapped to a bright finish. A soft iron wheel impregnated with a fine grit of powdered diamond is often used. Diamond tools are usually designed like the standard carbide tools described above, except that for turning nylon a sharp-pointed tool with a 20-degree positive rake is used.

A circular tool is economical in quantity production wherever it can be used.

With most thermoplastics, surface speeds may run as high as 600 fpm with feeds of from 0.002 to 0.005 in. The speed and feed must be determined largely by the finish desired and the kind of tool used. It is difficult to make fine finish cuts on nylon except with a carbide or diamond-tipped tool.

Good results are obtained without lubricating the cutting and turning tools if feed pressure is relaxed often enough to avoid overheating. On turret lathes, good lubrication is obtained, if necessary, by flowing a mixture of equal parts of soluble oil and water, or a soap solution, over the tools.

Milling operations must be held to very light cuts, not over 0.010 in. at normal feeds. End milling with a centercutting tool may be performed on an ordinary drill press, as long as the work is rigidly supported and the feed is steady and slow. Feeds may be increased with the aid of a suitable cooling agent. It is entirely practicable to use hand feed, but for the best finishes, power feed is advisable.

Vibration of the machine will cause a poor finish.

Tool design. In turning *polycarbonate* parts, a very sharp tool of high-speed steel or carbide is best suited for clean smooth cuts. Suggested tool angles for this particular use are as follows: back rake—5 to 10 degrees; side rake—15 degrees; side clearance—15 degrees; and front clearance—5 to 10 degrees. Poly-carbonate parts can be cut without coolant at turning speeds of 1500 to 2500 in./min. A coolant, preferably water, should be used with higher speeds to minimize strains and reduce surface galling.

Cutting depths range from 1 to 100 mils. When finish or dimensions are critical, the depth should not exceed 15 mils. The cross-feed rate, which is dependent on the cutting depth, varies from 1 to 3 mils per revolution.

The finest surfaces are obtained using a round-tip cutter, a high turning speed, a shallow cut, and a low cross-feed rate. Radii of 15 to 30 mils are suggested for round-tip cutters.

In turning *polystyrene* parts, tools should be ground to produce a straight non-twisting ribbon cut. The tool should have a slight negative rake (about 2 degrees), a clearance angle of from 10 to 15 degrees, and a side rake of about 10 degrees. By setting the cutting edge at 1 to 2 degrees above the center of the work and not directly at the center, the tendency of the work

Table 25-6. Turning, single-point and box tools*

Material	Hardness	Condition	Depth of Cut (in.)	High-Speed Steel Tool			Carbide Tool		
				Speed (fpm)	Feed (ipr)	Tool Material	Speed (fpm)	Feed (ipr)	Tool Material
Thermoplastics									
Polyethylene Polypropylene TFE-fluorocarbon Butyrates	$31R_R$ to $116R_R$	Extruded, molded or cast	0.150	250-350	0.010	M2, T5	400-450	0.010	C-2
			0.025	300-400	0.002	M2, T5	450-500	0.002	C-2
High impact styrene; acrylonitrile-butadiene-styrene; modified acrylic	$83R_R$ to $107R_R$	Extruded, molded or cast	0.150	250-350	0.015	M2 T5	400-450	0.015	C-2
			0.025	300-400	0.005	M2 T5	450-500	0.005	C-2
Nylon Acetals	$79R_M$ to $100R_M$	Molded	0.150	300-400	0.010	M2, T5	500-600	0.015	C-2
Polycarbonate	$80R_M$ to $103R_M$	Extruded, molded or cast	0.150	400-500	0.002	M2, T5	600-700	0.005	C-2
Acrylics			0.150	250-300	0.008	M2, T5	450-500	0.010	C-2
			0.025	300-400	0.005	M2, T5	500-600	0.005	C-2
Polystyrenes, low and medium impact	$70R_M$ to $95R_M$	Molded or extruded	0.150	75-100	0.005	M2, T5	200-300	0.010	C-2
			0.025	150-200	0.001	M2 T5	350-400	0.002	C-2
Thermosets Paper and cotton base	$50R_M$ to $125R_M$	Cast, molded or filled	0.150	500-1000	0.012	M2 T5	750-2000	0.012	C-2
Fiber glass and graphite base	$50R_M$ to $125R_M$	Cast, molded or filled	0.025	1000-2000	0.005	M2, T5	1000-3000	0.005	C-2
			0.150	400-500	0.012	M2, T5	500-1000	0.012	C-2
Asbestos base	$50R_M$ to $125R_M$	Molded	0.025	500-1000	0.005	M2 T5	750-1500	0.005	C-2
			0.150	650-750	0.012	M2, T5	700-1000	0.010	C-2
			0.025	750-1000	0.005	M2, T5	750-2000	0.005	C-2
Silica fiber-reinforced phenolic resin (Refrasil)	55† to 75	Molded	0.050	—	—	—	200	0.010	C-2
			—	—	—	—	—	—	—

* Reprinted from "Machining Data Handbook."
† Barcol hardness.

Table 25-7. Turning, Cutoff and Form Tools*

Material	Hardness BHN	Condition	Speed (fpm)	Feed–Inches Per Revolution								Tool Material
				Cutoff Tool Width–Inches				Form Tool Width–Inches				
				.062	.125	.250	.500	.750	1.00	1.50	2.00	
Thermoplastics† Polyethylene Polypropylene Fluorocarbons Butyrates	$31R_R$ to $116R_R$	Extruded, molded or cast	250 to 350	.003	.003	.003	.002	.002	.001	.001	.001	M2, M5 HSS or C-2 Carbide
High-impact styrene Acrylonitrile-butadiene-styrene Modified Acrylic	$83R_R$ to $107R_R$	Extruded, molded, or cast	250 to 350	.003	.003	.003	.002	.002	.001	.001	.001	M2, M5 HSS or C-2 Carbide
Nylon Acetals Polycarbonates	$79R_M$ to $100R_M$	Molded	300 to 400	.005	.005	.005	.005	.004	.003	.002	.002	M2, M5 HSS or C-2 Carbide
Acrylics	$80R_M$ to $103R_M$	Extruded, molded, or cast	250 to 350	.003	.003	.003	.002	.002	.001	.001	.001	M2, M5 HSS or C-2 Carbide
Polystyrenes medium and low impact	$70R_M$ to $95R_M$	Molded or extruded	150 to 200	.002	.002	.002	.001	.001	.0005	.0005	.0005	M2, M5 HSS or C-2 Carbide
Thermosets Paper and cotton base	$50R_M$ to $125R_M$	Cast, molded, or filled	400 to 500	.003	.003	.003	.002	.002	.002	.002	.002	M2, M5 HSS or C-2 Carbide
Fiber glass and graphitized base	$50R_M$ to $125R_M$	Cast, molded, or filled	400 to 500	.003	.003	.003	.002	.002	.002	.002	.002	M2, M5 HSS or C-2 Carbide
Asbestos Base	$50R_M$ to $125R_M$	Molded	400 to 500	.003	.003	.003	.002	.002	.002	.002	.002	M2, M5 HSS or C-2 Carbide
Silica fiber-reinforced phenolic resin (Refrasil)	55‡ to 75	Molded	100 to 200	.003	.003	.003	.002	.002	.002	.002	.002	C-2 Carbide

* Reprinted from "Machining Data Handbook."
† The width of the form tool should not be greater than the minimum diameter of the part unless the part is supported to prevent any deflection.
‡ Barcol Hardness.

Table 25-8.   Face milling*

Material	Hardness	Condition	Depth of Cut,† (in.)	High-Speed Steel Tool			Carbide Tool		
				Speed (fpm)	Feed (in./ tooth)	Tool Material	Speed (fpm)	Feed, (in./ tooth)	Tool Material
Thermoplastics	31R$_R$ to 125R$_R$	Extruded, molded, or cast	0.150 0.060	500-750 750-1000	0.016 0.004	M2, M7 M2, M7	1300-1500 1500-2000	0.020 0.005	C-2 C-2
Thermosets	50R$_M$ to 125R$_M$	Extruded, molded, or cast	0.150 0.060	200-300 400-500	0.015 0.005	M2, M7 M2, M7	1300-1500 1500-2000	0.015 0.005	C-2 C-2
Reinforced plastics Silica fiber- reinforced phenolic resin (Refrasil)	55‡ to 75	Molded	– 0.060	– –	– –	– –	– 1300	– 0.009	– C-2

* Reprinted from "Machining Data Handbook."
† Depth of cut measured parallel to axis of cutter.
‡ Barcol hardness.

to climb is minimized. In all turning and most especially the facing of large diameters, coolants are essential.

Pieces should be run at speeds at which there is no vibration. This can only be determined by trial and error. Speeds up to 1300 rpm, if the piece is held rigidly, are practicable for light cuts. Either a liquid or air blast coolant will do.

For inside turning or boring, if the diameter is large, comparable speeds may be used. But in feeding coolant to small hole operations, surface speeds may have to be decreased. To avoid crazing of the surface, the surface heat should never be raised above 175°F. Accordingly, the tools must be kept sharp and cool.

In turning *ABS*, high-speed steel tools are satisfactory, but for longer service life, carbide-tipped tools are recommended. A general-purpose tool design is shown in Fig. 25-5, with recommended angles as follows: (a) Positive rake angle—0 to 5 degrees (less frictional heat build-up occurs using a positive rake, since work is cut rather than scraped); (b) Front or end clearance angle—10 to 15 degrees (this angle must be great enough to prevent tool heel from contacting the work); (c) Nose or end radius—1/16-3/16 in. (in general, the greater the nose radius, the better the surface finish at a given speed or feed rate; cutting may be reduced, however, and heat build-up increased

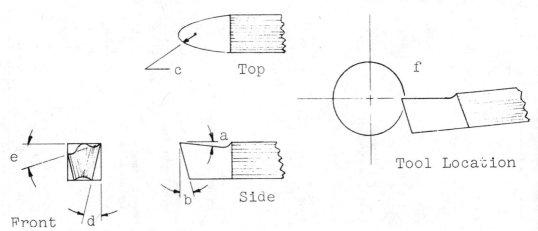

Fig. 25-5. General-purpose tool recommendations for ABS. (*Courtesy Borg-Warner Chemicals*)

Table 25-9. End Milling—Slotting*

Material	Hardness BHN	Condition	Depth* of cut (in.)	Speed (fpm)	Feed—inches per Tooth Width of Slot—Inches				HSS Tool Material except as noted
					$\frac{1}{4}$	$\frac{1}{2}$	$\frac{3}{4}$	1 to 2	
Thermoplastics	31R$_R$	Extruded,	.250	270 to 450	.002	.003	.005	.008	M2, M7
	to	molded,							
	125R$_R$	or cast	.050	300 to 500	.001	.002	.004	.006	M2, M7
Thermosets	50R$_M$	Cast,	.250	125 to 180	.002	.003	.005	.008	M2, M7
	to	molded,							
	125R$_M$	or filled	.050	140 to 200	.001	.002	.004	.006	M2, M7
Silica fiber-reinforced phenolic resin (Refrasil)	55		.250	760	–	–	.004	.005	C-2 Carbide
	to	Molded							
	75†		.050	850	–	–	.003	.004	C-2 Carbide

* Reprinted from "Machining Data Handbook."
† Barcol Hardness.

as the radius increases); (d) Side clearance angle—10 to 15 degrees (this dimension is not critical, excepting that a clearance must be allowed to minimize frictional heat build-up); (e) Side rake angle—0 to 15 degrees (to aid in removal of curls); and (f) Tool angle in relation to work—approximately 2 degrees above center-line of work (this will minimize tendency of work to climb or chatter).

Thermosetting Plastics. Molded phenolics and ureas should not be machined unless it is impossible to form the desired shape by molding. Since machining of molded articles destroys their lustrous surface, machining is restricted to articles made for utility rather than appearance. Only with a few special materials can the machined surface be satisfactorily polished.

Fillers cause difficulties in machining in the order (1) wood flour (least), (2) fabric, (3) cotton, (4) mineral, and (5) glass (most).

One of the greatest problems in machining phenolics is their inability to dissipate heat. Another is their tendency to chip.

The procedures for machining phenolics are similar to those for machining brass. The cutting action is more a scraping than a peeling. The speed in feet per minute is high and compares with those for brass. The principal difference is in the abrasive action. Phenolics dull the regular steel tool very quickly.

Most turning and boring of the phenolics on a lathe is done dry. A jet of air at the point of the cutting tool reduces the rate at which the tool is dulled, and clears away the chips.

Cast Phenolics. Regular machine-shop equipment is used in producing articles from cast phenolics. In quantity production on screw machines, operations similar to those applied to metal, such as drilling, turning, threading, tapping and milling, require only differently ground tools, and ranges of feed and speeds adjusted to the requirements of the materials.

There are a few hard-and-fast rules governing the machining of cast phenolics, but in general high speeds and light cuts are preferable. Nearly all the work is done dry. Non-alkaline cooling agents may be employed, but are rarely re-

Table 25-10. End Milling—Peripheral*

Material	Hardness BHN	Condition	Depth of Cut† (in.)	High Speed Steel Tool						Carbide Tool					
				Speed (fpm)	Feed—Inches Per Tooth Cutter Diameter Inches				Tool Mtl.	Speed (fpm)	Feed—Inches Per Tooth Cutter Diameter Inches				Tool Mtl.
					$\frac{1}{4}$	$\frac{1}{2}$	$\frac{3}{4}$	1 to 2			$\frac{1}{4}$	$\frac{1}{2}$	$\frac{3}{4}$	1 to 2	
Thermoplastics	31R$_R$ to 125R$_R$	Extruded, molded,	.050	500 to 750	.004	.006	.010	.015	M2 M7	1300 to 1500	.004	.006	.010	.020	C-2
		or Cast	.015	750 to 1000	.003	.005	.008	.010	M2 M7	1500 to 2000	.003	.005	.008	.010	C-2
Thermosets	50R$_M$ to 125R$_M$	Cast, molded,	.050	200 to 300	.005	.006	.010	.015	M2 M7	1300 to 1500	.005	.006	.010	.015	C-2
		or filled	.015	400 to 500	.004	.005	.008	.010	M2 M7	1500 to 2000	.004	.005	.008	.010	C-2
Silica fiber-reinforced phenolic resin (Refrasil)	55‡ to 75	Molded	.015	—	—	—	—	—	—	1300	.004	.005	.008	.010	C-2

* Reprinted from "Machining Data Handbook."
† The depth of cut is measured perpendicular to the axis of the cutter. The width of cut parallel to the axis can be equal to the cutter diameter, up to 1 inch maximum.
‡ Barcol Hardness.

quired. In producing articles on an automatic screw machine, where the taper of cast phenolic rod may hamper the feeding of it, the rod may first be centerless-ground accurately to uniform diameter.

Turning tools are sharpened very much as for brass. There should be plenty of clearance, 10 to 20 degrees, and a slightly negative or zero rake is desirable. The tool should be set 1 to 2 degrees above the center of the material. Cutting edges should be in condition to produce long ribbon shavings or chips. Honed tools produce the smoothest cuts. Diamond, carbide, or high-speed steel may be used to get the longest runs without resharpening or resetting.

For maximum efficiency and best working conditions, dust and shavings should be removed by a blower.

Spindle speeds in turning operations range from 450 to 6000 rpm, depending on the specific work being done and the diameter of the material, and should be regulated to give a surface speed of about 600 fpm.

Sawing

Tables 25-11 and 25-12 give general recommendations for the power band sawing of plastics. Band sawing is recommended because it is more versatile for making irregular or curve contours as well as straight cuts. Best results are obtained with a skip tooth or buttress-type tooth having zero front rake and a raker set to the teeth.

The greatest single problem in the sawing of most plastics is the dissipation of heat. Since many plastics. especially thermoplastics, have very low softening temperatures, and all plastics are poor thermal conductors, a cooling agent (liquid or compressed air) is needed, unless the cut is very short, such as in the removal of a gate.

Round saws should be hollow-ground, with burrs from sharpening removed by stoning, and band and jig saws should have enough set to give adequate clearance to the back of the blade. This set should be greater than is usual for cutting steel. It is always best to relieve the feed pressure near the end of a cut, to avoid chipping.

The proper rate of feed is important and, since most sawing operations are hand-fed, it can be learned only through experience. Attempts to force the feed will result in the heating of the blade, gumming of the material, loading of the saw teeth, and an excessively rough cut. Before the next cut is made, the saw must be cleaned. Chromium plating of the blade reduces friction and tends to give better cuts.

Above all, the saw, whether band or circular, must be kept sharp. Frequent sharpening of circular saws pays for itself. Dull bandsaws should be replaced.

Circular saws are usually from $1/32$ to $1/8$ in. thick. The width of bandsaws is usually $3/16$ to $1/2$ in.

Both thermoplastic and thermosetting resins can be sawed also by use of cut-off machines having abrasive wheels. This equipment is used for cutting rods, tubes, etc., into lengths, for slicing profiled bars of cast phenolics, and for removing large gates. Narrow-faced wheels are used (0.02 to 0.125 in.), containing usually a silicon-carbide abrasive in the range of No. 36 to No. 50. With an appropriate wheel properly used, clean cuts can be made. If necessary, water is used to prevent overheating.

Thermoplastics. As already indicated, the sawing of thermoplastics will involve a wide range of conditions, depending upon the type of resin involved. To illustrate, following are the recommendations for sawing styrene and for sawing polycarbonate, one of the so-called engineering plastics with greater toughness and heat resistance.

(1) *Polystyrene.* Of the three types of sawing operations possible on styrene, namely circular jig or band sawing, the most preferable technique is that of band sawing. This method lends itself most readily to sawing styrene since the heat build-up can be most easily dissipated.

Band Sawing–The use of coolant (usually water) is usually advisable here. Nevertheless, it has been found that for styrene sections from $1/8$ in. up, it is possible to increase the cutting efficiency and reduce the extent of lubrication by using a skip-tooth band, 4-6 teeth per in., instead of the usual metal cutting band. No coolant is required for $1/8$ in. thicknesses of

Table 25-11. Power Band Sawing, Vertical, High Speed Steel Blades*

Material	Hardness BHN	Condition	Material Thickness (in.)	Tooth Form†	Pitch (teeth per in.)	Band Speed (fpm)
Thermoplastics						
Polyethylene	$31R_R$	Extruded,	$<\frac{1}{2}$	P	10-14	4000‡
Polypropylene	to	molded,	$\frac{1}{2}$-1	P	6	3000‡
Fluorocarbons	$116R_R$	or cast	1-3	B	3	2500‡
Butyrates			>3	B	3	1500‡
High-impact styrene	$83R_R$	Extruded	$<\frac{1}{2}$	P	10-14	2300‡
Acrylonitrile-butadiene-	to	molded,	$\frac{1}{2}$-1	P	8	2000‡
styrene	$107R_R$	or cast	1-3	P	6	1500‡
Modified acrylic			>3	B	3	1000‡
Nylon	$79R_M$	Molded	$<\frac{1}{2}$	P	10-14	3000‡
Acetals	to		$\frac{1}{2}$-1	P	6	2500‡
Polycarbonates	$100R_M$		1-3	B	3	2000‡
			>3	B	3	1500‡
			$<\frac{1}{2}$	P	10-14	4000‡
Acrylics	$80R_M$	Extruded,	$\frac{1}{2}$-1	P	6	3000‡
	to	molded,	1-3	B	3	2500‡
	$103R_M$	or cast	>3	B	3	1500‡
			$<\frac{1}{2}$	P	8-10	2300‡
Polystyrenes,	$70R_M$	Molded	$\frac{1}{2}$-1	P	6	2000‡
medium and low impact	to	or	1-3	B	3	1500‡
	$95R_M$	extruded	>3	B	3	1000‡
			$<\frac{1}{2}$	P	8-10	2300‡
Fluorocarbons	$31R_R$	Extruded,	$\frac{1}{2}$-1	P	6	2000‡
	to	molded,	1-3	B	3	1500‡
	$116R_R$	or cast	>3	B	3	1000‡
			$<\frac{1}{2}$	P	10-14	4000‡
Cellulose acetate	$49R_R$	Molded	$\frac{1}{2}$-1	P	6	3000‡
	to		1-3	B	3	2500‡
	$112R_R$		>3	B	3	1500‡
			$<\frac{1}{2}$	P	10-14	3000‡
Nylon	$79R_M$	Molded	$\frac{1}{2}$-1	P	6	2500‡
	to		1-3	B	3	2000‡
	$100R_M$		>3	B	3	1500‡
			$<\frac{1}{2}$	P	10-14	5500‡
Thermosetting Plastics	$114R_M$	Unfilled,	$\frac{1}{2}$-1	P	8	5000‡
Melamine-	to	molded	1-3	B	3	4000‡
formaldehyde	$119R_M$		>3	B	3	3000‡
	$50R_M$	Cast, molded,	$<\frac{1}{2}$	P	10-14	5500‡
	to	laminated-	$\frac{1}{2}$-1	P	6	5000‡
	$93R_M$	asbestos or	1-3	B	3	4000‡
		fabric filler	>3	B	3	3000‡
	$100R_M$	Cast or	$<\frac{1}{2}$	P	10-14	5500‡
	to	molded,	$\frac{1}{2}$-1	P	6	5000‡
	$105R_R$	no filler	1-3	B	3	4000‡
			>3	B	3	3000‡
Phenolics	$108R_M$	Cast, molded	$<\frac{1}{2}$	P	10-18	75‡
	to	or laminated	$\frac{1}{2}$-1	P	8	50‡
	$115R_M$	with glass	1-3	P	6	50‡
		or mineral filler	>3	–	–	–
			$<\frac{1}{2}$	P	10-14	5000‡
			$\frac{1}{2}$-1	P	6	4000‡
Urea-formaldehyde	$94R_E$	Molded	1-3	B	3	3500‡
			>3	B	3	2200‡

* Reprinted from "Machining Data Handbook."
† P—Precision; C—Claw; B—Buttress.
‡ Recommended speed for high carbon blades.

Table 25-12.   Power band sawing, high carbon saw blades*

Material	Hardness	Condition	Pitch, teeth/in. Minimum thickness of material, in.			Speed (fpm)
			$\frac{1}{4}$ and under	$\frac{1}{4}$-$1\frac{1}{2}$	$1\frac{1}{2}$ and over	
Thermoplastics						
Cellulose acetate	$49R_R$ to $112R_R$	Molded	10	6-10	3-6	3000
. Acrylics	$80R_M$ to $103R_M$	Extruded, molded, or cast	14	6-10	3-6	3000
Nylon	$79R_M$ to $100R_M$	Molded	14	4-10	3-4	2500
Polystyrenes	$70R_M$ to $95R_M$	Molded or extruded	10	4-6	3-4	2000
Thermosets						
Melamine-formaldehyde	$114R_M$ to $119R_M$	Unfilled molded	14	6-10	3-4	4500
Phenolics	$50R_M$ to $93R_M$	Cast, molded laminated asbestos or fabric filler	14	6-10	3-6	4500
	$100R_M$ to $105R_M$	Cast or molded, no filler	14	6-10	3-6	4500
	$108R_M$ to $115R_M$	Cast, molded or laminated with glass fiber or graphitized cloth	14-18	8-10	6-8	2000
Urea-formaldehyde	$94R_E$	Molded	14	6-10	3-6	4500

* Reprinted from "Machining Data Handbook," Research Associates Inc:

styrene. Metal cutting band saws can be employed for cuts in sections of thickness smaller than $\frac{1}{8}$ in. Ample coolant should be used for the latter. Otherwise the band will be seized by gumming and cause chipping at the cut. The operating speeds for band sawing styrene should be from 465-550 fpm.

In sawing thick sections of styrene (4-5 in.) it is desirable to use a wide band slot to permit clearing of waste and also to remove the slot disc. A helpful device in sawing thick sections is to support the work on 2 wooden cleats which straddle the blade so that the specimen is held about $\frac{3}{4}$ in. from the base of the saw. For thick pieces, operating speeds of 1100 rpm and a band saw of 36 in. are recommended.

Circular Sawing. I It is adantageous to use hollow ground saw for this purpose, although this is not always necessary if all the other precautions are taken. Blade thicknesses may vary from 0.040 in. for sections up to $\frac{1}{8}$ in. and increasing to $\frac{3}{32}$ in. for heavier pieces. For general work a $\frac{1}{16}$ in. blade is a good compromise. Saw diameters may be from 4 to 10 inches. Smaller saws derive their advantages from the fact that they have lower peripheral speeds at any given spindle speed. Consequently, they have less tendency to heat themselves and the work. The thickness of the work determines, for the most part, the number of teeth per inch. Another consideration is the diameter of the saw. For a 4 in. blade, a good compromise is from 8-10 teeth per inch; for thicker work sections and larger diameter blades the number of teeth may increase to 15 per inch.

The use of coolant is almost always recommended for this operation. The saw

Saw Type	Teeth/In.	Thickness of Cut (in.)	Speed (ft./min)
Band	10 to 18	Up to $\frac{1}{4}$	2500-3000
		$\frac{1}{4}$ up to 1	2000-2500
		1 up to $1\frac{1}{2}$	1500-2000
		$1\frac{1}{2}$ and greater	800-1200
Circular	10 to 12	$\frac{1}{4}$ to $\frac{1}{8}$	6000-8000
	8	$\frac{1}{4}$ to 4	6000-8000

should be flooded with coolant directly above and at the point of operation. Customary operating speeds are from 1800-2000 rpm. Nevertheless, if an uninterrupted flow of coolant is fed to the saw, speeds as high as 3450 rpm can be reached without damaging the material or the tool. Only when small cuts (of 2 in. or less) are taken can coolant be dispensed with in circular saw operations using normal type blades. In so doing, the feed must be unforced. A helpful device in circular sawing is to have the clearance between the blade and saw table as narrow as possible to prevent the material from pulling into the gap.

Jig Sawing. The operation is not a recommended one for styrene. The jig saw blade, because of its reciprocal action, cannot handle the liberal quantity of coolant required and consequently tends to splash it up toward the operator on the upward stroke of the blade. It is generally preferable to rout holes out rather than resort to jig sawing. For sharp cornered holes, band filing without coolant is recommended, provided the feed is light. In those instances where it is absolutely necessary to use a jig saw, standard blades may be used at a speed of about 875 cycles per minute.

(2) *Polycarbonate.* Parts molded of polycarbonate can be cut with common band and circular saws as well as with hand or power hack saws. Special blades are rarely required. Blade speeds and cutting rates are not as critical as with other thermoplastics due to the high heat resistance of polycarbonate.

Special attention to blades and cutting rates is needed only when sawing is the final or only machining operation, or when very thick sections are involved. For material 1 in. or more thick, blade and feed speeds should be carefully chosen, just as they would be for soft metals. Band saw blades with 10 to 18 teeth per inch

are satisfactory. When the blades have a set, the preferred range is 20 to 30 mils. Tooth spacing for circular saw blades ranges from large (for cutting thick sections) to very small (for cutting thin sections) The best procedure is to use the smallest spacing which gives satisfactory performance.

The recommended clearance angle for circular saw teeth is between 20 and 30 degrees. The rake angle should be 15 degrees.

High-speed steel or carbide-tipped circular saws with alternate bevel and straight mill teeth are preferred. The use of designs which eliminate the possibility of the blade body rubbing the stock during sawing improves the results, particularly when working with thick sections.

The table above gives recommended sawing conditions for polycarbonate:

Thermosetting Plastics. The sawing of these plastics is confined to bandsaws (solid or inserted-segment), carbide circular saws, and abrasive cutoff discs.

Nearly all of these plastics are extremely abrasive. Some of them, notably the glass-filled alkyds, will dull a circular saw of high-speed steel in a cut of a few inches in $\frac{1}{4}$-in. material. In production runs of this material, it is probably best to use an abrasive cutoff disc, 0.040 to $\frac{1}{16}$ in. thick, 6 to 20 in. in diameter, run at from 3500 to 6000 rpm.

Phenolics, ureas and melamines are usually cut by bandsaws, which have the advantage of dissipating the heat.

An air jet should be used at the point of contact, and a suction hose to remove the dust.

Piercing, Trimming and Routing

Thermoplastics. Small holes may be punched through thin sheets of thermoplastics on standard hand-operated arbor presses, but an ordinary punch press or shearing machine is generally used. Thin sheets may be processed cold, but thicker sheets should be heated. For acrylics, about 185°F will be required for the heavier sections, while the other materials should be heated not above about 120°F, in order to prevent damaging the surface and finish.

It is common practice to remove ring gates from injection molded articles by shearing, even to the extent of making multiple-cavity punching dies for multiple-cavity moldings. It is important that the punch and die fit closely, in order to avoid producing ragged edges.

Blanking or shaving dies are frequently used for removing parting lines and flash lines. The punch and die are sometimes heated so as to leave the best possible finish. Routing and shaping are done on standard woodworking equipment. For fine cuts at high speed, single- or multiple-bladed fly cutters should be used.

Routing is a grooving or milling operation. It is done by means of cutters of small diameter having three or more teeth of the desired profile, and mounted on vertical spindles which project above a metal table top. These cutters revolve at high speeds, e.g., 12,600 rpm, so as to produce smooth cuts.

The work is held in a special fixture equipped with a metal master guideplate which is kept in contact with a metal collar on the table, directly below the cutter. The fixture is moved by hand.

For bringing rough-sawed articles of flat stock to size and shape, edge-molders are used, of the type used for finishing articles of wood. A cutter of desired profile is mounted on a vertical spindle, which revolves at 7000 to 10,000 rpm, depending on the size of the cutter and the amount of stock to be removed. The work is held in a fixture equipped with a metal master guideplate, which is kept in contact with a collar on the spindle below the cutter. The fixture is moved by hand. Fixtures of semi-automatic type have been developed to accomplish this same result; the operator loads and unloads the fixtures as they advance to and from the cutters. The cut is made rapidly to prevent burning the material.

Thermosetting Plastics. If the quantities are sufficiently large to warrant the cost of piercing tools, they provide the best method for removing flash from holes. A simple pad die with stripper plate may be made to pierce all holes and trim the outside shape of an article, at the rate of about 700 per hr. Small articles with several holes may be pierced in a dial press at the rate of about 1200 per hr. The piercing punches for removal of flash should be round-nosed and should have two or three helical grooves around the working end. These grooves should pass by the flash line in the hole, to rake the flash out of the hole. If the flash is on the surface of the article, it should be punched into the hole rather than out of it in order to prevent chipping.

Although thin sheets ($\frac{1}{4}$ to $\frac{3}{16}$ in.) of cast phenolic may be blanked with a steel-rule die, this is not recommended, because of the difficulty in obtaining accurate cuts. Blanks are better and more economically obtained by casting a bar of the desired profile and slicing or cutting it with standard slicing equipment or abrasive cutoff equipment.

Blanking and Die-Cutting

Many thermoplastics sheets are thermoformed into end-products like luggage, housings, toys, trays, etc. Most of these are easily trimmed or punched with hardened steel dies.

Other types of trimming operations may include the use of steel rules, punch dies, clicker dies, and hi-dies, depending on the gauge of the material. Punch presses, kick presses, clicker or dinking machines are used in actuating the different die-cutting methods.

Dies should be kept as sharp as possible. If they are not kept sharp, ragged or burred edges will result, possibly allowing cracks to propagate when the stock is under stress. Dies must also be kept free of grease and oils, some of which may create stress crazing or cracking in various plastics.

Many light-gauge plastics sheets may be trimmed with dies made of steel rules formed to the desired shape and placed in a wooden or metal frame. Steel rule is generally used to trim shallow drawn parts, but it can be built up economically into dies for cutting parts with draws more than two inches deep. Back-up surfaces in many cases (e.g., for the styrene

Table 25-13. Counterboring and Spotfacing*

Material	Hardness BHN	Condition	Speed (fpm)	Feed – Inches Per Revolution								Tool Material
				Nominal Hole Diameter – Inches								
				1/4	1/2	3/4	1	1.5	2	2.5	3	
THERMOPLASTICS												
Polyethylene Polypropylene Fluorocarbons Butyrates	31R$_R$ to 116R$_R$	Extruded, Molded, or cast	175	.001	.0015	.002	.002	.0025	.003	.0035	.0045	M2, M3 HSS
			400	.002	.002	.003	.003	.003	.004	.004	.005	C-2 Carbide
High-impact styrene Acrylonitrile-butadiene-styrene	83R$_R$ to 107R$_R$	Extruded, molded, or cast	175	.001	.0015	.002	.002	.0025	.003	.0035	.0045	M2, M3 HSS
			400	.002	.003	.003	.003	.003	.004	.004	.005	C-2 Carbide
Modified acrylic Nylon Acetals	79R$_M$ to 100R$_M$	Molded	200	.002	.003	.004	.006	.008	.009	.010	.012	M2, M3 HSS
			450	.002	.004	.005	.007	.009	.010	.011	.012	C-2 Carbide
Polycarbonates Acrylics	80R$_M$ to 103R$_M$	Extruded, molded, or cast	200	.002	.003	.004	.006	.008	.009	.010	.012	M2, M3 HSS
			450	.003	.004	.005	.007	.009	.010	.011	.012	C-2 Carbide
Polystyrenes, medium and low impact	70R$_M$ to 95R$_M$	Molded or extruded	150	.001	.0015	.002	.002	.0025	.003	.0035	.004	M2, M3 HSS
			300	.002	.002	.003	.003	.003	.004	.004	.005	C-2 Carbide

THERMOSETS

Material	Hardness	Form	Speed									Tool
Paper and cotton base reinforced	$50R_M$ to $125R_M$	Cast, molded, or filled	300	.002	.003	.004	.006	.008	.009	.010	.012	T15, M33 M41 Thru M47 HSS
			500	.003	.004	.005	.007	.009	.010	.011	.012	C-2 Carbide
Silica fiber-reinforced phenolic resin (Refrasil)	55† to 75†	Molded	25	.003	.004	.005	.007	.010	.012	.014	.016	T15, M33, M41 Thru M47 HSS
			150	.004	.005	.006	.008	.011	.013	.015	.017	C-2 Carbide
Fiber glass, graphitized, and asbestos base	$50R_M$ to $125R_M$	Cast, molded, or filled	200	.002	.003	.004	.006	.008	.009	.010	.012	T15, M33, M41 Thru M47 HSS
			450	.002	.004	.005	.007	.009	.010	.011	.012	C-2 Carbide

* Reprinted from "Machinery Data Handbook."
† Barcol Hardness.

sheets) may be hard rubber or wood blocks oriented to use the end grains.

Heavier-gauge sheet may require stronger dies. Tool steel up to $\frac{1}{4}$ in. thick can be bent or forged to shape to make clicker dies. The hi-die is also suitable for heavy-gauge material and long runs. Parts with draws as deep as 7 in. may be trimmed without building up these dies.

Counterboring and Spotfacing

Counterboring involves removal of material to enlarge a hole for part of its depth with a rotary, pilot guided, end cutting tool having two or more cutting lips and usually having straight or helical flutes for the passage of chips and the admission of a cutting fluid.

In spotfacing, one uses a rotary, hole piloted, end facing tool to produce a flat surface normal to the axis of rotation of the tool on or slightly below the workpiece surface.

Recommendations for counterboring or spotfacing plastics are indicated in Table 25-13.

Laser Machining

A relatively new technique, which may some day hold importance in the machining of plastics, is the use of the laser, particularly (at this point in time) the carbon dioxide laser. The carbon dioxide laser emits radiation at a wavelength of 10.6 microns. This is in the infrared region and is strongly absorbed by most plastics. In essence, then, the laser concentrates this invisible infrared beam through focusing lenses to create a point in space with tremendous power density. If a workpiece is held stationary under the laser, it will drill a hole; if the piece is moved, it slits the material. And the induced heat is so intense and the action takes place so quickly, that little heating of adjacent areas takes place.

Thus far, laser machining has been evaluated for cutting such plastics as acrylic, polyethylene film, polystyrene foam, acetal, and nylon. Apparently, most plastics can be machined with the technique (epoxies and phenolics do not perform as well and it is felt that vinyl may also have problems). But activity in this area is still experimental, rather than commercial.

Jigs, Fixtures, and Automatic Feeding Devices

The primary purpose of jigs and fixtures is to facilitate machining, so as to reduce manufacturing costs and to make practicable various operations which cannot be performed by hand. With their aid it is possible to produce more pieces per hour, to use less skilled labor, to improve accuracy of dimensions, and to reduce the fitting necessary in assembling.

It is generally agreed that a fixture is a device which holds the article while the cutting tool is performing the work, whereas a jig is a device which not only holds the object but also incorporates special arrangements for guiding the tool to the proper position. It would follow, therefore, that jigs are used principally for drilling, boring, etc., while fixtures are used in milling and grinding. Regardless of the nomenclature, the most important feature of any jig or fixture is its ability to perform the work for which it was made.

Simplicity of design makes for low cost, long life, and inexpensive maintenance. Keeping this in mind, the following rules may be used as a guide to proper design of jigs and fixtures.

(1) First, it should be shown on the basis of cost studies, that cost of designing and making a proposed tool will be less than the gross savings which its use will yield.
(2) All clamping devices should be quick-acting, easily accessible, and so placed as to give maximum resistance to the direction of the force of the cutting tool.
(3) The clamping pressure must not crack or distort the article of plastic.
(4) Locating points should be visible to the operator when positioning the article, in order to minimize loading time.
(5) All bearings should be of sealed type, for protection against the highly abrasive plastic chips and dust.
(6) Proper consideration should be given to cooling the cutting tool.
(7) The fixture must be foolproof, so that the operator cannot position the article in any way but the correct one.

(8) Lubrication must be provided for moving parts.

(9) The tool should be made as light as possible by using light materials or coring out unnecessary metal.

(10) Holes should be provided for the escape of chips or dirt.

When designing jigs or fixtures, it is well to keep in mind the possibilities of converting to automatic operation if production runs will be large enough to make this worth the cost. By coupling electrical or air-powered devices and automatic feeders to the fixture, it is possible to eliminate the need of an operator except to load the hopper and remove the finished pieces. Since one operator can service a number of automatic fixtures, this may reduce the manufacturing cost per piece to a very small fraction of the original cost.

A more intensive discussion of automation as it applies to machining, assembly, and decoration can be found on p. 809.

26
DECORATING PLASTICS*

Decorating or marking, whether for aesthetic purposes or for functional purposes (e.g., to improve resistance to wear, scratching, marring, light or heat, etc.) is frequently required in the manufacture of articles from any plastic, whether thermoplastic or thermoset. Each presents its own characteristics and hence its own problems; thus we caution the reader to investigate the various techniques and their applicability to the particular plastic that is involved in any given job.

Before attempting to decorate a part, it is necessary to:

(1) Determine the decorating technique that best suits the desired end effect.

(2) Make sure that the plastic being used is suitable for the technique selected. The choice can have a significant effect on the entire finishing line. Some plastics, like the polyolefins and acetals, for example, may require pretreatment. These pretreatment systems ideally should be hooked into the production line—either just prior to decorating or just after molding.

(3) Be certain that the design of the product does not place any limitations on the successful decoration of the part.

Good product design is a key to ease of

decorating and for keeping costs in line. Reinforcing ribs, marks caused by welding or knitting of the material as it enters the mold, and even knock-out pin marks may show through, and cause poor adhesion of the coating or other decorative surface to the part, or otherwise interfere with the method of decoration. If this is the case, the part may have to be redesigned, or post-finishing operations such as sanding may be required.

Typically, hot-stamped decorations, for example, are much easier to apply to a part when the area to be decorated is nearly flush with the surface than when it is located in a deep recess. Or, when working with sharply tapered parts, the user should be aware that such a design tends to cause wrinkling or distortion because the parts roll in an arc relative to the decorating head. To prevent trouble later, part design concepts should be a major input to early decision-making on the selection of a decorating technique.

A further caution: Make sure that the part is clean prior to decorating and rid of all mold lubricants and possible contamination. In some instances it is recommended that certain mold release agents (i.e., silicones) not be used at all if subsequent finishing operations are required.

A brief comparative review of the various techniques available for decorating of plastics can be found in Table 26-1.

* Reviewed and revised by J. O'Rinda Trauernicht, Assoc. Editor, Plastics Focus.

Table 26.1. Most Widely Used Techniques for Printing and Decorating on Molded Plastic Parts.*

The Process	What It's About	Equipment	Applications	Effect
Painting[a]				
1. Conventional Spray	Paint's sprayed by air or airless gun(s) for functional or decorative coatings. Especially good for large areas, uneven surfaces or relief designs. Masking used to achieve special effects.	Spray guns, spray booths,[b] mask washers often required; conveying and drying apparatus needed for high production.	Can be used on all materials (some require surface treatment[c]).	Solids, multi-color, overall or partial decoration, special effects such as woodgraining: possible.
2. Electrostatic Spray	Charged particles are sprayed on electrically conductive parts; process gives high paint utilization; more expensive than conventional spray.	Spray gun, high-voltage power supply; pumps; dryers. Pretreating station for parts (coated or preheated to make conductive).	All plastics can be decorated.	Generally for one-color, overall coating.
3. Wiping	Paint is applied conventionally, then paint is wiped off. Paint is either totally removed, remaining only in recessed areas, or is partially removed for special effects such as woodgraining.	Standard spray-paint set-up with a wipe station following. For low production, wipe can be manual. Very high-speed, automated equipment available.	Can be used for most materials. Products range from medical containers to furniture.	One color per pass; multi-color achieved in multi-station units.
4. Roller Coating	Raised surfaces can be painted without masking. Special effects like stripes.	Roller applicator, either manual or automatic. Special paint feed system required for automatic work. Dryers.	Can be used for most materials.	Generally one-color painting, though multicolor possible with side-by-side rollers.

(continued)

* Reprinted from *Plastics Technology, Manufacturing Handbook and Buyer's Guide,* Vol. 19, No. 6, Mid-April 1973, with the permission of the publishers, Bill Communications. [a]Another painting technique is Ford's Electrocure. It involves high-speed cure via electron beam. There's also a good deal of work going on in UV curing paints. [b]Spray booth apparatus must comply with OSHA requirements. [c]Fire-resistant paints may be required for some applications.

Most Widely Used Techniques for Printing and Decorating on Molded Plastic Parts *(continued)*

The Process	What It's About	Equipment	Applications	Effect
Screen Printing	Ink is applied to part through a finely woven screen. Screen is masked in those areas which won't be painted. Economical means for decorating flat or curved surfaces, especially in relatively short runs.	Screens, fixture, squeegee, conveyorized press setup (for any kind of volume). Dryers. Manual screen printing possible for very low-volume items.	Most materials. Widely used for bottles; also finds big applications in areas like tv and computer dials.	Single or multiple colors (one station per color).
Hot Stamping	Involves transferring coating from a flexible foil to the part by pressure and heat. Impression is made by metal or silicone die. Process is dry.	Rotary or reciprocating hot stamp press. Dies. High-speed equipment handles up to 6000 parts/hr.	Most thermoplastics can be printed; some thermosets. Handles flat, concave or convex surfaces, including round or tubular shapes.	Metallics, wood grains or multicolor, depending on foil. Foil can be specially formulated (e.g., chemical resistance).
Heat Transfers	Similar to hot stamp but preprinted coating (with a release paper backing) is applied to part by heat and pressure.	Ranges from relatively simple to highly automated with multiple stations for, say, front and back decoration.	Can handle most thermoplastics. A big application area is bottles. Flat, concave or cylindrical surfaces.	Multi-color or single color; metallics (not as good as hot stamp).
Electroplating	Gives a functional metallic finish (matte or shiny) via electrodeposition process.	Preplate etch and rinse tanks; tanks for plating steps; preplating and plating chemicals; automated systems available.	Can handle special plating grades of ABS, PP, polysulfone, filled Noryl, filled polyesters, some nylons.	Very durable metallic finishes.
Metallizing 1. Vacuum	Depositing, in a vacuum, a thin layer of vaporized metal (generally aluminum) on a surface prepared by a base coat.	Metallizer, base- and top-coating equipment (spray, dip or flow), metallizing racks.	Most plastics, especially PS, acrylic, phenolics, PC, unplasticized PVC. Decorative finishes (e.g., on toys), or functional (e.g., as a conductive coating).	Metallic finish, generally silver but can be others (e.g., gold, copper).

Method	Description	Equipment	Used on	Colors
Cathode sputtering	Uniform metallic coatings by using electrodes.	Discharge systems—to provide close control of metal buildup.	High-temperature materials. Uniform and precise coatings for applications like microminiature circuits.	Metallic finish. Silver and copper generally used. Also gold, platinum, palladium.
3. Spray	Deposition of a metallic finish by chemical reaction of water-based solutions.	Activator, water-clean and applicator guns; spray booths, top- and base-coating equipment if required.	Most plastics. For decorative items.	Metallic (silver and bronze).
Tamp Printing	Special process using a soft transfer pad to pick up image from etched plate and tamping it onto a part.	Metal plate, squeegee to remove excess ink, conical-shaped transfer pad, indexing device to move parts into printing area, dryers, depending on type of operation.	All plastics. Specially recommended for odd-shaped or delicate parts (e.g., drinking cups, dolls' eyes).	Single- or multi-color—one printing station per color.
In-the-Mold Decorating	Film or foil inserted in mold is transferred to molten plastics as it enters the mold. Decoration becomes integral part of product.	Automatic or manual feed system for the transfers. Static charge may be required to hold foil in mold.	Most plastics, especially polyolefins and melamines. For parts where decoration must withstand extremely high wear.	Single- or multi-color decoration.
Flexography	Printing of a surface directly from a rubber or other synthetic plate.	Manual, semi- or automatic press, dryers.	Most plastics. Used on such areas as coding pipe and extruded profiles.	Single- or multi-color.
Offset Printing	Roll-transfer method of decorating. In most cases less expensive than other multicolor printing methods.	Ranges from low-cost hand presses to very expensive automated units. Drying, destaticizers, feeding devices.	Most plastics. Used in applications like coding pipe.	Multi-color print or decoration.
Valley Printing	Uses embossing rollers to print in depressed areas of a product.	Embosser with inking attachment or special package system.	Used largely with PVC, PE, for such areas as floor tiles, upholstery.	Generally two-color maximum.
Labeling	From simple paper labels to multicolor decals and new preprinted plastic sleeve labels.	Equipment runs the gamut from hand dispensers to relatively high-speed machines.	Can be used on all plastics. Used mostly for containers and for price marking.	All sorts of colors and types.

Integral Coloring

One decorating concept that will only be briefly covered in this section (because it does not involve in-mold or post-molding decoration) is an important characteristic of plastics—the ability to impart color to plastics at the resin manufacturing or compounding stage. Various pigments or dyes can be added to the base plastic resin so that when the resin is molded, a solid through-and-through integral color results (see p. 843). No matter how deep you scratch the plastic part, it will always show the same color. These integrally colored parts can either be used as is or post-decorated with any of the techniques described in this chapter.

A more sophisticated approach to decorating via integral coloring is the so-called *two-color* or *two-shot* injection molding technique. This produces a single part incorporating two different colors. Best known among the products made in this way are one-piece typewriter keys in which the letter on the key and the background are in contrasting colors. This technique is based on the use of two injection molding units feeding into a single specially designed mold. One part in one color (e.g., the outline of the letter on the typewriter key) is molded in the first stage. This part is then sequenced to a second stage (e.g., another mold cavity) where it serves either as an insert around which the second color is applied (as in the typewriter keys) or, in other applications, as a shell into which the second color is injected.

More recently, injection molding machines (with multiple injection units) have been designed that can inject three shots in sequence to produce products like one-piece acrylic tail light lenses in three different colors (red, amber, and white).

Similar to this method is co-extrusion technology (see p. 197), which involves the extrusion of two different colored plastics rather than the injection molding, as above. Co-extrusion techniques produce a single sheet incorporating two (or more) layers of contrasting colors. Thermoforming such sheet produces products like decorative cups with a white interior and a colored exterior.

It is also possible to add various decorative fillers (e.g., glitter, tinsel, etc.) to the base resin at the compounding level so that the products produced display unusual visual effects. Typical is a clear vinyl beach ball with tinsel suspended in the wall sections of the ball.

Finally, injection molding operating techniques are available to create a swirl or mottled effect in the finished part. Some plastics, like many structural foams, will normally come out of a mold with a wood-like grained pattern. The ability to mold-in or form-in textured surfaces on plastics adds still another dimension to the possibilities for integral coloring.

Aside from these integral coloring systems, however, the decorating of plastics usually involves the kind of post-molding finishing systems described in this section—painting, hot-stamping, electroplating, metallizing, etc. More recently, the decorating concept has backed up to the molding machine itself, via the in-mold decorating systems described on p. 751.

It should be noted too that although each method is described individually, it is possible to combine several decorating methods in sequence on one part to create even more unusual effects (e.g., spray painting, hot-stamping, and metallizing a single automotive cluster or medallion). It is important, however, when using such combinations, to make sure that the various techniques are compatible.

CONVENTIONAL SPRAY PAINTING

Painting is still one of the more popular techniques for decorating plastics, and today it is being used as much for a functional finish as a decorative finish. ABS exterior trim and styrene structural foam exterior window shutters, for example, are painted not only for aesthetic purposes but to increase their resistance to UV exposure.

There are a number of different ways to apply paint to plastic. The most common is conventional spray painting, ranging from the simple hand gun system to highly automated systems with automatic screen washers and elaborate masks for multiple-color decoration. It should be noted, however, that not all parts are suited to an automated operation by virtue of size (e.g., large complex parts) or because the volume is not large enough to warrant the expense of such an operation.

In a spray painting system, a key factor is the type of gun used. Various methods are available: (1) rotary guns (usually two guns mounted so that they rotate around a part to get at hard-to-reach spots); (2) reciprocating guns that have a back-and-forth action and travel on a curved transverse to paint the sides of complex parts; (3) spindle machines in which the parts rotate on a spindle while the guns are stationary (and can paint small parts from the top and bottom at the same time); and (4) combined motions (like rotating guns and reciprocating parts or rotating parts and reciprocating guns).

Also essential to most spray operations is the use of a mask to shield off those sections of a part which are not meant to be painted. These form-fitting masks are generally made by electroforming and while they vary considerably in complexity, they usually fall into one of these four categories:

The lip-type mask (Fig. 26-1) is used for painting a depressed name or design. In this mask a lip of metal extends down the vertical side wall of the depressed design, all the way or only part way down the side wall, depending on the result desired. This lip must be thin, yet strong. The centers of letters and numbers such as O, A, 6, 8 must be securely held in place by bridges. The fit and the lip of the mask ensure a clean sharp paint line. The draft angle of the depressed design should be at least 5 degrees.

The cap type (Fig. 26-2) is the reverse of the lip. It is used where the embossed name or design is to be kept clean while painting the

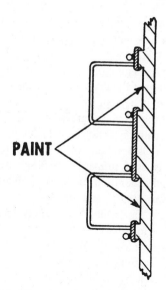

Fig. 26-2. The cap mask for capping raised design.

background. The lip of metal must cover the vertical side walls all the way to the bottom, so as to protect the embossing completely.

The plug mask (Fig. 26-3) is used for protecting a depressed design while painting the background. Its principal use is with tranparent articles such as automobile horn buttons and doors of refrigerator evaporators, and where vacuum plating is required. Here again the positive fitting essential to prevent fogging on vertical side walls and bottoms of such designs is accomplished by electroforming directly into the design. The plugs are cut out, and finished with the proper radii and draft angles to facilitate painting. They are then suspended by fine

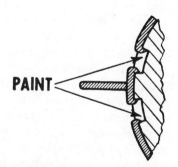

Fig. 26-1. The lip mask for sunken design.

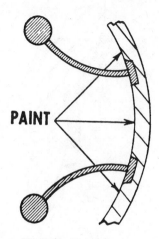

Fig. 26-3. The plug mask.

wires, usually attached to a frame so as to provide a unit which can be handled in production.

In designing molds, the engineer must remember that the draft angles of the depressed design should be kept to a minimum, preferably not more than 5 or 7 degrees.

Spray painting through a block cutout mask (Fig. 26-4) is a variant of fill-in painting. The method is used primarily in filling depressed letters and calibration marks on such articles as knobs for stoves and radios. (For articles which have fine calibrations, spray painting is not satisfactory.) As a rule, it is not practicable to mask perfectly; for masking small letters and pips, the openings in the mask would be so small that they would immediately become filled with paint. Instead, a cutout is made to enclose each character or group of characters, to confine the paint to the immediate area. Then the excess on the surface is removed by wiping or buffing, leaving the depressed characters filled with paint. Articles made in multiple-cavity molds are not fully uniform in dimensions, but the block cutout mask can usually be designed to accommodate all of them.

The mold engineer should design the mold with thought to the problems of masking. Slight changes in the demarcation of colors which will not essentially change the design of the article may definitely facilitate the masking.

In operating a spray painting set-up, it is important that the masks be washed on a

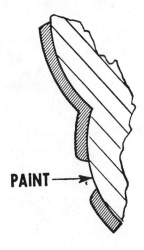

Fig. 26-4. Block cutout plane surface mask.

regular basis. Most set-ups today involve some type of automatic washing system. To achieve best production efficiency, whether the mask is on or off the machine, three or four masks are often required for each part. This gives at least one mask in the painting operation, one being washed, and one being dried.

Until recently, most masks were washed with solvents, Government regulations concerning the discharge of solvents into the atmosphere, however, has encouraged users to look in other directions. Typical: cleaning paint masks with water-based materials that are nontoxic, nonflammable, and lower cost. In one such system, a water-soluble protective film is applied to the mask surface. When the mask needs cleaning, it is immersed in hot water or a heated solution and the protective film dissolves, taking the paint along with it. In essence, the paint is not dissolved, but taken off mechanically.

Another key element in a spray painting operation is the spray booth itself (although in some automated and semi-automated operations, a closed chamber is used). Selection of the right type of booth will depend on the plastic being sprayed, the method of application, and the rate of production desired. There are three basic types on the market today:

(1) The water-wash booth incorporates a continuous waterfall in the back of the booth that literally washes paint out of the exhaust air. It is especially useful in continuous spray production lines.

(2) In a filter booth, a filtering device traps over-spray particles. It is most suitable for long runs when using slow drying or light viscosity materials or for intermittent or short runs no matter what type of paint is used. Since filters must be replaced regularly, roll-type dispensing units have been developed.

(3) Baffle booths are used where exhaust air to the outside does not have to be free of paint particles. This type is suitable for intermittent production with quick drying materials. Baffles assure an even air flow distribution through the work area of the spray booth.

Removing Mold Release Residues[1]

Use mold releases sparingly where parts are to be subsequently painted. If a mold release has to be used, zinc stearates are the least harmful to pain adhesion. Silicones should always be avoided.

Lack of adhesion, due to excessive mold release, can be quickly spotted if the adhesion problem is generally restricted to the same area on a number of parts. Look particularly at corners, or other sharp areas, where the part might be difficult to release from the mold. Mold release might be the problem if adhesion is spotty, for example, on every fifth or eighth part. This may be the interval in which the operator is having trouble releasing the part and uses excessive mold release.

To correct a mold release problem, the parts should be washed. Several methods of washing may be used. A 1% solution of Joy does a good job of removing mold release, dirt, and static charge. Alcohols, such as isopropanol or butanol, do not attack solvent-sensitive plastics and are suitable for washing parts. Do not use methanol, as it can be absorbed through the skin into the blood stream. An equal parts blend of isopropanol and VM&P Naphtha is effective, particularly where oil is on the parts. Blended solvents that are used to reduce the lacquers may be used to wash the parts, provided they do not attack the plastic. *One word of caution!* Be sure that the washing tanks do not become contaminated with mold release, thus spreading an even film of mold release over the entire part. Change solvents frequently.

Pretreatment[2]

Factors such as molecular weight, method of resin manufacture, processing conditions, and

the form and shape of the mold all affect paint adhesion and thus the coatability of plastics. Most plastics require some type of surface treatment prior to finishing. In most cases, this involves nothing more than cleaning the part of dirt or other residue (see above). For some plastics, however, particularly those in the polyolefins family, more involved treatment is required. The polyolefins, for example, have a wax-like surface that is difficult to wet.

Recommended pretreatments for various plastics are given in Table 26-2 and are described, in brief, below:

Solvent Treatment (hot or cold). This softens the substrate, but not to the extent of causing surface deformation, crazing or cracking.

Etching. Strong oxidizing agents etch the plastic surface to improve adhesion.

Flame or Heat Treatment. This also provides an oxidized surface, but does so without the use of liquid agents.

Corona or Arc Discharge. This is another technique for producing an oxidized surface, using an ozone field.

Mechanical Abrasion. Abrasive belts or grit roughen the surface to give it "tooth" and improve adhesion.

Prime Coat. The base coatings are formulated for two properties: to adhere to the plastic substrate and provide a good bond for the topcoat.

Because so many factors affect the ability of coatings to adhere to plastics, it is difficult to generalize on which surface treatment and type of coating to use. It is almost essential to develop the coating system based on samples from a preproduction run.

A number of advanced techniques for treating plastics are currently under investigation. Gas plasma surface treatment (a dry procedure that uses no acids) is being studied as a means of preparing the polyolefins (polyethylene and polypropylene) and TFE-fluoroplastics for painting. The effect of this treatment is to either cross-link the surface molecules to create a tight, coherent skin that permits stronger adhesive bonds, or to form free radicals on the polymer surface that can provide strong chemical bonds to coatings.

---

[1] By C. D. Storms, Executive Vice-President, Red Spot Paint and Varnish Co., Inc.; from a paper delivered at New York University's First National Annual Decorating Plastics Conference.

[2] Adapted from an article "Don't Trust to Luck when Painting Plastics," by T. P. Kinsella, F. J. Steslow, and J. R. Weschler, Sherwin-Williams Co., appearing in *Plastics World*, April 1973 and reprinted with the permission of the publishers, Cahners Publishing Co.

Table 26-2. Selection guide to paints for plastics*

Plastic	Coatability	Pretreatment	Recommended coating types
ABS	Good	Solvent wipe	Vinyl, modified vinyl, polyurethane
Acrylic	Good	Solvent wipe	Acrylic, modified acrylic, nitrocellulose, polyurethane, epoxy
Cellulose acetate and butyrate	Good	Solvent wipe	Acrylic, vinyl, nitrocellulose
Nylon	Fair	Detergent wash	Acrylic, vinyl, polyurethane
Phenolic and melamine	Difficult	Solvent wipe	Alkyds, polyurethane, epoxy, acrylic
Polycarbonate	Fair	Primer	Polyurethane, acrylic, modified acrylic
Polyester	Good	Solvent wipe	Polyurethane, epoxy
Polyolefins	Difficult	Flame solvent or primer	Polyurethane, nitrocellulose, modified acrylic
Polyphenylene oxide-based	Fair	Primer	Polyurethane
Polyurethane	Good	Primer	Polyurethane
Polyvinyl chloride	Fair	Solvent wipe or primer	Polyurethane, nitrocellulose

* Sherwin-Williams Co.

Selecting a Paint [2]

Just as with metal or wood, getting a tightly adhering coating depends on more than mere good luck. It requires carefully matching the paint system and the plastic, plus close attention to the nature of the plastic. Some of the guidelines to be followed are listed below and recommended paints for the various plastics are given in Table 26-2.

Heat-distortion Point and Heat Resistance. This determines if a bake-type paint can be used and, if so, the maximum baking temperature the plastic can tolerate.

Solvent Resistance. The susceptibility of the plastic to solvent attack dictates the choice of paint system. Some softening of the substrate is desirable to improve adhesion, but a

solvent that aggressively attacks the surface and results in cracking or crazing obviously must be avoided.

Resiaual Stress. Molding operations often produce parts with localized areas of stress. Application of coatings to these areas may swell the plastic and cause crazing. Annealing of the part before coating will minimize or eliminate the problem. Often it can be avoided entirely by careful design of the molded part to prevent locked in stress.

Mold-Release Residues. Excessive amounts of mold-release agents often cause adhesion problems. To assure satisfactory adhesion, the plastic surface must be rinsed or otherwise cleaned to remove the release agents.

Plasticizers and Other Additives. Most plastics are compounded with plasticizers and chemical additives. These materials usually migrate to the surface and may eventually soften the coating, destroying adhesion. A coating should be checked for short- and long-term

[2] Adapted from an article "Don't Trust to Luck when Painting Plastics," by T. P. Kinsella, F. J. Steslow, and J. R. Weschler, Sherwin-Williams Co., appearing in *Plastics World*, April 1973 and reprinted with the permission of the publishers, Cahners Publishing Co.

softening or adhesion problems for the specific plastic formulation on which it will be used.

Other Factors. Stiffness or rigidity, dimensional stability and coefficient of expansion of the plastic are factors that affect the long-term adhesion of the coating. The physical properties of the paint film must accommodate those of the plastic substrate.

Water-Based and High-Solids Coatings

In the early 1970's, the government enacted various air pollution laws that restricted the emission of photochemically reactive solvents into the atmosphere. Included in this classification were the aromatic hydrocarbons like toluol and the branched chain ketones. As a result, a great deal of effort has been expended on creating water-based coatings that do not involve the use of solvents. These paints are based on resins such as alkyds and acrylics and can be applied by the same methods used for solvent-based coatings. Water-based paints are advantageous for use on plastics because the absence of a solvent eliminates attack on the plastic substrate. However, such systems are higher in cost than solvent-based coatings and, at this writing, there are still limitations in terms of the system's failures in humidity or water soak. Water-based coatings are available today for use on acrylic, ABS, polystyrene, and polycarbonate.

Another development to meet air pollution regulations is the high-solids coatings (greater than 80% non-volatile content). These do not eliminate solvent emission completely, but greatly reduce the amount of solvent given off. Coatings based on polyester, epoxies, polyurethanes, and alkyds are available in these formulations. Because of their high solids content, special equipment is sometimes required for their application.

Ultraviolet or Radiation Curing

One of the newer and more important developments in the painting of plastics is the use of ultraviolet or electron beam radiation to cure the coating (usually based on unsaturated polyesters or acrylics). These techniques are adapt-

able to the air pollution laws described above since the coatings involved contain little or no solvent. Moreover, they do not require heat for drying and curing. Curing cycles are virtually instantaneous and the finished coatings show extreme hardness and abrasion resistance.

The first operational plant for radiation curing of coatings on plastics (mostly instrument panels) was set up by the Ford Motor Co. in the early seventies. In this set-up, the molded parts are first sensitized for electrostatic painting and a pigmented acrylic monomer coating is then applied by reciprocating electrostatic disc guns. The coated moldings pass into a concrete-enclosed radiation chamber, where a nitrogen atmosphere prevents oxidation and improves color control. In the curing chambers, the moldings pass under twin banks of electron accelerators which polymerize the coatings. The parts may be stacked immediately after painting. The system is called Electrocure and has been made available for licensing to other companies.

In the Ford operation, cure speeds are on the order of 1 to 2 sec. The coatings used are 85 to 90% solids to minimize pollution. It is also possible with this type of system to paint untreated or unprimed polypropylene.

Special Effects[3]

To meet the needs of the office machine industry, the dead front control panel was created. This consists of a control panel that looks black when the machine is off but gives colorful instructions when on. Many variations to create this effect exist. One technique follows:

1. Windows, numbers, or letters are silk screened second surface (or first surface) and left open.
2. These openings are then sprayed or silk screened with various transparent colors on the second surface.
3. Other instructions or dials that are to

[3] Adapted from a paper on painting plastics delivered at New York University's First National Annual Decorating Plastics Conference by C. D. Storms, Executive Vice-President, Red Spot Paint and Varnish Co.

remain visible permanently are applied first surface.

4. Finally, the part is generally sprayed first surface with a clear flat lacquer or catalyzed material to further hide the windows.

The use of transparent colored plastic and/or a transparent colored topcoat also enhance the dead front effect.

An innovative method combining hot-stamping and painting has also been developed for back-lighted controls such as automotive heater and air conditioning controls. The part is molded of clear or tinted acrylic with raised letters to indicate the control instructions such as temperature, etc. It is then hot-stamped with a clear (transparent) or pigmented (opaque or translucent) hot-stamp tape commonly called a *put-on foil*. The part is generally painted black (or other colors if desired) to eliminate light leaks and to achieve the proper color and gloss. The part is then re-hot-stamped (commonly called *take-off foil*) to remove the paint from the raised letters. Thus, when the part is back-lighted the only area that light is transmitted through is the letters. Often the part is sonically welded into an ABS housing.

Painting Structural Foams. Structural foam plastics, such as polystyrene foam or ABS foam, are commonly used to increase the strength of a part by producing a thicker piece with the same amount of plastic. This is accomplished by foaming the plastic (usually with nitrogen). Examples of such parts are chair backs, exterior house shutters, television cabinet bases, and automotive seat backs.

Foamed plastics are a problem to finish due to swirl marks and varying densities inherent in the molding process. This can cause soak-in of coatings and result in an uneven gloss-glossy and dull spots.

Swirl marks are particularly objectionable in wood-grain applications. Following barrier coating, the stain tends to hang in the swirl depressions just as it hangs in the deeper molded-in grain pattern. This causes a "dirty" wood-grained effect.

To combat these problems, a high solids catalyzed primer has been developed. The coating does a good job of filling in the swirl marks and providing a base of uniform density for further finishing. In some cases where extremely smooth, high gloss surfaces are required, the part is sanded. The primer accepts most lacquer type topcoats or barrier coats for further finishing.

Wood Graining. There are a number of painting techniques available to achieve a wood-like effect on plastics, particularly when working with structural foam plastics that offer molded-in grains. A typical system is as follows for achieving a wood grain effect on polystyrene structural foam parts:

The part is first sprayed with a barrier coat to shield the plastic from being attacked by the solvents in subsequent stains and topcoats. A wiping stain is then sprayed on and wiped with an absorbent cloth to give the basic tone effect. This stain is force dried so that it will not dissolve when the next glaze is applied. This subsequent glaze is a toning glaze that is sprayed on the part, then brushed into the raised portions to give a complimentary dark and light hue within the grained portion. Finally, a clear topcoat is sprayed over the entire piece to provide gloss and abrasion resistance.

This is considered a four-coat system. Others are available, such as a one-coat system, in which the part is sprayed with a stain, then brushed off to achieve the grain effect, or a two-coat system in which parts are sprayed with different color barrier coats to achieve a wood-like appearance, then covered with a wiping stain to fill in the grain.

OTHER PAINTING TECHNIQUES

Electrostatic Painting. One of the newer techniques for painting plastics involves creating an electrical attraction between paint and part so as to minimize overspray and paint waste. In electrostatic painting, the plastic part is first rendered conductive by applying salts such as hygroscopic ionized salt solutions. Methods of application for this pretreatment can range from conveyorized dip to rotation of the part on spindles.

The paint and the part are then oppositely

charged to create the electrical attraction. In essence, a high-voltage charge is placed on the paint particles while the part to be painted is grounded. Becuse of this attraction, paint that would normally be lost as over-spray in a conventional spray painting operation will wrap around the part and be attracted to the sides and back. Some systems claim up to 100% paint utilization.

One limitation of electrostatic coating is that it may not lend itself to mask painting. It is usually most applicable where the entire surface of the part is to be painted. Typical of the applications that have found use for electrostatic painting are ABS grilles and other automotive parts which incorporate surfaces at many angles from the gun. In choosing a solvent system for electrostatic painting it is important that the system be formulated to have the proper polarity to wrap around the part and to be slow enough to prevent dry spray in deep recessed areas and on the rear of the part.

Wiping. One modification of spray painting is the paint-wipe operation. While there are no clear-cut borders in the selection of paint/wipe *vs* mask painting, paint/wipe should be seriously considered when dealing with a product where, with a fairly flat surface, a lower surface is to be painted. The wipe set-up itself is an automatic operation. Its use is probably more expensive than the cost of a mask, but a much higher volume can be handled, and in some cases, where very fine lettering or design is involved, it is almost impossible to design masks to do the job.

Wiping can be either manual or automatic. However, unless labor cost is very low, it is rarely feasible to consider manual wipe. In automatic wipe, product indexes from the paint station to a wipe station where an absorbent fabric—generally a disposable soft paper, but some processors use a reusable fabric—passes over the top of the product taking paint off the flat surface and leaving the paint in the indented parts. A light solvent spray can be used to aid in paint removal.

Roller Coating. Roller coating is just the reverse of wiping. This coats a raised surface of a part, leaving recessed parts free of paint. No secondary operations are required; the process does not paint lower surfaces even when only slightly raised sections are used.

Flow Coating. This is another painting option that involves paint applied from overhead nozzles onto racked parts. The excess coating drains off the parts after painting and is recirculated for use in the next cycle, resulting in very little waste. The process is not continuous, but many parts can be handled in a single cycle. In flow coating, part geometry plays a key role, since features like sharp corners could lead to excessive paint build-up.

Dip Coating. In this technique, the part is literally dipped into a paint bath, the idea being again to limit waste through overspray. Automatic systems for raising and lowering parts are used to control the uniformity of the coating thickness. However, parts that might trap excessive air cannot be dip coated because they will require prolonged drain time.

Flocking. This decorating technique is also classified as a variation on painting. In essence, it involves coating a plastic part with adhesive lacquers tinted to match the flock color; then, the flock (i.e., fibers) is applied with a special flock spray gun. Flock can give a plastic part the appearance and feel of such materials as felt, suede, or short-napped mohair.

SCREEN PRINTING[4]

The screen printing process can print one or more colors in a pass in a machine with multiple printing heads. It is used to print bottles, molded items such as cups, and can be used on flat film and extruded shapes.

Formerly called silk screening, the preferred term is now *screen printing.* Of all the screen materials, silk is now the least popular media used. The most widely used screen media are now nylon, polyester, and stainless-steel mesh—you get longer life and better reproducibility.

One advantage cited for the process is quick color change-over, making it especially suitable

[4] Adapted from "Printing and Decorating Plastics" by J. O'Rinda Trauernicht, *Plastics Technology*, September 1969. Reprinted by permission of the publisher, Bill Communications.

for containers where ingredients change frequently or for private labeling. With multiple heads, the machines will print four colors in one pass—even more if more heads are added on the machine.

Among special ink developments are: flame-dry inks that dry in seconds and do not require long dryers; hot inks that dry instantly—no dryers are needed at all. While hot inks are now commercial, they are still undergoing development. They are used most widely in the dairy industry, and are pretty well limited to polyethylene. The flame-dry inks are also largely limited to use on polyethylene.

Standard inks are so-called *cold inks*—they do not have to be heated before use—and are composed of quite a variety of materials—e.g., PVC products require a vinyl-base ink—based on type of material, gloss, product resistance, and drying time.

Adhesion of most inks in screen printing is excellent because the ink actually fuses to the product in many cases.

Both polyethylene and polypropylene must be treated, of course, before being printed. The most effective way is in an integrated machine where surface treatment takes place right before printing. A time lapse will mean that the treatment will lose some of its effect. Three methods are used: (1) flame treatment—considered the most practical, this method is the most widely used; (2) corona discharge; (3) chemical—this method is not widely used, but does have the advantage of not destroying the gloss on polypropylene.

Basic components of a screen printing system are: (1) screen (prepared from a stencil); (2) frame to hold the screen; (3) squeegee. In operation, the squeegee is pulled across the screen, forcing the ink through the mesh to make the print. Either the screen frame or the squeegee can be moved (a top-notch system should be set up so that both move). This, of course, is a highly simplified description of the process.

(1) Screens. Nylon screens are reportedly preferable for plastics, but with hot inks stainless steel must be used. Average life for a screen is on the order of 50,000 impressions although much higher life has been reported.

Split screen printing is available for simultaneous printing of two colors.

(2) Frames. Most frames are wood, although screens framed on metal are desirable because the metal frames last longer and will not warp.

Specially operated frames are available for screens to stretch all four sides of the screen at the same time, lengthening mesh life.

(3) Squeegees. The "simple" squeegee makes the difference between top-notch screen printing and a less than second rate job. The squeegee blades are best if they are made of resilient rubber or plastic. The shape of the printing edge of the squeegee determines the sharpness and deposit of ink. A square edge, for example, is for printing on flat objects; a square edge with rounded corners is for extra-heavy deposits and for printing light colors on dark backgrounds. A double-sided squeegee with a beveled edge is for direct printing on uneven surfaces, for bottles, or containers.

The hardness of the squeegee is also an important factor. Blades vary from 45-50 to 75-80 Durometer. Softer blades are recommended for rough surfaces and for depositing more color. One other guideline: when buying or cutting a squeegee, the blade should be at least 2 in. longer than the print.

The squeegee should print from either direction, especially in high-speed machines, to prevent build-up of ink on one side of the screen. However, in certain jobs, because of the effect on registration, a one-way sweep is necessary.

When buying a screen printing unit, also consider automation possibilities. For example, a fully automatic system with conveyor feed to handle up to 8000 parts/hr orients parts, flame treats if necessary, cleans the part, destaticizes it, prints, and dries.

A rule of thumb for automation is 5000 plus items/day. But, this does not mean 5000 parts of a mixed variety of products. If product demands dictate, you should have separate screening equipment for flats, ovals, and rounds and conical shapes. While it is possible to convert from one type to another, change-over times involved usually make it impractical, especially if jobs are frequently run.

Optionals–A wide variety of optionals is available with screen printing equipment.

Nest carriers hold objects which must be printed in a special manner. This permits printing right up to the edge of the part; some operations may benefit from a vacuum system that holds parts securely throughout printing and feeding and releases them at take-off stage.

Most static eliminators are placed after the feeding mechanism and ahead of the printing head. Tinsel or tinsel bars are commonly used although much more expensive and efficient mechanical equipment such as electric static eliminators and remotely controlled ionized air units are available. (Also, permanent and temporary antistatic compounds can be used either in the resin itself or as a coating on the product.)

Dryers–Selection of dryer will depend on type of equipment used. For semi-automatic units, a flat-bed oven with circulating air, heated by gas or electricity is usually used. In automatic equipment, an oven with special nests that carry products through an up and down configuration are desirable. Drying time for regular cold inks can be anywhere from $\frac{1}{2}$ to 15 min.

Unscramblers are offered to orient bottles prior to putting them in the decorator.

Automatic registration in the printing station can be built into equipment. With this type of attachment there must be a notch in the part to match the registration mark in the printing station.

Other optionals: automatic feed and take-off; expandable mandrels to hold pieces with irregular inside dimensions and soft plastics or open tubes; stepless variable speed drive.

HOT STAMPING[5]

Of the various commercially accepted methods of decorating plastics, the application of roll leaf and heat transfers are the only "dry" processes. When the decorator purchases roll leaf or transfers, he is purchasing a film that has been pre-coated with a metallic or pigmented

[5] By Frank J. Olsen, Kensol-Olsenmark, Inc., Melville, N.Y.

finish (in the case of roll leaf) or a film that has been preprinted with a multi-color design (in the case of heat transfers). When he receives this material, which is sold in roll form, it is completely dry.

The process of transferring the decoration from roll leaf to the plastic surface is known as *hot stamping*. It is a versatile and inexpensive technique and is applicable to all thermoplastics and some thermosets. Ordinarily, thermosets are not decorated in this way since melting of the surface is required for adhesion.

Hot stamping involves the use of heat, pressure, and dwell time to effect the transfer of the color or design from the carrier film to the plastic. Basically, a heated metal or rubber die is brought into contact with the plastic under pressure. The coated film is mounted between the die and the plastic. After a dwell time of normally less than 1 sec, the coating is released from the carrier film and adheres itself to the plastic surface. When the die is removed, the part is completely dry and can be handled immediately.

There are three basic effects that can be obtained by the use of roll leaf and heat transfers:

(1) An inlaid or debossed effect (roll leaf only) where a heated metal die is used to deposit a roll leaf coating below the surface of the plastic to be decorated. A ladies' compact would be a good example of this effect.

(2) A surface coating of roll leaf or heat transfer. In this instance, a heated silicone rubber die or roller is used to transfer the coating onto a smooth or slightly textured surface. The sides of a wood-grained plastic television or radio cabinet are decorated by this technique.

(3) An embossed effect is obtained when a roll leaf or transfer coating is deposited onto the surface of a raised section in the plastic. This is normally applied through the use of a silicone rubber die. The lettering and chrome border highlighting on automobile instrument panels is an example of this effect.

Roll Leaf

Any roll leaf (or heat transfer, as described below) consists of a coating or several layers of coatings which are released from a carrier film and deposited on a substrate. In a typical roll leaf, we start with a carrier which, in most instances, is a thin film (approximately 0.0005 to 0.001 in.) of cellulose acetate or polyester film. In the manufacture of roll leaf, this carrier would be coated on one side with a release agent which would melt when heat is applied to the uncoated side of the carrier. A wax is the most common release coating. The carrier would then pass through a color or metallic coating process. The third coat would be a heat-activated sizing. Naturally, to obtain acceptable adhesion, the sizing must be formulated to bond onto the particular material being decorated.

When extreme abrasion and solvent resistance are required, a fourth protective topcoat is added. This coating is applied between the release coat and the color coat in pigment foils. In dyed metallics, the dye is added normally to the protective topcoat. Protective topcoats have been developed to protect the roll leaf from alcohol (common in the cosmetic industry), salt spray (for automotive applications), detergents (for appliances and housewares), etc.

Simulated wood-grain roll leaf is a relatively new development. These foils are currently in popular demand for wood graining radio and TV cabinets.

More complex roll leaf constructions are used in the manufacture of these wood-grain foils. In some wood grains, five coatings are applied. The first is the release, second is a protective topcoat, and the third coat is a grain. This grain is applied by a textured roller which is engraved to simulate the particular grain desired. The fourth is a base coating which is a solid pigment color to complete the two-tone effect. The size coating is then applied to complete the process. Wood-grain roll leaf is available with a gloss ("lacquered") finish, or with a matté ("hand rubbed") finish. When a matté finish is required, an etched polyester carrier strip is used.

In general, most roll leafs are formulated to release at from 250 to 300°F. Lately, manufacturers have been developing foils that will release at lower temperatures for several technical reasons. Lower die heats will reduce the distortion when hot stamping thin wall parts, and also will lengthen the life of silicone rubber dies and the soft metal dies (brass, zinc, etc.).

Heat Transfers

Multi-color heat transfers consist of a carrier strip which is printed with a specific design, pattern, picture, etc., in several colors, rather than continuously coated, as in roll leaf. The carrier strip is normally a thicker material than used in roll leaf. In most cases, it is a Kraft paper or polyester film.

Heat transfers have certain advantages when two or more colors must be applied in register to an item. Since the carrier is printed with the design already in register by the supplier, exact color registration is guaranteed. When applying multi-colors by the roll leaf process, the item must normally be passed through the press several times. The number of passes equals the number of colors to be applied. Multi-color hot stampers have been developed—normally a turntable system feeding several presses. However, many objects cannot be decorated on such a press due to limitations in size and shape.

Most transfers are formulated with a sizing so that they adhere to the plastic surface. However, the latest development is the formulation of inks with the same plastic as the object to be decorated. This results in fusion of the design when heat is applied. The superior abrasion resistance of this type transfer eliminates the need for a protective topcoat.

Application Equipment

As mentioned previously, heat, pressure, and dwell time are required to release a roll leaf or transfer coating from the carrier onto the substrate.

There are basically two types of machines used in the process (see Fig. 26-5). The reciprocating, or upright, press is recommended when applying a design onto a flat, convex, or

RECIPROCATING PRESS

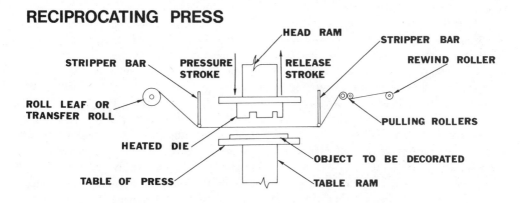

ROTARY PRESS — CYLINDRICAL DIE

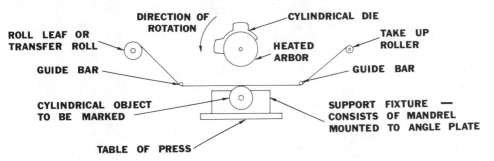

ROTARY PRESS — FLAT DIE

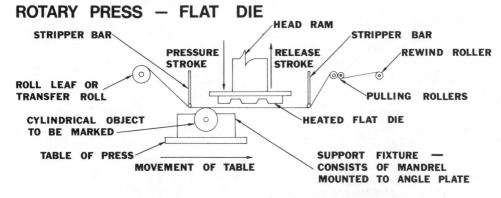

Fig. 26-5. Hot-stamping equipment. (*Courtesy Kensol-Olsenmark*)

concave surface. A reciprocating press is also used to mark cylindrical shapes where the marking area does not exceed 90° or ¼ of the circumference. The heated platen of a reciprocating press has been made as large as 24 x 36 in. Machines of this size have been used to apply a decorative bright chrome or simulated gold border on the bezel of a television set.

Rotary-type machines are recommended when decorating more than 90° of a cylindrically shaped object. They are also recommended when you desire to coat, using silicone rubber, more than 50 sq in. in a solid concentrated area. This rule of thumb applies for both roll leaf and heat transfer applications. It is difficult to eliminate air pockets which are trapped between the silicone rubber die, roll leaf (or transfer), and the part when reciprocating presses are used. In many instances, trapped air can be squeezed out when using a reciprocating press by doming the silicone rubber die. Doming is usually accomplished by adding a .005 in. shim beneath the center of the die. However, the rolling action of a rotating silicone-coated roller pushes air and gasses ahead of it as it passes across a broad area. Rotary presses also eliminate the need for extremely high pressures on broad area coverage since only line contact is made between the roller and the plastic surface.

The reciprocating press basically consists of a head ram which moves up and down, a table ram which can be adjusted up and down according to height of the plastic part, and a frame which contains the drive mechanisms, supports the pressures developed and maintains alignment between the two rams. Attached to the head ram is a heating head which supplies and maintains heat to the die and an automatic roll feed attachment which advances the roll leaf or transfers the required amount after each impression. The head ram can be driven by a direct air cylinder (for pressures up to 2 tons), an air-driven toggle system (for pressures of 2 to 10 tons), or a hydraulic or air hydraulic system for pressures over 10 tons. Straight mechanical presses have been made; however, it is difficult to vary the dwell time. The heated die should remain in contact with certain plastics such as PVC for a very short dwell time. Harder plastics such as polystyrene or acrylic require longer die contact.

Figure 26-5 (center diagram) illustrates a rotary machine which has a cylindrical-shaped hot stamping die. This die can be of etched or engraved metal or molded silicone rubber when a design or lettering is to be applied. It also shows an engraved die rotating across the surface of a cylindrical plastic object. In order to back up the impression, the item must be placed over a support mandrel which must be cut to accurately fit. The mandrel must be mechanically driven at the same speed as the rotating stamping die when designs or lettering are applied. If a smooth silicone pad is used to release a solid foil coating or a transfer design, we normally do not have to mechanically drive

the mandrel. In this case, the mandrel would be designed to turn as freely as possible. This type rotary can also be used for applying roll leaf or transfers to a flat surface. The object to be coated is mounted onto a fixture and is transported, usually by a hydrocheck table, beneath the roller. The speed of rotation of the roller must equal the speed of movement of the table. Since there is a tremendous heat drain from the roller, it must normally be both internally and externally heated. For this reason, the roller is constantly rotating even when the object to be marked is not in contact. If the roller is stopped, the external heaters could cause overheating and rapid decomposition of the rubber coating. This type rotary is presently being used for wood-graining the sides of TV cabinets and other mass areas. On some rotaries, several rollers are used to assure complete transfer of coating. Another use of the rotary shown in Fig. 26-5 (with cylindrical die) is the decorating and marking of extrusions such as hose and picture frame moldings.

The bottom diagram in Fig. 26-5 illustrates a rotary machine design, whereby a cylindrical object is rotated on a hydrocheck table beneath a flat die. Since a flat die is far less costly than a cylindrically shaped die, the main advantage of this system is low tooling costs on short runs.

Metal and Rubber Dies

A hot stamping die can be of metal or silicone rubber. Several types are shown in Fig. 26-6. Naturally, it is important that the die have the ability to conduct heat. When the designer desires a decorative design or functional marking applied to a smooth or slightly textured surface, an engraved metal or molded rubber die must be used. A quality metal die is almost always hand engraved (by pantograph) onto a steel or brass blank. The blank must first be shaped to the contour of the part (if the surface is other than flat). In order to produce a quality die, black and white artwork (to scale) must be supplied. Because of the skilled labor required, engraved metal dies are quite expensive. Many times, photo-engraved zinc, magnesium, or copper dies may be used if the

METAL DIES — (CROSS SECTION)

ENGRAVED BRASS
OR STEEL DIE

ETCHED ZINC, MAGNESIUM
OR COPPER DIE

DIE FACE

SMOOTH, DEEP
SHOULDER

DIE FACE

RAGGED, SHALLOW
SHOULDER

METLCONE® —"SANDWICH" DIE — (CROSS SECTION)

DIE FACE

.032 COPPER OR STEEL
.062 SILICONE RUBBER
.062 ALUMINUM

SILICONE RUBBER

DIE FACE

SILICONE RUBBER — 50 TO 90 DUROMETER 1/32 – 1/16 – 1/8 THICK
ALUMINUM OR STEEL BACKING 1/32 – 1/16 – 1/8 THICK

PROPER RUBBER WIDTH — (WHEN COATING A RAISED AREA)

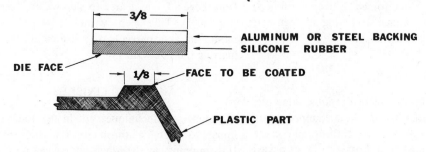

3/8

ALUMINUM OR STEEL BACKING
SILICONE RUBBER

DIE FACE

1/8

FACE TO BE COATED

PLASTIC PART

Fig. 26-6. Metal and rubber dies. (*Courtesy Kensol-Olsenmark*)

length of run prohibits the investment in an engraved die. These inexpensive dies, which are made by an acid-etching process, are also used in prototype work. It is difficult to estimate how long an etched die will last in production. This depends upon several factors: (1) If the artwork calls for fine line detail, the die will break down sooner. (2) Soft plastics, such as PVC or polyethylene will cause less die wear than the harder thermoplastics such as polystyrene and the acrylics. (3) If a plastic surface is not uniform, we must apply more pressure and dwell to the die so that the low spots can be reached. This can destroy an etched die after just a few impressions.

Molded silicone rubber dies have been successfully used to apply decorative logos, identifying marks, etc., onto plastic surfaces which have surface imperfections, variations, and surfaces which could not take the stress applied by a metal die. The surface of a molded rubber die will flex at low pressures and conform to irregularities in the part. However, if too much pressure is applied, the stretch in the rubber will cause distortions of the mark. A new die material called Metlcone (patented) has been developed to reproduce with the fidelity of a metal die, yet have the ability to flex like a rubber die. The artwork is reproduced by etching on a thin sheet of copper or steel. This is bonded to rubber which, in turn, is bonded to aluminum. The aluminum sheet is required so that the die can be easily attached to the heated head of the press through the use of heat-activated, die-bonding film or screws.

Much development work is being done at this time, to improve silicone rubber die material. Silicone rubber is available impregnated with metal filings to improve heat conductivity. Teflon coatings are being applied to rubber surfaces to reduce sticking when such dies are used to apply appliqués. Textured silicone dies are also available. A rubber die with a ticking pattern can be obtained for applying wood grain roll leaf to a smooth surface. This pattern applies a realistic porous look to the smooth surface.

Fibrous glass-reinforced silicone rubbers, which eliminate fatigue stretching as a cause of failure, are now also available.

Design Guides:

(1) Avoid positioning bosses, ribs, gates, etc. beneath an area to be hot stamped. Any imperfection will act as a shim and cause uneven stamping. If a boss or rib is placed beneath the stamping area, you will obtain sink marks over this heavy area (see Fig. 26-7).

(2) Use proper molding cycles. If the molder attempts to speed up the molding cycle beyond reasonable limits and does not allow the part to cool properly before it is ejected, excessive shrinkage and warp will result.

(3) Use the same hob for all molds if several molds are being built by different mold makers.

(4) When designing multi-cavity molds, try to position runners and gates so that all parts

will eject from the mold at the same temperature.

(5) If you have a choice of substrate, keep in mind its compatibility with the hot stamping and heat transfer process as well as its end use. The least expensive material may prove more costly after decorating.

(6) Allow sufficient draft on parts to prevent distortion during ejection. However, try to keep taper on cylindrical parts to under $\frac{1}{2}$ of a degree, if a roll-on decoration is to be performed.

(7) Polish out all tool marks in areas to be decorated. Any imperfections will be magnified when high luster metallics or colors are used.

(8) When designing parts utilizing raised lettering or patterns, try to raise these areas .030 in. or more (Fig. 26-7). We have coated raised areas of .020 in.; however, overstamping on adjacent areas could cause rejects on non-uniform or warped parts.

(9) Break the sharp corners on raised areas to prevent cutting of rubber dies (Fig. 26-7).

(10) When angular or spherical borders are to be coated, do not exceed more than a 40° angle from the horizontal. Step surface at least .050 in. from adjacent area (Fig. 26-7).

(11) When raised areas are located in a deep recess, provide a minimum clearance of .060 in. from edge of raised area to side wall of recess (Fig. 26-7). This keeps wrinkling of leaf to a minimum and prevents the heated die from marring the side wall.

METALLIC FINISHES

A number of techniques are in use today to impart a metallic finish to a plastics part. Hot-stamping, as described above, can be used to put a gold, silver, or other metallic decoration onto plastics. More recently, high luster metallized silver foils or roll leaf have been developed for exterior use (e.g., on automotive trim).

Spray plating is another method for giving plastics a metallic surface. Here, a silver or copper plating is deposited by chemical reaction of water-based solutions on the substrate. In most cases, the system requires a base coat and a topcoat.

1

POORLY POSITIONED RIB

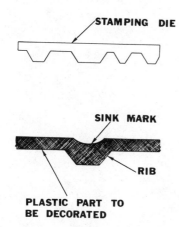

STAMPING DIE

SINK MARK

RIB

PLASTIC PART TO
BE DECORATED

2

RAISED AREAS SHOULD BE .030
OR HIGHER — BREAK SHARP EDGES

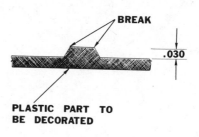

BREAK

.030

PLASTIC PART TO
BE DECORATED

3

WHEN COATING ANGULAR OR
SPHERICAL BORDERS DO NOT
EXCEED 40° FROM THE HORIZONTAL

.050 40° 40° .050

PLASTIC PART TO
BE DECORATED

4

WHEN COATING RAISED AREAS IN
A DEEP RECESS, RAISED AREA
SHOULD BE A MINIMUM OF .060
FROM SIDE WALL

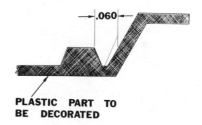

.060

PLASTIC PART TO
BE DECORATED

Fig. 26-7. Part design for hot-stamping. (*Courtesy Kensol-Olsenmark*)

In the system, silver-containing salts are deposited on a product through a dual nozzle gun. In cases where copper is to be applied, the silver is applied first and the copper is applied over the silver coat. A typical system would consist of four guns—one for water to clean the part after each processing step, one for the activator to treat the surface of the part, a dual nozzle gun for silver and a dual nozzle gun for copper—a spray booth with a special bottom to recover the silver, mixing vessels and demineralizer for water purification. Automated systems are available on a custom-made basis.

The two most popular techniques, however, for the metallic finishing of plastics are vacuum metallizing and electroplating. These are discussed below.

Metallizing. In vacuum deposition, the first step is to jig up the article so that when in the vacuum chamber no part of it will be shielded, and thus fail to be coated (see Fig. 26-8). Jigging is different for every shape and size to be metallized. Experience, or trial and error, sets the final pattern for jigging. Jigs should be made of metal.

Next, it is usually necessary to lacquer the

Fig. 26-8. Vacuum metallizing.

plastic to provide a base sufficiently shiny to give the deposited aluminum a reflective surface. Only rarely do the shape of the article and the high polish of the mold make this lacquering unnecessary. The lacquering is done by a dip with slow removal, by spraying, or by dipping and spinning, whichever will best provide a coating free from runs.

The lacquered articles are dried in an oven. All solvents should be removed, since residual solvent will interfere with attaining a vacuum in the chamber.

The jigged-up pieces of plastic are placed on rotating stations inside the vacuum machine. In the chamber, on tungsten wire filaments carried by fixtures, are hanging $\frac{3}{4}$-in. pieces of aluminum wire. The door of the chamber is closed and the vacuum pump started. When the gauge indicates the necessary vacuum, a transformer is turned on to heat the tungsten filaments so that the aluminum evaporates and coats the rotating pieces of plastic.

The thin coat of aluminum thus deposited is protected by a coat of lacquer, either waterwhite or colored, as required.

Adhesion of the lacquer to the plastics is very important to the quality of the product. For every undercoat there must be a top coat

that will not attack the undercoat, and both coats must be developed simultaneously.

It is essential that the plastic be free from mold lubricants, especially external ones. With polystyrene, no solvent will remove mold lubricants without attacking the plastic itself, and scrubbing each piece is not practicable. Hence, external mold lubricants should be avoided for articles which are to be metallized.

Scratches or flaws on the surface of the plastic will be exaggerated by metallizing. The undercoat of lacquer follows the contour of the plastic itself, and its mirror-like surface magnifies the defects underneath. Similarly, an orange-peel surface is likely to be exaggerated by metallizing.

Good results in metallizing can be promoted by the designer and maker of the mold. A highly polished mold that does not require an external lubricant will provide a favorable surface on the molded article, and simplify the process of metallizing. The problem of jigging the article for metallizing, without leaving a mark in a prominent place, can sometimes be solved by the designer of the mold, by providing a hole or a plug, or the like, by which the piece can be attached to a screw, spring, or taper pin, to facilitate jigging. Some very small articles are most economically metallized while still attached to the sprue. In such a case the gates are enlarged to ensure that the pieces will remain on the sprue until the metallizing has been done.

Aluminum does not tarnish, but the coating of aluminum is protected against abrasion by a coating of air-drying lacquer; a baked lacquer cannot be used, because the plastics themselves cannot tolerate baking temperatures. Hence, the metallized finish is appropriate only on expendable items, or other items that do not require abrasion-resistance.

Almost all plastics can be treated via vacuum metallizing. Among those considered especially receptive, however, are polystyrene, acrylic, phenolic, unplasticized vinyls, and polycarbonates. Some materials, especially the polyolefins, have posed serious problems in metallizing and development work is underway to improve their metallizing ability (see below).

Vacuum metallizing may be divided into first surface and second surface metallizing. First surface means that the decorating is done from the front side (or first surface) of the substrate. Second surface means that the decorating is done on transparent plastics from the rear (or second surface) of the substrate.

Recent Developments. While no universal basecoat (the coating that is applied prior to the deposition of the aluminum) is yet available, several formulations have been developed that work on a variety of substrates. Basecoats are available that are suitable for use on ABS, acrylic, polycarbonate, and phenylene oxide-based plastics. Others are suitable for ABS and polystyrene. In the past, polystyrene basecoats were not suitable for ABS.

A number of "fast cure" basecoats have also been developed to meet the needs of high-volume conveyorized shops. These basecoats are being baked at a minimum of 20 mins at $140°F$ and a maximum of 1 hr at $175°F$. Because of the short bake, production line speeds are faster.

Special vacuum metallizing basecoats are also available that can be applied to primerless polypropylene. Heretofore, polypropylene had to be primed or treated before basecoating.

Barrel Plating

Prior to the development of the electroplating systems described below, barrel plating was used to impart a metallic finish to small plastic parts. It is still in use today.

To prepare plastics parts for barrel plating, a conductive coat of silver is precipitated on them from silver nitrate solution, in order to make them conductive. They are then plated in rotating plating barrels that contain acid copper-plating solutions, and copper anodes. In from 12 to 24 hr, a thickness of copper from 0.002 to 0.005 in. is deposited. The articles are then polished with a steel shot in a burnishing barrel. A plating of gold, silver, or nickel is deposited on them by another barrel-plating operation.

Electroplating

Electrodeposition of a metallic coating on plastics, commonly called *plating*, can be successfully performed on a number of plastics. The two most popular substrates at this time are ABS and polypropylene, with ABS currently predominating (see Figs. 26-9 and 26-10). Polysulfone, phenylene-oxide-based resins (e.g., General Electric's Noryl), and

Fig. 26-9. Chrome-plated ABS shower head housing. (*Courtesy Borg-Warner Chemicalls*)

Fig. 26-10 Electroplated ABS housing for barometer. (*Courtesy Borg-Warner Chemicals*)

polyaryl ether are also considered candidates for electroplating.

The general procedure for electroplating plastics involves a series of preplating steps, followed by the electroplating operation (see Figs. 26-11 and 26-12). In most cases, the sequence is as follows:

In preplating, the parts are first cleaned thoroughly to remove soil, particles, fingerprints, etc., and racked.

Normally, an etching process follows, but a number of systems have been evolved that take advantage of conditioning or pre-etch steps to make the plastic more receptive to the critical etching procedure or to overcome the thermal problems involved with exposure to the heated etch baths. In one system for electroplating ABS, for example, an organic solvent pre-etch step is used to pre-condition the surface of the ABS part and permit a more uniform and controlled attack on the plastics substrate by the etchant treatment. In another proprietary system for electroplating polypropylene (a special silica-filled polypropylene is required for this system), two-stage etching is used to cut production time and to reduce the levels of heat (and thereby prevent warpage in some parts). The first etch tank contains an aqueous chromic acid solution and is run at 130°F.; the part is rinsed, then immersed in a second aqueous etching solution, this one at room temperature.

Following either cleaning and/or conditioning or pre-etching, the plastic part passes into an etch bath. This is an extremely critical part of the preplating operation, since it is here that the etch penetrates the plastic to create a micro-roughened or micro-porous surface. The roughness or porosity imparted to the plastic surface at this stage has a direct bearing on the strength of the metal-to-plastic adhesion.

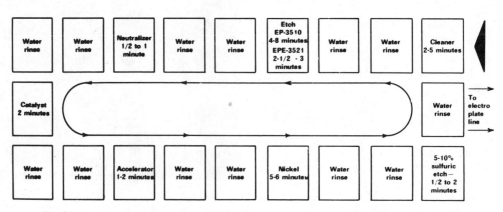

Fig. 26-11. Flow chart: electroless nickel plating process. (*Courtesy Borg-Warner Chemicals*)

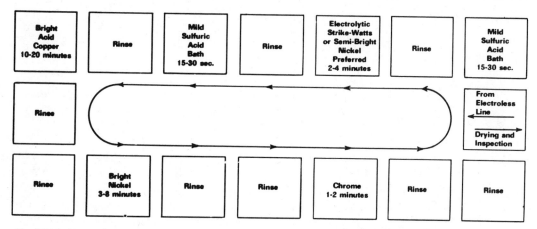

Fig. 26-12. Flow chart: typical electrolytic plating line for decorative plating. (*Courtesy Borg-Warner Chemicals*)

The plastic parts then move through rinsing and neutralizing baths (to neutralize the residues from the etch bath) and on to a catalyst treatment. In this step, colloidal seeds of palladium and tin are introduced to the plastic surface to make it more receptive to the subsequent electroless metal deposition. Accelerators are also used to enhance the effectiveness of the catalyst particles.

A thin conductive metallic coating is then deposited on the surface of the treated plastic to provide the electrical conductivity necessary for the electroplate operation. This step is known as *electroless metal deposition*. Generally, electroless metal is chemically reduced from solution onto the catalyzed surface. Either nickel or copper has been used as the metallic coating.

The electroplating procedure that follows the above pre-plating procedure is much the same as that used for plating metals. A typical sequence is as follows:

Following pre-plate, the plastic part moves into a mild sulfuric acid etch bath. Next, the first electroplate layer, copper, is deposited on the surface of the plastic. This copper layer constitutes the bulk of the electrodeposit and performs two key functions. First, it levels the etched surface to enhance the visual appeal of the finished part. Second, and more important, it acts as a cushion between the plastic and the subsequent layers of metal (which are not as ductile as copper), thereby absorbing the stresses produced by the differences in the thermal expansion of the plastic and the metal. The temperature coefficient of expansion for most plastics is three to five times greater than that of the metal plate structure.

Once electroplated copper has been put on the plastic substrate, nickel is generally deposited over the copper, followed by chromium electrodeposition. In some cases, gold can be deposited over the copper or over bright nickel; it is also possible to deposit tin or silver should the functional requirements of the application dictate.

The more common procedure, however, is to use nickel/chrome. It should also be noted that the choice of the nickel used is dictated by the end-uses of the electroplated plastic part. In exterior or outdoor applications (e.g., an automotive grille), where more corrosion protection is needed, a semi-bright nickel would be chosen over a bright nickel. It is also possible to use dual nickel systems (a sulfur-free, semi-bright nickel and a bright nickel) or even triple nickel systems (semi-bright nickel, bright nickel, and a thin deposit of a stress-inducing nickel to provide additional protection between the nickel and the chromium plate).

Part Design: For maximum efficiency in electroplating plastics, consideration should be given to designing the molded part specifically to take advantage of the plating technique. Some of the fundamental design principles for ABS are shown in Fig. 26-13.

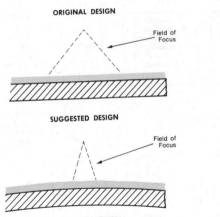

Crowned surfaces will tend to hide minor surface irregularities.

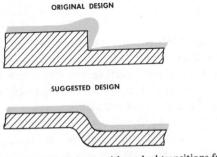

A uniform wall thickness with gradual transitions from one wall section to another will enhance the plated part performance.

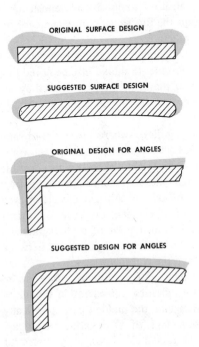

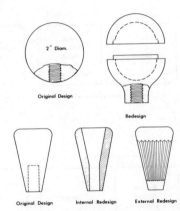

The heavy mass electroplated spherical ball (top left) failed three cycles of 160°/0°F. After internal mass was removed and part was designed into two parts using a snap fit (top right), the required three cycles of 180°/-20°F was met.

Removing excess internal bulk (bottom center) provides a more uniform cross section; adding external serrations (bottom right) acts as expansion joints which accommodate plastics thermal expansion.

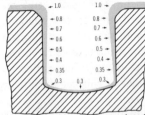

Thickness variations for electroplated nickel (in mils) in a groove with a width-to-depth ratio of 0.85.

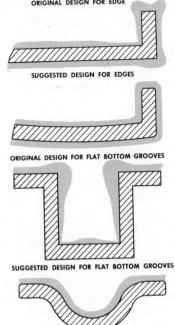

Fig. 26-13. Designing ABS for electroplating. (*Courtesy Borg-Warner Chemicals*)

Other design points to keep in mind are:

(1) Parts should be designed for molding in one piece (mechanical welds are difficult to plate).
(2) Gates should be hidden on non-critical surfaces or disguised.
(3) Deep recesses or blind holes should be avoided since plating solutions might build up in the cavity.
(4) Non-critical surface areas should be designated and designed for cathode contacts (since plating will not be as uniform or as bright in these areas).
(5) Sufficient rigidity should be allowed to prevent part warpage under elevated temperatures during plating.

PRINTING

Printing presses used to decorate plastic film and sheeting can also be used with some modification for printing or decorating molded parts. Three types are widely used, as follows: flexography, offset, and valley printers. Flexography and offset are used for identification printing on pipe and other profiles, and decorative printing on containers and other molded parts. Valley printers have found application in the decoration of vinyl floor tiles.

Here's a brief look at each.

Flexography. Flexography involves printing on a surface directly from a rubber or other synthetic plate. Each color to be printed via flexography will require a separate printing station. The press is generally a rotary type with one or more color stations. Flexography is widely used in identification printing because of its low cost plates, and because changeover is relatively simple. The process, however, is only recommended for parts of $1/2$ in. diameter and larger. Smaller parts must be handled by offset. Flexography is also most widely used on pipe and other continuous profiles, with offset being used for the bulk of the decorative applications.

Dry-offset Printing. In dry offset printing, a dry, pasty ink is transferred via inking rollers to image plates. From the image plates the ink is transferred to a printing surface and from

here to a product. In multi-color operations, all the colors are printed on the surface of the printing roll before being transferred to the product. (see Fig. 26-14).

For cylindrical and conical products, rotary-type machines are used. For flat items or hollow items with an oval cross section, flat-bed type units are used.

Offset printing is a high-speed method of decorating cylindrical containers, and it is estimated that 90% of all plastic tubes are decorated by this method, as well as many jars, cups, and product tubs.

Equipment for dry offset ranges from hand to completely automated; the demand for hand-operated units is quite small since some degree of automation is desirable even for runs as low as 300-400/hr.

The operation requires a dryer after the printing step unless air dry is preferred. Drying times for typical polyethylene parts range from 3 to 5 min at $150°F$ vs 15 min for air dry.

Valley Printing. This process, often called *inlaid printing*, actually uses an embossing roll as the transfer roll. Ink is picked up on the raised portions of the embossing rolls and therefore the print is in the recessed or valleyed part of the product.

In the case of one-color valley printing, the process can be hooked right into an existing embossing press. For two-color valley printing a separate press is required. Three-color, and to a very limited extent, four-color presses have been worked with, but the market for these presses is not large.

Transfer Printing. A new decorating technique recently introduced could probably best be described as *transfer printing* or *tamp printing*, as it is also called. The system involves the use of a debossed metal plate and a soft conical-shaped transfer pad. The design or decoration is acid etched in the metal plate. This etched area is filled with pigment by a moving brush which continually carries the pigment from a tray-type reservoir. A doctor blade cleans the surface of the plastic to assure that no pigment remains in the non-decorating areas. The soft transfer pad comes in contact with the plate and picks up the pigment from the etched area. The transfer pad then comes in

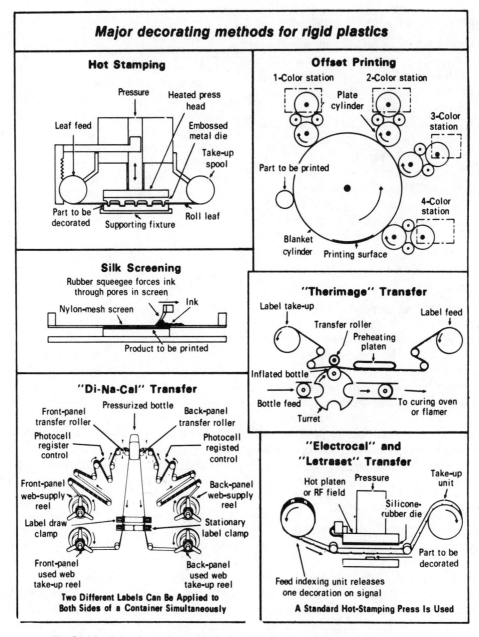

Fig. 26-14. Major decorating methods for rigid plastics. (*Courtesy Art Decorating Co.*)

contact with the object to be printed and transfers the design to the decorating surface.

The system does not require the heavy application pressures common with other decorating systems and hence can be used on sensitive and thin-walled products. The shape and soft flexibility of the transfer pad also allows printing over completely irregular surfaces.

Heat Transfers[6]

As described in the section on hot-stamping, heat transfers are a decorating variation, involving the printing of a design on a release-coated paper or plastic film carrier and then transferring the design from the carrier to

---

[6] By Art Decorating Co., Inc., Cedar Grove, N. J.; reprinted from the June 1969 issue of *Plastics World*, with the permission of the Cahners Publishing Co.

a part in a one-step operation. The advantage, of course, is that it allows the use of multi-colored designs at a relatively low cost. Described below are several of the more important heat-transfer systems available. Since they are known in the trade by their registered commercial names, they will be described as such in this discussion.

"Therimage" transfer (see Fig. 26-14) involves printing a design on a patented release-coated paper. Five-color rotogravure printing is used, which allows halftone and process art. Transfer of the design to the part is accomplished on a special machine that preheats the carrier and then presses it against the plastic part with a heated roller. After transfer, the part is oven- or flame-cured to improve adhesion and add gloss. The system is especially effective on plastic containers. Decorating speeds of the equipment are geared for high-volume production rates.

"Di-Na-Cal" transfer (see Fig. 26-14), similar to "Therimage", is basically a plastics-bottle decorating medium. A "Di-Na-Cal" machine transfers rotogravure-printed designs from a carrier using internally applied compressed air and a preheat principle. After the container and the carrier are preheated, the decoration is pressed against the container, and a permanent heat fusion takes place. Oven- or flame-curing is not required. Internal compressed air is maintained throughout the cycle and can be used to correct bottle deformities, such as concavity. The process has maximum advantages with oval-, flat- and square-shaped bottles. Two completely different designs can be applied to both sides of a container simultaneously. Up to six colors can be used on each side. Process art and halftones in unlimited color choice are possible.

In "Electrocal" transfer (see Fig. 26-14), designs are printed on a paper carrier by a web-type, multi-color silk-screen machine. Transfer of the design from the carrier to the plastic part is accomplished with standard hot-stamping equipment modified to handle the special transfer roll. Transfer takes place in a normal stamping cycle using heat and pressure. Rubber-silicone stamping dies, shaped to conform to the decorating surface, are used.

The transfer roll comes in various sizes (large, single sheets are also available for large parts). Production speeds are slower than other transfer methods because of the longer dwell time involved. The process works best on relatively flat surfaces.

IN-MOLD DECORATING

In-mold decorating involves the decorating of the plastic part while it is in the process of being compression molded or injection molded (more recently it has been adapted to blow molding; see below). It is applicable both to thermosets and thermoplastics and involves the use of a printed film or foil inserted in the mold manually or automatically at some time during the molding cycle.

The original application for the development was in the decorating of molded melamine dinnerware. In this system, a melamine-impregnated paper or foil is introduced into the mold after a part is partially cured. The mold is then closed and the cycle continued; after pressure has been applied for a given time, the mold is allowed to "breathe" (i.e., opened slightly to allow trapped gases to escape) and closed once more until the cure is complete. The printed foil becomes an integral part of the finished melamine piece and is actually under the surface of the piece so that, while it is fully visible, it cannot be damaged in any way. Other thermosets can also be decorated in this way (Fig. 26-15).

A similar system has also been developed for decorating reinforced plastic parts (Fig. 26-16). Briefly, the procedure is to place a charge of reinforced plastics bulk molding compound in a compression mold together with a sheet of unimpregnated paper, close the press and fully cure the material, reopen the press, pour in a quantity of catalyzed polyester resin, close the press again, cure the polyester resin, and then extract the finished molding (with the decoration again as an integral part). In essence, during the first stage, some of the resin in the compound partially impregnates the paper, firmly attaching it to the surface of the molding; when the glazing resin is applied in the second stage, it completes the impregnation of

Fig. 26-15. In-mold decorating used on compression molded phenolic handles and inserts. (*Courtesy Commercial Decal, Inc.*)

the paper and provides a protective glaze over the whole surface.

A third procedure of in-mold decorating is intended for injection molded thermoplastics and is applied as follows: A printed film or foil is inserted into the injection mold before the material enters. The insert is put in with the printed side away from the mold surface. Since the insert is always printed on a material compatible with the plastic being molded, when the molten plastic enters the mold, flows over the sheet, and cures, the decoration actually becomes an integral part of the finished molding. The part is thus removed from the mold completely decorated (Fig. 26-17).

The technique has been used successfully with polystyrene, polypropylene, polyethylene, polycarbonate, polyallomer, styrene-acrylonitrile, and ABS.

In-mold decoration as applied to blow-molded products involves feeding from a roll of printed plastic film labels (i.e., of polypropylene film, etc.) into the blow mold. The bottle is then blown against the label, the label adheres to the bottle, and when the mold opens, a finished decorated bottle is ejected.

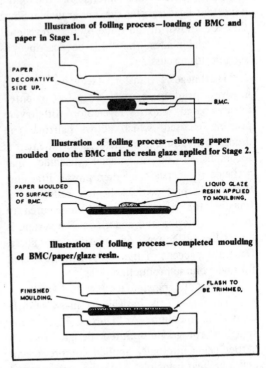

Fig. 26-16. In-mold foil decorating of reinforced plastics.

Fig. 26-17. In-mold decoration of injection molded polystyrene cannisters. (*Courtesy Commercial Decal, Inc.*)

27

JOINING AND ASSEMBLING PLASTICS

This chapter is concerned with basic techniques for joining plastics to themselves, to other plastics, and to other materials.

The techniques, of course, vary considerably in terms of the plastic involved and the application for which it is intended. Adhesives, for example, are widely used in plastics assembly. But they, in turn, are subdivided into solvent or dope cements which are suitable for most thermoplastics (not thermosets) and monomeric or polymerizable cements which can be used for most thermoplastics and thermosets. It should also be noted that there are some plastics with outstanding chemical resistance, like the polyolefins, which preclude the use of many cements, and generally require some form of surface treatment prior to adhesion.

To accommodate these distinctions, this chapter is subdivided into sections on (1) solvent cementing of thermoplastics; (2) adhesives for thermoplastics and thermosets; (3) thermal welding techniques for thermoplastics (including dielectric or heat sealing); (4) ultrasonic assembly techniques for thermoplastics and thermosets; and (5) mechanical fasteners (e.g., rivets, bolts, screws, etc.) and mechanical fastening techniques for thermoplastics and thermosets.

A good starting point is Table 27-1, which summarizes available bonding and joining techniques in terms of basic advantages and limitations. The costs indicated are as of the early 1970's and are included to put a degree of economic perspective on the various techniques available.

ADHESIVE BONDING OF PLASTICS

Adhesive bonding has found wide use in the assembly of plastics by virtue of low cost and adaptability to high-speed production. In addition, adhesives provide a more uniform distribution of stresses over the assembled areas and a high strength/weight ratio.

Three different types of cements are commonly used in bonding plastics:

(1) Solvent cements and (2) dope cements are used for most thermoplastics and function by attacking the surfaces of the adherends so that they soften and, on evaporation of the solvent, will join together. The dope cements, or bodied cements, differ from the straight solvents in that they also contain, in solution, a quantity of the same plastic which is being bonded. In drying, these cements leave a film of plastic which contributes to the bond between the surfaces to be joined.

(3) Monomeric or polymerizable cements consist of a reactive monomer, identical with or compatible with the plastic to be bonded, together with a suitable system of catalyst and

Table 27-1.   Bonding and Joining Techniques for Plastics Materials*

Technique	Description	Advantages	Limitations	Processing Considerations
SOLVENT CEMENT AND DOPES	Solvent softens the surface of an amorphous thermoplastic; mating takes place when the solvent has completely evaporated. Bodied cement with small percentage of parent material can give more workable cement, fill in voids in bond area. Cannot be used for polyolefins and acetal homopolymers.	Strength, up to 100% of parent materials, easily and economically obtained with minimum equipment requirements.	Long evaporation times required; solvent may be hazardous; may cause crazing in some resins.	Equipment ranges from hypodermic needle or just a wiping media to tanks for dip and soak. Clamping devices are necessary and air dryer is usually required. Solvent recovery apparatus may be necessary or required. Processing speeds are relatively slow because of drying times. Equipment costs are low to medium.
THERMAL BONDING 1. Ultrasonics	High-frequency sound vibrations transmitted by a metal horn generate friction at the bond area of a thermoplastic part, melting plastics just enough to permit a bond. Materials most readily weldable are acetal, ABS, acrylic, nylon, PC, polyimide, PS, SAN, phenoxy.	Strong bonds for most thermoplastics; fast, often less than one sec.	Size and shape limited. Limited applications to PVCs, polyolefins.	Converter to change 20 KHz electrical into 20 KHz mechanical energy is required along with stand and horn to transmit energy to part. Rotary tables and high-speed feeder can be incorporated.
2. Hot Plate and Hot Tool Welding	Mating surfaces are heated against a hot surface, allowed to soften sufficiently to produce a good bond, then clamped together while bond sets. Applicable to rigid thermoplastics.	Can be very fast, e.g., 4-10 sec. in some cases; strong bonds.	Stresses may occur in bond area.	Use simple soldering guns and hot irons, relatively simple hot plates attached to heating elements up to semi-automatic hot plate equipment. Clamps needed in all cases. Costs run gamut from $4.99 gun to $2000-$20,000 range for commercial hotplate units.

3. Hot Gas Welding	Welding rod of the same material being joined (largest application is vinyl) is softened by hot air or nitrogen as it's fed through a gun that's softening part surface simultaneously. Rod fills in joint area and cools to effect a bond.	Strong bonds, especially for large structural shapes.	Relatively slow; not an "appearance" weld.	Requires a hand gun, special welding tips, an air source and welding rod. Regular hand gun speeds run 6 in./min; high-speed hand-held tool boosts this to 48-60 in./min. Costs begins at under $100.
4. Spin Welding	Parts to be bonded are spun at high speed developing friction at the bond area; when spinning stops, parts cool in fixture under pressure to set bond. Applicable to most rigid thermoplastics.	Very fast (as low as 1-2 sec); strong bonds.	Bond area must be circular.	Basic apparatus is a spinning device but sophisticated feeding and handling devices are generally incorporated to take advantage of high-speed operation.
5. Dielectrics	High-frequency voltage applied to film or sheet causes material to melt at bonding surfaces. Material cools rapidly to effect a bond. Most widely used with vinyls.	Fast seal with minimum heat applied.	Only for film and sheet.	Requires rf generator, dies and press. Operation can range from hand fed to semi-automatic with speeds depending on thickness and type of product being handled. 3-25 kw units are most common.
6. Induction	A metal insert or screen is placed between the parts to be welded, and energized with an electromagnetic field. As the insert heats up, the parts around it melt, and when cooled form a bond. For most thermoplastics.	Provides rapid heating of solid sections to reduce chance of degradation.	Since metal is embedded in plastic stress may be caused at bond.	High-frequency generator, heating coil and inserts (generally 0.02-0.04 in. thick). Hooked up to automated devices, speeds are high. Cost: in the $1000/kw range. (1-5 kw's used) work coils, water cooling for electronics, automatic timers, multiple position stations may also be required.

Table 27-1. Bonding and Joining Techniques for Plastics Materials* (continued)

Technique	Description	Advantages	Limitations	Processing Considerations
ADHESIVES[a] 1. Liquids Solvent, Water Base, Anaerobics	Solvent-and-water-based liquid adhesives, available in a wide number of bases—e.g., polyester, vinyl—in one or two part form fill bonding needs ranging from high-speed lamination to one-of-a-kind joining of dissimilar plastics parts. Solvents provide more bite, but cost much more than similar base water-type adhesive. Anaerobics are a group of adhesives that cure in the absence of air, with a minimum amount of pressure required to effect the initial bond. Adhesives are used for almost every type of plastic.	Easy to apply; adhesives available to fit most applications.	Shelf and pot life often limited. Solvents may cause pollution problems; water-base not as strong; anaerobics toxic.	Application techniques range from simply brushing on to spraying and roller coating-lamination for very high production. Adhesive application techniques, often similar to decorating equipment, from hundreds (e.g., a glue pump for $400) to thousands of dollars with sophisticated laminating equipment costing in the tens of thousands of dollars. Anaerobics are generally applied a drop at a time from a special bottle or dispenser.
2. Mastics	Highly viscous single- or two-component materials which cure to a very hard or flexible joint depending on adhesive type.	Does not run when applied.	Shelf and pot life often limited.	Often applied via a trowel, knife or gun-type dispenser; one-component systems can be applied directly from a tube. Various types of roller coaters are also used. Metering-type dispensing equipment in the $2500 range has been used to some extent.
3. Hot Melts	100% solids adhesives that become flowable when heat is applied. Often used to bond continuous flat surfaces.	Fast application; clean operation.	Virtually no structural hot melts for plastics.	Hot melts are applied at high-speeds via heating the adhesive, then extruding (actually squirting) it onto a substrate, roller coating, using a special dispenser or roll to apply dots or simply dipping.

4. Film	Available in several forms including hot melts, these are sheets of solid adhesive. Mostly used to bond film or sheet to a substrate.	Clean, efficient.	High cost.	Film adhesive is reactivated by a heat source; production costs are in the medium-high range depending on heat source used.
5. Pressure Sensitive	Tacky adhesives used in a variety of commercial applications (e.g., cellophane too). Often used with polyolefins.	Flexible.	Bonds not very strong.	Generally applied by spray with bonding effected by light pressure.
MECHANICAL FASTENERS (Staples, screws, molded-in inserts, snap fits and variety of proprietary fasteners.)	Typical mechanical fasteners are listed on the left. Devices are made of metal or plastic. Type selected will depend on how strong the end product must be, appearance factors. Often used to join dissimilar plastics or plastics to non-plastics.	Adaptable to many materials; low-to-medium costs; can be used for parts that must be disassembled.	Some have limited pull-out strength; molded-in inserts may result in stresses.	Nails and staples are applied by simply hammering or stapling. Other fasteners may be inserted by drill press, ultrasonics, air or electric gun, hand tool. Special molding—i.e., molded-in-hole—may be required.

a Because of the thousands of formulations available within the various adhesive categories, it is not practical to attempt an analysis here of which adhesives will satisfy a particular application's need. However, typical adhesives in each class are:

Liquids: 1. Solvent—Polyester, vinyl, phenolics, acrylics, rubbers, epoxies, polyamide; 2. Water—acrylics, rubber, casein; 3. Anaerobics—cyan-acrylate; Mastics—rubbers, epoxies; Hot Melts—polyamides, PE, PS, PVA; Film—epoxies, polyamide, phenolics; Pressure Sensitive—rubbers.

* Reprinted from the August 1970 issue of *Plastics Technology*, published by Bill Communications. "Bonding and Joining Plastics," by J. O'Rinda Trauernicht.

promoter. The mixture will polymerize either at room temperature or at a temperature below the softening point of the thermoplastic. In order to hasten setting and to reduce shrinkage, a quantity of the solid plastic may be dissolved in the monomer. In addition to chemical bonding (specific adhesion), there may or may not be a degree of solvent attack by the monomer on the plastic. Adhesives of this type may be of an entirely different chemical type than the plastics being bonded; for example, many thermoplastics may be bonded by means of a liquid mixture of resin and hardener based upon epoxy resins, where the chemical reactivity and hydrogen bonding available from the epoxy adhesive contribute to excellent specific adhesion to many materials.

Techniques

Regardless of the composition of the cement employed, the following general rules should be observed in cementing plastic materials:

(1) The surfaces to be cemented must be clean; a slight film of oil, mold-release agent, water or polishing compound will cause poor bonding;

(2) the surfaces must be smooth, and aligned as nearly perfectly as practicable;

(3) where solvent or dope is used, it must be sufficiently active to soften the surfaces to such a depth that when pressure is applied a slight flow occurs at every point in the softened area;

(4) the solvent or cement must be of such composition that it will dry completely without blushing;

(5) light pressure must be applied to the cemented joint until it has hardened to the extent that there is no movement when released. (If the required clamping process actually deforms the parts, the stress due to spring-back of the flexible parts must be relieved before the joint hardens thoroughly, or else the joint may subsequently fail. This may be accomplished by releasing the clamps while the adhesive is just slightly "wet.");

(6) subsequent finishing operations must be postponed until the cement has hardened;

(7) care must be taken that the vapor from solvent cements is not confined, in order to prevent the surface of the molded piece from becoming etched;

(8) adequate ventilation and attention to fire hazards are further required for the protection of personnel.

In the application of adhesives, it is very important that the surfaces of the joint be clean and well matched. Poor contact of mating surfaces can cause many troubles. The problem of getting proper contact is aggravated by warpage, shrinkage, flash, marks from ejector pins, and non-flat surfaces.

Care must be taken to prevent application of adhesive to surfaces other than those to be joined, in order to avoid disfigurement of the surface. The adhesives should be applied evenly over the entire joint surface in sufficient quantity to ensure against voids. The assembly should be made as soon as the surfaces have become tacky, which usually means within a few seconds after application. Enough pressure should be applied to ensure good contact until initial bond strength has been achieved. Stronger bonds result when the adhesive is applied to both pieces of an assembly.

Joints

The design of the joint plays a large part in the effectiveness of the cement, the appearance of the joint, and the ease and cost of assembly. Butt, lap, and tongue-and-groove joints are in most frequent use, but angled or scarf joints and V-joints are used in some assemblies (see Fig. 27-1). If the strength of the joint is to be equivalent to that of the adjacent wall, the area of the joint should be increased by at least 50% over that of the original contiguous edge. A joint which combines both shear and tensile strength is most effective.

Lap Joint. The lap joint has some advantages over the others, in regard to both appearance and, to some extent, strength. For best appearance, the solvent should be applied by a felt pad to the half of the joint which fits

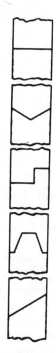

Fig. 27-1. Joint designs. From top to bottom: butt, "V", lap, tongue and groove, scarfed. (*Courtesy Borg-Warner Chemicals*)

on the inside of the article. Then no exudation of adhesive will appear at the outer parting line.

When two pieces of annular cross-section are to be cemented with a lap joint, it is advantageous to make the outer lip thin in proportion to the inner (e.g., in a ratio of 1 : 4), since the greater periphery thus gained adds to the area, and hence the strength, of the joint. Lap joints are generally used for adhesive joining while butt joints are preferred for solvent cementing.

Butt Joint. A butt joint gives less contact area than most lap joints. The butt joint often has the disadvantage of unsightly bond lines, but it is probably the easiest type to provide for in molding. The butt joint is not a self-locating joint, and locating pins or fixtures are often required to prevent slipping during clamping.

Tongue-and-Groove Joint. A tongue-and-groove joint is self-positioning, but unless it is very shallow it will require application of adhesive by other than felt-pad methods. If the adhesive is applied by a flow gun into the groove, the amount must be controlled so that after assembly the fluid will come to the edge

of the joint, but not flow out to mar the outer appearance of the finished article.

Other Joints. The V-joint and the scarfed joint are variations of the preceding joints. They have the disadvantage of requiring a somewhat more complex method of application of cement. The V-joint is self-positioning, and the scarfed bond is self-positioning on certain types of articles. It is difficult to mold the component parts to the close tolerance required by these joints.

SOLVENT CEMENTING THERMOPLASTICS

Structural bonds of up to 100% of the strength of the parent material are possible with solvent bonding, a technique suitable for the amorphous plastics such as polystyrene and acrylic. It cannot be used for the polyolefins and most acetal (although acetal copolymers have been solvent cemented with some success, and a development program is looking at bonding of the acetal homopolymers). The technique is also usable on some nylons and some vinyls.

Solvent cementing basically involves softening the bonding area with a solvent or a solvent containing small quantities of the parent material, referred to as *dope* or *bodied cement*, generally containing less than 15% resin. A bodied cement serves several purposes: it can slow down the solvent evaporation rate, it can provide a more workable solution, and it can fill in spaces where mating surfaces are not perfectly matched. (This last use should not be considered a final solution to mismatched parts that match up as closely as possible.)

A resin dissolves best in a solvent whose solubility most closely approaches its own. See Table 27-2 for a list of solvents for various plastics. Solvent bonding of dissimilar materials is possible where the materials can be bonded with the same solvents.

The suitability of solvents or solution-dispersed adhesives may depend upon the mechanical structure of the plastic; the importance of optical clarity in the bond; the presence or absence of molding stresses which could result in crazing when released by solvent; the composition of the solvent; and the

Table 27-2.   Typical Solvents for Solvent Cementing of Plastics*

ABS	Methyl ethyl ketone, methyl isobutyl ketone, tetrahydrofuran, methylene chloride
Acetate	Methylene chloride, acetone, chloroform, methyl ethyl ketone, ethyl acetate
Acrylic	Methylene chloride, ethylene dichloride
Cellulosics	Methyl ethyl ketone, acetone
Nylon	Aqueous phenol, solutions of resorcinol in alcohol, solutions of calcium chloride in alcohol
PPO	Trichloroethylene, ethylene dichloride, chloroform, methylene chloride
PVC	Cyclohexane, tetrahydrofuran, dichlorobenzene
Polycarbonate	Methylene chloride, ethylene dichloride
Polystyrene	Methylene chloride, ethylene ketone, ethylene dichloride, trichloroethylene, toluene, xylene.
Polysulfone	Methylene chloride

These are solvents recommended by the various resin suppliers. A key to the selection of solvents is how fast they evaporate: a fast evaporating product may not last long enough for some assemblies; too slow evaporation could hold up production.

* Reprinted from the August 1970 issue of *Plastics Technology*, published by Bill Communications, Inc.

presence of highly soluble compounding ingredients in the plastic.

Built-in molding stresses must be relieved, or else solvent-crazing is likely to occur. The avoidance of solvents of a low boiling point, or diluting the solvent through the addition of polymer, is generally desirable, and at the very least delays the appearance of crazing cracks. To minimize strains and resulting stresses which may be caused by flexing or by change of temperature, the layer of adhesive should always be as thin as possible, and no more rigid than the adherends.

Application

The solvent is usually applied by one of the following devices or techniques:

Dip Method.   In the dip method, one of the two parts to be joined is dipped into the cement just enough so that the solvent will act on the desired area for the maximum time. After dipping, the parts should be assembled immediately and held under light pressure for a short time. Bonding takes place by the formation of a "cushion" of solvent-swollen plastic surface. To avoid squeezing out this cushioned material from the dipped plastic surface before the solvent can act on the mating dry surface, about 15 to 30 sec should be allowed before pressure is increased.

Specially designed dipping fixtures, clamps and conveyors should be considered for each job. The area of attack of solvent on the piece can be controlled by laying a roll of glass rods in the cementing tray, on which the piece to be cemented can be rested. The level of the cement should be kept equal to the diameter of the rods. Skillful operators can assemble pieces by this method without spilling or wasting cement. The areas of contact of cement and solvent can be controlled also by lining the bottom of the cement tray with a felt pad. Pieces to be cemented are then placed on the felt pad, which is kept thoroughly wet with the liquid solvent cement.

Capillary Method.   For some applications where the surfaces to be cemented fit very closely, it is possible to introduce the cement by brush, eyedropper or hypodermic needle into the edges of the joint. The cement is allowed to spread to the rest of the joint by capillary action. In other cases it is possible to insert fine wires into the joint when the parts are assembled in the jig. The cement is then introduced into the joint. After the cement has reached all parts of the joint, the wires are removed. This procedure is useful also in removing air bubbles and in filling voids in joints made in other ways.

Soak techniques are also available in which pieces are immersed in the solvent, softened, removed and quickly brought together. Areas adjoining the joint area should be masked to prevent them from being etched. Cellophane tape is commonly used. After soak, parts should be assembled rapidly.

No matter which technique is selected,

surfaces to be joined should be clean (use of a cleaning solvent, e.g., acetone, is often recommended), not below 65°F, and fit together as flat surfaces. If the parts do not fit well, the time for a part to soften after application of cement must be increased to aid in obtaining satisfactory fit.

Once the parts are mated, they must be fastened together by spring clamps, C clamps, or toggle clamps so uniform pressure is exerted on bonded areas. Pressure should be low—100-200 psi is often recommended—so that the part will not be flexed or stressed. The pressure should be sufficient to squeeze all air bubbles from the joint and to assure good contact of the mating surfaces.

When the clamps are removed, an elevated temperature cure is often called for, e.g., warm acrylic to 120°F, heat ABS to 130-150°F, PPO over 200°F, polycarbonate up to 175°F.

Precautions. Cementing trays should be made of materials that are inert to the cementing solvents. Care should be taken to avoid excessive contact of the hands or skin with cement solvents, since most of them will remove natural oils from the skin, and may cause chapping or even mild dermatitis after too frequent exposure. Where glacial acetic acid is used, contact should be carefully avoided, because of its corrosive nature.

Working areas must be well ventilated. Breathing of concentrated vapors of ethylene dichloride, methylene chloride, glacial acetic acid, etc., may cause toxic effects. Cementing tanks, trays, and storage containers should be equipped with tight covers to reduce evaporation when not in use.

Since solvent vapors as well as solvents themselves can cause crazing, assemblies within enclosed spaces should be adequately ventilated. An air line can be inserted to sweep the vapors out, or vacuum can be used, along with a tube to bleed fresh air to the side opposite the vacuum takeoff. Where it is impossible to remove solvent vapors from the joint itself, after the bond has been made, the joint should be designed so that solvent vapors can be excluded from the adjacent area.

Different materials have somewhat different solvent bonding characteristics as noted below:

ABS. A thin layer of solvent cement provides the greatest bond strength and should be applied as uniformly as possible to one or both ABS surfaces, and joined under pressure. Excessive amounts of bonding agent do not develop strong bonds, require longer cure times, and may drip from joints, marring surrounding surfaces.

For applications where the mating surfaces cannot be coated, the cement may be introduced into the joint by brush, eyedropper, or a hypodermic needle. This should allow the solvent to spread throughout the joint. Other methods for applying cements may be oil cans, squeeze bottles, dip tanks, plunger pencils, and rollers.

For optimum bonding results, the surface must be smooth, free of all oil or grease film, mold-release agents and water or polishing compounds. The cement used should be quick drying to prevent moisture absorption, yet slow enough to allow assemblage of parts. The recommended cure time is 12-24 hr at room temperature (73°F). The time may be reduced by curing at 130-150°F.

The solvents that are recommended for ABS are methyl ethyl ketone (MEK), methyl isobutyl ketone (MIBK), tetrahydrofuran (THF), and methylene chloride.

Acetal. The solvent resistance of acetal (du Pont's recommendations for its Delrin acetal are covered in this section) makes it virtually impossible to cement this plastic with solvent—type cements. However, a number of commercial adhesives can be used to give satisfactory bonds, provided the surface of the part is prepared as described below. Optimum bond strength can only be obtained on acetal with a prepared surface. The preferred method is by the "Satinizing" procedure (see Fig. 27-2).

In this method, parts to be adhered are immersed for 10 to 30 sec in a special treating bath at 175 to 250°F. A typical bath formulation is made up of "Dicalite" bulk aid (0.5 wt %); para-toluene sulfonic acid (0.3); "Dioxane" (3.0); and perchlorethylene (96.2). After immersion, the parts are transferred immediately to an air oven maintained at 250°F, and maintained in the oven for 1 min. Parts are then rinsed in warm water and air dried thoroughly prior to cementing.

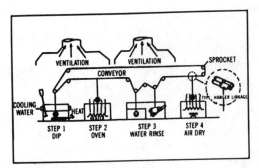

Fig. 27-2. "Satinizing" process for pre-treating acetal resins. (*Courtesy E. I. du Pont de Nemours & Co., Inc.*)

The bath should be stirred continuously to maintain a uniform mixture, and the area above the treating bath and other equipment should be well ventilated to remove formaldehyde fumes and perchlorethylene vapors. Galvanized iron or zinc alloys should not be used as a vessel for the bath mixture since the *para*-toluene sulfonic acid will react with zinc.

If "Satinizing" cannot be carried out, then the mating surfaces should be abraded with emery cloth with a 280A grit.

Acrylic. There is a difference between the bonding characteristics of molded and cast acrylics. The major problem encountered with molded or extruded parts is the molded-in strains and stresses which can cause crazing during the solvent bonding operation. Annealing is generally required on such parts where the bond must be good. Sheet can be annealed by heating it in a forced-air oven at about 10°F below its distortion temperature.

Cellulosics. Fast-evaporating adhesives must generally be used in combination with solvents that evaporate more slowly since rapid evaporation may cause a "moisture blush," a white, frosty appearance of the cemented joint. When making a bodied cement, a 10% resin solution is often recommended.

Nylon. While both aqueous phenol and resorcinol-ethanol are recommended for nylon, resorcinol-ethanol is safer to work with, so is generally preferred. A bodied cement using calcium chloride and ethanol combination is used for food-contact or potable water applications.

PPO. PPO can use a wide variety of solvents. For bodied cements, solution containing very small amounts, 1-7%, is suggested.

PVC. Tetrahydrofuran is especially recommended for PVC because it diffuses rapidly through the resin layers, minimizing set time. In one application, a PVC pipe, a THF solution with a small amount of resin and co-solvent enables joints to sustain 500 psi hydrostatic pressure. (The co-solvent is generally an aliphatic ketone, cyclohexanone or dimethyl formamide.)

Polycarbonate. Solvent bonding will produce bond strengths of about 6500 psi. When handling large parts or when a high working temperature is involved, a solution of methylene chloride and ethylene dichloride (up to 40%) is recommended. Bodied cements may have from 1 to 10% resin. Recommended clamping pressure is 100 to 600 psi.

Polystyrene. Excellent bond strengths—up to 100% that of the parent material—have been reported when solvent cementing PS. But the solvent must be selected carefully. The fast evaporating solvents may cause crazing in the joints and may also give the weakest bonds. Dopes generally contain 10-15% resin.

Polysulfone. Higher pressure, around 500 psi for 5 min, is recommended for solvent welded polysulfone parts than most other plastic parts. Solvent suggested is a 5% solution of the resin in methylene chloride.

ADHESIVES*

Adhesive bonding is a process in which the adhesive acts as an agent to hold two substrates together (as opposed to solvent welding where the parent materials actually become an integral part of the bond). Because of the smooth surfaces of plastics, adhesion must be chemical rather than mechanical.

Table 27-3 gives a list of various adhesives and typical applications in plastics. While this list is not complete, it does give a general idea of what types of adhesives are used where. It should be remembered, of course, that thousands and thousands of variations of standard adhesives are available off the shelf.

* Adopted from "Bonding and joining plastics," by J. O'Rinda Trauernicht, *Plastics Technology*, August 1970. Reprinted with permission from the publishers.

The computer may shape up as an excellent selection aid for adhesives. To date, however, its use has been limited. But one major chemical company offers a computerized selection service to compounders of urethane adhesives. Formulation selection is made according to the combination of properties desired, tack time, strength, method of application, and economics. The program narrows the selection process down to only those urethane formulations that fit all the given parameters. A major limitation: Substrate requirements have not yet been quantified, so selection cannot be on the basis of the end product.

Among the most important criteria in selecting an adhesive are:

1. Substrates involved, including additives used in the plastic.
2. Joint configuration and size.
3. Stresses the joint will be subjected to in use.
4. Environmental elements the joint can be expected to encounter.
5. Economics (performance/cost ratio).

Application

Just about anything from a finger to high-speed spray and laminating equipment is used to apply adhesives.

Before bonding, most plastics require surface preparation. Common surface treatment could include: (1) chemical cleaning with a solvent such as acetone; (2) abrasive cleaning by sandblasting, vapor honing, sandpaper or wire brush; (3) degreasing to remove residual contaminants from surfaces to be joined. In cases where a mold release agent has been used, make certain that no traces remain.

Some key application techniques are listed below:

Tube, Knife, Spatula, Brush, Roller, Squeeze Bottle. These hand-type techniques are used for the liquids and mastics. Equipment is simple and costs are low. Besides the actual application device, equipment should consist of mixing vessels (for multi-component materials), clamps where required, dryers where elevated cure is necessary. Depending on production require-

ments, labor costs are generally high. However, for some adhesives, e.g., some epoxies and silicones, this is the only practical method of application. Some of the less viscous adhesives can be applied via a self-feeding brush that speeds hand application somewhat. One epoxy supplier offers a two-part adhesive in a two-component pouch. The seal between the components is broken, the components mixed, then dispensed right from the pouch.

Hot Press Bonding. Use of a hot press is common for reactivating sheet-type adhesives. It is also used to reactivate dual-component adhesives that work by applying one-component of the adhesive to a substrate and the second to the second substrate, then allowing the adhesives to dry. This is a widely used lamination technique. When the substrates are brought together under the heat and pressure of the press, reactivation and bonding take place. A hot tool, e.g., an iron, can be used for some field and low-volume production.

Meter/Mix/Dispense. This type of equipment is widely used for accurate handling of multi-component adhesives, especially those with long cure times or requiring elevated cures. Some of the newer fast-cure adhesives must be applied by one of the hand techniques to ensure that they don't set in the dispensing equipment.

Spray Application. Spraying is fast and provides a means of reaching inaccessible areas quickly. It is often used for high-speed production application of the liquids, and for application of the rubber-type contact adhesives. Spraying is largely confined to one-component types. It is almost identical to the process used for spray painting, except that a simplified nozzle, called a *siphon nozzle*, is used and the process is slower than painting. When spraying it is essential that checks be made to assure that proper coverage is reached.

Roll Coaters. Essentially, roll coating consists of a glue reservoir, a roll which automatically picks up the glue from the reservoir, and a second roll which picks up the glue from the first roll and transfers it to a moving substrate. A doctor blade can be used to control the amount of glue on the roll. Other variations include a single-roll set-up, a system

	ABS	Acetal	Acrylic	Cellulosics	Fluoro-carbons	Nylon	PPO	PVC	Poly-carbonate
Metals	23	3,23	2,3	2,3	22,23	2,23	2,4,23	2,3,15, 23,36, 43	23
Paper		3,23	42	42	22,23	3,41	6,23	42	36
Wood	23	23	2,3,42	3	23	2,3	2,4,23	3,23, 36,42	23,36
Rubber	23	3	1-4	1-4	23	2	2,4,23	3,4, 15	4,36
Ceramics		23	2,3	3	23	3,23	4,23	3,4	23,36
ABS	23,43	2,4, 11,23				21,23		23	4,23
Acetal	2,4, 11,23	3,23	23		23	23		23	
Acrylic	2,4, 11,23	2,4,11, 23	S		23	2,3, 15,22		3,4, 11,23	
Cellulosics				3,4 14,36	23	2,3, 15,22		23	
Fluorocarbons		23			22,23			23	
Nylon	21,23	21,23	2,3, 15,22	2,3, 15,22		2,22, 23,36	21,23		
PPO							4,23, 43		
PVC	23	23	3,4, 11		23			3,4,11, 36,42	
Polycarbonate	23							3,4,11, 36,42	S
Polyethylene	23				23				
Polypropylene									
Polystyrene					23				
Polysulfone									
Alkyds							4,23, 43	23	
Epoxy							4,23, 43	23	
Melamines and Ureas							4,23, 43		
Phenolics	23,43	43	4,21, 23	4,21, 23	23		4,23, 43	4,15	
Polyesters			23	23			4,23, 43	23	
Polyurethanes			4,23		23		4		

Table 27-3. Typical adhesives for bonding plastics*

Table 27-3. Typical adhesives for bonding plastics* (continued)

Poly-ethylene	Poly-propylene	Poly-styrene	Poly-sulfone	Alkyds	Epoxy	Melamines & Ureas	Phenolics	Polyesters	Poly-urethanes
2,31, 41	1	31	4,23		4	3,43	2	4	3,4
41	1,41	4,31, 36			3	41,42	42	41	4,36
2,41	1,41	31,36			23,31	2,3	2,42	2	36
2,41	1,41	5			3	2,3,43	2,3	1-4	4,36
2,41	1,41	41,42			23,31	2	2	2	3
23									
							23		
							4,21, 23	23	4,23
							23		
23		23							23
				21,23	21,23	21,23	21,23, 43	21,23	
				4,23, 43	4,23, 43	4,23, 43	4,23, 43	4,23, 43	4
			23				4,15	23	
23,31 41			4						
	23,31 41								
		4,5,13, 23,31, 36					4,23, 43		
4			4,23						
					2,3,23, 31,36				
						2,3,23, 31,36			
		4,23				3,21, 23,24	2-4,23, 31,43		
								3,23, 31,36	
									3,4, 23,36

(Key and footnotes overleaf)

Key and footnotes for Table 27-3.

Elastomeric

1. Natural Rubber	2. Neoprene	3. Nitrile
4. Urethane	5. Styrene-Butadiene	6. Silicones

Thermoplastic

11. Polyvinyl acetate	12. Polyvinyl alcohol	13. Acrylic
14. Cellulose nitrate	15. Polyaminde	

Thermosetting

21. Phenol Formaldehyde (Phenolic)	22. Rescorcinol, Phenol-Rescorcinol	23. Epoxy
24. Urea-Formaldehyde		

Resin

31. Phenolic-Polyvinyl Butyral	32. Phenolic-Polyvinyl Formal	33. Phenolic-Nylon
36. Polyester	37. Acrylic	

Other

41. Rubber Latices	42. Resin Emulsions	43. Cyanoacrylate
S. Solvent only recommended		

NOTE: This information contains a compilation of suggestions and guidelines offered by various adhesive and materials and manufacturers and plastics molders. It is intended only to show typical adhesives used in various applications. Lack of a suggested adhesive for two materials does not mean that these materials cannot be bonded, only that suppliers do not commonly indicate which adhesive to use. Before making a final selection, consult both materials supplier and materials manufacturer. Key for Table and part of the data was supplied by USM Corp., Chemical Div.

* Reprinted from "Bonding and joining plastics," by J. O'Rinda Trauernicht, *Plastics Technology*, August 1970.

in which the substrate is dipped into the glue and the roll used to remove the excess adhesive. An engraved roll can be used to deposit a noncontinuous pattern of glue, generally in an even pattern of dots.

The actual lamination, using either water- or solvent-based adhesives can be done in one of two ways: wet combining or thermoplastic mounting. In wet combining, an adhesive is applied to either substrate and the materials are joined immediately and stacked for drying. In mounting, an adhesive is applied to one substrate (the board in panel lamination), allowed to dry, then reactivated to a tacky state and mated. Thermoplastic mounting is said to be faster and more versatile than wet combining, although it is more expensive.

Curtain and Screen Coating. Both curtain (applying a molten sheet of adhesive to a substrate passing underneath the flow) and screen (forcing adhesive through a specially prepared screen onto the substrate below) techniques can be adapted for adhesive processing. Curtain coating is commonly used for packaging, screen processing for high-quality printing. Both, when used with adhesive instead of molten plastic or printing inks, provide a closely controlled deposition process.

Hot Melt Application. Over a half-dozen techniques can be used to apply hot melts, with some suppliers requiring that their equipment be used, others recommending a wide range of commercial equipment. The job of a hot-melt applicator is to melt the material (most melt from 300-500°F), and apply it at a controlled rate to the substrate.

Two basic types of application equipment are melt reservoir and progressive feed. In melt-reservoir, the adhesive is put into the melting pot and heated to a predetermined temperature; in progressive feed, a rope of hot melt is fed through the heating system, keeping the amount of adhesive molten to a minimum at any one time.

Some application methods are: (1) extrusion—actually squirting the molten resin onto the substrate via a pistol-type or automated, air-operated dispenser; (2) hot-drop application with either manual or automatic dispensing in which a wide variety of droplet patterns can be applied; (3) hot-melt coating using a roller.

Adhesive Forms

The form that the adhesive takes (i.e., liquids, mastics, hot melts, etc.) can also have a bearing on how and where they are used, as follows:

Liquids. Liquid adhesives lend themselves very well to high-speed processing techniques such as spray and automatic roller coating. They may be solvent or water based; the anaerobics (cyanoacrylate-based bonding agents that cure in the absence of air) also fall into this class, as well as a number of epoxies.

An extensive area of activity in liquids is the increased use of wood-grain printed vinyl sheet bonded to particle board or other inexpensive solid substrate. Both solvent-based and water-based adhesives are used. The solvent-type gives a better bite (14-20 psi peel strength compared to 4-6 psi for water base), but also will be more expensive. Application is by direct roll coaters with speeds of 50-80 ft/min being typical.

The anaerobics, which can give some very high bond strengths, and are usable with all materials except PE and fluorocarbons, are dispensed by the drop. A thin application of the anaerobics is said to give better bond strength than a heavy application.

Mastics. The more viscous, mastic-type cements include some, although not all, of the epoxies, urethanes, and silicones. Mastics can normally be applied at low-to-medium production speeds, but in some cases can be heated for higher production speeds.

Epoxies adhere well to both thermosets and thermoplastics. But epoxies are not recommended for most polyolefin bonding.

Developments in epoxies include the introduction of more fast-cure types (pot life of 5 min or less), and lower temperature cure requirements for some of the elevated-cure materials.

Urethane adhesives are making inroads into flexible packaging, the shoe industry, vinyl bonding. Polyester-based polyurethanes are emerging as the frontrunner over polyether systems because of their higher cohesive and adhesive properties.

Silicones are especially recommended where both bonding and sealing are desired.

Hot Melts. Hot melts, 100% solids adhesives that are heated to produce a workable material, are based on polyethylene, saturated polyester or polyamide in chunk, granule, pellet, rope, or slug form. Saturated polyesters are the primary hot melt for plastics; the PE's are largely used in packaging; polyamides are used most widely in the shoe industry. Application speeds are high, and it probably does not pay to consider the hot melts if production requirements are not correspondingly high.

A big potential growth area foreseen for hot melts in plastics is the furniture industry for bonding non-structural parts. In one application, a urethane foam shell is attached to a vinyl covering. Hot melt is applied, a vacuum is pulled through the foam to get pressure to hold the vinyl against the urethane, and in a fairly short time, the chair is ready for shipping. A previous bonding technique used on the same chair required a 5-week cure.

Pressure Sensitives. Contact-bond adhesives—usually rubber-base—provide a low-strength, permanently tacky bond. They have a number of consumer applications, e.g., cellophane tape; but, they are also used in industrial applications: (1) where a permanent bond is not desirable and (2) where a strong bond may not be necessary. The adhesive itself is applied rapidly by spray. Assembly is a mere matter of pressing the parts together.

Film Adhesives. Film adhesives overlap several areas, but exist as a separate group because of the processing similarity. All of the materials in this group require an outside means such as heat, water, or solvent to reactivate them to a tacky state. Among the film types are some hot melts, epoxies, phenolics, elastomers, and polyamides. Production rates for film adhesives are generally slow but very precise.

Film adhesives can be die cut into complicated shapes to ensure precision bonding of unusual shapes.

Applications for this type of adhesive include bonding plastic bezels onto automobiles, attaching trim to both interiors and exteriors, attaching nameplates on luggage.

Pretreatment

While most materials bond without trouble once the proper adhesive has been selected, a few, notably polyolefins, fluorocarbons and acetals, require special treatment prior to bonding.

Untreated polyolefins adhere to very few substrates; this is why polyethylene is such a popular material for packaging adhesives. Treating methods include electronic or corona-discharge, flame treating (especially recommended for large, irregular-shaped articles), acid treating by dipping the articles in a solution of potassium dichromate and sulfuric acid; solvent treating. A rubber-based adhesive that remains permanently tacky is generally recommended.

Fluorocarbons must be cleaned with a solvent such as acetone, then treated with a special etching solution.

Acetals can be prepared for bonding by several techniques. One method involves immersing articles in a special solution composed mainly of perchloroethylene, drying at 250°F, rinsing, then air drying.

CEMENTING OF SPECIFIC PLASTICS

The same basic handling techniques apply to almost all thermoplastic materials. In the discussions which follow, each thermoplastic will be treated separately, with mention of the specific cements most suitable for each. However, the first subsections which follow (on the cementing of acrylic) include discussions of bonding jigs and bonding techniques which are equally applicable, in most cases, to all other thermoplastics. To avoid duplication, this detail has been largely omitted from the subsections dealing with other thermoplastics. These sections on acrylics should be read in connection with the cementing of any of the other thermoplastics.

Cast Acrylic Sheeting. Articles of considerable size and complexity can be fabricated from methyl methacrylate plastics by cementing sections together. The technique described in this subsection applies to cast sheeting. Cementing of articles made from methyl methacrylate molding powders or extruded rod, tubing or other shapes is more specialized, and reference should be made to the instructions applicable to these materials, in the subsection which follows.

With care and practice, the transparency of acrylic resin can be retained in cemented joints; the joint will be clear and sound as the result of complete union of the two surfaces brought into contact. In order to accomplish this, the surfaces to be cemented are thoroughly softened by means of a solvent, to the extent that a soft layer or cushion is produced on each; then the uniting and hardening will effect a homogeneous bond. This principle underlies all solvent-cementing of acrylic resins.

Usually one of the two surfaces to be joined is soaked in the cement until a soft layer has been formed upon it. This soft surface is then pressed against the surface to be attached and, in contact with it, softens it also, by means of the excess of cement contained in the soaked area. (See Fig. 27-3).

For some purposes it may be desirable to dissolve clean shavings of methyl methacrylate resin in the cement, in order to raise its viscosity, so that it may be handled like glue. However, even with a thickened cement it is still necessary to establish a sound bond.

Most of the conventional types of joint-construction can be employed, such as overlap, butt, rabbet, miter, scarf, etc., depending on

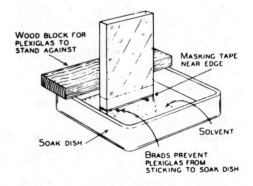

Fig. 27-3. Cementing acrylic. Edges are soaked until swollen into a cushion. Edges are then joined and pressure is applied. (*Courtesy Cadillac Plastic & Chemical Co.*)

the service requirements of the article. The area of the joint must be large enough to develop sufficient strength, and must be so designed as to give even distribution of stresses.

An accurate fit of mating pieces is essential in cementing acrylic resins. This need is primarily due to their rigidity, and the inability of a solvent cement to compensate for discrepancies in fit. In butt joints, it is necessary that both edges be true and square before cementing. Flat surfaces are more easily cemented than curved ones, and hence, wherever possible, it is desirable to rout or sand curved surfaces to form flat ones. When two curved surfaces are to be cemented, both must have the same radius of curvature. It is not good practice to force either piece in order to bring the surfaces into complete contact. The only exception to this is the case of very thin sheeting, such as 0.060 or 0.080 in., and even here the deviation must not be great.

The surfaces to be cemented should be smooth and clean, but need not be polished. A smooth machined surface is most satisfactory for cementing.

When ribs are to be cemented to curved pieces, such as panels for airplane enclosures, each rib should be machined from flat sheet stock, heated in an oven to shaping temperature, and then, without cement, and with the aid of a jig, shaped against the surface to which it is later to be cemented, and allowed to remain in contact until cool. Then the curved rib is soaked in cement, and cemented to the curved panel with the aid of the same jig.

In soaking the plastic in the cement, the softening action should be confined to the immediate region of the joint, by masking the rest of the surface with a tough paper or cellophane tape coated with a pressure-sensitive adhesive. The tape should be applied firmly, and with special care at the edges. The tape used should be impervious to the cement, and should not be applied long in advance of cementing, since its adhesive may loosen if it is allowed to stand too long.

The role of the cushion formed by soaking the surface in the cement is solely to enable the two surfaces to be brought into complete conformity, so as to exclude air bubbles and to compensate for lack of perfect fit between the two surfaces. The thickness of the cushion must be adequate to provide complete contact, but preferably no greater than this, since an excess reduces the strength of the joint and prolongs unnecessarily the setting of the bond. With solvent cements and unplasticized sheets, a soaking time of 15 min is usually enough to form an adequate cushion. Inadequacy of cushion is one cause of bubbles or uncemented areas in the finished joint.

The soaked surface, after removal from the cement, should be brought rapidly, while still wet, into contact with the surface to which it is to be joined, since soaked surfaces that have become dry do not wet the mating surface adequately. The softened surface, however, should not be dripping, since superfluous cement will run off and cause smears. The tray of cement, and the work, should be so situated that the transfer from cement to jig can be made conveniently and rapidly. A soaked surface that has become dry can be wetted again by brushing additional solvent cement on it.

Too early application of pressure on the joint will result in squeezing out some of the solvent needed to soften the dry surface. After the two surfaces have been brought into contact, only very slight pressure should be applied at first. Then, after about 15 to 30 sec, the cementing jig can be tightened, to apply pressure.

Small pieces can often be weighted with bags of shot, tied with cord, or taped to supply the required pressure (Fig. 27-4). However, assembly is best accomplished in well-designed cementing jigs (Figs. 27-5, 27-6, and 27-7) capable of holding the pieces firmly together until the joint is hard, without forcing either of them out of shape. Excessive pressure should be avoided, since it is likely to cause stress crazing. The pressure should be adequate to squeeze out bubbles or pockets of air from the joint, should be evenly applied over the entire joint, and should be maintained throughout the period of setting. Those requirements are met through the use of spring clips or clamps, either alone, in the case of simple assemblies, or in conjunction with cementing jigs of wood or metal in more complicated cases. For most

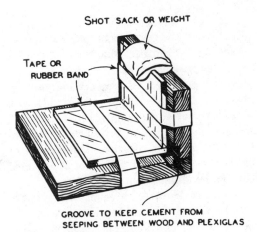

Fig. 27-4. In adhering small acrylic pieces, parts can be weighted with bags of shot, tied with cord, or taped. (*Courtesy Cadillac Plastic & Chemical Co.*)

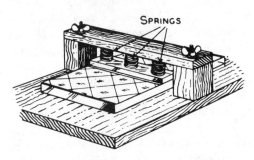

CEMENTING A RIB ON A SHEET

Fig. 27-5. Typical clamping method for cementing a rib on acrylic sheet. (*Courtesy Cadillac Plastic & Chemical Co.*)

joints, a pressure in the neighborhood of 10 psi is suitable, provided that it does not force either of the parts appreciably out of shape.

After assembly of the work in the jig, any excess cement that has been extruded from the joint should be removed by scraping onto the masking paper, which is subsequently to be removed. Prompt cleaning up of excess of soft cement will save time in finishing after the bond has set.

The assembly should be allowed to remain in the jig for from 1 to 2 hr, and an additional 4 hr should elapse before the work is subjected to heat treatment or to finishing operations. These times are approximate only, depending on the complexity of the job.

Methylene chloride, a solvent which produces joints of medium strength, requires a soaking time of 3 to 10 min and sets very

quickly. Ethylene dichloride is a trifle slower and is less apt to produce cloudy joints.

Acetic acid cement consists of glacial (100%) acetic acid (not to be confused with the 28% acid used in photographic work). This cement requires a soaking time of about 1 hr at room temperature, or 2 to 5 min at $140°F$ ($60°C$). The cemented joints can be handled in 3 to 5 hr, and machined in 16 to 24 hr. The fumes of glacial acetic acid, especially when it is used hot, are irritating; adequate ventilation is required, and this cement is accordingly waning in popularity.

The most universally applicable type of solvent cement is the polymerizable type, comprising a mixture of solvent and catalyzed monomer. These are mobile liquids, volatile, rapid in action, and capable of yielding strong sound bonds. They should be used with adequate ventilation.

An example of these is a 40-60 mixture of catalyzed methyl methacrylate monomer and methylene chloride. To ensure formation of the correct depth of cushion, the ratio of these two solvents should be maintained, within reasonable limits; this can be readily checked by measuring the specific gravity of the cement. Permissible ranges of specific gravities at various temperatures are as follows:

| Temperature | | Permissible Range of |
($°C$)	($°F$)	Specific Gravity
20	68	1.17-1.21
25	77	1.16-1.20
30	86	1.15-1.19
35	95	1.14-1.18
40	104	1.13-1.17

If the specific gravity is found to be low, it should be adjusted by adding methylene chloride.

Before using this cement, there is added, per pint of solvent, 2.4 grams of a 50/50 mixture of benzoyl peroxide and camphor (which serves as stabilizer for the benzoyl peroxide). After addition of this catalyst, the cement will remain usable for up to 45 days if the container is tightly closed and stored at not over $77°F$; or for up to 5 months if stored at $41°F$.

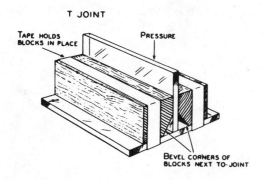

T JOINT

TAPE HOLDS
BLOCKS IN PLACE

PRESSURE

BEVEL CORNERS OF
BLOCKS NEXT TO-JOINT

Fig. 27-6. Another typical clamping method. (*Courtesy Cadillac Plastic & Chemical Co.*)

Additional cement formulations, and their adaptability to specific types of acrylic resin, are detailed in the wealth of literature available directly from the resin manufacturers.

Heat treatment or annealing of joints made with solvent cements is highly desirable because it greatly increases the strength of the joint, but it is frequently not necessary because the joints are adequately strong without it. Solvent joints never become completely dry; that is, they are never entirely freed of solvent through evaporation and diffusion. If the cemented piece is placed in an oven, the penetration of the diffusing solvent progresses farther away from the bond, with some resulting increase in the

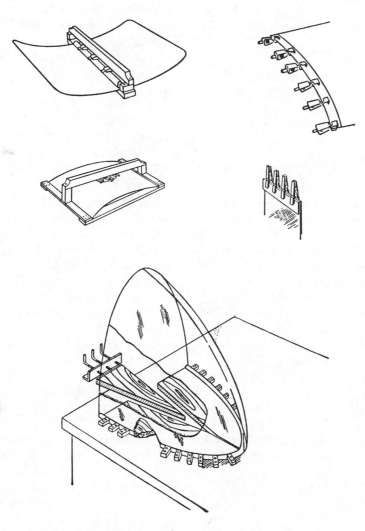

Fig. 27-7. Clamping set-ups for various contoured acrylic parts.

strength of the cemented joint. Heat treatment of joints made with solvent cement should be done carefully, so as not to warp the assembly or to approach the boiling point of the cement and thus produce bubbles.

The optimum and maximum heat-treating (annealing) temperatures depend upon the specific plastic. The recommendations of the resin manufacturer should be consulted.

Hardening of the joint, with or without heat treatment, should be completed before carrying out finishing operations such as sanding, machining, or polishing. Otherwise the softened material in the joint will subsequently recede into the joint.

Molded Acrylic Pieces. With molded acrylics, cemented joints are generally not as satisfactory as with cast sheeting. Considerable difficulty is frequently encountered, and many precautions must be taken in order to obtain satisfactory results. It is preferable to design molded articles so as to employ mechanical fasteners or closures, rather than cemented joints, but this is not always feasible, and recourse must sometimes be had to solvent cement.

Cemented joints in molded methyl methacrylate are frequently unsightly as a result of stress-crazing. Hence the recommended procedure is intended to avoid stress as much as possible, so as to keep it well below the "working stress."* In the presence of solvents, the working stress of thermoplastics is lowered appreciably. Stresses built in during fabrication, combined with the effect of exposure to solvent during the bonding operation, may exceed the working stress, and cause cracking or crazing. Mechanically built-in stresses may be greatly reduced by annealing.

If crazing appears, the cracks may be of such a size as to be easily visible to the eye, or so minute as to appear merely as surface clouding. In either case, crazing spoils the appearance of the article, promotes mechanical failure, and in some cases degrades electrical properties.

---

* Working stress is defined as that stress which can be applied, internally or externally, to an article for a period of time of the order of years without causing failure. Working stress is a function of the environment, and will be lowered by exposure to most solvents and chemicals or by thermal strain.

Crazing may occur either locally or generally. It may be caused by excessive local mechanical stresses (internally or externally applied), or it may be the result of localized loss of volatile constituents. The crazed cross-section must withstand tensile loads and elongations beyond its ability, and it finally fails.

When articles are to be cemented, the avoidance of stress should be considered in the design of the dies, since the conditions of molding and the design and location of the gates determine to a large extent the degree of stress that will be present in the molded articles. The gate should be located away from the area to be cemented, since frozen stresses are usually greatest near the gate. In designing two pieces to be cemented, the exact dimensions should be very carefully considered, so that the pieces will fit together easily and accurately without forcing, since stresses developed by forcing during assembly will promote solvent crazing.

When a cover is to be cemented into a recess or groove, right-angled corners should, if possible, be replaced by bevels. This promotes the escape of air, when the surfaces are brought together, and reduces the excess of cement which must be wiped away.

A cover which must be cemented should be made circular, if possible, so as to permit it to be twisted in place, immediately after cementing, to remove air bubbles.

The finishing operations that will be required should be anticipated when the original design is made. The design of the gate should be such as to minimize chipping during its removal, since surface fractures occasion solvent crazing. For the same reason, degating of the molded article should be very carefully done.

The conditions of molding also are important. To minimize stress, injection molded articles should be run with the mold as hot as possible and with as short a cycle as will permit ready removal of the shot from the mold.

The selection of a cement is of less importance than the foregoing precautions to minimize strain. A minimum amount of cement should be used. Cement is generally applied to well-fitting molded pieces by a brush, but in some cases the use of a hypodermic needle is advantageous. The solvent-type cements may be employed.

Cellulose Acetate Plastic. Sheeting and molded articles of cellulose acetate can be cemented readily to pieces of the same plastic with a bond practically as strong as the material itself. The usual precautions necessary for best results in cementing of plastics must be observed.

The cements used with cellulose acetate plastics are of two types: (1) solvent-type, consisting only of a solvent or a mixture of solvents; (2) dope-type, consisting of a solution of cellulose acetate plastic in a solvent or mixture of solvents.

The solvent-type cement is generally employed when the surfaces to be cemented are in a single plane and simple in nature, and when the surfaces can be readily held to a perfect fit. The dope-type cement is used when the surfaces are irregular or not easily accessible.

Acetone and mixtures of acetone and methyl "Cellosolve"* are commonly used as solvent cements for cellulose acetate. Acetone is a strong solvent for the plastic, but evaporates rapidly. The addition of methyl "Cellosolve" retards the evaporation, prevents blushing, and permits more time for handling the parts after application of the cement.

A cement of the dope type for cellulose acetate plastic, by virtue of its containing plastic in solution, leaves upon drying a film of plastic that forms the bond between the surfaces to be joined. These cements are generally used when an imperfect fit of the parts requires filling. Dope cements, satisfactory for many purposes, can be made by dissolving cellulose acetate plastic in solvents. A typical formula is:

	parts by weight
cellulose acetate plastic	130
acetone	400
methyl "Cellosolve"	150
methyl "Cellosolve" acetate	50

A general formula, suitable for use with a

* "Cellosolve," in this chapter, denotes ethylene glycol monoethyl ether (2-ethoxyethanol).

wide variety of ingredients, would be the following:

	parts by weight
cellulose acetate (low viscosity)	8-12
low- and medium-boiling solvents and diluents (under 100°C)	45-75
high-boiling solvents (over 100°C)	20-50

After being cemented, the pieces being united should be held under light pressure for 1 to 10 min, depending on the nature of the bond and the type of cement used. The assembly should be allowed to stand at least 24 hr before subsequent operations are performed, such as sanding, polishing, testing, and packing.

Cellulose Acetate Butyrate and Propionate. These plastics are cemented in accordance with the technique described for cellulose acetate. Solvent cements may be formulated from the following:

	Boiling Points (°C)
Methylene dichloride	39.8
Acetone	56.2
Methyl acetate	57.2
Chloroform	61.7
Ethyl acetate	76.7
Methyl ethyl ketone	79.6
Ethylene dichloride	83.5
Isopropyl acetate	88.7
Nitromethane	101.2
Dioxane	101.3
Nitroethane	114.0
Methyl "Cellosolve"	124.6
Butyl acetate	126.6
"Cellosolve"	135.1
Methyl "Cellosolve" acetate	145.1
Ethyl lactate	154.0
Cyclohexanone	155.4
"Cellosolve" acetate	156.4
Diacetone alcohol	169.2
Butyl lactate	188.0

In the case of dope cements, the plastic to be dissolved in solvents is cellulose propionate.

Cellulose Nitrate Plastic. This thermoplastic may be readily joined to itself with acetone or other solvents for cellulose nitrate. The joint hardens rapidly, the low-boiling solvents evaporate readily, and the bond is strong. The techniques for cellulose acetate are applicable to cellulose nitrate.

Ethyl Cellulose Plastic. The strongest bonds between pieces of ethyl cellulose plastic are made by solvents or by solvents bodied with ethyl cellulose plastic.

These solvents can be applied by felt pad, by dipping, or by a syringe. For brushing or flow-gun, 5 to 10% of finely ground plastic is dissolved in the solvent. Parts should be in contact with the solvents for 1 to 3 min prior to assembly. Usually better bonds are obtained if both of the surfaces to be joined are thus wetted with the solvent. The mating surfaces are then brought into contact and held under light pressure (5 psi) for about 10 min. The joints can be air-dried or oven-dried. Oven temperatures of 120 to 130°F are usually satisfactory; 150°F is the upper limit. Drying time will depend on the type and amount of solvent used. Bodied solvents are slower in drying than straight solvents.

Ethyl cellulose plastic can be joined also by heat-sealing techniques such as friction spinning and welding. Films and sheeting can be joined by use of dielectric heaters.

Nylon. The recommended adhesives for bonding nylon-to-nylon are generally solvents, such as aqueous phenol, solutions of resorcinol in alcohol, and solutions of calcium chloride in alcohol, sometimes "bodied" by the inclusion of nylon in small percentages.

Aqueous phenol containing 10 to 15% water is the most generally used cement for bonding nylon to itself. The cement is prepared by melting phenol (available from chemical supply houses) and stirring in the necessary water. The bond achieved by use of this cement is water-resistant, flexible, and has strength approaching the nylon.

The surfaces to be cemented should be cleaned thoroughly with a conventional degreasing agent such as acetone, or trichlorethylene. The cement is applied with a brush or cotton swab to the mating surfaces, which should be wetted completely using one or more coats as needed.

If the parts fit well, they can be assembled immediately. However, if the fit is poor or loose at the interface, a waiting period of 2 to 3 min to soften the surface will help in obtaining a satisfactory fit. A few experiments at this point will determine the time to form an adequately softened surface for the formulation of the best joint.

The parts are fastened together by means of spring clamps, "C" clamps, or toggle clamps in such a way that a uniform pressure is exerted on the bonded areas.

If the phenol is exposed to moist air, the cemented joint may turn white as it cures. A pre-cure after the parts are fastened for 1 hr at 100°F, or overnight in a dry atmosphere at ambient temperature, will prevent the joint from whitening.

The assembly is next immersed in boiling water in order to remove the excess phenol and to ensure a strong bond. Immersion time is dependent on the area of the bonded section. Five minutes immersion is used for $\frac{1}{8}$ in.-wide bonds; for 1 in. wide surfaces, boiling as long as 1 hr may be necessary. As a general guide, the assembly should be boiled until the odor of phenol has disappeared from the joint.

In cases where it is objectionable to immerse the sample in boiling water, the sample may be cured by prolonged air exposure at ambient temperature, or oven temperature of 150°F. Adequate bond strength for handling will be reached in 48 hr, and maximum strength in 3 or 4 days. The curing time is dependent on the quantity of phenol used.

Calcium chloride-ethanol bodied with nylon is recommended for nylon-to-nylon joints where there is potential contact with foods, or where phenol or resorcinol would be otherwise objectionable. Such solvents produce bonds which are equivalent in strength to those obtained with phenol or resorcinol, except when used with high molecular weight nylons. This adhesive also has the added advantages of forming a non-brittle bond, of being less corrosive, and of being easy to apply.

Bonding nylon to metal and other materials: Various commercial adhesives, especially those based on phenol-formaldehyde and epoxy resins, are sometimes used for bonding nylon to other materials. Although these commercial adhesives can also be used for bonding nylon-to-nylon, they are usually considered inferior to the solvent-type bonding because they result in a brittle joint.

Epoxy adhesives, for example, have been used to produce satisfactory joints between nylon and metal, wood, glass and leather.

These adhesives are two-part systems and are prepared before each use by mixing a curing agent with the adhesive resin according to the manufacturer's description. Thorough mixing is essential to obtain optimum performance.

Surfaces to be bonded must be thoroughly cleaned with a suitable degreasing agent, such as Perclene, acetone, etc. For best results, the nylon should be roughened with an emery cloth and the metal should be sandblasted.

The catalyzed epoxy adhesives are best applied with a wooden paddle or spatula. They will cure at room temperature in 24 to 48 hr. To accelerate the cure, they may be heated in a circulating air oven for 1 hr at 150°F.

Polycarbonate. Solvent cementing of parts molded of polycarbonate may be effected by the use of a variety of solvents or light solutions of resin in solvents.

The selection of the solvent system best suited for a particular application depends upon the area to be bonded, the speed with which parts can be assembled, and the practical limitation of necessary drying conditions.

The following solvents are recommended for solvent cementing parts molded of polycarbonate:

(1) Methylene chloride, which has a low boiling point of 40.1°C., and an extremely fast evaporation rate. This solvent is recommended for most temperate climate zones.

(2) A 1 to 5% solution of polycarbonate in methylene chloride can be used in extreme cases to obtain smooth, completely filled joints where perfectly mated bonding areas are impossible to obtain. It also has the advantage of a decreased evaporation rate. Higher con-

centrations of this solution are not recommended because of the great difficulty in obtaining completely bubble-free joints.

(3) A mixture of methylene chloride and ethylene dichloride, with a maximum of 40% ethylene dichloride, may be used where it is difficult to join both halves fast enough to prevent complete evaporation of methylene chloride.

The evaporation rate of methylene chloride to ethylene dichloride is 6.7 to 1. This is important since methylene chloride after bonding will dry completely in a much shorter time than ethylene dichloride.

When solvent cementing parts molded of polycarbonate, use the minimum amount of solvent necessary for good adhesion. This is the opposite of the procedure recommended for other thermoplastics where one or both halves are soaked in solvent for a considerable period. Solvent should be applied to only one of the bonding surfaces. The other half remains dry and ready in the clamping fixture.

Parts must fit precisely without pressure or any contact inhibiting irregularities. Locating pins, tongue-and-groove shapes or flanges may be used to align bonding areas and to effect rapid part matching after the application of solvent. These devices should be kept shallow in order to avoid trapping solvent in the mating surface.

As soon as the two parts have been put together, pressure should be applied immediately. Pressure between 200 and 600 psi is suggested for best results. Holding time in the pressure fixture should be approximately 1 to 5 min, depending on the size of the bonding area.

Although the cemented parts may be handled without damage after the holding time in the pressure fixture, sufficient bond strength for room temperature use is normally developed by drying the cemented parts for 24 to 48 hr at room temperature. If the solvent-cemented part is to be used at high temperatures, a longer solvent removal program is necessary.

Adhesive Bonding of Polycarbonate. Parts

molded of polycarbonate can be bonded to other plastics, glass, aluminum, brass, steel, wood, and other materials using a wide variety of adhesives. Generally, the best results are obtained with solventless materials, such as epoxies and urethanes. However, each application has unique requirements for flexibility, temperature resistance, ease of application, and appearance, requiring careful adhesive selection. Room temperature-curing products such as RTV silicone elastomers work well, as do a number of air-curing and baking-type products.

Parts should be cleaned vigorously before adhesive bonding. All oil, grease, paint, mold releases, rust, and oxides must be removed by washing with compatible solvents, such as petroleum ether, isopropyl alcohol, heptane, VM and P naphtha, or a light solution of detergents, such as Joy. Bond strength may be improved by sanding, sand blasting, or vapor honing the bonding surfaces.

Polyethylene. The good solvent resistance of polyethylene and other polyolefins precludes the use of solvent-type cements and maximum bonds can be made only when the surfaces of the parts are prepared to give anchor points for selected adhesives.

One technique for surface preparation is to dip polyethylene in a chromic acid bath (made up of concentrated sulfuric acid, 150 parts by weight; water 12 parts; and potassium dichromate 7.5 parts) for about 30 sec at 160°F. The parts are rinsed with cold water after this treatment.

Still another effective surface treatment for producing cementable surfaces on polyethylene is electrical discharge. The discharge and arc resulting from the electrical breakdown of air are an effective source of energy. Several types of commercial electrical discharge treating equipment are available. The open oxidizing flame method is also used extensively to give cementable surfaces of molded and extruded polyethylene.

Several commercial rubber-base adhesives have produced moderate adhesion with polyethylene which has been surface treated. Cements were applied with a cotton swab. The coated areas were air dried for 3 min before assembly. After assembly the samples were allowed to condition overnight at room temperature.

Polystyrene. Complex assemblies of polystyrene, usually molded in sections, may be joined by means of solvents and adhesives. Solvent action on the surface is the usual means by which parts are held together. Polystyrene plastics are soluble in a wide variety of organic solvents.

Care must be exercised in the fabrication (especially by injection molding) of polystyrene pieces which are subsequently to be joined by the use of solvents, so as to keep internal stress as low as possible.

External loads should not be applied until the residual solvent in the joint has evaporated completely. This may require drying at room temperature for a week, or as much as a month, before the article is placed in service. The drying time required depends, of course, upon the amount of solvent used and the opportunity for escape.

Typical solvents listed in Table 27-4 may be divided into three groups according to their relative volatilities. Those with low boiling point, and thus high volatility, are "fast-drying." They are low in cost, are readily available, and dry rapidly. But they are unsatisfactory for

Table 27-4. Solvent Cements for Polystyrene

Solvent	Boiling Point (°C)	Tensile Strength of Joint (psi)
Fast-drying		
Methylene chloride	39.8	1800
Carbon tetrachloride	76.5	1350
Ethyl acetate	76.7	1500
Benzol	80.1	—
Methyl ethyl ketone	79.6	1600
Ethylene dichloride	83.5	1800
Trichloroethylene	87.1	1800
Medium-drying		
Toluol	110.6	1700
Perchlorethylene	121.2	1700
Ethyl benzene	136.2	1650
Xylols	138.4-144.4	1450
p-Diethyl benzol	183.7	1400
Slow-drying		
Amylbenzol	202.1	1300
2-Ethylnaphthalene	251	1300

transparent articles of polystyrene because they cause rapid crazing.

Solvents of the second type are classified as "medium-drying," and have higher boiling temperatures.

High-boiling or "slow-drying" solvents often require excessive time for development of sufficient bond strength to permit handling. These solvents, however, will not cause crazing to appear so quickly, and thus may be useful where a "clean" joint is essential. It is desirable to add up to 65% of a fast- or medium-drying solvent, to speed up the development of initial tack without greatly reducing the time before crazing appears.

A bodied, or more viscous, solvent may be required by certain joint designs and for producing airtight or watertight seals. These can be easily made by dissolving polystyrene in a solvent. Usually 5 to 15% of polystyrene by weight is adequate. Grinding the plastic to a powder will aid in making a solution. The choice of solvent depends upon the properties desired, such as setting time, flammability, safety hazard, etc.

Solvents may be applied by dipping, felt pad, syringe, etc. Bodied solvents are best applied by brush, flow-gun, doctor-knife, etc. The assembly should be made as soon as the surfaces have become tacky, usually within a few seconds after application. Enough pressure should be applied to ensure good contact until initial bond strength has been achieved. Stronger bonds result when solvent is applied to both surfaces to be joined. Experience will show how long the joint must remain clamped in order to develop enough strength to be handled. Considerable additional time will be required before the bond can be expected to withstand service loads.

The impact grades of polystyrene are not as readily soluble as general-purpose polystyrene. Hence good bonds are more difficult to make. Most of the impact polystyrenes can be successfully joined by using the medium-drying solvents and those solvents with the highest boiling points in the fast-drying range.

Numerous commercial adhesives, mostly of the solvent-dispersed, lacquer or rubber-based type, are available for polystyrene plastics.

Polystyrene plastics can be joined also by conventional heat-sealing techniques such as friction spinning and welding.

Polystyrene-to-wood. As polystyrene moved into contention in the late 1960's as a material of construction for the furniture industry, considerable research went into the subject of "adhesives for bonding polystyrene furniture components to wood." One report (from Shell Chemical Co.) came up with the following evaluation:

Solvent-based contact cements provide the strongest bond between polystyrene and wood. These adhesives all have a neoprene (polychloroprene) base and a ketonic-aromatic solvent system. The significant difference among solvent-based contact cements is the solids content. The resistance of the joint to the cleavage type stresses increase with an increase in the solids content of the adhesive.

Solvent-based contact cements should also exhibit excellent stability during exposure to the temperatures of a finish drying oven. The use of this type of adhesive can present the furniture manufacturer with one potential problem, i.e., excessive amounts of the adhesive can induce crazing of the polystyrene since they contain polystyrene-active solvents: ketones and aromatics. This crazing could propagate into cracks and thereby produce premature failure of the polystyrene component during end use.

Water-based contact cements which do not present this problem were also tested. These adhesives produce joint strengths between polystyrene and wood only slightly inferior to that obtained with solvent-based contact cement having similar solids contents. Despite the slightly lower bond strength values, a comparable amount of wood failure was noted during testing of joints with the water-based contact cements.

A definite disadvantage of these adhesives is that they do not wet the surface of the polystyrene, and the water must be totally absorbed by the wood. (Polystyrene samples joined with a water-based contact cement were still wet after 24 hours drying at room temperature.) Consequently, the bond strengths of joints made with water based contact cement shortly after assembly (three hours) are considerably less than that exhibited by joints made with solvent-based contact cements of similar solids contents. In assembly situations where the speed of adhesive set is relatively unimportant and the degree of solvent crazing is critical, water-based contact cements with high solids contents (40%), should produce a satisfactory bond between polystyrene and wood.

Other miscellaneous adhesive types, including a polystyrene cement, an epoxy adhesive, two PVA-based adhesives, a casein glue, a urea-based glue, an aliphatic resin glue, and a rubber paper cement, were also evaluated. Of these, only the polystyrene cement was observed to give good bond strengths between polystyrene and wood. The bond strengths of joints made with the polystyrene cement (polystyrene in toluene) were actually comparable to those obtained with the best of the solvent-based contact cements, and superior to those obtained with the water-based contact cements. The speed of set, however, is slow compared to solvent-based contact cements because of the different solvent systems employed. The polystyrene cement also crazes polystyrene considerably more than the solvent-based contact cements and its use for joining thin overlays to wood should be limited. This type of cement is being used successfully in a number of furniture plants for pre-doweling with wood parts. In this situation, the speed of set and the extent of polystyrene crazing are usually not critical.

PPO resins. Modified PPO (Noryl, General Electric) may be solvent cemented to itself or a dissimilar plastic using a number of commercially available solvents, solvent mixtures, and solvent solutions containing 1 to 7% PPO resin.

The choice of the cementing technique best suited to each application is dependent upon the evaporation rate of the solvent (or assembly time), the type of joint to be cemented, and the area to be bonded. Follow these recommended cementing procedures to obtain maximum bond strength:

(1) Remove all traces of grease, dust and other foreign matter using a clean cloth dampened with isopropyl alcohol.

(2) Lightly abrade the surface with fine sandpaper or treat the surface with chromic acid. (E-20 etchant, Marbon Chemical Company). For best results, immerse the area to be bonded in an 80°C acid bath for 30 to 60 sec depending upon part design.

(3) Wipe the surface a second time with a cloth dampened with isopropyl alcohol.

(4) Apply solvent to bond surfaces and quickly assemble the two parts. The rapid connection of bonded surfaces will prevent excessive solvent evaporation.

(5) Apply pressure as soon as the parts have been assembled. The amount of pressure required will generally depend upon part geometry; however, moderate pressure (100 to 600 psi) is usually sufficient. Clamping pressure should be sufficient to ensure good interface contact, but not be so high that the materials are deformed or that solvent is extruded from the joint.

Maintain pressure for 30 to 60 sec depending upon part design, bonding area, and part geometry. Bonded parts may be handled safely after the original hold time, although maximum bond strength is not usually attained for 48 hr at room temperature or 2 hr at 100°C. The cure time necessary to achieve maximum bond strength will vary with the shape of the part.

Apply solvent in an area of good ventilation and avoid direct solvent contact. Isopropyl alcohol may be used to clean clothing and equipment.

Adhesive Bonding of PPO. Parts molded of PPO resins may be bonded to one another as well as to dissimilar materials using a wide range of commercially available adhesives. Because adhesive bonding involves the application of a chemically different substance between two molded parts, the end-use environment of the assembled unit is of major importance in selecting an adhesive. Operating temperatures, environments, bond appearance, and shape and flexural and tensile properties must all be taken into consideration. Epoxy adhesives, because of their versatility, are generally recommended.

The following factors should be considered when selecting an adhesive:

(1) The cure temperature of the adhesive must not exceed the heat deflection temperature of the resin.
(2) Avoid adhesives containing solvents or catalysts which are incompatible with the resins.
(3) Test adhesives for compatibility and bond strength under expected operating conditions.

With the exception of holding pressure and cure cycle, the bonding procedures used for solvents can also be used with adhesives. Be sure part surfaces are free of dirt, grease, dust, oil, or mold release agents. The surface of the part should be sanded or chromic acid etched before bonding for maximum strength. To ensure against misalignment during the cure cycle, apply only "finger-tight" pressure. Follow recommended cure times and temperatures outlined by the adhesive manufacturer.

Polyvinyl Chloride and Vinyl Chloride-Acetate Copolymers. Greater diversity of composition, and, correspondingly, of ability to be cemented by solvents, exists among the vinyl chloride-acetate copolymer resins than in most of the other plastics that are cemented. This is due to the relative insolubility of the polyvinyl chloride component of the copolymer.

As the percentage of vinyl acetate is increased in the copolymer resins, the effect of solvents is markedly increased. For this reason, the cements which depend upon solvent action for strength will be the more effective with copolymers containing the greater percentage of vinyl acetate. In general, cementing by use of solvent is less satisfactory with the vinyl copolymers than with the more soluble plastics, such as those of the cellulose esters.

Copolymer resins are available also in plasticized forms, and these, particularly in the more highly plasticized formulations, are more rapidly cemented than are the unplasticized resins.

In the case of cements depending on their tackiness alone for adhesion, little difference between the resins will be encountered. In heat-sealing operations, also, there will be little difference in weldability, providing the optimum temperature for the particular composition is used.

Cements for copolymer resins are usually of two types: (1) solvent cements (also called *laminating thinners*); (2) dope-type cements.

Cementing with Solvent. Where smooth, rigid surfaces are to be joined, the solvent adhesives can be readily used. The adhesive is applied to the edges of the two pieces, which are held closely together, and flows between them by capillary action. Initial bonding takes place rapidly, usually within a few seconds, but the full strength of the joint is not reached until the solvent has completely evaporated. The ultimate strength of properly prepared bonds is practically as high as that of the original plastic.

The vinyl chloride-acetate copolymer resins are most rapidly dissolved by the ketone solvents, such as acetone, methyl ethyl ketone, and methyl isobutyl ketone. The cyclic ketones, such as cyclohexanone, form solutions of the highest solids content, but they evaporate slowly and are ordinarily used only for copolymers of high molecular weight and straight polyvinyl-chloride polymers. Propylene oxide also is a very useful solvent in hastening solution of copolymer resins, especially those of high molecular weight, and of straight polyvinyl chloride. Mixtures of ketones and aromatic hydrocarbons have a more rapid softening action than do the ketones alone. Mixtures of solvents and nonsolvents are preferable to solvents alone, and additions of aliphatic hydrocarbons and alcohols are sometimes advantageous. Two per cent of glacial acetic acid, added to the solvent cement, improves wetting and speeds the capillary flow of the cement between the two surfaces being joined. It should not be used, however, in formulations containing propylene oxide.

Embrittlement by solvent is one of the most troublesome factors in the bonding of rigid sheets by the use of laminating thinners. This is probably caused by molecular orientation and release of stress when solvent is applied. Two procedures which minimize this embrittlement have been used in solvent bonding, with considerable success. In the first, mixtures based on the less powerful solvents are used;

these apparently do not penetrate the sheet at a rate rapid enough to bring about embrittlement. In the second method, resin solutions are used for the bonding; their viscosity apparently minimizes penetration by the solvent.

The formulations of a number of typical laminating thinners are shown in Table 27-5. The solvents are arranged in order of increasing solvent power and decreasing rate of setting of the joint.

Table 27-5. Laminating Thinners.

	A	B	C	D
		(parts by weight)		
Dioctyl phthalate	5	2.5	2.5	
Methyl acetate (82%)	58			
Ethyl acetate (85%)	10			
Butyl acetate	10			
Methyl ethyl ketone		63		40
Dioxane		20		
Isophorone		2.5	2.5	
Methylene chloride			50	
Ethylene dichloride			43	
Cyclohexanone				40
Propylene oxide				20
Petroleum solvent (p.p. 94.4-121.7°C)	15			
Acetic acid	2	2	2	
Methanol		10		
	100	100	100	100

Propylene oxide penetrates vinyl chloride-acetate copolymer resins very rapidly and, in amounts up to about 20 to 25%, improves the "bite" into the resin. Under conditions where propylene oxide evaporates too rapidly, acetone may be substituted. Solutions containing propylene oxide should be stored with care and well stoppered to prevent evaporation, since its boiling point is only 34°C.

The chlorinated hydrocarbons also are excellent solvents for the vinyl chloride-acetate copolymer resins, and are suitable for use in cements.

Cyclohexanone and isophorone are extremely high-boiling solvents which impart slow drying and prolonged tackiness. By themselves they are very slow penetrants, and hence they are usually used in combination with low-boiling solvents.

Since many of the solvents discussed above present possible toxicity hazards, they should always be used with adequate ventilation.

Dope-type Cements. These are prepared by the addition of small amounts of resin to solvent cements to thicken them. Dope-type cements are used where the surfaces to be cemented are in contact only over a very small area or are mated so inaccurately that a thin solvent cement will not fill the gap.

Small percentages of vinyl chloride-acetate resin, in the form of shavings, turnings, or chips, may be dissolved in the previously mentioned solvent cements to impart viscosity. In conjunction with the vinyl resin there should be included 0.2 to 0.5% of propylene oxide to help stabilize the solution against discoloration by light and heat when stored. Adhesion may be increased by the addition of about 5% of a plasticizer, such as tricresyl phosphate.

In general, the dope-type adhesives set slowly with vinyl chloride-acetate copolymer resins, and have little strength immediately after application. The strength of the bond develops as the solvent evaporates.

The vinyl chloride-acetate resin cements do not possess a high degree of tackiness when wet. However, this can be greatly increased by incorporation of certain other compatible resins, to yield cements of improved adhesion. Data covering resins suitable for this purpose may be obtained from suppliers.

Vinylidene Chloride Copolymers. Vinylidene chloride plastics are more inert chemically than most other thermoplastics. Hence the choice of good solvents for bonding purposes is greatly limited. However, the solvents listed in Table 27-6 can be used with success in making solvent-type bonds. These solvents will also dissolve small amounts of vinylidene chloride plastic to make a more viscous solution for application by flow-gun and brush on poorly fitted mating surfaces. Solution of the plastic is greatly aided by grinding the polymer to a powder and by warming the solvent slightly. Great care should be used, in heating solvents, to keep fire away and to provide adequate ventilation.

Because of the chemical resistance of vinylidene chloride plastics, the exposure to the

Table 27-6.

Solvent	Boiling Point (°C)	Flammability	Maximum Allowable Concentration for Continuous Breathing (ppm)	Hazard of Toxicity in Use
Cyclohexanone	155.4	Yes	100	moderate
Tetrahydrofuran	65.4	Yes	75	slight
o-Dichlorobenzol	180.4	Yes	75	moderate
Dioxane	101.3	Yes	100	slight

action of the solvents before assembly must necessarily be longer. When the surface of the plastic has become tacky, assembly is made with a small clamping pressure to assure intimate contact of the mating surfaces. Clamps can be removed when the solvent has dried enough to permit handling of the pieces. Better results are obtained if the solvent, or bodied solvent, is applied to both surfaces to be joined.

Some of the thermosetting adhesive resins, such as the epoxies, bond very well with vinylidene chloride resins, and produce bond strengths approaching that of the material. Their main disadvantages are limited pot life, lack of initial tack, and the sometimes excessive time required for initial bond strength to develop and permit handling.

Heat-sealing techniques work well with the vinylidene plastics. They can be bonded by friction welding, and films and sheets can be dielectrically welded.

TFE-fluoroplastic. Since there are no solvents for TFE-fluoroplastic (Teflon, du Pont), techniques other than conventional solvent cementing must be used. The surfaces of parts fabricated of TFE are first modified with strong etching solutions making them receptive to conventional self-curing adhesive systems. Actually, self-curing adhesives of the epoxy and phenolic types are the most suitable means of bonding modified TFE to itself and to other materials.

CEMENTING OF THERMOSETTING PLASTICS

The cementing of thermosetting materials to themselves, or to other materials, poses problems which are not inherent in the cementing of thermoplastics. The insolubility of thermosetting materials makes it impossible to apply the solvent techniques used with thermoplastics, and the smoothness of surface of molded thermosetting plastics adds to the difficulty of cementing them.

The surfaces to be joined must mate perfectly, unless a gap-filling cement can be used. The smooth surface must be roughened; if machining is not required for mating, then the surfaces should be sanded. This removes the gloss and the mold-release agent. However, where the specific adhesive strength to the plastic is high, sanding occasionally reduces net bond strength by providing nuclei for cohesive failure of the plastic.

Cementing of Dissimilar Materials

The great strength of reinforced and thermosetting plastics, and their usefulness under a wide range of environments and conditions, have led to many new applications, frequently in combination with metals and other rigid materials.

The new synthetic resin and elastomer adhesives are chemically related to the new plastic materials, particularly those used in "reinforced" applications, which also are being adapted for structural (load-bearing) applications. Adhesive bonding is, for various reasons, the logical method of fastening for the production of structures composed of reinforced or rigid plastics. Such reasons are given in the following paragraphs.

In service where exposed metal must not be used, for reason of corrosion, obviously metallic fasteners must not be used, with the plastic. Similar is the case where a plastic is used to cover or protect a metal.

The plastic article may not be able to accept the piercing and the concentration of stress involved with a metallic fastener.

Many plastics, and particularly the cross-linked and thermoset reinforced plastics, cannot be bonded to themselves, or to other materials, by thermal welding or heat sealing.

Adhesives are available for joining dissimilar materials, and for bonding plastics to very thin metal structures. Their use provides means of achieving smooth contours, free from projecting fasteners, and also of ensuring uniform distribution of stress over the area bonded. The adhesive contributes sealing action, insulation, and damping of vibration. In many cases, the use of adhesive, rather than metallic fasteners, reduces the cost of production, reduces the weight of the assembly, or provides longer life in service.

Adhesive Formulations. The general chemical similarity of adhesives to plastics not only contributes to their good specific adhesion to plastics, but also assures that properly formulated adhesives closely match the plastics with regard to modulus, flexibility, toughness, resistance to thermal degradation, resistance to solvents and chemicals, and cohesive strength.

In cemented assemblies of reinforced plastics and metals, where structural strength is generally desired, the adhesive must be more rigid than those used for bonding plastic to plastic; i.e., one with modulus, cohesive strength, and coefficient of thermal expansion between those of the plastic and the metal. In many cases, such adhesives are stronger than the plastic itself.

Since most reinforced plastics are quite resistant to solvents and heat, heat-curing solvent-dispersed adhesives may be used; obviously, these could not be considered for bonding many thermoplastics. Such adhesives consist of reactive or thermosetting resins (e.g., phenolics, epoxies, urea-formaldehydes, alkyds, and combinations of these), together with compatible film-formers such as elastomers or vinyl-aldehyde condensation resins. Isocyanates are frequently added as modifiers to improve specific adhesion to surfaces which are difficult to bond. These adhesives may be applied not only in solvent-dispersed form, but also in the form of film, either unsupported, or supported on fabric, glass mat, etc.

One type of solvent-dispersed formulation includes a monomer (such as vinyl or acrylic) as the solvent; a dissolved polymer of the monomer as the film-former, to increase viscosity and decrease shrinkage; and a catalyst, packaged separately and added just before use, to polymerize the monomer. Cross-linking monomers in small proportion, such as divinylbenzene and triallylcyanurate, improve the bond strength of adhesives of this type, and also their resistance to heat and solvents. It will be recognized that these are related to the monomer cements previously discussed.

A great many of the outstanding new formulations are based on epoxy resins. The use of polymeric hardeners such as polyamides and polyamines, phenolics, isocyanates, alkyds, combinations of amines with polysulfide elastomers, etc., and the "alloying" of the epoxy with compatible polymeric film-formers, such as polyvinyl acetate and certain elastomers, has led to the development of a broad spectrum of adhesive formulations with a wide range of available properties. In general, these adhesives are characterized by toughness, high cohesiveness, and specific adhesion to a great many materials, including most plastics. In addition, many of these adhesives contain no solvent.

By the selection of the proper hardener (and several chemical families of various degrees of reactivity are available: monomeric polymeric, or mixed)), the adhesive can be formulated to cure at room temperature, "low" temperature, or elevated temperature. Generally speaking, the phenolic-epoxy and the elastomer-phenolic-epoxy blends are characterized by outstanding resistance to high temperature.

WELDING OF THERMOPLASTICS

Welding by heat, or thermal welding, provides an advantageous means of joining most thermoplastics. All such techniques use heat to soften (or melt) the bonding areas of the mating surfaces to fusion temperature. After joining, the area is cooled, causing the plastic to harden (resolidify) and forming the joint or weld.

There are, however, several techniques available today for welding thermoplastics, including: (1) dielectric heat sealing; (2) thermal heat sealing (non-high-frequency); (3) hot gas welding; (4) hot plate welding; (5) hot tool welding; (6) induction heating; (7) spin welding; and (8) ultrasonic assembly.

In determining the techniques best suited for a job, several factors must be considered: strength required; shape and size of the parts to be joined; the plastics involved; the equipment to be used (i.e., whether to use available machines or invest in new facilities); and whether or not the joint must be concealed. By careful part design, it should be possible to weld most thermoplastics with strong joints.

With each type of weld, however, it is important to establish time, temperature, and pressure combinations which melt the polymers at the mating surfaces and hold them firmly in contact while cooling. Excessive heating in any type welding can result in melt flow, and in some instances, even resin degradation. Insufficient heating, on the other hand, can produce weak joints.

High-frequency or Dielectric Heat Sealing

In its simplest explanation, dielectric sealing involves the use of electricity or electronic waves from an AC generator passing through a plastic sheet or film. These waves cause a high level of molecular "friction" within the plastic, so that (given a plastic with sufficient dielectric loss) heat will be generated within the plastic and the plastic will melt at the point of contact (i.e., where the energy field is concentrated). By applying pressure at this point and allowing sufficient cooling time, a permament bond is formed.

Dielectric sealing can handle just about any of the thermoplastics: ABS, acetate, polyester film, PVC, polyurethane, acrylics, to mention several; it is even possible to dielectrically weld some of the tougher engineering plastics, like acetal film, but in many such instances, where the plastic has a relatively low loss factor, fairly high frequencies must be used, necessitating more expensive equipment. The biggest use for this technique today is in sealing vinyl film and sheeting into such products as rainwear, inflatables, shower curtains, swimming pool liners, and upholstery.

The most usual line for a high-frequency sealing operation consists of an rf generator (with an oscillator tube), a sealing press to apply pressure to the bonding area, and heat sealing dies set up in the press to determine the shape of the sealing area (commonly fabricated from brass strips bent into the desired pattern). Operating ranges from 1 to 100 kw are available, but the most commonly used range is from 3 to 25 kw. More recently, as production demands for sealed goods have increased, especially in the automotive industry, there has been an interest in the 35 to 100 kw range. The higher the frequency, the faster a material with a given dielectric constant, can be fused or bonded.

Another design trend directed toward increasing production has been the development of rotary table machines with automated indexing work areas. A typical unit of this type might incorporate a rotary table with four work stations: loading, sealing, cooling, unloading. Such machines can seal as many as 400 to 500 units an hour.

Materials handling systems for feeding the sealing machines also vary widely. The most simple, of course, is hand feeding the part to be sealed into the press. More complex set-ups involve mounting the sealing equipment on trolleys which moves on a track alongside the film to be sealed or use conveyors or moving tables to automatically feed the film into the press.

Over the years, the dielectric sealing technique has become an extremely versatile process. Although originally developed for sealing plastic to itself (i.e., PVC to PVC), the technique is now available for sealing different materials together. In the automotive industry, for example, door panels based on vinyl overlays welded to hardboard sheets with a urethane foam core in-between, are being dielectrically sealed together. In many of these applications, vinyl chloride latex adhesives are involved. These adhesives are bonded directly to a solid skin or impregnated into a porous material like urethane foam. Once dried, the treated parts can be bonded dielectrically.

Dielectric sealing is also used as a manufacturing technique in assembling shoe uppers. This process involves placing pre-cemented and assembled parts of a shoe in the press. Upon application of high-frequency current, heat-sensitive cement on the shoe upper absorbs energy and heats. The heat activates the cement which penetrates the adjacent surfaces of the various components of the upper and bonds them together. This reportedly permits assembly of the shoe in under a day as opposed to up to a week required with some other methods.

Finally, dielectric sealing has also come into use as a method of decorating plastic products. For example, by sealing two sheets of vinyl together in a predetermined pattern, it is possible to achieve a textured quilted effect. Another method is known as *applique bonding*. Basically, this is achieved by applying the "tear seal" system. In a tear seal, the sealing die is forced further into the plastic than is normally the case, causing it to thin out (as the melted plastic is displaced by the die) to the point where excess material can easily be torn off or separated by hand; the shape of the die, of course, determines which areas will be torn off. Thus, if a separate film is laid on top of the base film and the tear seal technique is applied (i.e., cutting through the top film and not the base film), it is possible to create an applique pattern by tearing off the excess material from around the sealing area and leaving a pattern defined by the sealing die.

Thermal Sealing of Film and Sheet

Non-high-frequency thermal sealing of film and sheet is also available and does not require the high-frequency energy that goes into dielectric sealing. This method usually involves a press with at least one moving platen bar. Bars are heated by low voltage, heavy electric current. Heat is applied to the outer surface of the product and must travel through the material to the interfaces to make the weld. A variation of this process, similar to high-energy induction bonding, involves putting a metal wire or other insert between the two sheets to be sealed to make the heating step more efficient.

Some of the most common types of heat sealing equipment is listed below:

Jaw-type bar sealers. Sealing takes place between an electrically heated bar and stationary base; TFE or other high-temperature resistant coverings should be used to cover the hot bar to prevent molten material from sticking to the bar.

Rotary sealers. For continuous operation, rotaries operate by passing the film over a revolving drum which acts as the base for the heated sealing bar to press against. In an alternative method, the material is carried on belts between thin heated rollers or wheels revolving in opposite directions.

Band rotary sealers. In this process, the film moves between metal bands, traveling first between heated jaws and then between pressure rollers, and finally between cooling jaws.

Impulse-type sealers. As mentioned above, impulse sealing involves the use of a metal wire. Heat is supplied on an intermittent basis so that heat exposure time is reduced.

Hot knife or side weld sealers. A seal is made by pressing a continuously heated knife-edge bar across the film.

Multi-point sealing. Heat is transmitted to the product at closely spaced points so that only a small portion of the surface of the film is actually sealed. The result is a peelable seal that will not open under normal conditions.

As with most sealing operations, temperature and pressure are the important factors. Low- and medium-density PE melt at about 230-240°F and linear PE at about 25°F higher.

Hot Gas Welding

Most thermoplastics can be joined via hot gas welding, but PVC is still the major material being assembled by this technique into such products as tank liners and ducting. Other candidates for hot gas welding are polyethylene, polypropylene, acrylics, some ABS blends, polystyrene, and polycarbonate. The process is not good for filled materials and is used only to join materials thicker than $\frac{1}{16}$ in. Parts thicker than $\frac{3}{8}$ in. are not generally gas welded, but they have been. Applications are usually in large, structural parts (like signs, piping, fans, ducts, fume scrubbers) and equipment manufacturers note that the weld is not an "appearance" joint, but a functional bond.

Basically, heated gas (generally air, but nitrogen is used for PE and acetals, which oxidize in air, and is recommended for polypropylene that will be subjected to UV exposure) is used to soften both the mating surfaces and the welding rod. This welding rod is made of the same material as the parts being joined and as the welding tool moves down the joint, the softened rod is deposited into the mechanically beveled joint area. On cooling, the mating surfaces and the rod filling the joint between are fused into an integral weld.

The following is a more detailed description of the hot gas welding process.*

Joints. The types of welded joints used in hot gas thermoplastic welding (Fig. 27-8) are

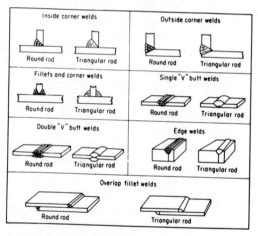

Fig. 27-8. Joint designs produced by hot-gas welding. (*Courtesy Kamweld Products Co.*)

similar to those used in metal welding. The same material preparations are required in plastic welding as in metal welding, except that beveling of thermoplastics edges is *essential*. Flux is not required; however, some materials

* Adapted from an article by S. J. Kaminsky, Kamweld Products Co., Norwood, Mass. 02062, appearing in the 1971-1972 issue of *Modern Plastics Encyclopedia*, reprinted with permission of the publishers, McGraw-Hill, Inc., New York, N.Y.

weld more satisfactorily in an inert gas atmosphere.

In thermoplastics, the materials do not melt and flow; they simply soften, and the welder must apply pressure to the welding rod to force the softened surface into the joint to create the required permanent bond.

Recently, plastics sheet and rod makers introduced triangular welding rods, in widths up to $3/8$ in., in PVC, PE and PP. Triangular rod can be used in "V" welds or fillet welds up to $3/8$ in. wide, which can be filled with one pass. Thus the weld can be completed in less time and with better appearance, especially inside tank corners. When triangular rod is used, the chances of porosity (or leaks) occurring between one round rod and another one are eliminated.

Triangular rod can be hand welded the same way as hand welding round rod; however, it can be applied more evenly with the properly designed high-speed tool. The speed weld also produces a much neater appearance than the hand weld. It must be understood that the use of heavy triangular rod, especially in horizontal runs, can be more tiring to the welder. But at the same time, the weld can be accomplished in one pass rather than in a series of passes, as with round rod. In some cases, a small-diameter round rod is welded in the bottom of the "V" to join the pieces together and prepare the joint for the larger triangular rod.

Welding guns contain a heating unit (electric or gas) through which is passed a stream of compressed air or inert gas that reaches the nozzle or tip at a temperature of between 400 and 900°F. For production work, a gun of the type having its air hose also enclosing its electrical cable is more efficient and flexible in operation. The wattage of the heater determines the effective range of available heat; the air pressure determines the actual amount of heat attained at the tip.

Tack welding is used to hold pieces in place for final welding. It can be effective in reducing or eliminating the need for clamps, jigs or additional manpower during assembly. Tack welds are produced simply by running a heated tacker tip along the joint at intervals or in a continuous line.

In hand welding, the welder uses both hands, one to hold the welding gun and the other the welding rod. Using a round tip on the gun, he directs a hot air stream with a fanning motion onto both sides of the joint to be welded, as well as onto the welding rod. After the weld is started, the welding rod is bent back slightly and enough pressure is exerted uniformly to fuse the heated surfaces.

Only experience and careful observation can assure good joints with maximum strength and minimum degradation of the material through overheating. Underheating makes a "cold weld" with poor tensile strength. The aim is always to produce adequate pressure at the joint to permit complete fusion, yet not so much as to cause the welding rod to "catch up" with the welding tip. This will cause the rod to stretch, resulting in a poor weld.

High-speed hand welding. Speed of making welds can be greatly increased through the use of welding gun tips that hold the welding rod in correct position and simplify applying optimum pressure while making the weld.

Flat strip may be used instead of round rod in a tip for welding tank linings and similar applications. Only a single pass is made, the wide strip providing adequate surface to give strength in the finished joint.

Hot Tool Welding

Generally, this description applies to any technique in which the edges to be joined are heated to fusion and then brought into contact and allowed to cool under pressure. Among the tools generally included in this category are electrical strip heaters, hot irons, and soldering guns. Hot plates are theoretically also included in this category, but since hot plate welding has taken on special attributes for joining plastics, it is covered separately and in more detail below.

Although work continues on adapting hot tools (i.e., strip heaters, soldering irons, etc.) to large-volume and high-speed production, the technique is still limited to low-volume production or experimental operations. Because temperature/pressure controls are difficult to maintain, most processors prefer to use techniques especially tailored for plastics.

Hot Plate Welding*

Hot plate welding will reportedly give strong, leak-free joints rated at 80-100% of the parent material. It's workable with HDPE, LDPE, PP, PS, PVC, acetals and other rigid thermoplastics. The technique has proved especially valuable for joining pipe, and is recommended for joining molded or extruded parts of the same material.

In hot plate welding, a plate—generally made of aluminum, although brass and stainless steel are also used—is heated to the melting point of the material to be joined. Two types of hot-plate welding are butt and groove welding.

In butt welding, the edges of the part are pressed against the opposite sides of the heated plate. When the plastic becomes pasty and a flash about half the thickness of the material is visible, pressure is removed. However, to avoid strains, the parts should remain in contact with the plate for some time. Parts are then removed from the heating element (or in many instances the plate is retracted from the parts) and the parts are pressed together. Flash can be removed with a peel knife.

Groove welding involves two heating elements, one to melt a groove the exact shape of the part to be bonded, and the other to soften the edge that will be bonded. The heated part is quickly placed into the heated groove and allowed to cool.

Processing times are fast. For example, acetal to be hot plate welded is pressed against a plate heated to 550°F with a contact time between 2-10 sec, depending upon joint geometry, then mated. If welds are wider than $\frac{1}{4}$ in., the heated parts should be sheared across each other during the contacting movement to prevent air entrapment.

While much hot-plate equipment is still hand operated, and often hand made, semiautomatic units have come on the scene to speed production. The welding cycle is automatic; feed and ejection are manual, although automatic ejection can also be incorporated. In a few custom jobs, automatic feed has also been

* Adapted from "Bonding and joining plastics," by J. O'Rinda Trauernicht, *Plastics Technology*, August 1970, reprinted by courtesy of the publisher, Bill Communications.

installed for automatic operation. However, use of a loading tool where an operator loads one tool while a second is in the machine and in a welding cycle is more common.

To better understand the operation of a hot plate welding system, following is production-line recommendations (from Celanese Corp.) for hot-plate welding Celcon acetal copolymer:

The hot plate welding method schematically represented in Fig. 27-9 has been found, both in laboratory studies and in production, to produce exceptionally strong bonds for acetal. These welds approach and, in many cases, equal the tensile strength of acetal.

The equipment used for hot plate welding consists, very simply, of a temperature controlled element to simultaneously heat both surfaces to be melted. During heat-up of the polymer on the heating tool, a pressure just adequate to insure intimate contact of the mating part surfaces on the heating tool should be used. As both surfaces reach the molten state, the surface material will become transparent. Sufficient melting for subsequent joining normally occurs when the surfaces in contact with the hot plate become transparent to a height of approximately $\frac{1}{16}$-$\frac{1}{8}$ in.

During joining, the molten surfaces should be brought carefully and quickly into contact and held with minimum pressure necessary. If

too much pressure is used during joining, starving of molten material from the weld will occur and result in poor weld strength. Normally, if weld flash greater than $\frac{1}{8}$ in. occurs, too much joining pressure is being used.

In welding two parts by the hot plate method, means should be provided to bring together the mating surfaces and hold them with light pressure sufficient to assure a bond without deformation.

The hot plate welding procedure is a two-step operation of placing the surfaces against the hot plate until molten and then holding them in contact until cooled. Thicker or thinner sections will require longer or shorter heat-up time on the hot plate.

Both time and temperature ranges are sufficiently broad, however, so that many jobs of welding by this method may be accomplished with minimum investigation to determine optimum conditions. Annealing of the weld joint will further enhance its strength. Annealing can be done by immersing the part in 300°F oil.

It is recommended that hot plate temperatures within the 430-560°F range be employed. Hot plate temperatures below 430°F will result in sticking of the material to the hot plate, while temperatures above 560°F increase the risk of polymer degradation.

The hot plate method of welding is ideally suited to straight, flat welds. It is adaptable to welding curved or angular joints also, however, when a heating tool is devised to provide uniform heat throughout the mating surfaces. The heated surfaces of such joints should also mate well without needing excessive, i.e. deforming, pressures. One means of welding irregularly shaped joints by this method is shown in Fig. 27-10.

This illustrates a circular or oval type joint with the ends of one piece flanged to conceal flash. The flange itself can add significant strength to a joint if heated and made part of the weld area.

Also shown is an angular joint. The flange in this joint also might help control and conceal flash while contributing finished weld strength. Such a flange may also help align the pieces as they are brought together after heating.

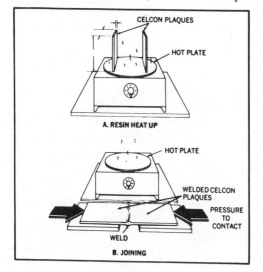

Fig. 27-9. Hot plate welding of acetal copolymer. (*Courtesy Celanese Plastics Co.*)

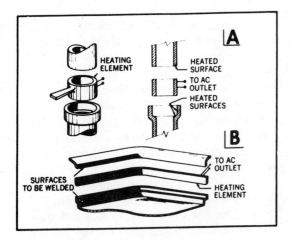

Fig. 27-10. Hot plate welding of irregular joints. (*Courtesy Celanese Plastics Co.*)

Induction Welding

Induction welding is exceptionally fast and versatile. Heat is induced by a high-frequency electrodynamic field in a metallic insert placed in the interface of the areas to be joined, and this heat brings the surrounding material to fusion temperature. Pressure is maintained as the electromagnetic field is turned off and the joint solidifies. Weld strength depends on size and geometry of the metal insert.

Induction welding is one of the fastest methods of joining plastics. Some applications require as little as 1 sec of welding time. Typical applications require from 3 to 10 sec. The rate of production is generally limited only by the speed with which the parts may be assembled in and removed from the welding jig.

Successful welds can be obtained by using stamped foil inserts, standard metallic shapes, such as wire screen, or various other configurations of conductive metal. The shape of the insert is not limited to the normal pattern of a closed resistance circuit; thus it may be made in a shape that provides a decorative effect in the finished article, e.g., stars or letters. Wire can be used, preferably of diameter from 0.010 to 0.030 in. Printed or metallized inserts may be feasible, but have not yet been fully evaluated.

The higher the pressure applied on the joint during welding, the higher the permissible temperature. Feasible pressure appears to be limited only by the strength and stiffness of the materials being welded. In most applications a pressure of 100 psi of joint is required, but greater pressure may be needed with some plastics, to minimize development of bubbles from thermal decomposition at the joint. The joint should be designed so that pressure is distributed uniformly throughout, and over the metal insert. The insert should be located as closely as possible to the generator coil. It should be centered accurately within the coil, to avoid its being pulled out of position by the attraction of the coil. A tongue-and-groove joint or a similar configuration is desirable, to locate and hold the metal insert. In any case, the insert must be located in the interface so that no portion of it is exposed to air, since this could cause rapid heating and subsequent disintegration of the insert.

Strengths of induction welds are limited by the relative smallness of the area in which the actual weld is formed. Welding occurs only in the area immediately adjacent to the metallic insert. In most cases this is a band, $\frac{1}{16}$ to $\frac{1}{8}$ in. wide, along the periphery of the insert Very little if any strength is obtained by metal-to-plastic bonding over the area of the insert.

Weld strengths can be determined by multiplying the unit weld strength by the area over which actual welding takes place. With wire screen as the insert between polyethylene slabs, weld strengths of better than 50% of that of the parent material can be consistently produced. Acrylics have been processed equally well, and from all indications welds of significant strength can be obtained in almost all thermoplastics.

A more recent adaptation of induction bonding involves replacing the metal insert with a metal-filled plastic. In essence, magnetic materials, in the form of micron size particles, are dispersed within a thermoplastic matrix which is compatible with the material to be joined (e.g., a polypropylene matrix when joining polypropylene parts together). The matrix can be supplied either in the form of a liquid or it can be molded into a shape (e.g., a gasket) that will conform to the shape of the weld area.

The metal-filled plastic bonding agent is simply applied to the parts to be joined and then subjected to the high-frequency magnetic field of an induction heating coil. The magnetic flux reaches through the work and generates heat in the bonding. Heating is the result of internal energy losses within the material that cause the temperature to rise. It is, of course, the metal particles (with magnetic properties) within the plastic that generate these losses through hysteresis and eddy current. (See Figs. 27-11 and 17-12).

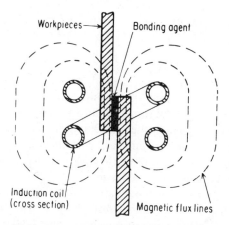

Fig. 27-11. Electrical induction heat is produced by the high-frequency magnetic field of an induction heating coil. The magnetic flux (dashed lines) reaches through the work and produces heat in the bonding agent. (*Courtesy Modern Plastics Magazine*)

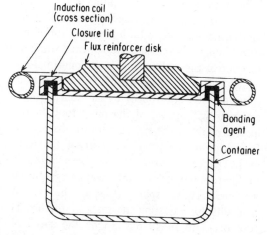

Fig. 27-12. Plastics lids are bonded to plastics containers using the induction bonding method. (*Courtesy Modern Plastics Magazine*)

As a consequence, the thermoplastic matrix in which the particles are confined melts and is subsequently fused to the two abutting surfaces, forming a structural bond on cooling.

To date, the process is available for polypropylene, polyethylene, nylon, ABS, polystyrene, and PVC. Its advantages include: adaptability to bonding most of the thermoplastics; speed of bonding; the ability to make structural bonds in either butt, peel, or shear; adaptability to hermetic seals; flexibility in bonding area in terms of size (can be a small spot weld or a long, continuous weld) and shape (irregular as well as symmetrical); and the fact that the physical and chemcial properties of the bond area are similar to those of the material being joined. It is also claimed that by using small particles of metal as opposed to larger pieces of metal or wire, stresses are not molded into the bond area.

Spin or Friction Welding

In this process, the two surfaces of thermoplastic are rubbed together until the friction develops enough heat to fuse and weld them. The strength of the welded joint approaches that of the parent material.

Typical welding cycles are: 1-2 sec. High-speed feeding devices, e.g., vibratory feeders, can be hooked into the operation for continuous production.

Nearly all the rigid thermoplastics can be spin welded, although there are some problems with the softer rigids, like low-density polyethylene or ethylene vinyl acetate.

The advantages of the process are: (1) the strength and good appearance of the joint; (2) the exclusion of oxidizing effect, in that the heated surfaces are directly in contact with each other throughout the process; (3) the economy resulting from adaptability to standard shop equipment, such as drill presses and lathes.

Its disadvantages are: (1) limitation to a circular or annular area (but sometimes a circular area of weld can be made part of the design of a non-circular article); (2) the squeezing out ("flashing") of soft material beyond the area of the weld before the weld is completed (but this can frequently be made unobjectionable by designing so that the flashing is directed to an internal area of the article).

Another drawback has generally been considered to be the technique's limitation to fairly small parts. More recently, however, work has been continuing in spin welding parts up to 20 in. in diameter.

Spin welding involves the rotating of one of the pieces against the other, which is held stationary. Rubbing contact is maintained at speed and pressure sufficient to generate frictional heat and melt the adjacent surfaces. The frictional heat is sufficient to effect almost immediate melting of the two surfaces, without substantially affecting the temperature of the material immediately beneath the surfaces. When sufficient melt is obtained, the spinning is stopped and pressure is increased to squeeze out all bubbles and to distribute the melt uniformly between the two surfaces. Pressure is maintained until the weld solidifies. In many instances, a single pressure setting has been found adequate. For thermoplastics having sharp melting points, braking is sometimes required, in order to stop the rotation of the parts quickly, so as to avoid tearing of the partially solidified weld.

Spin Welding Equipment.* Prototype or low production volume spin welding can be performed on a standard drill press with a driving tool and chuck. High volume applications require either a drill press equipped with a spin-welding tool, air cylinder, valves, and timer, or specially designed spin-welding equipment.

The basic equipment requirements for a spin welder are diagrammed in Fig. 27-13. Parts to be welded are shown held in a chuck. A spinning tool is attached to an electric motor by a drive shaft and V-belt. An air cylinder advances the tool and applies force to the parts during welding. A timer and air valve are used to control the air cylinder. A motor speed control, motor brake, and clutch may be required.

Variables of Spin Welding. Basic variables of spin-welding are rotational speed, joint pressure, and spin time.

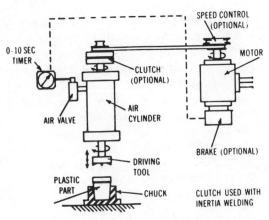

Fig. 27-13. Basic equipment requirements for spin welding. (*Courtesy E. I. du Pont de Nemours & Co., Inc.*)

Rotational speed depends on joint diameter.

Joint pressure is a convenient means for establishing the thrust required to generate heat between spinning parts. During welding, joint pressure forces bubbles, contamination, and excess material from the weld. Actual thrust will depend on joint design.

Spin time is the duration of relative motion between the rotating surfaces before melting takes place. Once melting has occurred, relative motion must be stopped to allow the weld to solidify under pressure since failure to arrest motion will cause tearing of the weld and give rise to low weld-strength.

Spin Welding Methods.* There are two basic methods for spin welding which differ only in the type of driving tool used: pivot or inertia methods. With both methods, rotational speed and joint pressure are adjusted similarly to control the welding process. Spin-times are controlled differently. With the pivot method, spin-time is controlled by a timer which activates withdrawal of the pivot tool. In the inertia method, the spinning tool is disengaged from the motor after the start of the weld cycle. Kinetic energy of the freely spinning inertia tool is converted into heat energy during welding. Thus, spin-time may be varied markedly by adjusting the mass of the inertia tool or more finely by varying the speed of rotation. Pivot tools are generally used for joints

* Adapted from a du Pont Technical Data Brochure on Machining Delrin acetal.

having a projected area of $\frac{1}{2}$ in^2 (3 cm^2) or less. Inertia tools are better for joints of large area because they offer fine control of input energy.

Essential parts of the pivot tool are the driving element, such as teeth or rubber facing, and the pivot pin. The pivot tool turns at constant speed throughout the weld cycle. A load is applied to parts by the spring loaded pivot pin before the driving element engages the parts. At the end of the cycle, the tool retracts and the driving element disengages while joint pressure is maintained by the pivot pin. Relative motion between parts stops almost instantly and the weld solidifies.

When part design does not permit use of a pivot pin, a thrust bearing may be used. A rubber facing may be used on the driving elements if no tooth indentations can be tolerated on the welded part.

Essential elements of the inertia tool are the driving element and the rotating mass or fly-wheel. The clutch for disconnecting the inertia tool from the motor may be part of the main welder drive system, or may be built into the inertia tool. With this inertia tool design, the rotating mass is supported by needle and ball bearings around a central shaft. A conical clutch connects shaft and mass when no vertical load is applied to the tool. When the spinning tool engages the upper part to be welded, the clutch disengages the rotating mass. Joint pressure is maintained during welding by a vertical load applied through tool shaft and lower thrust ball bearing to the rotating mass. The tool is raised after the rotating mass comes to rest.

Vibration Welding

A newer variation on frictional welding (to which "spin welding" also belongs) is called *vibration welding*. In vibration welding, frictional heat is generated by pressing the surfaces of two plastics parts together and vibrating the parts through a small relative displacement, which can be either linear or angular.

The major advantage of this technique lies in its application to non-circular parts (spin welding is suitable mostly for circular parts),

provided that a small relative motion between the parts in the welding plane is possible. Vibration welding, however, can also overcome one of the serious limitations of spin welding by its ability to join circular parts in a specific orientation.

Whatever the configuration, when vibrations cease at the end of the welding cycle, parts are in the exact desired position relative to one another. After a short cooling time under pressure, the welded parts are released.

Vibration welding machines operate at relatively low frequencies, 90 to 120 Hz. The amplitude of displacement can be varied in a range of 0.120 to 0.240 in. (3-6 mm). Joint pressure should be in the range of 200 to 250 psi (14-18 Kg/sq cm). Weld time will generally be between 2 to 3 seconds plus approximately 1 second hold time. Though slightly longer than typical spin welding and ultrasonic welding cycles, this is much shorter than hot plate welding and solvent cementing cycles.

Until now, it has been almost impossible to weld very large plastic parts, especially those with complex shapes. With this newest technique, parts with welding surfaces as long as 20 in. (50 cm) have been successfully joined. The linear method is applicable for all parts, while angular welding is limited to parts with a length to breadth ratio not exceeding 1.5:1. Both methods offer the capability of welding more than one part at a time.

Joint design. The basic joint is a butt joint. Unless parts have thick walls, a flange is generally required to provide rigidity and an adequate welding surface. As in any melt bonding technique, there is some displacement of melt out of the joint. For circular parts, angular welding with tongue-in-groove joints produces the best results.

Typical applications on which vibration welding techniques have been evaluated include automobile emission control cannisters, fuel pumps, expansion tanks, heater valves, water pump housings, motorcycle fuel tanks, pump impellers, etc. The technique is suited for use with parts molded or extruded of acetal, nylon, polyethylene, ionomer, acrylic, FEP fluorocarbon resins, Tefzel fluoropolymer, and Hytrel

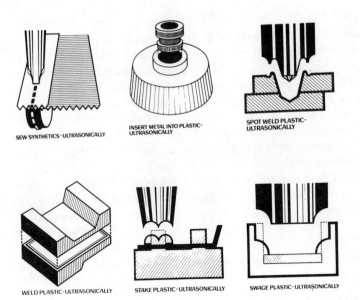

SEW SYNTHETICS-ULTRASONICALLY

INSERT METAL INTO PLASTIC-ULTRASONICALLY

SPOT WELD PLASTIC-ULTRASONICALLY

WELD PLASTIC-ULTRASONICALLY

STAKE PLASTIC-ULTRASONICALLY

SWAGE PLASTIC-ULTRASONICALLY

Fig. 27-14. Six ways in which ultrasonics can be used as an assembly technique for plastics. (*Courtesy Branson Sonic Power Co.*)

polyester elastomer. The technique is patented by du Pont and has been licensed to Branson Sonic Power Co.

Ultrasonic Assembly

The use of ultrasonic vibrations to assemble plastic components has become an increasingly important technique in recent years. Basically, the principle involves converting standard electrical energy at 50 or 60 cps into 20,000 Hz (cps). This is fed into a converter and is transformed into a 20,000 Hz mechanical signal. This mechanical signal, in turn, is transferred to a horn—a metal element that closely conforms to the top half of the product to be sealed. The action results in vibration of the part (while the bottom is held firmly in a fixture). The friction thus caused by the vibration melts the plastic at the joining surfaces. When the desired melting has occurred, the energy flow is cut and the parts are allowed to remain briefly in the press so the joint can set.

The process is quick, often taking less than 1 sec up to 4 sec for a complete cycle. Bond strengths are reported at 85 to 90% or more of that of the parent material.

There are four basic methods of ultrasonic assembly (Fig. 27-14), as follows:*

Welding. In welding, the horn contacts the plastic at a predetermined pressure and moves up and down, in phase with the top part being assembled. Mechanical vibrations at ultrasonic frequencies are transmitted from the horn into the plastic for a controlled period of time. The vibrations travel until they meet the joining surfaces. At this joint, or interface, the intense vibration of one plastic surface moving against the other causes sufficient frictional heat to melt and weld the two surfaces, producing a molecular bond.

Correct joint design is one of the most important requirements for a good ultrasonic weld. To provide sufficient material flow to fill high and low spots along the joint, a triangular projection called an *energy director* is added to one surface of a butt joint. The melted material acts as an adhesive, and flows uniformly around the periphery of the part to make an excellent weld.

Another important factor in successful

* Adapted from an article by Don J. Kolb, Branson Sonic Power Co., Danbury, Conn., appearing in the 1971-1972 issue of *Modern Plastics Encyclopedia* and reprinted by courtesy of the publishers, McGraw-Hill, Inc.

ultrasonic welding is the type of plastic used. Each plastic, like metal, has a different modulus of elasticity and will transmit vibrations differently. The more rigid the material, the better its acoustical properties. For example, lead is a very poor transmitter of vibrations. A bell made of lead would simply absorb the shock of a blow rather than transmit it to produce a tone. Materials such as general-purpose styrene, acrylic, polycarbonate, and ABS are good materials for transmitting vibrations, whereas the softer materials like polyethylene, polypropylene, and soft vinyl would react like a lead bell and absorb vibrations so no movement at the joint produced friction and melted plastic.

Staking. Most ultrasonic staking applications involve the assembly of metal and plastic. In staking, out-of-phase vibrations are required between the horn and the stud being staked. A hole in the metal part receives a plastic stud. A specially contoured horn contacts the stud, melts it, and reforms the plastic to form a locking head. Since frequency, pressure, and time are always the same for every production cycle, frictional heat generated between the horn and stud is consistent and faster than heat applied with a hot iron. Heads are formed rapidly, and the reformed plastic is permitted to cool for a fraction of a second before horn pressure is released. Thus, plastic recovery due to material "memory" is prevented, and tight assemblies are obtained. The final shape of the staked head conforms to that of the contoured horn.

Some typical ultrasonic staking applications include automobile instrument clusters and tail light assemblies, a broad range of appliances including radio and television panels, and electronic components.

Insertion. Ultrasonic insertion can, in many cases, replace the conventional and costly method of insert molding. A hole slightly smaller in diameter than the insert it is to receive guides the insert into the plastic under pressure applied by the ultrasonic horn. As with welding, out-of-phase vibrations between insert and plastic create heat at the plastic/metal interface, melting the plastic momentarily to permit the inserts to be driven into place.

Insertion can be accomplished by contacting either metal or plastic with the horn.

Ultrasonic exposure time is usually less than 1 sec, but during this brief contact the plastic reforms itself around knurls, flutes, undercuts, or threads to encapsulate the insert. Parts normally remain relatively cool because heat is generated only at the plastic/metal interface. Sizes of metal components and the number being inserted at one time will determine required ultrasonic exposure time for a given operation. Small inserts require ultrasonic exposure times as short as 0.1 sec.

Ultrasonic insertion applications include bushings, hubs, ferrules, terminals, pivots, retainers, and fasteners. Multiple inserts required on connectors or other electronic plug devices can be made simultaneously.

Spot Welding. The newest application for ultrasonic assembly is spot welding of heavy plastic sheets, some as thick as 0.250 in. Spot welding of plastic is possible because of recently developed, higher powered equipment and new pistol grip hand tools whose tip is designed to penetrate the first sheet, and enter the second sheet to a depth half the thickness of the first sheet. Displaced material then flows between the sheets, and forms a head created by the cavity in the face of the tool.

Ultrasonic Equipment. A typical ultrasonic assembly system is made up of five elements: the power supply, converter, stand, horn, and programmer.

The power supply converts conventional 60 cps of electrical energy to 20,000 cps. Existing power supplies produce power outputs at various levels, up to 2500 watts.

The converter transforms 20,000 cps of electrical energy as received from the power supply into mechanical energy vibrating at 20,000 cps by means of a piezoelectrical element made of lead zinconate titanate.

Converter and horn are mounted in a pneumatically operated assembly stand which contains the programming equipment. An air cylinder in the stand, operating under regulated pressure, brings the horn into contact with the work surface. Upon contact, the horn builds up a preset amount of pressure to clamp the part.

The ultrasonic energy dissipated at the joint welds the part. The timed "weld" sequence is followed by the timed "hold" sequence. At the end of the "hold" sequence, the stand withdraws the horn and converter assembly.

The ultrasonic horn, a one-half wave resonant metal section, is tuned to vibrate at the frequency of the system. The horn transmits the vibratory energy into the workpiece. Depending upon the power supply used, outputs of 2200 watts.

Horns are designed to meet the requirements of specific applications. Since the mass and shape of a horn determine the length at which it will oscillate at the required frequency, no two horns will have the same length. Horns may vary in length by as much as $1\frac{1}{4}$ in. They are constructed from a special titanium alloy which has an exceptionally high strength-to-wieght ratio, and is an efficient transmitter of vibrations.

The electronic programmer controls the duration of ultrasonic exposure and pressure applied to plastic parts by the horn and stand mechanism. When operating more than one converter from one power supply, the use of sequential timers permits multiple points of contact to be welded in sequence.

Ultrasonic equipment is also available for continuously welding film and sheet. One of the newer units is a "sewing machine" capable of "stitching" synthetics at 48 to 50 fpm.

MECHANICAL JOINTS

There are a variety of methods of providing mechanical fastenings and various types of joints in all kinds of plastics. This subsection covers: (1) fasteners, both in metal and in plastics; (2) such techniques as swaging, press-fitting, cold-heading, etc., which represent mechanical means of fastening; and (3) unique plastic fastening concepts such as the integral or "living" hinge.

Mechanical Fasteners

The type of fasteners that can be used with plastics covers a wide range: clips, nails, screws, assorted inserts, bolts, rivets, hinges, pins, to name a few. However, it is important when working with plastics to consider the special characteristics of the material as compared to those involved in joining metals.

For example, the coefficient of thermal expansion is one of the more important considerations in selecting a fastener or designing a product that will be using mechanical fasteners. Although a phenolic or other thermoset with a coefficient of thermal expansion similar to metal may not be severely affected, a real problem can be encountered with such thermoplastics as polystyrene with a coefficient of thermal expansion some 6 to 7 times greater than steel. This problem is especially magnified when using molded-in inserts, and some suppliers recommend using these inserts only with filled materials which have less expansion.

Plastics also vary in their ability to sustain loading (e.g., some are high in compression, others in tension) and plastics deflect more than metals under the same loading (therefore, metal and plastic should not share loads in parallel).

Following are more specific guidelines relating to the various types of mechanical fasteners in use today. However, because of the many variables among different plastic materials, it is suggested that users contact materials manufacturers for precise assembly data. Also, for a complete catalog of mechanical fasteners, a useful reference is the "Fasteners Handbook", by J. Soled (Van Nostand Reinhold Co., 1957).

Among the advantages offered by mechanical fasteners are: (1) ultimate bond strength is attained immediately; (2) a moveable joint can be attained, if desired; (3) disassembly is possible, with some types of fasteners; and (4) capital equipment costs are low.

It should also be noted that mechanical fasteners are available in plastics as well as in metals. Fasteners made of plastic are particularly useful in problems of fastening involving (1) corrosion-resistance, (2) color-matching, (3) sealing, (4) protection of painted or porcelained finishes, and (5) electrical insulation. They should not normally be considered where service temperatures are high; joints are subjected to high tensile or shear stresses; or the cost of the fastener itself is critical. A separate discussion on plastic fasteners can be found on p. 804.

Inserts

In using some of the mechanical fasteners described in this section, many applications will use threaded metal inserts to provide threaded holes in the plastic parts. These inserts are discussed in greater detail in the chapter on "Design Standards for Inserts." As a basic review, however, the major types of inserts include:

(1) Molded-in inserts
(2) Post-molding inserts. This category is subdivided into:

(a) Self-tapping inserts. There are two types of such inserts—thread-cutting and thread-forming. Thread-cutting inserts cut their own mating threads in the plastic material. They can be used in both thermoplastic and thermosetting materials. For most applications in plastics, the slotted type can be used. These inserts have coarse external threads which reduce installation torque and provide strong threads even in brittle plastics (Fig. 27-15).

In very hard plastics, or those with abrasive fillers, a case-hardened steel insert is used. This is the same type that is used in many metal applications and is available with internal locking features.

Thread-cutting inserts are also available with male threads. This self-tapping stud or male insert is useful where alignment or ease of assembly is a factor to be considered.

Installation of thread-cutting inserts consists of two steps: drilling or coring the holes and driving the insert. Because only two steps are needed and tapping is not required, thread-cutting inserts can be economically installed. Equipment available in most shops can be readily adapted for volume production.

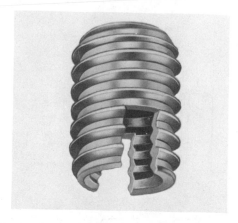

Fig. 27-15. Coarse external thread brass insert is ideal for both thermoplastics and thermoset materials. The wider thread spacing reduces the installation torque and provides stronger threads in weak or brittle plastics. Regular slotted casehardened steel inserts are recommended if the plastic contains highly abrasive fillers. (*Courtesy Groov-Pin Corp.*)

Thread-forming inserts create threads by mechanically displacing material as the insert is driven.

(b) Press-fit inserts (generally knurled).
(c) Expansion inserts which are placed in pre-drilled or pre-molded holes and expanded against the walls by insertion of a screw. Torsion resistance and pull-out retention are obtained by embedding the knurls into the material.
(d) Inserts installed ultrasonically. Friction, caused by ultrasonic vibrations, melts a thin film of resin at the metal-plastic interface. Pressure from the ultrasonic tool directs the insert into the cored or machined hole. When the energy source is removed, the molten area surrounding the flutes rapidly freezes securing the insert. Residual stresses in the boss are minimized because only a very thin film of the resin is melted. Because friction energy is concentrated at a point, melting is rapid and insertion time is relatively fast. (See Fig. 27-16).

Standard Machine Screws

For applications where assembly and disassembly are expected to be frequent, it is

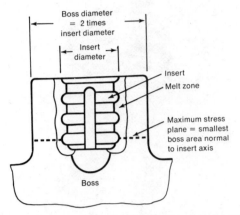

Fig. 27-16. Ultrasonic insert design for polycarbonate parts. (*Courtesy General Electric Co.*)

generally recommended that the part design incorporate metal threaded inserts (as described above) to accept a mating threaded screw.

Where the application involves infrequent assembly (less than six times), a hole in the plastic part with machined or molded-in threads to receive the mating threaded screws can be used. Most thermoplastics can be readily machined after molding with standard metal-working tools and this procedure is generally recommended for achieving threads below $\frac{1}{4}$ in. in diameter.

Larger internal threads may be formed by a threaded core pin which is unscrewed either manually or automatically. External threads may be formed in the mold by splitting the thread along its axis, or by unscrewing the part from the mold. When a parting line or the slightest flash on the threads cannot be tolerated, the second method (unscrewing the part from the mold) must be used. Mold designs with automatic unscrewing operations are feasible.

Coarse threads can be molded easier than fine threads and therefore are preferable. Threads finer than 28 pitch or closer than class 2 should not be specified. The roots and crests of all type threads should be rounded with a 0.005-0.010 in. radious to reduce stress concentration and provide increased strength.

When inserting a screw-type fastener, a hand-held or automatic drill is preferred, although it can be a simple hand insertion.

Many screw insertions make use of commonly available hand-held drills.

Standard taps can be used with phenolic parts. The pre-drilled or pre-molded hole should be chamfered to the maximum diameter of the thread. When threading or tapping a thermoplastic to receive a metal fastener, a slight increase over normal metal clearance is required for the difference in thermal expansion coefficients.

Self-Tapping Screws

These screws are available only in metal and are designed to produce their own mating threads when driven into a material. The unthreaded hole for a self-tapping screw can either be molded or drilled into the molding. The entering edge is usually suitably chamfered to prevent spalling.

Self-tapping screws provide a substantial cost savings by simplifying the molding operation and reducing assembly costs (molding or machining threads into the part is not necessary, since the self-tapping screw forms its own). They are not recommended, however, when frequent disassembly is required. Reassembly is generally limited to about 5 or 6 times.

There are two types of such screws: thread-cutting and thread-forming. Thread-cutting screws tap or cut a mating thread as the screw is driven. They are slotted to provide a channel for disposal of chips and therefore the depth of the hole should be slightly deeper than the screw (generally around $\frac{1}{32}$ in.) to provide a depository or reservoir for chips. Thread-cutting screws are usually suggested for use with plastics with higher flexural modulus (over 2×10^5 psi). Thread-cutting screws are used with thermosets and various high-performance engineering plastics, such as the acetals, nylon, ABS, etc. The stress factor with thread-cutting screws is relatively low.

Thread-forming screws create threads by mechanically displacing material as the screw is driven. They are used with the more ductile thermoplastics.

When using self-tapping screws, it is best to check with the screw manufacturer and the resin supplier as to installation specifications (i.e., how close to the edge of the part to place the fastener, how wide and how deep a hole should be, etc.). These specifications vary according to the type of fasteners and the plastic involved. For example, with Type BF or BT screws (see Fig. 27-17) in nylon or acetal, a hole size equal to the pitch diameter is recommended. For Type A, a hole diameter equal to 0.8 times the screw diameter is suggested.

Modified PPO that will receive a self-tapping screw should have a receiving hole diameter equal to the pitch of the screw. Minimum torque to keep screw assembly stress within the design limit of the material is also suggested. ASA Type T and BT screws perform well with this material.

For ABS, a minimum screw hole size for thread-forming is a hole diameter equal to the maximum root diameter, plus one-half the difference between the maximum root diameter and the maximum thread diameter.

Overall, the optimum hole diameter for self-tapping screws in plastic parts is the pitch diameter of the screw. Smaller hole sizes increase the stripping torque but decrease the ratio of stripping to driving torque. Both the stripping torque and the strip-drive ratio are important factors in fastener design. Larger holes reduce drive torque, strip torque, and vibration resistance.

Depth of Hole. In general, the number of screw theads engaged in the plastic is determined by the depth of the hole. The threads on the taper of a self-tapping screw provide little, if any, holding power. Therefore, the strength of a self-tapping screw is dependent upon the length of the clyinder of plastic which extends from the lowest full thread engaged in the plastic, up to the top of the hole.

In designing for a self-tapping screw in thermoplastics, the taper length and length tolerance of the screw must be subtracted from the depth of penetration available in the plastic.

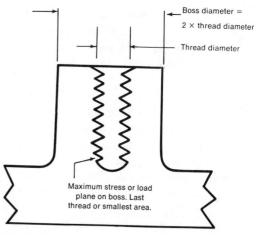

Fig. 27-18. In applications requiring the use of a metal screw and a threaded polycarbonate boss, the boss diameter must be equal to twice the thread diameter. This recommendation should generally be followed for screws up to 3/8 in. (*Courtesy General Electric Co.*)

As a rule, the hole should be deep enough to provide a minimum of two or three full threads of engagement. The practical upper limit of hole depth is a hole deep enough to provide a stripping torque equal to the torsional strength of the screw. For fine pitch threads, this is approximately eight to twelve threads. For coarse threads, three to eight full threads are required.

Bosses. The use of bosses may be desirable in many applications and they are quite commonly used in plastics. Bosses are protruding studs or pads used for the reinforcement of holes or for mounting and assembly. In general,

TYPE	DESIGNATION		
	ASA	MFG	FED'R'L.
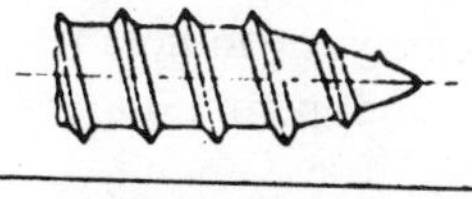	A	A	A
	B	B OR Z	B
	BF	FZ	BF
	BT	25	BG

Fig. 27-17. Self-tapping screws for plastics. (*Sketch courtesy E. I. du Pont de Nemours & Co., Inc.*)

when working with thermoplastics, the boss diameter is usually between two to three times the outside diameter of the screw (the average is $2\frac{1}{2}$ times). This is generally sufficient to take the possible hoop stresses developed due to screw insertion.

Trouble-Shooting Tips. The three most common problems associated with the use of self-tapping screws in plastic parts are: (1) fracture of the part as the screw is inserted; (2) stripping of the plastic threads as the screw is inserted; (3) loosening of the screw after a period of time.

Part Fracture. The basic cause of this problem is a high stress as the screw is inserted. The user should determine whether he is employing the right type of screw (e.g., thread-forming screws, as noted above, develop more stresses in the plastic than thread-cutting screws).

A second major cause of part fracture is the use of small diameter holes. With a small hole, the tap of the screw takes a deeper bite into the plastic material, and this may again produce excessive stress in the boss.

Stripped Threads in the Plastic. Often, the cause of this problem is an insufficient strip-drive ratio. The initial step in correcting this difficulty is to determine the hole size in the plastic. Larger diameter holes yield lower value of stripping torque but increase the strip-drive ratio.

If the hole depth in the plastic is too shallow, the screw may bottom out before the screw head seats. The hole should be deep enough to allow for the length tolerance of the screw.

If the screws are inserted with a power-driven screw driver, high driver speed may cause failure of the plastic thread. Slowing down the driver tends to increase the accuracy of torque setting and decrease the rate of loading of the threads in the plastic.

For some applications, a finer pitch screw may correct thread stripping as the finer pitch is less efficient in converting applied torque to compressive loadings of the threads in the plastic. Hole size should be checked when screw pitch is changed.

If all of the above have failed to correct the problem, redesign may be necessary. The shear area in the plastic can be increased with a larger diameter screw, a longer screw with a deeper hole, or a larger diameter head on the screw.

Screw Loosening. The common cause of this problem is insufficient thread engagement. The area of thread engagement may be increased by any or all of the following: smaller diameter holes; longer screws and deeper holes; or screws with finer pitch. Also, a careful re-evaluation of the application environment should be made to determine that estimates of temperature and moisture conditions are valid.

Drive Screws

These screws are designed to be driven into place with an impact and are used for permanent fastening. A distinct advantage with this type of screw is alignment. Drive screws are ideal for conduit installation or other operations where installation space or alignment may be a problem. It can also be used to join thin-guage materials (Fig. 27-19).

Bolts and Nuts

Bolts and nuts of standard configuration, made of metal or of plastic, can be used where mechanical requirements of the joined area permit.

Rivets

Rivets, made of metal or of plastic, can be used to hold parts together securely, or to hold two or more parts together and permit motion between them.

Rivets are noted for their low cost and simple mechanical installation that lends itself to automation. High clamp load or tension load is limited, and accuracy of location is not as good as other mechanical fasteners. Rivets can be installed to allow for dimensional change from temperature. Rivets are particularly applicable for use with plastics of high-impact strength. To join very thin sections of plastics, metallic eyelets may be used instead of commercial rivets. Eyelets are particularly advantageous in providing a bushing for a rivet, or a

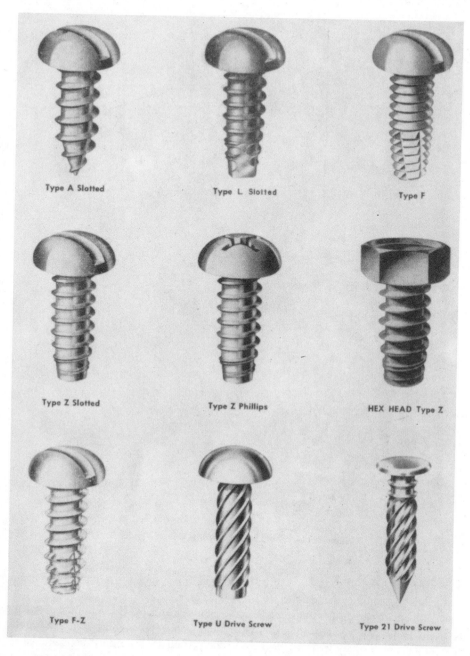

Fig. 27-19. Some typical self-tapping and drive screws. (*Courtesy Parker-Kalon*)

metal shaft which must rotate in service. The eyelet minimizes wear of the plastic.

Of the rivets used in joining plastics, the blind rivet is probably the most common. Blind rivets are available both in metal and plastics, and are designed for installation from one side only. Though there are a variety of proprietory designs for such rivets, essentially they consist of a hollow body and solid pin. The setting of the rivet is accomplished by driving or pulling (depending on the specific rivet being considered) the solid pin through the hollow shank and thus flaring the shank on the blind side of the rivet, to effect a positive locking action (Fig. 27-20).

Chemically expanded rivets have hollow

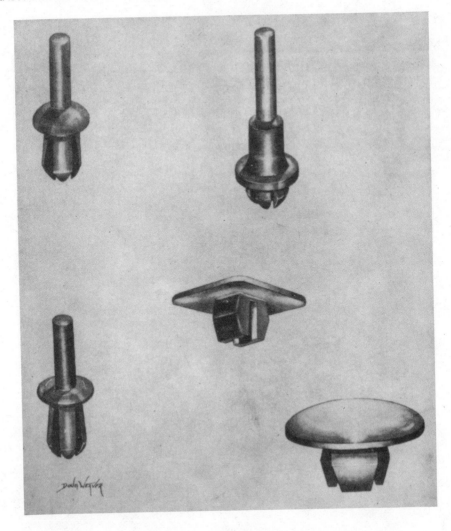

Fig. 27-20. Blind rivets. (*Courtesy Shakeproof, Div. of Illinois Tool Works*)

shanks in which a chemical charge is reacted to expand the hollow shank. Heat is applied to the rivet head with an approved soldering gun.

Chemically expanded metal rivets can be used for joining plastics as well as the light metals. In joining materials of relatively low elastic moduli, such as polyethylene and polytetrafluoroethylene resin, applications are limited to those where joints of low strength are permissible, since the expansion of the exploding rivet shank distorts most soft plastics.

Design considerations again vary with material and type of fastener. The following are examples of the manufacturer's recom-

mendations for riveting two engineering plastics:

Polycarbonate. Conventional riveting equipment and procedures can be used with polycarbonate. Care must be exercised to minimize stresses induced during the fastening operation. To do this, the rivet head should be $2\frac{1}{2}$ to 3 times the shank diameter. Also, rivets should be backed with either plates or washers to avoid high localized stresses (Fig. 27-21). Fastening procedures should be established to control compressive stress under the fastener head to less than 2000 psi for unfilled grades of polycarbonate.

ABS. Semi-tubular rivets are the most

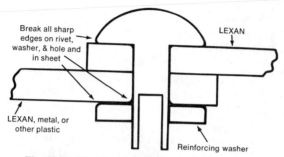

Break all sharp edges on rivet, washer, & hole and in sheet

LEXAN

LEXAN, metal, or other plastic

Reinforcing washer

Fig. 27-21. Riveting polycarbonate. (*Courtesy General Electric Co.*)

commonly used; however, the self-piercing rivets, such as the split or bifurcated types, are often useful in minimizing assembly cost. These self-piercing rivets may be used with flexible ABS sheet because of the relatively low clinching pressure required.

The bearing strength at rivet holes is more important than the strength of the rivet itself, particularly since the shear and pull-apart loads are focused at this point. (Shear failure is breaking of rivet shank; pull-apart is the material giving way.) Higher pull-apart retention and improved shear strengths can be attained by: specifying a large shank and head diameter; by placing the head on the plastic side of a plastic-to-metal joint, and by locating the rivet head on the thinner side of the plastic-to-plastic joint.

Several factors to remember when joining ABS to similar or dissimilar materials are:

(1) the distance between the rivet hole center line and the part edge should be at least equal to three times the diameter of the rivet shank (edge distance), but if self-piercing rivets are used, this distance should be extended to five rivet shank diameters. In riveting, the thickness, strength, and characteristics of the material being used should be considered.

(2) The interval between the centerlines of adjacent rivets is called the *pitch distance* and should be at least five times the rivet shank diameter. The clinch allowance recommended by most rivet manufacturers is between 50-70% of the rivet shank diameter.

(3) The proper size rivet holes are extremely important in joints involving plastic parts. If the rivet hole is too small, the rivet may bind in the hole and buckle. If the hole is too large, the rivet clinch will be off center and produce a loose joint.

In order to minimize buckling of the ABS part during the clinching process, a D/T ratio of 1 (rivet shank diameter/material thickness) is recommended for riveted joints in plastics.

Spring Clips and Nuts

A wide variety of proprietary metallic spring clips and nuts provides inexpensive methods of rapidly fastening plastics. These range from simple spring-type fasteners, which are forced over a molded stud, to multi-perforated rings, tubular devices, and irregular shapes. These can be applied unattached, or attached to one of the parts to be joined. If the spring device is attached to one of the parts, rapid assembly can later be completed from one side only. (See Fig. 27-22).

Speed nut stampings are used in combination with screws. The mating screw is usually of the coarse thread sheet metal variety. The speed nut acts to provide reinforcement and load distribution, and resists vibration. The speed clip provides push-on attachment over a molded boss or stud. Load capacity is usually low and re-assembly limited.

Pins can be effectively used in plastic. The physical characteristics of plastics related to loading and residual stress propogation must be considered in design. Spring pins with their inherent resilience combine well with most plastic materials. Pins have a variety of usages such as locating and locking devices, bearing surfaces, and hinges.

A number of means of providing movable assemblies, such as hinges, latches, snap locks

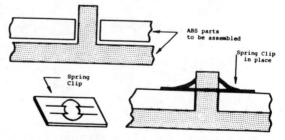

ABS parts to be assembled

Spring Clip in place

Spring Clip

Fig. 27-22. Tinnerman spring clips to lock ABS parts together. (*Courtesy Borg-Warner Chemicals*)

and bead-chain attachments, are largely derived from standard or proprietary devices used with common materials of construction other than plastics.

Other Devices

Hinges. There are several types of hinges available for use with plastics products. These include integral hinges such as those described on p. 000 (in which the two plastics parts are joined together by a strip of plastic). Another type of integral hinge consists of a molded ball and socket type hinge assembled through a snap-fit (Fig. 27-23). As an alternative, a socket may be molded for a drilled-through hole. In terms of more conventional mechanical fasteners, however, hinges that can be used include lug and pin and conventional hinge assemblies such as piano hinges.

The lug-pin-rivet type consists of a molded male lug (generally found in the cover of a box) and a molded female slot. The cover and bottom of the box are assembled and drilled as one complete unit. The hinge is formed by using a drive-pin or hinge-pin.

To reduce cost, holes may be drilled from both sides of the box using two drills diametrically opposed. The drill unit should have a kinematic linkage which controls the depth and feed of the hole.

The two drills may also be operated by small air cylinders instead of a mechanical linkage. Drilling fixtures with hardened bushings should always be used for best quality.

For the rivet type, the design may be such that no lugs and slots are required. A hole is drilled in the vertical wall of the cover and bottom and fastened by a rivet.

Hinge assemblies can be divided into hinges and snap-hinges. In some cases the usual hinge is a piano hinge screwed to the plastic cover with drive screws or swaged to molded plastic lugs. In designing the box, however, adequate wall thickness must be allowed around the molded or drilled hole. Otherwise, assemblies may crack when the drive screws are forced into the plastic.

In addition to the piano hinge, there are shorter type hinges assembled by riveting; also certain specialty designs are available whereby molded undercuts hold the hinge assembly to the plastic component.

Nails. As plastics continue to show up in applications originally intended for wood (e.g., furniture, shoe heels, etc.), there has been an increasing use of conventional nails as an assembly device. Many styrene structural forms used in furniture, for example, have been specially formulated to accept and hold nails. For ABS plastics used in heels, conventional woodworking equipment (i.e., a nailing machine) has been successfully used to attach the heels to the shoe. It is recommended in this latter case, however, that nails be located at least 5 diameters from an edge so that strains may be readily distributed.

Before using nails, however, it would be best to check first with your resin supplier. The stresses involved and the nail holding ability of the plastic must be taken into consideration.

Swaging or Peening

Swaging or peening is commonly used to fasten or connect molded pieces or metal parts to molded thermoplastics. It can also be used to provide an upset type head (like a common nail) on a molded shaft, pin or vee-slot.

Swaging is useful to assemble metal to plastic, or plastic to plastic, where motion

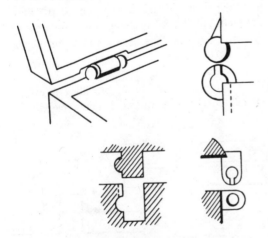

Fig. 27-23. Molded ball and socket type hinge assembled through a snap-fit. As an alternative, a socket may be a molded or drilled-through hole. (*Courtesy Monsanto Co.*)

between the parts is required. This operation is best performed when the plastic is heated to a softening temperature and formed under high pressure. If the temperature is high, less pressure required is needed, but the quality of the swage is reduced.

The simplest form of tool used to swage is a soldering iron. For greatest production, automation can be employed by using a kick-press or air-operated cylinder.

In addition, an air-operated indexing table can be used to feed the work and position it under the heated tips. The use of air will provide the necessary pressure and minimize operator fatigue.

The most commonly used soldering iron tips are $\frac{1}{4}$ to $\frac{5}{16}$ in. diameter copper. However, these are often soft and deform or bend under continued usage. Hardened brass, which has been nickel-plated and then flash chrome-plated, is most satisfactory. The brass is easy to machine into a concave tip to control the flow. Chrome improves the life and mold release.

During swaging operations a thin film of polymer may stick to the tip. This can be removed by a wire brush or a solvent.

Standard type mold release agents on the heated tips will minimize sticking.

Press or Shrink Fitting

These techniques are universally applicable to joining similar and dissimilar thermoplastics, and require no foreign elements such as cement or metal inserts in the finished joints. Properly applied, they produce serviceable joints with good strength at a minimum cost.

Plastics are press-fitted in the same manner as metals and other materials, but interfaces are generally increased to compensate for the relatively low elastic modulus of most plastics. For maximum joint strength, interferences should be made as large as possible without restricting assembly or stressing a piece beyond its yield point. Theoretical relationships of interference and stress level are based on geometry, and the properties of the materials. Interference can be calculated by standard stress-analysis procedures.

The relationship of maximum stress, caused by press-fitting a shaft or insert in a plastic hub, to diametral interference (ΔD) is expressed as:

$$\Delta D = \frac{S_d D_s}{L}\left[\frac{L + \mu_h}{E_h} + \frac{L - \mu_s}{E_s}\right]$$

and

$$L = \frac{1 + \left(\dfrac{D_s}{D_h}\right)^2}{1 - \left(\dfrac{D_s}{D_h}\right)^2}$$

where

D = Diametral interference, in.
S_d = Design stress, psi. Refer to AWS under "Design Parameters"
D_h = Outside diameter of hub, in.
D_s = Diameter of shaft, in.
E_h = Tensile modulus of elasticity of hub, psi
E = Modulus of elasticity of shaft, psi
μ_h = Poisson's ratio of hub material
μ_s = Poisson's ratio of shaft material
L = Geometry factor

Where large interferences are not required, shrink fitting may be suitable. Interferences for shrink fitting are determined by adding shrinkage of the hub to expansion of the shaft. In some applications it may be practicable to shrink-fit immediately after molding, while the article is still hot, so as to eliminate the necessity for reheating the hub.

Residual joint strengths in press or shrink-fitted articles are affected by complex variables, such as apparent modulus and coefficient of friction. For most thermoplastics, variation in apparent modulus becomes negligible after a year, so that joint strength becomes constant. Since the coefficient of friction is affected by variables such as lubrication, moisture, temperature and stress level, the coefficient under each of these conditions must be known in order to calculate accurately the strength of the joints. Axial strength and torsional strength of joints can then be calculated by standard equations. When torsional strength is critical, a ribbed shaft should be used; when axial strength is critical, rings or threads should be used. When both torsional and axial strength are critical, a knurled shaft or combinations of rings and ribs provide a good balance of properties.

When pieces are to be press-fitted for maximim holding power immediately after molding, they should be free from internal stresses; these can be reduced by annealing. Also in designing such joints, environmental conditions should be carefully considered. Expansion by heat and moisture can be compensated by designing for expected growth at the worst conditions, by considering expected expansion as an addition to the interferences selected for the desired joint strength. In plastics, a press-fitted assembly experiences its highest level of stress immediately after fittings; subsequent environmental dimensional shrinkage is usually more than compensated by creep, and thus can generally be neglected.

Internal stresses in a press-fitted article may tend to promote crazing in some plastics, such as acrylics and polystyrenes, and may reduce impact strength. Where a pressed article is expected to withstand impact it should be tested under actual conditions to determine the feasibility of this type of joint.

The force required to press-fit two parts may be approximated using the following equation:

$$F = \pi f P D_s T$$

and

$$P = \frac{S_d}{L}$$

where

F = Assembly force, lb
f = Coefficient of friction
P = Joint pressure, psi
D_s = Diameter of shaft, in.
T = Length of press-fit surfaces, in.
S_d = Design stress, psi
L = Geometry factor

Plastics Fasteners*

Designers can dramatically reduce production and component costs in fastening and assembly

*Adapted from an article, "With simple, integral design, 'hardware' joins the plastics revolution", by C. R. Baechtle, Systems Dev. Div., IBM Corp., from the September 1972 issue of *Plastics World.*

of both plastic and nonplastic parts by taking advantage of the design possibilities of plastics fasteners and hardware.

The flexibility and "springiness" of plastics permit simple, efficient designs. Furthermore, the ability to mold fasteners or joint members integrally with a plastic part saves both hardware and extra assembly operations.

These hardware items can be grouped into five major types: snap-on, snap-in, clasps, drive-pin, and hinges. Because of their inherent properties, plastics are the prime choice for these applications. No one material meets all requirements, however. The material must be carefully matched to the configuration, function, and cost requirements. The Chart in Fig. 27-24 lists recommended plastics for these five types of applications. Some basic design considerations are discussed in the general overview below, (see also Fig. 27-25). More details can be found elsewhere in this Chapter.

Materials for Plastic 'Hardware'							
	Application						
						Hinges	
Material	Snap-In	Snap-On	Clasp	Drive-Pin	Knuckle and Pin	Ball-Grip	Integral
ABS	✓	✓	✓		✓	✓	
Acetal	✓	✓	✓	✓	✓		
Acrylic			✓		✓		
Cellulosic		✓					
Fluorocarbon	✓		✓				
Polycarbonate	✓	✓		✓	✓	✓	
Polyethylene	✓	✓	✓				*✓
Polyamide	✓				✓	✓	
Polypropylene		✓	✓			✓	✓
Polystyrene				✓		✓	
Polyurethane					✓		
Vinyl	✓	✓	✓		✓		

*for strap hinge only

Fig. 27-24. Selecting plastics for fasteners. (*Courtesy Plastics World Magazine*)

Snap-on Fittings. Snap-on items are usually molded from polypropylene, polyethylene, fluorocarbons, or the flexible vinyls. These materials are particularly useful because, being somewhat elastic, they permit tighter, more secure interference fits.

Snap-on joints depend on friction. Changes in temperature and moisture content affect dimensions and therefore affect the holding friction at the interface. This factor must be

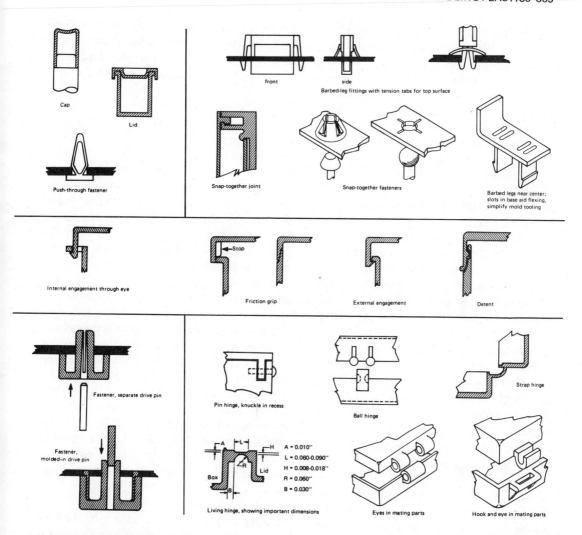

Fig. 27-25. Plastics fasteners: snap-on (upper left); snap-in (upper right); clasps (center); drive pins (lower left); and hinges (lower right). (*Courtesy Plastics World Magazine*)

carefully considered in assemblies of dissimilar materials, especially plastic-nonplastic. A more detailed discussion of snap fits can be found on p. 806.

Applications for snap-on joints include pen tops, container lids, and automotive trim.

Snap-in Fittings. Snap-in fittings are a simple, economical way to assemble plastic parts. They usually have a barb or projection to engage the part or, in the case of self-fasteners, some type of undercut in the mating element to engage the barb. Snap-in fittings provide a strong, reliable fastening. Their resistance to pull-out does not diminish with age.

In designing these fittings, care must be taken so that the elements will not be over-stressed as they move through the interference fit to the relatively stress-free seated position. The amount of stress that can be tolerated varies with the configuration, wall thickness, and yield strength of the plastic. Stiff materials, such as nylon and acetal, are recommended.

Clasps. Clasps are commonly used to secure tops and lids on display boxes, cosmetic cases, tool cases, and utility boxes.

Their holding power can be obtained from simple friction between the joint surfaces or from positive mechanical engagment. Material flexure is important because the parts must flex in order to release the clasp and spring back to their original position.

Drive-Pin Fasteners. Drive-pin devices, used to secure clamps, brackets, machine feet and similar items, provide an effective approach to mechanical fastening.

During assembly, the pin is driven inward to spread and wedge the tabs in the mounting hole. Either the pin or the space between the tabs must be tapered to produce this spreading action. The clearances between the tabs and mounting hole must be carefully selected to assure that the tabs will grip tightly.

Normally the drive pin is a separate piece. However, one design shown in Fig. 27-25 is molded with an integral pin. The pin is retained in the extended position by a controlled amount of molding flash. After the fitting is inserted, a blow at the end of the pin breaks the flash and drives the pin into the tabs.

Hinges. Integral hinges offer a great potential for savings. Since they are molded into the item, they cost almost nothing to produce and eliminate much of the assembly cost with standard hinges.

Many approaches can be used. One basic style is the knuckle-and-pin design, in which the set of knuckles that make up the hinge pivot on an axial pin. The knuckles can be either hooks or eyes or a combination of the two. The hook-and-eye combination does not require a pin, but the hinge can disengage if the cover is swung back too far.

A popular hinge, widely used on small boxes, is the ball-grip design. One half of the hinge consists of two balls. These snap into depressions in a projection molded into the mating part.

Still another approach is the intergral strap hinge. The top and base are molded simultaneously, connected by straps spaced along one edge. A typical strap is about 0.25 in. wide and 0.35 in. thick.

The living hinge, made possible by polypropylene and other plastics, has become extremely popular. This design is much neater in appearance than the strap hinge and is much longer-lived. Properly designed, a living hinge will flex over 300,000 times without failure.

Dimensions must be carefully designed, however, to make the hinge bend at the proper point (see Fig. 27-25). For additional information on the integral hinge, see below.

Snap Fittings

Snap fitting is a simple and rapid means of assembly, since all the elements for assembly (i.e., undercuts and corresponding lips on the mating part) are molded directly into the products so that they are ready to be joined together as they come out of the mold. Basically, snap fits involve an undercut on one part engaging a molded lip on the other to retain the assembly.

Joints of this type are strong, but are usually not pressure tight unless other features such as an O-ring are incorporated in the joint design. The critical factor in any snap fit is the amount of undercut that can be molded or machined into the part. Generally, undercuts can be snapped out of the mold by a standard ejection system; the critical depth of the undercut will depend upon the material chosen, the wall thickness, part configuration, etc. Where undercuts are too deep to be ejected conventionally, collapsible cores can sometimes be used. It is also possible to machine a deep undercut into a part.

In terms of design, undercuts and mating lips on snap-fit parts may be fully cylindrical or consist of flexible centilevered lugs. Cylindrical snap-fits are generally stronger but require greater assembly force than cantilevered-lug, snap-fits. For complex parts, use of centilevered lugs may simplify molding the parts.

The two general types of snap-fits are the snap-on and the snap-in. The snap-on is accomplished by molding an undercut on one part and a corresponding lip on the mating part (Fig. 27-26). The snap-in fittings generally incorporate molded-in prongs on one part that are pressed through holes in the mating part.

The force required to assemble parts by snap-fitting depends upon part geometry and on the coefficient of friction between the materials. This force may be divided arbitrarily into two elements—the force required initially to expand the hub, and the force required to

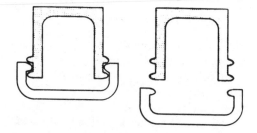

Fig. 27-26. Snap-on fit is accomplished by molding an undercut on one part and a corresponding lip in the mating part. (*Courtesy Borg-Warner Chemicals*)

overcome friction during the press fit stage. As the beveled edges slide past each other, the maximum force for the expansion occurs at the point of maximum hub expansion and is approximated by:

$$F_e = \frac{(1+f)\,\text{Tan}\,(\alpha)\,S_d \pi D_s L_h}{W}$$

where

F_e = Expansion force, lb (Kg)
f = Coefficient of friction
α = Angle of beveled surfaces
S_d = Stress due to interference, psi (Kg/cm^2)
D_s = Shaft diameter, in. (cm)
L_h = Length of hub expanded, in. (cm)
W = Geometry factor

For blind hubs, the length of hub expanded L_h may be approximated by twice the shaft diameter. For short open hubs such as in pulleys, the length of hub expanded will equal the axial hub length. For other part geometries or where greater accuracy is required, tests on actual parts are recommended. The force required to overcome friction during the press-fit stage is approximated by:

$$F_f = \frac{fS_d D_s L_s}{W}$$

where

F_f = Friction force, lb (Kg)
f = Coefficient of friction
S_d = Stress due to interference, psi (Kg/cm^2)
D_s = Shaft diameter, in. (cm)
L_s = Length of interference sliding surface, in. (cm)
W = Geometry factor

Generally, the friction force is less than the force for hub expansion for most assemblies.

The Integral Hinge

Although integral hinges are feasible with a number of plastics (e.g., the polyolefins, acetal, nylon, styrene-butadiene, etc.), the concept is generally associated with polypropylene. This section will discuss the integral hinge as it is generally applicable to polypropylene.

There are three techniques used to fabricate integral hinges: molded-in, cold-worked, and extruded. These are descibed below:*

Molded-in. The integral hinge can be injection molded by conventional techniques, providing certain factors are kept in mind.

The desirable molecular orientation is transverse to the hinge axis. This can be best achieved by fast flow through the hinge section using high melt temperatures. Since these requirements are consistent with good molding practices, optimum production rates can be maintained.

The main concern in integral-hinge molding is to avoid conditions that can lead to delamination of the hinge section. These include filling too slowly; too low a melt temperature; non-uniform flow-front through the hinge section; material contamination such as pigment agglomerates; and excessively high mold temperatures near the hinge area.

The integral hinge can also be produced by post-mold flexing. The hinge section is molded, then subjected to stresses beyond the yield point immediately after molding by closing the hinge. This creates a "necking-down" effect. Stretching the oriented polymer molecules on the outer surface of the hinge radius provides the remarkable flex strength of the thinned-down hinge section.

Flexing of the molded hinge must be done while it is still hot, through an angle sufficient to stress its outer fibers. This post-molding step provides maximum and uniform orientation in the hinge area with a minimum of applied stresses. The thinness of the hinge area requires that pigments be well dispersed so that agglo-

* Information supplied by Exxon Chemical Co.

merates will not provide focal points of weakness in the hinge structure.

Cold-worked. Where parts are heavy or complex, it may be impractical to force the necessary quantity of resin through the hinge sections. In these cases, integral hinges can be obtained by cold-working the molded parts. With this process, since the molecules are properly oriented during forming, the direction of polymer flow when molding is not critical.

A press, home-made toggle job, or hot-stamping machine can be used to perform the cold-working operation. The male forming die should be about 270 to 280°F. Pressure is maintained for about 10 sec. This time can be reduced if the part still retains residual molding heat or is pre-heated. Recommended pre-heating temperature is from 175 to 230°F.

Die backings may be either hard or flexible. With hard backing such as steel, the softened polypropylene is die-formed into the desired hinge contour. Thinner hinges are usually made using flexible backing such as stiff rubber. The deformation of this type of backing produces the hinge contour by stretching the softened plastic and generally results in thinner cross sections.

Extruded. Formation of the hinge cross section by use of an extruder die results in a hinge with poor flex-life. Because hinges are formed in the direction of polymer flow, they cannot be sufficiently oriented when flexed. However, if the extruded hinge is formed by the take-off mechanism while the polypropylene retains internal heat, the hinge will have properties approaching those of cold working.

Cold-Forming or Coining Hinges. More recently it has become possible to create hinges in some of the tougher engineering thermoplastics by *coining* techniques. A molded or extruded part is placed in a fixture between two coining bars. Pressure is applied to the bars, the part is compressed to the desired thickness elongating the plastic. Coining is effective only when the material is elongated beyond the tensile yield point. The process is usually used for such materials as acetal and nylon which cannot normally be molded in a sufficiently thin section for a strong, durable hinge.

Cold or Hot Heading

This technique is also useful for joining thermoplastics to similar or dissimilar materials.

In cold heading or staking, a compression load is applied to the end of a plastic shaft (while holding and containing the shaft body). As the load exceeds the shaft's yield strength, the shaft (or stake) will cold flow, forming a rivet-like head. The shaft, of course, can be molded right onto the part itself to facilitate assembly or a separate shaft can be used to join two components together. In this case, double heading of the shaft (at each end) is required.

Cold staking is normally done at room temperature (73°F). However, if a head is formed at a higher temperature than is expected in use, it will not recover unless the temperature at which it was formed is exceeded.

Cold staking may be accomplished by using equipment ranging from a simple arbor press and hand vise, to rivet-setting machines (Fig. 27-27).

Hot staking is similar to the cold staking method. It is accomplished by using a heated

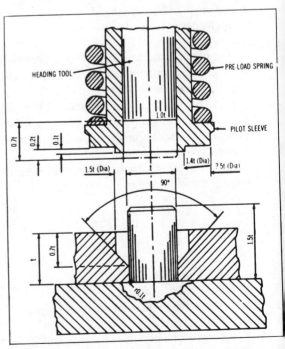

Fig. 27-27. Cold-heading of acetal. (*Courtesy E. I. du Pont de Nemours & Co., Inc.*)

(400-500°F) flaring tool. This tool is frequently concave in shape, and will effectively form a rivet-like head (Fig. 27-28).

Ultrasonic Riveting

In riveting or staking a part ultrasonically, a stud or projection is molded into the plastic part. In assembly, this protrudes through a hole in a second part that is not a thermoplastic. The ultrasonic tool is brought into contact with the stud and melts it, locking the assembly together.

AUTOMATION

As must be evidenced by the coverage of secondary processing operations (machining, finishing, decorating, joining) in the last three chapters, automated procedures are beginning to receive greater consideration. It is appropriate, therefore, to conclude these chapters with a few words on the subject.

In each of the preceding chapters on secondary processing, we have indicated the availability of automatic equipment to mark, decorate, seal, join, drill, etc. To automate these operations, it thus becomes necessary to devise automatic systems for removing parts from the mold, orienting them so that they will be in the proper physical position to be fed to the machine, mounting the parts appropriately, and then feeding or transferring them to the automatic secondary processing unit.

Robots

One technique for keeping parts under control as they are taken from the mold is an automatic extractor or, more recently, a robot (if parts are allowed to fall free or are otherwise ejected, some means must be added to unscramble and orient them for proper feeding). Although at this writing, it is still too early to assess the full role of the robot, it is not too early to predict that it will influence many segments of the plastics industry. To date, the main function of the robot has been in loading and unloading molding machines, in orienting the molded parts for secondary operations, in applying paints and coatings, and even in spraying up fibrous glass for reinforced plastic parts.

Using hydraulically controlled "hands" or grippers, the robot can perform most of the highly repetitive simple operations that humans can. Some of the more advanced types can handle a number of different articulations or movements.

Orienting Parts for Automatic Assembly*

To use automatic assembly effectively, it is important that parts be designed and molded so that they can be fed, oriented, and inserted by mechanical devices.

The parts must be not only dimensionally stable, but also free of protruding gate tabs and flash. If the part is not totally symmetrical, it generally must be oriented before it can be mated with the part to which it will be assembled.

Orientation requires a "detail" on the part that can be sensed by the orienting device. This detail could be one or several shoulders, a flat or a slot on the part.

Usually the detail must have a minimum size of 0.030 in. Preferably it should be larger, because the small dimensions require a precision orienting mechanism, which increases equipment costs. Undersized details also increase the possibility of jams.

When the designer of plastic parts has acquainted himself with modern automatic assembly methods, he is in a position to take full advantage of the cost-cutting opportunities

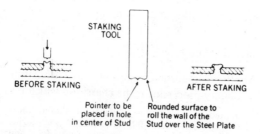

STAKING TOOL

BEFORE STAKING

AFTER STAKING

Pointer to be placed in hole in center of Stud

Rounded surface to roll the wall of the Stud over the Steel Plate

Fig. 27-28. Modification of hot-heading procedure. (*Courtesy Celanese Plastics Co.*)

* Adapted from an article, "Automatic Assembly is on the Way," as published in the January 1973 issue of *Plastics World*, with the permission of Cahners Publishing Co.

that exist between design engineering and manufacturing engineering.

Here are some specific suggestions on part design:

(1) *Try to design parts that are symmetrical and thus eliminate the need for orientation.* If a pin must have a chamfer on one end to function properly, put one on the other end also.

(2) *Use chamfers on pins and their mating holes.* This is useful in manual as well as automatic assembly.

(3) *Specify limits on burr or flash.* Make sure parts come to the machine clean and as free from molding flaws and foreign materials as possible.

(4) *If a part must be held in a specific orientation for placement, provide a suitable detail on the part.* The detail should enable the part to be selected in a feeder and held in orientation through the tracking and placement operations.

(5) *Choose a plastic tough enough to meet handling requirements.* When dealing with parts that are easily broken or deformed, specify presorting to eliminate scrap that can jam the machine.

(6) *In the basic design, give special consideration to springs.* Springs are difficult to handle and usually reduce assembly efficiency.

(7) *The more rigid the part, the fewer problems in the assembly operation.* The handling of soft parts, like rubber, can often be improved with the use of graphite or Teflon powder to aid tracking characteristics.

Common Feeding Methods. Usually there is more than one way to feed and orient a part. Some methods are more reliable and economical than others, but often the part design dictates the use of a specific method.

Probably the most common methods of feeding parts through a track to the assembly point are "end-to-end", "side-by-side" and "diameter-to-diameter".

If the part is very thin or has flanges or orienting details that can overlap, a horizontal vibrating track is preferable to a gravity track.

The "stack over a loose rod" is another approach. It is recommended for feeding very thin parts (0.030 in. or less), such as washers,

retainers and O-rings that have a hole in the center. In this method, the parts are shuttled off the bottom of the stack for placement, while the shuttle mechanism supports the loose rod vertically.

Another method of handling round parts is through a confined tube. This method is recommended for cylindrical parts that must be fed end-to-end or for discs with no hole in the center. Typical parts are pins, springs, and some long-headed fasteners.

Most headed fasteners and some flanged parts are tracked by the "hanging by a shoulder" method.

Common Placement Methods. After the parts are fed from the bulk supply, they are "picked-up" and positioned into a mating part on the assembly machine. There are three common placement methods: "vertical placement", "pick and place" and "horizontal shuttle with vertical placement".

For vertical placement, the parts may be fed end-to-end, side-by-side or diameter-to-diameter. The part must have some design detail, such as a hole in the center, so that the placement tool can guide the part for loading.

Pick and place requires a versatile piece of hardware. This approach, which may be used for all three methods of feeding, will accommodate parts that do not have a piloting detail such as a hole in the center.

It is usually tooled with a pair of clamp jaws that grip the part on its periphery. A variation of this method can be used to place parts that may have flanges that will overlap in the feeding track.

The third method is the horizontal shuttle and vertical placement. It can be used with any of the feeding methods, generally for complicated requirements, such as torquing fasteners or expanding and placing O-rings.

Indexing Units*

Another important element in automating secondary processing is the chassis on which the

* Adapted from an article, "What you should know about automating secondary processing," as appearing in the May 1972 issue of *Plastics Technology* and reprinted with the permission of the publisher, Bill Communications, Inc.

fixtured parts are indexed from one work station to the next, so that successive work steps can be performed.

Rotary indexing tables, sometimes called *dials*, are one of the most widely used devices for multiple-step operations. They are produced with circular tables ranging from a few inches to at least 10 ft in diameter. The size will basically be determined by the number of work stations and the size of the units doing the work that are mounted around the circumference of the table. The 360° of one revolution is divided by the number of stations to determine the size of the indexing step. Six stations would require the table to index 60°. One station would usually have a loading and

unloading function; the other five would be for work assembly.

The rotaries are often air-powered, using low-cost, readily available air cylinders to index the table. Air to power the cylinders is admitted through solenoid-actuated valves controlled by a timing mechanism. One of the main criticisms of this power source is that it is difficult to modulate the motion; the cylinder abruptly sets the table in motion and it comes to an equally abrupt halt.

If all workpieces are firmly held on the table, this may cause no inconvenience. If, however, a lightweight part is placed on the main workpiece at one station, and to be fastened in place at the next, the jerky action

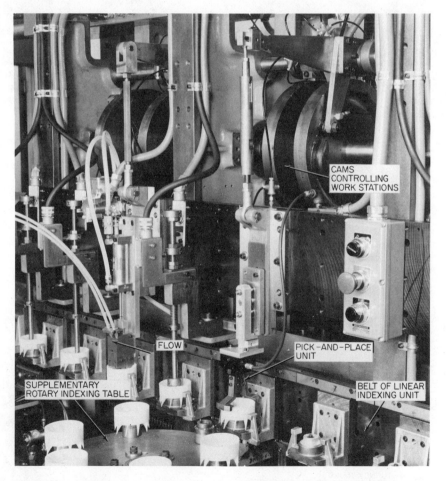

Fig. 27-29. Where parts are difficult to feed automatically, operator can be assisted by use of small dial or rotary indexing table. Parts are placed on table which rotates under a pick-and-place unit that, in turn, transfers parts to fixtures on main assembly machine. Fixtures are mounted on steel belt for precision transfer from station to station. (*Courtesy Bodine Corp. and Plastics Technology Magazine*)

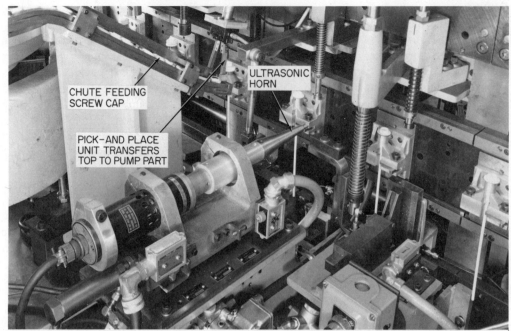

Fig. 27-30. Thirteen plastic and metal parts (for a pump) are assembled by ultrasonic welding at the rate of 62 parts/min. Note that one plastic part feeds in from upper left to a small pick-and-place unit which picks the part from the chute and positions it on top of the assembly held in the fixture. (*Courtesy Bodine Corp. and Plastics Technology Magazine*)

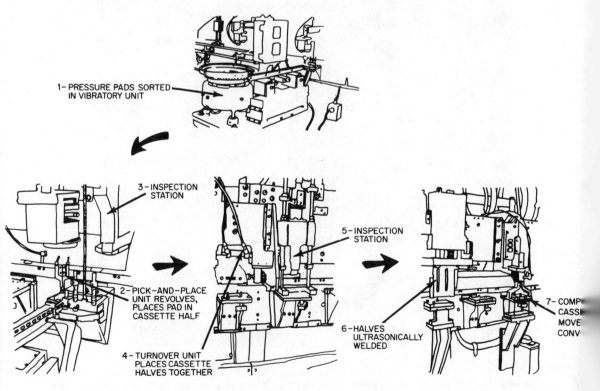

Fig. 27-31. Typical set-up for assembling tape cassettes automatically, using sorting devices, pick-and-place units, etc. (*Courtesy Plastics Technology Magazine*)

may promptly dislodge the part, unless some automatic finger attachment holds it in place.

More sophisticated operation of the rotary tables is achieved by a mechanical drive that uses a cam to index in a controlled pattern of acceleration-deceleration. The cam can be shaped to slowly accelerate the table up to some set top speed, maintain this speed for some fraction of the cycle, then decelerate to a soft, bumpless stop.

The work area available on a rotary is sometimes a problem and sometimes a virtue. For each station, the area available on the table is a pie-shaped sector, and space can become crowded if work or assembly devices extend too far toward the center of the table. On the other hand, the operating units at each station are unlikely to be of the same size, so that a large unit may be able to encroach on the space of a small adjacent unit. On the plus side, the pie-shaped sector broadens with distance from the center of the table, allowing a very large work unit to be mounted, just as long as the working head is sized within bounds.

Linear indexing units use endless belts on which the fixtured parts are mounted to pass in a straight line past the work stations. These can vary greatly in precision and sophistication. For some rough operations the belt could be of canvas, the indexing motion imparted by a pawl and rachet, or other simple mechanism.

At the other extreme, the belts are single-piece or segmented steel, mounted in massive frames, and under high tension. Very high precision and repeatability are claimed for these units. They are available in various standard lengths to provide a linear work surface of as much as 20 ft along each of the two sides of the frame. A very large number of work stations can be accommodated, the number depending, as with the rotaries, on the size of the parts and the work devices; and some types of work stations can even be located at the ends of the frames where the belts pass around drums.

One of the big advantages advanced in favor of the linear indexing units is that power can be transmitted along a shaft extending the length of the unit. Cams, gears or other motion-transmitting devices can be mounted on this shaft for perfect synchronization of all operations, either by directly transmitting the power required to operate a work station or by tripping a switch to operate a secondary power source. In comparison, direct, precision-transfer of power to stations with a rotary table is near-impossible.

As would be expected, indexing drives are sophisticated, provide controlled acceleration and deceleration, and position the workpieces at each station with high precision.

The work area at each station is rectangular, not pie-shaped, and more freedom is claimed with respect to designing and mounting work stations. Where the rotary tables are most often horizontal, with most work motions being performed vertically, the linear units have vertical mounting surfaces and work motions tend to be horizontal. Some linear units are tipped over 90° so that the belt surface is horizontal, but this limits the capacity to mount work stations, it being generally impractical to mount them under the belt. Gravity can assist in holding parts on the rotary, but usually not on the linear units.

Typical Set-ups. Figures 27-29 though 27-31 illustrate some typical operations involving automated secondary processing.

28

DESIGN STANDARDS FOR INSERTS

INTRODUCTION*

While the subject of inserts has been touched on in previous chapters (e.g., Design of Molded Products; Joining and Assembling Plastics), this section will deal specifically with design standards for inserts.

It would be well, however, at this point, to review the various mechanical fastening methods. The available fastening methods are enumerated as follows:

1. Cored or drilled hole with either thread cutting or thread forming self-tapping type screws.
2. Molded or tapped threads used in conjunction with regular machine screws.
3. Molded-in inserts of either the internal threaded bushing or external threaded stud varieties.
4. Post-molding inserts which are self-tapped, pressed, cemented or expanded into the parent material. Also included in this group would be those inserts which are installed into tapped or molded threads as well as those which are installed in thermoplastics by remelting using ultrasonics, spin friction, or other heat-generating principle.

* As prepared by Donald P. Viscio, Marketing Manager, Heli-Coil Products, Div. of Mite Corp.

Thread Cutting or Thread Forming Self-Tapping Screws

The basic difference between the two types of screws is that the thread cutting type cuts through the material, whereas the thread forming screw displaces the material to form the thread. Thread cutting screws are generally used with thermosetting plastics, while thread forming screws are generally used with thermoplastic materials.

Tap holes and clearance hole sizes and lengths are important variables. The three factors which must be analyzed in determining the proper hole sizes are ease of driving the screw, cracking of the boss (especially important is the long-term effect of stresses created), and adequate assembly strength. The optimum hole size must be thoroughly tested under all conditions, and must be controlled within close limits for the more friable materials. As is typical with interference fit systems, relatively small changes in size will have a drastic effect on the torque produced and stresses created in driving the screw into the material.

Advantages:
1. Least expensive.
2. Fastest to assemble.

Limitations (especially applicable to most thermosetting plastics):
1. Questionable holding power in compari-

son to standard machine screw used with a female imbedded insert in the molded part.

2. Noticeable loss of holding power after removal from the self-tapped hole and reassembly.
3. May cause chipping and cracking of the molded part unless hole and screw dimensions are closely held.
4. Re-entry of screw can easily damage threads by improper entry.

Molded or Tapped Threads
With Machine Screws

The standard sizes and types of machine screws (either metal or plastic) can be used.

The internal threads can be prepared by either tapping or by molding. Tapping using conventional methods can be achieved by having the proper tap drill size molded into the part or drilled. Factors such as complexity of mold, number of holes, and difficulty of machining by drilling, should be carefully evaluated. Many plastics are abrasive in nature, and both drilling and tapping may require special tooling. Oversize taps, chrome-plated and even carbide taps have proven to be well worthwhile for large volume production or when the material is extremely abrasive, such as a glass-filled plastic.

Molding threads into the part is only considered when the resultant thread class of fit, surface finish, and strength are critical, or if the threaded connection forms the essential function of the part. Generally, molded threads are limited to one to three per part, because of the increased complexity of the mold, and the extra operations required to unscrew the threaded core pins for each cycle. Another factor which must be considered is that, if split mold sections are used, a parting line or flash develops, which may have to be tapped out. Molded internal threads no finer than 32 pitch, or longer than $1/2$ in., should be used; otherwise, too much operator time will be required to disassemble the part from the mold.

Advantages:
1. Relatively inexpensive; screw cost is less than that for self-tapping style.

2. Holding power somewhat better than that of self-tapping drive screws.
3. Stress in material, due to thread forming, is minimal.
4. Easy assembly of screw.

Limitations:
1. Separate drilling and tapping operations are costly, due to tap wear and labor involved.
2. Strength is not equal to mold-in or post-mold insert.
3. Threads may chip and are easily damaged by improper entry of the screw.
4. Molding in threads increases mold cost significantly. Also, cycle time is increased, especially when more than one thread is molded.

Molded-In Inserts

Where repeated disassembly, or where strength of assembly is important, mold-in type inserts can be specified (Fig. 28-1). Careful attention must be given to design with mold-in inserts for the following reasons:

1. Unpredictable variations in molding.
2. Possibility of developing excessive strains on the part, due to differential cooling rates of the plastic and the insert.
3. Some plastic materials will crack around the insert after they have aged. Other materials will creep in aging, or, if two inserts are rigidly located together in a mating part, one or the other of the inserts will pull out as the plastic ages and shrinks.
4. Delicate or light inserts may be damaged or dislodged during the molding cycle.

Female inserts molded through the part frequently are not threaded when compression molding is used because the plastic will flow into the threads, and a retapping operation will be required. If injection or transfer methods are used, the female insert may be tapped prior to molding for most materials, because the mold is closed on the insert before the compound is forced into the mold.

Inserts which are molded in and intended to provide a sealed leakproof connection should

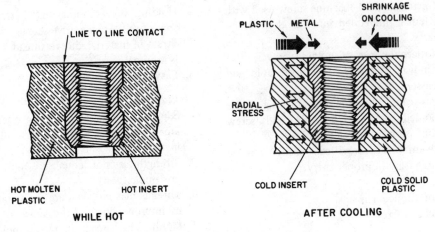

Fig. 28-1. Molded-in insert. (*Courtesy Heli-Coil Products*)

incorporate grooves and gaskets. Dipping the insert in liquid latex will give the same results as grooving with gaskets.

Location of the insert and boss configuration in the part are also important. The minimum wall thickness around the metal insert is as follows:

1. Steel inserts—boss should be equal to insert outer radius.
2. Brass inserts—0.9 times insert outer radius.
3. Aluminum inserts—0.8 times insert outer radius.

For those materials prone to cracking after cooling or aging, the following suggestions are offered:

1. Keep knurls to a minimum.
2. Knurls should not be too sharp.
3. Blind inserts should have a round head.
4. Remove all sharp corners.

For optimum location of inserts within the part itself, the following factors should be taken into account:

1. Floor under the insert should be at least $\frac{1}{6}$ of the diameter of the insert to prevent a sink mark.
2. Inserts should be longer or shorter than the boss itself.
3. Mold recess for male inserts should be at least 0.020 in. from the edge of the mold if cracking of the mold is to be avoided.

4. Inserts molded in opposite sides of a thermosetting plastic should be no closer than $\frac{1}{8}$ in.
5. Avoid bringing a boss down to a narrow fin around the insert.

Advantages:
1. Very good holding power of screw and insert is possible.
2. Subsequent reassembly may be achieved without the loss of screw holding power.

Limitations:
1. It is difficult to charge the mold with inserts on a automatic cycle, without a costly feeding, loading, and indexing mechanism.
2. If a part is produced with insert missing or misplaced, the entire part must be rejected.
3. Inserts must be cleaned of all flash, especially in threaded inserts. This can be costly.
4. May create residual stresses in the plastic after cooling and aging. Also, creep and shrinkage affect insert retention.
5. Floating inserts can cause extensive mold damage.
6. Large inserts require preheating above mold temperature to pre-expand them and improve flow and cure-in.

Post-Molding Inserts

The inserts which fall into this category have

gained acceptance in recent years. There are six basic types of inserts in this category:

A. Press-in inserts, which are assembled while the parts are still hot, after removal from the mold.
B. Inserts which are cemented in place.
C. Inserts which are self-tapped into the material.
D. Inserts which are installed into tapped holes.
E. Expansion-type inserts.
F. Inserts for ultrasonic installation.

Group A and B inserts can be considered together, because they have practically the same advantages and limitations.

Advantages:
1. Can be used with parts produced by automatic molding.
2. No cleaning of flash from inserts.
3. Installation mechanics are simple, requiring no investment for special tools.

Limitations:
1. Insert holding power not equal to mold-in inserts.
2. Insert retention reliability may be questionable, because of the reliance on close tolerances in the case of the press-in insert, and reliance on clean surfaces properly bonding to the cement under all conditions for the cemented insert.
3. Press-in bushings may crack the material if holes are slightly undersize, or if the bushing is slightly oversize.

Group C inserts are those which are of either the solid bushing or carbon steel coiled wire types. They are driven into a plain hole which has been prepared by either molding or drilling to the proper diameter. In both cases, the insert taps its own thread, and, when assembled, the internal threads form the desired female thread. Self-tapping inserts have somewhat greater holding strength than expansion types.

Advantages:
1. Strong assembly strength (resistance to pull-out), because of shearing of a large area.
2. Good quality internal threads are achieved.

Limitations:
1. Insert cost is usually quite high in comparison to expansion types.
2. Installation is the slowest on a per piece basis for all types of fasteners.
3. High driving torque, or poor insert retention, can result from extremes in hole-insert tolerance variations.
4. Size range generally restricted to #8 thru $\frac{1}{2}$ in.

Group D type inserts, which are installed into threaded holes, offer high quality and precision assembly. The tapped holes are either molded or tapped to close tolerances (Class 3B limits may be achieved if necessary). Also, improved strength of assembly will result from this group, because of the increase in shear area over the self-tapping group mentioned above. Generally available with inserts in this group is a screw locking device to prevent loosening, or for adjustment screws. Insert retention principles vary with each manufacturer, but the two basic principles are to "lock-in" the insert with a collar, keys, prongs or rings, nylon pellets, pins, or the spring expansion of the stainless steel coiled wire type insert.

Advantages:
1. Highest quality of internal thread.
2. Strongest insert strength of all types.
3. Screw locking feature can be utilized.
4. Hard, wear resistant internal threads.

Limitations:
1. Highest cost of all other types of inserts.
2. Installation slow, usually requiring expensive equipment if automatic installation is desired.

Group E inserts are relatively recent. Expansion-type inserts are installed into cored or drilled holes, and are secured by expanding a knurled portion of the insert. The expansion is achieved by either the entry of the screw or a spreader plate, the distinction being that the former relies on the full penetration through the insert by the screw to expand the knurls, while the latter are retained by integral expansion feature of the insert itself. Both types distribute the locking stresses evenly in the hole and provide ample resistance to torsional and tensile loads. They are easily installed, either manually or automatically.

The strength of the assembled expansion-type insert, while not as high as that for self-tapping inserts, is sufficient to cause the screw to shear before pulling out the insert. The strength of the expansion types is approximately halfway between that of the press-in and self-tapping inserts. Loading values for a given insert will vary with the properties of the material.

When working with post-mold inserts, the effect of hole size on torsional and tensile assembly strength varies greatly among plastics, and rather wide variations in hole dimensions result in less loss of assembly strength than might be expected. In short, the plastic used has far more effect on the assembly strength than has the hole to insert fit. See Fig. 28-2.

The boss diameters of the plastic should be at least two times the diameter of the insert, and, in many cases, 2.5 to 3 times the insert diameter is required. When designing clearance diameters of component parts, it is extremely important to see that the insert and not the plastic carry the load. See Fig. 28-3.

Several examples of expansion-type inserts are shown in Figs. 28-4 through 28-8.

Advantages:

1. Lower insert installation cost, by maximum utilization of press time.
2. Rework to remove flash in threads is eliminated.

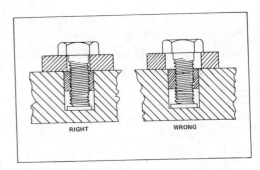

Fig. 28-3. Boss diameter should be 2 to 3 times insert diameter and design of the component should allow the insert, not the plastic, to bear the load. (*Courtesy Heli-Coil Products*)

Fig. 28-4. Standard insert provides brass threads in plastics parts of all types after molding. (*Illustrations of insert types courtesy Heli-Coil Products*)

3. Eliminates floating inserts, damage to mold, breaking of core pins.
4. Adaptable to fast-cycling automatic installation.

Limitations:

1. Strength not as high as mold-in types but higher than press-in types.
2. Cost slightly higher than mold-in types.

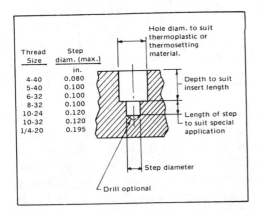

Thread Size	Step diam. (max.)
	in.
4-40	0.080
5-40	0.100
6-32	0.100
8-32	0.100
10-24	0.120
10-32	0.120
1/4-20	0.195

Hole diam. to suit thermoplastic or thermosetting material.

Depth to suit insert length

Length of step to suit special application

Step diameter

Drill optional

Fig. 28-2. Recommended design for stepped holes for use with post-molding inserts. Stepped holes are used to avoid sink marks opposite insert holes. (*Courtesy Heli-Coil Products*)

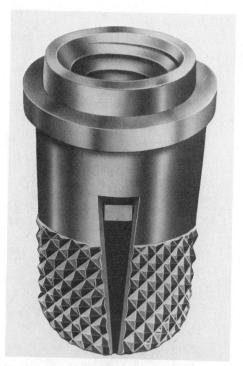

Fig. 28-5. Clinch insert features a pilot (that may be clinched over a terminal, eyelet fashion) and a flange (that offers a broad surface for electrical contact).

Fig. 28-7. Cone-spread insert is designed for use in miniaturized parts where No. 0 size screws and larger are used. In installation, downward pressure on the insert causes the cone vortex to shear. The cone enters the insert, expanding the knurled body into the wall of the hole and anchoring it in place.

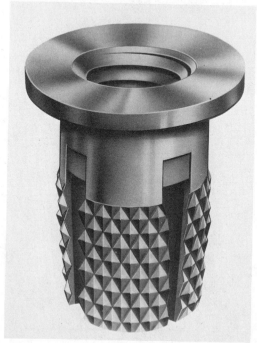

Fig. 28-6. Flange insert features a flange with large surface for effective electrical contact or for holding down mating parts.

Group F—Recently many manufacturers have been turning to inserts for ultrasonic installation into thermoplastics. The theory behind this method is quite simple: high frequency vibration develops frictional heat at the insert/plastic interface. The heat thus created remelts the plastic in a narrow zone around the insert. As the plastic cools, it shrinks away from the warm metal outward toward the cooler plastic. After shrinking, the diameter of the plastic hole increases, creating a microscopic relief zone which in no way deters the holding strength of the insert. This relief zone, however, does prevent the compressive stresses normally created by the mold-in method. Another advantage of the ultrasonic installation of specially engineered inserts is the avoidance of flash in the insert threads. The hole which is to receive the insert may be either tapered (which is easier to mold) or straight. The tapered hole, which is best for ease of insert installation, can be either

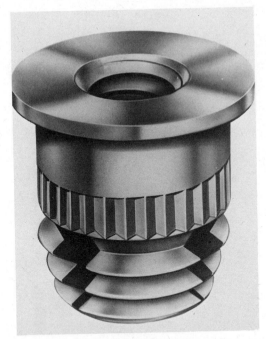

Fig. 28-8. In the wedge insert, the large wedge configurations of the insert body are forced deeply into the wall of the drilled hole. Especially applicable for use in soft plastics.

cored or drilled with comparatively liberal tolerances.

The ultrasonic installation cycle time for inserts in the #2 through #10 insert sizes is usually less than one second in many of the common thermoplastics.

A metal insert designed specifically for ultrasonic installation is shown in Fig. 28-9 and 28-10.

Advantages:
1. Achieves a high strength stress-free assembly without the problems normally encountered with mold-in or the high cost of inserts installed into tapped threads.
2. Can be installed quickly and easily into cored or drilled, straight or tapered hole. Tolerances are quite liberal.
3. Can be used in small diameter bosses or where wall thickness is limited. All other inserts require a much thicker wall to prevent bulging and cracking. This is especially significant in stress-sensitive polymers such as polycarbonates or acrylics.

Limitations:
1. Insert cost is slightly higher than mold-in or expansion type.
2. Original cost of installation equipment is comparatively high but it can be amortized in a relatively short run.

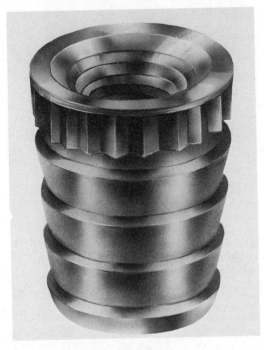

Fig. 28-9. Metal insert engineered for ultrasonic installation. (*Courtesy Heli-Coil Products Co.*)

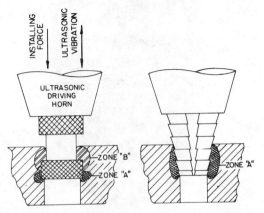

Fig. 28-10. Differences in ultrasonic installation of conventional insert (left) and inserts designed for ultrasonic use (right). (*Courtesy Heli-Coil Products Co.*)

DESIGN STANDARDS
FOR INSERTS

Inserts of many types are used: those made on screw machines and those made by cold forging, stamping, and drawing. The discussion in this chapter is divided, rather loosely, according to these types, but general instructions and precautions given under one heading are largely applicable to the others also. Inserts of various designs are shown in Fig. 28-11.

Maintaining a proper accuracy in various dimensions of inserts has always been a problem in the plastics industry as with the insert manufacturers, mainly because of the lack of information on design and standardization of dimensions. The technicians who were selected to prepare this engineering standard have endeavored to compile their own knowledge as well as that of the entire plastics industry. Engineers having a reasonable knowledge of plastics and an acquaintance with inserts and their use will find this standard of value in the proper design and selection of inserts.

Screw-Machine Inserts

Dimensions and Tolerances. Dimensions and tolerances for the usual types of male and female inserts in Fig. 28-12 and Table 28-1 are compiled with the cooperation of the National

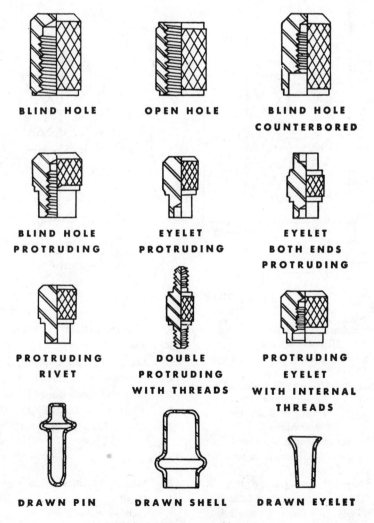

BLIND HOLE OPEN HOLE BLIND HOLE COUNTERBORED

BLIND HOLE PROTRUDING EYELET PROTRUDING EYELET BOTH ENDS PROTRUDING

PROTRUDING RIVET DOUBLE PROTRUDING WITH THREADS PROTRUDING EYELET WITH INTERNAL THREADS

DRAWN PIN DRAWN SHELL DRAWN EYELET

Fig. 28-11. Usual types of inserts.

Screw Machine Products Association as being practicable for machining as a single operation on an automatic screw machine, and hence are most economical.

Note that the dimensions given for tapped inserts apply only to nonferrous metals where the depth of usable tapping is not more than $1\frac{1}{2}$ times the tap diameter. On A-2 (minor diameter) and C (length of tapped inserts) the maximum "standard" tolerance should be specified whenever possible. However, for closer tolerances, "precision" can be specified when necessary. To maintain the "precision" tolerance, reaming and other additional operations will be necessary, at additional cost. In certain cases, the thread on a stud can be rolled,

reducing length B-1 (Fig. 28-12) but generally not increasing the cost.

If steel inserts are required, Fig. 28-12 and Table 28-1 cannot be used in design without several modifications which will increase the cost over that of inserts made of brass or, in special cases, of aluminum.

Minimum wall thickness of metal in the inserts depends entirely upon the desired accuracy of the inside dimensions of the insert. If too thin a wall of metal is used, the combination of stress caused by shrinkage of the plastic and by molding pressure may collapse the wall of the insert, so that the inside diameter will be out of the range of specified tolerances. Table 28-1 shows the minimum

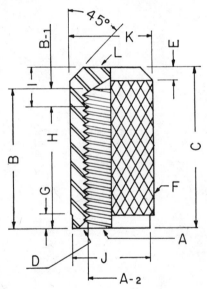
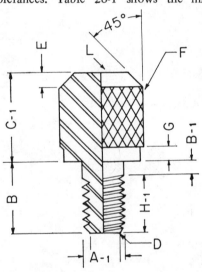

Fig. 28-12. Tolerance index.*

A	Thread, "Unified and American National" (ASA B1.1) Classes 2A/2B
A-1	Major diameter, Unified Thread Limits (ASA B1.1)
A-2	Minor diameter, "Regular" Tolerance, Unified Thread Limits (ASA B1.1) "Precision" Tolerance ± 0.0005 in.
B	Depth of minor and length of major diameter
B-1	Number of unusable threads from bottom and top
C	Length, "Regular" Tolerance ± 0.010 in. "Precision" Tolerance ± 0.001 in.
C-1	Length of body, male insert, ± 0.010 in.
D*	Thread chamfer, 45° ± 0.005 in.

E*	Body chamfer, 45° ± 0.010 in.
F	Knurl
G	Length of sealing diameter, minimum 1/32 in.
H	Length of usable thread, 1.5 × diameter
H-1	Length of usable thread H-1 + B-1 = B
I	Amount to add to H to obtain C, H + I = C
J	Sealing diameter, "Regular" Tolerance ± 0.003 in. "Precision" Tolerance ± 0.001 in.
K	Minimum diameter of bar stock
L	Small cut-off burrs are acceptable unless purchaser states otherwise.

* See Table 28-1.

Table 28-1. Dimensions and Tolerances*

Nonferrous inserts which have a usable thread length not more than 1½ times the tap diameter

Coarse	A Fine	K Minimum	J Maximum	Tap Drill	A-2	A-1 Maximum	A-1 Minimum	B-1	I	D and E	Knurl
2-56		3/16	9/64	#50	0.0700	0.0860	0.0820	3	3/32	1/64	Fine
	2-64	3/16	9/64	#49	.0730	.0860	.0822	3	3/32	1/64	Fine
3-48		7/32	5/32	#45	.0820	.0990	.0946	3	7/64	1/64	Fine
	3-56	7/32	5/32	#45	.0820	.0990	.0950	3	3/32	1/64	Fine
4-40		7/32	11/64	#43	.0890	.1120	.1072	2½	7/64	1/64	Fine
	4-48	7/32	11/64	#42	.0935	.1120	.1076	2½	7/64	1/64	Fine
5-40		1/4	3/16	#37	.1040	.1250	.1202	2½	7/64	1/32	Med.
	5-44	1/4	3/16	#37	.1040	.1250	.1204	2½	7/64	1/32	Med.
6-32		1/4	13/64	#33	.1130	.1380	.1326	2½	5/32	1/32	Med.
	6-40	1/4	13/64	#32	.1160	.1380	.1332	2½	9/64	1/32	Med.
8-32		9/32	7/32	#29	.1360	.1640	.1586	2½	5/32	1/32	Med.
	8-36	9/32	7/32	#28	.1405	.1640	.1590	2½	9/64	1/32	Med.
10-24		5/16	1/4	#23	.1540	.1900	.1834	2½	3/16	1/32	Med.
	10-32	5/16	1/4	#20	.1610	.1900	.1846	2½	5/32	1/32	Med.
12-24		3/8	5/16	#16	.1770	.2160	.2094	2½	13/64	3/64	Med.
	12-28	3/8	5/16	#13	.1850	.2160	.2098	2½	11/64	3/64	Med.
1/4-20		13/32	11/32	#6	.2040	.2500	.2428	2	13/64	3/64	Coarse
	1/4-28	13/32	11/32	7/32	.2187	.2500	.2438	2	11/64	3/64	Coarse
5/16-18		15/32	13/32	G	.2610	.3125	.3043	2	7/32	3/64	Coarse
	5/16-24	15/32	13/32	I	.2720	.3125	.3059	2	13/64	3/64	Coarse
3/8-16		9/16	15/32	O	.3160	.3750	.3660	2	1/4	3/64	Coarse
	3/8-24	9/16	15/32	Q	.3320	.3750	.3684	2	7/32	3/64	Coarse
7/16-14		5/8	17/32	U	.3680	.4375	.4277	2	9/32	3/64	Coarse
	7/16-20	5/8	17/32	25/64	.3906	.4375	.4303	2	1/4	3/64	Coarse
1/2-13		11/16	19/32	27/64	.4218	.5000	.4896	2	5/16	1/16	Coarse
	1/2-20	11/16	19/32	29/64	.4531	.5000	.4928	2	17/64	1/16	Coarse
9/16-12		3/4	21/32	31/64	.4843	.5625	.5513	2	11/32	1/16	Coarse
	9/16-18	3/4	21/32	33/64	.5156	.5625	.5543	2	9/32	1/16	Coarse
5/8-11		13/16	23/32	35/64	.5469	.6250	.6132	2	3/8	1/16	Coarse
	5/8-18	13/16	23/32	37/64	.5781	.6250	.6168	2	5/16	1/16	Coarse

* See Fig. 28-12.

recommended diameters of bar stock for various sizes of inserts.

Cold-Forged Inserts

In general, the volume or quantity needed to ensure economical production by cold-forging is about the same as that needed by other processes. When second operations are required, such as turning, drilling, tapping and others, larger quantities are needed.

In the discussion below, of specific problems, the utility of cold-forged inserts will become apparent. None of the cold-forged inserts shown has been machined, but they could be, of course, by either automatic or single-purpose equipment. If machining, drilling, reaming or tapping is involved, the tolerances are the same as those for screw-machine inserts given in Table 28-1.

There are no specific formulas controlling the individual relationships of the diameters and widths of collars to the shank or the kind and variety of shapes, like ribbed, finned, pinchneck, hexagon and so on, which may be combined with other symmetrical or unsymmetrical shapes in one piece. For each problem, therefore, the solution should be reached through cooperation between the designer of the molded piece and the manufacturer of the insert.

Materials. Almost any metal can be cold-worked, but cold-working grades of the following are preferred in the order named:

1. aluminum and aluminum alloys
2. brass
3. copper and copper alloys
4. carbon steels
5. alloy steels

6. stainless steel
7. silver and other precious metals

Tolerances without Finishing Operations. The tolerances given for the inserts in the layouts of Figs. 28-13 and 28-14 are those ordered, although closer ones could be met if necessary. Tolerances for any element such as length or diameter vary with the material and with the sizes and proportions of the piece, since they in turn determine the equipment or method of heading to be used.

In general, the following tolerances can be considered as commercial without finishing operations, although in some cases special care must be exercised to meet them:

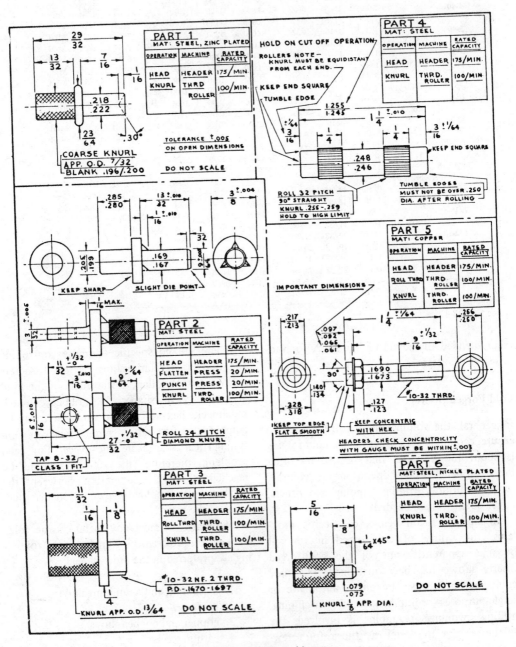

Fig. 28-13. Cold-forged inserts.

length   ±0.010 in. (maximum)
fillets   sharp or rounded, as
      specified
diameter   0.002 in. (minimum)
squareness, shoulders
  or collars with
  shank   ±1° maximum

Tolerances with Finishing Operations.
Whatever tolerance is needed can be met by

adding finishing operations. For example, aircraft studs, bolts and specials are commonly made today to tolerances as close as 0.0005 in. and even less.

Special Inserts

It would be an endless task to cover the entire field of special inserts. Some of the more

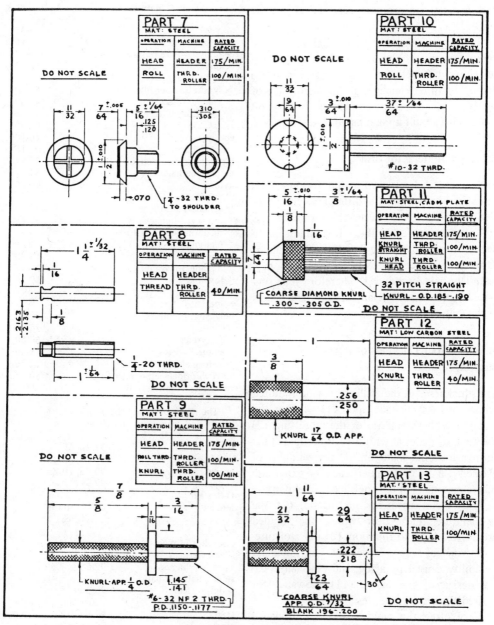

Fig. 28-14. Cold-forged inserts.

important phases of design will be covered in the succeeding paragraphs.

The design of special inserts for various applications requires as much engineering as other phases of preliminary work, if not more. In many cases too little significance is attached to planning special inserts. The design engineer, the manufacturer of the insert, and the molder must cooperate to obtain simplicity of design, which will result in the production of satisfactory articles and promote economical production.

Typical applications of special inserts are commutators, wire-and-insert connections on telephone handsets, and radio resistors where carbon or other elements are molded inside of the plastic. A radio condenser is a good example of a built-up laminated insert.

Selection of Metal for Inserts

The correct selection of metal for inserts is essential because of the differences in coefficient of expansion between the various metals and plastics (see Table 28-2). It is impossible to keep the plastics listing in Table 28-2 up to date, because of the wide variety of new materials being produced. Therefore it is suggested that the technician refer to data sheets of the material suppliers.

Minimum Wall Thickness of Material Around Inserts

The thickness of the wall of plastic required around inserts depends upon (1) whether the material is thermoplastic or thermosetting, (2) the type of material within each group, (3) the shrinkage of the material, (4) the modulus of elasticity and (5) the coefficient of expansion of the material, (6) the coefficient of expansion of the metal used in the inserts, (7) the temperature range over which the molded article will have to function, (8) the moisture-sensitivity of the plastic, (9) any loss of flexibility caused by aging, and especially (10) the design of the insert, (11) allowance for desired electrical properties.

Very often the molded article is designed first and the necessary inserts then fitted into

the remaining space. If inserts are required, they should be considered first and then the molded article designed around them. The shape and form of the insert govern the wall thickness of the plastic to a great degree, especially when the inserts are of irregular

Table 28-2. Coefficient of Thermal Expansion (30 to 60°C) per Degree Centigrade

Typical Material	Coefficient $\times 10^6$
alkyds	25-35
cellulose acetate	80-160
cellulose acetate butyrate	110-170
cellulose propionate	110-170
diallyl phthalate:	
synthetic-fiber-filled	50-60
mineral-filled	40-42
glass-fiber-filled	32-36
epoxy, filled	20-60
epoxy, unfilled	40-100
ethyl cellulose	100-160
melamine-formaldehyde	20-57
methyl methacrylate	54-110
nylon	90-108
phenolics:	
general-purpose	30-45
improved-impact	30-45
medium-impact—CFI-10	29
high-impact—CFI-20	22
medium-heat resistant	15-30
high-heat-resistant	20-35
low-loss	19-26
arc-resistant	49
polyester, colored	48
polyester, premixes	30-80
polyethylene	110-250
polystyrene	60-80
silicones	8-50
urea-formaldehyde	22-36
vinyl chloride-acetate resin	50-185
vinylidene chloride resin	190

The range of values shown reflects the variation in fillers and variations in resins.

aluminum 2S	99.2% Al	23.5
brass, ordinary	67 Cu, 33 Zn	18.8
bronze, commercial	90 Cu, 10 Zn	17.4
copper	99.9+	16.7
C.R. steel		11.7
Monel	67 Ni, 30 Cu, 1.4 Fe, 1 Mn	14.
nickel		13.3
phosphor bronze		17.
silver, German		18.
silver, sterling	92.5 Ag, 7.5 Cu	18.
solder, half-and-half		24.
stainless steel	90-2 Fe, 8 Cr, 0.4 Mn, 0.12 C	11.
steel	99 Fe, 1 C	10.8
zinc	95 Zn, 5 Al	28.

shape (rectangular, square, star, or any other shape having sharp corners).

The two main factors in the properties, especially of phenolic, urea and melamine materials, are modulus of elasticity and the ability of the material to cold-flow after curing so that it can stretch slightly without cracking. No one property of the material will solve the problem. For instance, a material having a low shrinkage of 0.002 in./in. but having a very rigid character will crack. Other materials which have a shrinkage of 0.010 in./in. but are capable of being stretched will not crack despite a minimum thickness of wall. It is impossible to set up comprehensive standards of wall thickness of material in relationship to diameters of inserts, particularly for some of the special designs. An insert $\frac{1}{4}$ in. in diameter requires a $\frac{1}{8}$ in. wall, while an insert 6 in. in diameter might require a $1\frac{1}{2}$ in. wall of material, depending upon the factors mentioned. Each individual article presents different problems and must be engineered according to the design of the insert and the material used. Table 28-3 shows recommended minimum wall thicknesses with plain round inserts for various plastics.

Anchorage

Firm and permanent anchorage of inserts is essential, and since there is no chemical or natural adherence between plastics and metal inserts, anchorage must be obtained by mechanical means. The slight anchorage that is obtained by the shrinkage of plastic around the insert is never sufficient.

Inserts must be anchored sufficiently to prevent turning when torque is applied and to prevent pulling out of the plastic when subjected to tension. However, internal stresses in the molded plastic must be kept to a minimum.

In the early days of plastics, it was customary to use hexagonal stock for inserts (Fig. 28-15). This is mechanically incorrect except in some special applications. Hexagonal stock provides torsional anchorage only. Grooves must be machined to obtain sufficient anchorage in tension. Combinations of sharp corners and grooves on hexagonal stock set up certain stresses in the plastic which often result in

cracking. In practically all instances, round stock is recommended, so that diamond knurling can be obtained. Diamond knurling provides the most satisfactory anchorage from the standpoint of torque and tension, and minimizes possible cracking around the insert.

Knurling of inserts is best accomplished in screw machines with end-knurling tools. The stock sizes given in Table 28-1 are ample to allow end knurling and to leave sufficient stock for a proper sealing diameter free of knurling at the open end of the insert. Cold-forged inserts are knurled on reciprocating or rotary rollers.

Grooves can be used in conjunction with diamond knurl (Fig. 28-16). Sharp corners must be avoided when machining the grooves. When using grooves, provide one wide groove in the center of the insert rather than two grooves, one on each end. The center groove allows the material to shrink or creep toward the center and minimizes strain within the piece, and thus possible cracking. Right and wrong designs are illustrated in Figs. 28-17 and 28-18.

See p. 831 for a discussion of the anchorage of special inserts.

Insert Testing

There has been considerable comment and evaluation of various testing procedures on inserts. While to date, these have not been standardized in the industry, several companies have suggested the following terminology and procedures:

Tensile Strength: Axial force (in pounds) required to pull the insert out of the material at least 0.020 in.

Jack Out Torque: Rotational force (in inch/ounces, inch/pounds, or foot/pounds) applied to a mating screw which pulls the insert out of the material through a washer with adequate clearance for the insert outside diameter.

Insert Rotation Torque: Rotational force required to turn the insert in the parent material. It is a good comparative measure of overall strength of the assembly.

Clamping Torque: Rotational force applied to a mating screw when the insert is allowed to contact a non-rotating plate and the insert

Table 28-3. Minimum Wall Thickness of Material (In.)

Diameter of Inserts (in.):	1/8	1/4	3/8	1/2	3/4	1	1-1/4	1-1/2	1-3/4	2
phenolics:										
general-purpose	3/32	5/32	3/16	7/32	5/16	11/32	3/8	13/32	7/16	15/32
medium-impact	5/64	9/64	5/32	13/64	9/32	5/16	11/32	3/8	13/32	7/16
high-impact (rag)	1/16	1/8	9/64	3/16	1/4	9/32	5/16	11/32	3/8	13/32
high-impact (sisal)	5/64	9/64	5/32	3/16	1/4	9/32	5/16	11/32	3/8	13/32
high-impact (glass)	1/16	3/32	1/8	1/8	3/16	3/16	1/4	1/4	5/16	5/16
high-heat-resistant, general-purpose type	1/8	3/16	7/32	1/4	11/32	3/8	13/32	3/16	15/32	1/2
high-heat-resistant, impact type	5/64	9/64	5/32	13/64	9/32	5/16	11/32	3/8	13/32	7/16
low-loss	5/32	7/32	1/4	9/32	3/8	13/32	7/16	15/32	1/2	17/32
special for large inserts	3/64	7/64	1/8	5/32	7/32	1/4	9/32	5/16	11/32	3/8
polyester, colors	3/32	5/32	3/16	7/32	5/16	11/32	3/8	13/32	7/16	15/32
polyester, sisal-filled	5/64	9/64	5/32	3/16	1/4	9/32	5/16	11/32	3/8	13/32
polyester, glass-filled	1/16	1/8	9/64	3/16	1/4	9/32	5/16	11/32	3/8	13/32
diallyl phthalate:										
(a) "Orlon"-filled	1/8	3/16	7/32	5/16	11/32	3/8	13/32	7/16	15/32	1/2
(b) mineral-filled	3/32	5/32	3/16	7/32	5/16	11/32	3/8	13/32	7/16	15/32
(c) glass-filled	5/64	9/64	5/32	3/16	1/4	9/32	5/16	11/32	3/8	13/32
cellulose acetate	1/8	1/4	3/8	1/2	3/4	1	1-1/4	1-1/2	1-3/4	2
cellulose acetate butyrate	1/8	1/4	3/8	1/2	3/4	1	1-1/4	1-1/2	1-3/4	2
ethyl cellulose	1/16	3/32	1/8	5/32	3/16	7/32	1/4	9/32	5/16	11/32
urea formaldehyde	3/32	5/32	3/16	7/32	5/16	11/32	3/8	13/32	7/16	15/32
*melamine formaldehyde (a)	3/32	5/32	3/16	7/32	5/16	11/32	3/8	13/32	7/16	15/32
(b)	1/8	3/16	7/32	5/16	11/32	3/8	13/32	7/16	15/32	1/2
vinylidene chloride resin	3/32	1/8	3/16	1/4	3/8	1/2	1/4	9/32	5/16	11/32
methyl methacrylate resin	3/32	1/8	3/16	3/16	7/32	1/4	5/8	3/4	7/8	1
polystyrene	3/16	3/8	9/16	3/4	1-1/8	1-1/2	1-7/8	2-1/4	2-5/8	3
polyethylene	1/16	3/32	1/8	5/32	3/16	7/32	1/4	9/32	5/16	11/32
nylon:										
"Zytel" 101 or equiv.	1/16	3/32	1/8	5/32	3/16	7/32	1/4	9/32	5/16	11/32
" 31 " "	3/32	1/8	5/32	7/32	1/4	5/16	11/32	13/32	7/16	15/32
" 63 " "	3/32	5/32	3/16	1/4	5/16	11/32	13/32	7/16	1/2	9/16
" 69 " "	1/8	7/32	9/32	11/32	7/16	1/2	19/32	21/32	23/32	13/16
" 105 " "	1/16	3/32	1/8	5/32	3/16	7/32	1/4	9/32	5/16	11/32
" 211 " "	3/32	1/8	5/32	7/32	1/4	5/16	11/32	13/32	7/16	15/32
" 42 " "	1/16	3/32	1/8	5/32	3/16	7/32	1/4	9/32	5/16	11/32
vinyl chloride-acetate resin	3/32	1/8	3/16	1/4	3/8	1/2	5/8	3/4	7/8	1

* Melamine formaldehyde   (a) mineral-filled melamine ignition material
                            (b) cellulose-filled melamine, electrical grade

diameter is larger than the clearance hole in the plate. This simulates the usual application condition. It is important to know what material the plate will be for accurate values.

Sketches of these various procedures are indicated in Fig. 28-19.

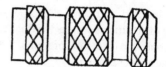

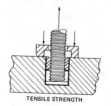

Fig. 28-18.
Wrong

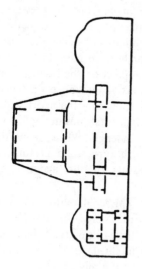

Fig. 28-15.

TENSILE STRENGTH   JACK OUT TORQUE

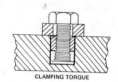

ROTATIONAL TORQUE   CLAMPING TORQUE

Fig. 28-19. Some suggested procedures for insert testing. (*Courtesy Heli-Coil Products Co.*)

Problems in Molding with the Usual Types of Inserts

Most of the data that follows relates to the use of inserts in thermoset molding. Portions, however, are applicable to thermoplastics.

Floating of Inserts. Floating of inserts can be controlled or prevented by several methods:

(1) The retaining pins may be tapered slightly, starting the taper at the fillet and carrying it up to one-third of the length of the pin. If too much taper is allowed, making the insert too tight on the retaining pin, the insert may pull out of the material.

(2) It has been found that a straight knurl on the retaining pin provides sufficient holding surface.

(3) Square retaining pins can be used.

(4) Split pins are practical for blind-hole inserts.

(5) Spring tension pins can be used, in which the retaining pin is slotted and music

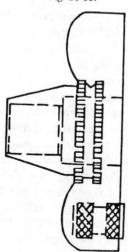

Fig. 28-16.

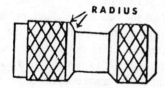

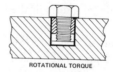

RADIUS

Fig. 28-17.
Right

wire is inserted into the slot. This method presents difficulties because the slightest flow of material into the slot prevents the spring from functioning properly.

(6) An extended shoulder can be provided on the insert, shown in Fig. 28-12 as J (sealing), and this shoulder allowed to enter into the mold proper. This is an ideal method of preventing the insert from floating, although it is not permissible when inserts must be flush with the surface of the material.

(7) On male inserts, a tapered hole can be provided for a drive fit if close accuracy of inserts is maintained. In such a case, a taper of 0.0005 to 0.001 in. for the depth of the hole is sufficient. If the insert is long enough, a small side hole can be drilled in the pin and music wire inserted to provide spring action. This spring action prevents the insert from floating in practically all methods of molding. The same method can be used for holding inserts in the top half of the mold.

(8) When precise location of the insert is essential, removable threaded pins are provided in the mold. Inserts are screwed to these pins. However, this procedure increases the cost of production. Subsequent removal of the flash from the thread is avoided in most cases.

(9) In some instances, location of a male or female insert may be affected in the top half of the mold by use of a ball-and-socket arrangement. There are commercially available the necessary ball, spring, and plug used in this arrangement. The insert is normally pushed into the hole, engaging the ball, and the side pressure exerted by the ball will hold the insert in place during molding.

(10) Male inserts can sometimes be held in the upper half of the mold by use of a magnetized section in the mold. Of course, the size and the type of material used for the insert determine the practicability of this method.

Crushing of Inserts. In transfer molding, there are very few difficulties with crushing of inserts if close tolerances on the length are maintained. In compression molding, however, when the insert must show on both sides and is molded vertically or in line with the press motion, crushing of inserts can be prevented by the use of preforms with holes to allow the preform to slip over the insert. Sliding pins are provided in the force plug, operated by spring, air, or hydraulic action. These pins are in a down position when the mold is being closed and they contact the surface of the inserts before the flow of material takes place. Since the pins are under constant pressure, no material can enter the insert. This method can be applied to blind- or open-hole inserts, and either top or bottom pins. Considerable pressure can be applied on the inserts. Actual tests on a brass insert $\frac{1}{2}$ in. long, 6 x 32 thread, with $\frac{1}{16}$ in. wall, show that the insert withstands 6 cycles of 500 lb total pressure with a reduction of 0.0005 in. in length. When the inserts are not of the through type, solid preforms may be used. Preheating of the material is recommended.

Flow of Material into an Open-Hole Through-Type Insert. In transfer molding, there is very little difficulty, if the length of the insert is maintained from 0.002 to 0.004 in. oversize. When the mold is closed, the insert is pinched in the mold, and it is impossible for the plastic to enter. In compression molding, however, it is impossible to prevent material from entering the hole unless pressure-type pins are used, as above, for prevention of crushing of inserts. If pressure pins cannot be used, it is advisable, especially on larger inserts, to tap the inserts undersize before molding, and then retap to proper size after molding. Extreme care should be taken in retapping to prevent stripping of threads, especially if considerable material has flowed into the thread. Small inserts are most economically molded with a drilled hole and tapped after molding.

Flow of Material into a Blind-Hole Insert. Difficulties with flow of material into blind-hole inserts are not as numerous as with the open-hole types. In most cases such flow is caused by loose retaining pins which allow the insert to float with the flow of material, uneven machining on the face of the insert, or knurling on the entire outside diameter of the insert, leaving extended burrs on the face which do not permit the insert to rest flat on the surface of the mold or the surface of the retaining pin. In all cases, it is good practice to provide a slight recess in the mold, accommodating the outside diameter of the insert. When the "J" diameter (Fig. 28-12) of the retaining pin is the same as that of the insert, and sharp corners can be retained in the hole, an 0.005-in. depth is sufficient to prevent the plastic from flowing in. This method allows the insert to protrude above the surface of the molded article, and this is desirable, especially when electrical contacts are to be made.

Protruding Inserts. Protruding inserts are frequently required and are molded in place for specific purposes. In most cases, the protruding section is used for assembly or for bearing points where mechanical action is required. In special cases, especially of large inserts where the molded article is subjected to considerable torque in order to obtain a tight connection, it is advisable to allow a hexagonal section of the insert to protrude above the molded surface for a wrench grip. Thus strain is applied on the insert rather than on the plastic.

Perfect anchorage also is necessary. Where the wall of material is limited, the anchorage section of the insert is turned and coarsely diamond-knurled. A groove can be added to increase anchorage for tension. However, sharp corners must be avoided. In the event that the hexagonal shape is used for anchorage, sharp corners must be reduced by turning, and grooves provided for tension anchorage. Figure 28-20 shows a recommended design.

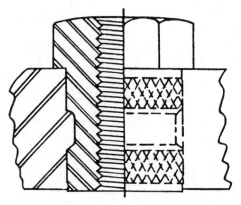

Fig. 28-20.

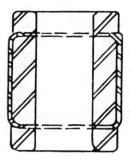

Fig. 28-21.

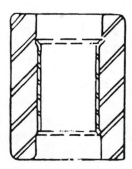

Fig. 28-22.

Anchorage of Special Inserts

Most of the following data relates to molded thermosets; in some cases, it is also applicable to molded thermoplastics.

Thin Tubular Inserts. These inserts are extremely difficult to anchor properly, if a tubular insert is molded part way up a molded article, it is possible to invert a bead which will act as a satisfactory anchorage. The bead can be used on outside or inside inserts as shown in Figs. 28-21 and 28-22, respectively. A perforated surface around the circumference also can be used where permissible. When molding an outside tubular insert, it is often necessary to coat the inside of the insert with neoprene or vinyl to improve the bonding.

Flat Plate-Type Inserts. These can be anchored by means of countersunk holes wherever it is permissible. Bevel all edges of the insert or, if certain sections of the insert are not required for the functioning of the article, the section can be partially cut out and bent over to provide anchorage. This method is illustrated in Fig. 28-23. If metal inserts must be thick, bosses can be extruded and slightly flared to provide satisfactory anchorage. Anchorage may be obtained also by spot-welding lugs to the underside of the insert.

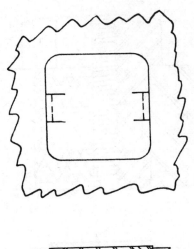

Fig. 28-23.

Drawn Shell-Type Inserts. Where a minimum wall thickness of plastic is specified and an insert of this type is used, extreme caution must be exercised to provide proper anchorage. Figure 28-24 shows unsatisfactory anchorage because it allows insufficient wall thickness of plastic to avoid cracking. Figure 28-25 shows an insert which is fairly well designed, and could be used to good advantage. In an insert of this type, the plastic has a chance to slide over the insert. However, to provide the best possible anchorage, the insert should be flared in slightly, as shown in Fig. 28-26. With this design, the plastic actually has a chance to anchor the insert and to creep while shrinking.

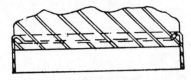

Fig. 28-24.

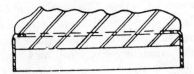

Fig. 28-25.

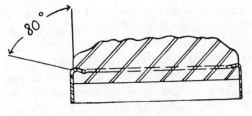

Fig. 28-26.

Drawn Pin-Type Inserts. Very often an insert of this type is molded into a plastic and then countersunk after molding, as illustrated in Fig. 28-27. A slight bead provided as an undercut for anchorage on an insert of this type is entirely insufficient to hold the insert properly. Wherever possible, when an insert of this type is used, piercing pins should be provided in the mold, so that the insert can be pierced during the molding operation and the necessary countersink molded into the plastic. During this piercing operation, the insert is flared out to provide proper anchorage, as shown in Fig. 28-28.

Figure 28-29 shows a drawn-type pin with an open end. Partial anchorage is obtained by shearing and folding two segments during the molding operation. Floating pressure-type

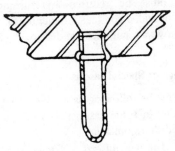

Fig. 28-27.

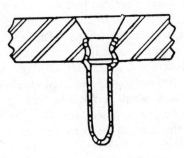

Fig. 28-28.

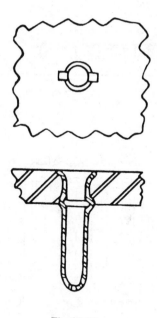

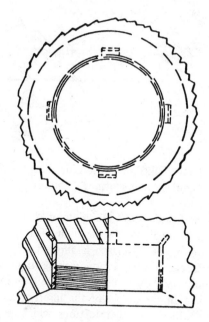

Fig. 28-29.

Fig. 28-30.

piercing pins in the mold are recommended to minimize flow of plastic into the insert.

Drawn-Shell Threaded Inserts. As illustrated in Fig. 28-30, these are often used in large molded articles where it is not necessary to have 75 per cent of thread, or where insert space is limited. Because the shell is usually thin, approximately 50 per cent of the depth of thread is obtained. The four flared lugs provide a satisfactory anchorage in every respect. It is impossible to provide sealing points on an insert of this type, and hence flow of material into the thread must be expected. Tapping after molding is recommended for most satisfactory results.

Intricate Inserts. An intricate insert is shown in Fig. 28-31. Considerable difficulty

with cracking of the plastic was encountered until aluminum inserts were selected. Actually there were two factors in favor of aluminum, i.e., its coefficient of expansion and its ability to yield or spring slightly when the plastic was shrinking.

Large-Surface Inserts. It is often necessary to mold one or more large-surface inserts on one side of the plastic, as illustrated in Fig. 28-32. Inserts of this type cause nonuniform shrinkage of plastic and considerable warpage. Even when shrinkage or cooling fixtures are used, it is certain that surface A will be convex and B concave after the piece is allowed to cool and age. If a flat surface is required, the surface must be machined. Best results will be obtained when the articles are allowed to age or, if

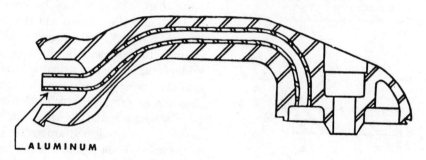

ALUMINUM

Fig. 28-31.

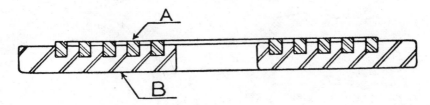

Fig. 28-32.

possible, are baked in an oven for at least 72 hr, at suitable temperatures, before being machined.

Large Inserts with a Minimum of Wall Thickness of Material. Where a minimum thickness of a thermosetting plastic is allowed around a large insert, a special noncracking type will generally have to be used. Extreme care must be taken in the design of the insert to avoid sharp corners or other features which might create local stresses.

Irregular-Shaped Inserts. These inserts cause the greatest difficulty. Figure 28-33 shows a U-shaped insert approximately $1\frac{1}{2}$ in. long, on which two rib projections are required. From the standpoint of economy in forming this insert and loading it into the mold, it can be made in one piece, but it will cause difficulty with cracking of the plastic. It would be more economical in the long run to make two separate inserts, as shown in Fig. 28-34. If electrical contact is required, a wire can be fastened between the two inserts; or, if a more solid connection is desired, the insert can be made solid with cutout slots, as shown in Fig. 28-35. Provision must be made in one half of the mold to prevent these slots from being

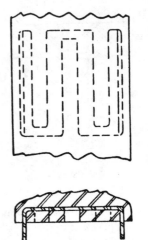

Fig. 28-35.

filled with plastic. When these slots are open, there will be a slight give in the insert when the plastic shrinks. This will consequently reduce or eliminate the possibility of cracking the plastic.

When a long bar-type insert is used, it is always advisable to provide an anchorage in the middle of the bar by means of grooves or slots, or coarse diamond knurl for round bars (see Anchorage, p. 827). The center anchorage will allow the plastic to creep along the surface of the insert while it is shrinking toward the center. If additional anchorage is desired on round bars, the ends can be knurled with straight knurl (Fig. 28-36) and still retain the creeping action. Where dimensional accuracy is required, full allowance for shrinkage should be made. If the article is of cylindrical shape, then, instead of using a knurl for anchorage, circular rings can be provided, which will give satisfactory anchorage and at the same time allow the plastic to creep uniformly around the periphery of the insert. When a knurl is used on a piece of this type and the plastic begins to shrink, it has

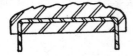

Fig. 28-33.

Fig. 28-34.

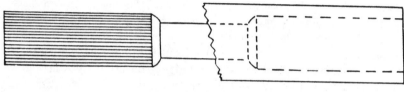

Fig. 28-36.

a tendency to climb up on the knurl, producing stress on the plastic and causing it to crack.

Leakproof Inserts

Because of the difference in the behavior between plastics and metals (e.g., difference in coefficient of thermal expansion), the characteristics of some plastics, and the problems of providing proper adequate anchorage, it is usually impossible to make an insert remain airtight within the plastic even under small pressures. If inserts are used in articles that must withstand high internal pressures, special methods must ordinarily be used to make them airtight.

To retain an airtight joint between the plastic and a metal insert it is necessary to provide a flexible wall of other material between the two. When the molded article and the insert expand and contract, this flexible material, although it consists of only a very thin coating, will compensate for the difference in coefficient of thermal expansion between metal and plastic.

A few successful methods are recommended. The insert is knurled the same as for normal anchorage and is provided with at least two grooves, about $\frac{1}{32}$ in. wide and 0.020 in. deep. The head or anchorage part is dipped in neoprene, polyvinyl chloride-acetate, or other rubbery synthetic material, and then oven-dried before using. This will supply sufficient coating on the insert to give it the necessary cushioning action.

It is possible also, especially on round inserts, to provide a groove in the anchorage head of the insert large enough so that a neoprene washer can be used. Under normal molding conditions, the washer will produce satisfactory results. On some applications, a retaining groove is molded or machined between the insert and the plastic. The groove is filled with alkyd resin and allowed to dry at room temperature, or is oven-baked.

Special Inserts for Reinforcement

It is often necessary to mold inserts into plastics as reinforcements to provide greater strength, greater rigidity, greater safety (as in automobile steering wheels) or greater dimensional accuracy.

In molding a thermoplastic housing, for example, instead of molding a thick wall to obtain rigidity, a sheet-metal reinforcement can be molded on the inside of the housing. This will not only produce greater rigidity with a minimum of wall thickness, but it will also assist in maintaining better dimensional accuracy. Various materials can be used as reinforcement—molding board, laminated phenolics, perforated metal, metal screens.

Nonmetallic Inserts

Inserts of various nonmetallic materials are used successfully. The use of wooden inserts in applications such as doorknobs or automobile gearshift knobs saves considerable material and shortens the molding cycle.

Glass inserts are being successfully molded into thermoplastics by injection and into thermosetting materials by transfer. Difficulties can be reduced during the initial engineering of mold design. The most difficult problems are caused by the nonuniformity of contours and dimensions. Glass, being of a brittle nature, does not lend itself to the application of full clamping pressures during molding. It is necessary to provide a cushion, by means of springs or rubber, to compensate for the normal irregularities in dimensions of glass inserts. In some cases, paper is glued to the surface of glass to provide additional cushion, and also to protect the surface from scratching during handling.

Locating the insert in the mold is difficult. Figure 28-37 illustrates the use of a sleeve-type ejector. The inside diameter of the sleeve is the same as the outside diameter of the glass insert. The insert is located by placing it inside of the ejector sleeve when it is protruding in ejected position. Figure 28-38 illustrates a step-molded article. The diameter of one of the steps must be the same as the outside diameter of the insert. A sleeve-type ejector is used on this step, to locate the insert, as in the preceding case.

Preparation of Inserts Before Molding

Cleaning of Inserts. Considerable significance should be attached to the cleaning or washing of inserts prior to molding, especially screw-machine inserts. If inserts are improperly washed, even though they appear clean, there may be loose metal chips hanging on to the threads, or fine metal dust in the knurls. This latter is often rolled into the surface by the process of knurling, and it is not easily washed off, but it will be loosened by the flow of the plastic. These metal chips may flow up to the surface and impair the appearance of the molded article. The most serious difficulty, however, is in electrical applications, where a small particle or a slight amount of metal dust may cause a total breakdown electrically. Grease and oil also are detrimental to molded articles from the standpoint of appearance, and should be thoroughly washed off.

Processes of cleaning are divided into three types:

1. Mechanical, including hand polishing, tumbling, shot- or sandblasting, or washing with solvent or alkali;
2. Chemical, such as removal of iron rust and silver tarnish by an acid bath;
3. Use of electrolytic cleaners.

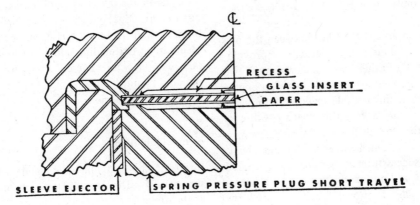

Fig. 28-37.

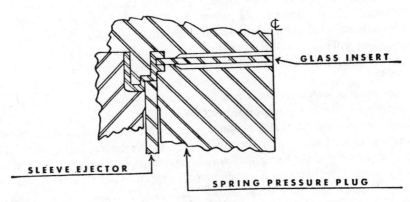

Fig. 28-38.

Oil and machining chips can best be removed by a well-stirred alkali bath followed by a rinse with hot water, except where the nature of the metal, such as aluminum, rules out the alkali in favor of degreasing with a solvent.

In many cases a reasonable amount of tarnish can do no harm, but where the function or the appearance of the piece demands chemically clean inserts, an acid dip is necessary. For brass and bronze, a mixture of nitric acid and sulfuric acids or nitric alone is commonly used.

Silver tarnish can be removed with nitric acid or a diluted solution of one of the cyanides. Trisodium phosphate has been found to be an efficient remover of iron rust.

Preheating of Inserts. Large inserts should be preheated (above the mold temperature if possible) prior to molding. This will allow the maximum expansion and improve the flow and cure of the plastic. With thermoplastic materials, preheating of inserts will reduce the likelihood of weld marks, which often result in cracking of the plastic after molding.

Cleaning Flash from Inserts

Most of the difficulty with flash can be avoided in the design of the article and the insert by providing sealing points so that the flow of plastic is cut off or at least minimized. However, even with the best design there will be some material on the inserts, especially when the mold becomes worn or close tolerance on inserts is not maintained. Several methods are recommended to minimize this, particularly lubricating the insert, prior to molding, with wax, soap, grease, or oil. For thermoplastics it is recommended that a mineral oil of viscosity SAE 100 or greater be used. Plating and polishing the inserts minimize the adherence of flash.

To remove the flash, cut it close to the molded article and peel it off. In most cases a mild solution of caustic soda will loosen the flash so that it can be easily removed. This method, however, requires extreme caution because too long contact or too strong a solution will harm the surface of the article, and may even loosen the insert in its anchorage.

Salvage of Inserts

When the inserts are of the through type, they can be knocked out by means of a foot press and fixture. When they are anchored part way in the material, a strong solution of caustic soda will loosen the inserts in thermosetting material so that they can be picked out. For thermoplastic material use suitable solvents for the plastic, or soften the articles in an oven and pull out the inserts. Reclaimed inserts should be inspected before reuse.

Relieving Molding Stresses Around Inserts

Considerable stresses are set up in molded articles or irregular design, such as those having both thin and thick sections, and especially those with metal inserts. The best method to relieve stresses is to allow the article to cool slowly. The ideal condition would be to carry the articles on a conveyor through an oven which has various stages of temperatures, starting at $50°F$ below the molding temperature, then gradually decreasing until the article is cooled to room temperature. This method, however, requires special equipment.

The next best method requires two ovens, one at approximately $225°F$ and one at $150°F$. The molded article remains in each oven successively until its temperature is reduced to oven temperature.

The final step is cooling to room temperature. In case of thermoplastic materials, molding stresses are relieved by annealing in an oven or a water bath at suitable recommended temperatures.

29

COMPOUNDING AND MATERIALS HANDLING

Up to this point, we have referred to plastics in their basic resin form. Obviously, in actual practice, many resins involve the addition of various chemicals and additives to make them tougher (reinforcing fibers), more flexible (plasticizers), more resistant to ultraviolet (stabilizers), colorful (colorants), and so on. Most of these additives are added to the resin by various compounding techniques, either at the resin supplier level or by specialists in compounding or, as has been the more recent case, by plastics processors themselves. In this chapter, we will present a brief review of compounding.

Another aspect of the manufacturing operation that has taken on added importance for the processor is the manner in which the plastic moves from point of delivery through premolding operations (e.g., drying, coloring, etc.) into the processing machine and, after processing, on to assembly operations, regrind, etc. These subjects are covered in the sections on materials handling (p. 858) and plant layout (p. 864).

COMPOUNDING*

Few resins are useful in their natural form and are therefore mixed with other materials to

* Adapted from an article on "Compounding," by A. A. Schoengood, Editor, *Plastics Engineering*, as it appeared in the February 1973 issue of the publication, then known as *SPE Journal*. With the permission of the Society of Plastics Engineers.

improve and enhance their properties and thereby make them more useful for a variety of applications. The process by which ingredients are intimately mixed together into as nearly a homogeneous mass as is possible is known as *compounding*. And because of the nature of both the resin and the other ingredients, compounding requires a wide range of mixes—dry powders, slurries, pastes, doughy consistencies, for example, and a corresponding range of mixing operations.

The task of mixing becomes one of changing the original distribution of two or more nonrandom or segregated masses, so that an acceptable probability distribution of one mass throughout the others is achieved. Thus the problem becomes one of deforming or redistributing masses in order to achieve this desired probability distribution, in the absence of diffusion or other random molecular motions. The problem becomes more complicated if the ultimate particles are not independent of one another, but instead exert interparticulate forces leading to particle agglomerations. Thus, external force must be exerted on such agglomerates to allow them to mix. Dealing with these forces, or stresses, is at the heart of the problem of dispersion processes.

In mixing and dispersing processes involving thermoplastic melts, the thermoplastic material is regarded as essentially a fluid subjected only to laminar flow, capable of being deformed. Thus, the problem of mixing in thermoplastics is that of subjecting such materials to laminar shear deformations in such a manner that an initially non-random distribution of ingredients approaches some arbitrary scale of randomness. The problem is usually complicated by the fact that the ingredients do exhibit interparticulate forces, so that the stresses accompanying the deformation must be considered, as well as the deformation process itself.

Mixing Theory

The general theory of mixing usually considers a non-random or segregated mass of two components, and their deformation by a laminar or shearing deformation process. The object of shearing is to mix the mass in such a way that samples taken from the mass exhibit minimal variations, ultimately tending to zero. The shearing process is generalized, i.e., it is not limited to any particular kind of shearing action or mixing device, but applies to all such devices.

Three basic principles are given by the general theory of mixing:

1. The interfacial area between different components must be greatly increased in the mixing process because it results in a decrease in average striation thickness.
2. Elements of the interface must be distributed uniformly throughout the mass being mixed. Uniformity can be defined by the intensity of segregation, or by the average deviation of particles from a mean concentration.
3. The mixture components must be distributed so that for any unit of volume, the ratio of components within the unit is the same as that of the whole system.

The first and second principles obviously require, for any particular mixing process or device, that the dynamics of fluid or particulate flow within the device ideally obey the three requirements of efficient mixing, namely, that the shear be at right angles to the direction of flow, that the shear rate be controlled by varying the gap between the stationary and rotating elements, or by varying rotational speed, or both, and that each particle follow the same flow path to experience the same

shear history (Fig. 29-1). They also suggest that the optimum initial orientation of the interface between components should be normal to the flow lines or streamlines within the device and that all the streamlines of flow should lead to a region of maximum shear. The last principle implies that the scale of sampling must be large in comparison to the ultimate particle size (Fig. 29-2).

Limitations of Mixing Theory

Being generalized, the mixing theory suffers from limitations which must be recognized

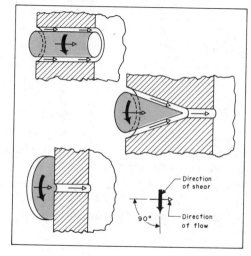

Fig. 29-1. Three ideal mixing devices. A. Shear is applied at right angles to the direction of flow, the shear rate is varied by varying the speed of the rotating element, and the particles follow a spiral path to the exit. B and C. In both, shear rate can be regulated by controlling gap size or rotational speed. Gap size is varied by axially displacing rotating element.

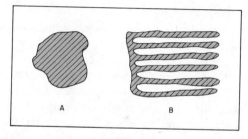

Fig. 29-2. The interfacial area in (A) is increased to that of (B) to pass through each volume element of the mixture. The material in each volume element is in the same ratio as in the entire mixture.

when it is applied to actual plastic materials and present mixing devices:

1. It is limited to purely viscous (Newtonian) materials, while actual plastic materials are viscoelastic, plastic, or otherwise non-Newtonian. However, the qualitative concepts which hold for simple viscous bodies are of great value in furthering the understanding of the mixing of real materials.
2. It is assumed that there are no van der Waals or other forces between particles.
3. Initial orientation of components is assumed to be known although this is almost never true in practice.
4. Mixing process is assumed to be isothermal. Yet the heat generated in mixing most plastic materials is sufficiently high so that the energy dissipation during mixing is considerable.

The major gap between theory and practice lies in our inability to fully describe the rheological behavior of complex plastic melts and the dispersion of powdered materials, particularly since the flow within actual mixing devices is very complex. Still, the application of qualitative and semi-quantitative concepts has thrown considerable light on the process of mixing.

Mixing Evaluation

A mixture is described in terms of the statistical deviation of a suitable number of samples from a mean, the sample sizes being dependent upon some length, volume, or area characteristics of the mixture or its properties. For example, if color is imparted by adding pigment and homogeneity is measured by visual impression, the characteristic length is the resolving power of the eye, say 0.001 in. A completely mixed compound exhibits pigment streaks no greater than 0.001 in. On the other hand, the color value for any given series of samples would appear uniform to a spectrophotometer that integrates over a 1-in.-diameter circle even if the streaks were 0.1 in. thick. Likewise, the intensity of color difference between streaks would affect the resolving power of the eye—or

the spectrophotometer—and, hence, the characteristic length.

The means for measuring the degree of mixing are varied. In commercial practice, inspection for color homogeneity, color comparison for specks, streaks, or spots of unmixed filler or resin is visual. Frequently, changes in physical properties, such as tensile strength, modulus, or density, are used to evaluate the degree of mixing. Since these properties may be affected during polymer degradation and thermal effects on the mixture, property changes do not strictly evaluate the mixing as such.

Some of the more fundamental measures evaluate changes in rehological properties, chemical reactions, and electrical conductivity. These most nearly approach the ideal measurement of the basic criteria of mixing.

The only direct measure of a basic criterion of mixing is "striation thickness," though this property is measurable at present only under certain conditions (Fig. 29-3).

Because of the many variables of mixing and the scantiness of criteria for measuring mixing effectiveness, current theory of mixing of heavy viscous materials is less than satisfactory. Often practical and extensive tests are the only means of determining the suitability of a particular

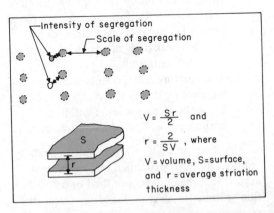

$$V = \frac{Sr}{2} \quad \text{and}$$

$$r = \frac{2}{SV} \text{ , where}$$

V = volume, S = surface, and r = average striation thickness

Fig. 29-3. Average striation thickness (r) is computed from the ratio of the interfacial surface area between the components and the total volume of the mixture. Average striation thickness in effect measures the scale of segregation, which is the average distance between clumps of the same component. Intensity of segregation is the average deviation of particles from a mean concentration.

mixing machine for a specified plastics mixture. Most compounding can probably be classified as an art.

The factors that contribute most to the ultimate properties of plastic compounds are many, depending on the type of basic resin, type of compound, the types and amounts of other ingredients, and the desirable degree of homogeneity. The basic resin may vary considerably chemically, that is, as to molecular weight and configuration and particle size. The compound may range from an adhesive or coating solution to a molding powder. In addition, there are a large number of secondary plastics materials which may be incorporated with the plastics resin in the mixture, such as plasticizers or solvents, solids such as fillers or pigments, and others such as stabilizers, dyestuffs, and lubricants. Finally, the uniformity of the mixture contributes greatly to the properties of a plastic, for usually the more homogeneous the mixture is, the better the properties, or at least the more uniform, they will be.

Some mixtures are quite simple, such as those used for clear plastic film, containing 90-95% of the basic resin and only small amounts of additives. On the other hand, other mixtures are quite complicated with a dozen or more ingredients mixed with as little as 20-30% of the basic resin. These are referred to as the "filled plastics" from which many of the molded articles are formed.

The major additives or ingredients compounded with basic resins are: fillers, plasticizers, colorants, heat stabilizers, antioxidants, ultraviolet light absorbers, antistatic agents, flame retardants, blowing agents, and lubricants.

Fillers

Fillers are classified as inorganic, organic, mineral, natural, and synthetic. They are more commonly used with the thermosetting resins such as the phenolics, ureas, and melamines, although they fill some thermoplastics as well. Large amounts of fillers are commonly referred to as extenders, because they increase the bulkiness and likewise decrease the cost of a plastic. Since the properties of an extended plastic often suffer, the use of fillers is limited to less critical applications. Fillers normally endow plastics with specific mechanical, physical, and electrical properties such as strength characteristics, hardness, density and dielectric strength. Fillers also tend to increase the resistance of a plastic to one or more of the various service conditions. A filler usually comprises between 10 and 50% of the weight of the mix. Some of the common fillers and their functions are listed below:

Purpose	Filler
Bulk	Wood flour
	Sawdust
	Wood pulp
	Sisal–jute
	Purified cellulose
	Mica–rock
Reinforcement	Glass fibers and spheres
	Asbestos fibers
	Cellulosic fibers
	Cotton fabric
	Paper
	Synthetic fibers
Hardness	Inorganic pigments
	Mineral powders
	Metallic oxides
	Powdered metals
	Graphite
	Silica
Thermal insulation	Asbestos
	Diatomaceous earths
	Ceramic oxides
	Silica
Chemical resistance	Glass fibers and fabrics
	Synthetic fibers and fabrics
	Graphite
	Metallic oxides
Appearance	Color pigments
	Dyestuffs
	Carbon blacks
	Powdered metals
	Phosphorescent minerals
	Woven fabrics

One of the most widely used fillers, especially for the thermosetting plastics, is wood flour, which consists of a finely ground powder of one or more of the hardwoods or sometimes of nut shells. Wood flours are readily available, cheap, lightweight, strong (due to their fibrous nature), and easy to compound, being easily wet by the resin. But they also exhibit low thermal resistance, low dimensional stability (due to high moisture content), and poor electrical characteristics.

Mineral fillers are varied and comprise a large number of natural minerals as well as refined minerals and inorganic pigments. Virtually every type of rock has been used in powder form. Many are fairly inexpensive and easily available, but more expensive minerals are also used, such as the refined metal oxides, sulfates, and other inorganic pigments, to impart both color and hardness to the plastic. The purer forms of silica, such as mica and quartz, also provide good heat and electrical insulation. However, the inorganic and mineral fillers used in substantial amounts result in increased brittleness. Asbestos, a mineral fiber, combines the advantages of an inorganic filler with those of a fiber, namely, imparting good strength and heat resistance, but recently asbestos has come under stringent safety regulations. The factors for choosing a filler are:

- Cost, availability and uniformity
- Compatibility or wettability with the resin
- Moisture absorption
- Physical properties
- Thermal stability to mold temperatures
- Resistance to chemicals
- Abrasiveness
- Effect on plastic flow characteristics

Plasticizers

Plastics may need to be plasticized to enhance flexibility, resiliency, and melt flow. Without the addition of a plasticizer, it would not be possible to make plastics sheeting, tubing, film, and other flexible forms of plastics. In theory plasticizers enable the molecular chains of polymers to move freely with respect to one another, with a minimum of entanglement or internal friction. A plasticizer therefore acts as an internal lubricant, overcoming attractive forces between the chains and separating them

to prevent intermeshing. The higher the temperature, the greater the penetration of plasticizer between chains, and the greater is the melt flow or moldability.

A plasticizer can be anything incorporated into plastic, but not chemically linked to it. Any solvent that would dissolve PVC, for example, could be called a plasticizer. But in reality, a plasticizer must meet much more exacting requirements. As a matter of fact, the use of more than one plasticizer in any one formulation is not uncommon in today's applications.

Although nearly 80% of all plasticizers are used in PVC, they are also often used in cellulosics, nylon, ABS, polystyrene, among other resins.

Most plasticizers are liquids, although a few are solids that melt at compounding temperatures. All must exhibit good compatibility with the resin they modify. They are usually colorless and odorless, and have low vapor pressure and good thermal stability. Unfortunately, they decrease the strength characteristics, heat resistance, dimensional stability, and solvent resistance of resins.

The most popular general-purpose plasticizers are the phthalates, although epoxies, phosphates, adipate diesters, sebacates, polyesters are also in common use. But with some 500 different plasticizers to choose from and combine with a virtually unlimited number of formulations, the choice of a plasticizer becomes exacting, particularly as already stated, because it affects the physical properties of the end product as well as processability.

A good starting point is to decide on the properties required in the end product. The choice of plasticizer often comes down to finding one that satisfies the end product property requirements, that is compatible with the resin and that is least expensive.

As a general rule, the more plasticizer added, or the higher the ratio of plasticizer to resin, the more flexibility is achieved. Plasticizers might compose up to 50-60% of a final PVC product.

Epoxy plasticizers, for example, add heat and light stabilization. The phosphates improve flame resistance, but may impair heat and light stability. The polyesters and trimellitates are used where durability is vital, but where low temperature properties are less important. Aliphatic diesters, on the other hand, impart good low temperature properties.

Other factors to be considered include Federal approval for food packaging, specific gravity, compatibility with other additives in the formulation.

Secondary plasticizers can be used in smaller quantities to impart properties not achieved with primary plasticizers.

Colorants

The use of colorants makes it possible to produce a great variety of materials in colors varying from pastels to deep hues besides the varicolors and marblelike shades. Broadly speaking, there are two types of colorants used in plastics, namely, dyes and pigments, both organic and inorganic, the essential difference between them being that of solubility. Dyes are fairly soluble in plastics, while the pigments, being insoluble, are dispersed throughout the mass. The choice of either depends on resin compatibility or the need for solubility. Of almost equal importance is color stability, which means that a dyestuff or a pigment is stable at molding temperatures and on exposure to light, moisture, and air expected in the end use. Colorants are also chosen for strength, electrical properties, specific gravity, clarity, and resistance to migration (bleeding). The darker phenolic resins need bright pigments to hide them, whereas the lighter colored ureas and melamines call for pastel shades to take advantage of their attractive translucency and glasslike transparency. Typical are the acrylics requiring soluble dyes, which will not dim their clarity.

With some exceptions, colorants are supplied as dry powders of various specific gravities and bulking values. Soluble dyes are the easiest to use. Powders are simply tumbled or rolled in a drum with preweighed quantities of resin powder and distributed evenly throughout the mix. As the resin liquefies in an extruder or injection molding machine, the dye dissolves in the resin.

All dyes are transparent. Most have relatively poor light fastness and limited heat stability, but give bright shades in transmitted light, and are widely used, therefore, where transparency is desired. Dyed plastics often hold their color much better against fading than pigmented plastics, a fact attributed to the greater depth of color imparted with the dyes. Surface color may actually fade but the more deeply placed undamaged dye will show through and obscure the faded dye.

Dyes have much lower thermal stability than pigments, particularly inorganic pigments. Thus they are much more sensitive to molding temperatures and use conditions, and may easily become discolored under the higher heats. With some of the newer heat resistant plastics requiring higher molding temperatures and longer mold cycles for the larger parts, the problem of thermal stability of organic dyes and pigments becomes most significant. With molding temperatures of 400-600°F not uncommon, choosing a colorant becomes limited to few organic pigments and a great number of inorganic pigments, which can withstand high heat without yellowing or browning to some extent.

Dyes are also subject to color migration or bleeding, a condition that not only lessens the beauty of a color but requires selection of government certified colors for toys, cosmetic containers, nursing bottles, plastics film for food packaging, and similar consumer items. In the same way, color migration may be a serious problem in costume jewelry, buttons, or tableware, for example, where the dye can stain the skin, a dress, or table linen, or cause an allergic reaction.

Pigmented plastics are opaque and light sensitive, fading on the surface and obscuring the unchanged pigment beneath. Coloring with pigments is more difficult, particularly with organic pigment. Incorporating pigment into the various resins often requires special technology and equipment.

A pigment must first of all meet given end-use requirements, such as lightfastness, transparency or opacity, brilliance, or color, down to the specific shade. It must also disperse well. The dispersion of a pigment is a process by which pigment particles are "wetted" down by the resin in the liquid or molten stage. How well a pigment disperses depends on the temperature at which the two materials are mixed, the particle size of the pigment and the molecular weight of the polymer. Mixing time and equipment are very important.

Improper dispersion will affect color shade since color strength is lost. Often the undispersed particles will detrimentally affect the physical properties of the plastic. It is possible to compensate for loss of color strength by using more pigment but this substantially increases cost.

Moreover, a large number of undispersed pigment agglomerates can block extruder screens and eventually cause them to burst under the pressure. In these circumstances, the surfaces of extruded material may lose smoothness and in films these particles will also cause pin holes. Dispersion is the key to successful coloring.

Pigments can only broadly be classified in terms of dispersibility. Large particle size materials such as titanium whites are easiest to handle. Generally inorganic pigments have fairly large particle sizes but only a few are as easy to work with as titanium pigments.

By contrast, organic pigments are usually the most difficult to disperse. Many can be dry blended with resins and extruded or injection molded on conventional equipment but preparation prior to the dry blending often requires skill. The blending sequence as well as the expert use of dispersing aids can be important.

Some organic pigments are so difficult to disperse that only the most efficient equipment can be effective. Organic pigments are very light and fluffy and carry electrostatic charges, all of which makes dry blending very difficult.

Smaller particle size pigments carry greater coloring strength over the entire powder, but their specific gravity creates problems in metering or weighing out. Too, these colorants also require proper handling facilities and procedures to ensure color uniformity and prevent contamination of other plant equipment and material.

One of the recent developments has been the color concentrates which are furnished with the natural plastics molding compounds, enabling the molder to color his plastics by mixing some of the color with the molding powder in a blender, or by adding it directly in the hopper of an extruder press. The use of color concentrates provides better color control and greatly facilitates handling. Color concentrates are particularly useful in wire coverings, where a large variety in color is desirable, especially for identification in coaxial cables.

How can colorants be used most economically? Should preblended or predispersed pigments be used, or dry powder purchased directly from pigment manufacturers and blended and dispersed in plant?

The choice is obvious when large volumes of single pigment or single color are involved, particularly if the pigment is inorganic. There are, unfortunately, only a few areas where this is the case. The resin supplier himself, however, is often willing to supply precolored resin at only nominal extra cost if volume is large.

Most color manufacturers supply a variety of colors. But since only very seldom will a single pigment produce a desired shade, blending of several pigments is necessary. An experienced color matcher must be on hand and accurate weighing of colorants the rule to ensure batch to batch uniformity of shade.

For frequent short runs, clean-up, and the increased frequency of risks, and an inventory of many different colors cannot be ignored when calculating costs. Even infrequent mistakes can be very costly. Mistakes can occur as a result of switching production from one machine to another, variations in equipment size, accuracy of temperature-measuring instruments, and differences in speed, dwell time, and the size of a part. These can produce different shades of color even for the same resin and pigment formulation.

Only very large operations can afford a colorant specialist, fairly large capital investment in special or modified equipment, and a practical colorant inventory. Smaller or even medium size operations find it very often more economical to consult specialists in color blending, color compounding, or master batching. Color compounders maintain close contact with suppliers and fairly extensive information files on all pertinent materials. The larger compounders also maintain well-equipped testing laboratories.

Heat Stabilizers

Stabilizers are used to prevent the degradation of resins during processing, when melts are subjected to high temperatures, or used to extend the life of end products of which they become a part. Polyvinyl chloride is particularly vulnerable to degradation during processing, and therefore a prime consumer of heat stabilizers, although other resins requiring stabilization are chlorinated polyethylene and blends of ABS and polyvinyl chloride.

In polyvinyl chloride, stabilizers are chosen for their ability to prevent changes in color during processing. Lead, organotin, cadmium/barium, and other systems are chosen for their specific effects on end products. Generally, liquid stabilizers are used for flexible polyvinyl chloride.

Since rigid polyvinyl chloride is processed at higher temperatures, the stabilizer choice is much more critical. Commercial systems are based on solid barium/cadmium, lead, and the organotins. High-cadmium systems contain secondary stabilizers and require the addition of an organic phosphite. Lead stabilizers and dibutyl tin salts are used when toxicity is not a factor.

Barium/cadmium/zinc liquids are the work horses in the vinyl industry, particularly in the area of packaging and handling. Over the past two years, the trend has been toward the bulk handling of plastic additives, both liquids and solids.

The first widespread move toward bulk handling has involved use of the tote bin, a packaging system regarded as an important first step toward the bulk handling of materials ranging anywhere from straight stabilizers to multicomponent blends containing a stabilizer, epoxy plasticizer, and UV absorbers. Additive mixtures are particularly suitable for operations involving large quantities and constant ratios, as, for example, in vinyl flooring compositions and plastisol foam constructions.

The tote bin is just a small step away from the handling of these materials in bulk. Liquid barium/cadmium/zinc stabilizers, as is well recognized, are antioxidants. Many of them are also subject to hydrolysis. When blended with epoxy plasticizer, however, they can be handled by the same kind of storage tanks, piping, and other equipment used for straight epoxy plasticizer. But when it comes to the bulk handling of straight liquid stabilizers, special precautions are important, such as the use of stainless steel tanks, pipes, pumps, and other handling devices, and the provision of an inert atmosphere, since liquid barium/cadmium/zinc stabilizers, as antioxidants essentially are prone to degradation through the action of oxygen and moisture.

Antioxidants

Antioxidants are used to protect materials from deterioration through oxidation brought on by heat, light, or chemically induced mechanisms. Deterioration is evidenced by embrittlement, melt flow instability, loss of tensile properties, and discoloration.

The three main preventive mechanisms to control deterioration of polymers operate by absorbing or screening ultraviolet light, deactivating metal ions, and decomposing hydroperoxides to nonradical products. The mechanisms can very often be compounded when two or more antioxidants are used in one system. For example, since many chain terminating antioxidants are also metal decomposers and some peroxide decomposers the radical terminators, the same result can be achieved in more than one way.

Although antioxidants are not ultraviolet deactivators, the decomposition byproducts of antioxidants are effective hydrogen conjugators, that is, capable of hydrogen bonding—a reaction that permits degrading the electromagnetic energy of incoming light and redistributing it as thermal dissipation in the polymer without the formation of free radicals. In this aspect, it compares with the mechanism of ultraviolet absorbers such as the benzotriazoles and benzophenones.

To counteract the oxidizing effect of metal

ions, through the formation of unstable complexes with hydroperoxides, followed by an electron transfer and the emission of free radicals, antioxidants remove the ability of metal ions to complex with free radicals, thereby effectively nullifying their detrimental reaction.

Materials subjected to heat, as they are during processing, undergo chain scission and crosslinking and the formation of hydroperoxides and free radicals, the rates of each being dependent on the polymer and processing conditions. Antioxidants deactivate these sites by decomposing the hydroperoxide or terminating the free radical by hydrogen abstraction and radical conjugation. The free radicals remaining within the antioxidant do not react toward other hydroperoxides or free radicals to continue the chain reaction.

Peroxide decomposing antioxidants are normally used as secondary additives. They differ from primary phenols or amines in that they undergo decomposition when reacted with hydroperoxide rather than contain the hydroperoxide. Secondary antioxidants do not form free radicals when they induce peroxide decomposition. Thus the products of decomposition differ from those produced by free radical trapping.

Synergism, or the cooperative effect of two antioxidants which may react in the same or alternative methods, often produces more effective results than one system at the same concentration. One theory has it that the synergist regenerates the primary antioxidant; the other holds that the secondary antioxidant is destroyed. Both are accepted.

The total effect of synergism in polymeric materials, particularly in polyethylene and polypropylene, can be such as to produce improved antioxidant efficiencies to as much as 200%.

Ultraviolet Light Absorbers

Virtually every plastic degrades in sunlight in a number of ways, the most common being discoloration and the loss of physical properties. Particularly susceptible to this type degradation are the polyolefins, polystyrene, PVC, ABS, polyesters, and polyurethanes. The job of stabilizing color and lengthening the life of a product falls to ultraviolet light absorbers.

Black, in any form, whether carbon black, black paint, or black dye, is the most effective UV absorber, but cannot be universally used. Hence a variety of chemicals are used instead.

Benzophenones are used in clear polyolefin systems. The benzophenones are in fact good general-purpose UV absorbers and can also be used in pigmented systems, although they normally defer to a nickel complex of an alkylated hindered phenol. The nickel complex functions as a UV absorber, an antioxidant, and, in polypropylene, as a dye acceptor. The benzotriazoles are used almost exclusively for polystyrene.

Both the benzotriazoles and the benzophenones are used as UV absorbers in the polyesters. Formulations of both the benzotriazoles and the benzophenones are varied.

The choice of UV absorber depends on the particular application, the basic resin, the ultimate effect on color, and the required durability of the end product, since UV absorbers are used up in time. Only a small number have Government approval for applications involving food and drugs. Concentrations in any formulation run in the order of 0.25-1%.

Antistatic Agents

Antistatic agents, sometimes called destaticizers, are used to reduce the buildup of electrostatic charges on the surface of plastics materials by increasing surface conductivity.

Plastics particularly susceptible to the accumulation of electrostatic charge are polyethylene, polypropylene, polystyrene, nylon, polyesters, urethanes, cellulosics, acrylics and acrylonitriles.

The most common antistatic agents for plastics are amines, quaternary ammonium compounds, phosphate esters, and polyethylene glycol esters.

Antistatic agents are selected on the basis of application, durability, necessary concentration, FDA approval, where applicable, and effectiveness at low humidity.

Flame Retardants

Flame retardants are used to affect combustion in plastics. There are many flame retardants and their choice depends primarily on the resin to which it is being added. For example, the flame retardant might be added to keep temperatures below a given combustion level, or smother a reaction between the material and oxygen or other combustion-aiding gases, or finally, work on combustion through various types of vaporization.

Flame retardants therefore work on four basic principles: either they insulate, create an endothermic cooling reaction, coat the product thereby excluding oxygen, or actually influence combustion through a reaction with materials that have different physicals.

The most common flame retardants are antimony, boron, halogens, nitrogen, and phosphorus. The most common in PVC are phosphate esters.

Flame retardants can be organics like phosphate esters, inorganics like antimony oxide and zinc borate and dispersions, and reactives like polyols with phosphorus. Reactive flame retardants are mainly used in epoxies, polyesters and urethanes. Organic and inorganic flame retardants can be used in a variety of plastics.

Blowing Agents

A blowing or foaming agent is used alone or in combination with other substances to produce a cellular structure in a plastic mass. The term covers a wide variety of products and techniques, but in compounding we are limiting ourselves to chemical agents, which decompose or react under the influence of heat to form a gas. Chemical blowing agents range from simple salts such as ammonium or sodium bicarbonate to complex nitrogen-releasing agents.

The basic requirements for an ideal chemical blowing agent are:

- Gas-release temperature must be within a narrow range.
- Release rate must be controllable but rapid.
- Gas must not be corrosive.
- Compound must be stable in storage.
- Residue should be colorless, nonstaining, and free of unpleasant odors.
- Compound and residue must be nontoxic.
- Residue must be compatible with the plastic to be foamed and have no effect on its properties.

Nitrogen-releasing compounds dominate the field of chemical blowing agents. Some of the products commercially available are the azo compounds, such as azodicarbonamide (azobisformamide), the N-nitroso compounds, and sulfonyl hydrazides.

Azodicarbonamide has been for years the principal choice of the chemical blowing agent family.

Since the temperature range in which azodicarbonamide liberates coincides with the temperature at which a number of polymers have a melt viscosity suitable for foaming, it has been used extensively in vinyl plastisols and calendered vinyl, as well as during the early stages of extrusion and injection molding of structural foams based on polyethylene, polypropylene, vinyl, styrene, and ABS.

New blowing agents and new blowing agent systems are being developed both to meet the challenge of foaming engineering polymer systems or to solve such problems as plateout, screw, and barrel corrosion, corrosion of the mold, and uneven cell structure.

With emphasis on new engineering thermoplastics for high-temperature applications, the requirements for blowing agents with high decomposition temperature are increasing. With a decomposition temperature between 500 and 550°F, and a gas yield of approximately 175 ml/g at 536°F, the THT type of blowing agent is being commercially used to expand nylons, thermoplastic polyesters, and polycarbonates.

The blowing agent or system used in foamable glassfiber reinforced thermoplastics is generally a proprietory mixture.

Lubricants

Lubricants are used to enhance resin processibility and the appearance of end products. Effective lubricants are compatible with the resins in which they are used, do not adversely affect the properties of end products, are easily combined, and have FDA approval, where applicable.

Lubricants fall into five categories: metallic stearates, fatty acid amides and esters, fatty acids, and hydrocarbon waxes, and low molecular weight polyethylenes.

Clear grades of polystyrene require a clear-melt grade of zinc stearate, although bis-stearamides are generally adequate. ABS systems require metallic stearates in combination with glycerol monostrearate; styrene acrylonitrile requires only fatty acid amides. Polyvinyl chloride needs large concentrations of calcium stearate and hydrocarbon waxes, one serving as an internal lubricant and the other as an external lubricant. Low molecular weight polyethylenes are reported to be one of the most efficient external lubricants available. Most systems incorporate polymeric processing aids for faster fusion and higher gloss.

In polyolefins, lubricants tie up catalyst residues, generally calcium stearate. Stearates and ethylene bis-stearamide waxes are sometimes used in the processing of fine powdered polyolefins.

For phenolics and melamines, the choices are ethylene bis-stearamide, calcium, and zinc stearate, the last particularly for molding grades. Zinc stearate is also widely used as a lubricant or release agent for reinforced polyester compounds, although calcium stearate is used in certain applications.

Underlubrication will cause degradation and frequently higher melt viscosities, but overlubrication can cause too much slippage and lower outputs. An imbalance of lubricant and stabilizer can cause plateout or the migration of pigment from a melt system.

Methods of Compounding

Strictly speaking, compounding involves the fusion of different materials into a homogeneous mass, uniform in composition and structure. The actions in this type of mixing consist of smearing, folding, stretching, wiping, compressing, and shearing. However, component materials may be dry-blended either in preparation for compounding or for direct product fabrication.

In compounding, the order of addition of ingredients is most important in order to maintain as liquid a mix as possible at the beginning, progressing stepwise to a stiff plastic mixture. The completeness of the mixing is limited by the toughest and most viscous ingredient, which supplies a supporting structure and folds about the other ingredients as the mixing proceeds. Thus the more mobile materials are locked up inside a tough viscous doughy envelope.

In other instances, extensive exposure of the mixing surfaces produces a drying or hardening effect which causes striations. The use of heat either internally generated by shear or externally applied is needed to obtain compound uniformity. Since it is always necessary to minimize heat history during compounding, it is often necessary to supplement heat transfer through machine elements with a cooling system. This is particularly true for thermosetting compounds, so as to avoid overpolymerization of the resin through excessive heating.

With this in mind, let us consider the major types of plastics compounding equipment in current use—batch and continuous internal mixers, two-roll mills, and extruders.

Intensive Dry Mixers

Intensive dry mixers are used for dry-blending powdered resins, such as PVC, with plasticizers and other additives. A typical mixer consists essentially of a high-speed propeller-like impeller located at the bottom of a container (Fig. 29-4). Heat generated during the blending cycle is continuously removed to stabilize the dry blend and improve flow characteristics. In one design, the blend is cooled by impeller action. Heat is transferred from the constantly moving fluidized bed of material as it makes continuous and intimate contact with a cooling jacket wall. Other designs make use of cooling plate coils.

Impellers operate at the relatively high speed of 80-160 rpm. Moisture volatilized during the blending cycle is sucked into a vortex formed in

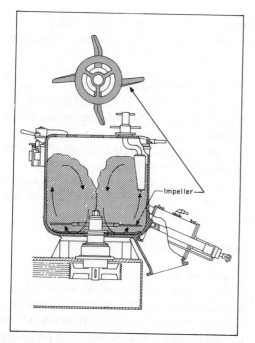

Fig. 29-4. High-speed impeller at the bottom of container housing is used to dry-blend powdered resins, such as PVC, with plasticizers and other additives. Heat and volatiles are removed during blending.

the center of the mixer and vented from there. High-speed chopper blades operating at about 3600 rpm can be used to break up agglomerates and lumps and thereby assure the rapid incorporation of fluids.

Large-capacity systems generally combine high-intensity mixing units with low-intensity cooling units. The cooling units usually exceed the mixing units in capacity to maximize operating efficiency. Mixing cycles are short, averaging 5-10 min.

Internal Intensive Batch Mixer

Internal batch mixers have been widely used in the production of vinyl film and sheeting since the introduction of these products. In fact, every large manufacturer of vinyl film and sheet uses internal mixers for dispersion and fluxing. High-speed operation, which is used extensively, results in overall cycles of 2-4 min. Internal mixers are also used for processing plastics such as vinyl, polyolefins, ABS, and styrene, along with thermosets such as mela-

mines and ureas because they can hold materials at a constant temperature.

Essentially, internal mixers consist of cylindrical chambers or shells within which materials to be mixed are deformed by rotating blades or rotors (Fig. 29-5). In most cases a shell actually consists of two adjacent cylindrical shells with a rotor describing an arc concentric with each shell section. Because the rotor-blade tips clear the cylindrical segment of the shell by a small amount, the blade motion causes the mixture to be sheared between blade tip and shell. The blades also interact between themselves so as to cause folding or "shuffling" of the mass. The shuffling is further accentuated by the helical arrangement of the blade along the axis of the rotor, thereby imparting motion to the mass in the third, or axial, direction. Frequently, the blade is divided into two helices of opposite direction of pitch in order to further the shuffling of components within the mixture. In some cases, the rotor blades are Z-shaped arms

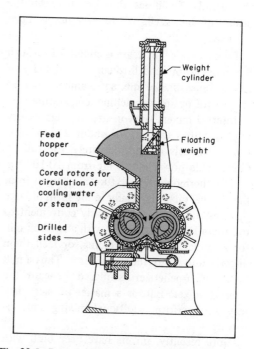

Fig. 29-5. Rotors within cylindrical chambers deform material in the tight clearances between rotor blade tips and chamber walls. Rotated at fairly high speeds, the blades provide kneading, shearing, smearing, pulling, and compressing action. Cored rotors provide channels for cooling water or steam. The batch is held confined by a ram in the feeding neck.

instead of elliptical. Still further variations are haben and fishtail blades. The shell and rotor are cored or otherwise provided with means for heating or cooling for the purpose of controlling the batch temperature.

The principle of internal batch mixing was first introduced in 1916 with the development of the Banbury mixer. The Banbury mixer consists of a completely enclosed mixing chamber in which two spiral-shaped rotors operate, a hopper to receive the material for mixing, and a door for discharging the batch. Revolving in opposite directions and at slightly different speeds , the two rotors keep the stock in constant circulation. The ridge between the two cylindrical chamber sections helps force mixing; and the acute convergence of the rotors and the chamber walls imparts shearing. This combination of intensive working produces a highly homogeneous mix. The batch is confined within the sphere of mixing action by an air-operated ram in the feeding neck. The ram also greatly facilitates the flow between the blade tip and shell, where shearing is most extensive.

Cooling water or steam is circulated through the cored rotors and through a multiplicity of drilled holes in the mixing chamber liner to provide the optimum machine temperature.

Internal mixers range in size from laboratory models with a capacity of about 2 lb of mixture to models of approximately 100-150 lb. Comparable power ranges are from 10-100 hp.

The thermoplastic mixture discharged from the internal mixer is usually in the form of large shapeless lumps; it is frequently convenient to roll these lumps into sheets with a two-roll mill for eventual grinding or dicing, or feed them directly into an extruder-pelletizer. Thus a mill or extruder-pelletizer is placed beneath the internal mixer. But, as a matter of fact, they may also perform additional mixing and dispersing.

More recently, mixers have been built with the capability of dewatering and devolatizing of a variety of plastics. Slurries containing more than 70% moisture can be dewatered in 3-4 min. Most of the liquid is drained or pumped out of the machine; the remainder is vaporized by heat produced by mechanical action of the

stock and contact with the mixer's heated working elements. Steam escapes through an exhaust system.

Color and formula changes can be made easily in internal mixers because mixing chambers are self-cleaning. Thus a variety of stocks of different color or formulas can be handled in succession. Many internal mixers in operation today are used for repeated changes of color in the same machine—from white to green, to gray, to red, to brown, to black, all in the course of one day. The same adaptability applies to formula changes.

Continuous Mixer

A continuous mixer is one whose rotors and mixing action are similar to those of the Banbury. Raw material is automatically fed from feed hoppers into the first section of the rotor which acts as a screw conveyor, propelling the material to the mixing section where it undergoes intensive shear between rotor and chamber wall, kneading between the rotors, and a rolling action (Fig. 29-6). Interchange of material between the two bores of the mixing section is an inherent feature of rotor design. The amount and quality of mixing can be

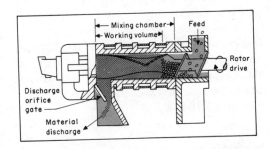

Fig. 29-6. The working volume is that portion of the mixer volume between the point where fluxing occurs and the discharge end. The working volume of the material in the mixer body is affected by production rate and other operating conditions. Raw materials coming into the mixing chamber through the feed screws are converted to a homogeneous mass somewhere between the end of the last feed screw flight and the rotor apex (change in direction of pitch). Intensive shear occurs between rotor and chamber wall. Like the Banbury batch mixer, material is interchanged between the two bores of the mixing section.

controlled accurately by adjustment of speed, feed rates, and orifice opening. Since the feed screw is constantly starved and the mixing action is rotary, there is little thrust or extruder action involved. Stock discharge temperature is controlled by the rotor speed and the discharge orifice; production rates are controlled primarily by feed rates.

The continuous mixer can maintain optimum dispersion or mixing over a wide range of production rates. On most materials, a production rate range of 5 to 1 or even 10 to 1 is possible. For example, a mixer designed to produce over 10,000 lb/hr PVC sheeting compound has proved capable of handling the same material, with equivalent quality, at rates of 1000 or 2000 lb/hr. Such operation is desirable in a calendering line where gage changes in the sheet require different production rates.

In practically all mixing, discharge temperature is critical, either because it is a limiting factor, a measure of the amount of work done, or a goal. To maintain versatility, it is desirable for discharge temperature to be independent of production rates. The temperature of a continuous mixer can be adjusted within a range of $100°F$ at any given production rate and then maintained at the set temperature throughout the production run.

Generally, the mixer is fed—either with a preblend or with two, three, four, or more ingredients—on a continuous basis. But when the form of the materials or the proportions preclude accurate continuous-flow weighing and feeding, one or all of the ingredients can be fed in batch weighings. When these are fed at regular time intervals, there is no loss in mixing or dispersion quality and a continuous discharge is maintained. Sometimes minute quantities of an additive may require preblending with the primary ingredient before being fed to the mixer. Sometimes a masterbatch may have to be prepared.

The continuous mixer is not completely self-purging, but one stock can be followed by another, without cleaning, with contamination of only a small amount of material. Still, before following with a different compound, it is generally desirable to clean the orifice unit, which is removable as a complete assembly.

Two-Roll Mills and Pelletizers

As stated earlier, after being processed in a mixer, material may be dropped directly onto a mill for sheeting and dicing or into an extruder or pelletizer. Arrangements vary. Where a mixer is located directly above the sheeting mill, the discharged stock drops into the bite of the rolls and can be taken off in a continuous strip. In other installations, where the sheeting mill is at a distance from the mixer, conveyors of various types are used to carry the mixed batch to the mill. High-production mixers require mechanical means of handling the large quantities of materials passing through them, both before and after mixing.

The two-roll mill consists of two opposite-rotating parallel rollers placed close to one another with the roll axes lying in a horizontal plane, so that a relatively small space or nip between the cylindrical surfaces exists. Material reaching the nip is deformed by friction forces between itself and the rollers and made to flow through the nip in the direction of roll motion (Fig. 29-7). Liquid plasticizers or finely divided solid ingredients are also placed in the nip and are incorporated into the resin through shearing action. Strip cutters are used to cut ribbons of stock from the roll for feeding to a cooling tank and dicer.

Usually, by adjusting roll temperature, the compound can be made to adhere to one of the rolls as a relatively thin sheet. The rolls are heated or cooled by a heating or cooling media introduced into their hollow cores. They are usually rotated at different speeds to facilitate the formation of a sheet or band on one of them. During mixing, the band is frequently cut and manually pulled loose from the roller. The gap between the rollers generally may be adjusted manually by means of hand-driven or motor-driven screws.

Roll mills vary greatly in size from very small laboratory machines with rollers of about 1 in. in diameter and driven by fractional-horsepower motors to very large mills with rollers of nearly 3 ft in diameter and 7 or 8 ft in length and driven by motors over 100 hp.

Special-purpose roll mills are available with three, four, or five rolls, in which the material is

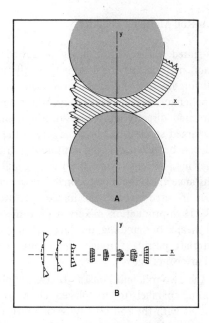

Fig. 29-7. A. Rolls act on thin, flowing wedge of material simultaneously compressed and forced to flow through the roll nip. Rolls rotate in opposite directions. Material leaving the nip adheres to either roll if rolls rotate at equal speeds. B. Velocity profiles of flows are symmetrical about the X-axis within the nip area for equal roll speeds.

caused to pass from one nip to the next, in succession. Likewise, rolls may be arranged in pairs within a single frame, each independently adjustable in clearance. Material is led through the various pairs in tandem in a cascade arrangement for continuous mixing.

The underwater pelletizer adapts to a variety of materials, such as polyethylene, linear polyethylene, polystyrene, polypropylene, color masterbatch, and plastic blends. The pellets can be easily handled, accurately weighed and conveniently stored. In operation, the mixer discharges directly into the pelletizer, where the material, after being worked by the extruder screw, is forced through a straining screen and then pelletized in the head. Water circulated through the head separates the pellets and conveys them to the dryer. For light materials, the pellets are carried upward by flowing water to a dewatering system. For heavier materials, spray nozzles are used to flush the pellets to the bottom of the chamber and out through a side discharge port.

Single-Screw Extruders

Although developed to form thermoplastics, the single-screw extruder also functions as a mixing device because it subjects materials to laminar-flow deformation. Thus, although the extruder is not primarily used as a mixer, it is frequently used to add ingredients to a resin during melt extrusion and thereby to take advantage of their inherent mixing action. It is quite common for colorants to be added to extruded products in this way. Both the colorant and resin are metered into a feed hopper and a suitably colored extrudate frequently obtained.

Often, reclaimed scrap stocks of different colors are mixed satisfactorily in the extruder provided that the materials are suitably metered in the feed hopper. Again, a masterbatch or premixed concentrate of colored thermoplastics may be fed to the extruder simultaneously with the resin to be colored. Here the extruder becomes an auxiliary mixer.

The amount of mixing a given volume of resin receives may be expressed in terms of the total amount of shear to which the resin is exposed. It is the product of shear rate and the extruder residence time. It is a measure of the relative displacement of one particle with respect to its neighbor.

The total amount of shear in an operation can be increased in one of several ways: by diminishing the channel depth of the screw, by letting the screw helix angle approach 0 or 90 deg, or by increasing the amount of pressure flow and leakage flow through increased die restriction (Fig. 29-8).

The total amount of shear to which a particular element of liquid resin is exposed depends upon its initial position in the screw channel, since flow path, and thus both residence time and shear rate vary according to the initial position of the material in the cross section of the screw channel. Material near the center of the channel receives less mixing than material near the screw surface or the barrel wall. This is because of the shorter residence time of the material near the center of the channel. It is also important to note that no motion of material backward along the screw

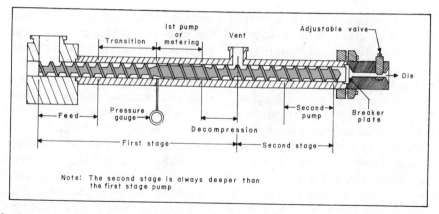

Fig. 29-8. Two stages of a vented extruder. Shear can be varied by varying screw channel depth, screw helix angle, or the amount of pressure flow and leakage flow through die restriction. The second-stage pump is always deeper than the first-stage pump.

axis ever takes place. Hence, single-screw extruders provide no "turnover" and little bulk mixing in the axial direction.

Many types of special mixing heads or torpedoes have been designed as attachments to the end of extruder screws. All have the primary purpose of further increasing the amount of shear to which the extrudate is exposed and of crossblending the somewhat nonuniformly mixed material leaving the screw channel.

There are different kinds of conventional single-screw extruders for different operations. In the area of compounding, they can be conveniently grouped under the following headings: mixer-fed extruders, hot-melt extruders, and cold-feed extruders. Available head arrangements for each type include strip, strand, slug, cone, pelletizing, and underwater pelletizing.

Mixer-fed extruders are usually located directly under a batch or continuous mixer and fed by gravity.

Operation is automatic; a variable speed drive is used to synchronize the extruder speed with the production rate of the mixer in order to keep the screw always full of stock and the discharge uniform and unbroken.

Extruder cylinders are generally jacketed for steam or liquid heating, but they may be electrically heated, as required. Extruder size is dependent on the production rate of the continuous mixer or the size and mixing cycle duration of the batch mixer. Drive require-

ments are determined to a great extent by the compound to be handled.

Hot-melt extruders take the resin from materials suppliers' reactors or receivers and deliver it in flat-strip or pelletized form. A great many hot-melt extruders have been built in recent years, most of them for handling polymers within the range of 0.2-50 melt index.

Extruder size is determined by the hourly production rate desired. Horsepower requirements, throughput rate, and screw design are dependent on material characteristics.

Cold-feed extruders are available for working high-density and low-density polyethylene, plasticized PVC, and other compounds. The machines are generally cube or pellet fed and deliver extrudate to pelletizing heads. Most are completely jacketed construction, designed for use with pressurized water circulating systems. They are started up with circulating hot water and then, as material flow is established, water temperature is gradually reduced to obtain the best operating conditions for the resin being handled.

Compounder-Extruder

One modification of the single screw is a single-step "enforced order" variable-intensity compounder and extruder operating over a stepless gradient from extensive to intensive mixing, with a running cutoff on the intensive mixing end at any desired shear rate between 30 and 3000 reciprocal seconds.

The unusual design incorporates a "rotor," or single rotating screw, inside a series of "stators," or stationary screws, which replace the smooth barrel of a conventional extruder. The stator and rotor grooves are opposite handed and alternatively increase or decrease in depth. The grooves in both rotor and stator are shaped to help the material vortexing.

The cross-sectional diagram of the rotor and stator shows six extensive mixing stages, a vent section for removing volatiles and entrapped air, and a wide-landed, conical, high-intensity, variable-shear section (Fig. 29-9). The material starts to *vortex* in the grooves of the rotor as a solid as soon as it reaches the barrel. Since it is immediately transferred to the stators where it makes contact with the heat transfer walls of the stator grooves, it starts to melt and immediately is mixed with the bulk of the solid material, melting further resin.

This transfer of material from rotor to stator and vice versa, can be broken down into different actions, namely, vortexing, in which the vortex of material in the giver screw diminishes and increases, because in the taker screw the outside layer of the vortex of the giver screw *unpeels* and then becomes the outside of the vortex in the taker screw. In this way, all of the material comes in contact with the heat transfer walls of both rotor and stator, once per stage. (A stage is defined as one transfer from rotor to stator, and back again.)

Besides the vortexing in the grooves, the material in the rotor moves forward, more or less in a straight line or gentle helix. But as material is transferred into the stator, it must also follow the exact helical path of the stator. Thus there is a *change in direction or flow* with every pass from rotor to stator and back.

It is this change of direction of flow that is responsible for the pumping or extruding capability of the machine regardless of the nature of the material. If the material tends to turn with the rotor, it is nevertheless pumped forward by grooves of the stator. If it does not turn with the rotor, it is propelled forward like a nut on a bolt.

Twin-Screw Extruder

In twin-screw extruders, two screws are arranged side by side. One design incorporates co-rotating screws that are intermeshing and self-wiping. Since the screws rotate in the same direction, material moves helically along the inside barrel wall in a figure-8 path from the feed section to the discharge point (Fig. 29-10).

The geometry of the screw components is such that the root of one screw is constantly wiped by the flight tip of the second screw with a uniformly small clearance between them at every point. Thus dead spots are eliminated, the residence time of each melt particle is uniform, and purging times are shortened. Because of the positive conveyance of material, all types of material—molten polymers, pastes, flakes, pow-

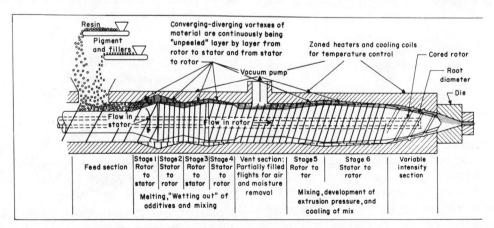

Fig. 29-9. Material taken up at the inlet by rotor helix is peeled off by stator helix and alternately by rotor helix, once each stage. Vortexing action of rotor and stator transfers material to and from rotor and stator heat transfer surfaces. Pumping is achieved by enforced-order interaction of material in opposite-handed screws.

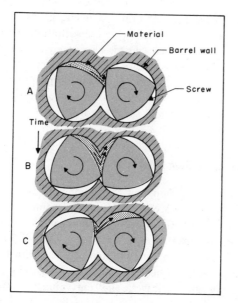

Fig. 29-10. Twin-screw geometry of a co-rotating system. A. Screw transports material to a point of intermeshing. B. Majority of material is in process of being transferred to second screw. Very little material passes between screws. C. Transfer is complete. Material follows a figure-8 path around the screws as it moves down the length of the barrel.

ders of low bulk density, and fibrous materials—are efficiently conveyed irrespective of friction coefficients. With positive conveying, operations with partially filled screw flights is possible while degasing is going on in some zones or additional components are introduced into others.

Screw configurations are used to vary conveying efficiency, throughput rate, the degree of filling, and pressure buildup. Screws with reversed flights are used to generate localized high pressure. Hydraulically operated dynamic valves between barrel sections can be used to allow pressure variations during operation. Staggered stepped screws provide intensive transversal mixing by shear forces of varying intensity. Various mixing and kneading effects are obtained by suitable screw design.

Residence time is determined by barrel length, screw lead, screw speed, and throughput rate. Residence time may vary between 20 sec and 10 min, depending on the process and operating conditions. Residence time distribution may be influenced by changing the

screw geometry, by which residence time distribution can be widened to handle considerable longitudinal mixing.

Selective removal of heat from the melt by cooled screws and barrel walls results in increased melt viscosity and, consequently, in higher shear rates. The need for external heating can be minimized by appropriate screw geometry. Depending on product and process the specific input energy may be varied between 0.05 and 1.2 kwh/kg.

Material in the barrel is worked into thin layers by relatively narrow screw flights. The design increases processing intensity, particularly at the barrel crest, where the material transfers from one screw to the other and the layers are continuously mixed and inverted.

Volatiles are removed through vent ports at different locations along the barrel length. Fluids or melts of high or low viscosity or free-flowing solids may be introduced into the barrel at various points by means of suitable metering devices.

Twin-screw extruders with counter-rotating screws are also available and, despite similarities, there are some significant differences in the way they handle the melt.

With counter-rotating screws, screw flights carry material by friction in such direction that all of it is forced toward the point where the two screws meet, there forming a "bank" of material similar to that of a two-roll mill, although some material slips through the gap between the two screws. Like the two-roll mill, the theory is to feed material through the nip from the bank on top of the nip gradually and statistically, in order that each particle is processed equally over a period of time.

Counter-rotating twin screws can put the material passing through the nip under an extremely high degree of shear, in the manner of a two-roll mill. Varying the clearance between the screws varies the portion of material fed between the screws against the portion of material accumulated in the bank and simply moving down the barrel. Material passing through the nip can be subjected to great shear forces by narrowing the clearance, and also the portion of material in the bank is greater.

Statistically, however, only a portion of

the material is subjected to the high shear condition of the nip, and a portion only to the low shear condition of the bank. Counter-rotating twin screws are not totally self-cleaning.

In co-rotating twin screws, one screw transports the material around up to the point of intermeshing, where, because of the existence of two opposing and equal velocity gradients, a great majority of the material is transferred from one screw to the other along the entire barrel length in the figure-8 path mentioned earlier. Because the figure-8 path is relatively long, the chances of controlling melt temperature are much better.

The amount of shear energy developed at the point of deflection can be regulated within very wide limits by choosing the depths of the screw flights.

Self-cleaning screws, aside from the obvious advantage of facilitating color change, provide control over residence time distribution. Control of residence time is also of great importance for heat-sensitive resins and pigments, or for operating at higher processing temperatures. Short and uniform residence time is essential to minimize heat and shear history, and, thereby, maximum quality.

Compounding Lines

By now it is apparent that a compounding line involves the careful selection, arrangement, and location of the foregoing equipment, if it is to achieve maximum effectiveness (Fig. 29-11). Ideally, a compounding line will deal with the questions of shear, temperature sensitivity of resins and additives, and the volume of the operation.

Compounding operations vary from

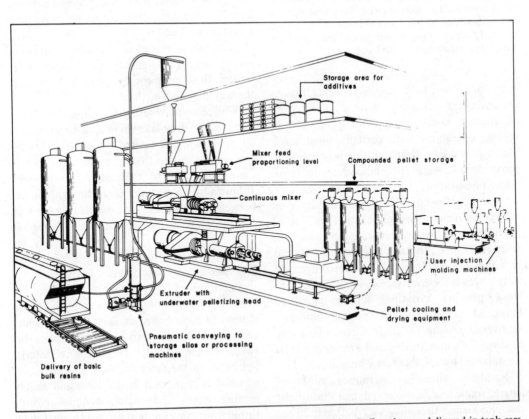

Fig. 29-11. Typical in-plant compounding system for an injection molder. Bulk resins are delivered in tank cars and pneumatically conveyed to silos for storage or directly to machines for processing. Resins and additives are proportioned and fed to continuous mixer. Output of mixer goes to extruder and underwater pelletizer and then to pellet storage silos for subsequent use in injection molding machine or fabricating extruder.

500-10,000 lb/hr in throughput capacity, whereas the capacities of in-line compounding systems rarely exceed 1000 lb/hr. Generally, the compounder preblends the material in drum tumbling systems, ribbon mixers, paddle mixers, or double-cone twin-shell blenders, but he may also use devices set up on his extruders. The throughput rate of such systems is generally only 50-500 lb/hr.

Most PVC compounds can be processed from a dry blend, that is, one already containing plasticizer, and fed into cable coating extruders, provided they are large, with throughput rates in excess of 250 lb/hr. When coating extruders are small, handling all kinds of thin wire, separate compounding systems are preferred.

Single-screw compounding extruders are adequate in most applications, particularly when heat history is less critical. With advances in screw design and in new devices that localize and control the introduction of shear, they will do the job probably 70% of the time. For medium shear operation, the transfer mix represents a relatively new approach.

Twin screw compounding extruders are one answer to the need for very good control of shear and melt temperature, and the removal of relatively large quantities of volatiles. In applications involving the use of heat-sensitive material, as in fluxing PVC, twin-screw compounding extruders may be used to perform different compounding operations, that is, provide different levels of shear, in different sections of the same machine. Twin-screw compounding machines are generally smaller than single-screw extruders, requiring less floor space. For example, the overall length of a 2000 lb/hr twin-screw compounding extruder is around 12 ft, whereas the length of a comparable single-screw extruder would probably exceed 20 ft.

Feeding and preblending equipment have achieved high levels of performance. But commercially, volumetric or gravimetric feeders, while able to satisfy any number of requirements, are usually subject to short-term inaccuracy, being unable to maintain minimal acceptable percent variations for the now common residence times of 20-30 sec. Short-term feed rate fluctuations can be overcome, however, through continuous preblending, by which volumetric or gravimetric feeders feed their output into an intermediate device, from which a continuous preblended stream of material is then fed to a compounding extruder. Preblenders also perform intimate mixing at ambient temperatures, thereby minimizing heat history.

The majority of pelletizing operations still involve the practice of extruding strands of melt into a water bath for cold cutting. For throughput rates of 1500 lb/hr or more, underwater pelletizers tend to be the rule. Underwater cutters must contend, however, with delicate heat balance between the heated die plate and the body of water into which the melt is extruded, especially during startup and shutdown and with higher temperature melting polymers, where the temperature difference between die plate and water exceeds 200°F.

In-line compounding is an alternative to separate compounding systems. The idea, of course, is to obtain the benefit of separate compounding and extrusion while avoiding the operational and inventory complications of a separate compounding operations. Typical examples of in-line compounding operations are to be found in PVC pipe extrusion, foamed polystyrene sheet extrusion, and in the production of siding material from certain plastic alloys. It is generally desirable to combine preblending for multiple in-line compounding lines to a single unit in order to make best use possible of the rather expensive metering and blending equipment. Here the problem is one of designing a distribution and hold system to inventory preblended materials.

A second alternative to separate compounding is on-the-machine compounding, where the plasticating section of an extruder or injection molding machine performs the compounding. On-the-machine blenders are used to meter in correct proportions of ingredients, but are limited in their capability of introducing a controlled amount of shear into a polymer mixture.

Another approach to on-the-machine compounding is that of mounting a continuous mixer on top of a large injection molding machine. Such a step has been used successfully in compounding a thermoplastic and an inorganic filler for the large-scale production of high quality parts.

A compounding plant can be set up on the ground floor of a conventionally built warehouse, provided the ceilings are at least 16-18 ft high. The compounding machine can be fed by gravity from a preblending and mixing room located on a mezzanine. Because of its general height the building can be used to store incoming raw materials in silos and outgoing materials in bags and containers. Only one operator is needed. He can usually supervise two lines. Manpower is needed more in formula preparation and product bagging. The use of continuously fed systems electrically monitored can eliminate operator mistakes. The same holds true downstream. Automated metering and blending, continuous systems, and automated packaging are more economical in the long run.

An overhead rail parallel to the axis of the compounding machine will facilitate screw and die changes. The rail can be equipped with multiple die arrangements to stay abreast of color changes

Water-cooled machines generally operate better in closed-circuit cooling systems. The use of untreated plant water in the cooling channels will eventually lead to a formation of hard-to-remove scales and possibly to plugging.

Control cabinets should be away from the main extruder, protected from heat generated and mechanical vibrations. They should also be purged and kept under positive pressure. To conserve space, they should be designed with front access and placed against walls.

The amount of space required for a compounding operation depends largely on how feed stocks arrive—whether by bulk-shipment or in bags or other containers—and how much of the product mix has to be kept in inventory. A typical product mix of a commercial compounding unit producing 2000 lb/hr of compound will require a floor space of 10,000 sq ft. The figure includes laboratory space, but not offices and off-site facilities.

AUTOMATIC MATERIAL FLOW AND PRE-CONDITIONING*

With increased demands for higher production rates and material consumption, the volume of material moving through a typical processing plant has grown tremendously. Hand loading machines to keep up with this demand is often well beyond both physical and practical limits, thereby creating a need for automation.

It is also a critical economic factor that the processor be able to maximize product quality during every cycle of every machine, so that the amount of scrap being generated is minimized. One condition that plays a role in quality control is the proper pre-conditioning of moisture-sensitive materials before they enter the processing machine.

These two elements—automation in material flow and pre-conditioning—offer many secondary benefits (in addition to saving material, labor, and time) for the processor. Prime among these for those processors who have storage facilities (i.e., outdoor silos) and in-plant coloring capabilities, is the economics inherent in purchasing uncolored materials in bulk quantities at lower prices, in saving storage space, and in reducing inventory problems.

Pre-heating and drying materials automatically at the processing machine can also be more effective than oven drying, which involves hand loading and unloading of trays and may expose material to ambient moisture on its trip from the oven to the machine. Automatically conveying regrinds back into the processing operation as they are generated offers secondary benefits by eliminating the need for hand-emptying grinder bins, extra storage space, and regrind inventory problems.

The choice of materials handling equipment depends on (1) the type of material—pellets, powders, etc., (2) the amount of material needed to keep up with processing, (3) the vertical and horizontal distances over which material is moved, and (4) the special functions required of the equipment to meet processing requirements—meter in colorants, proportion regrinds with virgin, etc.

* By David C. Whitlock, Conair, Inc., Franklin, Pa.

Vacuum Loaders

Vacuum conveyors are by far the most popular units for moving plastic pellets and powders from shipping containers to processing. Small integral motor/pump units, which convey 250 to 1200 lb of pellets or regrinds per hour, require no floor space and operate on standard 115 v. They are fully automatic and load on demand (Fig. 29-12).

Larger, integral motor hopper loaders are available for loading at rates to 2000 lb/hr. These are recommended for carrying pellets over short distances up to 50 ft.

When considerable regrind is generated, virgin-regrind ratio loaders will automatically keep regrind used up, while the processor maintains control over the amount of the regrind mixed with virgin. Such loaders have two feed tubes, one for the virgin material, and one that mounts in the grinder or regrind source (Fig. 29-13).

Fig. 29-12. Low-cost integral motor type hopper loader. (*Illustrations courtesy Conair, Inc.*)

Positive Displacement Vacuum Loaders

Processors who require high capacity loaders, must convey materials over long distances, or must move free-flowing powders (such as PVC) will find that positive-displacement vacuum loaders may best satisfy their needs. Vacuum is supplied by a positive-displacement pump driven by a three-phase motor. Such pumps handle up to 15,000 lb/hr and distances up to 600 ft. These larger units are frequently used for unloading bulk railcars to silo storage.

The positive-displacement vacuum unit can also power a multiple machine loading installation with the cost of the pump being spread over a number of presses.

Control is centralized, and a common vacuum line joins the stations to the central pump. Individual feed lines eliminate the danger of cross-contamination, so that different colors and/or materials can be used in each press. Common material feed lines can also be used where operations involve only one material.

Most recently, solid-state controls have appeared on several types of hopper loaders and some manufacturers are now incorporating printed circuits and solid-state relays in both their small units and central conveying systems.

Coloring Loaders

Hopper mounted coloring loaders combine virgin materials, regrinds, one or two colorants or additives (such as slip agents, inhibitors, etc.), mix the materials by tumbling or gravity, and drop the mixture into the process hopper. Some coloring loaders allow use of dry, powdered colorants, color concentrates or new liquid colorants through the same unit without major equipment alterations.

The coloring units are self-loading and mount directly over processing machines so there is no need to manually handle virgin or regrind materials and risk contamination and waste (Fig. 29-14). When powdered dry pigments are used, they are placed in a canister in a separate color room and the filled canister is then mounted on the coloring loader.

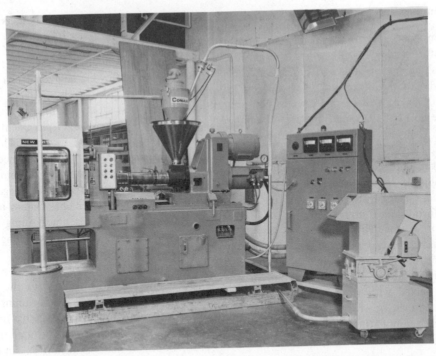

Fig. 29-13. Some hopper loaders are designed to automatically bring a pre-set amount of regrind into the hopper during each loading cycle. The most efficient method is to pick up the regrind right from the grinder box.

Fig. 29-14. Set-up for dry-coloring uses automatic coloring loaders located on a balcony over the injection molding room. Mixed materials flow by gravity through tubes connected to the machine throats.

Color concentrates are usually conveyed automatically into a color storage bin on the coloring loader. The colorant metering screws meter in the desired percentage of color for each fixed volume batch of natural virgin or virgin ratio loaded with regrinds. Of course, the regrinds are already colored, so they aren't considered in the color calibration.

Liquid coloring is a more recent technique. It has the advantage of eliminating the dustiness of dry coloring. If the liquid is of a constant density and viscosity, very accurate color metering may be achieved. Cost of coloring is reported to be slightly higher than dry color, but less than custom color concentrates.

One method of using liquid colorants is simply to mount an interchangeable metering assembly onto the coloring loader previously described. It meters a precise amount of the colorant into a precalibrated charge of virgin and/or regrind in the unit's mixing drum. All materials, additives and colorants are thoroughly mixed together before discharging into the machine hopper.

Another method involves a pressure pump which feeds a controlled amount of colorant through a metering valve and then directly into the screw of the injection molding machine as virgin materials are flowing into the throat. When machine screws don't give adequate mixing and dispersion because of short L/D ratios, a small mechanical mixer may be placed above the throat and the liquid injected into the pellets as they flow through this mixer.

Freeze-dried colorants, which are conglomerations of powdered dry colors in a cleaner and more easily handled particle form, can be successfully metered through standard coloring equipment.

Consideration of regrinds and thorough incorporation of colorants with both virgin and regrind materials are important to successful on-the-machine coloring, regardless of the colorant type. It is important too, that an accurate, constant volume of the basic resin be delivered into the system if colorants are to be added in pre-set amounts. The color calibration depends on it.

Metering accuracy and the ability to handle a wide range of colorants and material types should be provided in the unit design. It is usually advisable to pre-test materials through these units before deciding if they will meet processing requirements.

Disposable Machine Hoppers

While the coloring loaders minimize labor and time costs during color changes, the machine hopper has always presented a major down-time factor because it had to be wiped clean before the next color was used. To get at the machine hopper it was usually necessary to move the loader out of the way, which slowed the process further.

A properly sized loading unit will meet material usage demands of any processing machine so all that is needed is a funnel to direct the material into the machine throat as it falls from the loader. Large supplies of material over the machines, whether colored or not, can be a disadvantage when changes are frequent.

A clear plastic cone held in a steel frame is available on which either a regular hopper loader, or a coloring loader, is mounted (Fig. 29-15). The cone is low cost, so it can either be thrown away or be saved for the color the next time it is run. It can usually be changed from floor level, eliminating climbing to clean large machine hoppers.

Bulk Vacuum Flow Systems

Since the first shipments of plastics in bulk railcar were sent to a wire coating company in 1958, there have been hundreds of conveying installations erected to satisfy bulk material buying, storage, and automatic in-plant distribution (Fig. 29-16).

Properly designed, a bulk conveying system enables processors to realize the lowest price for raw materials and add savings from efficient space utilization, elimination of hand trucking material and hand loading, and minimize material waste through spillage, breakage, and contamination. Systems range from only one, truck-filled silo to a "tank farm" of many silos sized to hold railcar shipments.

Because nearly all free flowing plastic materials can be shipped in bulk quantities,

Fig. 29-15. Clear hopper provides instant visual check on material being used. Cone is sufficiently low in cost to be disposable, if desired.

most processors using the 100,000 lb/month quantity of one type can justify such a program. Even materials usually shipped in smaller containers, such as moisture-sensitive acrylics can be moved in bulk truck quantities by special arrangements with the supplier. They can be delivered dry, kept dry in storage silos by adding a dehumidifying dryer to maintain dry air in the silo void, and brought to the machine in a closed system where final drying and pre-heating can be accomplished.

While the bulk trucks mentioned above are self-unloading with their own pressure-type blowers, railcar shipments are usually unloaded by the positive-displacement vacuum pumps. Similar equipment then distributes the stored materials within the plant as needed, whether directly to processing or into batch weighing and blending stations.

However, a bulk system can require rates which exceed the practical limits of vacuum-type conveyors. Pressure systems have capacity advantages because of their higher power supply and continuous, rather than cyclic, operation, but they also cost considerably more because of their higher horsepower, inclusion of cyclone separators (or similar air/dust separation), and rotary feeder valves.

One of the major deterrents to a processor taking advantage of bulk conveying is in the instance where the material he is using is not

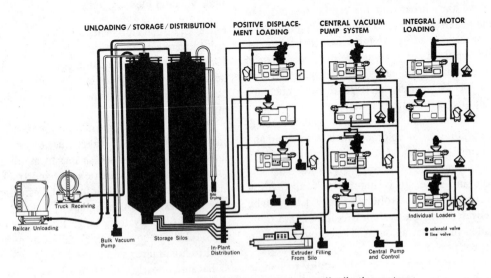

Fig. 29-16. Total bulk unloading, storage, and in-plant distribution system.

free-flowing, especially the powdered materials which may include fillers or additives which make them cake when compressed in storage and nearly impossible to get moving through a conveying system.

Powdered PVC resins, for example, can fall into the "hard-to-handle" category. Although many dry blends flow and convey readily from the manufacturer, they may require in-plant blending where their characteristics are changed when lubricants and additives are put into the blend. Such materials will require special equipment to allow their smooth handling into, and through, the processing plant.

Plastics Material Drying

Some plastics (e.g., nylon, polycarbonate, etc.) are usually hygroscopic in nature and sensitive to moisture, which means they need controlled pre-heating and thorough drying prior to processing to ensure the surface and strength qualities for which they were selected.

While non-hygroscopic materials may not require dehumidified drying, they will carry surface moisture that should be removed before processing by use of hot air dryers. This preheating also removes one variable from processing, i.e., the materials are maintained at a constant temperature year-round. This can mean improved cycling and increased production.

The most widely accepted types of plastics material dryers use a molecular sieve desiccant through which air is passed for dehumidification. The sieve traps moisture molecules, but allows air molecules to pass. Then the dry air is heated to a predetermined temperature and delivered to the plastics material, usually held in a special drying hopper. The dry air picks up moisture from the plastics, and is carried back into the dryer for dehumidifying and reheating. The desiccant is periodically taken off the drying stream for high heat regeneration, which purges it of any moisture picked up.

Good performance in dryers depends on exclusion of ambient air. Just as important are the air through-put and the proper sizing of units or systems to do the job needed at each machine.

A typical dehumidifier might use four continuously rotating desiccant cartridges (Fig. 29-17). The beds are rotated slowly through a stationary valving arrangment. The regeneration blower and heater are completely separate from the process air. The manifold arrangement is such that as the freshly regenerated bed is brought on-stream, a small amount of process air is bled into it for cooling. Since the bed cooling process air must be reheated before being sent to the drying hopper, part of the heat put into the desiccant during regenerations is saved and sent to the process.

Also available are twin-tower units with two alternating desiccant beds. These units switch from a wet desiccant bed to a regenerated one on a timed basis.

For the processor with several smaller machines, some of which might involve a need for drying, compact individual dryers that mount on the side of the drying hopper have been introduced (Fig. 29-18).

Fig. 29-17. Large dehumidifying dryer has continuously rotating desiccant cartridges so that desiccant is automatically moved into regeneration before saturation. Fresh desiccant is continuously introduced to the process air stream.

Fig. 29-18. Compact pre-heating/dehumidifying dryers can mount on the side of the drying hopper. The removable desiccant cartridge is regenerated on a separate small dryer. Such units are generally used on machines running under 100 lb/hr.

Central Dry Air System

However, providing a single dryer for each machine may still be too expensive for the processors who needs drying facilities only on occasion. As an alternative, he can install a central dry air system that provides automatic regeneration convenience for all machines, and spread the cost of this across the number of machines being serviced. Drying is used only at those machines that need it, without shifting equipment or congesting the molding area.

The heart of the central system is a single source of low dewpoint, dehumidified air, unheated but continuously available for delivery to any machine in the system. A

special drying hopper, usually insulated, and sized for the recommended exposure time for thorough drying, replaces normal machine hoppers. On the side of each drying hopper, or a frame close by, is an individual pre-heater that draws in dry air from the central source through a common line, heats the air to the temperature selected for that station's material, and delivers it to the material. The pre-heater's blower pushes the dry air up through the material and into the return air line, which carries moisture laden air from all stations back to the central dehumidifier.

PLANT LAYOUT

The subject of plant layout is complex and involved and could not possibly be covered in detail here. Obviously, plant layouts differ in terms of the process used (e.g., rotational molding generally requires more space because of the swing of the arms), as well as in terms of the specific end-product being made, the degree of secondary processing (e.g., decorating) required, the degree of automation needed, and so on. As a basic guide to the subject, however, and as an indication of the various elements that must be taken into consideration in planning plant layout, this section will concentrate specifically on plant layout for injection molding.*

Injection Molding Plant Layout

Planning is the most important part of building a new injection-molding operation, or modernizing an existing one. The purpose of planning is to minimize operating costs, and thus maximize profits. Obviously, planning starts by market analysis and definition of product requirements (quality, quantity, characteristics). Following this, the planning of the manufacturing operation can begin.

Mold selection is the first step in planning injection-molding operations. It is the mold that determines ultimate product quality, production rate, and profit. Mold designs are

---

* Adapted from a publication of Husky of America, Inc., Bolton, Ontario, Canada. Reprinted through the courtesy of Husky of America, Inc.

becoming more sophisticated. Stack molds, hot-runner molds, split-cavity molds, cold-runner molds, hot-runner reflector molds, and so forth, all can provide top quality parts on fully automatic cycles.

Major considerations in mold selection are (1) raw materials, (2) number of cavities, (3) runner system, (4) cooling/heating system, (5) method of part ejection, (6) quality of part desired, and (7) shots/minute rate. Service connections to the molds—electrical, chilled water, compressed air, hot water, hot oil—must be considered, as must accessories such as blow-off nozzles, fans, mold wipers and so forth.

The second step involves the selection of the injection-molding machine. Planning must include consideration of service connections to the molding machines, taking into account what is needed for your particular products.

The next step is to select product-handling equipment to save hand labor and control product quality. As much as 40% of the total *processing* cost (not including materials, transportation, taxes, R&D, etc.), of injection-molding operations can come from product handling. Consideration must be given to such operations as conveying, sorting, orienting, stacking, and so forth. The product-handling equipment chosen must match molding productivity and be compatible with the injection-molding machine and mold. Conveyors can be the in-press type, take-out conveyors, or special stacking and orienting machines designed for a particular product. Other considerations are automatic bulk-handling systems, in-press grinders, and other such pieces of equipment.

Special systems are those in which the mold, injection-molding machine, and product-handling equipment are designed to function together as a system rather than a collection of interchangeable machinery. On long-run products, this can often improve the efficiency of the operation. Some such special applications are injection-blow-molding systems, or other systems perhaps including parts-removal robots, product handling conveyors, automatic take-off equipment, parts-and-runner separation systems or equipment for automatic assembly of two component parts.

Once the basic hardware has been selected, efficient storage and handling of raw materials is the next major consideration in planning. The design of the overall raw materials system can have a major impact on the ability of the plant to achieve low manufacturing costs and good housekeeping.

Many injection-molding plants start operations without serious planning of the total raw materials system. For small plants, where growth and expansion are unlikely, this may have little impact on manufacturing performance. But for growing operations with many machines, major cost increases, efficiency reductions, and poor housekeeping often result from poor materials-flow planning. To avoid these problems, the raw materials system must be designed with expansion requirements, future automation, and flexibility in mind (see Fig. 29-19).

Plant Layout

Plant layout has a major impact on every phase of manufacturing performance. For this reason, it is important that the plant layout be designed according to guidelines that have proven successful and profitable in other injection-molding operations. An efficient plant layout provides maximum plant capacity (lb/year) and maximum productivity (lb/year/person) at minimum operating cost. The objectives of plant layout are these:

1. Optimum material flow
2. Expansion without disturbing existing production
3. Maximum operating efficiency through maximum use of automation
4. Efficient use of floor space.

Material Flow. A well-planned materials flow will minimize the amount of time the product spends in the plant, and will reduce confusion, congestion and scrap. Arrange the main plant operations next to each other according to a flow chart. Receive materials at one end of the plant; ship from the opposite end. Materials should move in as straight a line as possible. Service equipment and related departments should blend into the operation,

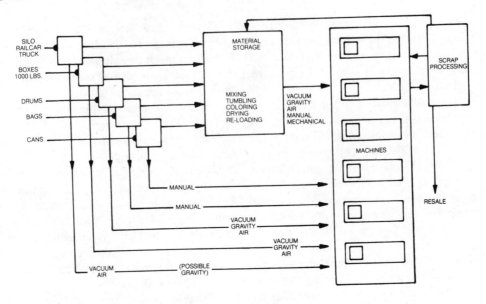

Fig. 29-19. Flow chart of typical injection molding operation. (*Illustrations courtesy Husky of America, Inc.*)

not disrupt it. Plan single-purpose aisles and large storage areas to reduce congestion.

Service Equipment. Next, equipment must be chosen to supply all plant services. Reliable supplies of power, water and air are absolutely essential. Consider plant lighting, ventilation, air-conditioning, compressed-air supply, an oil- or water-heating and circulation unit, chillers, water treatment, cooling towers, and an electrical substation.

Expansion. The cost of plant expansion includes the capital cost of buildings and equipment plus potential costs of lost production. Plan for expansion by including long-range requirements in the original plant design. Plan two permanent walls. Put extra capacity into your substation, cooling tower and silos, and locate them along permanent walls. Plan services distribution for expansion.

Automation. A logical solution to the problem of increasing direct labour costs is automation. Provide space at production machines for automatic loading and unloading systems. Use automatic material-handling between operations where possible.

Efficient Floor Space. Keep areas between injection-molding machines clear of stored materials, supplies and unused auxiliary equipment.

System design is based on these variables:

number of different materials, annual volume of each raw material, number of different colors, number of injection-molding machines and average length of production run. Purchase of raw materials in bulk quantities, by truck or rail car, usually provides major savings in the cost of raw materials. If annual consumption of a single raw material exceeds one million pounds, a silo system for bulk storage can usually be justified by the savings in raw material costs. In some cases, use of the rail car itself for bulk storage can be less costly than silos. Investigate this with raw material suppliers. Loading and unloading raw materials from silos is usually done by air blower or by combination air blower/vacuum systems. Remember that there are different service requirements for different kinds of unloading and storage systems.

Figures 29-20 through 29-23 indicate four methods for the pneumatic conveying of raw materials. Figure 29-20 illustrates the use of an air blower. Raw materials are gravity fed into rapidly moving air streams and conveyed with air to be desired destination (e.g., silo, surge bin, machine hopper, etc.). This system is best for large flow rates of materials required over long distances to a large number of widely separated receivers.

Figure 29-21 shows a vacuum pump system.

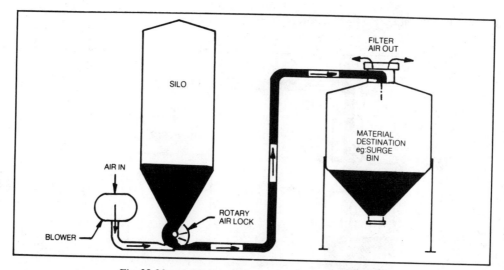

Fig. 29-20. Air blower method of conveying raw materials.

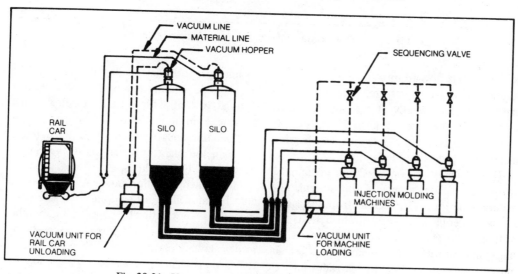

Fig. 29-21. Vacuum pump method of conveying raw materials.

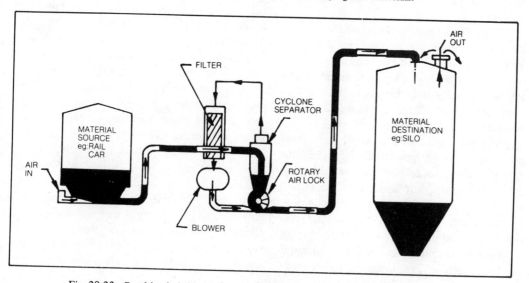

Fig. 29-22. Combined air blower/vacuum pump method of conveying raw materials.

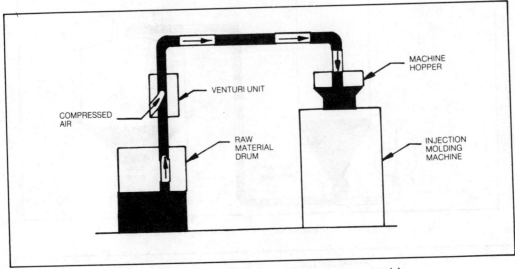

Fig. 29-23. Compressed air method of conveying raw materials.

Raw materials flow from a storage unit toward location of negative pressure. A vacuum pump can be installed directly at the point of destination (e.g., machine hopper). This system is used for smaller flow rates of free-flowing materials to a variety of locations.

The schematic in Fig. 29-22 is a combined air/blower vacuum pump. In this system, raw materials are moved by vacuum from storage to a small intermediate receiver which feeds the material into rapidly moving air streams. A single unit provides both vacuum and fast air movement. This combination provides highest flow rates, farthest delivery, and the largest number of receiving points.

Figure 29-23 shows a compressed air system. Plant compressed air is directed through a Venturi, creating a vacuum which is used to convey materials from storage to receiver. This system is good mainly for relatively low flow rates over short distances. Clean, dry compressed air is essential, however.

If colored products are required, precolored materials may be purchased from material suppliers. This provides best color matching and color consistency, but also means the highest cost/lb of material. Coloring may also be done in the plant, either with dry color, color concentrate, liquid color or solid-bar coloring.

Loading requirements at the injection-molding machine can involve single-material loading, proportional loading of raw material and regrind, blending and loading color and raw material, or central blending and loading from a distance. Careful cost estimates comparing central blending and the use of precolored resins can help the plant engineers make the choice between the two methods.

Miscellaneous Equipment. All plants must plan for miscellaneous equipment to carry out operations not directly involved in the injection-molding process. Miscellaneous equipment includes machines for secondary operations, maintenance, scrap processing, and so forth. Secondary equipment would be printers, labelers, assembling equipment and packing equipment, such as carton unfolders, folders, wrappers, etc.; a scrap grinder may be required for central grinding of large volumes of plastic parts. Mold maintenance equipment depends upon the size of the plant. Small plants with few molds normally require equipment and facilities for minor maintenance only. Larger plants with many molds usually perform major maintenance to molds requiring machine tools such as surface grinders, lathes, and milling machines. As far as machine maintenance goes, most will be done right at the machine. However, all plants require a certain space to be allocated for a mechanic's work area, for storage of spare parts, tools and lubricants. In addition, offices for plant supervision should be provided as close as possible to the production area. Administration offices should be in a separate building attached to the manufacturing facility.

Distribution of Services. Design and layout of plant service distribution systems should take place after machine and plant layout have been done. Service lines should be planned for easy accessibility for maintenance, and up-to-date schematic and installation drawings will simplify plant engineering and maintenance

tasks. Three basic service layouts—service tunnel, overhead installation and service trench—are available. The service tunnel is recommended for new plants with multiple machines. It is the most efficient. Overhead installation is the least expensive alternate with the most flexibility. The service trench is best for existing plants and relatively small operations of less than ten machines. The planning should start by listing all services required for the specific plant layout.

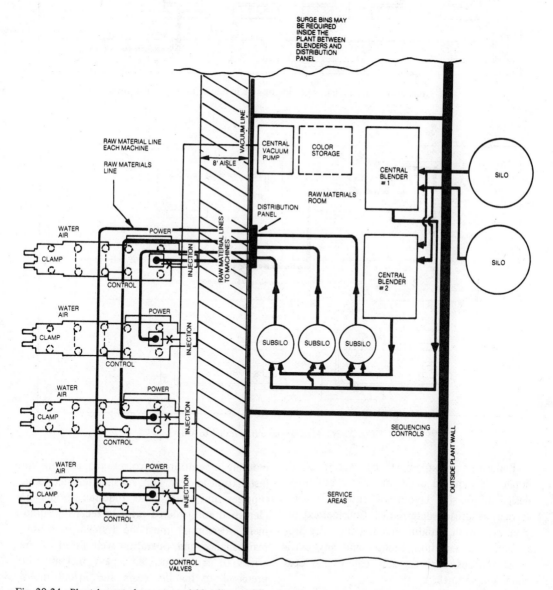

Fig. 29-24. Plant layout shows central blending with a central vacuum system for distribution to each machine. This is the most automated method of machine loading.

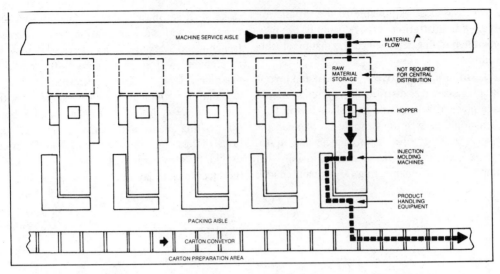

Fig. 29-25. Machine layout, side by side in one or more parallel rows.

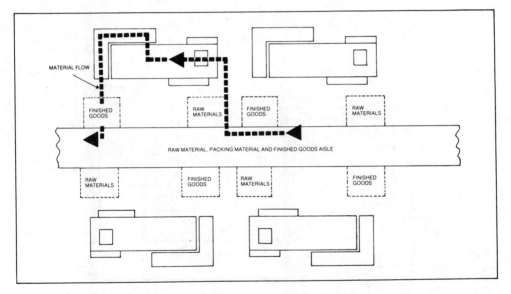

Fig. 29-26. Machine layout, parallel to aisles.

Buildings. Next must be specified all details of required plant and office buildings, designed to suit layout of equipment and services as already determined. Consider expansion, insulation, building maintenance, fire protection, site selection, builder and any extra considerations such as soil condition, local codes and so forth.

Production and Maintenance. The final step in the technical planning of injection-molding operations involves preparation for installation and start-up. All maintenance tools, supplies and spare parts required should be identified and ordered prior to receiving equipment. Personnel must be trained in set-up, maintenance and operation well ahead of the arrival of equipment. Mold personnel must be prepared to handle, store and install molds safely and properly. Machines must be unloaded and installed, and essential production

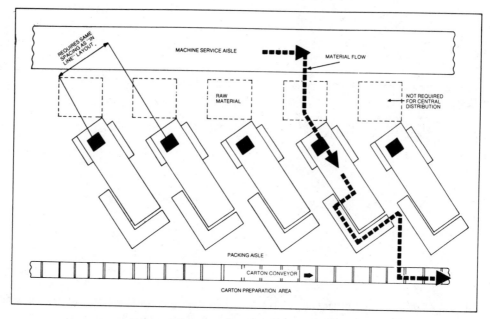

Fig. 29-27. Machine layout, at an angle to aisles.

tools and supplies—purge bar, mold releases, etc.—should be on hand. Good housekeeping practices should be set up from the beginning. Finally, production and inventory control systems, preventive maintenance and other operating procedures must be planned right along with the technical layout, in order to achieve the optimum production efficiency.

Typical Layouts

Figs. 29-24 through 29-27 shows several typical layouts for an injection molding plant.

Fig. 29-24 is a layout involving central blending with a central vacuum system for distribution to each machine. It is the most automated system for machine loading, requires minimum floor space, is least costly to operate, and is best for housekeeping and plant appearance. It is recommended under the following conditions: more than three machines; few material and color combinations (color concentrate) run at the same time; relatively long production runs.

Fig. 29-25 shows a layout in which machines are placed side-by-side in one or more parallel rows and 'in-line' with material flow. This represents a very good use of floor space and also allows a single packer to service more than one machine. Single purpose aisles relieve congestion and over-all, the layout facilitates supervision of raw materials needs.

In Fig. 29-26, machines are placed parallel to aisles. This layout is good for large machines with large products, slow cycles, and palletizing at the machine. It has some disadvantages as a multiple-machine layout for small units, namely: multi-purpose aisles add to congestion; machines are too far apart for one packer to service more than one machine; there is a waste of floor space.

In Fig. 29-27, the machines are at an angle to aisles. This can create confusion in material flow and congestion around machines. It also uses more floor space than the 'in-line' concept.

For multiple machine layouts, the side-by-side system seems to offer a number of advantages.

30
PERFORMANCE TESTING OF PLASTICS PRODUCTS

Evaluation of physical and other characteristics of objects made from plastics is a function which the plastics engineer is frequently required to perform. The purpose may be control of quality in production, acceptance testing against specifications, establishment of data for engineering and design, or other ends of substantial economic importance.

Whatever the purpose, the question of the reliability of the evaluation cannot be evaded, and when the economic importance of the evaluation is large, the reliability of the evaluation must be in proportion. Neither the authority of the engineer nor the computations performed on his data can produce reliable predictions from unreliable data; there is no substitute for a valid method of test, which must be selected before evaluation begins.

Evaluation of articles made from plastics, perhaps more than most other classes of articles, requires specialized testing methods, because of the characteristics of the resins and compounds, and of the production processes used to manufacture the articles economically. When the engineer tests an object made from plastic, he tests not only its material, but also the way the object was made. There are inherent variations in the articles, from lot to lot, and within each lot, and also from place to place within each individual article. This place-to-place variation establishes the rule that the characteristics of each article made from a

plastic are governed not only by the material of which it is made, but also by its shape and dimensions, i.e., its design. These variations are rooted in the nature of the production process itself.

The very word "plastic," implying ability to be shaped, implies the economic basis of the processes of fabrication. The manufacture of articles from plastics must be a repetitive process of maximum rapidity, involving rapid changes in temperature and in pressure, and hence necessarily abrupt transitions between fluid (or plastic) and solid condition. The quality of the finished product is notably sensitive to any irregularity in temperature, pressure or time in the production cycle. Hence, uniformity of performance of a finished product, from lot to lot, and even within a lot, depends not only upon uniformity of the plastic used but also very considerably upon uniformity of the operation of manufacturing the article, i.e., upon precision of control of all factors in the operation.

To establish the degree of uniformity to be expected, or to determine the reproducibility of a method of test, may require the testing of all, or a large percentage, of the articles in a lot. But, particularly because many performance tests are destructive, considerations of cost usually restrict testing, for control purposes, to a very small number of samples from a lot. In this connection, it must be emphasized that the results of testing only a few samples from a lot can be reliable only when the lot as a whole has been manufactured under constant conditions—uninterrupted production, with effective instrument control over all variables.

Performance tests, realistically pertinent to the conditions and hazards of service of the article, may become a very potent tool for

(1) choosing the kind, and specific grade, of plastic for the job, or an acceptable alternative material,

(2) establishing, at the start, the proper operating conditions for production (preferably at the time when samples are made for approval),

(3) proving the soundness of design of the product,

(4) checking the uniformity of the material, and of the article as made,

(5) checking the effects of subsequent changes in tooling or machinery or operating conditions (e.g., enlarging of gates; change to a larger machine; change in any factor in the cycle of manufacture).

There is considerable and useful literature on methods of test for measuring the physical and other characteristics of the various plastics as materials. Almost without exception, these methods utilize specimens of standard dimensions and shapes, prepared specifically for the purpose. The resulting data are valuable in establishing identification and uniformity of the materials themselves, but usually the published standard data derived from such test specimens are not reliably applicable directly to design calculations.

Hence it is difficult to predict from such standard test data the performance of a specific article, made from a specific plastic, when it is subjected to the stresses and exposures which it is expected to withstand in service.

Few commercial articles will resemble, even remotely, in shape or dimensions, the standard test specimens. And even when it is feasible to cut specimens of standard dimensions from the larger object which is to be evaluated, the results of tests on such specimens rarely resemble those obtained on standard molded

specimens; neither can such results be relied upon to indicate precisely the probable performance of the whole object itself.

The plastics engineer, then, recognizing the difficulty of evaluating with reliability the probable performance of the article, on the basis of either the standard test data on the material, or tests of standard-sized specimens cut from the article itself, must arrive at a method of test which is adaptable to the specific case at hand, and he may have to design one. His prime consideration must be the relevance of the test to conditions of service of the article. Ideally, the result of the performance test should correlate perfectly with the actual performance of the article in service.

The literature on methods of testing objects of nonstandard shapes and dimensions is not— nor, by definition, can it ever be—anything like the orderly and compact, yet comprehensive, set of guides which covers the testing of plastics simply as materials. Not only are there as many or more properties to be measured, but there is an infinity of possible different shapes and, for each possible shape, a wide range of possible dimensions. Given the nature of the material used, and the process by which the article is made, each article is likely to have its own unique set of characteristics. Hence, in the testing of articles made from plastics there can be no complete codification of methods, apparatus or acceptable levels of performance.

This chapter represents an attempt to describe tests and equipment, in several categories, which have been effectively used for testing specific articles, and for measuring certain important characteristics of plastics in the form of finished articles. Although the methods outlined may not meet the requirements of any given case, it is suggested that the general principles set forth herein be used in devising variations or new methods which will be suited to the particular article to be tested.

PERFORMANCE TESTING*

One test may measure a single property or several properties at once. In every case the test

* Based on "Standard Tests on Plastics, Bulletin GIC" issued by Celanese Plastics Co., Div. of Celanese Corp.

has been devised to be as accurate as possible. After many years of work by thousands of technical specialists in the plastics industry, the tests presently used are generally regarded as suitable. Nevertheless, further improvement is constantly sought. The American Society for Testing and Materials regards testing as a dynamic science always receptive to further improvement. To make tests more accurate, more reproducible and more meaningful, the plastics industry and interested university and independent groups are continually perfecting present tests and developing better ones for the future.

Of course, tests are not ends in themselves, but rather means of extracting knowledge about materials. The real test of a material comes with actual service. Once a plastic product is taken home and used by the consumer, it no longer matters whether tensile strength is 6500 or 6600 psi. The product succeeds entirely, or it fails. To assure success of toys, housewares, industrial products, and automotive components, the properties of likely materials are studied by design engineers who, through experience and judgment, balance material characteristics and service requirements against the amount of material needed in a part to give an adequate safety margin. It is in this area of product design and specifications that the tests themselves are tested.

In certain cases, service requirements may be so complex that suitability of a material for that service can be determined only in actual service. In one such case, plastic for pipe is tested by making pipe of it, attaching it to a pressured water line, and watching what happens (U.S. Department of Commerce Commercial Standard CS 255-63). A plastic material considered for large-scale use, as in automobiles, is likewise fabricated into finished parts for in-test service. A maker of plastic dishes may drop them on concrete, run them through dishwashers repeatedly, and otherwise put them through informal "practical" tests.

The tests included here are grouped as they are in the ASTM book. They are described as briefly as possible (far too briefly to serve as a laboratory guide) in order to give interested persons a general idea of what the tests are

about. Fully detailed procedures for all ASTM tests on plastics are available from the American Society for Testing and Materials, 1916 Race Street, Philadelphia, Pa. Commercial Standard CS 255-63 is available from the Supt. of Documents, U.S. Government Printing Office, Washington, D.C. 20025.

Tensile Properties (ASTM D638)

Specimen: Specimens can be injection molded or machined from compression molded plaques. They are given standard conditioning.* Typically ⅛ inch thick, their size can vary; their shape is exemplified in the figure below.†

TENSILE TEST SPECIMEN

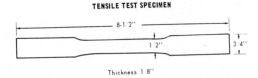

Thickness 1 8''

Procedure: Both ends of the specimen are firmly clamped in the jaws of an Instron testing machine. The jaws may move apart at rates of 0.2, 0.5, 2, or 20 inches a minute, pulling the sample from both ends. The stress is automatically plotted against strain (elongation) on graph paper.

Significance: Tensile properties are the most important single indication of strength in a material. The force necessary to pull the specimen apart is determined, along with how much the material stretches before breaking.

The elastic modulus ("modulus of elasticity" or "tensile modulus") is the ratio of stress to strain below the proportional limit of the material. It is the most useful tensile data because parts should be designed to accommodate stresses to a degree well below this.

For some applications where almost rubbery elasticity is desirable, a high ultimate elongation may be an asset. For rigid parts, on the other

* See ASTM D618-"Conditioning Plastics."
† For polyethylene, the Standard (D-1248-72) requires the use of specimens described in ASTM D638 as Type IV. This specimen is smaller and allows for the much greater elongation of PE.

hand, there is little benefit in the fact that they can be stretched extremely long.

There is great benefit in moderate elongation, however, since this quality permits absorbing rapid impact and shock. Thus the total area under a stress-strain curve is indicative of overall toughness. A material of very high tensile strength and little elongation would tend to be brittle in service.

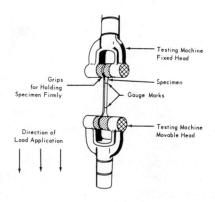

Flexural Properties of Plastics (ASTM D790)

Specimen: Usually $\frac{1}{8}$ x $\frac{1}{2}$ x 5 inches. Sheet or plaques as thin as $\frac{1}{16}$ inch may be used. The span and width depend upon thickness.

Specimens are conditioned according to Procedure A, ASTM D618.

Procedure: The specimen is placed on two supports spaced 4 inches apart. A load is applied in the center of a specified rate and the loading at failure (psi) is the flexural strength. For materials which do not break, the flexural property usually given is Flexural Stress at 5% strain.

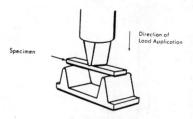

FLEXURAL PROPERTIES OF PLASTICS

Significance: In bending, a beam is subject to both tensile and compressive stresses, as indicated in the following sketch:

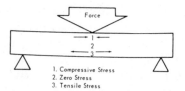

STRESSES IN FLEXED SAMPLE

1. Compressive Stress
2. Zero Stress
3. Tensile Stress

Since most thermoplastics do not break in this test even after being greatly deflected, the flexural strength can not be calculated. Instead, stress at 5% strain is calculated—that is, the loading in psi necessary to stretch the outer surface 5%.

Stiffness in Flexure (ASTM D747)

Specimen: The specimens must have rectangular cross section, but dimensions may vary with the kind of material.

Specimens are conditioned according to Procedure A, ASTM D618.

Procedure: The specimen is clamped into the apparatus (sketch at right) and a 1% load is applied manually. The deflection scale is set at zero. The motor is engaged and the loading increased, with deflection and loading figures recorded at intervals. A curve is drawn of deflection versus load, and from this is calculated stiffness in flexure in pounds per square inch.

STIFFNESS IN FLEXURE APPARATUS

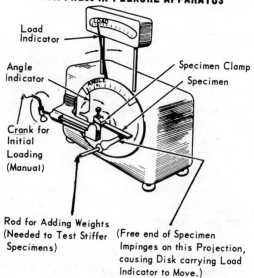

Load Indicator

Angle Indicator

Crank for Initial Loading (Manual)

Specimen Clamp

Specimen

Rod for Adding Weights (Needed to Test Stiffer Specimens)

(Free end of Specimen Impinges on this Projection, causing Disk carrying Load Indicator to Move.)

Significance: This test does not distinguish the plastic and elastic elements involved in the measurement and therefore a true elastic modulus is not calculable. Instead, an apparent value is obtained and called "stiffness in flexure." It is a measure of the relative stiffness of various plastics and taken with other pertinent property data is useful in material selection.

STIFFNESS IN FLEXURE

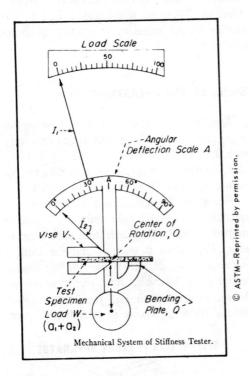

Mechanical System of Stiffness Tester.

© ASTM–Reprinted by permission.

Izod Impact (ASTM D256)

Specimen: Usually $\frac{1}{8} \times \frac{1}{2} \times 2$ inches.

Specimens of other thicknesses can be used (up to $\frac{1}{2}$ inch) but $\frac{1}{8}$ inch is frequently used for molding materials because it is representative of average part thickness.

A notch is cut on the narrow face of the specimen, which is conditioned according to Procedure A of ASTM D618.

Procedure: A sample is clamped in the base of a pendulum testing machine (see figure) so that it is cantilevered upward with the notch facing the direction of impact. The pendulum is released, and the force consumed in breaking the sample is calculated from the height the pendulum reaches on the follow-through.

IZOD IMPACT

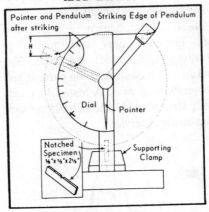

Significance: The Izod Impact test indicates the energy required to break notched specimens under standard conditions. It is calculated as ft lb per inch of notch and is usually calculated on the basis of a one inch specimen although the specimen used may be thinner in the lateral direction. (This is indicated in the sketch of the Izod Specimen).

The Izod value is useful in comparing various types or grades of a plastic. In comparing one plastic with another, however, the Izod impact test should not be considered a reliable indicator of overall toughness or impact strength. Some materials are notch-sensitive and derive greater concentrations of stress from the notching operation. The Izod impact test may indicate the need for avoiding sharp corners in parts made of such materials. For example nylon and acetal-type plastics, which in molded parts are among the toughest materials, are notch sensitive and register relatively low values on the notched Izod impact test.

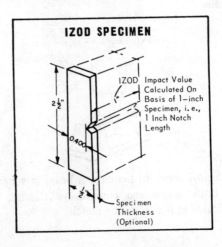

Tensile Impact (ASTM D1822)

Specimen: Small standard specimens of "tensile bar" shape measuring $2\frac{1}{2}$ inches long.

Specimens are conditioned according to Procedure A, ASTM D618.

Procedure: The specimen is mounted between a pendulum head and crosshead clamp on the pendulum of an impact tester (see sketch). The pendulum is released and it swings past a fixed anvil which halts the cross head clamp. The pendulum head continues forward, carrying the forward portion of the ruptured specimen.

The energy loss (tensile impact energy) is recorded, as well as whether the failure appeared to be a brittle or ductile type.

Significance: This test is comparatively new, having been adopted by ASTM in 1961, and there is relatively little industry data available to adequately assess its accuracy and utility. However, possible advantages over the notched Izod test are immediately apparent: the notch sensitivity factor is eliminated, and energy is not used in pushing aside the broken portion of the specimen.

TENSILE IMPACT

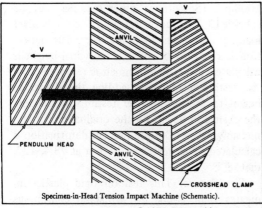

Specimen-in-Head Tension Impact Machine (Schematic).

© ASTM—Reprinted by permission.

Rockwell Hardness (ASTM D785)

Specimen: Sheets or plaques at least $\frac{1}{4}$ in. thick. This thickness may be built up of thinner pieces, if necessary. Normally specimens are conditioned according to Procedure A, ASTM D618.

A steel ball under a minor load is applied to the surface of the specimen. This indents slightly and assures good contact. The gauge is then set at zero. The major load is applied for 15 seconds and removed, leaving the minor load still applied. The indentation remaining after 15 seconds is read directly off the dial. This value is preceded by a letter representing the Rockwell hardness scale used.

The size of the balls used and loadings vary, and values obtained with one set cannot be correlated with values from another set.

Significance: Rockwell hardness can differentiate relative hardness of different types of a given plastic. But since elastic recovery is involved as well as hardness, it is not valid to compare hardness of various kinds of plastic entirely on the basis of this test.

Rockwell hardness is not an index of wear qualities or abrasion resistance. For instance, polystyrenes have high Rockwell hardness values but poor scratch resistance.

ROCKWELL HARDNESS

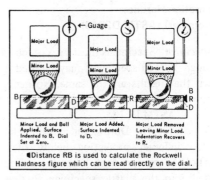

◀Distance RB is used to calculate the Rockwell Hardness figure which can be read directly on the dial.

Compressive Properties of Rigid Plastics (ASTM D695)

Specimen: Prisms $\frac{1}{2} \times \frac{1}{2} \times 1$ inch or cylinders $\frac{1}{2}$ inch diameter x 1 inch.

Procedure: The specimen is mounted in a compression tool between testing machine heads which exert constant rate of movement. Indicator registers loading.

Specimens are usually conditioned according to Procedure A, ASTM D618.

The compressive strength of a material is calculated as the psi required to rupture the specimen or deform the specimen a given percentage of its height. It can be expressed as psi either at rupture or at a given percentage of deformation.

Significance: The compressive strength of plastics is of limited design value, since plastic products (except foams) seldom fail from compressive loading alone. The compressive strength figures, however, may be useful in specifications for distinguishing between different grades of a material, and also for assessing, along with other property data, the over-all strength of different kinds of materials.

COMPRESSIVE PROPERTIES OF RIGID PLASTICS

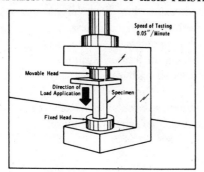

Indentation Hardness (Durometer) (ASTM D2240)

Specimen: Specimens must be $\frac{1}{4}$-inch thick, although $\frac{1}{8}$-inch specimens may be used if it is shown that results are the same. Normally the specimens are conditioned according to Procedure A, ASTM D618.

Procedure: The Durometer instrument has a pointed indenter projecting below the base (face) of the pressure foot. When the indenter is pressed into the plastic specimen so that the base rests on the plastic surface, the amount of indentation registers directly on the dial indicator.

Significance: This test measures the indentation into the plastic of the indenter under load, according to a scale of 0 to 100. There is no unit of measurement. Readings taken immediately after application may vary from those taken after pressure has been held for a time, because of creep. This test is preferred for polyethylene, because the Rockwell test loses meaning when excessive creep is encountered. For other materials (acetate, acetal, etc.) the Rockwell hardness test is still the standard.

Shear Strength (ASTM D732)

Specimen: Sheets or molded discs from 0.005 to 0.500 inch thick are used. Conditioning method is Procedure A, ASTM D618.

Procedure: The specimen is mounted in a punch-type shear fixture and the punch (1 in. D) is pushed down at a rate of 0.05 in./min. until the moving portion of the sample clears the stationary portion.

Shear strength is calculated as the force/area sheared.

Significance: Shear strength is particularly important in film and sheet products where failures from this type load may often occur.

For the design of molded and extruded products it would seldom be a factor.

Deformation Under Load (ASTM D621)

Specimen: $\frac{1}{2}$-inch cube, either solid or composite.

Conditioned according to Procedure A, ASTM D618.

Procedure: The specimen is placed between the anvils of the testing machine, and loaded at 1000 psi. (Sometimes other loadings may be specified.) The gauge is read 10 seconds after loading, and again 24 hours later. The deflection is recorded in mils. The original height is calculated after the specimen is removed from the testing machine by adding the change in height to the height after testing. By dividing the change in height by the original height and multiplying by 100, the percent deformation is calculated. This test may be run at 73.4, 122, or 158°F.

Significance: This test on rigid plastics indicates their ability to withstand continuous short-term compression without yielding and loosening when fastened as in insulators or other assemblies by bolts, rivets, etc. It does not indicate the creep resistance of a particular plastic for long periods of time.

It is also a measure of rigidity at service temperatures and can be used as identification for procurement.

DEFORMATION UNDER LOAD

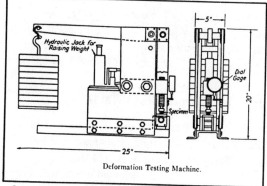

Deformation Testing Machine.

© ASTM—Reprinted by permission.

Vicat Softening Point
(ASTM D1525)

Specimen: Flat specimens must be at least $\frac{3}{4}$ inch wide and $\frac{1}{8}$ inch thick. Two specimens may be stacked, if necessary, to get the thickness, and the specimens may be compression or injection molded.

Procedure: The apparatus for testing Vicat softening point consists of a temperature regulated oil bath with a flat ended needle penetrator so mounted as to register degree of penetration on a gauge.

A specimen is placed with the needle resting on it. The temperature of the bath (preheated to about 50°C lower than anticipated Vicat softening point) is raised at the rate of 50°C/hr or 120°C/hr. The temperature at which the needle penetrates 1 mm. is the Vicat Softening Point.

VICAT SOFTENING POINT

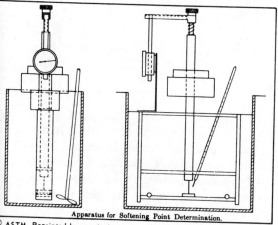

Apparatus for Softening Point Determination.
© ASTM—Reprinted by permission.

Significance: The Vicat softening temperature is a good way of comparing the heat softening characteristics of polyethylenes.

It also may be used with other thermoplastics.

Deflection Temperature*
(ASTM D648)

Specimen: Specimens measure 5 x $\frac{1}{2}$ inch x any thickness from $\frac{1}{8}$ to $\frac{1}{2}$ inch. Conditioned according to Procedure A, ASTM D618.

Procedure: The specimen is placed on supports 4 inches apart and a load of 66 or 264 psi is placed on the center. The temperature in the chamber is raised at the rate of 2° ± 0.2°C per minute. The temperature at which the bar has deflected 0.010 inch is reported as "deflection temperature at 66 (or 264) psi fiber stress."

DEFLECTION TEMPERATURE

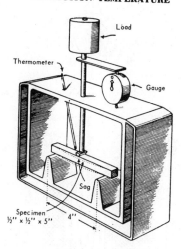

Significance: This test shows the temperature at which an arbitrary amount of deflection occurs under established loads. It is not intended to be a direct guide to high temperature limits for specific applications. It may be useful in comparing the relative behavior of various materials in these test conditions, but it is primarily useful for control and development purposes.

* This property used to be called "Heat Distortion Temperature."

Flow Rate (Melt Index) by Extrusion Plastometer (ASTM D1238)

Specimen: Any form which can be introduced into the cylinder bore may be used, e.g., powder, granules, strips of film, etc.

The conditioning required varies, being listed in each material specification.

Procedure: The apparatus (shown) is preheated to 190°C for polyethylene. Material is put into the cylinder and the loaded piston (approx. 43.25 psi) is put into place. After 5 minutes, the extrude issuing from the orifice is cut off flush, and again one minute later. These cuts are discarded. Cuts for the test are taken at 1, 2, 3, or 6 minutes, depending on the material or its flow rate. The melt index is calculated and given as grams/10 minutes.

FLOW RATE (MELT INDEX) BY EXTRUSION PLASTOMETER

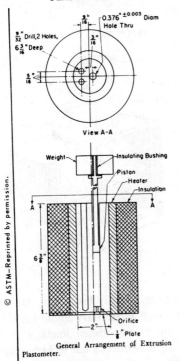

General Arrangement of Extrusion Plastometer.

Significance: The melt index test is primarily useful to raw material manufacturers as a method of controlling material uniformity. While the data from this test is not directly translatable into relative end-use processing characteristics, the melt index value is nonetheless strongly indicative of relative "flowability" of various kinds and grades of PE.

The "property" measured by this test is basically melt viscosity or "rate of shear." In general, the materials which are more resistant to flow are those with higher molecular weight.

Brittleness Temperature (ASTM D746)

Specimen: Pieces $\frac{1}{4}$ inch wide, 0.075 inch thick, and $1\frac{1}{4}$ inch long. They are conditioned according to Procedure A, ASTM D618.

Procedure: The conditioned specimens are cantilevered from the sample holder in the test apparatus (shown) which has been brought to a low temperature (that at which specimens would be expected to fail). When the specimens have been in the test medium for 3 minutes, a single impact is administered and the samples are examined for failure. Failures are total breaks, partial breaks, or any visible cracks. The test is conducted at a range of temperatures producing varying percentages of breaks. From these data, the temperature at which 50% failure would occur is calculated or plotted and reported as the brittleness temperature of the material according to this test.

BRITTLENESS TEMPERATURE

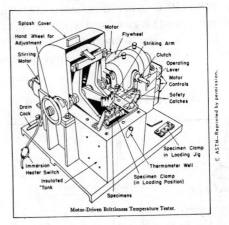

Motor-Driven Brittleness Temperature Tester.

Significance: This test is of some use in judging the relative merits of various materials for low temperature flexing or impact. However, it is specifically relevant only for materials and conditions specified in the test, and the values cannot be directly applied to other shapes and conditions.

The brittleness temperature does not put any lower limit on service temperature for end-use products. The brittleness temperature is sometimes used in specifications.

Haze and Luminous Transmittance of Transparent Plastics (ASTM D1003)

Specimen: Transparent film and sheet samples are used, or molded samples which have parallel plane surfaces. A disc, 2 inches in diameter, is recommended. No conditioning is required.

Procedure: Procedure A is followed when a hazemeter is used in the determinations.

Procedure B is followed when a recording spectrophotometer is used.

Significance: In this test, haze of a specimen is defined as the percentage of transmitted light which, in passing through the specimen, deviates more than 2.5 deg. from the incident beam by forward scattering.

Luminous transmittance is defined as the ratio of transmitted to incident light.

These qualities are considered in most applications for transparent plastics. They form a basis for directly comparing the transparency of various grades and types of plastic.

From the raw material manufacturing standpoint they are important control tests in the various stages of production.

Luminous Reflectance, Transmittance and Color (ASTM E308)

Specimen: Opaque specimens should have at least one plane surface. Translucent and transparent specimens must have two surfaces which are plane and parallel. The piece must be at least 2 inches in diameter.

Procedure: The sample is mounted in the instrument and along with it a comparison surface (white chalk). The samples are placed in the instrument and light of different wavelength intervals is impinged against the surface. Reflected or transmitted light is then measured to obtain property values listed below.

Significance: This test is the primary method

of obtaining colorimetric data. Properties determined include the following:

1. Total luminous reflectance or the luminous directional reflectance.
2. Luminous transmittance.
3. The chromaticity coordinates x and y (color).

LUMINOUS REFLECTANCE, TRANSMITTANCE AND COLOR

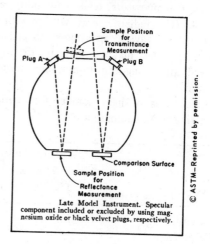

Late Model Instrument. Specular component included or excluded by using magnesium oxide or black velvet plugs, respectively.

© ASTM—Reprinted by permission.

Outdoor Weathering (ASTM D1435)

Specimen: No specified size. Specimens for this test may consist of any standard molded test specimen or cut pieces of sheet or machined samples.

Procedure: Specimens are mounted outdoors on racks slanted at a 45° angle and facing south. It is recommended that concurrent exposure be carried out in many varied climates to obtain the broadest, most representative total body of data. Sample specimens are kept indoors as controls and for comparison.

Reports of weathering describe all changes noted, areas of exposure, and period of time.

Significance: Outdoor testing is the most accurate method of obtaining a true picture of weather-resistance. The only drawback of this test is the time required for several years exposure.

A large number of specimens are usually required to allow periodic removal and to run representative laboratory tests after exposure.

Accelerated Weathering (ASTM G23)- (Recommended Practice)

Specimen: Any shape; size up to 5 x 7 x 2 inches.

Procedure: Artificial weathering has been defined by ASTM as "The exposure of plastic to cyclic laboratory conditions involving changes in temperature, relative humidity, and ultraviolet (UV) radient energy, with or without direct water spray, in an attempt to produce changes in the material similar to those observed after long-term continuous outdoor exposure."

Three types of light sources for artificial weathering are in common use:

Source	UV Energy Output, Approx. (X Sunlight)
Enclosed UV Carbon Arc	7.5
Open-flame Sunshine Carbon Arc	3
Water-cooled Xenon Arc	1*

* (ASTM G26).

Selection of light source involves many conditions and circumstances, such as what material is being tested, the proposed end-use, previous testing experience, the type of information desired, etc.

Significance: Since weather varies from day to day, year to year, and place to place, no precise correlation exists between artificial laboratory weathering and natural outdoor weathering. However, standard laboratory test conditions produce results with acceptable reproducibility and which are in *general* agreement with data obtained from outdoor exposures. Fairly rapid indications of weatherability are therefore obtainable on samples of known materials which through testing experience over a period of time, have general correlations established. There is no artificial substitute for *precisely* predicting outdoor weatherability on materials with no previous weathering history.

Water Cooled Xenon Arc Type (ASTM D2565)

Specimen: This recommended practice covers procedures applicable when operating Light and Water Exposure Apparatus (Water-Cooled Xenon-Arc Type) for artificial weathering.

Procedure: Xenon arcs have been shown to have a spectral energy distribution when properly filtered, which closely simulates the spectral distribution of sunlight at the surface of the earth.

Significance: Since the emitted energy from the Xenon lamps decays with time, and since the parameters of temperature and water do not represent a specific known climatic condition, the results of laboratory exposure are not necessarily intended to correlate with data obtained by outdoor weathering. There may be no positive correlation of exposure results between the Xenon arc and other laboratory weathering devices.

Accelerated Exposure to Sunlight using the Atlas Type 18FR Fade-Ometer® (ASTM G23)

The Atlas Type 18FR Fade-Ometer is used primarily to check and compare color stability. Besides determining the ability of various pigments needed to provide both standard and custom colors, the Fade-Ometer is helpful in preliminary studies of various stabilizers, dyes, and pigments compounded in plastics to prolong their useful life. It is primarily for testing materials to be used in articles subject to indoor exposure to sunlight.

The Fade-Ometer was extensively used in the development of UV absorbing acetate film for store windows to protect merchandise displayed in direct sunlight.

Exposure in the Fade-Ometer cannot be directly related to exposure in direct sunlight, partially because other weather factors are always present outdoors.

Pipe Tests (Commercial Standard CS 255*)

Specimen: Sections of pipe are used in these tests. The lengths are specified for each type, grade and dimension of pipe.

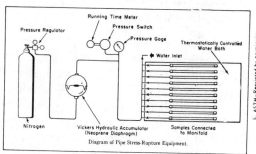

PIPE TESTS

Diagram of Pipe Stress-Rupture Equipment.

Test	Procedure	Significance
Sustained Pressure	Specimens are filled with water and put under pressure and temperatures indicated for each type of pipe and held for at least 1000 hours.	Service installations of plastic pipe may be expected to give many years of service. This test, by using relatively high pressures, indicates whether the materials will hold up in normal service. The test advises continuation for a year or longer for further reliability.
Incremental Pressure	Specimens are filled with water and brought to pressure level indicated. Pressure is raised in steps until maximum indicated pressure is attained. No more than 2 of 6 specimens can fail.	Most pipe in service bears pressure of varying amounts sometimes abruptly changed. This can place a greater strain on pipe than gradual changes. The incremental pressure tests show ability of pipe to withstand relatively sudden increases in pressure to levels above "normal" service ratings.
Environmental Stress Cracking	Lengths of pipe are put under pressure and coated with Igepal CO 630. After 3 hours, 4 of 6 specimens must retain full pressure.	Susceptibility to ESC is checked as a quality control measure, to assure that excessive levels of internal strain in the pipe will not contribute to failure in service contacts with such agents as detergents.
Density	Density of base resin is calculated by the density gradient technique. (Amount of carbon black is given as percentage.)	Many properties are density related. The base resin density in pipe determines classification and size requirements.

* This Standard lists pipe wall thicknesses required for various materials and kinds of service, general appearance and finish requirements, and tolerances, as well as the tests described here. (The pressure tests apparatus are described in ASTM D1598-63T.) Passing all tests is a requirement for marketing pipe as meeting this commercial standard.

Environmental Stress Cracking (ASTM D1693)

Specimen: This test is limited to Type 1 (low density) polyethylenes. Specimens measure $\frac{1}{8} \times \frac{1}{2} \times 1\frac{1}{2}$ inch. These are annealed in water or steam at $100°C$. (boiling) for one hour and then equilibrated at room temperature for 5 to 24 hours. After conditioning the specimens are nicked according to directions given.

Procedure: The specimens are placed in an air circulating oven and next inserted into a test tube which is then filled with fresh reagent (Igepal). The tube is stoppered with an aluminum covered cork (as shown) and placed in a constant temperature bath at $50°C$. These are inspected periodically and any visible crack is considered a failure. The duration of the test is reported along with the percentage of failures.

Significance: The cracking obtained in this test is indicative of what may be expected from a wide variety of other stress cracking agents. The information cannot be translated directly into end-use service prediction, but serves to rank various types and grades of PE in categories of resistance to ESC. Though restricted to Type 1 polyethylene, this test can be used on high and medium density materials as well, in which case it would be considered a "modified" test.

ENVIRONMENTAL STRESS CRACKING

Test Equipment.

Permanent Effect of Heat (ASTM D794)

Specimen: Any piece of plastic or molded plastic part.

Procedure: The specimens are placed in an air circulating oven at a temperature (multiple of $25°C$) which is thought or known to be near the temperature limit of the material. If, after four hours, there is no change observed, the temperature is increased in increments of $25°C$ at four hour intervals until a change does occur.

The change might be any property or properties of special interest—mechanical, visual, dimensional, color, etc. The test is written so that many effects of heat can be studied and specification requirements can be individually agreed upon by parties concerned.

Significance: This test is of particular value in connection with established or potential applications which involve service at elevated temperatures. It permits comparison of various plastics and grades of one plastic in the form of test specimens, as well as molded parts in finished form.

Water Absorption (ASTM D570)

Specimen: For molding materials the specimens are discs 2 inches in diameter and $\frac{1}{8}$ inch thick. For sheet materials the specimens are bars 3 inches x 1 inch x thickness of the material.

The specimens are dried 24 hours in an oven at $50°C$, cooled in a desiccator, and immediately weighed.

Procedure: Water absorption data may be obtained by immersion for 24 hours or longer in water at $73.4°F$. Upon removal, the specimens are wiped dry with a cloth and immediately weighed. The increase in weight is reported as percentage gained.

For materials which lose some soluble matter during immersion—such as cellulosics—the sample must be re-dried, re-weighed, and reported as "percent soluble matter lost." The % gain in weight + % soluble matter lost = % water absorption.

Significance: The various plastics absorb varying amounts of water, and the presence of absorbed water may affect plastics in different ways.

Electrical properties change most noticeably with water absorption, and this is one of the reasons that polyethylene, since it absorbs almost no water, is highly favored as a dielectric.

Materials which absorb relatively larger amounts of water tend to change dimension in the process. When dimensional stability is required in products made of such materials, grades with less tendency to absorb water are chosen.

The water absorption rate of acetal type plastics is so low as to have a negligible effect on properties.

Density by Density Gradient Technique (ASTM D1505)

Specimen: Any small piece, so long as its shape does not encourage entrapment of air bubbles, it doesn't contain voids, and it permits accurate determination of center of volume. Conditioning is required only if specimens might change in density more than the limits of accuracy would allow.

Procedure: A density gradient column contains denser liquid at the bottom and the gradient is decreasingly dense at higher levels. A group of tubes can contain a range of densities from as low as 0.80 and may range as high as 2.89. Into such columns, glass floats of various known densities are inserted. When they reach equilibrium levels, the density gradient within the tube is established. Plastic specimens may then be placed into the column and, at equilibrium level, their density may be read from the center line of their volume. The whole system is kept at constant temperature of 73.4°F.

Significance: Density determinations by this method are very accurate and quick, and it is a widely used technique. It requires very careful preparation and handling.

Specific Gravity and Density (ASTM D792)

Specimen: May be determined on moldings, sheet, rod or tubes (Method A) as well as on molding powders, pellets, flake, etc. (Method B). Conditioning of specimens is optional. Extrudate or molded specimens are preferred over molding pellets, as these may contain voids. Molded specimens are not tested for specific gravity until after post-molding shrinkage has been accomplished—usually 24 hours.

Procedure: In Method A, a piece of the article is held by a fine wire, weighed and submerged in water. While it is in the water it is weighed again. From the weight difference the density can be calculated.

In Method B, 5 grams of the pellets or powdered material are added to a measured volume in a pyconometer and the specific gravity is calculated from the weight and volume change at 73.4°F. This method requires that all air be evacuated, and with such specimens as film this can be troublesome and there is a risk of error from bubbles which may remain.

Significance: Specific gravity is a strong element in the price factor and thus has great importance. Beyond the price/volume relationship, however, specific gravity is used in production control, both in raw material production and molding and extrusion. Polyethylenes, for instance, may have density variation, depending upon the degree of "packing" during molding, or the rate of quench during extrusion.

While *specific gravity* and *density* are frequently used interchangeably, there is a very slight difference in their meaning.

Specific gravity is the ratio of the weight of a given volume of material at 73.4°F (23°C) to that of an equal volume of water at the same temperature. It is properly expressed as "Specific Gravity, 23/23°C."

Density is the *weight per unit volume* of material at 23°C and is expressed as "D23C, g per cm^3."

The discrepancy enters from the fact that water at 23°C has a density slightly less than *one*. To convert specific gravity to density, the following factor can be used:

D23C, g per cm^3 = specific gravity, 23/23°C x 0.99756.

Tests for Electrical Resistance (ASTM D257)

Specimen: Specimens for these tests may be any practical form, such as flat plates, sheets, and tubes.

Procedure: These tests describe methods for determining the several properties defined below. Two electrodes are placed on or imbedded in the surface of a test specimen. The following properties are calculated.

Property	Definition	Significance
Insulation Resistance	Ratio of direct voltage applied to the electrodes to the total current between them; dependent upon both volume and surface resistance of the specimen.	In materials used to insulate and support components of an electrical network, it is generally desirable to have insulation resistance as high as possible.
Volume Resistivity	Ratio of the potential gradient parallel to the current in the material to the current density.	Knowing the volume and surface resistivity of an insulating material makes it possible to design an insulator for a specific application.
Surface Resistivity	Ratio of the potential gradient parallel to the current along its surface to the current per unit width of the surface.	
Volume Resistance	Ratio of direct voltage applied to the electrodes to that portion of current between them that is distributed through the volume of the specimen.	High volume and surface resistance are desirable in order to limit the current leakage of the conductor which is being insulated.
Surface Resistance	Ratio of the direct voltage applied to the electrodes to that portion of the current between them which is in a thin layer of moisture or other semi-conducting material that may be deposited on the surface.	

Dielectric Constant and Dissipation Factor (ASTM D150)

Specimen: The specimen may be a sheet of any size convenient to test, but should have uniform thickness. The test may be run at standard room temperatures and humidity, or in special sets of conditions as desired. In any case, the specimens should be pre-conditioned to the set of conditions used.

Procedure: Electrodes are applied to opposite faces of the test specimen. The capacitance and dielectric loss are then measured by comparison or substitution methods in an electric bridge circuit. From these measurements and the dimensions of the specimen, *dielectric constant* and *loss factor* are computed.

Significance: *Dissipation factor* is a ratio of the real power (in phase power) to the reactive power (power 90° out of phase). It is defined also in other ways:

Dissipation factor is the ratio of conductance of a capacitor in which the material is the dielectric to its susceptance.

Dissipation factor is the ratio of its parallel reactance to its parallel resistance. It is the tangent of the loss angle and the cotangent of the phase angle.

The dissipation factor is a measure of the conversion of the reactive power to real power, showing as heat.

Dielectric Constant is the ratio of the capacity of a condenser made with a particular dielectric to the capacity of the same condenser with air as the dielectric. For a material used to support and insulate components of an electrical network from each other and ground, it is generally desirable to have a *low* level of dielectric constant. For a material to function as the dielectric of a capacitor, on the other hand, it is desirable to have a high value of dielectric constant, so the capacitor may be physically as small as possible.

Loss Factor is the product of the dielectric constant and the power factor, and is a measure of total losses in the dielectric material.

Arc Resistance (ASTM D495)

Significance: This test shows ability of a material to resist the action of an arc of high voltage and low current close to the surface of the insulation in tending to form a conducting path therein.

The arc resistance values are of relative value only, in distinguishing materials of nearly identical composition, such as for quality control, development, and identification.

Dielectric Strength (ASTM D149)

Specimen: Specimens are thin sheets or plates having parallel plane surfaces and of a size sufficient to prevent flashing over. Dielectric strength varies with thickness and therefore specimen thickness must be reported.

Since temperature and humidity affect results, it is necessary to condition each type material as directed in the specification for that material. The test for dielectric strength must be run in the conditioning chamber or immediately after removal of the specimen from the chamber.

Procedure: The specimen is placed between heavy cylindrical brass electrodes which carry electrical current during the test. There are two ways of running this test for dielectric strength:

1. *Short-Time Test*: The voltage is increased from zero to breakdown at a uniform rate—0.5 to 1.0 kv/sec. The precise rate of voltage rise is specified in governing material specifications.
2. *Step-By-Step Test*: The initial voltage

applied is 50% of breakdown voltage shown by the short-time test. It is increased at rates specified for each type of material and the breakdown level noted.

Breakdown by these tests means passage of sudden excessive current through the specimen and can be verified by instruments and visible damage to the specimen.

Significance: This test is an indication of the electrical strength of a material as an insulator. The dielectric strength of an insulating material is the voltage gradient at which electric failure or breakdown occurs as a continuous arc (the electrical property analogous to tensile strength in mechanical properties). The dielectric strength of materials varies greatly with several conditions, such as humidity and geometry, and it is not possible to directly apply the standard test values to field use unless all conditions, including specimen dimension, are the same. Because of this, the dielectric strength test results are of relative rather than absolute value as a specification guide.

The dielectric strength of polyethylenes is usually around 500 volts/mil. The value will drop sharply if holes, bubbles, or contaminants are present in the specimen being tested.

The dielectric strength varies inversely with the thickness of the specimen.

Conditioning Procedures (ASTM D618)

Procedure: Procedure A for conditioning test specimens calls for the following periods in standard laboratory atmosphere (50 ± 2% RH, 73.4 ± 1.8°F):

Specimen Thickness, inch	Time, hr.
0.25 or under	40
Over 0.25	88

Adequate air circulation around all specimens must be provided.

Significance: The temperature and moisture content of plastics affects physical and electrical properties. To get comparable test results at different times and in different laboratories, this standard has been established.

In addition to Procedure A descibed above, there are other conditions set forth to provide for testing at higher of lower levels of temperature and humidity.

NONDESTRUCTIVE TESTING

The need for nondestructive testing presents a continuing challenge to every member of the plastics industry concerned with the production of high-quality merchandise at lowest possible cost. The present practice of testing to destruction a few pieces out of perhaps many thousands not only creates a false sense of security but may represent an unsound compromise between the testing of a large number of pieces in the interests of maintaining good quality and, on the other hand, the natural reluctance to destroy salable merchandise.

The problem is receiving attention in many quarters, and is being diligently studied by professional groups particularly expert in this field, two examples being the American Society for Testing Materials Committee E-7 on Nondestructive Testing, organized in 1938, and the more recently established Society for Nondestructive Testing.

Most nondestructive methods of test involve far more than external visual inspection of the surfaces. Nearly every basic principle of physics has been used to obtain, nondestructively, the necessary information concerning the properties of finished articles. The great majority of methods rely upon mechanical measurements or upon a flow or transfer of energy. The energy must usually be supplied from an external source such as an x-ray tube, magnetizing coil, ultrasonic generator or mechanical force, and must be chosen so that distribution of energy within the test piece is modified by the presence of defects. Among the recognized agencies used for nondestructive testing are conduction of electric current; electromagnetic induction; magnetic field; electric (potential) field; conduction of heat; penetrating radiation (x-rays, gamma rays, etc.); mechanical vibration; luminous energy; chemical (spot tests); and static electricity.

Many of these methods originated in the technology of metals and some of them are obviously not applicable to plastics. However, there is little doubt that future testing of articles made from plastics will depend to an increasing extent on modifications of existing techniques or on new concepts arising from the unique needs of organic materials.

INDEX